D1351450

Iron and Steel Works of the World

1988 9th Edition

Price £80.00 plus postage and packing

Edited by Richard Serjeantson
Raymond Cordero
and Henry Cooke

Compiled by Sharon Sexton
and Sheridan Jordain

Published by Metal Bulletin Books Ltd
Park House, Park Terrace, Worcester Park,
Surrey, KT4 7HY, England.

MEMBER OF THE EUROPEAN
ASSOCIATION OF DIRECTORY PUBLISHERS

ASSOCIATION OF BRITISH
DIRECTORY PUBLISHERS

British Library Cataloguing in Publication Data
Iron and steel works of the world.—9th ed.
1. Iron industry and trade—Directories
2. Steel industry and trade—Directories
I. Serjeantson, Richard II. Cordero, Raymond III. Cooke, Henry, *1952*
338.7'6691'025 HD9510.3

ISBN 0–94767–108–0 ISSN 0075 0875

Copies are also available from:

Metal Bulletin Books Ltd,
16 Lower Marsh,
London SE1 7RJ,
England.

Tel: 01 633 0525
Telex: 917706
Telefax: 01 928 6539

Metal Bulletin Inc,
220 Fifth Ave.,
New York,
NY 10001, USA

Tel: 212 213 6202
Telex: 640213

Foreword

THE world's steel industries have undergone another period of immense change in the four years since the last edition of Iron & Steel Works of the World was published, particularly in Europe and North America. Closures, management buyouts, takeovers, mergers, divestments and in some countries, privatisation, have all contributed to a much altered scene.

With a few notable exceptions, like South Korea, there has been little in the way of major investment. The accent has been on rationalisation, reconstruction and revitalisation to meet the still difficult economic conditions and many former giants have been slimmed down drastically in the process. Yet the uncrushable vigour of the industry is reflected in the number of entries – 1,581 – in this ninth edition.

In this new edition we have again aimed to include all major producers of iron, raw steel and rolled steel products, as well as many of the world's re-rollers, tube makers, iron powder producers, strip coaters and cold rolled section makers. In deciding eligibility for inclusion we have continued our established policy of omitting companies which produce forgings, castings and wire except where they are also engaged in the production of iron, raw and rolled steel, etc. Again, we have not included companies producing solely ferro-alloys – these are the subject of a separate international directory published by Metal Bulletin Books.

Information from the Communist countries continues to be relatively sparse, but in the case of China there are welcome signs of a better flow of data. In Russia, unfortunately, *glasnost* has not yet been embraced by the steel authorities.

We have not altered the basic layout of the book – alphabetical arrangement of countries with alphabetical listing of companies within each country – buyers' guides arranged by product. All tonnages are metric unless otherwise stated.

There is also a comprehensive index which includes cross-references to relevant subsidiaries which do not have separate entries, but are included under their parent companies, and to companies now operating under a new name but which are better known under an old name. In indexing we have adopted the general principle of using the key word: for example Sté des Aciéries de Montereau will be found under Montereau; and whenever possible we have used the abbreviated title – e.g. Sidor–CVG Siderúrgica del Orinoco CA – where the abbreviation is reasonably well-known.

Foreword (continued)

We wish to acknowledge the help of all those companies which supplied details for their entries in this book, especially as they had this time to grapple valiantly with a form designed to make feeding the computer easier. We thank, too, steel industry associations all over the world for assistance in various ways. The photograph on the front cover is reproduced by courtesy of British Steel Corp.

This is the last edition in which Raymond Cordero will figure as an editor, as he retires this year from full-time employment with Metal Bulletin PLC. He has been involved with every edition since the book was launched by his father, the late H. G. Cordero, in 1952.

Richard Serjeantson
Henry Cooke

SIGRI

Contents

Abbreviations

ABNT	Associação Brasileira de Normas Técnicas
ACSR	Aluminium conductor, steel reinforced
AEF	American Electric Fusion (welded tube mills)
a/f	Across flats
Afnor	Association Française de Normalisation
AISI	American Iron & Steel Institute
AOD	Argon-oxygen decarburisation (converter for refining stainless steel, developed by Union Carbide Corp, USA)
API	American Petroleum Institute
AREA	American Railway Engineering Association
ASTM	American Society for Testing & Materials
BAP	Bath agitation procedure (combined top and bottom blowing practice)
BBC	Brown Boveri & Co
Best	Böhler electro-slag topping (refining process)
BG	British gauge
BOF	Basic oxygen furnace
BS	British Standard
BWG	Birmingham wire gauge
CAB	Capped argon bubbling (secondary steelmaking process), or calcium argon bubbling (secondary steelmaking process involving desulphurisation)
CDW	Cold drawn welded tubes
CEM	Cie Electro-Mécanique
CLU	Creusot-Loire Uddeholm (converter for stainless steel refining)
CPE	Cross roll piercing and elongating
CRCA	Cold rolled close annealed
CTD	Cold twisted deformed
DH	Dortmund-Hörder (vacuum degassing process)
dia	Diameter
DIN	Deutsche Industrie-Norm
DOM	Drawn over mandrel
DR	Direct reduction, or double reduced
DRI	Direct reduced iron
DSAW	Double submerged arc weld (tubes)
EBT	Eccentric bottom tapping (arc furnace)
EOF	Energy optimising furnace
ERW	Electric resistance weld
ESR	Electro-slag refining
fpm	Feet per minute
g	Gauge
GC	Galvanized corrugated
GFM	Gesellschaft für Fertigungstechnik & Maschinenbau (forging machines)
HBB	Hindustan Brown Boveri
HF	High-frequency
HH	Huntington Heberlein (sinter plants)
H/V	Horizontal/vertical (rolling mills)
HYL	Direct reduction process (from Hojalata y Lámina)
i.d.	Inside diameter

Abbreviations (continued)

IPN	I (joist) profil normal
IPS	International pipe size
JIS	Japan Industrial Standards
JRS	Japanese National Railway Standards
K-Bop	Basic oxygen converter (combined top and bottom blowing practice developed by Klöckner-Werke AG, West Germany)
kg	Kilogram
kj	Kilojoule
KMS	Klöckner-Maxhütte Stahlverfahren (combined top and bottom blowing oxygen converter – can be referred to as OBM-KMS)
KS	Klöckner Stahlverfahren (bottom blowing with oxygen and hydrocarbons, melting 100% solid charge)
kVA	Kilovolt-Ampère
KW	Kilowatt
KYS	Klöckner Youngstown Steel (shaft furnace feeding electric arc furnace)
lb	Pounds (weight)
LBE	Lance bubbling equilibrium (oxygen steelmaking process involving combined top and bottom blowing)
LD	Linz-Donawitz (oxygen steelmaking process)
LD-AC	Linz-Donawitz Arbed-Centre (oxygen steelmaking process for high-phosphoric irons, using powdered lime injection)
LF	Low-frequency, or ladle furnace
LNG	Liquefied natural gas
LPG	Liquid petroleum gas
LWS	Loire-Wendel-Sprunck (oxygen steelmaking process)
m	Metre
max	Maximum
MF	Medium-frequency
MIG	Metal (electrode) inert gas (welding)
min	Minimum
MKW	Type of Schloemann cold strip mill
mm	Millimetre
MN	Mega-Newton (press power)
mpm	Metres per minute
MVA	Megavolt-Ampère
n.b.	Nominal bore
nom	Nominal
NPS	Nominal pipe size
OBM	Oxygen Boden Maxhütte (bottom blown oxygen steelmaking process)
OCTG	Oil country tubular goods
o.d.	Outside diameter
OLP	Oxygen lime powder (steelmaking process)
PIW	Per inch of width
Q-Bop	Quick/quiet basic oxygen process
RCS	Round-cornered square
RF	Radio-frequency
RH	Rheinstahl Heraeus (vacuum degassing process)
RHS	Rectangular hollow section
SAE	Society of Automotive Engineers
SAW	Submerged arc weld

TRIMTEC® – The Economical Perfection of Secondary Metallurgy Process Technology. Easy, Safe, Reliable.

Current steelmaking practices are characterized by continuous efforts towards the most economical production methods with simultaneous improvement of the material properties.
Secondary metallurgy has already substantially contributed to the achievement of these objectives and, on account of its integration into this part of the steelmaking process, the wire injection technique has now acquired additional importance.
The main features of the technique are the marked improvements in the treatment technology and the economics. It has never been so economical and easy to safely and objectively introduce powdered substances or alloys into the melt. TRIMTEC® cored wire is the optimal way of treating melts, regardless of whether the operation involves calcium treatment – the current major field of application – or alloying with sensitive elements. Our TRIMMASTER wire injection machines are equally suitable for the introduction of solid and cored wires and offer you numerous advantageous facilities. For instance, it is possible to feed wires with cross-sections of 5–18 mm at infinitely variable rates of 5–300 m/min.

And the Odermath Stahlwerkstechnik GmbH service by no means ends there. We will advise you on the installation of wire injection plants, provide designs incorporating relevant know-how and delegate our application engineers to commission the plant. Worldwide – without limitation.

Ask us. Our capabilities are confirmed by more than 100 cases. You too can gain the benefit of our experience.

ODERMATH STAHLWERKSTECHNIK GMBH
Am Kielsgraben 2–4 · D-4019 Monheim/Rhld. · Tel. (0 21 73) 59 51 · Telex 8 515 775

Abbreviations (continued)

SBQ	Special bar quality
SCH	Schedule
SG	Spheroidal graphite (ductile iron)
SIS	Standardiseringskommissionen i Sverige
STB	Top and bottom blown converter
swg	Standard wire gauge
TFS	Tin-free steel
TIG	Tungsten (electrode) inert gas (welding)
TN	Thyssen Niederrhein (secondary steelmaking process involving desulphurisation)
tpd	Tons per day
tpm	Tons per month
tpy	Tons per year
UHP	Ultra-high power
UIC	Union Internationale des Chemins de Fer
UNI	Ente Nazionale Italiano di Unificazione
UOE	U-press, O-press, electric weld (large-diameter heavy-wall welded pipe production)
UPA	U (channel) profil américain
UPE	U (channel) profil européen
UPN	U (channel) profil normal
VAD	Vacuum arc degassing
VAR	Vacuum arc refining
VIM	Vacuum induction melting
VOD	Vacuum oxygen decarburising (secondary steelmaking process for stainless steel)
VOR	Vacuum oxygen refining

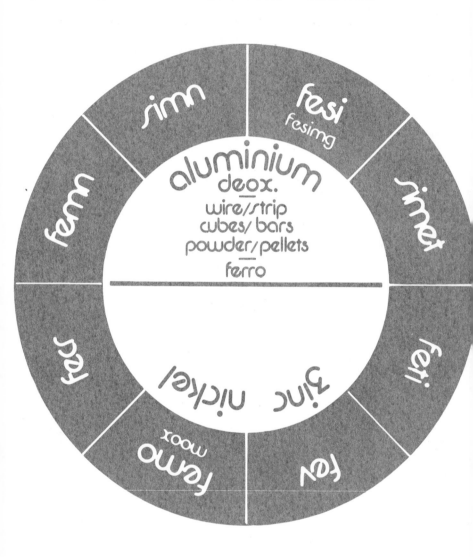

FIRST INDUSTRIAL S.A.
P.O. Box 1052, 40 Boulevard Napoleon
Luxembourg City, LUXEMBOURG
Tel: 44 65 66 Telex: 60134
Cable: ICM ALLOYS
Telefax: 44 63 27

Index to Countries

Index to Countries

tability, competitiveness and
ess in the international iron,
and non-ferrous metal industry
etermined by advanced technol-
s and plant concepts aimed at
bility. More than ever before, the
et is looking for ideas to improve
ty and efficiency.
ag's specialists know what's
ded:
m of project and design
neers who are experts in steel
uction, continuous casting,
g mill engineering and tube-
pipe-making technology.

team designs and builds
hinery and systems for capaci-
to meet market demands
d on techniques requiring less
gy and less raw materials. This provides the stimulus for optimum linking of individual process stages to arrive at an integrated total technical solution, improve and expand the fields of application and extend the product range.

Hand in hand with industry Mannesmann Demag help find solutions for revamping, modernizing and rationalizing existing plants, for rational use of energy, residual energy recovery, paying even greater attention to the environment.
In this way, plant projects are optimally implemented using the most modern manufacturing facilities for very large components of highest quality.

mannesmann *technology*

Mannesmann Demag Hüttentechnik

Metallgewinnung
Postfach 100141,
Wolfgang-Reuter-Platz,
D-4100 Duisburg 1,
Tel.: (203) 605-1, Telex: 855855

MEER
Postfach 365, Ohlerkirchweg 66,
D-4050 Moenchengladbach 1
Tel.: (2161) 350-1, Telex: 852525

MDS Mannesmann Demag Sack GmbH
Postfach 330370, Wahlerstrasse 2,
D-4000 Duesseldorf 30,
Tel.: (211) 6504-01, Telex: 8586849

2-1674-021

Index to Countries

Alphabetical Index to Companies

Alphabetical Index to Companies

ANDRITZ

Alphabetical Index to Companies

Alphabetical Index to Companies

Recent carryings-on at Port Talbot are enough to put a glint in a steelman's eye.

Our kilometre-long hot strip mill in Port Talbot has been complet rebuilt, within budget and on target. And for 95 per cent of the time it too do the job, it remained fully operational.

A world record, we believe. And not our only one last year.

At Redcar, we rebuilt and relined the largest blast furnace in Europ just 135 days. (It's somewhat taller than St Paul's Cathedral.) The m objective of such investments is to improve quality and customer servi So we can compete even more strongly at home and abroad.

We're also cutting costs. Take energy, for example: In our busines

er cent cut in energy consumption represents a saving of £6 million. ce 1980, energy consumption per tonne is down 17 per cent.

We invest millions in research and development of new production hniques and new applications.

And we give high priority to marketing skills: not just to respond ckly to demand; to develop new markets all over the world.

If you'd like to know more about the new, profitable tish Steel, write to: British Steel Information Services, lbert Embankment, London SE1 7SN and we'll send our colour brochure.

British Steel
In shape for things to come.

Alphabetical Index to Companies

Alphabetical Index to Companies

Alphabetical Index to Companies

Alphabetical Index to Companies

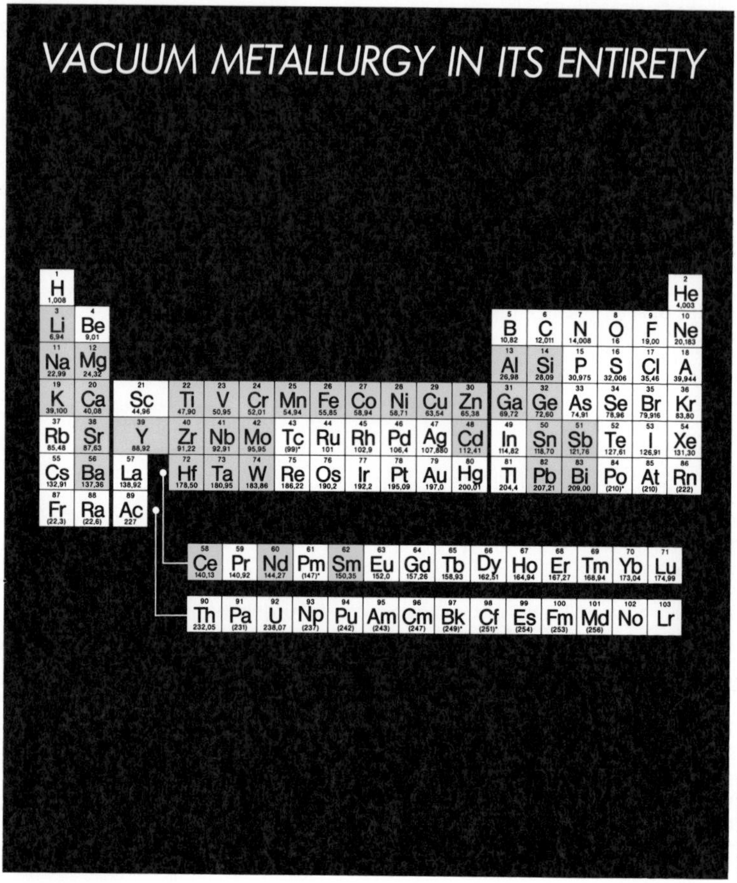

Alphabetical Index to Companies

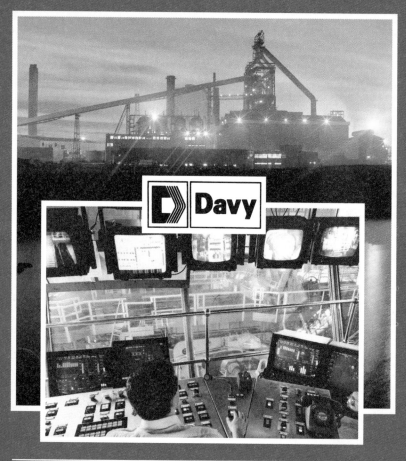

Alphabetical Index to Companies

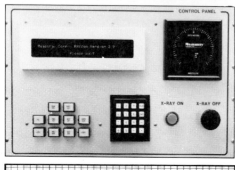

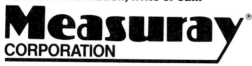

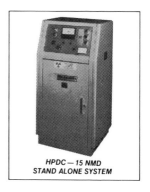

Alphabetical Index to Companies

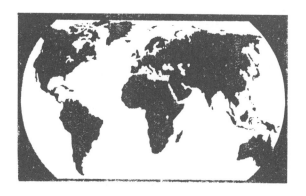

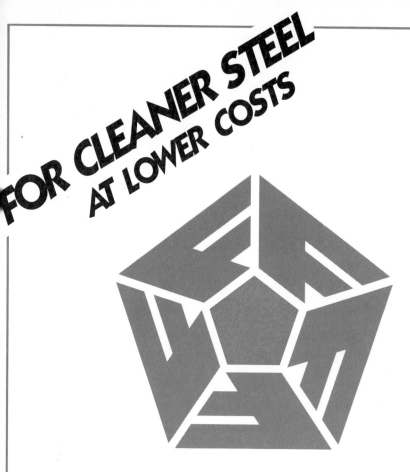

Alphabetical Index to Companies

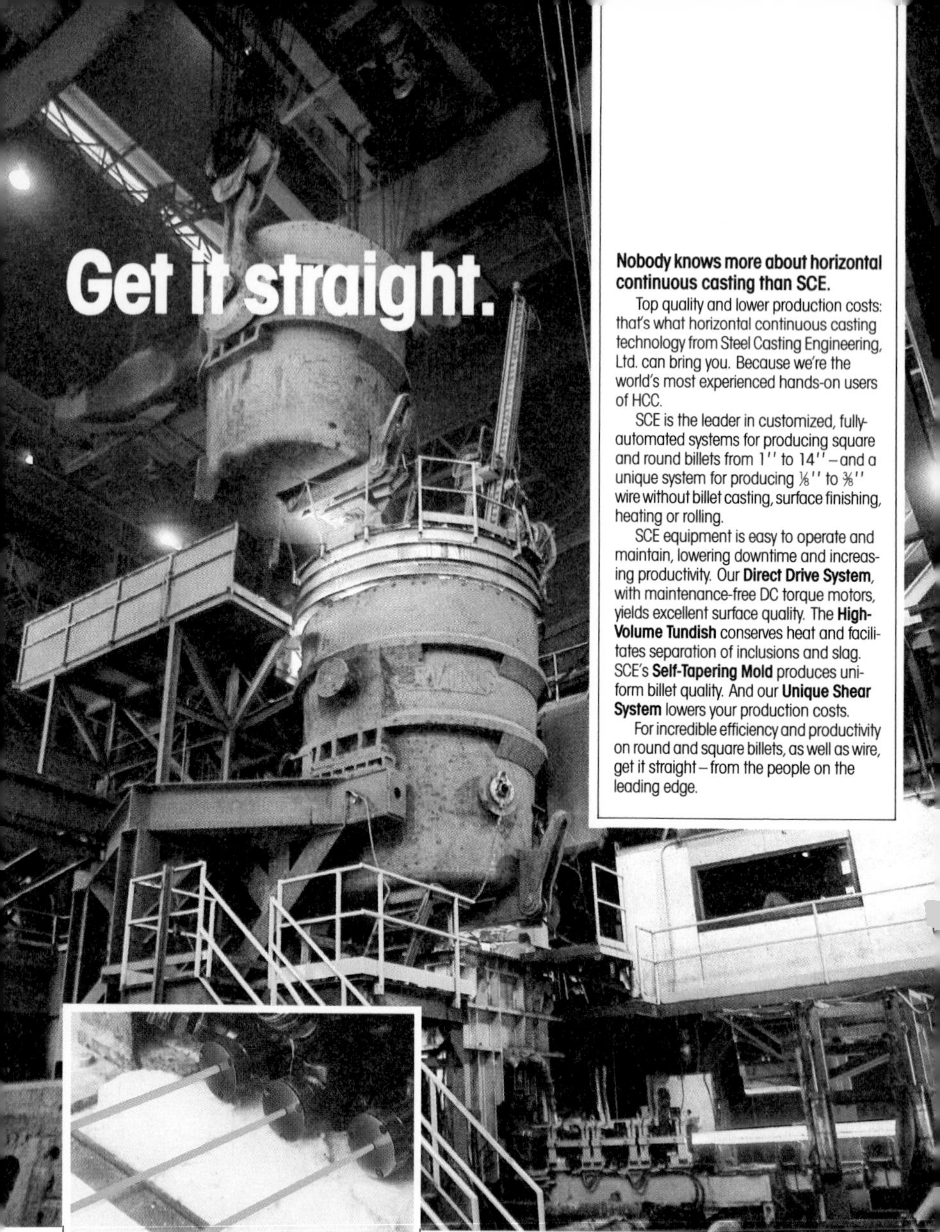

SCE technology produces horizontal continuously cast wire in large or small lots – eliminating billet inventory, rolling cost, and long delivery times.

Alphabetical Index to Companies

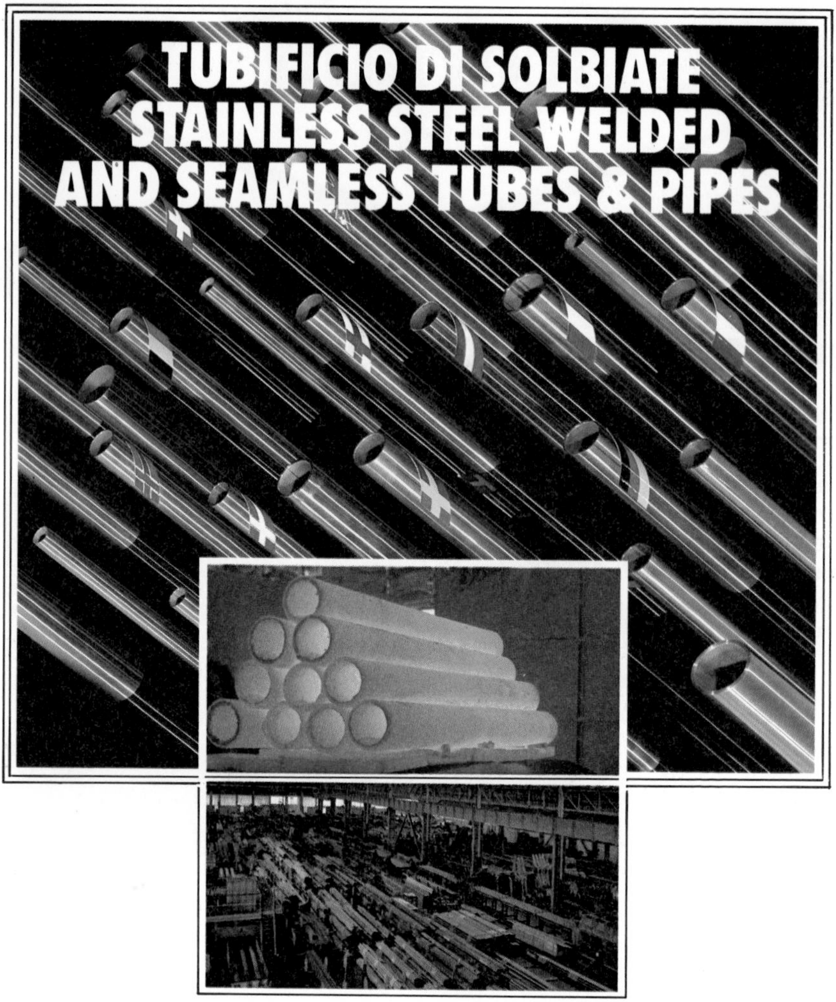
TUBIFICIO DI SOLBIATE
STAINLESS STEEL WELDED
AND SEAMLESS TUBES & PIPES

Alphabetical Index to Companies

dms
dujardin montbard somenor
Équipements pour la sidérurgie et la métallurgie.
Equipment for the iron and steel industry.
COULEE CONTINUE ROTATIVE
ROTARY CONTINUOUS CASTER
LIGNE DE TRAITEMENT
PROCESSING LINE
LAMINOIR TRANSVAL
TRANSVAL MILL
DRESSEUSE
STRAIGHTENER
L.A.T. SOUDE
WELDED TUBE MILL
dujardin montbard somenor
Z.I. SECLIN - BP 219 - 59472 SECLIN CEDEX - FRANCE - Tél. 20.90.92.26
TELEX. DMS LIL A/B 130321 F

Alphabetical Index to Companies

Alphabetical Index to Companies

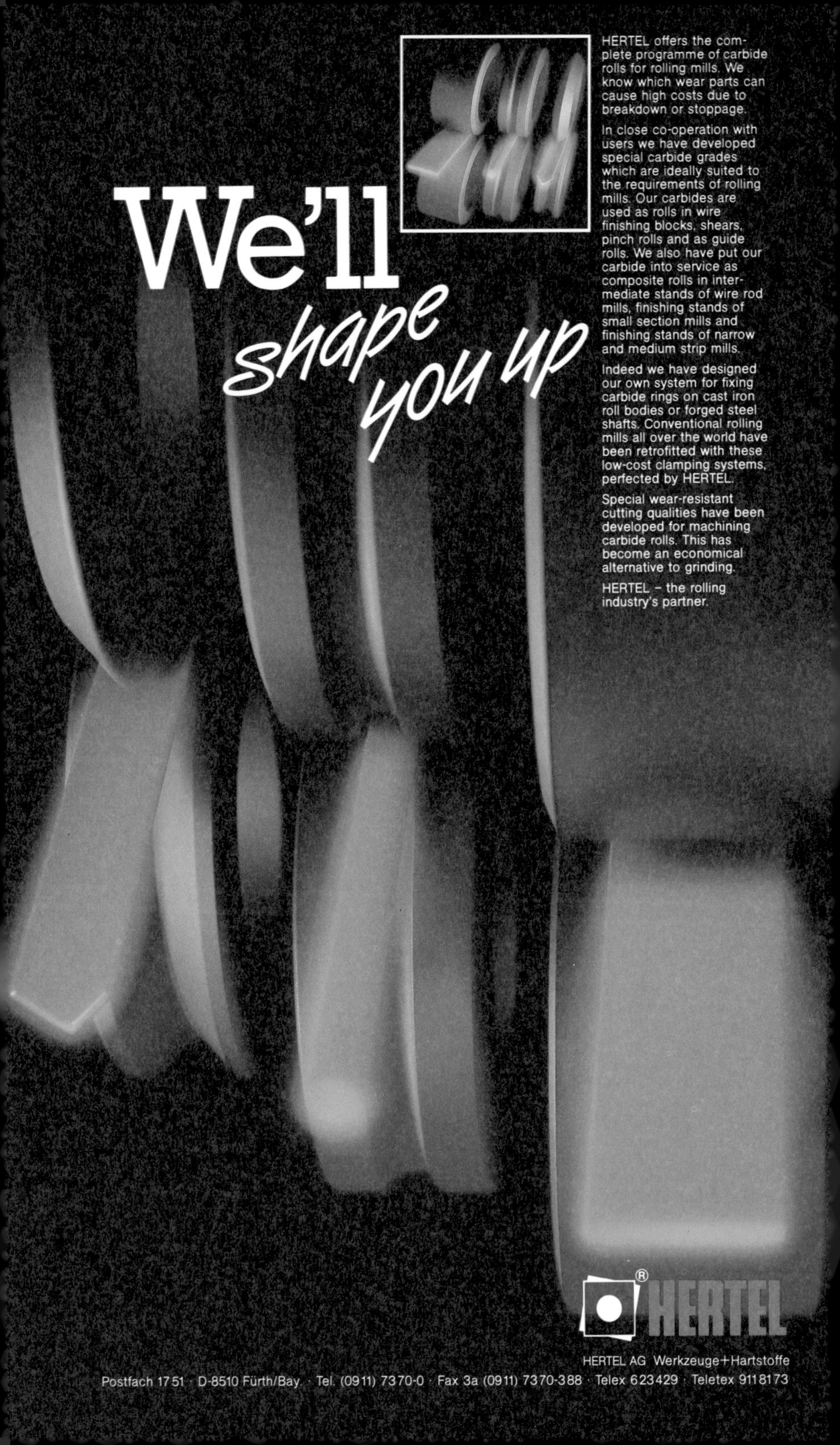
We'll
shape
you up
HERTEL offers the complete programme of carbide rolls for rolling mills. We know which wear parts can cause high costs due to breakdown or stoppage.
In close co-operation with users we have developed special carbide grades which are ideally suited to the requirements of rolling mills. Our carbides are used as rolls in wire finishing blocks, shears, pinch rolls and as guide rolls. We also have put our carbide into service as composite rolls in intermediate stands of wire rod mills, finishing stands of small section mills and finishing stands of narrow and medium strip mills.
Indeed we have designed our own system for fixing carbide rings on cast iron roll bodies or forged steel shafts. Conventional rolling mills all over the world have been retrofitted with these low-cost clamping systems, perfected by HERTEL.
Special wear-resistant cutting qualities have been developed for machining carbide rolls. This has become an economical alternative to grinding.
HERTEL – the rolling industry's partner.
HERTEL®
HERTEL AG Werkzeuge+Hartstoffe
Postfach 1751 · D-8510 Fürth/Bay. · Tel. (0911) 7370-0 · Fax 3a (0911) 7370-388 · Telex 623429 · Teletex 9118173

Alphabetical Index to Companies

Slag Reduction

Leaders in Slag Handling Processing & Utilisation.

Converting waste products into commercial opportunities.

Recovering guaranteed quality scrap.

Processing slag into high quality roadstone aggregate.

Meadowbank House Meadowbank Road Rotherham South Yorkshire S61 2DY
Telephone: (0709) 740666 (INTL +44 709 740666) Telex: 547714 Facsimile No.: (0709) 561013

Alphabetical Index to Companies

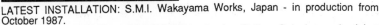

Alphabetical Index to Companies

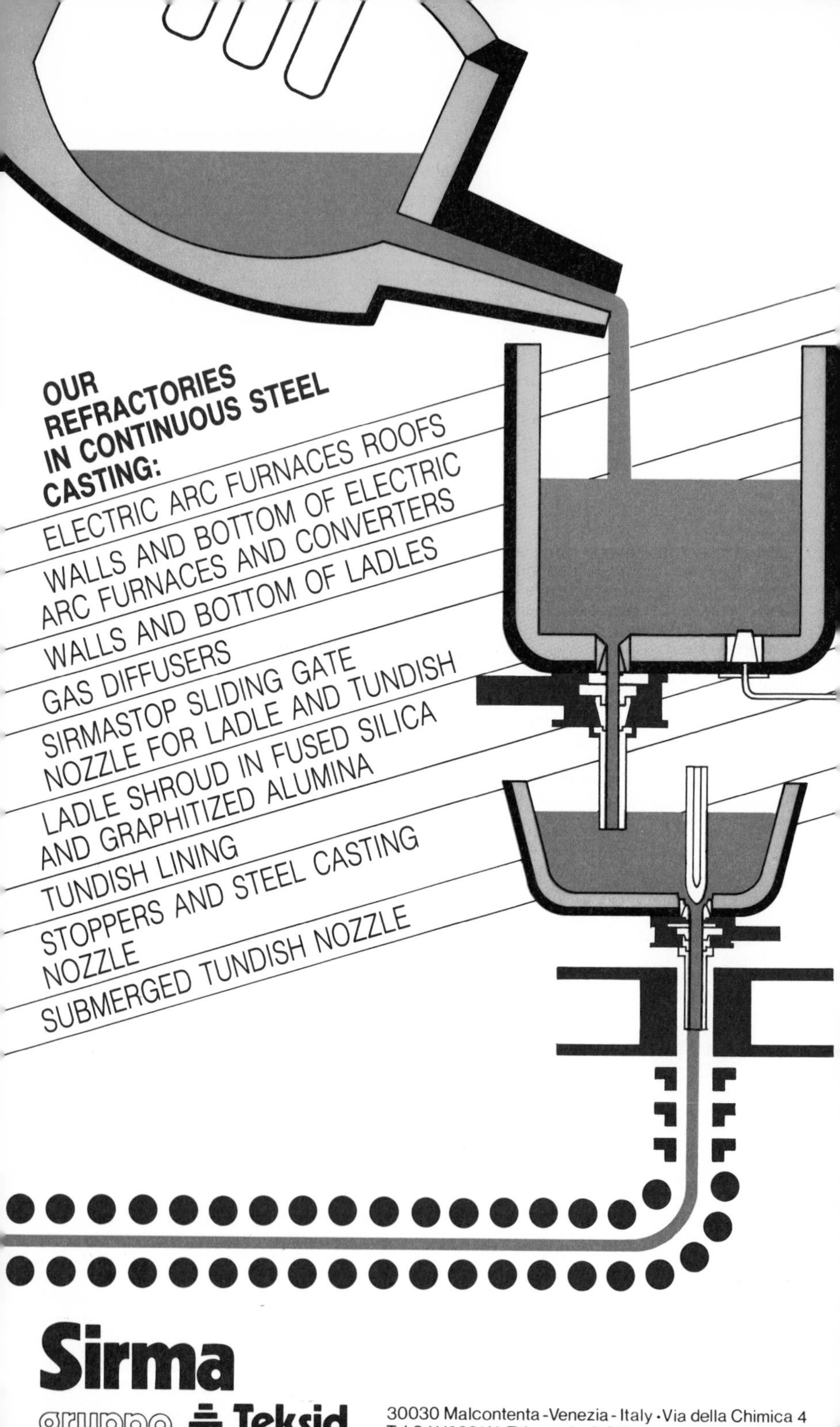
OUR
REFRACTORIES
IN CONTINUOUS STEEL
CASTING:
ELECTRIC ARC FURNACES ROOFS
WALLS AND BOTTOM OF ELECTRIC
ARC FURNACES AND CONVERTERS
WALLS AND BOTTOM OF LADLES
GAS DIFFUSERS
SIRMASTOP SLIDING GATE
NOZZLE FOR LADLE AND TUNDISH
LADLE SHROUD IN FUSED SILICA
AND GRAPHITIZED ALUMINA
TUNDISH LINING
STOPPERS AND STEEL CASTING
NOZZLE
SUBMERGED TUNDISH NOZZLE
Sirma
gruppo Teksid
30030 Malcontenta - Venezia - Italy • Via della Chimica 4
Tel. 041/663111 • Telex 410119 SIRMAR I • Telefax 041/663220

Albania

No recent Albanian iron and steel output figures are available. Rolled steel output increased "Two-fold" between 1981 and 1986. In 1987 pig iron output was due to be 34% more than in 1986, raw steel 15% more, and rolled steel 22% more. Rolled steel output in 1990 is planned to be 30-32% more than in 1985. A steel wire plant is included in the current 5-year plan.

Enver Hoxha Tractor Plant

Works:
■ Tirana. **Steelmaking plant:** HF electric furnace making carbon and special steel.

Steel of the Party Metallurgical Combine

Works:
■ Elbasan. ***Coke ovens:*** (two batteries). ***Blast furnaces:*** two (annual capacity 250,000 tonnes). ***Steelmaking plant:*** basic oxygen converters; three electric arc furnaces. ***Continuous casting machines:*** Two 2-strand bloom (210-220mm). ***Rolling mills:*** rail and medium section; bar and wire rod; heavy plate and two sheet mills.
Products: Foundry pig iron. Carbon steel – round bars, square bars, flats, hexagons, light angles, light angles, light tees, light channels, light rails, medium plates, heavy plates. Stainless steel – round bars, square bars, flats, medium plates, heavy plates. Alloy steel (excl stainless) – round bars, square bars, flats, medium plates, heavy plates.
Expansion programme: Construction of a tube mill started in late 1982. Another wire rod mill is planned under the current five-year plan. By the end of 1986 it was planned that most of the steel requirements of the country's nail and bolt would be supplied from Elbasan instead of from imports.

Algeria

SNS – Sté Nationale de Sidérurgie

Head office: 23 Avenue de l'Aln la Glaciere, Algiers. **Telex:** 53693.
Established: 1964.
Works:
■ El Hadjar, Annaba. ***Coke ovens:*** Two batteries (85 ovens each). ***Sinter plant:*** One 240 sq m strand. ***Blast furnaces:*** One 2,000 cu m (annual capacity 1,200,000 tonnes). ***Steelmaking plant:*** Six basic oxygen converters – three 90-tonne, and three 60-tonne; one 85-tonne electric arc furnace. ***Continuous casting machines:*** Three 4-strand Demag for billet; two 1-strand Demag for slab. ***Rolling mills:*** 4-strand bar and wire rod ***Coil coating lines:*** One electrolytic tinning line; one hot-dip galvanizing line.
■ 2 Avenue des Martyrs de la Révolution, Oran. ***Steelmaking plant:*** One open hearth furnace. ***Rolling mills:*** One 14-stand bar.
■ Reghaia. ***Tube and pipe mills:*** welded – Spiral and longitudinal. ***Other plant:*** Pipe galvanizing plant; cold-formed section plant.
■ Ghardaia. ***Tube and pipe mills:*** welded – Spiral.
Products: Steelmaking pig iron. Carbon steel – slabs, wire rod, reinforcing bars, cold roll-formed sections (from strip), medium plates, hot rolled sheet/coil (uncoated), cold rolled sheet/coil (uncoated), hot-dip galvanized sheet/coil, electrolytic tinplate (single-reduced), bright wire, longitudinal weld tubes and pipes, spiral-weld tubes and pipes, large-diameter tubes and pipes, galvanized tubes and pipes.

Angola

Siderurgia Nacional (Sina UEE)

Head office: Zona Industrial do Farol das Lagostas, Luanda. **Telex:** 3178. **Established:** 1968.
Works:
■ Luanda. ***Steelmaking plant:*** One Tagliaferri electric arc furnace. ***Rolling mills:*** bar mill.
Products: Carbon steel – reinforcing bars.

Argentina

Acindar Industria Argentina de Aceros SA

Head office: Paseo Colón 357, 1063 Buenos Aires. **Tel:** 331 8431-9, 30 0121-9. **Telex:** 22352, 21270, 9111 and 9112.
Management: Board directors – Alcides López Aufranc (chairman), Juan Carlos Manuel Gurmendi (vice chairman), Arturo F.A. Acevedo, Jorge E. Acevedo, Arturo T. Acevedo, Oscar Rodolfo García, Marta Zulema Gurmendi, Carlos A. Leone, Alberto F. Méndez Cañás, Alberto F. Ibáñez (board secretary); managers – Oscar A. Pigazzi (investigation and commercial development), Julio Carozzi (sales, merchant steels), Arnoldo Pfister (sales, industrial steels), Marcos Carranza Velez (sales, foreign markets), Juan E. MacCulloch (plant 1 manager), Ricardo López Mayorga (plant 2 manager), Gustavo A. Pittaluga (industrial relations), Hugo A. Untersander (purchasing), Carlos E. Soete (systems), Julian P. Rooney (control), Miguel O. Socas (institutional relations), Jorge N. Vid ela (legal), Ricardo M. Rivero Haedo (finance), Guillermo H. Sorondo (planning and budget), Cesar A. Deymonnaz (international finance relations), Walter J. Villar (relations with associated companies). **Established:** 1942. **Capital:** Pesos 61,108,521. **No. of employees:** Works – 4,288. Office – 3,090. **Annual capacity:** Pig iron: (DRL) 700,000 tonnes. Raw steel: 1,338,000 tonnes. Finished steel: 1,630,000 tonnes.
Works:
■ Planta No 2, Ruta Prov 21, Km 247, Villa Constitucion, Pcia Santa Fe. (Tel: 74099, 74306, 74485). Works director: Arturo Acevedo. ***Direct reduction plant:*** One Midrex 400 series (annual capacity 700,000 tonnes). ***Steelmaking plant:*** Three 110-120-tonne Tagliaferri electric arc furnaces (1,050,000 tonnes). ***Refining plant:*** One 110-120-tonne ladle furnace (1,050,000 tonnes). ***Continuous casting machines:*** Two 10-strand (combined) Concast billet (100 x 100 to 150 x 150) and one 2-strand Concast bloom (127mm sq) (combined capacity 1,050,000 tonnes). ***Rolling mills:*** One Morgan No 2 bar and wire rod (640,000 tonnes); one Morgan No 1 Wire Rod and hot strip and hoop (430,000 tonnes). ***Tube and pipe mills:*** welded – Aetna Standard, butt welding (81,000 tonnes); Yoder, electric welding (84,000 tonnes). ***Other plant:*** 39 drawing machines (205,000 tonnes).
■ Planta No 1, Dr Ignacio Arieta 4936, Tablada, Pcia Buenos Aires. Tel: 651 0051-9. Works director: Juan MacCulloch. ***Steelmaking plant:*** Two electric arc furnaces – one 25-tonne Whiting, one 50-tonne Lectromelt (annual capacity 288,000 tonnes). ***Refining plant:*** One 50-tonne ladle furnace (250,000 tonne). ***Rolling mills:*** One Birdsboro blooming (240,000 tonnes); one Birdsboro bar (120,000 tonnes); one Schoelmann-Birdsboro wire rod (155,000 tonnes). ***Other plant:*** Drawing machines (90,000 tonne) and Davy forging presses (24,000 tonnes).
Products: Direct-reduced iron (DRI). Carbon steel – ingots, slabs, blooms, billets, wire rod, reinforcing bars, round bars, square bars, flats, hexagons, bright wire, black annealed wire, galvanized wire (plain), galvanized wire (barbed), longitudinal weld tubes and pipes, galvanized tubes and pipes. Stainless steel – round bars, square bars, flats, hexagons. Alloy steel (excl stainless) – ingots, slabs, blooms, billets, round bars, square bars, flats, hexagons.
Sales offices/agents: Paseo Colón 357, 1063 Buenos Aires (Tel: 331 4091-9);

Acindar Industria Argentina de Aceros SA (continued)

Balcarce 358, 1063 Buenos Aires (331 4071-7).
Expansion programme: 1988 is the expected completion date of the continuous casting machine in Planta No 1 (2-strand Concast, product dimensions – 150 x 150 billets, 220 x 290 bloom; annual capacity 150,000 tonnes).

Armco Argentina SA

Head office: Av. Corrientes 330, 1378 Buenos Aires. **Tel:** 311 6215, 311 4947. **Telex:** 22195.
Established: 1938.
Works:
■ Valentín G6mez 210, Haedo, BA. ***Tube and pipe mills:*** welded – One Yoder. ***Coil coating lines:*** One Armco-Sendzimir hot-dip galvanizing (and aluminizing) line; one Armco-Sendzimir terne/lead coating (and galvanising).
Products: Carbon steel – hot-dip galvanized hoop and strip; terne-plate (lead coated) sheet/coil (narrow strip); longitudinal weld tubes and pipes; galvanized tubes and pipes.

Aceros Bragado Sacif

Head office: Bernardo de Irogoyen 190, 9 y 10 pisos, Buenos Aires. **Postal address:** Codigo Postal 1072. **Telex:** 17414 acebra ar.
Established: 1969.
Works:
■ Rosario, SF (formerly operated by Cura Hnos). ***Steelmaking plant:*** electric arc furnaces. ***Rolling mills:*** bar and wire rod mills.
■ Bragado, BA. ***Steelmaking plant:*** electric arc furnaces. ***Continuous casting machines:*** 3-strand for (75 to 140mm) billet . ***Rolling mills:*** bar and wire rod mills.
Products: Carbon steel – wire rod, reinforcing bars.

La Cantábrica SA Metalúrgica, Industrial y Comercial

Head office: Moreno 755, 1092 Buenos Aires.
Established: 1902.
Works:
■ Tres Arroyos 329 Haedi 1706, Buenos Aires. ***Rolling mills:*** One medium section; one bar.
Products: Carbon steel – round bars, square bars, flats, light angles, special light sections, medium angles, medium joists, light rails.

Comesi Saic

Head office: Avda. Belgrano 1225, Buenos Aires. **Tel:** 38 9016/8. **Telex:** 17783 globe ar.
Management: Managers – Atilio Paglino (general), Ricardo Ali (plant), Mario Belchior (financial). **Established:** 1962. **Capital:** Australes 115,000. **No. of employees:** Works – 349. Office – 68. **Ownership:** Private. **Subsidiaries:** None.
Works:
■ ***Coil coating lines:*** hot-dip galvanizing (non-ox Sendzimir) (annual capacity 120,000 tonnes); zinc-aluminium (Galvalume) (proposed); colour coating (60,000 tonnes).
Products: Carbon steel – hot-dip galvanized sheet/coil 0.30 x 1,000-2.5-1,220mm (1986 output 108,041 tonnes), galvanized corrugated sheet/coil, colour coated sheet/coil 0.30 x 1,000-1.6 x 1,220mm (6,587 tonnes), zinc-aluminium alloy coated sheet/coil 0.30 x 1,000-2.5-1,220mm.
Expansion programme: Planned increase in capacity of hot-dip coating line to 200,000 tonnes and conversion to zinc-aluminium coating (early 1990).

Laminfer SA

Head office: Provincias Unidas y Junin, Rosario 2000. **Telex:** 41820 lafer ar. **Established:** 1960.
Works:
■ ***Rolling mills:*** One RWF cold strip. ***Tube and pipe mills:*** welded – $\frac{5}{8}$ to 3".
Products: Carbon steel – cold rolled hoop and strip (uncoated), longitudinal weld tubes and pipes.

Ostrilion SA Comercial e Industrial

Head office: Av Don Pedro de Mendoza 3875, 1294 Buenos Aires. **Postal address:** Casilla de Correo 1627, Correo Central, 1000 Buenos Aires. **Tel:** 21 4621-8. **Telex:** 21058 artib ar.
Management: President – Eric G. Campbell; vice-president – Hector A. Magro; directors – Leonardo C. Campbell, Daniel A. Magro, R.A. Robinson, R.M.R. de H. Ohlsson, Carlos A. Solano (personnel manager), Patricio J. Wilson (general manager). **Established:** 1908. **Capital:** US$ 5,000,000. **No. of employees:** Works – 125. Office – 63. **Ownership:** Pemenza SA de Finanzas. **Subsidiaries:** Industrias El Rosario SAIC y A, Conquistador SA. **Annual capacity:** Finished steel: 41,000 tonnes.
Works:
■ At head office. Works manager: Fernando M. Canosa. ***Coil coating lines:*** One Sendzimir hot-dip galvanizing (oxidation-reduction) (annual capacity 41,000 tonnes).
Products: Carbon steel – hot-dip galvanized sheet/coil 0.30-1.90mm thickness, 1000mm width (output 35,000 tonnes).
Brands: Ostrilion – galvanized corrugated sheet. Ostriform – roll-formed galvanized sheet. Ostritub – spiral, lock form tubes made from galvanized steel.

Propulsora Siderurgica Saic

Head office: Leandro N Alem 1067, piso 22°-23°, Buenos Aires. **Tel:** 313 6777. **Telex:** 22016 props ar. **Telecopier (Fax):** 313 6165.
Management: President – Agustín Rocca; vice president – Carlos D. Tramutola; managers – Fernando Freytes (general), Carlos Ormachea (financial), Daniel Novegil (commercial), Reinaldo Castilla (works). **Established:** 1961. **Capital:** Pesos 60,000,000. **No. of employees:** Works – 856. Office – 548. **Annual capacity:** Finished steel: 735,000 tonnes.
Works:
■ Avenida Costanera Almirante Brown, 1925 Ensenada, BA. (Tel: 21002-8. Telex: 31142 prosi ar). ***Rolling mills:*** One 4-stand tandem cold reduction (572 x 1,524 x 1,675) and one 1-stand temper/skin pass (572 x 1,524 x 1,675) (combined annual capacity 735,000 tonnes).
Products: Carbon steel – cold rolled sheet/coil (uncoated) (1986 output 379,406 tonnes).

Aceros Revestidos SA

Head office: Cmno Gral Belgrano km 31,500, 1888 Florencio Varela, Pcia Buenos Aires. **Postal address:** Casilla Correo 6. **Tel:** 313 8528, 313 8578, 313 8539. **Telex:** 23776 servi ar.
Established: 1979. **Ownership:** Techint. **Annual capacity:** Finished steel: 36,000 tonnes.
Works:
■ At head office. Works manager: M. Chimienti. ***Coil coating lines:*** One continuous soluble anode Lavezzari electro-galvanizing (annual capacity 36,000 tonnes); colour coating One Lavezzari (sheets) (12,000 tonnes).
Products: Carbon steel – electro-galvanized sheet/coil thickness 0.20-2.5mm, width up to 12.0mm (1986 output 12,000 tonnes); colour coated sheet/coil thickness 0.40-1.25mm, width up to 12.50mm (4,000 tonnes).
Brands: Sidercolor – electrogalvanized and colour coated sheets; Siderzinc – electrogalvanized sheets; Siderskin – electrogalvanized and PVC coated.

Rosati y Cristófaro Saic

Head office: Sarmiento 930, 3° piso, Buenos Aires. **Telex:** 18760 rocri ar.
Works:
■ San Nicolás, BA. ***Rolling mills:*** light section and bar mills.
■ Don Bosco, BA. ***Rolling mills:*** light section and bar mills.
Products: Carbon steel – reinforcing bars, round bars, square bars, flats, special light sections, light rails.

Siam SA – Division Siat

Head office: Tucuman 633, Buenos Aires. **Telex:** 24181.
Established: 1948.
Works:
■ Lamús. ***Tube and pipe mills:*** welded – Spiral and longitudinal.
Products: Carbon steel – longitudinal weld tubes and pipes, spiral-weld tubes and pipes, large-diameter tubes and pipes.

Siderca Saic

Head office: Leandro N Alem 1067, 1001 Buenos Aires. **Tel:** 311 1091, 312 7041. **Telex:** 339134 ar dalsi. **Telecopier (Fax):** 331 6165.
Management: President – Roberto Rocca; vice presidents – Agustin O.F. Rocca, Federico A. Pefia, Hilario Testa; executive vice president – Carlos Tramutola; general manager – Javier O. Tizado; directors – Federico A. Pefia jr (financial), Alberto Valsecchi (commercial), Ricardo Lammertyn (manufacturing), Mario Giadorou (development).
Established: 1948. **Capital:** Pesos 1,513,114,000. **No. of employees:** 5,115.
Subsidiaries: Cometarsa Saic (structural fabrication, vessels etc); Tubos y Perfiles Saic (seamless tube distribution, light pole production and sales); Siderca International SA (Panama) (steel distribution and sales company); Bernal SA (service company); Comatter SA (welded tube production and sales); Propulsora Siderurgica Saic (cold rolled sheet production); Serviacero SA (steel service center).
Works:
■ Simini and Route 12, 2804 Campana, BA. ***Direct reduction plant:*** one Midrex (annual capacity 360,000 tonnes). ***Steelmaking plant:*** Three electric arc furnaces – two 75-tonne and one 60-tonne (585,000 tonnes). ***Continuous casting machines:*** One 4-strand Demag round (336,000 tonnes). ***Tube and pipe mills:*** seamless – 3-stand Innocenti Pilger ($5\frac{1}{2}$"-$9\frac{5}{8}$") (130,000 tonnes); Aetna Standard mandrel ($\frac{1}{2}$"-$5\frac{1}{2}$") (195,000 tonnes). ***Other plant:*** Cold drawn and cold Pilger finishing plant, and coupling plant.
Products: Carbon steel – seamless tubes and pipes $\frac{1}{2}$-$9\frac{5}{8}$" (1986/87 fiscal output 311,253 tonnes) of which oil country tubular goods 1.033-$9\frac{5}{8}$" (193,808 tonnes); cold drawn tubes 8-114mm (3,738 tonnes) and standard and linepipe, etc (113,707 tonnes).
Expansion programme: The expansion plan includes a new mandrell mill (seamless) $5\frac{1}{2}$-$9\frac{5}{8}$" and a new continuous casting machine.

Sidinsa – Siderurgica Integrada Saic

Head office: Avda. Corrientes 1516, Buenos Aires. **Telex:** 122918 sidsa ar.
Established: 1975.
Works:
■ Project for an integrated steelworks based on direct reduction, electric steelmaking and cold rolling.

Somisa – Sdad Mixta Siderurgia Argentina

Head office: Avda Belgrano 737, 1092 Buenos Aires. **Tel:** 30 0181-9. **Telex:** 23135 smsa ar.
Management: President – Carlos A. Magliano; vice-president – Jorge Otto Diaz;

Argentina

Somisa – Sdad Mixta Siderurgia Argentina (continued)

directors – Miguel Angel Alvarez, Juan Carlos A. Azzarri, Osvaldo J. Etchepareborda, Miguel Angel Nicodemo; sindicos – Marcelo Fernandez Romero, Jorge A. Martinsen, Rodolfo O. Puente. **Established:** 1947. **Capital:** Pesos 517,361,200. **No. of employees:** Works – 11,594. Office – 554. **Ownership:** Dirección General de Fabricaciones Militares (Estatal) (99.9%); private capital (0.1%). **Annual capacity:** Pig iron: 2,190,000 tonnes. Raw steel: 2,500,000 tonnes. Finished steel: 2,011,000 tonnes.
Works:
■ At head office. ***Coke ovens:*** Two compound batteries (89 ovens) (annual capacity 494,000 tonnes) and two compound batteries (80 ovens) (900,000 tonnes). ***Sinter plant:*** One 2.5 x 29m Dwight Lloyd (800,000 tonnes). ***Blast furnaces:*** Two – No 1 8.8m hearth dia, 1,428m cu m (912,500 tonnes) and No 2 10.3 hearth dia 2,000 cu m (1,277,500 tonnes). ***Steelmaking plant:*** Three 200-tonne basic oxygen converters (2,200,000 tonnes); electric arc; five 235-tonne (two operating) open hearth (400,000 tonnes). ***Continuous casting machines:*** Two 6-strand bloom (178 x 178 to 254 x 254mm) (1,250,000 tonnes) and one 2-strand slab (165/250 x 750/1600mm) (1,000,000). ***Rolling mills:*** slabbing and blooming reversing high lift (1168 x 2921mm) (1,300,000 tonnes); 9-stand 554mm continuous billet (1,300,000 tonnes); 2-stand 3-high and 1-stand 2-high cross country rail (673 and 813mm) (250,000 tonnes); one 11-stand fully continuous wide hot strip (1,676mm) (coil weight 11.3 tonnes) (1,600,000 tonnes – coils); one 4-stand 4-high tandem cold reduction (1626mm) (coil weight 20.4 tonnes) (700,000 tonnes); two temper/skin pass – No 1 1-stand 4-high (1,626mm) (coil weight 20.4 tonnes) and No 2 2-stand 4-high (1,219mm) (coil weight 20.4 tonnes). ***Coil coating lines:*** One Ferrostan electrolytic tinning (110,000 tonnes).
Products: Carbon steel – slabs 165 to 180mm x 500 to 1,525mm (1986 output 987,116 tonnes); blooms thickness and width – 126mm (min) 315mm (max) (588,287 tonnes); billets 60 x 60mm to 120 x 120mm (492,084 tonnes); medium channels 127/151/178/203/254mm and wide-flange beams (medium) 178/203/254mm (22,169 tonnes); heavy rails 50.88/60.34 kg/metre (14,522 tonnes); medium plates 6.5 to 25.4mm thickness (11,276 tonnes); heavy plates 25.5 to 100mm thickness (6,707 tonnes); hot rolled sheet/coil (uncoated) 1.4 to 15.5 x 560 to 1550mm (960,889 tonnes); cold rolled sheet/coil (uncoated) 0.20 to 2.5 x 560 to 1500mm (397,376 tonnes); electrolytic tinplate (single-reduced) 0.15 to 0.38mm thickness (69,520 tonnes).

Tamet – SA Talleres Metalurgicos San Martin

Head office: Diogenes Taborda 1533, 1437 Buenos Aires. **Postal address:** Casilla de Correo 84, Sucursal 37, Buenos Aires. **Telex:** 18287.
Established: 1909.
Works:
■ Juan Manuel de Rosas 4000, 3503 Puerto Vilelas, Chaco. **Blast furnaces**.
■ Ribera Sud Riachuelo, 1773 Puente La Noria, Ing. Budge, BA. **Spun grey iron pipe plant**.
Products: Foundry pig iron. Cast iron – pipes.

Establecimiento Altos Hornos Zapla

Head office: Cuidad de Palpala, Juyuy. **Telex:** 66106 ahzap ar.
Established: 1943.
Works:
■ At head office. ***Coke ovens:*** Two batteries (32 ovens). ***Sinter plant:*** Four 4.8sq m AIB pans. ***Blast furnaces:*** No III 3 4.2m hearth dia, 194.7 cu m; No IV 4 4.1m hearth dia, 191.1 cu m; No V 5m hearth dia, 279.2 cu m. ***Steelmaking plant:*** Two 25-tonne OBM basic oxygen converters; two electric arc furnaces. ***Refining plant:*** One vacuum degassing unit. ***Rolling mills:*** One 3-high blooming (2,000 x 730mm) mill; one 3-high cross-country medium section (1,400/1,200 x 475mm) mill; one wire rod. ***Other***

Establecimiento Altos Hornos Zapla (continued)

plant: Ferro-alloy plant. Foundry.
Products: Steelmaking pig iron. Carbon steel – ingots, billets, wire rod, reinforcing bars, round bars, square bars, flats, light angles, medium angles, medium tees, medium joists, medium channels. Stainless steel – ingots, billets. Alloy steel (excl stainless) – ingots, billets.

Australia

Australian Tube Mills Pty Ltd (ATM)

Head office: Tube Street, North Sunshine, Vic 3020. **Tel:** 311 6666. **Telex:** 32617. **Telecopier (Fax):** Austube Melborne.
Management: H.R. Fowler (general manager). **Established:** 1971. **Capital:** A$4,400,000. **No. of employees:** Works – 140.
Works:
■ At head office. ***Tube and pipe mills:*** welded – One ERW (annual capacity 50,000 tonnes). ***Other plant:*** Pipe galvanizing plant.
Products: Carbon steel – longitudinal weld tubes and pipes commercial grade to AS 1074, BS 1387, $\frac{3}{4}$-6" nb; oil country tubular goods – line pipe to (API) 15L grades A & B, ASTM A53 grades A & B, 1.906-6.625" od; hollow sections to ASA 177 from $\frac{3}{4}$" to 5" sq and up to 6 x 4" rectangular, wall thickness 0.072-0.250"; special tube for automotive axles, etc.

The Broken Hill Pty Co Ltd

Head office: 140 William Street, Melbourne, Vic 3000. **Postal address:** GPO Box 86A, Melbourne, Vic 3001. **Tel:** 03 609 3333. **Telex:** 30408. **Cables:** Hematite. **Telecopier (Fax):** 03 609 3015.
Management: Senior executives: D.M. Rice (chief executive officer), P.J. Laver (general manager), R.A. Nevison (general manager, engineering & projects), J.E.C. McKenzie (general manager, production), G.S. Boileau (group manager, marketing), W.J. Moss (group manager, planning), M. Pirola (group manager, information analysis), R.A.N McCall um (group manager, human resources), A.J. Castleman (general manager finance), P.E. Jacobs (manager, accounting), C.W. Wirth (manager, financial evaluation & development), K.D. Thurgood (manager, taxation). **Established:** 1885. **Capital:** A$1,400,000. **No. of employees:** 61,349. **Subsidiaries:** Australia including – Associated Airlines Pty Ltd (executive air transport), Australian Industrial Refractories Ltd (refractories plants), Australian Iron & Steel Pty Ltd (iron and steel production and coal mining), Queensland BHP Steel Ltd (construction of rolling mills), AWI Holdings Pty Ltd (holding company), BHP Finance Ltd (finance), BHP Minerals Holdings Pty Ltd (investments in mining and allied projects), Mt Newman Ore Sales Pty Ltd (sales representation), Mt Newman Iron Ore Co Ltd (investment), Pibara Iron Ltd (Mining & Marketing), Hunter Valley Aluminium Pty Ltd), BHP Nominees Pty Ltd (investments and overseas representation), Coca-Cola Bottlers (soft drink bottler), North Queensland Pty Ltd (soft drink deliveries), BHP Petroleum Pty Ltd (hydrocarbons exploration), JLA Holdings Ltd (manufacturing and distribution of steel products, steel office furniture) Mt. Newman Mining Co Pt Ltd (management), Orbital Engine Co Pty Ltd (engine research), Queensland Coal Mining Co Ltd (inactive), Tasmanian Electro Metallurgical Co Pty Ltd (alloy production), The Port Waratah Stevedoring Co Pty Ltd, (stevedoring services), Groote Eyland Mining Co Pty Ltd (manganese ore mining).
Overseas including – USA – BHP Finance (USA) Inc (finance), RAL Pacific Holdings, Rheem Pacific Packaging Corp (metal containers), BHP Holdings (USA) Inc (holding company), March Pipe Co Inc (pipe distribution), Fathom Management Corp (ship management), Mt Newman Iron Ore Sales Inc (sales representation), Mt Newman Iron Ore Sales (Europe) Inc (sales representation), Utah International Inc (mining), San Juan Coal Co (mining), Utah Exploration Inc, Utah Marcona Corp (holding company),

Australia

The Broken Hill Pty Co Ltd (continued)

Marcona Conveyor Corp (shipping), Utah Mines Ltd, Utah Pacific Inc (minerals exploration), Utah Queensland Coal Ltd (coal mining), Utah Sulawes Inc (metal exploration). Indonesia – PT Lamipak Primula Indonesia (plastic tubes), PT Rheem Indonesia (metal containers). Hong Kong – BHP International Holdings Ltd (investment and overseas represent ation), John Lysaght Hong Kong Ltd (trading), Mt Newman Iron Ore Sales (Asia) Ltd. New Zealand – BHP New Zealand Ltd (representative and steel marketing). Singapore – Resources Insurances Pte Ltd (insurance underwriters). UK – BHP Financial Services Pty Ltd (investment and finance), Utah Minerals (UK) Ltd (minerals marketing). Brazil – Mineracao Marex Ltda (holding company), Participacoes Ltda (holding company). Panama – John Lysaght Int Holdings SA (investments). Malaysia – John Lysaght (Malaysia) (rollformed steel). Papua New Guinea – John Lysaght (Papua New Guinea) Pty Ltd (rollformed steel). New Caledonia – Sté Industrielle Pour La Co-operation Technique (rollformed steel).

Slab & Plate Products Division
Works:

■ Port Kembla Steelworks, PO Box 1854, Wollongong, NSW 2500. (Tel: 042 74 5011.Telex: 29083. Fax: 042 751 235). Divisional general manager: J.K. Ellis. General manager: G.J. Greenlees. ***Coke ovens:*** No 3A A battery (48 ovens) (annual capacity 280,000 tonnes); No 4 battery (66 ovens) (501,000 tonnes); No 5 battery (66 ovens) (501,000 tonnes) and No 6 battery (101 ovens) (769,000 tonnes). ***Sinter plant:*** Two Dwight Lloyd (Lurgi) 208 sq m strand (1,800,000 tonnes) and three Dwight Lloyd (Lurgi) 420 sq m strand (4,500,000 tonnes). ***Blast furnaces:*** No 2 7.60m hearth dia, 1,148 cu m (880,000 tonnes); No 3 8.53m hearth dia, 1,509 cu m (980,000 tonnes); No 4 9.07m hearth dia, 1,815 cu m (1,060,000 tonnes) and No 5 12.15m hearth dia, 3,045 cu m (2,080,000 tonnes). ***Steelmaking plant:*** Three 275-tonne basic oxygen (4,200,000 tonnes); one 50-tonne electric arc. ***Refining plant:*** One 50-tonne AOD; one 275-tonne RH vacuum degassing; one ladle injection station; two ladle stirring, alloying and re-heating stations (CAS-OB). ***Continuous casting machines:*** Two slab – one 2-strand Demag 10.5m radius (850-1,750 x 212-230mm (1,600,000 tonnes) and one 2-strand SMS/Concast (1-strand 2 x 750-950 x 230mm; 1 strand 1,100-2,200 x 200-300mm) (2,500,000 tonnes). ***Rolling mills:*** One universal slabbing (1,220 x 2,285mm (horizontal), 1,015 x 2,240mm (vertical)); one heavy plate – one 4-high 1,090 x 3,500mm roughing stand and one 4-high 1,090 x 3,500mm finishing stand; one wide hot strip – 4-high 1,200 x 1,800mm reversing with edger, Coilbox, two 4-high 676 x 1,676mm and four 4-high 676 x 2,120mm; two cold reduction – one 5-stand 4-high 1,370 x 600 x 1,220 (470,000 tonnes) and 1-stand 4-high 1,380 x 550 x 1,220mm reversing; temper/skin pass – No 1 2-stand 4-high 1,240 x 470 x 1,120mm and No 2 2-stand 4-high 1,350 x 550 x 1,350mm. ***Coil coating lines:*** Two electrolytic tinning – No 1 Halogen (250,000 tonnes) and No 2 Halogen (250,000 tonnes). ***Other plant:*** Continuous pickling line (max strip width 1,600m , max strip thickness 6.4mm(530,000 tonnes); continuous annealing line (tempers T4 and T5) (220,000 tonnes); batch annealing bases (T1, T2 and T3 tempers) (280,000tonnes); electrolytic cleaning line (240,000 tonnes); foundries casting ingot mould and stools, slag pots, etc; plain carbon and alloy steel castings up to 50 tonnes; copper and copper alloy castings.

Rod & Bar Products Division
Works:

■ Newcastle Steelworks, PO Box 196B, Newcastle, NSW 2300. (Tel: 049 69 0411. Telex: 28006. Fax: 049 69 0416). Divisional general manager: W.Farrands. ***Coke ovens:*** No 5 battery (78 ovens) (capacity 2,208 tonnes per day). ***Sinter plant:*** Lurgi 4m wide 208 sq m grate area (annual capacity 2,000,000 tonnes). ***Blast furnaces:*** No 3 7.32m hearth dia, 1,075 cu m (675,000 tonnes) and No 4 7.31m hearth dia, 1,268 vu m (850,000 tonnes). ***Steelmaking plant:*** Two 200-tonne basic oxygen converter (1,900,000 tonnes). ***Rolling mills:*** No 2 2-high reversing, 1,250mm blooming (1,800,000 tonnes); continuous billet – one 4-stand 800mm vertical/horizontal roughing, 4-stand 660mm vertical/horizontal finishing; one medium section and bar (No 2 Merchant mill) with 10 primary stands and two parallel 4-stand finishing trains (280/330mm) (500,000 tonnes); one light section and bar (No 1 Merchant mill)

The Broken Hill Pty Co Ltd (continued)

with 13-stand, 250/330mm rolling flats, angles, Y's and channels (300,000 tonnes); one wire rod with 4-stand, thirteen primary stands and four parallel 10-stand No-twist finishing trains, producing rod 5.5-12.5mm dia (670,000 tonnes); 4-strand 13 primary stands and four parallel 10-stand No-Twist finishing trains, producing rod 5.5-12.5mm dia (670,000 tonnes); hot strip and hoop with 460mm 2-high continuous, 10 horizontal and vertical stands, producing skelp 100-457mm wide narrow strip and flats (500,000 tonnes); two cold strip – one 4-high 460mm reversing breakdown and one 200mm reversing Steckel; one temper/skin pass – 4-high 460mm reversing. ***Other plant:*** Steel foundry with two 50-tonne mains frequency coreless induction furnaces and one 6-tonne VIP induction furnace producing high-alloy iron and steel for rolls and castings; one vertical centrifugal casting machine for rollmaking; strapping line (18,000 tonnes/year).

■ Kwinana Works, PO Box 160, Kwinana, WA 6167.(Tel: 09 410 1111. Telex: 92074). Works manager: M.F. Baker. ***Sinter plant:*** Dwight Lloyd (Lurgi) 2.5 metres wide grate area, 100 m sq (annual capacity 1,000,000 tonnes) (out of service since April 1982). ***Blast furnaces:*** 7.8m hearth dia, 1,385 cu m (750,000 tonnes) (out of service since April 1982). ***Rolling mills:*** One light section comprising 1-stand 3-high 410mm breakdown and 8-stand 310mm continuous-rolling angles, flats, Y', channels and fence posts (250,000 tonnes).

■ Geelong Rod Mill, PO Box 803, Geelong, Vic 3220. (Tel: 052 78 9333. Telex: 30772. Fax: 052 78 3525). Works superintendent: Geoff Miller. ***Rolling mills:*** One wire rod comprising Kocks 1-strand with 6-stand roughing train, 10-stand intermediate train and 10-stand finishing train producing 5.5-14mm dia coiled rod and rebar at a max finishing speed of 60m/sec (annual capacity 220,000 tonnes).

■ Queensland BHP Steel Ltd, PO Box 300, Acacia Ridge, Brisbane, Qld 4110. (Tel: 07 273 5500). Works manager: Grahame Milton. ***Rolling mills:*** One 15-stand continuous bar producing rounds, Tempcore reinforcing bar and flats (commissioned mid 1987) (annual capacity 225,000 tonnes).

Long Product Division
Works:

■ PO Box 21, Whyalla, SA 5600. Tel: 086 45 7000, 086 45 9444. (Telex: 80270. Fax: 086 45 3017). General manager: I. Bartholomew. ***Coke ovens:*** No 1 (72 ovens) and No 2 (36 ovens) (combined annual capacity 900,000 tonnes). ***Blast furnaces:*** No 2 8.6m hearth dia, 1,543 cu m (1,000,000 tonnes). ***Steelmaking plant:*** Two 125-tonne basic oxygen (1,200,000 tonnes); one 19-tonne electric arc furnace (not in use). ***Rolling mills:*** 2-high reversing 1,000 x 2,450mm blooming (1,240,000 tonnes); two structural/ heavy section – 4-stand 2-high producing rails, steel sleepers, sleeper plates and billets. Universal mill – 1,270mm roughing and 1,140mm finishing, producing universal beams, columns, channels, bearing piles and heavy rails. ***Other plant:*** Direct metal foundry producing ingot moulds and shots.

Coated Products Division GPO Box 196, Sydney, NSW 2001. ***Tel:*** 02 239 4444. ***Telex:*** 20355. ***Telecopier (Fax):*** 02 239 4458.
Works:

■ Westernport. Works manager: I.C. Thomas. ***Rolling mills:*** One 2,050mm semi-continuous wide hot strip comprising 1-stand 4-high reversing rougher, Coilbox, 5-stand 4-high finishing train; One Hitachi cold reduction; one Hitachi temper/skin pass. ***Coil coating lines:*** One 1,500mm Wean hot-dip galvanizing; Three zinc-aluminium one 1,220mm Wean continuous Zincalume, one 1,220, and one 1,600mm.

■ Port Kembla. Works manager: J.M. O'Donnell. ***Rolling mills:*** Two cold reduction – one 5-stand 1,330mm Davy United and one 1,830mm Davy reversing one 1,420mm Sueco temper/skin pass. ***Coil coating lines:*** Two Bethlehem/Mesta continuous hot-dip galvanizing; one Bliss electro-galvanizing; one Bethlehem/Mesta Zincalume zinc-aluminium; two colour coating.

■ Unanderra Works, Five Islands Road, Unanderrra, NW 2526. Works manager: M. Chapman. ***Rolling mills:*** Two Sendzimir cold reduction – one ZR22 1,067mm and one

Australia

The Broken Hill Pty Co Ltd (continued)

ZR21AA 1,677mm; 2-high 1,575mm temper/skin pass. ***Other plant:*** Two softening and descaling lines (1,575mm and 1,067mm); one bright annealing line (1,280mm) and slitting, cutting-to-length and polishing facilities.
■ Alexandria, Sydney, NSW. Stainless steel tube plant.

Products: Steelmaking pig iron; foundry pig iron. Carbon steel – ingots; slabs; blooms; billets; hot rolled coil (for re-rolling); wire rod; reinforcing bars; round bars; square bars; flats; hexagons; light angles; light channels; special light sections Ys; medium angles; medium joists; medium channels; wide-flange beams (medium); universals (medium); heavy angles; heavy joists; heavy channels; wide-flange beams (heavy); universals (heavy); heavy rails; light rails; rail accessories (fishplates, sleepers, sleeper plates); cold roll-formed sections (from strip); sheet piling and universal bearing piles, piling beams; pipe piling octaganol; hot rolled hoop and strip (uncoated); skelp (tube strip); cold rolled hoop and strip (uncoated); medium plates; heavy plates; universal plates; floor plates; hot rolled sheet/coil (uncoated); cold rolled sheet/coil (uncoated); hot-dip galvanized sheet/coil; electro-galvanized sheet/coil; enamelling grade sheet/coil; terne-plate (lead coated) sheet/coil; colour coated sheet/coil; zinc-aluminium alloy coated sheet/coil; electrical grain oriented silicon sheet/coil; electrical non-oriented silicon sheet/coil; electrical non-oriented non-silicon sheet/coil; blackplate; electrolytic tinplate (single-reduced); electrolytic tinplate (double-reduced); bright wire; black annealed wire; galvanized wire (plain); galvanized wire (barbed). Stainless steel – ingots; slabs; blooms; billets; medium plates; heavy plates; Alloy steel (excl stainless) – ingots; slabs; blooms; billets; wire rod.

Bunge Industrial Steels Pty Ltd

Head office: Resolution Drive, Unanderra, NSW 2526. **Postal address:** PO Box 231. **Tel:** 042 714944. **Telex:** 29227. **Telecopier (Fax):** 042 718470.
Management: Managers – R. Wade (general), J. Croll (technical), D. Muggleton (administration), J. Mannix (sales), J. Ryall (quality), L. McBride (operations). **Established:** 1979. **No. of employees:** Works – 25. Office – 9. **Ownership:** Bunge Industrial Ltd. **Annual capacity:** Finished steel: 60,000 tonnes (plate).
Works:
■ At head office. Heat treatment facilities capable of normalising, quenching and tempering plate in thickness 5-100mm, width to 3,100mm (maximum annual capacity 60,000 tonnes).
Products: Carbon steel – medium plates and heavy plates (also low alloy and alloy grades). Alloy steel (excl stainless) – medium plates, heavy plates.
Brands: Bisalloy – high-strength and abrasion-resistant quenched and tempered steel plates. HY-80/100 grades to military specifications-16216J.
Sales offices/agents: At head office. BMP Trading Inc, Commerce, CA, USA, (Tel 213 725 1177); Balfour Guthrie & Co, Vancouver, BC, Canada, (Tel 604 685 0211).
Additional information: As-rolled plates are supplied to the company by the nearby Port Kembla works of BHP.

Commonwealth Steel Co Ltd

Head office: Maud Street, Waratah, NSW 2298. **Postal address:** PO Box 14. **Tel:** 049 680411. **Telex:** 28115. **Cables:** Comsteel Newcastle. **Telecopier (Fax):** 049 676815.
Management: Chief executive – E. Rees; managers – W. Ainsworth (manufacturing), J. Pattenden (group marketing), R. Farley (technical service); group controller – M. Payne. **Established:** 1918. **Capital:** Authorised – A$ 10,000,000, Issued – A$ 6,000,000. **No. of employees:** Works – 768. Office – 268. **Ownership:** Australian National Industries Ltd (100%). **Annual capacity:** Raw steel: 130,000 tonnes. Finished steel: 100,000 tonnes.
Works:
■ At head office. ***Steelmaking plant:*** one 50-tonne electric arc furnace (annual capacity 130,000 tonnes). ***Refining plant:*** One 1,100mm ESR (3,000 tonnes); one

Commonwealth Steel Co Ltd (continued)

50-tonne ladle furnace (130,000 tonnes); one DH vacuum degassing unit and one LM stream degasser (130,000 tonnes); one 30-tonne VOD (5,000 tonnes). ***Continuous casting machines:*** 2-strand Technica-Guss horizontal billet (90-205mm) (70,000 tonnes). ***Rolling mills:*** One 2-high reversing blooming (750mm) (10,000 tonnes); one 3-high breakdown billet (500mm) (10,000 tonnes); bar – 5-stand parallel open train with independent drive for finishing stand (300mm) (10,000 tonnes). ***Other plant:*** Plant producing assembled wheel sets, locomotive wheels, carriage and wagon wheels, tyres and axles, all types of open-die forgings, large castings over 20 tonnes, hardened steel rolls, grinding balls and rods, cast steel abrasives, railway castings (couplers and draft gear, etc).
Products: Carbon steel – ingots 2-6-tonnes (1985-86 output 500 tonnes); billets 90 x 90-50 x 150mm (200 tonnes); round bars 100-215mm (5,000 tonnes). Stainless steel – ingots 2-tonnes (output 100 tonnes); blooms 200 x 200mm & 150 x 150mm (50 tonnes); billets 50 x 50mm-125 x 125mm (50 tonnes); round bars 20-200mm (1,000 tonnes); bright (cold finished) bars 20-70mm. Alloy steel (excl stainless) – ingots 2-6-tonnes (3,000 tonnes); blooms 150 and 200mm (300 tonnes); billets 50-125mm (200 tonnes); round bars 20-200mm (10,000 tonnes); bright (cold finished) bars 20-70mm.
Brands: Comsteel – Crucible Steel's range of tool steel brands.
Sales offices/agents: Sydney – 75 George Street, Parramatta, NSW; Melbourne – 780 Elizabeth Street, Melbourne, Vic; Brisbane – 17 Hayling Street, Salisbury, Qld; Adelaide – Cromwell Road, Kilburn, SA; Perth – 46-48 Kings Park Road, West Perth, WA.

Email Tube Division, Email Ltd

Head office: 31 Ashmore Street, Alexandria, NSW 2015. **Postal address:** PO Box 124, Alexandria, NSW 2015.
Established: 1961. **Ownership:** Email Ltd.
Works:
■ Alexandria. ***Tube and pipe mills:*** seamless. Welded.
Products: Carbon steel – seamless tubes and pipes, longitudinal weld tubes and pipes, cold drawn tubes and pipes. Stainless steel – seamless tubes and pipes, longitudinal weld tubes and pipes, cold drawn tubes and pipes, precision tubes and pipes. Alloy steel (excl stainless) – seamless tubes and pipes, longitudinal weld tubes and pipes, cold drawn tubes and pipes, precision tubes and pipes. Incoloy, titanium and cupro-nickel tube. Element sheathing. Copper-coated mild steel tube.

Hills Industries Ltd

Head office: 944-956 South Road, Edwardstown, SA 5039. **Tel:** 08 297 3888. **Telex:** 82110.
Established: 1955.
Works:
■ 29 Morrow Road, O'Sullivan Beach, SA 5166. (Tel: 08 382 2022). Works directors: L.K. Venus. ***Tube and pipe mills:*** welded – RF
Products: Carbon steel – longitudinal weld tubes and pipes.

Martin Bright Steels Ltd

Head office: Clifford Road, Campbellfields, Vic. **Telex:** 31934 marson aa.
Management: Tom Mulcairn (general manager). **Established:** 1923. **Ownership:** Atlas Steels Ltd (Ultimate holding company – Rio Algom Ltd, Canada (RTZ). **Subsidiaries:** Advance Steel Co, Hornsby, NSW (producer of chrome plated bar); Castlebright Steels, Melbourne (bright bar producer), Gilbert Lodge (distribution).
Associates: Atlas Steels of Australia (wire drawing).
Works:
■ Campbellfield, Vic. Bar drawing, precision grinding, turning and polishing.
Products: Carbon steel – bright (cold finished) bars, cold drawn tubes and pipes.

Australia

Martin Bright Steels Ltd (continued)

Stainless steel – bright (cold finished) bars, cold drawn tubes and pipes. Alloy steel (excl stainless) – bright (cold finished) bars, cold drawn tubes and pipes.

Palmer Tube Mills Ltd

Head office: 146 Ingram Road, Acacia Ridge, Qld. **Postal address:** PO Box 246, Sunnybank, Qld 4109. **Tel:** 61 7 3455566. **Telex:** 42056. **Telecopier (Fax):** 61 7 345 5318.
Management: Directors – Ross L. Palmer (chairman and managing), Terrence G. O'Haire (finance), Frederick R. Gaydon (marketing). **Established:** 1976. **Capital:** A$ 128,000,000. **No. of employees:** Works – 100. Office – 200. **Subsidiaries:** Welded Tube Co of America, Chicago, IL (100%) (producers of welded steel tubing). **Annual capacity:** Finished steel: 400,000 tonnes.
Works:
■ Palmer Tubes Mill (Australia) Pty Ltd. Works director: Graham Miller. ***Tube and pipe mills:*** welded – Two Kusakabe 15 to 100 nb pipe, 13 x 13 nb to 100 x 100 nb RHS (annual capacity 90,000 tonnes).
■ Palmer Tubetek (Tel: 7 343 4366). Works director: Mike Brook. ***Tube and pipe mills:*** welded – One Abbey Etna 20 to 90 nb tube (annual capacity 25,000 tonnes), one Mackay 15 to 45 nb tube (5,000 tonnes).
Products: Carbon steel – longitudinal weld tubes and pipes, precision tubes and pipes, hollow sections. Alloy steel (excl stainless) – longitudinal weld tubes and pipes, precision tubes and pipes ½ through to 3½" round, square, and rectangle hollow sections ½ through to 16" square and rectangle, ½ through to 20" dia round. (output in 1986/87 including Palmer Tubetek and Welded Tube Co of America – precision 12,000 tonnes, hollow section 220,000 tonnes).
Brands: Painted tubular products – Tru-Blue (tubular hollow sections); Red-Roo (round pipe), Greens-Tuf 450 (tubular hollow sections); Tru-Blu II (round pipe).
Sales offices/agents: Palmer Tube Mills (Aust) Pty Ltd, 70B Harley Crescent, Condell Park, NSW 2200 (PO Box 880, Bankstown, NSW 2200) (Tel: 02 708 4688. Telex: 027081283); PO Box 472, 55 St Vincents Street, Port Adelaide, SA 5015 (Tel: 08 341 1412. Fax: 08 341 1499); PO Box 54, 8 Dynon Road, Flemington, Vic 3031 (Tel: 03 376 4480. Fax: 03 376 7498); Suite 2B, 16 Kearns Crescent, Applecross, WA 6153 (Tel: 09 364 7840. Fax: 09 364 8103).

Rydal Steel – Division of LNC Industrial Products Pty Ltd

Head office: 34 South Street, Rydalmere, NSW. **Postal address:** PO Box 29, Lidcombe, NSW 2141. **Telex:** 73963 incho aa.
Established: 1925.
Works:
■ At head office. ***Rolling mills:*** Three bar
Products: Carbon steel – round bars, square bars, flats, hexagons, special light sections.

Sandvik Australia Pty Ltd

Head office: Cnr Percival and Warren Roads, Smithfield, NSW. **Telex:** 121268 sandvik aa.
Ownership: Sandvik AB, Sweden.
Works:
■ ***Tube and pipe mills:*** welded .
Products: Stainless steel – longitudinal weld tubes and pipes.

Siddons Steel Mills

Head office: 1 Culverlands Street, Heidelberg West, Vic 3081. **Telex:** 34358 ssm aa.
Established: 1955. **Ownership:** Wholly-owned division of the Siddons Industries group of companies.
Works:
■ McEwens Road, Heidelberg West. ***Rolling mills:*** Two medium section – 4-stand 3-high 18" cross-country (annual capacity 15,000 tonnes) and 5-stand 3-high 14" cross-country (5,000 tonnes).
Products: Carbon steel – special light sections (truck rim sections), light rails up to 15kg/m, rail accessories. Cutting edges for earthmoving up to 250mm wide, light merchant sections.

Smorgon Steel

Head office: 105 Dohertys Road, Laverton North, Vic 3026. **Tel:** 61 3 3692311. **Telex:** 151102. **Telecopier (Fax):** 61 3 3696829.
Management: Directors – Charlie Holckner, David Holckner, Robert Smorgon. **Established:** 1982. **No. of employees:** 370. **Ownership:** Humes Ltd. **Annual capacity:** Raw steel: 330,000 tonnes. Finished steel: 310,000 tonnes.
Works:
■ At head office. ***Steelmaking plant:*** 80-tonne Fuchs electric arc furnace (5.5 x 6.4m oval) (annual capacity 330,000 tonnes). ***Refining plant:*** 80-tonne Fuchs ladle furnace (330,000 tonnes). ***Continuous casting machines:*** 3-strand Rokop billet (150 x 150mm) (330,000 tonnes). ***Rolling mills:*** bar and rod – 28-stand continuous in-line combination (max coil weight 1,300kg) (310,000 tonnes).
Products: Carbon steel – billets 150 x 150mm (1986 output 211,217 tonnes); hot rolled coil (for re-rolling); wire rod 5.5mm upwards; reinforcing bars 12-36mm, round bars 12-65mm, flats 32-100mm wide, and light angles 50 x 50mm to 100 x 100mm and some unequal (combined output 184,576 tonnes).
Brands: Welbend (reinforcing bar).
Expansion programme: New scrap handling facilities, scrap preheater, 80-tonne arc furnace, ladle furnace and rod outlet (1987/88).

Tubemakers of Australia Ltd

Head office: 1 York Street, Sydney, NSW 2000. **Postal address:** GPO Box 536, Sydney, NSW 2001. **Tel:** 239 6666. **Telex:** 22849. **Telecopier (Fax):** 251 3042.
Management: Chairman – J.A.L. Hooke; managing director – J.M. Griggs; executive general managers – K.T. Cocks (subsidiaries), A.B. Daniels (steel pipe division), R.F. Griffiths (personnel), R.R. Johnson (finance and corporate affairs), K.F. Kelly (BTM division), C.R. Stubbs (merchandising divisions). **Established:** 1946. **Capital:** A$ 350,444,000. **No. of employees:** Works – 3,387. Office – 2,197. **Ownership:** BHP Petroleum Pty Ltd (49.75%). **Subsidiaries:** Wholly-owned – William Adams Tractors Pty Ltd, Amcast Engineering Co Pty Ltd, Coastline Foundry (Qld) Pty Ltd, P.J. O'Connor & Co Pty Ltd, Steel Mains Pty Ltd, Voca Communications Pty Ltd; 60% – Bundy Tubing (Aust) Pty Ltd. **Associates:** Vinidex Tubemakers Pty Ltd (37½%) (manufacture and marketing of PVC pipe and fittings).

Steel Pipe Division
Works:
■ Newcastle, NSW and Kembla Grange, NSW.

BTM Division
Works:
■ Kilburn, SA; Sunshine, Vic and Granville, NSW.

Products: Carbon steel – seamless tubes and pipes longitudinal weld tubes and pipes large-diameter tubes and pipes galvanized tubes and pipes plastic coated tubes and pipes cold drawn tubes and pipes precision tubes and pipes hollow sections

® INDUSTRIEANLAGENBAU GESELLSCHAFT M.B.H. & CO. KG

TEL. 0043/222 65 89 17-0
65 89 18
65 89 19
65 89 10
TELEFAX 0043-222-65891731
65891831
TELEX 136046
112591

A-1040 WIEN, WOHLLEBENGASSE 16 AUSTRIA – EUROPA

PRODUKTE & LEISTUNGEN

ANLAGEN FÜR

- COIL-VERPACKUNG
- ENERGIERÜCKGEWINNUNG
- GUSS BRECHKEILE
- HOCHMODERNE ADJUSTAGEN
- ROHSTOFFVORBEREITUNG
- SCHLEUDERGUSS
- ROHSTOFFRÜCKGEWINNUNG
- SFEROKLAV (SPHÄROGUSS)
- UMWELTSCHUTZ
- WASSERAUFBEREITUNG
- ZINK- & ZINNRÜCKGEWINNUNG

EINRICHTUNGEN FÜR
STAHL- UND WALZWERKE,
GIESSEREIEN

PROJEKTE FÜR

- GASANLAGEN
- FEUERUNGSANLAGEN

LIZENZEN

PRODUCTS & SERVICES

PLANTS FOR

- COIL PACKING
- ENERGY RECOVERY
- FEEDERHEAD BREAKER
- TOPMODERN ADJUSTING
- RAW MATERIALPREPARATION
- ROLLER CASTING
- RAW MATERIAL RECYCLING
- SFEROKLAV (SPHEROLITIC CAST IRON)
- POLLUTION CONTROL
- WATER TREATMENT
- ZINC & TIN RECOVERY

EQUIPMENT FOR
STEELWORKS, ROLLING MILLS
FOUNDRIES

PROJECTS FOR

- GAS PLANTS
- HEATING PLANTS

LICENCES

SFEROKLAV

FEEDERHEADBREAKER

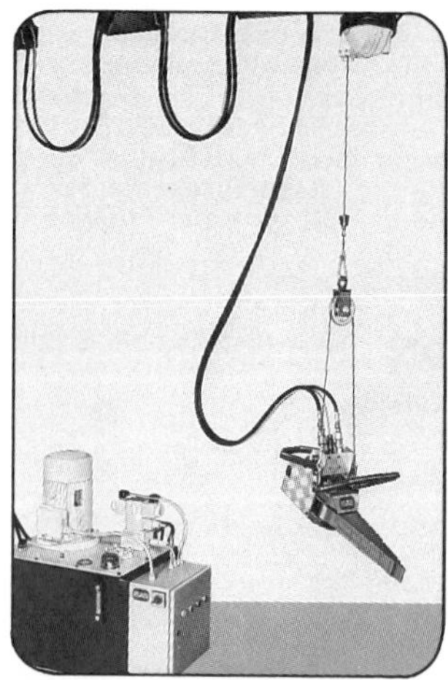

Austria

Eisenwerk Breitenfeld Ges.mbH Nachf KG

Head office: A 8662 Mitterdorf, Stmk. **Tel:** 03858 2511. **Telex:** 036667. **Cables:** Breitenfeld Mitterdorf Murztal.
Management: Gottfried Pengg. **Established:** 1942. **No. of employees:** Works – 449. Office – 67. **Ownership:** Joh Pengg Draht- und Walzwerke. **Annual capacity:** Raw steel: 80,000 tonnes.
Works:
■ At head office. Works director: Otto Gross. ***Steelmaking plant:*** One 30-tonne electric arc furnace (annual capacity 80,000 tonnes). ***Continuous casting machines:*** One 4-strand billet (own construction). ***Other plant:*** Forging plant, plant for insulated cables.
Products: Carbon steel – ingots (1986 output 2,800 tonnes); billets 80, 90, 100, 115 and 140mm sq (80,000 tonnes). Forgings (blanks and rings) (623 tonnes), bars (3,117 tonnes), insulated electric cables (3,557 tonnes).
Brands: Comet Stahl.
Sales offices/agents: At head office, also at Tonfabrikgasse 4, 1210 Vienna and Unionstrasse 73 , 4020 Linz.

Stahl- und Walzwerk Marienhütte GmbH

Head office: Südbahnstrasse 11, A 8021 Graz. **Tel:** 0316 52 1 97. **Telex:** 03 11828.
Management: Helmut Grossschädl, Herifried Hornich. **Established:** 1960. **Capital:** Sch 15,200,000. **No. of employees:** Works – 240. Office – 30. **Annual capacity:** Raw steel: 120,000 tonnes. Finished steel: 120,000 tonnes.
Works:
■ At head office. ***Steelmaking plant:*** One 25-tonne electric arc furnace (annual capacity 120,000 tonnes). ***Refining plant:*** One 25-tonne ladle furnace (120,000 tonne). ***Continuous casting machines:*** One 2-strand Marienhütte billet (120 x 120mm) (120,000 tonnes); one Korf Engineering rotary (110 x 110 and 130 x 130mm) (120,000 tonnes). ***Rolling mills:*** One 3-high roughing (420mm) bar and wire rod 15-stand 2-high (320mm/260mm) (coil weight 350kg) (80,000 tonnes).
Products: Carbon steel – billets 120 x 120mm (output 120,000 tonnes); wire rod 5.5 to 14mm dia (5,000 tonnes); reinforcing bars (8 to 30mm dia) (70,000 tonnes); round bars (8 to 30mm dia) (4,500 tonnes); square bars (12 to 24mm sq) (500 tonnes).

Martin Miller KG

Head office: A 3133 Traismauer. **Tel:** 027 83 302. **Telex:** 15566 miller.
Products: Carbon steel – cold rolled hoop and strip (uncoated). Sawblabes, machine knives.

Joh. Pengg, Draht- & Walzwerke

Head office: A 8621 Thörl, Stmk. **Tel:** 03861 2302. **Telex:** 36668. **Telecopier (Fax):** 03861 2318.
Subsidiaries: Eisenwerk Breitenfeld Ges.mbH, Mitterdorf (steel production); Stahlcord GmbH & Co KG, Fürstenfeld (tyre cording).
Works:
■ At head office. ***Rolling mills:*** wire rod mill; wire drawing plant.
Products: Carbon steel – wire rod. Stainless steel – wire rod, wire. Alloy steel (excl stainless) – wire (low-alloy).

Vereinigte Edelstahlwerke AG

Head office: Elisabethstrasse 12, Vienna. **Tel:** 0222 58833. **Telex:** 111683 rewe a, 111109 vewe a. **Telecopier (Fax):** 0222 564246.
Management: Executive board – Alexander Martinowsky, Ernest Bachner, Karl Krobath, Hans Wehsely. **Established:** 1975. **No. of employees:** 9,600.
Ownership: Voest-Alpine, Austria (100%). **Annual capacity:** Raw steel: 175,000 tonnes.
Works:
■ Mariazellerstrasse 25, Postfach 96, A 8605 Kapfenberg (Tel: 03862 20/0. Telex: 36612 vek a). ***Steelmaking plant:*** Three electric arc furnaces – 50-tonne, 35-tonne, and 15-tonne (annual capacity 195,000 tonnes). ***Refining plant:*** Four ESR units – 1,000/550/550/1,150mm dia (32-tonne) (10,000 tonnes); one 50-tonne ladle furnace furnace; two vacuum degassing units – 50-tonne and 40-tonne; one 800mm dia (12-tonne) VAR unit (1,000 tonnes); 40-tonne VOD unit; one vacuum melting unit with permanent rotating electrode "Rotel" (600 tonnes). ***Rolling mills:*** 1-stand 2-high 880 x 2,200mm reversing blooming (100,000 tonnes); 3-stand 2-high 720 x 2,000mm dia reversing heavy section (80,000 tonnes); medium section, light section, bar, and wire rod comprising 15+5 stands (being rebuilt) (45,000 tonnes). ***Other plant:*** Open die forge; closed die forge; bright drawing shop; heat treatment shop, etc.
■ Bleckmanngasse 10, Postfach 28, A 8680 Mürzzuschlag (Tel: 03852 2060. Telex: 36676 vem a). ***Rolling mills:*** heavy plate, wide hot strip and sheet comprising 3-high 800/650/800 x 2,000mm dia, 2-high 800 x 1,350/1,600/2,000mm dia; 2-high 800 x 1,250mm dia (high speed steel) only (combined annual capacity 20,000 tonnes).
■ Hauptstrasse 2, Postfach 31, A 2630 Ternitz (Tel: 02630 8351. Telex: 16660 vet a). ***Tube and pipe mills:*** seamless – 33MN extrustion press, 14-stand reducing mill, cold Pilger mill (combined annual capacity 12,000 tonnes).
Products: Stainless steel – round bars 6 to 200mm dia; square bars 16 to 130mm sq;

Austria

Vereinigte Edelstahlwerke AG (continued)

flats 25 to 240mm x 6 to 100mm; hexagons 17 to 60mm; seamless tubes and pipes 50 to 250mm dia (output 12,000 tonnes). Alloy steel (excl stainless) – wire rod 6 to 20mm dia (10,000 tonnes); round bars 6 to 200mm dia, square bars 16 to 130mm sq, flats 25 to 240mm x 6 to 100mm, and hexagons 17 to 60mm (combined output, 60,000 tonnes); medium plates 1 to 60mm thick (20,000 tonnes).
Brands: Tool steels – Isobloc 2000; Isodisc Superior; Regulit; Isomatrix PM. Pre-machined bar IBO; IBO-Ecomax.

Voest-Alpine AG

Head office: Postfach 2, A 4031 Linz. **Tel:** 0732 5851. **Telex:** 21788 va a.
Management: Chairman and president – Herbert C. Lewinsky; steel division – Ludwig von Bogdandy (executive vice president and board member); flat products – Heribert Kreulitsch (executive vice president and deputy board member). **Ownership:** Oesterreichische Industrie-Holding AG (100%).
Works:
■ Postfach 2, A 4031 Linz. (Tel: 0732 585-0. Telex: 21115). Works director: Harald Mayer. ***Coke ovens:*** Five Otto, four Didier and one Still, including five batteries of 191 ovens (annual capacity 1,483,000 tonnes) and three of 120 ovens (889,000 tonnes). ***Sinter plant:*** One 2-strand 75 sq m each (1,950,000 tonnes); one 168 sq m strand (1,660,000 tonnes). ***Blast furnaces:*** Three 8m hearth dia, 1,153 cu m each (620,000 tonnes respectively); one 7m hearth dia, 850m (450,000 tonnes); one 10.5m hearth dia, 2,386 cu m (1,920,000 tonnes). ***Steelmaking plant:*** Three 60-tonne and two 130-tonne LD basic oxygen converters (3,350,000 tonnes); one 20-tonne electric arc furnace and one 5-tonne induction (27,000 tonnes); ***Refining plant:*** 24/62-tonne ladle vacuum degassing unit (100,000 tonnes). ***Continuous casting machines:*** Four 1-strand Voest-Alpine bow type slab (740-1,600mm x 165-300mm) (3,270,000 tonnes). ***Rolling mills:*** 4-high 1-stand heavy plate (max width 4,200mm) (600,000 tonnes); 4-high 7-stand wide hot strip (max width 1,600mm, 1.5-18mm thick) (coil weight 28.8 tonnes) (2,570,000 tonnes); cold reduction – two reversing stands (max width 1,550mm wide) (470,000 tonnes), and one 4-stand tandem (max width 1,650mm) (906,800 tonnes); 4-high 2-stand temper/skin pass (max widths 1,500mm and 1,650mm) (567,000 tonnes and 674,800 tonnes); one 20-high Sendzimir reversing (max width 1,350mm) (45,000 tonnes). ***Coil coating lines:*** One hot-dip galvanizing (220,000 tonnes); one electro-galvanizing (150,000 tonnes).
■ Donawitz Works, A 8704 Leoben. (Tel: 03842 21521 0. Telex: 33331 vad a). Works director: Gerhard Mitter. ***Sinter plant:*** One 120 sq m strand Voest-Lurgi (annual capacity 1,500,000 tonnes). ***Blast furnaces:*** One 8m hearth dia, 1,111 cu m (550,000 tonnes); one 7m hearth dia, 1,130 cu m (480,000 tonnes). ***Steelmaking plant:*** Three 60-tonne basic oxygen converters (1,200,000 tonnes). ***Refining plant:*** One 12 MW ESR (400,000 tonnes); ladle furnace; VOD (planned). ***Continuous casting machines:*** One 6-strand Voest billet (130 x 130mm) (570,000 tonnes); one 3-strand Voest bloom (360 x 360mm, 250 x 250m and 225 x 225mm) (590,000 tonnes). ***Rolling mills:*** Lewis/United heavy section and medium section – 3-high with universal stand (12-96kg/m) (120,000 tonnes); Blaw Knox bar – roughing train with nine horizontal and three vertical stands, intermediate train with four horizontal and one vertical stand and four 2-high finishing train (91,000 tonnes); Morgan/Voest wire rod – 2-stand continuous with eight roughing stands, two 10-stand intermediate train and two 10-stand finishing train (coil weight 1,300kg) (450,000 tonnes).
■ Kindberg/Krieglach Tube Works, A 8652 Kindberg/Aumühl. (Tel: 03865 2215. Telex: 36647 vaki a). Works directors: Robert Mitter, Hubert Wastl. ***Tube and pipe mills:*** seamless – one push-bench and stretch reducing mill (½-7") (annual capacity 250,000 tonnes) (four shifts, according to product mix). Welded – two HF welding units (12.7-177.6mm) (60,000 tonnes plus 11,500 tonnes (two shifts)).
Products: Carbon steel – blooms (1986 output 485,341 tonnes); billets (433,743 tonnes); hot rolled coil (for re-rolling); wire rod (407,082 tonnes); reinforcing bars (48,612 tonnes); round bars (12,593 tonnes); square bars (896 tonnes); flats (28,625

Voest-Alpine AG (continued)

tonnes); light angles (21,002 tonnes); light channels (2,090 tonnes); medium joists (4,460 tonnes); medium channels (9,007 tonnes); wide-flange beams (medium) (4,710 tonnes); heavy angles (2,394 tonnes); heavy joists (7,328 tonnes); heavy channels (5,905 tonnes); universals (heavy) (2,627 tonnes); heavy rails (81,718 tonnes); rail accessories (13,497 tonnes); medium plates (90,600 tonnes); heavy plates (478,600 tonnes); floor plates (24,300 tonnes); clad plates (two and three-ply) (2,000 tonnes); hot rolled sheet/coil (uncoated) (698,200 tonnes); cold rolled sheet/coil (uncoated) (941,300 tonnes); hot-dip galvanized sheet/coil (220,700 tonnes); electro-galvanized sheet/coil (114,620 tonnes); electrical non-oriented non-silicon sheet/coil (51,500 tonnes); blackplate; seamless tubes and pipes ($\frac{1}{2}$-7″) (201,800 tonnes); longitudinal weld tubes and pipes ($\frac{3}{8}$-7″) (49,700 tonnes); oil country tubular goods (1.9-7″) (199,500 tonnes); cold drawn tubes and pipes (6-180mm) (14,600 tonnes); precision tubes and pipes (12.7-62mm) (9,000 tonnes).

Bangladesh

Bangladesh Steel & Engineering Corp—see Chittagong Steel Mills Ltd

Bengal Steel Works Ltd

Head office: Amin Court (6th Floor), 62/63 Motijheel CA, Dacca 2. **Tel:** 231061-3. **Telex:** 65613 bomb bj. **Cables:** Bomb.
Management: Mohammad Bhai (chairman), Mubarak Ali (executive director), Aziz Mohammad Bhai (executive director). **Established:** 1964. **Capital:** Tk 15,637,500. **Ownership:** Public limited company. **Associates:** Bengal Carbide Ltd (33.33%) (manufacture of dry cell batteries). **Annual capacity:** Rolled steel: 15,000 long tons. Tor-steel: 15,000 long tons.
Works:
■ 4 Fouzderhat Industrial Estate, Chittagong. (Tel: 51408-9. Telex: 65613 bomb bj). Works director: Mubarak Ali. ***Rolling mills:*** One bar wire rod – 18″ roughing train, 14″ intermediate train and two 10″ finishing trains (annual capacity 15,000 long tons). ***Other plant:*** Tor-steel plant (15,000 long tons).
■ 218 Tejguon Industrial Area, Dacca 8. (Tel: 602396, 600768. Telex: 65613 bomb bj). Works director: Mubarak Ali. **Wire drawing plant** (annual capacity 10,800 long tons).
Products: Carbon steel – wire rod 6 to 25mm dia; reinforcing bars 10 to 40mm dia; square bars up to 50mm; flats 25mm x 6mm to 75mm x 100mm; galvanized wire (plain) 8 to 30 gauge; nail wire 4 to 14 gauge, roofing screws.
Additional information: There is also an aluminium conductor plant at Tejgaon Industrial Area, Dacca with an annual capacity of 6,000 long tons.

Chittagong Steel Mills Ltd

Head office: Bangladesh Steel House, Airport Road, Karwan Bazar, Dhaka. **Postal address:** Post Box 457, Dhaka. **Tel:** 32 62 76, 32 72 41. **Telex:** 65880 bj da. **Cables:** Steelhouse Dhaka.
Management: Chairman – Nefaur Rahman; directors – K.M. Badiul Alam (production and engineering), Waseq-Al-Azad (commercial and planning and development), Uttam Ali Miah (finance); managrs – A.N.M. Mahfuzar Rahman (general, steel). **Established:** 1967. **Capital:** Tk 580,000,000. **No. of employees:** Works – 2,552. Office – 933. **Ownership:** Bangladesh Steel & Engineering Corp. **Annual capacity:** Raw steel: 250,000 tonnes. Finished steel: 194,000 tonnes.
Works:
■ PO Box 429, Chittagong (Tel: 501223. Telex: 66243 csm bj). Works director:

Bangladesh

Chittagong Steel Mills Ltd (continued)

Golam Muqtader. ***Steelmaking plant:*** Four 60-tonne open hearth furnaces (annual capacity 250,000 tonnes). ***Rolling mills:*** One blooming (146,000 tonnes); one billet – two 3-high and one 2-high stands (800 x 2,000mm, 700 x 1,700mm) (100,000 tonnes); one 1-stand 3-high heavy plate (800/520/800 x 2,000mm) (57,000 tonnes). ***Coil coating lines:*** Three hot-dip galvanizing (60,000 tonnes).
Products: Carbon steel – billets 110 x 110mm, 85 x 85mm, 65 x 65mm, 50 x 50mm (1986 output, 57,000 tonnes); round bars 28-50mm (1,000 tonnes); medium angles 100 x 100 x 9mm (200 tonnes); medium plates 2.38-4.76mm (3,000 tonnes); heavy plates 6.35-25.40mm (4,000 tonnes); hot-dip galvanized sheet/coil 24-30 gauge (43,000 tonnes), and galvanized corrugated sheet/coil 24-30 gauge 40,000 tonnes (produced from imported sheet)

Dhaka Steel Group of Industries

Head office: Bangladesh Steel House, Airport Road, Karwan Bazar, Dhaka. **Postal address:** Post Box 457, Dhaka. **Tel:** 32 62 76, 32 72 41. **Telex:** 65880 bj da. **Cables:** Steelhouse Dhaka.
Management: Chairman – Nefaur Rahman; directors – K.M. Badiur Alam (production and engineering), Waseq-Al-Azad (commercial planning, and development), Uttam Ali Miah (finance); manager – A.N.M. Mahfuzar Rahman (general, steel). **Established:** 1964. **Capital:** Tk 2,500,000. **No. of employees:** Works – 450. Office – 180. **Ownership:** Bangladesh Steel & Engineering Corp. **Annual capacity:** Finished steel: 7,500 tonnes.
Works:
■ Tel: 40 75 05. Works director: K.A.M. Kamaluddin. ***Rolling mills:*** 3-stand 3-high bar mill (annual capacity 7,500 tonnes).
Products: Carbon steel – reinforcing bars 6.36-31.75mm (annual output 4,500 tonnes).

National Tubes Ltd

Head office: Bangladesh Steel House, Airport Road, Karwan Bazar, Dhaka. **Postal address:** Post Box 457, Dhaka. **Tel:** 32 62 76, 32 72 41. **Telex:** 65880 bj da. **Cables:** Steelhouse Dhaka.
Management: Chairman – Nefaur Rahman; directors – K.M. Badiul Alam (production and engineering), Waseq-Al-Azad (commercial, planning, and development), Uttam Ali Miah (finance); managers – A.E.A. Munim (general). **Established:** 1964. **Capital:** Tk 25,000,000. **No. of employees:** Works – 221. Office – 382. **Ownership:** Bangladesh Steel & Engineering Corp. **Annual capacity:** Finished steel: 45,000 tonnes.
Works:
■ Tongi Industrial Area, Dhaka (Tel: 69 12 62). Works director: A.M.N.F. Zaman. ***Tube and pipe mills:*** welded – one ERW/HFW 12.5mm-204mm dia (annual capacity 45,000 tonnes).
Products: Carbon steel – longitudinal weld tubes and pipes 12.5 (½") to 204mm (8") dia (1986 output, 7,100 tonnes).

Panther Steel Ltd

Head office: 62-63 Motijheel Commercial Area, Dacca. **Postal address:** PO Box 3026. **Tel:** 255158-9, 231272, 231286. **Telex:** 65613 bomb bj, 642835 ambee bj. **Cables:** Pansteel.
Management: Directors – Raja Mohammad Bhai (managing), Mrs Nurjehan Hudda, Mrs Sakina Miraly, Mubarak Ali; company secretary – MD. Nazimuddin. **Established:** 1979. **Capital:** Tk 10,000,000. **No. of employees:** Works – 114. Office – 31. **Subsidiaries:** Panther Engineers Ltd (civil and mechanical engineers and consultants); Bengal Steel Works Ltd (manufacturer of iron and steel products and steel re-enforced aluminium conductors); Bengal Carbide Ltd (dry-cell battery

Panther Steel Ltd (continued)

manufacturing); Bengalad Ltd (advertising and sales promotion agents); Bengal Food Ltd (vegetable oil refining); Panther International Ltd (imports, exports and general trading); Venus International Ltd (imports, exports and general trading); Ambee Pharmaceuticals Ltd (pharmaceutical products manufacturer); Ambee Ltd (imports, exports and general trading). **Annual capacity:** Finished steel: 15,000 tonnes.
Works:
■ Block B, Plot 4, Tongi Industrial Area, Dacca. (Tel: 691105, 691362. Telex: 65613 bomb bj, 642835 ambee bj). Works director: Raja Mohammad Bhai. ***Rolling mills:*** light section and bar – one 10" and one 9" (annual capacity 15,000 tonnes).
Products: Stainless steel – round bars ($\frac{3}{8}$" and $\frac{1}{2}$" dia); light angles (2 x 2 x $\frac{1}{4}$", $2\frac{1}{2}$ x $2\frac{1}{2}$ x 5/16", $1\frac{1}{2}$ x $1\frac{1}{2}$ x $\frac{3}{16}$", $1\frac{1}{2}$ x $1\frac{1}{2}$ x $\frac{1}{8}$", $1\frac{1}{2}$ x $1\frac{1}{2}$ x $\frac{1}{4}$", $2\frac{1}{2}$ x $2\frac{1}{2}$ x $\frac{1}{4}$", 1 x 1 x $\frac{1}{8}$", $\frac{3}{4}$ x $\frac{3}{4}$ x $\frac{1}{8}$"); light channels ($\frac{3}{4}$ x $\frac{3}{4}$ x $\frac{1}{8}$"); zeds ($\frac{3}{4}$ x $\frac{3}{4}$ x 1/8").

Belgium

ALZ NV

Head office: Industrieterrein Rechteroever, B 3600 Genk-Zuid. **Tel:** 011 35 39 81. **Telex:** 39058 aldozg.
Established: 1961.
Works:
■ At head office. ***Steelmaking plant:*** One 35-tonne electric arc furnace (annual capacity 120,000 tonnes). ***Refining plant:*** One 60-tonne VAD (140,000 tonnes). ***Continuous casting machines:*** One 1-strand Concast slab (360,000 tonnes). ***Rolling mills:*** Two cold reduction – one Schloemann MKW 1200 (105,000 tonnes), one SKP 1500 (105,000 tonnes).
Products: Stainless steel – ingots, slabs, hot rolled band (for re-rolling), cold rolled hoop and strip (uncoated), medium plates, hot rolled sheet/coil (uncoated), cold rolled sheet/coil (uncoated).
Sales offices/agents: Avenue Cortenberg 79-81, B 1040 Brussels (Tel: 02 733 9893).
Expansion programme: New 90-tonne electric furnace, 100-tonne MRP converter, and VOD, raising rawsteel annual capacity to 360,000 tonnes in 1988. New 2m wide CR mill (June 1989).

Armco SA, Belgium

Head office: Ave Werihet 55, B 4520 Wandre-Liège.
Products: Carbon steel – cold rolled hoop and strip (uncoated); copper-coated

Belgium

Armco SA, Belgium (continued)

strip.
Additional information: In January 1987 TI Group (UK) acquired most of Armco's European steel operations.

Usines Gustave Boël

Head office: B 7100 La Louvière. **Tel:** 64 27 27 11. **Telex:** 57 228. **Cables:** Boel La Louviere. **Telecopier (Fax):** 64 27 27 48.
Management: Pol Boël (president), Jacques Boël (administrateur délégué), Henri Zunsheim (administrative director), Robert Nachtergaele (technical director), Charles Staquet (commercial director). **Established:** 1851. **Capital:** BFr 1,440,000,000. **No. of employees:** Works – 3,171. Office – 446. **Associates:** Fabrique de Fer de Maubeuge, France (galvanized sheets); Umitol, France (slit strips, coils). Drahtwerk Eberesbach, West Germany (wire mesh). **Annual capacity:** Pig iron: 1,600,000 tonnes. Raw steel: 2,016,000 tonnes. Finished steel: 1,380,000 tonnes.
Works:
■ At head office. ***Coke ovens:*** four Koppers batteries (179 ovens) (annual capacity 898,000 tonnes). ***Sinter plant:*** one 210 sq m strand Dwight Lloyd (1,560,000 tonnes). ***Blast furnaces:*** Two 4.5m hearth dia, 444.4 cu m; two 4.5m hearth dia, 489.8 cu m; one 5.5m hearth dia, 583.9 cu m; and one 6.5m hearth dia, 837.0 cu m (combined annual capacity 1,600,000 tonnes). ***Steelmaking plant:*** Three 85-tonne basic oxygen (2,000,000 tonnes); one 8-tonne electric arc furnace (16,000 tonnes). ***Refining plant:*** One ladle furnace. ***Continuous casting machines:*** One 6-strand billet (36,000 tonnes); two 1-strand slab (5,203.000 tonnes). ***Rolling mills:*** One 27-stand, 5-12.7mm wire rod (coil weight 2,600kg) (300,000 tonnes). one 1-stand heavy plate; one 6-stand wide hot strip; one 1-stand cold reduction; one 1-stand temper/skin pass.
Products: Carbon steel – billets, hot rolled coil (for re-rolling), wire rod, cold rolled hoop and strip (uncoated), hot rolled sheet/coil (uncoated), cold rolled sheet/coil (uncoated), bright wire, black annealed wire, galvanized wire (plain).
Sales offices/agents: Ets Tourey, Paris, France; Real & Co KG, Düsseldorf, West Germany; Queenborough Steel Co Ltd, Flackwell Hearth, Bucks, UK, Duo-Staal, Im Zoetemeer, Netherlands; Fusco, Milan, Italy.

SA Fabrique de Fer de Charleroi

Head office: Rue de Châtelet 266, B 6030 Charleroi, Marchienne-au-Pont. **Postal address:** BP 239, B 6000 Charleroi. **Tel:** 071 36 21 90. **Telex:** 051234. **Cables:** Fabferchar. **Telecopier (Fax):** 071 43 51 11.
Management: K. Choquet (administrateur-directeur général), G. Lefebvre (ingénieur en chef, directeur), J.P. Gerard (ingénieur principal, directeur). **Established:** 1863. **Capital:** Bfr 452,000,000. **No. of employees:** Works – 900. Office – 110. **Subsidiaries:** Charleroi (USA) (51%); Tolco SA (49%).
Works:
■ At head office. Works director: G. Lefebvre. ***Steelmaking plant:*** Two electric arc furnaces – one 180-tonne and 160-tonne (annual capacity 650,000 tonnes). ***Refining plant:*** One 180/200-tonne VAD, one 160/180-tonne VOD. ***Continuous casting machines:*** One 1-strand Mannesmann Demag slab (1,700-2,150mm wide, 180-300mm thick) (650,000 tonnes). ***Rolling mills:*** One 1-stand 4-high heavy plate (4,250mm wide) (600,000 tonnes).
Products: Carbon steel – ingots, 10 to 120 tonnes; slabs up to 1,200 x 2,100 x 300mm; heavy plates, 4 to 400mm thick and 4.1m wide. Stainless steel – ingots, 10 to 40 tonnes; slabs up to 1,200 x 2,100 x 300mm; heavy plates, 4 to 400mm thick and 4.1m wide. Alloy steel (excl stainless) – ingots, 10 to 120 tonnes; slabs up to 12,000 x 2,100 x 300mm; heavy plates, 4 to 400mm thick and 4.1m wide.
Sales offices/agents: Charleroi (USA), 88 Danbury Road, Wilton, CT 06897, USA.

Forges de Clabecq SA

Head office: Rue de la Déportation 218, B 1361 Tubize, Clabecq. **Tel:** 02 355 77 55. **Telex:** 21253. **Cables:** Forges de Clabecq. **Telecopier (Fax):** 02 355 90 36. **Management:** Administrateurs – Pierre Dessy (président, délégué), José Petit (directeur général), Harry Landgraf (conseiller à la prësidence); directeurs – Jacques Leclercq (général-adjoint), Hubert Galant (technique), Edgard Danneau (financier), José Engelbeen (commercial), Michel Levita (commercial). **Established:** 1888. **Capital:** Bfr 3,850,000,000. **No. of employees:** 2,716 (as at 31st March 1987). **Ownership:** Sococlabecq (21.34%); Sté Nationale pour la Restructuration des Secteurs Nationaux (SNSN) (38.96%); Cockerill Sambre (11.13%) (includes non voting shares). **Annual capacity:** Pig iron: 1,500,000 tonnes. Raw steel: 1,750,000 tonnes. Finished steel: 980,000 tonnes.
Works:
■ At head office. Works director: Hubert Galant. ***Iron ore pelletizing plant:*** Polysius (annual capacity 800,000 tonnes). ***Blast furnaces:*** Four – three 5,200m hearth dia, 620 cu metres each (250,000 tonnes respectively), one 7,900m hearth dia, 1,200 cu metres (750,000 tonnes). ***Steelmaking plant:*** Three 80-tonne LDAC-LBE basic oxygen converters (1,500,000 tonnes). **Refining plant:** Three TN ladle inert gas desulphurising units (1,500,000 tonnes). ***Continuous casting machines:*** 2-strand FCB slab (max 2,100 x 300mm) (1,350,000 tonnes). ***Rolling mills:*** heavy plate – single-stand 4-high 3,000 x 900mm reversing roughing and 4-stand 4-high 3,000 x 790mm. ***Other plant:*** Shotblasting and painting plant.
Products: Steelmaking pig iron (1986 output 693,850 tonnes); Carbon steel -. Slabs (843,283 tonnes), medium plates 2 to 15m long, heavy plates 700 to 2,800mm wide and floor plates 3 to 1,500mm thick (585,000 tonnes).
Brands: SA Forges de Clabecq.
Sales offices/agents: J.N. Barker & Co Ltd, 20 Aldermans Hill, Palmers Green, London N13 4PN, UK (Tel: 23882 barker); A.C. Lemvigh Muller, Box 2029, DK 1306 Copenhagen K, Denmark (Telex: 22338 aclm dk); Fuglesangs Ltd A/S, Postboks 9040, Vaterland, Oslo, Norway (Telex: 71348 fugli n); Oy Algol Ab, PO Box 170, SF 00131 Helsinki 13, Finland (Telex: 121430 algol sf); Davum Stahl AG, Leonhardgraben 8, CH 4000 Basle 3, Switzerland (Telex: 65103 dav ch); SA Forges de Clabecq, Postbus 545, Sir Winston Churchilllaan 813, 2280 AM Riswijk ZH, Netherlands (Fax: 70 948 957).

Cockerill Sambre SA

Head office: Chaussée de la Hulpe 187, B 1170 Brussels. **Tel:** 32 2 674 02 11. **Telex:** 23268 cs bru. **Cables:** Cockerill Sambre Brussels. **Telecopier (Fax):** 32 2 660 36 40.
Management: Jean Gandois (chairman); Philippe Delaunois (president and chief executive officer). **Established:** 1981. **Capital:** Bfr 20.400,000,000. **No. of**

Belgium

Cockerill Sambre SA (continued)

employees: 13,000. **Ownership:** Belgian state and other state-owned companies (98%). **Subsidiaries:** Financial companies: Belgium – SA Cockerill Sambre Finances et Services (100%); France – SA Sté Industrielle de Souvigny (100%), Sté Civile Immobiliére (SCI) (50%).
Commercial and trading companies: Belgium – SA Disteel (100%), SA DLC (100%), SA Dema-Jouret (79.60%), SA Metalutil Steel (100%), NV Metaalhandel Vermeersch (100%), NV New Scaffo (100%), NV Craco Steel (100%), SA Cassart Echafaudages (95%), SA Sté Belge d'Oxycoupage (Oxybel) (74.93%), SA Parametal (99.92%), SA Steelinter (100%), SA Cie des Produits Sidérurgiques (Coprosid) France – Sarl Frère-Bourgeois France (100%), SA Cie des Priduits Siderurgiqus – Coprosid (50%), SA Produits d'Usines Métallurgiques (PUM) Station Service Acier (100%), SA Chaillous, Station Service Acier (PUM) (91%), SA Deville (50%), SA Boules à Jouer (84%), SA Jean D'Huart & Cie (100%), SA Eucosider France (100%), SA Exma (100%), SA CPI (100%), SA Prosimo (100%), SA Le Fer à Béton (50%), SA Guille (51%), SA Thionville Acier-Lorac (100%), Sté Civile Immobilière (SCI) (50 %), SA Fermatec (100%), SA Guillot (50%), SA Laminoirs et Ateliers de Jeumont (100%), SA Lille-Aciers (100%), Sarl Produits Métallurgiques des Ardennes (100%), SA Produits Métallurgiques du Sud-Ouest – Station Service Acier (PUM) (100%), SA Produits Métallurgiques de l'Aisne (Prometaisne) (100%), SA Produits Métallurgiques de l'Orléanais (Prometo) (100%), SA Plastiques Développement (100%), SNC PUM Plastiques et Cie (100%), SA Anciens Etablissements Quinchon (100%), SA Robert & Cie (50%), Sté Civile Immobiliére (SCI) (97.50%), SA Savoie Métal (75%), SA Sté Industrielle Métallurgique et d'Entreprise (SIME) (50%), SA Sté Industrielle et Métallurgique de l'Est (Simest) (50%), SA Cie de Récupération des Métaux (Cimeret) (90%), SA Cirec (100%), SA Sté de Transports de Produits d'Usines Métallurgiques (Transpum) (100%), SA Lopez (100%), SA Letierce (67%), SA Mirouze-Novacier (65%), SA Hardy-Tortuaux (50%), SA Berdin (100%), SA Bizalion Fers (100%), SA Centre Acier Service (100%), SA Charpe Métal (100%), SA Ecoma (100%), SA Fertube (100%), SA Hardy BA (56.14%), SA Isère M étal Service (100%), SA Lapauze (100%), SA Loeb Hardy (100%), SA Mahe Caillard (100%), SA Metalsitt (100%), SA Moreau (100%), SA Promet (100%), SA Quincaillerie Régionale de la Brie (75%), SA Sovelam (100%), SA Sté de Vente de Produits Métallurgiqu es (100%), SA Cofrafer (50%), SA Sogap (100%); West Germany – Frère Bourgeois Deutschland GmbH (100%), Eisenstahl GmbH (100%), Cockerill Stahl GmbH (74.90%), Transcometal Gmbh (75.51%), SA Eucostahl (100%); Luxembourg – SA Frecolux (100%), SA Eucosider (50%), SA Euromill (100%); UK – United Continental Steels Ltd and Frère-Bourgeois UK (100%); USA – Frère-Bourgeois USA (60%); Italy – Steelinter SpA (80%); Netherlands – BV Cockerill Nederland Holding (100%), BV Dikema en Chabot's Handelmaatschappij (100%), BV Dikema Staal (100%), BV Financieringsmaatschappij Dikema (100%), BV Oxyned (100%), BV Werle Staal (100%).
Steel and other industrial companies: Belgium – SA Sté Carolorégienne de Cokéfaction (Carcoke) (78.10%), SA Centrale Commune d'Oxygène (CCO) (95%), SC Fabrique d'Oxygène de Charleroi (FOC) (61.48%), SC L'Oxygène Métallurgique (100%), SA Sté Carolorégienne de Laminage (Carlam) (75.31%), SA Laminoirs du Ruau (100%), SA Tréfileries de Fontaine-L'Evêque (66.66%), SA Cockerill Industries (94.46%), SA Cockerill Mechanical Industries (CMI) (100%), SA Heurbel (80%), SA International Countertrading (90.89%), SA New Denaeyer Thermal Industries (51%), SA Babcock Services (70%), SA OSB Co (100%), SA Forges de la Meuse, SA Cockerill Forges and Ringmill (100%), SA Polypal (100%); West Germany – Schwarwälder Röhrenwerk GmbH (SRW) (100%), Stahl-Profilier-und Schneidebetrieb GmbH (100%).
Coated Products: Belgium – SA Phénix Works (100%), SA Sté Européenne pour le Commerce International (Eurinter) (100%), SA Eurinbel (100%), SA Galvacenter (100%), SC FP Homes Belgium (100%), SA Saviar (64.71%), SA Polytuil (69.60%),SA Metal Profil Belgique (100%); France – Sarl Eurinter France (100%), Sarl Polytuil France (100%), SA Sté des Forges d'Haironville (92.99%), Sté Portexter (100%), SA Couvracier (100%), Sarl Cobim (100%), Sarl Comoest (100%), Sarl Couvrapose (100%), SA Monopanel (50%), SA Aciérie et Laminoirs de Beautor (81.27%), SA Sté des Forges de Froncles (99.91%), SA Sté Meusienne de Galvanisation de Produits d'Usines Métallurgiques (Galvameuse) (100%); West Germany – Eurinter Metalhandelgesellschaft GmbH (100%), SFH Deutschland GmbH (100%),

Cockerill Sambre SA (continued)

Profilvertrieb GmbH (100%); UK – Phenix Works UK Ltd (100%), TAC Metal Forming Ltd (50%); Netherlands – BV Eurinter Nederland (100%), BV Metal Profil Nederland (100%); Sweden – AB Eurinter Svenska (100%), AB Ekonomiplat (70%); Luxembourg – SA Galvalange (50%); Italy – Srl Eurinter Italie (100%); Switzerland – SA Socometaux (100%).
Other: SA Sté d'Opérations Maritimes et Fluviales (Somef) (100%), SA Service et de Conseil en Informatique, Bureautique, Télématique (IBT) (100%). **Associates:** Sté Cie Fours à Chaux de Ben-Ahin (49.60%), SA Centrale Zeebrugge (50%), SA Sté Carolorégienne pour la Production d'Air Liquide (Caral) (49.94%), Thymarmon Eisen und Stahl GmbH (75%), Sarl Sté de Distribution des Produits Sidérurgiques (Districid) (100%), SC Groupement Immobilier (Groupimo) (84%), SA Sté de Participations Minières et Industrielles (Soparmi) (94%), SC Immobilière de l'Association des Maîtres de Forges du Hainaut (Forgimo) (92%), Sté des Mines de Valleroy (100%), SA Tilmetal (100%), Ferimpex GmbH (100%). **Annual capacity:** Pig iron: 5,340,000 tonnes. Raw steel: 5,400,000 tonnes. Finished steel: 5,000,000 tonnes.
Works:
■ Marcinelle-Marchienne, Charleroi. Works director: H. Jacobs. ***Coke ovens:*** One Koppers battery (50 ovens), one Didier battery (20 ovens) and two Coppee (52 ovens) (total annual capacity 770,000 tonnes). ***Sinter plant:*** One 102 cu m strand Basse Sambre and one 152 cu m strand Delattre Levivier (total annual capacity 3,000,000 tonnes). ***Blast furnaces:*** Two – No 4 (Marcinelle) 9m hearth dia, 1,490 cu m (1,300,000 tonnes) and No 5 (Marchienne) 8.6m hearth dia, 1,361 cu m (960,000 tonnes). ***Steelmaking plant:*** Three 170-tonne K-OBM basic oxygen converters (2,500,000 tonnes). ***Refining plant:*** Two ALT injection ladle furnace. ***Continuous casting machines:*** One 6-strand Concast S-type (Radius 4m) billet (120 x 120mm, 130 x 130mm) (650,000 tonnes); one Voest slab (220 x 915/1,605mm) (1,850,000 tonnes). ***Rolling mills:*** bar comprising roughing mill – 9-stand (530-395mm dia); intermediate mill – 2-strand 5-stand (four horizontal/one vertical) (355-340mm dia); bar mill – 2-strand 4-stand (two horizontal/two vertical) (320-340mm dia) (rolling programme: rounds – 6-40mm plain and crenellated, squares 10-30mm, flats 20-80mm x 4.7-20(25)mm. Wire rod comprising 2-strand Morgan No-Twist/SMS – 10-stand (two 8" and eight 6") max speed 88m/sec, with Stelmor cooling (coil weight 500-1,650kg) (rolling programme: 5.5-12 (14)mm) Total annual capacity 750,000 tonnes.
■ Division Ruau, Train de 600, Charleroi. Works director: M. Hamaide. ***Rolling mills:*** One semi-continuous Schloemann/Cockerill medium section comprising one 2-high 780mm reversing roughing stand and a continuous finishing mill with two 2-high 610mm horizontal stands, seven universal stands 950mm dia horizontal and 600mm dia vertical, and three 500mm tilting stands, (rolling programme: IPN, IPE 80-270mm, HE 100 to 160mm, American sections 4" to any size, WF 4 x 4" to 6 x 6", UPN 80 to 320mm, L 100 to 150mm and special sections SLS, SLP, PPL etc) (annual capacity 525,000 tonnes).
Laminoirs du Ruau SA (Monceau sur Sambre): Continuous merchant bar mill comprising one 6-stand 530mm dia Pomini roughing mill (three horizontal and three vertical), one 6-stand horizontal 350/450mm dia Banning roughing mill, one 4-stand horizontal 365mm dia Thiriau intermediate mill, one 4-stand horizontal 365mm dia Thiriau finishing mill (two horizontal, two vertical) (rolling programme: equal and unequal angles U's, T's, flats and special sections on request (300,000 tonnes).
■ Division Coke-Fonte-Energie-Transports, Liege. Works director: P. Bruyere. ***Coke ovens:*** one Coppee battery (24 ovens), two Koppers batteries (44 ovens) and two Krupp-Koppers battery (95 ovens) (annual capacity 986,000 tonnes). ***Sinter plant:*** one 321 sq m strand DL-HH-Lurgi (3,600,000 tonnes). ***Blast furnaces:*** Two – HF6 (Seraing) 9.30m hearth dia, 1,680 cu m (1,080,000 tonnes), HFB (Ougrée) 9.75m, 1,666 cu m (1,300,000 tonnes).
■ Division Chertal, Liege. Works director: P. Delvaux. ***Steelmaking plant:*** Three LD/LD-HC 200-tonne basic oxygen converters (annual capacity 2,900,000 tonnes). ***Refining plant:*** Two injection ladle furnace, one RH vacuum degassing unit (900,000 tonnes). ***Continuous casting machines:*** One 2-strand Mannesmann-Demag (radius 20.5m) slab (215 x 900/2200mm) (2,000,000 tonnes). ***Rolling mills:*** Sack 102" universal slabbing with one 2-high 1,220mm dia horizontal stand and one vertical

Belgium

Cockerill Sambre SA (continued)

stand 1,000mm dia (coil weight 30 tonnes) (600,000 tonnes). One Sack 88" semi-continuous wide hot strip comprising roughing mill – with one 4-high horizontal 1,050mm dia before one vertical 1,000mm dia and behind one vertical 1,100mm dia; finishing mill with six 4-high continuous stands 780mm dia (max speed 14m/sec) (coil weight 30-tonne max) (rolling programme: carbon steel from 1.50-13mm thick and 600-2,050mm wide) (2,600,000 tonnes).

■ Division Siderurgie à Froid, Rue des Martyrs 143, B 4210 Saint Nicolas (Liege). (Tel: 041 351731. Telex: 42455). Works director: J. Arnolis. ***Rolling mills:*** cold reduction Jemeppe – one tandem four 4-high 66" Bliss (due to close in 1990) (coil weight 30-tonne) (annual capacity 920,000 tonnes). Tileur – one tandem four 4-high 66" Mesta Sack (will become continuous in 1988) (coil weight 27-tonne) (1,000,000 tonnes); one tandem five 4-high 48" Mesta Sack (coil weight 27-tonne) (550,000 tonnes); one reversing 4-high 42" United (50,000 tonnes). One 2-stand 4-high and one 1-stand 4-high temper/skin pass for cold rolled. ***Coil coating lines:*** Ferblatil, Tilleur – one electrolytic tinning width 1,020mm (290,000 tonnes) and one tin-free steel (ECCS) (width 1,020mm) (85,000 tonnes). ***Other plant:*** Continuous annealing line, Howaq type for 600-1,550mm CR coil (600,000 tonnes); two lines for grain-oriented and non-oriented electrical sheet (110,000 tonnes).

■ Division Carlam Decapage, Charleroi. Works director: F. Desoheemaker. ***Rolling mills:*** One 88" semi-continuous Schloeman Siemag wide hot strip comprising roughing train with one 4-high 1,080/1,200mm dia reversing stand preceded by one 950/1,050mm dia vertical edger, finishing train with six 4-675/760mm dia high continuous stands (max speed 16m/sec) (coil weight 40 tonnes max) (rolling programme: carbon steel 1.50-20mm thick, and 600-2,030mm wide, stainless steel 2.5-20mm thick and 600-1,800mm wide) (annual capacity 2,600,000 tonnes). ***Other plant:*** Hydrochloric and pickling line.

Products: Carbon steel – hot rolled coil (for re-rolling) 1.5-25mm x 16-2,050mm; wire rod 5.5-16mm, high, low and medium carbon for drawing, welding, cold heading, etc; reinforcing bars 6-40mm, plain and ribbed; flats 20-120mm; light angles 16-200mm equal, 30 x 20-100 x 50mm unequal; light tees 20-80mm; light channels 30-70mm; medium angles 16-200mm equal, 30 x 20-100 x 50mm unequal; medium tees 20-80mm; medium joists 80-550mm IPN, 80-600mm IPE; medium channels 30-70mm; wide-flange beams (medium) 4-21", 100-600mm HE; universals (medium); heavy channels 80-320mm UPN; wide-flange beams (heavy) 4-21", 100-600mm HE; hot rolled hoop and strip (uncoated) 1.5-25mm x 16-2,050mm; cold rolled hoop and strip (uncoated) 0.4-3mm x 12-1,520mm; coated hoop and strip – electro-galvanised; medium plates up to 25mm; hot rolled sheet/coil (uncoated) 1.5-25mm x 16-2,050mm; cold rolled sheet/coil (uncoated) 0.4-3mm x 12-1,520mm; electro-galvanized sheet/coil 0.4-3mm x 12-1,500mm, coating 2.5-7.5 microns both sides; enamelling grade sheet/coil; electrical grain oriented silicon sheet/coil up to 790mm wide; electrical non-oriented silicon sheet/coil up to 1,000mm wide; electrical non-oriented non-silicon sheet/coil, blackplate, electrolytic tinplate (single-reduced) and electrolytic tinplate (double-reduced) 0.18-0.49mm wide; tin-free steel (ECCS). Alloy steel (excl stainless) – wire.

Brands: Hot-rolled flat products (structural and deep drawing): Supertenax, Soudotenax, Cor-Ten, Paten, Pole-alter. Cold-rolled flat products (TC, XE, XES, ZC, ZE, ZES sheets): Pole-alter, Cor-Ten, Magnetil BC. Electrozinc-coated sheet/coil: see under subsidiaries (Zincor I, II and III). Concrete reinforcing bar: Tempcore.

Coldstream

Head office: Ruelle Gros Pierre 10, B 7800 Ath. **Tel:** 068 280161. **Telex:** 57578. **Telecopier (Fax):** 068 285775.

Management: Pierre Boisot (directeur, sales), Jean-Pierre Trachte (directeur-adjoint, purchasing, administration, finance), Françis Houbion (sous-directeur, production). **Established:** 1968. **Capital:** Bfr 156,000,000. **No. of employees:** 91. **Ownership:** Kanthal-Höganäs AB, Sweden, (100%). **Associates:** Höganas Gadelius KK, Tokyo, Japan; Höganäs GmbH, Düsseldorf, West Germany; Höganäs (Great Britain)

Coldstream (continued)

Ltd, Aylesbury, UK; Höganäs Meppi SA, Villefranche S/Saône; Höganäs Italia Srl, Como, Italy.
Works:
■ At head office. Atomization towers, zinc furnaces, Coldstream processes, annealing furnaces, reduction furnaces. Powders (atomized) – iron, nickel, cobalt bearing alloys, high speed steels, soft magnetic alloys, wearsurfacing and hardfacing alloys, alloys to customers' specifications. Recycled pre-alloyed tungsten carbide cobalt powder and grit. Toll-crushing and micronization.

Tôleries Delloye-Matthieu SA

Head office: Les Forges, B 5270 Marchin. **Tel:** 085 23 32 23. **Telex:** 59622. **Cables:** Delloy Huy. **Telecopier (Fax):** 085 23 42 20.
Management: Chairman and managing director – C. Delloye; directors – P. Gelise (commercial), S. Marchant (financial), J. Labye (technical). **Established:** 1896. **Capital:** Bfr 152,000,000. **No. of employees:** 480. **Associates:** SA Fonderies N. Porta, Huy, Belgium; Deutsche Zincor, Rosbach, West Germany. **Annual capacity:** Finished steel: 300,000 tonnes.
Works:
■ At head office. ***Coil coating lines:*** Four electro-galvanizing (annual capacity 300,000 tonnes).
Products: Carbon steel – electro-galvanized sheet/coil (1986 output 180,000 tonnes).
Brands: Zincor – electorgalvanized steel (phosphated and chromate rinse); Zincrox – electrogalvanized steel with chromium and chromium oxyde.

SA Laminoirs de Longtain

Head office: Rue Emile Vandervelde, 1-3, B 7170 Manage. **Tel:** 064 22 11 71. **Telex:** 57271. **Cables:** Lamilong, La Croyere.
Management: Chairman – Pierre Haillez; directors – Jacques Boël, Patrice Becker, Bernard de Closset, Philippe Delaunois, Michel Wylenmann; senior executive Albert D'Hernoncourt (manager). **Established:** 1925. **Capital:** Bfr 450,000,000. **No. of employees:** 172. **Ownership:** Boelinves and Cockerill-Sambre (principal shareholders).
Works:
■ At head office. ***Tube and pipe mills:*** welded – square (40 x 40 x 2 up to 250 x 250 x 6mm) and rectangular (50 x 20 x 2 up to 300 x 200 x 8mm) (annual capacity 90,000 tonnes).
Products: Carbon steel – longitudinal weld tubes and pipes, hollow sections.

Belgium

Métal Profil Belgium SA

Head office: Parc Industriel des Hauts-Sarts, B 4400 Herstal. **Telex:** 41240
Ownership: Phénix Works.
Products: Carbon steel – cold roll-formed sections (from strip).

Phenix Works SA

Head office: 10 Quai du Halage, B 4110 Flémalle. **Tel:** 041 33 78 19. **Telex:** 41 210.
Management: Board – J. Dubois (president), R. Jemasse (executive director), V. Moreau, (general manager). **Established:** 1910. **Capital:** BFr 1,963,000. **No. of employees:** Works – 1648. Office – 485. **Ownership:** Cockerill Sambre SA (100%) (Ultimate holding company – Belgian state). **Subsidiaries:** Metal Profil SA (100%) (profiles); Monopanel (50%) (profiles, panels, composites), Ekonomiplat AB (70%) (profiles); Tac Metal Forming, Manchester (50%) (profiles); Couvracier SA (100%) (roofing, wallworks); Saviar (65%) (holding), Galvalange SA (50%) (Alusuisse sheet/coil); Galvacenter (100%) (distribution centre). **Annual capacity:** Finished steel: 995,000 tonnes.
Works:
■ At head office. ***Coil coating lines:*** electrolytic tinning (annual capacity 170,000 tonne), hot-dip galvanizing (681,000 tonnes), colour coating Skinplate PVC (153,000 tonnes); Skinplate (PVC coating line) (18,000 tonnes).
Products: Carbon steel – hot-dip galvanized sheet/coil thickness 0.35-6.35mm, width up to 1,825mm (1986 output 640,000 tonnes), colour coated sheet/coil thickness 0.30-1.50mm, width up to 1,500mm (153,000 tonnes), zinc-aluminium alloy coated sheet/coil, coated sheet/coil (18,000 tonnes), electrolytic tinplate (double-reduced) thickness 0.17-0.40mm, width up to 1,100mm (128,000 tonnes).
Brands: Galbest – coil/sheet, Skinplate – PVC coated coil/sheet, Estetic -prepainted coil/sheet.
Sales offices/agents: Eurinter SA, B 4110 Flémalle, Belgium. (Tel: 041 33 49 30, Telex: 041 210 evrin B).

Laminoirs du Ruau SA

Head office: 147 Rue de Trazegnies, Monceau-sur-Sambre, B 6031 Charleroi. **Tel:** 071 32 00 87. **Telex:** 51237. **Telecopier (Fax):** 071 32 42 47.
Management: Jean Lecomte (président du conseil d'administration); Michel Hamaide (directeur); Michel Duriau (secrétaire général). **Established:** 1905 (as Laminoirs et Usines du Ruau). **Capital:** Bfr 650,000,000. **No. of employees:** Works – 443. Office – 51. **Ownership:** Cockerill Sambre (100%). **Annual capacity:** Finished steel: 300,000 tonnes.
Works:
■ At head office. ***Rolling mills:*** One 6-stand cantilever light section (530 x 315mm (H-V-H-V-H-V) and 14-stand 2-high; bar (six 400 x 900mm (H) four 360 x 750mm (H) and four 360 x 750m (V-H-V-H)) (combined annual capacity 300,000 tonnes).
Products: Carbon steel – round bars 8-40mm; square bars 8-32mm; flats 20 x 4 to 120 x 20mm (1986 output 51,500 tonnes); light angles 20 x 20 to 80 x 80/30 x 20 to 100 x 50mm (54,500 tonnes (equal), 20,200 tonnes (unequal)); light tees 20 x 20 to 80 x 80 (12,500 tonnes); light channels 40 x 20 to 100 x 50mm (incl American sections 1-4") (32,500 tonnes); special light sections – hand rails, window sections, half rounds anchoring rails, etc. PNs 80-100mm (including American sections 3") (10,000 tonnes).
Sales offices/agents: Steelinter SA, B 6090 Couillet, Belgium.

Segal SC – Sté Européenne de Galvanisation

Head office: Chaussée de Ramioul 50, B 4120 Ivoz-Ramet. **Tel:** 32 41 75 34 99. **Telex:** 42282.
Management: Board – V. Moreau (president), M. Muller, S. Hoenselaar.

Segal SC – Sté Européenne de Galvanisation (continued)

Established: 1983. **Capital:** BFr 990,000,000. **Ownership:** Sidmar SA (33⅓%), Hoogovens NV (33⅓%), Phenix Works (33⅓%). **Annual capacity:** Finished steel: 260,000 tonnes.
Works:
■ At head office. ***Coil coating lines:*** One hot-dip galvanizing (annual capacity 240,000 tonnes).
Products: Carbon steel – hot-dip galvanized sheet/coil 0.4-1.6mm, 750-1850mm.
Brands: Segal.
Sales offices/agents: Eurinter SA, B 4110 Flémalle; Segal Commercial, Rue Baugnée, 4120 Ivoz-Ramet.

Sidmar NV

Head office: John Kennedylaan 51, B 9020 Ghent. **Tel:** 091 42 31 11. **Telex:** 11491. **Cables:** Sidmarstaal. **Telecopier (Fax):** 91 42 49 07.
Management: General directors – Henri Muller, Norbert von Kunitzki; executive directors – Jaques Delori (purchasing and personnel), Roger Flammang (EDP and accounting), Albert Hamilius (works director), Paul Matthys (finance and external relations), Pierre Wilmes (sales). **Established:** 1962. **Capital:** Bfr 14,508,305,000. **No. of employees:** Works – 4,726. Office – 1,733. **Ownership:** Arbed Group (51.01%); Belgian government (28.11%); Luxembourg government (15.96%); Falck Group (4.92%). **Subsidiaries:** NV Sidarfin, Gent (97.30%) (holding company); NV Esp, Oevel (99.99%) (steel slitting and cutting centre); Ass Coop Zélandaise de Carbonisation, Sluyski (50%) (coke ovens). **Associates:** Sté Coop Segal, Ivoz-Ramet (33.33%) (steel galvanizing). **Annual capacity:** Pig iron: 2,705,000 tonnes. Raw steel: 3,200,000 tonnes. Finished steel: 2,100,000 tonnes.
Works:
■ At head office. Works director: A. Hamilius. ***Coke ovens:*** Two Otto batteries (100 ovens) (annual capacity 1,300,000 tonnes). ***Sinter plant:*** Dwight Lloyd – one 176 sq m strand and one 304 sq m strand (5,420,000 tonnes). ***Blast furnaces:*** One 9.20 m hearth dia, 1,632 cu m (1,30,000 tonnes) and one 10 m hearth dia, 1,764 cu m (1,600,000 tonnes). ***Steelmaking plant:*** Two 300-tonne basic oxygen LD converters (3,500,000 tonnes). ***Continuous casting machines:*** One 2-strand SMS slab (2,400 x 2,300 x 10mm) (2,000,000 tonnes). ***Rolling mills:*** One slabbing (2,500mm wide) (2,900,000 tonnes) one wide hot strip (2,800,000 tonnes); two Siemag cold reduction (1,930,000 tonnes); temper/skin pass (1,800,000 tonnes). ***Other plant:*** Continuous annealing line (500,000 tonnes).
Products: Carbon steel – slabs hot rolled coil (for re-rolling) medium plates; heavy plates; hot rolled sheet/coil (uncoated) – 1.50-12.7mm thick, 600-1900mm wide (1,880mm with sheared edges) 1.2-16m long (773,608): Cold rolled sheet/coil (uncoated) – 0.20-3mm thick, 600-1880mm wide, up to 7m long (1,325,360 tonnes).

Belgium

Sidmar NV (continued)

Blackplate
Brands: Sidca (Sidmar continuous annealed cold rolled steel), Sidlec (lamination steel), Incrasteel, Profilar (high strength low alloy steel); Lasertex.
Sales offices/agents: TradeArbed, Avenue de la Liberté, Case Postale 1802, Luxembourg; also agencies. Acciaierie & Ferriere Lombarde Falck SpA, Via Mazzini 23, I 20099 Sesto San Giovanni (Milan). Europese Staal Prefabrikatie (ESP), Tweemolensstraat 8, B 2440 Geel (Tel: 014 21 64 64. Telex: 34743 espoevb) (steel service centre for cold rolled and hot rolled pickled sheet, slit strip and sheets.

Tubel SA

Head office: Avenue Maréchal Foch 867, B 7310 Jemappes. **Tel:** 065 88 57 96, 065 82 45 20. **Telex:** 57103.
Management: Managers – P. Becker (general), R. Dispersyn (sales, gas pipes), J. Blondiau(sales). **Established:** 1927. **Ownership:** Boelinvest (89%); Laminoirs de Longtain (11%).
Works:
■ Gentbrugge. ***Tube and pipe mills:*** ERW precision welded
■ Nimy. ***Tube and pipe mills:*** welded ERW gas pipe plant.
■ Jemappes. ***Tube and pipe mills:*** ERW precision welded
Products: Carbon steel – longitudinal weld tubes and pipes 9.5-114mm; galvanized tubes and pipes 9.5-60mm; precision tubes and pipes 9.5-30mm; hollow sections 10x10 to 100x100mm; blast furnace burners, electrical tubing, oxygen lance pipes.

Brazil

Acesita – Cia Aços Especiais Itabira

Head office: Rua Tupis 38, 13° andar, 30000 Belo Horizonte, MG. **Telex:** 311030 caei br.
Established: 1944.
Works:
■ Praça 1° de Maio 9, 35180 Timóteo, MG. ***Sinter plant:*** One Greenawalt non-continuous. ***Blast furnaces:*** One 294 cu m and one 558 cu m. ***Other ironmaking plant:*** One electric reduction furnace. ***Steelmaking plant:*** Two basic oxygen converters – one 30-tonne, and one 75-tonne; three electric arc converters. ***Refining plant:*** One 35-tonne AOD and one 75-tonne VOR ***Continuous casting machines:*** Two 1-strand Concast curved mould (900 to 1,600 x 150/200mm) slab machine. ***Rolling mills:*** One slabbing mill; one blooming three bar (hand mills); one Steckel wide hot strip two Sendzimir cold reduction one temper/skin pass ***Other plant:*** Bight bar plant; ferro-silicon furnace; non-ferrous foundry.
Products: Foundry pig iron Carbon steel – ingots, slabs, blooms, billets, round bars, square bars, flats, bright (cold finished) bars, hot rolled sheet/coil (uncoated), cold rolled sheet/coil (uncoated), electrical grain oriented silicon sheet/coil, electrical non-oriented silicon sheet/coil. Stainless steel – ingots, slabs, blooms, billets, round bars, square bars, flats. Alloy steel (excl stainless) – ingots, slabs, blooms, billets, round bars, square bars, flats, hot rolled hoop and strip (uncoated), cold rolled hoop and strip (uncoated), hot rolled sheet/coil (uncoated), cold rolled sheet/coil (uncoated). High-speed steels, leaded steels, free-cutting steels. Iron, steel and non-ferrous castings.

Açominas – Aço Minas Gerais SA

Head office: Rua dos Inconfidentes 1001, Belo Horizonte, Minas Gerais. **Postal address:** PO Box 30000. **Tel:** 031 212 5055. **Telex:** 0311269.
Management: President – Manoel Braga de Paula Ferreira; directors – Antônio

Açominas – Aço Minas Gerais SA (continued)

Augusto de Andrade Oliveira (industrial), Delson de Miranda Tolentino (financial), Eloy Heraldo dos Santos Lima (foreign affairs), Luiz Carlos Ferraz, José Maurício Cota (commercial), José Augusto Ferreira Filho (administrative). **Established:** 1961. **Capital:** Cz$13,231,861,745. **No. of employees:** 6,091. **Ownership:** Siderbrás – Siderurgia Brasileira SA (96.14%); Government of the State of Minas Gerais (1.98%).
Works:
■ Usina Presidente Arthur Bernardes, Km. 09 Rodovia MG-030, Ouro Branco, Minas Gerais (Tel: 031 721 1388. Telex 031 1738). ***Coke ovens:*** Two direct injection batteries with recirculating gas (106 ovens) (annual capacity 1,154,000 tonnes). ***Blast furnaces:*** One 11.5m hearth dia, 2,761cu m (1,857,120 tonnes). ***Steelmaking plant:*** Two 200-tonne LD basic oxygen converters (2,000,000 tonnes). ***Refining plant:*** One 200-tonne vacuum degassing (600,000 tonnes). ***Rolling mills:*** One slabbing / blooming 2-high reversing 1,350 x 3,000mm (3,000,000); and one semi-continuous billet /blooming – 1,060 x 2,200mm blooming stand and four continuous billets stands (2,000,000 tonnes).
Products: Carbon steel – slabs 140-300 thick, 850-1,900mm wide, max length 12m (1986 output 61,417 tonnes); blooms 163 x 163mm, 185 x 185mm, 210 x 210mm, 350 x 420mm, max length 16 m (30,565 tonnes); billets 80mm sq, 100mm sq, 120mm sq, 130mm sq, max length 16m (654,162 tonnes);
Expansion programme: Heavy section and rail mills – two 2-high 1,070mm reversing roughing stands, three universal reversing stands, replaceable by three 2-high reversing stands, three edgers (capacity was scheduled to be 440,000 tonnes in the first phase); medium section mill – one 2-high 780mm reversing stand, two 610mm roughing stands, 12-stand tandem train with seven universal stands, four horizontal/vertical stands and one 2-high stand (capacity 710,000 tonnes in the first phase). This new plant will produce wide-flange beams 160 x 160mm to 400 x 300mm (heavy mill); and 100 x 100mm to 160 x 160mm (medium mill); joists 200 x 100mm to 600 x 220mm (heavy mill) and 80 x 46mm to 220 x 110mm (medium mill); rails TR37, TR45, TR50, TR68.

Siderúrgica Aconorte SA

Head office: BR 232 Km 12.7, Distrito Industrial do Curado, 50791 Recife, PE. **Tel:** 081 251 3488. **Telex:** 811329, 811344.
Established: 1958. **Ownership:** Gerdau Group. **Annual capacity:** Raw steel: 260,000 tonnes. Finished steel: 240,000 tonnes (rolled); 114,000 tonnes (drawn).
Works:
■ Recife, PE. ***Steelmaking plant:*** Two 25-tonne electric arc furnaces. ***Refining plant:*** One ladle furnace . ***Continuous casting machines:*** One 2-strand Demag for billet (120mm sq). ***Rolling mills:*** One Schloemann bar and wire rod mill. ***Other plant:*** 34 wire-drawing machines. Hot-dip galvanizing.
■ Igarassu, PE. Barbed wire plant, nail and staple machines, wire rope and strand making facilities.
Products: Carbon steel – wire rod, reinforcing bars, bright wire, black annealed wire, galvanized wire (plain), galvanized wire (barbed). ACSR wire, wire strand, wire rope, nails and staples.

Açopalma – Cia Industrial de Aços Várzea de Palma

Head office: Rua Carlos Gomes 40, Bairro Santo Antonio, 30000 Belo Horizonte, MG. **Telex:** 311764 acvp br.
Works:
■ Alameda Lourival Boechat 550, Várzea de Palma, MG. ***Steelmaking plant:*** One 10-tonne electric arc furnace. ***Refining plant:*** One 8-tonne ESR unit. ***Rolling mills:*** One 3-high 480/mm billet and bar one 3-high bar 350/300mm for bar in coil.
Products: Alloy steel (excl stainless) – billets, round bars (high-speed steel, hot and cold die steel).

Siderúrgica J.L. Aliperti SA

Head office: Rua Affonso Aliperti 180, 04156 São Paulo, SP. **Tel:** 011 275 2211, 011 275 1777. **Telex:** 1122200 sjla br.
Established: 1924.
Works:
■ Rua Dalila Magalhães 180, São Paulo, SP. ***Blast furnaces:*** Two furnaces (charcoal) – one 4.1m hearth dia, 200 cu m, and one 4.8m hearth dia, 240 cu m. ***Steelmaking plant:*** electric arc and open hearth furnaces (combined annual capacity 300,000 tonnes). ***Rolling mills:*** One blooming mill; one billet one 19-stand light section ; one 8-stand bar and wire rod mill.
Products: Carbon steel – wire rod, reinforcing bars, square bars, flats, light angles.

Siderúrgica Alterosa Ltda

Head office: Rua Pequi 189, 35660 Pará de Minas, Minas Gerais. **Tel:** 037 231 3304. **Telex:** 031 3436 sial br.
Management: Managers – Djalma Marçal Rezende (works), Rinaldo Assunção Meireles (export), Geraldo Duarte (financial). **Established:** 1959. **Capital:** Cz$120,991,284. **No. of employees:** Works – 242. Office – 26. **Annual capacity:** Pig iron: 100,000 tonnes.
Works:
■ At head office. ***Blast furnaces:*** Four – 1.60m hearth dia, 34.5 cu m (annual capacity 9,900 tonnes); 1.90m hearth dia, 45.5 cu m (14,850 tonnes); 2.10m hearth dia 74.0 cu m and 3.23m hearth dia 144.0 cu m (49,500 tonnes).
Products: Foundry pig iron (output 1986 69,623 tonnes).

Siderurgica Amaral SA

Head office: Rua Pernambuco 635, Belo Horizonte, Minas Gerais. **Postal address:** PO Box 30130. **Tel:** 55 31 201 90 11. **Telex:** 0391193.
Management: Rodrigo Valladares (financial director). **Established:** 1959. **Capital:** US$3,000,000. **No. of employees:** Works – 220. Office – 30. **Ownership:** Viena Administração e Participações Ltda. **Subsidiaries:** Viena Agro Pecuaris (85%) (agriculture and cattle raising); Andrape Valladapes Ltda (50%) (construction). **Associates:** Rofari Transportes (99%) (transport); Refloral Ltda (25%) (reafforestation); Viena Sid do Maranhão (70%) (pig iron production, under construction). **Annual capacity:** Pig iron: 65,000 tonnes.
Works:
■ Av. Amazonas 2585, Betim, MG. (Tel: 531 11 66). Works director: Jenner Valladares. ***Blast furnaces:*** Three – No 1 (annual capacity 16,250 tonnes), No 2 (16,250 tonnes), and No 3 (32,500 tonnes).
Products: Steelmaking pig iron 2-6kg (output 1,600 tonnes); foundry pig iron 2-6kg (51,400 tonnes).
Sales offices/agents: Hofflinghouse; M & M Ferrous Corp; Marc Rich.
Expansion programme: Viena Sid do Maranhão (near Carajas iron ore deposit) will have a capacity of 54,000 tonnes of pig iron in its first stage (1988), with plans to double this by 1990.

Aços Anhanguera SA (Acoansa)

Head office: Avenida Paulista 2037, 19° andar, Edifiío Horsa 11, CZ 03 Conjunto Nacional, 01395 São Paulo. **Tel:** 011 288 4455. **Telex:** 1123805 anha br.
Established: 1962.
Works:
■ Estrada São Paulo, Salesópdis, Km 56, PO Box 354, 08700 Mogi das Cruzes, SP (Tel: 011 469 7711. Telex: 01133448 anha br). ***Blast furnaces:*** Two furnaces (charcoal). ***Steelmaking plant:*** Three electric arc furnaces – two 40-tonne, and one 70-tonne (annual capacity 360,000 tonnes). ***Refining plant:*** One 42-tonne vacuum degassing ladle unit. ***Rolling mills:*** One 2-high reversing 850 x 2,400mm blooming

Brazil

Aços Anhanguera SA (Acoansa) (continued)

mill; one 2-high reversing 700 x 2,200mm billet mill; one 4-stand open bar (530 x 1500) and one 6-stand continuous bar mill (350 x 630mm).
Products: steelmaking pig iron; foundry pig iron. Carbon steel – ingots, blooms, billets, round bars, square bars, flats, bright (cold finished) bars. Stainless steel – ingots, blooms, billets, round bars, square bars, flats, bright (cold finished) bars. Alloy steel (excl stainless) – ingots, blooms, billets, round bars, square bars, flats, bright (cold finished) bars.

Siderúrgica Nossa Senhora Aparecida SA

Head office: Rua Beija Flor 87, 04739 São Paulo, SP. **Tel:** 011 548 0037. **Telex:** 01144441.
Established: 1944.
Works:
■ Rua Padre Madureira 431, 18100 Sorocaba, SP. Tel: 0152 325100. Telex: 152154. ***Steelmaking plant:*** electric arc furnaces. ***Rolling mills:*** One 2-stand 3-high billet three bar – one 4-stand 3-high open, one 5-stand 3-high open, and one 14-stand. ***Other plant:*** Forge.
■ Av Fagundes de Oliveira 510, 09900 Diadema, SP. **Drawing machines, centreless grinders, peeling machines**.
■ Av Dom Pedro 1 734/790, 09000 Santo André, SP. **Wire works**.
Products: Carbon steel – billets, wire rod, round bars, square bars, flats, bright (cold finished) bars (centreless ground and peeled). Stainless steel -. Billets, wire rod, round bars, square bars, flats, bright (cold finished) bars (centreless ground and peeled). Alloy steel (excl stainless) – billets, wire rod, round bars (including high-speed steel), square bars, flats, bright (cold finished) bars (centreless ground and peeled), wire (including high-speed steel). Welding wire, needle wire, wire products; free-cutting steel; tool steels; forged bars; electric resistance wire.

Apolo Produtos de Aço SA

Head office: Rua do Passeio 70-8th floor, Rio de Janeiro, RJ. **Tel:** 220 8196. **Telex:** 2122069 apac br.
Management: President – Antônio Joaquim Peixoto de Castro Palhares; vice president – Paulo Cesar Peixoto de Castro Palhares; directors – Cid Gabriel Ferreira Sampaio (executive), Dimas Correa Filho (sales and purchasi, domestic), Carlos Eduardo de Sá Baptista (industrial), Célio Rodolfo Pary (sales and purchasi, – international), Emílio Salgado Filho (financing and accounting). **Established:** 1939. **Capital:** Crz 139,916,000. **No. of employees:** Works – 415. Office – 165. **Ownership:** GPC Indústria e Comércio (87.88%). **Annual capacity:** Finished steel: 72,000 tonnes.
Works:
■ Estrada Rio do Pau, Cep 20021, 2651 Pavuna, Rio de Janeiro, RJ. Tel: 371 2321. Telex: 02122069. Works director: Carlos Eduardo de Sá Baptita. ***Tube and pipe mills:*** welded – Five ERW 1/2 to 3'' (annual capacity 72,000 tonnes). ***Other plant:*** Tube galvanizing plant (60,000 tonnes).
Products: Carbon steel – longitudinal weld tubes and pipes ½ to 3" (1986 output, 43,869 tonnes, including galvanized); galvanized tubes and pipes ½ to 3" (37,294 tonnes).

Armco do Brasil SA Indústria e Comércio

Head office: Rua Dr Francisco Mesquita 1575, São Paulo, SP. **Telex:** 1123277 armc br.
Established: 1911.
Works:
■ ***Rolling mills:*** cold strip – Sendzimir, Schloemann MKW, Sundwig and other – 2-high and 4-high. **Coil coating lines** Universal electrolytic line for zinc, terne, brass, copper,

Armco do Brasil SA Indústria e Comércio (continued)

and nickel coating.
Products: Carbon steel – cold rolled hoop and strip (uncoated). Alloy steel (excl stainless) – cold rolled hoop and strip (uncoated). Hardened and tempered strip.

Cia Metalúrgica Barbará

Head office: Av Almirante Barroso 72/12, Rio de Janeiro, RJ 20031. **Postal address:** Caixa Postal 1509. **Tel:** 021 292 0058. **Telex:** 21 478. **Telecopier (Fax):** 021 240 0616.
Management: Directors – Jean Michel (presidente), Eugenio Palomino (general), Jean Jacques Faust, Luiz Renato Bueno (juridico), Carlos Alberto Rosito (commercial), João Sergio Marinho Nunes, Antonio Guilherme Dias (industrial), Elias Rodriques Costa. **Established:** 1937. **No. of employees:** Works – 1,500. Office – 150. **Ownership:** Cie de Saint Gobain (68%). **Subsidiaries:** Válvulas Barbará SA (100%) (water and industrial valves). **Annual capacity:** Pig iron: 160,000 tonnes.
Works:
■ Via Dr Sergio Braga 452, 27400 Barra Mansa, RJ. (Tel: 243 22 1022. Telex: 021 22 876). Works director: Antonio Guilherme Dias. ***Blast furnaces:*** Two 3.3m hearth dia (168 m3) each (annual capacity 45,000 tonnes respectively). ***Other ironmaking plant:*** One 33 MVA Elkem elctric reduction furnace (70,000 tonnes).
Products: Foundry pig iron. Cast iron – pipes 3"-48" (output 100,000 tonnes); pipe fittings (push on and flanged and mechanical joints) (8,500 tonnes).

Siderúrgica Barra Mansa SA

Head office: Av José Cesar de Oliveira 21, 9° andar 05317 São Paulo, SP. **Tel:** 831 3422. **Telex:** 01131641 usbm br
Established: 1937. **Ownership:** Votorantim Group.
Works:
■ ***Sinter plant:*** Semi-continuous (annual capacity 100,000 tonnes). ***Blast furnaces:*** Two charcoal furnaces – No.1 (61,000 tonnes); No.2 (39,000 tonnes). ***Steelmaking plant:*** Two 15-tonne LD basic oxygen converters (120,000 tonnes); two 50-tonne electric arc furnaces. ***Continuous casting machines:*** Two Demag billet machines – one 4-strand (120 x 120mm to 130 x 130mm) (150,000 tonnes); one 2-strand (80 x 80mm to 130 x 130mm) (90,000 tonnes). ***Rolling mills:*** blooming; medium section and bar (Demag) ***Other plant:*** Wire plant.
Products: Carbon steel – wire rod, reinforcing bars, round bars, square bars, and light angles (total output in 1986, 181,182 tonnes); black annealed wire
Expansion programme: Revamp of Demag merchant mill will increase annual capacity of the rolling mills from 210,000 tonnes to 325,000 tonnes.

Cia Siderúrgica Belgo-Mineira

Head office: Avenida Carandaí 1115, 19°/26° andares, 30134 Belo Horizonte, MG. **Postal address:** Caixa Postal 15. **Tel:** 031 219 1122. **Telex:** 0311154.
Established: 1921.
Works:
■ 35930 João Monlevade, MG (Tel: 031 851 1101. Telex: 0311398). ***Sinter plant:*** Dwight Lloyd 120 sq m strand. ***Blast furnaces:*** Nos. I, II, and III 4.6m hearth dia, 214 cu m; IV 3.9m hearth dia, 164 cu m; V 6m hearth dia, 406 cu m. ***Steelmaking plant:*** Two 100-tonne basic oxygen converters (annual capacity 1,000,000 tonnes). ***Continuous casting machines:*** billet (400,000 tonnes). ***Rolling mills:*** One 2-stand reversing 1,000 x 2,100mm blooming 3-stand continuous billet; one bar and wire rod Krupp semi-continuous; one Morgan no-twist wire rod (coil weight 500 kg)
■ Rua da Ponte 102, 34100 Sabará, MG. ***Steelmaking plant:*** Two 15-tonne open hearth (probably closed). ***Rolling mills:*** One 3-high 650 x 1,900mm blooming and two 3-high 450 x 1,450mm billet mill; one 5-stand 300 x 900mm light section ***Other plant:*** Drawing plant for wire and reinforcement.
■ Av General David Sarnoff 909, 32210 Contagem, MG (Tel: 031 333 1244. Telex:

Cia Siderúrgica Belgo-Mineira (continued)

0311142, 0311263). **Wire drawing plant.**
Products: Carbon steel – wire rod (including high-carbon) up to 16mm dia; reinforcing bars; bright wire; black annealed wire; galvanized wire (plain); galvanized wire (barbed).
Expansion programme: Morgan wire rod mill at Monlevade (annual capacity 450,000 tonnes of rod 5 to 32mm diameter) in 1989.

Cia Brasileira de Aço

Head office: Ria Antonio Frederico 267, 04244 São Paulo, SP. **Tel:** 011 272 8522. **Telex:** 1122603 chba br.
Works:
■ ***Steelmaking plant:*** Three electric arc furnaces (annual capacity 96,000 tonnes). ***Rolling mills:*** light section
Products: Carbon steel – light angles, light channels.

Cia Brasileira de Ferro

Head office: Rodovia Br 262, Km 8 Guarita, Viana, ES.
Ownership: Gerdau group.
Works:
■ ***Blast furnaces:*** One (capacity 104 tpd).

Siderúrgica Cearense SA

Head office: Av. Parque Oeste 1400, Distrito Industrial de Fortaleza, 61900 Maranguape, CE. **Telex:** 0852023.
Ownership: Gerdau group.
Works:
■ Maranguape, CE. ***Steelmaking plant:*** One 10-tonne electric arc furnace. ***Continuous casting machines:*** One 2-strand for billet. ***Rolling mills:*** One bar
Products: Carbon steel – reinforcing bars, round bars, square bars, flats.

Cimetal Siderurgia SA

Head office: Rua Padre Marinho 455, Belo Horizonte, Minas Gerais. **Tel:** 201 4655. **Telex:** 311246, 316053 cisi br.
Management: President – Ricardo Guedes Pinto Ferreira; directors – Mauro Santos Ferreira (commercial), Sérgio Prates Octaviani Bernes (financial), Alceu de Castro Parreiras (industrial). **Established:** 1987. **Capital:** Cz$128,477,340. **No. of employees:** Works – 4,351. Office – 404. **Subsidiaries:** Cimetal Florestas SA (99.9997%) (reafforestation); Cimetal Carvão Ltda (0.013%) (charcoal transport and trading); Fazenda São Francisco SA (62.80%) (agriculture and farming). **Annual capacity:** Pig iron: 565,000 tonnes. Raw steel: 100,000 tonnes. Finsihed steel: 140,000 tonnes.
Works:
■ Barão de Cocais, Av. Getúlio, Vargas 1555, Barão de Cocais, MG. (Tel: 031 837 1231. Telex 031137 teim br). Works head manager: Márcio Moreira. ***Blast furnaces:*** Three – No 2 4.70m hearth dia, 278 cu m (annual capacity 135,000 tonnes); No 3 3.17m hearth dia, 101 cu m (50,000 tonnes); No 4 2.00m hearth dia, 98 cu m (40,000 tonnes). ***Steelmaking plant:*** One 21-tonne basic oxygen LD converter (230,000 tonnes). ***Continuous casting machines:*** One 4-strand billet (100mm sq and 120mm sq) (240,000 tonnes).
■ Sete Lagoas Plant, BR 040 Km. 635, Sete Lagoas, MG. (Tel: 031 921 5088). Works manager: Geraldo Martins Coura. ***Blast furnaces:*** Three – 2.30m hearth dia, 65 cu m (annual capacity 40,000 tonnes); 3.55m hearth dia, 158 cu m (75,000 tonnes); 3.87m hearth dia, 183 cu m (90,000 tonnes).
■ João Neiva Plant, BR 101 Km. 204 Ibiraçu, João Neiva, ES. (Tel: 027 258 1343.

Cimetal Siderurgia SA (continued)

Telex 0272772 cisi br). Works head manager: Walter Kopperschimidt. ***Blast furnaces:*** Two – 3.50m hearth dia, 139 cu m (annual capacity 65,000 tonnes); 3.80m hearth dia, 175 cu m (75,000 tonnes).
■ Itaúna Plant, Av. Lenhita 1752, Itaúna, MG. (Tel: 037 241 1412. Telex 0371569 teim br). Works head manager: Maurício Santiago dos Santos. ***Blast furnaces:*** Six – No I-1 2.00m hearth dia, 45 cu m (annual capacity 24,000 tonnes); No II-1 2.60m hearth dia, 70 cu m (38,000 tonnes); No II-2 2.70m hearth dia, 92 cu m (46,000 tonnes); No V-1 2.40m hearth dia, 72 cu m (35,000 tonnes); No V-2 2.60m hearth dia, 73 cu m (37,000 tonnes); No VI-1 2.40m hearth dia, 80 cu m (40,000 tonnes).
Products: Steelmaking pig iron 0.5-7.0kg pigs (1986 output 161,465 tonnes); foundry pig iron 0.5-7.0kg pigs (205,435 tonnes). Carbon steel – reinforcing bars $\frac{1}{2}$"-1$\frac{1}{4}$" (49,052 tonnes); round bars $\frac{5}{8}$-1$\frac{3}{8}$" (23,708 tonnes); flats (1,346 tonnes); light angles (equal angles) (39,386 tonnes). Granulated iron (205,435 tonnes); ingot molds (15,740 tonnes); SG iron (45,609 tonnes); cast iron counterweights (12,390 tonnes); special grade SG iron (28,624 tonnes); high purity nodular SG iron (30,540 tonnes); Cylpebs (6,040 tonnes); malleable iron (25,730 tonnes).
Brands: Cimetal.

Cobrasma SA

Head office: Rua da Estação 523, 06001 Csasco, São Paulo. **Postal address:** Caixa Postal 969. **Tel:** 011 704 6122. **Telex:** 011 71545, 011 71589. **Cables:** Cobrasma. **Telecopier (Fax):** 11 704 6856.
Management: President – Luis Eulálio de Bueno Vidigal; vice presidents – Marcos V. Xavier da Silveira, Luis Eulálio de Bueno Vidigal Filho; directors – José Teixeira Beraldo (managing), Pedro Paulo Leite de Barros (finance), Alberto Martinez (marketing), Geraldo da Rocha Menezes; assistant directors – Norberto de Souza Aranha (sales), Eduardo Hubert, K. Monteiro (export). **Established:** 1944. **Capital:** Cz$1,717,565,225. **No. of employees:** Works – 2,039. Office – 133. **Subsidiaries:** Fornasa SA (100%) (manufacturer of longitudinal ERW steel tubes and pipes).
Works:
■ At head office. Works director: Geraldo Rocha Menezes. ***Steelmaking plant:*** Several electric arc furnaces (annual capacity 42,400 tonnes). ***Other plant:*** Steel foundry (44,400 tonnes castings).
Products: Carbon steel – ingots (1986 output 11,827 tonnes), castings (24,982 tonnes).
Expansion programme: An expansion and re-equipment programme, to be completed early in 1991, will raise steel melting capacity from 120,000 to 128,400 tonnes a year.

Cofavi – Cia Ferro e Aço de Vitória

Head office: Av Espírito Santo, 216 Jardim América, Cariacica, Espírito Santo. **Tel:** 027 226 3211. **Telex:** 027 2168 cfav br. **Cables:** Ferroaco.
Management: President – Edward Thomaz Merlo; directors – Mário Victor Cardoso Monteiro (financial), Carlos Alberto Ferrari Ferreira (industrial), Olinto Alvarez Villas Boas (commercial), Marcus Alexandres Fundão Pessoa (administration). **Established:** 1942. **Capital:** Cz$1,234,718,438. **No. of employees:** 2,067. **Ownership:** Siderbrás – Siderurgia Brasileira SA (94.38%).
Works:
■ At head office. ***Steelmaking plant:*** Four electric arc furnaces – one 18-tonne, two 20-tonne, and one 70-tonne (annual capacity 500,000 tonnes). ***Continuous casting machines:*** One 6-strand billet (80, 100 and 130mm) (465,000 tonnes); one 2-strand bloom (160 and 170 x 110mm) (35,000 tonnes). ***Rolling mills:*** One blooming (155 and 170 x 110mm) (36,000 tonnes); one billet (80, 100 and 130mm) (324,000 tonnes); one medium section (63.50 and 152.40mm) (180,000 tonnes); one light

Cofavi – Cia Ferro e Aço de Vitória (continued)

section (38.10 and 76.20mm) (180,000 tonnes); one bar (12 and 32mm) (30,000 tonnes).
Products: Carbon steel – 265mm (output 856 tonnes); ingots; slabs; blooms 160 and 170 x 110mm (11,072 tonnes); billets 80, 100 and 130 (168,630 tonnes); semis for seamless tubemaking; hot rolled coil (for re-rolling); wire rod; round bars 12 and 32/31.71 and 101.60mm (42,611 tonnes); square bars 65 and 130mm; flats; light angles 76.20mm (4,565 tonnes); light tees; light channels 38.10 and 63.50mm (53,392 tonnes); special light sections; medium angles 76.20, 101.60, 127.00 and 152.40mm (47,601 tonnes); medium tees 88.90 and 127.00mm (8,642 tonnes); medium channels 63.50 and 101.60mm (43, 015 tonnes); wide-flange beams (medium).

Comesa – Cia Siderúrgica de Alagoas

Head office: Av Des Dr José Jerônimo de Albuquerque s/nº, 57690 Atalaia, AL.
Ownership: Gerdau group.
Works:
■ Atalaia, AL. ***Steelmaking plant:*** One electric arc furnace. ***Rolling mills:*** One light section and bar mill.
Products: Carbon steel – reinforcing bars, round bars, light angles.

Confab Industrial SA

Head office: Al. Rio Negro 433, Alphaville, 06400 Barueri, São Paulo. **Postal address:** Caixa Postal 137. **Tel:** 011 421 1222. **Telex:** 011 71423, 011 71424 confbr. **Telecopier (Fax):** 011 421 4297.
Management: President – Roberto Caiuby Vidigal; vice president – Antônio Carlos Vidigal; directors – Adrianus Franciscus ten Wolde (finance and administration), Aldo Passos Batista (pipe division), Ayrton Bassani (equipment division), Caio Luiz Barboza Ferraz (materials), Emílio Wainer; manager- Guilherme Muylaert Antunes (export).
Established: 1943. **Capital:** Cz$2,259,689,000. **No. of employees:** Works – 3,281. Office – 601. **Subsidiaries:** Confab Montagens e Equipamentos (89.26%) (erection); Confab Revestimentos SA (100%) (pipe coating); Confab Trading SA (100%) (exporting); Confol NV (100%) (sales office); Oca Mineração (50%) (mining); Conave-Reparos e Construção Naval SA (100%) (shipbuilding and repair); Nippon Steel Confab Ltd (55%) (engineering); Confab Industrial Corp (100%) (sales office).
Annual capacity: Finished steel: 480,000 tonnes (welded steel pipes).
Works:
■ Estrada Municipal, Bairro do Feital, Pindamonhangaba, São Paulo. (Tel: 0122 42 33 11. Telex 0122 172). Works director: Mário Ferraz de Arruda Filho. ***Tube and pipe mills:*** welded – No 1 ERW 6⅝ to 18" (annual capacity 120,000 tonnes); No 2 SAW-UOE 12¾ to 48" (238,000 tonnes). ***Other plant:*** Plant for parts and accessories (1,000 tons); threaded pipe plant (8⅝ to 20") (53,000 tonnes).
■ Av. Prosperidade 374, São Caetano do Sul, SP. (Tel: 011 441 4455. Telex 011 4039). Works director: Sílvio Vidigal Monteiro de Barros. ***Tube and pipe mills*** welded – No 1 SAW-Bending 10.¾ to 32" (annual capacity 30,000 tonnes); No 2 SAW-Spiral 24 to 72" (18,000 tonnes); No 3 SAW-Large dia 32 to 100" (20,000 tonnes).
Products: Carbon steel – longitudinal weld tubes and pipes 12¾", 16", 20", 24", 30" 36" (output 1986 246,488 tonnes); spiral-weld tubes and pipes 40", 42", 48", 60' (4,395 tonnes); large-diameter tubes and pipes 80", 84", 92", 100" (12,222 tonnes) oil country tubular goods 13⅜", 20" (53,393 tonnes).
Sales offices/agents: Confab Industrial Corp, 5959 Westheimer, Suite 214 Houston, TX 77057, USA (Tel: 713 780 2070. Telex: 794281. Fax: 713 780 8450).

Copalam – Cia Paulista de Laminação

Head office: Av Dr Alberto Soares Sampaio, 1880, 09380 Mauá, SP. **Tel:** 011 450 14 77. **Telex:** 011 44755 pdla br. **Cables:** Copalam.
Management: Directors – José Roberto Patricio, Adib Joaquim Mendes. **Established:** 1950. **Capital:** Cz$11,928,252. **No. of employees:** Works – 100. Office – 41. **Annual capacity:** Finished steel: 24,000 tonnes.
Works:
■ At head office. ***Rolling mills:*** One 1-stand 350mm billet (annual capacity 24,000 tonnes); 3-stand 280mm light section (24,000 tonnes).
Products: Carbon steel – light angles :FR::5/8 x $\frac{1}{8}$" to 1.$\frac{1}{2}$" x $\frac{1}{8}$ (output 5,224 tonnes); light tees $\frac{3}{4}$" x $\frac{1}{8}$" to 1.$\frac{1}{4}$ x $\frac{1}{8}$" (5,723 tonnes); special light sections (3,112 tonnes).
Brands: Copalam.

Cosigua – Cia Siderurgia da Guanabara

Head office: Avenida João XXIII 6777, Zona Industrial, 23568 Santa Cruz, RJ. **Tel:** 021 305 1515. **Telex:** 2123422, 2122535.
Established: 1961. **No. of employees:** 4,887. **Ownership:** Gerdau Group. **Annual capacity:** Pig iron: 360,000 tonnes. Raw steel: 1,030,000 tonnes. Finished steel: 1,020,000 tonnes (rolled), 265,000 tonnes (drawn).
Works:
■ Contagem, MG. ***Blast furnaces:*** Two continuous flow furnaces – one relined December 1984, and the other relined March 1985.
■ Pará de Minas, MG. ***Blast furnaces:*** One continuous flow – relined February 1987. ***Other plant:*** Foundry and finishing shop.
■ Pedro Leopoldo, MG. ***Blast furnaces:*** One continuous flow – relined October 1983.
■ Prudente de Morais, MG. ***Blast furnaces:*** One continuous flow – relined November 1982, and one small continuous – relined February 1986. ***Other plant:*** Small foundry.
■ Santa Cruz, RJ. ***Steelmaking plant:*** Two 100-tonne electric arc furnaces. ***Refining plant:*** One ladle furnace . ***Continuous casting machines:*** Two billet machines – one 5-strands, and one 6-strand. ***Rolling mills:*** One single-strand semi-continuous quick-change bar mill; wire rod – 2-strand continuous with two Morgan No-Twist finishing blocks (finishing speed 90 metres/sec) mill. ***Other plant:*** 76 drawing machines; annealing furnaces; hot-dip galvanizing lines; wire strand machines; barbed wire, nail and staple making equiment.
■ Cotia, SP.
■ Mauá, SP.
Products: Carbon steel – billets, wire rod, reinforcing bars (bars and coils), round bars, square bars, flats, light angles, light tees, light channels, bright wire, black annealed wire, galvanized wire (plain), galvanized wire (barbed). Copper-coated wire, welding wire, wire strand, nails and staples. Manhole covers, frames, grates; valve and meter boxes.
Brands: Elefante, and Urso (barbed wire).

Cosim – Cia Siderúrgica de Mogi das Cruzes

Head office: Av Cavalheiro Nami Jafet 265, Vila Industrial, Mogi das Cruzes, São Paulo. **Tel:** 011 469 8844. **Telex:** 011 33815 csmc.
Management: President – Jorge da Costa Lino; directors – Altamiro de Oliveira Netto (administration), Haley Castanho (industrial), Laurival Gabrieli jr (financial). **Established:** 1968. **Capital:** Cz$805,545,250. (April $987). **No. of employees:** 773. **Ownership:** Siderbrás – Siderurgia Brasileira SA (98.63%).
Works:
■ At head office. Works director: Altamiro de Oliveira Netto. ***Blast furnaces:*** One 4.10m hearth dia, 185 cu m (annual capacity 100,000 tonnes). ***Steelmaking plant:*** Five 60-tonne open hearth (not in operation) (288,000 tonnes). ***Rolling mills:*** 2-stand 3-high blooming / billet and bar (26$\frac{1}{2}$ x 114") (266,000 tonnes). ***Tube and pipe mills:***

Cosim – Cia Siderúrgica de Mogi das Cruzes (continued)

One Ehrhardt piercing process hot finishing seamless, 1½ NPS upto 4" NPS SCH-40 and SCH-80 (36,000 tonnes).
Products: Carbon steel – blooms billets and semis for seamless tubemaking, (1986 output – blooms 3,256 tonnes, billets 9,076 tonnes); square bars (14,073 tonnes); seamless tubes and pipes (19,478 tonnes).

Cosinor – Cia Siderúrgica do Nordeste

Head office: BR 101 Sul, Km 21.8 Pontezinha, Cabo, Pernambuco. **Postal address:** Caixa Postal 1071, Recife 50,000, PE. **Tel:** 081 341 3711. **Telex:** 0811342.
Management: Directors – Paulo Roberto de Freitas Araujo (superintendente), Pedro de Nigris, Durval Vieira de Freitas, Raul Alves Kalckmann, Mário Jorge de Menezes Andrade. **Established:** 1939. **Capital:** Cz$480,542,537. **No. of employees:** Works – 342. Office – 1,021.
Works:
■ At head office. ***Steelmaking plant:*** electric arc – furnaces (5-tonne, 10-tonne and 12-tonne) (annual capcity 80,000 tonnes). ***Continuous casting machines:*** One 2-strand Concast billet (100mm sq) (120,000 tonnes). ***Rolling mills:*** bar (10-25mm dia) (96,000 tonnes).
Products: Carbon steel – reinforcing bars 10, 12, 16, 20 and 20mm dia (output 75,000 tonnes).

Cosipa – Cia Siderúrgica Paulista

Head office: Avenida São João 473, 11° andar, 01035 São Paulo, SP. **Postal address:** Caixa Postal 8090, São Paulo, SP. **Tel:** 011 223 8844, 011 223 7755. **Telex:** 1121196 cspa br. **Cables:** Cosiderpa Spo. **Telecopier (Fax):** 011 223 4960.
Management: President – Antonio Maria Claret Reis de Andrade; directors – Carlos Alberto Azevedo do Amaral (financial), Benedito Nicotero Filho (commercial), Sergio Matheus Antunes Mattos (industrial), Israel Aron Zilberman (engineering), Gunther Bantel (environment), Osvaldo Vasconcelos Malmegrin (administration).
Established: 1953. **Capital:** Cz$18,105,066,549. (April 1987). **No. of employees:** Works – 12,500. Office – 3,000. **Ownership:** Siderbrás – Siderurgica Brasileira SA (98.5%). **Annual capacity:** Pig iron: 2,100,000 tonnes. Raw steel: 2,700,000 tonnes. Finished steel: 2,500,000 tonnes.
Works:
■ Usina José Bonifácio de Andrada e Silva, Estrada de Piaçaguera Km 6, 11500 Cubatão, SP. Tel: 0132 61 1333. Telex: 013 1291. Works director: Sergio Matheus Antunes Mattos. ***Coke ovens:*** Five Otto (modified) underjet batteries (203 ovens) (annual capacity 1,800,000 tonnes). ***Sinter plant:*** Three Dwight Lloyd – (1) 2.5 x 32.5m strand (820,000 tonnes), (2) 3.7 x 39.9m strand (1,445,000 tonnes), (3) 4.0 x 67.0m strand (301,000 tonnes). ***Blast furnaces:*** (1) 9.8m hearth dia, 1,829 cu m (1,155,000 tonnes), (2) 10.9m hearth dia, 2,565 cu m (2,180,000 tonnes). ***Steelmaking plant:*** Two 130-tonne and four 100-tonne LD basic oxygen converters (3,900,000 tonnes). ***Refining plant:*** Two RH vacuum degassing units (200,000 tonnes); one ladle refining – CaSi injection (2,600,000 tonnes). ***Continuous casting machines:*** Three 1-strand Clesid/Concast curved mould slab (185/215/260 x 1,000mm-1,900 x 9,000mm) (1,200,000 tonnes). ***Rolling mills:*** One 2-high reversing slabbing (2,600,000 tonnes); one 1-stand heavy plate and medium plate (800,000 tonnes); one 6-stand wide hot strip and hot strip and hoop (coil weight 4/15 tonnes) (2,500,000 tonnes); one finishing hot mill (coil weight 4/15 tonnes) (300,000 tonnes); one 4-stand cold reduction and cold strip (coil weight 8/24 tonnes (900,000 tonnes); one temper/skin pass (coil weight 8/24 tonnes) (900,000 tonnes); two finishing cold mills (coil weight 8/24 tonnes) (700,000 tonnes). ***Other plant:*** Foundry.
Products: Carbon steel – ingots (1986 output 2,323,200 tonnes); slabs (output 7,000 tonnes); hot rolled hoop and strip (uncoated) 2.65 x 1,200mm (857,000 tonnes); cold rolled hoop and strip (uncoated) 1.20 x 1,200mm (674,000 tonnes);

Cosipa – Cia Siderúrgica Paulista (continued)

medium plates, heavy plates and universal plates 16 x 2,440 x 1,200mm, 50 x 2,000 x 6,000mm (657,000 tonnes); floor plates 4.75 x 1,200mm and hot rolled sheet/coil (uncoated) 2.65 x 1,200mm (130,000 tonnes); cold rolled sheet/coil (uncoated) 1.20 x 1,200mm and enamelling grade sheet/coil 1.20 x 1,200mm (197,000 tonnes).
Brands: COS-AR-COR (hot and cold sheets and plates).
Sales offices/agents: 26 Rue de La Pepiniére, F 75008 Paris, France (Tel: 00331 43871321. Telex: 290756 cosipa. Cables: Apisoc); Shinkokusai Bldg, 5th Floor, Room 536, 4-1 Marunouchi 3-chome, Chiyoda-ku, Tokyo 110, Japan (Tel: 00813 214 0988, 214 0980. Telex: 2226522 cosipa. Cables: Cosiderpa).

CSN – Cia Siderúrgica Nacional

Head office: Av 13 de Maio, 13 8° andar, 20003 Rio de Janeiro, RJ. **Postal address:** Caixa Postal 2736. **Tel:** 021 297 7177. **Telex:** 021 23025.
Management: President – Juvenal Osório Gomes; directors – André Martins de Andrade Jr (adminstrative), José de Gusmão Campelo Lima (development), Francisco Ari Souto (operations), Alexandre Henriques Leal Filho (finance), Oduwaldo Antonio Ferreira (social services), José Maria Carvalho Jr (property). **Established:** 1941. **Capital:** Cz$13,192,537,455 (Dec 31 1986). **No. of employees:** Works – 17,858. Office – 3,985. **Ownership:** Siderbrás – Siderúrgia Brasileira SA. **Subsidiaries:** Fem – Fábrica de Estruturas Metálicas SA, Próspera – Sdade Carbonífera Próspera SA. **Annual capacity:** Pig iron: 4,219,000 tonnes. Raw steel: 4,600,000 tonnes. Finished steel: 3,950,000.
Works:
■ Usina Presidente Vargas, Rua 21 No 10, 27180 Volta Redonda, RJ. (Tel: 0243 42 9955. Telex: 0223 123). Works director: Francisco Ari Souto. ***Coke ovens:*** Four underjet batteries – one (68 ovens) (annual capacity 384,000 tonnnes), one (45 ovens) (464,500 tonnes), one (60 ovens) (619,000) and one (45 ovens) (464,500 tonnes). ***Sinter plant:*** Four Dwight-Lloyd – 93.65 sq m (778,800 tonnes), 165.84 sq m (1,393.920 tonnes), 165.84 sq m (1,526,910) one 190.00 sq m (2,257,530 tonnes).. ***Blast furnaces:*** One 8.38 m hearth dia, 1,335 cu m (841,000 tonnes), one 8.53 m hearth dia, 1,556 cu m (885,000 tonnes) and one 13.00 m hearth dia, 3,815 cu m (2,493,000 tonnes). ***Steelmaking plant:*** Three 220-tonne basic oxygen LD converter (4,600,000 tonnes). ***Continuous casting machines:*** One 2-strand straight mould for slab (208 x 835/1,500mm) and two 2-strand curved mould (250 x 810/1,630mm) (combined capacity 3,700,000 tonnes). ***Rolling mills:*** slabbing one 2-high reversing 44" (80/1,370mm) (slabs – 1,526,000 tonnes; billets – 652,000 tonnes); one heavy section / rail and medium section – 1-stand 2-high reversible, 2-stand 3-high open, 2-stand 3-high open and 2-stand 2-high open (330/280mm max) (290,000 tonnes). No 1 continuous wide hot strip – comprising 1 edger, 1-stand 2-high roughing, 1-stand 4-high reversing, seven 4-high finishing (slab 257/105) 1,270/710, 4,900/3,100; coil 12 7/1.47, 1,200/600mm (coil weight 8.3) (1,350,000 tonnes); No 2 continuous hot strip mill over 1m – 1-stand 2-high roughing, 23-stand 4-high roughing and 7-stand 4-high finishing (slab – 305/152, 1,575/610mm; coil – 12.7/1.20, 1,575/610) (coil weight 40.86kg) (3,200,000 tonne.; cold reduction 5-stand 4-high continuous (finished coil 2.67/0.24, 1,575/610mm) (coil weight 40.8kg) (1,500,000 tonnes). cold strip 5-stand 4-high continuous (finished coil 0.45/0.18, 1000/600mm) (coil weight 18.9) (1,200,000 tonnes); temper/skin pass 1-stand 4-high continuous (finished coil 1.9/0.3, 1,220/610mm) (coil weight 18.1) (285,000 tonnes); Four double reduction mills (finished coil 0.61/0.15, 1,067/600mm) (coil weight 13 and 40) (1,631,000 tonnes). ***Coil coating lines:*** Five Ferrostan electrolytic tinning (850,000 tonnes); one Ferrostan/CSN tin-free steel (ECCS) (tin/chromium combination) (180,000 tonnes); three continuous NOF-Armco hot-dip galvanizing (two horizontal and one vertical furnace) (500,000 tonnes). ***Other plant:*** Forge, foundry (two) (6,000 tonnes), five continuous annealing lines and 55 batch annealing furnaces.
Products: Carbon steel – hot rolled coil (for re-rolling) 300 x 1,200mm (output 1,226,000 tonnes); round bars 80.0mm and square bars 100 mm (37,000 tonnes); light angles 63.5 x 63.5mm (1,000 tonnes); medium angles 152.4mm (18,000 tonnes); medium channels 152.4mm (5,000 tonnes); heavy angles 203.2mm (2,000

Brazil

CSN – Cia Siderúrgica Nacional (continued)

tonnes); heavy channels 203.2/304.8mm (66,000 tonnes); heavy rails TR-57/TR-68 (26,000 tonnes); light rails TR-25/TR-37/TR-45 (57,000 tonnes); rail accessories TR-45/PA-45 (4,000 tonnes); hot rolled sheet/coil (uncoated) 3.0 x 1,200mm (1,600,000 tonnes); cold rolled sheet/coil (uncoated) 0.75 x 1,000 (494,000 tonnes); hot-dip galvanized sheet/coil 0.60 x 1,000 (366,000 tonnes); blackplate 0.24 x 825mm (103,000 tonnes); electrolytic tinplate (single-reduced) 0.24 x 825mm (378,000 tonnes); electrolytic tinplate (double-reduced) 0.16 x 825mm (59,000 tonnes); tin-free steel (ECCS) 0.24 x 825mm (127,000 tonnes).
Sales offices/agents: Brazilian National Steel Corp, NY, USA (Tel: 001 212 486); Pell Hnos SACIFA, Buenos Aires, Argentina (Tel: 0054-1 11-1223); Juan M. Gonzalez Representaciones Ltda, Montevideo, Uruguay (Tel:00598-2 406.020, 402.571, 402.540).
Expansion programme: Stage III of expansion plans, due for completion in 1987 – one slab reheating furnace, one continuous annealing furnace, two electrolytic tinplate lines, one continuous hot-dip galvanizing line. The compnay plans to increase liquid steel and slab capacity to 5,300,000 tonne/year by 1990.

Dedini SA Sideúrgica

Head office: Rua Itaddock Lobo 585, 1° andar, 01414 São Paulo, SP. **Tel:** 011 883 0411, 011 64 0451. **Telex:** 1130964, 1131666 dedi br.
Established: 1955.
Works:
■ Avenida Marechal Castelo Branco 101, 13400 Piracicaba, SP. Tel: 0194 33 1122. Telex: 191912, 192736 dedi br. ***Steelmaking plant:*** electric arc furnaces. ***Continuous casting machines:*** Two 3-strand machines for billet – one Concast and one Voest. ***Rolling mills:*** Three bar – one 2-stand 3-high 485mm; one 2-stand 3-high 395mm and 5-stand 2-high; one 2-stand 3-high 435mm, 5-stand 2-high 300mm and 8-stand 270mm. ***Other plant:*** Foundry.
Products: Carbon steel – billets, reinforcing bars, round bars, square bars. Castings.

Eletrometal SA – Metais Especiais

Head office: Via Anhanguera, Km 113, 13170 Sumaré, SP. **Tel:** 55192 641 800. **Telex:** 39191963 elsm br. **Telecopier (Fax):** 5592 6059.
Management: José Diniz de Souza (president), Edmundo Alcoforado (production director), Naila Diniz, R. Croft, Marcos Diniz Ribeiro (board members), Nigel R. Croft (international business manager). **Established:** 1961.
Works:
■ Via Anhanguera Km 113, 13100 Sumaré, SP. ***Steelmaking plant:*** Two electric arc furnaces – one 22-tonne (annual capacity 40,000 tonnes), one 11-tonne (30,000 tonnes). ***Refining plant:*** One 35-tonne and one 10-tonne ESR unit; one 26-tonne VAD /VOD unit; one 8-tonne VAR unit. ***Rolling mills:*** bar and wire rod. ***Other plant:*** Forging presses and GFM forging machines. Wire drawing plant.
Products: Bright wire, Stainless steel – ingots, blooms, billets, wire rod (5.5-101.6mm dia), round bars, square bars. Alloy steel (excl stainless) – ingots, blooms, billets, wire rod (5.5-101.6mm dia), round bars and square bars (including high-speed steel, hot and cold work tool steels). Forged bars, blanks, discs, rings, mill rolls, valve steel, nickel and cobalt superalloys, aerospace alloys, electrical and electronic alloys, high-purity niobium, molybdenum and tungstem metal and their alloys.
Brands: Cold strip mill for alloy steel and ring rolling mill (1988), two horizontal extrusion presses and cold Pilgen mill (1989), ladle steelmaking facilities (1990), electron beam facilities (1990).

Siderúrgica Fi-El SA

Head office: Rua Guaicurus 225, 05033 São Paulo, SP. **Telex:** 1122168, 1130074.
Established: 1926. **Ownership:** Mannesmann SA (Ultimate holding company – Mannesmann AG, West Germany).
Works:
■ Praça Cariri 303, S. José dos Campos, SP. ***Steelmaking plant:*** Two 15/17-tonne electric arc furnaces. ***Continuous casting machines:*** One 2-strand for billet (80 to 160mm sq). ***Rolling mills:*** Schloeman/Ashlow wire rod (coil weight 300 kg). ***Other plant:*** Cold drawing plant; wire products plant.
Products: Carbon steel – billets, wire rod, bright wire, black annealed wire, galvanized wire (barbed). Wire nails, baling and fencing wire.

Fornasa SA

Head office: Av João Batista, 824-6° andar, Osasco, SP 06090. **Tel:** 011 704 6408. **Telex:** 1172027 fovr br. **Cables:** Fornasa. **Telecopier (Fax):** 011 702 8656.
Management: President – Luis Eulalio de Bueno Vidigal; vice presidents – Marcos V. Xavier da Silveira, Luis Eulalio Bueno Vidigal Filho; directors – Rodolfo de Almeida Prado, Alberto Martinez, João de Arruda Meyer Filho (assistant), Arycélio Barroso Silva (assistant). **Established:** 1945. **Capital:** US$5,030,837. (Dec 31 1986). **No. of employees:** Works: 570. Office – 65 (Dec 31 1986). **Ownership:** Cobrasma SA (99.97%). **Annual capacity:** Finished steel: 72,000 tonnes.
Works:
■ Rua Vice Prefeito Wilson de Paiva 20, Bairro do Conforto, 27265 Volta Redonda, RJ. (Tel: 0243 42 3055. Telex: 223163 fovr br). Works director: Arycélio Barroso Silva. ***Tube and pipe mills:*** welded – one Yoder W3½ with Thermatool welder 21.3 to 88.9mm od (annual capcity 25,000 tonnes) and one Etna 4K 4.5 with Tocco welder, 33.4 to 114.3mm od (47,000 tonnes).
Products: Carbon steel – longitudinal weld tubes and pipes (output 65,000 tonnes); oil country tubular goods; galvanized tubes and pipes; cold drawn tubes and pipes; precision tubes and pipes, boiler and heat-exchanger tubes 8 to 100mm nominal size (21.3 to 114.3mm od); wall thickness 2-6.02mm; standard lengths 6 and 6.4m special lengths – black 12.2m, galvanized 6.5m (output 65,000 tonnes).
Brands: Fornasa (black and galvanized pipes, rigid galvanized steel conduits, mechanical tubes, precision tubes, boilder and heat exchanger tubes, line pipe).

Gerdau Group

Head office: Av. Farrapos 1811, 1° andar, Porto Alegre, RS.
Management: Directors – Jorge Gerdau Johannpeter (presidente), Germano H. Gerdau Johannpeter (conselheiro, vice presidente), Klaus Gerdau Johannpeter (conselheiro, vice presidente), Frederico C. Gerdau Johannpeter (conselheiro, vice presidente), Carlos Leoni R. Siqueiraq (conselheiro, vice presidente, consultor juridico), Carlos J. Petry (conselheiro, Planejamento e Recursos Humanos), Edgar K. Oliveira (produção), Claus J. Süffert (engenharia), Ruben Rohde (sistemas), Ruy Lopes Filho (financeiro), Peter Wilm Rosenfeld (relações externas), Gert Funcke (mercado externo), Artur Cesar Brenner Peixoto (holdings), Luiz Celestino Pedó (superintendente geral, siderurgia), Ery J. Bernardes (superintendente geral, informática), Rudolfo T. Tanscheit (superintendente geral, comerciais), Pedro F. Hoerde (superintendente geral, florestal). **Subsidiaries:** Metalúrgica Gerdau SA, Siderúrgica Riograndense SA, Siderúrgica Açonorte SA, Cia Siderúrgica da Guanabara – Cosigua, Siderúrgica Guaíra SA, Cia Siderúrgica de Alagoas – Comesa, Siderúrgica Cearense SA, Comercial Gerdau Ltda, Cifsul – Cia de Indústrias Florestais do Rio Grande do Sul, Seiva SA – Florestas e Indústrias. **Activities:** Holding company.

Siderúrgica Guaíra SA

Head office: Avenida Almirante Barroso 22, 19° andar, 20031 Rio de Janeiro, RJ. **Tel:** 021 262 5022. **Telex:** 021 23574, 021 22853 csgb br.
Established: 1958. **Ownership:** Gerdau group.
Works:
■ Rua Mato Grosso 889, 80000 Vila Guaira, PR (Tel: 041 242 4611. Telex: 041 5019 ggsg br. ***Steelmaking plant:*** electric arc furnace. ***Rolling mills:*** One bar mill.
■ BR 423 Km 24.5, 83700 Araucária, PR (Tel: 041 842 2237. Telex: 041 6374 ggsg br). ***Steelmaking plant:*** One 80-tonne electric arc furnaces (annual capacity 240,000 tonnes). ***Continuous casting machines:*** One 4-strand for billet.
Products: Carbon steel – reinforcing bars.
Expansion programme: Wire rod, reinforcing bar and section mill (annual capacity 270,000 tonnes) to be installed in 1988, increasing rolled steel capacity by 60%.

Siderúrgica Hime SA

Head office: Avendida Almirante Barroso 22, 19° andar, 20031 Rio de Janeiro, RJ. **Tel:** 021 262 5022. **Telex:** 23574, 22853 csgb br.
Established: 1973/1982.
Works:
■ Rodovia – Presidente Dutra Km 197, Quadra 4, 21720 Nova Iguaçu, RJ. (Tel: 021 767 4137. Telex: 32335 hime br). ***Steelmaking plant:*** electric arc furnace. ***Continuous casting machines:*** One Demag machine for billet .
■ Rua Barão de São Gonçalo 196-A, 24500 São Gonçalo. ***Rolling mills:*** One 3-stand 3-high 500mm blooming; one 5-stand 2-high 360mm bar. ***Other plant:*** Forge. Iron foundry.
Products: Carbon steel – billets, round bars, square bars, flats, light angles, rail accessories. Forgings. Castings.

Cia Industrial Itaunense

Head office: Rua São Paulo 409, 13° andar, 30000 Belo Horizonte, MG. **Tel:** 031 201 9155. **Telex:** 312829 cii br.
Established: 1963.
Works:
■ Rua Godofredo Gonçalves 150, 35680 Itaúne, MG. (Tel: 037 241 2333. Telex: 313670 cii br). ***Steelmaking plant:*** electric arc furnace. ***Continuous casting machines:*** One 2-strand Concast for billet . ***Rolling mills:*** Two bar.
Products: Carbon steel – reinforcing bars, round bars.

Lafersa – Laminação de Ferro SA

Head office: Avda General David Sarnoff 270, Cidade Industrial Cel Juventino Dias, 32000 Contagem, MG. **Postal address:** Caixa Postal 2.
Works:
■ Contagem, MA. ***Blast furnaces:*** One (charcoal) furnace. ***Steelmaking plant:*** One 10-tonne basic oxygen converter (annual capacity 40,000 tonnes); open hearth furnace. ***Continuous casting machines:*** One 2-strand billet ***Rolling mills:*** bar and wire rod.
Products: Carbon steel – wire rod, reinforcing bars, round bars.

Mangels São Bernardo SA

Head office: Rua Max Mangels Senior 777, São Bernardo do Campo, São Paulo. **Postal address:** PO Box 561. **Tel:** 011 448 6622. **Telex:** 011 44709, 011 44463. **Cables:** Animax.
Management: President – Max Ernst Mangels; directors – Ariovaldo Carmignani (vice president, works), Jair Moggi (financial), Marcus Benedito Teixeira, Rubens Galhardo,

Mangels São Bernardo SA (continued)

Hans Gerhard Dislich, Arnold Krayer Krauss, Flavio Edson del Soldato, Mário Eduardo Barra, Aloísio Frazão. **Established:** 1980. **Capital:** Cz$114,100,000. **Ownership:** Mangels Industrial SA (100%). **Subsidiaries:** Maxitec SA (13.3%) (computerised numerical controls programmable controls). **Associates:** Mangels S Bernardo SA (100%) (low-carbon cold rolled steel strip, high-carbon and heat treated high-carbon cold rolled strip, steel service centre); Mangels Minas Industrial SA (100%) (LPG cylinders, fuel tanks, passenger car wheels), Mangels Mians Industrial SA – Filial São Paulo (hot-dip galvanizing); Cryometal SA – Metais Especiais e Equips Criogênicos (99.0%) (cryogenic equipment and cryogenic vessels); Maxitrade SA (11.3%) (international trade). **Annual capacity:** Finished steel: 60,000 tonnes (low carbon) and 20,000 (high carbon).
Works:
■ At head office. ***Rolling mills:*** One cold strip – cold rolling mill: Schloemann Mkw-40 reversing, Schmitz 4-high reversing, various 4-high and 2-high non-reversing, Sendzimir 12-roll reversing (annual capacity (low carbon) 60,000 tonnes; (high carbon) 20,000 tonnes). ***Other plant:*** – high convection bell-type annealing furnaces, continuous hardening and tempering lines, grinding, polishing, edge treating lines, steel strapping for paint line.
Products: Carbon steel – cold rolled hoop and strip (uncoated) (low-carbon, high cardon and heat treated high-carbon).
Additional information: Laminação Baukus SA has been incorporated with Mangels São Bernardo.

Mannesmann SA

Head office: Rua Conde Francisco Matarazzo 9500, 09500 São Caetano do Sul, SP. **Tel:** 011 453 3722. **Telex:** 1144135, 1144177, 1144130.
Management: Peter Ulrich Schmidthals (chairman), Dieter Althoff, Joseph Doll, Friedrich-Hams Grandin, Paul Josef Günther. **Established:** 1952. **Ownership:** Mannesmann AG, West Germany (76%). **Subsidiaries:** Siderúrgica Fi-El, São José dos Campos.
Works:
■ Usina do Barreiro, Caixa Postal 2153/2154, 30161 Belo Horizonte, MG (Tel: 031 333 2244. Telex: 311170 mann br). ***Blast furnaces:*** One 5.5m hearth dia, 478 cu m (annual capacity 300,000 tonnes). ***Other ironmaking plant:*** Two 17MW submerged arc furnaces (150,000 tonnes). ***Steelmaking plant:*** One 70-tonne basic oxygen converter (520,000 tonnes); three 40-tonne electric arc furnaces (230,000 tonnes). ***Rolling mills:*** One 3-stand 3-high 750 x 1,200mm blooming and bar (560,000 tonnes); 3-stand 3-high 430 x 1,400mm bar and wire rod (coil weight 100kg) (150,000 tonnes). ***Tube and pipe mills:*** seamless -one 6 – 10$\frac{3}{4}$" Mannesmann plug (170,000 tonnes); one 1$\frac{1}{2}$" 6" Pilger (110,000 tonnes); two $\frac{1}{2}$ to 3" extrusion presses (70,000 tonnes); seamless cold rolling and drawing mill – three cold Pilger and twelve drawbenches. Welded – two (50,000 tonnes).$\frac{3}{8}$ to 3" continuous. ***Other plant:*** OCTG plant. Forging plant for tubulars. Hot-dip tube galvanizing plant.
■ Guarulhos, SP. ***Rolling mills:*** Three cold strip mills – one 2-high, and two 4-high. ***Tube and pipe mills:*** welded – Seven continuous. ***Other plant:*** Drawbenches.
■ Sae Caetano, SP. ***Rolling mills:*** One bar and wire rod. ***Other plant:*** Bright bar plant.
Products: Carbon steel – ingots, blooms, billets, semis for seamless tubemaking, wire rod, round bars, square bars, flats, hexagons, bright (cold finished) bars, black annealed wire, seamless tubes and pipes (1986 output, 370,000 tonnes), longitudinal weld tubes and pipes (71,000 tonnes), large-diameter tubes and pipes, oil country tubular goods, galvanized tubes and pipes, cold drawn tubes and pipes, hollow bars, precision tubes and pipes. Alloy steel (excl stainless) – wire rod, wire, seamless tubes and pipes, longitudinal weld tubes and pipes, oil country tubular goods, cold drawn tubes and pipes, precision tubes and pipes, hollow sections. High speed and bearing steel.
Additional information: Raw steel output in 1986: 842,000 tonnes, of which oxygen steel 700,000 tonnes.

Siderúrgica Mendes Junior SA (SMJ)

Head office: BR-040, km 769, Distrito de Dias Tavares, 36105 Juiz de Fora, Minas Gerais. **Postal address:** Caixa Postal 232. **Tel:** 032 229 1000. **Telex:** 032 2115. **Telecopier (Fax):** 032 229 1717.
Management: President – J. Murillo Mendes; vice president – Fernando Zenobio Carvalho; directors – Sérgio de Moraes (marketing), Newton Cotrim (works), José Carlos Nogueira (financial). **Established:** 1984. **Capital:** US$260,000,000. **No. of employees:** Works – 2,080. Office – 1130. **Annual capacity:** Raw steel: 400,000 tonnes. Finished steel: 840,000 tonnes.
Works:
■ At head office. Works director: Newton Cotrim. ***Steelmaking plant:*** 70/80-tonne electric arc (annual capacity 400,000 tonnes). ***Continuous casting machines:*** 4-strand Concast billet (130 x 130 x 15,700mm) (400,000 tonnes). ***Rolling mills:*** Two combined bar and wire rod – No 1 producing 12.5-40mm dia material and No 2 5.5-12.5mm (coil weight 2,000kg) (combined annual capacity 840,000 tonnes).
Products: Carbon steel – wire rod 5.5. to 12.5mm (output 263,000 tonnes); reinforcing bars 5.5 to 40.0mm (302,000 tonnes); round bars 5.5 to 40.0mm (9,000 tonnes); square bars 12.5 to 28.8mm (3,000 tonnes); bright wire 1.2 to 11.0mm (20,000 tonnes); black annealed wire 1.24 to 6.0mm (4,000 tonnes); galvanized wire (plain) 1.5 to 5.0mm (10,000 tonnes); galvanized wire (barbed) 1.6 to 2.2.mm (12,000 tonnes); nails and fence staples (2,500 tonnes).
Sales offices/agents: Av Almirante Barroso 52 20031 Rio de Janiero, Brazil. (Tel: 021 220 9910. Telex: 021 35853. Telefax: 021 240 0246).

Cia Siderúrgica Pains

Head office: Av Gabriel Passos 102, 35500 Divinópolis, Minas Gerais. **Postal address:** Caixa Postal 140. **Tel:** 037 221 1000. **Telex:** 031 2553. **Telecopier (Fax):** 037 221 0358.
Management: Directors – Ralph Weber (managing), Dalton Nosé (technical), Antônio Carlos Cascão (sales), Gunther H. Porst (financial), Antônio A. Rodrigues (adminstrative). **Established:** 1953. **Capital:** Cz$103,524,000. **No. of employees:** Works – 1,539. Office – 344. **Ownership:** Korf Group (33%), João Jabour (33%), Deutsche Finanzierungsgesellschaft für Beteiligungen in Entwicklungsländern GmbH (33%). **Subsidiaries:** Expresso Rio Mar SA (100%) (transport), Pains Florestal SA (100%) (reafforestation). **Annual capacity:** Pig iron: 270,000 tonnes. Raw steel: 400,000 tonnes. Finished steel: 430,000 tonnes.
Works:
■ At head office. Works director: Dalton Nosé. ***Blast furnaces:*** One 2,750mm hearth dia, 110 cu m (annual capacity 70,000 tonnes); one 3,730mm hearth dia, 192 cu m (120,000 tonnes); one 2,900mm hearth dia 125 cu m, (80,000 tonnes). ***Steelmaking plant:*** Three 28-tonne open hearth (160,000 tonnes); one 30-tonne EOF (260,000 tonnes). ***Continuous casting machines:*** Two 2-strand Concast billet (120mm sq) (400,000 tonnes). ***Rolling mills:*** bar – 3-stand, (8.0 to 40.0mm) (annual capacity 300,000 tonnes); wire rod – 1-stand, (6.35 and 8.0mm) (coil weight 370kg) (60,000 tonnes).
■ Rua Jonas Barcelos Corrêa, 1,340, Contagem, MG ***Rolling mills:*** bar – 1-stand, (6.35 to 12.5mm) (70,000 tonnes). ***Other plant:*** Wire drawing mill (capacity 24,000 tonnes).
Products: Carbon steel – billets 120mm sq (1986 output 6,000); wire rod 6.35 and 8mm; reinforcing bars 8-40mm (245,000 tonnes); round bars 6.35-12mm (40,000 tonnes); flats (2,000 tonnes); bright wire (18,000 tonnes); black annealed wire (2,500) tonnes; galvanized wire (plain) (511 tonnes); galvanized wire (barbed) (1,000 tonnes).
Sales offices/agents: Main sales office – Rua Jonas Barcelos Corrêa, 1,340, Contagem, MG.
Expansion programme: Steelmaking: Second EOF (300,000 tpy) (1988) Continuous casting: (300,000 tpy) machine (1988), Rolling mill modernisation (1987/1991).

Persico Pizzamiglio SA

Head office: Rodovia Presidente Dutra km 219, Guarulhos 07270, São Paulo. **Tel:** 912 4711. **Telex:** 011 33703.
Management: Vittorio Vignoli (commercial director), Alberto Seródio (commercial superintendent), Luiz Roberto H. Pinto (international sales manager). **Established:** 1952. **Capital:** US$84,000,000. **No. of employees:** Works – 2,300. Office – 400. **Annual capacity:** Finished steel: 240,000 tonnes (tube and pipe).
Works:
■ At head office. Works director Salvatore Di Mino. ***Tube and pipe mills:*** welded – ERW and TIG tube mills, $\frac{1}{4}$-7" od (annual capacity 300,000 tonnes).
Products: Carbon steel – pipe piling, longitudinal weld tubes and pipes, oil country tubular goods, galvanized tubes and pipes, cold drawn tubes and pipes, precision tubes and pipes. Stainless steel – seamless tubes and pipes, longitudinal weld tubes and pipes, oil country tubular goods, cold drawn tubes and pipes, precision tubes and pipes.

Aços Finos Piratini SA

Head office: Rua Câncio Gomes 127, Porto Alegre, 90000 Rio Grande do Sul. **Tel:** 051 222 5611.
Management: President – Walter Jobim Filho; directors – Waldir Comerlato (financial), Carlos Olivério Arnt (commercial), Francisco Guilherme Alcântara Oliveira (industrial), José Manoel Gonzalez de Souza (administrative). **Established:** 1961. **Capital:** Cz$ 2,326,370,645. **No. of employees:** 2,694. **Ownership:** Siderbrás – Siderurgica Brasileira SA (83.87%).
Works:
■ Av Getúlio Vargas, Charqueadas, 96705 Rio Grande do Sul (Tel: 051 221 9800. Telex: 512404). ***Direct reduction plant:*** One 50m SL/RN (annual capacity 55,000 tonnes). ***Steelmaking plant:*** Two electric arc furnaces – one 12-tonne, and one 50-tonne (285,000 tonnes). ***Refining plant:*** One 50-tonne Asea-SKF unit (240,000 tonnes). ***Rolling mills:*** One 2-high 750/600mm reversing billet (236,000 tonnes); 6-stand 3-high/alternate 2-high open bar (260,000 tonnes); one 6-stand 600/280mm continuous wire rod (coil weight 600kg) (84,000 tonnes).
Products: Carbon steel – billets 100 – 175mm (output 3,435 tonnes); wire rod 6.00-30.16mm (9,311 tonnes); round bars 6.00 – 152.40mm dia (66,407 tonnes); square bars 30.00 – 113.00mm sq (3,621 tonnes); flats 50.80 – 279.40mm wide, 7.94 x 63.50mm thick (1,731 tonnes). Stainless steel – billets 100 – 175mm; wire rod 6.00 – 30.16mm (2,578 tonnes); round bars 6.00 – 147.63mm (2,816 tonnes); square bars 30.00 – 113.00mm; flats. Alloy steel (excl stainless) – billets 100 to 175mm (1,851 tonnes); wire rod 6.00 to 30.16mm (16,431 tonnes); round bars 6.00 to 152.40mm dia (38,945 tonnes); square bars 30.00 to 113.00mm sq (3,018 tonnes); flats 50.80 x 279.40 wide, 7.94 x 63.50 thick (1,294 tonnes).

Usina Queiroz Júnior SA Indústria Siderúrgica

Head office: Rodovia do Inconfidentes km 51, 35450 Itabirito, MG. **Telex:** 031 1493 uqsa.
Works:
■ At head office. ***Blast furnaces:*** One (annual capacity approx 60,000 tonnes); foundry with electric furnaces.
Products: Foundry pig iron, iron and steel castings.

Siderúrgica Riograndense SA

Head office: Avenida Borges de Medeiros 650, 93200 Sapucaia do Sul, RS. **Tel:** 0512 73 1166, 0512 73 1288. **Telex:** 051 1456.
Established: 1938. **No. of employees:** 2,422. **Ownership:** Gerdau Group. **Annual capacity:** Pig iron: 80,000 tonnes. Raw steel: 280,000 tonnes. Finished steel: 605,000 tonnes (rolled), 170,000 tonnes (drawn).
Works:
■ Sapucaia do Sul, RS and Porto Alegre, RS. ***Steelmaking plant:*** Three electric arc

Siderúrgica Riograndense SA (continued)

furnaces – one 10-tonne and two 20-tonne. ***Refining plant:*** Two ladle furnace. ***Continuous casting machines:*** Three 2-strand Demag billet (100 and 120mm sq). ***Rolling mills:*** Two bar – one continuous and one semi-continuous; one wire rod – high speed semi-continuous single-stand with Morgan No-twist finishing block (max finishing speed steel. ***Other plant:*** 32 drawing machines, hot-dip wire galvanizing line, barbed wire machine, wire mesh plant and chain-making plant.
■ Antonina, PR. ***Blast furnaces:*** Two.
Products: Carbon steel – billets, wire rod, reinforcing bars, round bars, bright (cold finished) bars, light angles, light channels, bright wire, black annealed wire, galvanized wire (plain), galvanized wire (barbed). Wire mesh, chains, rivets, nails, fence staples and pig iron.
Brands: Elefante, Urso, Potro (barbed wire); Hoefel Sander (chains).

Sandvik do Brasil SA

Head office: Av das Nacões 1130, São Paulo, SP.
Ownership: Sandvik AB, Sweden.
Products: Stainless steel – wire and tubes.

Laminação Santa Maria SA – Indústria e Comércio

Head office: Avenida Ugo Fumagalli 770, Cumbica Guarulhos, SP.
Established: 1940.
Works:
■ At head office. ***Rolling mills:*** One 500/300mm wire rod mill. ***Other plant:*** Bright drawing plant.
Products: Carbon steel – wire rod, round bars, square bars, flats, hexagons, special light sections, cold drawn tubes and pipes. Alloy steel (excl stainless) – round bars, square bars, flats, hexagons, special light sections, cold drawn tubes and pipes.

Siderúrgica São Cristóvão Ltda

Head office: Rua Ipatinga 597, Divinópolis, Minas Gerais. **Postal address:** PB 35. **Tel:** 037 221 3077. **Telex:** 039 1079.
Management: Jaime Martins Filho (general manager). **Established:** 1959. **Capital:** CZ$16,800,000. **No. of employees:** Works – 200. Office – 20. **Annual capacity:** Pig iron: 120,000 tonnes.
Works:
■ At head office. Works director: Paulo César. ***Blast furnaces:*** Three – No 1 60 cu m (annual capacity 32,000 tonnes); No 2 35 cu m (18,000 tonnes); No 3 130 cu m (70,000 tonnes).
Products: Steelmaking pig iron (output 24,918,530 tonnes); foundry pig iron (7,469,120 tonnes).

Siderama – Cia Siderúrgica da Amazônia

Head office: Estrada do Paredão s/n, Distrito Industrial, Manaus 69000, AM. **Tel:** 237 2121. **Cables:** Siderama.
Ownership: State-owned.
Works:
■ At head office.
Products: Carbon steel – wire rod, reinforcing bars.

Siderbrás – Siderurgia Brasileira SA

Head office: SAS, Quadra 2, Bloco E, Ed Siderbrás, Brasília 70070, Distrito Federal. **Tel:** 061 216 7171. **Telex:** 061 1542.
Management: President – Amaro Lanari Jr; directors – José Barros Cota (technical), Antônio Roberto de Oliveira Zappia (planning), Gabriel Mousinho Furtado Gomes (financial), Claudio Cesar de Avellar (foreign affairs). **Established:** 1973. **Capital:** Cz$39,963,160,241. (May 1987). **No. of employees:** 244. **Ownership:** State owned. **Subsidiaries:** Aço Minas Gerais SA (Aćominas); Cia Siderúrgica Nacional – CSN; Cia Siderúrgica de Mogi das Cruzes – Cosim; Cia Ferro e Aço de Vitória – Cofavi; Cia Siderúgica Paulista – Cosipa; Usinas Siderúrgicas de Minas Gerais SA – Usiminas; Usina Siderúrgica da Bahia SA – Usiba; Cia Siderúrgica de Tubarão – CST; Cia Brasileira de Projetos Indutriais – Cobrapi (engineering company). **Associates:** Siderúrgica Mendes jr. **Annual capacity:** Raw steel: 14,000,000 tonnes. Finished steel: 13,000,000 tonnes.
Activites: Holding company for Brazilian state-owned steel interests.

Sidersete – Siderúrgica Sete Lagoas Ltda

Head office: Street Sítio da Abadia 21, Sete Lagoas, Minas Gerais. **Postal address:** PO Box 57. **Tel:** 031 921 6911. **Telex:** 0312976.
Management: Paulo Vicente Hernandez (manager), José Gonçalves Filho (manager), Silvio João Bosco (superintendent). **Established:** 1959. **Capital:** Cz$16,500,000. **No. of employees:** Works – 369. Office – 43. **Ownership:** Aços Anhanguera SA (99.97%), Caemi Internacional SA (0.03%) (Ultimate holding company – Cia Auxiliar de Empresas de Mineração (Caemi)). **Annual capacity:** Pig iron: 120,000 tonnes.
Works:
■ At head office. ***Blast furnaces:*** Two (charcoal) 80-100 tonnes per day.
Products: Foundry pig iron.

Tekno SA – Construcções Industria e Comércio

Head office: Eugênio de Freitas 130, 02060 São Paulo, SP. **Tel:** 011 292 1411. **Telex:** 1122680.
Management: President – José Lyra Madeira; vice president – João Alberto A. Borges; superintedent director – Guilherme Luis do Val; directors – Antonio Monte Costa, Enio Perillo, Sebastião Paulino, Valter Sassaki, Fábio Giannini, Claudio Lanari. **Established:** 1939. **Capital:** US$3,300,000. **No. of employees:** Works – 500. Office – 150. **Subsidiaries:** Perkrom – Construcção, Industria e Comercio Ltda (100%) (building panels from hot dip galvanized steel coil). **Annual capacity:** Finished steel: 60,000 tonnes.
Works:
■ Rodovia Washington Luis 181 2 km, Caixa Postal 3, 12500 Guaratinguetá, SP. (Tel: 125 22 4700. Telex: 125541). Works director: Claudio Lanari. ***Coil coating lines:*** One tandem line colour coating (Hunter Engineering Co, USA) (annual capacity 48,000 tonnes). ***Other plant:*** For pre-painted sheets and special profiles for roofs and sidings (large spans) (Tekno) and for pre-painted profiles for roofs and sidings (Perkrom).
Products: Carbon steel – colour coated sheet/coil 0.20-2.00mm, 1220mm wide (output 36,000 tonnes).
Brands: Kalha – pre-painted HDG steel profiles for large spans, Kroma – pre-painted metals in coils and cut-to-length sheets.
Sales offices/agents: Caixa postal 3489, 01000 Sao Paulo, Brazil.

Cia Siderúrgica de Tubarão

Head office: BR 101 Norte, km 8.5, Planalto de Carapina, Serra, Espirito Santo. **Postal address:** Caixa Postal 211. **Tel:** 027 228 2111. **Telex:** 027 2225, 027 2235.
Management: Directors – Arthur Carlos Gerhardt Santos (president), Antiocho Carneiro de Mendonça (financial), Fernando A. Paschoal Guerra (development), Ichiro

Cia Siderúrgica de Tubarão (continued)

Fukunaga (technical and production), Giorgio Dogliotti (control); José Wellington G. Rezende (general superintendent, purchasing), Leopoldino de Oliveira (sales superintendent), Celio Marcio Diniz Gomes (head of corporation planning staff). **Established:** 1983 (start-up). **Capital:** Cz$9,712,940,202. **No. of employees:** Works – 4,810. Office – 1,550. **Ownership:** Shareholders – Siderbrás – Siderúrgia Brasileira SA (51%), Kawasaki Steel Corp (24.5%), Stà Finanziaria Siderúrgica Finsider pAz (24.5%). **Annual capacity:** 1986 – Pig iron: 3,340,000 tonnes. Raw steel: 3,404,000 tonnes. Finished steel: 3,008,000 tonnes (slabs).
Works:
■ At head office. ***Coke ovens:*** Three Carl Still batteries (147 ovens) (annual capacity 1,372,000 tonnes). ***Sinter plant:*** One Lurgi-Dwight Lloyd 440 sq m strand (width 5m and length 88m) (4,857,000 tonnes). ***Blast furnaces:*** One 14m hearth dia, 3,707 cu m (3,285,000 tonnes). ***Steelmaking plant:*** Two 280-tonne/heat basic oxygen LD (3,370,000 tonnes). ***Rolling mills:*** Universal 1-stand slabbing 1,270 x 2,425mm (3,000,000 tonnes).
Products: Carbon steel – slabs – thickness 100 to 305mm, width 650/1,911mm, length 4,450 – 12,500mm (3,008,000 tonnes).
Sales offices/agents: 477 Madison Avenue, Room 73, NY 100022, USA (Tel 011 212 421071). Shinkokusai Bldg 5th Floor, Room 536, 4-1-3-chome, Marunouchi Chyioda-ku, Tokyo, Japan. (Tel 214 0988-9).
Expansion programme: Continuous casting machine, (annual capacity 2,000,000 tonnes) – 1990. Second stage (submitted to Federal Government) includes a hot strip mill (capacity 4,000,000 tonnes) -1992.

Usiba – Usina Siderúrgica da Bahia SA

Head office: BR 324, km 16, Centro Industrial de Aratu, Simões Filho, Bahia. **Tel:** 071 594 8311. **Telex:** 071 1521 usib br.
Management: President – Helder Parente Prudente; directors – Guilherme Furtado (financial), Matthias Jödicke (commercial), Edson Pitta Lima (administration). **Established:** 1963. **Capital:** Cz$2,643,026,568.(April 1987). **No. of employees:** 1,426. **Ownership:** Siderbrás – Siderurgia Brasileira SA (99.42%).
Works:
■ At head office. ***Direct reduction plant:*** One 600mm HYL I (annual capacity 240,000 tonnes). ***Steelmaking plant:*** One 90-tonne, 42 MVA electric arc furnace (330,000 tonnes). ***Continuous casting machines:*** One 6-strand billet (straight mould with bending) (120mm) (320,000 tonnes). ***Rolling mills:*** One 17-stand continuous light section (up to 3") and one 17-stand continuous bar (10-50mm) (400,000 tonnes); one 8-stand continuous wire rod (6.35-10mm) (coil weight 1.3 tonne) (400,000 tonnes).
Products: Direct-reduced iron (DRI) (1986 output 243,635 tonnes). Carbon steel – billets 120mm (313,900 tonnes); wire rod 6.35m and 8mm (14,000 tonnes); reinforcing bars 6.35m and 32mm (264,800 tonnes).
Sales offices/agents: Overseas agents: Ferrostaal, Interbrás, Coutinho Caro.
Expansion programme: Arc furnace transformer uprating to 76 MVA and increase in works annual capacity to 400,000 tonnes.

Usinas – Usinas Siderúrgicas de Minas Gerais SA

Head office: Rua Prof Vieira de Mendonça 3011, Pampulha, Belo Horizonte 30000, Minas Gerais. **Tel:** 031 441 4222. **Telex:** 031 1261.
Management: President – Ademar de Carvalho Barbosa; directors – Uajará Rodrigues (financial), Luiz Eduardo Kikinger de Abreu (development), Rinaldo Campos Soares (operations), Lauro Cesar de Abreu (administrative), Kazumasa Nagasaki (Japanese Affairs). **Established:** 1957. **Capital:** Cz$8,879,113,243. **No. of employees:** 14,555. **Ownership:** Siderbrás – Siderurgia Brasileira SA (87.25%). **Annual capacity:** Pig iron: 3,026,000 tonnes. Raw steel: 4,300,000 tonnes. Finished steel 2,952,000 tonnes.
Works:
■ Usina Intendente Câmara, Caixa Postal 22, 35160 Ipatinga, Minas Gerais (Tel: 031

Usinas – Usinas Siderúrgicas de Minas Gerais SA (continued)

821 2444. Telex: 031 1334). Works director: Rinaldo Campos Soares. ***Coke ovens:*** Two Nittetsu Duplo-S underjet batteries (100 ovens) (annual capacity 545,000 tonnes), and two Nittetsu Duplo-M underjet batteries (110 ovens) (1,116,000 tonnes). ***Sinter plant:*** Three Dwight Lloyd – one 90 sq m strand, and two 180 sq m strand (5,140,000 tonnes). ***Blast furnaces:*** Two 7,000mm hearth dia, 885 cu m (535,000 tonnes each); one 11,500mm, 2,700 cu m (1,956,000 tonnes). ***Steelmaking plant:*** LD basic oxygen converters – No 1 three 80-tonne (800,000 tonnes), No 2 two 180-tonne (2,400 tonnes). ***Refining plant:*** One 192-tonne Confab Industrial/ Lectromelt ladle furnace (360,000 tonnes). ***Continuous casting machines:*** Three 2-strand Hitachi/Demag 10.5 m radius slab (250-300 x 940-1,910 x 6,000mm) (2,500,000 tonnes). ***Rolling mills:*** One 2-high reversing slabbing (1,150 x 2,930mm max slab weight) (coil weight 19 tonnes) (1,800,000 tonnes); one 4-high stand quarto-reversing heavy plate comprising work roll 1,100mm x 4,100mm dia, back-up roll 2,000 x 4,000mm dia (960,000 tonnes); six-stand 4-high wide hot strip comprising work roll 700 x 2,100mm dia, back-up roll 1,400 x 1,950mm (max coil weight 16 tonnes) (2,400,000 tonnes); 5-stand 4-high continuous cold reduction comprising work roll 584 x 1,780mm dia, back-up roll 1,423 x 1,680mm dia (1,331,000 tonnes); two temper/skin pass – one 6-high (work roll 533 x 1,680mm, back-up roll 1,420 x 1,680mm and int dia 538 x 1,420mm) (720,000 tonnes), and one 4-high (work roll 535 x 1,690mm and back-up roll 1,420 x 1,690mm) (600,000 tonnes).

Products: Steelmaking pig iron and foundry pig iron (1986 output, 2,862,610 tonnes); liquid steel (3,138,789 tonnes); raw steel (3,073,482 tonnes). Carbon steel – slabs; hot rolled coil (for re-rolling) 1.50 to 5.00mm and 700 to 1,900mm; medium plates 1.50 to 5.00mm and 700 to 1,900mm x 1,829 to 12,192mm, and heavy plates 5.00 to 150mm x 760 to 3,400mm x 2,400 to 18,000mm (combined 1986 output, 793,393 tonnes); hot rolled sheet/coil (uncoated) 1.50 to 5.00mm x 700 to 1,900mm x 1,829 to 12.192mm (sheet – 237,356 tonnes, coil 817,125 tonnes); cold rolled sheet/coil (uncoated) 0.40 to 3.00mm x 700 to 1,700mm x 1,215 to 1,600mm (or in coil form) (cold rolled sheet – 205,947 tonnes, cold rolled coil – 842,108 tonnes, uncoated coil – 56,361 tonnes); electrical non-oriented silicon sheet/coil 0.45 to 0.75mm x 1,200mm x coil; electrical non-oriented non-silicon sheet/coil 0.45 to 0.75mm x 1,200mm x coils; blackplate 0.21 to 0.38mm x 500 to 1,250mm (33,414 tonnes); ingot equivalent (3,098,123 tonnes); semis for sale (15,062 tonnes). Total 1986 output of finished products – 3,000,766 tonnes. Also coke (1,659,407 tonnes); sinter (4,631,650 tonnes).

Sales offices/agents: Brazil – 60 Rua da Candelária, Rio de Janeiro, RJ (Tel: 021 233 6622. Telex: 021 22169); 433 Alameda Campinas, São Paulo, SP (Tel: 011 288 0122. Telex: 011 23098); 236 Rua General Câmara, Porto Alegre, RS (Tel: 0512 24 8444. Telex: 051 1235); 143 Rua Comendador Araújo, Curitiba, PR (Tel: 041 232 7144. Telex: 041 5810); 4,777 Av. Conselheiro Aguiar, Recife, RE (Tel: 081 326 4667. Telex: 081 4766). Japan – Room 536, Shin Kokusai Bldg 3-4-1, Chiyodai-Ku, Marunouchi (Tel: 00813 214 1567. Telex: 2226522).

Usipa – Usina Siderúrgica Paraense SA

Head office: Av João Cesar de Oliveira 5365, Contagem, MG. **Telex:** 316261 usip br, 316260 usip br.

Established: 1959. **Ownership:** Gerdau Group.

Works:

■ Para de Minas. ***Blast furnaces:*** – one 2.2m hearth dia, 82 cu m, one 1.6m hearth dia, 38 cu m.

■ Contagem. ***Blast furnaces:*** two 3.45m hearth dia, 157 cu m.

Products: Steelmaking pig iron. Foundry pig iron.

Vibasa – Villares Industrias de Base – See Villares Group

Brazil

Villares Group

Head office: Av. do Estado 6116, São Paulo, SP. **Postal address:** PO Box 01516. **Tel:** 011 278 7322. **Telex:** 011 22857. **Telecopier (Fax):** 2797015. **Management:** President – Paulo D. Villares; vice presidents – André Musetti (steel and industrial group), Iwan Oleg von Hertwig (steel), Theodoro Niemeyer (development); directors – Humberto H. Parreiras Henriques (commercial), Werner Viertler (industrial). **Established:** 1944. **Capital:** Cz$ 707,794,000. **No. of employees:** Works – 4,667. Office – 409. **Ownership:** Industrias Villares SA, Villares Family, Catena Participações public shareholding, BNDES. **Subsidiaries:** Villares Overseas Corp, USA; Alvorada; Florestradora Perdizes; Técnica Villares; Curiango; Ascensores Atlas, Uruguay. **Annual capacity:** Raw steel: 464,000 tonnes. Finished steel: 372,000 tonnes.

Aços Villares SA
Works:

■ Av. Dr Ramos de Azevedo 133, São Caetano do Sul, SP 09520 (Tel: 011 453 2788. Telex: 011 44010). Works director: Werner Viertler. ***Steelmaking plant:*** Four electric arc furnaces – two 25-tonne, one 15-tonne, and one 5-tonne (annual capacity 96,000 tonnes); two induction furnaces – 2.5-tonne and 0.3-tonne (4,000 tonnes). ***Refining plant:*** One 25-tonne Asea-SKF. ***Rolling mills:*** One 3-stand billet 600mm (58,000 tonnes); two medium section – one 3-stand 600mm, and one 3-stand 400mm; one 6-stand bar 380mm and one 7-stand wire rod 320mm (coil weight 350kg) (combined capacity 29,800 tonnes). ***Other plant:*** Forging shop with a 2,000-tonne hydraulic press for parts up to 15-tonnes; 2,000kg drop forging hammer for parts up 400kg. Heat treating facilities including two induction hardening machines for rolls, and two continuous furnaces. Heat treatment plant. Foundry for casting up to 90-tonnes with two centrifugal casting machines for tubes. Bar finishing and pickling lines.

Products: Carbon steel – billets; round bars 6-203mm (1986 output 325 tonnes); square bars 8-50.80mm; flats 3 to 38mm thickness, 15 to 102 width (60 tonnes). Stainless steel – billets; wire rod (coils) 8mm up to 21mm (2,150 tonnes); round bars 6-203mm (9,500 tonnes); square bars 8-50.80mm (400 tonnes); flats 3-38mm thickness, 15 to 102 width (5,000 tonnes); hexagons 9.5-50.80mm (850 tonnes); bright (cold finished) bars drawn squares, rounds and hexagons up to 50mm and ground bars up to 165mm; light angles 3-10mm thickness, 12 to 76mm leg length (1,000 tonnes). Alloy steel (excl stainless) – round bars 6mm to 203mm (12,000 tonnes); square bars 8mm to 50.80mm (750 tonnes); flats 3-38mm thickness, 15 to 102mm width (4,300 tonnes); hexagons 95mm to 50.80mm (60 tonnes). Bright (cold finished) bars tool steels 1.00mm to 8.90mm and ground bars up to 165mm. Forged special steel bars, shafts, rings, discs, forged steel rolls, pressing die blocks and other parts. Main sizes up to 15 tonnes and 2,000mm dia. Castings: centrifugally cast steel tubes, cast steel rolls, return bends and fittings, light and heavy castings in carbon steel, alloy steel, heat resisting steel, corrosion-resistant stainless steel, high-alloy iron casting of the "Ni-resist" type and others. Sizes range from 10kg to 90 tonnes for castingsand up to 54 tonnes for rolls.
Brands: Carbon and low-alloy steel, hot/cold work tool steels, high-speed tool steels, stainless steels, engine valve steels, electric resistance alloys and nickel base alloys.
Sales offices/agents: Villares Overseas Corp (USA), 630 Fifth Avenue, Suite 1750, New York NY 5544 (Tel: 212 307 1011. Telex: 423483).
Expansion programme: Electroslag remelting and vacuum induction melting (1989). Plasma vapour deposition (July 1987).

Vibasa – Villares Industria de Base SA
Works:

■ Rodovia Luis Dumont Villares Km 2 s/n°, Moreira Cesar, Pindamonhangaba, São Paulo, 12,400. (Tel: 0122 40 1120. Telex: 0122 177). Works director: Werner Viertler. ***Steelmaking plant:*** Two 80-tonne and one 40-tonne electric arc furnaces (combined annual capcity 364,000 tonnes). ***Refining plant:*** Tap degassing system for each EAF vacuum degassing unit; Vacuum tank for ladle degassing or ingot making under vacuum. ***Rolling mills:*** One 2-high 900mm dia reversing blooming (300,000

Villares Group (continued)

tonnes); one 2-high 750mm dia reversing billet (300,000 tonnes); one 21-stand 1-strand continuous bar and wire rod, 560 to 315mm dia (coil weight 1,700kg) (180,000 tonnes). ***Other plant:*** Steel Foundry for parts up to 125 tonnes finished weight. Forging shop with a hydraulic press of 6,300/8,000 tonnes for parts up to 125 tonnes finished weight. Iron foundry for castings and mill rolls up to 45 tonnes finished weight with induction furnaces of 5 tonne, 8 tonne and 25 tonne nominal capacities.Four 4 centrifugal casting machines for mill rolls to 10 tonne finished weight. Total annual capacity 24,000 tonnes.
Products: Carbon steel – blooms round or square 203.2mm to 254mm (1986 output 3,023 tonnes); billets RCS 66.6 to 177.8mm (26,000 tonnes); wire rod 6.15 to 38.10mm (9,100 tonnes); round bars 12.70 to 76.20mm (102,600 tonnes); square bars 50.80 to 63.50mm (5,000 tonnes); flats 5.4 to 27.0 thick x 50.8 to 102.mm wide (500 tonnes). Alloy steel (excl stainless) – blooms 203 to 254mm (9,030 tonnes); billets RCS 66.6 to 203mm (30,014 tonnes); wire rod (coils) 6.35 to 38.10mm (21,204 tonnes); round bars 12.70 to 254mm (48,637 tonnes); square bars 50.80 to 63.50mm (4,640 tonnes); flats 5.4 to 27.0mm thick x 50.8 to 102mm wide (35,145 tonnes); Forging up to 125 Mt finished weight including forged rolls. Castings up to 125 Mt finished weight, including rolling mill rolls static cast up to 40 Mt finished weight and centrifugally cast up to 10 Mt finished weight.
Brands: Carbon and low alloy steels, including grades for spring, bearing and cold heading grades and steels for the forging and automotive industries.
Sales offices/agents: Villares Overseas Corp (USA), 630 – Fifth Avenue, Suite 1750, New York, NY 1011.

Bulgaria

Burgas Steelworks

Head office: Debelt, Near Burgas.
Works:
■ ***Rolling mills:*** One light section comprising Sket single-strand 4-stand vertical/horizontal roughing train, 2-strand 6-stand roughing and 6-stand intermediate trains, two single-strand finishing trains of eight vertical/horizontal stands each (annual capacity 800,000 tonnes).
Products: Carbon steel – reinforcing bars 12 to 45mm; round bars 12 to 60mm; flats 5 to 20mm x 30 to 70mm; light angles 20 x 20 x 3mm to 75 x 75 x 8mm and 25 x 16 x 3mm to 80 x 50 x 5mm.
Expansion programme: A direct reduction plant and up to five electric arc furnaces are planned.
Additional information: Due to start-up by end of 1987. Billets will initially be supplied from Kremikovtsi and the Lenin Works at Pernik.

Kremikovtsi Iron & Steel Works

Head office: Near Sofia.
Established: 1963. **Annual capacity:** Finished steel: 4,000,000 tonnes.
Works:
■ ***Coke ovens:*** Four batteries (annual capacity 643,000 tonnes – No 3 battery, 650,000 tonnes – No 4 battery). ***Sinter plant:*** 4-strand (300 sq m) (3,500,000 tonnes). ***Blast furnaces:*** Three furnaces – two 1,033 cu metres (560,000 tonnes), one other (520,000 tonnes). ***Steelmaking plant:*** Four basic oxygen converters –

Bulgaria

Kremikovtsi Iron & Steel Works (continued)

three 100-tonnes (1,265,000 tonnes) and one bottom blown; three 100-tonne electric arc furnaces (one furnace has a capacity of 135,000 tonnes). ***Refining plant:*** One 120-tonne Asea-SKF unit (500,000 tonnes). **Continuous casting machines**. ***Rolling mills:*** One 1,500mm slabbing and blooming (1,265,000 tonnes); one continuous billet (850/700/500mmm); 250mm continuous bar and wire rod; one 6-stand 1,700mm semi-continuous wide hot strip (530,000 tonnes); four cold reduction – 5-stand 4-high 585/1,350 x 1,200mm (450,000-550,000 tonnes), 1,700mm and 20-roll cluster mill (600,000 tonnes combined), and 1,200mm (500,000 tonnes). ***Tube and pipe mills:*** seamless – One (50,000 tonnes). Welded – One longitudinal and one spiral (50,000 tonnes). ***Coil coating lines:*** One electrolytic tinning (120,000 tonnes); one hot-dip galvanizing (135,000 tonnes); one colour coating aluminizing line (105,000 tonnes); ***Other plant:*** Cold forming line for sections. Tube galvanizing plant.
Products: Carbon steel – ingots, slabs, blooms, billets, semis for seamless tubemaking, wire rod, reinforcing bars, light angles, light tees, light channels, cold roll-formed sections (from strip), hot rolled hoop and strip (uncoated), medium plates, heavy plates, hot rolled sheet/coil (uncoated), cold rolled sheet/coil (uncoated), hot-dip galvanized sheet/coil, aluminized sheet/coil, electrical non-oriented silicon sheet/coil, electrolytic tinplate (single-reduced), seamless tubes and pipes, longitudinal weld tubes and pipes 20 to 120mm dia, spiral-weld tubes and pipes 530 to 1,020mm dia, galvanized tubes and pipes up to 140mm dia. Stainless steel – ingots, slabs, blooms, billets, hot rolled sheet/coil (uncoated), cold rolled sheet/coil (uncoated). Alloy steel (excl stainless) – ingots, slabs, blooms, billets. Ferro-manganese. Bearing steels.
Expansion programme: Three new stands were being added to the billet mill during 1984.

Lenin Iron & Steel Works

Head office: Pernik.
Established: 1953.
Works:
■ ***Sinter plant:*** 50 sq metres, ***Blast furnaces:*** Two 230 cu m (annual capacity 235,000 tonnes). ***Steelmaking plant:*** Eight electric arc furnaces – four 10-tonne and four 100-tonne (1,150,000 tonnes); seven open hearth furnaces – one 140-tonne, six 70-tonne. **Continuous casting machines:** Three one-strands. ***Rolling mills:*** One 550mm heavy section and medium section, one 300mm medium section, one 250/500mm light section and bar; one reversing medium plate; one 1,100mm sheet. ***Coil coating lines:*** One hot-dip galvanizing line.
Products: Carbon steel – ingots, wire rod, reinforcing bars, round bars, light angles, medium joists, medium channels, medium plates, hot-dip galvanized sheet/coil.

Burma

No.3 Mining Corp

Head office: Kanbe Road, Yankin, Rangoon. **Telex:** 21511.
Works:
■ ***Direct reduction plant:*** Two Kinglor Metor (annual capacity 40,000 tonnes). ***Steelmaking plant:*** One electric arc furnace. ***Other plant:*** Grinding ball plant.

Ywama Steel Mill

Head office: Insein, Rangoon.
Established: 1957.
Works:
■ ***Steelmaking plant:*** One 15-tonne 5MVA electric arc furnace (annual capacity 12,192 tonnes). ***Continuous casting machines:*** 2-strand Kobe Steel curved mould

Ywama Steel Mill (continued)

billet and slab (100mm sq and 90 x 230mm). ***Rolling mills:*** medium section, light section, and wire rod (48,768 tonnes); hot sheet mill (7,000 tonnes). ***Other plant:*** Wire-drawing, nail-making and welded mesh plant.
Products: Carbon steel – wire rod 5.5mm dia, flats, light channels, galvanized wire (plain), sheet bars, nails, wire mesh.
Expansion programme: The plant was modernised by Kobe Steel's Engineering Division in 1985-87.

Canada

The Algoma Steel Corp Ltd

Head office: 503 Queen Street East, Sault Marie, Ont P6A 5P2. **Tel:** 705 945 2351. **Telex:** 02777169.
Management: Board of directors – Russell S. Allison, Robert W. Campbell, Stuart E. Eagles, John Macnamara (chairman and chief executive officer) , Arthur H. Mingay, James F. Hankinson, Charles H. Hantho, Peter M. Nixon (president and chief operating officer), Leonard N. Savoie, Robert J. Theis, Adam H. Zimmerman; officers – R.N. Robertson (senior vice president, commercial), R.H. Cutmore (vice president, finance and accounting), S.H. Ellens (vice president, employee relations), G.S. Lucenti (vice president, manufacturing), G.B. Hudson (vice president, sales), J.T. Melville (vice president, treasurer and general counsel), P.J. Paciocco (vice president, engineering and technical services), P.C. Finley (secretary), W.J. Reed (controller, steel and iron ore operations), R.A. Foran (assistant secretary), R.J. Greenwood (assistant treasurer), R.J. Backstrom. **Established:** 1934. **No. of employees:** 8,942. **Ownership:** Canadian Pacific Enterprises Ltd (53.82% of common shares). **Subsidiaries:** Cannelton Industries Inc (coal and ore mining). **Annual capacity:** Pig iron: 3,150,000 short tons. Raw steel: 2,800,000 tons.
Works:
■ At head office. ***Coke ovens:*** One Koppers battery (86 ovens) (annual capacity 352,000 tons). ***Sinter plant:*** one travelling grate (1,794,000 tons). ***Blast furnaces:*** Three – No 4 17′ 3″ hearth dia, 12,025 cu ft (450,000 tons), No 6 27′ 45,605 cu ft (900,000 tons), No 7 35′, 88,169 cu ft (1,800,000 tons). ***Steelmaking plant:*** Five basic oxygen LD converters – three 105-ton (1,000,000 tons), and two 260-ton (2,600,000 tons). ***Continuous casting machines:*** One 4-strand bloom (600,000 tons) (with beamblank machine); one 2-strand slab (1,800,000 tons); one 2-strand beam blank. ***Rolling mills:*** One slabbing (46″) (1,200,000 tons); one blooming (45″) (1,200,000 tons); one rail and structural (30″/50″ structural) (600,000 tons); one heavy plate (52″ x 126″/40″ x 67″ x 166″) (346,000 tons); one wide hot strip (30½″ x 61″ x 106″); one cold reduction (20½″ x 61″ x 80″) (350,000 tons); one temper/skin pass (23″ x 61″ x 80″) (550,000 tons). ***Tube and pipe mills:*** One seamless (2⅜ to 12¾″) (400,000 tons). ***Other plant:*** Welded beam division (38,000 tons).
Products: Steelmaking pig iron. Carbon steel – ingots; slabs; blooms; billets; wide-flange beams (medium) (rolled and welded) and H/bearing piles; heavy angles (for elevators) and zeds; heavy joists; heavy channels; heavy rails; light rails; rail accessories (tie plates and splice bars); medium plates; heavy plates; floor plates; hot rolled sheet/coil (uncoated); cold rolled sheet/coil (uncoated); seamless tubes and pipes (including mechanical tubing, and compiling stock); oil country tubular goods.
Sales offices/agents: Mississauga, Ont; Calgary, Alta; Algoma Tube Corp, Houston, TX and Denver, Colorado, USA.

Arc Tube Inc

Head office: 59 Industrial Park Crescent, Sault Ste Marie, Ont P6B 5P3. **Tel:** 705 949 8402. **Telecopier (Fax):** 705 949 6098.
Management: A. Rea (president), Don Esson (vice president), John Caul (general

Canada

Arc Tube Inc (continued)

manager). **Established:** 1976. **Capital:** C$4,000,000. **No. of employees:** Works – 30. Office – 5. **Associates:** Form Rite Ltd (tube forming).
Works:
■ At head office. ***Tube and pipe mills:*** welded – two Yoder, $\frac{3}{16}$-$\frac{1}{2}$" (annual capacity 8,000 short ton). ***Other plant:*** Two Yoder tube terne coating lines (8,000 tons).
Products: Carbon steel – longitudinal weld tubes and pipes – $\frac{3}{16}$", $\frac{1}{4}$", 5/16", $\frac{3}{8}$", $\frac{1}{2}$" (output 8,000 tons); terne coated tubes.
Brands: Arcote (plain); Terne Arcfan (terne coated).
Sales offices/agents: 6735 Telegraph Road, Suite 30, Birmingham, MI 48010, USA (Tel: 313 540 3271).

Associated Tube Industries Ltd (ATI)

Head office: 7455 Woodbine Avenue, Markham, Ont L3R 1A7. **Tel:** 416 475 6464. **Telex:** 06 986871. **Cables:** Aticanco.
Management: Vice presidents – Scott Sweatman (general manager), Ron Sinclair (operations); joint sales managers – Ian O'Connor (stainless steel and fabricated parts), Al Krempulek (tube products). **Established:** 1956. **No. of employees:** Works – 200.
Works:
■ Markham, Ont. ***Tube and pipe mills:*** various welded. ***Other plant:*** Draw benches.
Products: Carbon steel – longitudinal weld tubes and pipes $\frac{3}{16}$-6$\frac{5}{8}$" od; cold drawn tubes and pipes. Stainless steel – longitudinal weld tubes and pipes 300 and 400 grades; cold drawn tubes and pipes $\frac{1}{8}$-2" od. Alloy steel (excl stainless) – longitudinal weld tubes and pipes (including high nickel).

Atlas Specialty Steels Division of Rio Algom Ltd

Head office: Centre Street, Welland, Ont 3B 5R7. **Tel:** 416 735 5661. **Telex:** 615114.
Annual capacity: Raw steel: Welland – 325,000 short tons; Tracy 88,000 tons (semis).
Works:
■ Welland, Ont. ***Steelmaking plant:*** Three electric arc furnaces – one 30-ton, two 65-ton. ***Refining plant:*** One VAD /VOD unit; two VAR units – one 10 – 30", one 18 – 40". ***Rolling mills:*** One 2-high reversing 26" blooming; one 3-high 22" billet; two bar mills – one 16" and one 10". ***Other plant:*** Forging hammer and presses; cold drawing facilities for bar and wire; etc.
■ Atlas Stainless Steels Division, 1675 Marie-Victoria Road, Tracy, Que J3R 4R4 (Tel: 514 746 5000). ***Steelmaking plant:*** One 60-ton electric arc furnace. ***Refining plant:*** One VOD unit. ***Continuous casting machines:*** One 1-strand Concast-Mesta curved mould for slab . ***Rolling mills:*** One Sendzimir planetary; two 50" Sendzimir reversing cold reduction; one 2-high temper/skin pass. ***Other plant:*** Bright annealing line.
Products: Carbon steel – (special quality) ingots, slabs, blooms, billets, round bars, medium plates. Stainless steel – ingots, slabs, blooms, billets, hot rolled band (for re-rolling), round bars, square bars, flats, cold rolled hoop and strip (uncoated) up to 18", medium plates, hot rolled sheet/coil (uncoated), cold rolled sheet/coil (uncoated). Alloy steel (excl stainless) – ingots, slabs, blooms, billets. Tool steels, high speed steels, rock drill steels, forgings in stainless steel.

Atlas Tube

Head office: 200 Clark Street, Harrow, Ont N0R 1G0. **Postal address:** PO Box 970. **Tel:** 519 726 5061, 519 738 2267. **Telecopier (Fax):** 519 738 3115.
Management: Alan Zekelman (president), Barry Zekelman (vice president), Joeseph Ponic (purchasing and sales manager). **Established:** 1984. **Capital:**

Atlas Tube (continued)

C$12,000,000. **No. of employees:** Works – 45. Office – 5. **Ownership:** Division of 538750 Ontario Ltd **Annual capacity:** Finished steel: 32,000 tons.
Works:
■ At head office. ***Tube and pipe mills:*** welded – one 5" Wean McKay (annual capacity 25,000 tons) and one $2\frac{1}{2}$" Yoder (8,000 tons).
Products: Carbon steel – longitudinal weld tubes and pipes, galvanized tubes and pipes and hollow sections (Squares: 1 x 1", $1\frac{1}{4}$ x $1\frac{1}{4}$", $1\frac{1}{2}$ x $1\frac{1}{2}$", (in gauges 0.083", 0.100" and 0.125"); 2 x 2", $2\frac{1}{2}$ x $2\frac{1}{2}$, 3 x 3", $3\frac{1}{2}$ x $3\frac{1}{2}$", 4 x 4" (in gauges 0.100", 0.125", 0.188" and 0.250") all in lengths of 20ft to 48ft. Rectangular: 1 x $1\frac{1}{2}$", 2 x 1", $2\frac{1}{2}$ x $1\frac{1}{4}$" (in gauges 0.083", 0.100" and 0.125" and lengths of 24ft); 3 x 2", 4 x 2", 4 x 3", 5 x 2", 5 x 3" and 6 x 2" (in gauges 0.100", 0.125", 0.188" and 0.250" and lengths of 20ft to 48ft). Round: $1\frac{1}{4}$, $1\frac{1}{2}$ and $1\frac{7}{8}$ (in gauges 0.083", 0.100" and 0.125" 0.188: and lenths of 4ft to 24ft); 2"$\frac{1}{2}$, 3 and 5" (in gauges of 0.083", 0.100", 0.125", and 0.250" in lengths of 20ft to 48ft); etc.

Barton Tubes Ltd

Head office: 2170 Queens Way, Burlington, Ont L7R 3Y2. **Postal address:** PO Box 277. **Tel:** 416 637 8261. **Telex:** 0618738 barton bur.
Established: 1952.
Works:
■ At head office. ***Tube and pipe mills:*** welded – induction $\frac{1}{2}$-4".
Products: Carbon steel – longitudinal weld tubes and pipes, galvanized tubes and pipes. Alloy steel (excl stainless) – longitudinal weld tubes and pipes.

Baycoat Ltd

Head office: 244 Lanark Street, Hamilton, Ont L8N 3K7. **Postal address:** PO Box 624. **Tel:** 416 561 0965. **Telex:** 618300.
Management: Ross Craig (president), Jack Paines (vice president and general manager). **Ownership:** Dofasco and Stelco (50% each).
Products: Carbon steel – colour coated sheet/coil.

Brockhouse Canada Ltd

Head office: 275 Orenda Road, Bramalea, Ont L6T 3T7. **Tel:** 416 457 9200.
Cables: Brocktor Toronto.
Management: N.C. Eiloart (president and general manager), M.K. McLachlan (vice president, marketing), L.W. Rhodenizer (treasurer and assistant secretary), Robert Howchin (manager, operations), K. Stacey (vice president, sales). **Established:** 1952. **No. of employees:** Works – 75. Office – 38. **Ownership:** Brockhouse Ltd, West Bromwich, UK (100%).
Works:
■ At head office. Range of cold forming mills with capacities to 0.375" thick, 48" wide, finished depth 9". Brake presses for up to 16ft x 250 tons.
Products: Carbon steel – cold roll-formed sections (from strip).

Brooks Tube Ltd—see Ipsco Inc

Burlington Steel—see Slater Steels, Hamilton Specialty Bar Division

Canada

Canadian Steel Wheel, Division of Hawker Siddeley Inc

Head office: 1900 Dickson Street, Montreal, Que H1N 2H9. **Tel:** 514 255 3605. **Telex:** 05828603.
Management: M.J. Colman (vice president), M. Gamble (general manager), R.D. Jones (director, sales and development), J.H. Allan (comptroller). **Established:** 1957. **No. of employees:** 400. **Ownership:** Hawker Siddeley Canada Inc, Toronto, Ont.
Works:
■ At head office. General manager: M. Gamble. ***Steelmaking plant:*** Two 60-short ton electric arc furnaces (annual capacity 160,000 short tons). ***Other plant:*** Wrought steel railway wheel production line with three forging presses (1,000-ton, 3,000-ton, and 6,000-ton).
Products: Carbon steel – ingots. Alloy steel (excl stainless) – ingots. Wrought steel railway wheels and industrial wheels.

Co-Steel Inc

Head office: 1601 Hopkins Street, Whitby, Ont L1N SR6. **Tel:** 416 686 2500. **Telex:** 06981358. **Cables:** Co-Steel Toronto.
Management: G.R. Heffernan (chairman), W.J. Shields (president).
Activities Co-Steel Inc is the parent company of the wholly-owned Raritan River Steel Co, USA (see separate entry) and Lake Ontario Steel Co (Lasco), Canada (see separate entry) and owns two thirds of Sheerness Steel Co Ltd, UK (see separate entry).

Courtice Steel Ltd

Head office: 160 Orion Place, Cambridge, Ont N1R 7G8. **Postal address:** PO Box 1526. **Tel:** 519 740 2488. **Telex:** 06959438. **Telecopier (Fax):** 519 623 2062.
Management: A. Kerney (board-chairman), J.B. Kelly (president), W. Oerlemans (vice president, operations), P. Kelly (vice president, marketing and sales). **Established:** 1976. **No. of employees:** Works – 199. Office – 38. **Ownership:** Harris Steel Group Inc, Toronto, Canada. **Annual capacity:** Raw steel: 330,000 short tons. Finished steel: 300,000 short tons.
Works:
■ At head office. ***Steelmaking plant:*** Two electric arc furnaces – one 25-ton Empco (annual capacity 120,000 tons), and one 40-ton Empco (on order) (250,000 tons). ***Continuous casting machines:*** One 2-strand Rokop billet (4 x 4") (250,000 tons).
■ Bowmanville, Ont. ***Rolling mills:*** One bar and section – 2-stand roughing 14", 5-stand intermediate 10", and 2-stand finishing 9" (75,000 tons).
Products: Carbon steel – billets 4 x 4" (100mm) (1986 output 102,351 tons), reinforcing bars 3-11 (10 – 35mm), round bars $\frac{3}{8}$-2", square bars $\frac{3}{8}$-2", flats $\frac{1}{8}$" x $\frac{3}{4}$" to $\frac{1}{2}$" x 2$\frac{1}{2}$", light angles $\frac{3}{4}$" x $\frac{3}{4}$" to 3" x 3" and light channels $\frac{3}{4}$ to 3" (combined output 85,000 tons).

Dofasco Inc

Head office: 1330 Burlington Street East, Hamilton, Ont. **Postal address:** PO Box 460, Hamilton, Ont L8N 3J5. **Tel:** 416 544 3761. **Telex:** 0618682. **Telecopier (Fax):** 416 544 3761 Ext 2331.
Management: Officers: Frank H. Sherman (chairman and chief executive), R. Ross Craig (vice chairman), John G. Sheppard (vice chairman), Paul J. Phoenix (president and chief operating), William L. Wallace (executive vice president), William D. Simon (senior vice president, commercial), Thomas Van Zuiden (senior vice president, finance and administration), John H. McAllister (vice president, raw materials, purchases and traffic), Noel G. Thomas (vice president, technology), David H. Samson (vice president, engineering), H. Graham Wilson (vice president and secretary), Robert J. Swenor (vice president, personnel), Bill P. Solski (treasurer), Robert W. Grunow (comptroller), Jack W. Craven (vice president, quality), William H. Mulveney (vice president, sales),

Dofasco Inc (continued)

Lawrence V. Walsh (vice president, works manager), L. Allen Root (assistant treasurer), R. Eric Moore (assistant comptroller), Urmas Soomet (assistant secretary), Peter A. Yakiwchuk (assistant comptroller). Directors: George H. Blumenauer, R. Ross Craig, Roger G. Doe, Robert C. Dowsett, John R. Evans, Howard J. Lang, John D. Leitch, Frank H. Logan, Paul J. Phoenix, John G. Sheppard, Frank H. Sherman. **Established:** 1912. **No. of employees:** 11,600. **Subsidiaries:** National Steel Car Corp Ltd, Hamilton, Ont (100%); Prudential Steel Ltd, Calgary, Alb (100%); BeachviLime Ltd, Beachville, Ont (100%); Whittar Steel Strip, Detroit, MI (100%). **Associates:** Iron ore mining & pelletizing interests: Adams Mine, Kirkland Lake (100%); Sherman Mine, Temagami, Ont (90%); Wabush Mines comprising Skully Mine, Wabush, Nfld and Arnaud Pellets, Point Noire, Que (16.4%); Iron Ore Co of Canada, Labrador City, Sept-Iles, Que (6.1%). Other investments: Baycoat Ltd, Hamilton, Ont (50%); ITL Industries Ltd, Windsor, Ont (20%); Arnaud Railway Co, Que (16.4%); Wabush Lake Railway Co Ltd, Nfld (16.4%) Aberford Resources Ltd, Alb (12.3%); Abermin Corp, Vancouver, BC (10%); Knoll Lake Minerals Ltd, Nfld (9.5%); Northern Land Co Ltd, Nfld (8.2%); Twin Falls Power Corp Ltd, Nfld (2.8%). **Annual capacity:** Pig iron: 3,822,000 short tons. Finished steel: 4,500,000 short tons.

Works:

■ Hamilton, Ont. ***Coke ovens:*** Five Koppers-Becker batteries (211 ovens) (annual capacity 1,265,000 tonnes),one Gun Flue battery (35 ovens) (435,000 tonnes), six Didier underjet batteries (246 ovens) (1,700,000 tonnes). ***Blast furnaces:*** Four – (No.1) 6.33m hearth dia, 901.6 cu m (771,680 tons), (No.2) 6.33m hearth dia, 923.5 cu m (771,680 short tons); (No.3) 6.55m hearth dia, 904.8 cu m (862,680 short tons), (No.4) 8.53m hearth dia, 1,573 cu m (1,415,960 short tons). ***Steelmaking plant:*** Three 140-tonne and one 250-tonne basic oxygen LD converters (combined annual capacity 4,500,000 tonnes); three electric arc furnaces – two 11' dia Volta (annual capacity 4 tonnes/hour) and one 7' dia Lectromelt (1.4 tonne/hour). ***Continuous casting machines:*** One 2-strand slab (annual capacity 2,200,000 tonnes). ***Rolling mills:*** Two wide hot strip mills – No 1 hot strip mill: 1-stand 2-high 45 x 88" reversing slabber with close coupler 42 x 42" attached edgers, 1-stand 2-high 39⅛ x 68" reversing rougher with attached edgers, 7-stand 4-high 28¾ × 68/58 1/2 × 68'' finishing (3,500,000 tonnes); No.2 hot strip mill : existing 2-high 54 x 72" reversing roughing mill with attached edger 42 x 43½", 5-stand 4-high finishing 30¾ x 60½ x 68" (1,600,000 tonnes) (in 1987 No.2 hot strip mill was being updated by addition of a sizing mill, modernisation of five finishing stands and addition of two others, also additional ancillary equipment). Two cold reduction mills – one single-stand 4-high 16½ x 66" and one 5-stand 4-high 24 x 72" (1,680,000 tonnes). Two temper/skin pass mills – one single-stand 4-high 66", and one 2-stand 4-high 56" (734,000 tonnes). ***Coil coating lines:*** electrolytic tinning – one 40" Aetna (1956) Ferrostan tinning line (0.006 – 0.0149" product), one 48" Wean (1972) Ferrostan T.F.S./tin line (0.005 – 0.023" product) (total tinplate production 420,000 tpy); three hot-dip galvanizing lines - one 60" Aetna (1960) Sendzimir (0.013 – 0.168") (276,000 tons), one 52" Wean (1968) Sendzimir (0.01 to 0.064") (216,000 tons), one 60" wide Wean (1984) Sendzimir (0.012 to 0.080") (240,000 tons); one zinc-aluminium line – 48" wide Aetna (1955) Sendzimir, converted to Galvalume (1984) (0.012 to 0.066" products) (152,000 tons).

Products: Carbon steel – skelp (tube strip) cold rolled hoop and strip (uncoated), medium plates, heavy plates, floor plates, hot rolled sheet/coil (uncoated), cold rolled sheet/coil (uncoated), hot-dip galvanized sheet/coil, enamelling grade sheet/coil, zinc-aluminium alloy coated sheet/coil, coated sheet/coil, electrical grain oriented silicon sheet/coil, blackplate, electrolytic tinplate (single-reduced), electrolytic tinplate (double-reduced), tin-free steel (ECCS). Alloy steel (excl stainless) – hot rolled sheet/coil (uncoated). Plates up to 60" wide and 0.500" thick; floor plate; hot rolled sheet in commercial, drawing and structural quality and high-strength low-alloy (Dofascoloy); hot rolled skelp up to 60" wide; cold rolled sheet 2 – 60" wide and 0.015 – 0.142" thick; cold rolled strip ½ to 23 5/16" wide and 0.010 to 0.142" thick; cold rolled skelp; electrical steel; porcelain enamelling sheets; tin mill blackplate; coated steels; galvanised sheets (see Brands); culvert (CSP) sheets; Galvalume sheets; electrolytic

Canada

Dofasco Inc (continued)

tinplate (including double-reduced); electrolytic chromium coated steel; Dofasco Pre-Coat; carbon, alloy and stainless steel castings up to 25,000lb.
Brands: Dofascoloy (high strength low-alloy hot rolled, cold rolled and galvanized sheet); Dofasco Premier (regular galvanized sheet); Satincoat (wipedcoat galvanized sheet); Galvalume (aluminium/zinc-coated sheet); Pre-Coat (prepainted galvanized, Galvalume, cold rolled or tinplate coils).
Expansion programme: No.2 steelmaking shop (No.4 BOF) is being converted to the K-OBM process. A ladle metallurgy station, two additional hot strip mill finishing stands and two extra coilers are being installed.

Domfer Metal Powders Ltd

Head office: PO Box 970, Station H, Montreal, Que H3G 2M9. **Tel:** 514 933 3178. **Telecopier (Fax):** 514 933 4080.
Management: Managers – L.G. Roy (president general), R. Benoit (manufacturing), E.R. Magnusson (marketing), R.R. Piperni (sales); controller – M. Janelle.
Established: 1976. **No. of employees:** Works – 80. Office – 9.
Works:
■ 6090 Newman Boulevard, La Salle, Quebec. (Tel: 514 365 8254). Plant manager: P. Bertrand. **Melting and Shotting plant** (using steel scrap and pig iron). **Iron powder plant**.
Products: Iron powder, Domfer MP (sponge iron powder). Domfer MP (sponge iron powder).

Fischer Stainless Steel Tube Plant

Fischer Edelstahirohre, Aachen, West Germany, plans to set up a stainless steel tube plant at North York, Ontario, with production due to begin early in 1988.

Ipsco Inc (formerly Interprovincial Steel & Pipe Corp Ltd)

Head office: Armour Road, Regina, Sask S4P 3C7. **Postal address:** PO Box 1670. **Tel:** 306 949 3530. **Telex:** 071 2269. **Telecopier (Fax):** 306 949 3500.
Management: Chairman of the board – W.M. Elliott; president and chief executive officer – Roger Phillips; executive vice president and chief operating officer – Milan Kosanovich; senior vice presidents – G.C. Backman (chief administrative officer), B.E.A. Yeo (chief financial officer); vice presidents – R.A. Acquaviva (manufacturing), W.D. Bailey (director of product development), L.D. Cooke, J.R. Gough, J.K. Kwong, R.A. Rzonca (personnel), J.R. Tulloch; treasurer – M.J. Dalla-Vicenza; secretary – H.H. MacKay, J.W. Comrie (assistant); controller – E.J. Tiefenbach. **Established:** 1956.
Capital: C$90,413,000 (1986). **No. of employees:** Works – 853. Office – 283.
Subsidiaries: Lambton Steel Ltd, North Vancouver, BC (cut-to-length lines); Brooks Tube Ltd, Brooks, Alta (pipe producer – inoperative); Ipsco Enterprises Inc, Aurora, CO (US sales office). **Annual capacity:** Raw steel: 680,000 tonnes. Finished steel: 515,000 tonnes.
Works:
■ Red Deer Works, PO Box 593, Central Park Road, Red Deer, Alta. Tel: 403 342 1717. Fax: 403 342 1797. Works director: Don Lines. ***Tube and pipe mills:*** welded – one 3-12" ERW (annual capacity 120,000 tonnes).
■ Regina Works, PO Box 1670, Regina, Sask. (Tel: 306 949 3530. Telex: 071 2269). Works directors: John Wild (steel), Dieter Schaeble (pipe). ***Steelmaking plant:*** Four electric arc furnaces – one 132-tonne, one 120-tonne, one 70-tonne and one 34-tonne (annual capacity 680,000 tonnes). ***Continuous casting machines:*** One 1-strand Mesta slab (35-85" wide, 8-10" thick, up to 40' long) (680,000 tonnes). ***Rolling mills:*** Three wide hot strip – one 2-high 48 x 114" slabbing, one 4-high reversing roughing 28" and 49 x 80" and one 4-high reversing finishing 28" and 60 x 80" (875,000 tonnes). ***Tube and pipe mills:*** Four welded – one ERW $2\frac{3}{8}$" (25,000 tonnes), one ERW 3-16" (60,000 tonnes), two DSA spiral weld 16-80" (115,000 tonnes).
■ Calgary Works, 7201 Ogdendale Road SW, Calgary, Alta. (Tel: 403 279 3351. Fax:

Ipsco Inc (formerly Interprovincial Steel & Pipe Corp Ltd) (continued)

403 236 8166). Works director: Bernie Brauer. ***Tube and pipe mills:*** welded – one ERW 4½-10¾" (annual capacity 140,000 tonnes).
■ Edmonton Works, 6735 75th Street, Edmonton, Alta. (Tel: 403 460 1321. Fax: 403 469 7660). Works director: Paul Hibbeln. ***Tube and pipe mills:*** Three welded – one ERW 3½-16" (annual capacity 95,000 tonnes), two 1 DSA spiral 16-80" (105,000 tonnes).
■ Port Moody Works, 3190 Murray Street, Port Moody, BC. (Tel: 604 461 1131. Fax: 604 461 2441). Works director: Dave Dahlgren. ***Tube and pipe mills:*** Two welded – one ERW ½-1½" and one ERW 2-4½" (combined annual capacity 60,000 tonnes).
Products: Carbon steel – ingots (not for sale); slabs (for sale starting in 1987); medium plates (output 11,000 tonnes); hot rolled sheet/coil (uncoated) (15,104 tonnes); longitudinal weld tubes and pipes (13,541 tonnes); spiral-weld tubes and pipes (5,712 tonnes). Alloy steel (excl stainless) – medium plates (25,667 tonnes); hot rolled sheet/coil (uncoated) (35,244 tonnes); longitudinal weld tubes and pipes (54,163 tonnes); spiral-weld tubes and pipes (51,408 tonnes); oil country tubular goods (60,681 tonnes); hollow sections (26,216 tonnes).

Ivaco Inc

Head office: Place Mercantile, 770 W Sherbrooke Street, Montreal, Que H3A 1G1. **Tel:** 514 288 4545. **Telex:** 055 60305.
Management: Chairman – Isin Ivanier; president – Paul Ivanier (chief executive officer); senior vice presidents – Sydney Ivanier, Michael Herling (secretary), Jack Klein; vice presidents – John Loveridge, M.L. Cairns, Albert A. Kassab (chief financial officer), George G. Goldstein; director – John G. Metrakos (marketing, raw materials and materials management). **Annual capacity:** Finished steel: 375,000 short tons.
Works:
■ Ivaco Rolling Mills, Division of Ivaco Inc, PO Box 322, L'Original, Ont KOB 1KO. (Tel: 613 675 4671). Managers: G.D. Silverman (general), T. Perlus (melt shop), D.S. Goldsmith (rod mill), B. Retty (commercial); finance director: – G. Guillon; superintendents – M. Coburn (melt shop), M. Goldsmith. ***Steelmaking plant:*** Two 65-ton electric arc furnaces. ***Continuous casting machines:*** One 4-strand billet. ***Rolling mills:*** One 2-strand wire rod.
Products: Carbon steel – billets (high and low carbon), wire rod (high and low carbon), bright wire, black annealed wire (high and low carbon), galvanized wire (plain), galvanized wire (barbed), nails, fencing, fasteners, fourdrinier fabric, machined parts.

Lake Ontario Steel Co (Lasco Steel)

Head office: Hopkins Street South, Whitby, Ont L1N 5T1 **Tel:** 416 668 8811. **Telex:** 06 981265. **Cables:** Lasteel. **Telecopier (Fax):** 416 668 4536.
Management: C.G. Schmelzle (chairman), L.C. Hutchinson (president and chief executive officer), D.J. Blucher (executive vice president, sales), R.C. Smith (executive vice president, finance). **Established:** 1963. **Capital:** C$50,000,000. **No. of employees:** Works – 880. Office – 338. **Ownership:** Division of Co-Steel Inc. **Annual capacity:** Finished steel: 854,000 short tons.
Works:
■ At head office. ***Steelmaking plant:*** Two electric arc furnaces – one 65-ton, 45,000 kVA with ladle arc refiner (annual capacity 300,000 tons), and one 130-ton, 95,000 KVA (610,000 tons). ***Refining plant:*** One arc ladle furnace. ***Continuous casting machines:*** Two 3-strand high-head straight mould billet and one 4-strand low-head curved mould bloom (5¼" and 5 x 7⅜") (combined capacity 910,000 tons). ***Rolling mills:*** One continuous bar – four 2-high 21" horizontal roughing stands, six 2-high 16" intermediate stands, and four 2-high 13" finishing stands (470,000); one structural mill – four 2-high 16" roughing stands, two 2-high 14" intermediate stands and four 2-high 16" finishing stands (384,000 tons).
Products: Carbon steel – reinforcing bars; round bars; square bars and other merchant

Canada

Lake Ontario Steel Co (Lasco Steel) (continued)

mill products, and special quality bars; light angles; light channels; heavy angles; heavy channels. Alloy steel (excl stainless) – round bars (special quality low alloy).
Brands: Grader blades.

Manitoba Rolling Mills, Amca International

Head office: Lot 27, Main Street, Selkirk, Manitoba R1A 2B4. **Postal address:** PO Box 2500. **Tel:** 204 284 5395. **Telex:** 07 57506. **Telecopier (Fax):** 204 785 2193.
Management: President – R.R. Robertson (general manager); managers – H.B. Irvine (marketing and sales), R.P. Grove (employee relations), A. Critchley (manufacturing services), T. Jarvis (melt shop), J. Greaves (rolling mill); controller – R.C. Warrentin.
Established: 1906. **No. of employees:** Works – 525. Office – 125. **Ownership:** Amca International (Ultimate holding company – Canadian Pacific Railways). **Annual capacity:** Finished steel: 300,000 short tons.
Works:
■ At head office. ***Steelmaking plant:*** Two 50-ton U.H.P. Electric arc furnaces (annual capacity 300,000 tons). ***Continuous casting machines:*** One 4-strand Concast billet (8 x 8", $6\frac{3}{4}$ x $6\frac{3}{4}$", $5\frac{1}{4}$ x $5\frac{1}{4}$) (300,000 short tons). ***Rolling mills:*** light section – 24" reversing rougher and 14 continuous finishing stands (300,000 tons).
Products: Carbon steel – billets, reinforcing bars, round bars, square bars, flats, light angles, light channels, special light sections, medium angles, medium joists, medium channels, heavy angles, heavy channels. Alloy steel (excl stainless) – round bars, flats, light angles, light channels, special light sections.

Nor-Sand Metal Inc

Head office: 425 McCartney, Arnprior, Ont. **Tel:** 613 623 6501. **Telex:** 534576 nor smd arpr.
Management: B.S. Boneham (president), L. Roslund (vice president, operations), K. Burgess (marketing manager), T. Moylan (comptroller). **Ownership:** Sandvik (100%).
Products: Stainless steel – longitudinal weld tubes and pipes. Nickel alloy tubing.

QIT – Fer et Titane Inc

Head office: 770 Sherbrooke Street West, Suite 1800, Montreal, Que H3A 1G1. **Tel:** 514 288 8400. **Telex:** 05 268575.
Management: Bruce Grierson (chairman and chief executive officer), Gilles Charette (president and chief operating officer). **Established:** 1948. **Ownership:** BP. **Subsidiaries:** Quebec Metal Powders (iron powder); Richards Bay Minerals (titanium minerals, etc). **Annual capacity:** molten iron: 900,000 tonnes of molten iron.
Works:
■ Sorel, Que J3P 5P6. **Iron making plant:** Nine electric furnaces (annual capacity 900,000 tonnes). ***Steelmaking plant:*** One KOBM basic oxygen converter. ***Continuous casting machines:*** One 4-strand billet machine.
Products: Foundry pig iron. Iron powder Carbon steel – billets. Titania slag.
Brands: Sorel (titania slag); Sorelmetal (high quality pig iron); Atomet (iron powder); Sorelsteel (continously cast billets).

Sidbec-Dosco Inc

Head office: 300 Léo-Pariseau, Montreal, Que. **Postal address:** PO Box 2000, Succursale Place du Parc, Montreal, Que H2W 2S7. **Tel:** 514 286 8600. **Telex:** 05 24515. **Telecopier (Fax):** 514 286 4071.
Management: Chairman of the board – Pierre Laurin; president and chief executive officer – John LeBoutillier; vice presidents – Georges H. Laferrière (supplies and

Sidbec-Dosco Inc (continued)

engineering), Paul E. Landry (finance, administration, and treasurer), J.-Pierre Picard (marketing and sales), Serge Wagner (human resources), Pierre Lacroix (secretary and general counsel). **Established:** 1969. **No. of employees:** Works – 215. Office – 2,985. **Ownership:** Sidbec (100%). **Associates:** Sidbec-Feruni Inc (100%) (metal recycling). **Annual capacity:** Raw steel: 1,500,000 short tons. Finished steel: 1,100,000 short tons.

Works:

■ Contrecoeur Works, 1900 Montée de la Pomme d'Or, Contrecoeur, Que J0L 1C0. (Tel: 514 587 2091. Telex: 05 25498). Works director: R. Lévesque. ***Direct reduction plant:*** Two Midrex (annual capacity 500,000 long tons and 700,000 long tons). ***Steelmaking plant:*** Three electric arc furnaces – one 100-ton and two 150-ton (1,300,000 short tons); one ladle furnace. ***Continuous casting machines:*** Two 6-strand Concast billet (3½-7½") (650,000 short tons); one 1-strand Voest Alpine slab (7" x 28-60") (650,000 short tons). ***Rolling mills:*** slabbing – 45 x 109" roughing combination 2-high United with 42 x 42" vertical edger; one 25-stand bar and wire rod comprising 9-stand 17 x 42", 4-stand 12½ x 24", 4-stand 14 x 33", and 8-stand 10½ x 20" (325,000 short tons); one 4-high United Steckel wide hot strip (27" and 49 x 66") (600,000 short tons); one Sendzimir cluster cold reduction and cold strip (39-50", 26-39") (250,000 short tons); one 2-high Waterbury Farrel temper/skin pass (31 x 56").

■ Montreal Works, 5870 Rue Saint-Patrick, Montreal, Que H4E 1B3. (Tel: 514 766 7761. Telex: 05 566331). Works director: M. Lalanne. ***Steelmaking plant:*** One 60-ton electric arc furnace (annual capacity 200,000 short tons). ***Continuous casting machines:*** One 4-strand Concast billet (3½-6¼") (200,000 short tons). ***Rolling mills:*** medium section and merchant bar comprising 4-stand 3-high and 1-stand 2-high (130,000 short tons). ***Tube and pipe mills:*** welded – one Aetna Standard butt-weld (¼-4"). ***Other plant:*** Pipe galvanizing lines, hot-dip wire galvanizing lines, high and low-carbon wire drawing department.

■ Longueuil Works, 2555 Chemin du Lac, Longueuil, Que J4N 1C1. (Tel: 514 651 1500. Telex: 05 268767). Works director: R. Beaudin. ***Rolling mills:*** One Birdsboro 16-stand continuous medium section comprising 3-stand 21", 3-stand 17", 4-stand 14", and 6-stand 12" (annual capacity 300,000 short tons).

■ Etobicoke Works, 1020 Martin Grove Road, Rexdale, Ont M9W 4W2. (Tel: 416 247 2121. Telex: 06 989247). Works director: F. Palmer. Hot-dip wire galvanizing lines, high and low-carbon wire drawing department, nail department.

Products: Carbon steel – slabs, billets, wire rod, reinforcing bars, round bars, square bars, flats, light angles, medium angles, medium channels, skelp (tube strip), hot rolled sheet/coil (uncoated), cold rolled sheet/coil (uncoated), bright wire, black annealed wire, galvanized wire (plain), longitudinal weld tubes and pipes, galvanized tubes and pipes.

Slater Steels – Hamilton Speciality Bar Division

Head office: 319 Sherman Avenue North Hamilton, Ont L8N 3P9. **Postal address:** PO Box 943. **Tel:** 416 549 4774. **Telex:** 0618480. **Telecopier (Fax):** 416 549 3785.

Management: President – J.F. Miles; vice presidents – A.P. Hyde (manufacturing), B.E. Naber (sales and marketing). **Established:** 1911. **No. of employees:** Works – 552. Office – 148. **Ownership:** Slater Industries Inc (Ultimate holding company – Fobasco Ltd, Toronto (controlling interest)). **Subsidiaries:** Norforge – Ball Forging, Sept-Iles, PQ (100%). **Associates:** Fort Wayne Specialty Alloys Division, Fort Wayne, IN (steel mill); Sorel Forge Division, Sorel, Que and Slacan Division, Brantford, Ont (all operating divisions of Slater Industries Inc (100%)). **Annual capacity:** Raw steel: 300,000 tons. Finished steel: 250,000 tons.

Works:

■ At head office. Works director: A.P. Hyde. ***Steelmaking plant:*** One 45-ton and one 65-ton electric arc furnaces (annual capacity 300,000 tons). ***Refining plant:*** One 60-ton ladle furnace. ***Continuous casting machines:*** Two 3-strand Concast billet (5

Slater Steels – Hamilton Speciality Bar Division (continued)

x 7") (300,000 tons). ***Rolling mills:*** One bar – one 2-high 30" stand, one 3-high 21" roughing stand, two 2-high 16" stands, two 2-high 14" stands, four 14"/12" vertical-horizontal stands (250,000 tons). ***Other plant:*** Hamilton Works – Two grinding ball forging machines, Norforge Division – one grinding ball forging machine.
Products: Carbon steel – reinforcing bars 10mm through 45mm incl; round bars ¾" to 2¼" dia; square bars 1"; flats ¼ to ¾" thick and 1½ to 6" wide; light channels 2"; and light channels 2" and special light sections (tyre ring/plough share/jack bar/sign posts) (1986 combined output approx 200,000 tonnes, including alloy). Alloy steel (excl stainless) – round bars ¾ to 2¼" dia; flats ¼ to ¾ thick and 1½ to 6" wide.
Brands: Burlington – forged Steel Grinding Balls.

Slater Steels, Sorel Forge Division

Head office: 201 Montcalm Street, St Joseph de Sorel, Que J3P 5P2. **Postal address:** PO Box 520. **Tel:** 514 746 4030. **Telex:** 05268603. **Telecopier (Fax):** 514 746 4092.
Management: Raymond Pallen (president), J.C. Raimondi (vice president sales and marketing), Michel Cardin (director of marketing). **Established:** 1953. **No. of employees:** Works – 111. Office – 140. **Ownership:** Slater Industries Inc. **Annual capacity:** Finished steel: 20,000 short tons of forged products.
Works:
■ At head office. ***Steelmaking plant:*** One 42-tonne electric arc furnace (annual capacity 30,000 short tons). ***Refining plant:*** One 42-tonne vacuum degassing unit (ladle to ladle) (30,000 tons).
Products: Carbon steel – ingots, slabs, blooms, billets, round bars, square bars, flats, hexagons. Stainless steel – ingots, slabs, blooms, billets, round bars, square bars, flats, hexagons. Alloy steel (excl stainless) – ingots, slabs, blooms, billets, round bars, square bars, flats, hexagons. Combined 1986 output for carbon, alloy & stainless steel – slabs (9,500 tons), ingots (1,500 tons), billets and blooms (1,500 tons), bars (1,500 tons), custom forgings (3,500 tons).
Brands: Nudie (AISI H-13, (tool steel); Nudie XL, (AISI H-13, ESR, tool steel); CSM-2 (AISI P20, mould steel); Holder block, (AISI 4150 Modified, mould steel); Crudie, (AISI 4350 Modified, forging steel); Airkool, (AISI A2, tool steel); Airdi 150, (AISI D2, tool steel); Ketos, (AISI 01, tool steel), Custom forgings to any shape.

Sonco Steel Tube, Division of Ferrum Inc

Head office: 14 Holtby Avenue, Brampton, Ont L6X 2M1. **Tel:** 416 451 2400. **Telex:** 06 97542. **Telecopier (Fax):** 416 451 2795.
Management: Vice presidents – R.G. Stewart (general manager), B.W. Smith (finance); managers – D.W. Boughner (sales), K. Peacock (materials); controller – J. Teolis; manufacturing manager – D. Thomas (Holtby), M. Katz (Van Kirk). **Established:** 1954. **No. of employees:** Works – 255. Office – 87. **Ownership:** Ferrum Inc (Ultimate holding company – Jannock Ltd). **Associates:** Lyman Tubeco.
Works:
■ At head office. Works director: Dave Thomas. ***Tube and pipe mills:*** welded :ERW mechanical – No 1 ¾-3" round (annual capacity 14,000 short tons); No 2 ⅜-1" round (7,000 tons); No 3 ¾-3" round (14,000 tons); No 4 ¾-3" round (10,500 tons). ERW structural: No 5 1½-4"sq (54,000 tons); No 6 3-7"sq (82,800 tons) and No 9 1-4"round (24,480 tons).
■ Van Kirk, Van Kirk Drive, Brampton, Ont. (Tel: 416 451 2400). Works director: Marvin Katz. ***Tube and pipe mills:*** welded – No 1 ERW structural 4-12"sq (annual capacity 100,000 short tons); ERW pipe 2 ⅜-7"od (40,000 tons).
Products: Carbon steel – pipe piling, oil country tubular goods, galvanized tubes and pipes, hollow sections.

Standard Tube Canada, Division of TI Canada Inc

Head office: 193 Givins Street, Woodstock, Ont N4S 7Y6. **Postal address:** PO Box 430. **Tel:** 519 537 6671. **Telex:** 064 74115. **Telecopier (Fax):** 519 539 6804. **Management:** S. Taylor (chairman), L. Buganto (president), C. Ohlson (vice president and general manager), W. James (controller), H. Davis (sales and marketing manager). **Established:** 1904. **No. of employees:** Works – 300. Office – 150. **Ownership:** Canada Inc, London, Ont (Ultimate holding company – TI Group PLC, London, UK). **Annual capacity:** Finished steel: 150,000 short tons.
Works:
■ At head office. Works director: W. Shaw.. ***Tube and pipe mills:*** welded – Two Yoder 3½", one McKay 3½", one McKay 5¾", one Wean 5¾" and one Abby Etna 8¾" (combined annual capacity 120,000 tons).
■ 5700 Pare Street, Montreal, Ont H4P 2M2 (Tel: 514 731 3451. Telex: 055 60799). Works director: P. Ledoux. ***Tube and pipe mills:*** welded – One Etna 2", and one Etna 3" (annual capacity 20,000 tons).
■ 47 Milner Street, Winnipeg, Man R2X 2P7 (Tel: 204 633 2510. Telex: 07 57834). Works director John E. Wren. ***Tube and pipe mills:*** welded – One AEF 3½" (annual capacity 10,000 tons).
Products: Carbon steel – longitudinal weld tubes and pipes, galvanized tubes and pipes, cold drawn tubes and pipes, hollow sections laminated. Stainless steel – longitudinal weld tubes and pipes, cold drawn tubes and pipes. Alloy steel (excl stainless) – longitudinal weld tubes and pipes, cold drawn tubes and pipes.

Stanley Steel Co Ltd

Head office: Imperial Street, Hamilton, Ont L8L 7V2. **Postal address:** PO Box 66. **Tel:** 416 544 2803. **Telex:** 021712.
Management: R. Weir (president and general manager), J.W. Watson (manager, sales and marketing), J. Butcher (controller), P.C. Brown (vice president, operations). **Established:** 1916. **No. of employees:** 245. **Ownership:** Wholly-owned by The Stanley Works, New Britain, CT, USA. **Subsidiaries:** Stanley Strapping Systems Division.
Works:
■ At head office. ***Rolling mills:*** Three cold strip – one 4-high 20", one 2-high 18", and one 6-high 920mm; one 18" temper/skin pass. ***Other plant:*** Slitting and annealing facilities.
Products: Carbon steel – cold rolled hoop and strip (uncoated) (precision) and oscillated strip coils.

Stelco Inc

Head office: IBM Tower, Toronto-Dominion Centre, Toronto, Ont M5K 1J4. **Tel:** 416 362 2161.
Management: Chairman, president and chief executive officer – J.D. Allan; executive vice presidents – G.W.R. Bowlby, R.E. Heneault, J.E. Hood; vice presidents – G. Binnie (comptroller), P.D. Matthews (treasurer), L.M. Killaly (secretary and general counsel). **Established:** 1910. **No. of employees:** 17,768 (average 1986). **Subsidiaries:** All wholly-owned – Cloture Frost Inc-Frost Fence Inc, Montreal, Que; Stelco Holding Co, Cleveland, OH, USA; Stelco Coal Co, Pittsburgh, PA, USA; Pikeville Coal Co, Louisville, KY, USA (Chisholm Mine); Kanawha Coal Co, Ashford, WV, USA (Madison Mine); Ontario Eveleth Co, Minneapolis, MN, USA; Ontario Hibbing Co, Minneapolis, MN, USA; Stelco USA Inc, Troy, MI, USA. **Annual capacity:** Pig iron: 5,332,050 short tons. Raw steel: 5,200,500 tons.
Works:
■ Hilton Works. ***Coke ovens:*** Four underjet batteries (274 ovens) (annual capacity 1,974,000 tons). ***Sinter plant:*** One 6'7" strand (617,000 tons). ***Blast furnaces:*** Three – 25' hearth dia, 37,700 cu ft; 29'6" hearth dia, 52,586 cu ft; 32' hearth dia, 73,940 cu ft. ***Steelmaking plant:*** Three 140-ton basic oxygen converters (2,800,000 tons). ***Continuous casting machines:*** Combination bloom and slab; one

Canada

Stelco Inc (continued)

slab. ***Rolling mills:*** One 46″/36″ universal slabbing (3,150,000 tons); one 54″ blooming (1,500,000 tons); one 30″/24″ billet (1,112,000 tons); two bar – one 20″ (118,000 tons) and No 1 (435,000 tons); one 18″ wire rod (1,600lb coil) (725,000 tons); one 148″ heavy plate (750,000 tons); one 56″ wide hot strip (2,320,000 tons); three cold reduction – one 80″ continuous (1,115,000 tons), one 48″ continuous (460,000 tons), and one 56″ reversing (153,000 tons); four temper/skin pass – one 56″ continuous (405,000 tons), one 60″ (273,000 tons), one 56″ (tandem) continuous (420,000 tons), and one 80″ continuous (650,000 tons). ***Coil coating lines:*** Three electrolytic tinning – 35-½″ (coil and sheets), 42″ (coil and sheets) and 42″ (coil only); three continuous hot-dip galvanizing – two 48″ and one 60″.

■ Lake Erie Works. ***Coke ovens:*** One battery (45 ovens) (annual capacity 540,000 tonnes). ***Blast furnaces:*** One 33′ hearth dia (1,330,000 tonnes). ***Steelmaking plant:*** Two 230-tonne basic oxygen converters (1,570,000 tonnes). ***Continuous casting machines:*** One 2-strand slab (1,570,000 tonnes). ***Rolling mills:*** One 2,050mm wide hot strip (1,165,000 tonnes).

■ McMaster Works, Contrecoeur, Que. ***Steelmaking plant:*** One 85-ton electric arc furnace. ***Continuous casting machines:*** One 4-strand low-head curved mould billet (3½″, 4″ and 6″ square) (annual capacity 345,000 short tons). ***Rolling mills:*** One 18″/12″ bar (235,000 tons).

■ Edmonton Works, Edmonton, Alta. ***Steelmaking plant:*** Two 75-ton electric arc furnaces. ***Continuous casting machines:*** One 3-strand high-head straight mould billet (4¼″-8″ square) (annual capacity 325,000 short tons). ***Rolling mills:*** One 12″/14″ bar (250,000 tons), for grader blades, sucker rods, cultivator shanks, studs and rock bolts.

■ Camrose Works, Camrose, Alta. ***Tube and pipe mills:*** welded – one ERW (4½-16″) (annual capacity 80,000 short tons), and one arc fusion electric U & O mill 18-42″ (150,000 tons).

■ Welland Tube Works, Welland, Ont. ***Tube and pipe mills:*** welded – one U&O fusion electric 20-36″ (annual capacity 190,000 short tons); one Stelform spiral forming 36-60″ (300,000 tons).

■ Page Hersey Works, Welland, Ont. ***Tube and pipe mills:*** welded – one ERW 2⅜-8⅝″ (60,000 short tons), one ERW 6⅝-16″ (90,000 tons), one continuous weld pipe 0.540-4½″ (120,000 tons), one electric weld tubing 0.706-4″ (40,000 tons), one seamless 1.315-4½″ (30,000 short tons).

■ Other works: There are wire drawing and finishing facilities at Dominion Works (Lachine, Que) and Parkdale Works (Hamilton, Ont); forging hammers and presses at Gananoque Works (Gananoque, Ont); bolts, nuts and rivets at Swansea Works (Toronto, Ont); bolts, rivets and lockwashers at Brantford Works (Brantford, Ont); barbed wire production facilities at Frost Works (Hamilton, Ont) and Dominion Works (Lachine, Que); equipment for cold drawing and finishing bars at Canadian Drawn Steel Co (Hamilton, Ont). There are iron mines and pellet plants in Quebec and Labrador, Canada and in Minnesota and Michigan, USA; coal mines in Pennsylvania, West Virginia and Kentucky, USA, and a limestone quarry and lime kiln in Ingersoll, Ont.

Products: steelmaking pig iron. Carbon steel – ingots, slabs, blooms, billets, wire rod (1,600lb coils), reinforcing bars, round bars, bright (cold finished) bars, light angles, light tees, light channels, heavy angles, medium plates (sheared and torch cut), heavy plates (sheared and torch cut), hot rolled sheet/coil (uncoated), cold rolled sheet/coil (uncoated), hot-dip galvanized sheet/coil, coated sheet/coil (pre-painted), blackplate, electrolytic tinplate (single-reduced), electrolytic tinplate (double-reduced), tin-free steel (ECCS), bright wire, black annealed wire, galvanized wire (plain), galvanized wire (barbed), longitudinal weld tubes and pipes, spiral-weld tubes and pipes, oil country tubular goods. Alloy steel (excl stainless) – ingots (incl high-strength low-alloy), slabs, blooms, billets. Mechanical tubing; pressure tubing; twisted prestressed and strand wire; baling and bale ties; staples; reinforcing fabric; fencing, fence posts and gates; bolts, nuts, rivets, washers, screws; pole line hardware; forgings; cultivator shanks, grader blades, grinding rods, sucker rods; coke and coal chemicals.

Sydney Steel Corp (Sysco)

Head office: Inglis Street, Sydney, Cape Breton, NS B1P 6K5. **Postal address:** PO Box 1450. **Tel:** 902 564 5471. **Telex:** 019 35197. **Cables:** Systco. **Telecopier (Fax):** 902 539 6009.
Management: Chairman – M.H. Cochrane; president – E.A. Boutilier; vice presidents – R.J. Skinner (finance and secretary), J.A. Strasser (marketing); engineering director – L.A. Chiasson; works manager – J.B. Currie. **Established:** 1967. **No. of employees:** 1,350. **Ownership:** Crown Corp wholly-owned by the Province of Nova Scotia. **Annual capacity:** Raw steel: 1,000,000 short tons.
Works:
■ At head office. ***Coke ovens:*** One Koppers-Becker underjet battery (61 ovens) (annual capacity 220,000 tons). ***Blast furnaces:*** Two – (No 3) 21′ 6″ hearth dia (24,000 cu ft of stack) (540,000 tons) and (No 1) (idled, will not be reactivated). ***Steelmaking plant:*** Four 225-ton (heat) open hearth furnaces (1,000,000 tons). ***Refining plant:*** One 225-ton (heat) Lectromelt DH vacuum degassing unit. ***Continuous casting machines:*** Voest-Alpine combination bloom slab – 3-strand bloom and 1-strand slab (blooms – 406.4 x 304.8mm, slabs – up to 305mm x 2,134mm) (750,000 tons). ***Rolling mills:*** 1-stand 2-high (40″ reversing blooming (900,000 tons); 7-stand 2-high 16″ continuous billet (450,000 tons); 3-stand 3-high 28″ cross-country rail (400,000 tons).
Products: Carbon steel – ingots re-rolling and forging, slabs up to 305mm thick x 2,134mm wide, blooms 152.4 mm sq and up, billets, heavy rails 37 kg/m to 67kg/m (75 lb/yd to 140 lb/yd), rail accessories, tie plates (standard sizes).
Brands: Sydney (railroad rail).
Sales offices/agents:
Expansion programme: C$157,000,000; 3-year programme includes installation of an electric arc furnace and ladle refining facilities, modification of continuous casting facilities and rolling mill improvements.

Tubular Steels Products Div of Chartcan

Head office: 62 Donshire Court, Scarborough, Ont MIL 2EL. **Tel:** 416 752 4477.
Management: Vice presidents – E.J. Horton (executive), B.P. Watson (finance), R.D. McKinnon (sales). **No. of employees:** 100. **Ownership:** Chartcan Inc.
Works:
■ At head office. ***Tube and pipe mills:*** six welded (3/8-3″)
Products: Carbon steel – longitudinal weld tubes and pipes.

Union Drawn Steel Co Ltd

Head office: 1350 Burlington St East, Hamilton, Ont L8N 3A2. **Postal address:** PO Box 98. **Tel:** 416 547 4480. **Telex:** 061 8314. **Telecopier (Fax):** 416 544 3852.
Management: President – R.F. Hawkins; vice presidents – J.J. Yellana (sales and marketing), D.I. Williams (finance and administration), J.P. Paliwoda (manufacturing).
Established: 1905. **No. of employees:** Works – 110. Office – 20. **Ownership:** Shares privately held by above and other investors. **Annual capacity:** Finished steel: 76,500 short tons.
Works:
■ At head office. One continuous coil drawing line, one coil bull block, three straightening and cutting machines, three bar drawnbenches, one centreless turning machine and four centreless grinders, etc.
Products: Carbon steel – bright (cold finished) bars – rounds to 4½″, squares and hexagons to 2½, flats ¼ to 2″ x 1 to 5″. Alloy steel (excl stainless) – bright (cold finished) bars – rounds to 4″, squares, hexagons, flats on enquiry. Precision ground shafting up to 6″.
Brands: Century – 100,000 psi minimum yield (1045, 1050, 1141, 1144); Supreme 100 – precision ground shafting – 100,000 psi minimum yield.

Canada

Western Canada Steel Ltd

Head office: 450 South East Marine Drive, Vancouver, BC V5X 2T2. **Tel:** 604 325 2271. **Telex:** 04 51178. **Telecopier (Fax):** 325 2226.
Management: J.M. Willson (president and chief executive officer), G.W. McGibney (vice president new business development), P.A. Hoffman (vice president and sales).
Established: 1947. **No. of employees:** Works – 442. Office – 82. **Ownership:** Cominco Ltd (100%). **Subsidiaries:** Richmond Steel Recycling Ltd (100%) (scrap merchants). Hawaiian Western Ltd (51%) (steel bar production). Pacific Western Steel, Inc (100%) (steel sales). **Annual capacity:** Raw steel: 270,000 short tons. Finished steel: 350,000 tons.
Works:
■ At head office. Works director: S.G. Gilbertson. ***Steelmaking plant:*** One 80-ton electric arc furnace (annual capacity 150,000 tons). ***Continuous casting machines:*** One 4-strand Rokop billet (5½" and 8" sq) (250,000 tons). ***Rolling mills:*** One bar – 4-stand 18" continuous roughing and 16-stand 16" and 12" continuous finishing (200,000 tons). ***Other plant:*** Hot forging for carbon and low alloy steel fasteners (12,000 tons).
■ 2601 52nd Street SE, Calgary, Alta T2P 2J1 (Tel: 403 272 4056. Telex: 038 22712). Works director: D.A. Lamont. ***Steelmaking plant:*** One 40-ton electric arc furnace (annual capacity 120,000 tons). ***Rolling mills:*** One bar – 2-stand 18" roughing and 5-stand 12" cross-country (100,000 tons).
Products: Carbon steel – ingots 5½" square, reinforcing bars, round bars, square bars, rail accessories fasteners.

Chile

Cintac – Cĭa Industrial de Tubos de Acero SA

Head office: Camino a Melipilla 8920, Maipú, Santiago. **Postal address:** Casilla 14294, Correo 21. **Tel:** 575070. **Telex:** 240256 cinta cl.
Management: Managers – Patricio de Groote Pérez (general), Patricio Giglio Gazzana (marketing), Neil C. Peebles (production). **Established:** 1956. **Capital:** Pesos 667,927. **No. of employees:** Works – 83. Office – 126. **Subsidiaries:** Siderúrgica Aza SA (99.33%) (steelworks).
Works:
■ At head office. ***Tube and pipe mills:*** welded – one MEP ⅛-2" (annual capacity 4,000 tonnes), one Abbey Etna 3XU ½-3" (10,000 tonnes), one own manufacture ½-4" (11,000 tonnes).
Products: Carbon steel – hot-dip galvanized hoop and strip, longitudinal weld tubes and pipes, galvanized tubes and pipes.

Famae – Fabrica y Maestranzas del Ejercito

Head office: Pedro Montt 1606, Casilla de Correo 4100, Santiago.
Works:
■ ***Rolling mills:*** 7-stand 3-high bar. ***Other plant:*** Foundry and forge.
Products: Carbon steel – reinforcing bars.

Cia Siderúrgica Huachipato SA

Head office: Casilla 1-C, Talcahuano.
Established: 1946. **Ownership:** Cía de Acero del Pacífico (Cap) (state-owned).
Works:
■ Av Las Industrias s/n, Talcahuano. ***Coke ovens:*** Six Koppars-Becker underjet batteries – five (13 ovens each) and one (5 ovens). ***Blast furnaces:*** One 6.3m hearth dia, 712 cu m and one 7m hearth dia, 818 cu m (combined annual capacity 950,000 tonnes). ***Steelmaking plant:*** Two 100-tonne basic oxygen furnaces (1,000,000 tonnes). ***Continuous casting machines:*** One 1-strand concast slab. ***Rolling mills:*** One 2-high reversing 813mm slabbing and blooming; one 3-high 660mm billet;

Cia Siderúrgica Huachipato SA (continued)

21-stand bar and wire rod; one 3-high roughing heavy plate mill; one 4-high Steckel reversing and 4-stand wide hot strip; one cold reduction; one temper/skin pass. ***Tube and pipe mills:*** welded . ***Coil coating lines:*** One electrolytic tinning line. ***Other plant:*** Galvinizing pots.
Products: Foundry pig iron. Carbon steel – slabs, blooms, wire rod, reinforcing bars, round bars, medium plates, heavy plates, hot rolled sheet/coil (uncoated), cold rolled sheet/coil (uncoated), hot-dip galvanized sheet/coil, electrolytic tinplate (single-reduced), longitudinal weld tubes and pipes.

Cia de Acero Rengo SA

Head office: Los Migueles 600, Casilla 16-D, Rengo, Colchagua.
Works:
■ ***Steelmaking plant:*** One 10-tonne electric arc furnace. ***Rolling mills:*** bar and wire rod mills comprising 1-stand 3-high 500mm, 6-stand 2-high 400mm, 5-stand 2-high 380mm cross-country, and 6-stand 2-high 250mm finishing block.
Products: Carbon steel – wire rod, reinforcing bars.
Additional information: Formerly operated as Indac – Industries del Acero Ltda.

China

Anshan Iron & Steel Co

Works:
■ Anshan City, Liaoning Province. ***Sinter plant:*** several (annual capacity 12,000,000 tonnes). ***Blast furnaces:*** No 1 568 cu m, No 2 826 cu m, No 3 821 cu m, No 4 1,002 cu m, No 5 939 cu m, No 6 1,050 cu m, No 7 2,580 cu m, No 9 983 cu m, No 10 1,627 and No 11 2,025 cu m. ***Steelmaking plant:*** Two 150-tonne and one 180-tonne basic oxygen converters; one 5-tonne, one 1.5-tonne and one 0.5-tonne electric arc furnaces; 300-tonne open hearth fourteen furnaces (eight in No 1 plant, six in No 2 plant). ***Continuous casting machines:*** One bloom ***Rolling mills:*** one slabbing (4,000,000 tonnes; two blooming (3,000,000 tonnes); two heavy section (3,000,000 tonnes); bar; wire rod; 1,550mm wide hot strip (1,000,000 tonnes); one 1,050 hot sheet. ***Tube and pipe mills:*** one 57-168mm dia seamless. ***Coil coating lines:*** hot-dip tinning three sheet galvanising;
Products: Steelmaking pig iron. Carbon steel – ingots, blooms, billets, heavy angles, heavy joists, heavy rails, medium plates, hot rolled sheet/coil (uncoated), hot-dip galvanized sheet/coil, hot-dip tinplate, seamless tubes and pipes, large-diameter tubes and pipes. Production in 1985 was 24,100,000 tonnes of iron ore, 6,600,000 tonnes of pig iron, 7,250,000 tonnes of raw steel, 5,040,000 tonnes of rolled steel, and 4,145,000 tonnes of coke.
Expansion programme: It is planned to increase raw steel capacity to 8,000,000 tonnes and to produce CR sheet including enamelling grade. New plant would include a wide hot strip mill and a 2,000,000 tonnes a year slab caster; two old blast furnaces would be modernised.

Anyang Iron and Steel Works

Head office: Heran Province.
Works:
■ At head office. **Blast furnaces. *Steelmaking plant:*** Three 15-tonne basic oxygen converters. ***Continuous casting machines:*** One billet **Rolling mills and seamless tube mills**.
Products: Carbon steel – products include electrical non-oriented silicon sheet/coil, seamless tubes and pipes. Production in 1985 was pig iron 675,000 tonnes, raw steel 610,000 tonnes, and rolled products 418,000 tonnes.

China

Baoji Steel Tube Works

Works:
■ Shanxi Province. ***Tube and pipe mills:*** HF welded.
Products: Carbon steel – spiral-weld tubes and pipes.

Baoshan Iron & Steel Works

Works:
■ Shanghai Municipality. **Coke ovens**. ***Sinter plant:*** One (annual capacity 4,500,000 tonnes). ***Blast furnaces:*** one 4,063 cu m (10,000 tonnes/day). ***Steelmaking plant:*** Two 300-tonne basic oxygen converters (3,000,000 tonnes). ***Tube and pipe mills:*** seamless (500,000 tonnes).
Products: Steelmaking pig iron. Alloy steel (excl stainless) – seamless tubes and pipes 21-140mm dia. Completion of the first stage was delayed, so production in 1985 reached only 540,000 tonnes of pig iron and 400,000 tonnes of steel.
Expansion programme: The expansion programme foresees a doubling of iron and steel capacity and the extension of products to HR coil and plate; CR sheet and galvanized and colour coated sheets. A second blast furnace is being added for operation in 1991.

Baotau Iron & Steel Co

Head office: Baotou, Inner Mongolia.
Works:
■ At head office. **Coke ovens and sinter plant.** ***Iron ore pelletizing plant:*** Dravo-Lurgi (annual capacity 1,200,000 tonnes). ***Blast furnaces:*** Three – No 1 1,513 cu m, No 2 1,513 cu m and No 3 1,800 cu m. ***Steelmaking plant:*** Three 50-tonne basic oxygen converters; five 500-tonne open hearth furnaces. ***Rolling mills:*** heavy section; rail. **Seamless tube mills.**
Products: Carbon steel – ingots, heavy joists, wide-flange beams (heavy), heavy rails, seamless tubes and pipes. Production in 1985 was 6,200,000 tonnes of iron ore, 1,670,000 tonnes of pig iron, 1,555,000 tonnes of raw steel and 770,000 tonnes of rolled products.

Benxi Iron & Steel Co

Head office: Renmin Road 2, Pingshan District, Benxi City, Liaoning Province. **Tel:** 3191.
Works:
■ Liaoning Province. ***Blast furnaces:*** Five – No 1 380 cu m, No 2 330 cu m, No 3 1,070 cu m, No 4 1,070 cu m, No 5 2,000 cu m. ***Steelmaking plant:*** Three 120-tonne basic oxygen converters (annual capacity 1,700,000 tonnes); electric arc furnace. ***Rolling mills:*** 2-high blooming/ slabbing (2,000,000 tonnes); one 7-stand 1,700mm wide hot strip (1,600,000 tonnes).
Products: Alloy steel (excl stainless) – round bars flats light angles High speed steel; ball bearing steel; heat resistant and stainless steel. Production in 1985 was 12,250 tonnes of iron ore, 3,050,000 tonnes of pig iron, 1,480,000 tonnes of steel and 890,000 tonnes of rolled products.

Betai Iron & Steel Works

Small plant which produced about 200,000 tonnes of pig iron, 10,000 tonnes of raw steel and 9,000 tonnes of rolled steel in 1988.

Changcheng Steel Works

Head office: Zhongba, Jiangyou County Sichuan
Established: 1964. **No. of employees:** Works – 16,000. Office – 8,000.

Changcheng Steel Works (continued)

Ownership: Ministry of metallurgical industry.
Works:
■ No 1 plant (head office). ***Steelmaking plant:*** Four electric arc furnaces – 5 and 15-tonnes; one medium frequency induction 0.5-tonne. ***Refining plant:*** Five 0.5-2-tonne ESR; one 0.2-tonne VIM; one 2.5-tonne consumable electrode furnace. ***Rolling mills:*** One medium section – 2-stand 500mm, 5-stand 300mm. ***Other plant:*** Five cold drawing and seven cold-rolling mills for tubes; four 1.5-tonne steam hammers.
■ No 2 Plant, Houba, Jiangyou County, Sichuan. (Moved in 1965 on merger of Shanghai Strip Plant, Shanghai Cold Rolling Strip Plant and Shanghai Tube Plant). ***Rolling mills:*** slabbing – two trains with three 400mm stands and three 300mm stands; hot strip and hoop – three trains with 300 and 400mm stands; fourteen cold strip – 4-high and 2-high. ***Tube and pipe mills:*** welded Four 32-76mm HF. ***Other plant:*** Fifteen wire drawing machines.
■ No 3 Plant, Tongkou, Jiangyou County, Sichuan. ***Steelmaking plant:*** One 5-tonne and one 15-tonne electric arc furnaces; one 0.6-tonne induction furnace. ***Refining plant:*** Five 0.5-2.5-tonne ESR; two 3,000 lb and 5-tonne VIM; one 7-tonne consumable electrode furnace and two electron beam furnace – one 200k and one 1,200kw. ***Rolling mills:*** Two bar – one 3-high with three 650mm stands and one 3-high with four 400mm stands; one 3-high Lauth heavy plate; one 2-high pullover hot sheet; two 4-high cold sheet. ***Other plant:*** One 300-tonne steam hammer and two 2,000-tonne rapid presses.
■ No 4 Plant, Wudu, Jiangyou County, Sichuan. ***Steelmaking plant:*** Five 10-30-tonne electric arc furnaces. ***Refining plant:*** Two 2.5-tonne ESR. ***Rolling mills:*** One 2-high reversing 825mm blooming One Light/ medium section – semi continuous with four 550mm stands, nine 400-280 stands and eight 320-280mm stands; one hot strip and hoop with 750mm edging mill and 2-high roughing 660 x 700mm and 4-high finishing 400/800 x 700mm; one 4-high reversing 200/500 x 600mm cold strip. ***Tube and pipe mills:*** one seamless with 3,150-tonne hot extrusion, 1,100-tonne vertical piercer and 18-stand reducing mill. ***Other plant:*** Cold drawing machines.
Products: Carbon steel – wire rod 6-24mm; round bars 12-275mm; bright (cold finished) bars; hot rolled hoop and strip (uncoated); cold rolled hoop and strip (uncoated) 80-500mm wide, 0.3-3mm thick; medium plates; heavy plates; seamless tubes and pipes hot extruded 80-194mm od; longitudinal weld tubes and pipes. Stainless steel -; wire rod round bars; seamless tubes and pipes; cold drawn tubes and pipes. Alloy steel (excl stainless) – bright (cold finished) bars; hot rolled hoop and strip (uncoated); cold rolled hoop and strip (uncoated); medium plates; heavy plates; hot rolled sheet/coil (uncoated); wire; cold drawn tubes and pipes. Carbon and alloy engineering steels, carbon and alloy tool steels, spring steels, ball bearing steels. Heat resistant steels, super alloys. Forged bars.
Brands: Chuanzi (stainless and nickel-chrome seamless tubes).
Expansion programme: A 50,000 tonnes-a-year continuous billet caster for start-up in early 1989 is being supplied by Krupp Industrietechnik GmbH.

Changzhi Iron & Steel Works

Head office: Changzhi City, Shanxi Province. **Tel:** 2938.
Established: 1947.
Works:
■ No 1 Works. **Coke ovens, sinter plant and blast furnaces.** ***Steelmaking plant:*** basic oxygen converter; two 15-tonne electric arc furnaces (annual capacity 40,000-60,000 tonnes). ***Rolling mills:*** One 3-stand 500mm blooming (250,000 tonnes); one 5-stand light section (70,000 tonnes); one 300mm Planetary hot strip and hoop (20,000 tonnes); one 4-high 380/190/430mm cold strip (500 tonnes) **Seamless tube and pipe mills**. ***Other plant:*** Cold drawing plant (3,000 tonnes); forge with one 1-tonne and one 3-tonne steam hammer (4,000 tonnes) tube drawing plant.
Products: Carbon steel – billets 60 and 65mm sq; wire rod 14-28mm; bright (cold finished) bars; hot rolled hoop and strip (uncoated); cold rolled hoop and strip (uncoated) up to 160mm gauges 0.4-1.2mm; seamless tubes and pipes including boiler and thin wall; cold drawn tubes and pipes. Raw steel output in 1985 was 239,000 tonnes, rolled steel 100,000 tonnes and pig iron 260,000 tonnes.

China

Chengde Iron & Steel Works

Raw steel production in 1985 was 120,000 tonnes, rolled steel 107,000 tonnes and pig iron 127,000 tonnes.

Chengdu Metallurgical Experimental Plant

Head office: No 2 Erhuan Road (North Section), Chengdu, Sichuan Province.
Established: 1958.
Works:
■ At head office. ***Steelmaking plant:*** Three 5-tonne electric arc furnaces. ***Rolling mills:*** One 2-stand 500mm blooming; medium section / bar. ***Other plant:*** Cold drawing machines; two steam hammers; wire drawing machines for high strength wire.
Products: Carbon steel – reinforcing bars 12-15mm; round bars 10-28mm; flats including spring steel; bright (cold finished) bars. Alloy steel (excl stainless) – round bars; flats; bright (cold finished) bars. Carbon and alloy engineering steels; carbon and alloy tool steels; high speed steel. Production in 1985 was: pig iron 40,000 tonnes, raw steel 80,000 tonnes and rolled steel 200,000 tonnes.
Expansion programme: It is planned to increase annual raw steel capacity to 100,000 tonnes and to introduce argon stirring and continuous casting. New mills would be for bars and for pipes (piercing mill) with a 20,000 tonne capacity.

Chengdu Seamless Steel Tube Plant

Head office: Tel: 43412 504.
Works:
■ At head office. ***Steelmaking plant:*** Two 5-tonne electric arc furnaces. ***Tube and pipe mills:*** seamless three hot rolling (4,7, and 12") (annual capacity 200,000 tonnes).
Products: Carbon steel – seamless tubes and pipes. Stainless steel – seamless tubes and pipes. Alloy steel (excl stainless) – seamless tubes and pipes. Production in 1985 was 332,000 tonnes of raw steel and 275,000 tonnes of finished products.

Chongqing Iron & Steel Co

Head office: Sechuan Province.
Works:
■ ***Blast furnaces:*** Three – No 1 100 cu m, No 2 620 cu m, No 3 620 cu m. ***Steelmaking plant:*** basic oxygen -two 30-tonne, two 50-tonne and two 80-tonne. **Rolling mills**.
Products: Carbon steel – round bars, light angles, heavy plates, clad plates copper. Production in 1985 was pig iron 66,000 tonnes, raw steel 735,000 tonnes and rolled finished products 630,000 tonnes.

Chongqing Special Steel Works

Head office: Chongqing City, Sichuan. **Tel:** 661961.
Established: 1937. **Ownership:** Ministry of Metallurgical Industry.
Works:
■ At head office. ***Sinter plant:*** one 3-strand. ***Blast furnaces:*** Two 620 cu m. ***Steelmaking plant:*** Eleven 5-20-tonne electric arc furnaces. ***Refining plant:*** Five 0.5-2-tonne ESR; one 10-tonne vacuum degassing unit; four VIM including atmosphere induction; one 15-tonne VHD/ VOD . ***Rolling mills:*** Three blooming (750, 600 and 500mm); one medium section; two bar (250mm); two 3-stand heavy plate (650 x 1,230mm); one 4-high reversing hot strip and hoop; four other hot mills; twelve 2 high and 4-high and one 12-high cold strip; one 4-high other cold mill two temper/skin pass; one 4-high sheet. ***Other plant:*** One GFM 800-tonne precision forging machine; air and steam hammers; eight cold wire drawing blocks.
Products: Stainless steel – cold rolled hoop and strip (uncoated) 0.10-2 x 400mm max; wire. Alloy steel (excl stainless) – round bars; square bars; hexagons; bright (cold

Chongqing Special Steel Works (continued)

finished) bars, medium plates; hot rolled sheet/coil (uncoated) Cold finished alloy; forgings; spring steel, high speed steels, high quality carbon steels, constructional steels; heat resistants steels; alloy tool and die steels; raw steel output in 1985 was 250,000 tonnes, rolled products 185,000 tonnes.
Expansion programme: A continuous billet caster is planned.

Dalian Steel Plant

Head office: Ganjingzi District, Dalian City, Liaoning Province. **Tel:** 62112.
Established: 1948. **No. of employees:** About 14,000.
Works:
■ ***Steelmaking plant:*** One 20-tonne, two 10-tonne, four 7-tonne and three 5-tonne electric arc furnaces. ***Refining plant:*** One VOD. Dalian is a leading speciality steel plant producing high speed, bearing, spring and stainless steels. Production of raw steel in 1985 was 220,000 tonnes.
Expansion programme: The existing rolling mills are to be replaced and capacity raised from 130,000 tonnes-200,000 tonnes. A new continuous rod mill with an annual capacity of 200,000 tonnes will also be installed; construction is due to start in 1985 and be completed three years after. This Birdsboro mill will comprise 3-stand 15" cartridge roughing mill, three 13" intermediate stands and six 11" finishing stands.

Dalian Steel Rolling Mill

Head office: 249 Fuguo Street, Shahekou District, Dalian City, Liaoning Province. **Tel:** 41121 and 41193.
Works:
■ At head office.
Products: Carbon steel – wire rod 6,5,8 and 9mm; round bars 6,8,9 and 10mm; flats 4 x 16mm, 4 x 20mm, 4 x 25mm.

Daxian Iron & Steel Works

Some 57,000 tonnes of raw steel and 60,000 tonnes of rolled finished steel were produced in 1985.

Daye Steel Works

Head office: Huangshi, Hubei Province. **Tel:** 4911, 4921.
Established: Founded nearly 100 years ago; the longest established steel plant in China.
Works:
■ ***Blast furnaces:*** Two small furnaces; ***Steelmaking plant:*** three 90-tonne open hearth furnaces; nine electric furnaces (one 20-tonne, five 9-tonne, two 3-tonne and one 1-tonne). ***Rolling mills:*** blooming; several light section and bar. ***Other plant:*** Cold drawing plant for bars.
Products: Carbon steel – round bars, Production in 1985 was 465,000 tonnes of raw steel and 320,000 tonnes of finished rolled steel.
Expansion programme: A small continuous bar/rod mill (annual capacity 250,000 tonnes) is to be installed for the production of bearing and other steels and high carbon steels in rods and rounds up to 120mm, and spring steel flats (from 16-60mm). A continuous billet caster (annual capacity 100,000 tonnes) will start up in early 1989 – it is being supplied by Krupp Industrietechnik.

Echeng Iron & Steel Works

Works:
■ Echeng Hubei Province. ***Sinter plant:*** Two. ***Blast furnaces:*** Four – one 554 cu m

China

Echeng Iron & Steel Works (continued)

and three 100 cu m. ***Steelmaking plant:*** One 7-tonne basic oxygen; three 5-tonne electric arc furnaces. ***Rolling mills:*** blooming; bar; wire rod; hot strip and hoop; cold strip;
Products: Carbon steel – wire rod, round bars, light angles, hot rolled hoop and strip (uncoated), cold rolled hoop and strip (uncoated). Production 1985 was: pig iron 247,000 tonnes, raw steel 240,000 tonnes and rolled steel 270,000 tonnes.

Fushun Steel Plant

Head office: Fushun City, Liaoning Province. **Tel:** 89589.
Established: 1937
Works:
■ ***Steelmaking plant:*** One 15-tonne and one 30-tonne electric arc furnaces; ***Refining plant:*** Several ESR 1-30-tonne, one 30-60-tonne VOD /VHD; 1-7 tonne consumable electrode furnaces. ***Rolling mills:*** Four light section – 250mm, 420mm, 500mm and 650mm; hot sheet mill (1,400mm). ***Tube and pipe mills:*** One 100mm seamless. ***Other plant:*** 2,000-tonne hydraulic press; 1,000-tonne precision forging machine; forging hammer; cold drawbenches. Stainless and heat resistant steels; high speed steels; ball bearing steels; carbon and alloy engineering steels; carbon tool steels; spring steels; titanium alloys; superalloys. Production in 1985 was 277,000 tonnes of raw steel and 210,000 tonnes of rolled steel.

Fushun Xinfu Steel Plant

Head office: Gongnong Street, Wanghua District, Fushun City, Liaoning Province. **Tel:** 89251.
Established: 1958. **No. of employees:** 12,000.
Works:
■ At head office. **Sinter plant**. ***Blast furnaces:*** two 300 cu m (annual capacity 400,000 tonnes). ***Steelmaking plant:*** Three 6-tonne basic oxygen converters (200,000 tonnes); four 5-tonne electric arc furnaces (100,000 tonnes). ***Refining plant:*** Power injection SL unit. ***Rolling mills:*** blooming one 650mm stand, three 650mm stands (270,000 tonnes); four bar and wire rod (70,000 tonnes, 50,000 tonnes, 45,000 tonnes and 5,000 tonnes); one wire rod 50,000 tonnes. ***Other plant:*** Drill rod plant with rotary piercing mill, reducing mill, drawing machine, etc.
Products: Steelmaking pig iron; foundry pig iron. Carbon steel – billets 50, 55, 60, 75, 90, 100 and 110mm sq; semis for seamless tubemaking; wire rod in coils; reinforcing bars 10, 12, 16, 20, 25, 32, 40, 50mm (including coils); round bars 10-50mm; flats (including spring steel); hexagons 25-6mm x 60-16mm and 50-100mm x over 16mm; light angles Free-cutting steels and hollow drill bars. Production in 1985 was 186,000 tonnes of pig iron, 205,000 tonnes of raw steel and 185,000 tonnes of finished products.

Guangzhou Alloy Steel Plant

Head office: Lianhe, North Suburb of Guangzhou, Guangdong Province. **Tel:** 61082.
Established: 1960's.
Works:
■ At head office. ***Steelmaking plant:*** Two 5-tonne, two 15-tonne and one 3-tonne electric arc furnaces. ***Rolling mills:*** Several bar. ***Other plant:*** Forging hammers; roll foundry with 5-tonnes per hour cupola and two 5-tonne electric furnaces; precision and magnet castings shop with medium frequency furnaces. A VIM furnace and a consumable electrode furnace.
Products: Carbon steel – reinforcing bars 25 and 28mm; round bars 12-32mm; square bars 9-20mm; flats 4-7 x 12-20mm. Low alloy bars; spring steels; chilled cast iron rolls, alloy nodular iron rolls, alloy chilled rolls, etc; carbon and alloy steel rolls under 600mm; precision castings including alloy, stainless, acid and wear resistant; high speed steel; magnet steels; steel balls.

Guangzhou Iron & Steel Works

Head office: Baihedong, Guangzhou City, Guangdong Province. **Tel:** 81212.
Established: 1958.
Works:
■ At head office. **Coke ovens and sinter plant.** ***Steelmaking plant:*** One basic oxygen converter; several electric arc furnaces. ***Rolling mills:*** One 650mm blooming; two light section 250 and 400mm. ***Tube and pipe mills:*** one seamless
Products: Steelmaking pig iron. Carbon steel – reinforcing bars; round bars (low carbon); seamless tubes and pipes. High manganese steel castings; metallurgical coke; chemicals. Production in 1985 was pig iron 235,000 tonnes, raw steel 280,000 tonnes and rolled steel 230,000 tonnes.

Guiyang Steel Mill

Head office: Youzha Steel, Guiyang. **Tel:** 27901.
Established: 1958.
Works:
■ At head office. ***Steelmaking plant:*** Five 10-tonne electric arc furnaces. ***Refining plant:*** One ESR. ***Rolling mills:*** One 650mm blooming; four bar; two double duo ***Other plant:*** Forging equipment; plant for cold drawing seamless tubes.
Products: Carbon steel – wire rod, round bars, square bars, seamless tubes and pipes, cold drawn tubes and pipes. Stainless steel – wire rod, round bars, square bars. Alloy steel (excl stainless) – round bars, square bars, seamless tubes and pipes, cold drawn tubes and pipes. Tool steel; bearing steel; rock drill steels. Production in 1985 was 130,000 tonnes of raw steel and 105,000 tonnes of rolled products.

Haerbin Steel Rolling Mill

Some 200,000 tonnes of finished steel were produced in 1985.

Handan Iron & Steel Works

Works:
■ Hopeh Province. ***Blast furnaces:*** three medium-size (annual capacity about 1,000,000 tonnes). ***Steelmaking plant:*** Three 20-tonne basic oxygen converters (500,000 tonnes). ***Continuous casting machines:*** one billet. ***Rolling mills:*** Including one wire rod with Ashlow twist-free finishing block and controlled cooling system (200,000 tonnes).
Products: Carbon steel – including wire rod. Production in 1985 was 775,000 tonnes of pig iron, 746,000 tonnes of raw steel and 450,000 of finished rolled products.

Hangyang Iron & Steel Works

Production in 1985 was 122,000 tonnes of raw steel and 155,000 tonnes of rolled finished products.

Hangzhou Iron & Steel Works

Works:
■ Zhejiang Province. **Coke ovens; ironmaking plant; steelmaking plant** (including a 5-tonne converter); **rolling mills**. These include sheets and light rails. Production in 1985 was coke 27,000 tonnes, pig iron 325,000 tonnes and raw steel nearly 390,000 tonnes rolled finished products 350,000 tonnes.

Hefei Iron & Steel Co

Head office: Heping Road, Hefei City, Anhui Province. **Tel:** 82889.
Established: 1956 as the Hefei Special Steel Plant. **No. of employees:** 15,000.
Works:
■ At head office. **Blast furnaces**. ***Steelmaking plant:*** One basic oxygen converter;

Hefei Iron & Steel Co (continued)

five 10-tonne (special steel section) electric arc furnaces. ***Refining plant:*** One ladle furnace. ***Continuous casting machines:*** One billet. ***Rolling mills:*** One 630mm blooming; one 500mm medium section; two light section 350 and 250mm; one hot strip and hoop; one hot sheet; one cold strip. ***Tube and pipe mills:*** One seamless. One welded.
Products: Carbon steel – round bars, square bars, flats, hexagons, bright (cold finished) bars, seamless tubes and pipes, longitudinal weld tubes and pipes. Alloy steel (excl stainless) – bright (cold finished) bars. Alloy engineering steel; carbon and alloy tool steels; spring steels; bearing steels; silicon steels; boiler tubes. Production in 1985 was 190,000 tonnes, raw steel 250,000 tonnes and rolled products 170,000 tonnes.
Expansion programme: The plant is undergoing expansion, including the conversion of the side blown converter to LD.

Hengyang Steel Tube Plant

Head office: Hengyang, Hunan Province. **Tel:** 22151.
No. of employees: 3,000.
Works:
■ At head office. ***Tube and pipe mills:*** seamless – one 108mm trio mill (700-114mm od, 7-20mm wall thickness) (piercing and plug rolling) (annual capacity 30,000 tonnes); one 100mm automatic (32-114mm od, 2.8-15mm wall thickness) (50,000 tonnes); one 76mm automatic with cold drawing (5-76mm od, 1-8m wall thickness (20,000-25,000 tonnes). Welded – two HF 50.8 and 63.5mm wall thickness. ***Other plant:*** Hot dip tube galvanizing plant.
Products: Carbon steel – seamless tubes and pipes including boiler, high pressure, bearing; longitudinal weld tubes and pipes 19.05-65.5mm od, 1.5-3.75mm wall thickness; galvanized tubes and pipes; cold drawn tubes and pipes. Stainless steel – seamless tubes and pipes; cold drawn tubes and pipes.
Expansion programme: It is planned to install a rotary sizing mill and a three-roll planetary mill.

Huhehaote Iron & Steel Works

Works:
■ Inner Mongolia. **Plant details:** Steelmaking, including oxygen steelmaking. Raw steel production in 1985 was 104,000 tonnes and pig iron 44,000 tonnes.

Jiangxi Steel Works

Head office: Xingu, Jiangxi Province.
Established: 1965.
Works:
■ At head office. ***Steelmaking plant:*** Two 25-tonne basic oxygen converters; four 5-tonne electric arc furnaces. ***Refining plant:*** Two 1-tonne and 3-tonne ESR. ***Continuous casting machines:*** Two curved billet. ***Rolling mills:*** One 650mm blooming with; light section two 400mm stands, three 360mm stands, five 250mm stands, two 360mm stands and five 250mm stands; one wire rod; two 760 and 1,200mm pack sheet; four 4-high cold strip with one 4-stand, one 10 stand, two 1-stand. ***Tube and pipe mills:*** seamless – one 3-high cold rolling (RD30) and one 3-high cold rolling (RD50). ***Other plant:*** Two steam forging hammers; cold drawbenches; wire drawing machines, etc; one 5-tonne cupola; steel foundry with two 5-tonne electric arc furnaces and two other electric furnaces.
Products: Carbon steel – ingots; slabs; blooms; wire rod; reinforcing bars 12-18mm; flats; light angles 5 x 50mm-8 x 75mm; special light sections (rim, front axle, plough shares); cold rolled hoop and strip (uncoated) 10-150mm wide, 0.05-2.5mm thick; hot rolled sheet/coil (uncoated); electrical non-oriented silicon sheet/coil; bright wire; seamless tubes and pipes; cold drawn tubes and pipes; hollow sections. Forgings; iron and steel castings including cast steel rolls; carbon and alloy engineering and tool

Jiangxi Steel Works (continued)

steels; high speed steels; spring steels; bearing steels; stainless, silicon and free-cutting steels. Production in 1985 was 470,000 tonnes of raw steel and 300,000 tonnes of rolled steel.
Brands: Shanjeng.

Jiangyou Changzheng Steel Plant

Raw steel production was 320,000 tonnes in 1985 and rolled finished products 320,000 tonnes.

Jinan Iron Plant

Produced 200,000 tonnes of pig iron in 1985.

Jinan Iron & Steel Works

Works:
■ Shandong Province. **Plant details:** Medium-sized works producing pig iron (600,000 tonnes in 1985), raw steel (about 500,000 tonnes) and rolled finished steel (335,000 tonnes).

Jinxi Steel Tube Plant

Production of finished steel was about 44,000 tonnes in 1985.

Jiuquan Iron & Steel Co

Works:
■ At head office. ***Blast furnaces:*** One 1,513 cu m. Production in 1985 was 2,460,000 tonnes of iron ore and 620,000 tonnes of pig iron.
Additional information: Production in 1985 was 2,460,000 tonnes of iron ore and 620,000 tonnes of pig iron.

Kunming Iron & Steel Works

Works:
■ Yunnan Province. **Coke ovens, sinter plant and four blast furnaces.** ***Steelmaking plant:*** Three 15-tonne basic oxygen converters. ***Continuous casting machines:*** One billet. ***Rolling mills:*** heavy plate. ***Tube and pipe mills:*** seamless.
Products: Carbon steel – ingots, medium plates, heavy plates, seamless tubes and pipes. Production in 1985 was: iron ore 1,260,000 tonnes; pig iron 520,000 tonnes; raw steel 490,000 tonnes and rolled products 380,000 tonnes.

Laiwu Iron & Steel Plant

Head office: Laiwu City, Shandong Province.
Established: Special steel plant established in 1965 (production 1969) became No 1 steel plant when integrated works was established in 1970.
Works:
■ No 1 Steel Plant. ***Steelmaking plant:*** One basic oxygen converter; three 5-tonne (special) electric arc furnaces. ***Rolling mills:*** One 2-stand 550mm blooming; one 5-stand 300mm medium section; one 3-stand 330mm and 5-stand 280mm light section; several cold strip. ***Other plant:*** Cold drawing bench.
Products: Carbon steel – bright (cold finished) bars; light angles; light tees; medium angles; medium joists; heavy angles; heavy joists; cold rolled hoop and strip (uncoated) 80-150mm wide, 0.65-3mm thick. Carbon and alloy engineering steels; ball bearing steels; carbon and alloy tool steels; spring steels, stainless steels. Production for the whole works in 1985 was 446,000 tonnes of pig iron, 158,000 tonnes of raw steel and 73,000 tonnes of rolled steel.

China

Lanzhou Steel Plant

Raw steel output in 1985 was 227,000 tonnes, rolled finished products 163,000 tonnes.

Liayuan Iron & Steel Works

Works:
■ Hunan Province. **Coke ovens** (two batteries); **one sinter plant; blast furnaces.** ***Steelmaking plant:*** Two 15-tonne basic oxygen converters. ***Continuous casting machines:*** One billet (annual capacity up to 300,000 tonnes). **Rolling mills.** Pig iron production in 1985 was 40,000 tonnes.

Linfen Iron & Steel Co

Pig iron output in 1985 was 170,000 tonnes, and that of finished rolled steel 55,000 tonnes.

Lingyuan Iron & Steel Works

Head office: Chaoyang City, Lingyuan County, Liaoning Province. **Tel:** 3212.
Products: Carbon steel – round bars, cold rolled hoop and strip (uncoated), longitudinal weld tubes and pipes. Production of raw steel in 1985 was 142,000 tonnes, rolled steel 120,000 tonnes and pig iron 115,000 tonnes.

Liuzhou Works

Works:
■ Guangxi Province. **Coke ovens**. ***Blast furnaces:*** two 255 cu m. ***Steelmaking plant:*** Three Bessemer converters. ***Continuous casting machines:*** One billet. **Rolling mills**. Production in 1985 was 250,000 tonnes of pig iron, 285,000 tonnes of steel and 240,000 tonnes of rolled products.

Lueyang Iron & Steel Works

Pig iron production in 1985 was 120,000 tonnes, raw steel 52,000 tonnes and rolled steel 33,000 tonnes.

Luoyang Steel Plant

Produced 90,000 tonnes of raw steel and 118,000 tonnes of rolled products in 1985.

Maanshan Iron & Steel Co

Works:
■ Anhui Province. **Coke ovens**. ***Blast furnaces:*** Ten – No 1 255 cu m; No 2 294 cu m; No 3 294 cu m; No 4 294 cu m; No 7 100 cu m; No 9 300 cu m; No 10 300 cu m; No 11 300 cu m; No 12 300 cu m; No 13 250 cu m. ***Steelmaking plant:*** Three 50-tonne and three 9-tonne basic oxygen converters; four 5-tonne and one 1.5-tonne electric arc two 185-tonne open hearth furnaces. two 185-tonne. ***Continuous casting machines:*** One 2-strand billet (annual capacity 200,000 tonnes). ***Rolling mills:*** Including one blooming (up to 1,500,000 tonnes); small mill for special steels; one single stand 2,300mm heavy plate. ***Other plant:*** Wheel and tyre shop. Production in 1985 totalled 1,550,000 tonnes of coke, 6,770,000 tonnes of iron ore, 1,990,000 tonnes of pig iron, 1,645,000 tonnes of raw steel and 1,115,000 tonnes of rolled products.

Expansion programme: A 1,200 cu m blast furnace (9.15 m hearth dia) with an annual capacity of 800,000 tonnes, and additional rolling capacity including a No-Twist rod are planned and may be under construction.

Meishan Metallurgical Corp

Works:
■ Near Nanjing, Jiangsu Province. ***Sinter plant:*** two 130 sq m grate area. ***Blast furnaces:*** two 1,060 cu m (annual capacity 1,500,000 tonnes). Production in 1985 was 1,290,000 tonnes of steelmaking pig iron.
Kaiser Engineers & Constructors Inc is carrying out a study into the feasibility of integrating Meishan works by the installation of a third blast furnace, a coke oven battery, sinter plant, basic oxygen steel plant, continuous caster and hot and cold rolling mills. Raw steel capacity would be 2,000,000 tonnes/year. Currently the wors, pig iron goes to Shanghai melting shops.

Nanchang Iron & Steel Works

A very small plant which produced about 30,000 tonnes of pig iron, 70,000 tonnes of raw steel and 175,000 tonnes of rolled steel in 1985.

Nanjing Iron & Steel Works

Works:
■ Jiangxi Province. **Coke ovens and sinter plant.** ***Blast furnaces:*** one. ***Steelmaking plant:*** Two 15-tonne basic oxygen converters. **Rolling mills.**
Products: Carbon steel – light rails, hot rolled sheet/coil (uncoated), electrical non-oriented silicon sheet/coil. Output in 1985 was: pig iron 335,000 tonnes; raw steel 302,000 tonnes; rolled steel 219,000 tonnes.

Ningxia Shizuishan Iron & Steel Works

Works:
■ At head office. **General information:** Very small iron and steel plant with rolled steel production in 1985 of about 40,000 tonnes. ***Blast furnaces:*** one 100 cu m.

Panzhihua Works

Head office: Dukou, Sichuan Province.
Works:
■ At head office. Four 90 sq m grate area ***Sinter plant:***. ***Blast furnaces:*** No 1 1,000 cu m; No 2 1,000 cu m; No 3 1,200 cu m. ***Steelmaking plant:*** Three 120-tonne basic oxygen converters. ***Rolling mills:*** One Rail/ heavy section.
Products: Carbon steel – heavy angles, heavy joists, heavy channels, heavy rails. Production in 1985 was 1,490,000 tonnes of iron ore, 1,940,000 tonnes of pig iron, 1,665,000 tonnes of raw steel and 920,000 tonnes of rolled products.
Brands: Zeds.
It was announced in August 1987 that production would be doubled to 3,000,000 tonnes annually with the installation of a new blast furnace, continuous casting plant and hot and cold rolling mills.

Pingxiang Iron & Steel Works

Output in 1985: pig iron 240,000 tonnes, raw steel 145,000 tonnes.

Qingdao Steel Plant

Produced 350,000 tonnes of raw steel and 320,000 tonnes of rolled products in 1985.

China

Qiqihar Steel Plant

Head office: Fularji, Heilongjiang Province. **Tel:** 3931-3937.
Established: 1954 (formerly North Manchurin Steel Plant). **No. of employees:** 15,000.
Works:
■ At head office. ***Steelmaking plant:*** Two 20-tonne, one 10-tonne, one 5-tonne and one 3-tonne electric arc furnaces; three 90-tonne basic open hearth. ***Refining plant:*** Four ESR; ladle furnace. several VIM; one SL powder injection, one CAB (capped argon bubbling). ***Rolling mills:*** blooming; medium section; ***Other plant:*** Cold drawing machines; hydraulic forging presses; foundry. High quality carbon steels; spring steels; bearing steels; tool steels; die steels; high speed steels; titanium alloys and super alloys. Production in 1985 was 380,0000 tonnes of raw steel and 280,000 tonnes of rolled steel.

Sanming Iron & Steel Works

Head office: Liexi, Sanming City, Fujian Province. **Tel:** 22075.
Works:
■ At head office. **Blast furnaces**. ***Steelmaking plant:*** Two 15-tonne basic oxygen converters. **Rolling mills.**
Products: Carbon steel – wire rod, light angles, light channels, medium angles, medium joists, seamless tubes and pipes. Production 1985 was 275,000 tonnes of pig iron, 260,000 tonnes of raw steel and 200,000 tonnes of finished steel.

Shaanxi Precision Alloy Plant

Head office: Xian, Shaanxi.
Established: 1958 (moved from Dalien to Xian in 1965). **No. of employees:** Over 2,000.
Works:
■ At head office. ***Steelmaking plant:*** One electric arc furnace; induction furnace. ***Refining plant:*** One ESR; VIM; consumable electrode. ***Rolling mills:*** Including one 4-high reversing bar/ hot strip and hoop; cold strip – 2-high, 4-high, 12-high and 20-high. ***Tube and pipe mills:*** seamless – 2-high hot piercer. ***Other plant:*** Steam hammers; dry and wet wire drawing machines; single and multistrand tube rolling mills and tube spinning machines.
Products: Stainless steel – round bars, flats, hot rolled hoop and strip (uncoated), cold rolled hoop and strip (uncoated), wire, seamless tubes and pipes, cold drawn tubes and pipes. Alloy steel (excl stainless) – round bars, flats, hot rolled hoop and strip (uncoated), cold rolled hoop and strip (uncoated), wire, seamless tubes and pipes, cold drawn tubes and pipes. Precision alloys (magnetic, thermometal, expansion, etc); corrosion resistant and stainless steels; heat resistant steels; high strength steels; spun thin walled tubes; ground rods; fine wires.
Additional information: A new 4-high reversing cold strip mill (215/560/460mm) built by Josef Fröhling GmbH started up in late 1986/early 1987. It is for rolling chrome and nickel alloy steels and nickel iron.

Shaanxi Steel Works

Head office: Xing Fu Road, Dongjiao, Xian, Shaanxi. **Tel:** 3 1726.
Established: Moved to Xian in 1965. **No. of employees:** Over 6,000.
Works:
■ At head office. ***Steelmaking plant:*** Four or five electric arc furnaces (size reported as 5 or 10-tonne); one 150kg induction furnace. ***Refining plant:*** Two 1-tonne ESR. ***Rolling mills:*** Two blooming – one 650mm and two 500mm stands; light section – five 300mm and two 400mm stands; wire rod – three 300mm, five 250mm and three 250mm and two double-duo stands. ***Other plant:*** Two and three-tonne forging hammers; cold drawing facilites; wire drawing plants (single and double die); powder and sintering equipment for hard alloys. Carbon and alloy tool steels; ball bearing steels;

Shaanxi Steel Works (continued)

spring steels; high speed steels; stainless steels; resistance alloys etc, mostly as small sections, wire and forgings. Raw steel production in 1985 was 100,000 tonnes and rolled finished products 78,000 tonnes.
Expansion programme: A billet caster (Krupp) with a capacity of 240,000 tonnes/year is due for completion in early 1989.

Shanghai Iron & Steel Works

Works:
■ No 1 Iron & Steel Works. ***Blast furnaces:*** two 255 cu m each. ***Steelmaking plant:*** Three 30-tonne, three 15-tonne basic oxygen converters; two 85-tonne open hearth. **Rolling mills.**
■ No 2 Iron & Steel Works. ***Steelmaking plant:*** Two 3-tonne electric arc furnaces.
■ No 3 Iron & Steel Works, 300 Shangan Road, Pudong, Shangai. (Tel: 839180). ***Steelmaking plant:*** Three 25-tonne basic oxygen converters; six 15-tonne electric arc furnaces; two 100-tonne open hearth furnaces. ***Refining plant:*** One VOD unit (planned). ***Continuous casting machines:*** One 1/2-strand for bloom /slab (blooms – 130, 150, and 200mm sq; slabs – 600 to 1,000mm x 130 to 200mm) (annual capacity 400,000 tonnes) ***Rolling mills:*** One 630mm blooming; one light/ medium section; one 3,300mm heavy plate one hot and one cold sheet mill;
■ No 5 Iron & Steel Works, Shui Chan Lu, Wu Song, Shanghai. (Tel: 671494). ***Steelmaking plant:*** Three 25-tonne basic oxygen converters; two 30-tonne, three 10-tonne, five 5-tonne and one 3-tonne; electric arc furnaces, medium frequency furnace. ***Refining plant:*** One 1-tonne and one 0.2-tonne ESR; one ladle furnace; one 7-tonne VAR; one 1.5-tonne VIM; vacuum consumable electrode furnace. ***Rolling mills:*** slabbing; one 750mm blooming; billet; bar; wire rod; heavy plate; wide hot strip. ***Coil coating lines:*** hot-dip tinning. ***Other plant:*** Hot extrusion press; forging hammers; forging press; cold rolled tube plant; cold drawing plant; wire drawing equipment.
■ Shanghai Eight-Five Steelworks (Special Steel) plant, Post Box 419, Guichi County, Anhui Province. ***Steelmaking plant:*** Two 5-tonne electric arc furnaces. ***Rolling mills:*** One medium section; one light section. ***Other plant:*** Two steam hammers; wire drawing facilities; seamless tube plant with piercer and cold drawbenches.
Products: Carbon steel – reinforcing bars, round bars, flats, bright (cold finished) bars, light angles, light channels, medium plates, heavy plates, hot rolled sheet/coil (uncoated), hot-dip galvanized sheet/coil, colour coated sheet/coil, cold drawn tubes and pipes. Stainless steel – medium plates, heavy plates. Alloy steel (excl stainless) – round bars bright (cold finished) bars, seamless tubes and pipes, precision tubes and pipes, Carbon and alloy engineering steels; alloy tool steels; high speed steel; ball bearing steels, boiler tubes; needle wire; high temperature alloys; heat resistant steels; magnetic alloys; titanium alloys. Production in 1985 was 320,000 tonnes of pig iron, 5,050,000 tonnes of raw steel and 4,950,000 tonnes of finished rolled steel.
Expansion programme: Krupp Industrie Technik GmbH is to supply a bar mill for high alloy steels to No 5 Steel works for operation in 1990. The size range will be 8-40mm dia. The mill configuration is semi-continuous roughing, a 4-stand horizontal/vertical roughing, two intermediate trains with eight horizontal/vertical stands, and a finishing mill.

Shanghai Xihu Iron & Steel Works

Head office: Tangshan Road 1034, Shanghai. **Tel:** 456860.
Products: Carbon steel – reinforcing bars 10-18mm; special light sections for windows. Chilled alloy iron rolls.

Shaogan Iron & Steel Works

Pig iron output in 1985 was 285,000 tonnes, raw steel 300,000 tonnes and rolled steel 285,000 tonnes.

China

Shenyang Steel Rolling Mill & Shenyang Wire Rod Mill

Output in 1985 of finished rolled steel was 110,000 tonnes and 217,000 tonnes respectively.

Shoudu Iron & Steel Co

Works:

■ Peking Suburbs. ***Coke ovens:*** Four batteries. ***Blast furnaces:*** No.1 576 cu m, No.2 1,327 cu m, No.3 1,036 cu m, No.4 1,200 cu m; there is also a 23 cu m experimental furnace. ***Steelmaking plant:*** Five basic oxygen converters – three 30-tonne, one 5-tonne top blown, one 5-tonne bottom blown (annual capacity 1,400,000 tonnes); three electric arc furnaces. ***Refining plant:*** One ladle furnace. ***Continuous casting machines:*** Two 8-strand for billet. ***Rolling mills:*** slabbing, blooming, and billet (combined capacity 1,800,000 tonnes); one 500/400mm heavy section (100,000 tonnes); several light section; one bar; wire rod mill. ***Tube and pipe mills:*** welded – One 20-120mm mill (70,000 tonnes). ***Other plant:*** Cast iron pipe plant.

■ Metallurgical Research Institute, Special Steel Division. **Steelmaking plant:** High frequency and intermediate. ***Refining plant:*** VIM unit. ***Rolling mills:*** One wire rod; one hot strip and hoop; several cold strip including Sundwig 380mm 20-roll. ***Other plant:*** Cold drawing plant.

Products: Steelmaking pig iron. Production in 1985 was 14,725,000 tonnes of iron ore, 3,260,000 tonnes of pig iron, 2,580,000 tonnes of raw steel and 2,200 tonnes of rolled products. Carbon steel – ingots, slabs, blooms, billets, reinforcing bars, round bars, flats, light angles, heavy angles, heavy channels, longitudinal weld tubes and pipes, Stainless steel – wire Alloy steel (excl stainless) -, wire, Cast iron -, pipes. Free cutting steel bars.

Additional information: The complex of works called Shondon was previously known as Peking or Capital Works, and incorporates the former Shihchingshan Works.

Shuicheng Works

Works:

■ Guizhou Province. ***Blast furnaces:*** one 1,200 cu m (annual capacity 650,000 tonnes) and one 690 cu m. **Steelmaking plant. Rolling mills**. Production in 1985 was: pig iron 500,000 tonnes, raw steel 35,000 tonnes and rolled steel 68,000 tonnes, but capacity is believed to have been expanded.

Siping Steel Rolling Plant

Head office: Pingdong Road 80, Siping City, Jilin Province. **Tel:** 3212.

Works:

■ At head office. **Rolling mills:** Two 3-high (DS 200, DS 400) open train conical mills. Steel shafting (17-80 x 50 x 8mm).

Special Steel Co of Shougang

Head office: Guchengu, Shijingshan, Peking. **Tel:** 872531.

Established: 1983 through merger of Beijing Steel Works and Beijing Special Steel Works and other establishments. **No. of employees:** 19,000.

Works:

■ At head office. ***Steelmaking plant:*** One 10-tonne, ten 5-tonne, three 3-tonne and one 1:FR:1./2-tonne electric arc furnaces; eight induction furnaces. ***Refining plant:*** Twenty-five ESR; one 15-tonne ladle furnace; one 15-tonne VOD. ***Rolling mills:*** Including two 3-stand 650mm and 2-stand 500mm billet; three light section; one wire

Special Steel Co of Shougang (continued)

rod (Swedish built); two 4-stand hot sheet; Four cold strip – two 2-high, one 4-high and one 20-high. ***Other plant:*** Eight forging hammers; six cold drawing machines; 111 wire drawing machines; atomising steel shot plant; steel foundry.
Products: Alloy steel (excl stainless) – round bars, bright (cold finished) bars; hot rolled sheet/coil (uncoated); wire; seamless tubes and pipes. Carbon and alloy engineering steels (including gear steels); carbon tool and die steels; high speed steels; bearing steels; silicon steels, stainless steels; steel shot; steel castings under 2 tonnes; forgings.

Suzhou Iron & Steel Works

Output in 1985 was 165,000 tonnes of pig iron. 108,000 tons of finished steel and 65,000 tonnes of raw steel.

Taiyuan Iron & Steel Co

Head office: Jian Cao Ping, Taiyuan, Shanxi Province. **Tel:** 312277. **Cables:** 6922 taiyuan.
Established: 1934.
Works:
■ At head office. **Coke ovens**. **Sinter plant**. ***Blast furnaces:*** No 1 149 cu m, No 2 296 cu m, No 3 1,200 cu m. ***Steelmaking plant:*** Two 50-tonne basic oxygen converters; one 50-tonne Asea electric arc furnace; three 100-tonne open hearth. ***Refining plant:*** One 18-tonne AOD; one DH vacuum degassing unit. ***Rolling mills:*** One 650mm slabbing; one blooming; One Medium/ light section; one heavy plate; one 4-high reversing Steckel wide hot strip; one 4-high universal hot mill; 8-high and 20-high cold reduction.
Products: Steelmaking pig iron. Carbon steel – ingots, wire rod, light angles, light tees, light channels, medium angles, medium tees, medium joists, heavy angles, heavy tees, heavy joists, heavy rails, medium plates, heavy plates, electrical non-oriented silicon sheet/coil; Stainless steel – hot rolled sheet/coil (uncoated), cold rolled sheet/coil (uncoated). Boiler and ship plates; Armco iron; die steels; cold heading steels; drum sheets; high alloy chill rolls. Production in 1985 was iron ore 2,560,000 tonnes; pig iron 1,025,000 tonnes; raw steel 1,415,000 tonnes and rolled steel 950,000 tonnes.
Expansion programme: A chinese-built converter to operate the K-OBM process under licence with the Klöckner-CRA-Technology group will be operating from mid 1989 producing carbon and high grade steels, including stainless. Taiyuan is said to be the largest producer of special steels in China. A continuous caster is also believed to be operating.

Tangshan Iron & Steel Co

Head office: Tel: 23911.
Works:
■ Tangshan City, Hebei Province. ***Blast furnaces:*** four 100 cu m. **Steelmaking plant and rolling mills**.
Products: Carbon steel – reinforcing bars, light channels (including U sections for mines. High alumina bricks. Production in 1985 was 310,000 tonnes of pig iron; 1,230,000 tonnes of raw steel; and 850 tonnes of rolled products.

Tianjin Sbexian County Iron Plant

Works:
■ At head office. ***Blast furnaces:*** No 2, No 3 and No 4 – 500 cu m each.

Tianjin Seamless Steel Pipe Plant

The first stage in modernising the plug mill was completed in late 1986 with the commissioning of a seamless rolling mill supplied by Friedrich Kocks GmbH & Co. The mill produces finished tubes from 25-83mm dia, with 2.5-11mm wall thickness. Throughput is about 20 tonnes per hour.

Tianjin Steel Plant

Works:
■ At head office. ***Steelmaking plant:*** Three 75-tonne basic oxygen converters; one 3-tonne and three 5-tonne electric arc furnaces; two 100-tonne open hearth.
Products: Carbon steel – wire rod, light angles etc.

Tonghua Iron & Steel Plant

Works:
■ Jilin Province. Raw steel output was 230,000 tonnes in 1985, rolled steel 145,000 tonnes and pig iron 300,000 tonnes.

Weiyuan Iron & Steel Works

Pig iron production in 1985 was 74,000 tonnes, raw steel 100,000 tonnes, rolled steel 75,000 tonnes.

Wuhan Iron & Steel (Wisco)

Head office: Hubei Province.
Works:
■ At head office. **Coke ovens**. ***Sinter plant:*** No 1 plant 4 strand 75 sq m grate area and No 2 plant 4 strand 90 sq m grate area. ***Blast furnaces:*** No 1 1,386 cu m, No 2 1,536 cu m, No 3 1,513 cu m and No 4 2,516 cu m. ***Steelmaking plant:*** Three 70-tonne basic oxygen converters (annual capacity 1,500,000 tonnes); two 250-tonne and six 500-tonne open hearth. ***Refining plant:*** Two RH vacuum degassing including one 80-tonne. ***Continuous casting machines:*** Three 1-strand slab (1,500,000 tonnes). ***Rolling mills:*** One blooming (annual capacity 2,000,000 tonnes); one billet; one heavy section; one rail; one heavy plate (2,800mm max width 8mm and up thick) (550,000 tonnes); one 1,700mm wide hot strip (3,000,000 tonnes); one 1,700mm cold reduction (1,000,000 tonnes); one temper/skin pass; one silicon sheet (70,000 tonnes). ***Coil coating lines:*** One electrolytic tinning (300,000 tonnes); one hot-dip galvanizing (100,000 tonnes).
Products: Steelmaking pig iron. Carbon steel – ingots, slabs, billets, medium angles, medium tees, medium joists, medium channels, wide-flange beams (medium), heavy angles, heavy tees, heavy joists, heavy channels, wide-flange beams (heavy), heavy rails, medium plates, heavy plates, hot rolled sheet/coil (uncoated), cold rolled sheet/coil (uncoated), electrical non-oriented silicon sheet/coil, electrolytic tinplate (single-reduced). Production in 1985 was: iron ore 4,550,000 tonnes; pig iron 4,060,000 tonnes; raw steel 3,980,000 tonnes; rolled steel 3,335,000 tonnes.
Expansion programme: Mannesmann Demag Sack is supplying a 6-high CR mill for strip 700-1,100mm wide and 0.19-0.55mm thick for start up in 1989. Annual capacity will be 130,000 tonnes.

Wuhu Iron & Steel Works

A small plant which produced 10,000 tonnes of pig iron in 1985.

Wuxi Iron & Steel Works

Production in 1985 totalled 130,000 tonnes of raw steel and 230,000 tonnes of rolled steel.

Wuyang Iron & Steel Co

Rolled steel output in 1985, 155,000 tonnes.

Xialu Iron & Steel Works

Production in 1985: pig iron 80,000 tonnes, raw steel 122,000 tonnes, rolled steel 78,000 tonnes.

Xiangtan Iron & Steel Plant

Pig iron output in 1985 was 530,000 tonnes, raw steel 545,000 tonnes and rolled steel 445,000 tonnes.
Works:
■ Hunan Province. Blast furnace, open hearth furnaces and sheet mill.

Xilin Iron & Steel Works

Ingot production in 1985 totalled 106,000 tonnes, rolled steel 70,000 tonnes and pig iron 94,000 tonnes.

Xinfu Steel Plant – See Fushun Xinfu Steel Works

Xingtai Iron & Steel Works

Works:
■ Hebei Province. Blast furnaces and iron foundry pig iron. Output in 1985 was 327,000 tonnes.

Xining Plant

Pig iron output in 1985 reached 620,000 tonnes.

Xinjiang Iron & Steel Works

Production in 1985 was follows: pig iron 150,000 tonnes, raw steel 175,000 tonnes, rolled steel 155,000 tonnes.

Xinyu Iron & Steel Works

Works:
■ Xinyu City, Jiangxi Province. The works supplied some of its own iron ore (550,000 tonnes in 1985) and produces pig iron (300,000 tonnes), raw steel (100,000 tonnes) and rolled steel (55,000 tonnes). A key ferro alloy producer.

Xuanhau Iron & Steel Co

Works:
■ At head office. ***Blast furnaces:*** No 1, No 2 and No 5 297 cu m each and No 6 300 cu m. **Steelmaking plant. Rolling mills**.
Additional information: Production in 1985 was 620,000 tonnes of pig iron, 58,000 tonnes of raw steel and 44,000 tonnes of rolled steel.

Xuzhou Iron & Steel Works

Pig iron output in 1985 was 128,000 tonnes, raw steel and finished rolled products 25,000 tonnes.

Zhangdian Iron & Steel Works

Small plant with output of 145,000 tonnes of pig iron and 44,000 tonnes of rolled steel in 1985.

Colombia

Acesco – Acerias de Colombia SA

Head office: Carrera 68, 19-24, Apartado Aereo 27133, Bogota, DE. **Tel:** 2616066. **Telex:** 45451. **Cables:** Acesco.
Management: Managers – Mario Escobar (general), Oscar Zuluaga (finance), Santiago d'Apena (sales), Carlos Zuluaga (operations); director – Alfonso Rodriguez (technical). **Established:** 1970. **Capital:** Col Pesos 750,000,000. **No. of employees:** Works – 180. Office – 100. **Annual capacity:** Finished steel: 40,000 tonnes.
Works:
■ Barranquilla Works, Apartado Aereo 6498. (Tel: 423266. Telex: 33532). Works director: Alfonso Rodriguez. ***Coil coating lines:*** One Kawasaki hot-dip galvanizing (annual capacity 40,000 tonnes).
Products: Carbon steel – hot-dip galvanized sheet/coil 0.16-1.90mm thick (output 33,000 tonnes).
Expansion programme: Increase in capacity 1988, colour-coating line 1989.

Colmena – Consorcio Metalúrgico Nacional SA

Head office: Caretera al Sur 61-95, Apartado Aereo 9180, Bogotá. **Tel:** 2383980. **Telex:** 44531.
Established: 1957.
Works:
■ ***Tube and pipe mills:*** welded – Yoder and Mannesmann Meer.
Products: Carbon steel – longitudinal weld tubes and pipes, galvanized tubes and pipes.

Futec – Fundiciones Technicas SA

Head office: Calle 17 No 43 F-23, Medellin, Antioquia. **Postal address:** Apartado Postal 3459.
Established: 1960.
Works:
■ ***Steelmaking plant:*** electric arc furnaces. ***Rolling mills:*** bar.
Products: Carbon steel – reinforcing bars.

Holasa – Hojalata y Laminados SA

Head office: Calle 17 43F-122, Medellin. **Postal address:** Apartado Aereo 51865. **Tel:** 66 66 11. **Telex:** 66640, 65017. **Telecopier (Fax):** 66 22 41.
Management: Presidente – Pablo Pelaez Gonzalez; vice presidente – Jorge Gallego Montaño (operaciones), Alvaro Hernandez Bonnet (inversiones), Hector Hoyos Velez (financiero), Carlos Lega Rueda (mercadeo), Carlos Enrique Monreno M (planeacion). **Established:** 1968. **Capital:** Col$60,000,000. **No. of employees:** Works – 90. Office – 94. **Annual capacity:** Finished steel: 63,500 tonnes.
Works:
■ At head office. ***Coil coating lines:*** One Wean United Halogen electrolytic tinning (annual capacity 63,500 tonnes).
Products: Carbon steel – electrolytic tinplate (single-reduced), electrolytic tinplate (double-reduced).
Expansion programme: It is planned to increase capacity to 90,000 tonnes annually.

Acerías Paz del Rio SA

Head office: Carretra 8a No 13-31, pisos 7-11, Bogatá, Cundinamarca. **Postal address:** Apartado Postal 4260. **Tel:** 2828111, 2411570. **Telex:** 044693. **Cables:** Siderrio.
Management: President – Jaime Garcia Parra; vice presidents – Francisco Triana Murcia (senior executive), Maria del Rosario Sintes (executive, administration and finance), Jorge Freese Torres (operations), Juan Pablo Vargas Gallo (sales); director – Dario Naranjo Dousdebes (purchasing). **Established:** 1947. **Capital:** Pesos 1,556,200. (April 1987). **No. of employees:** Works – 3,800. Office – 1,460. **Annual capacity:** Pig iron: 340,000 tonnes. Raw steel: 380,000 tonnes. Finished steel: 285,000 tonnes.
Works:
■ At head office. ***Coke ovens:*** One gun-flue battery (57 ovens) (annual capacity 356,000 tonnes). ***Sinter plant:*** One 68 sq m strand Lurgi (400,000 tonnes). ***Blast furnaces:*** One 6.3m hearth dia, 689 cu m (330,000 tonnes). ***Steelmaking plant:*** Two 28-tonne basic oxygen converters (340,000 tonnes); one 22-tonne open hearth furnace (40,000 tonnes). ***Rolling mills:*** 1-stand 1,100mm slabbing (700,000 tonnes); 3-stand 710/660mm billet (230,000 tonnes); 3-stand 710/660mm medium section; 14-stand bar – 3-stand 450mm, 5-stand 300mm and 6-stand 250mm (165,000 tonnes); Steckel, 690 and 1,350 x 1,425mm (coil weight 6.5 tonnes) (400,000 tonnes).
Products: Carbon steel – billets 70 x 70 x 5,000mm (output 45,000 tonnes); hot rolled coil (for re-rolling) 317mm (25,000 tonnes); wire rod $\frac{3}{8}$", $\frac{1}{4}$" (50,000 tonnes); reinforcing bars $\frac{1}{4}$-1$\frac{1}{2}$" (95,000 tonnes); medium angles 4 x $\frac{1}{4}$" to 6 x $\frac{1}{2}$" (3,000 tonnes); medium channels 4", 6", 8" (1,000 tonnes); hot rolled sheet/coil (uncoated) 8-3.17mm (45,000 tonnes).

Sideboyaca – Siderurgica Boyaca SA

Head office: Cl 75, No 12-12, Bogota. **Telex:** 43270.
Works:
■ ***Steelmaking plant:*** electric arc furnaces. ***Continuous casting machines:*** 2-strand billet machine. ***Rolling mills:*** 6-stand 380/270 light section. Light sections.

Sidelpa – Siderúrgica del Pacifico SA

Head office: Apartado Aereo 1986, Urb Industrial Acopi, Cali. **Tel:** 644474. **Telex:** 55695. **Cables:** Sidelpa Cali. **Telecopier (Fax):** 93 644474.
Management: President – Roberto Corredor; vice president – Hernando Murcia (technical); managers – Victor Garcia (quality control), Augusto Molina (engineering), Mirtha de Villegas (sales), Hector A. Parra (administration). **Established:** 1961. **Capital:** Authorised: $500,000. Subscribed: Pesos 331 417. **No. of employees:** 500. **Ownership:** Ingenio Mayaguez SA (23%). Central Dona Ana SA (11%). Others (66%). **Associates:** Siderúrgica de Boyacá, Siderúrgica del Muna and Siderúrgica de Caribe (all mini-mills). **Annual capacity:** Raw steel: 120,000 tonnes. Finished steel: 110,000 tonnes.
Works:
■ At head office. ***Steelmaking plant:*** One 30-tonne and one 18-tonne electric arc furnaces (annual capacity 120,000 tonnes). ***Refining plant:*** One 30-tonne ladle furnace (100,000 tonnes). ***Continuous casting machines:*** One 3-strand Danieli billet (160 x 160) (120,000 tonnes). ***Rolling mills:*** One 11-stand 450mm roughing and 300mm finishing bar (90,000 tonnes); one 5-stand 320mm cross country (30,000 tonnes).
Products: Carbon steel – blooms (output 60,000 tonnes); reinforcing bars (20,000 tonnes); round bars (12,000 tonnes); flats (10,000 tonnes); hexagons (500 tonnes); medium angles (6,000 tonnes).

Colombia

Sidemuña – Siderúrgica de Muña SA

Head office: Calle 78 No 9-57, piso 13, Bogotá. **Telex:** 44874.
Established: 1948.
Works:
■ ***Steelmaking plant:*** electric arc furnaces. ***Continuous casting machines:*** One 2-strand Rokop billet (100-150mm sq). ***Rolling mills:*** Danieli wire rod.
Products: Carbon steel – wire rod, reinforcing bars.

Simesa – Siderúrgica de Medellín

Head office: Carrera 48, 17-226, Medellín. **Tel:** 32 27 00. **Telex:** 66708. **Cables:** Simesa. **Telecopier (Fax):** 232 22 70.
Management: Members of the board – Pedro Jose Camargo Bohorquez, Enrique Andrade Rodriguez, Fabio Echeverri Correa, Jose Antonio Paternostro Matera, Jorge Guzman Moreno. President – Alberto Leon Mejia Zuluaga; vice presidents – Emilio Mira Vasquez (operations), Oliverio Henao Marin (administrative), José Luis Arango Cañas (finance manager), Hernan Jaramillo Gomez (sales manager); head technical and planning – Jairo Gomez Gomez; mill manager – Antonio Giron Higuita. **Established:** 1938. **No. of employees:** Works – 1,019. Office – 357. **Subsidiaries:** Siminera (coal mining and trading); Minerales Industriales (clays mining and trading). **Associates:** Erecos (manufacture of refractory products). **Annual capacity:** Raw steel: 130,000 tonnes. Finished steel: 150,000 tonnes.
Works:
■ At head office. ***Steelmaking plant:*** One Lectromelt 12′ 6″ electric arc furnace (annual capacity 130,000 tonnes). ***Refining plant:*** One ladle furnace (130,000 tonnes). ***Continuous casting machines:*** One 3-strand Concast billet (100 x 100, 130 x 130mm) (150,000 tonnes). ***Rolling mills:*** One light section and bar – cross-country with two 3-high roughing stands and 2-stand 250mm double-duo finishing train (70,000 tonnes); wire rod with four 2-high 475mm roughing stands and six monoblock finishing stands (coil weight 300kg) (80,000 tonnes). ***Tube and pipe mills:*** Two Yoder seamless – 3½″ and 2½″ (combined capacity 18,000 tonnes). ***Other plant:*** Forging plant producing grinding balls (5,000 tonnes/year). Foundries producing grey, malleable and SG iron and steel (5,000 tonnes).
Products: Carbon steel – billets 100 x 100 and 130 x 130mm; wire rod (low-carbon) ¼″ and ⅜″ (1986 output 29,373 tonnes); reinforcing bars (deformed and plain) 12-40mm (36,893 tonnes deformed; 13,471 tonnes plain; flats 1½-4″ (1,000 tonnes); galvanized wire (plain) (2,121 tonnes); galvanized wire (barbed) (2,357 tonnes); longitudinal weld tubes and pipes and galvanized tubes and pipes ⅜-4″ (6,245 tonnes); hollow sections (square 30 x 30-90 x 90mm, rectangular 20 x 40-50 x 130mm, round 38-114mm dia – estimated combined output in 1987 1,000 tonnes). Alloy steel (excl stainless) – billets 100 x 100 and 130 x 130mm. Grinding balls (1986 output 2,705 tonnes); Iron and Steel castings and fittings (2,426 tonnes).
Brands: Simesa.
Expansion programme: Expansion of melting shop to annual capacity of 150,000 tonnes (1992). Ferro-alloy plant – 80,000 tonnes (1992). Rolling mill revamping and automation: (1988). New mini steel mill – 150,000 tonnes (1995). Direct reduction plant 80,000 tonnes (1995).

Costa Rica

Metalco SA

Head office: Colima de Tibas, 1000 San José. **Postal address:** PO Box 1131. **Tel:** 35 4366. **Telex:** 2593 mtalco cr.
Management: Managers – Jorge Enrique Gómez (general), Manrique Arrea (sales), Alfonso Binda (production); comptroller – Juan A. Montoya. **Established:** 1962. **No. of employees:** Works – 90. Office – 62. **Associates:** Acesco, Colombia.
Works:

Metalco SA (continued)

■ At head office. ***Coil coating lines:*** One continuous B & K hot-dip galvanizing (annual capacity 40,000 tonnes).
Products: Carbon steel – hot-dip galvanized sheet/coil gauges 20-34 (output 6,000 tonnes); galvanized corrugated sheet/coil gauges 26-34 (5,000 tonnes); colour coated sheet/coil gauges 20-26 (3,000 tonnes);

Cuba

Empresa Metalúrgica José Marti

Head office: Cotorro, Havana.
Ownership: State-owned.
Works:
■ El Cotorro, 30 Km from Havana. ***Steelmaking plant:*** open hearth furnaces. ***Rolling mills:*** One medium section, light section mill; one bar, one wire rod,
Products: Carbon steel – wire rod, reinforcing bars, light angles, medium angles.

Cyprus

BMS Metal Pipes Industries Ltd

Head office: PO Box 210, Anatolikon, Paphos, **Tel:** 061 33578. **Telex:** 4503 bms cy.
Management: Costas Koutsos (managing director). **Established:** 1976. **Capital:** C£1,000,000. **No. of employees:** Works – 47. Office – 8. **Associates:** Anatolikon Steel Industries, Paphos (slitting of hot and cold rolled coil into strip). **Annual capacity:** Finished steel: 15,000 tonnes.
Works:
■ At head office. ***Tube and pipe mills:*** welded – one Voest-Alpine ½-4", (annual capacity 15,000 tonnes).
Products: Carbon steel – longitudinal weld tubes and pipes.

Czechoslovakia

Export sales of Czechoslovak iron and steel products are handled by Ferromet, 27 Opletalova PO Box 779, 11181 Prague 1. (Tel: 2141. Telex: 121411. Fax: 228517).

Chomutov Tube Works

Works:
■ Chomutov. ***Tube and pipe mills:*** seamless – including Mannesmann-Meer 1,650mm extrusion press with stretch reduction facilites (annual capacity 400,000-500,000 tonnes), Loewy-Robertson 3-roll piercer for carbon, alloy and stainless seamless tubes; multi-strand cold tube reduction mill; and two plants producing special steel tubes.
■ Podbrezova. ***Tube and pipe mills:*** seamless for precision tubes (including mill for tubes 57-76mm (annual capacity 10,000 tonnes); extrusion plant; two drawing plants;

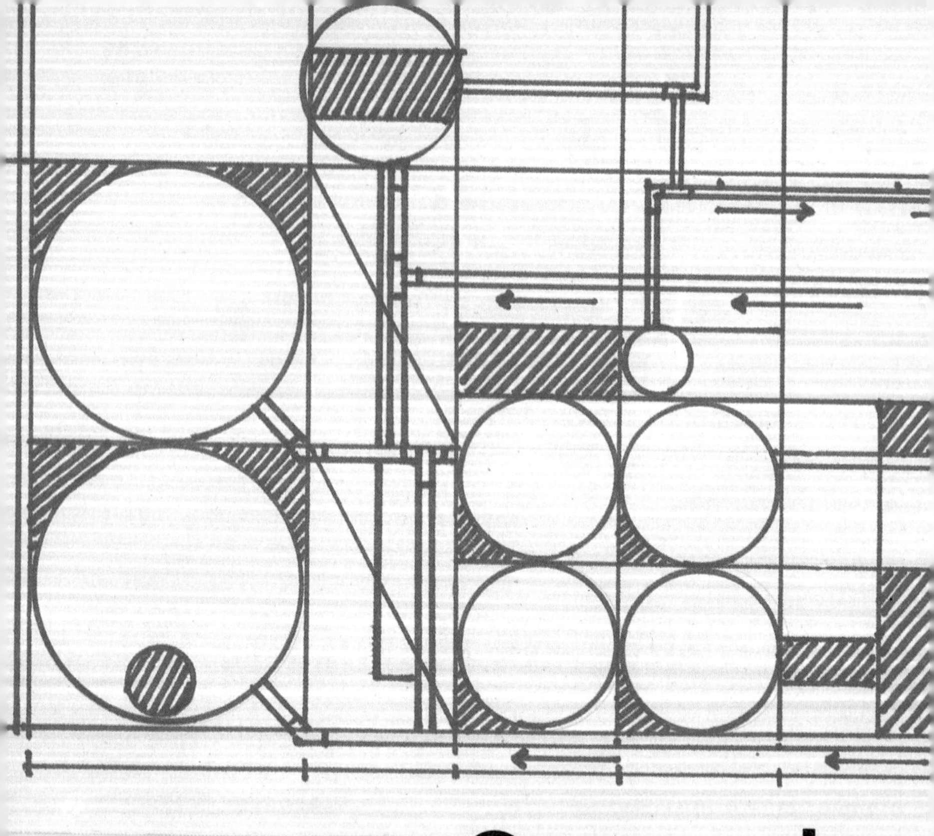

Chomutov Tube Works (continued)

amd ancillary equipment.
Products: Carbon steel – seamless tubes and pipes, longitudinal weld tubes and pipes, large-diameter tubes and pipes, precision tubes and pipes, Stainless steel – seamless tubes and pipes, precision tubes and pipes, Alloy steel (excl stainless) – seamless tubes and pipes, special tubes for nuclear reactors.
Additional information: There is also a special steel works at Chomutov with a capacity of about 5,000 tonnes a year.

East Slovak Iron & Steel Works

Established: 1960.
Works:
■ ***Coke ovens:*** Three batteries (annual capacity 2,550,000 tonnes). ***Sinter plant:*** 4-strand (capacity 8,000-10,000 tonnes per day). ***Blast furnaces:*** Three furnaces including two 1,719 cu m. ***Steelmaking plant:*** Two melting shops comprising LD basic oxygen converters – No 1 three 150-tonne (annual capacity 3,000,000 tonnes), and No 2 two 160-tonne (1,800,000 tonnes). ***Refining plant:*** One 150-tonne vacuum degassing unit. ***Continuous casting machines:*** One 2-strand for slab (190 x 960 to 1,550mm (1,000,000 tonnes). ***Rolling mills:*** One 1,150mm universal slabbing; one wide hot strip – 5-stand roughing and 6 stand finishing; two cold reduction comprising of one 5-stand tandem for coil 0.18 to 1mm thick and 1,050mm wide, and one 4-stand 1,700mm for coil 0.4 to 2mm x 1,500mm; temper/skin pass. ***Tube and pipe mills:*** welded – One spiral (60,000 tonnes); one longitudinal (100,000 tonnes). ***Coil coating lines:*** One Ferrostan electrolytic tinning (160,000 tonnes); five Dubnica hot-dip tinning (40,000 tonnes); one continuous hot-dip galvanizing aluminizing (100,000-70,000 tonnes); two colour coating – one for plastic (60,000 tonnes), and one for strip (20,000-30,000 tonnes).
Products: Carbon steel – medium plates; heavy plates 1.8 to 10mm thick, 700 to 1,500mm wide; hot rolled sheet/coil (uncoated); cold rolled sheet/coil (uncoated) (including high-tensile and deep-drawing, 0.18 to 2mm thick, 650 to 1,500mm wide); electro-galvanized sheet/coil; aluminized sheet/coil; colour coated sheet/coil; electrical non-oriented non-silicon sheet/coil; electrolytic tinplate (single-reduced); hot-dip tinplate.

Poldi Works, Kladno – put under Poland by mistake

Třinec Iron & Steel Works

Established: 1839.
Works:
■ **Coke ovens**. ***Blast furnaces:*** Four. ***Steelmaking plant:*** Two 200-tonne basic oxygen converters and electric arc furnaces (combined annual capacity 2,500,000 tonnes). ***Refining plant:*** One 75-tonne RH vacuum degassing unit (No 2 shop). ***Continuous casting machines:*** bloom (730,000 tonnes). ***Rolling mills:*** Two slabbing and blooming (950 and 1,150mm); one 700mm continuous billet; one 520mm continuous sheet bar; 880/900mm reversing heavy section; billet and rail; one 400mm medium section; light section and medium section; wire rod, including one producing 5.5-12mm dia (600,000 tonnes).
Products: Carbon steel – slabs, 180-1,040mm x 100-200mm; blooms, 160-280mm sq; sheet bars, 250-330mm wide; billets, 80-132mm sq; wire rod, 5.5-11mm dia; reinforcing bars, 5.5-32mm dia (including Tor-Steel and deformed); round bars 5.5-70mm dia; square bars 35-65mm sq; flats, 25-150mm x 5-40mm; light angles, 32-50mm x 3-6mm; medium joists, 280-360mm wide; heavy channels, 100-300mm; heavy rails; rail accessories.

Lenin Works

Works:
■ Plzen. ***Steelmaking plant:*** electric arc and open hearth furnaces.
Additional information: The works is part of the Skoda engineering works.

Czechoslovakia

NHKG – Nová Hut Klementa Gottwalda

Works:

■ Kuncice, Near Ostrava. ***Coke ovens:*** Including one battery with an annual capacity of 1,000,000 tonnes. **Sinter plant**. ***Blast furnaces:*** Several of 1,082 cu m furnaces; and one of 1,260 cu m (capacity 2,000 tonnes/day). ***Steelmaking plant:*** Two Oxyvit bottom-blown basic oxygen converters, one 60-tonne electric arc furnace (105,000 tonnes), and open hearth including two double furnaces (combined annual capacity 4,800,000 tonnes). ***Rolling mills:*** One blooming; one billet; one heavy section; one medium section (900,000 tonnes); one bar and wire rod; one bar (1,000,000 tonnes); one medium plate for plates 6 to 10mm thick and 1 to 3mm wide (over 500,000 tonnes); one wide hot strip. ***Tube and pipe mills:*** seamless. Spiral welded.
Products: Carbon steel – slabs, blooms, semis for seamless tubemaking, wire rod, reinforcing bars (deformed), round bars (heavy), heavy angles, heavy joists, heavy channels, light rails, medium plates and heavy plates (6-10mm thick and 1-3m wide), hot rolled sheet/coil (uncoated), seamless tubes and pipes, longitudinal weld tubes and pipes, spiral-weld tubes and pipes; car and railway wheels.

Sverma Steel Works

Works:

■ Podbrezova. **Steelmaking plant**. ***Continuous casting machines:*** One 4-strand curved mould for bloom.
Products: Carbon steel – semis for seamless tubemaking.

Veselinad Moravou Works

Products: Carbon steel – longitudinal weld tubes and pipes, galvanized tubes and pipes.

Vitkovice Steelworks

Established: 1826.
Works:

■ Ostrava. **Coke ovens**. **Sinter plant**. ***Blast furnaces:*** Three including one of 1,095 cu m (capacity 1,400 tonnes/day). ***Steelmaking plant:*** One 70-tonne Oxyvit bottom-blown basic oxygen converter; five electric arc furnaces including 75-tonne Brown Boveri; and open hearth furnaces including three tandem units (combined annual capacity 2,250,000 tonnes). ***Refining plant:*** One 60-tonne DH vacuum degassing unit; Vakuvit vacuum degassing unit; Duvavit argon-oxygen decarborisation vessel. ***Rolling mills:*** blooming; billet; and two cogging; heavy section; rail; 450mm medium section; two light section; plate mills including 4-high 3,500mm heavy plate (annual capacity 450,000 tonnes); hot strip and hoop (100 to 900m wide and 6 to 60mm thick); 5 to 200mm wide and 0.1 to 5mm thick cold strip (5 to 200mm wide and 0.1 to 5mm thick). ***Tube and pipe mills:*** seamless – Three Mannesmann and one Stiefel. One welded. ***Other plant:*** Steel foundry; grey iron foundry; forges; machine shops; plant building shops.
Products: Steelmaking pig iron and foundry pig iron. Carbon steel -. Ingots; slabs 270 x 80 to 1,300 x 220mm; blooms 140 to 300mm sq, 170 x 140mm to 450 x 300mm; billets 88 to 290mm dia, 125 x 80 to 235 x 85mm; semis for seamless tubemaking; round bars 35 to 60mm dia; square bars 30 to 60mm; flats 40 x 12mm to 120 x 60mm; light angles; light tees; light channels; medium angles; medium tees; medium channels; heavy angles; heavy joists; heavy channels; heavy rails; sheet piling; hot rolled hoop and strip (uncoated) 100 to 900mm wide and 6 to 60mm thick; skelp (tube strip); cold rolled hoop and strip (uncoated) 3 to 200mm wide and 0.1 to 5mm thick; heavy plates 1,000-3,500mm wide and 8-120mm thick; seamless tubes and pipes; longitudinal weld tubes and pipes. Stainless steel – ingots; slabs; billets 300 x 10mm to 300 x 36mm; semis for seamless tubemaking; heavy plates; seamless tubes and pipes. Alloy steel (excl stainless) – ingots; blooms; billets; round bars; square bars; flats; light angles; light channels. Steel and grey iron castings; non-ferrous castings; forgings; structural steelwork; steelworks plant, etc.
Additional information: Products are mainly for own use.

Denmark

Danish Steel Works Ltd (Det Danske Stålvalseværk A/S)

Head office: Stålværksvej, D 3300 Frederiksværk. **Tel:** 45 2 12 03 33. **Telex:** 40191 staal dk. **Telecopier (Fax):** 45 2 12 46 66.
Management: Directors – K. Stausholm-Pedersen (managing), F. Knudsen (financial), Per Krogh Jensen (sales). **Established:** 1940. **Capital:** Dkr 180,000,000. **No. of employees:** Works – 1,300. Office – 400. **Subsidiaries:** Dansteel Ltd, UK (100%) (sales company).
Works:
■ At head office. ***Steelmaking plant:*** Two 110-tonne electric arc furnaces (annual capacity 850,000 tonnes). ***Refining plant:*** One 110-tonne 15 MVA ladle furnace (600,000 tonne). ***Continuous casting machines:*** One 6-strand Demag billet (142 mm sq); one 1-strand Demag slab (210-260 mm sq). ***Rolling mills:*** One Continuous bar – 560/400/360/345/260 mm dia (275,000); one 4-high 3,300 x 900 x 1,500mm heavy plate (600,000).
Products: Carbon steel – reinforcing bars 8-35mm, round bars 10-60mm, square bars 10-40mm, flats 5 x 20 to 20 x 120, heavy plates 6-100mm x 3,000mm. DSK – bars. Dansteel – bars and heavy plates.

Nordisk Simplex AS

Head office: Vejlegaardsvej 34, DK 2665 Vallensbaek Strand. **Tel:** 45 2 73 19 20. **Telex:** 33400 tubes dk. **Cables:** Dantubes. **Telecopier (Fax):** 45 2 54 64 17.
Management: Directors – Pekka Sipilä (managing), Cyril Soerensen (sales); managers – Paul Christensen (production), Jens Vestergaard (financial), Niels Bolt (export). **Established:** 1910. **Capital:** Dkr 10,000,000. **Ownership:** Rautaruukki Oy, Finland (100%).
Works:
■ At head office. ***Tube and pipe mills:*** Four HF 16-64mm dia welded.
Products: Carbon steel – longitudinal weld tubes and pipes and precision tubes and pipes 16-64mm dia.

Dominican Republic

Metaldom – Complejo Metalúrgico Dominicano C por A

Head office: Carretera Sanchez Km 7, Santo Domingo.
Works:
■ ***Steelmaking plant:*** electric arc furnace. ***Continuous casting machines:*** billet machine. ***Rolling mills:*** One light section; one bar and wire rod. ***Tube and pipe mills:*** welded .
Products: Carbon steel – wire rod, reinforcing bars, round bars, square bars, light angles, longitudinal weld tubes and pipes, galvanized tubes and pipes.

Dubai

Ahli Steel Co

Head office: Al-Ramool, Rashidiya. **Postal address:** PO Box 2302. **Tel:** 257457-8. **Telex:** 45754 steel em. **Cables:** Steel Dubai.
Management: Director – Abdulrehman Mohd Kayed (managing); managers – Iqbal Ahmed (factory/sales), P.V. Ashokan (purchasing). **Established:** 1975. **Capital:** Dh 30,000,000. **No. of employees:** Works – 160. Office – 10. **Ownership:** Abdulla

Dubai

Ahli Steel Co (continued)

Kayed Ahli. **Annual capacity:** Raw steel: 60,000 tonnes. Finished steel: 55,000 tonnes.
Works:
■ ***Steelmaking plant:*** Two 10/14-tonne electric arc furnaces (annual capacity 60,000 tonnes). ***Continuous casting machines:*** One 2-strand 75 x 75mm billet machine (55,000 tonnes). ***Rolling mills:*** 2-strand 10-20mm bar (50,000 tonnes).
Products: Carbon steel – reinforcing bars (deformed), round bars.

Ecuador

Adelca – Acería del Ecuador CA

Head office: Avenida Orellana 1791, Quito. **Postal address:** PO Box 2209, Quito. **Telex:** 22407 adelca ed.
Established: 1963.
Works:
■ Quito. ***Rolling mills:*** One cross-country 18"/10" light section and bar; one Danieli 17-stand continuous bar (coil weight 400kg).
Products: Carbon steel – reinforcing bars, round bars, flats, light angles.

Andec – Acerías Nacionales del Ecuador SA

Head office: Av Juan Tanca, Marengo Km 51/2, Guayaquil. **Telex:** 043265 andec ec.
Established: 1964.
Works:
■ ***Rolling mills:*** bar and wire rod.
Products: Carbon steel – wire rod, reinforcing bars.

Funasa – Fundiciones Nacionales SA

Head office: via Puerto Marítimio Camino a Las Esclusas, Guayaquil. **Postal address:** PO Box 6146. **Tel:** 430835. **Telex:** 43265.
Management: Ricardo Avendaño (presidente); Fausto Banderas R (gerente general), Francisco Duque C (gerente técnico), Gustavo Palma Y (gerente financiero). **Established:** 1976. **Capital:** $376,000,000. **No. of employees:** Works – 160. Office – 40. **Annual capacity:** Raw steel: 25,000 tonnes.
Works:
■ At head office. ***Steelmaking plant:*** One 10-tonne, 9ft dia electric arc (annual capacity 25,000 tonnes). ***Continuous casting machines:*** One 2-strand Concast billet (100mm sq) (25,000 tonnes).
Products: Carbon steel – billets 100mm sq, 2,800mm-4,500mm long; wire rod.
Expansion programme: Capacity increase to 200,000 tonnes/year under study.

Egypt

Alexandria National Iron & Steel Co

Head office: Dikheila-Alexandria. **Postal address:** PO Bag, Alexandria. **Tel:** 4300 220, 4300 031, 4300 132. **Telex:** 45715 ansdk un, 54278 ansdk un. **Telecopier (Fax):** 4300 667.
Management: Directors – Ibrahim Salem Mohammadain (chairman and managing), Mohamed Abdel Aziz Khattab (joint managing); managers – M. Ijuin (general), Mr Isoi (sales), H. Saleh (sales), Mr Ohgi (purchasing), Mr Soufi (purchasing). **Established:**

Alexandria National Iron & Steel Co (continued)

1982. **Capital:** E£235,000,000. **No. of employees:** 1,800. **Annual capacity:** Sponge iron: 716,000 tonnes. Raw steel: 798,500 tonnes (billets). Finished steel: 745,000 tonnes (re-bars).
Works:
■ At head office. ***Direct reduction plant:*** One Midrex series 600 (annual capacity 716,000 tonnes DRI). ***Steelmaking plant:*** Four 70-tonne NKK 5800 MH type electric arc furnaces (840,500 tonnes). ***Continuous casting machines:*** Three 4-strand V-B type billet 130sq mm (798,500 tonnes). ***Rolling mills:*** One continuous 16-stand bar (coil weight 2 tonnes) (425,000 tonnes); one continuous 25-stand wire rod (coil weight 2 tonnes) (320,000 tonnes).
Products: Carbon steel – wire rod – 6, 8, 10, 12, 13mm (start up date April 1987); reinforcing bars – 10, 13, 16, 19, 22, 25, 28mm (output in July-December 1986, 47,000 tonnes).
Additional information: Cairo office at: 21 Ahmed Orabi Street, Elmohandessein.

Delta Steel Mill SAE

Head office: Mostorod, Kalyoubiah. **Telex:** 94220.
Established: 1952.
Works:
■ At head office. ***Steelmaking plant:*** Two 25-tonne electric arc furnaces (annual capacity 100,000 tonnes). ***Continuous casting machines:*** One Concast billet (130mm sq). ***Rolling mills:*** bar. ***Other plant:*** Foundry.
Products: Carbon steel – reinforcing bars. Cast iron – pipes.
Expansion programme: The arc furnaces are being modernised and fitted with water-cooled panels to double capacity by 1988.

The Egyptian Copper Works

Head office: Hagar El Nawatia, Alexandria. **Telex:** 54293.
Established: 1935. **Ownership:** State-owned.
Works:
■ At head office. ***Steelmaking plant:*** Two electric arc furnaces – one 25-tonne, and one 50-tonne; one open hearth furnace. ***Continuous casting machines:*** One 4-strand Concast for billet (100 to 130mm sq) (annual capacity 100,000 tonnes). ***Rolling mills:*** One bar – modernised 1974 with addition of Danieli cantilever intermediate stands and new finishing stands (70,000 tonnes). ***Other plant:*** Steel foundry; wire plant; baling hoop plant.
Products: Carbon steel – reinforcing bars, black annealed wire. Baling hoop; steel castings.

The Egyptian Iron & Steel Co (Hadisolb)

Head office: 54 Abdel Khalek Sarwat Street, Cairo. **Postal address:** PO Box 746, Cairo. **Tel:** 911980, 911852, 913040. **Telex:** 92007 solb un. **Cables:** Hadisolb Cairo.
Management: M.D. Tantawi (chairman). **Established:** 1955. **Capital:** E£350,000,000. **No. of employees:** Works – 20,671. Office – 2,016. **Ownership:** General sector. **Annual capacity:** Pig iron: 1,750,000 tonnes. Raw steel: 1,250,000 tonnes. Finished steel: 750,000 tonnes.
Works:
■ El-Tibbin, Helwan (Tel: 790242. Telex: 92007). ***Sinter plant:*** Russian type – two 50m sq strand (annual capacity 880,000 tonnes), four 75m sq strand (2,650,000 tonnes). ***Blast furnaces:*** Two 5,100mm hearth dia, 575 cu m (410,000 tonnes); two 7,200mm hearth dia, 1,033 cu m (1,340,000 tonnes). ***Steelmaking plant:*** Three 80-tonne basic oxygen converters (1,200,000 tonnes); two 12-tonne electric arc furnaces (50,000 tonnes). ***Continuous casting machines:*** Three 6-strand vertical (Russian) billet (150-200mm) (605,000 tonnes); three 2-strand vertical (Russian) slab

Egypt

The Egyptian Iron & Steel Co (Hadisolb) (continued)

machine (150 x 150 x 480mm) (595,000 tonnes). ***Rolling mills:*** 1-stand 950mm blooming (200,000 tonnes); 3-stand and edger 750mm heavy section (170,000 tonnes); 8-stand 450mm medium section (200,000 tonnes); 9-stand 380mm light section mill (72,000 tonnes); wide hot strip – semi-continuous with one 2-high 1,200 x 1,000mm rougher and 6-stand 4-high 1,200 x 600mm finishing (750,000 tonnes); two reversing 4-high 1,200 x 500mm cold reduction mill (coil weight 15 tonnes) (360,000 tonnes); one temper/skin pass (coil weight 15 tonnes).
Products: Carbon steel – slabs 150 x 150 x 480 (annual output 595,000 tonnes); billets 150-200mm (605,000 tonnes); hot rolled coil (for re-rolling) 2-4mm thick x 1,020mm wide (71,816 tonnes); reinforcing bars 16mm, 19mm (23,700 tonnes); light angles 30mm, 40mm, 50mm (19,000 tonnes); medium angles 60mm, 70mm, 80mm (10,700 tonnes); medium joists 120mm (1,100 tonnes); medium channels 80mm, 100mm (3,300 tonnes); heavy angles 90mm, 100mm, 120mm, 150mm (2,900 tonnes); heavy joists 160mm, 200mm, 260mm (10,400 tonnes); heavy channels 120mm, 140mm, 160mm, 200mm, 260mm (14,300 tonnes); heavy rails (19,200 tonnes); light rails 18 Kg/m (1,600 tonnes); medium plates 10-60mm (38,600 tonnes); hot rolled sheet/coil (uncoated) 2-7mm thick x 1,000mm wide (4,327 tonnes); cold rolled sheet/coil (uncoated) 0.5-1.5mm thick x 1,000mm wide (75,453 tonnes); hot-dip galvanized sheet/coil 0.8-1.5mm thick x 1,000mm wide (1,949 tonnes).
Expansion programme: Blooming, section and plate mills being modernised by Krupp Industrietechnik – 1986/87. Expansion to annual raw steel capacity of 2,600,000 tonnes under study with the USSR, involving a new sinter plant, a third blast furnace, a new LD shop (two 125/135-tonne converters) a curved mould continuous slab caster, and a plate mill of 2,000 or 2,500mm (1,300,000 tonnes). A new 200,000-tonne electrolytic tinning line is also under study with the Russians.

National Metal Industries Co

Head office: 18 Emad El Din Street, Cairo. **Postal address:** PO box 1673. **Tel:** 748721, 753440. **Telex:** 21624 sfnmt un. **Cables:** Nametin. **Telecopier (Fax):** 698192.
Management: Board of directors – Said El Hadidy (chairman), Abdel Monem Shatilla (technical), Mohamed El Khalifa (commercial), Abdel Halim Eissa (financial), Rizkalla Shokralla (works). **Established:** 1946/1947. **Capital:** E£30,000,000. **No. of employees:** Works – 2,400. Office – 300. **Ownership:** Public sector co (Ultimate holding company – Metallurgical Industries Corp). **Annual capacity:** Raw steel: 90,000 tonnes. Finished steel: 150,000 tonnes.
Works:
■ Abou Zaabal, Kaliobia. (Tel: 698718, 698404. Telex: 21624 sfnmt un). Works director: Rizkalla Shokralla. ***Rolling mills:*** Three bar : semi-mechanised rolling mill No 1 – 1-stand 3-high 470mm (annual capacity 75,000 tonnes), 5-stand 2-high 280mm and 1-stand 3-high 280mm (60,000 tonnes); semi-mechanised rolling mill No 2 – 1-stand 3-high and 6-stand 2-high 230mm (22,000 tonnes); mechanised rolling mill – 2-stand 3-high 260mm (95,000 tonnes) and 8-stand 2-high 320mm (90,000 tonnes).
Products: Carbon steel – ingots 138 x 138, 158 x 158mm (output 1985/86 90,000 tonnes); reinforcing bars 8,10,13,16,19,22,25 mm dia (150,000 tonnes).
Sales offices/agents: At head office; Alexandria office – 23 El Tahrir Square, Alexandria.
Expansion programme: Modernisation of the automatic rolling mill to increase the capacity from 90,000 to 125,000 tonnes. New rolling mill with capacity of 125,000 tonnes. Modification of the existing open hearth furnaces and addition of ladle furnace. Addition of 3-strand continuous casting machine (capacity 100,000 tonnes).

El Salvador

Acero SA—see Siderúrgica Centroamericana del Pacífico SA

Corinca SA – Corporación Industrial Centroamericana SA

Head office: Edificio Su Casa, 2° No Local N° 13, Avenida Olimpica y Pasaje 3, Colonia Flor Blanca, San Salvador. **Postal address:** PO Box 1142. **Tel:** 23 1519. **Telex:** 20065. **Cables:** Corinca.
Established: 1966. **Capital:** Colon 6,400,000. **No. of employees:** Works – 250. Office – 50. **Annual capacity:** Finished steel: 24,000 tonnes.
Works:
■ Steelmaking and steel rolling plant, Quezalteque (Tel: 28 1728. Telex: 20065). Works director: Francisco Alvarado. ***Steelmaking plant:*** One 5-tonne 2.500 KVA electric arc furnace (annual capacity 12,000 tonnes). ***Rolling mills:*** One 9-stand 12" bar (24,000 tonnes).
■ Wire Works, km 5 1-2 Boulevard del Ejercito, Soyapango (Tel: 27 0512. Telex: 20065). Works director: Francisco Alvarado. **Wire Works**: Seven drawing machines, 20 nail making machines, four barbed wire machines; wire galvanizing plant (24-strands); electric pot furnace for wire annealing; two wire staightening machines (wire rod is imported).
Products: Carbon steel – ingots (200lb) (output 12,000 tonnes); reinforcing bars No 3 – No 8 (12,000 tonnes); bright wire 17 BWG-4 BWG (6,000 tonnes); black annealed wire 16 BWG (1,200 tonnes); galvanized wire (plain) 17 BWG-9 BWG (1,200 tonnes); galvanized wire (barbed) 16 BWG (600 tonnes); nails (1,200 tonnes).

Siderúrgica Centroamericana del Pacífico SA (Sicepasa)

Head office: Edificio Borgonovo, Ciudad Merliot, Calles L-1 y L-2, Carretera a Santa Tecle. **Tel:** 28 1133. **Telex:** 20163 borgo. **Cables:** Borgo.
Management: Antonio Borgonovo (president), Alfredo Borgonovo Avila (director and general manager). **Established:** 1976. **Capital:** US$6,000,000. **No. of employees:** Works – 165. Office – 15. **Subsidiaries:** Acero SA (hot rolled bars and angles, bright and galvanized wire, barbed wire nails) – see below.
Works:
■ Km 110-111 Carretera del Litorel, Sonsonate. Works director: Santiago Lopez. ***Steelmaking plant:*** One 50-tonne electric arc furnace (annual capacity 100,000 tonnes). ***Continuous casting machines:*** One 4-strand Danieli/Continua billet (90-130mm sq) (100,000 tonnes). ***Rolling mills:*** One semi-continuous bar – 3-high 450mm stand, four 280mm open stands, four continuous 280mm cantilever stands, and two groups of four 250mm stands each (80,000 tonnes).
■ Acero SA, Km 58, Carretera Zacatecolouce (Tel: 340002). Works director: Antonio Borgonovo. ***Steelmaking plant:*** One 3-tonne induction. ***Rolling mills:*** bar and wire rod – 3-high 400mm and five open 250mm stand hand mill, and 3-high 320mm semi-continuous, 3-high 320mm, four cross-country 320mm stands, two groups of four 280mm open stands and two cantilever stands (30,000 tonnes).
Products: Carbon steel – billets 90mm sq (SAE 1008 to 1040); wire rod (180kg coil weight); reinforcing bars; light angles $\frac{3}{4}$ x $\frac{3}{4}$ x $\frac{1}{8}$" – $2\frac{1}{2}$ x $1\frac{1}{2}$ x $\frac{1}{4}$"; bright wire; galvanized wire (plain); nails and wire mesh.

Finland

Dalsbruk Oy Ab

Head office: Rådhustorget, 10601 Ekenäs. **Postal address:** PB 66. **Tel:** 91162400. **Telex:** 13190. **Telecopier (Fax):** 91115053.
Established: 1987. **Ownership:** Rautaruukki Oy (80%), Ovako Steel Oy AB (20%).
Works:
■ Koverhar, 10820 Lappvik, Lappohja. (Tel: 91143100. Telex: 13122). General manager: Klas-Göran Eriksson. Works manager: Erkki Ristimäki. ***Sinter plant:*** Four 24 sq m Greenawalt pans (annual capacity 750,000 tonnes). ***Blast furnaces:*** One 6.5m hearth dia, 573 cu m (500,000 tonnes). ***Steelmaking plant:*** Two 50-tonne LD basic oxygen converters. **Refining plant:** One ladle treatment station. ***Continuous casting machines:*** Two 4-strand Concast S type billet (110-130 mm sq) (500,000 tonnes). ***Other plant:*** Pig casting plant.
■ Dalsbruk, 25900 Dalsbruk, Taalintehdas. (Tel: 95261211. Telex: 62143). General manager: Klas-Göran Eriksson. ***Rolling mills:*** One blooming and bar for bars 8-32mm; one wire rod 5.5-16.0mm. ***Other plant:*** Wire drawing and galvanizing plant. Pre-stressed concrete plant.
■ 10410 Åminnefors. (Tel: 911 30755. Telex: 13155). ***Rolling mills:*** One bar.
■ Tehdasraudoite division: Reinforcement fabricating works at Helsinki, Kuopio and Tampere.
Products: Foundry pig iron. Carbon steel – billets for rebar, wire rod, spring steel and cold heading steel; also forging billets; wire rod including high carbon; reinforcing bars 6-32mm; galvanized wire (plain), spring wire, welding wire, etc. Alloy steel (excl stainless) – billets, wire rod, wire. Alloy and non-alloy welding rod. Cut and bent reinforcement, slag products.

Haato-Tuote Oy

Head office: Pohjantähdentie 17, SF 01450 Vantaa 45. **Tel:** 358 0 8726011. **Telex:** 124593. **Telecopier (Fax):** 358 0 8722148.
Products: Stainless steel – longitudinal weld tubes and pipes 50-800mm, and pipe fittings.

Oy Hackman Ab

Head office: Olavinkatu 1 A, Helsinki. **Postal address:** Box 955. **Tel:** 358 0 6946055. **Telex:** 124757. **Telecopier (Fax):** 358 0 6941575.
Management: Curt Lindbom (managing director), Ilkka Brotherus (consumer goods), Matti Glöersen (metal), Rabbe Kihlman (economy and administration), Kai Lillandt (wood), Esa Niemi (tools). **Established:** 1790. **Capital:** Fmk 321,000,000. **No. of employees:** 2,255.
Works:
■ Hackman Metal, F 79130 Sorsakoski. (Tel: 358 72 43111. Telex: 4327). Works director: Matti Glöersen. ***Tube and pipe mills:*** welded – one for 53-219.1mm, one for 48.3-114.3mm and one for up to 406mm (combined annual capacity 4,500 tonnes).
Products: Stainless steel – longitudinal weld tubes and pipes 53-1,200mm; welded pipe fittings (bends, tees, reducers, collars, etc).

Oy Ja-Ro Ab

Head office: Rautatienkatu 24, SF 68601 Pietarsaari. **Postal address:** PO Box 15. **Telex:** 7511, 7528 jaro sf.
Established: 1949. **Ownership:** Outokumpu Oy, Stainless Steel Division.
Works:
■ At head office. Works director: Christer Asp.
Products: Stainless steel – longitudinal weld tubes and pipes.

Finland

Outokumpu Oy

Head office: Töölönkatu 4, SF 00100 Helsinki. **Postal address:** PO Box 280, SF 00101 Helsinki. **Tel:** 358 04031. **Telex:** 124441 okhi sf. **Cables:** Kumpu Helsinki. **Telecopier (Fax):** 90 403 25600.
Management: Board of directors – P. Voutilainen (chairman and president); J. Hakkarainen (vice chairman and executive vice president); V. Lehtinen, O. Siltari, O. Virolainen, R. Virrankoski (members and executive vice presidents). **Established:** 1932. **Capital:** Fmk 493,000,000. **No. of employees:** 15,000. **Subsidiaries:** Wirsbo Bruks Ab, Oy Ja-Ro Ab, Viscaria Ab, A/S Bidjovagge Gruber, Tara Mines Ltd, Outokumpu Mines Ltd, Tietokumpu Oy, The Nippert Co, Valleycast Inc, Metallverken Ab, Finska Stenindustri Ab, Okmetic Oy, Princeton Gamma-Tech Inc, Rammer Oy, Roxon Oy, Outokumpu Equipmentos Industriais e Participacoes Ltda, Outokumpu Equipment Canada Ltd, Outokumpu Técnica-Chile Ltda, Outokumpu Mexicana SA de CV, Outokumpu Sucursal del Peru, Outokumpu Engineering Inc, Transamine Service Ltd, Outokumpu Benelux BV, Outokumpu Danmark A/S, Outokumpu Deutschland GmbH & Co, Outokumpu France SA, Outokumpu Italia Srl, Outokumpu Japan KK, Outokumpu Metals (USA) Inc, Outokumpu Norge A/S, Outokumpu Representative Office, (Moscow), Outokumpu (SEA) Pte Ltd, Outokumpu (Schweiz) AG, Outokumpu Stål & Metall Ab, Outokumpu (UK) Ltd, Outokumpo (USA) Inc. **Associates:** Løkken Gruber A/S, Trout Lake Joint Venture, Micronas Oy, AOR Industries Oy, Ab Etiproducts Oy, Okphil Inc. **Annual capacity:** Raw steel: 200,000 tonnes.
Works:
■ Stainless Steel Division, Tornio. ***Steelmaking plant:*** One 60-tonne electric arc furnace (annual capacity 200,000 tonnes). ***Refining plant:*** One 65-tonne AOD (200,000 tonnes). ***Continuous casting machines:*** One 1-strand Voest-Alpine slab (170 x 1,600mm max). ***Rolling mills:*** One Sendzimir type ZR 21 BGGI cold reduction (0.5-6.0mm x 1,600mm max) (coil weight 26 tonnes) (125,000 tonnes); one temper/ skin pass (coil weight 26 tonnes) (125,000 tonnes). ***Other plant:*** Two submerged arc furnaces for ferro-chrome production (180,000 tonnes).
Products: Stainless steel – slabs, hot rolled band (for re-rolling), hot rolled sheet/coil (uncoated) and cold rolled sheet/coil (uncoated) (1986 output – of rolled products 153,000 tonnes.
Brands: Polarit (Outokumpu stainless steel).
Expansion programme: Steckel hot rolling mill (end-1987).
Additional information: Other Outokumpu divisions produce copper, nickel, zinc, cobalt and lead.

Ovako Steel Oy Ab

Head office: Bulevarden 7, SF 00101 Helsinki. **Postal address:** PB 790. **Tel:** 90 61621. **Telex:** 12747. **Cables:** Ovako Helsinki. **Telecopier (Fax):** 7531565.
Management: Matti Sundberg (managing director), Esa Sardlainen, Reijo Antola, Gösta Engman. **Established:** 1915. **No. of employees:** 2,800. **Ownership:** Ovako Steel AB, Sweden (100%). **Associates:** Dalsbruk Oy AB (20%). **Annual capacity:** Raw steel: 800,000 tonnes. Finished steel: 600,000 tonnes.
Works:
■ 55100 Imatra. (Tel: 954 636 88. Telex: 5711 ovai). Works director: Reijo Antola. ***Steelmaking plant:*** Two 60-tonne electric arc furnaces. ***Refining plant:*** One ladle furnace (installation in 1988). ***Continuous casting machines:*** One 3-strand S-type Concast billet (100-125mm sq); one bloom (280 x 350mm (installation in 1988)). ***Rolling mills:*** One medium section (33-85mm dia); one light section (12-32mm dia); one wire rod (7-23mm dia). ***Other plant:*** Turning, drawing, grinding, heat treatment, chemical surface treatment.
■ Works at Billnäs, Loimaa and Jokioinen. Plants for production of springs, chains, wire, wire products, etc.
Products: Carbon steel – wire rod, round bars, flats, bright (cold finished) bars, light rails, rail accessories. Alloy steel (excl stainless) – wire rod 5.5-14mm (coil weight 1,000 kg), round bars, flats. Bright (cold finished) bars. Special steels including boron steels, case-hardening, spring steel, wear-resisting and cold heading steel.

Rautaruukki Oy

Head office: Kiilakiventie 1, SF 90101 Oulu. **Postal address:** PO Box 217. **Tel:** 358 81 327711. **Telex:** 32109 steel sf. **Telecopier (Fax):** 358 81 327506. **Management:** Chairman and president – Mikko Kivimäki; executive vice presidents – Holger Sweins (senior, marketing), Reino Mäkelä (senior, operations), Matti Haltia (steel upgrading), Esko Erkkilä (general manager, Raahe Steel Works), Pekka Einamo (engineering), Aulis Saarinen (research and development). **Established:** 1960. **Capital:** Fmk 602,000,000. **No. of employees:** 7,205. **Ownership:** The Finnish State (99.01%); Outokumpu Oy (0.3%); Valmet Oy (0.2%); Oy Fiskars Ab (0.1%); Rauma-Repola Oy (0.2%); Oy Wärtsilä Ab (0.1%). **Subsidiaries:** (these include associated companies) Hedpipe AB, Sweden (100%) (steel tube works); Nordisk Simplex A/S, Denmark (100%) (steel tube works); Dalsbruk Oy Ab (80%) (rebars, wire rod and wire products); Structo D.O.M. Europe AB (33.3%) (precision tubes); Suomen Kuonajaloste Oy (38%) (slag products). **Annual capacity:** Pig iron: 1,700,000 tonnes. Raw steel: 2,000,000 tonnes. Rolled steel: 1,500,000 tonnes.

Works:

■ Raahe Steel Works, 92170 Raahensalo. (Tel: 358 82 301. Telex: 32162). Works director: Esko Erkkilä. ***Coke ovens:*** One twin-flue battery (35 ovens) (annual capacity 470,000 tonnes). ***Sinter plant:*** Three 75 sq m strand Dwight Lloyd (106 tph each (combined 2,250,000 tonnes)). ***Blast furnaces:*** Two 7.2m hearth dia, 900 cu m (annual capacity 850,000 tpy each). ***Steelmaking plant:*** Three 90/100-tonne basic oxygen converters (2,000,000 tonnes). ***Continuous casting machines:*** Five slab machines – two 1-strand bending (210 x 950mm to 210 x 1,800mm), and three 1-strand vertical type (150 x 970mm to 290 x 1,800mm) (total annual capacity 2,000,000 tonnes). ***Rolling mills:*** One 4-high reversing 1,045/1,825 x 3,600mm heavy plate mill (600,000 tonnes); one semi-continuous wide hot strip comprising 4-high 990/1,700 x 3,600mm reversing roughing stand and six 4-high 724/1,524 x 2,040mm continuous finishing stands (1,500,000 tonnes).

■ Hämeenlinna Works, 13300 Hämeenlinna. (Tel: 358 172061. Telex: 2327). Works director: Erkki Kutilainen. ***Rolling mills:*** One 4-stand tandem 650/1,525 x 1,680mm cold reduction (700,000 tonnes); one single-stand 650/1,525 x 1,688mm temper/skin pass (1,000,000 tonnes). ***Coil coating lines:*** Two hot-dip galvanizing lines – one 0.4 to 2.0mm x 650 to 1,400mm (200,000 tonnes), and one 0.4 to 1.2mm x 650 to 1,250mm (120,000 tonnes); one colour coating plastic-coating line 0.5 to 1.5mm x 650 to 1,400mm (100,000 tonnes).

■ Tubular products division. Works: Hämeenlinna, Lappohja, Oulainen, Pulkkila, Nordisk Simplex A/S (Denmark) and Hedpipe AB (Sweden). Tel: 358172061. Telex: 2327. General manager Raimo Raassina (works). ***Tube and pipe mills:*** welded – eighteen longitudinal-weld (combined annual capacity 300,000 tonnes); four spiral-weld (50,000 tonnes); one tapered tube mill (60,000 pieces).

■ Toijala Works, Hämcentie 100, PO Box 69, 37801 Toijala. (Tel: 358 3724800. Telex: 22408). Cold formed section lines (annual capacity 30,000 tonnes).

■ Ylivieska Works, Koskipuhdontie, PO Box 89, 84101 Ylivieska. (Tel: 358 8323221. Telex: 3427). Two lines for welded sections (annual capacity 15,000 tonnes).

■ Transport Systems Division, Kiilakiventie 1, PO Box 217, 90107 Oulu (Tel: 358 81327711. Telex: 32109 steel sf). Vice president: Seppo Sahlman. Works at Oulu, Otanmäki, Taivalkoski, and Halikko making freight cars, wagons, containers, pressure vessels, tanks, etc.

Products: Carbon steel – welded sections (heavy) height up to 3,000mm, max length 30,000mm (deliveries in 1986, 12,000 tonnes); cold roll-formed sections (from strip) L, U, Z, C and other shapes, max wall thickness 10mm (deliveries in 1986, 27m,000 tonnes); pipe piling; medium plates and heavy plates 5 to 100mm thick, 5.5 to 50m long, up to 3,200mm wide; floor plates with chequer and tear pattern; hot rolled sheet/coil (uncoated) up to 1,750mm wide, 1.8 x 16mm thick in steels for structural purposes, pressure vessel steels, shipbuilding steels, and steels with good formability (deliveries of hot rolled products in 1986, 650,000 tonnes); cold rolled sheet/coil (uncoated) 0.4 to 3mm thick, and max 1,525mm wide in steels with good formability and steels for structural purposes (deliveries in 1986, 350,000 tonnes); hot-dip galvanized sheet/coil 0.5 to 2mm thick, max 1,440mm wide, zinc coating 100 to 350 grams/sq metre; colour coated sheet/coil 0.4 to 1.5mm thick, max width 1,440mm,

Rautaruukki Oy (continued)

coated with PVC, polyester, PVF, laminates (deliveries of galvanized and colour coated sheets in 1986, 200,000 tonnes); longitudinal weld tubes and pipes , line pipe, precision, and hollow sections – round 8 to 323.9mm dia, square and rectangular 10 x 10mm to 400 x 200mm, max wall thickness 12.5mm from hot rolled, cold rolled, and zinc-coated steels in structural steel grades, and grades with good formability, also API grades (deliveries in 1986, 215,000 tonnes); spiral-weld tubes and pipes and pipe piling 200 to 1,200mm dia, max wall thickness 16mm, max length 18,000mm with structural steel grades and pressure vessel steels, also API line pipe steels (deliveries in 1986, 30,000 tonnes); galvanized tubes and pipes ; tapered tubes max 300mm dia, min 60mm dia, up to 15,000mm long, max wall thickness 6mm in steels for structural purposes (deliveries in 1986, 40,000 pieces).
Sales offices/agents: Helsinki Office – Fredrikinkatu 51-53, PO Box 860, SF 00101 Helsinki; Rautaruukki AB, Pyramidvägen 7, S 17136 Solna, Sweden; Rautaruukki (UK) Ltd, Merevale House, Parkshot, Richmond, Surrey TW9 2RW, UK; Finnstål A/S, Postboks 211, N 1322 Høovik, Norway; Nordisk Simplex A/S, Vejlegårdsvej 34, DK 2665 Vallensbaek Strand, Denmark; Rautaruukki (Deutschland) GmbH, Grafenberger Allee 68, D 4000 Düsseldorf 1, West Germany.

France

Aciers d'Allevard

Head office: BP 17, Le Cheylas, F 38570 Goncelin. **Tel:** 76 45 45 45. **Telex:** 320 602. **Telecopier (Fax):** 76 45 45 33.
Management: Chairman – Pierre Assayag; managers – Patrick Bandrier (staff), Bernard Chandesris (engineering), Daniel Françon (financial), Denis Sauvagnargues (works); Roland Gueschir (commercial and marketing director). **Capital:** Fr 29,000,000. **No. of employees:** 570.
Works:
■ At head office. ***Steelmaking plant:*** One 30-tonne electric arc furnace (annual capacity 100,000 tonnes). ***Refining plant:*** One 30-tonne ladle furnace (100,000 tonnes). ***Continuous casting machines:*** 3-strand billet (105, 120, 135) (120,000 tonnes). ***Rolling mills:*** Two bar – 550mm (75,000 tonnes) and 500mm (60,000 tonnes).
Products: Alloy steel (excl stainless) – round bars 11-35mm dia (output 40,000 tonnes); flats – 40 x 70 – 140mm (47,000 tonnes).

Alpa – Aciéries et Laminoirs de Paris—see Usinor Sacilor, Long Products

Aciéries et Forges d'Anor

Head office: 40 Rue du Maréchal Foch, F 59186 Anor. **Tel:** 27 59 57 55. **Telex:** 820665. **Telecopier (Fax):** 27 59 61 64.
Management: Jean-Claude Poitte (gérant), Richard Poitte (adjoint directeur générale), Jean-Claude Desreumaux (service administratif et financier), Pierre Line (service commercial France), Michel Letoret (service commercial export). **Established:** 1902. **Capital:** Fr 6,151,500. **No. of employees:** Works – 190. Office – 40.
Works:
■ At head office. ***Steelmaking plant:*** Induction furnaces (annual capacity 700 tonnes). ***Rolling mills:*** bar – Universal reversing (3,500 tonnes).
Products: Alloy steel (excl stainless) – flats 8 x 70mm to 60 x 300mm. Compound steel bars.
Brands: Anor.

Armco SA, France

Head office: 4 Rue de l'Abreuvoir, F 92400 Courbevoie. **Telex:** 602306.
Products: Carbon steel – longitudinal weld tubes and pipes.
Additional information: In January 1987, TI Group (UK) acquired most of Armco's steel works in Europe.

Ascométal—see Usinor Sacilor, Long Products

Aubert et Duval

Head office: 41 Rue de Villiers, F 92200 Neuilly-sur-Seine. **Postal address:** BP 120. **Tel:** 1 47 58 11 72. **Telex:** 620072. **Telecopier (Fax):** 47 57 69 39.
No. of employees: Works – 2,700. Office – 200.
Works:
■ F 63770 Les Ancizes Comps. (Tel: 73 86 04 21. Telex: 390779). ***Steelmaking plant:*** electric arc furnace. ***Refining plant:*** AOD; ESR; vacuum degassing unit; VAR and VIM. ***Rolling mills:*** blooming and billet.
Products: Stainless steel – billets, round bars, square bars, flats, hexagons, bright (cold finished) bars, cold rolled sheet/coil (uncoated). Alloy steel (excl stainless) – billets, round bars, square bars, hexagons, bright (cold finished) bars, medium plates, hot rolled sheet/coil (uncoated), oil country tubular goods, hollow bars. Forgings.

Laminoirs de Beautor SA

Head office: F 02800 La Fère, Beautor. **Tel:** 23 57 21 11. **Telex:** 140500.
Products: Carbon steel – cold rolled sheet/coil (uncoated) and electro-coated coil/ sheets.

Biraghi SA

Head office: 20/22 Rue Jean Blanchard, F 74200 Thonon-Les-Bains. **Postal address:** BP 55. **Tel:** 50 71 01 12. **Telex:** 385655F. **Telecopier (Fax):** 50 26 55 08.
Management: J.L. Benard (chairman), S. Bienfait (plant manager). **Established:** 1986. **Capital:** Fr 250,000. **No. of employees:** Works – 13. Office – 7. **Ownership:** Spiro-Gills France (Ultimate holding company – Hamon-Sobelco, Belgium). **Annual capacity:** Finished steel: finned tubes – 1,100 tonnes, studded tubes – 500 tonnes.
Works:
■ At head office.
Products: Carbon steel – extended surface tubes (high frequency welded, finned and studded pipes) 25-168mm dia. (output 750 tonnes). Stainless steel – extended surface tubes (high frequency welded, finned and studded pipes) 25-168mm dia. (25 tonnes). Alloy steel (excl stainless) – extended surface tubes (high frequency welded, finned and studded pipes) 25-168mm dia. (300 tonnes).

Aciéries de Bonpertuis

Head office: F 38140 Rives sur Fure. **Tel:** 076 051754. **Telex:** 320632.
Products: Carbon steel – round bars, square bars, flats. Stainless steel – round bars, square bars, flats. Ovals; cutlery bevels.

Laminoirs de Bretagne—see Usinor Sacilor, Long Products

Sté Métallurgique de Brévilly—see Usinor Sacilor, Long Products

BSL Tubes et Raccords SA

Head office: 108 Route de Reims, F 02202 Soisson. **Postal address:** PO Box 40.
Ownership: Sitindustrie SpA. **Associates:** Stainless steel tube and pipe – Sitai SpA and Nuova CMC SpA (Italy) and Zwahlen & Mayr (Switzerland). **Annual capacity:** Finished steel: 15,000 tonnes (welded stainless steel pipe).
Products: Stainless steel – longitudinal weld tubes and pipes, especially large diameter.
Sales offices/agents: Sitai International BV, Rotterdam, Netherlands.

Sté Calibracier

Head office: F 80800 Corbie. **Tel:** 224 82044. **Telex:** 140748. Bright bars.

Camus & Cie

Head office: 35 Rue Vaucorbe, F 89700 Tonnerre. **Postal address:** BP 24. **Tel:** 86 550145. Steel tubes.

Carnaud Basse-Indre

Head office: 65 Avenue Edouard Vaillant F 92100 Boulogne sur Seine. **Tel:** 1 47 61 22 84. **Telex:** 270 425.
Management: F. de Wendel (president directeur général), P.F. Milleron (directeur général adjoint), C. Desnoes (directeur usine), G. Rotger (directeur commercial). **Established:** 1984. **Capital:** Fr 78,350,000. **No. of employees:** Works – 1,400. Office – 33. **Ownership:** Sté des Forges de Basse Indre (Ultimate holding company – Carnaud SA (96%)). **Annual capacity:** Finished steel: 400,000 tonnes.
Works:
■ F 44610 Indre. (Tel: 40 86 02 00). Works director: C. Desnoes. ***Rolling mills:*** One 5-stand cold reduction (annual capacity 540,000 tonnes); one 2-stand temper/skin pass. ***Coil coating lines:*** One electrolytic tinning (Ferrostan) and one dual tin-free steel (ECCS) (combined annual capacity 400,000 tonnes).
Products: Carbon steel – blackplate, electrolytic tinplate (single-reduced), electrolytic tinplate (double-reduced), tin-free steel (ECCS) (combined output 310,000 tonnes).
Expansion programme: Continuous annealing line to be completed beginning of 1988.

Aciers de Champagnole

Head office: 23 Rue G. Clemenceau, F 39300 Champagnole. **Postal address:** BP 104. **Tel:** 084 52 64 44. **Telex:** 360190. **Telecopier (Fax):** 084 52 61 25.
Management: P. Barme (managing director). **Established:** 1916. **Capital:** Fr 15,000,000. **No. of employees:** Works – 90. Office – 25. **Ownership:** Kloster Speedsteel AB, Sweden (100%). **Annual capacity:** Finished steel: 2,500 tonnes.
Works:
■ At head office. Works director: J.C. Breuil. ***Rolling mills:*** medium plate – 3-high 550/320mm (annual capacity 1,500 tonnes). ***Other plant:*** Forge with one press (1,000-tonne); two hammers (4 tonne and 1.5 tonne); wire works; finishing shop; heat treatment for tool bits. High speed steel.
Sales offices/agents: Zone Industrielle du Coudray, 23 Av Albert Einstein, F 93150 Le Blanc Mesnil; 5 Rue du Moulin Perrault, F 42100 St Etienne Terre Noire.

Aciers de Châtillon et Gueugnon—see Usinor Sacilor, Stainless & Special Products

Tubes de Chevillon

Head office: F 52170 Chevillon. **Tel:** 25 04 40 21. **Telex:** 840602, 840630 tubchev.
Products: Carbon steel – longitudinal weld tubes and pipes and cold drawn tubes and pipes 8-160mm dia.

Forges de Clairvaux

Head office: 6 Rue de Léningrad, F 75008 Paris. **Tel:** 42 936551. **Telex:** 280454.
Products: Carbon steel – special light sections.

Commentryenne des Aciers Fins Vanadium Alloys

Head office: 33 Rue d'Alsace, F 92531 Levallois Perret. **Postal address: Tel:** 47 59 67 30. **Telex:** 61374 cochi. **Telecopier (Fax):** 47 59 67 72.
Management: President – Jean Rousseau; managers – Michel Mercier (general), Bernard Laurent (sales). **Established:** 1956. **Capital:** Fr 17,325,000. **No. of employees:** Works – 470. Office – 30. **Ownership:** Chiers-Châtillon. **Associates:** Forecreu (high speed oil well drilling rods), TAT (heat treatment). **Annual capacity:** Raw steel: 25,000 tonnes. Finished steel: 15,000 tonnes.
Works:
■ F 03600 Commentry. Tel: 070 646222. Works director: Jean Carré. ***Steelmaking plant:*** One 30-tonne and one 18-tonne electric arc furnaces (annual capacity 15,000 and 10,000 tonnes). ***Refining plant:*** One 5-tonne ESR (500 tonnes). ***Rolling mills:*** 3-high 2-stand billet (650mm) (40,000 tonnes); two bar (300 and 250mm) (10,000 tonnes); 2-high 800 and 650mm sheet mills (4,000 tonnes and 6,000 tonnes); one Sendzimir mill (2,000 tonnes); ***Other plant:*** One 1,000-tonne press for forging billets; drawing and grinding facilities for high speed steel; electron-beam welding equipment for hi-metal strip.
Products: Alloy steel (excl stainless) – wire rod (high speed) and high speed steels bars, sheets (output 1985: 9,000 tonnes); aircraft steels; refractory metals – sheets. Hire rolling of special alloys.
Brands: Rhino (high speed steels); Jet (aircraft steels).
Sales offices/agents: Commentryenne Ltd, Holmwood House, Courtworth Road, Sheffield S11 9LP, UK; Commentryenne Inc, 920 Lloyd Avenue, Latrobe, PA 15650, USA.

Creusot Loire Industrie—see Usinor Sacilor, Metallurgy

Etilam-Gravigny (EG)—see Usinor Sacilor, Transformation & Other Activities

Exma SA

Head office: 1 Rue de l'Industrie, F 57331 Yutz. **Tel:** 82 56 22 07. **Telex:** 930165. **Telecopier (Fax):** 82 51 91 12.
Management: Charles d'Huart (président directeur général), Patrick Hirsch (directeur général). **Established:** 1981. **Capital:** Fr 10,900,000. **No. of employees:** 65. **Ownership:** Jean D'Huart et Cie. **Annual capacity:** Finished steel: 50,000 tonnes.
Works:
■ At head office. ***Tube and pipe mills:*** welded (annual capacity l25,000 tonnes).
Products: Carbon steel – longitudinal weld tubes and pipes, hollow sections.

Forges de Froncles

Head office: BP 1, F 52320 Froncles. **Tel:** 010 33 25 02 3005. **Telex:** 840617.
Products: Carbon steel – cold rolled sheet/coil (uncoated).

Galvameuse SA – Sté Meusienne de Produits d'Usines Métallurgiques

Head office: Revigny-sur-Ornain, F 55800 Contrisson. **Postal address:** BP 1. **Tel:** 029 797700. **Telex:** 961353. **Telecopier (Fax):** 029 751944.
Ownership: Cockerill-Sambre, Belgium (99%).
Works:
■ At head office. ***Rolling mills:*** cold reduction. ***Coil coating lines:*** hot-dip galvanizing.
Products: Carbon steel – cold rolled sheet/coil (uncoated), hot-dip galvanized sheet/coil.

Galvanor/Coloracier—see Usinor Sacilor, Flat Products

Grousset

Head office: Boulevard de l'Industrie, F 42176 St Just-St Rambert. **Tel:** 66 36 50 61. **Telex:** 330989 f. **Telecopier (Fax):** 66 55 49 70.
Management: Mme de Chemellier (gérante), J. Monistrol (directeur général), J.P. Cote (directeur production). **Established:** 1904. **Capital:** Fr 1,500,000. **No. of employees:** Works – 200. Office – 20. **Annual capacity:** Rolled products: 15,000 tonnes.
Works:
■ At head office. ***Rolling mills:*** 2-stand (17-102mm) bar (annual capacity 15,000 tonnes).
Products: Carbon steel – round bars 18-62mm (annual capacity 2,000 tonnes); hexagons 17-102mm (13,000 tonnes); bright (cold finished) bars.

GTS Industries—see Usinor Sacilor, Flat Products

Imphy SA—see Usinor Sacilor, Stainless & Special Products

Iton Seine SA

Head office: Quai de Seine, F 78270 Bonnières sur Seine. **Postal address:** BP 13. **Tel:** 30 93 03 41. **Telex:** 695555. **Telecopier (Fax):** 30 93 31 14.
Management: Managers – Valério Bisio (general, works director), Dominique Wilbois (purchasing), Jean Prudhomme (sales). **Established:** 1973. **Capital:** Fr 46,500,000. **Annual capacity:** Raw steel: 250,000 tonnes. Finished steel: 250,000 tonnes.
Works:
■ At head office. Works director: Valério Bisio. ***Steelmaking plant:*** One 4.65m dia Tagliaferri electric arc furnace (annual capacity 250,000 tonnes). ***Continuous casting machines:*** One 4-strand Danieli billet 100 to 140mm sq (250,000 tonnes).
Products: Carbon steel – billets 100-140mm sq; reinforcing bars and round bars 6-32mm dia; square bars.

Les Usines Laprade SA—see Usinor Sacilor, Transformation & Other Activities

Sté Lorraine de Vernissage et d'Imprimerie—see Usinor Sacilor, Flat Products.

Manufer SA

Head office: Le Pizou, F 24700 Montpon-Menesterol (Dordogne). **Tel:** 53 81 81 63. **Telex:** 570422.
Products: Carbon steel – reinforcing bars.

Fabrique de Fer de Maubeuge

Head office: Avenue de Beco, F 59720 Louvroil. **Tel:** 27 64 60 01. **Telex:** 160 866. **Telecopier (Fax):** 27 65 87 70.
Management: Michel Corpet (président). **Established:** 1884. **Capital:** Fr 59,799,960. **No. of employees:** 480. **Ownership:** Usines Gustave Boël, Belgium. **Subsidiaries:** S.F.P.M., Louvroil (non-prime galvanized and coated sheets). **Annual capacity:** Finished steel: 300,000 tonnes.
Works:
■ At head office. ***Rolling mills:*** cold strip – Sendzimir (annual capacity 200,000 tonnes). ***Coil coating lines:*** hot-dip galvanizing (300,000 tonnes); colour coating (60,000 tonnes).
Products: Carbon steel – hot-dip galvanized sheet/coil (1986 output 250,000 tonnes); colour coated sheet/coil (30,000 tonnes).
Brands: Galvazot, Galvaliss, Algazot, Algaliss, Agricolor.
Sales offices/agents: At head office. Germany – Real, Postfach 300844, 4000 Düsseldorf 30; UK – Queenborough, Aries House, Flackwell Heath, Bucks; Netherlands – Duo Staal, Postbus 6011, 2702 AA Zoetermeer; Italy – Fusco, Via Salasco, 20136 Milan.

Sté Meusienne de Constructions Mécaniques

Head office: 32 Rue La Boëtie, F 75008 Paris. **Tel:** 1 43 59 79 72. **Telex:** 290024 smcm f. **Cables:** Socomeca Paris. **Telecopier (Fax):** 1 45 63 55 47.
Management: Chairman – René Giros; managers – Claude Gabet (general), Bernard Boutillot (commercial and export). **Established:** 1905. **Capital:** Fr 12,000,000. **No. of employees:** Works – 527. Office – 25. **Ownership:** Private. **Annual capacity:** Finished welded tubing: 45,000 tonnes.
Works:
■ F 55170 Ancerville. (Tel: 33 29 75 21 21. Telex: 961468 F). Works director: Claude Gabet. ***Tube and pipe mills:*** welded – five HF 12-114mm od (annual capacity 40,000 tonnes, for carbon steel tubing); two HF 12-51mm od (5,500 tonnes, for stainless steel tubing); six TIG 10-129mm od (6,000 tonnes, for stainless steel

Sté Meusienne de Constructions Mécaniques (continued)

tubing).
Products: Carbon steel – longitudinal weld tubes and pipes, round and square sections (1986 output 35,000 tonnes). Stainless steel – longitudinal weld tubes and pipes, 10 to 129mm and profiles (12,000 tonnes).

Sté des Aciéries de Montereau (SAM)—see Usinor Sacilor, Long Products

Sté des Tubes de Montreuil

Head office: 7-9 Rue du Parc, BP 17, F 93151 Le Blanc-Mesnil. **Tel:** 010 331 48671536. **Telex:** 212170 bardini f. Steel tubes.

Sté Métallurgique de Normandie—see Usinor Sacilor, Long Products

Pont-à-Mousson SA

Head office: 91 Avenue de la Libération, F 5400 Nancy. **Tel:** 08 396 81 21. **Telex:** 850003 Pamsa F.
Management: Directeur général – Bernard Novel (président), Pierre Blayau; directeur général adjoint – Michel Bousquet, Philippe Lenoir. **Capital:** Fr 629,129,400. **No. of employees:** 6,344 (average 1986). **Ownership:** St-Gobain group. **Subsidiaries:** These include – Stanton PLC, UK (100%), Halbergerhütte GmbH, West Germany (100%), Funditubo, Spain (100%), Tubi-Ghisa, Italy (25%), Barbara, Brazil (56.2%).
Works:
■ Pont-à-Mousson, Foug, Toul, Liverdun (Meurthe-et-Moselle), Bayard (Haut-Marne) and Fumel (Lot-et-Garone).
Products: Carbon steel – seamless tubes and pipes. Cast iron – pipes, pipe fittings. Iron castings.

Produits d'Acier SA

Head office: Z.I. La Silardiere, F 42501 Le Chambon Feugerolles Cedex, St. Etienne. **Tel:** 77 894282. **Telex:** 307075.
Management: Lidio Neirotti (president directeur générale), Claudio Neirotti (directeur générale), Bernard Raffin (directeur). **Established:** 1986. **Capital:** Fr 6,000,000. **Ownership:** Neirotti Tubi SpA, Italy.
Works:
■ At head office. ***Tube and pipe mills:*** welded.
Products: Carbon steel – longitudinal weld tubes and pipes and galvanized tubes and pipes (rounds 20-152mm, square 20-120mm, rectangular 20 x 10mm to 180 x 60mm, special sections 1.5-5mm thick).

Profilafroid—see Usinor Sacilor, Transformation & Other Activities

Profilés et Tubes de l'Est—see Usinor Sacilor, Transformation & Other Activities

Sté Métallurgique de Revigny

Head office: Av du XXe Corps, Revigny sur Ornain. **Telex:** 961475 smrrevi. Bright bars.

Aciéries et Laminoirs de Rives

Head office: Renage, F 38140 Rives sur Fure. **Tel:** 76 91 42 44. **Telex:** 320277 alr.
Products: Carbon steel – billets; reinforcing bars merchant bars including round bars, flats; light angles; light tees; special light sections. Alloy steel (excl stainless) – billets; round bars; flats; light angles; light tees; special light sections including agricultural sections. Fence posts.

Sacilor – Aciéries et Laminoirs de Lorraine—see Usinor Sacilor

Sat – Sté des Aciéries de Thionville—see Usinor Sacilor, Long Products

Hauts Fourneaux Réunis de Saulnes & Uckange—see Usinor Sacilor, Transformation Activities

Aciéries et Laminoirs du Saut du Tarn (ALST)

Head office: F 81160 St Juéry. **Postal address:** BP 10. **Tel:** 63 55 15 00. **Telex:** 520789.
Management: Chairman – M. Faivre; managers – J. Desvaux (general), M. Thomasset, J. Balleriaud (sales). **Established:** 1983. **Capital:** Fr 2,500,000. **No. of employees:** Works – 105. Office – 15. **Ownership:** Immobiliere Bujon (50%); Sté des Forges de Clairvaux (50%).
Works:
■ At head office. ***Steelmaking plant:*** Two electric arc furnaces – one 15-tonne and one 10-tonne (annual capacity 30,000 tonnes). ***Rolling mills:*** billet Three 3-high 600mm stands (40,000 tonnes); bar – two 3-high stands and one 2-high stand, 450mm (15,000 tonnes); 7-stand 2-high 250-450mm mill (20,000 tonnes).
Products: Carbon steel – (Special carbon), hot rolled bars – including round bars, square bars, flats, hexagons. Stainless steel – round bars, square bars, flats, hexagons. Alloy steel (excl stainless) – round bars, square bars, flats, hexagons.

SME – Sté Métallurgique de l'Escaut—see Usinor Sacilor, Long Products

Sollac – Sté Lorraine de Laminage Continu—see Usinor Sacilor, Flat Products.

Solmer – Sté Lorraine et Méridionale de Laminage Continu——see Usinor Sacilor, Flat Products.

Laminoirs de Strasbourg—see Usinor Sacilor, Flat Products

Forges de Syam

Head office: F 39300 Champagnole, Syam (Jura). **Tel:** 84 52 20 42. **Telex:** 360744 forsyam f.
Products: Carbon steel – flats, special light sections.

Laminoirs à Froid de Thionville—see Usinor Sacilor, Transformation & Other Activities

Trancel—see Usinor Sacilor, Long Products

Ugine Savoie—see Usinor Sacilor, Stainless & Special Products

Unimétal, Sté Française des Aciers Longs—see Usinor Sacilor, Long Products

Usinor Aciers—see Usinor Sacilor, Flat products

Usinor Sacilor

Head office: Immeuble Ile-de-France, 4 Place de la Pyramide, F 92070 Paris, La Défense. **Tel:** 1 49 00 60 10. **Telex:** 614730 usinr.
Management: Chairman – Francis Mer (chief executive officer); directors – Robert Hudry (finance), François Labadens (communications), Gerard Belorgey (human affairs), Jean-Claude Georges-François (economic affairs and regional development), Jean Jacquet (Long Products branch), Philippe Choppin de Janvry (international affairs and stainless and special products), Edmond Pachura (flat products branch). **Established:** 1987. **No. of employees:** 80,000. **Subsidiaries:** Principal subsidiaries and holdings: Flat products – Usinor Aciers, Solmétal- Sollac, Solmer. Long products – Unimétal, Ascométal. Stainless and special products – Ugine, Ugine-Savoie, Imphy. Metallurgy – Creuot-Loire Industrie, C3F. Large-diameter welded tubes – GTS Industries. International/Exports – Daval, IMS.

Flat Products Division

Usinor Aciers
Management: Bernard Rogy, Jacques Français, Guy Dollé (directeurs généraux adjoints). **Established:** 1985. **Capital:** F 4,600,250,000. **No. of employees: 3,000. Subsidiaries:** Galvanor (99%), Laminoiors de Strasbourg (99%), GTS Industries (95%), Solmer (50%). **Annual capacity:** Pig iron: 7,700,000 tonnes. Raw steel: 7,500,000 tonnes. Finished steel: 2,500,000 (sheets/coils) (4,500,000 hot coils and 1,200,000 plates).
Works:
■ Dunkirk, BP 2 508 59 381, Dunkirk Cédex 1. (Tel: 28 29 30 00. Telex: 820070 usidunk f). Works director: M. Frimat. ***Coke ovens:*** Five batteries (214 ovens) (annual capacity 2,000,000 tonnes). ***Sinter plant:*** Three Dwight-Lloyd – 140 sq m, 160 sq m and 460 sq m (8,000,000 tonnes). ***Blast furnaces:*** one 9.5m hearth dia, 1,608 cu m, (1,400,000 tonnes); one 9.5m hearth dia, 1,601 cu m (1,400,000 tonnes); one 10.2m hearth dia, 1,994 cu m, (1,700,000 tonnes) and one 13.9m hearth dia, 3,626 cu m (3,200,000 tonnes). ***Steelmaking plant:*** Three 140-tonne and three 230-tonnes basic oxygen converters (7,500,000 tonnes). ***Refining plant:*** One RHOB 220-tonne vacuum degassing unit (1,400,000 tonnes). ***Continuous casting machines:*** One 10-strand slab (950-2,100mm, 250-300mm) (6,000,000 tonnes). ***Rolling mills:*** heavy plate (width 5.0m max, length 33m max) (1,200,000 tonnes) and wide hot strip (600-1,930mm) (coil weight 32 tonnes max) (4,500,000 tonnes).
■ Biache Saint Vaast, Rue du Général de Gaulle, 62118, Biache St Vaast. (Tel: 21 50 18 30. Telex: 810 649). Works director: Jean Chaon. ***Rolling mills:*** One 4-high tandem cold reduction (width 650-1,525mm, thickness 0.3-2.5mm) (coil weight 28 tonnes) (annual capacity 626,000 tonnes) and one 1-stand 4-high temper/skin pass (coil weight 28 tonnes) (410,000 tonnes).
■ Mardyck, 1 Route de Spycker, PO Box 129, 59760 Grande Synthe. (Tel: 28 27 88 88. Telex: 820087). Works director: Roger Patou. ***Rolling mills:*** One 5-stand 4-high tandem cold reduction (width 2,030mm, thickness entry 1.8-6mm and exit 0.17-3mm) (coil weight 42 tonnes) (annual capacity 950,000 tonnes); two temper/skin pass – one 2-stand 4-high (width 2,030mm, thickness exit 0.14-3) (coil weight 42 tonnes) (400,000 tonnes) and one 1-stand 4-high (width 1,450mm, thickness 0.3-3) (coil weight 42 tonnes) (750,000 tonnes). ***Coil coating lines:*** One electrolytic tinning (12 baths) (width 1,200mm) (300,000 tonnes).
■ Montataire, BP 2, 60160. (Tel: 44 55 75 01. Telex: 140487 mtaer). Works director:

France

Usinor Sacilor (continued)

Jean Cenac. ***Rolling mills:*** One 5-stand 66" cold reduction (coil weight 42 tonnes) (annual capacity 1,400,000 tonnes); temper/skin pass – two 4-high 66" (coil weight 27 tonne) and one continuous annealing line with one 6-high temper 1,650mm (coil weight 42 tonnes) (600,000 tonnes). ***Other plant:*** Open coil annealing facilities (66").
Products: Steelmaking pig iron (1986 output 143,000 tonnes). Carbon steel – slabs – 207 and 247 x 1,040, 1,250, 1540mm (304,000 tonnes); hot rolled coil (for re-rolling) – 1.7-2.5 x 940, 1,250, 1,540mm (2,450,000 tonnes); flats 4-5 x 200-400 and 2 x 330mm (28,000 tonnes); hot rolled hoop and strip (uncoated) 4-5 x 200-400mm (70,000 tonnes); skelp (tube strip) 1.5-3 x 200-400mm (115,000 tonnes); heavy plates 10-40 x 2,000-3,000mm (550,000 tonnes); hot rolled sheet/coil (uncoated) – 940-1,540mm wide (Dunkirk 3,300,000 tonnes including HR coil for re-rolling); (Biache Saint Vaast 43,000 tonnes); cold rolled sheet/coil (uncoated) – Biache Saint Vaast (510,000 tonnes), Mardyck 3.37 tonnes/km (767,049 tonnes); Montataire 0.3-3mm thick, 600-1,600mm wide (1,000,000 tonnes); electrical non-oriented non-silicon sheet/coil – Mardyck semi-processes 5.0 tonnes/km (62,524 tonnes); blackplate – Mardyck 1.66 tonnes/km (232,240 tonnes); electrolytic tinplate (single-reduced) – Mardyck 1.61 tonnes/km (197,000 tonnes); electrolytic tinplate (double-reduced) – Mardyck 1.18 tonnes/km (32,000 tonnes). Alloy steel (excl stainless) – hot rolled coil (for re-rolling) 2.2.5 x 900 x 1,300mm (3,000 tonnes); flats 4 x 70-220mm (3,000 tonnes); hot rolled hoop and strip (uncoated) 1.5-3 x 250mm (6,000 tonnes); skelp (tube strip) 1.5-3 x 250mm (1,000 tonnes); heavy plates 10-40 x 2,000-3,000mm (2,500 tonnes); hot rolled sheet/coil (uncoated) 2-8 x900-1,300mm (40,000 tonnes).
Brands: Usinor Acier – Usiten (plates), Normeca (sheet and coils) and Indaten (sheet and coils); Montataire – Usifer (for enamelling), Usical (continuously annealed steels).

Sollac – Sté Lorraine de Laminage Continu 17 Avenue des Tilleuls,F 57191 Florange,Moselle. ***Tel:*** 82 51 56 00. ***Telex:*** 86 00 02 florange. ***Cables:*** Sollacos Florange. ***Telecopier (Fax):*** 82 58 24 05.
Management: President – Edmond Pachura; direction commerciale – Pierre Jacque; direction financière – François Faijean and direction des usines – Gilbert Heim. **Established:** 1948. **Capital:** F 1,215,000,000. **No. of employees:** 8,000. **Ownership:** Solmetal (Sacilor subsidiary) (72%), AG der Dillinger Hüttenwerke, West Germany (25%), Carnaud SA (3%). **Subsidiaries:** Solmer (50%), Marcoke (72%) and Sidfos (50%). **Annual capacity:** Pig iron: 2,300,000 tonnes. Raw steel: 3,000,000 tonnes. Finished steel: 2,800,000 tonnes.
Works:
■ Suzange. (Tel: 8 3514437. Telex: 860002). Works director: Jacques Saillour. ***Sinter plant:*** One 106 sq m strand and one 138 sq m strand (combined annual capacity 2,650,000 tonnes).
■ Fensch. (Tel: 8 3514437. Telex: 860002). Works director: Jacques Saillour. ***Blast furnaces:*** One (P3) 8.70m hearth dia, 1,422 cu m and two (P4 and P6) 8.50m hearth dia, 1,275 cu m (combined capacity 2,300,000 tonnes).
■ Serémange. (Tel: 8 3514437. Telex: 860002). Works director: Jacques Saillour. ***Sinter plant:*** One battery (64 ovens) (annual capacity 700,000 tonnes) (including 650,000 tonnes of 25mm and above for blast furnaces). ***Steelmaking plant:*** Two 240-tonne LWS basic oxygen converters (2,900,000 tonnes). ***Refining plant:*** Two 240-tonne Asea-SKF (2,900,000 tonnes). ***Continuous casting machines:*** One 2-strand Clesid (190 x 620-1,260mm) and one 2-strand Fives-Babcock (190 x 1,100-1,860mm) slab (combined capacity 2,850,000 tonnes). ***Rolling mills:*** One 80" wide hot strip comprising one SMS reversing horizontal/vertical roughing stand, one 'M'-stand and six 4-high finishing stands (coil weight 20-28 tonnes) (2,800,000 tonnes).
■ Ebange-Florange. (Tel: 8 251 56 36. Telex: 860002). Works director: Jean-Marie Paul-Dauphin. ***Rolling mills:*** One 5-stand tandem cold reduction (1,200,000 tonnes); four temper/skin pass – two 48" 2-stand for tinplate, one 57" single-stand and one

Usinor Sacilor (continued)

80" single-stand (combined capacity 2,200,000 tonnes). ***Coil coating lines:*** Two electrolytic tinning (400,000 tonnes); one tin-free steel (ECCS) (200,000 tonnes) one electro-galvanizing for sheets (Solzinc). ***Other plant:*** One 57" pickling line; one continuous annealing line for tinplate; open-coil annealing.
■ St-Agathe. Tel: 8 251 56 36. Telex: 860002. Works director: Jean-Marie Paul-Dauphin. ***Rolling mills:*** One 74" pickling line coupled to one 74" 4-stand tandem cold reduction (annual capacity 1,600,000 tonnes). ***Coil coating lines:*** One Carosel (Solcan) electro-galvanizing (300,000 tonnes).
Products: Steelmaking pig iron (1986 output 1,876,000 tonnes). Carbon steel – slabs – 600-1,860mm wide, 190mm thick (2,224,000 tonnes); hot rolled coil (for re-rolling) – 600-1,850mm wide – 1.5-4.76mm thick; hot rolled hoop and strip (uncoated) – 50-600mm wide; 1.5-4.76mm thick (66,000 tonnes); cold rolled hoop and strip (uncoated) – 16-500mm wide; 0.2-3mm thick (51,000 tonnes); hot rolled sheet/coil (uncoated) – (2,267,000 tonnes); cold rolled sheet/coil (uncoated) – 600-1,850mm wide, 0.4-3mm thick (2,128,000 tonnes); electrolytically galvanized sheet/coil – coils 825-1,830mm wide, 0.3-2mm thick, sheets 600-1,570mm wide, 0.5-3mm thick (240,000 tonnes); electrical non-oriented silicon sheet/coil – (42,000 tonnes); electrolytic tinplate (single-reduced) – 600-975mm wide, 0.17-0.49mm thick and electrolytic tinplate (double-reduced) – 600-975mm wide, 0.15-0.30mm thick (combined output 306,000 tonnes); tin-free steel (ECCS) – 600-975mm wide, 0.17-0.49mm thick (140,000 tonnes).
Brands: Sollege (double reduced tinplate); Solchrome (TFS – chrome-plated steel sheets); Soldur (high-strength thin hot and/or cold rolled steel sheets); Solfer (enameling steel); Soldecor (textured steel); Solzinc (electro-galvanized steel sheets); Solcar (electro-galvanized steel sheets in coil).
Sales offices/agents: Le Fer Blanc (tinplate sales), Paris (Tel: 1 47 67 92 00); Valor Acier (domestic market), Paris (Tel: 1 47 67 85 00); Daval (export sales), Paris (Tel: 1 47 67 83 83).
Expansion programme: Continuous annealing line and skinpass mill – capacity 400,000 tonnes/year. Start-up mid-1988.

Solmer – Sté Lorraine et Méridionale de Laminage Continu Tour Franklin 100-101, Quartier Boieldieu,La Défense 8, F 92800 Puteaux,Hauts de Seine. ***Tel:*** 16 1 47 76 42 21. ***Telex:*** 620498. ***Telecopier (Fax):*** 16 1 47 76 42 21 poste 56 33.
Management: Président du Conseil de Surveillance – Edmond Pachura; directoire – Bernard Rogy (président), Claude Diot, Michel Boucraut, Henri Faure; directeur administratif – Bernard Serpentier; directeur financier – Gérard Millet. **Established:** 1971. **Capital:** F 2,338,726,000. **No. of employees:** 5,750. **Ownership:** Usinor Acier (50%), Sollac (50%).
Works:
■ 13776 Fos Sur Mer Cedex. (Tel: 42 47 33 33. Telex: 430396). Works director: H. Faure. ***Coke ovens:*** Two Compound underjet batteries (108 ovens) (annual capacity 1,580,000 tonnes). ***Sinter plant:*** One 400 sq m strand (5,380,000 tonnes). ***Blast furnaces:*** Two 11.2m hearth dia, 2,850 cu m (capacity 2,000,000 tonnes each). ***Steelmaking plant:*** Two 310-tonne LBE basic oxygen (combined capacity 4,200,000 tonnes). ***Refining plant:*** AOB. ***Continuous casting machines:*** Two slab – No 1 2-strand (850-1,550 x 225 (2,000,000 tonnes) and No 2 2-strand (1,050-2,200 x 220 (2,000,000 tonnes). ***Rolling mills:*** One wide hot strip – five roughing and seven finishing stands (width 650-2,133mm, thickness 1.2-16mm) (max coil weight 38 tonnes) (4,600,000 tonnes).
Products: Carbon steel – slabs; hot rolled coil (for re-rolling); hot rolled hoop and strip (uncoated); hot rolled sheet/coil (uncoated). Stainless steel – hot rolled band (for re-rolling); hot rolled sheet/coil (uncoated). Alloy steel (excl stainless) – slab; hot rolled coil (for re-rolling); hot rolled hoop and strip (uncoated); hot rolled sheet/coil (uncoated).

Sté Lorraine de Vernissage et d'Imprimerie, 1 Rue d'Alsace,F 57190 Florange.

Usinor Sacilor (continued)

Postal address: BP 14. F 57192 Florange Cedex. ***Tel:*** 82 58 41 23. ***Telex:*** 860 028.
Management: Pierre Couveninhes (président directeur général) and Jacques Demange (directeur de l'usine de Florange). **Established:** 1954. **Capital:** Fr 8,100,000. **No. of employees:** Works – 177. Office – 47. **Ownership:** Solmetal (75%) and Dilling (25%).
Works:
■ At head office (Florange). Nine sheet coating lines, printing and spot coating lines. Printed sheets and coated sheets.
Sales offices/agents: Sté Le Fer Blanc SA, Paris.

Laminoirs de Strasbourg, 1 Rue du Bassin de l'Industrie,F 67016 Strasbourg Cedex. ***Tel:*** 88 61 48 64. ***Telex:*** 870 739 strasid. ***Cables:*** Lastral Strasbourg.
Management: Jacques Français (président directeur général); Léon Tordoir (directeur général adjoint); Maurice Calleau (secrétaire général); Lucien Spendov (directeur des usines); Michel Birckel (directeur commercial). **Established:** 1966. **Capital:** Fr 65,000,000. **No. of employees:** 914. **Ownership:** Usinor Aciers (99%). **Subsidiaries:** Eurinco (99%). **Associates:** Cofrastra AG, Switzerland (50%) (trade in building materials); Helsitra Maroc (49%) (ventilation and air conditioning equipment). **Annual capacity:** Finished steel: 470,000 tonnes.
Works:
■ Strasbourg (head office). ***Rolling mills:*** One Moeller and Neumann 4-high reversing cold reduction (annual capacity 230,000 tonnes); One Demag temper/skin pass (400,000 tonnes). ***Coil coating lines:*** One Sendzimir hot-dip galvanizing (240,000 tonnes). ***Other plant:*** Insulating panel manufacturing line (2,000,000 sq m).
Products: Carbon steel – cold rolled hoop and strip (uncoated) and hot-dip galvanized hoop and strip – 20-599mm wide, 0.30-4.00mm thick (1986 output 55,060 tonnes); cold rolled sheet/coil (uncoated) – 500-1,250mm wide, 0.5-3.00mm thick (39,630 tonnes); hot-dip galvanized sheet/coil – 600-1,250mm wide, 0.4-3.30mm thick (135,020 tonnes). Building products – steel profiles for roofing, boarding and concrete floors (18,790 tonnes); insulating panels (832,000 sq m).
Brands: Strabande, Stralisse, Monogal, Ondatherm, Cofrastra, Cofradal.

Ziegler SA, Cedex 35,F 92072 Paris la Defense. ***Tel:*** 1 47 67 90 00. ***Telex:*** 612552 ziegler. ***Telecopier (Fax):*** 1 47 67 99 78.
Management: Président – Jacques Français; managers – Michel François (general), Jean Bataille (sales), Roger Poussin (financial), Michel Silvert (production). **Established:** 1919. **Capital:** Fr 86,400,000. **No. of employees:** 705. **Ownership:** Sacilor (96%). **Annual capacity:** Finished steel: 413,000 tonnes.
Works:
■ Chemin de la Demie-Lune , F 08210 Mouzon. (Tel: 16 24 26 10 34. Telex: 840123). Works director: François Prat. ***Coil coating lines:*** Two Sendzimir hot-dip galvanizing and aluminizing lines (1,020 and 1,500mm) (annual capacity 200,000 tonnes).
■ Rue Bidet, F 62240 Desvres. (Tel: 16 21 91 66 33. Telex: 810 814 ziegler). Works director: Jean Pierre Lahouste. ***Coil coating lines:*** One 1,500mm Sendzimir hot-dip galvanizing (annual capacity 103,000 tonnes).
■ 32 Rue Gambetta, BP 62, F 59264 Onnaing. (Tel: 16 27 46 81 53. Telex: 110722 ziegler). Works director: Bruno Fontaine. ***Coil coating lines:*** colour coating (1,500mm) (annual capacity 110,000 tonnes).
Products: Carbon steel – hot-dip galvanized sheet/coil 0.25-5mm thick, 1,540mm wide; galvanized corrugated sheet/coil (145,000 tonnes) (1986 output 608,000 tonnes) aluminized sheet/coil – 0.40-2.5mm thick, 1540mm wide; colour coated sheet/coil 0.25-2mm thick, 1,540mm wide.
Brands: Galvabande, Galvallia, Extragal (galvanized sheets); Alusi type 1 and Alupur 2, Extratherm (aluminized sheets); Galvacolor and Lacmetall (pre-painted sheets); Placicolor (plastic-coated sheets); Nergal, Nergal Isol and Galvabac (corrugated sheets); Modulisol (sandwich panels).
Sales offices/agents: At head office address. Overseas agents – Daval, Paris. Service

Usinor Sacilor (continued)

Center, Saint-Ouen (Tel: 1 46 06 88 01. Telex: 650 530 ziegler).

Galvanor/ Coloracier – SA de Construction et de Galvanisation de Montataire, Route de St Leu,F 60160 Montataire. ***Postal address:*** BP 1. ***Tel:*** 16 44 55 35 12. ***Telex:*** 140330 ganor f. ***Telecopier (Fax):*** 16 44 25 10 20.
Management: J. Français (président directeur général); J. Portier (directeur général); R. Parent (directeur industriel); J.M. Rozey (directeur commercial). **Established:** 1957. **Capital:** Fr 166,360,000. **No. of employees:** Works – 793. Office – 144. **Ownership:** Usinor Group (99%). **Annual capacity:** Finished steel: 750,000 tonnes.
Works:
■ Montataire (head office). ***Coil coating lines:*** Three continuous hot-dip galvanizing (annual capacity 750,000 tonnes); two continuous colour coating (200,000 tonnes); five continuous cold roll-forming lines for sections (60,000 tonnes) and finishing facilities (slitting lines).
Products: Carbon steel – hot-dip galvanized sheet/coil and galvanized corrugated sheet/coil – 0.25-4mm thick, 600-1,650mm wide (1986 output 608,000 tonnes); colour coated sheet/coil – 0.30-2mm thick, 610-1,500mm wide (145,000 tonnes). Nervesco, Muresco, Toitesco (galvanized or galvanized and pre-painted cold roll-formed sections for the building industry) (45,000 tonnes).
Brands: Galvanor, Coloracier, Nervesco, Muresco and Toitesco.

GTS Industries 4 Place de la Pyramide,F 92070 Paris La Defense,Puteaux (Hauts de Seine). ***Tel:*** 1 49 00 67 17. ***Telex:*** 614730 usinr gts. ***Telecopier (Fax):*** 1 49 00 68 28.
Management: Guy Dollé (chairman); Bernard Pinan-Lucarre (executive vice president); Gérad Monchicourt (marketing vice president). **Established:** 1985. **Capital:** Fr 100,000,000. **No. of employees:** 700. **Ownership:** Usinor-Sacilor (95%).
Works:
■ Dunkirk, Belleville and Joeuf. ***Tube and pipe mills:*** welded three 8-56" DSAW (annual capacity 1,050,000 tonnes). Welded tube and pipe.

Long Products Division

Unimétal – Sté Francaise des Aciers Longs 47 Rue Haute-Seille,Metz (Moselle),Lorraine. ***Postal address:*** BP 4019, 57040 Metz Cedex 1. ***Tel:*** 87 37 78 00. ***Telex:*** 860017 f. ***Telecopier (Fax):*** 87 36 19 09.
Management: Executive board: Jean Jacquet (président), Guy Dethier, Jean-Didier Dujardin, Jean Dupuy, Claude Loppinet, Gérard Picard, Henri Rigo, Pierre Rivière, Jean-Marie Schaack, Paul Tordjman, François Wasservogel. **Established:** 1984. **Capital:** Fr 800,000,000. **No. of employees:** 10,073. **Ownership:** Sacilor (100%) (ultimate holding company – Usinor Sacilor). **Subsidiaries:** Steel making and rolling – Sté des Aciéries de Thionville (65%), Trancel (99.9%), Sté Métallurgique de Normandie (100%), Alpa (80%), Sté des Aciéries de Montereau (50%), Metalescaut (98.2%); steel rolling – Sté Métallurgique de Brévilly (99.9%), Laminoirs de Bretagne (99%); Sales of railway material – Matériel de Voie (74.84%). **Annual capacity:** Raw steel: 8,432,000 tonnes. Finished steel 6,495,000 tonnes.
Works:
■ Gaudroupe, Rombas, Joeuf, Hayuspe. BP 3, F 57360 Amneville. (Tel: 87 67 90 00. Telex: 860266). Works director: M. Mangeot. ***Sinter plant:*** Three Lurgi – one 196 sq m strand and two 400 sq m strand (total annual capacity 7,300,000 tonnes). ***Blast furnaces:*** one 8.6m hearth dia, 1,273 cu m, one 8.4m hearth dia, 1,317 cu m and one 8.5m hearth dia, 1,934 cu m (3,600,000 tonnes). ***Steelmaking plant:*** Two 260-tonne OLP/LBE basic oxygen converters (3,400,000 tonnes). ***Refining plant:*** One 260-tonne Asea-SKF (1,200,000 tonnes) and one 260-tonne vacuum degassing unit. ***Continuous casting machines:*** Two 6-strand Demag bloom (255 x 320 to 360 x 480) (combined capacity 2,200,000 tonnes). ***Rolling mills:*** One 1,350mm blooming; one 650mm continuous billet; one 1,176mm heavy section (300,000

• Gandrange + Hayange

France

Usinor Sacilor (continued)

tonnes); one 950mm rail (650,000 tonnes); one 1-stand bar (12-60mm dia) (coil weight 1,150kg) (200,000 tonnes); one 2-strand wire rod (5-12mm dia) (coil weight 1,150kg) (500,000 tonnes).

■ BP 29, F 54402, Longwy. (Tel: 82 24 67 90. Telex: 860 924). Works director: Mr Mangeot. ***Rolling mills:*** One universal heavy section (annual capacity 510,000 tonnes); one 3-strand wire rod (5-18mm dia) (coil weight 1,300kg) (760,000 tonnes).

■ BP 1, F 54230 Neuves Maisons. (Tel: 83 47 20 10. Telex: 850030). Works director: M. Broggi. ***Steelmaking plant:*** One 117-tonne Clecim electric arc furnace (annual capacity 530,000 tonnes). ***Refining plant:*** One 100-140-tonne Heurtey-Safe ladle furnace (530,000 tonnes). ***Continuous casting machines:*** One 4-strand Fives Cail Babcock billet (120-155 sq m) (530,000 tonnes). ***Rolling mills:*** One 2-strand wire rod (5.5-13mm) (coil weight 1,400kg) (460,000 tonnes).

■ SAT – Sté des Aciéries de Thionville, BP 200, F 57104 Thionville. Tel: 82 88 24 24. Works director: M. Jacob. ***Steelmaking plant:*** One 65-tonne CEM electric arc furnace (annual capacity 280,000 tonnes). ***Refining plant:*** One 65-tonne CEM vacuum degassing (280,000 tonnes). ***Continuous casting machines:*** One 4-strand Clecim billet (120-135 sq m) (260,000 tonnes).

■ Trancel, 31 Rue St Marcel, F 59330, Haumont. (Tel: 27 64 26 16. Telex: 160736). Works director: M. Blanc. ***Rolling mills:*** Bar and light section comprising three 470mm (annual capacity 30,000 tonnes), four continuous and five finishing stands (25,000 tonnes).

■ Sté Métallurgique de Brévilly, F 08140, Douzy. (Tel: 24 26 30 30. Telex: 840 166). Works director: M. Charnet. ***Rolling mills:*** light section comprising one 380mm, three 350mm and six 280mm stands (annual capacity 22,000 tonnes).

■ Sté Métallurgiqe de Normandie, BP 52, F 14125 Mondeville. (Tel: 31 84 81 37. Telex: 17040). Works director: P. Rivière. ***Coke ovens:*** 120 ovens (annual capacity 520,000 tonnes). ***Sinter plant:*** One 187 sq m strand Dwight Lloyd (1,200,000 tonnes). ***Blast furnaces:*** one 6.7m hearth dia, 805 cu m and one 8.0m hearth dia, 1,180 cu m (combined capacity 1,090,000 tonnes). ***Steelmaking plant:*** Two 100-tonne LDAC/LBE basic oxygen converters (1,200,000 tonnes). ***Refining plant:*** One 100-tonne Heurtey ladle furnace (670,000 tonnes). ***Continuous casting machines:*** Two 4-strand Clecim/Concast billet (105-120 sq m) (670,000 tonnes). ***Rolling mills:*** One 3-strand wire rod (5-13mm dia) (coil weight 1,100kg) (650,000 tonnes).

■ SAM – Sté des Aciéries de Montereau, BP 5, F 77130 Montereau. (Tel:1 62. Telex: 691951). Works director: J de Miscault. ***Steelmaking plant:*** One 80-tonne CEM electric arc furnace (annual capacity 550,000 tonnes). ***Refining plant:*** One 85-tonne ladle furnace (500,000 tonnes). ***Continuous casting machines:*** One 6-strand Concast Rokop SAM billet (120 x 120) (550,000 tonnes). ***Rolling mills:*** wire rod – 1-strand, fifteen horizontal stands and ten-stand No-Twist finishing block (5.5-12mm dia) (coil weight 1,500kg) (463,000 tonnes). ***Other plant:*** Sté d'Etudes et de Rélisations Hydrométallurgiques (18.7% owned) – zinc and lead recovery.

■ Alpa – Aciéries et Laminoirs de Paris, Zone Industrielle Limay-Porcheville, BP 39, F 78440 Garcenville. (Tel: 34 77 59 37. Telex: 696903). Works director: C. Cornier. ***Steelmaking plant:*** One 60-tonne CEM/BBC electric arc (annual capacity 300,000 tonnes). ***Continuous casting machines:*** One 4-strand Concast billet (120 sq m) (300,000 tonnes). ***Rolling mills:*** One 2-strand bar (8-40mm dia) (275,000 tonnes).

■ SME – Sté Métallurgique de l'Escaut, F 59125, Trith-Saint-Léper. (Tel: 27 45 08 50. Telex: 110711). Works director: A. Vigneron. ***Steelmaking plant:*** One 74-tonne DC (Clecim) electric arc furnace (annual capacity 250,000 tonnes). ***Continuous casting machines:*** One 4-strand Concast billet (120 x 120mm) (250,000 tonnes). ***Rolling mills:*** One bar – 4-stand 500mm and 16 continuous stands (230,000 tonnes).

■ Laminoirs de Bretagne, BP 46, F 22440 Ploufragan. (Tel: 99 94 21 65. Telex: 950762). Works director: M. Geffriaud. ***Rolling mills:*** bar comprising one 500mm, two 300mm and 10 continuous stands (60,000 tonnes).

Products: Carbon steel – ingots; blooms and billets – (combined 1986 output 280,000 tonnes); wire rod – (1,970,000 tonnes); reinforcing bar – (430,000 tonnes); beams – (540,000 tonnes); merchant bars – (550,000 tonnes); underground roadway

Usinor Sacilor (continued)

supports – (30,000 tonnes); heavy rails and light rails (combined output 230,000 tonnes); sheet piling – (110,000 tonnes).
Sales offices/agents: Daval – Immeuble Elysées La Défense, F 92072 Paris La Défense, Cedex 35 (Tel: 1 47 67 83 83. Telex: 614850); Le Matériel de Voie, Immeuble Elysées La Défense, F 92072 Paris La Défense, Cedex 35 (Tel: 1 47 67 99 78. Telex: 612550).

Ascométal, Immeuble Elysées La Défense, 29 Le Parvis,F 92072 Paris La Défense, Cedex 35. ***Tel:*** 47 67 95 00. ***Telex:*** 614623 ***Telecopier (Fax):*** 47 67 95 99.
Management: Executive board – Jean Jacquet (président), Guy Dethier, Jean-Didier Dujardin, Jean Dupuy, Claude Loppinet, Gérard Picard, Henri Rigo, Pierre Rivière, Jean-Marie Schaak, Paul Tordjman, François Wasservogel. **Established:** 1984. **Capital:** Fr 300,000,000. **No. of employees:** 51,979 **Ownership:** Sacilor (100%), (ultimate holding company – Usinor Sacilor). **Subsidiaries:** Estamfor (100%) (forgings); Sté Métallurgique de Revigny (100%), Calibracier (100%), Sté Industrielle de Pré-à-Varois (100%) (cold finishing). **Annual capacity:** Raw steel: 765,000 tonnes. Finished steel: 860,000 plus 235,000 tonnes of tube rounds.
Works:
■ Les Dunes, BP 41, F 59441, Dunkirk Cedex 2. (Tel: 28 29 60 00. Telex: 820071). Works director: J.M. Chappe. ***Steelmaking plant:*** One 82-tonne Clesid electric arc furnace (annual capacity 330,000 tonnes). ***Refining plant:*** One 80-tonne ladle furnace; one 80-tonne vacuum degassing unit; one 80-tonne VAD; one RH (combined capacity 350,000 tonnes). ***Continuous casting machines:*** One 4-strand DMS centrifugal round (770-325mm dia) (300,000 tonnes). ***Rolling mills:*** One 37" blooming, one 37" billet and one 37" heavy section (combined capacity 240,000 tonnes). ***Other plant:*** Forging shop with one 6,000-tonne press, two 1,500-tonne presses and one 800-tonne press. Machining shop.
■ Fos, BP 30,F 13771 Fos S/Mer Cedex. (Tel: 42 47 93 00. Telex: 420465). Works director: J.J. de Cadenet. ***Steelmaking plant:*** One 115-tonne Tagliaferri electric arc furnace (annual capacity 300,000 tonnes). ***Rolling mills:*** blooming and billet (78-325mm round and 60-350mm sq) (combined capacity 130,000 tonnes); 1-strand wire rod (5.5-32mm dia) (55,000 tonnes).
■ Hagondange, BP 38 F 57301 Hagondange, Cedex. (Tel: 87 70 51 21. Telex: 860216). Works director: A. Gevanudan. ***Steelmaking plant:*** One 100-tonne FCB-Demag electric arc furnace (475,000 tonnes). ***Refining plant:*** One 100-tonne Heurtey-Safe ladle furnace (415,000 tonnes). ***Continuous casting machines:*** One 4-strand vertical bloom (240 x 240) (415,000 tonnes). ***Rolling mills:*** bar and wire rod – one 3-high roughing and eighteen continuous stands (325,000 tonnes). ***Other plant:*** Cold forging (10,000 tonnes); hot forging (40,000 tonnes) for automobile parts.
■ Le Marais, BP 515, F 42007, St Etienne Cedex. (Tel: 77 74 91 66. Telex: 330631). Works director: Y. Combescure. ***Steelmaking plant:*** One 80-tonne Clesid electric arc furnace (annual capacity 250,000 tonnes). ***Refining plant:*** One 80-tonne Heurtey-Safe VAD. ***Rolling mills:*** blooming with one 850mm and one 550mm finishing stand (110,000 tonnes).
■ Valenciennes. BP 12, F 59125 Trith Saint Leger. (Tel: 27 30 92 65. Telex: 820544). Works director: C. Brochard. Machining Shop. ***Other plant:*** Machining shop.
Products: Carbon steel – ingots, blooms and billets – (combined 1986 output 86,000 tonnes); semis for seamless tubemaking – (82,000 tonnes); wire rod – (8,000 tonnes); round bars – (230,000 tonnes); flats and hexagons – (3,000 tonnes). Alloy steel (excl stainless) – Ingots,blooms and billets (86,000 tonnes); semis for seamless tubemaking (122,000 tonnes); round bars (251,000 tonnes); flats, etc (1,000 tonnes). Asfor (structural steels); Ugifos (bearing steels).
Sales offices/agents: Ascométal GmbH, Erkrath, West Germany (Tel: 49 210 43 00 40. Telex: 8589362); Ascométal SpA, Milan, Italy (Tel: 39 260 72 644/5. Telex: 322221); Asfor Steel Corp, New Jersey, USA (Tel: 1 201 893 0001. Telex: 138685).

France

Usinor Sacilor (continued)

Transformation & Other Activities Division

Les Usines Laprade SA F 64260 Arudy. ***Tel:*** 59 05 61 54. ***Telex:*** 570953.
Management: Xavier Martin Laprade (président directeur général), Bernard Martin Laprade (sales management). **Established:** 1931. **Capital:** Fr 9,450,000. **No. of employees:** 242. **Ownership:** Usinor (52.8%). **Annual capacity:** Finished steel: 15,000 tonnes (strip).
Works:
■ Arudy. Works director: Jean Marc Capdeboscq.
Products: Carbon steel – (and alloy steel) cold rolled hoop and strip (uncoated) – drawn and stamped parts.

Etilam-Gravigny (EG), 52 Avenue du Général Sarrail,F 52101 Saint Dizier. ***Tel:*** 25 56 11 45. ***Telex:*** 840628.
Management: Jacques Wallon (président directeur général), Georges Ferrer (responsable service commercial), Robert Spiridonoff (responsable fabrication). **Established:** 1924. **Capital:** Fr 14,000,000. **No. of employees:** Works – 200. Office – 40. **Ownership:** Usinor (91%). **Annual capacity:** Rolled steel: 28,000 tonnes.
Works:
■ Saint Dizier. Works director: R. Spiridonoff. ***Rolling mills:*** cold strip – one 660mm Sendzimir (annual capacity 15,000 tonnes), one 400mm Mackintosh (8,000 tonnes) and one 250mm Chavanne (2,000 tonnes); one 660mm temper/skin pass (25,000 tonnes). ***Coil coating lines:*** Three hot-dip tinning (6,000 tonnes); one hot-dip galvanizing (2,500 tonnes); four colour coating – two reprinting (6,000 tonnes) and two printing (2,000 tonnes).
Products: Carbon steel – cold rolled hoop and srip – (1986 output 13,000 tonnes); coated cold rolled strip – hot-dip galvanized hoop and strip – (2,500 tonnes), hot dip tinned and leaded strip (5,000 tonnes); Etilac lacquered strip (5,500 tonnes); lacquered and printed strip with a peelable varnish (1,500 tonnes).
Sales offices/agents: UK agent – Production Steel Supply, Chelford, Macclesfield, Cheshire. SK11 9BD.

Profilafroid, 29 Avenue de Laumière,F 75926 Paris Cedex 19. ***Postal address:*** BP 26. ***Tel:*** 42 02 43 03. ***Telex:*** 216273 filapra (France), 216274 filaexp (export). ***Telecopier (Fax):*** 42 02 82 80.
Ownership: Usinor.
Works:
■ Usine de l'Epinette, F 60930 Bailleul-sur-Therain.
Products: Carbon steel – cold roll-formed sections (from strip) – including stainless, highway guard rails; light sheet piling; slotted angles.

Profilés et Tubes de l'Est Usine di Messempre,Pure,F 08110 Carignan. ***Tel:*** 24 22 01 11. ***Telex:*** 840126 pteme. ***Cables:*** Profilest Carignan. ***Telecopier (Fax):*** 24 22 21 20.
Management: M. Benoist (chairman of the board), J. de Vreese (general manager), B. Libert (sales manager). **Established:** 1973. **Capital:** Fr 37,800,000. **No. of employees:** Works – 135. Office – 15. **Ownership:** GPRI (100%); (Ultimate holding company – Usinor Sacilor). **Annual capacity:** Finished steel: 55,000 tonnes.
Works:
■ Carignan. Works director: S. Gualtierotti.
Products: Carbon steel – cold roll-formed sections (from strip) – (output 48,000 tonnes); sheet piling.
Sales offices/agents: 29 Avenue de Laumière, BP 7, F 75019 (Tel: 42 02 43 43. Telex: 213265 f).

Hauts Fourneaux Réunis de Saulnes & Uckange, 28 Rue de Lisbonne,F75008

Usinor Sacilor (continued)

Paris. ***Postal address:*** BP 151-08, F 75363 Paris Cedex 08. ***Tel:*** 1 45 63 06 30. ***Telex:*** 642614 sufonte. ***Cables:*** Sonuckange Paris 37.
Management: Xavier Lauras (président directeur général), Claude-Marie Granier (secrétaire général), Claude Galloy (directeur des usines), Jean Maitre (directeur commercial). **No. of employees:** Works – 478. Office – 21. **Ownership:** Usinor. **Associates:** Centrale Sidérurgique de Richemont (11.4%), Sté pour l'Exploitation du Port Commun d'Illange – Sopcillange (99.98%), Chambre de la Métallurgie de Longwy (11.67%), Transest (10%) (transport). **Annual capacity:** Pig iron: 1,200,000 tonnes.
Works:
■ Route de Thionville, F 57270 Uckange. (Tel: 82 58 92 25. Telex: 860096). Works director: Claude Galloy. ***Sinter plant:*** Two 75 sq m strand Dwight Lloyd (annual capacity 1,700,000 tonnes). ***Blast furnaces:*** one 7m hearth dia, 1,002 cu m, one 6.5m hearth dia, 669 cu m and one 6.5m hearth dia, 1,015 cu m (1,200,000 tonnes).
Products: Foundry pig iron 10kg ingots, 25 x 15 x 10 cm (output 410,400 tonnes).

Laminoirs à Froid de Thionville, Route de Manom,F 57101 Thionville. ***Postal address:*** BP 50. ***Tel:*** 82 53 87 17. ***Telex:*** 860029. ***Cables:*** Lamfroid Thionville.
Established: 1925. **Capital:** Fr 16,000,000. **No. of employees:** Works – 190. Office – 60. **Ownership:** Etilam-Gravigny (99.9%).
Works:
■ Thionville. Cold rolling facilities, electrolytic, hot-dip and plastic coating lines.
Products: Carbon steel – cold rolled hoop and strip; electrolytic coated cold rolled steel strip (galvanized hoop and strip; copper, nickel, chromium and brass coated); hot dipped tinned strip; cold rolled steel strapping strip (blue and waxed); laquer, plastic and film coated cold rolled steel sheets.

Valexy, 29 bis, Avenue Laumière,F 75019 Paris. ***Tel:*** 1 42 03 77 78. ***Telex:*** 211518 Valxy. ***Telecopier (Fax):*** 1 42 03 67 38.
Established: 1979. **Ownership:** Usinor (66%), Vallourec (34%).
Works:
■ BP 1, F 54720 Lexy. (Tel: 82233391. Telex: 650478). ***Tube and pipe mills:*** welded.
■ BP 14, F 02230 Fresnoy-le-Grand. (Tel: 23660474. Telex: 150798). ***Tube and pipe mills:*** welded.
■ Sierck-les-Bains. (Tel: 87 50 31 38). ***Tube and pipe mills:*** welded for rectangular and square tubes.
■ Rue des Usines, F 59330 Hautmont. (Tel: 27622345. Telex: 160942). ***Tube and pipe mills:*** welded.
■ 141 Rue Bataille, F 69356 Lyons. (Tel: 78745431. Telex: 300417). ***Tube and pipe mills:*** welded.
Products: Carbon steel – small dia and thin-wall welded tubes, longitudingal-weld tube and pipe, spiral-weld tube and pipe, oil country tubular goods, galvanized tube and pipe, hollow sections including square and rectangular.

Stainless & Special Products Division

Imphy SA, 19 Le Parvis, Elysees,F 92072 Paris La Défense. ***Tel:*** 1 47 67 98 00. ***Telex:*** 614846. ***Telecopier (Fax):*** 1 47 67 98 85.
Management: President and chief executive officer – Thierry Chereau; vice president – Pierre Legendre; managers – Gerarde de Liege (sales), Jacques Renaud (production), Dominique Drieux (finance), Marcel Hua (purchasing); non-executive directors – Jean-Paul Courtier, Jean Morlet. **Established:** 1979. **Capital:** Fr 189,414,960. **No. of employees:** Works – 2,000. Office – 100. **Ownership:** Sacilor. **Subsidiaries:** Industrial subsidiaries – Sprint Metal (100%) (stainless and alloy wires); Mecagis (100%) (magnetic parts); Mecalev (100%) (forks for lift trucks). Marketing subsidiaries

France

Usinor Sacilor (continued)

– Metalimphy Alloy Corp (USA); Imphy Deutschland GmbH; Imphy Italiana; Imphy UK Ltd; Imphy Holland BV; Imphy SA Switzerland; Imphy Far East Co Ltd (Hong Kong); Imphy SA Korea (Seoul). Warehouse – Imphy Service. **Associates:** Hood (USA) (28%) (thermostatic bi-metals). **Annual capacity:** Finished steels: 60,000 tonnes.
Works:
■ Imphy Works, F 58160 Imphy. (Tel: 86 57 80 00. Telex: 800420). Works director: Jacques Renaud. ***Steelmaking plant:*** One 30-tonne UHP electric arc furnace (annual capacity 65,000 tonnes). ***Refining plant:*** One 10-tonne ESR (2,500 tonnes); one 33-tonne ladle furnace; 3-tonne, 10-tonne and 11-tonne VAR (5,000 tonnes); one 6-tonne VIM (4,500 tonnes); one 33-tonne VOD. ***Continuous casting machines:*** 2-strand SCEC rotary round (120-180mm dia) (40,000 tonnes). ***Rolling mills:*** 2-high 700mm reversing blooming (25,000 tonnes); continuous wire rod (5.5-26mm dia) (coil weight 450kg) (45,000 tonnes); six precision multi-roll type cold strip (80-600mm wide) (9,000 tonnes). ***Other plant:*** Forge – 2 hydraulic presses 900-tonne and 700-tonne. Hammers. Wire drawing works. Stainless and alloy wires for cold heading, spring, welding, profile wire. Metal powder plant with inert gas atomisation and hot isostatic presses.
Products: Stainless steel – wire rod – (1986 output 30,000 tonnes); wire (Sprint Metal subsidiary) – (22,000 tonnes). Alloy steel (excl stainless) – wire rod; cold-rolled hoop and strip (uncoated); wire. Bi-metals – controlled expansion alloys (strip and wires). Soft magnetic alloys (strip and parts). High temperature superalloys (bars and parts). Pre-alloyed powder (parts). Nickel and cobalt base.
Sales offices/agents: USA – Metalimphy Alloys Corp and Sprint Metal Corp, PA (Tel: 215 8221348. Telex: 5106614838); West Germany – Imphy Deutschland GmbH, Dusseldorf (Tel: 0211 594055. Telex: 8586328); Italy – Imphy Italiana, Turin (Tel: 011 510701. Telex: 224131); UK Imphy UK Ltd, – Jacksons Industrial Park, Buckingham (Tel: 6285 29241. Telex: 847876); Netherlands – Imphy Holland BV, Tilburg (Tel: 013 636065. Telex: 52317); Switzerland – Imphy SA, Nyon/VD (Tel: 022 619556. Telex: 27259); Hong Kong – Imphy Far East Co Ltd, Hong Kong (Tel: 5833 5767. Telex: 74279); South Korea – Imphy SA Korea, Seoul (Tel 2742 2611. Telex: 25251).

Ugine Aciers de Châtillon et Gueugnon, Immeuble Ile de France, Cedex 33,F 92070 Paris La Défense. ***Tel:*** 49 00 60 10.
Management: Président directeur général – Ph. Choppin de Janvry; directeur généraux adjoints – M. Bouvard, M. Coudert, D. Georges-Picot, SM. Monthiers; directeurs – D. Franchot (politique sociale générale), J. Lefèvre (recherche et développement), JF. Magnan (investissements, stratégie, marketing), G. Martel (moyens logistigques, approvisionnements), JP. Thévenin (exploitation des sites industrielles). **Capital:** Fr 312,078,825. **No. of employees:** 6,600. **Ownership:** Usinor Sacilor (92%). **Subsidiaries:** Sté Métallurgique de Saint Chély (100%); Uginox SAE (Spain); Uginox Turin and Uginox Milan (Italy); Inoxium Luxembourg; Ugine Edelstahl GmbH (Erkrath and Renningen, West Germany); Inoxium UK; Ugine Stahl AG (Switzerland).
Works:
■ L'Ardoise. ***Steelmaking plant:*** basic oxygen converter; electric arc furnaces – two 40-tonne and one 60-tonne (one 18 MVA and two 32 MVA). ***Refining plant:*** One 110-tonne AOD. ***Continuous casting machines:*** One 150-250 x 1,620mm slab.
■ Firminy. ***Rolling mills:*** One 670mm wide multi-roll cold strip; one 670mm wide temper/skin pass. ***Other plant:*** One 1,350mm wide slitter; one 670mm wide polishing line; three slitters; one cut-to-length line; one conditioning line; one 670mm wide bright annealing line. Double-bevelled martensitic steels plant comprising three mills and one heat treatment line.
■ Gueugnon. ***Rolling mills:*** Five Sendzimir reversing cold reduction (0.25-5mm thick) and one 0.35-3mm thick and 600-1,280mm wide. Three 2-high temper/skin pass. ***Other plant:*** One coil preparation line; pickling and annealing facilities including three bright annealing lines; finishing plant including polishing, slitting and cutting to length; non-oriented electrical sheet plant comprising decarburising line and two slitting lines.
■ Isbergues. ***Steelmaking plant:*** Two 100-tonne 46 MVA electric arc furnaces.

Usinor Sacilor (continued)

Refining plant: One 100-tonne AOD; one 70-tonne vacuum refining unit. ***Continuous casting machines:*** One 1-550mm wide and 150-200mm thick slab. ***Rolling mills:*** cold reduction – two Sendzimir reversing ZR 22 B 52 (0.3-4mm thick); one 2-high temper/skin pass; electrical sheet plant comprising five coil preparation lines and two Sendzimir reversing cold reduction mills – one ZR 22 B 42 and one ZR 22-50 (0.23-0.50mm dia). ***Coil coating lines:*** One terne/lead coating (0.4-2mm thick and 600-1,250mm wide). ***Other plant:*** Annealing and pickling plant; finishing plant including polishing machines, slitting and cutting to length; heat treatment decarburising and insulation coating lines.
■ Pont de Roide.
Stainless steel sheet plant. ***Rolling mills:*** cold reduction – one 1,350mm wide multi-roll; two 670mm wide multi-roll; temper/skin pass – one 1,350mm wide and one 670mm wide. ***Other plant:*** Two bright annealing lines 670mm wide and one pickling and annealing line 670mm wide.
Carbon steel strip plant. Rolling mills: Cold strip mills – one 4-high 510mm wide; one 4-high 400mm wide and one 4-high 250mm wide. One 650mm wide temper/skin pass.
■ Sté Métallurgique de Saint Chely.
Electrical sheet plant: ***Rolling mills:*** One 4-high cold reduction. ***Other plant:*** One pickling line, coating line and four slitting lines.
Ferro-Alloy plant.
Products: Stainless steel – slabs, sheet/coil, strip. Grain-oriented and non-oriented electrical sheet; special carbon and alloy steel strip; terne/lead coated sheet; ferro-alloys; precision metal recovery plant.

Ugine Savoie, Immeuble Ile de France, Cedex 33,F 92070 Paris La Défense, *Tel:* 49 00 60 10.
Management: Ph. Choppin de Janvry (chairman), M. Coudert (general director), M. Louichon (works director). **Capital:** Fr 180,000,000. **No. of employees:** 1,900.
Ownership: Usinor Sacilor (100%).
Works:
■ Ugine Savoie, F 73400. (Tel: 79 89 30 30). ***Steelmaking plant:*** Two 40-tonne electric arc furnaces (annual capacity 160,000 tonnes). ***Refining plant:*** Two 40-tonne AOD (160,000 tonnes). ***Continuous casting machines:*** bloom. ***Rolling mills:*** One 600mm blooming with finishing stand for blooms 70-180mm; bar and wire rod for bars 23-80mm and wire rods 5.5-35mm. ***Other plant:*** Cold finishing shop (pickling, grinding, lathe).
Products: Stainless steel – blooms, billets, bars and wire rod. Alloy steel (excl stainless) – blooms, billets, bars and wire rod.

Metallurgy Division

Creusot Loire Industrie, Division Creusot Marrel, Immeuble Ile de France, 4 Place de la Pyramide,F 92070 Paris la Defense. ***Postal address:*** Cedex 33. ***Tel:*** 33 1 49 00 60 50. ***Telex:*** 620132. ***Telecopier (Fax):*** 33 1 49 00 57 30.
Management: F. Di Pace (chairman), A. Demus (president, Division Creusot-Marrel). **Established:** 1985. **Capital:** Fr 150,000,000. **No. of employees:** 3,392.
Ownership: Usinor.
Works:
■ Chateauneuf, BP 46, F 42800 Rive de Gier. (Tel: 33 77 75 38 00. Telex: 310839). Works director: A. Van Honacker. ***Steelmaking plant:*** One 90-tonne electric arc furnace (annual capacity 100,000 tonnes); ***Refining plant:*** ladle furnace and 90-tonne Dortmund Hörder vacuum degassing unit. ***Rolling mills:*** One 4-high heavy plate (4,700mm wide, 40-950mm thick) (coil weight 130 tonnes) (80,000 tonnes). ***Other plant:*** For ultra heavy plates – 12,000-tonne press; cold leveller; grinding machines; heat treatment furnaces.
■ Le Creusot, 56 Rue Clémenceau, BP 56, F 71202 Le Creusot Cedex. (Tel: 33 85 80 55 55. Telex: 801500). Works director: J. Caillard. ***Steelmaking plant:*** Two 100-tonne electric arc furnaces (annual capacity 145,000 tonnes). ***Refining plant:*** ladle furnace and 100-tonne Finkl System vacuum degassing unit (145,000 tonnes);

France

Usinor Sacilor (continued)

100-tonne VOD (for stainless); tap degassing for forgings. ***Rolling mills:*** 2-high 4,000mm heavy plate and 4-high 3,600mm medium plate (combined annual capacity 105,000 tonnes); Sendzimir cold mill (2-20mm thick, 2,500mm wide). 2-20mm thick, 2,500mm wide Sendzimir. ***Other plant:*** Heavy forge 7,500-tonne and 9,000-tonne presses (ingots up to 240-tonne); heavy foundry (castings up to 300-tonnes). Le Creusot works – carbon steel ingots, alloy steel ingots, stainless steel ingots, carbon steel slabs, alloy steel slabs, stainless steel slabs, heavy plates, medium plates and stainless steel plates: 2-200mm thick; special cold rolled plates 1.2-20mm thick; clad plates. Steel grades: carbon and alloy engineering steels, heat-treated steels, structural and fabrication steels (high-yield weldable steels, low temperature and cryogenic steels), abrasion-resistant steels, steels for hot working, tool steels, stainless and heat-resisting steels, titanium and alloys. Other products: Forgings (especially rotors for large turbines and generators, vessel components for nuclear power plants and petrochemical industry, etc). Steel castings (unit weight up to 200 tonnes and more). Chafeauneuf works – heavy plates 40-950mm up to 4,500mm wide and unit weight up to 100 tonnes in carbon and alloy engineering steels. Output in 1986 – stainless and clad plates 25,000 tonnes; carbon and alloy steel plates 75,000 tonnes.
Sales offices/agents: Western Germany – Creusot-Loire Industrie, 4000 Düsseldorf 1 (Tel: 49 211 37 03 09. Telex: 8588220); USA – Creusot-Marrel Inc, PA 19087 (Tel: 1 215 254 99 00. Telex: 420388 ur); France and Export – Creusot-Loire Industrie, Paris (Tel: 33 1 49 00 60 50. Telex: 620132); Italy – Usinor Italia, Milan (Tel: 39 2 78 39 57.)

Products: Steelmaking pig iron. Foundry pig iron. Carbon steel – ingots, slabs, blooms, billets, semis for seamless tubemaking, hot rolled coil (for re-rolling), wire rod, reinforcing bars, round bars, square bars, flats, hexagons, light angles, light tees, light channels, special light sections, medium angles, medium tees, medium joists, medium channels, wide-flange beams (medium), universals (medium), welded sections (medium), heavy angles, heavy tees, heavy joists, heavy channels, wide-flange beams (heavy), universals (heavy), welded sections (heavy), heavy rails, light rails, cold roll-formed sections (from strip), sheet piling, hot rolled hoop and strip (uncoated), skelp (tube strip), cold rolled hoop and strip (uncoated), hot-dip galvanized hoop and strip, coated hoop and strip, medium plates, heavy plates, hot rolled sheet/coil (uncoated), cold rolled sheet/coil (uncoated), hot-dip galvanized sheet/coil, electro-galvanized sheet/coil, galvanized corrugated sheet/coil, aluminized sheet/coil, enamelling grade sheet/coil, terne-plate (lead coated) sheet/coil, colour coated sheet/coil, coated sheet/coil, electrical grain oriented silicon sheet/coil, electrical non-oriented silicon sheet/coil, electrical non-oriented non-silicon sheet/coil, blackplate, electrolytic tinplate (single-reduced), electrolytic tinplate (double-reduced), tin-free steel (ECCS), longitudinal weld tubes and pipes, spiral-weld tubes and pipes, large-diameter tubes and pipes, oil country tubular goods, galvanized tubes and pipes, hollow sections. Stainless steel – ingots, slabs, blooms, billets, hot rolled band (for re-rolling), wire rod, round bars, square bars, flats, hexagons, cold rolled hoop and strip (uncoated), medium plates, heavy plates, clad plates, hot rolled sheet/coil (uncoated), cold rolled sheet/coil (uncoated), wire. Alloy steel (excl stainless) – ingots, slabs, blooms, billets, semis for seamless tubemaking, hot rolled coil (for re-rolling), wire rod, round bars, square bars, flats, hexagons, hot rolled hoop and strip (uncoated), skelp (tube strip), cold rolled hoop and strip (uncoated), medium plates, heavy plates, hot rolled sheet/coil (uncoated), wire.
Sales offices/agents: Export: Daval, Immeuble Elysées La Défense, Cedex 35, F92072 Paris La Défense (Tel: 1 47 67 83 83. Telex: 614850 Daval). Subsidiaries: USA, Canada, Denmark, Netherlands, Norway, Belgium, Greece, United Kingdom, Sweden, West Germany, Switzerland, Italy, Spain. Daval is represented in more than 100 countries. Home Sales: Valor, Immeuble Elysées, La Défense, Cedex 35, F 92072 Paris La Défense (Tel: 1 47 67 85 00).

Valexy—see Usinor Sacilor, Transformation & Other Activities

Vallourec Industries

Head office: 7 Place du Chancelier Adenauer, F 75116 Paris. **Postal address:** BP 180, F 75764 Paris Cedex 16. **Tel:** 1 45 02 30 00. **Telex:** 611906 vlre paris. **Telecopier (Fax):** 1 45 02 31 91.
Management: Officers – Arnaud Leenhardt (chairman and chief operating), Jean-Claude Cabre (president and chief operating); vice presidents – René Patrier (executive, marketing), Jean-Claude Verdière (senior, finance controller), Alain A. Honnart (executive, operations manufacturing, research and development), Maurice Rentier (senior, investments, research and quality), Jean-Claude Prouhèze (research and development), François Coyon (manufacturing operations); director – François Fabre (personnel). **Established:** 1899 (Vallourec). **Capital:** Fr 508,964,200 (Vallourec). **No. of employees:** Office – 400. Works – 8,000. **Ownership:** Vallourec (100%). **Subsidiaries:** Interfit (100%) (fittings, elbows, tees, caps); Specitubes (100%) (small-diameter welded, welded redrawn and seamless redrawn tubing in stainless steel, special alloys, titanium and titanium alloys, high-nickel alloys); Starval (100%) (distribution of tubes, fittings and accessories); Valti (80%) joint venture with Tube Investments, UK (seamless pipe for bearings); Escofier Technologie SA (65%) (engineering and manufacture of equipment for cold rolling facilities); Hughes DPA (51%) joint venture with Baker-Hughes, USA (well completion equipment, rock bits and tool joints, manufacturing and welding of tool joints on to drill pipe). **Associates:** Sopretac (100%) (other metallurgical activities) (main subsidiaries of Sopretac: Affival (100%) (cold rolling of blackplate and tinplate, Affival cored wire for ladle injection); Cerec (100%) (pressed products for the mechanical, automobile and public works industries); Tubauto (100%) (seats and equipment for cars, garage doors and parking barriers)). Valinco (51%) (contracting activities) (main subsidiaries of Valinco: GTM-Entrepose (43%) (all areas of building, civil engineering and contracting – offshore, power plants, engineering and contracting)).
Works:
■ 54 Rue Anatole France, F 59620 Aulnoye-Aymeries (Tel: 27 66 66 00. Telex: 160793 vlrau). Works manager: Philippe Malicet. ***Tube and pipe mills:*** seamless – including one plug mill for tubes 114 to 419mm dia and finishing facilities including coating, threading and heat treatment for OCTG.
■ 75 bis Rue Jean-Jaurès, F 59410 Anzin (Tel: 27 30 92 45. Telex: 110834 vlraz). Works manager: Jacques Tellier. ***Steelmaking plant:*** electric arc furnace.
■ Rue Haverssière, F 76250 Deville-les-Rouen (Tel: 35 74 78 78. Telex: 770942 vlrdev). Works manager; Christian Demuynck. ***Tube and pipe mills:*** seamless – including one plug mill for tubes 60 to 273mm dia and finishing facilities including coating and heat treatment.
■ Place de la Gare, F 60290 Laigneville Rantigny (Tel: 44 71 04 44. Telex: 140126 vlrlai). Works manager: Jean-Pierre Ducoup. Cold drawing facilities for tubes from 10 to 100mm dia.
■ 21 Rue Marthe Paris, F 21150 Venarey-lès-Laumes (Tel: 80 96 00 13. Telex: 350465 vlrlau). Works manager: Jean-Claude Journée. ***Tube and pipe mills:*** welded – Four argon weld mills for stainless steel tubes 15 to 35mm od and up to 50m long max. Heat treatment and bending facilities. Four argon weld mills for titanium tubes 15 to 35mm od and up to 24m long.
■ Saint-Saulve Steel Plant, Zone Industrielle, F 59880 Saint-Saulve (Tel: 27 41 72 88. Telex: 110784 acierie). Works manager: Pierre Boussard. ***Steelmaking plant:*** One 80-tonne electric arc furnace. ***Refining plant:*** One vacuum degassing unit. ***Continuous casting machines:*** 4-strand rotary SCEC for round (120 to 220mm dia).
■ Saint-Saulve Tube Plant, Zone Industrielle, F 59880 Saint-Saulve (Tel: 27 46 17 50. Telex: 110887 vlrsolv). Works manager: Bertrand Cantegrit. ***Tube and pipe mills:*** seamless – including one Neuval continuous mandrel mill for tubes 27 to 146mm dia. Cold drawing facilities for tubes 25 to 273mm.
■ ZI Vitry Marolles, F 51303 Vitry-le-François (Tel: 26 74 02 20). Works manager: Jean Brecheteau. Cold drawing facilities for tubes 4 to 250mm dia.
Products: Carbon steel – seamless tubes and pipes – general purpose; boiler tube (smooth-bore or rifled); structural; mechanical; steam generator tubes, straight or bent; tubes for production of components for armament, naval and aerospace industries; longitudinal weld tubes and pipes; oil country tubular goods including seamless tubes for oil-well drilling equipped with Hughes tool joints, lined or unlined; seamless

Vallourec Industries (continued)

threaded tubes for casing for oil and gas exploration, API, Buttress and VAM joints; cold drawn tubes and pipes welded and seamless precision for mechanical applications, motor industry, hydraulics, gas cylinders; cold finished seamless tubes for heat exchangers, straight feedwater heaters, or bent, smooth or finned tubes; precision tubes and pipes; hollow sections. Stainless steel – seamless tubes and pipes for oxidation resistance at high temperature; nuclear fuel cladding tubes; round or hexagonal for nuclear applications such as structural parts and control rods; longitudinal weld tubes and pipes – condenser tubes; nuclear fuel cladding tubes; oil country tubular goods; cold drawn tubes and pipes; precision tubes and pipes; hollow sections – extruded or extruded and cold drawn, including solids.

Valti – SA pour la Fabrication de Tubes à Roulements

Head office: 7 Place du Chancelier Adenauer, F 75116 Paris. **Tel:** 4502 19 00. **Telex:** 611906. **Cables:** Tubroulement Paris.
Products: Alloy steel (excl stainless) – Bearing tube.

Sté des Tubes de Vincey

Head office: F 88450 Vincey. **Tel:** 48 389213. **Telex:** 230643 tubvincey brget.
Products: Carbon steel – longitudinal weld tubes and pipes.

Ziegler SA—see Usinor Sacilor, Flat Products.

East Germany

Export sales of East German iron and steel products are handled by Metallurgiehandel, Brunnenstrasse 188-190, 1054 Berlin (Tel: 0 28 920. Telex: 115123. Cables: Metallurgiehandel).

VEB Bandstahlkombinat Hermann Matern

Works:
■ This group of steelworks includes: Eisenhüttenkombinat Ost, Eisenhüttenstadt; VEB Walzwerk Finow, Eberswalde-Finow 2; VEB Walzwerk Burg, Burg; VEB Kaltwalzwerk Salzungen, Bad Salzungen; VEB Blechwalzwerk Olbernhau, Olbernhau; VEB Kaltwalzwerk Oranienburg, Oranienburg.

VEB Rohrwerke Bitterfeld

Works:
■ Bitterfeld. ***Tube and pipe mills:*** welded – Longitudinal and spiral.
Products: Carbon steel – longitudinal weld tubes and pipes, spiral-weld tubes and pipes 200 to 800mm dia.

VEB Stahl- und Walzwerk Brandenburg

Head office: 1800 Brandenburg. **Tel:** 550. **Telex:** 0157622.
Works:
■ Brandenburg. ***Steelmaking plant:*** Four 20-tonne QEK basic oxygen converters (annual capacity 500,000 tonnes); two 120-tonne Asea UHP electric arc furnaces (550,000 tonnes); open hearth furnace. **Refining plant:** One 180-tonne Scandinavian

VEB Stahl- und Walzwerk Brandenburg (continued)

Lancers unit. ***Continuous casting machines:*** Two 8-strand Danieli for billet (100 to 160mm sq) (500,000 tonnes). ***Rolling mills:*** One 6,500mm slabbing (3,000,000 tonnes); one blooming; bar; wire rod (including one with max finishing speed of 70 metres/seconds for rod 5 to 9mm dia; heavy plate. ***Other plant:*** Wire drawing plant (770,000 tonnes).
Products: Carbon steel – slabs, blooms, billets, wire rod, square bars, hexagons, heavy plates 6 to 16mm thick, bright wire. Alloy steel (excl stainless) – billets.
Additional information: This works is part of VEB Qualitäts- und Edelstahl-Kombinat.

VEB Walzwerk Hermann Matern, Burg

Head office: 3270 Burg. **Tel:** 470. **Telex:** 087425.
Works:
■ Burg. ***Rolling mills:*** medium plate.
Products: Carbon steel – medium plates 1.5 to 6.0mm. Alloy steel (excl stainless) – medium plates 1.5 to 6.0mm.
Additional information: This works is part of VEB Bandstahlkombinat.

VEB Edelstahlwerk 8 Mai 1945

Works:
■ 8210 Freital (Tel: 3751 8860. Telex: 028021). ***Steelmaking plant:*** electric arc furnace (annual capacity 240,000 tonnes); two plasma furnaces – one 15-tonne (8-10 tonnes/hour), and one 40-tonne (20-25 tonnes/hour). ***Refining plant:*** One ladle furnace; one vacuum degassing unit; one Leybold Heraeus VOD unit. ***Rolling mills:*** One blooming; one bar. ***Other plant:*** Forge.
Products: Stainless steel – blooms; round bars. Alloy steel (excl stainless) – (including engineering steels, high-speed steel, bearing steel, spring steel) blooms; round bars; square bars; flats; hexagons; wire. Forgings.

VEB Walzwerk Finow

Head office: 1302 Eberswalde-Finow 2. **Tel:** 550. **Telex:** 0168332.
Works:
■ Finow. ***Rolling mills:*** light section and bar; hot strip and hoop. ***Tube and pipe mills:*** welded.
Products: Carbon steel – including hot rolled hoop and strip (uncoated), longitudinal weld tubes and pipes 18 to 80mm dia, merchant bars, sections including box sections.
Additional information: This works is part of VEB Bandstahlkombinat.

Freital Works—see VEB Edelstahlwerk 8 Mai 1945

VEB Stahl- und Walzwerk Gröditz

Head office: 8402 Gröditz. **Tel:** 20. **Telex:** 028821.
Works:
■ Gröditz. ***Steelmaking plant:*** electric arc furnace. ***Refining plant:*** 20/40-tonne RH vacuum degassing unit. Rolling mills. Forging, engineering and bearing steels; pipe joints; tyres and flanges 500 to 4,000mm.

VEB Stahl- und Walzwerk Wilhelm Florin, Hennigsdorf

Head office: 1422 Hennigsdorf. **Tel:** 372 4829996. **Telex:** 0158522.
Works:
■ Hennigsdorf. ***Steelmaking plant:*** Three electric arc furnaces including a Krupp

VEB Stahl- und Walzwerk Wilhelm Florin, Hennigsdorf (continued)

furnace with an annual capacity of 175,000 tonnes. ***Refining plant:*** One 50-tonne RH vacuum degassing unit. ***Continuous casting machines:*** Two 4-strand for billet (93 to 145mm sq and 90 to 160mm sq) (annual capacity 300,000 tonnes). ***Rolling mills:*** light section and bar (250mm) (200,000 tonnes); one Schloemann single-pass for special sections.
Products: Carbon steel – billets, reinforcing bars, round bars 13 to 100mm, square bars 14 to 50mm, flats 16 x 8mm to 80 x 16mm, hexagons 13 to 55mm, special light sections.

Hennigsdorf Works—see VEB Stahl- und Walzwerk Wilhelm Florin

VEB Walzwerk Michael Niederkirchner, Ilsenburg

Head office: 3705 Ilsenburg. **Tel:** 850. **Telex:** 088432.
Works:
■ Ilsenburg. ***Rolling mills:*** One 4-high SMS/Voest-Alpine 3,200mm reversing heavy plate (annual capacity 500,000 tonnes). ***Other plant:*** Hot plate leveller, hot dividing shear for plates 40 x 3,000mm; curved knife cross-cut shear for plates up to 2,800mm wide and 40mm thick.
Products: Carbon steel – medium plates; heavy plates 5 to 160mm thick and 1,000 to 3,000mm wide (including plates 5 to 40mm thick and 3,000mm wide with hot yield point between 220 and 100 N/mm sq). High-strength and controlled-rolling plate grades.

VEB Rohr- und Kaltwalzwerk Karl-Marx-Stadt

Head office: 9057 Karl-Marx-Stadt. **Tel:** 5670. **Telex:** 07326.
Works:
■ Karl-Marx-Stadt.
Products: Carbon steel – Lead-coated cold rolled strip; longitudinal weld tubes and pipes 30 to 76mm (welded); precision tubes and pipes 7 to 16mm.

VEB Maxhütte Unterwellenborn

Head office: 6806 Unterwellenborn. **Tel:** Saalfeld 460. **Telex:** 05887141.
Works:
■ Unterwellenborn. **Blast furnaces**. **Steelmaking plant:** Basic Bessemer. ***Rolling mills:*** medium section – built by Cockerill Sambre and SMS (annual capacity 770,000 tonnes).
Products: Carbon steel – flats, bulb flats, medium angles, medium joists, medium channels, light rails, sheet piling up to 400mm.

VEB Blechwalzwerk Olbernhau

Head office: 9332 Olbernhau-Grünthal. **Tel:** 4141. **Telex:** 078331.
Works:
■ Olbernhau. ***Rolling mills:*** medium plate.
Products: Carbon steel – medium plates 1.5 to 6.0mm; floor plates (chequer plate).
Additional information: This works is part of VEB Bandstahlkombinat.

VEB Kaltwalzwerk Oranienburg

Head office: 1400 Oranienburg. **Tel:** 70. **Telex:** 0158528.
Works:
■ Oranienburg. ***Rolling mills:*** cold strip.
Products: Carbon steel – cold rolled hoop and strip (uncoated) 0.2 to 3.0mm, up to 400mm wide. Alloy steel (excl stainless) – cold rolled hoop and strip (uncoated) 0.2 to 3.0mm, up to 400mm wide.
Additional information: This works is part of VEB Bandstahlkombinat.

VEB Eisenhüttenkombinat Ost

Head office: 1220 Eisenhüttenstadt. **Tel:** 530. **Telex:** 0168421.
Works:
■ Eisenhüttenstadt. ***Sinter plant:*** Rumanian (annual capacity 4,000,000 tonnes). ***Blast furnaces:*** Six (2,000,000 tonnes). ***Steelmaking plant:*** Two 230-tonne basic oxygen converters (2,200,000 tonnes). ***Continuous casting machines:*** One 6-strand Voest-Alpine bloom (170 x 300 to 200 x 380mm (1,000,000 tonnes); one 2-strand Voest-Alpine slab (850 to 1,850mm x 150 to 350mm) (2,000,000 tonnes). ***Rolling mills:*** cold reduction comprising 4-high 1,700mm (1,000,000 tonnes), and 6-high (400,000 tonnes); 4-high temper/skin pass. ***Coil coating lines:*** One Heurtey hot-dip galvanizing (210,000 tonnes); one colour coating. ***Other plant:*** Annealing facilities. Cold forming line for sections. Pickling line for coil 2 to 6mm thick and up to 1,850mm wide.
■ Zweigbetrieb Bandstahlveredlung Porschdorf, 8321 Porschdorf (Tel: Bad Schandau 2556. Telex: 02593). **Coil coating lines**.
Products: Carbon steel – cold roll-formed sections (from strip); cold rolled sheet/coil (uncoated) 0.3 to 2mm thick; hot-dip galvanized sheet/coil and colour coated sheet/coil up to 1,500mm wide and 2.25mm thick; coated sheet/coil (zinc, brass, nickel, copper, copper-brass, copper-nickel). Stainless steel – cold rolled sheet/coil (uncoated). Electrical sheet.
Brands: Ekotal.
Additional information: This works is part of VEB Bandstahlkombinat.

Zweigbetrieb Bandstahlveredlung Porschdorf—see VEB Eisenhüttenkombinat Ost

VEB Qualitäts- und Edelstahl-Kombinat

Works:
■ This group of steelworks includes Stahl- und Walzwerk Brandenburg, Brandenburg; VEB Stahl- und Walzwerk Wilhelm Florin Hennigsdorf, Hennigsdorf; VEB Walzwerk Michael Niederkirchner Ilsenburg, Ilsenburg; VEB Maxhütte Unterwellenborn, Unterwellenborn; VEB Draht- und Seilwerk Rothenburg, Rothenburg.

VEB Stahl- und Walzwerk Riesa

Head office: 8400 Riesa. **Tel:** 880. **Telex:** 028825.
Works:
■ Riesa/Elbe. ***Steelmaking plant:*** electric arc furnace and open hearth furnace

VEB Stahl- und Walzwerk Riesa (continued)

(combined annual capacity 1,000,000 tonnes). ***Continuous casting machines:*** Two 4-strand bloom (203mm sq). ***Rolling mills:*** blooming; light section and bar; wire rod; hot strip and hoop. ***Tube and pipe mills:*** seamless – including alloy (180,000 tonnes). Welded.
Products: Alloy steel (excl stainless) – seamless tubes and pipes (20-114.3mm od, wall thickness 2-6.5mm) in bearing steel; welded tubes; sections and bars.

VEB Rohrkombinat

Works:
■ This group of works includes: Stahl- und Walzwerk Riesa; VEB Edelstahlwerk "8. Mai 1945" Freital; VEB Stahl- und Walzwerk Gröditz; VEB Rohr- und Kaltwalzwerk Karl-Marx-Stadt.

VEB Kaltwalzwerk Salzungen

Head office: 6200 Bad Salzungen. **Tel:** 570. **Telex:** 0628931.
Works:
■ Bad Salzungen. ***Rolling mills:*** cold strip.
Products: Carbon steel – cold rolled hoop and strip (uncoated) 0.2 to 4.0mm, up to 530mm wide.
Additional information: This works is part of VEB Bandstahlkombinat.

VEB Eisen- und Hüttenwerke Thale

Head office: 4308 Thale. **Tel:** 70. **Telex:** 048521.
Works:
■ Thale. Iron powder plant, forging equipment.
Products: Iron powder. Steel plates, sintered parts from non-alloy and alloy iron powder.
Expansion programme: 120,000,000-130,000,000 Ostmark expansion programme for both powder production and forging equiment scheduled for completion in 1988/89.
Additional information: This works is part of VEB Mansfeld Kombinat Wilhelm Pieck Eisleben, which is a grouping of non-ferrous metal works.

West Germany

Max Aicher KG Stahlwerk Annahütte

Head office: D 8229 Hammerauobb. **Tel:** 08654 8001. **Telex:** 56610.
Established: 1975.
Products: Carbon steel – reinforcing bars and mesh.

Alte & Schröder GmbH & Co

Head office: Bandstahlstrasse 14, D 5800 Hagen-Halden. **Tel:** 02331 693-1. **Telex:** 823152.
Established: 1959. **Ownership:** Affiliated to Freidr. Gustav Theis Kalkwalzwerke GmbH.
Works:
■ **Strip coating plant**.
Products: Carbon steel – coated hoop and strip – electrolytically zinc – and nickel-plated, phosphate treated and patterned steel and non-ferrous strip.

Andernach & Bleck KG

Head office: Lennestrasse 92, D 5800 Hagen-Halden. **Tel:** 02331 3530. **Telex:** 823846 aub d.
Products: Carbon steel – cold rolled hoop and strip (uncoated).

Badische Stahlwerke AG

Head office: Weststrasse 31, D 7640 Kehl/Rhein. **Tel:** 07851 83 0. **Telex:** 753358. **Cables:** Stahlwerke Kehl. **Telecopier (Fax):** 07851 73674.
Management: Directors – Horst Weitzmann (managing), Wilhelm Dening (technical), Hanns-Joachim Pape (sales and purchasing). **Established:** 1968. **Capital:** DM 30,000,000. **No. of employees:** 1,168 (incl Badische Drahtwerke GmbH). **Subsidiaries:** Badische Drahtwerke GmbH, Kehl (wire products); Neckar Drahtwerke GmbH, Eberbach (wire products); Besta Beton stahl GmbH & Co KG, Lübbecke (wire products); Badische Stahl-Engineering GmbH, Kehl (engineering and consulting); Frachtservice Kehl GmbH, Kehl (transport). **Annual capacity:** Raw steel: 826,000 tonnes. Finished steel: 785,000 tonnes.
Works:
■ At head office. ***Steelmaking plant:*** Two 65-tonne electric arc furnaces (annual capacity 826,000 tonnes). ***Refining plant:*** Two 65-tonne ladle furnace (826,000 tonnes). ***Continuous casting machines:*** 9-strand curved mould billet (120mm sq) (826,000 tonnes). ***Rolling mills:*** bar (12-28mm) and 5.5-14.0mm wire rod (coil weight 1,500kg) (785,000 tonnes).
Products: Carbon steel – billets (1986 output 728,000 tonnes); wire rod and reinforcing bars (757,000 tonnes).

Walzwerk Becker (Wabec) GmbH & Co

Head office: Johannesstr 2, D 5350 Euskirchen. **Tel:** 02251 52051-3. **Telex:** 8869131 wbek. **Cables:** Metallwerk Euskirchen.
Products: Carbon steel – reinforcing bars 8-28mm dia; cold rolled hoop and strip (uncoated) to DIN 1544 and 1624, 0.30-3mm thick, 10-360mm wide.

Walzwerk Becker-Berlin Verwaltungsges.mbH & Co Strangguss KG

Head office: Berlinerstrasse 19-37, D 1000 Berlin 27. **Tel:** 030 44380030. **Telex:** 8869131 wbek.
Products: Carbon steel – billets and Alloy steel (excl stainless) – billets 90-130mm square, 2-11m long.
Additional information: This company was in receivership as we went to press.

Benteler AG

Head office: Residenzstrasse 1, D 4790 Paderborn, North Rhine Westfalia. **Postal address: Tel:** 05254 81 0. **Telex:** 936866. **Telecopier (Fax):** 05254 13666.
Management: Executive board – Peter Adams (managing director), Erich Mager (vice managing director, finance and administration), Hubertus Benteler (technical), Rainer Fuess (sales), Heinrich Kürpick (personnel), Gert Vaubel (quality assurance), W. Dickhoff (sales director, automotive industries), G. Friederich (sales director, steel and hot-finished pipe), D. Kretschmer (sales director, cold-drawn pipe). **Established:** 1876. **Ownership:** Erich Benteler KG; Dr. Helmut Benteler KG. **Subsidiaries:** Benteler Industries Inc, Grand Rapids, MI, USA.
Works:
■ D 4450 Lingen. ***Steelmaking plant:*** One 80-tonne electric arc furnace. ***Refining plant:*** One ladle furnace. ***Continuous casting machines:*** billet (160mm sq); bloom (230 x 120mm, 200 x 160mm and 240 x 200mm); slab (420 x 200-670 x 200mm); round (150mm).
■ D 4790 Paderborn. ***Rolling mills:*** One cold strip (32.0 x 0.5mm to 670.0 x 4.0mm).

Benteler AG (continued)

Tube and pipe mills: seamless – one 21.3-114.3mm Erhardt; five Pilger; four draw-benches for seamless precision tubes 4.0-140.0mm. Welded – two for precison tubes and ferritic/austenitic alloy pipe 4.0-80.0mm.
■ D 4220 Dinslaken. ***Tube and pipe mills:*** seamless – one 21.3-146.0mm Erhardt.
■ D 5900 Siegen/Weidenau. ***Tube and pipe mills:*** seamless – one 21.3-88.9mm CPE.
■ D 4800 Bielefeld. Plant for hydraulic and telescopic cylinders and cylinder tubes.
Products: Carbon steel – slabs, 420 x 200mm to 670 x 200mm; blooms, 230 x 120mm, 200 x 160mm and 240 x 200mm rectangular; billets 160 x 160mm sq; semis for seamless tubemaking, 150mm round, 210 and 270mm octagonal diagonal; pipe piling; skelp (tube strip); cold rolled hoop and strip (uncoated); seamless tubes and pipes; longitudinal weld tubes and pipes; oil country tubular goods; galvanized tubes and pipes; plastic coated tubes and pipes; cold drawn tubes and pipes; hollow bars; precision tubes and pipes. Stainless steel – longitudinal weld tubes and pipes, precision tubes and pipes. Alloy steel (excl stainless) – seamless tubes and pipes, oil country tubular goods, cold drawn tubes and pipes, hollow bars, precision tubes and pipes. Automotive products – pressed, drawn, stamped and other parts for the automotive industry; mechanical engineering products.
Brands: Zista (tubes); Istatherm (tubes).
Sales offices/agents: Associated tube stockholding and trading companies: Röhren- und Stahllager GmbH, Bad Vilbel, Berlin, Echterdingen, Munich and Ratingen. Helens Roer A/B, Halmstad, Sweden. Helens Roer A/B, Oslo, Norway. Helens Roer A/S, Copenhagen, Denmark. Helens Italiana Srl, Milan, Italy. Kindlimann AG, Wil, Switzerland.

Bergrohr GmbH Herne

Head office: Bochumer Strasse 229, D 4690 Herne 1. **Postal address:** Postfach 1340. **Tel:** 2323 4950. **Telex:** 8229855. **Telecopier (Fax):** 2323 495253. **Management:** President – Hans Berg; managing directors – Wilhelm Kirchhoff (technical), Manfred Lohfink (sales), Bernd Mönkemöller (finance); non-executive directors – Hans-Gerd Jodorf (sales, line pipe), Arnold Scheibe (sales, structural pipe). **Established:** 1899. **Capital:** DM 30,000,000. **No. of employees:** Works – 600. Office – 200. **Ownership:** AG der Dillinger Hüttenwerke (50%), Berg family (50%). **Associates:** Berg Steel Pipe Corp, Panama City, FL, USA (large diameter pipe mill – longitudinal, DSAW).
Works:
■ At head office. ***Tube and pipe mills:*** Three welded – 20-120" od DSAW (annual capacity 350,000 tonnes approx); 30-120" od DSAW (50,000 tonnes approx); 30-64" od DSAW (600,000 tonnes approx).
Products: Carbon steel – pipe piling 30-48" od (1986 output 15,000 tonnes approx); longitudinal weld tubes and pipes; large-diameter tubes and pipes 56" od (650,000 tonnes approx); plastic coated tubes and pipes 44" od (40,000 tonnes approx).
Sales offices/agents: Line pipe sales – Königsallee 98, PO Box 200107, D 4000 Düsseldorf (Tel: 211 133029. Telex: 8587110. Fax: 211 324972).

Bergrohr GmbH Siegen

Head office: Siegstrasse 70, D 5900 Siegen. **Postal address:** Postfach 210205. **Tel:** 0271 7070. **Telex:** 271310 berg.
Products: Carbon steel – longitudinal weld tubes and pipes 20-120" od up to 1½: wall thickness; large-diameter tubes and pipes, plastic coated tubes and pipes.

Berliner Stahlwerk KG Gerd Becker GmbH & Co

Head office: Postfach 1527, D 5350 Euskirchen-Euenheim. **Tel:** 02251 52051-3. **Telex:** 8 869131 wbek.
Works:
■ Berliner Strasse 19-37, D 1000 Berlin 27. (Tel: 030 4341011. Telex: 181234). ***Steelmaking plant:*** 45-tonne electric arc furnace.
Additional information: The steel is cast into billets on the continuous caster of Walzwerk Becker-Berlin Verwaltungsges.mbH & Co Strangguss KG. This company, and Berliner Stahlwerk KG Gerd Becker, were in receivership as we went to press.

Stahlwerke Bochum AG (SWB)

Head office: Castroper Strasse 228, D 4630 Bochum 1. **Tel:** 0234 5080. **Telex:** 825210. **Cables:** SWB Bochum. **Telecopier (Fax):** 0234 508111.
Management: Rolf Müller (commercial), Franz Josef Lenze (technical), Friedhelm Rappard (social), Manfred Espenhahn (deputy manager). **Established:** 1947. **Capital:** DM 24,200,000. **No. of employees:** Works – 1,281. Office – 493. **Ownership:** Thyssen AG, Duisburg (48.5%); Eisen- und Hüttenwerke AG, Cologne (48.5%).
Works:
■ At head office. ***Rolling mills:*** cold reduction , special cold reduction for electrical sheet. ***Other plant:*** Continuous pickling line, annealing and coating lines for electrical sheets, slitting and cut-to-length lines, steel foundry and machine shop.
Products: Carbon steel – cold rolled sheet/coil (uncoated) (1986 output 173,000 tonnes); electrical non-oriented silicon sheet/coil (216,000 tonnes). Steel castings (output 6,900 tonnes).
Sales offices/agents: Electrical sheets – EBG Elektroblech GmbH, Castroper Strasse 228, D 4630 Bochum 1; Cold rolled sheets – Otto Wolff Flachstahl GmbH, Zeughausstrasse 2, D 5000 Cologne; steel castings – Stahlwerke Bochum AG, Castroper Strasse, D 4630 Bochum 1.

Böhler AG, Edelstahlwerke

Head office: Hansaallee 321, D 4000 Düsseldorf 11. **Postal address:** Postfach 110246. **Tel:** 0211 587 0. **Telex:** 8584489. **Cables:** Boelerag Dusseldorf. **Telecopier (Fax):** 0211 587 2657.
Management: Board members – T. Schumacher (finance), W. Süssmann (commercial), F. Fuhrmann (personnel); director – C. Weiser (mill products). **Established:** 1870. **Capital:** DM 57,000,000. **No. of employees:** Works – 1,800. Office – 420. **Ownership:** Vereinigte Edelstahlwerke (VEW), Vienna, Austria. **Annual capacity:** Raw steel: 45,000 tonnes. Finished steel: 40,000 tonnes.
Works:
■ At head office. Works director: Clemens Weiser. ***Steelmaking plant:*** 25-tonne electric arc furnace (annual capacity 45,000 tonnes). ***Refining plant:*** ESR (max 800mm dia) (2,500 tonnes); 25-tonne vacuum degassing unit (45,000 tonnes); 25-tonne VOD (6,000 tonnes). ***Other plant:*** Forging plant (25,000 tonnes). Foundry (3,000 tonnes). Welding wire and electrode works (20,000 tonnes).
Products: Carbon steel – ingots 1-20 tonne; round bars up to 700mm; square bars up to 600mm; flats up to 1,060 x 510mm. Stainless steel – ingots 1-20 tonne; round bars up to 700mm; square bars up to 600mm; flats up to 1,060 x 510mm. Alloy steel (excl stainless) – ingots 1-20 tonne; round bars up to 700mm; square bars up to 600mm; flats up to 1,060 x 510mm.

Remscheider Walz- und Hammerwerke, Böllinghaus & Co

Head office: Neuenkamper Strasse 12-20, D 5630 Remscheid 1. **Postal address:** PO Box 10 01 48. **Tel:** 0 21 91 34 00 57-9. **Telex:** 8513486. **Cables:** Bollinghaus Walzwerk. **Telecopier (Fax):** 0 21 91 3 90 09.
Management: Director – Alfred Härtel (owner); sales managers – Hartwig Härtel (purchasing), Adolf Schulte (tool steel – domestic), Joachim Werner (tool steel – export). **Established:** 1889. **Capital:** DM 400,000. **No. of employees:** Works – 35. Office – 15. **Subsidiaries:** Stabstahlgesellschaft Remscheid mbH, Remscheid (sales organisation for file steel in all profiles, e.g. flat, round and square bars). **Annual capacity:** Finished steel: 4,000 tonnes approx.
Works:
■ At head office. ***Rolling mills:*** Double-duo bar (rounds 8-40mm; squares 8-40mm; hexagons 8-40mm; octagons 10-35mm; flats 3-30mm thick, 8-100mm wide; special shapes; boring drill rods for mining industry) (combined annual capacity 4,000 tonnes).
Products: Carbon steel – round bars, square bars, flats, hexagons, light angles, light tees, special light sections. Stainless steel – round bars, square bars, flats, hexagons, light angles, light tees, light channels, special light sections. Alloy steel (excl stainless) – round bars, square bars, flats, hexagons, light angles, light tees, special light sections.
Additional information: The company also sells file steel, special profile shapes, clicker die steel, boring drill rods, stainless steel sheets/plates/angles/merchant bars, seamless tubes.

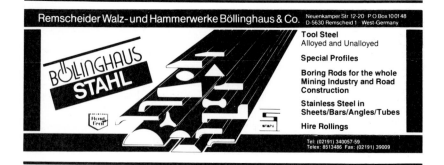

Röhrenwerke Bous/Saar GmbH (RBS)

Head office: D 6626 Bous. **Postal address:** Postfach 160. **Tel:** 06834 811. **Telex:** 445528. **Cables:** Rohrwerk Bous/Saar.
Management: Managers – Adolf Binz (technical), Heribert Becher (commercial). **Established:** 1887/1959. **Capital:** DM 40,000,000. **No. of employees:** Works – 800. Office – 215. **Ownership:** Mannesmannröhren-Werke AG (100%). **Annual capacity:** Raw steel: 250,000 tonnes. Finished steel: 180,000 tonnes.
Works:
■ At head office. ***Steelmaking plant:*** One 60-tonne EBT electric arc furnace. ***Tube and pipe mills:*** seamless – One Pilger mill (annual capacity 180,000 tonnes). ***Other plant:*** Heat treating and plastic coating facilities.
Products: Carbon steel – seamless tubes and pipes and line pipe upto 406mm dia. Stainless steel – seamless tubes and pipes. Alloy steel (excl stainless) – seamless tubes and pipes.
Brands: RBS.
Sales offices/agents: Mannesmannröhren-Werke AG, Postfach 1104, D 4000 Düsseldorf (Tel: 0211 8751. Telex: 08581421).

Boschgotthardshütte O. Breyer GmbH

Head office: Industriestrasse 9, D 5900 Siegen. **Postal address:** Postfach 21 01 63. **Tel:** 0271 7040. **Telex:** 872771. **Cables:** Boschhutte. Special steel billets, bars and forgings.

Kaltwalzwerk Brockhaus GmbH (KWB)

Head office: D 5970 Plettenberg 2. **Postal address:** POB 3220. **Telex:** 8201800 kwbp d.
Works:
■ At head office. ***Rolling mills:*** cold strip.
Products: Carbon steel – cold rolled hoop and strip (uncoated). Spring steels.

Edelstahlwerke Buderus AG

Head office: Buderusstrasse 25, D 6330 Wetzlar. **Tel:** 06441 374 0. **Telex:** 0483843 ebu w. **Telecopier (Fax):** 06441 374 882.
Management: Holger Flieth, Hans-Ulrich Plaul. **Established:** 1920. **Capital:** DM 32,500,000. **No. of employees:** Works – 1,660. Office – 407. **Ownership:** Feldmühle Nobel AG (38.5); Buderus AG (61.5%). **Annual capacity:** Raw steel: 350,000 tonnes.
Works:
■ At head office. ***Steelmaking plant:*** Two electric arc furnaces – one 85-tonne and one 45-tonne (annual capacity 350,000 tonnes). ***Refining plant:*** One 85-tonne vacuum degassing unit (350,000 tonnes); one TN plant. ***Rolling mills:*** One blooming

Edelstahlwerke Buderus AG (continued)

– one 2-high stand 900 x 2,000 mm (250,000 tonnes); one billet – two 3-high stands 650 x 2,000mm (120,000 tonnes); one hot strip and hoop – eight 4-high stands 315 x 450mm (coil weight 2.8 tonnes) (120,000 tonnes); one cold strip 4-high stand/2-high stand (coil weight 2.8 tonnes) (30,000 tonnes). ***Other plant:*** One forging shop, one drop forging shop.
Products: Carbon steel – billets, round bars, square bars, flats, hexagons, hot rolled hoop and strip (uncoated) up to 350mm wide and 13mm thick, cold rolled hoop and strip (uncoated) up to 320mm wide and 6.0mm thick, medium plates. Stainless steel – billets, round bars, square bars, flats, hexagons, hot rolled hoop and strip (uncoated) up to 350mm wide and 13mm thick, cold rolled hoop and strip (uncoated) up to 320mm wide and 6.0mm thick, medium plates. Alloy steel (excl stainless) – billets, round bars, square bars, flats, hexagons, hot rolled hoop and strip (uncoated) up to 350mm wide and 13mm thick, cold rolled hoop and strip (uncoated) up to 320mm wide and 6.0mm thick, medium plates. Bright steel. Open die forgings up to 110 tonnes ingot weight, weld filler materials, precision flats. Alloy and non-alloy tool and engineering steels, high-speed steels, wear-resistant steels, anti-friction bearing steels, stainless and chemically resistant steels, heat-resistant steels.

Karl Diederichs Stahl-, Walz- und Hammerwerk

Head office: Luckhauser Strasse 1-5, D 5630 Remscheid 11. **Postal address:** POB 12 01 20. **Tel:** 2191 593-0. **Telex:** 8512882. **Telecopier (Fax):** 2191 593 165.
Management: Managers – Manfred Diederichs (general), Rolf Hülsenbeck (sales), Herbert Neuhaus (buyer), Hans Radekopp (works), Volkmar Schulz (quality). **Established:** 1931. **No. of employees:** Works – 370. Office – 80. **Annual capacity:** Finished steel: 60,000 tonnes (approx). Forged bars – heat-treated or as forged, black or machined (up to approx 15,000mm long, up to approx 800mm dia, up to approx 750mm square, up to approx 1,200 x 400mm flat); hammer forgings – heat-treated or as forged, black or machined (shafts up to approx 25,000kg piece weight, discs up to approx 2,000mm outside dia, hollow forged parts up to approx 2,000mm od, collar and flange shafts up to approx 1,500mm collar or flange dia); seamless rolled rings – forged or rolled, heat-treated or as forged, black or machined (up to approx 3,200mm od, up to approx 650mm high, up to approx 6,000kg piece weight).

AG der Dillinger Hüttenwerke

Head office: D 6638 Dillingen, Saar. **Postal address:** Postfach 1580. **Tel:** 068 31 470. **Telex:** 4437110 dh d. **Cables:** Dillingerhutte 6638 Dillingensaar.
Ownership: Sacilor is the majority shareholder. **Subsidiaries:** Sollac (25%).
Works:
■ At head office. **Sinter plant. Blast furnaces.** ***Steelmaking plant:*** Two 200-tonne LD basic oxygen converters. ***Continuous casting machines:*** Two 2-strand slab – one 1,600mm and one 2,200mm. ***Rolling mills:*** One 4-high 4,800mm heavy plate .
Products: Carbon steel – medium plates, heavy plates.

Dörrenberg Edelstahl GmbH

Head office: Hammerweg 7, D 5250 Engelskirchen. **Postal address:** PO Box 21 64. **Tel:** 02263 79 1. **Telex:** 884545. **Cables:** 49 2263 79205.
Management: Managing directors – Werner Fuss, Helmut Laczkovich, Eckhard Müller; sales managers – Joachim Holzapfel (export), Horst Borges (castings), H.-J. Hielscher (precision castings), Joachim Neumann (steels); purchasing manager – Gebhard Heischeid. **Established:** 1860. **Capital:** DM 9,000,000. **No. of employees:** 650.
Works:
■ At head office. ***Steelmaking plant:*** One 10-tonne electric arc furnace; five 1-5-

Dörrenberg Edelstahl GmbH (continued)

tonne induction furnaces. ***Refining plant:*** 10-tonne Detem vacuum degassing unit (Dörrenberg patents). ***Other plant:*** Forge. Foundry incl. investment casting.
Products: Stainless steel – ingots 0.5-9 tonnes; round bars. Alloy steel (excl stainless) – ingots 0.5-9 tonnes; round bars; square bars; flats; hexagons; special light sections.

Duisburger Kupferhütte GmbH

Head office: Werthauser Strasse 182, D 4100 Duisburg 1. **Postal address:** PO Box 10 01 03. **Tel:** 0203 601 0. **Telex:** 855 863. **Cables:** Kupferhutte Duisburg. **Telecopier (Fax):** 0203 665173.
Management: Managing director – Christopher M. Moore; managers – Jürgen Redhardt (sales), Karl Heinz Born (purchasing). **Established:** 1876. **Capital:** DM 15,000,000. **No. of employees:** Works – 226. Office – 71. **Annual capacity:** Pig iron: 200,000 tonnes.
Works:
■ At head office. ***Sinter plant:*** One 80 sq m strand Lurgi (annual capacity 500,000 tonnes). ***Blast furnaces:*** Two – No 3 5.5m hearth dia, 580 cu m (200,000 tonnes); No 4 4.5m hearth dia, 425 cu m (150,000 tonnes). ***Other ironmaking plant:*** One 30-tonne Demag electric pig iron furnace (100,000 tonnes).
Products: Steelmaking pig iron (output 1,500 tonnes), foundry pig iron (output 195,000 tonnes), granulated iron.

J.N. Eberle & Cie GmbH

Head office: Eberlestrasse 28, D 8900 Augsburg. **Postal address:** Postfach 101507. **Tel:** 0821 5212-1. **Telex:** 533756 esaeg d. **Cables:** Eberle Augsburg.
Works:
■ At head office. ***Rolling mills:*** cold reduction. **Saw factory**.
Products: Carbon steel – Hardened and tempered precison cold rolled hoop and strip (uncoated) 0.02-2.0mm thick to 4,50mm wide. Bimetal strip – 0.6-2.0mm thick, up to 100mm wide.

Stahlwerke Ergste GmbH & Co KG

Head office: Letmather Strasse 69, D 5840 Schwerte. **Tel:** 02304 791. **Telex:** 8229646 aerg d.
Established: 1918. Special steel strip, bar and wire.

Eisen- und Metallwerke Ferndorf GmbH

Head office: D 5910 Kreuztal-Ferndorf. **Postal address:** Postfach 1140. **Tel:** 02732 2940. **Telex:** 875552 benf. **Cables:** Eisenmetallerk Ferndorfkrsiegen.
Products: Carbon steel – longitudinal weld tubes and pipes DN 508 5.0 to 11.0mm, DN 609.6 and 660 5.0 to 12.7mm, DN 711 6.0 to 16.0mm, DN 812.0 to 1420 7.0 to 16.0mm; spiral-weld tubes and pipes DN 300.0 and 350 4.0 to 8.0mm, DN 406.4 and 457.2 4.0 to 9.5mm, DN 508 5.0 to 12.7mm, DN 610 5.0 to 14.0mm, DN 711 6.3 to 20.0mm, DN 812 7.1 to 20.0mm, DN 914.4 7.1 to 20.0mm, DN 1016 to 1620 8.0 to 25.0mm; qualities according to DIN 17100, 17155, API, ASTM. Also tube fittings.

Rudolf Flender GmbH & Co KG

Head office: Eiserfelder Strasse 110, D 5900 Siegen. **Postal address:** POB 101043. **Tel:** 0271 33050. **Telex:** 872785 frsg.
Products: Carbon steel – longitudinal weld tubes and pipes.

Fürstlich Hohenzollernsche Hüttenverwaltung Laucherthal (FHH)

Head office: D 7480 Sigmaringen. **Postal address:** PO Box 220.
Works:
■ ***Rolling mills:*** light section. Plant for bright drawing, turning and grinding bars.
Products: Alloy steel (excl stainless) – round bars, square bars, flats, hexagons, bright (cold finished) bars.

Ewald Giebel KG

Head office: D 5860 Iserlohn 7. **Postal address:** Postfach 7324. **Tel:** 02374 560. **Telex:** 827217 gibl d.
Established: 1932. **Ownership:** Giebel family. **Subsidiaries:** Ewald Giebel Luxembourg GmbH.
Products: Carbon steel – cold rolled hoop and strip (uncoated); hot-dip galvanized hoop and strip; coated hoop and strip (tinned non-ferrous metal strip); cold rolled sheet/coil (uncoated) 0.06 to 4mm, up to 1,000mm wide; hot-dip galvanized sheet/coil (0.2-2mm, up to 1,000mm wide); electro-galvanized sheet/coil (0.2-2mm, up to 1,000mm wide).

Blech- und Eisenverarbeitung Grösgen GmbH & Co KG

Head office: D 4100 Duisburg 1. **Postal address:** Postfach 210319. **Tel:** 02 03 3120 38. **Telex:** 0855429 groes d. Steel tubes and pipes.

Halbergerhütte GmbH

Head office: Saarbückerstrasse 51, D 6604 Saarbrücken-Brebach. **Postal address:** Postfach 1170. **Telex:** 4428830.
Established: 1756. **Ownership:** Pont-à-Mousson SA, France.
Products: Cast iron – pipes; pig iron, castings.

Hamburger Stahlwerke GmbH

Head office: Dradenaustrasse 33, D 2103 Hamburg 95. **Tel:** 0 40 74 08 0. **Telex:** 217657 hsw d. **Telecopier (Fax):** 740 1432.
Management: Wolf-Dietrich Grosse, Eckehard Förster, Gerd G. Weiland. **Capital:** DM 31,500,000. **No. of employees:** Works – 651. Office – 215. **Ownership:** Protei GmbH & Co KG, Hamburg (100%). **Subsidiaries:** Salzgitter Draht GmbH, Salzgitter; Müller Bewehrungstechnik GmbH Co KG, Garbsen (cutting and bending); Weserdraht Friedrich Klinke GmbH & Co KG, Minden (wire drawing); Böco Baustahl-Matten GmbH, Biebesheim (mesh welding); HSC Hamburger Stahlwerke Consulting GmbH, Hamburg (technology transfer and consulting); Hamburger Transport Unternehmung, Hamburg (forwarding and shipping agent). **Annual capacity:** Raw steel: 700,000 tonnes. Finished steel: 540,000 tonnes.
Works:
■ At head office. ***Direct reduction plant:*** One 4.8m stack dia gas reduction (annual capacity 550,000 tonnes). ***Steelmaking plant:*** Two 110-tonne electric arc furnaces (700,000 tonnes). ***Refining plant:*** 110-tonne ladle furnace (700,000 tonnes). ***Continuous casting machines:*** 6-strand Concast billet (100-130mm) (700,000 tonnes). ***Rolling mills:*** Combined bar and wire rod – 1-strand 14-35mm and 2-strand 5.5-12.5mm (coil weight 1.5 tonnes) (540,000 tonnes).
Products: Direct-reduced iron (DRI) (1986 output 168,761 tonnes). Carbon steel – billets 100-130mm; wire rod 5.5-12.0mm; reinforcing bars 6-28mm.

Hille & Müller Kaltwalzwerk

Head office: Am Trippelsberg 48, D 4000 Düsseldorf 13-Reisholz. **Tel:** 211 79500. **Telex:** 8582103 hmd g.
Subsidiaries: Trierer Walzwerke GmbH, Trier (strip plating and coating); Thomas Steel Strip Corp, Warren, OH, USA (cold rolling and strip coating); Rafferty Brown, Waterbury, CT and Greensboro, NC, USA (steel stockholding).
Works:
■ ***Rolling mills:*** cold strip. Strip coating lines (electro plating).
Products: Carbon steel – cold rolled hoop and strip (uncoated); coated hoop and strip – electro-plated strip with nickel, brass, copper, chromium, etc; galvanized strip.

Hindrichs-Auffermann AG

Head office: Heckinghauser Strasse 116-126, D 5600 Wuppertal 2. **Postal address:** Postbox 24 02 23. **Tel:** 0202 627061. **Telex:** 8591734. **Cables:** Pressplatt.
Products: Stainless steel – hot rolled hoop and strip (uncoated), cold rolled hoop and strip (uncoated), medium plates, hot rolled sheet/coil (uncoated), cold rolled sheet/coil (uncoated). Stainless steel press-plates for the laminate industry; metal-clad steel strip and sheets.

Hoesch AG

Head office: Eberhardstrasse 12, D 4600 Dortmund 1. **Tel:** 0231 8441. **Telex:** 822 123. **Cables:** Hoesch.
Management: Board – Detlev Rohwedder (chairman); Hero Brahms (finance); Günter Flohr (sales); Günter Sieber (personnel). **Established:** 1871. **Capital:** DM 355,600,000. **No. of employees:** 32,400. **Subsidiaries:** These include associated companies: Steel – Hoesch Stahl AG, Dortmund (see below). Manufacturing operations – Hoesch Hohenlimburg AG, Hagen (see below), Hoesch Rohr AG, Hamm (see below), Hoesch Rothe Erde-Schmiedag AG, Dortmund (see below), O & K Orenstein & Koppel AG, Berlin/Dortmund (75%) (construction and mining machinery, plant and systems, escalators; Faun division – Dittmann & Neuhaus AG, Witten (springs); Vacmetal, Dortmund (50%) (units and process technology for treatment of liquid metals, especially steel and iron); Hoesch Verpackungssysteme GmbH, Schwelm (steel and non-metallic strapping); Hoesch Maschinenfabrik Deutschland AG, Dortmund (heavy machine tools); Schwinn AG, Homburg/Saar (drop forgings); Blefa GmbH, Kreuztal (stainless steel kegs); Blefa-Felser GmbH, Ateendorn (steel-barrels, special containers). Trading and service operations – Eisen und Metall AG, Gelsenkirchen (see below); Herzog Coilex GmbH, Stuttgart (SSC); Schrottver wertung Celler GmbH, Dortmund (scrap trading and treating); Dortmunder Eisenhandel GmbH, Dortmund (secondary steel products); Hoesch Export AG, Dortmund (see below); MBP Software & Systems GmbH, Dortmund (analysis, consulting service). Also subsidiaries and associated companies in France, Italy, Netherlands, Great Britain, Spain, Switzerland, Austria, Sweden, Liberia, USA, Mexico, Argentina, Brazil, Japan etc.

Hoesch Stahl AG, Rheinische Strasse 173,D 4600 Dortmund 1. ***Tel:*** 0231 8441. ***Telex:*** 822141. ***Cables:*** Hoesch Hutten.
Management: Board – Hans Wilhelm Grasshoff (chairman); Knut Consemüller (technical); Alfred F. Heese (personnel); Jens Wagner (sales); Heinz-Peter Klenz (controlling).; **Established:** 1984. **Capital:** DM 210,000,000. **No. of employees:** 15,800. **Annual capacity:** Pig iron: 5,300,000 tonnes. Raw steel: 4,700,000 tonnes.
Works:
■ Phoenix Works, Dortmund. ***Blast furnaces:*** Two – No 3 9.8m hearth dia, 1,611 cu m (annual capacity 1,380,000 tonnes) and No 5 9.5m hearth dia, 1,419 cu m (1,080,000 tonnes). ***Steelmaking plant:*** Three 180-tonne LD basic oxygen converters (4,700,000 tonnes). ***Refining plant:*** One 180-tonne DH vacuum degassing unit. ***Continuous casting machines:*** Three – slab one 2-strand Concast, one 2-strand Demag and 1-strand SMS (combined annual capacity 4,700,000 tonnes).
■ Westfalen Works, Dortmund. ***Sinter plant:*** Two Dwight Lloyd/Lurgi – one 150sq

West Germany

Hoesch AG (continued)

m strand (annual capacity 2,010,000 tonnes) and one 175 sq m strand (2,430,000 tonnes). ***Blast furnaces:*** Two – No 4 9.7m hearth dia, 1,573 cu m (annual capacity 1,320,000 tonnes) and No 7 9.7m hearth dia, 1,834 cu m (1,500,000 tonnes). ***Rolling mills:*** One wide hot strip – three roughing and seven finishing stands (3,605,000 tonnes); two cold strip – one 5-stand tandem (460,000 tonnes) and one 5-stand tandem (1,620,000 tonnes). ***Coil coating lines:*** One electrolytic tinning and tin-free steel (ECCS) (1,200mm); one electro-galvanizing (1,920mm) and one galvanised electrolytic and one terne/lead coating (1,550mm). ***Other plant:*** Slag processing plant.

■ Union Works, Dortmund. ***Rolling mills:*** heavy section – one roughing stand and 3-stand (950mm) (annual capacity 325,0000 tonnes).

■ Hamm Works. ***Coil coating lines:*** One colour coating (1,850mm).

■ Kreuztal-Eichen Works. ***Coil coating lines:*** One hot-dip tinning (1,650mm) and one colour coating. ***Other plant:*** Cold roll-forming lines; plant for steel sheet sandwich panels.

■ Kreuztal-Ferndorf Works. ***Coil coating lines:*** One hot-dip galvanizing (1,550mm) and one colour coating (1,350mm).

■ Wissen Works. ***Coil coating lines:*** One electrolytic tinning (995mm). ***Other plant:*** Tinplate coating line.

■ Hagen/Hohenlimburg Works. ***Coil coating lines:*** One electrolytic tinning and one hot-dip tinning; one hot-dip galvanizing and terne/lead coating.

Products: Carbon steel – slabs; blooms; heavy angles; heavy channels; cold roll-formed sections (from strip); sheet piling and mining sections; hot rolled sheet/coil (uncoated); cold rolled sheet/coil (uncoated); hot-dip galvanized sheet/coil; electro-galvanized sheet/coil; terne-plate (lead coated) sheet/coil; colour coated sheet/coil; blackplate; electrolytic tinplate (single-reduced); electrolytic tinplate (double-reduced); tin-free steel (ECCS). Sandwich panels for roofing and walling, trapezoidal sheets, steel structures.

Brands: Zincal (electro-galvanized sheet/coil); Ternal (lead-tin alloy coated sheet/coil); Platal/Pladur/Zincometal (hot-dip galvanized and coil coated sheet/coil); Isowand/Isopaneel/Isodach (sandwich panels).

Hoesch Hohenlimburg AG, Langenkampstrasse 14.D 5800 Hager 5,Hohenlimburg. ***Tel:*** 02334 881. ***Telex:*** 821891 11 ho d. ***Cables:*** Hoesch.
Management: Board – Gerhard Bergmann (sales); Götz-Peter Blumbach (technical); Karl Borggräffe (finance, personnel). **Established:** 1969 (in present form). **No. of employees:** 5,000. **Ownership: Subsidiaries:** Döhner AG, Hagen-Hohenlimburg (sales company); Dittmann & Neuhaus AG, Witten (springs); Luhn & Pulvermacher, Hagen (springs); Industria Española de Suspensiones SA, Madrid, Spain (springs); Hoesch Argentina Saiyc, Buenos Aires, Argentina (springs); Hoesch Industria de Molas Ltda, São Paulo, Brazil (springs); Suspensiones Automotrices SA de CV, Tlalnepantla, Mexico (springs); Hoesch-Isocar, Eisenstadt, Austria (plastic springs).

Works:

■ Hagen/Hohenlimburg. ***Rolling mills:*** One continuous hot strip and hoop with one roughing and nine finishing stands (annual capacity 845,000 tonnes); one 4-stand cold strip with slitting line (160,000 tonnes).

■ Schwerte. ***Rolling mills:*** Two light section (annual capacity 246,000 tonnes).

Products: Carbon steel – special light sections hot rolled, hot extruded, bright drawn, and cold roll-formed sections (from strip) including plastic-coated; hot rolled hoop and strip (uncoated) 25-650mm x 1.5-16mm in special steels; cold rolled hoop and strip (uncoated) 20-650mm x 0.3-5mm thick, unhardened or hardened, annealed in special steel, particularly free cutting qualities, key steel, soft magnetic iron strip and deep drawing quality carbon steel strip; universal plates up to 650mm wide and 16mm thick; electrical non-oriented silicon sheet/coil and electrical non-oriented non-silicon sheet/coil 30-1,250mm wide, 0.35, 0.50 and 0.65mm thick, insulated or non-insulated. Stainless steel – special light sections and cold roll-formed sections (from strip). Springs (leaf and coil springs), torsion bars, stabilisers; bright machine parts (spindles, bars, pistons etc), hardened and chrome plated.

Hoesch AG (continued)

Hoesch Rohr AG, Kissinger Weg,D 4700 Hamm. ***Tel:*** 02381 4201. ***Telex:*** 828661. ***Cables:*** Hoesch Hamm/West.
Management: Board – Jochen Schroer (sales); Hans-Jürgen Tolkemit (technical); Wilhelm Vogel (commercial). **Established:** 1931. **No. of employees:** 2,000. **Annual capacity:** Finished steel: 500,000 tonnes.
Works:
■ Hamm. ***Tube and pipe mills:*** welded – Two HFI (up to 18"); one laser longitudinal weld (1:FR:3/12-2").
■ Dortmund. ***Tube and pipe mills:*** one 20-80" spiral welded.
Products: Carbon steel – longitudinal weld tubes and pipes; spiral-weld tubes and pipes; large-diameter tubes and pipes; oil country tubular goods; plastic coated tubes and pipes (polyethylene and epoxy coated); hollow sections mechanical tubes, laser-welded special steel pipes, boiler and exchanger tubes, line pipes and line pipe fittings, parts and components made of steel tubing.

Hoesch Rothe Erde-Schmiedag AG, Tremoniastrasse 5-11,D 4600 Dortmund 1. ***Tel:*** 0231 1860. ***Telex:*** 822245. ***Cables:*** Rotheerde.
Management: Board – Karl-Friedrich Golücke (chairman); Heinz Hofmann (technical); Jürgen Remmerbach (sales). **Established:** 1970 (in present form). **No. of employees:** 3,000. Seamless rolled rings in steel and non-ferrous metals; drop forgings, spheroidal graphite iron castings; drive system components including large anti-friction bearings and turntables; pit support structures and railway track fastening systems.

Sales offices/agents: West Germany – Eisen und Metall AG, Gelsenkirchen (Tel: 0209 8011. Telex: 8241000) (Trade/Services: Steel – semis, structural shapes and wide flange beams, mining and shipbuilding sections, merchant bars, hot-rolled wire rods, special sections – hot rolled, cold formed, cold drawn and hot extruded, steel sheet piling systems, light-weight sections, rail track material; coated and uncoated strip/sheets, medium and heavy plates, hot-dip and electrolytically galvanized strip/sheets, blackplate, tinplate, tin free steel (ECCS). Steel tubes and pipes – mechanical steel tubes, hollow sections, water and gas pipes, boiler, condenser and superheater tubing, line pipe longitudinal and spiral-weld, steel pipes with polyethelene and epoxy coatig, pile-driving pipes. Processed and finished products – steel wire and wire products, steel strapping, plastic strapping, strapping tools and machines, building components – sandwich panels for roofing and walling, trapezoidal sheets, sectional door elements. Raw materials – steel scrap and pig iron for steelworks and foundries; ferro-alloys, alloyed steel scrap; metal scrap and primary metals, industrial demolition. Non-ferrous semis – rolled products, extrusions and forgings of aluminium, aluminium-alloys, copper, copper-alloys and other metals. Cutting, saw cutting, slitting to length and width. Hoesch Export AG, Dortmund (Tel: 0231 5450. Telex: 17231500) (Activities – Hoesch Export is the export trade organisation of the Hoesch group of companies. Sales programme – Flat rolled products (hot- and cold- rolled); steel strip, steel coils, steel sheets, medium plates, heavy plats, universal mill plates, shipbuilding steel, electrical steel strip/sheets, hot-dip and electrolytically galvanized sheets, plastic coated sheets and coils, blackplate, electrolytic tinplate, tin free steel (ECCS), steel strip (electrolytically metal-coated or clad with non-ferrous metals), sound-absorbing composite sheets. Profiles – semis, structural shapes and wide flange beams, mining sections, merchant bars, hot rolled wire rods, special sections, steel sheet mining systems, trench sheet piling, rail track material. Steel tubes and pipes – mechanical steel tubes, cylinder tubes, square and rectangular tubes, hollow sections, water and gas pipes, boiler, condenser and superheater tubing, line pipe longitudinal and spiral weld, steel pipes with polyethelene and epoxy coating, pile-driving pipes, oil country tubular goods, pipeline accessories. Processed and finished products – steel wire, wire products, strapping tools and machines, steel strapping, plastic strapping, steel castings, drop forgings, blanks. Special products – non-ferrous metals (ingots, slabs, semis, wire, foils), cables, ferroalloys, Blefa roof windows, Blefa drums, tyres and tubes, general merchandise products. Netherlands – Hoesch Nederlands, Den Haag; France – Hoesch France Sarl, Paris; Italy – Hoesch Italia Srl, Milan; UK – Hoesch Ltd, Surrey;

Hoesch AG (continued)

Denmark – Hoesch Danmark, Rostilde; Spain – Hoesch Ibérica SA Madrid; Austria – Hoesch Handelsgesellschaft mbH, Vienna; Switzerland – Hoesch Basel AG, Basel; Sweden (Finland and Norway) – Skandinaviska Hoesch AB, Danderyd; Iran – Hoesch Export AG/Hoesch Handel AG (West Germany), (representative office – Tehran); Japan – Nippon Roballo Co Ltd, Tokyo; Hong Kong – Hoesch Export (Hong Kong) Ltd, Central Hong Kong; China – Hoesch Export AG, Peking; USA – Hoesch America Inc, Atlanta.

J.P. Hüsecken & Co GmbH

Head office: D 5800 Hagen 5. **Postal address:** Postfach 5306. **Tel:** 023 34 48 66-68. **Telex:** 821888.
Products: Carbon steel – cold rolled hoop and strip (uncoated).

Kammerich-Reisholz GmbH—see Mannesmannröhren-Werke AG

Kind & Co Edelstahlwerk

Head office: D 5267 Wiehl. **Telex:** 884291 kdbn. Special steels including high-speed steel, hot and cold work tool steel, die steel, plastic mould steel, stainless steel.

Klöckner Stahl GmbH

Head office: Klöcknerstrasse 29, D 4100 Duisburg 1. **Postal address:** POB 100 248. **Tel:** 0203 396 1. **Telex:** 855817. **Telecopier (Fax):** 396 3535.
Management: Herbert Gienow (chairman), Karl Brotzmann, Günter Büker, Friedrich Haffner, Josef Mennen, Karl Sinkovic. **Established:** 1985. **Capital:** DM 200,000,000. **No. of employees:** 21,000. **Ownership:** Klöckner-Werke AG (100%). **Subsidiaries:** Schmiedewerke Krupp-Klöckner GmbH (50%); Klöckner Draht GmbH (98%); Klöckner CRA Technologie GmbH (50%); Klöckner Stahltechnik GmbH (50%); Klöckner Stahlforschung GmbH (50%).
Works:
■ Hütte Bremen, PO Box 21 02 20, D 2800 Bremen 21. (Tel: 0421 6481. Telex: 240240 kwbd). Works director: D. Terlaak. ***Sinter plant:*** One 150 sq m strand Dwight Lloyd (annual capacity 1,980,000 tonnes). ***Blast furnaces:*** No 3 9.2m hearth dia, 1,561 cu m (1,230,000 tonnes); No 2 12.0m hearth dia, 2,787 cu m (2,520,000 tonnes). ***Steelmaking plant:*** Two 280-tonne basic oxygen converters (5,220,000 tonnes). ***Refining plant:*** One 280-tonne ladle vacuum degassing unit. ***Continuous casting machines:*** 2-strand Voest/Klöckner slab (2,150 x 220mm max) (3,060,000 tonnes). ***Rolling mills:*** wide hot strip – four continuous roughing and finishing stands 90" (max coil weight 45 tonne) (5,508,000 tonnes); hot strip skin pass mill 90" (max coil weight 45 tonne) (480,000 tonnes); two cold reduction – 4-stand 80" (max coil weight 36 tonnes) and 1-stand 90" (max coil weight 36 tonnes) (1,488,000 tonnes); two temper/skin pass – 1-stand 80" (max coil weight 36 tonnes) and 1-stand 56" (max coil weight 23 tonnes) (1,488,000 tonnes).
■ Georgsmarienhütte, PO Box 27 80, D 4500 Osnabrück. (Tel: 0541 3221. Telex: 94742 gmw d). Works director: K. Schäfer. ***Blast furnaces:*** No 3 6.0m hearth dia, 540 cu m (annual capacity 511,000 tonnes). ***Steelmaking plant:*** One blowing station, two 125-tonne KS reactors used alternately (720,000 tonnes). ***Refining plant:*** One 130-tonne ladle vacuum degassing unit. ***Continuous casting machines:*** 6-strand bloom (200 x 240mm and 165 x 165mm) (720,000 tonnes). ***Rolling mills:*** 1-stand blooming and 2-stand billet (480,000 tonnes); 10-stand 700/600/450mm medium section and 15-stand 350/300mm light section (408,000 tonnes).
■ Mannstaedt-Werke, PO Box 11 89, D 5210 Troisdorf. (Tel: 02241 841. Telex: 889551 kmt d). Works director: H. Billen. ***Rolling mills:*** 6-stand 900/450mm medium section (annual capacity 200,000 tonnes); 9-stand 650/330mm light section

Klöckner Stahl GmbH (continued)

(100,000 tonnes). ***Other plant:*** Strip forming lines for cold roll-formed and hollow sections; cold drawing plant for hot rolled special sections; fabrication plant for finished parts including elevator guide rails, wheel rims, turntables, and parts for the automotive industry.
Products: Steelmaking pig iron. Carbon steel – ingots, slabs, blooms, billets, hot rolled coil (for re-rolling), round bars, square bars, flats, special light sections, hot rolled hoop and strip (uncoated), skelp (tube strip), cold rolled hoop and strip (uncoated), medium plates, heavy plates, universal plates, floor plates, clad plates, hot rolled sheet/coil (uncoated), cold rolled sheet/coil (uncoated), electro-galvanized sheet/coil, electrical non-oriented non-silicon sheet/coil, hollow sections. Alloy steel (excl stainless) – ingots, billets, hot rolled coil (for re-rolling), round bars, square bars, flats, special light sections.

Eisenbau Krämer mbH

Head office: An der Bahn 52, D 5912 Hilchenbach-Dahlbruch. **Postal address:** PO Box 4020. **Tel:** 02732 204 0. **Telex:** 875554. **Cables:** Eisenbaukramer Dahlbruch. **Telecopier (Fax):** 02732 20420.
Management: Managing directors – F. Burat (technical), W. Mikulla (purchasing and sales). **Established:** 1921. **Capital:** DM 990,000 (registered). **No. of employees:** 280. **Ownership:** Shareholders: Krämer, Czwalinna.
Works:
■ At head office. Plant manager: Oswald Martin. ***Tube and pipe mills:*** welded – mill for line pipe 20-80" (annual capacity 120,000 tonnes); two hot forming lines for thick-wall pipe (83,000 tonnes).
■ Hellbachstrasse 84-86, D 4350 Recklinghausen-Sud. (Tel: 02361 63025. Telex: 829664). Plant manager: Hans Fuchs. ***Tube and pipe mills:*** welded – three cold forming lines for thick-wall line pipe 30-150" (annual capacity 80,000 tonnes).
Products: Carbon steel – longitudinal weld tubes and pipes, large-diameter tubes and pipes and plastic coated tubes and pipes (thick-wall and line pipe). Stainless steel – longitudinal weld tubes and pipes (thick-wall and line pipe). Alloy steel (excl stainless) – longitudinal weld tubes and pipes (thick-wall and line pipe).
Brands: EKD.

Kronenberg GmbH & Co KG

Head office: Hochstrasse 2, D 5653 Leichlingen 1. **Postal address:** Postfach 249. **Tel:** 2175 9970. **Telex:** 8515895. **Cables:** Kronenberg Leichlingen.
Established: 1900.
Products: Carbon steel – cold roll-formed sections (from strip).

Krupp Stahl AG (KS)

Head office: Alleestrasse 165, D 4630 Bochum 1. **Tel:** 0234 6271. **Telex:** 8258310 ksd. **Telecopier (Fax):** 0234 627 4088.
Management: Gerhard Cromme (chairman), Fritz Fischer (flat products), Hans Graf (sections), Günter Fleckenstein (finance), Karl Meyerwisch (personnel). **Established:** 1965. **Capital:** DM 573,000,000. **No. of employees:** Works – 14,467 (13.12.1986). Office – 5,254 (13.12.1986). (Group: Works – 18,239. Office – 6,693). **Ownership:** Fried. Krupp GmbH, Essen (70.4%); National Iranian Steel Co, Isfahan (25.1%). **Subsidiaries:** Krupp Brüninghaus GmbH, Werdohl (98%); Krupp Stahl Vertriebsgesellschaft mbH, Bochum (100%); Gerlach-Werke GmbH, Homburg/Saar (58%); Schmiedewerke Krupp-Klöckner GmbH, Bochum (50%); Vacmetal Ges. für Vakuum-Metallurgie mbH, Dortmund (50%); Krupp Stahltechnik GmbH, Duisburg (50%). **Annual capacity:** Pig iron: 5,340,000 tonnes. Raw steel: 5,376,000 tonnes. Finished steel: 3,438,000 tonnes.
Works:
■ Rheinhausen. ***Coke ovens:*** Two Didier batteries (48 ovens each) (annual capacity 570,000 tonnes). ***Sinter plant:*** One Lurgi 400m sq strand (4,800,000 tonnes). ***Blast furnaces:*** Four – one 11.5m hearth dia, 2,355 cu m (1,800,000 tonnes); one 11.5m

STOMMEL & VOOS

P.O. Box 10 13 87
Schaberger Straße 39–47
D-5650 Solingen 1 (W.-Germany)

Telephone (02 12) 4 20 09
Telex 8 514 524
Telefax 02 12/4 60 00

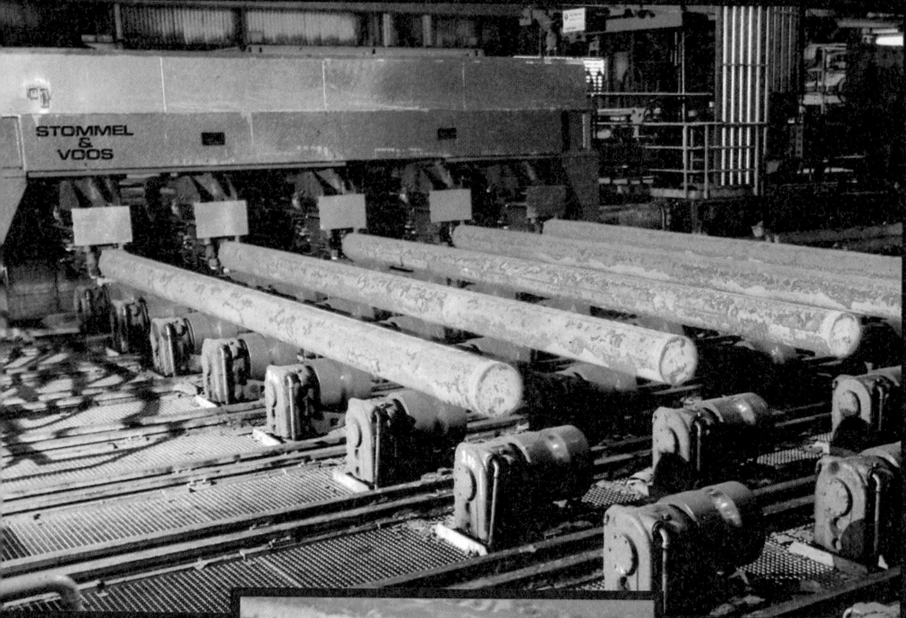

Steelworks all over the world demand the highest standards and reliability in the marking of their products. They trust in the 50 years of experience gained by Stommel & Voos in the construction and development of sophisticated marking machines for steelworks and rolling mills. Stommel & Voos machines stamp slabs, blooms, billets, plates, pipes and sections, at temperatures up to and in excess of 1 000 degrees celsius.

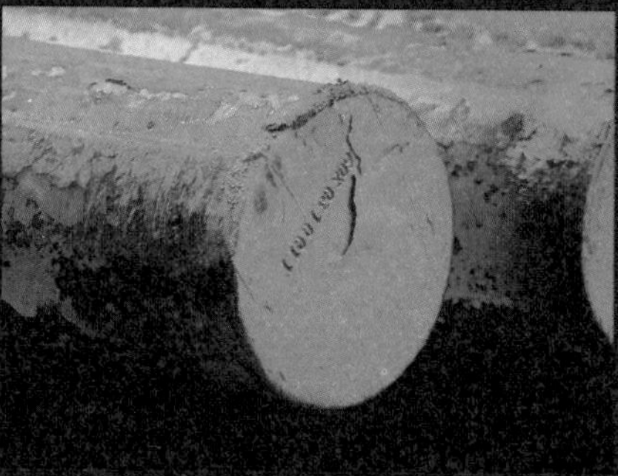

These machines are microprocessor controlled with 100% acknowledgement of the stamped data. They are interfaced to the computer in the customer's own material tracking system. In addition to our comprehensive range of standard machines which cover most applications, we also develop special machines for any specific requirements. For the marking of steelworks and rolling mill products, always specify Stommel & Voos.

Hydraulically operated stamping machine for tubes, bars and sections

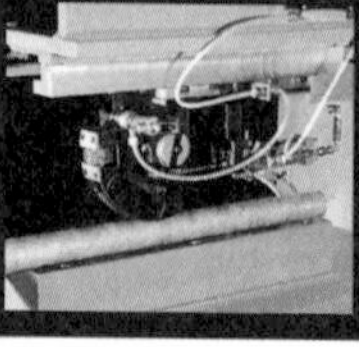

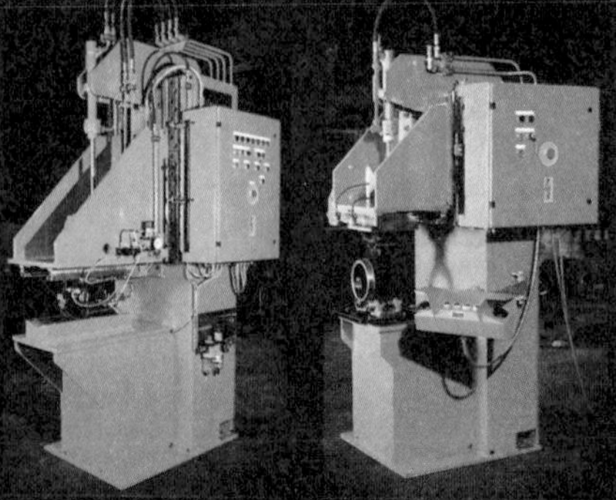

Hydraulically operated universal roll-off stamping machine for flanges, rings and other round products

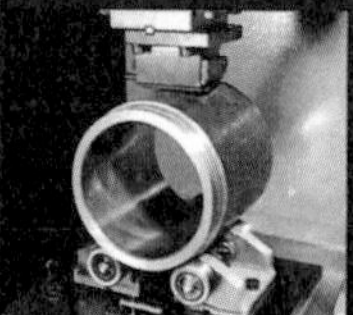

Krupp Stahl AG (KS) (continued)

hearth dia 2,355 cu m (1,800,000 tonnes); one 9.5m hearth dia, 1,693 cu m (900,000 tonnes); one 8.9m hearth dia, 1,436 cu m (800,000 tonnes). ***Steelmaking plant:*** Two 300-tonne basic oxygen converters (3,600,000 tonnes). ***Refining plant:*** ladle furnace – 300-tonne and 120-tonne (1,800,000 tonnes); vacuum degassing – 300-tonne and 120-tonne (1,500,000 tonnes). ***Continuous casting machines:*** One combination bloom and slab – 4-strand for blooms 600 x 260 (1,800,000 tonnes) and 2-strand for slabs 850-1,650 x 260mm (2,200,000 tonnes). ***Rolling mills:*** One 2-high blooming 1,230mm dia (coil weight 3 to 11 tonnes) (1,560,000 tonnes); billet – continuous 10-stand 800-600mm (2,400,000 tonnes (2 x 1,200,000 tonnes)); heavy section and rail – 2-stand 2/3-high 850mm (384,000 tonnes).

■ Bochum. ***Steelmaking plant:*** 140-tonne 85 MVA electric arc furnace (annual capacity 670,000 tonnes). ***Refining plant:*** Two 75-tonne AOD converters (600,000 tonnes); one 75-tonne ladle furnace (600,000 tonnes); two 75-tonne vacuum degassing (600,000 tonnes). ***Continuous casting machines:*** 1-strand curved mould slab (240 x 865-1,650mm) (480,000 tonnes). ***Rolling mills:*** wide hot strip – 4-high reversing roughing stand and seven 4-high finishing stands, 1,800mm roll length (coil weight 19kg/mm, maximum 30 tonnes) (3,200,000 tonnes); cold reduction – 4-stand 4-high 1,600 width, 0.35-3.5mm thick (coil weight 38 tonnes) (1,100,000 tonnes); temper/skin pass – 1-stand 4-high 1,600mm/0.35-3.5mm (coil weight 38 tonnes) (900,000 tonnes). ***Coil coating lines:*** electro-galvanizing insoluble anodes, horizontal Krupp/Sundwig (180,000 tonnes).

■ Siegen. ***Steelmaking plant:*** Two electric arc furnaces – one 105-tonne (annual capacity 720,000 tonnes), one 80-tonne (216,000 tonnes). ***Refining plant:*** One 2.6-13.2-tonne ESR unit (5,640 tonnes); one 140-tonne ladle furnace (600,000 tonnes); one 140-tonne vacuum degassing (600,000 tonnes); one 80-tonne VOD (240,000 tonnes). ***Continuous casting machines:*** 6-strand Krupp billet (135/150/265mm sq) (480,000 tonnes (1987)). ***Rolling mills:*** One 3-stand billet 50-150mm sq/75-160mm round (576,000 tonnes); one 21-stand, bar 15-75mm round (288,000 tonnes); one 24-stand wire rod 5.5-27mm dia (174,000 tonnes); one merchant mill for 35/22-300/64mm flats, 42-98mm hexagons and 38-45mm squares (192,000 tonnes); one 4-high reversing medium plate 800-2,500mm x 2-100mm (28,200 tonnes).

■ Düsseldorf. ***Rolling mills:*** cold reduction : Four Sendzimir – two 1,300mm, one 1,550mm, and one 1,600mm (coil weight 17kg/mm) (annual capacity 306,000 tonnes); one 4-high 1,150mm. Three 2-high temper/skin pass – one 1,300mm, one 1,500mm, and one 1,600mm (coil weight 17kg/mm).

Products: Carbon steel – ingots, slabs, blooms, billets, wire rod, round bars, square bars, flats, hexagons, heavy joists, heavy channels, heavy rails, light rails, rail accessories, hot rolled hoop and strip (uncoated), cold rolled hoop and strip (uncoated), medium plates, heavy plates, hot rolled sheet/coil (uncoated), cold rolled sheet/coil (uncoated), hot-dip galvanized sheet/coil, electro-galvanized sheet/coil, bright wire. Stainless steel – ingots, slabs, blooms, billets, round bars, square bars, flats, hexagons, hot rolled hoop and strip (uncoated), cold rolled hoop and strip (uncoated), medium plates, heavy plates, hot rolled sheet/coil (uncoated), cold rolled sheet/coil (uncoated), bright annealed sheet/coil, wire, Alloy steel (excl stainless) – ingots, slabs, blooms, billets, round bars, square bars, flats, hexagons, hot rolled hoop and strip (uncoated), cold rolled hoop and strip (uncoated), medium plates, heavy plates, hot rolled sheet/coil (uncoated), cold rolled sheet/coil (uncoated), wire.

Lech-Stahlwerke GmbH

Head office: Industriestrasse 1, D 8901 Meitingen-Herbertshofen. **Tel:** 0 82 71 820. **Telex:** 533180.

Associates: Saarstahl Völklingen.

Works:

■ Meitingen-Herbertshofen. ***Steelmaking plant:*** Three electric arc furnaces (annual capacity 500,000 tonnes). ***Continuous casting machines:*** One 4-strand Danieli billet. ***Rolling mills:*** One Danieli continuous bar.

Products: Carbon steel – billets, wire rod, reinforcing bars, round bars, square bars, flats, light angles, light tees, light channels.

Lemmerz-Werke KGaA

Head office: D 5330 Königswinter 1. **Postal address:** Postfach 1120. **Tel:** 0 22 23 71-0. **Telex:** 885243. Teletext: 222350. **Telecopier (Fax):** 02223 71620.
Works:
■ Königswinter. ***Steelmaking plant:*** Two 30-tonne electric arc furnaces. **Rolling mills:** 2-high reversing universal plate mill.
Products: Carbon steel – (Engineering steel) ingots; universal plates. Alloy steel (excl stainless) – ingots; universal plates.

Mannesmann Edelstahlrohr GmbH

Head office: Industriestrasse 8, D 4018 Langenfeld 2.
Established: 1985 (to combine special steel tube operations of Mannesmann and VDM). **Ownership:** Mannesmann Group.
Products: Stainless steel – seamless tubes and pipes, longitudinal weld tubes and pipes. Tubes in nickel, nickel-base alloys and other materials.
Sales offices/agents: Mannesmann Edelstahl GmbH, Weyerstrasse 41-43, Postfach 11 07 40, 5650 Solingen 11.

Mannesmannröhren-Werke AG

Head office: Mannesmannufer 3, D 4000 Düsseldorf. **Postal address:** Postfach 1104. **Tel:** 0211 8750. **Telex:** 8581421. **Telecopier (Fax):** 0211 875 3245.
Management: Board directors – Wulf Dietrich Liestmann (chairman), Karl-Heinz Kaufmann, Gerd Pfeiffer, Reinhold Schreiner, Wolfgang Wiedenhoff. **Established:** 1970. **Capital:** DM 400,000,000. **No. of employees:** 32,000 (1985 total). **Ownership:** Mannesmann AG, Düsseldorf (75%); Thyssen AG, Duisburg (25%). **Subsidiaries:** (Including Associates) Röhrenwerke Bous/Saar GmbH, Mannesmann Edelstahlrohr GmbH; Mecano-Bundy GmbH; Wälzlagerrohr GmbH; Robur Buizenfabriek BV (Netherlands) and Mannesmann-Sümerbank/Boru Endųstrisi TAS, Turkey (all steel producers); Mannesmann Oilfield Tubulars Corp, Houston, TX, USA (sales of tubes in North America); Mannesmann Kronprinz AG (automotive wheels); Mannesmann Rohstoffwerke GmbH (limestone and refractories); Mannesmann Reederei GmbH (freight transport and forwarding); HWT Gesellschaft für Hydrid- und Wasserstofftechnik mbH (hydride and hydrogen technology). Mannesmannröhren-Werke is part of the Mannesmann AG group which also includes Mannesmann Demag AG (plant and equipment for the metallurgical, mining, materials handling and other industries); Mannesmann Rexroth GmbH (hydraulic systems); Mannesmann Anlagenbau AG (general contracting including pipelines, piping systems and industrial plant); Mannesmann Hartmann & Braun AG (instrumentation and process control systems); Mannesmann Kienzle GmbH (computers and information systems); Mannesmann Handel AG (trading in pipe and tube, steel products, raw materials, etc); Mannesmann SA, Brazil (iron and steel products including seamless tubes, rolled steel products and wire). **Annual capacity:** Raw steel: 3,700,000 tonnes.
Works:
■ Huckingen (Duisburg). **Coke ovens and sinter plant**. ***Blast furnaces:*** one (A) 10.3m hearth dia, 2,081 cu m (annual capacity 2,000,000 tonnes); one (B) 10.3m hearth dia, 2,226 cu m (1,600,000 tonnes). ***Steelmaking plant:*** Three 225-tonne LD-CB basic oxygen converters (change vessel system with two blowing stations) (3,700,000 tonnes). ***Continuous casting machines:*** Two 2-strand Mannesmann Demag slab (150-250 x 800-2,050mm) (2,000,000 tonnes); round – one 6-strand Mannesmann Demag (177 and 220mm) (1,400,000 tonnes) and one 5-strand Mannesmann Demag (177mm, 220mm, 270mm, 310mm and 406mm) (1,200,000 tonnes).
■ Mülheim. ***Rolling mills:*** One 4-high Mannesmann Demag Sack heavy plate (5m). ***Tube and pipe mills:*** seamless – one 27-140mm x 25mm Mannesmann Demag continuous with stretch reducing; one 21-178mm x 25mm Mannesmann Demag continuous with stretch reducing; one 89-226mm x 70mm Mannesmann Demag/Kocks Pilger. Welded – one 13-89mm x 7mm Aetna Fretz-Moon; one 159-508mm x

West Germany

Mannesmannröhren-Werke AG (continued)

15mm Mckay high-frequency induction. ***Other plant:*** Plant for pipe bends.
■ Düsseldorf-Rath. ***Tube and pipe mills:*** seamless – one 178-346mm x 25mm Mannesmann Demag plug and sizing; one 216-660mm x 120mm Mannesmann Demag plug and sizing. Welded – one 32-168mm x 9mm Mannesmann Demag high frequency induction.
■ Remscheid. ***Tube and pipe mills:*** seamless – cold Pilger mills for roller bearing tube. ***Other plant:*** Hydraulic extrusion press for high alloy tubes and hollow billets of 60-220mm for cold finishing at Mannesmann Edelstahlrohr GmbH.
■ Düsseldorf-Reisholz. Erhardt press for heavy wall seamless pipe up to 1,440mm.
■ Wickede. ***Tube and pipe mills:*** welded – for precision tubes.
■ Brackwede. ***Tube and pipe mills:*** seamless – plant for precision tubes.
■ Holzhausen. ***Tube and pipe mills:*** seamless – plant for precision tubes.
Products: Carbon steel – pipe piling; heavy plates 7-150mm thick, up to 4,800mm wide, up to 25m long; seamless tubes and pipes 21-660mm dia, up to 120mm wall thickness; longitudinal weld tubes and pipes; large-diameter tubes and pipes; oil country tubular goods; galvanized tubes and pipes; cold drawn tubes and pipes; precision tubes and pipes (seamless 1.5-380mm dia, 0.2-25mm dia; also welded); pipe bends, axle parts, cylinders and pressure vessels.
Sales offices/agents: Mannesmann Handel AG, Düsseldorf, with representation worldwide; Mannesmann Edelstahl GmbH, Solingen; Stahlkontor Hahn GmbH, Ratingen; and Mannesmann Oilfield Tubulars Corp, Houston, TX, USA.

Maxhütte – Eisenwerke-Gesellschaft Maximilianshütte mbH

Head office: Hauptstrasse 51, D 8458 Sulzbach-Rosenberg. **Tel:** 09661 60 1. **Telex:** 63837 0. **Cables:** Maxhutte Sulzbach-Rosenberg. **Telecopier (Fax):** 09661 60 570.
Management: Joachim Oberländer (chairman), Winfrid Groetschel, Friedrich Höfer, Manfred Leiss. **Established:** 1853. **Capital:** DM 192,000,000. **No. of employees:** Works – 3,910. Office – 840. **Ownership:** Klöckner-Stahl GmbH (49.9%), Eschweiler Bergwerksverein (15%). **Associates:** Salmax, Gesellschaft für Oberflächenveredelte Feinbleche mbH, Salzgitter (49%) (galvanized sheet). **Annual capacity:** Pig iron: 1,100,000. Raw steel: 1,100,000.
Works:
■ At head office. ***Blast furnaces:*** Five 5.0m hearth dia, 413 cu m; 6.8m hearth dia, 709 cu m; 6.3m hearth dia, 675 cu m; 6.8m hearth dia, 709 cu m; 6.3m hearth dia, 697 cu m. ***Steelmaking plant:*** Three 70-tonne OBM basic oxygen converters. ***Refining plant:*** One GHH ladle furnace; one DH and one RH vacuum degassing units; argon rinsing stations. ***Continuous casting machines:*** Two 3-strand Concast-Schloemann billet (one 80 and 108mm sq; and one 109, 140, 160 and 180mm sq, (annual capacity 600,000 tonnes); one 3-strand Schloemann-Siemag bloom (320 x

Maxhütte – Eisenwerke-Gesellschaft Maximilianshütte mbH (continued)

430mm) (600,000 tonnes). ***Rolling mills:*** One 2-high reversing 1,100mm blooming; one 4-stand 850/900mm billet and section (500,000 tonnes); One 18-stand merchant and reinforcing bar (500,000 tonnes approx). ***Tube and pipe mills:*** seamless – one Erhardt with cold finishing plant (100,000 tonnes), seven drawing and two cold Pilger machines; one planetary (600,000 tonnes).
Products: steelmaking pig iron (for own use). Carbon steel – blooms (rolled and continuous cast) 320mm sq, 320 x 430mm; billets 50-190mm sq, 100-180mm dia; semis for seamless tubemaking 50-190mm sq, 100-180mm dia; wire rod 12-37mm dia; reinforcing bars 8-28mm dia; round bars 12-37mm dia; square bars 12-30mm sq; flats 5 x 20mm to 30 x 55mm (12 x 70); light angles 20 x 3mm to 45 x 7mm; medium joists 240-400mm; medium channels 180-400mm; heavy rails (S 49, S 54; UIC 54, UIC 60; AREA 115, 132, 133, 136); seamless tubes and pipes 1.8 x 17.2mm to 36 x 177.8mm (60.0 x 165.1). Alloy steel (excl stainless) – blooms (rolled and continuous cast) – 320mm sq, 320 x 430mm); billets 50-190mm sq, 100-180mm dia; semis for seamless tubemaking 50-190mm sq, 100-180mm dia; wire rod 12-37mm dia; round bars 12-37mm dia; square bars 12-30mm sq; flats 5 x 20mm to 30 x 55mm (12 x 70); light angles 20 x 3mm to 45 x 7mm; light channels 180-400mm.
Additional information: Alloy steel accounts for about 40% of total production. Maxhütte filed for bankruptcy in April 1987. A number of plans to continue production under new arrangements were being considered as we went to press.

Mecano-Bundy GmbH

Head office: Dischinger Strasse 11, D 6900 Heidelberg. **Postal address:** Postfach 103940. **Tel:** 06221 702 0. **Telex:** 461826 bundy d. **Telecopier (Fax):** 06221 702255.
Ownership: Mannesmann Group.
Products: Carbon steel – precision tubes and pipes Bundy tubing – double-walled tube made from from copper-plated strip. Plastic tubing and control valves.

Moselstahlwerk GmbH & Co KG

Head office: Hafenstrasse, D 5500 Trier, Rheinland-Pfalz. **Postal address:** PO Box 181 229. **Tel:** 555 02. **Telex:** 472839.
Management: Paul G. Lauer (managing director), Bertram Wagner (technical manager). **Established:** 1970. **No. of employees:** Works – 260. Office – 27. **Ownership:** Alfred Rass, Walter Rass (Ultimate holding company – Rass Stahl AG). **Associates:** Hochwald Drahtwerk GmbH & Co (reinforcement mesh, wire, etc); Kaufmann & Lindgens GmbH (reinforcement mesh). **Annual capacity:** Finished steel: 220,000 tonnes (billets); 200,000 tonnes (wire rod).
Works:
■ At head office. ***Steelmaking plant:*** Two 45-tonne electric arc furnaces (annual capacity 250,000 tonnes). ***Continuous casting machines:*** One 3-strand billet (130 x 130 x 4,000mm, 115 x 115 x 4,000mm) (250,000 tonnes). ***Rolling mills:*** One semi-continuous wire rod (coil weight 420kg) (200,000 tonnes).
Products: Carbon steel – billets, wire rod.
Sales offices/agents: Stinnes Montanhandel GmbH & Co KG, PO Box 5870, D 6800 Mannheim 1.

Stahlwerke Peine-Salzgitter AG (P&S)

Head office: Eisenhüttenstrasse 99, D 3320 Salzgitter 41, Niedersachsen. **Tel:** 05341 21 1. **Telex:** 954481 0 sg d, 92665 spspe d.
Management: Kurt Stähler (Vorsitzender), Günter Geisler, Peter Kehl, Jürgen Kolb, Eberhard Luckan, Jürgen Meyer. **Established:** 1970. **Capital:** DM 874,000,000 (incl reserves). **No. of employees:** 12,288 (30 Sept 1986). **Ownership:** Salzgitter AG (99.46%). **Subsidiaries:** Salzgitter Stahl GmbH, Düsseldorf (steel and pipe trading); Deumu Deutsche Erz- und Metall-Union GmbH, Hannover (steel, scrap and metal

Stahlwerke Peine-Salzgitter AG (P&S) (continued)

trading); Salmax Gesellschaft für oberflächenveredelte Feinbleche mbH, Salzgitter (surface treated sheets).

Works:

■ At head office. ***Coke ovens:*** Two Otto batteries (54 ovens each). ***Sinter plant:*** One 180 sq m strand and four 75 sq m strands. ***Blast furnaces:*** Five – No A 10.8m hearth dia, 2,330 cu m; No7 9.5m hearth dia, 1,606 cu m; No 6 and 5 8.2m hearth dia, 1,164 cu m each; No 1 7.6m hearth dia, 1,065 cu m. ***Steelmaking plant:*** Three 200-tonne LD basic oxygen converters. ***Refining plant:*** Ladle degassing plant, TN-desulphurisation plant. ***Continuous casting machines:*** One 2-strand slab (bow type) (210-250mm x 1,000-2,000mm); one 1-strand round (bow type) (210-250mm x 1,950-2,600mm). ***Rolling mills:*** One 3,200mm heavy plate; one 7-stand 80" wide hot strip (coil weight 32 tonnes); cold reduction – one 4-stand 4-high and 1-stand 6-high 80" tandem (coil weight 32 tonnes); two 4-high temper/skin pass. ***Tube and pipe mills:*** welded – large diameter spiral pipe. ***Coil coating lines:*** One hot-dip galvanizing; one electro-galvanizing; one colour coating (plastic). ***Other plant:*** Hot strip slitting line; coil slitting lines and cut-to-length lines; profiling line; pickling line; foundry.

■ Peine, Ilsede. ***Coke ovens:*** Two Koppers batteries (56 ovens and 23 ovens). ***Steelmaking plant:*** Three 100-tonne LD/AC basic oxygen converters. ***Refining plant:*** Ladle degassing plant. ***Continuous casting machines:*** One 6-strand bloom (bow type) (165 x 165mm, 200 x 200mm, 230 x 230mm, 230 x 340mm). ***Rolling mills:*** One universal heavy section (up to 1,000mm web); one continuous universal medium section (up to 400mm web). ***Other plant:*** Sheet piling plant.

Products: Steelmaking pig iron, foundry pig iron and special pig iron. Carbon steel – slabs; blooms; billets; special light sections; medium joists; medium channels; wide-flange beams (medium); universals (medium); heavy joists, heavy channels, wide-flange beams (heavy) and universals (heavy) (up to 1,000mm web); sheet piling, box piling and bearing piles; medium plates and heavy plates (up to 3,000mm wide, surface treated plate); hot rolled sheet/coil (uncoated) up to 1,880mm wide; cold rolled sheet/coil (uncoated) up to 1,850mm wide; hot-dip galvanized sheet/coil; electro-galvanized sheet/coil; colour coated sheet/coil; spiral-weld tubes and pipes; large-diameter tubes and pipes. Grey and spheroidal cast iron, cast steel, trapezoidal sections made from galvanized and plastic-coated sheets, ground basic slag, Thomas lime.

Brands: Folastal (plastic coated sheet).

Sales offices/agents: Sales office – at head office. Agents – Salzgitter Stahl GmbH, Düsseldorf, West Germany; Salzgitter Belgique SA, Brussels, Belgium; Salzgitter France SA, Saint-Mandé, France; Salzgitter Italiana Srl, Milan, Italy; Salzgitter Española SA, Madrid, Spain; Salzgitter Stahl GmbH, Vienna, Austria; Salzgitter (London) Ltd, London, UK; Salzgitter de México SA de CV, Mexico, DF; Delta Steel Inc, Houston, TX, and Feralloy Corp, Chicago, IL, USA.

Stahlwerke Plate GmbH & Co KG

Head office: D 5880 Lüdenscheid-Platehof. **Tel:** 02351 4390. **Telex:** 826869 plat d.

Works:

■ At head office. Alloy and special steel bars, rings and forgings.

Poppe & Potthoff GmbH & Co

Head office: Feldweg 21-23, D 4806 Werther. **Postal address:** Postfach 1150. **Tel:** 05203 7010. **Telex:** 931904 rohr d.

Established: 1928.

Products: Stainless steel – longitudinal weld tubes and pipes, cold drawn tubes and pipes, precision tubes and pipes. Alloy steel (excl stainless) – seamless tubes and pipes, cold drawn tubes and pipes, precision tubes and pipes.

West Germany

Rasselstein AG

Head office: Engerser Landstrasse 17, D 5450 Neuwied. **Postal address:** Postfach 2020. **Tel:** 02631 810. **Telex:** 867841 a ras d. **Cables:** Rasselstein Neuweid. **Telecopier (Fax):** 812 903.
Management: Claus Freiling, Heinrich Hütten, Franz Weber. **Established:** 1760. **Capital:** DM 146,000,000. **No. of employees:** Works – 3,250. Office – 350. **Ownership:** Thyssen AG (50%), Eisen- und Hüttenwerke AG (50%). **Annual capacity:** Rolled steel: 2,502,000 tonnes.
Works:
■ Neuwied. ***Rolling mills:*** One 4-high reversing cold reduction (annual capacity 432,000 tonnes). ***Coil coating lines:*** Electrolytic (150,000 tonnes).
■ Andernach (Tel: 02631 810. Telex: 867841 ras d). ***Rolling mills:*** Two cold reduction – 6-stand tandem (1,300,000 tonnes), 5-stand tandem (770,000 tonnes). ***Coil coating lines:*** Four combined electrolytic tinning and tin-free steel (ECCS) (combined capacity 940,000 tonnes), three other coil coating lines (56,000 tonnes).
Products: Carbon steel – cold rolled hoop and strip (uncoated) (output in 1986, 5,000 tonnes), cold rolled sheet/coil (uncoated) (295,000 tonnes), terne-plate (lead coated) sheet/coil (102,000 tonnes), blackplate (108,000 tonnes), electrolytic tinplate (single-reduced) and electrolytic tinplate (double-reduced) (520,000 tonnes), tin-free steel (ECCS) (146,000 tonnes).
Brands: Andralyt (electrolytic tinplate), Ancrolyt (tin-free steel), Raternet and Neuratern (terne-plate), Neuralyt (corrosion-protected sheet).

Hagener Gussstahlwerke Remy GmbH (Remystahl)

Head office: Eckeseyerstrasse 112-116, D 5800 Hagen 1. **Postal address:** PO Box 1340. **Tel:** 02331 25051. **Telex:** 823786 remy d. **Telecopier (Fax):** 02331 25059.
Established: 1856. Rolled and forged bars. High-speed steel, tool steel, stainless steel. Also nickel-base alloys.

Risse & Wilke GmbH & Co

Head office: Oeger Strasse 8, D 5800 Hagen. **Postal address:** Postfach 54 09. **Tel:** 02334 4941-5. **Telex:** 0821856. **Cables:** R & W.
Products: Carbon steel – cold rolled hoop and strip (uncoated) 0.004-0.08% carbon, 20-440mm wide, 0.30-4.00mm thick.

Rogesa Roheisengesellschaft Saar mbH

Head office: Postfach 16 07, D 6638 Dillingen, Saarland. **Tel:** 0 68 31 47 0. **Telex:** 443711 0 dhd. **Telecopier (Fax):** 0 68 31 47 22 12.
Management: Jean Lang (chairman), Rudolf Judith (deputy chairman), Roland de Bonneville, Peter Hartz. **Established:** 1981. **Capital:** DM 30,000,000.
Ownership: AG der Dillinger Hüttenwerke (50%); Saarstahl Völkingen GmbH (50%). **Annual capacity:** Pig iron: 4,626,000 tonnes.
Works:
■ At head office. Works director: H. Ehl. ***Sinter plant:*** One 180 sq m strand (annual capacity 2,000,000 tonnes), one 258 sq m strand (3,200,000 tonnes). ***Blast furnaces:*** No 1 8.5m hearth dia, 1,270 cu m (in reserve); No 3 8.5m hearth dia, 1,270 cu m (798,000 tonnes); No.4 10.0m hearth dia, 1,790 cu m (1,730,000 tonnes); No 5 11.0m hearth dia, 2,220 cu m (2,098,000 tonnes).
Products: Steelmaking pig iron.

Rothrist Rohr GmbH

Head office: An der Knippenburg, D 4250 Bottrop. **Telex:** 8579408 roro.
Products: Carbon steel – longitudinal weld tubes and pipes.

Eisen- und Stahlwalzwerke Rötzel GmbH

Head office: Josefstrasse 82, D 4054 Nettetal 1/Breyell. **Postal address:** Postfach 3160. **Tel:** 02153 7350. **Telex:** 854212 waroe d.
Works:
■ Nettetal. ***Rolling mills:*** Two hot strip and hoop; one cold strip.
Products: Carbon steel – hot rolled hoop and strip (uncoated), cold rolled hoop and strip (uncoated).

Saarstahl Völklingen GmbH

Head office: Bismarckstrasse 57-59, D 6620 Völkligen, Saarland. **Postal address:** PO Box 10 19 80. **Tel:** 0 68 98 101. **Telex:** 4424110 svk. **Cables:** Saarstahl Völklingen. **Telecopier (Fax):** 0 68 98 10 40 01.
Management: Board members – Kurt K.E. Kühn (chairman, sales), Peter Hartz (personnel), Hans-Günter Herfurth (technical). Supervisory board – Manfred Schäfer (chairman). Chief representatives – Olaf-Roman Baron von Engelhardt (sales), Guido Scheer (controller). **Established:** 1881. **Capital:** DM 330,000,000. **No. of employees:** Works – 8,705. Office – 2,406 (31.12.86). **Ownership:** Manfred Schäfer (in trust for the Government of Saarland) (76%), Aciéries Réunis de Burbach-Eich-Dudelange SA (Arbed), Luxemburg (24%). **Subsidiaries:** TechnoSaarstahl GmbH (100%); Burbach-Homburger Stahl-und Waggonbau GmbH (49.99%); Lech-Stahlwerke GmbH (77.18%); IWKA Stahlflaschen GmbH (90%). **Associates:** Rogesa Roheisengesellschaft Saar mbH (50%); ZKS Zentralkokerei Saar GmbH (25.5%); Kraftwerk Wehrden (33.33%). **Annual capacity:** Raw steel: 3,672,000 tonnes. Finished steel: 3,500,000 tonnes.
Works:
■ At head office. ***Steelmaking plant:*** Three 160-tonne basic oxygen (annual capacity 3,240,000 tonnes); one 125-tonne electric arc furnace (428,000 tonnes); one 3.5-tonne vacuum induction furnace (4,000 tonnes). ***Refining plant:*** Two VAR – No 1 13-tonne, No 2 165-tonne. ***Continuous casting machines:*** billet – one 6-strand for special steel (240 x 240) (384,000 tonnes); and four 6-strand for LD steel (240 x 320) (3,240,000 tonnes). ***Rolling mills:*** Two blooming – one reversing, 2-stand 3-high and 1-stand 2-high (1,150mm, 700mm) (1,520,000 tonnes); 2-stand 3-high 750mm semis mill (1,200,000 tonnes); one heavy section – three 3-high, one 2-high and one universal (1,340mm) (450,000 tonnes); one medium section – ten 2-high, two universal, and one edging stand (460mm; 650mm; 980mm) (600,000 tonnes); one light section – four 2-high, and seventeen 2-high stands (340mm) (420,000 tonnes); one wire rod – one 3/2-high, and twenty 2-high stands (280mm) (coil weight 660/

Saarstahl Völklingen GmbH (continued)

1,000kg) (222,000 tonnes). ***Other plant:*** Plant for ready-to-use short bars; plant for wire rod, quenched and tempered in coil; bar peeling plant; heat treatment plant; forging plant; special steel processing plant.
■ Saarbrücken-Burbach. ***Rolling mills:*** One wire rod – 15-stand and 4 x 10-stand 2-high (coil weight 2,500kg) (annual capacity 920,000 tonnes). ***Other plant:*** Tor-steel plant.
■ Neunkirchen. (Tel: 0 68 21 161. Telex: 444813). Works director: Jörg Haschke. ***Rolling mills:*** One 18-stand continuous light section (365-600mm) (coil weight 2,200kg) (annual capacity 516,000 tonnes); one 28-stand continuous wire rod (coil weight 2,000kg) (372,000 tonnes). ***Other plant:*** Mining frames plant.
■ Homburg. (Tel: 0 68 41 1 931. Telex: 044613). Works director: Aribert Becker. ***Tube and pipe mills:*** welded ***Other plant:*** Three galvanizing lines; bright drawing plant; plant for screws, bolts, nuts etc; light-weight steel cylinder plant.
Products: Carbon steel – ingots; blooms; billets; wire rod 5.5-47mm; reinforcing bars 6-20mm; round bars 10-110mm; square bars 16-80; flats 18 x 5-155 x 80; hexagons 16-67mm; bright (cold finished) bars; light angles 45-100 and 30 x 20 – 150 x 75mm; light tees 30-120; light channels 30 x 15-70 x 40mm; medium angles; medium tees; medium tees; medium joists; medium channels; wide-flange beams (medium); universals (medium); welded sections (medium); heavy angles; heavy tees; heavy joists; heavy channels; wide-flange beams (heavy); universals (heavy); universal plates; bright wire; longitudinal weld tubes and pipes; galvanized tubes and pipes. Stainless steel – ingots, wire rod, round bars, square bars, flats, hexagons, light angles, light tees, light channels, special light sections. Alloy steel (excl stainless) – ingots, billets, wire rod, round bars, square bars, flats, hexagons, bright (cold finished) bars, light angles, light tees, light channels, special light sections, universal plates.
Sales offices/agents: Saarstahl Vertriebsges mbH; Saarstahl Export GmbH; Secosar SA, Paris., France; Les Aciers Fins de la Sarre SA, Liège, Belgium; Saarsteel Inc, Whitestone, USA; Acciai della Saar SpA, Milan, Italy; Saarstahl AG, Zürich, Switzerland; Saarstahl Iberica SA, Barcelona.

Schmidt & Clemens GmbH & Co

Head office: PO Box 11 40, D 5253 Lindlar, Nordrhein-Westfalen. **Tel:** 02266 920. **Telex:** 884547. **Cables:** Schmidtclemens. **Telecopier (Fax):** 02266 92 355.
Management: President – Christoph Schmidt-Krayer; vice presidents – Franz Erdmann (technical), Peter M. Greissmayr (finance). **Established:** 1879. **Capital:** DM 12,000,000. **No. of employees:** Works – 900. Office – 200. **Annual capacity:** Finished steel: 15,000 tonnes.
Works:
■ At head office. ***Steelmaking plant:*** One 8-tonne electric arc furnace (annual

West Germany

Schmidt & Clemens GmbH & Co (continued)

capacity 10,000 tonnes). Refining plant: One 8-tonne AOD (10,000 tonnes). ***Other plant:*** Alloy steel – forge with 1,900-tonne press and hammers; static castings foundry up to 5,000-kg; investment castings foundry up to 150-kg; centrifugal castings foundry up to 2,000mm od and single lengths up to 6,400mm.
Products: Stainless steel – billets; round bars and square bars (forged); forgings. Alloy steel (excl stainless) – billets; round bars and square bars (forged); forgings. Finish-machined metal extrusion tools, and rolls for various applications.
Brands: Märker (steel for all Schmidt & Clemens brands); Irrubigo (steel for all Schmidt & Clemens stainless steel brands); Euzonit (alloys for all Schmidt & Clemens highly corrosion resistant superalloys).

Schoeller Werk GmbH & Co KG

Head office: D 5374 Hellenthal/Eifel. **Postal address:** Postfach 67. **Tel:** 0 24 82 810. **Telex:** 833622. **Cables:** Schoellerwerk Hellenthal. **Telecopier (Fax):** 02482 81109.
Products: Stainless steel – longitudinal weld tubes and pipes, cold drawn tubes and pipes.

Schwarzwälder Röhrenwerk

Head office: Freudenstädter Strasse 58, D 7272 Altenstieg 3. **Telex:** 765396 srfw d.
Ownership: Cockerill Sambre (Belgium) (99.52%).
Products: Carbon steel – longitudinal weld tubes and pipes.

Friedrich Gustav Theis Kaltwalzwerke GmbH

Head office: Bandstahlstrasse 14, D 5800 Hagen-Halden. **Tel:** 02331 6930. **Telex:** 823152.
Products: Carbon steel – cold rolled hoop and strip (uncoated), nickel and zinc plated strips, flat wire.

Thyssen AG vorm. August Thyssen-Hütte

Head office: Kaiser-Wilhelm-Strasse 100, D 4100 Duisburg 11. **Postal address:** Postfach 11 05 61. **Tel:** 0203 521. **Telex:** 855401 thy d. **Telecopier (Fax):** 0203 52 25102.
Management: Dieter Spethmann (chairman), Werner Bartels, Heinz Kriwet, Karlheinz Rösener, Heinz-Gerd Stein, Dieter H. Vogel, Hans Gert Woelke, Karl-August Zimmermann. **Subsidiaries:** These include Thyssen Industrie AG (90%); Thyssen Handelsunion AG (100%); Thyssen Edelstahlwerke AG (100%); Thyssen Stahl AG (100%). **Activities** Holding company of the Thyssen Group.

Thyssen Bandstahl Berlin GmbH

Head office: Berliner Strasse 19, D 1000 Berlin 27. **Postal address:** Postfach 27. **Telex:** 181829 band d.
Ownership: Thyssen AG.
Works:
■ At head office. ***Rolling mills:*** hot strip and hoop.
Products: Carbon steel – hot rolled hoop and strip (uncoated).

Thyssen Edelstahlwerke AG

Head office: Oberschlesienstrasse 16, D 4150 Krefeld. **Postal address:** Postfach 730. **Tel:** 02151 83 1. **Telex:** 85312 0 te d.
Management: Karlheinz Rösener (chairman), Tilmann Kömpel, Wilhelm Küpper, Siegfried Robert, Gerd Topp. **Established:** 1927/1975. **Capital:** DM 150,000,000. **No. of employees:** 13,338. **Ownership:** Thyssen AG, Duisburg (100%). **Subsidiaries:** These include – Deutsche Edelstahlwerke GmbH, Krefeld; Essener Stahl- und Metallhandelsgesellschaft mbH, Essen; Wälzlagerrohr GmbH, Krefeld.
Works:
■ At head office. ***Steelmaking plant:*** Two electric arc furnaces – one 80-tonne and one 30-tonne; 4-tonne medium frequency furnace. ***Refining plant:*** Two 25-tonne AOD; two ESR; two vacuum arc furnaces. ***Continuous casting machines:*** One 1/2-strand Sack slab (150 x 800mm to 280 x 1,600mm and 200 x 260mm to 300 x 500mm). ***Rolling mills:*** light section and wire rod; three Sendzimir cold reduction (one 1,144mm, one 1,444 and one 1,650mm). ***Other plant:*** Forging machine, forging presses, annealing and heat treatment facilities; bright drawing plant;

Thyssen Edelstahlwerke AG (continued)

mechanical workshop; department for production of carbide tool material; titanium plant.
■ Auestrasse 4, Postfach 1369, D 5810 Witten. ***Steelmaking plant:*** One 110-tonne electric arc (bottom tapping). ***Rolling mills:*** Combined blooming and billet; light section. ***Other plant:*** Mechanical workshops; department for heating conductors and resistance alloys; bright drawing plant; heat treatment facilities.
■ Postfach 21 40, D 5885 Schalksmühle 2 (Dahlerbrück). Plant for production of precision strip.
■ Robertstrasse 88, Postfach 529, D 4630 Bochum. Welding electrode plant.
■ Ostkirchstrasse 117, Postfach 41 01 44, D 4600 Dortmund 41. Plant for production of magnets, magnet systems and porous sintered materials.
Products: Carbon steel – slabs, blooms, billets, wire rod, round bars, square bars, flats, hexagons, bright (cold finished) bars, special light sections, cold rolled hoop and strip (uncoated) (precision), medium plates, heavy plates, hot rolled sheet/coil (uncoated), cold rolled sheet/coil (uncoated). Stainless steel – slabs, blooms, billets, semis for seamless tubemaking, wire rod, round bars, square bars, flats, hexagons, bright (cold finished) bars, special light sections, cold rolled hoop and strip (uncoated) (precision), medium plates, heavy plates, hot rolled sheet/coil (uncoated), cold rolled sheet/coil (uncoated), wire. Alloy steel (excl stainless) – slabs, blooms, billets, wire rod, round bars, square bars, flats, hexagons, bright (cold finished) bars, special light sections, cold rolled hoop and strip (uncoated), medium plates, heavy plates, hot rolled sheet/coil (uncoated), cold rolled sheet/coil (uncoated), wire. Special steels: High-speed steels; tool steels; high-grade structural steels including for special applications; ball and roller bearing steels; aircraft steels and steels for nuclear reactor engineering; high-temperature steels and alloys; stainless steels; heat-resistant steels and alloys; steels and alloys with special physical properties. Special materials: Titanium and titanium alloys; high-alloy welding materials; machinable and hardenable carbide tool material; porous sintered materials; heat conductors and resistance alloys; high-melting-point metals; hard copper alloys; materials for resistance-welding electrodes; contact spring material; permanent magnet alloys; dental alloys; cemented carbide tool material. Semi-finished products: Drawn wire in coils, on spools or in barrels; cold work rolls; die blocks; precision ground flats; pre-machined tool steel; pre-machined moulds; tools in tool and high-speed steels, hardenable carbide tool material and cemented carbide material.

Thyssen Grillo Funke GmbH

Head office: Kurt-Schumacher-Strasse 95, D 4650 Gelsenkirchen. **Postal address:** Postfach 10 01 04. **Tel:** 0209 4071. **Telex:** 824848 grifu d.
Established: 1969. **Ownership:** Thyssen AG.
Works:
■ At head office. **Cold strip**. ***Other plant:*** Annealing furnaces, etc for electrical sheet production.
Products: Carbon steel – electrical grain oriented silicon sheet/coil.

Thyssen Henrichshütte—see Thyssen Stahl AG

Thyssen Niederrhein AG—see Thyssen Stahl AG

Thyssen Stahl AG

Head office: Kaiser-Wilhelm-Strasse 100, D 4100 Duisburg 11. **Postal address:** Postfach 11 05 61. **Tel:** 0203 521. **Telex:** 855483 tst d. **Telecopier (Fax):** 0203 52 25 102.
Management: Heinz Kriwet (chairman), Karl-August Zimmermann (deputy chairman), Werner Hartung, Karlheinz Sandhöfer, Ekkehard Schulz, Helmut Wilps, Hans Gert Woelke. **Established:** 1983 (dates from 1891/1953). **Capital:** DM 750,000,000. **No. of employees:** Works – 27,300. Office – 9,536. **Ownership:** Thysen AG, Duisburg (100%). **Subsidiaries:** Thyssen Bandstahl Berlin GmbH, Berlin (100%);

West Germany

Thyssen Stahl AG (continued)

Thyssen Grillo Funke GmbH, Gelsenkirchen (100%); Thyssen Heinrichshütte AG, Hattingen (100%); Thyssen Niederrhein AG, Oberhausen (100%); Thyssen Draht AG, Hamm (100%); Nedstaal BV, Alblasserdam, Netherlands (100%); Veerhaven BV, Rotterdam, Netherlands (100%); Baustahlgewebe GmbH, Düsseldorf (34%); Gemeinschaftsbetrieb Eisenbahn und Häfen GmbH, Duisburg-Hamborn (69%); Exploration und Bergbau GmbH, Düsseldorf (54%); Rohstoffhandel GmbH, Düsseldorf (50%); Ruhrkohle AG, Essen (12.7%); Seereederei "Frigga" GmbH, Hamburg (44.4%); Bong Mining Co Inc, Monrovia, Liberia (21.4%) (total Thyssen group); Ferteco Mineração SA, Rio de Janeiro, Brazil (57.7%); Ertsoverslagbedrijf Europoort CV, Rotterdam, Netherlands (50%). **Annual capacity:** Pig iron: 12,600,000 tonnes. Raw steel: 10,800,000 tonnes.

Works:

■ Hochofenwerk, Hamborn/Schwelgern. ***Coke ovens:*** Koppers batteries Nos. 1, 3 and 4 (190 ovens); Still batteries Nos. 2, 6a and 6b (164 ovens). ***Sinter plant:*** Two 150 sq m strand GHH/McKee, three 444 sq m strand Krupp/Koppers, four 250 sq m strand Krupp/Koppers, four 94 sq m strand GHH/McKee Walsum. ***Blast furnaces:*** Six – (No.3) 8.1m hearth dia, 1,046 cu m; (No.4) 10.7m hearth dia, 1,862 cu m; (No.6) 7.8m hearth dia, 870 cu m; (No.8) 9.3m hearth dia, 1,303 cu m; (No.9) 9.4m hearth dia, 1,362 cu m; Schwelgern 13.6m hearth dia, 3,594 cu m.

■ Bruckhausen. ***Steelmaking plant:*** Two 380-tonne basic oxygen converters. ***Refining plant:*** One 400-tonne DH vacuum degassing unit. ***Continuous casting machines:*** 2-strand Sack slab 215 x 1,800-2,600mm. ***Rolling mills:*** One 2-stand 1,240mm reversing slabbing mill; two heavy section – 3-stand reversing (1,270mm universal beam); and 5-stand open (980mm beam); 7-stand 59" wide hot strip (coil weight 23.4-tonnes); 4-stand 66" tandem cold reduction (coil weight 25-tonnes); two temper/skin pass – 56" and 66" (coil weight 25-tonnes). ***Coil coating lines:*** one combined hot-dip galvanizing /aluminizing line; one continuous electro-galvanizing line 0.4-2.5 x 600-1,250mm.

■ Beeckerwerth. ***Steelmaking plant:*** Three 240-tonne basic oxygen converters. ***Continuous casting machines:*** slab – 2-strand Demag curved mould for 255 x 1,1180mm-2,070mm, and 2-strand Sack curved mould for 255 x 1,420mm-2,400. ***Rolling mills:*** 7-stand 88" wide hot strip (coil weight 36-tonne); 5-stand tandem 80" cold reduction mill (coil weight 36-tonnes); two 4-high temper/skin pass – 80" and 66" (coil weights 32-tonne and 36-tonne). ***Coil coating lines:*** One continuous hot-dip galvanizing line; one continuous electro-galvanizing line; one plastic coating line.

■ Ruhrort. ***Blast furnaces:*** Three – (No.6) 10.8m hearth dia, 2,151 cu m; (No.7) 9.0m hearth dia, 1,346 cu m; (No.8) 9.0m hearth dia, 1,754 cu m. ***Steelmaking plant:*** Two 140-tonne basic oxygen converters; one 140-tonne furnace for ferro manganese. ***Refining plant:*** One 140-ton RH vacuum degassing unit. ***Continuous casting machines:*** bloom one 3-strand Demag for 160x160-265 x 380mm; one 6-strand Siemag/Schloemann for 260 x 330mm. ***Rolling mills:*** billet mills – Three open stands (870mm).

■ Hochfeld. ***Rolling mills:*** one 7-stand 420mm dia medium section mill; one 13-stand 280mm dia light section mill; 23-stand (4-strand) wire rod .

■ Hüttenheim. ***Rolling mills:*** One 4-high 3,700mm heavy plate. ***Other plant:*** One hot shear line and one hot shear slitting line.

■ Finnentrop. ***Coil coating lines:*** One continous Heurtey Sendzimir hot-dip galvanizing line, and one continuous hot-dip galvanized line for narrow strip.

■ Thyssen Niederrhein AG, Oberhausen. ***Steelmaking plant:*** Two 120-tonne GHH electric arc furnaces. ***Refining plant:*** One TN ladle furnace. ***Continuous casting machines:*** 6-strand SMS/Concast for billet (105-160mm). ***Rolling mills:*** One 5-stand 550mm medium section mill; 24-four-stand (2-strand) wire rod mill.

■ Thyssen Henrichshütte, Hüttenstrasse 45, D 4320 Hattingen. ***Blast furnaces:*** Two – (No.2) 7.5m hearth dia, 984 cu m; (No.3) 7.5m hearth dia, 879 cu m. ***Steelmaking plant:*** One 150-tonne basic oxygen converter (one blowing station, two vessels); two electric arc furnaces – one 150-tonne and one 40-tonne. ***Refining plant:*** One 150-tonne ESR unit. ***Continuous casting machines:*** slab – 1-strand Concast/ Schloemann 200 x 1,700, 300 x 2,100mm; and 1-strand Concast/Schloemann

Thyssen Stahl AG (continued)

adjustable mould, 160 x 1,350-1,700 and 200 x 1,750-2,000. ***Rolling mills:*** One 4-high 4,200mm heavy plate. ***Other plant:*** Steel foundry; forge; machine shops; plant for making dished products; welding and cutting facilities.
Products: Foundry pig iron (1985/86 output, 529,000 tonnes); blast furnace ferro-manganese (85,000 tonnes); Carbon steel – (includes some alloy) and ingots; slabs (694,000 tonnes); blooms (498,000 tonnes); billets (234,000 tonnes); hot rolled coil (for re-rolling) (3,365,000 tonnes); wire rod (927,000 tonnes); reinforcing bars (43,000 tonnes); round bars (50,000 tonnes); square bars (2,000 tonnes); flats (36,000 tonnes); light angles (56,000 tonnes); light tees (4,000 tonnes); light channels (5,000 tonnes); special light sections (23,000 tonnes); heavy angles (3,000 tonnes); heavy tees (12,000 tonnes); heavy joists (79,000 tonnes); heavy channels (9,000 tonnes); wide-flange beams (heavy) (154,000 tonnes); heavy rails (112,000 tonnes); light rails (5,000 tonnes); rail accessories (4,000 tonnes); heavy plates (876,000 tonnes); universal plates (72,000 tonnes); hot rolled sheet/coil (uncoated) (831,000 tonnes); cold rolled sheet/coil (uncoated) (859,000 tonnes); hot-dip galvanized sheet/coil (602,000 tonnes); electro-galvanized sheet/coil (120,000 tonnes); aluminized sheet/coil (193,000 tonnes); colour coated sheet/coil (136,000 tonnes);

Tillmann Kaltprofilwalzwerk GmbH & Co KG

Head office: Stockumer Strasse 42, D 5768 Sundern. **Postal address:** Postfach 80 20. **Tel:** 02933 2008. **Telex:** 84206. **Telecopier (Fax):** 02933 6796.
Works:
■ At head office.
Products: Carbon steel – cold roll-formed sections (from strip), Stainless steel – cold

West Germany

Tillmann Kaltprofilwalzwerk GmbH & Co KG (continued)

roll-formed sections (from strip), also roll-formed sections made of non-ferrous metals, enamelled strip steel, etc; strip thickness up to 5mm, max depth of section 115mm; strip width up to 500mm max length of section 12,000mm – plastic coated, galvanized and electro-plated sections.

TPS-Technitube Röhrenwerke GmbH

Head office: Industriegebiet, D 5568 Daun. **Postal address:** PO Box 1360. **Tel:** 0 65 92 7120. **Telex:** 4729937. **Cables:** Technitube Daun/Eifel. **Telecopier (Fax):** 065 92 1305.
Management: Managing directors – Peter Lepper and Heinz Nägel.
Products: stainless – seamless tubes and pipes, longitudinal weld tubes and pipes, oil country tubular goods, cold drawn tubes and pipes, precision tubes and pipes; fittings, joints, flanges; thin walled tubing.

Trierer Walzwerke GmbH

Head office: Brühlstrasse 14-15, D 5500 Trier. **Tel:** 0651 2031. **Telex:** 472873.
Ownership: Hille & Müller, Düsseldorf.
Works:
■ Electro-plating lines.

VDM – Vereinigte Deutsche Metallwerke AG

Head office: Nickel Technology Division, Plettenberger Strasse 2, D 5980 Werdohl. **Postal address:** PO Box 1820. **Tel:** 02392 550. **Telex:** 826433. **Telecopier (Fax):** 0 23 92 55 2217.
Management: Hansjürgen Hauck, Volker Groth, Heinz Koch, Josef Weiler. **Established:** 1930. **Capital:** DM 91,000,000. **No. of employees:** 1,700.
Ownership: Metallgesellschaft AG, Frankfurt (98.9%).
Works:
■ Altena. ***Rolling mills:*** Sheet and medium plate; forge.
■ Unna. Foundry.
■ Werdohl. Strip mill.
■ Werdohl-Bärenstein. Wire works.
Products: Stainless steel – cold rolled hoop and strip (uncoated); medium plates; wire; longitudinal weld tubes and pipes; forgings and castings.
Brands: Nicrofer (nickel-chromium-iron alloys); Cronifer (chromium-nickel-iron alloys); Crofer (chromium-iron alloys);
Sales offices/agents: Australia – VDM Australia Pty Ltd, North Clayton, Vic (Tel: 00 61 3 5 61 13 11. Telex: 134903 vdm); Hurstville, NSW 2220 (Tel: 00 61 2 570

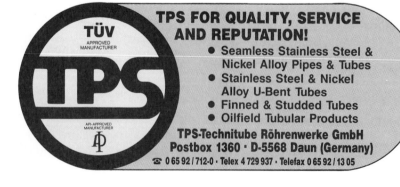

VDM – Vereinigte Deutsche Metallwerke AG (continued)

9444. Telex: 71588 vdm). Belgium – SA VDM Belgium NV, Waterloo (Tel: 00 32 2 3 54 29 00. Telex: 62704 vdmbel b. Fax: 00 32 2 354 36 26). Brazil – VDM do Brasil Ltda, Santo André, Sao Paulo (Tel: 00 55 11 444 70 63. Telex: 114005 kspl br). France – VDM Sarl, Saint-Cloud Cédex (Tel: 00 331 46 02 87 89. Telex: 201125f); Saint Genis Laval (Tel: 00 33 78 56 66 40. Telex: 305622 f vdm lyon). UK – VDM (UK) Ltd, Claygate-Esher, Surrey (Tel: 00 44 372 6 71 37, Telex: 929601 vdmuk g. Fax: 00 44 372 6 63 88). Italy – VDM Italia Srl, Milan (Tel: 00 39 2 8 354 895. Fax: 00 392 8 321 990). Japan – VDM Division of Nihon Metallgesellschaft KK, Tokyo (Tel: 0081 3 284 7341. Telex: 26684 mgtok j. Fax: 0081 3 281 7379). Yugoslavia – VDM AG c/o Joint Representative Office Metallgesellschaft-Degussa, Zagreb (Tel: 0038 41 22 08 54. Telex: 22519 mgzag yu). Canada – Vereinigte Deutsche Metallwerke Canada Ltd, Toronto, Ont (Tel: 001 416 961 9363. Telex: 06218366 vdm canada tor); Vereinigte Deutsche Metallwerke Canada Ltd, Calgary, Alb (Tel: 001 403 252 7751. Telex: 03826750 vdm canada cgy). Mexico – VDM AG Representative Office, Mexico, DF (Tel: 00 525 250 51 77. Telex: 1773078). Netherlands – Vereinigte Deutsche Metallwerke Nederland BV, Rotterdam (Tel: 0031 10 4 33 22 77. Telex: 26294 vdm nl). Austria – VDM Gesellschaft mbH, Vienna (Tel: 0043 222 6 54 78 60. Telex: 136610 vdm a. Fax: 0043 222 65 26 85). Rumania – VDM AG c/o Metallgesellschaft AG, Bucaresti/SR (Tel: 151 199. Telex: 10940 mg bukr). Singapore – VDM Division, Metallgesellschaft (Singapore) Pte Ltd, Singapore (Tel: 0065 223 8322. Telex: 29039 mg met. Fax: 0065 224 0566 de). South Africa – VDM AG South Africa Division, Ansfrere (Tel: 002 711 802 4130. Telex: 422273 sa. Fax: 011802 3018). USA – VDM Technoliges Corp, Parsippany, NJ (Tel: 001 201 267 8545. Telex: 257523 vdmt ur. Fax: 001 20 129 249 19); Arlington Heights, IL (Tel: 001 312 577 4660. Telex: TWX 910 687 3986a vdmt); Paoli, PA (Tel: 001 215 647 5860. Telex: ITT 4762206 vdmt pha); VDM Technologies Corp, Houston, TX (Tel: 001 713 682 2670. Telex: TWX 910 881 3796. Fax: 001 713 682 8186); Irvine, CA (Tel: 011 714 476 0912. Telex: 160788 vdmt la).

C.D. Wälzhölz KG

Head office: Feldmuelhen Strasse 55, D 5800 Hagen. **Tel:** 02331 3641. **Telex:** 823571. **Telecopier (Fax):** 02331 364-337.
Products: Carbon steel – cold rolled hoop and strip (uncoated) hardened and tempered, electrical sheets, wire.

Wälzlagerrohr GmbH

Head office: Krützpoort 1, D 4150 Krefeld. **Postal address:** Postfach 4160. **Tel:** 02151 716 1. **Telex:** 853362. **Cables:** Walzrohr. **Telecopier (Fax):** 02151 716211.
Management: Managing directors – Hartmut Küssner (sales), Franz Rauner (manufacturing). **Established:** 1970. **Capital:** DM 20,000,000. **No. of employees:** 430. **Ownership:** Thyssen Edelstahlwerke AG, Krefeld (50%); Mannesmannröhren-Werke AG (50%).
Works:
■ At head office. ***Tube and pipe mills:*** seamless – two Assel mills, each consisting of Assel type mill, sizing mill, rotary sizer and a number of cold reducing machines.
Products: Alloy steel (excl stainless) – seamless tubes and pipes, precision tubes and pipes. Seamless steel tubes for bearing races. Tubes are available as follows: Hot rolled and spheroidized, turned, max Brinell hardness 207; spheroidized, cold finished, Brinell hardness 260-320; spheroidized, cold finished, final annealed, max Brinell hardness 250. Main sizes: Hot rolled 50-230mm od, cold finished 5-140mm od.

Karl Krauss

Masters Red Hot Problems in both Professional and Private Life

For almost 25 years Karl Krauss (46) has been responsible for the removal of liquid slag in the steel mill. For many long years the handling and disposal of the hot slag was a very troublesome and extremely dangerous operation. During

one of the many slag transports, a small splash of the red hot slag struck Karl on the leg leaving a life-long scar. This scar serves as a per- manent reminder of the real dangers involved in what Karl viewed as an adventurous experiment. It was always Karl, the old fox, that prevented the more serious injuries or damage to equipment. Through the foresight of the management at Karl's steel plant, a KAMAG slag pot carrier was purchased and placed in service. The KAMAG carrier masters the red hot slag problems so well that Karl's only problems, if any, are his strong appetite for red hot foods.

Higher Tonnages

In today's slag pot operations not only must there be the highest regard for safety but an ever-increasing emphasis is being placed on capacities. These increased capacities are a direct result of the modified techniques of steel production and places high demands on slag transport equipment. Today's equipment must be able to transport ever-increasing capacities as well as provide maximum availability.

Solutions for slag removal by the KAMAG product range include using the standard KAMAG transporter system or the dedicated single-purpose transporters. Within the proven standard KAMAG pallet system, special pallets are used in direct connection with the KAMAG elevating transporters for loads up to 60 tons. Should your requirements exceed this range, KAMAG offers a complete line of dedicated single-purpose transporters with ladle weight capacities of 160 tons or more. These dedicated transporters are custom manufactured to customer's ladle specifications and are available in both the platform style and U-frame design. The transporters in either design are available with mechanical drive trains.

All transport problems presented to KAMAG are solved in the most efficient and economical method possible. This alone could be the most important reason to talk to KAMAG. Challenge us. Our specialists are at your disposal.

KAMAG

Transporttechnik GmbH & Co.
Daimlerstraße 18, 7900 Ulm
Telefon: 0731/40 32 02
Telex: 712367
Telefax: 0731/40 33 31

Westig GmbH

Head office: Hochstrasse 32, D 4750 Unna, Westphalia. **Postal address:** PO Box 2040. **Tel:** 02303 203 0. **Telex:** 8229255. **Cables:** Westigstahl. **Telecopier (Fax):** 02303 20 32 03.
Management: Managing directors – Ernst-Dieter Aufenberg (chairman), Werner Buss (finance and administration), Willy Berends (marketing). Board chairman – Klaus G. Heger. **Established:** 1911. **Capital:** DM 9,100,000. **No. of employees:** Works – 430. Office – 142. **Ownership:** Ernst-Dieter Aufenberg (40%), Werner Buss (30%), Willy Berends (30%). **Subsidiaries:** Westig of America Inc, Pine Brook, NJ, USA (100%); Westig (UK) Ltd, Sheffield, UK (98%); Westig GmbH, Niederlassung Wien, Vienna, Austria (99%). **Annual capacity:** Finished steel: 15,000 tonnes.
Works:
■ At head office. Managing director: E.D. Aufenberg. ***Rolling mills:*** One cold reduction. ***Other plant:*** Drawing plant, hardening department, bi-metal shop.
Products: Carbon steel – bright (cold finished) bars 0.80-30mm (0.315-1.811"); cold rolled hoop and strip (uncoated) (unhardened) 5-160mm wide (0.197-6.304") and 0.05-2mm thick (0.002-0.079"); bright wire (cold drawn) in coils 0.20-16mm (0.0079-0.6299"). Stainless steel – bright (cold finished) bars 0.80-30mm (0.315-1.811"); cold rolled hoop and strip (uncoated) (unhardened) 5-160mm wide (0.197-6.304") and 0.05-2mm thick (0.002-0.079"); wire (cold drawn) in coils 0.20-16mm (0.0079-0.6299"). Alloy steel (excl stainless) – bright (cold finished) bars 0.80-30mm (0.315-1.811"); cold rolled hoop and strip (uncoated) (unhardened) 5-160mm wide (0.197-6.304") and 0.05-2mm thick (0.002-0.079"), including tool, alloy and non-alloy engineering steel; wire (cold drawn) in coils 0.20-16mm (0.0079-0.6299"). Ground and polished tool steel up to 50mm (1.9680"); high-speed steels, stainless and engineering steels; needle wire in carbon and stainless qualities; hardenable free-cutting steels, etc. Applications include cutting tools, parts for textile machinery and office machines, electrical and surgical instruments, measuring tools, etc. Hardened special steel strip in coil, non-alloy (0.70-1.30% carbon), alloy (Cr/Si, Cr/V, Mn/Si), and stainless; applications include loom droppers, healds, sinkers, measuring tapes, feeler gauges, springs, valves, saws, knives, etc; electron beam welded bi-metal strip for metal cutting saws.
Brands: Westig; Westig-Duro-Biflex (bi-metal strip).
Sales offices/agents: Westig GmbH Trading, Goethestrasse 11, D 4000 Düsseldorf 1 (Tel: 0211 683020. Telex: 8586514. Fax: 675950). Overseas agents: Argentina and Brazil – Notlim Steel Representacoes S/C Ltda, São Bernardo do Campo, SP, Brazil. Belgium and Netherlands – Roestvaststaal & RVS Armaturen Hollimex Handelsonderneming BV, Krimpen a/d IJissel, Netherlands. Canada – Fagersta Secoroc Ltd, Burlington, Ont. Denmark – AST Stainless ApS, Glostrup. Finland – Oy Gronblom Ab, Helsingfors. France – Avesta SA, Paris. Hong Kong and China – Axel Johnson Corp (HK) Ltd, Wanchai, Hong Kong. India – Sri Padmanabha & Co, Madras. Italy - Guido Fasoli, Sesto Giovanni. Japan – Doitsu Kinzoku Co Ltd, Osaka. Korea – Sandvik Korea Co Ltd, Seoul, South Korea. Mexico, Venezuela and Ecuador – Rescandi SA de SN, Mexico, DF. Norway – Fagersta Haak A/S, Oslo. Spain – Aceros Nobles SA, Barcelona. Sweden – Stellan Brufors, Gothenburg. Switzerland – L. Klein AG, Biel; Pfenninger Stahl AG, Oberengstringen ZH. Taiwan – Polytech International Ltd, Taipei. UK – Westig (UK) Ltd, Sheffield. USA – Westig of America Inc, Pine Brook, NJ.

Wickeder Eisen- und Stahlwerk GmbH

Head office: Hauptstr 6, D 5757 Wickede-Ruhr. **Tel:** 0 23 77 860. **Telex:** 8202889. **Cables:** Wickedereisen.
Management: Joint managing directors – Hans-Werner Löhr, Klaus-Peter Helmetag. **Established:** 1913. **Capital:** DM 10,000,000. **Ownership:** Private limited company. **Subsidiaries:** Westfalenstahl GmbH, D 5850 Hohenlimburg (rolling mill); Wissner GmbH, Werl/Westfalen (metal works); Feran Steel Ltd, UK (sales office, UK and export).
Works:
■ At head office. Cold rolling mills, cold bonding lines and cold forming mills for

Wickeder Eisen- und Stahlwerk GmbH (continued)

sections.
Products: Carbon steel – cold roll-formed sections (from strip), cold rolled hoop and strip (uncoated), coated hoop and strip aluminium-clad steel strip, nickel-clad steel strip, multi-layer strip, nickel-clad copper strip, trimetals copper-clad steel strip, brass-clad steel strip.
Sales offices/agents: Feran Steel Ltd, Priory Gatehouse, Church Hill, Orpington, Kent (Tel: 0689 36579. Telex: 896916. Fax: 0689 76881).

Greece

Metallurgiki Athinon SA

Head office: Papada Girokornion, Athens. **Telex:** 215121 met gr.
Works:
■ Inofita. ***Rolling mills:*** One cold strip mill. ***Tube and pipe mills:*** welded – Seuthe.
Products: Carbon steel – cold rolled hoop and strip (uncoated), longitudinal weld tubes and pipes, galvanized tubes and pipes.

Metallurgiki Halyps SA

Head office: Adrianiou 2, Athens. **Telex:** 221014.
Established: 1973.
Works:
■ Almyros, Near Volos (Tel: 282112). ***Steelmaking plant:*** Two 50-tonne Asea electric arc furnaces. ***Continuous casting machines:*** One 3-strand Concast billet machine (75 to 130mm sq); one 2-strand Concast slab machine. ***Rolling mills:*** wire rod and hot strip and hoop (combination) mill.
Products: Carbon steel – wire rod, hot rolled hoop and strip (uncoated).

Halyvourgia Thessalias SA

Head office: Loudovikou 1, Piraeus. **Telex:** 212883.
Established: 1963/1972.
Works:
■ 4 Km, Volos-Larissa Road, Volos. ***Steelmaking plant:*** Two 45-tonne electric arc furnaces. ***Continuous casting machines:*** One 3-strand billet machine. ***Rolling mills:*** One wire rod mill.
Products: Carbon steel – wire rod, reinforcing bars, round bars.

Halyvourgiki Inc

Head office: 8 Dragatsaniou Street, 10559 Athens. **Tel:** 3237811. **Telex:** 216631-2. **Cables:** Ferohal. **Telecopier (Fax):** 3222392.
Management: D.S. Economou (managing director), G. Halkias (domestic sales), G. Kalogeras (export sales). **Established:** 1948. **Capital:** Drs 1,350,000,000. **No. of employees:** Works – 1,100. Office – 50. **Annual capacity:** Pig iron: 1,000,000 tonnes. Raw steel: 2,200,000 tonnes. Finished steel: 2,000,000 tonnes.
Works:
■ Eleusis. (Tel: 5546351. Telex: 216631). Works director: G. Dalatsis. ***Coke ovens:*** 52 ovens (annual capacity 500,000 tonnes). ***Blast furnaces:*** Two – 7m hearth dia, 843 cu m each (500,000 tonnes each respectively). ***Steelmaking plant:*** basic oxygen converter (1,000,000 tonnes); electric arc furnaces (1,200,000 tonnes). ***Continuous casting machines:*** Five 4-strand billet (80-150mm x 80-150mm) (annual capacity 1,500,000 tonnes); three 1-strand slab (600-1,550 x 140-220mm) (1,200,000 tonnes). ***Rolling mills:*** bar and wire rod mills (1,300,000 tonnes); wide

Halyvourgiki Inc (continued)

hot strip (1,500,000 tonnes); cold reduction and cold strip cold mill (700,000 tonnes);
Products: Carbon steel – slabs, billets, hot rolled coil (for re-rolling), wire rod, reinforcing bars, hot rolled hoop and strip (uncoated), cold rolled hoop and strip (uncoated), medium plates, heavy plates, hot rolled sheet/coil (uncoated), cold rolled sheet/coil (uncoated).

Hellenic Steel Co

Head office: 1 Mitropoleos Street, GR 105 57 Athens. **Postal address:** PO Box 8432, GR 100 10 Athens. **Tel:** 32 30 585, 32 37 345. **Telex:** 216459, 210416. **Cables:** Hellenicsteel. **Telecopier (Fax):** 01 32 32 277.
Management: Managing director – Alexandros Tiktopoulos; deputy managing directors – Damianos Hadjikokkinos, Horst Jungk, Toshio Kimura. **Established:** 1963. **Capital:** Drs 12,897,644,000. **No. of employees:** Works – 980. Office – 50. **Annual capacity:** Finished steel: over 1,000,000 tonnes.
Works:
■ Salonica. ***Rolling mills:*** cold reduction – one 4-stand tandem cold 56", one 4-high reversing 54"; One 2-stand 56" temper/skin pass. ***Coil coating lines:*** One 42" Wean (Halogen) electrolytic tinning (annual capacity 60,000 tonnes); hot-dip galvanizing – One Clecim 48" continuous line (80,000 tonnes); one 48" Republic Steel Corp sheet (25,000 tonnes).
Products: Carbon steel – cold rolled sheet/coil (uncoated) (1986 output 291,000 tonnes); hot-dip galvanized sheet/coil (61,000 tonnes); galvanized corrugated sheet/coil (4,000 tonnes); electrolytic tinplate (single-reduced) (43,000 tonnes).

Helliniki Halyvourgia SA

Head office: 40-42 Syngrou Avenue, GR 11742 Athens. **Tel:** 01 923 5932. **Telex:** 221756 elha gr. **Cables:** Elhalvourgia.
Management: Board chairman – Stavros Salapatas; managing directors – Manos Efstratiou, Vasilios Salapatas. **Established:** 1938. **Capital:** Drs 788,870,000. **No. of employees:** Works – 360. Office – 30. **Annual capacity:** Raw steel: 400,000 tonnes. Finished steel: 480,000 tonnes.
Works:
■ GR 19300 Aspropyrgos. (Tel: 01 557 3415. Telex: 216890 elha gr). Works director: V. Salapatas. ***Steelmaking plant:*** Two 55-tonne electric arc furnaces (annual capacity 400,000 tonnes). ***Continuous casting machines:*** One 4-strand curved mould billet (120 x 120mm sq) (400,000 tonnes). ***Rolling mills:*** light section – two 3-high and five 2-high stands (500 x 1,500/1,000mm, 350 x 800mm) (140,000 tonnes); bar wire rod – twenty-two-stand 2-high (500/400/350 x 800m, 270 x 500mm); sixteen-stand 2-high; finishing block (500/400/350 x 800mm); 8-stand (two 2,110mm and six 1,660mm) (coil weight 1,300kg) (340,000 tonnes).
Products: Carbon steel – billets 120mm sq; wire rod 5.5-12mm dia; reinforcing bars (plain and deformed) 8-26mm dia.

Profil SA

Head office: Kallithea 17610, Athens. **Postal address:** PO Box 75093. **Tel:** 9587552-4. **Telex:** 219851 metr gr.
Works:
■ ***Tube and pipe mills:*** welded.
Products: Carbon steel – longitudinal weld tubes and pipes.

Greece

Sidenor SA (formerly Steel Works of Northern Greece SA)

Head office: 115-119 Kifissias Avenue, Athens 606. **Telex:** 215881.
Established: 1962.
Works:
■ 12 Km from Salonica on Athens National Highway. ***Steelmaking plant:*** Two electric arc furnaces. ***Continuous casting machines:*** Two billet machines – one 2-strand, and one 4-strand. ***Rolling mills:*** One bar mill; one hot strip and hoop mill.
Products: Carbon steel – wire rod, reinforcing bars, hot rolled hoop and strip (uncoated).

Guatemala

Hornos SA

Head office: 20 C17-62 Zona 1, Guatemala City. **Telex:** 9302 distun gu.
Established: 1965.
Works:
■ At head office. ***Steelmaking plant:*** Three electric arc furnaces. ***Continuous casting machines:*** One 2-strand Danieli billet machine. ***Rolling mills:*** One Danieli/ Banning wire rod mill (coil weight 1,000kg). ***Other plant:*** Wire drawing and galvanizing; nail plant.
Products: Carbon steel – wire rod, bright wire, galvanized wire (plain), galvanized wire (barbed). Nails.

Indeta SA

Head office: 11 Avenida 10-23, Zona 1, Guatemala City.
Works:
■ San Ignacio, Mixco. ***Rolling mills:*** Pomini Farrel bar and wire rod – 450mm roughing train, 350mm intermediate, 320mm finishing and 260mm monobloc (annual capacity 65,000 tonnes).

Haiti

Aciérie d'Haiti SA

Head office: Route Nationale 1, Port au Prince. **Tel:** 2 2163, 2 5718. **Telex:** 2030171. **Telecopier (Fax):** 509 1 23767.
Established: 1974. **Capital:** US$150,000. **No. of employees:** Works – 250. Office – 40. **Annual capacity:** Raw steel: 25,000 tonnes. Finished steel: 24,000 tonnes.

Honduras

Industria Nacional Del Acero SA

Head office: Quebrada Seca, Choloma, Cortes. **Postal address:** PO Box 797, San Pedro Suia. **Tel:** 52 7275. **Telex:** 1192 conadi.
Management: Managers – Carlos Caraccioli C (general), Mario Zeiaya (finance); comptroller – Marcos Ferrera. **Established:** 1984. **Capital:** US$ 2,500,000. **No. of employees:** Works – 225. Office – 25. **Ownership:** Corporacion Nacional De Inversiones (95%). **Annual capacity:** Raw steel: 14,000 tonnes.
Works:
■ At head office. ***Steelmaking plant:*** electric arc furnace (annual capacity 14,000

Industria Nacional Del Acero SA (continued)

tonnes). Continuous casting machines: 1-strand billet (90 x 90) (20,000 tonnes). ***Rolling mills:*** One 17-stand bar ($\frac{3}{8}$ to 1$\frac{3}{8}$") (20,000 tonnes).
Products: Carbon steel – billets (Output 12,000 tonnes); reinforcing bars (10,000 tonnes).
Brands: Inacero.

Hong Kong

Shiu Wing Steel Ltd

Head office: 1209 Connaught Centre, Hong Kong. **Tel:** 5 255391. **Telex:** 83018 swing hx. **Cables:** Swingco. **Telecopier (Fax):** 852 5 8611454.
Management: Chairman – D.Y. Pong; directors – Hong Siu-Chu (managing), P.Y. Cheung, J.B. Lam, H. Pong, F. Pong. **Established:** 1954. **No. of employees:** Works – 330. Office – 60. **Subsidiaries:** Mack & Co (FF) Ltd (inland freight forwarder, containerised cargo); Easiatic Warehouse & Forwarding Ltd (general cargo). **Annual capacity:** Finished steel: 290,000 tonnes.
Works:
■ S.D. 3, Lot 1066, Junk Bay, Kowloon. (Tel: 3 7755171). ***Steelmaking plant:*** One 45-tonne Brown Boveri electric arc furnace (annual capacity 160,000 tonnes). ***Continuous casting machines:*** One 4-strand Concast for billet 115mm sq (270,000 tonnes). ***Rolling mills:*** One 16-stand Tandem bar mill (290,000 tonnes).
Products: Carbon steel – reinforcing bars 16-40mm (1986 output, 260,000 tonnes).
Brands: S.W. Steel.
Additional information: Addition of second arc furnace, revamping of rolling mill to annual capacity of 350,000 tonnes.

Shun Fung Iron Works Ltd

Head office: 301a New World Tower, Queen's Road, Central. **Telex:** 63024.
Works:
■ Junk Bay, New Territories, Kowloon. ***Steelmaking plant:*** electric arc furnace. ***Continuous casting machines:*** One 4-strand (140mm sq) for billet . ***Rolling mills:*** One bar.
Products: Carbon steel – reinforcing bars.

Hungary

Export sales of Hungarian iron and steel products are handled by Metalimpex, Május 1.u.17, H 1146 Budapest XIV (Tel: 187 611. Telex: 225251. Cables: Mineralimpex Budapest) and Pannonia, Bajza u.26, H 1062 Budapest VI (Tel: 228 850. Telex: 225128. Cables: Panexport Budapest).

Borsodnádasd Sheet Metal Works

Works:
■ **Rolling mills**.
Products: Carbon steel – Sheets, including electrical non-oriented non-silicon sheet/ coil.

Hungary

Csepel Steel & Tube Works

Works:

■ Csepel Island, Near Budapest. ***Steelmaking plant:*** Three electric arc and four 40-tonne open hearth furnaces (combined annual capacity 250,000 tonnes). **Refining plant:** One 60-tonne Scandinavian Lancers. ***Continuous casting machines:*** bloom. ***Rolling mills:*** bar; heavy plate; hot strip and hoop. ***Tube and pipe mills:*** seamless (over 80,000 tonnes). Welded. ***Other plant:*** Cold drawing plant. Foundry.

Products: Carbon steel – ingots, semis for seamless tubemaking, round bars, hot rolled hoop and strip (uncoated), steel cable, heavy plates, electrical non-oriented non-silicon sheet/coil. Seamless tubes and pipes, longitudinal weld tubes and pipes. Castings. "Glassy" metal (0.003 to 0.004mm thick).

Danube Works

Works:

■ Dunaujváros. ***Coke ovens:*** Four batteries (including one Indian-built unit with an annual capacity of 620,000 tonnes). ***Sinter plant:*** Annual capacity 120,000-130,000 tonnes). ***Blast furnaces:*** Two 770 cu m (annual capacity 700,000-750,000 tonnes each). ***Steelmaking plant:*** Two 130-tonne USSR basic oxygen converters (1,150,000 tonnes); two 5-tonne electric arc furnaces; four 18-tonne (oxygen blown) open hearth furances (1,000,000 tonnes). ***Refining plant:*** One vacuum degassing unit; one Scandinavian Lancers unit. ***Continuous casting machines:*** Three 2-strand for slab (180 x 1,050 to 240 x 1,550mm). ***Rolling mills:*** One 1,000mm billet; one 150mm roughing; one 3-high Lauth heavy plate; one 5-stand wide hot strip (800,000 tonnes); cold reduction comprising 4-high 1,700mm and 4-high 1,200mm (over 400,000 tonnes). ***Tube and pipe mills:*** welded (three, including one spiral). ***Coil coating lines:*** One electrolytic tinning; one hot-dip tinning (10,000 tonnes); one hot-dip galvanizing (22,000 tonnes). ***Other plant:*** Cold forming line for sections (130,000 tonnes).

Products: Steelmaking pig iron. Carbon steel – slabs, cold roll-formed sections (from strip), medium plates, hot rolled sheet/coil (uncoated), cold rolled sheet/coil (uncoated), electrical non-oriented silicon sheet/coil, electrolytic tinplate (single-reduced), hot-dip tinplate, longitudinal weld tubes and pipes, spiral-weld tubes and pipes.

Diosgyör Works (Lenin Metallurgical Works)

Head office: PO Box 567, 3540 Miskolc. **Tel:** 4651411. **Telex:** 62242, 62326.

Works:

■ Diosgyör, Near Miskolc. ***Blast furnaces:*** Three furnaces (capacity 450 tonnes per day). ***Steelmaking plant:*** One 80-tonne Demag basic oxygen converter (annual capacity 700,000 tonnes); electric arc furnaces including 50-tonne (100,000 tonnes),

Diosgyör Works (Lenin Metallurgical Works) (continued)

one 80-tonne, two 10-tonne and two 15-tonne (total capacity 1,100,000 tonnes). Refining plant: One 80-tonne Asea ladle furnace. ***Continuous casting machines:*** One 5-strand Kobe Steel for billet (120 to 180mm sq) (408,000 tonnes). ***Rolling mills:*** heavy section; medium section, light section and bar for special steel (73,000 tonnes); one light section and bar for special steel (147,000 tonnes).
Products: Steelmaking pig iron. Carbon steel – billets, heavy plates, sections and heavy rails. Stainless steel – billets, round bars, heavy plates, stainless steel for nuclear and chemical engineering. Alloy steel (excl stainless) – billets, round bars, sections, heavy plates. Forgings and screws.

Koebanya Tube Works

Products: Stainless steel – longitudinal weld tubes and pipes.

Ozdi Kohászati Uzemek (Okü)

Head office: Alkotmány ut 1, Ozd 1.
Established: 1845.
Works:
■ Ozd. ***Blast furnaces:*** Four 7.5m hearth dia, 600 cu m (annual capacity 937,600 tonnes). ***Steelmaking plant:*** Nine 100-tonne and 180-tonne open hearth furnaces (1,400,000 tonnes). ***Continuous casting machines:*** One 6-strand for billet (75 to 130mm sq) (1,400,000 tonnes). ***Rolling mills:*** one slabbing; one blooming; medium section and light section; bar and wire rod Schloemann-Siemag (5.5 to 40mm dia) (350,00 tonnes); one universal medium plate; one hot strip and hoop. ***Other plant:*** Wire drawing plant.
Products: Carbon steel – billets, reinforcing bars, round bars, sections, hot rolled hoop and strip (uncoated), medium plates, bright wire. Alloy steel (excl stainless) – billets, round bars. Free-cutting steel.
Expansion programme: The open hearths are being replaced by two Korf energy optimising furnaces, reducing annual capacity to about 900,000 to 1,000,000 tonnes.

Salgótarján Works

Works:
■ ***Rolling mills:*** cold strip.
Products: Carbon steel – cold rolled hoop and strip (uncoated), electrical non-oriented non-silicon sheet/coil, galvanized wire (plain); aluminium-coated wire.

India

Agarwal Foundry & Engineering Works

Head office: 1841 Tobacco Bazar, Secunderabad 500003, Andhra Pradesh. **Tel:** 76127, 63269. **Telex:** 04256333.
Management: Prabhu Dayal Agarwal (managing director). **Established:** 1980. **Capital:** R 50,000,000. **No. of employees:** Works – 300. Office – 60. **Ownership:** Agarwal group of companies. **Annual capacity:** Raw steel: 35,000 tonnes. Finished steel: 30,000 tonnes.
Works:
■ Survey No 66, 67 & 68, Pet Basheerabad, Pin 500 040 Taluq, Medchal, Dist Ranga Reddy, Andhra Pradesh. Works director: Prabhu Dayal Agarwal. ***Steelmaking plant:*** One 16 ft dia electric arc (annual capacity 35,000 tonnes). ***Rolling mills:*** medium section – 3-stand 3-high (16″ dia), and 5-stand 3-high (10″ dia); 6-stand 3-high light section (8″ dia); and 6-stand 3-high bar (8″ dia) (combined capacity 35,000 tonnes).

India

Agarwal Foundry & Engineering Works (continued)

Other plant: Small forge, iron foundry (including SG iron) and wire drawing machines.
Products: Carbon steel – ingots, slabs, blooms, billets, wire rod, reinforcing bars, round bars, square bars, flats, hexagons, light angles, light tees, light channels, special light sections, medium angles, medium tees, medium joists, bright wire, black annealed wire, galvanized wire (plain), galvanized wire (barbed). Cast iron – pipes, pipe fittings.
Expansion programme: Direct reduction plant (annual capacity 1,000,000 tonnes) is proposed.

Ahmedabad Steelcraft & Rolling Mills Pvt Ltd

Head office: Chinubhai Centre, 2nd Floor, Ashram Road, Ahmedabad 380009. **Telex:** 0121487 ascr in. **Cables:** Steelcraft Ahmedabad, India.
Management: S.M. Jhaveri (director, technical). **Established:** 1960. **Capital:** Rs 80,000,000. **No. of employees:** Works – 200. Office – 25. **Subsidiaries:** Gujarat Bright Bar Industries Pvt Ltd, Ahmedabad (manufacturer of mild steel bright bars and shaftings, textile, automobile and other engineering sections/shapes). **Annual capacity:** Rolled steel: 51,900 tonnes.
Works:
■ Odhav, Ahmedabad 382410. (Tel: 887551-2). Works director: S.M. Jhaveri. Works manager: S. Murthy. ***Rolling mills:*** Three light section and bar (8", 10" and 12").
Products: Carbon steel – round bars, square bars, flats, light angles, special light sections including door, window and ventilator sections; textile, automobile and engineering sections; triangular and trapezoidal sections.
Brands: ASCRM.
Sales offices/agents: At head office address. Consignment agents in India: United Steel Corp, New Delhi; Roopa Steels, Bangalore; Krishna & Co, Madras. Overseas agents: Mauritius Commercial Enterprises, PB 280, Port Louis, Mauritius; Shica & Co, Singapore; Tractor Trading Co, Cairo, Egypt.

Andhra Steel Corp Ltd

Head office: 29/2 K H Road, Bangalore 7. **Telex:** 845475.
Established: 1959.
Works:
■ Malkapuram, Visakhapatnam. ***Steelmaking plant:*** One electric arc furnace. ***Rolling mills:*** medium section and flat products mill; one wire rod mill.
■ Whitefield Road, Bangalore. ***Steelmaking plant:*** One electric arc furnace. ***Continuous casting machines:*** Two billet machines (own design).
Products: Carbon steel – reinforcing bars, round bars, square bars, flats.

Apollo Tubes Ltd

Head office: Ranipet 632403, Tamil Nadu. **Telex:** 402205 aplo in.
Established: 1973.
Works:
■ At head office. ***Tube and pipe mills:*** Longitudinal welded. ***Other plant:*** Galvanizing facilities.
Products: Carbon steel – longitudinal weld tubes and pipes, galvanized tubes and pipes.

BD Steel Castings Ltd

Head office: Poonam Chambers, Shivsagar Estate, Dr Annie Besant Road, Worli (South), Bombay 400018. **Telex:** 0114890.
Established: 1972.
Works:

BD Steel Castings Ltd (continued)

■ ***Steelmaking plant:*** One GEC 10/12-tonne electric arc furnace. ***Rolling mills:*** One light section; one bar.
Products: Carbon steel – ingots, reinforcing bars, light angles.

Bharat Steel Tubes Ltd—see BST Manufacturing Ltd

Bhartia Electric Steel Co Ltd

Head office: 8 Anil Maitra Road, Calcutta 700019. **Telex:** 0217174.
Established: 1922.
Works:
■ At head office. **Steel plant and foundry.**
Products: Carbon steel – ingots.

Bhoruka Steel Ltd

Head office: Whitefield Road, Mahadevapura Post, Bangalore 560048. **Tel:** 812 58595. **Telex:** 08452646 bsl in. **Cables:** Besteel.
Management: S.N. Agarwal (managing director), R.C. Purohit (vice president). **Established:** 1969. **Capital:** Rs10,480,000. **No. of employees:** Works – 400. Office – 100. **Ownership:** Public Limited Company **Associates:** Transport Corp of India Ltd (road transport), Bangalore Wire Rod Mill (wire rod mill – annual capacity 125,000 tonnes), Karnataka Oxygen Ltd (producer of industrial and special gases), Bhoruka Aluminium Ltd (producer of aluminium extrusions), Bhoruka Engineering Industries Ltd (civil engineering contractors and fabricators), Bhoruka Goldhoffer Ltd (manufacturers of heavy duty trailors and tractors), Bhoruka Textiles (manufacturers of viscose yarn), ABC (India) Ltd (road transport). **Annual capacity:** Finished steel: 110,000 tonnes.
Works:
■ At head office. ***Steelmaking plant:*** One 25-ton UHP electric arc furnace (annual capacity 110,000). ***Refining plant:*** One (Asea) ladle furnace with VD/ VOD station. Scandinavian Lancers powder injection systems. ***Continuous casting machines:*** One 2-strand Demag billet (50,000 tonnes); one 2-high Demag bloom (60,000 tonnes). ***Rolling mills:*** One wire rod – Ashlow non-twist 8-stand monobloc (coil weight 400kg) (125,000 tonnes).
Products: Carbon steel – blooms, billets, wire rod (including high carbon) 5.5mm to 22mm dia. Stainless steel – blooms, billets. Alloy steel (excl stainless) – billets, wire rod 5.5mm to 22mm.
Sales offices/agents: 6 Muthumari Chetty Street, Madras-1 (TN); 16th Floor, Nirmal, Nariman Point, Bombay-21; Minister Road, Ranigunj, Secunderabad (AP); 75-C, Park Street, Calcutta-16; Hindustan Times House, 7th Floor, Kasturba Gandhi Marg, New Delhi-1.
Expansion programme: Installation of scrap pre-heating, oxy fuel burner and two eccentric bottom tapping systems; installation of 2-high reversing mill.
Additional information: The company also makes refractory items used by mini-steel plants, including nozzles, ladle sleeves and magnesite bricks.

Bhushan Industrial Corp

Head office: Industrial Area, Chandigarh. **Telex:** 395262.
Works:
■ ***Steelmaking plant:*** One electric arc furnace. ***Continuous casting machines:*** One 2-strand Concast billet.
Products: Carbon steel – billets. Alloy steel (excl stainless) – billets.

Bihar Alloy Steels Ltd

Head office: Hadley House, Old Hazaribagh Road, Ranchi 834001 (Bihar). **Telex:** 625223 basl in.
Established: 1975.
Works:
■ Patrata (DT). ***Steelmaking plant:*** Two 32-tonne electric arc furnaces. ***Continuous casting machines:*** One 2-strand billet machine. ***Rolling mills:*** bar and wire rod mill. ***Other plant:*** Forge.
Products: Carbon steel – ingots, blooms, billets, wire rod, round bars, square bars, flats. Stainless steel – ingots, blooms, billets, wire rod, round bars, square bars, flats. Alloy steel (excl stainless) – ingots, blooms, billets, wire rod, round bars, square bars, flats.

BP Steel Industries Pvt Ltd

Head office: 39/44 Sion East, Bombay, Maharashtra. **Tel:** 473423, 476984, 482576, 486227. **Telex:** 11 71255 bpsi. **Cables:** Bipisteel.
Management: Directors – Rampaul Gupta, Vinodkumar Gupta, Rameshkumar Gupta. **Established:** 1959. **Capital:** Rs 2,250,000. **No. of employees:** Works – 60. Office – 4. **Annual capacity:** Finished steel: 4,500 tonnes.
Works:
■ Bombay. Bright bar plant.
Products: Carbon steel – bright (cold finished) bars. Stainless steel – bright (cold finished) bars. Alloy steel (excl stainless) – bright (cold finished) bars.

BST Manufacturing Ltd

Head office: Chiranjiv Tower, 43 Nehru Place, New Delhi 110019. **Tel:** 6431724-6. **Telex:** 03165676 bes in.
Management: Directors – Raunaq Singh (chairman), Onkar S. Kanwar (managing), S.S. Kanwar, Prem Pandhi, Jagjit Singh Aurora, V. Ramachandran, J.S. Maini, G.V. Kapadia, S.V. Bhave, V.N. Nadkarni, Stephen Stranahan, Nelson Daniel Abbey; executive – P.P. Daryani (president, tube division). **Established:** 1959. **Capital:** Authorised – Rs 100,000,000. Issued and subscribed – Rs 43,500,000. **No. of employees:** 1924. **Subsidiaries:** Reliance Overseas Ltd. **Associates:** Apollo Tubes Ltd, Apollo Tyres Ltd; Bharat Gears Ltd, Raunaq International Ltd, Universal Steel & Alloys Ltd. **Annual capacity:** Finished steel: 144,000 tonnes (steel pipes and tubes).
Works:
■ Ganuar, Distt. Sonepat, Haryana. Vice president (Works): J.P. Jain. ***Tube and pipe mills:*** welded – ERW. ***Other plant:*** Tube galvanizing plant. PVC pipe plant.
Products: Carbon steel – longitudinal weld tubes and pipes, galvanized tubes and pipes and precision tubes and pipes 15 to 200mm nb to ISI, BS, DIN, and API specifications (1985/86 output, 33,104 tonnes). Scaffolding; fabricated pipework for chemical, petrochemical and fertiliser industries, power stations, etc.
Sales offices/agents: Hoechst House, Nariman Point, Bombay; 17 Bipalbi Trailakya, Maharaj Saranee, Calcutta; Raji Building, 730 Mount Road, Madras, 5.8.328/9 Chapel Road, Nampally Hyderabad, Hyderabad.
Additional information: The company also has an office at – Allahabad Bank Building, 17 Sansad Marg, New Delhi 110001 (Tel: 310310, 310014, 310318. Telex: 031 66621 bst in. Cables: Tubemakers New Delhi).

Calcutta Steel Co Ltd

Head office: Steel House, 20 Hemanta Basu Sarani, Calcutta 700001.
Established: 1959.
Works:
■ BT Road, Khardah, Dist 24 Parganas, West Bengal. ***Rolling mills:*** One bar; one semi-continuous wire rod.
Products: Carbon steel – wire rod (low and medium carbon for electrodes; high-carbon), reinforcing bars, round bars, square bars, flats, hot rolled hoop and strip (uncoated). Octagons; spring steel flats.

Chase Bright Steel Ltd

Head office: Meher Chambers, Nicol Road, Ballard Estate, Bombay 400038. **Telex:** 1173572 cbs in.
Established: 1959.
Works:
■ Majiwada, Kapur Bawdi, Thana, Maharashtra. Cold drawing, turning and grinding plant for bars.
■ Aurangabad. Bright bar plant.
Products: Carbon steel – bright (cold finished) bars. Alloy steel (excl stainless) – bright (cold finished) bars.

Choksi Tube Co Pvt Ltd

Head office: Baajaj Bhavam, Backbay Reclamation, Bombay 40021. **Telex:** 011394.
Established: 1962.
Works:
■ Plot 60A, Vatva Industrial Area, Vatva, Ahmedabad (Tel: 877471-2. Telex: 012461). ***Tube and pipe mills:*** welded – for stainless steel. ***Other plant:*** Drawbenches.
Products: Stainless steel – longitudinal weld tubes and pipes, cold drawn tubes and pipes, precision tubes and pipes (capillary).
Brands: CT.

Damehra Steels & Forgings Pvt Ltd

Head office: P-35 India Exchange Place, Calcutta 700001, West Bengal. **Tel:** 26 3225-6. **Telex:** 021 5009. **Cables:** Damehra.
Management: Directors – R.P. Mehra (managing), Rakesh Mehra, Kanchan Mehra; managers – D. Krishnan (export), S.S. Sen (sales), B.N. Dhar (production).
Established: 1964. **Capital:** Rs 50,000,000. **No. of employees:** Works – 160. Office – 25. **Subsidiaries:** D.A. Mehra Steels Pvt Ltd, Damehra Investment Pvt Ltd, Ujjaini Rolling Mills Pvt Ltd, Apollo Warehousing Pvt Ltd, Apollo Electronics Corp.
Annual capacity: Finished steel: 23,000 tonnes.
Works:
■ Plant 1 – 15/64 G.T. Road, Lilluah, Howrah (Tel: 66 4695). Plant 2 – Timberpond, Foreshore Road, Shalimar, Howrah (Tel: 67 4164). Works director: Rakesh Mehra. Bright bar plant comprising drawing, centreless turning, and centreless grinding.
Products: Carbon steel – bright (cold finished) bars, Stainless steel – bright (cold finished) bars and Alloy steel (excl stainless) – bright (cold finished) bars – rounds, squares, flats and hexagons up to 3" dia.
Brands: Damehra.
Sales offices/agents: At head office. 4647 Bazar Ajmeri Gate, Delhi 6 (Tel: 52 3640).

Ellora Steels Ltd (ESL)

Head office: E26/27 Chikalthana Industrial Area, Aurangabad, Maharashtra. **Telex:** 745263.
Works:
■ At head office. ***Steelmaking plant:*** One electric arc furnace. ***Refining plant:*** One 5-tonne AOD unit. ***Rolling mills:*** One light section; one bar mill.
Products: Carbon steel – ingots, round bars, square bars, flats, light angles, light channels, special light sections, medium tees, medium joists. Stainless steel – ingots, round bars, square bars, flats. Alloy steel (excl stainless) – ingots, round bars, square bars, flats.

India

Ferro Alloys Corp Ltd (Facor) (Steel division), (Vidarbha Iron & Steel Corp Ltd)

Head office: Shreeram Bhavan, Ramdaspeth, Nagpur 440010, Maharashtra. **Tel:** 7814-5, 7614-5. **Telex:** 0715 347 NP. **Cables:** Facor Steel.
Management: Managing director – M.D. Saraf; president – Vinod Saraf; managers – A.C. Kapoor (works) and D.B. Moharil (development). **Established:** 1972. **Capital:** Rs 50,000,000. **No. of employees:** Works – 417. Office – 119. **Annual capacity:** Finished steel: 30,000 tonnes.
Works:
■ At head office. ***Steelmaking plant:*** One 10-12-tonne electric arc furnace (annual capacity 20,000 tonnes); One 3.5-tonne induction furnace (10,000 tonne). ***Refining plant:*** One 18-tonne AOD (30,000 tonnes). ***Continuous casting machines:*** One 2-strand S-type 8.5 metre radius bloom (140 x 140-200 x 200mm) (45,000 tonnes). ***Rolling mills:*** Two bar – one 24" 2-stand (50,000 tonnes) and one 16" 5-stand (36,000 tonnes) (2 shift basis).
Products: Carbon steel – blooms 140-200mm sq, round bars, square bars, flats. Stainless steel – blooms 140-200mm sq, round bars, square bars, flats. Alloy steel (excl stainless) – blooms (140-200mm sq), round bars, square bars, flats.
Additional information: Vidarbha Iron & Steel Corp has been taken over by Facor on a "leave and licence" basis until December 31st 1989.

Graham Firth Steel Products (India) Ltd

Head office: Firsteel Compound, Western Express Highway, Goregaon (East), Bombay 400 063, Maharashtra. **Tel:** 686731, 686734, 686735, 686755. **Telex:** 01171243 gfsp in. **Cables:** Gramfirth Bombay 62.
Management: Board of directors – Pranlal J. Patel (chairman and managing director), H.J. Nash, Mr Ardeshir, R. Wadia (alternate to H.J. Nash), D.S. Mulla, G.B. Newalkar, K.B. Rao, R.P. Patel, Mrs Chandralata P. Patel; secretary – G.A. Shah; advisor – Vasant N. Kamte. Sr executive administration – B.S. Irani; managers – A.J. Mistry (works), G.R. Desai (corporate – finance and personnel), P.D. Kanungo (marketing). **Established:** 1960. **Capital:** Rs 15,000,000. **No. of employees:** Works – 122. Office – 69. **Subsidiaries:** None. **Associates:** None. **Annual capacity:** Finished steel: 15,880 tonnes.
Works:
■ At head office. ***Rolling mills:*** Three Farmer Norton cold strip – 4-high 330mm with automatic gauge control (annual capacity 12,000 tonnes), 4-high 250mm (2,000 tonnes) and 2-high 360mm (1,880 tonnes); one 2-high Guest Keen Williams skin pass mill (15,880 tonnes).
Products: Carbon steel – cold rolled hoop and strip (uncoated) 12-340mm wide, 0.30-7mm thick (1986-87 output 10,900 tonnes).

Guest Keen Williams Ltd (GKW)

Head office: 97 Andul Road, Howrah 711103, West Bengal. **Telex:** 217319 gkw in.
Established: 1922. **Ownership:** GKN, UK (46.8%).
Works:
■ At head office. ***Steelmaking plant:*** electric arc furnace. ***Rolling mills:*** light section , bar , and wire rod. ***Other plant:*** Bright bar plant.
Products: Carbon steel – round bars, square bars, flats, hexagons, bright (cold finished) bars, special light sections. Stainless steel – round bars, square bars, flats, hexagons, bright (cold finished) bars; special light sections; Alloy steel (excl stainless) – round bars, square bars, flats, hexagons, bright (cold finished) bars, special light sections. Spring steels, leaded steels, free-cutting steels.
Additional information: In mid 1987 it was reported that GKN was negotiating to sell 7% of the company to Birla, subject to government approval.

Gujarat Steel Tubes Ltd (GST)

Head office: Bank of India Bldg, Bhadra, Ahmedabad 380001. **Tel:** 343841. **Telex:** 121699, 121266. **Cables:** Steelpipes.
Management: Board directors – Apoorva S. Shah (chairman and managing); Mrs Karuna Shah (managing), Navnitlal S. Shodhan, Kanchanlal C. Parikh, Nalin M. Shah, Dhanvantlal C. Gandhi, Amolukroy G. Patel, Ravindra M. Mehta, Ram M. Moghe, Champklal P. Mehta, Ravishankar S. Bhatt. **Established:** 1960. **Capital:** Rs 4,600,000. **No. of employees:** Works – 2,000. Office – 108. **Subsidiaries:** Neeka Tubes Ltd (100%) (manufacturers welded of stainless steel tubes 6 to 144mm o.d., tube-making machinery and machined components).
Works:
■ Kali, near Sabarmati, Ahmedabad 382470. (Tel: 476476-9. Telex: 121 395 8247). ***Tube and pipe mills:*** HF induction welded (annual capacity 185,500 tonnes).
Products: Carbon steel – longitudinal weld tubes and pipes (plain end, screwed and socketed) (15mm, 20mm 25mm 32mm, 40mm, 50mm, 65mm, 80mm, 100mm, 125mm, 150mm, 200mm nb); galvanized tubes and pipes; precision tubes and pipes. API line pipe, boiler tubes. Total output in 1986-87, 76,381 tonnes.
Sales offices/agents: 712-715 Tulsiani Chambers, 212 Nariman Point, Bombay 400021; Tinson House, 2nd Floor, 54 Janpath, New Delhi 110001; Agurchand Mansion, 2nd Floor, 151 Mount Road, Madras 600002; Commerce House, 6th Floor, 2 Ganeshchandra Avenue, Calcutta 700013; 52/2 Silver Jubilee Park Road, 1st Floor, Bangalore 560002; 5-5-89/32, 1st Floor, Sara Iron Market, Ranigunj, Seconderabad 5003; I 790 Industrial Area B, Ludhiana 141003; Mehta Chambers, 34 Siyaganj, Indore 4520. Agents – CM Nihalani, United Agencies, POB 2212, Dubai; Nishat Enterprises, PO Box 3524, Ruwi, Muscat, Oman; Haresh, Manaf Trading & Contracting Est, PO Box 4304, Riyadh, Saudi Arabia; Mohomed Ali Al-Audi, PO Box 4312, Hodeidah, Yemen; Harish Lalwani, Al-Anwar Modern Co, PO Box 1795, Doha, Qatar; Suresh Sawhney, Navshal Co Ltd, PO Box 41/2115, Tehran, Iran; Emile Srouji, Synco Jordan Services & Rep Co Ltd, PO Box 1168, Amman, Jordan; Harish Sawhney, Alpina International Co, 33 East 29 Street, New York, NY 10016, USA; Mohd Hassan, Salah Trading Agencies, PO Box 5287, Bahrain.

Guru Ispat Ltd

Head office: 18a Park Street, 16 Stephen Court, 4th Floor, Calcutta 700071. **Tel:** 28 8760-4. **Telex:** 0212365, 0217714, 0215460. **Cables:** Guru Eximp.
Management: Pawan K. Parolia (director). **Associates:** Toyo Ispat Ltd, Nissho Yodo Ispat Ltd, Parolia Metal Industries, Vinay Alloy Rolling Mills, East Coast Sponge Iron Ltd.

Indian Iron & Steel Co Ltd (IISCO)—see Sail – Steel Authority of India Ltd

Indore Steel & Iron Mills Pvt Ltd (Isim)

Head office: 63a Mittal Court, Backbay Reclamation Scheme, Nariman Point, Bombay 400021. **Telex:** 0113444 isim.
Established: 1949.
Works:
■ 314 Bhagirathpura, Indore 452003. ***Rolling mills:*** One wire rod.
■ Shri Ishar Alloy Steels Pvt Ltd, New Industrial Estate, Sector D, Sanwer Road, Indore 452003. ***Steelmaking plant:*** Three electric arc furnaces. ***Continuous casting machines:*** One 2-strand Concast billet machine. ***Rolling mills:*** Two bar mills.
■ National Metal Industries, 314 Bhagirathpura, Indore 452003. ***Rolling mills:*** One light section and bar.
Products: Carbon steel – ingots, billets, wire rod, reinforcing bars, round bars, square bars, flats, light angles, light channels. Alloy steel (excl stainless) – billets, wire rod.

India

Shri Ishar Alloy Steels Pvt Ltd – See Indore Steel & Iron Mills Pvt Ltd (Isim)

Jain Tube Co Ltd

Head office: D-20 Connaught Place, New Delhi 110001. **Tel:** 353217, 353267, 353317, 353367. **Telex:** 31 66902 jtc in. **Cables:** Jaintubes.
Management: Directors – R.N. Jain (managing), S.R. Jain (joint managing), J.R. Jian, N.L. Jain, O.P. Shroff, V.K. Chopra (nominated by PICUP), S.S. Jayarao (nominated by UPSIDC), A.N. Kacker (nominated by UPSIDC). **Established:** 1964. **Capital:** Rs 4800,000. **No. of employees:** Works – 1,100. Office – 150. **Subsidiaries:** Mahavir Investments Ltd.
Works:
■ 21 K.M. Meerut Road, Chaziabad, UP (Tel: 84 1173, 84 1316). ***Tube and pipe mills:*** welded – Two HF 15-150mm ($\frac{1}{2}$-6" dia) (annual capacity 108,000 tonnes). ***Other plant:*** Tube galvanizing plant.
Products: Carbon steel – longitudinal weld tubes and pipes, galvanized tubes and pipes.
Brands: JTC.
Expansion programme: The company is planning to manufacture API Grade Pipes up to 300mm (12" dia).

Jotindra Steel & Tubes Ltd

Head office: 151 Mittal Tower, Nariman Point, Bombay. **Telex:** 115887 jstb in.
Works:
■ 14/3 Mathura Road, Faridabad, Haryana. ***Tube and pipe mills:*** welded – ERW.
Products: Carbon steel – longitudinal weld tubes and pipes, galvanized tubes and pipes.
Brands: JST.

Kalyani Steels Ltd

Head office: PO Mundhwa, Poona 411036. **Telex:** 0145427.
Established: 1974.
Works:
■ At head office. ***Steelmaking plant:*** Two Hindustan Brown Boveri electric arc furnaces.
Products: Carbon steel – ingots, billets, round bars, square bars. Stainless steel – ingots, billets, round bars, square bars. Alloy steel (excl stainless) – ingots, billets, round bars, square bars.

Kap Steel Ltd

Head office: 53 Radha Bazar Lane, Calcutta. **Tel:** 58653-4. **Telex:** 08452559. **Cables:** Kapsteel.
Management: Director – J.C. Kapur (managing), N.C. Kapur (executive); managers – Sunil Kapur (general, purchasing), Dilip Kapur (general, sales), S.K. Pramanik (chief executive, operations), C. Nataraja Pillai (general, rolling mills). **Established:** 1977. **Capital:** Rs 89,950,000. **No. of employees:** Works – 234. Office – 96. **Subsidiaries:** Anekal Castings Pvt Ltd. **Associates:** Kap Chem Ltd, Anekal Castings Pvt Ltd, J.C. Metals Pvt Ltd, R.J. Metals Pvt Ltd, RKS Investments Pvt Ltd, RJ Investments Pvt Ltd, J.C. Investments Pvt Ltd. **Annual capacity:** Finished steel: 48,000 tonnes of rolled products.
Works:
■ Whitefield Road, Mahadevapura Post, Bangalore 560048 (Tel: 58691-3. Telex: 0845559). ***Steelmaking plant:*** Three electric arc furnaces – two 15/17-tonnes (GEC

Kap Steel Ltd (continued)

and HBB), and one 5-tonne (GEC) (annual capacity 48,000 tonnes). Continuous casting machines: 2-strand Concast India 4m radius for billet (85 to 125mm sq) (96,000 tonnes). ***Rolling mills:*** One 4-stand 20"/14" combination light section; one 14", 12", 10" combination bar mill.
Products: Carbon steel – ingots 100-76MM Sq x 1,500mm, billets 100 x 100mm, reinforcing bars, round bars, square bars, light angles, medium angles.
Sales offices/agents: Bangalore, Madras, Salem, Madurai, Tiruchy, and Cochin.

Khandewal Tubes, Division of Khandelwal Ferro Alloys Ltd

Head office: Nirmal (20th Floor), Nariman Point, Bombay 400021. **Telex:** 112410, 116222.
Established: 1958.
Works:
■ ***Tube and pipe mills:*** welded – One Mannesmann-Meer HF. ***Other plant:*** Galvanizing plant.
Products: Carbon steel – longitudinal weld tubes and pipes, galvanized tubes and pipes.

KMA Ltd

Head office: Whitefield Road, Mahadevapura, Bangalore 560048. **Tel:** 58901. **Telex:** 0845460. **Cables:** Kamalloys.
Management: Sundip Sen (general manager).
Works:
■ L.B. Shastri Marg, Kurla, Bombay 400470 (Tel: 512 9348. Telex: 01171156). ***Rolling mills:*** One cold strip.
Products: Carbon steel – cold rolled hoop and strip (uncoated) (mild steel and medium carbon steel).

KR Steelunion Pvt Ltd

Head office: Mittal Court, B Wing, 4th Floor, Nariman Point, Bombay 400021.
Works:
■ Kalwa (near Bombay). ***Coil coating lines:*** One electrolytic tinning (annual capacity 150,000 tonnes).
Products: Carbon steel – electrolytic tinplate (single-reduced).

Maharashtra Elektrosmelt Ltd

Head office: Nirmal, 2nd Floor, Nariman Point, Bombay 400021, Maharashtra. **Tel:** 2024285. **Telex:** 0115652. **Cables:** Eleksmelt.
Management: Executive director/board member – H.S. Aswath; deputy general managers – G. Rangarajan (operation), Nishant Tanksale (technical development and co-ordination), A.K. Podey (services) and R.K. Singh (steel). **Established:** 1974. **Capital:** Paid up – Rs 50,000,000. **No. of employees:** Works – 779. Office – 107. **Ownership:** Steel Authority of India Ltd. **Annual capacity:** Pig iron: 75,000 tonnes. Raw steel: 75,000 tonnes. Finished steel: 75,000 tonnes.
Works:
■ Chanda-Mul Road, Chandrapur 442401. (Tel: 3345, 3349. Telex: 716208). Executive director: H.S. Aswath. ***Steelmaking plant:*** Two x 15/20-tonne basic oxygen converter (annual capacity 75,000 tonnes). ***Refining plant:*** 1 x 20 tonne CLU (60,000 tonnes) (under installation). ***Continuous casting machines:*** 2-strand ISPL (Demag) billet (100 x 100 – 180 x 180mm) (75,000 tonnes). ***Other plant:*** 33 MVA electric smelting furnace for ferro-manganese (67,000 tonnes).
Products: Carbon steel – billets 120 x 120mm sq (output 1986/87 5,500 tonnes).

India

Mahindra Ugine Steel Co Ltd (Muscosteel)

Head office: Bakhtawar, Nariman Point, Bombay 400021. **Telex:** 0113599.
Established: 1962.
Works:
■ Jagdish Nagar, PO Khopoli 410203, Maharashtra. ***Steelmaking plant:*** Two electric arc furnaces. ***Rolling mills:*** One 2-high 800mm reversing blooming; one 4-stand 3-high Morgardshammar bar. ***Other plant:*** Forging press and hammers.
Products: Carbon steel – slabs, blooms, billets, round bars, square bars, flats, hexagons. Stainless steel – slabs, blooms, billets, round bars, square bars, flats, hexagons. Alloy steel (excl stainless) – slabs, blooms, billets, round bars, square bars, flats, hexagons. Forged round and blooms. Carbon and alloy engineering steels, free-cutting steels, spring steels and valve steels, stainless and heat-resisting steels, tool and die steels, bearing steels, etc.

Man Industrial Corp Ltd

Head office: Near Loco, Jaipur 6. **Postal address:** PB 131. **Telex:** 036236 man in.
Established: 1945.
Works:
■ ***Steelmaking plant:*** One 4.5-tonne electric arc furnace. ***Rolling mills:*** light section and wire rod. ***Other plant:*** Forge.
Products: Carbon steel – ingots, wire rod (including high-carbon), light angles, light tees, special light sections. Stainless steel – ingots, wire rod. Alloy steel (excl stainless) – ingots, wire rod. High-speed steels, leaded steels, free-cutting steels, spring steels, bearing steels. Forgings for the railway, defence, agriculture, and automobile industries; tools, dies, etc.

Metalman Pipe Mfg Co Ltd

Head office: Sanwer Road, Indore 452003, Madya Pradesh. **Postal address:** PO Box 212 **Tel:** 36581. **Telex:** 0735283 tube in. **Cables:** Metalman.
Management: Directors: J.R.D. Soni (chairman and managing), S.D. Varma, A. Ramachandra Rao, Mohan Jayakar, Rajiv Lochan Soni, Vijay Soni. **Established:** 1972. **Capital:** Rs 15,000,000. **No. of employees:** Works – 300. Office – 30.
Works:
■ At head office. ***Tube and pipe mills:*** welded Three HF – one ½ to 3″, one 2 to 6″, one for precision tube ¼ to 3″ (combined annual capacity 70,000 tonnes). ***Other plant:*** Continuous pipe galvanizing plant. Normalising and annealing facilities.
Products: Carbon steel – longitudinal weld tubes and pipes, and galvanized tubes and pipes 15 to 150mm nb; precision tubes and pipes up to 88.9mm od; hollow sections square and rectangular (periphery not exceeding 520mm).
Brands: Metalman.
Sales offices/agents: Head office and 409 Maker Chamber V, Nariman Point, Bombay 400021.

Modi Industries Ltd, Steel Division

Head office: Modinagar 201204, UP. **Telex:** 0313403.
Established: 1963.
Works:
■ At head office. ***Steelmaking plant:*** Six electric arc furnaces. ***Rolling mills:*** Two wire rod – one comprising 3-high 18″ and 14″ and 2-high 10″; and one comprising 2-high 18″, 3-high 14″, and 2-high 11″ and 12″. ***Other plant:*** Wire-drawing plant with galvanizing facilities.
Products: Carbon steel – ingots, billets, wire rod, round bars, square bars, bright wire, galvanized wire (plain). Alloy steel (excl stainless) – ingots, billets, wire rod, round bars, square bars.

Mohan Steels Ltd

Head office: 82/2 Cooper Ganj, Kanpur 208003, Uttar Pradesh. **Tel:** 52433. **Cables:** Mosteels.
Management: Directors – R.S. Gupta (managing), M.K. Kejriwal, G.K. Kejriwal, S.K. Kejriwal; chief executive officer – Madhav Kejriwal. **Established:** 1973. **No. of employees:** Works – 180 approx. Office – 19. **Annual capacity:** Raw steel: 28,000 tonnes. Finished steel: 24,000 tonnes.
Works:
■ Gazauli Industrial Area, Unnao-Lucknow Highway, Unnao (Tel: 82323/82571). ***Steelmaking plant:*** electric arc – 12-tonne HBBSSKD320 (annual capacity 27,000 tonnes). ***Rolling mills:*** bar / wire rod 4-stand 3-high (470mm), 8-stand 2-high (235mm) (coil weight 235mm) (combined capacity 24,000 tonnes).
Products: Carbon steel – ingots 125mm x 150mm x 1,500mm (Output 12,000 tonnes); wire rod and round bars 8mm, 10mm, 12mm, 16mm, 20mm, 25mm, 32mm, (16,000 tonnes).
Brands: MSL.
Expansion programme: New steelmaking furnaces (induction and arc furnace) (annual capacity 50,000 tonnes), new bar/rod mill (straight lengths and coils) (45,000 tonnes). Expected date of completion October 1988.

Mohta Electro Steel Ltd

Head office: Plot Nos. 84-86, Industrial Area, Bhiwani, Haryana. **Telex:** 344204 mes in.
Established: 1972.
Works:
■ Bhiwani. ***Rolling mills:*** Seven cold strip – up to 450mm, including 2-high Sendzimir.
Products: Carbon steel – cold rolled hoop and strip (uncoated).

Mukand Iron & Steel Works Ltd

Head office: Bajaj Bhavan, 226 Nariman Point, Bombay 400021. **Tel:** 202 10 25. **Telex:** 0114359. **Cables:** Mukandbom.
Management: Chairman and managing director – Viren J. Shah; chief executive – Rajesh V. Shah; deputy cheif executive – Vinod Shah; financial controller – S.B. Jhaveri; general managers – N.S. MohanRam (steel plant), T.P. Shah (machine building), R.H.G. Rau (research and development and foundry), T.S. Anand (management audit), Niraj Bajaj (marketing), Sukumar V. Shah (commercial). **Established:** 1937-38. **Capital:** Rs 710,000,000 (30.6.1986). **No. of employees:** 4,159. **Subsidiaries:** Beco Engineering Co Ltd (100%). **Annual capacity:** June 30 1987 – billet 270,000 tonnes (licenced), 270,000 tonnes (installed). Rolled steel 264,000 tonnes (licenced), 302,000 tonnes (installed). Steel castings 12,000 tonnes (licenced), 12,000 tonnes (installed).
Works:
■ Thane Belapur Road, Kalwe, Thane 400605. (Tel: 922 2170-9. Telex: 01177011).General manager: N.S. MohanRam (steel plant). ***Steelmaking plant:*** Three 30-tonne electric arc furnaces (annual capacity 270,000 tonnes). ***Refining plant:*** One 30/35-tonne ladle furnace (72,000 tonnes), and one 30/35-tonne VOD (72,000 tonnes) (based on 8 heats/day). ***Continuous casting machines:*** Two 3-strand billet (100 x 100mm) (288,000 tonnes); one 2-strand bloom (160 x 160mm) (144,000 tonnes). ***Rolling mills:*** One 9-stand heavy section (25-90mm) (80,000 tonnes); one 12-stand Morgardshammer wire rod (5.5-22mm) (coil weight 400kg) (222,000 tonnes).
■ Lal Bahadur Marg, Kurla, Bombay 400070. (Tel: 512 01 80. Telex: 01171499). General manager: R.H.G. Rau. **Steel foundry**: Three electric arc furnaces (annual capacity 12,000 tonnes of steel castings).
Products: Carbon steel – billets 100-120mm sq; wire rod (incl high-carbon) 5.5-12mm dia; reinforcing bars; round bars; flats. Bright (cold finished) bars. Stainless steel – billets 100-120mm sq; wire rod 5.5-12mm dia; round bars 25-60mm dia. Alloy steel (excl stainless) – billets 100-120mm sq; wire rod 5.5-12mm dia; round bars 25-60mm dia.

We're proud to announce a new steel.

Sorry Nippon, BSC and Bethlehem.

For a company that's not among the top 25 companies of the world, we've done pretty well.

We've just developed Mukand Supersteel. A dual phase high tensile steel bar that is already in use in the construction industry. And is being researched for further applications.

We've also invented a rapid electrolytic descaling process for stainless steel wire rods which shortens descaling time to an incredible 2 seconds!

Back home in India, we're the largest producer of steel through the Electric Arc Furnace route. And have a state-of-the-art plant with an annual capacity of 270,000 billet tonnes.

In fact, Mukand has introduced India to many innovations in the steel world. The continuous billet casting technology. Easy draw cooling system for wire rods. CHQ Boron steel. And many more.

What's more, Mukand is not a new-comer to the export world. We regularly supply Bright Bars, Predrawn & Finished Wires in high alloy grades such as stainless steel, leaded free cutting steel and spring steel to discerning buyers in the USA, Canada, West Germany, Finland and U.K.

Now after following the leaders for years, we've at last come of age to make our mark in the world of steel.

MUKAND IRON & STEEL WORKS LTD.
L.B. Shastri Marg, Kurla, Bombay 400 070, INDIA.

Rediffusion/M/55

Mukand Iron & Steel Works Ltd (continued)

Output of wire rods, bars, flats and rounds by category – carbon steel billets, alloy steel billets and stainless steel billets – 100/120mm sq (output in year ended June 30th, 1986 207,435 tonnes). Carbon steel wire rods, high carbon steel wire rods, alloy steel wire rods and stainless steel wire rods 5.5-12mm (127,725 tonnes). Reinforcing bars, round bars and alloy steel bars 25-60mm (57,146 tonnes). High-carbon (49,002 tonnes); stainless steel (25,619 tonnes); alloy steel (14,114); free-cutting (2,701); semi free-cutting (1,875 tonnes); silico-manganese steel (967 tonnes); ordinary carbon steel (28,649 tonnes); low-carbon (12,071 tonnes); mild steel (49,873 tonnes). Steel castings (8,411 tonnes).

Nagarjuna Steels Ltd (NSL)

Head office: Nagarjuna Hills, Panjagutta, Hyderabad 500004. **Telex:** 1556538 nsl in.
Established: 1974.
Works:
■ Patancheru, Medak Dist, AP. ***Rolling mills:*** cold strip.
Products: Carbon steel – cold rolled hoop and strip (uncoated).

National Metal Industries – See Indore Steel & Iron Mills Pvt Ltd

National Steel Industries

Head office: Village Sejwaya, Madhya Pradesh. **Telex:** 314665.
Works:
■ **Coil coating lines:** Galvanizing line (annual capacity 40,000 tonnes).
Additional information: The line, being built by Cockerill, was due to start operating in mid-1987.

Neeka Tubes Ltd

Head office: 4th Floor, Bank of India Building, Bhadra, Ahmedhbad 380001. **Postal address: Tel:** 877567-9. **Telex:** 0121595.
Management: M.N. Kothari (executive director), S.H. Prithyani (works manager), M.A. Kamalwala (assistant marketing and general manager). **Capital:** Rs 30,600,000. **No. of employees:** Works – 178. Office – 82. **Ownership:** Gujarat Steel Tubes Ltd (100%). **Annual capacity:** Finished steel: 1,000 tonnes.
Works:
■ 114-123A GIDC Vatwa, Ahmedabad 382445. Works director: S.H. Prithyani. ***Tube and pipe mills:*** welded – one TIG ($\frac{1}{2}$ to 4") (annual capacity 1,000 tonnes).
Products: Stainless steel – seamless tubes and pipes, longitudinal weld tubes and pipes cold drawn tubes and pipes.

Nippon Denro Ispat

Head office: Nagpur.
Works:
■ **Coil coating lines:** Galvanizing line.
Products: Carbon steel – Galvanized sheet.
Expansion programme: A 120,000 tpy cold strip mill was scheduled for start-up in 1988.

India

Panchmahal Steel Ltd

Head office: 3rd Floor, Kothari Chambers, Kothi Road, Baroda 390001, Gujarat. **Tel:** 540447, 557472. **Telex:** 175317. **Cables:** Panchsteel Baroda.
Management: Managing director – Ashok Malhotra; chairman – R.L. Malhotra; directors – Ravi Malhotra, Mr Lokhandwala; managers – N.K. Bharal (general), B.S. Shekhawat (sales), J.C. Panchal (purchase). **Established:** 1973. **Capital:** Rs 3,480,000 (at 30.9.85). **No. of employees:** Works – 465. Office – 37. **Annual capacity:** Finished steel: 54,000 tonnes (billets/blooms), 32,000 tonnes (rolled products).
Works:
■ GIDC Estate, Kalol, Panchmahal District, Gujarat 389330. (Tel: 392, 393, 402, 407, 407, 417 (Derol exchange). Telex: 177201). ***Steelmaking plant:*** Two electric arc furnaces – 10/12-tonne SSKD-320 and 15/17-tonne SSKD-340 (annual capacity 54,000 tonnes). ***Refining plant:*** 15/17-tonne AOD; 15/17-tonne ladle furnace. ***Continuous casting machines:*** Two billet – 2-strand (100 x 100mm and 120 x 120mm, 2 strand 180 x 160mm) (54,000 tonnes). ***Rolling mills:*** one bar comprising 3-stand 3-high 100-50mm round-cornered square, 1-stand 2-high 100-28mm dia, 100 x 20mm flats (combined capacity 24,000 tonnes).
Products: Carbon steel – blooms 180 x 160mm (1986 output 4,345 tonnes); billets 100 x 100mm and 120 x 120mm (8,092 tonnes); round bars 28-100mm dia (3,142 tonnes); square bars 50-100mm sq (5,987 tonnes). Stainless steel – blooms 180 x 160mm (31 tonnes); billets 100 x 100mm and 120 x 120mm (255 tonnes); round bars 28-100mm dia (81 tonnes); square bars 50-100mm sq; flats 100 x 20mm (141 tonnes). Alloy steel (excl stainless) – blooms 180 x 160mm (3,078 tonnes); billets 100 x 100mm and 120 x 120mm (3,896 tonnes); round bars 28-100mm dia (3,970 tonnes); square bars 50-100mm sq (2,335 tonnes).

Parikh Steel Pvt Ltd

Head office: 116 Stephen House, 4 Dalhousie Sqare East, Calcutta 700001. **Tel:** 205355. **Telex:** 0212695.
Established: 1971.
Works:
■ 188 Girish Ghosh Road, PO Belurmath, Howrah. ***Rolling mills:*** Two bar – one 6", one 9".
Products: Carbon steel – round bars, flats.

Partap Steel Rolling Mills (1935) Ltd

Head office: Partap Estates, Grand Trunk Road, Chheharta, Amritsar, Punjab. **Tel:** 41380, 8215, 8367, 51028. **Telex:** 0384 202, 0384 222. **Cables:** Partapmill, Chheharta, Stelroling, Amritsar.
Management: Board directors – Amir Chand Maheshwari (chairman); Partap Chand Maheshwari (managing director); Kailash Chand Maheshwari, Parkash Chand Maheshwari, Hari Singh (chief executive), Darshan Singh (general manager, commercial), Er Ajmer Singh (general manager, projects and research and development), P. Suri (material manager). **Established:** 1935. **Capital:** Subscribed – Rs 65,000,000. **No. of employees:** Works – 1800. Office – 150. **Subsidiaries:** Partap Steel Rolling Mills (1935) Ltd, Mandi Gobindgarh and Indore. **Associates:** Partap Steels Ltd, Ballabgarh and Pattsncheri. **Annual capacity:** Raw steel: 72,000 tonnes. Finished steel: 165,000 tonnes.
Works:
■ At head office. ***Steelmaking plant:*** Three electric arc furnaces – two 12-tonne 11 Birlec GEC furnaces and one 15-17 tonne (combined annual capacity 72,000 tonnes). ***Continuous casting machines:*** One 2-strand billet (100,120 and 160mm sq) (72,000 tonnes). ***Rolling mills:*** Two bar – one 14" 6-stand (for flats, profiles and rounds) and one 12" 6-stand (for flats, profiles); one 12-stand wire rod – six 14" and six 12" (coil weight 180kg) (30,000 tonnes). ***Other plant:*** Foundry with a capacity of 5,000 tonnes/yearly of grey iron castings.
■ Pithampur, District Dhar, near Indore. (Tel: 54, 55 and 99). ***Steelmaking plant:*** Two electric arc furnaces – one GEC Model-11 and one GEC Model 13S (annual capacity 85,000 tonnes). ***Continuous casting machines:*** One 3-strand 3/6m radius billet (100, 120 and 160mm sq) (85,000 tonnes). ***Rolling mills:*** medium section, light section and bar – 5 stand 22" (50,000 tonnes), 12" cassette stand, 6-stand 14" open-type (20,000 tonnes), 6-stand 8" fibre bearing (5,000 tonnes).
■ Mandi Gobindgarh, Punjab. (Tel: 54, 55, 99. Telex: 399-205). ***Steelmaking plant:*** One Birlec/GEC Model 8 electric arc furnace (annual capacity 15,000 tonnes). ***Continuous casting machines:*** One 3-strand 11/6m radius billet (100, 120 and 160 mm sq) (85,000 tonnes). ***Other plant:*** Steel foundry producing rolls, large ingots and special castings.
Products: Carbon steel – billets 100,120 and 160mm dia; wire rod 8-12mm dia; round bars 16-55mm dia; flats 15 x 4-282 x 8; light angles 50 x 50 x 60-75 x 75 x 8; light tees; special light sections (door hinges, lock rings, flanges, tank sections); medium tees 125 x 75; medium joists; wide-flange beams (medium) 125 x 100. Alloy steel (excl stainless) – billets 100, 120 and 160mm dia; flats 15 x 4-282 x 8.

Punjab Con-Cast Steels Ltd

Head office: Focal Point, Ludhiana 141010, Punjab. **Tel:** 20997, 33590. **Telex:** 0386311. **Cables:** Billets.
Management: Directors – Jawahar Lal Jain (managing), R.K. Singla (executive), A.C. Katiyar (executive). **Established:** 1970. **Capital:** Rs 11,785,000 (share capital), Rs 125,309,000 (fixed capital). **No. of employees:** Works – 950. Office – 200. **Annual capacity:** Raw steel: 100,000 tonnes. Finished steel: 85,400 tonnes.
Works:
■ At head office. Works director: A.C. Katiyar. ***Steelmaking plant:*** Four electric arc – Two 10-tonne, one 15-tonne, and one 5-tonne (annual capacity 100,000 tonnes). ***Refining plant:*** One 14-tonne ladle furnace (35,000 tonnes); one 14-tonne vacuum degassing unit (35,000 tonnes); one 14-tonne VOD (35,000 tonnes). ***Rolling mills:*** One 4-stand 3-high 22" blooming (35,500 tonnes); three bar – one 5-stand 3-high 14" (18,000 tonnes), one 6-stand 3-high 14" (18,000 tonnes), and one 8-stand 3-high 8" (14,000 tonnes).
Products: Carbon steel – ingots 8 x 9", 11½ x 10"; reinforcing bars 10mm and above; round bars 10-100mm; square bars 40-125mm; flats thickness 6-25mm, width 37mm-150mm; hexagons 25mm and above. Stainless steel – ingots 8 x 9", 11½ x 10"; round bars 10-100mm; square bars 40mm to 125mm; flats thickness 6-25mm, width 37-150mm. Alloy steel (excl stainless) – ingots 8 x 9", 11½ x 10"; round bars 10-100mm; square bars 40-125mm; flats thickness 6-25mm, width 37-150mm.

India

The Punjab Steel Rolling Mills

Head office: Old Station, Baroda 390002, Gujarat. **Telex:** 0175257.
Established: 1940.
Works:
■ At head office. ***Rolling mills:*** bar and wire rod comprising 3-stand 3-high 380mm, 4-stand 2-high 270mm, 5-stand 2-high 270mm, and 6-stand 2-high 260mm.
Products: Carbon steel – wire rod, round bars, flats, light angles, rail accessories.

Raipur Wires & Steel Ltd (RWSL)

Head office: P-49 Industrial Area, Raipur. **Telex:** 775263 rwsl in.
Established: 1977.
Works:
■ ***Steelmaking plant:*** Three 10/14-tonne electric arc furnaces. ***Continuous casting machines:*** One 2-strand ISPL billet.
Products: Carbon steel – ingots, billets. Alloy steel (excl stainless) – ingots, billets. Spring and bearing steels.

Rashtriya Ispat Nigam Ltd, Visakhapatnam Steel Project

Head office: RTC Complex, Visakhapatnam 530020, Andhra Pradesh.
Management: D.R. Ahuja (chairman and managing director). **Established:** 1982.
Works:
■ Visakhapatnam (planned). ***Coke ovens:*** Three 7m underjet compound batteries (67 ovens each) (annual capacity 3,598,600 tonnes of coal). ***Sinter plant:*** Two 312 sq m strand Dwight Lloyd (5,256,000 tonnes each). ***Blast furnaces:*** Two 12m hearth dia, 3,200 cu m (1,700,000 tonnes). ***Steelmaking plant:*** Five 130-tonne basic oxygen converters (3,400,000 tonnes). ***Continuous casting machines:*** Ten 4-strand for bloom (250mm sq to 300 x 550mm) (3,196,000 tonnes). ***Rolling mills:*** One 13-strand semi-continuous universal heavy section (800,000 tonnes); one 33-stand light and medium merchant bars bar (1,572,000 tonnes); one 4-strand 25-stand wire rod (coil weight 600kg) (600,000 tonnes); one medium merchant and structural (700,000 tonnes). ***Other plant:*** foundries and forge.
Additional information: The works is scheduled for commissioning in 1989.

Rathi Udyog Ltd

Head office: Rathi Katra, Nai Sarak, Delhi 110006. **Tel:** 2512776, 2519657.
Cables: Rathimill.
Management: P.R. Rathi (chairman), P.C. Rathi (director, technical), Pradeep Rathi (president). **Established:** 1973. **Capital:** Rs 7,040,050 (share capital). **No. of employees:** Works – 359. Office – 57. **Associates:** Gordhan Das Rathi Steels Pvt Ltd; Rathi Ispat Ltd (mini-steel works). **Annual capacity:** Finished steel: 77,100 tonnes.
Works:
■ South Side of GT Road, Ispat Nagar, Ghaziabad (UP) (Tel: 8 49241. Telex: 0592269 ru in). Works directors: P.R. Rathi, P.C. Rathi. ***Rolling mills:*** bar and wire rod mills comprising six 400mm stands, one 360mm stand, six 290mm stands, four 2-high 260mm stands (finishing speed 20 metres/second) (coil weight 140kg) (annual capacity 77,100 tonnes).
Products: Carbon steel – wire rod 6mm, 8mm, 10mm, 12mm (1986 output, 9,500 tonnes); reinforcing bars 8 to 45mm, round bars 6 to 45mm and square bars (34,500 tonnes).
Expansion programme: Roughing mill with two 600 mm stands and four 400mm continuous stands expected to be completed by June 1988. Second re-heating furnace (25-tonnes/hour) with ceramic recuperator, September 1987.

Sail – Steel Authority of India Ltd

Head office: Integrated Office Complex, Lodhi Road, New Delhi 110003. **Telex:** 312979 sail nd. **Cables:** Steelindia.
Subsidiaries: Hindustan Steel Ltd; Bokaro Steel Ltd; Salem Steel Ltd; Sail International Ltd; Indian Iron & Steel Co Ltd (IISCO); Visvesvaraya Iron & Steel Ltd; National Mineral Development Corp Ltd; Hindustan Steelworks Construction Ltd (construction of steel plants, etc); Metallurgical & Engineering Consultants (India) Ltd (Mecon) (consultancy and engineering services for the metallurgical industries); Bharat Refractories Ltd (refractories). **Associates:** Rashtriya Ispat Nigam Ltd; Visakhapatnam Steel Project.
Works:
■ Hindustan Steel (HSL), Bhilai Steel Plant, Madhya Pradesh. ***Coke ovens:*** Eight batteries. ***Sinter plant:*** 8 strands. ***Blast furnaces:*** Seven furnaces including three 7.2m hearth dia, 886 cu m (capacity 1,149 tonnes/day); and three 9.1m hearth dia, 1,489 cu m (1,738 tonnes/day). Combined annual capacity 3,948,000 tonnes. ***Steelmaking plant:*** Three 100/130-tonne LD basic oxygen converters (annual capacity 1,500,000 tonnes); ten open hearth furnaces comprising five 250-tonne, and five 500-tonne (2,500,000 tonnes). ***Continuous casting machines:*** One 4-strand USSR for bloom (250 x 280mm to 300 x 350mm); two 1-strand USSR for slab (180 x 1,500mm to 3,200 x 2,000mm). ***Rolling mills:*** One 2-high 1,150mm reversing blooming (2,500,000 tonnes); one 2-high H/V 1,000/700/500mm billet (1,263,000 tonnes); heavy section and rail comprising 1-stand 2-high reversing roughing, 2-stand 3-high 800mm finishing, and 1-stand 2-high 800mm finishing (750,000 tonnes); light section and bar comprising 4-stand 2-high 500mm roughing, 400mm edger, 4-stand 2-high 400mm intermediate, 400mm edger, 350mm edger and 1-stand 2-high 350mm finishing (500,000 tonnes). one wire rod comprising 39-stand with seven 450mm roughing and 380/320/280mm finishing (400,000 tonnes); one heavy plate (400,000 tonnes); one wide hot strip (1,000,000 tonnes).
■ Bokaro Steel Plant, Bokaro Steel City, Dhanbad, Bihar. ***Coke ovens:*** Seven underjet twin-flue compound regenerative batteries (69 ovens each) (annual capacity 2,397,000 tonnes). ***Sinter plant:*** Three. ***Blast furnaces:*** Five furnaces (4,585,000 tonnes). ***Steelmaking plant:*** Five 100-tonne LD basic oxygen converters (2,500,000 tonnes). ***Rolling mills:*** One 1,250mm universal slabbing (1,465,000 tonnes); one wide hot strip comprising 1-stand 2-high roughin, 4-stand 4-high universal and 6-stand 4-high finishing (1,430,000 tonnes); two cold reduction comprising 4-stand 2,000mm tandem, 5-stand 1,400mm tandem (475,000 tonnes). ***Coil coating lines:*** One 1,250mm hot-dip galvanizing. ***Other plant:*** Ingot mould foundry; steel foundry; iron and non-ferrous foundry. Forge.
■ Durgapur Steel Plant, Durgapur, West Bengal. ***Coke ovens:*** Five batteries including four with a total of 312 ovens and an anuual capacity of 1,740,000 tonnes. ***Sinter plant:*** 2-strand 142.7 sq m (annual capacity 1,100,000 tonnes). ***Blast furnaces:*** Three 8.23m hearth dia, 1,171 cu m (1,270 tonnes/day); one 8.84m hearth dia , 1,550 cu m (1,500 tonnes/day). Total annual capacity 1,700,000 tonnes. ***Steelmaking plant:*** Nine open hearth furnaces – one 120-tonne, eight 220-tonne (annual capacity 1,600,000 tonnes). ***Rolling mills:*** One blooming comprising 1-stand 2-high 1,067mm reversing and 1-stand 2-high 813mm reversing (1,600,000 tonnes); one billet comprising 2-stand 610mm vertical and 2-stand 2-high 706mm horizontal roughing and 4-stand 2-high finishing (957,000 tonnes); one medium section comprising 24" – 1-stand 2-high 660mm roughing, 2-stand 3-high 610mm intermediate and 1-stand 2-high 610mm finishing (211,000 tonnes); one light section and bar comprising 7-stand 2-high 406/372mm roughing and two edgers; 2-stand 2-high330mm intermediate with edger; 6-stand 2-high 330/279mm finishing with edger (240,000 tonnes); one hot strip and hoop (skelp) comprising 6-stand 2-high 406mm roughing with two edgers, 5-stand 330mm finishing with 3 edgers (250,000 tonnes). ***Other plant:*** Fishplate plant, sleeper plant, wheel and axle plant.
■ Rourkela Steel Plant, Rourkela, Orissa. ***Coke ovens:*** Five batteries including four with 290 ovens and and annual capacity of 1,810,000 tonnes. ***Sinter plant:*** Two 125 sq m strands (1,200,000 tonnes). ***Blast furnaces:*** Three 7.4m hearth dia, 990 cu m (1,000 tonnes per day); one 9m hearth dia, 1,448 cu m (1,500 tonnes per day) (total annual capacity 1,600,000 tonnes). ***Steelmaking plant:*** Five basic oxygen converters – three 50-tonne, two 60-tonne (1,800,000 tonnes); ***Rolling mills:*** One

India

Sail – Steel Authority of India Ltd (continued)

2-high 1,180mm high-lift slabbing and blooming (1,800,000 tonnes); one 4-high, 3,100mm heavy plate (280,000 tonnes); two wide hot strip – one semi-continuous 2-stand 4-high roughing and 6-stand 4-high 1,700mm continuous finishing (1,106,000 tonnes); 3-stand 2-high 1,300mm for electrical sheets; three cold reduction comprising two 4-high 1,700/1,200mm reversing and one 5-stand 4-high 1,420mm tandem (669,000 tonnes); five temper/skin pass – three 2-high 1,400mm (temper) (50,000 tonnes), two 1-stand 4-high 1,700/1,200mm and one 2-stand 4-high 1,200mm (skin pass) (260,000 tonnes). ***Tube and pipe mills:*** welded – 8⅝ to 20" od ERW (75,000 tonnes); 16 to 64" od spiral (55,000 tonnes). ***Coil coating lines:*** One 1,040mm electrolytic tinning (150,000 tonnes); two 1,220mm hot-dip galvanizing (160,000 tonnes).
■ Alloy Steels Plant, Durgapur, West Bengal (Indian Iron & Steel Co Ltd). ***Steelmaking plant:*** Four electric arc furnaces – three 50-tonne KT, one 10-tonne NOT; and two HF induction furnaces – one 2-tonne, one 500kg (combined annual capacity 160,000 tonnes). ***Rolling mills:*** One 2-high 900 x 2,200mm reversing blooming; one billet and sheet bar – 2-stand 2-high 700 x 1,800mm reversing and 2-high 650 x 1,800mm x 650 x 1,300mm finishing; one 15-stand bar; one sheet mill comprising 1-stand 3-high 1,825 x 800/640/800mm roughing, one 2-stand 1,525 x 750mm finishing; one Sendzimir cold reduction; and one 2-high 1,525 x 650mm temper/skin pass (combined annual capacity 17,400 tonnes). ***Other plant:*** Forge.
■ Salem Steel Plant, Salem. ***Rolling mills:*** One Sendzimir cold reduction (annual capacity 32,000 tonnes). ***Other plant:*** Coil build-up line. Annealing and pickling lines.
■ Burnpur, West Bengal. ***Coke ovens:*** Three Simon Carves regenerative underjet (one with 72 ovens, two 78 ovens). ***Blast furnaces:*** Two 5.2m hearth dia, 434 cu m; two 7.6m hearth dia, 1,041 cu m (combined annual capacity 1,300,000 tonnes). ***Steelmaking plant:*** Seven open hearth furnaces – six 200-tonne tilting, one 100-tonne fixed (1,000,000 tonnes); three 25-tonne acid Bessemer converters (800,000 tonnes). ***Rolling mills:*** One 2-high 40" reversing blooming (1,000,000 tonnes); one 10-stand billet and sheet bar comprising three vertical and 7 horizontal stands (1,000,000 tonnes); one 2-high 34" reversing heavy section and rail (300,000 tonnes); one 18" 2-stand 3-high roughing, 1-stand 3-high intermedite, 1-stand 2-high finishing bar (150,000 tonnes); one 23-stand wire rod and merchant (200,000 tonnes); one sheet mill comprising of 2-stand Lewis 3-high roughing and 4-stand 2-high finishing. ***Coil coating lines:*** One hot-dip galvanizing (100,000 tonnes).
■ Kulti, West Bengal (Indian Iron & Steel Co). Steel foundry with 4-tonne electric arc furnace, three centrifugal pipe foundries, other foundries.
Products: Steelmaking pig iron; foundry pig iron. Carbon steel – ingots; slabs; billets; semis for seamless tubemaking; wire rod; reinforcing bars; round bars; square bars; flats; light angles; light channels; special light sections; medium tees; medium joists; medium channels; heavy tees; heavy joists; heavy channels; wide-flange beams (heavy); heavy rails; light rails; rail accessories; sheet piling; medium plates; heavy plates; hot rolled sheet/coil (uncoated); cold rolled sheet/coil (uncoated); hot-dip galvanized sheet/coil; electrical grain oriented silicon sheet/coil; electrical non-oriented silicon sheet/coil; blackplate; electrolytic tinplate (single-reduced); hot-dip tinplate; longitudinal weld tubes and pipes; spiral-weld tubes and pipes. Stainless steel – ingots; billets; round bars; square bars; flats; hot rolled hoop and strip (uncoated); skelp (tube strip); cold rolled hoop and strip (uncoated); medium plates; heavy plates; hot rolled sheet/coil (uncoated); cold rolled sheet/coil (uncoated). Alloy steel (excl stainless) – ingots; billets.

The Sandur Manganese & Iron Ores Ltd (Smiore)

Head office: 20 Palace Road, Bangalore. **Telex:** 845427 smio in.
Established: 1954.
Works:
■ **Ironmaking plant:** One submerged arc 15 MVA furnace. ***Other plant:*** Ferro-silicon furnaces.
Products: Foundry pig iron. Ferro-alloys.

Sanghvi Steels Ltd

Head office: Mahavir Darshan, 4th Floor, M.N. Koli Marg, Bombay 400003, Maharashtra. **Tel:** 330271, 346794. **Telex:** 01175128 ssl in. **Cables:** Magansteel Bombay.
Management: C.M. Sanghvi (managing director), H.M. Sanghvi (joint managing director). **Established:** 1983. **Capital:** Rs 20,000,000. **No. of employees:** Works – 300. Office – 35. **Associates:** Hiralal Maganlal & Co (ship repair and import of dried fruits); Patel Engineering Works (ship repair and general engineering); Sanghvi Constructions Pvt Ltd (construction); Sanghvi Cargo Handling Pvt Ltd (bulk cargo /scrap handling). **Annual capacity:** Finished steel: 44,000 tonnes.
Works:
■ 12 Midc Area, Taloja, Dist. Raigad, Maharashtra State. (Tel: 377/379). Works director: H.M. Sanghvi (jt managing director). ***Steelmaking plant:*** One 17/19-tonne GEC Model 11-S electric arc furnace (annual capacity 45,000 tonnes). ***Continuous casting machines:*** One 2-strand ISPL (Mannesmann Demag) billet (100, 120, 150 and 160mm sq) (45,000 tonnes). ***Rolling mills:*** Two bar – one 5-stand 3-high and 5-stand 2-high (16, 12 and 10") and one 3-stand 3-high and 2-stand 2-high (10").
Products: Carbon steel – ingots 3 x 4", $3\frac{1}{2}$ x $4\frac{1}{2}$", 5 x 6"; billets 100, 120 and 150mm sq; reinforcing bars 8, 10, 12, 16, 20 and 25mm dia; round bars 8-50mm dia; flats 5-16mm thick, 30-100mm wide. Alloy steel (excl stainless) – ingots 3 x 4", $3\frac{1}{2}$ x $4\frac{1}{2}$", 5 x 6"; billets 100, 120 and 150mm sq; round bars; flats (spring steel, file steel). Carbon, alloy and stainless steel output Sept 85-Aug 86: Ingots – 32,055 tonnes; reinforcing bars 17,610 tonnes; flats – 9,300 tonnes.

Singh Electrosteel Ltd (SES)

Head office: BPT Plot 52 Victoria, (Overbridge) Road, Mazgaon, Bombay 10. **Tel:** 370912, 374126. **Telex:** 0112984 isc. **Cables:** Steelworth Bonbay.
Management: J. Singh (technical director), N. Singh, R. Singh. **Established:** 1979. **Capital:** Rs 40,000,000. **No. of employees:** Works – 70. Office – 10. **Subsidiaries:** Indian Steel Corp, Calcutta and Bombay; Singh Bros, Bombay and Bangalore; Devi Commercial Co, Calcutta; Singh Alloys & Steel Ltd, Calcutta; Indian Steel Equipment, Calcutta; ISC Processros (P) Ltd, Calcutta (manufacturing electro magnets and magnetic separators). **Annual capacity:** Raw steel: 100,000 tonnes.
Works:
■ F2 MIDC, Tarapur, Bombay. (Tel: 284. Telex: 0132207). Works director: Jitendra Singh. ***Steelmaking plant:*** Two 15-tonne electric arc furnaces (annual capacity 50,000 tonnes). ***Continuous casting machines:*** One 2-strand ISPL-Demag 6m radius billet (75,000 tonnes). ***Rolling mills:*** One medium section and bar (100,000 tonnes).
■ D 34 Industrial Area, Kalyani, Calcutta. (Tel: 599). Works director: Arabindra Singh. ***Steelmaking plant:*** One 15-tonne electric arc furnace (annual capacity 50,000 tonnes). ***Continuous casting machines:*** One 2-strand billet. ***Rolling mills:*** One wire rod (coil weight 500kg) (10,000 tonnes).
Products: Carbon steel – billets 100-180mm; reinforcing bars 12-20mm; round bars 16-20mm. Stainless steel – billets 100-180mm. Alloy steel (excl stainless) – billets 100-180mm. SES, SAS.

Sipta Coated Steels Ltd

Head office: 10th Floor, Chander Mukhi, Nariman Point, Bombay 400021. **Tel:** 2870746/48/49/53/60. **Telex:** 011 4613 sipt in.
Management: Chairman – H.G. Vartak; managing director – Mukesh Gupta; directors – V.B. Gadkari, G. Golapakrishnan, Miss Y.J. Tarachand, S.G. Subrahmanyan, Nandkishore Kagliwal, M.P. Poddar, B.L. Agarwal, S.P. Jain; vice presidents – P.C. Soni (commercial and company secretary), J.L. Zutshi (technical); works manager – A.K. Gupta. **Established:** 1984. **Capital:** Rs 7,080,000. **No. of employees:** Works – 260. Office – 40. **Annual capacity:** Finished steel: 40,000 (galvanized plain coils).
Works:
■ A/3 MIDC Industrial Area, Nanded 431603, Maharashtra. (Tel: 3542,3810,2816).

India

Sipta Coated Steels Ltd (continued)

Works director: A.K. Gupta. ***Coil coating lines:*** One CMI, Belgium/Stein Heurtery, France hot-dip galvanizing (furnace line) (40,000 tonnes).
Products: Carbon steel – hot-dip galvanized sheet/coil 0.16-0.5mm thickness, 70-1,200mm.
Brands: Sipta.

Somani Iron & Steels Ltd

Head office: Somani Bhawan, 51/27 Nayaganj, Naughara, Kapur 208001. **Tel:** 69824, 67582, 68227. **Telex:** 0325 322. **Cables:** Furnace.
Management: Directors – R.K. Somani (managing), K.K. Somani, N.K. Somani, D.K. Jaiswal, Ramji Agnihotri. **Established:** 1971. **Capital:** Issued, subscribed and paid up – Rs 5,000,000. **No. of employees:** Works – 168. Office – 20. **Annual capacity:** Raw steel: 36,000 tonnes.
Works:
■ Sonik, District Unnao (UP). (Tel: 330 and 331). ***Steelmaking plant:*** Two 12-tonne electric arc furnace (annual capacity 36,000 tonnes).
Products: Alloy steel (excl stainless) – ingots 100 x 75, 113 x 87, 125 x 100, 125 x 150, 400 x 450mm (1985/86 output 24,684 tonnes).

The South India Steel & Starch Industries

Head office: 110/1-A Nattamangalam Main Road, Maniyanoor, Salem 636010, Tamilnadu. **Postal address:** PB 207, Nethimedu, Salem 636002. **Tel:** 51255, 52303. **Cables:** Sahumills.
Management: Ravikumar Gupta (managing partner); P.R. Doshi (chief executive). **Established:** 1968. **Capital:** Rs 815,000. **No. of employees:** Works – 100. Office – 12. **Associates:** Cauvery Industries, Salem (bright bar producer); Sonal Vyapar Ltd, Salem (ingot producer). **Annual capacity:** Finished steel: 19,500 tonnes.
Works:
■ At head office. ***Rolling mills:*** bar (annual capacity 19,500 tonnes). ***Other plant:*** Cold drawing plant.
Products: Carbon steel – light angles 35 x 35 x 5mm, 40 x 40 x 6mm (output 3,000 tonnes); medium angles 50 x 50 x 6mm, 65 x 65 x 6mm (5,000 tonnes).

Special Steel Co

Head office: 211 Jawahar Marg, Indore, Madhya Pradesh. **Tel:** 34835.
Management: Narendra Kalani (proprietor), R.S. Yadav (sales manager), Mr Amarnath (stores purchase manager). **Established:** 1987. **Capital:** Rs 5,500,000. **No. of employees:** Works – 125. Office – 15. **Associates:** Prime Steel Co (manufacturers of Torsteel). **Annual capacity:** Finished steel: 12,000 tonnes.
Works:
■ 36-A, Sector III, Industrial Area, Pithampur (Dhar). Works director: Mr M. Pandey. ***Rolling mills:*** one 12″ single-stand light section and one 6-stand 8″ centre roll fix type bar (annual capacity 12,000 tonnes).
Products: Carbon steel – reinforcing bars 8 to 20mm; round bars 8 to 20mm; square bars 8 to 20mm; flats 25 x 3 to 50 x 12mm; light angles 25 x 25 x 3 to 50 x 50 x 6mm; light tees 25 x 25 x 3 to 40 x 40 x 6mm.
Expansion programme: Electric arc furnace.

Sponge Iron India Ltd

Head office: Khanij Bhavan, 6th Floor, Masab Tank, Hyderabad 500028, Andhra Pradesh. **Tel:** 227543, 227643. **Telex:** 04236640. **Cables:** Spongeiron Hyderabad.
Management: S. Vangala (chairman, managing director), V.L.N. Murthy (deputy

Sponge Iron India Ltd (continued)

generla manager, plant operations), K.P. Patnaik (deputy manager, engineering and projects), V. Subba Roa (manager, sales). **Established:** 1975. **Capital:** Rs 130,000,000 (authorised capital), Rs126,000,000 (paid-up capital). **No. of employees:** Works – 400. Office – 50. **Ownership:** The Government of India and The Government of Andhra Pradesh (100%).
Works:
■ PO Siil Campus 507 154, Khammam Dist, AP. (Tel: 08744 2326. Telex: 0153202). Works director: V.L.N. Murthy (deputy general manager, operations). ***Direct reduction plant:*** Two SL/RN Lurgi Chemie rotary kilns (annual capacity 60,000 tonnes).
Products: Direct-reduced iron (DRI) 3 to 25mm particle size (1986-7 output 52,500 tonnes).
Expansion programme: Production of low-phosphorus pig iron using DRI fines and other waste products from DR plant (June 1990), 100,000 tones/year direct production at Hospet, of Karnataka (1990).

Star Steel Pvt Ltd (Starsteel)

Head office: Opposite Makarpura Railway Station, Maneja, Baroda 390013.
Established: 1972.
Works:
■ ***Steelmaking plant:*** Two 10/14-tonne electric arc furnaces. ***Continuous casting machines:*** One 2-strand billet machine. ***Rolling mills:*** One 4-stand 3-high 14", 10", and 12" bar.
Products: Carbon steel – billets, reinforcing bars, round bars, square bars, flats, rail accessories. Stainless steel – billets. Alloy steel (excl stainless) – billets.

Steel Ingots Pvt Ltd

Head office: Trivedi Chambers, 2 Maharani Road, Indore 452007. MP. **Telex:** 0735331.
Established: 1973.
Works:
■ Indusrial Area, A.B. Road, Dewas. ***Steelmaking plant:*** One electric arc furnace.
■ A.B. Road, Hukamkhedi, Indore. ***Rolling mills:*** 10" light section and bar.
Products: Carbon steel – ingots, round bars, light angles.

Steel Tubes of India Ltd

Head office: Steel Tubes Road, Dewas 455001. **Telex:** 0732202.
Established: 1959.
Works:
■ ***Tube and pipe mills:*** welded – ERW. **Cold drawing plant**.
Products: Carbon steel – longitudinal weld tubes and pipes, cold drawn tubes and pipes.

Sunflag Iron & Steel Co

Head office: Bhandara, Near Nagpur, Maharashtra.
Works:
■ ***Steelmaking plant:*** One 50-tonne eccentric bottom tapping electric arc furnace. ***Refining plant:*** One 50-tonne ladle furnace . ***Continuous casting machines:*** One 3-strand billet (130mm sq). ***Rolling mills:*** 18-stand light section (annual capacity 200,000 tonnes).
Products: Reinforcing bars 10 to 40mm; round bars 10 to 50mm; flats 30 x 100mm to 5 x 20mm; light angles; light channels. Alloy steel (excl stainless) – round bars and flats (spring steels).
Additional information: Orders for the plant were placed with Mannesmann Demag in the second half of 1986.

India

Taloja Rolling Mills

Head office: 402-A Poonam Chambers, Bombay 400018. **Telex:** 1173799 trm in.
Established: 1975.
Works:
■ A-1, MIDC Area, Taloja, Dist. Kolaba. ***Rolling mills:*** 330/250mm light section and bar.
Products: Carbon steel – reinforcing bars, round bars, square bars, flats, light angles, light tees, light channels, special light sections.

The Tata Iron & Steel Co Ltd (Tisco)

Head office: Bombay House, 24 Homi Mody Street, Bombay 400001. **Tel:** 2049131, 2048453. **Telex:** 0112018, 0112731 tata in.
Management: Board of directors – R.H. Mody (chairman and managing), R.N. Tata (deputy chairman), S.A. Sabavala (vice chairman), J.R.D. Tata, N.A. Palkhivala, Keshub Mahindra, M.P. Wadhawan, Akbar Hydari, N.N. Wadia, Mantosh Sondhi, L.P. Singh, A.S. Gupta, R.P. Khosla, D.C. KothariManagement: President – J.J. Irani; vice presidents – K.C. Mehra (operations), R.N. Sharma (administration and raw materials), S.N. Pandey (industrial relations), A. Kashyap (corporate) (under the overall supervision of the Board of Directors headed by the chairman and managing director). **No. of employees:** 69,800. **Ownership:** Public Limited Company. **Subsidiaries:** Special Steels Ltd (production of wire rods and wire); Tata Refractories Ltd (production of refractories); Tata Pigments Ltd (production of ferrous sulphate and iron oxide); Kalimati Investment Co Ltd. **Annual capacity:** Hot metal: 2,000,000 tonnes. Raw steel: 2,400,000 tonnes. Finished steel: 2,000,000 tonnes (after Phase II of the modernisation programme nearing completion).
Works:
■ Jamshedpur 831001, Bihar State, India (Telex: 626201, 626280 tisc in). ***Coke ovens:*** Six batteries (total 324 ovens). ***Sinter plant:*** Two 76 sq ft Dwight Lloyd/Lurgi (annual capacity 1,200,000 tonnes). ***Blast furnaces:*** Six furnaces – (A) 7.2m hearth dia, 937 cu m (358,000 tonnes); (B) 6.24m hearth dia, 633 cu m (267,000 tonnes); (C) 6.24m hearth dia, 608 cu m (267,000 tonnes); (D) 7.20m hearth dia, 937 cu m (359,000 tonnes); (E) 5.64m hearth dia, 558 cu m (197,000 tonnes); (F) 8.63m hearth dia, 1,337 cu m (562,000 tonnes). ***Steelmaking plant:*** Two 130-tonne LD basic oxygen converters (1,013,000 tonnes); Two 10-tonne electric arc furnaces (31,000 tonnes); Melting Shop No 1 – four 80-tonne open hearth furnaces (60,000 tonnes); Melting Shop No 3 – eight 200-tonne open hearth furnaces (966,000 tonnes). ***Refining plant:*** One ESR unit (3,500 tonnes); one 130-tonne VAD unit (170,000 tonnes). ***Continuous casting machines:*** Two billet machines – one 6-strand curved mould (100 to 130mm sq) (348,000 tonnes), one 1-strand straight mould (16,000 tonnes). ***Rolling mills:*** Two blooming – No 1 40″ for blooms and slabs (900,000 tonnes), No 2 46″ for blooms and slabs (1,100,000 tonnes); two sheet bar and billet – No 1 32″ reversing; No 2 630/690mm continuous; one 28″ semi-continuous cross country medium section and light structural (187,000 tonnes); one 480/360mm continuous light section (320,000 tonnes); one bar and wire rod (300,000 tonnes); One 96″ heavy plate (100,000 tonnes); one 16 x 14″ continuous hot strip and hoop (180,000 tonnes); six hot sheet mills comprising three semi-mechanised with 3-high 838mm roughing, 2-high 813mm finishing, and three 2-high 813mm (150,000 tonnes). ***Tube and pipe mills:*** seamless – Two 51 to 219.10mm od Pilger (52,000 tonnes). Welded – One 15 to 80mm nb Fretz-Moon (175,000 tonnes); one 15.9 to 76.2mm od precision ERW (18,000 tonnes). ***Other plant:*** Batch type sheet galvanizing pots (66,000 tonnes); bar forging plant for axles and long bars (18,000 tonnes); wheel, tyre and axle plant (9,600 tonnes); plant for agricultural implements (picks and shovels).
Products: Steelmaking pig iron (1986-87 output, 1,940,467 tonnes); blast furnace ferro-manganese (24,004 tonnes). Carbon steel – ingots (2,250,360 tonnes including alloy); slabs 4 to 10″ thick by 24 to 30″ (156,800 tonnes including alloy and carbon, and alloy blooms); blooms 130 x 130mm to 350 x 350mm; billets 63mm, 75mm,

The Tata Iron & Steel Co Ltd (Tisco) (continued)

100mm, 125mm sq and RCS (813,100 tonnes including alloy); semis for seamless tubemaking (126,900 tonnes); wire rod (201,100 tonnes including bars); reinforcing bars; round bars; light angles, light tees, and light channels (combined output, 61,300 tonnes); medium angles, medium tees, medium joists, and medium channels (combined output, 70,800 tonnes); heavy angles, heavy tees, heavy joists, and heavy channels (combined output, 15,200 tonnes); hot rolled hoop and strip (uncoated) (151,600 tonnes); medium plates and heavy plates 8 to 63mm thick, 1,000mm, 1,250mm, and 1,500mm wide (combined output, 93,100 tonnes); hot rolled sheet/coil (uncoated) (83,100 tonnes); electrical non-oriented silicon sheet/coil (39,000 tonnes); seamless tubes and pipes 51 to 219.1mm od (32,825 tonnes); longitudinal weld tubes and pipes 15 to 80mm nb (62,683 tonnes); oil country tubular goods 114.3 to 177.8mm od (7,816 tonnes); cold drawn tubes and pipes 31.8 to 159mm od (4,628 tonnes); precision tubes and pipes 15.9 to 76.2mm od (10,348 tonnes); hollow sections (994 tonnes). Alloy steel (excl stainless) – ingots; slabs; blooms; billets. Forged bars (11,600 tonnes); hoe bars (8,000 tonnes); wheel, tyres and axles (13,000 tonnes); ferro-manganese (24,084 tonnes).
Brands: Tisten (high tensile steels); Tisstrip (medium and high carbon strip); Tisforge (forging quality steel); Tiscon (reinforcing bars); Tispure (products made from Tata Special A Grade Steel); Tisflow (close die forged steel blanks and rolled rings); Tisfree (free cutting steel); Tiscral (wear resistant steel); Tis-EFR (steel produced by electro-slag refining); Tis-BARFORGE (forged bars with superior dimensional tolerances); Tisring (rolled/forged rings); Tis-draw (deep drawing and extra deep drawing quality steels).
Expansion programme: Korf Energy Optimising furnace technology is being installed.

The Tinplate Co of India Ltd

Head office: 4 Bankshall Street, Calcutta 700001. **Tel:** 20 2401, 20 2407, 20 2410, 20 9880, 20 0663, 20 0461. **Telex:** 2945 tcil in. **Cables:** Tinplates Calcutta.
Management: R.K. Bhasin (managing director), M.M.L. Khullar (general manager).
Established: 1920. **Ownership:** Public limited company with major shares held by The Tata Iron & Steel Co Ltd and government financial institutions, banks and members of the public.
Works:
■ Golmuri, Jamshedpur, Bilhar. **Rolling mills:** Seven hot mills for pack rolling; cold mills. ***Coil coating lines:*** Wean United, Ferrostan and TFS III electrolytic tinning (annual capacity 90,000 tonnes); hot-dip galvanizing (70,000 tonnes black and galvanized sheets). ***Other plant:*** Annealing furnaces (coal fired), batch annealing units and corrugating machines.
Products: Carbon steel – hot rolled sheet/coil (uncoated), hot-dip galvanized sheet/coil, galvanized corrugated sheet/coil, electrical non-oriented non-silicon sheet/coil, blackplate, electrolytic tinplate (single-reduced), tin-free steel (ECCS).

Trichy Steel Rolling Mills Ltd

Head office: 290 West Masi Street, Madurai 1, Tamilnadu. **Tel:** 24385. **Telex:** 0445 205. **Cables:** Rolmill.
Management: Directors – S.B. Sankar (managing), V. Kandaswamy, S. Padmanabhan, S. Rajaratnam, V.T. Soma Sun Daram; secretary and general manager – N.G. Menon. **Established:** 1961. **Capital:** Rs 5,550,000. **No. of employees:** Works – 450. Office – 100. **Ownership:** Trichy Steel Rolling Mills Public Ltd (100%).
Associates: Shri Aravind Steel Pvt Ltd; Arun Ingots Pvt Ltd. **Annual capacity:** Raw steel: 32,000 tonnes.
Works:
■ PO Box 603, Senth Anni Puram, Trichy 620004. (Tel: 25301. Telex: 0455 204). General manager: S.P. Swamy. ***Steelmaking plant:*** Three electric arc furnaces – 10-tonne, 7.5-tonne, and 2.5-tonne (annual capacity 32,000 tonnes). ***Continuous***

India

Trichy Steel Rolling Mills Ltd (continued)

casting machines: One 2-strand ISPL curved mould billet (100mm sq) (60,000 tonnes). ***Rolling mills:*** light section -two roughing 3-high (14"), and five finishing 2-high (9") (6,000 tonnes); bar – one roughing 3-high (20"), one roughing 3-high (14"), three intermediate 2-high (14"), seven finishing 2-high (10") (36,000 tonnes). ***Other plant:*** Steel foundry (100 tonnes/month).
Products: Carbon steel – reinforcing bars 8mm, 10m, 12mm, 16mm, 20mm, 25mm; round bars 10m, 12mm, 16mm; square bars 10m, 12mm, 16mm 20mm; flats 20 x 6mm, 25 x 6mm, 32 x 6mm; light angles 50 x 50 x 6mm; light channels gate.
Sales offices/agents: At head office; Trichy Steel Rolling Mills Ltd, Senthanm Puram, Trichy 620004, Tamilnadu, India; TSRM Rosy Tower, 1st Floor, Branch Office,

Tube Products of India (TPI)

Head office: Tiam House, 28 Rajaji Road, Madras 600001. **Tel:** 511591. **Telex:** 7477. **Cables:** Teeaye.
Management: A.M.M. Arunachalam (chairman, TII Ltd), M.V. Arunachalam (managing director), S. Nagarajan (special director, Tube Products of India). **Established:** 1955. **No. of employees:** 1,550. **Ownership:** Tube Investments of India Ltd (TII).
Works:
■ Avadi, Madras 600054. (Tel: 680448). Special director: S. Nagarajan. ***Rolling mills:*** Two 4-high 12"/14" cold strip (annual capacity 43,500 tonnes); two 2-high 12"/17" temper/skin pass. ***Tube and pipe mills:*** Three RF welded – one ½ to 2½" o.d., one ½ to 3" o.d. and one 3 to 4½" o.d. (37,000 tonnes).
Products: Carbon steel – cold rolled hoop and strip (uncoated) low-carbon (drawing, deep drawing and extra deep drawing qualities), medium-carbon and high-carbon/ high-carbon low-alloy steels in coils or in cropped lengths up to 3,500mm and widths from 25mm to 350mm and thickness from 0.35mm to 4.90mm. Longitudinal weld tubes and pipes – mechanical, manipulated and pressure tested application in bright or semi-bright condition in round, square, rectangular. Cold drawn tubes and pipes – with closer dimensional tolerance and better internal finish, round, hexagonal and oval shapes to BS/IS/JIS/DIN/ASTM specification; in the ranges ERW 15.88mm to 114.3mm od in thickness 0.81mm to 6mm; CDW 4.76mm to 70mm od in thickness 0.46mm to 4.06mm.
Brands: Tru-wel (ERW and cold drawn welded steel tubes).
Sales offices/agents: Madras, Poonah, Hosur, Bombay, Calcutta, New Delhi and Ludhiana.
Additional information: The company has a co-operation agreement with TI Tube Products, Oldbury, Birmingham, UK.

Upper India Steel Manufacturing & Engineering Co Ltd

Head office: Dhandari Industrial Focal Point, Sherpur Kalan, Ludhiana 141010, Punjab. **Tel:** 27071, 27069, 31680, 31893. **Telex:** 0386277. **Cables:** Ispat Ludhiana.
Management: Directors – Indermohan Singh Grewal (managing), Pritpal Singh Grewal (joint managing), Gurlal Singh Grewal (works), Gurparshad Singh Grewal, S.S. Grewal. **Established:** 1962. **Capital:** Rs 40,000,000. **No. of employees:** Works – 700. Office – 100. **Annual capacity:** Finished steel: 40,000 tonnes.
Works:
■ At head office. ***Steelmaking plant:*** Two 15/17-tonne electric arc furnaces (annual capacity 40,000 tonnes). ***Continuous casting machines:*** One 2-strand Concast 6m radius billet (100-180mm sq) (40,000 tonnes). ***Rolling mills:*** Two billet – one 18" semi-automatic cogging for roughing 8x7" ingots; and one 14" roughing; three bar – one semi-automatic 16/12" for flats; one 10"; and one 14/9" cross-country for rolling file steel and special sections (combined capacity (124,000 tonnes).
Products: Carbon steel – ingots, billets, bars incl. triangular and other profiles. Alloy

Upper India Steel Manufacturing & Engineering Co Ltd (continued)

steel (excl stainless) – Triangular bars and other engineering profiles.
Expansion programme: Ladle injection system, ladle furnace, scrap preheating furnace, etc.

Usha Alloys & Steels Ltd (UASL)

Head office: Adityapur, Jamshedpur 831001.
Established: 1974.
Works:
■ Adityapur. ***Steelmaking plant:*** One 20/25-tonne electric arc furnace. ***Refining plant:*** Calcium injection plant. ***Continuous casting machines:*** One 2-strand Concast/India billet.
■ 12-16 Nawalganj, Agra 282006. ***Rolling mills:*** light section and wire rod.
Products: Carbon steel – billets, reinforcing bars, round bars, flats. Stainless steel – billets, round bars. Alloy steel (excl stainless) – billets.

Vidarbha Iron & Steel Corp Ltd—see Ferro Alloys Corp Ltd

Visvesvaraya Iron & Steel Ltd (formerly The Mysore Iron & Steel Ltd)

Head office: Bhadravati 577301, Shimoga Dist, Karnataka. **Tel:** 6322 (manging director), 6338 (general manager). **Telex:** 2366 (Bangalore office), 0831217 (Bhadrvati). **Cables:** Mysiron, Bhadravathi.
Management: Board of thirteen directors – six nominees of the State Government of Karnataka, four from Steel Authority of India Ltd (Indian Goverment holding company) and three nominees from financial institutions. The chief secretary, Government of Karataka, is the chairman of the board. The controlling ministry is the Ministry of Large Scale Industries, Government of Karnataka. **Established:** 1923. **Capital:** (31st Dec 1986) Authorised – Rs750,000,000. Paid up – Rs 46,400,000,000. **No. of employees:** 9,876. **Ownership:** State Government of Karnataka (60%); Sail – Steel Authority of India Ltd (40%). **Annual capacity:** Pig iron: 180,000 tonnes. Raw steel: 180,000 tonnes. Rolled steel: 125,000 tonnes.
Works:
■ At head office. Works director: A.A. Raju (vice-chairman and managing director). **Ironmaking plant:** Two Tysland-Hole electric pig iron furnaces (33', – 100 tonnes/day) (annual capacity 60,000 tonnes); two closed rotating electric pig iron furnaces (33', – 200 tonnes/day) (120,000 tonnes). ***Steelmaking plant:*** Two 12-tonne LD basic oxygen converters (73,200 tonnes); five electric arc furnaces – two 20-tonne (44,400 tonnes), one 8-tonne (15,600 tonnes), 1 6-tonne (6,900 tonnes), and one 3.5-tonne (4,000 tonnes); two 25-tonne open hearth (41,000 tonnes). ***Rolling mills:*** One blooming and heavy section four 3-high stands (78,000 tonnes); one light section (36,000 tonnes); one bar and wire rod 14-stand (36,000 tonnes); hot strip and hoop and rod (12,000 tonnes). ***Other plant:*** Five ferro-alloy furnaces including three ferro-silicon furnaces; plant producing centrifugally-cast iron pipes; steel foundry and general foundry; refractories plant; cement plant; forge; ferro-vanadium slag plant; plant for fabricated structures.
Products: Steelmaking pig iron (1986/87 output 4,532 tonnes). Carbon steel – round bars 12-50mm and flats 6 x 25-12 x 50mm (combined capacity 10,850 tonnes). Alloy steel (excl stainless) – round bars 10-140mm (forged rounds 140-500mm dia); square bars (round-cornered 40-140mm, forged squares 140-500mm), flats (50 x 20-20 x 170mm, forged flats 40 x 80-180 x 180 and 100-600mm x 180-600mm) and forged discs (300-1,200 x 100mm thick, 100mm min bore) (combined output 41,115 tonnes); ferro-silicon (4,065 tonnes); ferro-manganese, silico-manganese, silico-chrome, ferro-chrome (4,661 tonnes); slag cement (1,350 tonnes); steel castings (224 tonnes); iron castings (3,354 tonnes); fireclay refractories (2,334 tonnes).
Sales offices/agents: Branch offices in Bangalore, Madras, Bombay, New Delhi and Calcutta. Stockyard at Poona.
Expansion programme: New oxygen plant (1987-88).

Indonesia

PT Bakrie & Bros

Head office: Wisma Antara 7th Floor, Jalan Merdeka Selatan 17, Jakarta Pusat. **Telex:** 47368.
Established: 1942.
Works:
■ Jakarta. ***Tube and pipe mills:*** welded – Three Mannesmann-Meer.
Products: Carbon steel – longitudinal weld tubes and pipes, oil country tubular goods.

Bakrie Pipe Industries PT

Head office: Wisma Bakrie, Jalan HR Rasune Said Kar B-1, Jakarta 12940. **Telex:** 49339 bpi ia.
Works:
■ ***Tube and pipe mills:*** two welded. ***Other plant:*** Tube galvanizing plant.
Products: Carbon steel – longitudinal weld tubes and pipes, galvanized tubes and pipes.

PT Budidharma Jakarta

Head office: 7th Floor, Skyline Bldg, Jalan Thamrin 9, Jakarta 10340. **Tel:** 021 326762. **Telex:** 46427 sky jkt.
Management: Suhendro Notowidjojo (president and managing director), The Ning King (vice president), The Nien Khong (sales director), Pandu Atmadja (purchasing/ finance manager). **Established:** 1974. **Capital:** US$30,000,000. **No. of employees:** Works – 280. Office – 62. **Subsidiaries:** PT Sumber Sejahtera Jakarta (100%) (distributor of steel products). **Annual capacity:** Finished steel: 127,000 tonnes.
Works:
■ Semper, Cilincing, Jakarta Utara. (Tel: 492008. Telex: 49181 bdj ia). Works director: Leonardi Simadjaja. ***Rolling mills:*** bar – horizontal/vertical, 12-29mm (annual capacity 127,000 tonnes).
Products: Carbon steel – reinforcing bars, round bars.
Brands: Budicon.
Expansion programme: 50-tonne electric arc furnace; 4-strand continuous casting machine for billets 100, 110, and 120mm sq.

PT Fumira, Galvanized Sheet Works

Head office: Jalan Pintu Kecil 38/Atas, Jakarta Barat.
Works:
■ Jalan Dr Setia Budi 104, Srondol Timir, Semarang, Cental Java. ***Coil coating lines:*** One hot-dip tinning (sheet); one colour coating.
Products: Carbon steel – hot-dip galvanized sheet/coil, colour coated sheet/coil.

PT Gunung Gahapi Steel

Head office: Jalan Teuku Umar 6, Medan. **Telex:** 51316 gahapi mdh.
Works:
■ ***Steelmaking plant:*** electric arc furnace. ***Continuous casting machines:*** One 2-strand billet (100mm sq). ***Rolling mills:*** bar and wire rod.
Products: Carbon steel – wire rod, reinforcing bars, flats, galvanized wire (plain).

PT Inastu – Indonesian National Steel Tube Co

Head office: 47-A Jalan Bungur Besar, Jakarta. **Telex:** 49101 denso ia.
Established: 1970.
Works:
■ ***Tube and pipe mills:*** Longitudinal welded.
Products: Carbon steel – longitudinal weld tubes and pipes, hollow sections.

PT Industri Badja Garuda (IBG)

Head office: Jalan Ke Belawan Km 11, Medan 20243, Sumatra Utara. **Postal address:** PO Box 333, Medan 20001, Sumatera Utara. **Tel:** 22154, 24127. **Telex:** 51161 citoh.
Management: Directors – Nobuyoshi Higashi (president), Seitsuke Miyazaki (finance), David Pitoy (general affairs and sales). **Established:** 1968. **Capital:** US$1,575,000. **No. of employees:** Works – 186. Office – 13. **Ownership:** C. Itoh & Co Ltd (73%); Kawasaki Steel Corp (6.4%); and PT Zig-Zag Indonesia (20.0%). **Subsidiaries:** None. **Associates:** None. **Annual capacity:** Finished steel: 36,000 tonnes.
Works:
■ At head office. ***Coil coating lines:*** Two semi-continuous Kawasaki hot-dip galvanizing (annual capacity 36,000 tonnes); One Kawasaki colour coating (roll coating) (12,000 tonnes).
Products: Carbon steel – hot-dip galvanized sheet/coil 0.18-0.60mm (1986 output 20,650 tonnes); colour coated sheet/coil.
Brands: Swan.

PT Inti General Jaya Steel

Head office: Jl Benteng 116, Semarang. **Postal address:** PO Box 193.
Works:
■ ***Steelmaking plant:*** electric arc furnace. ***Rolling mills:*** light section and bar.
Products: Carbon steel – reinforcing bars, light angles.

PT Ispat Indo

Head office: Desa Kedungturi, Taman, Sidoarjo, East Java. **Postal address:** PO Box 83, Surabaya. **Tel:** 031 811424, 032 818493. **Telex:** 33281, 33112 ispat ia. **Cables:** Steelmill, Surabaya. **Telecopier (Fax):** 031 813202, 021 356988.
Management: Directors – M.L. Mittal (president), L.N. Mittal (managing). **Established:** 1975. **Capital:** US$13,800,000. **No. of employees:** Works – 750. Office – 250. **Ownership:** The Ispat Projects Ltd, India; Autumn Investments Ltd, Hongkong; Amalgamated Trading Ltd, Isle of Man; Mitco Fabrication Consultants Pvt Ltd, India (70%); Eddy Kowara & Gani Djemat SE (30%). **Annual capacity:** Raw steel: 390,000 tonnes. Finished steel: 234,000 tonnes.
Works:
■ At head office. ***Steelmaking plant:*** One 80-tonne NNK UHP electric arc furnace (annual capacity 390,000 tonnes). ***Refining plant:*** One GEC ladle furnace with Danieli eccentric bottom tapping. ***Continuous casting machines:*** One 4-strand Kobe billet (80-160mm sq) (390,000 tonnes). ***Rolling mills:*** One bar / wire rod – 8-stand non-twist finishing block (234,000).
Products: Carbon steel – billets (low-carbon) (1986 output 224,573 tonnes); wire rod (including high-carbon, cold-heading and electrode grade) 5.2-10.2mm.
Sales offices/agents: Bankok Bank Building, 6th Floor, Jalan MH Thamrin 3, Jakarta; Sinapore Rubber House 08-03, 14 Collyer Quay, Singapore 0104.

Indonesia

PT Jakarta Kyoei Steel Works Ltd

Head office: Jalan Rawa Terate II/1, Kawasan Industri, Pulogadung, Jakarta. **Postal address:** PO Box 49. **Telex:** 49167.
Works:
■ ***Rolling mills:*** bar and wire rod.
Products: Carbon steel – wire rod, reinforcing bars.

Krakatau Hoogovens International Pipe Industries Ltd PT

Head office: Wisma Baja, 6th Floor, Jalan Jenderal Gatot Suboroto Kav 54, Jakarta Selatan, DKI Jaya. **Postal address:** PO Box 3055. **Tel:** 514140. **Telex:** 45848 khipipe ia. **Cables:** Khipipe.
Management: Directors – S. Hutapea (president), R. Ijpenberg (financial), Abubakar Ayub (production). **Established:** 1972. **Capital:** US$ 15,000,000. **No. of employees:** Works – 800. Office – 50. **Ownership:** PT Krakatau Steel (66.66%); Hoogovens Group BV (16.66%); PT Indhasana (16.66%). **Annual capacity:** Finished steel: 60,000 tonnes.
Works:
■ Cilegon (Kota Baja), Banten, West Java. (Tel: Via Pertamina 367653-367542 (ext 346 or 329)). Works director: Pat de Guzman. ***Tube and pipe mills:*** Three spiral welded – SPX 480 (4-14"), SPX 780 (4-32"), SPM 2000 (16-80") (combined annual capacity 60,000 tonnes).
Products: Carbon steel – spiral-weld tubes and pipes 100-2,000mm dia and 3-19mm wall thickness; large-diameter tubes and pipes.
Brands: KHI.

PT Kratkatau Steel

Head office: Wisma Baja, Jalan Gatot Subroto Kav 54, Jakarta Selatan. **Postal address:** PO Box 174. **Tel:** 5204003. **Telex:** 45958-9 pt ks ia. **Cables:** PT KS Jakarta.
Management: Directors – T. Ariwibowo (president), S. Mangoensoewargo (marketing), Mr Kadarisman (personnel), J.L. Rombe (planning), Djoko Subagyo (production), Aidil Juzar (financial), Mr Moesdiono (junior, steel service centre); corporate secretary – A.B. Yusra. **Established:** 1971. **Capital:** US$ 2,165,000. **Ownership:** Government of Indonesia. **Subsidiaries:** PT Krakatau Hoogovens International Pipe Industries Ltd (66%); PT Cold Rolling Mill Indonesia Utama (40%); PT Pelat Timah Nusantara (15%); PT Krakatau Industrial Estate Cilegon (100%). **Annual capacity:** Pig iron: 1,500,000 tonnes. Raw steel: 1,500,000 tonnes. Finished steel: 1,413,000 tonnes.
Works:
■ Cilegon, West Java. ***Direct reduction plant:*** Four 500,000-tonne HYL I (annual capacity 1,500,000 tonnes). ***Steelmaking plant:*** Eight electric arc furnaces – four 65-tonne (500,000 tonnes), four 130-tonne (1,000,000 tonnes). ***Continuous casting machines:*** Two 4-strand Concast/Schloeman-Siemag billet (100 to 120mm sq) (500,000 tonnes); two 1-strand Concast/Schloeman-Siemag slab (950 x 2,080 x 200mm) (1,000,000 tonnes). ***Rolling mills:*** One medium section comprising three 2-high (two 550 x 1,200mm, one 650 x 1,500mm) (45,000 tonnes); one bar comprising fifteen x 2-high 11" (150,000 tonnes); one 2-strand 24-stand 5.5 to 12.5mm dia wire rod (coil weight 850kg) (220,000 tonnes); one roughing 5 finishing 88" wide hot strip (coil weight 30-tonnes) (1,000,000 tonnes).
Products: Direct-reduced iron (DRI) (1986 output 1,325,000 tonnes). Carbon steel – slabs 1,280 x 20mm (655,000 tonnes); billets 120 x 120mm (455,000 tonnes); hot rolled coil (for re-rolling) 2.0 x 1,200mm (635,000 tonnes); wire rod 5.5 to 6 to 8mm dia (180,000 tonnes); reinforcing bars 16 and 25mm dia (93,000 tonnes); round bars 16 and 25mm dia (7,000 tonnes); medium angles 60 x 60 x 6mm (11,500 tonnes); medium joists 100 x 50mm (4,500 tonnes); medium channels 100 x 50mm (6,500 tonnes); medium plates; heavy plates; hot rolled sheet/coil (uncoated) 1.8 to 25mm thick x 2 to 4m long; bright wire and galvanized wire (plain) 6 to 18g (18,000 tonnes).

PT Latinusa

Head office: Jl Let Jend S Parman N7, Jakarta. **Telex:** 45591 ptn jkt.
Management: Board members – M. Simatupang, Mohamad Hasan, Soetoro Mangoensoewargo; directors – Kasmir Batubara (president), Ibnu Bangsawan (financial), Mr. Sawardjiman (technical); manager – Karim Pohan (marketing).
Established: 1985. **Capital:** Rp 26,617,920,000. **No. of employees:** 431.
Ownership: PT Tambang Timah (61%); PT Krakatau Steel (15%); PT Nusamba (24%). **Annual capacity:** Finished steel: 130,000 tonnes (tinplate).
Works:
■ Kawasan Industri Berat Kav-D, Cilegon. (Telex: 48402 ptn cia). ***Coil coating lines:*** One Mannesmann/Demag/Sack Ferrostan electrolytic tinning (annual capacity 130,000 tonnes).
Products: Carbon steel – electrolytic tinplate (single-reduced).

PT Master Steel Mfg Co

Head office: Jalan Bekasi Raya Km 21, Pulogadung, Jakarta Timur. **Telex:** 49134 master ia.
Works:
■ ***Rolling mills:*** One bar. ***Other plant:*** Steelmaking plant.
Products: Carbon steel – billets, reinforcing bars.

PT National Union Steel Ltd

Head office: Jalan P. Jayakarta 40, Jakarta. **Tel:** 633011. **Telex:** 41350 nustel ia.
Cables: Natusteel.
Established: 1975.
Works:
■ 26 Jalan Babelan, Bekasi (Tel: 99 71333-4). ***Rolling mills:*** medium section and bar (combined annual capacity 9,000 tonnes).
Products: Carbon steel – reinforcing bars, flats, medium angles.

PT Pabrik Pipa Indonesia (The Indonesia Tube Mill Corp Ltd)

Head office: Jl H.R. Rasuna Said Kav B-29, Kuningan, Jakarta 12910. **Postal address:** PO Box 1016, Jakarta. **Tel:** 5780101. **Telex:** 49349 sst jkt. **Cables:** Tube Mill Jakarta. **Telecopier (Fax):** 021 5780827.
Management: Directors – Rachman Utan, Usmar Utan. **Established:** 1973.
Capital: US$1,000,000. **No. of employees:** Works – 200. Office – 50. **Annual capacity:** Finished steel: 24,000 tonnes.
Works:
■ Jl Raya Bekasi Km 21, Pulogadung, Jakarta 13920. (Tel: 4894893, 4895293. Telex: 49349 sst jkt). Works director: Armen Jaya Manan. ***Tube and pipe mills:*** welded – three ERW $\frac{1}{2}$-8" (annual capacity 24,000 tonnes). ***Other plant:*** Galvanizing plant, indirect firing, (Furnace Engineering, Australia (18,000 tonnes).
Products: Carbon steel – longitudinal weld tubes and pipes (output 24,000 tonnes); galvanized tubes and pipes.
Brands: PPI.

PT Radjin Steel Pipe Industry

Head office: Jl Kalibutuh 189-191, Surabaya. **Tel:** 031 479221. **Telex:** 33187.
Cables: Radjin Surabaya.
Management: President – Ibnu Susanto; directors – Solichin Salam, Wardhana Hudianto (marketing). **Established:** 1957. **Capital:** US$2,000,000. **No. of employees:** Works – 310. Office – 40. **Associates:** Sarana Steel (steel centre); Sarana Surya Sakti (bicycle rim and bailing hoop manufacture). **Annual capacity:** Finished steel: 30,000 tonnes.
Works:
■ Jl Rungkut Industry, I/28-30-32, Surabaya and II/10 Surabaya. (Tel: 031 819503.

Indonesia

PT Radjin Steel Pipe Industry (continued)

Telex: 33187). ***Tube and pipe mills:*** Three welded – $\frac{1}{2}$-1$\frac{1}{2}$", 1-4" and 3-8" (annual capacity 50,000 tonnes).
Products: Carbon steel – longitudinal weld tubes and pipes (output 24,000 tonnes).

PT Sermani Steel Corp

Head office: Jalan Gowa Jaya, Tello Baru, Ujung Pandang. **Telex:** 71139 sermani up.
Established: 1970.
Works:
■ ***Coil coating lines:*** One hot-dip galvanizing sheet. ***Other plant:*** Corrugating facilities.
Products: Carbon steel – hot-dip galvanized sheet/coil, galvanized corrugated sheet/coil.

PT Witikco

Head office: Jl Jembatan Tiga Blok F & G, Jakarta, DKI. **Postal address:** PO Box 2816 Jakarta, 10001. **Tel:** 6691522. **Telex:** 46020 sh jkt. **Telecopier (Fax):** 6696595.
Management: Gaotama Setiawan (vice president, business development); Muktar Widjaya (works director). **Established:** 1978. **Capital:** Rp 3,500,000,000. **No. of employees:** Works – 16. Office – 27. **Ownership:** PT Sinar Mas Inti Perkasa, Jakarta. **Associates:** PT Bimoli (coconut oil factory). **Annual capacity:** Finished steel: 12,000 tonnes.
Works:
■ Jl Walanda Maramis, Bitung. (Tel: 21032/21130). Works director: Muktar Widjaya. ***Coil coating lines:*** hot-dip galvanizing – Roller chain conveyor GL1500 Nakayama Steel (1978) (annual capacity 12,000 tonnes/year).
Products: Carbon steel – hot-dip galvanized hoop and strip 3 x 6 ft (output 12,000 tonnes); hot-dip galvanized sheet/coil 3 x 6 ft (12,000 tonnes); galvanized corrugated sheet/coil 3 x 6 ft (10,800 tonnes).
Brands: Gajah Berlian, Anoang, Mata, Lima Berlian.
Sales offices/agents: PT Witikco, Jl Sam Ratulangi 12, Manado, North Sulawesi.

Iran

Navard Va Luleh Ahwaz

Head office: 17 Shahide Mohammed Esmaeel Mohammadi Ave, Sepahboud Gharani St, Tehran. **Tel:** 827022-5.
Management: Chairman and managing director – M. Maysami; board members – A. Dialameh (director and commercial department manager), S.A. Zakeri (director).
Established: 1983. **Capital:** Rial 30,000,000. **Ownership:** State-owned. **Annual capacity:** Finished steel: 100,000 tonnes (hot strip), 55,000 tonnes (tube and pipe).
Works:
■ At head office. ***Rolling mills:*** hot strip and hoop – 520mm medium strip mill with one reversing roughing stand and six continuous finishing stands (annual capacity 100,000 tonnes). ***Tube and pipe mills:*** Three HF induction welded 6" (25,000 tonnes), 4" (20,000 tonnes), and 2" (10,000 tonnes).
Products: Carbon steel – hot rolled hoop and strip (uncoated) 400-520mm wide (1986 output 90,000 tonnes); longitudinal weld tubes and pipes 6", 5", 4", 3", 2", 1$\frac{1}{2}$", 1" (40,000 tonnes).

Kalup Corp

Head office: 17 20th St, Park Ave, Teheran. **Tel:** 623957-9, 622337, 626630. **Telex:** 222421 lasr ir.
Management: Board members – P. Toutounchi, M. Tamadon, R. Kianersi. **Established:** 1981. **Capital:** Rls 1,500,000,000. **No. of employees:** Works – 145. Office – 15. **Ownership:** Private. **Annual capacity:** Finished steel: 100,000 tonnes.
Works:
■ ***Tube and pipe mills:*** Five ASTM A53 A120 welded – 6", 8", 10" 12" and 14" (combined capacity 50,000 tonnes).
■ ***Tube and pipe mills:*** Five square welded – 134 x 134mm, 174 x 174", 215 x 215", 255 x 255" and 280 x 280" (combined capacity 50,000 tonnes).
Products: Carbon steel – longitudinal weld tubes and pipes 6-14" nom. dia, 4.37-9.52mm wall thickness (output 100,000 tonnes).

National Iranian Steel Corp (Nisc)

Head office: Vali-E Asr Avenue, Tehran. **Tel:** 212334 nisc ir.
Ownership: State-owned.
Works:
■ Isfahan Steelworks. ***Coke ovens:*** Two batteries (annual capacity 444,000 tonnes). ***Sinter plant:*** One Dwight Lloyd (75sq m) (700,000 tonnes). ***Blast furnaces:*** Two – one 7.2m hearth dia, 1033 cu m (550,000-700,000 tonnes); and one 2,000 cu m. ***Direct reduction plant:*** One HYL (1,000,000 tonnes). ***Steelmaking plant:*** Two 80-tonne basic oxygen converters (550,000 tonnes); three 150-tonne electric arc furnaces. ***Continuous casting machines:*** Three bloom – two 4-strand vertical USSR (360 x 250mm, 250mm sq, 200mm) and one 6-strand vertical USSR (200mm and 150mm sq). ***Rolling mills:*** One heavy section (650mm) (380,000 tonnes); light section and wire rod 320/250mm semi-continuous (120,000 tonnes).
■ Ahwaz Steelworks. ***Iron ore pelletizing plant:*** Straight grate (annual capacity 5,000,000 tonnes). ***Direct reduction plant:*** one Purofer (330,000 tonnes), three Midrex (1,000,000 tonnes) and three HYL (1,000,000 tonnes). ***Steelmaking plant:*** Six 150-tonne electric arc furnaces (1,600,000 tonnes). ***Continuous casting machines:*** One 6-strand Concast billet and bloom (150-260mm sq) (500,000 tonnes); one 2-strand Concast slab (150 x 600-250 x 1,200mm (1,000,000 tonnes).
■ Ahwaz. ***Steelmaking plant:*** One 60-tonne electric arc. ***Continuous casting machines:*** Two 4-strand Concast billet (80-130mm sq and 100-190mm sq). ***Rolling mills:*** One light section; one wire rod.
■ Ahwaz. ***Rolling mills:*** Schloemann-Siemag slabbing and blooming one horizontal stand, one vertical stand (annual capacity 1,000,000 tonnes).
■ Mobarekeh, Isfahan. ***Iron ore pelletizing plant:*** Straight grate (annual capacity 4,500,000 tonnes). ***Direct reduction plant:*** 3,200,000 tonnes. ***Steelmaking plant:*** Eight 180-tonne electric arc furnaces. ***Continuous casting machines:*** Four 8-strand slab (2,700,000 tonnes) (due to start up 1988). ***Rolling mills:*** One wide hot strip (2,500,000 tonnes). one cold reduction (1,000,000 tonnes). ***Coil coating lines:*** One hot-dip galvanizing.
Products: Steelmaking pig iron. Direct-reduced iron (DRI). Carbon steel – slabs, blooms, billets, wire rod, round bars, light angles, light channels, heavy angles, heavy channels, wide-flange beams (heavy), light rails.

Iraq

State Company for Steel Industries

Head office: Awkaf Building, Samawel Street, PO Box 3091, Hakemia, Basrah, Baghdad. **Telex:** 207046 scisi ik.
Works:
■ Khor al Zubair. ***Direct reduction plant:*** Two Midrex (annual capacity 400,000 and

State Company for Steel Industries (continued)

750,000 tonnes). ***Steelmaking plant:*** Four 70-tonne 35/42 MVA Lectromelt/Clesid electric arc furnaces. ***Continuous casting machines:*** Two 6-strand Concast/Clesid billet (80-150mm sq) (440,000 tonnes). ***Rolling mills:*** One 4-stand Wean United/Secim medium section; one 17-stand Wean United/Secim bar.
Products: Carbon steel – reinforcing bars, medium sections.

Irish Republic

Irish Steel Ltd

Head office: Haulbowline, Co Cork. **Tel:** 021 811731. **Telex:** 76123. **Cables:** Steelile Haulbowline. **Telecopier (Fax):** 021 811347.
Management: Board – W. Hugh O'Connor (chairman), J.J Byrne, G.T.P. Conlon, J.F. Coyne, J. Dorgan, M. Fahy, B.J. Foley, P.A. Howard, D.F. Quirke. Chief executive – L.S. Coughlan; managers – C.P. Vaughan (deputy general), B.W. Logan (deputy general), M.T. Walley (assistant general, production), G.S. Reynolds (technical), P.F. Dunlea (scrap procurement), M. Walsh (personnel); financial controller/secretary – T.B. Clifford. **Established:** 1947. **Capital:** IR£125,000,000. **No. of employees:** Works – 386. Office – 134. **Ownership:** State-owned. **Subsidiaries:** Irish Steel Sheet Ltd (100%) (fabricators and converters of coated and uncoated steel coils and sheets by re-rolling and other processes). **Annual capacity:** Raw steel: 435,000 tonnes. Finished steel: 333,000 tonnes.
Works:
■ At head office. ***Steelmaking plant:*** Two electric arc furnaces – 90-tonne and 30-tonne (annual capacity 435,000 tonnes). ***Continuous casting machines:*** 3-strand Concast billet (205 x 175mm) (345,000 tonnes). ***Rolling mills:*** medium section – thirteen Universal/horizontal and vertical stands, 90-280mm (333,000 tonnes).
Products: Carbon steel – reinforcing bars 16-40mm; round bars 16-40mm; light angles 50 x 50mm to 80 x 80mm (equal); light channels 102 x 51mm to 152 x 76mm, also UPN 100 to UPN 140; medium angles 90 x 90mm to 120 x 120mm (equal); medium channels 178 x 76mm to 203 x 76mm, also UPN 160 to UPN 200; wide-flange beams (medium) HEA/B 100 to HEA/B 160; universals (medium) 127 x 76mm to 203 x 133 x 30mm, also IPE 100 to IPE 220.
Sales offices/agents: At head office. Agents – Irish Steel Products Ltd, Steetley Industrial Estate, Bean Road, Coseley, Bilston, W Midlands, UK; Sté Métallurgique de Normandie, 42 Rue la Boétie, F 75008 Paris, France; Klaus Dee GmbH, Beethovenstrasse 12, D 4000 Düsseldorf 1, West Germany; Nillesen BV, Buizenwerf 123, 3063 AB Rotterdam, Netherlands; Ivo Daddi, Piazza Rossetti 3C/2, I 16129 Genoa, Italy; Helsider AG, Scheuchzerstrasse 4, PO Box 164, CH 8033 Zürich, Switzerland; Nordrum Trading A/S, Riddervoldsgt 4, PO Box 2319, Solli, N 0201 Oslo 2, Norway; SA Svendsen A/S, 6 Hammerensgade, DK 1267 Copenhagen, Denmark; Cevece Montan AB, Box 11171, S 40424 Gothenburg, Sweden.

Israel

Koor Steel

Head office: 5 Hama'apilim Str, Kfar-Shmaryahu 46910. **Postal address:** PO Box 316 Herzlia B 46103.
Management: Directors – U. Baratson (managing), A. Porat (deputy managing); assistant to managing – L. Segal; managers – E. Neumann (commercial), A. Ilan (financial), B. Schneeman (engineering); head of economics and data center – D. Ginzburg. **Subsidiaries:** United Steel Mills Ltd (USM) (U. Baratson, managing director),

Koor Steel (continued)

Middle East Tube Co Ltd (Metco) (A. Porat, managing director), Halum, Central Scrap Co Ltd (subsidiary of USM), Ma'as Wires & Meshes Ltd (subsidiary of USM). **Activities:** Holding Co.

Middle East Tube Co Ltd

Head office: 5 Hamaapilim Street, Kfar Shmaryahu 46910. **Postal address:** PO Box 316, Herzliah "B". **Tel:** 052 573284, 052 573188. **Telex:** 342569. **Telecopier (Fax):** 052 72278.
Management: U. Yellin (chairman), A. Porat (managing director), M. Porbocherai (works manager, Ramle works), D. Maliniak (works manager, Akko works). **Established:** 1950. **Capital:** US$21,000,000. **No. of employees:** 400.
Ownership: Koor Industries Ltd (93%); Mekoroth Water Co Ltd (7%) (Ultimate holding company – Hevrat Ovdim – General Cooperative Association of Jewish Labour in Eretz, Israel). **Subsidiaries:** Joint Pipe Industries Ltd (property).
Works:
■ Akko Works, PO Box 788, 31007 Haifa. (Tel: 04 910467. Telex: 46386. Fax: 04 912745). Works director: Mr Dan Maliniak. ***Tube and pipe mills:*** Two seamless – one plug (5½") (annual capacity 25,000 tonnes); and one stretch reducing (4½" (20,000 tonnes). Three high frequency longitudinal welded – two (½-2") (25,000 tonnes); and one (2⅜-6⅝") (25,000 tonnes). ***Other plant:*** One hot dip tube galvanizing plant (½-4½" dia), one tape coating plant (½/4-4½"). one internal concrete lining plant (1 1/4-4½"), and one cold section forming line.
■ Ramile Works, PO Box 62, Ramile. (Tel: 08 229555. Telex 33396. Fax: 08 229783). Works director: Mordechai Porbocherai. ***Tube and pipe mills:*** Three welded – two high frequency longitudinal weld (one 3-10" and one 6-20") (combined annual capacity 50,000 tonnes); and one DSAW spiral (18-80") (25,000 tonnes). ***Other plant:*** Outside coating – bitumen, plastic tape, concrete; inside lining – concrete.
Products: Carbon steel – seamless tubes and pipes ½ – 4½" (output 5,000 tonnes); longitudinal weld tubes and pipes ½ – 20" (47,000 tonnes); spiral-weld tubes and pipes 18 – 80" (13,000 tonnes); oil country tubular goods (7,500 tonnes); galvanized tubes and pipes (16,000 tonnes).
Sales offices/agents: Solcoor (Pipe and Tube Dept) Tel Aviv 67215. (Tel: 03 336068. Telex: 342384); Middle East Tube Co Ltd 46910 Kfar Shmaryahu; Solcoor Inc, NY 10016 USA (Tel: 212 5617200. Telex: ITT420761).
Additional information: Sales in 1986 were: Seamless pipe ½-4½" (6,000 tonnes); longitudinal welded pipe ½-20" (35,000 tonnes); spiral weled pipe 18-80" (13,000 tonnes); galvanized tube and pipe ½-4" (12,000 tonnes).

United Steel Mills Ltd (USM)

Head office: 5 Hama' Apilim St, Kfar-Shmaryahu 46910. **Postal address:** PO Box 316, Herzlia B 46103. **Tel:** 052 573188. **Telex:** 342569. **Telecopier (Fax):** 052 72278.
Management: M. Bala (chairman), U. Baratson (managing director), S.A. Dauber (manager, Kiryat Haplada works), A. Vaida (manager, Kiryat Gat works). **Established:** 1954. **No. of employees:** Works – 415. Office – 130. **Ownership:** Koor Industries Ltd (100%) (Ultimate holding company – Hevrat Ovdim (general Co-operative Association of Jewish Labour in Erez, Israel)). **Annual capacity:** Raw steel: 130,000. Finished steel: 250,000-300,000 tonnes.
Works:
■ Kiryat Haplada, Akko; PO Box 1708, Haifa 31016. (Tel: 04 917941. Telex 46393 irgul il). Works director: Samuel A. Dauber. ***Steelmaking plant:*** One 60-tonne 5.8m dia, 50 MVA electric arc furnace with EBT, water-cooled side panels and roof (annual capacity 130,000 tonnes). ***Continuous casting machines:*** One 4-stand Concast curved mold 4m billet (100-130mm sq) (130,000 tonnes). ***Rolling mills:*** light section and bar – one 12-stand 2-high horizontal/vertical for angles 40-60mm and round and

United Steel Mills Ltd (USM) (continued)

squares 20-55mm; and one 21-stand 2-high horizontal/verical bar – one 21-stand 2-high horizontal/vertical for bars 8-28mm dia (combined capacity 180,000 tonnes).
■ Kiryat Gat Rolling Mill, PO Box 185 Kiryat Gat 82000 (Tel: 051 885955. Telex: 26274). Works director: Andre Vaida. ***Rolling mills:*** bar and wire rod with 19-stand 2-high for bar 8-18mm dia and 8-stand Morgan finishing block for wire rod 5.5-12mm dia (coil weight 1,350kg) (combined annual capacity 180,000 tonnes).
Products: Carbon steel – billets 100-130mm sq and 8m long; wire rod 5-12mm dia; reinforcing bars 8-28mm dia; round bars 20-55mm; square bars 20-55mm; flats; hexagons 20-55mm; light angles 40-60mm.
Sales offices/agents: Kiryat Haplada-Akko, South Industrial Zone PO Box 1708 Haifa 31016. Solcoor Ltd, 74 Petach-Tikva Ay Maya House, Tel Aviv 67215.
Expansion programme: Meltshop addition of ladle furnace and modification of continuous casting machine for sequence casting, 12m long billets and higher quality steels (during 1987 and 1988). Rolling mill addition of vertical roughing stand and erection of line for sorting, weighing and bundling of finished products (during 1987).

Italy

Afim – Acciaierie Ferriere Industrie Metallurgiche

Head office: Via S.lle Minola 23, Nave, BS. **Tel:** 030 2632661. **Telex:** 301169.
Products: Carbon steel – reinforcing bars.

AFP – Acciaierie Ferriere del Po SpA

Head office: Viale Certosa 249, I 20151 Milan. **Tel:** 02 30700. **Telex:** 303098.
Works:
■ Villapoma (MN).
Products: Carbon steel – billets, wire rod, reinforcing bars, round bars, square bars, flats.

Albasider SpA

Head office: V. dei Conradi 18, I 17013 Albisola Superiore (SV). **Tel:** 019 482791. **Telex:** 270310 albsid i.
Products: Carbon steel – hot rolled hoop and strip (uncoated), cold rolled hoop and strip (uncoated).

Alfer – Azienda Laminazione Ferro SpA

Head office: Via Milano 146, I 25055 Pisogne, BS. **Tel:** 0364 8028. **Telex:** 300589 alferx i.
Products: Carbon steel – reinforcing bars.

Arvedi Group

Head office: Zona Porto Canale, I 26100 Cremona. **Postal address:** PO Box 160. **Tel:** 0372 4091. **Telex:** 312664 arvedi i. **Cables:** Arvedi. **Telecopier (Fax):** 0372 413170.
Management: President – G. Arvedi (Acciaierie Tubificio Arvedi SpA); managers – G. Gosio (general), F. Mazzolari (vice), W. Stella (sales). **Established:** 1972. **Capital:** Lire 4,000,000,000. **No. of employees:** Works – 519. Office – 47. **Subsidiaries:** The Arvedi group includes Acciaierie Tubificio Arvedi, Cremona; Ilta – Robecco d'Oglio,

Arvedi Group (continued)

Cremona and Tubificio Lombardo FT, Corbetta, Milan.

Acciaierie Tubificio Arvedi Spa
Works:
■ Zona Porto Canale, 26100 Cremona. Works director: G. Gosio. ***Steelmaking plant:*** One 46 MVA electric arc furnace (annual capacity 290,000 tonnes). ***Refining plant:*** One 7 MVA ladle furnace (290,000 tonnes). ***Continuous casting machines:*** One bloom (160 x 280mm) and one slab (320 x 160mm, 640 x 160mm) (combined annual capacity 290,000 tonnes). ***Tube and pipe mills:*** welded One Hot stretch reducer (17 by 88.9 dia) (120,000 tonnes); three ERW – 32 by 127 dia (55,000 tonnes), 60 by 168 dia (65,000 tonnes), 25 by 51 dia (15,000 tones). ***Other plant:*** Hot dip galvanizing plant for tubes (capacity 10 tonne/hour).
Products: Carbon steel – longitudinal weld tubes and pipes 17-168mm; galvanized tubes and pipes.

Ilta Ind. Lav. Tubi Acciaio SpA
Works:
■ Strada Statale 45 bis, I 26010 Robecco d'Oglio, Cremona. (Tel: 0372 91116. Telex: 320272 iltubi. Telefax: 0372 91538). Works director: Giancarlo Stringhini. ***Tube and pipe mills:*** welded.
Products: Carbon steel – longitudinal weld tubes and pipes 17.2-114.3mm x 1-8mm wall thickness. Stainless steel – longitudinal weld tubes and pipes 10-355.6mm x 1-8mm wall thickness.

Tubificio Lombardo FT
Works:
■ Corbetta, Milan. (Tel: 02 9777792. Telex: 335369). Works director: G. Canepa.

Arvedi Group (continued)

Tube and pipe mills: welded two ERW Abbey-Etna ($\frac{1}{2}$-4") (annual capacity 30-40,000 tonnes). ***Other plant:*** Dom plant – size $\frac{3}{8}$-4$\frac{1}{2}$" (30,000 tonnes); product line – Dom for mechanical applications, cylinder tubes, hydraulic circuit tubes, shock absorber tubes.
Products: Carbon steel – longitudinal weld tubes and pipes $\frac{1}{2}$-4" and $\frac{3}{8}$-4$\frac{1}{2}$" DOM.

Acciaierie Tubificio Arvedi SpA – See Arvedi Group

ASO SpA – Acciai Speciali Ospitaletto

Head office: Via Seriola 122, I 25035 Ospitaletto. **Tel:** 30 641261. **Telex:** 301667. **Cables:** Aso Ospitaletto. **Telecopier (Fax):** 30 641264.
Management: Aldo Artioli (president and general manager), Silano Panza (technical and sales manager), Luciano Baglioni (production manager). **Established:** 1972. **Capital:** Lire 2,020,000,000. **No. of employees:** Works – 40. Office – 6. **Annual capacity:** Raw steel: 35,000 tonnes.
Works:
■ At head office. Works director: Aldo Artioli. ***Steelmaking plant:*** 30-tonne TIBB electric arc furnace (annual capacity 35,000 tonnes). ***Refining plant:*** ASO/TIBB ladle furnace (35,000 tonnes); Leybold-Heraeus vacuum degassing unit (35,000 tonnes).
Products: Carbon steel – ingots (round, square and octagonal) (output 20,000 tonnes). Alloy steel (excl stainless) – ingots (round, square and octagonal) (15,000 tonnes).
Sales offices/agents: Sipro Siderprodukte AG, Seestrasse 33, CH 8702 Zollikon, Switzerland (Mr Goedersmann); SEPF, Place de l'Europe, F 38296 La Verpilliere, France (Mrs Berriot); PTS Ltd, Wester Moffat Cres, Airdrie, Scotland (Mr Paterson).

ATB – Acciaieria e Tubificio di Brescia SpA

Head office: Via F. Folonari 20, I 25100 Brescia. **Postal address:** Casella Postale 308. **Tel:** 030 53361. **Telex:** 25122, 305344.
Products: Carbon steel – large-diameter tubes and pipes 1-7m dia. Fabricated pipework, tanks and vessels, forgings.

Ferriere di Barghe SpA

Head office: Loc Ponte Re, I 25070 Barghe, BS. **Tel:** 0365 84134, 0365 84102. **Telex:** 300653 barghe i.
Products: Carbon steel – reinforcing bars.

Baroni SpA

Head office: V. Giovanni XXIII 1, I 46030 San Niccole Po, MN. **Tel:** 0376 414131. **Telex:** 301331.
Products: Carbon steel – longitudinal weld tubes and pipes.

A.F.V. Beltrame SpA

Head office: Viale della Scienza 81, I 36100 Vicenza. **Tel:** 0444 565544.
Management: Giancarlo Beltrame (amministratore delegato). **Established:** 1949. **Capital:** Lire 10,000,000,000. **No. of employees:** Works – 395. Office – 57. **Annual capacity:** Raw steel: 400,000 tonnes. Finished steel: 600,000 tonnes.
Works:
■ At head office. ***Steelmaking plant:*** One 105-tonne electric arc furnace (annual capacity 400,000 tonnes). ***Refining plant:*** One 105-tonne ladle furnace (400,000

A.F.V. Beltrame SpA (continued)

tonnes). ***Continuous casting machines:*** One 6-strand Danieli billet (140 x 140mm) (400,000 tonnes). ***Rolling mills:*** One Domini medium section 60-150mm (400,000 tonnes); one Simac light section 20-60mm (200,000 tonnes).
Products: Carbon steel – billets (1986 output 300,000 tonnes), square bars (23,000 tonnes), flats (142,000 tonnes), light angles (28,000 tonnes), light tees (24,000 tonnes), light channels (20,000 tonnes), medium angles (50,000 tonnes), medium tees (13,000 tonnes), medium joists (48,000 tonnes), medium channels (52,000 tonnes). Total rolled products 400,000 tonnes.

Metallurgica G. Berera SpA

Head office: Corso Promessi Sposi 7, I 22053 Lecco, CO. **Tel:** 0341 360 343. **Telex:** 380377 berera i.
Products: Carbon steel – longitudinal weld tubes and pipes, galvanized tubes and pipes.

SpA Officine Fratelli Bertoli fu Rodolfo

Head office: 39 Via Molin Nuovo, I 33100 Udine. **Postal address:** Casella Postale 278. **Tel:** 0432 42341-2. **Telex:** 450151 bertol i.
Products: Carbon steel – round bars 28-160mm; bright (cold finished) bars. Alloy steel (excl stainless) – round bars 28-160mm; bright (cold finished) bars. Castings – rough and machined.

Ferriere Bertoli SpA

Head office: Via Molin Nuovo 65, I 33100 Udine. **Tel:** 0432 42951.
Products: Carbon steel – billets; reinforcing bars 40-70mm.

Acciaierie di Bolzano SpA

Head office: Via A. Volta 4, I 39100 Bolzano. **Tel:** 0471 933341. **Telex:** 400065 accibz i. **Cables:** Acciaierie Bolzano.
Management: Managing director – V. Portanova; managers – F. Zuliani (general), M. Holzner (sales), G.P. Brioschi (financial). **Established:** 1938. **Capital:** Lire 22,000,000,000. **No. of employees:** Works – 1,080. Office – 220. **Ownership:** AFL Falck. **Annual capacity:** Raw steel: 200,000 tonnes. Finished steel: 180,000 tonnes.
Works:
■ At head office. ***Steelmaking plant:*** One 45-tonne electric arc furnace (annual capacity 200,000 tonnes). ***Refining plant:*** One 45-tonne Asea-SKF (160,000 tonnes); one 45-tonne ladle furnace (40,000 tonnes); one 45-tonne vacuum degassing unit (in combination with Asea-SKF) (110,000 tonnes); one 45-tonne VOD (in combination with Asea-SKF) (50,000 tonnes). ***Rolling mills:*** One 1-stand 3-high reversing 830 x 2,000mm blooming (200,000 tonnes); one 3-stand 3-high 630 x 1,800mm heavy section (65,000 tonnes); One continuous Pomini-Kocks bar (30,000 tonnes) and 2-strand continuous Pomini-Kocks-Morgan wire rod (coil weight 1,000kg) (100,000 tonnes). ***Other plant:*** Forges (one 1,000-tonne and one 630-tonne forging presses; one 2,500-kg forging hammer; one high speed hammer). Wire works for 1.5-27mm dia wire. Grinding and peeling works. Heat treatment facilities.
Products: Carbon steel – semis for seamless tubemaking 90-180mm (1986 output 1,000 tonnes); wire rod 5.3-32mm (13,000 tonnes); round bars 5.3-180mm (3,000 tonnes). Stainless steel – billets 40-150mm (1,000 tonnes); wire rod 5.3-32mm (10,000 tonnes); round bars 5.3-140mm (4,000 tonnes). Alloy steel (excl stainless) – blooms 160-200mm (1,000 tonnes); billets 65-150mm (2,000 tonnes); semis for seamless tubemaking 90-180mm (12,000 tonnes); wire rod (5.3-32mm) (54,000 tonnes); round bars 5.3-180mm (13,000 tonnes). Cold drawn round bars 1.5-27mm;

Acciaierie di Bolzano SpA (continued)

cold drawn wire 1.5-27mm; centreless ground bars 2-90mm; peeled bars 17.5-100mm; rolls for cold rolling; forged axles for underground coach.
Sales offices/agents: Domestic – Siau, Via Amp/ere 87, I 20131 Milan (Telex: 310648 siaumi i); Uginox, Corso Sommeiler 12, I 10125 Turin (Telex: 212095 uginto i). Export – Bozen Stahl, Am Kaiserberg 6-8, Postfach 101803, D 4100 Duisburg, West Germany (Telex: 8551212 boz d); Bolzano Steel Inc, 733 Summer Street, Suite 503, Stamford, CT 06901, USA (Telex: 0023643397 bolzano steel).

Giuseppe & Fratelli Bonaiti SpA

Head office: Via Cesare Battisti 11, I 24032 Calolziocorte. **Postal address:** Casella Postale 20. **Telex:** 380449 bona i.
Products: Carbon steel – cold roll-formed sections (from strip), cold rolled hoop and strip (uncoated).

Bredina Srl

Head office: 45 Via Campagnola, I 25076 Odolo, BS. **Tel:** 365 860103, 365 860781. **Telex:** 301461. **Cables:** Bredina Odolo.
Products: Carbon steel – reinforcing bars.

Profilati a Freddo Brollo SpA

Head office: Via F. da Desio 75, I 20033 Desio, MI. **Tel:** 0362 624351. **Telex:** 333022 brollo.
Products: Carbon steel – cold roll-formed sections (from strip) and welded tubes.

Acciaieria e Ferriera del Caleotto SpA (AFC)

Head office: Via Arlenico 22, I 22053 Lecco, Como. **Tel:** 0341 364517. **Telex:** 380357.
Products: Carbon steel – billets; wire rod 5.5-12mm. Alloy steel (excl stainless) – billets; wire rod 5.5-12mm.

Metallurgica Calvi SpA

Head office: Via 4 Novembre 2, I 22055 Merate, CO. **Tel:** 039 592985-7. **Telex:** 330146.
Products: Carbon steel – Bright drawn special light sections; Stainless steel – special light sections; and Alloy steel (excl stainless) – special light sections 5 x 3mm-150 x 150mm.
Sales offices/agents: UK agents – Ralph J. Batchelar Ltd.

Acciaierie di Calvisano SpA

Head office: Via Kennedy, Viadana, I 25012 Calvisano, BS. **Tel:** 030 968 6561. **Telex:** 300490.
Products: Carbon steel – billets.

Industrie Cantieri Metallurgici Italiani SpA

Head office: Piazza Municipio 84, I 80133 Naples. **Postal address:** Casella Postale 446, I 80100 Naples. **Telex:** 710054 cmi i.
Works:
■ At head office. ***Coil coating lines:*** Two electrolytic tinning and other coil coating lines.
Products: Carbon steel – electrolytic tinplate (single-reduced).

Acciaierie e Ferriere di Caronno SpA

Head office: Via Bergamo 1484, I 21042 Caronno Pertusella, VA. **Tel:** 02 2 30700. **Telex:** 3115051 accar i.
Products: Carbon steel – billets, wire rod. Alloy steel (excl stainless) – billets for forging and rolling (engineering steel).

Ferriera Acciaieria Casilina SpA

Head office: Via Casilina Km 21, Montecompatri (Rome). **Postal address:** Casella Postale 100. **Tel:** 06 9462292. **Telex:** 614244.
Management: Gian Antonio Bertoli (president), Gian Odolfo Bertoli (vice president), Giuseppe Lisandrelli (director). **Established:** 1963. **Capital:** Lire 1,260,000. **No. of employees:** Works – 30. Office – 7. **Annual capacity:** Finished steel: 30,000 tonnes.
Works:
■ At head office. ***Rolling mills:*** bar (400/360/280mm) (annual capacity 35,000 tonnes).
Products: Carbon steel – reinforcing bars – 6mm, 8mm, 10mm, 12mm, 14mm, 16mm, 18mm, 20mm (output 21,000 tonnes).

Ferriera Castellana SpA

Head office: Via Castellana 75, I 31030 Riese Pio X (Treviso). **Tel:** 0423 486001-2. **Telex:** 431395.
Management: Alessandro Tapergi (general manager). **Capital:** Lire 250,000,000. **No. of employees:** Works – 18. Office – 4. **Annual capacity:** Finished steel: 16,500 tonnes.
Products: Carbon steel – flats, 40-100mm wide, 3-6mm thick. Flats in grades Fe37, Fe50, Fe60, Fe70, C30, C40, C50, C60, 50S7, 52S8, 55S8. Saw-blades for granite.

Figli di E. Cavalli SpA

Head office: Via Tagliamento 61, I 20089 Quinto Stampi Rozzano (Milan). **Tel:** 02 8251141. **Telex:** 311675. **Cables:** Cavalli Nastri Rozzano.
Management: Managing directors – Edgardo Cavalli, Luigi Manzetti; director – Mr Lugana; sales managers – Mrs Griffini (export), Mrs Mongilli (Italy). **Established:** 1910. **Capital:** Lire 500,000,000. **No. of employees:** Works – 23. Office – 7. **Annual capacity:** Finished steel: 12,200 tonnes.
Works:
■ At head office. Works director: Adriano Lugana. **Continuous electrolytic coating lines:** Five – one line able to coat two 500mm strips (nickel) (annual capacity 3,500 tonnes), one 500mm (chrome) (500 tonnes), one 700mm (brass) (2,200 tonnes), one 700mm (copper) (1,000 tonnes, one 850mm (zinc) (5,000 tonnes).
Products: Carbon steel – electro-galvanized sheet/coil 1.2 x 850mm (output 5,000 tonnes). Coated sheet/coil – electro-nickel plated strip, 500mm wide (3,500 tonnes); electro-chrome plated strip, 500mm wide (500 tonnes); electro-copper plated strip, 500mm wide (1,000 tonnes), electro-brass plated strip, 700mm wide (2,200 tonnes).
Expansion programme: New polishing line.

Acciaieria di Cividate al Piano SpA

Head office: Via San Zeno 16, I 25076 Odolo, BS. **Tel:** 0365 860508.
Products: Carbon steel – billets.

Nuova CMC SpA

Head office: Via Asiago 244, I 21042 Caronno Pertusella (VA). **Tel:** 02 965675, 02 9650602. **Telex:** 320494 tub cmc. **Telecopier (Fax):** 0163 47783.
Management: Germano Bocciolone (president), Luciano Badin (director). **Established:** 1904. **Capital:** Lire 1,000,000,000. **Ownership:** Sitindustrie SpA.
Associates: Stainless steel pipe and tube – Sitai SpA (Italy), BSL Tube & Raccords (France), Zwahlen & Mayr (Switzerland).
Works:
■ At head office. Works director: Luciano Badin. ***Tube and pipe mills:*** welded – one ERW.
Products: Stainless steel – longitudinal weld tubes and pipes, dairy pipes, shaped pipes.
Sales offices/agents: Domestic market – Sitindustrie SpA, Prato Sesia. Exports – Sitai International BV, Rotterdam, Netherlands.

CoGeMe SpA – Cia Generale Metalli

Head office: I 26041 Casalmaggiore, CR. **Postal address:** Casella Postale 21. **Tel:** 0375 42596. **Telex:** 321376 cogeme i. **Cables:** CoGeMe Casalmaggiore. Merchant bars.

Profilati Leggeri Cogoleto Srl

Head office: 56 Via Recagno, I 16016 Cogoleto, Genoa. **Tel:** 39 10 9183338, 39 10 9183398. **Telex:** 286574. **Cables:** Profilati.
Management: Ernesto Bianchi (managing director). **Established:** 1961. **No. of employees:** Works – 25. Office – 4.
Works:
■ At head office. Cold roll-forming facilities for sections.
Products: Carbon steel – cold roll-formed sections (from strip).

Tubificio Dalmine Italsider SpA – TDI

Head office: Via per Statte 10, I 74100 Taranto. **Tel:** 099 4841. **Telex:** 860023 tdi. **Cables:** Tdi SpA.
Ownership: Dalmine SpA (51%).
Works:
■ At head office. ***Tube and pipe mills:*** welded – medium and large dia.
■ Piombino, I 57025 Piombino (Livarno) (Tel: 05652 6461. Telex: 5000050). ***Tube and pipe mills:*** welded – Fretz Moon continuous.
■ Torre Annunziata, I 80058 Torre Annunziata (Naples). (Tel: 081 8612155. Telex: 710088). ***Tube and pipe mills:*** welded – small and medium dia ERW. ***Other plant:*** Plant for lighting pole production.
Products: Carbon steel – longitudinal weld tubes and pipes, large-diameter tubes and pipes.

Dalmine SpA

Head office: I 24044 Dalmine (Bergamo). **Tel:** 035 560 111. **Telex:** 304463-5. **Telecopier (Fax):** 035 566 666.
Management: Fulvio Milano (chairman), Ilario Testa (managing director and general manager). **Established:** 1906. **Capital:** Lire 340,200,000,000. **No. of employees:** 9,097 (Dec 31st 1986). **Ownership:** IRI Finsider Group. **Associates:** Tubificio Dalmine-Italsider SpA (51%); Acciaierie Tubificio Arvedi (30%); Siderlandini Siderurgica delle Dolomiti SpA (29.76%); Rifinsider SpA (20%).
Works:
■ At head office. ***Steelmaking plant:*** Two electric arc furnaces. ***Refining plant:*** One VAD. ***Continuous casting machines:*** One, tube blanks. ***Tube and pipe mills:*** seamless – for carbon and alloy steels.
■ Massa, I 54100 Massa. (Tel: 0585 47871. Telex: 500049). Finishing plant for oil

Dalmine SpA (continued)

country tubular goods.
■ Costa Volpino, I 24062 Costa Volpino (Bergamo). (Tel: 035 970 220. Telex: 300233). ***Tube and pipe mills:*** seamless – for carbon, alloy and stainless steel. Welded – for stainless steel.
■ Pipe Fabrication Plant, I 24044 Sabbio Bergamasco (Bergamo). (Tel: 035 566191. Telex: 300101). Plant for manufacture of pipe fittings, fabrications, gas cylinders and pressure vessels.
Products: Carbon steel – seamless tubes and pipes, oil country tubular goods, cold drawn tubes and pipes. Stainless steel – seamless tubes and pipes, longitudinal weld tubes and pipes, cold drawn tubes and pipes. Alloy steel (excl stainless) – seamless tubes and pipes, oil country tubular goods, cold drawn tubes and pipes.

DeltaCogne SpA

Head office: Viale della Resistenza 2, I 57025 Piombina, Livorno. **Postal address:** General management – Viale Savea 336, I 20126 Milan (Tel: 02 64461. Telex: 313110). **Tel:** 0565 641.11. **Telex:** 500162.
Ownership: Deltasider (Ultimate holding company – Finsider Group).
Works:
■ Via Paravera 16, I 11100 Aosta (Tel: 0165 3021. Telex: 210357). ***Steelmaking plant:*** One electric arc and induction furnace. ***Continuous casting machines:*** One bloom. ***Rolling mills:*** One blooming; light section and bar; bar / wire rod. ***Other plant:*** Forge.
■ Via Glair 21, I 11029 Verrès, AO (Tel: 0125 929323. Telex: 211374). Press forming plant for coins, cutting tools etc, precision castings.
■ Via Cadorna 2, I 20010 Vittuone, MI (Tel: 02 9017241. Telex: 330122). Bright bar plant for stainless and valve steels.
Products: Stainless steel – blooms, billets, wire rod, round bars, square bars, flats, hexagons, bright (cold finished) bars. Alloy steel (excl stainless) – (tool and high-speed steel) blooms, billets, wire rod, round bars, square bars, flats, hexagons. Bright (cold finished) bars, Coin blanks, precision castings.

Nuova Deltasider SpA

Head office: Viale della Resistenza 2, I 57025 Piombino, Livorno. **Tel:** 0565 64111. **Telex:** 500162.
Ownership: Deltasider (Ultimate holding company – Finsider Group).
Works:
■ Piombino. **Coke ovens: Sinter plant:** – one. ***Blast furnaces:*** Three. ***Steelmaking plant:*** Three 100-tonne LD-LBE basic oxygen. ***Refining plant:*** Two ladle furnace. ***Continuous casting machines:*** One billet; one bloom; one slab and bloom. ***Rolling***

Nuova Deltasider SpA (continued)

mills: One reversing blooming; one primary mill for blooms, billets and round; one rail; one light section and bar; one wire rod. ***Other plant:*** Welded rail plant.
■ Serto San Giavanni, Viale Sarca 336, I 20126 Milan (Tel: 02 64461. Telex: 330311). ***Rolling mills:*** One bar. ***Other plant:*** Cold finishing plant.
■ Corso Mertava 7, I 10149 Turin (Tel: 011 57351. Telex: 221055). ***Steelmaking plant:*** One electric arc. ***Continuous casting machines:*** One billet. ***Rolling mills:*** One blooming; one continuous bar and wire rod.
■ Vertek Works, Via Torino 10, I 10155 Condove, TO (Tel: 011 964 3333. Telex: 210308). Bright bar plant – peeling, bright drawing, grinding and heat treatment plant.
■ Via Padua 5, I 84018 Seafati, SA (Tel: 081 8632723. Telex: 710459). Drawing plant for bar and wire – stainless, high-speed and tool steel.
Products: Carbon steel – blooms, billets, wire rod, round bars, square bars, flats, hexagons, bright (cold finished) bars, heavy rails, light rails, rail accessories. Stainless steel – bright (cold finished) bars, wire. Alloy steel (excl stainless) – blooms, billets, wire rod, round bars, square bars, flats, hexagons, bright (cold finished) bars, wire. Mild and special steel including engineering steel.

DeltaValdarno

Head office: Viale della Resistenza 2, I 57025 Piombino, Livorno. **Postal address:** General management – Viale Sarea 336, I 20126 Milan (Tel: 02 64461. Telex: 313110). **Tel:** 0565 64111. **Telex:** 500162.
Ownership: Deltasider (Ultimate holding company – Finsider Group).
Works:
■ Piazza Matteotti 13, I 52027 San Giovanni Valdarno, AR (Tel: 055 92041. Telex: 570119). ***Rolling mills:*** One continuous light section and bar. ***Other plant:*** Railway track fittings plant.
■ Via del Commercio 5, I 30155 Marghera, VE. ***Rolling mills:*** One light section and bar. ***Other plant:*** Welded beam plant, engineering workshop, steel pallet plant.
Products: Carbon steel – round bars, square bars, flats, light angles, light tees, light channels, medium angles, medium tees, medium joists, medium channels, welded sections (heavy), rail accessories.

Diano SpA

Head office: Via Nazionale 113, Lazzaro 962, RC. **Tel:** 0965 712021. **Telex:** 890031 diano i.
Products: Carbon steel – reinforcing bars.

Dora Acciaierie e Ferriere (formerly Acciaierie Ferriere Alpine SpA (AFB))

Head office: I 10050 Borgone di Susa, TO. **Tel:** 011 9645245. **Telex:** 210167.
Products: Carbon steel – flats, light angles.

Emiliana Tubi – Profilati Acciaio SpA (Setpa)

Head office: Via Cisa Ligure 25/A, I 42041 Brescello, RE. **Tel:** 0522 687171, 0522 687269. **Telex:** 531283 emtubi i. **Cables:** Setpa Brescello.
Products: Carbon steel – longitudinal weld tubes and pipes, galvanized tubes and pipes.

Eurocolfer Acciai SpA

Head office: I 28020 Palianzeno, NO. **Telex:** 200281.
Subsidiaries: (Ferdofer SpA Domegliara) (merchant bars); Acciaiera del Tirenno (Sicily) (beams).
Products: Carbon steel – medium joists, wide-flange beams (medium).

Falci SpA, Reparto Laminazione

Head office: Via Cuneo 7, I 12025 Dronero, CN. **Tel:** 0171 918106. **Telex:** 212451. **Cables:** Falci Dronero. **Telecopier (Fax):** 0171 918084.
Management: E. Conte (president), G. Simondi (managing director), G.P. Acchiardi (product manager, steel). **Established:** 1911. **Capital:** Lire 1,000,000,000. **No. of employees:** Works – 35. Office – 5. **Annual capacity:** Finished steel: 8,000 tonnes.
Works:
■ At head office. **Rolling mills:** SL special hot mill (annual capacity 8,000 tonnes).
Products: Carbon steel – square bars, flats, hexagons, light angles, special light sections, medium angles. Stainless steel – special light sections. Alloy steel (excl stainless) – square bars, flats, hexagons, light angles, special light sections. Special sections in coil.

Acciaierie e Ferriere Lombarde Falck SpA

Head office: Via G. Enrico Falck 63, I 20099 Sesto San Giovanni (MI). **Tel:** 02 24901. **Telex:** 321834 falck i. **Telecopier (Fax):** 02 2428830.
Management: Alberto Falck (president), Giorgio Enrico Falck (vice president and managing director), Paolo Scaroni (vice president), Vicenzo Portanova (managing director). **Established:** 1906. **Capital:** Lire 73,743,900,000. **No. of employees:** Works – 4,438. Office – 1,044. **Subsidiaries:** Stà Nordelettrica SpA (Sondel), Milan (50.02%); SpA Cantieri Metallurgici Italiani, Naples (77.19%); Acciaierie di Bolzano SpA, Bolzano (54.18%); AFL Acciaierie e Fonderie SpA, Dongo (99.99%); Transider SpA, Milan (50.10%); Ferrometalli Safem SpA, Bergamo (41.55%); Tubicar SpA, Carbonara Scrivia (76%); Sté des Raccords AFY SA, Lyon, France (55%); Falck France

Acciaierie e Ferriere Lombarde Falck SpA (continued)

SA, Lyon, France (52.34%); Finstahl SA, Luxembourg (99.99%); Falck Steel Inc, Stamford, USA (63.08%); Franco Tosi SpA, Milan (18.47%); Antonio Badoni SpA, Lecco (45.35%); Stà Italiana Acciai Utensili Siau SpA, Turin (12%). **Annual capacity:** Raw steel: 1,180,000 tonnes.
Works:
■ Unione, Via Mazzini 23, I 20099 Sesto San Giovanni (MI). (Tel: 02 24901. Telex: 310155). ***Steelmaking plant:*** electric arc furnaces – one 120-tonne and one 70-tonne Tagliaferri (combined annual capacity 630,000 tonnes). ***Refining plant:*** One 120-tonne ladle furnace; one 60/120-tonne VAD. ***Continuous casting machines:*** One 2-strand Concast curved mould slab (100-160mm x 320-640mm) (400,000 tonnes); one 4-strand DMS centrifugal round (115-215mm) (200,000 tonnes). ***Rolling mills:*** One semi-continuous 26" hot strip and hoop (730,000 tonnes).
■ Concordia, Via Mazzini 45/A, I 20099 Sesto San Giovanni (MI). (Tel: 02 24901. Telex: 352021). ***Steelmaking plant:*** One 130-tonne electric arc furnace (annual capacity 600,000 tonnes). **Refining plant:** one 140-tonne CAD. ***Refining plant:*** One 140-tonne CAD. ***Continuous casting machines:*** One 1-strand Concast curved mould slab (160-300 x 1,350-2,100mm) (400,000 tonnes). ***Rolling mills:*** One 4-high heavy plate (530,000 tonnes).
■ Vittoria, Viale Italia 234, I 20099 Sesto San Giovanni (MI). (Tel: 02 24901). ***Rolling mills:*** Two 4-high reversing cold strip (annual capacity 220,000 tonnes). ***Other plant:*** Steel wire works.
■ Arcore, Via C. Battisti 6, Arcore (MI). (Tel: 039 6650). ***Tube and pipe mills:*** seamless – two piercing and plug mills up to 9$\frac{5}{8}$", one Assel Transval up to 8",twelve cold Pilger 1$\frac{1}{2}$-4$\frac{1}{2}$", four drawbenches 2-4".
■ Vobarno, Via G. Enrico Falck, Vobarno. (Tel: 0365 61061. Telex: 30022). ***Rolling mills:*** One 1-high combined billet, light section and bar; 3-high reversing cold strip (annual capacity 130,000 tonnes). ***Tube and pipe mills:*** welded – seven 1-4"; four 1-4" drawbenches.
■ Dongo, Via Rubini 4, Dongo (CO). (Tel: 0344 81201). Malleable iron fittings and castings (annual capacity 18,000 tonnes).
■ Novate Mezzola, Via Nazionale 7, Novate Mezzola (SO). (Tel: 0343 4441). Ferro-chrome plant (annual capacity 13,000 tonnes).
Products: Carbon steel – slabs, billets, semis for seamless tubemaking, round bars, hot rolled hoop and strip (uncoated), skelp (tube strip), cold rolled hoop and strip (uncoated), medium plates, heavy plates, hot rolled sheet/coil (uncoated), cold rolled sheet/coil (uncoated), seamless tubes and pipes, longitudinal weld tubes and pipes, cold drawn tubes and pipes. Wire and strand for prestressed concrete, malleable iron fittings and castings, low carbon ferro-chrome.

Acciaierie e Ferriere Feralpi SpA

Head office: Via Industria 23, I 25017 Lonato (Brescia). **Tel:** 030 9131161. **Telex:** 300302 fera i. **Cables:** Feralpi SpA. **Telecopier (Fax):** 030 9132786.
Management: Camilla Pasini (president); Giovanni Tolettini Fu Andrea (vice president); Giuseppe Sansone Pasini (managing delegate); Ragionier Cassetti Giuseppe (general manager and administration); Armando Fantinelli (works manager and technical director). **Established:** 1955. **Capital:** Lire 15,000,000,000. **No. of employees:** Works – 452. Office – 34. **Ownership:** Family Pasini (Camilla and Giuseppe Sansone) (78%), others (22%). **Subsidiaries:** Cogedil (100%) (reinforcement services). **Annual capacity:** Raw steel: 390,000 tonnes. Finished steel: 383,000 tonnes.
Works:
■ At head office. (Tel: 030 913 1161. Telex: 300302). Works director: G. Armando Fantinella. ***Steelmaking plant:*** Three 35-tonne electric arc furnace (annual capacity 420,000 tonnes). ***Continuous casting machines:*** Two 3-strand Concast-S Type billet (115 x 115mm) (420,000 tonnes). ***Rolling mills:*** Continuous bar mill and wire rod (6-40mm rounds).
Products: Carbon steel – billets 115 x 115mm; wire rod 6-12mm; reinforcing bars

Acciaierie e Ferriere Feralpi SpA (continued)

6-40mm.
Sales offices/agents: Comsider, Lyon, France. (Tel: 78 83 41 41). Feralpi Deutschland, Munich, West Germany. (Tel: 82 54 15 80).
Expansion programme: Ladle furnace.

Ferdofin SpA

Head office: Via Pastrengo 29, I 10028 Turin. **Tel:** 011 593 668. **Telex:** 22596.
Products: Carbon steel – wire rod, reinforcing bars, square bars, flats, light angles, light tees, light channels.

Ferdofer SpA

Head office: Passaggio Napoleone 4, I 35017 Domegliara, VR. **Tel:** 045 773 0722. **Telex:** 480484.
Subsidiaries: Eurocolfer Pallanzero (NO) (sections); Acciaieria del Tirenno (Sicily) (beams and sections).
Products: Carbon steel – square bars, flats, light angles, light tees, light channels.

Acciaierie Ferrero SpA

Head office: Via G. Galileo 26, I 10036 Settimo Torinese, TO.
Products: Carbon steel – reinforcing bars, square bars, flats.

Ferrosider SpA

Head office: Via Domenico Ghidoni 169, I 25035 Ospitaletto Bresciano, Brescia. **Tel:** 030 640637. **Telex:** 301451. **Cables:** Ferrosider Ospitaletto B. **Telecopier (Fax):** 030 642628.
Management: Director – Giuseppe Masserdotti. **Established:** 1972. **Capital:** Lire 10,020,000,000. **No. of employees:** Works – 94. Office – 21.
Works:
■ At head office. ***Rolling mills:*** bar.
Products: Carbon steel – round bars 10-50mm; square bars 10-50mm; flats 14 x 3-150 x 25mm.

Finsider – Stà Finanziaria Siderurgica pA

Head office: Viale Castro Pretorio 122, I 00185 Rome. **Tel:** 06 4697. **Telex:** 610088 finsid.
Management: Mario Lupo (president), Giavanni Gainsardella (managing director).
Ownership: State-owned through IRI. **Subsidiaries:** These include Nuova Italsider, Deltasider, Terni, Dalmine, Siderexport.

Acciaieria Foroni SpA

Head office: Via A. Colombo 285, I 21055 Gorla Minore, VA. **Tel:** 0331 600460. **Telex:** 330570 foroni i.
Products: Stainless steel – ingots, billets, semis for seamless tubemaking, round bars. Nickel-base superalloys.

Forsidera SpA

Head office: Strada Battaglia 85, I 35020 Albignasego, Padua. **Tel:** 049 681013. **Telex:** 430411.
Products: Carbon steel – cold rolled hoop and strip (uncoated) – low-carbon steel for pressings 40-450mm wide.

Officine e Fonderie Galtarossa SpA

Head office: Lungadige A. Galtarossa 21, I 37100 Verona, VR. **Tel:** 45 594 944. **Telex:** 480062 galveri. **Cables:** Galtarossa Verona.
Products: Carbon steel – wire rod, reinforcing bars and welded mesh.

Gamma Serbatoi SpA

Head office: Località Torre Fessa, I 03013 Ferentino, FR. **Tel:** 0775 34121.
Products: Carbon steel – longitudinal weld tubes and pipes from hot rolled strip.

General Sider Europa SpA

Head office: Via E. Piaggio 29, I 66013 Chieti Scalo. **Tel:** 0871 59345. **Telex:** 600108. **Cables:** Generalsider Chieti Stazione.
Management: Board – Mario Gambini, Alberto Mazzoni, Graziano Serra; sales manager – Pasquale Frezza; directors – Aldo Cerulli (executive), Giuseppe Valloni (technical). **Established:** 1974. **Capital:** Lire 1,925,000,000. **No. of employees:** Works – 50. Office – 20. **Annual capacity:** Finished steel: 13,000 tonnes.
Works:
■ Via Custoza 28, I 66013 Chieti Scalo. (Tel: 0871 54630. Telex: 600108). Works director: Giuseppe Valloni. ***Tube and pipe mills:*** welded – one Wean Damiron WRS 800 Brown Boveri, 3-10″ (annual capacity 130,000 tonnes).
Products: Carbon steel – longitudinal weld tubes and pipes 3-10″; oil country tubular goods 3-10″; galvanized tubes and pipes 3-6″; plastic coated tubes and pipes 3-10″; bitumen coated tube and pipe 3-10″.

General Sider Italiana SpA

Head office: Via E. Piaggio 29, I 66013 Chieti Scalo. **Tel:** 0871 59345. **Telex:** 600108. **Cables:** Generalsider Chieti Stazione.
Management: Board – Mario Gambini, Alberto Mazzoni, Graziano Serra; sales manager – Pasquale Frezza; directors – Aldo Cerulli (executive), Giuseppe Valloni (technical). **Established:** 1961. **Capital:** Lire 1,535,000,000. **No. of employees:** Works – 35. Office – 5. **Annual capacity:** Finished steel: 40,000 tonnes.
Works:
■ At head office. Works director: Giuseppe Valloni. ***Tube and pipe mills:*** welded – one $\frac{1}{4}$-$1\frac{1}{2}$″ Yoder W 20 (annual capacity 15,000 tonnes); one $\frac{3}{8}$-$2\frac{1}{2}$″ Somenor M $2\frac{1}{2}$″ (25,000 tonnes).
Products: Carbon steel – longitudinal weld tubes and pipes $\frac{1}{4}$-$2\frac{1}{2}$″; galvanized tubes and pipes $\frac{1}{4}$-$2\frac{1}{2}$″; plastic coated tubes and pipes $\frac{1}{4}$-$2\frac{1}{2}$″; bitumen coated tube and pipe $\frac{1}{4}$-$2\frac{1}{2}$″. Tubular scaffolding.

Fratelli Goffi SpA

Head office: Via Legnago 47, I 25089 Villanuova, BS. **Tel:** 0365 31861. **Telex:** 301608.
Products: Carbon steel – longitudinal weld tubes and pipes.

IAI Cofermet Industria Acciai Inox SpA—see Terni Acciai Speciali SpA

Ilfo – Industria Laminati Ferrosi Odolesi Srl

Head office: Via Brescia 7/D, I 25076 Odolo, BS. **Tel:** 0365 860361-3. **Telex:** 300654 ilfo i. **Cables:** Ilfo Odolo.
Products: Carbon steel – reinforcing bars.

Ilta – Ind. Lav. Tubi Acciaio SpA—see Arvedi Group

Indumetal – Industrie Metallurgiche SpA

Head office: Via E. Fermi 30, Zona Industriale Aussa-Corno, San Giorgio di Nogaro, UD. **Tel:** 431 65581-4. **Telex:** 450132 indume i.
Products: Carbon steel – billets. Alloy steel (excl stainless) – billets.

Inossman – Inossidabili Maniago SpA

Head office: Via delle Industrie 4, I 33085 Maniago, PN. **Tel:** 0427 72772. **Telex:** 450679. Stainless and heat resisting steels.

Nuova Italsider SpA

Head office: Via Corsica 4, I 16128 Genoa. **Postal address:** Casella Postale 1727, I 16100 Genoa. **Tel:** 010 55181. **Telex:** 270039 itasid 1. **Cables:** Italsider Genoa. **Telecopier (Fax):** 010 541790.
Management: Presidente – Michele Civallero; vice presidente – Didimo Badile; amministratore delegato – Sergio Noce; direttori generali – Emilio Ainis, Attilio Angelini, Girolamo Morsillo, Pierluigi Renier, Vincenzo Rizza, Giovanni Battista Spallanzani. **Established:** 1949. **Capital:** Lire 3,059,475,000,000. **No. of employees:** 28,974 (December 31, 1986). **Ownership:** Finsider SpA (100%).
Subsidiaries: These include: ILVA (100%) (production of cold rolled flat products); ITA-Industria Trasformazione Acciaio (100%) (production of cold rolled flat products); Rivestubi (100%) (internal and external steel line pipe coating); Sidermontaggi (100%) (assembling industrial, mechanical and electrical plant; medium-heavy steel structural construction; tube and boiler work; special industrial plant maintenance); Lavemetal (94.3%) (production of sheet cut lengths aluminium and zinc-aluminium coated); Icrot-Lavorazioni Sussidiarie Finsider (60%) (iron recovery from steelworks slag, scrap preparation, repair of iron and steel equipments, industrial cleaning operations, subsidiary operations for steel industry, including maintenance); Sidercomit (51%) (selling steel products from warehouse stock and preworked steel products from service centres); Siderexport (51%) (Finsider export organization); Sidermar (49%) (shipping); Tubificio Dalmine Italsider (49%) (steel tubemaking). **Annual capacity:** Pig iron: 12,350,000 tonnes. Raw steel: 14,550,000 tonnes. Finished steel: HR flat products 12,400,000 tonnes, CR flat products 3,550,000 tonnes, coated flat products 770,000 tonnes, welded pipe 1,464,000 tonnes.
Works:
■ Taranto, Via Appia km 648, 74100 Taranto. (Tel: 099 4811. Telex: 860009 itasid 1). Works director: Girolamo Morsillo. ***Coke ovens:*** Six Still batteries (270 ovens) (annual capacity 2,500,000 tonnes); four Still batteries (172 ovens) and one ITI/Still battery (43 ovens) (3,000,000 tonnes); ***Sinter plant:*** Three 168mm strand Dwight Lloyd (4,850,000 tonnes) and two 400mm strand Dwight Lloyd (8,650,000 tonnes). ***Blast furnaces:*** Five – (No.1) 10.6m hearth dia, 2,471 cu m (1,700,000 tonnes); (No.2) 10.2m hearth dia, 2,406 cu m (1,665,000 tonnes); (No.3) 10.3m hearth dia, 2,448 cu m (1,665,000 tonnes); (No.4) 10.6m hearth dia, 2,471 cu m (1,700,000 tonnes) and (No.5) 14.0 m hearth dia, 4,335 cu m) (3,270,000 tonnes). ***Steelmaking plant:*** basic oxygen converters – three 330-tonnes LD (5,500,000 tonnes) and three 350-tonnes LD (6,000,000 tonnes). ***Refining plant:*** Three vacuum degassing units – two RH-OB 330-tonne and 350-tonne and one DH-OB 350-tonne. One 330-tonne IP; one 350-tonne CAB; one 330-tonne CAS/OB and one 350-tonne TAC-4. ***Continuous casting machines:*** Five slab – one 2-strand Demag (wide 1,000 x 2,200mm, thick 165/220/240/280mm, long 4,000 x 12,000mm) (1,660,000 tonnes); one 2-strand Concast (1,050 x 2,200mm, 200/240/270mm, 4,000 x 13,000mm) (1,700,000 tonnes); one 2-strand Concast (1,050 x 2,200mm, 200/240/270mm, 4,000 x 13,000mm) (1,700,000 tonnes); one 2-strand Innse/Italimp (650 x 1,360mm, 180/210/240mm, 4,000 x 12,600mm) (1,340,000 tonnes) and one 2-strand Innse/Italimp (650 x 1,350mm, 240mm, 4,500 x 12,600mm) (2,050,000 tonnes). ***Rolling mills:*** One 50 x 96" horizontal stand/46 x 81" vertical stand slabbing (550 tonnes/hour). one United/Innse heavy plate – one 4-high reversing/roughing stand (40/80 x 190") and one 4-high reversing/finishing stand

Italy

Nuova Italsider SpA (continued)

(40/80 x 190") (length 5,000 x 25,000mm, wide 1,000 x 4,500mm, thickness 7 x 150mm) (2,000,000 tonnes). Two wide hot strip – one Blaw Knox/United comprising one edger stand (45 x 17"); one 2-high roughing stand (46 x 68"); four 4-high roughing stand (44/56 x 68") and six 4-high finishing stand (28.5/60 x 68") (width 600 x 1,560mm, thickness 1.5 x 8mm) (coil weight 27 tonnes) (3,500,000 tonnes); one Blaw Knox/Innse comprising one edger stand (46 x 22"); three 2-high roughing stand (50.4 x 90.2"); three 4-high roughing stand (46.1/64.0 x 90.2") and seven 2-high finishing stand (32/64.0 x 90") (width 600 x 2,180mm, thickness 1.2 x 16mm) (coil weight 45 tonnes) (4,500,000 tonnes). one cold reduction – five 4-high stand (22.9/60 x 66") (width508 x 1,550mm, thickness 0.4 x 3.00mm) (coil weight 35 tonnes) (1,300,000 tonnes). Two temper/skin pass – one 4-high stand (24/60 x 66") (width 508 x 1,550mm, thickness 0.4 x 3.00mm) (coil weight 35 tonnes) and one 4-high stand (24/60 x 66") (width 508 x 1,550mm, thickness 0.4 x 3.00mm) (coil weight 35 tonnes) (combined annual capacity 1,300,000 tonnes). ***Tube and pipe mills:*** welded – four 18 x 42" longitudinal weld (thickness 5.56 x 27.8mm, length 12,300mm max) (340,000 tonnes); one 30 x 56" longitudinal weld (7.92 x 27.5mm, 17,300mm max) (880,000 tonnes); one 18 x 64" spiral weld (5.56 x 12.7mm, 16,000mm max) (44,000 tonnes); one 31.5 x 110.2" (6.35 x 25.4mm, 18,300mm max) (200,000 tonnes).

■ Via Nuova Bagnoli 435, 80124 Naples-Bagnoli. (Tel: 081 7231111. Telex: 710039 itasid i). Works director: Francesco Segreti. ***Coke ovens:*** Five Still batteries (150 ovens) (annual capacity 1,200,000 tonnes). ***Sinter plant:*** Two 168 sq m strand Lurgi (3,200,000 tonnes). ***Blast furnaces:*** One 8,306m hearth dia, 1,357 cu m (1,054,000 tonnes) and one 9,600m hearth dia, 1,915 cu m (1,296,000 tonnes). ***Steelmaking plant:*** Three 150-tonne LD basic oxygen converters (2,700,000 tonnes). ***Refining plant:*** Two 150-tonne ladle furnace. ***Continuous casting machines:*** Two 2-strand Innse slab (1,350 x 10,500 x 210mm) (1,950,000 tonnes). ***Rolling mills:*** hot strip and hoop – one vertical scale breaker (45.3 x 33.8"); one edger (39.3 x 31.9"); one 4-high reversing/roughing (39.3/57 x 56") (coil weight 24 tonnes); one edger (32.6 x 23.6") and six 4-high finishing stands (27/57 x 56") (annual capacity 2,000,000 tonnes).

■ Coated Products Division, Stabilimento Cornigliano, Via Pionieri e Aviatori d'Italia 8, 16154 Sestri Ponente (Genoa). (Tel: 010 41071. Telex: 270243 itasid i). Works director: Roberto Fabbri. ***Rolling mills:*** Two cold reduction – one 5-stand 4-high Continental (22/59 x 66") (coil weight 15 tonnes) (annual capacity 700,000 tonnes); one 5-stand 4-high Innse (27.8/43.3 x 48") (coil weight 20 tonnes) (300,000 tonnes); two temper/skin pass – one 2-stand 4-high Bliss (18/39.3 x 48") (coil weight 15 tonnes) and one 2-stand 4-high Innse (18/43.3 x 48") (coil weight 20 tonnes) (combined annual capacity 1,000,000 tonnes). ***Coil coating lines:*** Two Ferrostan electrolytic tinning - one 38" Wean and one 38" Wean Damiron and one HCD/TFS Clecim/Vec tin-free steel (ECCS) (combined annual capacity 420,000 tonnes (ECCS 170,000 tonnes)); two Sendzimir hot-dip galvanizing – one 60" and one 49" max width (combined annual capacity 350,000 tonnes).

■ Via Boscomarengo 1, 15067 Novi Ligure (Alessandria). (Tel: 0143 7751. Telex: 210084 itasid i). Works director: Ledo Gesi. ***Rolling mills:*** One 4-high Blaw Knox cold reduction (23/56 x 80") (coil weight 33.5 tonne) and two temper/skin pass – one 1-stand 4-high Blaw Knox (23/56 x 80") (coil weight 33.5 tonne) and one 1-stand 4-high Continental (21/56 x 66") (coil weight 18.0 tonne) (combined annual capacity 1,250,000 tonnes).

■ Campi, Corso FM Perrone 15, 16152 Cornigliano Ligure (Genoa). (Tel: 010 41071. Telex: 270072 itasid i). Works director: Pietro Meschi. ***Steelmaking plant:*** One 120-tonne Tagliaferri (with EBT System) electric arc furnace (annual capacity 350,00 tonnes). ***Refining plant:*** One 30-tonne ESR; one 120-tonne ladle furnace and one 120-tonne vacuum degassing unit. One Italimpianti (controlled pressure pouring machine for slabs) (2,300 x 10,200 x 75/150/600mm) (300,000 tonnes). **Continuous Casting Machines.** ***Rolling mills:*** One 4-high reversing heavy plate (950/1,450 x 3,750mm and vertical edging stand (plates – 1,100 x 3,750) 4,000 x 30,000 and 6 x 230 mm) (coil weight 30 tonne) (400,000 tonne). ***Other plant:*** One

Nuova Italsider SpA (continued)

steel forging plant for mill rolls; one clad plate plant, and normalizing, quenching and tempering heat-treating furnaces.
■ Corso Giuseppe Mazzini 3, 17100 Savona. (Tel: 019 83171. Telex: 270244 itasid i). Works director: Mario Miliardi. Workshop for railway wagons and components; special working with welding-surfacing for steel industry equipment; rollers and drums for conveyor belts and pre-machined products.
Products: Steelmaking pig iron (1986 output 9,218,000 tonnes); Carbon steel – slabs 140-2,200 x 100-500mm; hot rolled coil (for re-rolling) 600 to 2,080 x 1.5-16mm; hot rolled hoop and strip (uncoated) 80-599 x 1.5-16mm; skelp (tube strip); cold rolled hoop and strip (uncoated) 30-499 x 0.4-2.99mm; hot-dip galvanized hoop and strip 30-499 x 0.3-2mm; medium plates; heavy plates and plates for pipes 6-200mm thick x 4,500 max wide x 25,000mm max long; clad plates stainless, nickel, and Inconel clad 6-100mm thick 3,000mm max wide and 12,000mm max long; hot rolled sheet/coil (uncoated) 600-2,080 x 1.5-16mm; cold rolled sheet/coil (uncoated) 500-1,800 x 0.36-2.99mm; hot-dip galvanized sheet/coil 500-1,500 x 0.3-3mm; electrolytic tinplate (single-reduced) and tin-free steel (ECCS) 600-965mm x 0.17-0.49mm; longitudinal weld tubes and pipes; spiral-weld tubes and pipes and large-diameter tubes and pipes 457.2-2,743mm od, 5-56-30mm wall thickness. Total 1986 output – Pig iron (9,218,000 tonnes), steel (9,231,000 tonnes), hot rolled flat (8,396,000 tonnes), cold rolled flat (2,378,000 tonnes), welded pipes (844,000 tonnes), electrolytic tinplate (212,000 tonnes), galvanized flat-rolled (300,000 tonnes).
Sales offices/agents: Export sales organizations – Siderexport, Viale Brigata Bisagno, 2, I 16129 Genoa GE (Tel: 10 55171. Telex: 270202 sdx ge i). Nuova Italsider sales offices – mill rolls, Cornigliano Ligure; Affilliated companies – Sidernederland BV, Rotterdam, Netherlands (covering Belgium and Netherlands) (Tel: 4130666. Telex: 27360 sidne nl); Siderfrance SA, Paris, France (Tel: 47237756.Telex: 610347 siderfrance); Deutsche Siderexport GmbH, Düsseldorf, West Germany (Tel: 211 632031-5. Telex: 8586690 dsdx d); Ibersider SA, Barcelona, Spain (Tel: 2099322. Telex: 987401 b5de); Sidercom SA, Zürich, Switzerland (Tel 01 2423250-2. Telex: 812073 sider ch); Siderital Ltd, London, UK (Tel: 723 4002. Telex: 28697 siderital london); Siderius Canada Inc, Ontario, Canada (Tel: 416 9615777. Telex: 065 24397); Siderius Inc, New York (Tel: 212 4897470-3. Telex: 220061) and Texas (Tel: 713 6672222. Telex: 791177); Socomar, Addis Ababa, Ethiopia (Tel: 443440. Telex: 21030 A ababa). Representative offices – Siderexport SpA, Athens, Greece (Tel: 6925074. Telex: 2125057.); Siderexport SpA, Budapest, Hungary (Tel: 223284. Telex: 227501); Siderexport SpA, Zagreb, Yugoslavia (Tel: 275025. Telex: 21792 yu sider); Finsider-Siderexport, Moscow, USSR (Tel: 2301110. Telex: 413233 finsid u.) ; Siderexport SpA, Mexico (Tel: 52533920. Telex: 1777590 sideme) ; Siderexport SpA, Algiers, Algeria (Tel: 634868-9. Telex: 66511 sixal dz); Siderexport SpA, Cairo, Egypt (Tel: 3419596. Telex: 23349 ventur un); Siderexport SpA, Kowloon, Hong Kong (Tel: 37234686. Telex: 43919 ishko hk); Siderexport SpA, Tehran, Iran (Tel: 68507. Telex: 215242 sidex ir); Siderexport SpA, Jeddah, Saudi Arabia (Tel: 2 6431673. Telex: 603887 bared sj); Siderexport, Dubai, United Arab Emirates (Tel: 421515. Telex: 48968).

Lampre Srl

Head office: C Italia 40, I 20040 Usmate, MI. **Tel:** 039 672094-5. **Telex:** 330587 lampre i.
Products: Carbon steel – colour coated sheet/coil.

Las – Laminazione Acciai Speciali SpA

Head office: Via Buonarroti 5, I 25016 Ghedi, BS. **Tel:** 30 902 061-3. **Telex:** 300588 lasbs i.
Products: Carbon steel – round bars, square bars, flats, special light sections. Alloy steel (excl stainless) – round bars, square bars, flats, special light sections, 40-115mm. Engineering steel bars.

Italy

Acciaierie e Ferriere Luigi Leali SpA

Head office: Via Garibaldi 5, I 25076 Odolo, BS. **Tel:** 365 860401. **Telex:** 300040. **Telecopier (Fax):** 365 860423.
Established: 1954. **Capital:** Lire 10,000,000,000. **Subsidiaries:** Acciaieria Valsugana (100%) (see below); Ferriera di Roè Volciano (100%) (see below); Siderpotenza (50%); Cogea (22%). **Associates:** LAS-Ghedi. **Annual capacity:** Raw steel: 200,000 tonnes. Finished steel: 250,000 tonnes.
Works:
■ At head office. ***Steelmaking plant:*** Two electric arc furnaces (annual capacity 200,000 tonnes). ***Continuous casting machines:*** Two 2-strand curved mould billet (115-130mm sq) (300,000 tonnes). ***Rolling mills:*** One two 3-high and ten 2-high bar (310/360/490mm dia) (250,000 tonnes).
■ Acciaieria Valsugana SpA, Borgo Valsugana. ***Steelmaking plant:*** One electric arc (annual capacity 200,000 tonnes). ***Refining plant:*** One ladle furnace (200,000 tonnes). ***Continuous casting machines:*** One 4-strand straight mould billet (130-160mm sq) (200,000 tonnes), one 2-strand straight mould bloom (280-320mm sq) (200,000 tonnes).
■ Ferriera di Roè Volviano SpA, Roè Volciano. ***Rolling mills:*** bar comprising one 3-high and sixteen 2-high (260/330/380/500mm dia) (annual capacity 210,000 tonnes).
Products: Carbon steel – blooms 280-320mm; billets 115, 130, 140, 160mm; semis for seamless tubemaking 120mm dia; reinforcing bars 6-57mm; round bars 6-120mm; square bars 60-120mm; flats 80 x 30-160 x 60mm. Alloy steel (excl stainless) – blooms 280-320mm; billets 115, 130, 140, 160mm; semis for seamless tubemaking 120mm dia; round bars 6-120mm; square bars 60-120mm; flats 80 x 30-160 x 60mm.

Lombarda Tubi SpA

Head office: Via Milano 41, I 22050 Lomagna, CO. **Tel:** 530 0233, 530 0353-5. **Telex:** 333492 lomtub i.
Products: Carbon steel – longitudinal weld tubes and pipes from cold rolled strip.

Tubificio Lombardo—see Arvedi Group

Acciaierie di Lonato SpA

Head office: Via Salvo d'Acquisto 19, I 25017 Lonato, BS. **Tel:** 030 9130821. **Telex:** 300343 lonato i.
Products: Carbon steel – billets.

Lovere Sidermeccanica SpA

Head office: Viale Castro Pretorio 122, I 00185 Rome. **Tel:** 064 97951. **Telex:** 611312.
Capital: Lire 200,000,000. **Ownership:** Terni – Stà per l'Industria e l'Elettricità SpA (Ultimate holding company – Finsider Group).
Works:
■ Via G. Paglia 45, I 24065 Lovere, BG. ***Steelmaking plant:*** Three electric arc furnaces – one 60-tonne UHP, and two 6-tonne. ***Refining plant:*** One 52-tonne Asea-SKF. ***Other plant:*** Ring mill, wheel mill, four forging presses for axles and other forgings. Railway wheels, tyres and axles.

Lucchini Siderurgica SpA

Head office: Via Oberdan 6, I 25100 Brescia, BS. **Tel:** 03 39921. **Telex:** 300014, 301245.
Subsidiaries: Bisider SpA.
Products: Carbon steel – billets, wire rod, round bars, square bars, light angles (equal and unequal), light tees, light channels, medium joists, universals (medium), heavy joists. Alloy steel (excl stainless) – wire rod (for welded chaines).

La Magona d'Italia SpA

Head office: Via di Villamagna 92-94, I 50126 Florence. **Tel:** 055 2754-1. **Telex:** 570140 magona, 571571 magona. **Cables:** 570140 Magona Florence.
Products: Carbon steel – hot-dip galvanized sheet/coil, colour coated sheet/coil (painted and plastic-coated), electrolytic tinplate (single-reduced).

Fonderie Acciaierie Giovanni Mandelli

Head office: Via Torino 21, I 10197 Turin. **Postal address:** Casella Postale 300. **Tel:** 011 781901, 011 7800505. **Telex:** 210427 fmand i.
Products: Carbon steel – ingots.

Marcegaglia SpA

Head office: Via Bresciani 16, I 46040 Gazoldo Ippoliti, MN. **Tel:** 0376 6541. **Telex:** 300514.
Products: Carbon steel – longitudinal weld tubes and pipes.

Metallurgica Marcora SpA

Head office: Via Goito 19-20, I 21052 Busto Arsizio, VA. **Postal address:** Casilla Postale 429. **Tel:** 0331 631257, 0331 634477. **Telex:** 330596 ubibus i. **Cables:** Metalferro.
Products: Carbon steel – ingots, blooms, billets, semis for seamless tubemaking, round bars, square bars, flats.

Acciaierie e Tubificio Meridionali SpA

Head office: Corso Cavour 130, I 70121 Bari. **Postal address:** Casella Postale 268. **Tel:** 80 544 544. **Telex:** 810383. **Cables:** Tubificio Bari.
Products: Carbon steel – seamless tubes and pipes, galvanized tubes and pipes, cold drawn tubes and pipes.

Italy

Metalgoi SpA

Head office: Viale S. Eufemia 184, I 25080 Brescia. **Postal address:** PO Box 34, I 25100 Brescia. **Tel:** 30 366261. **Telex:** 300428 metgoi i. **Cables:** Metalgoi Brescia. **Telecopier (Fax):** 30 366365.
Management: Vittorio Goi (president). **Established:** 1947. **Capital:** Lire 1,020,000,000. **No. of employees:** Works – 35. Office – 15. **Annual capacity:** Finished steel: 60,000 tonnes.
Works:
■ At head office. ***Rolling mills:*** One 3-high 450mm light section (annual capacity 60,000 tonnes).
Products: Carbon steel – light angles 20-50mm; light tees 20-40mm; light channels 25-40mm; special light sections 20-50mm.

La Metallifera SpA

Head office: Via Provinciale, I 25070 Barghe, BS. **Tel:** 0365 84181. **Telex:** 303228 lamet.
Products: Carbon steel – longitudinal weld tubes and pipes.

Metalmanzoni SpA

Head office: Via S. Vecchia 15, I 22049 Valmadrera (CO). **Tel:** 0341 582165. Hot and cold rolled sheet.

Neirotti Tubi SpA

Head office: S.S. 10, Km 26, I 14019 Villanova d'Asti, AT. **Tel:** 141 946093. **Telex:** 210060.
Management: Lidio Neirotti (chairman), Claudio Neirotti (vice chairman). **Established:** 1973. **Capital:** Lire 1,000,000,000. **Subsidiaries:** Produits d'Acier SA, France.
Works:
■ At head office. ***Tube and pipe mills:*** welded.
Products: Carbon steel – longitudinal weld tubes and pipes and galvanized tubes and pipes (rounds 20-152mm, square 20-120mm, rectangular 20 x 10mm to 180 x 60mm, special sections 1.5-5mm thick).

Ferriere Nord SpA

Head office: Zona Industriale, I 33010 Osoppo, Udine. **Tel:** 0432 98 18 11. **Telex:** 450181 pitfer i. **Cables:** Pittini Osoppo. **Telecopier (Fax):** 0432 98 18 00.
Management: President – A. Pittini; managers – E. Masera (general), G. Cattapan (sales), B. Borgini (financial), I. Manzocco (plant), G. Tognarini (plant), A. Basso (plant), W. Ballandino (engineering). **Established:** 1972. **Capital:** Lire 20,000,000,000. **No. of employees:** Works – 670. Office – 198. **Ownership:** Pittini family (100%). **Subsidiaries:** Impianti Industriali SpA, Osoppo, UD (100%) (industrial equipment manufacturing); Siat SpA, Gemona, UD (100%) (mechanical drawn wire); Trafilerie Metallurgiche SpA, Catania (51%) (electro-weld mesh and lattice girders); Siderlazio Ilfe SpA, Pomezia, Rome (51%) (drawn wire and weld mesh); Norm Stahl AG, Papenburg, West Germany (97.5%) (drawn wire, weld mesh and lattice girders); Eurofer SpA, S. Giorgio di Nogaro, UD (100%) (scrap metal processing). **Annual capacity:** Raw steel: 450,000 tonnes. Finished steel: 500,000 tonnes (rolled products).
Works:
■ At head office. ***Steelmaking plant:*** One electric arc furnace (annual capacity 350,000 tonnes). ***Continuous casting machines:*** One 5-strand billet (450,000 tonnes). ***Rolling mills:*** One wire rod (coil weight 2,000kg) (500,000 tonnes).
Products: Carbon steel – billets 130 x 130mm, up to 16m long (1986 output 420,000 tonnes); wire rod 5.5-16mm dia (240,000 tonnes); electro-weld mesh (140,000 tonnes); lattice girders 12m long (40,000 tonnes); drawn wire 4-14mm dia (240,000 tonnes).

Ocsa – Officine di Crocetta SpA

Head office: Via delle Industrie 5, I 31035 Crocetta del Montello, TV. **Tel:** 0423 839225. **Telex:** 430107 ocsa i. **Cables:** Ocsa.
Products: Carbon steel – cold rolled hoop and strip (uncoated), longitudinal weld tubes and pipes.

Mini Acciaieria Odolese Srl

Head office: Via del Bosco, I 25076 Odolo, BS. **Tel:** 0365 860337, 0365 860481. **Telex:** 301433.
Products: Carbon steel – wire rod and mesh.

Olan – Officina Laminazione Nastri Srl

Head office: Via Sesia 2, I 20089 Quinto Stampi di Rozzano, MI. **Tel:** 02 8254473, 02 8254001.
Products: Carbon steel – cold rolled hoop and strip (uncoated), coated hoop and strip – nickel, copper, brass and chromium coated.

Ferriera Olifer Srl

Head office: Via G. Marconi 4, I 25076 Odolo, BS. **Tel:** 0365 860 175. **Telex:** 300638 olifer i.
Products: Carbon steel – merchant bars and light angles.

OMV – Off Metall. Ventura SpA

Head office: Via Caduti della Patria 87, Frazione Gerno, I 20050 Lesmo, MI. **Tel:** 039 6981371. **Telex:** 335895 ventur i.
Products: Carbon steel – longitudinal weld tubes and pipes.

Ori Martin SpA

Head office: Via Canovetti 13, I 25100 Brescia. **Tel:** 030 39991. **Telex:** 300037.
Subsidiaries: Ori Martin Sud, Via Martin, Ceprano, FR (reinforcing bars).
Products: Carbon steel – wire rod – low, medium and high carbon, for all general purposes including welding, pre-stressed concrete and springs; reinforcing bars.

Ferriera Padana SpA

Head office: Via Silvio Pellico, I 35100 Padua. **Postal address:** Castella Postale 555. **Tel:** 49 772115, 49 772213. **Telex:** 430370 ferpad i. **Cables:** Ferpadana Padua.
Products: Carbon steel – wire rod, reinforcing bars.

Padana Tubi & Profilati Acciaio SpA

Head office: Via Portamurata 14, I 42016 Villapieve di Guastalla, RE. **Tel:** 0522 825441-2, 0522 825857-9. **Cables:** Padana Tubi Guastalla.
Products: Carbon steel – longitudinal weld tubes and pipes.

Pasini Siderurgica SpA

Head office: Via Brescia 60, I 25076 Odolo Brescia. **Tel:** 0365 860531. **Telex:** 300027 pasini i.
Works:
■ At head office. ***Steelmaking plant:*** One Tagliaferri electric arc furnace.

Pasini Siderurgica SpA (continued)

Continuous casting machines: One 3-strand Concast billet (120mm sq). ***Rolling mills:*** One 6-30mm bar.
Products: Carbon steel – billets, reinforcing bars, round bars.

Pietra SpA – Acciaierie Ferriere e Tubifici

Head office: Orzinuovi 2, 25125 Brescia. **Tel:** 030 342161. **Telex:** 301529, 301530. **Cables:** Pietra-Brescia. **Telecopier (Fax):** 030 342046.
Management: Michele Salatini (sales manager). **Established:** 1952. **Capital:** Lire 5,000,000,000. **No. of employees:** Works – 451. Office – 83. **Annual capacity:** Raw steel: 115,000 tonnes. Finished steel: 97,000 tonnes (tubes).
Works:
■ At head office. Works director: Mario Finardi. ***Steelmaking plant:*** One 40-tonne electric arc furnace (annual capacity 115,000 tonnes). ***Continuous casting machines:*** One 3-strand Danieli billet (120-185mm sq); and two round – one 3-strand Danieli (180-205mm) and one 3-strand Demag (180-205m) (combined capacity 115,000 tonnes). ***Tube and pipe mills:*** one 6-89mm dia seamless (extrusion presses) (97,000 tonnes).
Products: Carbon steel – billets, and semis for seamless tubemaking (rounds 180-205mm, squares 120-185mm) (1986 combined output 115,000 tonnes); seamless tubes and pipes 8-89mm dia (76,000 tonnes); galvanized tubes and pipes ⅜-3" (11,000 tonnes); cold drawn tubes and pipes 6-60mm (10,000 tonnes). Gas and water seamless tubes API Std 5L-ASTM A53-73. Commercial tubes, black, galvanized, threaded and coupled.
Expansion programme: 1985/1987 – Budget for modernisation and up-dating plants 15 billion lira.

Plalam SpA

Head office: Zona Industriale Campolungo, Loc Villa S. Antonio, I 63100 Ascoli Piceno. **Telex:** 334617.
Products: Carbon steel – colour coated sheet/coil.

Acciaieria di Porto Nogaro Srl

Head office: Via E. Fermi, I 33058 S. Giorgio di Nogaro, UD. **Tel:** 0431 65843-4. **Telex:** 450229 apnoga i.
Products: Carbon steel – billets.

Predalva Acciaieria e Ferriera Srl

Head office: Via Provinciale, I 25050 Piancamuno, BS. **Tel:** 0364 55051-3. **Telex:** 300401 predal i. **Cables:** Predalva Piancamuno.
Products: Carbon steel – reinforcing bars.

Presider

Head office: Strada del Francese 13/17, I 10071 Borgaro Torinese, TO. **Tel:** 011 4702206.
Products: Carbon steel – reinforcing bars.

Profilati Nave SpA

Head office: Via Brescia Km 7, I 25075 Nave, BS. **Tel:** 030 2632091-3. **Telex:** 300427 pnnave i.
Products: Carbon steel – billets, light angles, light tees, light channels, special light sections.

Italy

Profilmec SpA

Head office: Cso F. Turati 15/H, I 10128 Turin. **Tel:** 50 52 22. **Telex:** 220438 promec i.
Products: Carbon steel – longitudinal weld tubes and pipes.

Profilnastro SpA

Head office: Corso Industria 20, Dusino San Michele, Asti. **Tel:** 0141 930153. **Telex:** 210108 pfntub i. **Telecopier (Fax):** 0141 930348.
Management: Steno Marcegaglia (president), Antonio Marcegaglia (vice president), Pierluigi Spagnol (general manager). **Established:** 1962. **Capital:** Lire 1,260,000. **No. of employees:** Works – 53. Office – 10. **Associates:** The Marcegaglia Group Metals – Steel Industries. **Annual capacity:** Finished steel: 80,000 tonnes.
Works:
■ At head office. ***Tube and pipe mills:*** welded – 4 high frequency induction machines (25-152mm dia and hollow sections 1.5 x 5mm thick) (annual capacity 80,000 tonnes).
Products: Carbon steel – longitudinal weld tubes and pipes – 20 to 120mm square, 20 x 10 to 180 x 60mm rectangular, 25 to 152mm round, plus special sections (55,000 tonnes); galvanized tubes and pipes; hollow sections.

Prolafer

Head office: Strada Statale 31 bis, I 13039 Trino Vercellese, VC. **Tel:** 161 828281. **Telex:** 200280 profer i.
Products: Carbon steel – billets, reinforcing bars.

ProReNa – Produttori Reggetta Nastri SpA

Head office: Via Provinciale 11, I 22040 Civate, CO. **Tel:** 0341 550296. **Telex:** 380091 ilexpo for Prorena. **Cables:** Prorena Civate.
Products: Carbon steel – cold rolled hoop and strip (uncoated) and strapping.

Raimondi SpA

Head office: Via G. Prati 9, I 20145 Milan. **Tel:** 02 314941. **Telex:** 331351 raival i.
Products: Carbon steel – billets. Alloy steel (excl stainless) – billets.

Redaelli Tecna SpA

Head office: Via Volta 16, I 20093 Cologno Monzese, MI. **Tel:** 02 253071. **Telex:** 333051 retec i.
Products: Carbon steel – cold rolled hoop and strip (uncoated). Wire, wire rope.

Acciaierie e Ferriere Riva SpA

Head office: Via Bergamo 1484, I 21042 Caronno Pertusella, VA. **Tel:** 029 659101.
Products: Carbon steel – billets, wire rod, reinforcing bars, square bars, flats, hexagons.

Cesare Rizzato & C SpA

Head office: Via Venezia 29, I 35131 Padova. **Postal address:** PO Box 988. **Tel:** 049 908 1722. **Telex:** 432137 rizzpd i. **Cables:** Rizzato Padova.
Management: Ing Zanardo (technical manager). **Capital:** Lire 3,645,000,000. **Annual capacity:** Finished steel: 40,000 tonnes.
Works:

Cesare Rizzato & C SpA (continued)

■ At head office. ***Rolling mills:*** Cold strip and temper/skin pass (340/400mm wide) (annual capacity 25,000 tonnes); cold reduction, cold strip and temper/skin pass (15,000 tonnes).
Products: Cold rolled hoop and strip (uncoated) thick 0.05-4mm; width 340-400mm (output 25,000 tonnes); cold rolled sheet/coil (uncoated) thick 0.4-3mm; width 20-1,500mm (output 25,000 tonnes); longitudinal weld tubes and pipes; cold drawn tubes and pipes. Low-carbon cold rolled strip, hardened spring steel strip 0.2-2mm x 10-250mm, hardened spring steel strip 0.05-4mm x 390mm.

Rodacciai Industria Trafilati SpA

Head office: Via Leopardi, I 22040 Bosisio Parini, CO. **Tel:** 31 854111. **Telex:** 380340 rodaco i.
Products: Carbon steel – bright (cold finished) bars and sections; bright wire.

Ferriera di Roè Volciano SpA—see Acciaierie e Ferriere Luigi Leali SpA

Fonderie Officine Rovelli—see Tubificio di Solbiate SpA

Acciaieria di Rubiera SpA

Head office: Via S. Cataldo 115, I 41100 Modena. **Tel:** 059 334195. **Telex:** 511179 accrub i.
Products: Carbon steel – ingots including forging ingots; billets, semis for seamless tubemaking.

Safas – Stà Azionaria Fonderia Acciai Speciali SpA

Head office: Via Verona 23, I 36077 Tavernelle di Altavilla, VI. **Tel:** 044 572100. **Telex:** 480344 safas.
Products: Stainless steel – billets, round bars, square bars.

Ferriere San Carlo SpA

Head office: Via Nazionale 1, I 25070 Caino, BS. **Tel:** 30 630024, 630051, 630007. **Telex:** 300485 scafer i.
Products: Carbon steel – reinforcing bars.

Sandvik Italia SpA

Head office: I 20100 Milan. **Postal address:** Casella Postale 3180. **Tel:** 02 30705. **Telex:** 331101 sandvk.
Ownership: Sandvik AB, Sweden.
Products: Stainless steel – seamless tubes and pipes.

Siderurgica Commerciale Santo Stefano SpA

Head office: Via Pisacane 54-56, I 20016 Pero, MI. **Tel:** 02 353 1200. **Telex:** 335287.
Products: Carbon steel – billets.

Italy

Seta Acciai SpA

Head office: 250 Strada Statale, 45 bis, km 41, S, Zeno Naviglio, Brescia. **Tel:** 030 2667321. **Telex:** 304154. **Telecopier (Fax):** 030 2667598.
Management: President – Germano Bocciolone; managers – Alberto Achini (general and managing director), Enrico Nolli (sales), Sergio Troglio (production). **Established:** 1986 (ex Seta SpA, 1980). **Capital:** Lire 30,000,000,000. **Ownership:** Sitai (51%), Seta SpA (49%).
Works:
■ At head office. ***Steelmaking plant:*** One 100 tonne, 50 MVA electric arc furnace (annual capacity 350,000 tonnes). ***Refining plant:*** One Seta 10 MVA ladle furnace (350,000 tonnes); one ASEA vacuum degassing unit (100,000 tonnes); and one ASEA VOD (100,000 tonnes). ***Continuous casting machines:*** Two bloom and round – 4-strand Danieli (tube rounds 270-420mm and squares 260-28mm), 5-strand Voest Alpine (rounds 120-250mm and squares 120-250mm) (150,000 tonnes).
Products: Carbon steel – billets, semis for seamless tubemaking. Stainless steel – billets, semis for seamless tubemaking 120-420mm round. High-chrome steels.

Seta Tubi SpA

Head office: Via Vittorio Emanuele II 39, Roncadelle 25030, Brescia. **Tel:** 030 2780401. **Telex:** 301446 seta i. **Telecopier (Fax):** 030 2782993.
Management: President – Germano Bocciolone; managers – Alberto Achini (general and managing director), Giancarlo Maugredotti (sales), Renato Scaldenzi (production); **Established:** 1985 (ex Seta SpA, 1980). **Capital:** Lire 20,000,000. **No. of employees:** 184. **Ownership:** Sitai (91%), Seta SpA (9%). **Annual capacity:** Finished steel: 100,000 tonnes (pipes).
Works:
■ At head office. Works director: Renato Scaldenzi. ***Tube and pipe mills:*** seamless – One 5,450-tonne Schloemann horizontal hot extrusion press; one 18-stand Kocks stretch reducing mill.
Products: Carbon steel – seamless tubes and pipes, and oil country tubular goods $2\frac{3}{8}$-$10\frac{3}{4}$" dia x 0.141-2" wall thickness. Stainless steel – seamless tubes and pipes, oil country tubular goods. High chrome steels.

Nuova Sidercamuna SpA

Head office: Via Vittoria Emanuelle II 46, I 25040 Berzo Inferiore, BS. **Tel:** 0364 40261. **Telex:** 300458 sidcam.
Products: Carbon steel – reinforcing bars, light channels, medium joists, medium channels.

Siderpo SpA

Head office: V. Carlo Pisacane 40, I 45010 Cavanella Po, RO. **Tel:** 0422 382 235, 0426 93191. **Telex:** 410604.
Products: Carbon steel – longitudinal weld tubes and pipes.

Siderpotenza SpA

Head office: Zona Industriale, Rione Betlemme, I 85100 Potenza. **Tel:** 0971 25911-4. **Telex:** 812357.
Products: Carbon steel – reinforcing bars.

Ferriera Siderscal SpA

Head office: Via Fossa Fresca 35, I 37050 Raldon, VR. **Postal address:** Casella Postale 4. **Tel:** 045 873 0105.
Products: Carbon steel – billets.

Silpa Tubi e Profilati SpA

Head office: Strada del Mare 19, I 10040 Piobesi, TO. **Tel:** 011 9 657 825. **Telex:** 211451 silpa i.
Products: Carbon steel – longitudinal weld tubes and pipes.

Simet – Stà Industriale Metallurgica di Napoli SpA

Head office: Via Stefano Barbato 16, I 80147 Naples. **Postal address:** Casella Postale 381, I 80100 Naples. **Tel:** 7523 233. **Telex:** 710151 simet i.
Products: Carbon steel – reinforcing bars.

Sisva Srl

Head office: Via Brescia 65, I 25012 Calvisano, BS. **Tel:** 030 968226.
Products: Carbon steel – billets.

Sitai SpA

Head office: Via Valsesia 2/8, I 28077 Prato Sesia (NO). **Postal address:** PO Box 9. **Tel:** 0163 850221. **Telex:** 200198 tubval. **Telecopier (Fax):** 0163 850221.
Management: Germano Bocciolone (president), Carlo Colombo (director).
Established: 1966. **Capital:** Lire 11,000,000,000. **Ownership:** Sitindustrie SpA.
Associates: Stainless steel tube and pipe – Nuova CMC (Italy), BSL Tube & Raccords (France) and Zwahlen & Mayr (Switzerland). **Annual capacity:** Finished steel: 35,000 tonnes (welded stainless steel pipe).
Works:
■ At head office. Works director: Carlo Colombo. ***Tube and pipe mills:*** seamless – 6-1,200mm dia. Welded – automatic pipe forming lines up to 10″ dia and 12mm thick; heavy presses for large-diameter heavy wall pipe 10-25mm.
Products: Stainless steel – seamless tubes and pipes up to 38mm dia, longitudinal weld tubes and pipes 6-1,200mm; Monel and titanium pipes.
Sales offices/agents: Domestic market – Sitindustrie SpA, Prato Sesia. Exports – Sitai International BV, Rotterdam, Netherlands.

Snar SpA

Head office: Via Badia 6/8, I 25060 Cellatica, Brescia. **Tel:** 030 312062. **Telex:** 303350 snar.
Management: Giordani Sperandio (director), Franzoni Aldo (sales manager).
Established: 1956. **Capital:** Lire 200,000,000. **No. of employees:** Works – 20. Office – 3. **Annual capacity:** Finished steel: 600 tonnes.
Products: Alloy steel (excl stainless) – hot rolled coil (for re-rolling), hot rolled hoop and strip (uncoated), cold rolled hoop and strip (uncoated).

Tubificio di Solbiate SpA (formerly Fonderie Officine Rovelli)

Head office: Via Rossini 5, I 21058 Solbiate Olona, Varese. **Tel:** 0331 641370. **Telex:** 325346 tubsol. **Telecopier (Fax):** 0331 642289.
Management: President – Guido de Haag; managers – Domenico Rinaldini (general), Elio Pirone (works), Claidio Maffia (export). **Ownership:** SIR Group.
Works:
■ At head office. Works director: Elio Pirone.
Products: Stainless steel – seamless tubes and pipes – 19.5-88.9mm dia, 1.65-7.62mm wall thickness; longitudinal weld tubes and pipes – 6-114.3mm dia, 1.0-4.0mm wall thickness; hollow bars (centrifugally cast); precision tubes and pipes (cold Pilgered).
Sales offices/agents: SIR France, Div. Tubif. Solbiate, 28 Rue Basfroi Bat 3, F 75011 Paris, France (Tel: 00331 4348613. Telex: 042215404 sirfran). General manager: Daniel Asselinne.

Italy

Acciaieria e Ferriere Stefana F.lli fu Girolamo SpA

Head office: Via Bologna 19/21, I 25075 Nave, Brescia. **Tel:** 030 2632061. **Telex:** 300065 stefna i. **Cables:** Ferrostefana Nave.
Management: Mauro Stefana (managing director). **Established:** 1949. **Capital:** Lire 39,330,000,000. **No. of employees:** Works – 545. Office – 55. **Annual capacity:** Raw steel: 500,000 tonnes. Finished steel: 350,000 tonnes.
Works:
■ Ospitaletto (Bs) (Tel: 030 640861. Telex: 300065 stefna). Works director: Osvaldo Bonetti. ***Steelmaking plant:*** Two 120-tonne electric arc furnaces (annual capacity 500,000 tonnes). **Refining plant:** CaSi injection. ***Continuous casting machines:*** One 6-strand Concast billet (100-120-130) (250,000 tonnes); one 4-strand Concast bloom (160-200-220-260) (250,000 tonnes).
■ Via Bologna 19 Nave (Bs). (Tel: 030 2632061, Telex: 300065 stefna). Works director: Camillo Recalcati. ***Rolling mills:*** One 3-high medium section (600 x 1,800-1,500) (annual capacity 200,000 tonnes); one wire rod with 10-stand non-twist finishing block (coil weight 1,300kg) (240,000 tonnes).
Products: Carbon steel – blooms 160-200-220-260mm (output 20,000 tonnes); billets 120-130mm (100,000 tonnes); wire rod 5.50 x 14mm dia (160,000 tonnes); heavy angles 80 x 120 (10,000 tonnes); heavy joists IPN 80 x 220 (60,000 tonnes); heavy channels UPN 80 x 220 (80,000 tonnes); wide-flange beams (heavy) IPE 120 x 200 (20,000 tonnes).

Stilma SpA

Head office: V. Emilia Ovest 96/A, I 41100 Modena. **Tel:** 059 330545. **Telex:** 511211.
Products: Carbon steel – round bars, square bars, flats, hexagons.

Acciaieria e Ferriera di Crema P. Stramezzi & C SpA

Head office: Viale S. Maria della Croce 9, I 26013 Crema, CR. **Postal address:** Casella Postale 71. **Tel:** 0373 59 022. **Telex:** 320617 fercrema. Merchant bars.

Tanga SpA

Head office: 90 Via Giotto da Bondone, I 33079 Sesto al Reghena, PN. **Tel:** 0434 689117. **Telex:** 450005.
Products: Carbon steel – longitudinal weld tubes and pipes.

Terni Acciai Speciali SpA

Head office: Viale Castro Pretorio 122, I 00185 Rome. **Tel:** 06 497951. **Telex:** 611312 terrom.
Capital: Lire 200,000,000. **Ownership:** Terni – Stà per l'Industria e l'Elettricità (Ultimate holding company – Finsider Group).
Works:
■ Viale B. Brin 218, I 05100 Terni, TR (Tel: 0744 4901. Telex: 660008). ***Steelmaking plant:*** Six electric arc furnaces including two 180-tonne. ***Refining plant:*** One AOD. ***Continuous casting machines:*** Two slab s. ***Rolling mills:*** One 4-high reversing heavy plate; one wide hot strip; four Sendimir cold reduction upto 1,300mm wide; two temper/skin pass. ***Other plant:*** Heat treatment plant for electrical steels.
■ Corso Regina Margerita 400, I 10143 Turin (Tel: 011 57351. Telex: 220667). ***Steelmaking plant:*** One 150-tonne electric arc. ***Refining plant:*** One 80-tonne AOD. ***Continuous casting machines:*** One 1-strand curved mould slab. ***Rolling mills:*** cold reduction; cold strip and temper/skin pass. ***Other plant:*** Bright annealing furnace.
Products: Stainless steel – (AISI 300 and 400 series) – slabs, hot rolled band (for re-rolling), hot rolled hoop and strip (uncoated), cold rolled hoop and strip (uncoated), medium plates, heavy plates, hot rolled sheet/coil (uncoated), cold rolled sheet/coil (uncoated), bright annealed sheet/coil.
Sales offices/agents: Stainless Steel Direzione Commerciale: Via C. Battisti 156, I

Terni Acciai Speciali SpA (continued)

20090 S. Maurizio al Lambro, Milan (Tel: 02 254 3661. Telex: 310366). Export sales subsidiaries in West Germany, France, UK and Switzerland. Electrical Sheets Direzione Commerciale: Viale E. Marelli 165, I 20099 Sesto San Giavanni, Milan (Tel: 02 249512. Telex: 310605). Exports through Siderexport, Viale B. Bisajno 2, I 16129 Genoa.

Terni – Stà per l'Industria e l'Elettricità SpA—see Terni Acciai Speciali SpA

Terninoss – Acciai Inossidabili SpA—see Terni Acciai Speciali SpA

Acciaierie del Tirreno SpA

Head office: Zona Industriale di Milazzo, I 98040 Pace del Mela, ME. **Tel:** 090 934 037/77. **Telex:** 980144 adt i.
Products: Carbon steel – heavy joists, wide-flange beams (heavy).

Metallurgica Tognetti SpA

Head office: Via Belfiore 32, I 22053 Lecco, CO. **Tel:** 0341 362622, 0341 369309.
Products: Carbon steel – cold rolled hoop and strip (uncoated), bright wire and wire products.

Profilerie Trentine SpA

Head office: V. del Garda 65, I 38068 Rovereto Trentino. **Tel:** 0464 25484-5. **Telex:** 400671.
Products: Carbon steel – cold roll-formed sections (from strip) up to 5mm.

Attività Industriale Triestine SpA

Head office: Viale Castro Pretono 122, I 00185 Rome. **Tel:** 06 497951. **Telex:** 611312.
Capital: Lire 200,000,000. **Ownership:** Terni – Stà per l'Industria e l'Elettricità SpA (Ultimate holding company – Finsider Group).
Works:
■ Via di Servola 1, I 34145 Trieste, TS. ***Coke ovens:*** one (52 ovens). ***Blast furnaces:*** One 6.15m hearth dia. ***Other plant:*** Two pig casting machines.
Products: Foundry pig iron (all types).

Tubicar SpA

Head office: S.S. Giovi 13, I 15050 Carbonara Scrivia, AL. **Tel:** 0131 872221-2. **Telex:** 210571 tubica i.
Products: Carbon steel – seamless tubes and pipes, cold drawn tubes and pipes. Alloy steel (excl stainless) – seamless tubes and pipes, cold drawn tubes and pipes.

Tubimar Ancona SpA

Head office: Via del Lavoro 6, I 60100 Ancona. **Tel:** 071 51861. **Telex:** 560083. **Telecopier (Fax):** 071 51864.
Management: Luciano Dori (presidente), Carlo Brambilla (amm. delegato), Giovanni Colombo (direttore generale). **Established:** 1985. **Capital:** Lire 2,000,000,000. **No.**

Italy

Tubimar Ancona SpA (continued)

of employees: Works – 135. Office – 15.
Works:
■ At head office. Works director: Lanfranco Mandolesi. ***Tube and pipe mills:*** welded – five $\frac{3}{8}$-6" ERW (annual capacity 100,000 tonnes).
Products: Carbon steel – longitudinal weld tubes and pipes $\frac{3}{8}$-5" (1986 output 26,000 tonnes); oil country tubular goods up to $4\frac{1}{2}$" (4,000 tonnes); galvanized tubes and pipes $\frac{3}{8}$-$6\frac{1}{2}$" (20,000 tonnes).
Sales offices/agents: Via Lentasio 1, I 20122 Milan (Tel: 02874116. Telex: 310553).

Tubinar SpA (formerly Maraldi SpA)

Head office: Via Lenta 1, I 20122 Milan. **Tel:** 02 874116. **Telex:** 310553.
Products: Carbon steel – longitudinal weld tubes and pipes.

Acciaierie Valbruna SpA

Head office: Viale della Scienza Zona Industriale, I 36100 Vicenza. **Tel:** 0444 563200, 0444 563312, 0444 563372. **Telex:** 480090 valvi i.
Products: Carbon steel – (High-speed steels, carbon, medium-alloy and stainless steel bars, forgings – stainless and tool steel) forgings and castings. Round bars, square bars, flats, hexagons, bright (cold finished) bars, light angles. Stainless steel – round bars, square bars, flats, hexagons, bright (cold finished) bars, light angles. Alloy steel (excl stainless) – round bars, square bars, flats, hexagons, bright (cold finished) bars.

Ferriera Valchiese SpA

Head office: Via Fiamme Verdi 20, I 25078 Vestone, BS. **Tel:** 0365 81 294, 0365 81 497. **Telex:** 300687 valkie i.
Products: Carbon steel – reinforcing bars.

Valducci SpA

Head office: Via Emilia 992, I 47032 Bertinoro, FO. **Tel:** 543 448460. **Telex:** 550282.
Products: Carbon steel – longitudinal weld tubes and pipes, galvanized tubes and pipes.

Ferriera Valsabbia SpA

Head office: Via Marconi 13, I 25076 Odolo, Brescia. **Postal address:** Casella Postale 13. **Tel:** 0365 860301-3. **Telex:** 300271. **Cables:** Valsabbia Odolo. **Telecopier (Fax):** 860376.
No. of employees: Works – 240. Office – 20. **Annual capacity:** Finished steel: 300,000 tonnes.
Works:
■ At head office. ***Steelmaking plant:*** One 40-60-tonne electric arc furnace (annual capacity 400,000 tonnes). ***Continuous casting machines:*** Two 3-strand each Concast and Sidercast billet (115 x 115) (500,000 tonnes). ***Rolling mills:*** One 1-stand bar (450mm dia) (250,000 tonnes).
Products: Carbon steel – reinforcing bars.

Acciaieria Valsugana SpA—see Acciaierie e Ferriere Luigi Leali SpA

Acciaierie Venete SpA

Head office: Z.I. Sud, Riviera Francia 9-11, I 35020 Camin, PD. **Tel:** 049 760 566. **Telex:** 431325 accven i.
Products: Carbon steel – billets.

Siderurgica Villalvernia SpA

Head office: Via Novi, I 15050 Villalvernia, AL. **Tel:** 019 482791. **Telex:** 210107.
Products: Carbon steel – medium plates, heavy plates.

Zincor Italia SpA

Head office: Varzi, PV. **Tel:** 383 52168, 383 52267, 383 53241. **Telex:** 315290 zincor.
Products: Carbon steel – electro-galvanized sheet/coil, colour coated sheet/coil.

Ivory Coast

Tôles Ivoire SA – Sté de Galvanisation de Tôles en Côte d'Ivoire

Head office: Rue des Pétroliers, Zone Industrielle de Vridi, Abidjan. **Postal address:** 15 BP 144, Abidjan 15. **Tel:** 35 53 38. **Telex:** 23371 tisa c.
Established: 1970.
Works:
■ At head office. ***Coil coating lines:*** hot-dip galvanizing for 1m wide steel strip (annual capacity 35,000 tonnes).
Products: Carbon steel – hot-dip galvanized sheet/coil, galvanized corrugated sheet/coil. Tisacier (roof cladding).

Japan

Aichi Steel Works Ltd

Head office: 1 Wanowari, Arao-machi, Tokai-shi, Aichi-ken. **Postal address:** PO Box 476. **Tel:** 052 604 1111. **Telex:** 446 9981 aswnag j. **Cables:** Aichi Steel Nagoya. **Telecopier (Fax):** 052 601 0301.
Management: President – Masuo Amano; executive vice president – Hirokazu Arai; senior managing director – Toru Kato; managing directors – Kazushi Nishiyama, Takashi Ito, Hitoshi Ono; directors – Eiji Toyoda, Motoo Matsubara, Hironaka Ito, Isamu Nomaru, Hideyuki Tanahashi, Michihiko Suzuki, Toshiro Yamamoto, Koichi Mori, Takashi Nishikawa. **Established:** 1940. **Capital:** Y 6,900,000,000. **No. of employees:** Works – 2,709. Office – 785. **Associates:** Toyota Motor Corp; Aiko Co Ltd; Tokai Special Steel Co Ltd. **Annual capacity:** Raw steel: 1,548,000 tonnes. Finished steel: 1,250,000 tonnes.
Works:
■ Chita Works (At head office). Works director: Hiroatsu Kizukuri. ***Steelmaking plant:*** Six electric arc furnaces – one 80-tonne, two 50-tonne, one 30-tonne and two 10-tonne (combined annual capacity 1,548,000 tonnes). ***Refining plant:*** One 10-tonne AOD (108,000 tonnes); two ladle furnace – one 80-tonne and one 10-tonne, two vacuum degassing units – one 50-tonne and one 10-tonne (ladle degassing) and one 2-tonne VIM (for R&D) (combined capacity 300,000 tonnes); one 80-tonne RH (780,000 tonnes). ***Continuous casting machines:*** One 1-strand curved mould billet (185 x 185mm) (96,000 tonnes); one 2-strand curved mould bloom (370 x 480mm)

Japan

Aichi Steel Works Ltd (continued)

(780,000 tonnes). ***Rolling mills:*** One 4-stand blooming and billet (bar 80-230mm dia, 95-350mm sq); one bar (40-75mm); one combined wire rod and bar (bar 10-40mm dia, wire rod 5.5-26mm); one light flat bar (width 40-101mm, thickness 4.36-36mm). ***Other plant:*** Forging press, forging machine, ring rolling mill etc (mainly for automobile parts).

■ Kariya Works, 3-chrome 6, Toyada-cho, Kariya-shi, Aichi-ken. ***Rolling mills:*** Two light section – rolling angles (20-100 x 2-13 thick) and stainless steel flats; one light section rolling stainless steel channels; one sheet bar.

Products: Carbon steel – round bars. Stainless steel – wire rod 5.5-26mm; round bars 10-230mm; flats width 12-150mm, thickness 3-25mm; bright (cold finished) bars; light angles width 20-100mm, thickness 3-25mm; light channels height 80-150mm, width 40-100mm, thickness 2-13mm; special light sections (small). Alloy steel (excl stainless) – wire rod 5.5-26mm; round bars; flats width 40-101, thickness 4.5-36mm; bright (cold finished) bars.

Sales offices/agents: Export Dept – Aichi Steel Works Ltd, Nippon Building, 6-2, 2-chrome, Ohte-machi, Chiyoda-ku, Tokyo, Japan. (Tel: 03 270 0851. Fax: 03 245 0649).

Asahi Industries Co Ltd

Head office: Sunshine 60 Bldg 1-1, Higashi-Ikebukuro 3-chome, Toshima-ku, Tokyo 170. **Tel:** 03 987 2165. **Telex:** 272 2662. **Telecopier (Fax):** 03 987 2170.

Management: Chairman – Seiji Tsutsumi; president – Akira Kato. Steel and construction material division – Hedeki Ikeda (managing director), Toshio Shimada (director). **Established:** 1935. **Capital:** Y 2,130,000,000. **No. of employees:** Works – 400. Office – 200. **Subsidiaries:** Affiliated companies – Mikasa Coca Cola Bottling Co Ltd, Mikasa Foods Co Ltd, Seiyu Farm Co Ltd, Japan Packaging Materials Co Ltd; Jobu Tetsudo Co Ltd; Jobu Industries Co Ltd; EAC Co Ltd (Environmental Analysis Center); Hakuro Shuzo Co Ltd; Shuho Echuya Ltd. **Annual capacity:** Raw steel: 300,000 tonnes. Finished steel: 300,000 tonnes.

Works:

■ Saitama Plant, 222 Watarase, Kamikawa-Mura, Kodama-gun, Saitama Pref. (Tel: 02745 2 2711). ***Steelmaking plant:*** 70-tonne electric arc furnace (annual capacity 300,000 tonnes). ***Continuous casting machines:*** 4-strand Mitsubishi billet (120 x 120mm) (300,000 tonnes). ***Rolling mills:*** bar 13-41mm dia (300,000 tonnes).

Products: Carbon steel – reinforcing bars 16-41mm (output 180,000 tonnes); round bars 13-38mm (120,000 tonnes).

Azuma Steel Co Ltd—see Toa Steel Co Ltd

Chubu Steel Plate Co Ltd

Head office: 5-1 Kousu-dori, Nakagawa-ku, Nagoya 454.

Works:

■ Nakagawa. ***Steelmaking plant:*** electric arc. ***Rolling mills:*** medium plate.

Products: Alloy steel (excl stainless) – medium plates.

Daido Steel Co Ltd

Head office: Kogin Bldg, 11-18 Nishiki 1-chome, Naka-ku, Nagoya 460. **Tel:** 052 201 5111. **Telex:** 4422243 daido j. **Cables:** Steel Nagoya. **Telecopier (Fax):** 052 211 1664.

Management: Representative director and chairman – Kunio Hakamada; representative director and president – Hiroji Kinoshita; representative directors and executive vice presidents – Yasuo Ohishi, Toshio Kishida, Tsutomu Horie, Seiji Suzuki; senior managing directors – Isamu Eguchi, Hideo Ohsawa, Takanori Yamamoto; managing directors – Hiroshi Hirata, Toshikazu Kurusu, Kanji Tomita, Shozo Watanabe,

Daido Steel Co Ltd (continued)

Jiro Miyamoto, Tetsuo Kato; directors and senior advisors – Kenichiro Ishii, Masaya Akita; directors – Tatsuo Yamada, Kohsuke Imaizumi, Nobu Ohba, Toshio Adachi, Shigeo Katsuta, Hiromi Ushiyama, Kousaku Maeda, Teikichi Yoshikawa, Kohtaro Yada, Yasukazu Tohdo, Jiro Ishida. **Established:** 1916. **Capital:** Y 17,610,000,000. **No. of employees:** Works – 5,538. Office – 2,542. **Subsidiaries:** Daido Machinery Co Ltd (metal fabricating machines); Daido Stainless Steel Wire Co Ltd (stainless steel wires); Fuji Valve Co Ltd (valves for engines and related parts); Daido-Sprag Ltd (couplings and clutches for industrial machines); Daido Steel (America) Inc; Daido Kogyo Co Ltd (sales of alloy steel bars and wire rod, castings and forgings, narrow strips, and brass products). **Associates:** Daido Inco Alloys Ltd (sales of nickel, nickel alloys and materials for electronic equipment materials); Japan Drop Forge Co Ltd (steel die-forgings); Ohji Steel Co Ltd (carbon steel bars); Tokai Special Steel Co Ltd (alloy steel ingots and billets); NHK Spring Co Ltd (springs of various types, mainly for automobiles); Tohoku Special Steel Works (high-grade special steel rolled products).
Works:
■ Hoshizaki Plant, 2-30 Daido-cho, Minami-ku, Nagoya 457 (Tel: 052 611 2511).Works director: Shohei Shibuya. ***Steelmaking plant:*** Three electric arc furnaces – two 15-tonne (annual capacity 120,000 tonnes) and one 30-tonne (100,000 tonnes). ***Refining plant:*** One 20-tonne AOD (120,000 tonnes); one ESR (1,500 tonnes); one ladle furnace (30-tonne Daido vacuum unit) (100,000 tonnes); one 2-tonne VAR (1,200 tonnes). ***Continuous casting machines:*** One 2-strand curved mould billet (120,000 tonnes). ***Rolling mills:*** blooming, heavy section, and bar comprising 1-stand 2-high 950mm, 1-stand 2-high 730mm and 2-stand 3-high 680mm (384,000 tonnes); one 4-stand 500mm medium section and bar (36,000 tonnes); one wire rod comprising 1-stand 600mm, 5-stand 400/360mm, 6-stand 320mm and monoblock finishing stand (coil weight 1 tonne max) (156,000 tonnes). ***Other plant:*** Bar and wire rod finishing facilities; heat treatment equipment.
■ Chita Plant, 30 Motohama-cho, Tokai-shi, Aichi-ken 477 (Tel: 0562 33 3111). Works director: Jiro Ishida. ***Steelmaking plant:*** Four 70-tonne electric arc furnaces (annual capacity 1,440,000 tonnes). ***Refining plant:*** Two 70-tonne Daido ladle furnace (1,440,000 tonnes); two 70-tonne RH vacuum degassing (1,440,000 tonnes). ***Continuous casting machines:*** One 2-strand curved mould bloom (370 x 480mm) (624,000 tonnes). ***Rolling mills:*** blooming and billet comprising 1-stand 2-high 1,100mm, 1-stand 2-high 820mm, 1-stand 2-high 780mm and 4-stand 700mm (1,680,000 tonnes); light section and bar comprising 2-stand 600mm, 6-stand 480mm, 4-stand 420mm and 4-stand 350mm (504,000 tonnes); wire rod comprising 6-stand 590/420mm roughing, 6-stand 440/360mm 1st intermediate, 8-stand 340/320mm 2nd intermediate and 8-stand 210/158mm finishing (coil weight 2 tonnes) (384,000 tonnes). ***Other plant:*** Die forging and extrusion facilities (72,000 tonnes).
■ Shibukawa Plant, 500 Ishihara, Shibukawa-shi, Gunma-ken 377 (Tel: 02792 2 1900). Works director: Toshiji Ishikawa. ***Steelmaking plant:*** Three electric arc furnaces – two 15-tonne (72,000 tonnes) and one 25-tonne (78,000 tonnes); one 2-tonne induction furnace (4,200 tonnes), one 2-tonne vacuum induction furnace (3,000 tonnes), one 2-tonne plasma induction furnace (1,500 tonnes). ***Refining plant:*** One 20-tonne AOD (72,000 tonnes); four 0.5-6-tonne ESR (9,000 tonnes); two ladle furnace – one 20-tonne (72,000 tonnes) and one 30-tonne Daido vacuum (78,000 tonnes); two 10-tonne VAR (12,000 tonnes). ***Continuous casting machines:*** One 1-strand curved mould billet and bloom (50,000 tonnes). ***Rolling mills:*** bar (medium) and hot strip and hoop comprising 1-stand 2-high 680mm reversing with edger, 2-stand 3-high 550mm,1-stand 2-high 550mm and 5-stand 4-high 320mm (60,000 tonnes). ***Other plant:*** Forging facilities (one 3,500-tonne, one 2,600-tonne and one 1,500-tonne press and a GFM rotary forging machine). Bar peeling and finishing and heat treatment equipment.
■ Kawasaki Plant, 4-1 Yako 2-chome, Kawasaki-ku, Kawasaki-shi 210 (Tel: 044 266 3751). Works director: Yoshitake Nakayama. ***Rolling mills:*** bar and wire rod comprising 3-stand 3-high 500mm, 3-stand 450mm, 6-stand 420/278mm and 6-stand 320/280mm (annual capacity 60,000 tonnes). ***Other plant:*** Bar finishing facilities; heat treatment furnaces and peeling machines.
■ Tsukiji Plant, 10 Ryugu-cho, Minato-ku, Nagoya 455 (Tel: 052 691 5181). Works

Japan

Daido Steel Co Ltd (continued)

director: Nobuo Suzuki. Foundry with one 10-tonne and two 4-tonne electric arc furnaces and one 5-tonne AOD, and eight induction furnaces (2-tonne, etc) (annual capacity 18,000 tonnes). Self-setting moulding and low-pressure moulding equipment (12,000 tonnes), centrifugal casting equipment (2,400 tonnes) and precision casting equipment (400 tonnes).

■ Chita Steel Strip Plant, 39 Motohama-cho, Tokai-shi, Aichi-ken 477 (Tel: 0562 33 3111). Works director: Motokazu Yamaguchi. ***Rolling mills:*** Five cold strip – one 12-high and two 20-high reversing (annual capacity 18,000 tonnes), 4-high and 6-high reversing (33,000 tonnes). ***Other plant:*** Heat treatment equipment and narrow strip finishing facilities.

■ Ohji Plant, 8-5 Kamiya 3-chome, Kita-ku, Tokyo 115 (Tel: 03 901 4161). Works director: Yutaka Okamoto. Strip hardening and tempering furnaces (7,000 tonnes). Polishing, punching and forming equipment.

Products: Carbon steel – wire rod 5.5-32mm; round bars 13-300mm; square bars 7-100mm; flats 3-100mm thick, 13-400mm wide; bright (cold finished) bars up to 250mm dia; cold finished wire rods up to 31.5mm dia; light angles 3-10mm thick, 13-100mm legs; rail accessories; hot rolled hoop and strip (uncoated) 6-18mm thick, 100-400mm wide; cold rolled hoop and strip (uncoated) 0.05-6mm thick, 3-430mm wide. Stainless steel – round bars; square bars; flats; bright (cold finished) bars; hot rolled hoop and strip (uncoated), cold rolled hoop and strip (uncoated). Alloy steel (excl stainless) – wire rod 5.5-32mm; round bars; square bars; flats; bright (cold finished) bars; hot rolled hoop and strip (uncoated); cold rolled hoop and strip (uncoated). Hollow drill steels – 19-38mm hexagonal, 32 and 38mm round. High-speed steel, bearing steel, spring steel, free-cutting steel. Die forgings, forged bars, castings incl precision castings. Output in 1986: Rolling mill products – carbon steels (205,000 tonnes), alloy steels (425,000 tonnes), tool steels (52,000 tonnes), spring steels (89,000 tonnes), bearing steels (118,000 tonnes); stainless steels (137,000 tonnes), heat resistant steels (5,000 tonnes); free-cutting steels (249,000 tonnes); welding electrodes (20,000 tonnes). Forgings – die forgings (65,000 tonnes), forged bars (67,000 tonnes). Castings (7,000 tonnes).

Additional information: The company also has a metal powder division, a titanium division and a magnet products division. The metal powder division produces a wide variety of powder products in special steels, such as stainless and high-speed steel. The main equipment used is water-atomising and gas-atomising equipment. The titanium division produces titanium and titanium alloy products for connecting rods, valves, valve retainers, bolts, glass frames, etc. PPC (Plasma Progressive Casting) equipment is used for electrode production. The magnet products division produces Alnico magnets, ferrite magnets, plastic magnets and magnet applied products.

Daido Steel Sheet Corp (Daido Kohan)

Head office: 2-1 Minamishinmachi 3-chome, Kuise, Amagasaki City, Hyogo Pref 660. **Tel:** 06 488 1181. **Telex:** 524 5498. **Cables:** Amagasaki Steel. **Telecopier (Fax):** 06 488 1365.

Management: President – Hayato Kunitake; senior managing directors – Masumi Shibuya, Yoshinaho Kakihara; managing directors – Michio Mabuchi, Kiyomi Horie, Teruaki Takahashi, Shohei Matsumoto; directors – Haruki Fujii, Yoshikazu Tagawa, Nobuyoshi Nozawa, Yukio Nakajima, Kazuo Migita. **Established:** 1950. **Capital:** Y 2,250,000,000. **No. of employees:** Works – 440. Office – 190. **Ownership:** Nippon Steel Corp (49%). **Subsidiaries:** Daido Steel Container Corp (100%). **Associates:** Daido Kenzai Kogyo Co Ltd (40%); Kansai-Okamura Mfg Co Ltd (22.4%). **Annual capacity:** Finished steel: 310,000 tonnes.

Works:

■ Amagasaki Works (at head office). Works director: Michio Mabuchi. ***Rolling mills:*** Three 4-high reversing cold strip (0.25-1.6mm thickness x 610-1,219mm width) (coil weight 13.6 tonnes) (annual capacity 480,000 tonnes). ***Coil coating lines:*** Two Sendzimir hot-dip galvanizing (360,000 tonnes); three continuous colour coating (210,000 tonnes).

■ Konan Works, 4-2 Takamatsu-cho, Kosei-Cho, Koga-gun, Shiga Pref. (Tel: 07487 5

Daido Steel Sheet Corp (Daido Kohan) (continued)

1311. Fax: 07487 2003). Works director: Michio Mabuchi. Two continuous sandwich panel production lines (annual capacity 2,280,000 sq m).
Products: Carbon steel – hot-dip galvanized sheet/coil 0.25-1.6mm thick, 610-1,219mm wide (output 124,000 tonnes); colour coated sheet/coil 0.27-1.2mm thick, 610-1,219mm wide (162,000 tonnes); zinc-aluminium alloy coated sheet/coil 0.25-1.6mm thick, 610-1,219mm wide (18,000 tonnes). Sandwich wall panels – 22-120mm thick, 910mm wide, and 1.8-10m long (680,000 sq m); sandwich roof panels – 25-45mm thick, 1,000mm wide, and 1.8-15m long (130,000 sq m).
Brands: Evergrip (hot-dip galvanized); Galvalume steel sheet (zinc-aluminium alloy coated); Colorgrip, Vinyever, Everlamiacryl (plastic coated); Isowand (sandwich wall panel); Isodach (sandwich roof panel).
Sales offices/agents: Nihon Bldg, 6-2 Otemachi 2-chome, Chiyoda-ku, Tokyo 100 (Tel: 03 279 0751).

Daiwa Steel Tube Industries Co Ltd

Head office: Nakanoshima Center Building, 22nd Floor 6-2-27 Nakanoshima, Kita-ku, Osaka 530. **Telex:** 63877.
Established: 1932.
Works:
■ Sakai. ***Tube and pipe mills:*** welded – RF incorporating in-line galvanizing.
Products: Carbon steel – longitudinal weld tubes and pipes, galvanized tubes and pipes.
Brands: Daiwa Z pipe.

Funabashi Steel Works Ltd (Funabashi Seito)

Head office: 2-2-1, Minamikaijin, Funabashi City, Chiba Pref. **Tel:** 0474 33 2251.
Management: M. Harada (president). **Established:** 1949.
Works:
■ ***Steelmaking plant:*** electric arc furnaces. ***Continuous casting machines:*** Mitsubishi/Olsson billet – 4-strand and 8-strand. ***Rolling mills:*** bar.
Products: Carbon steel – billets, reinforcing bars, flats.

Godo Steel Ltd (Godo Seitetsu)

Head office: Nishi-Hanshin Building, 3-24 Umeda 2-chome, Kita-ku, Osaka 520. **Tel:** 06 3461031. **Telex:** 5233860 godo j. **Telecopier (Fax):** 06 3461788.
Management: Chairman – Fujio Ohashi; deputy chairman – Seiji Takaishi; president – Takeshi Kato; senior managing directors – Yoshiyuki Kimura, Taro Nakano, Tamio Himuro, Kiyoji Sekine; managing directors – Fumitomo Okamoto, Takeshi Arai, Tarsuo Ohoya; directors and senior advisor directors – Katsuo Naito, Toshio Ninomi, Hiroshi Hirotani, Issei Osawa, Keisuke Mori, Takashi Komae, Kazuyoshi Kunishima, Senji Kanayama, Katsumi Takaishi, Hiroshi Tsukada, Tadami Itoigawa, Sumihiro Atomi.
Established: 1937. **Capital:** Y 3,407,065,250.
Works:
■ Osaka works, 1-2 Nishijima 1-chome, Nishiyodogawa-ku, Osaka 555. ***Sinter plant:*** One Dwight Lloyd. **Blast furnaces.** ***Steelmaking plant:*** Two 41-tonne basic oxygen furnaces. ***Refining plant:*** Two 41-tonne RH vacuum degassing units. ***Continuous casting machines:*** One 6-strand vertical with bending billet; one 4-strand Mannesmann-Demag bloom. ***Rolling mills:*** One universal heavy section; one continuous bar; one 2-strand 25-stand Morgan wire rod.
■ Tokyo works, 10-1 Haneda-Asahicho, Ohta-ku, Tokyo 144. ***Steelmaking plant:*** Two electric arc furnaces – one 60-tonne Lectromelt, and one 70-tonne. ***Continuous casting machines:*** One 4-strand Mannesmann-Demag bloom. ***Rolling mills:*** One semi-continuous bar.
■ Himeji works, 2946 Nakashima, Shikama-ku, Himeji 672. ***Steelmaking plant:*** Two electric arc furnaces – one 60-tonne Lectromelt, and one 70-tonne. ***Refining plant:***

Japan

Godo Steel Ltd (Godo Seitetsu) (continued)

One 70-tonne ladle furnace. ***Continuous casting machines:*** One 4-strand Hitachi bloom. ***Rolling mills:*** One 3-high heavy section.
Products: Steelmaking pig iron. Carbon steel – blooms, billets, wire rod, reinforcing bars, round bars, heavy angles, heavy joists, heavy channels, wide-flange beams (heavy), heavy rails, light rails. OH Bar (deformed bar).

Hitachi Metals Ltd (Hitachi Kinzoku)

Head office: 1-2 2-chrome, Marunouchi, Chiyoda-ku, Tokyo. **Tel:** 211 5311. **Telex:** 4494. **Cables:** Hitachimetal Tokyo.
Management: President – Koji Matsuno; vice presidents – Teiichi Suzumoto, Kenji Takitani; senior executive managing directors – Hiroshi Sakakibara, Takashi Kawai; executive managing directors – Sachiro Fujita, Yoshio Saida; directors – Katsushige Mita, Hideki Harada, Kantaro Nakamura, Yoshiteru Asai, Tomizo Tanaka, Yoshimasa Nagashima, Yoshimi Tokuda, Masazo Takegawa, Shigeru Kimura. **Subsidiaries:** Hitachi Metals International Ltd; Hitachi Metals America, Div of HMI; Hitachi Magnetic Corp; Singapore Foundry & Machinery Co Pte Ltd.
Works:
■ Kyushu Works, Moka Works, Yuasugi Works, Kumagaya Works, Electronic Devices Works, Kuwana Works, Wakanatsu Works, Plant & Equipment Works.
Products: Stainless steel – round bars, square bars, flats, bright (cold finished) bars, cold rolled hoop and strip (uncoated). Alloy steel (excl stainless) – round bars, square bars, flats, bright (cold finished) bars, cold rolled hoop and strip (uncoated). High-speed steels, die steel, and tool steel; rolls, super-high-tensile alloy steel, welding rods; electronic materials and heat-resistant and corrosion-resistant alloys; tools for extrusion, small cutting tools, cold forming rolls, alloy tubes and investment castings and forgings; cemented carbides. Magnetic and electronic materials – Alnico, ceramic and rare-earth cobalt magnets; single crystals; soft magnetic materials. Electronic parts – parts for copiers and office automation machines; parts for magnetic recording heads; switching mode power supplies. Steel castings and forgings.
Brands: YSS.
Sales offices/agents: Hitachi Metals America, Div of HMI – New York, New Jersey, Chicago, Detroit, Los Angeles, Cleveland, Atlanta Sales Office, Houston Sales Office, and Mexico. Hitachi Metals Europe GmbH – Düsseldorf, Milan, and London. Hitachi Metals Singapore Pte Ltd.

Igeta Steel Sheet Co Ltd

Head office: 2 Dejima Nishi-Machi, Sakai City. **Tel:** 0722 47 0111. **Telecopier (Fax):** 0722 47 0123.
Management: President – Akira Funachi; managing director – Sigeji Ueno; directors – Shigeaki Fujii, Toshihiko Nagai, Noboru Hase, Teruo Yasutake, Akira Tabana, Takashi Matsuoka, Shigenobu Matsui; auditors – Shigemitsu Kawabata, Toshikatsu Uesaki.
Established: 1951. **Capital:** Y 1,100,000,000. **No. of employees:** Works – 198. Office – 128. **Ownership:** Sumitomo Metal Industries Ltd (50%), Sumikin Weld Pipe Co Ltd (24%), Sumikin Bussan Kaisha Ltd (14%), etc (ultimate holding company – Sumitomo Metal Industries Ltd). **Subsidiaries:** Izumi Service Co Ltd (100%). **Associates:** Daiki Transportation Co Ltd (transport); Igeta Building Material Co Ltd (dealer); Shinwa Industries Co Ltd (construction of roofing and side walls); **Annual capacity:** Finished steel: cold rolled sheet 96,000 tonnes; colour-coated sheet (including aluminium) 204,000 tonnes.
Works:
■ At head office. ***Rolling mills:*** Single-stand reversing cold reduction (1,050 x 400 x 1,300mm) (max coil weight 17 tonnes) (annual capacity 96,000 tonnes). ***Coil coating lines:*** Three continuous colour coating – No 1 84,000 tonnes (for steel), No 2 12,000 tonnes (for aluminium), No 3 3,108,000 tonnes (for steel)).
Products: Carbon steel – cold rolled sheet/coil (uncoated) – steel foil 0.03-0.17mm thick x 610-1,000mm wide (output 4,200 tonnes); base steel for galvanising 0.15-0.4mm x 610-1,000mm (91,780 tonnes); colour coated sheet/coil 0.27-1.6mm x 610-1,219mm (output 160,000 tonnes (including 8,500 tonnes of aluminium)).

The Japan Steel Works Ltd

Head office: 1-2 Yurakucho, 1-chome, Chiyoda-ku, Toyko. **Tel:** 03 501 6111. **Telex:** 24256 jsw. **Cables:** Seikosho Tokyo. **Telecopier (Fax):** 03 504 0727.
Management: President – Naohiko Yagai; executive vice president – Shigeru Sawabe; directors – Tsuneo Yamaguchi (senior managing), Masaya Noji (senior managing), Yasuo Oshima (managing), Hiroshi Kawakami (managing), Tsutomu Kadowaki, Hiroyuki Tokushige, Yoshitaka Yoshida, Hitoshi Emachi, Nobuo Itho, Tatsuo Yoshida. **Established:** 1907. **Capital:** Y 18,405,000,000 (March 31 1987). **No. of employees:** 5,964. **Ownership:** Mitsui Group. **Subsidiaries:** Japan Steel Works America Inc (sales representative); JSW (Pte) Ltd (manufacturing works, Singapore).
Works:
■ Muroran. ***Steelmaking plant:*** Four electric arc furnaces, one 10-tonne, one 25-tonne, one 100-tonne, and one 120-tonne. ***Refining plant:*** One ESR; one ladle furnace; and one VOD. ***Rolling mills:*** One 4-high reserving 5,300mm heavy plate. ***Other plant:*** Amstrad pressure pouring unit for slabs up to 300mm thick, 2.2m wide, and 6.15m long. Forging and bending presses.
■ Hiroshima. ***Steelmaking plant:*** One electric arc and induction furnaces. ***Other plant:*** Forging hammers.
Products: Carbon steel – ingots, medium plates, heavy plates. Stainless steel – ingots, medium plates, heavy plates, clad plates. Steel castings and forgings.
Additional information: The company also has works at Yokohama and Tokyo.

Kansai Steel Corp (Kansai Seiko)

Head office: 5 Shiohama-cho, Sakai City, Osaka.
Management: T. Kimura (president). **Established:** 1951.
Works:
■ Sakai City. ***Steelmaking plant:*** electric arc. ***Continuous casting machines:*** One 2-strand Mitsubishi bloom. ***Rolling mills:*** bar. Bulb flats.

Kawasaki Steel Corp

Head office: Hibiya Kokusai Bldg, 2-3 Uchisaiwaicho 2-chome, Chiyoda-ku, Tokyo 100. **Tel:** 03 597 3111. **Telex:** 0222 3673, 0222 3833. **Cables:** Riverstcorp Tokyo. **Telecopier (Fax):** 03 595 4868, 03 595 4869.
Management: Chairman – Eiro Iwamura; president – Yasuhiro Yagi; executive vice presidents – Toyohiko Ohta (technical research), Kinji Ibaraki (sales, purchasing, physical distribution), Shinobu Tosaki (steel business, Tubarao Project, steel technology), Shigeru Koino (general affairs, personnel, overseas administration, engineering); senior managing directors – Einosuke Murakami (industrial relations, subcontract planning, corporate planning and accounting), Masashi Kawana (general superintendent, Chiba Works); managing directors – Kazutaka Tsutsumi (engineering), Toshio Kuchiki (finance), Shoji Iwaoka (general superintendent, Chita Works), Toshio Ueno (general superintendent, Hanshin Works), Eiji Shibata (sales), Nobuo Ohashi (technical research), Yasuhiko Takagi (industrial relations, subcontract planning), Tadaaki Yanazawa (general superintendent, Mizushima Works), Kenzo Monden (corporate administration, personnel, general affairs); directors – Toshio Kogure (assistant general superintendent, Mizushima Works), Eichi Komoto (sales, steel business planning), Takao Yamada (Tubarao Project, purchasing), Makoto Ohshima (systems planning and data processing), Hiroshi Narahara (chemicals), Haruaki Yanagisawa (steel technology), Zenichiro Yokota (finance), Nobutsune Hirai (new business), Sinji Komiyama (sales), Makoto Saigusa (steel business planning); Mitsunori Suzuki (engineering), Jun Nagai (assistant general superintendent, Mizushima Works), Hidehiko Kimishima (assistant general superintendent, Chiba Works). **Established:** 1950. **Capital:** At March 31, 1987: Y 146,664,000,000. **No. of employees:** At March 31, 1987: Works – 20,010. Office – 4,355. **Subsidiaries:** Trading services: Kawasho Corp. Raw materials and supplies: Kawasaki Refractories Co Ltd, Kawatetsu Mining Co Ltd, Mizushima Ferro-Alloy Co Ltd, Mizushima Joint Thermal Power Co Ltd, Philippine Sinter Corp. Steel fabrication: Chita Pipe Fitting Co Ltd, Daiwa Steel Corp,

Japan

Kawasaki Steel Corp (continued)

Hanshin Metal Machining Co Ltd, Kawaden Co Ltd, Kawasaki Thermal Systems Inc, Kawatetsu Container Co Ltd, Kawatetsu Electrical Steel Co Ltd, Kawatetsu Galvanizing Co Ltd, Kawatetsu Instruments Co Ltd, Kawatetsu Kizai Kogyo Co Ltd, Kawatetsu Kozai Kogyo Kaisha Ltd, Kawatetsu Metal Industry Co Ltd, Kawatetsu Steel Products Corp, Kawatetsu Steel Tube Co Ltd, Kawatetsu Wire Products Co Ltd, Kohnan Steel Center Co Ltd, River Building Materials Co Ltd, River Steel Co Ltd, Shikoku Iron Works Co Ltd, Tohoku Steel Corp, Toyohira Seiko Co Ltd. Others: California Steel Industries Inc, Chiba Riverment & Cement Corp, Kawasaki Enterprises Inc, Kawasaki Steel Systems R&D Corp, Kawasaki Steel Techno-Research Corp, Kawatetsu Civil Engineering Co Ltd, Kawatetsu Electric Engineering Co Ltd, Kawatetsu Engineering Ltd, Kawatetsu Real Estate Co Ltd, Kawatetsu Transportation Co Ltd, Kawatetsu Warehouse Co Ltd, KX Corp, Kobe Catering Co Ltd, Mizushima Riverment Corp, NBK Corp, Nihon Semiconductor Inc, Nihon Ugimag Corp, Nihon Yupro Co Ltd, Rifine Co Ltd, Saikai Industry Co Ltd. **Annual capacity:** At March 31, 1987 – Pig iron: 10,951,279 tonnes. Raw steel: 9,885,864 tonnes. Finished steel (rolled): 8,146,394 tonnes.

Works:

■ Chiba Works, 1 Kawasaki-cho, Chiba 260. Tel: 0472 64 2111. Telex: 3722212 kawatetsu. Works director: Masashi Kawana. ***Coke ovens:*** Three Otto (compound) batteries (208 ovens) (annual capacity 1,350,500 tonnes); three Carl Still batteries (260 ovens) (2,883,500 tonnes). ***Sinter plant:*** Three Dwight Lloyd – 80 sq m strand (1,095,000 tonnes), 203 sq m strand (2,555,000 tonnes), 210 sq m strand (2,555,000 tonnes). ***Blast furnaces:*** No 2 8.4m hearth dia (913,000 tonnes), No 3 9.5m hearth dia (1,278,000 tonnes), No 4 9.4m hearth dia (1,278,000 tonnes), No 5 11.1m hearth dia (1,898,000 tonnes), No 6 14.1m hearth dia (3,650,000 tonnes). ***Steelmaking plant:*** Seven basic oxygen converters – two 85-tonne K-Bop, three 150-tonne LD-KG and two 230-tonne Q-Bop (8,500,000 tonnes). ***Refining plant:*** Three RH vacuum degassing units – No 1 85-tonne, No 2 150-tonne, No 3 230-tonne;

Kawasaki Steel Corp (continued)

one 50-tonne Elo-Vac vacuum decarburisng and degassing unit. ***Continuous casting machines:*** Three slab – No 1 1-strand Concast (480,000 tonnes), No 2 2-strand Voest (1,800,000 tonnes), No 3 Mannesmann-Demag (2,400,000 tonnes). ***Rolling mills:*** Two universal slabbing – No 2 universal (3,960,000 tonnes), No 3 (3,600,000 tonnes); one 4-high 166" reversing heavy plate; two wide hot strip – No 1 6-stand 56" semi-continuous (1,800,000 tonnes), No 2 7-stand 80" continuous (3,600,000 tonnes); five cold reduction – No 1 5-stand 56" (600,000 tonnes), No 2 6-stand 56" (930,000 tonnes), No 3 4-stand 71" and 80" reversing (1,440,000 tonnes), 3-stand 4-high 49" double cold reduction (180,000 tonnes). ***Tube and pipe mills:*** welded – one 64" UOE (840,000 tonnes). ***Coil coating lines:*** Two electrolytic tinning – No 1 Halogen (0.1-0.6mm thick, up to 965mm wide) (120,000 tonnes), No 2 Halogen (0.08-0.4mm thick, up to 1,067mm wide) (300,000 tonnes); one tin-free steel (ECCS) (0.1-0.6mm thick, up to 1,067mm wide) (180,000 tonnes); one hot-dip galvanizing (0.25-2.3mm thick, up to 1,270mm wide) (150,000 tonnes); one electro-galvanizing (0.4-1.6mm thick, up to 1,550mm wide) (240,000 tonnes). ***Other plant:*** Calcined hard-ferrite powder plant (300 tonnes/month).

■ Mizushima Works, 1 Kawasaki-dori, Mizushima, Kurashiki City, Okayama 712. (Tel: 0864 44 3111. Telex: 5933432 kurashiki). Works director: Tadaaki Yanazawa. ***Coke ovens:*** Six Carl Still (compound) batteries (465 ovens) (annual capacity 5,505,000 tonnes). ***Sinter plant:*** Four Dwight Lloyd – 183 sq m strand (2,044,000 tonnes), 250 sq m strand (2,811,000 tonnes), 300 sq m strand (3,358,000 tonnes), 410 sq m strand (4,526,000 tonnes). ***Blast furnaces:*** No 1 10m hearth dia (1,643,000 tonnes), No 2 11.8m hearth dia (2,190,000 tonnes), No 3 12.4m hearth dia (2,665,000 tonnes), No 4 14.4m hearth dia, 3,650,000 tonnes). ***Steelmaking plant:*** Six basic oxygen furnaces – three 180-tonne LD-KG and three 250-tonne K-Bop (12,000,000 tonnes); one 30-tonne electric arc furnace (72,000 tonnes). ***Refining plant:*** 100-tonne Asea-SKF; 50-tonne ESR; three RH vacuum degassing units – No 1 180-tonne, No 2 250-tonne, No 3 180-tonne; one 100-tonne ESHT (electro-slag hot-topping unit). ***Continuous casting machines:*** 6-strand Concast bloom (1,770,000 tonnes); three slab – 2-strand Concast (1,200,000 tonnes), 2-strand Mannesmann-Demag (2,500,000 tonnes), 2-strand Concast (2,000,000 tonnes); one 4-strand Concast bloom/beam blanks (1,000,000 tonnes). ***Rolling mills:*** One universal slabbing (5,200,000 tonnes); one universal heavy section (wide flange beam) (640,000 tonnes); one continuous bar (360,000 tonnes); one wire rod and billet (1,440,000 tonnes); one bar/rod (30,000 tonnes/month); one 4-high 216" reversing heavy plate (1,880,000 tonnes); one 7-stand 90" continuous wide hot strip (4,800,000 tonnes); two cold reduction – No 1 5-stand 68" continuous (1,584,000 tonnes), No 2 3-stand (1,080,000 tonnes). ***Coil coating lines:*** One electro-galvanizing (30,000 tonnes/month). ***Other plant:*** Steel foundry and forge. Steel powder plant (5,000 tonnes/month). Ingot mould plant (168,000 tonnes).

■ Chita Works, 1-1 Kawasaki-cho, Honda City, Aichi 475. (Tel: 0569 21 5151. Telex: 4563 622 kwst cj). Works director: Shijo Iwaoka. ***Steelmaking plant:*** Three electric arc furnaces – two 15-tonne, one 10-tonne (annual capacity 108,000 tonnes). ***Tube and pipe mills:*** seamless – one small dia (21.3-168.3mm od) (444,000 tonnes), one medium dia (177.8-426.0mm od) (468,000 tonnes). Welded – No 1 small dia ERW (50.8-168.3mm od) (120,000 tonnes), No 2 small dia ERW (21.3-76.3mm od) (36,000 tonnes), No 3 small dia ERW (21.3-89.1mm od) (34,000 tonnes), No 1 medium dia ERW (165.2-508mm od) (240,000 tonnes), No 2 medium dia ERW (267.4-660.4mm od) (180,000 tonnes), No 1 spiral weld 406.4-1,524mm od) (48,000 tonnes), No 2 spiral weld (406.4-1,524mm od) (48,000 tonnes), one butt-weld (21.3-144.3mm od) (288,000 tonnes). ***Other plant:*** Cold drawing equipment. Steel foundry. Iron foundry. Small-dia polye-coated pipe plant (7,200 tonnes), medium-dia polye-coated pipe plant (2,400 tonnes).

■ Hanshin Works (Fukiai District), 2-88 Wakihamakaigan-dori, Kobe 651. (Tel: 078 221 4141. Telex: 5622 130 kwst kj). Works director: Toshio Ueno. ***Steelmaking plant:*** One 40-tonne electric arc furnace (annual capacity 100,000 tonnes). ***Rolling mills:*** One heavy section (medium flange beam) (360,000 tonnes); five Sendzimir cold reduction (432,000 tonnes). ***Coil coating lines:*** One electro-galvanizing (0.26-2.5mm

Kawasaki Steel Corp (continued)

thick, 1,829mm wide) (134,000 tonnes); one colour coating (0.24-1.6mm thick, 1,250mm wide) (60,000 tonnes). ***Other plant:*** Nishinomiya District: Rolling mills: Three Sendzimir mills.

Products: Steelmaking pig iron (output for fiscal year ended 31.3.87 – 10,951,279 tonnes); iron powder. Raw steel (9,885,864 tonnes). Carbon steel – wire rod 5.5-38mm dia; round bars 12-350 mm dia; heavy joists 150 x 75mm-900 x 300mm, heavy channels and wide-flange beams (heavy) (100 x 100mm-400 x 400mm and 150 x 100-500 x 300mm) (842,000 tonnes); sheet piling (U, Z and flat types); medium plates and heavy plates (4.5-200mm thick, up to 5,300mm wide, up to 35,000mm long); hot rolled sheet/coil (uncoated) (1.2-25.4mm thick, up to 2,200mm wide, max coil weight 45 tonnes), cold rolled sheet/coil (uncoated) (0.1-3.2mm thick, up to 1,880mm wide, max coil weight 40 tonnes), hot-dip galvanized sheet/coil and electro-galvanized sheet/coil (0.23-3.25mm thick, up to 1,829mm wide), colour coated sheet/coil, electrical grain oriented silicon sheet/coil and super oriented, electrical non-oriented non-silicon sheet/coil, electrolytic tinplate (single-reduced) and electrolytic tinplate (double-reduced) (0.12-0.60mm thick, up to 1,067mm wide), and tin-free steel (ECCS) (0.12-0.60mm thick, up to 965mm wide) (combined annual capacity 7,305,000 tonne); seamless tubes and pipes (21.3-426mm od, 2-40.5mm wall thickness), longitudinal weld tubes and pipes (ERW pipe 19-660.4mm od, 1.2-16mm wall thickness, and butt-weld pipe 21.3-114.3mm od, 2-8.6mm wall thickness), spiral-weld tubes and pipes (406.4-2,032mm od, 4-22mm wall thickness), large-diameter tubes and pipes and UOE pipe (508-1,626.6mm od, 6.35-38.1mm wall thickness) (1,273,801 tonnes). Stainless steel – hot rolled sheet/coil (uncoated) and cold rolled sheet/coil (uncoated) (0.2-8mm thick, up to 1,550mm wide). Heavy columns 400 x 400mm, 500 x 500mm. Castings and forgings. Corrugated pipe and liner plate. Welding electrodes.

Brands: River Ace (high tensile strength steel plates); River Ten (atmospheric corrosion resistant steels); Kap Clad (abrasion-proof clad steel plates); Mariwel (seawater corrosion resistant steels); River Lite (extra low-carbon stainless steels); River Zinc (electrogalvanized steel sheets); River Color (prepainted galvanized steel sheets); RG-H (high magnetic induction grain oriented silicon steels); KPP (polyethylene coated steel pipe); KLP (PVC-lined steel pipe); KFP (polyethylene powder-lined steel pipe); KSP (sheet piles); River Con (steel bars for concrete reinforcement); CHRX-BH (high strength, deep drawable automotive sheet steel); River Ex Zinc (corrosion resistant electroplated automotive steel sheet); Riverwelt (tin-coated steel for beverage cans); River Hi-Zinc Super (double layered electro-galvanized sheet steel for automotive applications); Phoenix (corrosion resistant high-performance paint); Laser Mirror (sheet steel with high surface reflectivity); Luminacolor (coloured stainless steel sheet).

Sales offices/agents: USA – Kawasaki Steel America Inc, Park Avenue Plaza, 55 East 52nd Street, New York, NY 10055 (Tel: 212 935 8710. Telex: 423786 kwny ui. Cables: Kawatetsu New York); 6375 Texas Commerce Tower, Houston, TX 77002 (Tel: 713 654 0031. Telex: 076 2672 kawatetsu hou); 444 South Flower Street, Suite 1590, Los Angeles, CA 90071-2991 (Tel: 213 629 9005. Telex: 067 4330 kawatetsu lsa. Cables: Kawatetsu Los Angeles). West Germany – Kreuzstrasse 20, D 4000 Düsseldorf 1 (Tel: 211 350411. Telex: 08582399 kwst d. Cables: Kawatetsu Dusseldorf). UK – 7th Floor, 1/2 Finsbury Square, London EC2A 1AJ (Tel: 01 638 1048; 01 920 0428. Cables: Kawatetsu London EC2). Thailand – Dusit Thani Office Bldg., Room 310, Rama 4 Road, Saladaeng Circle, Bangkok (Tel: 233 2915, 0705. Telex: 82878 kwst th. Cables: Kawatetsu Bangkok). Singapore – Unit 1303, 13th Floor, Hong Leong Bldg, 16 Raffles Quay, Singapore 1 (Tel: 2201174; 2204696. Telex: 21952 riversp rs). Philippines – 11th Floor, Allied Bank Center, 6754 Ayala Avenue, Makati, Metro Manila (Tel: 86 78 68; 87 57 96. Telex: 7564292 kwst pn. China – Beijing Minzu Hotel, Room 2801, Fuxingmen Nei Daijie 51, Beijing (Tel: 65 8541. Telex: 22745 kwstb cn). Brazil – Kawasaki Steel Comercio e Siderurgia Ltda, Av. Nilo Pecanha 50, Grupo 2109, Rio de Janeiro, RJ (Tel: 21 262 1232. Telex: 2121897 kste br. Cables: Kawatetsu Rio); Av Paulista 1159, 8 andar, conj 804, São Paulo, SP (Tel: 11 288 7728, 9631. Telex: 011 32842 kste br).

Kawatetsu Steel Tube Co

Head office: 1 Niihama-cho, Chiba-shi, Chiba 260.
Ownership: Kawasaki Steel Corp.
Products: Carbon steel – spiral-weld tubes and pipes.

Kobe Steel Ltd (Kobe Seikosho)

Head office: 3-18 Wakinohamacho 1-chome, Chuo-ku, Kobe 651. **Postal address:** Tokyo office – Tekko Bldg, 8-2 Marunouchi 1-chome, Chiyoda, Tokyo 100. **Tel:** Kobe – 078 251 1551; Tokyo – 03 218 7111. **Telex:** Kobe – 5622177 kobstl j; Tokyo – 2223601 kobstl j. **Cables:** Kobe – Kobe Steel Kobe; Tokyo – Kobe Steel Tokyo. **Telecopier (Fax):** Kobe – 078 232 3459; Tokyo – 03 287 2278.
Management: Chairman of the board – Yugoro Komatsu; director and president – Sokichi Kametaka; directors and executive vice presidents – Osamu Kataoka, Soichiro Yoshimura, Masato Watanabe; directors and senior executive officers – Chiyuki Honda, Tomokazu Godai, Toshimasa Saishoji, Chikahisa Nagai, Akio Suzuki, Masayuki Tsuchimoto; directors and executive officers – Tadashi Kawaguchi, Kennosuke Abe, Tadashi Moriyasu, Tetsuro Imai, Yoshihiro Izutani, Yukitoshi Hirose, Masumi Sato, Kazumasa Taguchi, Yoshiki Miyaji; director and senior advisor – Fuyuhiko Maki; directors – Takahiko Shiraishi, Shoichi Tokuda, Kaizo Okamoto, Yoshiaki Maeda, Makoto Nishizaki, Terumasa Ando, Kazuro Sasagawa, Mizune Sakato, Akihiro Sannomiya, Tsunehiko Okada, Shigeru Oda, Kotaro Shimo, Takuya Negami, Masahiro Kumamoto; auditors – Toshio Kobayashi, Shozo Uematsu, Michiaki Doi.
Established: 1911. **Capital:** Y 122,763,000,000 (March 31, 1987). **No. of employees:** 26,151 (March 31, 1987). **Subsidiaries:** (These include associated companies) Domestic – Ark System Co Ltd (60%) (systems engineering); Daido Light Metal Industry Co Ltd (74.7%) (aluminium alloy wire); Daiichi Taika Renga Co Ltd (95.2%) (fireclay bricks); Hanshin Yosetsukizai Co Ltd (75%) (welding consumables); International Training Service Co Ltd (66.5%) (training for plant operations); Japan Parallel Wire Co Ltd (50%) (parallel wire stranding); Jastic Kansai Co Ltd (50%) (building security); Kanmon Sohgo Service Ltd (100%) (employee housing, security service, internal mail); The Kansai Coke & Chemicals Co Ltd (39%) (coke, coke-oven gas); Kobe General Service Co Ltd (100%) (landscape gardening); Kobe Kotobuki Iron Co Ltd (31.4%) (planning and construction of plants); Kobelco Construction Machinery Co Ltd (100%) (sales of construction machinery); Kobelco PR Center Co Ltd (100%) (advertising, corporate publications, film and video production); Kobelco Research Institute Inc (100%) (material and physical analysis, environmental studies); Kobelco Telecommunications Technology Co Ltd (100%) (communication services and sales of communications equipment); Kokoku Steel Wire Ltd (18.6%) (wire ropes, wire strands, and wire); Leadmikk Ltd (50%) (plating of electronic parts); Moji Pipe Mfg Co Ltd (80%) (copper tubes); Niko-Aluminium Industries Ltd (70%) (aluminium painting); NHK Steel Co Ltd (25.1%) (special steel bars and sections); Nippon Air Brake Co Ltd (25.25%) (brake control systems, hydraulic equipment, remote control systems for main engines); Nippon Koshuha Steel Co Ltd (49.6%) (ball-bearing steel, tool steel, stainless steel, small tools); Nishi-Nippon Steel Ltd (38.1%) (round bars); Osaka Chain & Machinery Ltd (46.3%) (gears, gear reducers, gear couplings); Osaka Titanium Co Ltd (12.7%) (titanium sponge, semi-conductor grade silicon); Sakai Steel Sheet Works Ltd (62.5%) (cold rolled, galvanized and prepainted sheets); Sanwa Tekko Co Ltd (34.6%) (hot and cold rolled sheets, plates); Security Service Co Ltd (60%) (guards and maintenance for buildings); Shinagawa Rozai Co Ltd (40%) (fireclay bricks); Shinko Alfresh Co Ltd (40%) (repair of sashes and walls); Shinko Bolt Ltd (60%) (high-tensile bolts and nuts); Shinko Computer System Co Ltd (100%) (planning of computer systems); Shinko Electric Co Ltd (33.17%) (heavy electric machinery, material handling machinery, information control machinery); Shinko Engineering Co Ltd (48.8%) (diesel engines); Shinko Goods Service Co Ltd (70%) (stationery and books); Shinko Human Create Co Ltd (100%) (consulting and training); Shinko Industrial Co Ltd (83%) (gas cylinders); Shinko Inspection & Service Co Ltd (100%) (testing and inspection of metals, machinery and equipment); Shinko Kaiun Co Ltd (53.3%) (marine transport); Shinko Kakogawa Koun Co Ltd (40%) (marine transport); Shinko Kenzai Ltd (47.1%) (guard fences, gratings); Shinko Kohan Kako Ltd (40%) (hot and cold rolled sheets, plates); Shinko Kosan Ltd (44.5%) (real estate services, insurance agency); Shinko

Japan

Kobe Steel Ltd (Kobe Seikosho) (continued)

Lease Co Ltd (60%) (leasing); Shinko-North Co Ltd (95%) (aluminium sashes); Shinko Pfaudler Co Ltd (91.7%) (process equipment, cooling towers, sewage treatment, water treatment, waste treatment); Shinko Plant Construction Co Ltd (100%) (planning and construction of plants); Shinko Research Co Ltd (100%) (information research, technology transfer); Shinko Rikuun Co Ltd (48.8%) (overland transport); Shinko Sugita Seisen Ltd (19.7%) (cold-heading wires); Shinko Technology & Engineering Co Ltd (100%) (design and development of machinery); Shinko Travel Service Co Ltd (70%) (travel agency); Shinko Wire Co Ltd (52.6%) (PC wire, wire rope, stainless steel wire); Shinsho Corp (49.61%) (trading – iron and steel, non-ferrous metals, machinery, construction equipment); Shinwa Wood Works Ltd (50%) (packing with wood); Sun Aluminium Industries Ltd (65.4%) (aluminium foil); Sunrode Ltd (49.7%) (welding electrodes); Tokyo Copper Tube Co Ltd (52.3%) (copper tubes, aluminium alloy tubes); Yutani Heavy Industries Ltd (99.1%) (construction equipment).
Overseas – Chiaphua Shinko Copper Alloy Co Ltd, Hong Kong (30%) (brass strip); Daehan Chemical Machinery Manufacturing Co Ltd, South Korea (28.6%) (chemical equipment); Durakut International Corp, USA (66.67%) (trading – cutting tools); Earth Development (S) Pte Ltd, Singapore (100%) (aggregate for concrete structures); Harnischfeger Corp, USA (8.3%) (construction machinery, mining equipment); Harnischfeger of Australia Pty Ltd, Australia (25%) (import and servicing of construction machinery and mining equipment); Kobe Alumina Associates (Australia) Pty Ltd, Australia (40%) (bauxite mining, alumina refining, aluminium ingot marketing); Kobe Development Corp, USA (100%) (research and development in advances technologies); Kobe International (S) Co Pte Ltd, Singapore (100%) (trading – construction equipment, heavy machinery, air compressors, marine products and services); Kobe Leadframe Singapore Pvt Ltd, Singapore (100%) (production and marketing of leadframes); Kobe Steel America Inc, USA (100%) (marketing of Kobe Steel products); Kobe Steel Europe Ltd, UK (100%) (marketing of Kobe Steel products); Kobe Steel (Newcastle) Pty Ltd, Australia (99.9%) (technical consultation); Kobe Welding (Singapore) Pte Ltd, Singapore (62%) (welding electrodes); Kobelco America Inc, USA (100%) (trading – construction equipment); Kobelco Middle East (EC), Bahrain (100%) (business information research, marketing of hydraulic excavators and industrial machinery); Midrex Corp, USA (100%) (engineering and construction); Midrex International BV, Netherlands (100%) (technology licensing); Qatar Steel Co Ltd, Qatar (20%) (reinforcing bars); Singapore Kobe Pte Ltd, Singapore (70%) (copper tubes); Thai Kobe Welding Co Ltd, Thailand (33%) (welding electrodes); Titan Steel & Wire Co Ltd, Canada (28.9%) (steel wire, PC wire, nails); Zhuoshen Non-ferrous Metals Plant & Equipment Co Ltd, China (75%) (design engineering, manufacture and procurement, installation of equipment for non-ferrous metal industries). **Annual capacity:** 1985 fiscal – Pig iron: 9,321,000 tonnes. Raw steel: 8,763,000 tonnes. Finished steel: 9,705,000 tonnes.
Works:
■ Kakogawa Works, 1 Kanazawa-cho, Kakogawa. ***Sinter plant:*** One 65.5m 4-strand Dwight Lloyd (annual capacity 2,847,000 tonnes). ***Iron ore pelletizing plant:*** Grate kiln type – No 1 (2,190,000 tonnes), No 2 (2,190,000 tonnes). ***Blast furnaces:*** No 1 11.9m hearth dia, 3,090 cu m (2,500,000 tonnes); No 2 13.2m hearth dia, 3,850 cu m (3,120,000 tonnes); No 3 14.2m hearth dia, 4,500 cu m (3,650,000 tonnes). ***Steelmaking plant:*** Three 235-tonne LD-OTB basic oxygen converters (6,410,000 tonnes). ***Refining plant:*** One 240-tonne LF ladle furnace unit (max 60,000 tonnes/month); one 240-tonne DH vacuum degassing unit (max 60,000 tonnes/month), two 240-tonne RH degassing units (max 330,000 tonnes/month). ***Continuous casting machines:*** One 4-strand bloom, 380 x 600mm (max 125,000 tonnes/month); one 2-strand slab, 230 x 900mm to 250 x 1,900mm (max 130,000 tonnes/month); one 2-strand slab, 230 x 800mm to 280 x 2,100mm (max 150,000 tonnes/month). ***Rolling mills:*** No 1 universal slabbing, 80-450mm (3,600,000 tonnes); No 2 2-high reversing blooming and continuous billet, 115-155mm sq (175,000 tonnes/month); No 8 continuous wire rod, 5.5-16mm (115,000 tonnes/month); 4-high reversing heavy plate, 4.5-350mm (1,620,000 tonnes); 86" continuous wide hot strip, 1.0-25.4mm (3,000,000 tonnes); 5-stand tandem cold strip, 0.2-3.2mm (130,000 tonnes/month); reversing cold strip, 0.1-2.3mm (14,000 tonnes/month). ***Coil coating lines:*** Continuous hot-dip galvanizing , 0.3-3.2mm (144,000 tonnes); electro-

Kobe Steel Ltd (Kobe Seikosho) (continued)

galvanizing – No 1 0.3-3.2mm (108,000 tonnes), No 2 0.3-2.3mm (25,000 tonnes/month). ***Other plant:*** Hot metal pre-treatment plant – flux injection and oxygen top blowing in torpedo car (140,000 tonnes/month). Continuous annealing and pickling line (45,000 tonnes/month); Heroult-type ferro-alloy furnaces – No 1 (61,200 tonnes F Mn-H), No 2 (31,700 tonnes Si-Mn).

■ Kobe Works, 2 Nadahama-Higashi, Nada-ku, Kobe. ***Sinter plant:*** Two 2.3 x 35m strand Dwight Lloyd (annual capacity 905,000 tonnes each). ***Iron ore pelletizing plant:*** Grate kiln type (1,095,000 tonnes). ***Blast furnaces:*** No 1 7.2m hearth dia, 904 cu m (730,000 tonnes); No 2 9.1m hearth dia, 1,618 cu m (1,310,000 tonnes); No 3 9.5m hearth dia, 1,845 cu m (1,500,000 tonnes). ***Steelmaking plant:*** Three 80-tonne LD-OTB basic oxygen converters (2,300,000 tonnes); two Daido Lectromelt electric arc furnaces – one 60-tonne (88,000 tonnes), one 70-tonne (183,000 tonnes). ***Refining plant:*** One 90-tonne LR unit with arc heating and vacuum degassing (40,000 tonnes/month); two 90-tonne Asea-SKF units (50,000 tonnes/month). ***Continuous casting machines:*** One 4-strand billet, 155 x 155mm (40,000 tonnes/month); one 2-strand bloom, 300 x 430mm (50,000 tonnes/month). ***Rolling mills:*** No 3 2-high reversing blooming and continuous billet, 85-155mm (100,000 tonnes/month); continuous bar producing bar 12.7-55mm and wire rod 12.7-36mm (45,000 tonnes/month); No 7 continuous wire rod, 5.5-50mm (45,000 tonnes/month); continuous bar, bar 20-120mm and wire rod 20-55mm (75,000 tonnes/month). ***Other plant:*** Hot metal pre-treatment plant – flux injection and oxygen top blowing in LD converter (80,000 tonnes/month).

■ Amagasaki Works, 2-23 Ohama-cho, Amagasaki. ***Blast furnaces:*** No 2 7.2m hearth dia, 904 cu m (annual capacity 730,000 tonnes foundry pig iron).

■ Chofu-kita Plant, 2 Minato-machi, Chofu, Shimonoseki. ***Tube and pipe mills:*** seamless – Ugine-Séjournet (annual capacity 15,000 tonnes).

■ Takasago Steel Casting & Forging Plant, 3-1 2-chome, Niihana, Arai-cho, Takasago. ***Steelmaking plant:*** Five electric arc furnaces – one 100-tonne, one 60-tonne, two 20-tonne, one 15-tonne; four induction furnaces – one 3-tonne, one 1-tonne, one 0.75-tonne, one 0.5-tonne (annual capacity 272,000 tonnes).

Products: Steelmaking pig iron. Foundry pig iron. Iron powder. Carbon steel – ingots, slabs, blooms, billets, wire rod, reinforcing bars, round bars, square bars, flats, hexagons, hot rolled hoop and strip (uncoated), cold rolled hoop and strip (uncoated), hot-dip galvanized hoop and strip, medium plates, heavy plates, floor plates, clad plates, hot rolled sheet/coil (uncoated), cold rolled sheet/coil (uncoated), hot-dip galvanized sheet/coil, electro-galvanized sheet/coil, galvanized corrugated sheet/coil, colour coated sheet/coil, zinc-aluminium alloy coated sheet/coil, electrical non-oriented non-silicon sheet/coil, seamless tubes and pipes, cold drawn tubes and pipes. Stainless steel – ingots, billets, wire rod, seamless tubes and pipes, cold drawn tubes and pipes, hollow bars, precision tubes and pipes. Alloy steel (excl stainless) – ingots, slabs, blooms, billets, round bars, hot rolled hoop and strip (uncoated), cold rolled hoop and strip (uncoated), medium plates, heavy plates, floor plates, hot rolled sheet/coil (uncoated), cold rolled sheet/coil (uncoated), wire, seamless tubes and pipes, cold drawn tubes and pipes. Production (1985 fiscal year) – Pig iron 7,312,000 tonnes; raw steel 6,332,000 tonnes; bar and wire rod 2,349,000 tonnes; plate/sheet 3,288,000 tonnes; bloom and billet 118,000 tonnes.

Sales offices/agents: Overseas: USA – Kobe Steel America Inc, 299 Park Avenue, New York, NY 10171; 300 South Grand Avenue, Suite 2620, Los Angeles, CA 90071. West Germany – Postfach 240143, Immermann Strasse 10, D 4000 Düsseldorf. UK – Kobe Steel Europe Ltd, Alton House, 174/177 High Holborn, London WC1 7AA. Singapore – 22-02 DBS Bldg, 6 Shenton Way, Singapore 0106. Mexico – Rio Danubio 69, 9° piso, Col Cuauhtemoc, Delegacion Cuauhtemoc, 06500 Mexico DF. Bahrain – Kobelco Middle East (EC), PO Box 2309, 1st Floor, Zaayani House, Manama. Australia – ANZ Tower, 27th Floor, 55 Collins Street, Melbourne, Vic 3000. China – Room 1524, Beijing Hotel, Dong Chang An Jie, Peking. Domestic branch office: 1 Bingomachi 5-chome, Higashi-ku, Osaka 541. Domestic sales offices: 1-1 Kitasanjo-Nishi 4-chome, Chuo-ku, Sapporo, Hokkaido 060; 1-26 Ichibancho 3-chome, Sendai, Miyagi 980; 4-10 Higashiodori 2-chome, Niigata 950; 1-18

Japan

Kobe Steel Ltd (Kobe Seikosho) (continued)

Sakurabashidori, Toyama 930; 7-23 Naeki 4-chome, Nakamura-ku, Nagoya, Aichi 450; 5-1 Kameicho, Takamatsu, Kagwa 760; 16-11 Hatchobori, Naka-ku, Hiroshima 730; 1-1 Hakata-Ekimae 2-chome, Hakata-ku, Fukuoka 812.
Expansion programme: Kakogawa Works – installation of pulverised coal injection system (March 1988); rationalisation of slag disposal equipment (January 1989); installation of hot metal pre-treatment furnace (construction to begin in 1st half of 1987 fiscal year); installation of water cooling facilities for hot strip mill (first half of 1987 fiscal year).

Kokko Steel Works Ltd (Kokko Seiko)

Head office: 8-60 Shibatani, 2-chome, Suminoe-ku, Osaka 559. **Tel:** 06 685 1031. **Cables:** Kokkoseiko Osaka. **Telecopier (Fax):** 06 682 1031.
Management: Yukio Ishira (president). **Established:** 1934. **Capital:** Y 1,200,000,000.
Works:
■ ***Steelmaking plant:*** electric arc. ***Continuous casting machines:*** One Voest-Alpine billet (130mm sq). ***Rolling mills:*** bar.
Products: Carbon steel – reinforcing bars.

Kubota Ltd (Kubota Tekko)

Head office: 2-47, Shikitsuhigashi 1-chome, Naniwa-ku, Osaka 556-91. **Tel:** 06 648 2111. **Telex:** 5267785 kubota j. **Cables:** Ironkubota Osaka. **Telecopier (Fax):** 06 648 3862.
Management: President – Shigekazu Mino (and representative director), Masao Mikayama (executive vice president); executive managing directors – Kenji Kawakami, Kohsaku Uda, Yasoji Yasukawa, Mutsumi Sakata; managing directors – Eisuke Masuda, Tadashi Miyaoka, Kei Takada, Eiji Nakaya, Kazuo Ohkita, Ryoso Tanaka, Kazutaka Iseki, Shinji Tsuji, Yoshinobu Murayama. **Established:** 1890. **Capital:** Y 67,470 million (April 1986).
Works:
■ Oskaa, Hyogo, Shiya, Kanagawa, Aichi, Chiba, Ibaragi, Tochigi. Ductile iron pipe foundries. Other foundries producing centrifugally cast steel pipe, columns, mill rolls and ingot moulds. Engineering works. ***Tube and pipe mills:*** seamless. Welded – spiral.
Products: Carbon steel – spiral-weld tubes and pipes, hollow sections. Cast iron – pipes, pipe fittings. Ductile iron pipe, centrifugally cast steel cargo-oil pipe, centrifugally cast steel columns ("G-columns"), ingot moulds, mill rolls, PVC pipe, filament wound pipe, industrial castings, pumps, valves, building materials, industrial and agricultural machinery.
Sales offices/agents: In Bangkok, Düsseldorf, New York, Jakarta, Los Angeles, London, Singapore, Cairo, Peking.
Additional information: The company's Tokyo office is at 3-2, Nihonbashi-Muromachi 3-chome, Chuo-ku, Tokyo 103 (Tel: 03 245 3111. Telex: 2226068 kubota j. Cables: Ironkubota Tokyo. Fax: 03 245 3822).

Kurimoto Ltd

Head office: 1-9 Shinbashi, 4-chome, Minato-ku, Osaka. **Telex:** 24946.
Established: 1909.
Works:
■ Osaka area. **Ironmaking plant:** Induction furnaces, cupolas, etc. ***Tube and pipe mills:*** Spiral welded. ***Other plant:*** Ductile iron pipe plants. Foundries.
Products: Carbon steel – spiral-weld tubes and pipes. Ductile iron pipes, accessories and fittings. Wear-resistant and heat-resistant and corrosion-resistant steel castings.

Kyoei Steel Ltd

Head office: 3-1-1 Nakamiya-Ohike, Hirakata, Osaka 573. **Tel:** 0720 49 3225. **Telex:** 5347 215 skyoeij. **Cables:** Hirakata Tetukyoei. **Telecopier (Fax):** 0720 49 3334.
Management: President – Koichi Takashima; senior managing directors – Akihiko Takashima, Chikakazu Namiki; managing directors – Toshio Tanaka, Kazuo Kamo, Hisao Okada; directors – Gotaro Morita, Kunio Maeda, Kiyokatsu Nishiguchi, Yoshiaki Omori. **Established:** 1947. **Capital:** Y 100,000,000. **Subsidiaries:** Daiichi Seiko Co Ltd (88.5%) (production of steel bars); Kyoei Doboku Co Ltd (72%) (manufacture of sandmat, etc). **Associates:** Kyoei Iron Works Ltd (16.4%), Yamaguchi Kyoe Ind Co Ltd (23.3%), Auburn Steel Co Inc, USA (20%), Continental Iron & Steel Co Ltd (11.4%), Sudanese Products Co Ltd (9.8%) (and all/producers of steel bars).
Works:
■ Hirakata Plant (at head office). Works director: G. Morita. ***Steelmaking plant:*** 100-tonne electric arc furnace (annual capacity 400,000 tonnes). ***Continuous casting machines:*** 4-strand curved mould billet (115mm sq) (400,000 tonnes). ***Rolling mills:*** 20-stand bar (8-16) (400,000 tonnes).
■ Hanaten Plant, 1-8-37 Imazu-kita, Tsurumi-ku, Osaka. (Tel: 06 962 2162). Works director: T. Tanaka. ***Rolling mills:*** 8-stand bar for flats 4.5-9mm x 38-75mm (annual capacity 100,000 tonnes).
Products: Carbon steel – billets, reinforcing bars, round bars, square bars, flats.
Expansion programme: Installation of large transformer for electric arc furnace, and PSA oxygen generator.

Maruichi Steel Tube Ltd

Head office: 3-9-10 Kita Horie, Nishi-ku, Osaka 550. **Tel:** 06 531 0101. **Telex:** 64352 maruhan j. **Cables:** Oikokan Osaka.
Management: Chairman of the board – T. Horikawa; president – S. Yoshimura; managing directors – Y. Yoshino (senior, marketing), Y. Maruno (general manager, Sakai Plant), M. Goda (supervisor, Eastern Territory); directors – K. Shiota (president, Maruichi American Corp), H. Taniura (general manager, general affairs dept), Y. Okada (general manager, Nagoya office/supervisor Central Territory), H. Suzuki (general manager, Business dept, supervisor, Western Territory), S. Tamura (general manager, Nagoya plant), K. Ibata (general manager, Finance Dept), S. Juba (Fabricated Product Div), Y. Miné (general manager, Tokyo Plant No 2), D. Horikwas (general manager, Tokyo Office); auditors - K. Marutani (statutory), K. Shimazu. **Established:** 1926 (incorporated **Capital:** Authorised – Y10,000,000,000. Issued and fully paid – Y4,477,000,000. **No. of employees:** 679 (at March 31, 1987). **Ownership:** These include – Seiji Yoshimura (president) (8.93%); Kawasaki Steel Corp (3.79%); Nippon Steel Corp (2.40%); Kobe Steel Ltd (2.04%); Mitsubishi Corp (1.84%); Nippon Kokan KK (1.70%); Nissho Iwai Corp (1.29%). **Subsidiaries:** Domestic – Maruichi Kohan Ltd (40.98%); Hokkaido Maruichi Steel Tube Ltd (40%); Shikoku Maruichi Steel Tube Ltd (19.55%); Kyushu Maruichi Steel Tube Ltd (40%). Overseas – Masan Steel Tube Works Co Ltd, South Korea (55.09%); Maruichi Malaysia Steel Tube Bhd, Malaysia (6.62%); The Malaysia Steel Pipe Mfg Co Ltd, Singapore (22.50%); PT Indonesia Steel Tube Works Ltd, Indonesia (3.33%); Maruichi American Corp, USA (38%). **Annual capacity:** 900,000 tonnes.
Works:
■ 7-3-2 Kami Higashi, Hirano-ku, Osaka 547 (Tel: 06 791 5551). Works director: F. Ito. ***Tube and pipe mills:*** Six ERW welded (annual capacity 216,000 tonnes). ***Other plant:*** Two steel conduit finishing line.
■ 16 Ishizu Nishimachi, Sakai City, Osaka 592 (Tel: 0722 41 0301). Works director: Y. Maruno. ***Tube and pipe mills:*** Three ERW welded (annual capacity 408,000 tonnes). ***Other plant:*** Continuous galvanizing/pickling line; reversing cold reduction mill; stretch reducing mill.
■ 2-12-1 Tajiri, Ichikawa City, Chiba Pref 272. (Tel: 0473 79 1101). Works director: H. Takemaka. ***Tube and pipe mills:*** Eight ERW welded (annual capacity 102,000 tonnes).
■ 1-11 Shiohama, Ichikawa City, Chiba Pref 272 (Tel: 0473 95 1201). Works

Japan

Maruichi Steel Tube Ltd (continued)

director: Y. Mine. ***Tube and pipe mills:*** Four ERW welded (annual capacity 216,000 tonnes). ***Other plant:*** Steel conduit finishing line.
■ 14 Kanaoka, Tobishima-mura, Ama-gun, Aichi Pref 490-14. (Tel: 056755 1101). Works director: S. Tamura. ***Tube and pipe mills:*** Seven ERW welded (annual capacity 156,000 tonnes).
■ 2-125 Ishihara-cho, Sakai City, Osaka 591. (Tel: 0722 58 1858). Works director: S. Juba. ***Tube and pipe mills:*** welded – One 6,000-tonne UO press, one CO_2 shield arc welder and one taper shear (annual capacity 70,000 taper poles or about 6,000 tonnes).
Products: Carbon steel – longitudinal weld tubes and pipes; oil country tubular goods; galvanized tubes and pipes; hollow sections, mechanical tube (including furniture tube), standard (gas and water) pipe, linepipe, electrical conduit, taper pole products.
Brands: MKK (with circle).
Sales offices/agents: Osaka, Tokyo, Nagoya, Fukuoka, Hiroshima, Sapporo. Principal overseas agents: Nissho Iwai Corp. Mitsubishi Corp.

Mitsubishi Steel Mfg Ltd (Mitsubishi Seiko)

Head office: 2-22, Harumi 3-chome, Chuo-ku, Tokyo 104. **Tel:** 03 536 3111. **Telex:** 2523254 msmqj. **Cables:** Hishisteelexpo Tokyo. **Telecopier (Fax):** 03 533 3123.
Management: Chairman – Sadao Takeshita; president – Yoshihei Abe; executive vice president – Kazuki Yamanaka; managing directors – Shigeo Iwasaki, Masao Oka, Hajime Ushijima, Tsuneo Miura, Keishi Fukuma; directors – Masao Kanamori, Youhei Mimura, Kiyohiko Kizuki, Masafumi Inoue, Sanao Takeuchi, Taku Kinouchi, Tsuneo Yoshimura, Shuzo Otani. **Established:** 1949. **Capital:** Y 2,000,000. **Ownership:** Mitsubishi Group.
Works:
■ Tokyo, Utsunomiya, Hirota and Ichikawa. ***Steelmaking plant:*** electric arc – one 60-tonne, one 30-tonne. ***Refining plant:*** Two 60-tonne ladle furnace. ***Continuous casting machines:*** bloom (320 x 270mm-390mm). ***Rolling mills:*** blooming; bar and light section; bar.
Products: Alloy steel (excl stainless) – ingots, blooms, billets. Special steel bars, including spring steel and tool steel, hollow drill steel, alloy and carbon engineering steel. Forgings and castings. Leaf, coil and torsion bar springs for cars and trucks. Rolls for cold strip mills.
Sales offices/agents: 875 North Michigan Avenue, Chicago, IL 60611, USA (Tel: 312 943 0376, 312 943 0377).

Nakayama Steel Products Co Ltd (Nakayama Kogyo)

Head office: 2-4-12 Namamugi, Tsurumi-ku, Yokohama, Kanagawa Pref. **Telex:** 2222983.
Management: Asahi Kondo (president). **Established:** 1951.
Works:
■ At head office. ***Rolling mills:*** cold reduction. ***Coil coating lines:*** hot-dip galvanizing; colour coating.
■ Nishijimacho 1-chome, Nishiyodogawa-ku, Osaka. ***Continuous casting machines:*** One 4-strand Olsson billet. ***Rolling mills:*** bar. ***Coil coating lines:*** colour coating.
Products: Carbon steel – reinforcing bars (deformed); cold rolled hoop and strip (uncoated) and colour-coated.

Nakayama Steel Works Ltd (Nakayama Seikosho)

Head office: 1-66, 1-chome, Funamachi, Taisho-tu Osaka. **Tel:** 06 551 3331. **Telex:** 525325. **Telecopier (Fax):** 06 554 2111.
Management: Sanshiro Sumiya (director). **Established:** 1923. **Capital:** Y 6,270,000,000. **No. of employees:** Works – 1,709. Office – 595. **Annual capacity:** Pig iron: 1,530,000 tonnes. Raw steel: 2,220,000 tonnes. Finished steel:

Nakayama Steel Works Ltd (Nakayama Seikosho) (continued)

2,326,000 tonnes.
Works:
■ At head office. ***Coke ovens:*** Two Otto batteries (66 ovens) (annual capacity 460,000 tonnes). ***Sinter plant:*** One Dwight Lloyd 2.5 x 40m strand (1,570,000 tonnes). ***Blast furnaces:*** No 1 7.6m hearth dia, 1,000 cu m (950,000 tonnes), No 2 7.0m hearth dia, 757 cu m (580,000 tonnes). ***Steelmaking plant:*** Two 105-tonne LD-CB basic oxygen converters (1,800,000 tonnes); one 40-tonne electric arc furnace (420,000 tonnes). ***Refining plant:*** One 108-tonne RH vacuum degassing; one 108-tonne PM (Pulsating Mixing) process. ***Continuous casting machines:*** One 6-strand Concast billet (95-135mm sq) (5,586,000 tonnes); one 4-strand Olsson bloom (150 x 225, 190 x 235, 180mm sq) (864,000 tonnes); one 1-strand Concast slab (150 x 400-600mm (twin casting), 170 x 800-1,400mm (single casting)) (564,000 tonnes); one 4-strand Concast billet (95-135mm sq, 150 x 225mm) (5,137,000 tonnes); one 2-strand NSC bloom and slab (190 x 235mm, 150-250 x 400-800mm) (337,000 tonnes). ***Rolling mills:*** heavy section and medium section – 10-stand semi-continuous (angles 90-150mm, channels 100 x 50-250 x 90mm, H-beam 100 x 100-400 x 200mm, plate 600-1,300mm, flat bar 150-600m) (combined capacity 420,000 tonnes); bar and wire rod – 20-stand bar (12.7-50mm dia) (coil weight 2 tonnes), 40-stand rod (5.5-16mm) (coil weight 2 tonnes) (combined capacity 798,000 tonnes). one 1-stand Lauth type heavy plate (9 x 1,219-40 x 2,000mm) (276,000 tonnes); one 13-stand semi-continuous type hot strip and hoop (1.6 x 170-6 x 310mm) (coil weight 0.69 tonnes) (436,000 tonnes). ***Coil coating lines:*** One continuous type Selas furnace – hot-dip galvanizing (120,000 tonnes).
■ Shimizu, 13 4025 Banchi, Mihokaizima Shimizu, Shizuoka (Tel: 0543 34 1185). ***Rolling mills:*** One 3-stand Lauth type medium plate (3.2 x 800-12 x 1,700mm) (annual capacity 276,000 tonnes).
Products: Steelmaking pig iron. Carbon steel – slabs; blooms; billets; wire rod 5.5-16mm; round bars 12.7-50mm; flats 100-600mm; light angles, medium angles, medium channels, wide-flange beams (medium) and wide-flange beams (heavy) (angles – 90-150mm, channels – 100 x 50-250 x 90mm, wide-flange beams – 100 x 100-400 x 200mm); hot rolled hoop and strip (uncoated) 1.6 x 170-6 x 310mm; medium plates; heavy plates 9 x 1,219-40 x 2,000mm.

Nippon Kinzoku Co Ltd

Head office: Shin-Tokyo Bldg 3-1, 3-chome, Marunouchi, Chiyoda-ku, Tokyo 100. **Telex:** 2223549 nkl j.
Established: 1930. **Ownership:** Nippon Steel Corp.
Works:
■ Itabashi, Tokyo, and Oji, Tokyo. ***Rolling mills:*** cold reduction and cold strip – Sendzimir mills. ***Other plant:*** Cold-forming lines for sections. Bright annealing facilities.
Products: Carbon steel – cold rolled hoop and strip (uncoated) (high-carbon); electrical grain oriented silicon sheet/coil (extra-thin gauges). Stainless steel – flats 10 to 100mm x 2 to 6mm; light angles 20 x 20mm to 50 x 50mm x 1.5 to 5mm; cold roll-formed sections (from strip); cold rolled hoop and strip (uncoated); precision tubes and pipes. Coloured stainless steel strip. Stainless steel bicycle rims and other fabricated products.
Additional information: Stainless steel grades produced include SUS 201, SUS 202, SUS 301, SUS 304, SUS 304L, SUS 309S, SUS 310S, SUS 316, SUS 316L, SUS 321, SUS 347, SUS 405, SUS 430, SUS 434, SUS 444, SUS 410, SUS 410S, SUS 420J2, SUS 440A, SUS 631, SUS 301-CSP, SUS 304-CSP, SUS 631-CSP.

Nippon Koshuha Steel Co Ltd (Nippon Koshuha Kogyo)

Head office: 7-2 1-chome, Otemachi, Chiyoda-ku, Tokyo. **Tel:** 03 231 6761. **Telex:** 2223039. **Cables:** Koshuha Tokyo.
Established: 1950.
Works:

Japan

Nippon Koshuha Steel Co Ltd (Nippon Koshuha Kogyo) (continued)

■ Toyama. ***Steelmaking plant:*** electric arc. ***Refining plant:*** One Asea-SKF; one 1-tonne ESR; one vacuum degassing. ***Rolling mills:*** One 3-high blooming. 3-high and 2-high bar with kocks finishing block. ***Other plant:*** Forging presses. Drawing machines.
■ Hachinohe. Electric furnaces. Foundry.
■ Kitashinagawa. Tool plant.
Products: Alloy steel (excl stainless) – round bars including high-speed, bearing and tool steels.

Nippon Metal Industry Co Ltd (Nippon Kinzoku Kogyo)

Head office: Shinjuku Mitsui Building, 1-1 2-chome Nishishinjuku, Shinjuku, Tokyo 160. **Tel:** 03 344 2345. **Telex:** 2324155 nikkin j. **Cables:** Nikkinko Tokyo.
Management: Fujio Tsukamoto (president). **Established:** 1932.
Works:
■ Sagamihara. ***Steelmaking plant:*** electric arc. ***Refining plant:*** One 55-tonne AOD. ***Continuous casting machines:*** One 1-strand Olsson straight mould slab. ***Rolling mills:*** One heavy plate; wide hot strip – one 1-stand 4-high roughing; 1 Steckel mill. Two Sendzimir cold reduction.
■ Kinuura. ***Steelmaking plant:*** electric arc. ***Refining plant:*** One 75-tonne AOD. ***Continuous casting machines:*** One 1-strand Concast curved mould slab. ***Rolling mills:*** cold reduction Sendzimir.
■ Yokohama. Wire plant.
Products: Stainless steel – medium plates, heavy plates, hot rolled sheet/coil (uncoated), cold rolled sheet/coil (uncoated), wire.

Nippon Stainless Steel Co Ltd

Head office: 8-2 Honshio-cho, Shinjuku-ku, Tokyo 160. **Telex:** 2322057 narj.
Established: 1934. **Ownership:** Sumitomo Metal Industries Ltd.
Works:
■ Naoetsu Works, Ibaragi Pref. ***Steelmaking plant:*** One electric arc. ***Refining plant:*** One 1-tonne VAR; one 38-tonne VOD. ***Continuous casting machines:*** One 2 or 1-strand Concast bloom or slab. ***Rolling mills:*** One 2-high reversing slabbing. two bar – one 3-stand 3-high and one 2-high; one 3-high finishing wide hot strip; two cold reduction – 20-high Sendzimir and 4-high reversing; 6-high Sendzimir cold sheet. ***Other plant:*** Foundry and forge.
■ Kashima Works, Ibaragi Pref. ***Rolling mills:*** Two Sendzimir cold reduction.
Products: Stainless steel – ingots, slabs, blooms, round bars, square bars, light angles, hot rolled sheet/coil (uncoated), cold rolled sheet/coil (uncoated) castings and forgings. Titanium semis.

Nippon Steel Corp (Shin Nippon Seitetsu)

Head office: 6-3 Otemachi 2-chome, Chiyoda-ku, Tokyo 100. **Tel:** 03 242 4111. **Telex:** 22291 nsc j. **Cables:** Nipponsteel Tokyo.
Management: Representative director and chairman of the board – Yutaka Takeda; vice chairman – Akira Miki; representative director and president – Hiroshi Saito; executive vice presidents – Kensuke Koga (China projects co-operations), Hiroshi Tsukada, Makio Yamane (general administration, new materials, electronics and information systems, bio-technology business), Shuichi Kimura, Zensaku Yamamoto, Tatsuo Yamada (engineering business); managing directors – Takeshi Miyazaki (secretariat, personnel), Takashi Imai (corporate planning, corporate and economic research, budget and accounts, finance), Takao Katsumata (affiliate enterprises), Kiyoshi Kobayashi (labour relations, contracts planning), Toyohiko Ono (iron ore, coal and fuel, ferrous materials and machinery purchasing),Yoshiro Sasaki (sales, excluding exports), Tokuji Ohashi (exports), Hajime Nakagawa (technical administration bureau),

Nippon Steel Corp (Shin Nippon Seitetsu) (continued)

Hiroshi Kitanishi (central research and development bureau), Kazuo Sugiyama, Tatsuo Nagano, Minoru Sato, Kazuo Sugiyama (service business), Eiji Umene (supervision of business in North and Central America, president, Nippon Steel USA Inc). **Established:** 1970. **Capital:** US 2,275,181,000. **No. of employees:** 64,060 (as at March 31, 1987). **Subsidiaries:** Principal subsidiaries: Ordinary steel – Godo Stee Ltd (wire rods, reinforcing bars, structural bars, special steel bars, sections, rails and billets); Osaka Steel Co Ltd (sections and rails); Topy Industries Ltd (structural steels, automobile wheels, plate shearing and steel fabrication).
Special steel – Japan Casting and Forging Corp (coolant channel heads, bottom domes, Francis runners, roll housings, generator rotors, back-up rolls); Nippon Kinzoku Co Ltd (cold rolled stainless steel strip, cold rolled special (high-carbon) steel strip); Takasago Tekko KK (cold rolled stainless steel sheets and strip, cold rolled high-carbon and alloy steel strip, cold rolled mild steel sheet and strip, wound cores). Steel forming – Nittetsu Steel Drum Co Ltd (drums); Nippon Steel Metal Products Co Ltd (light-gauge steels, pre-painted sheets, rectangular hollow sections, corrugated pipe, guard rails, metal forms and other fabricated products, steel making flux and continuous casting powder); Sanko Metal Industrial Co Ltd (metal rooofing and siding, steelwork, metal swimming pools and ceilings); Daido Steel Sheet Corp (galvanized steel sheets, pre-painted sheets, laminated steel sheets, and Isowand (insulation wall) and Isodach (insulation roof)); Hokkai Koki Co Ltd (galvanized sheets, pre-painted sheets, heat insulating panels, steel wire and round wire nails); Taiyo Steel Co Ltd (cold rolled sheets, galvanized sheets and pre-painted sheets); Daiwa Can Co Ltd (sanitary cans, beer cans, metal containers, fibre containers, plastic containers, metal and plastic caps); Nippon Steel Wire Rope Co Ltd (wire rope, wire strand and wire); Nichia Steel Works Ltd (galvanized and aluminized wire, galvanized steel core wire, stranded steel cores for ACSR and high-tension bolts); Suzuki Metal Industry Co Ltd (piano wire, PC wire, spring wire, oil-tempered wire, tyre bead wire and stainless steel wire); Nippon Steel Bolten Co Ltd (high-tension bolts and nuts); Nippon Steel Welding Products and Engineering Co Ltd (welding electrodes).
Chemical, energy, non-ferrous and refractories – Japan Metals and Chemicals Co Ltd (ferro-manganese, chrome and silicon, fertilisers, geothermal power development and plant engineering); Nippon Denko Co Ltd (ferro-manganese, ferro-chromium, ferro-silicon, ferro-vanadium and other ferro-alloys, metallic silicon, chromium metal, and chromium compounds); Pacific Metals Co Ltd (ferro-nickel, ferro-manganese and ferro-silicon, stainless steel billets, carbon and alloy steel castings and forgings)); Nippon Steel Chemical Co Ltd (coke, benzene, styrene, naphthalene, phthalic anhydride, ammonia, methanol, polystyrene, blast furnace cement, mineral wool and other chemical products, pitch coke, carbon black, anti-corrosion paints, coumarone resin); Nittetsu Chemical Engineering Ltd (planning, design, manufacture and construction of chemical plants); Nittetsu Cement Co Ltd (Portland cement, blast furnace slag cement, Super Fine grouting material); Kurosaki Refractories Co Ltd (silica bricks, high-alumina bricks, fireclay bricks, basic bricks, dolomite bricks and monolithic refractories); Harima Refractories Co Ltd (silica bricks, high-alumina bricks, fireclay bricks, basic bricks, dolomite bricks and monolithic refractories); Kyushu Oil Co Ltd (oil products).
Engineering – Nittetsu Mining Co Ltd (limestone, copper ores, industrial machinery, iron ore and mining consultancy); Nippon Marine Construction Co Ltd (marine Construction); Nippon Tetrapod Co Ltd (Tetrapod, Igloo, Dolos, Tribars and Cyclonet);
Urban development, housing – Nippon Steel Site Planning Corp – Nippon Steel Real Estate Corp (real estate); Nippon Homes Corp (housing); Fudo Construction Co Ltd (general contractor for civil engineering works and building construction, especially in steelworks construction and weak soil stabilisation work (composer system)).
Transport, flow – Shinwa Kaiu Kaisha Ltd (marine transport, marine agency and marine brokerage); Nippo Kisen Co Ltd (marine transport, marine agency and marine brokerage); Nittetsu Shoji Co Ltd (import of ferrous and non-ferrous raw materials, export of steel products and general goods).
New business – Micron Co Ltd (manufacture and marketing of IC encapsulation material); NSC Electron Corp (manufacture and marketing of silicon products for the semi-conductor industry); Nippon Steel Techno-Research Corp (material analysis

Japan

Nippon Steel Corp (Shin Nippon Seitetsu) (continued)

service); Concurrent Nippon Corp (manufacture and marketing of mini-computers); Nippon Steel Technos Corp (plant and equipment diagnostic and maintenance services); PNN Corp (manufacturing and marketing of ceramic capacitors); Japan Technical Information Service (engineering and patent information research service); Nippon Micrometal Corp (manufacture and marketing of semi-conductor bonding wires); Colloid Research Institute (commercialisation of new sol and gel manufacturing method for fine ceramics).

Works:

■ Yawata Works, Yawata Higashi-ku, Kitakyushi City, Fukuoka Pref. (Tel: 093 662 8111). Managing director: Tomokatsu Kotani. ***Coke ovens:*** Tobata No 2 (annual capacity 420,000 tonnes), Tobata No 4 (391,000 tonnes) and Tobata No 5 (986,000 tonnes). ***Sinter plant:*** Dwight-Lloyd – Tobata No 3 320 sq m (3,285,000 tonnes) and Wakamatsu 6,570,000 tonnes. ***Iron ore pelletizing plant:*** (400,000 tonnes). ***Blast furnaces:*** Tobata No 1 13.4m hearth dia, 4,140 cu m (3,588,000 tonnes), Tobata No 4 13.8m hearth dia 4,250 cu m (3,610,000 tonnes). ***Steelmaking plant:*** Four basic oxygen converters No 1 shop two 170-tonne LD (4,655,000 tonnes); No 3 shop two 350-tonne LD (4,405,000 tonnes). ***Continuous casting machines:*** 4-strand vertical-bending bloom (840,000 tonnes); Four slab one-strand vertical (840,000 tonnes), one 2-strand bending (2,640,000), one 1-strand bending and one 1-strand bending bloom (combined capacity 22,280,000 tonnes). ***Rolling mills:*** Tobata No 2 Universal slabbing (3,350,000 tonnes) and Tobata No 1 1-stand 2-high reversing blooming. Slabbing (1,350,000 tonnes); one rail – two 2-high reversing and three universal (850,000 tonnes); one heavy plate one 2-high roughing and one 4-high finishing (1,244,000 tonnes); one wide hot strip one 2-high reversing, one 4-high reversing and two 4-high and four 6-high finishing (4,560,000 tonnes); cold reduction No 1 four 4-high tandem (1,219,000 tonnes), No 2 five 4-high tandem (600,000 tonnes), No 3 one 4-high reversing (294,000 tonnes), No 4 six 4-high tandem (1,050,000 tonnes) and one 6-high reversing (85,000 tonnes) and three for silicon steel in two 20-high Sendzimir and 1-stand 4-high reversing (48,000 tonnes). ***Tube and pipe mills:*** seamless One Mannesmann plug mill (600,000 tonnes) and one semi-floating mandrel mill (600,000 tonnes). Welded Three Spiral (224,000 tonnes). ***Coil coating lines:*** Three electrolytic tinning (522,000 tonnes); two tin-free steel (ECCS) (390,000 tonnes); four hot-dip galvanizing (492,000 tonnes); three electro-galvanizing (486,000 tonnes); one terne/lead coating (210,000 tonnes).

■ Muroran Works, Muroran City, Hokkaido. (Tel: 0143 45 3131). Managing director: Takuo Kitamura. ***Coke ovens:*** No 5 (annual capacity 708,000 tonnes) and No 6 (548,000 tonnes). ***Sinter plant:*** Dwight Lloyd No 6 460 sq m (5,694,000 tonnes). ***Blast furnaces:*** No 2 10.7m hearth dia, 2,296 cu m. ***Steelmaking plant:*** Two 270-tonne (No 1 shop) basic oxygen converters (4,265,000 tonnes). ***Continuous casting machines:*** One 4-strand bending bloom (680,000 tonnes), one 1-strand vertical slab (349,000 tonnes), one 2-strand bending blooming/ slab (1,824,000 tonnes). ***Rolling mills:*** One 1-stand 2-high reversing slabbing (2,150,000 tonnes); One blooming / billet 1-stand 2-high reversing, 6-stand 2-high and 4-stand 2-high (1,340,000 tonnes); one 6-stand 2-high roughing, 4-stand 2-high intermediate and 4-stand 3-high finishing bar (396,000 tonnes); one bar and wire rod – fully-continuous single strand 26-stand vertical/horizontal (490,000 tonnes); one 1-stand 4-high, 1-stand 4-high reversing and 6-stand 4-high tandem finishing train wide hot strip (1,810,000 tonnes); one 1-stand 4-high reversing cold reduction (156,000 tonnes).

■ Kamaishi Works, Kamaishi City, Iwate Pref. (Tel: 0193 24 3331). Managing directors: Hiroshi Asai. ***Coke ovens:*** No 1 (131,000 tonnes) and No 5 (427,000 tonnes). ***Sinter plant:*** Dwight Lloyd No 1 170 sq m (2,008,000 tonnes). ***Blast furnaces:*** No 1 8m hearth dia, 1,150 cu m (745,000 tonnes), No 2 9.35m hearth dia, 1,730 cu m (1,285,000 tonnes). ***Steelmaking plant:*** Two 90-tonne basic oxygen (1,220,000 tonnes). ***Continuous casting machines:*** One 4-strand bloom (570,000 tonnes). ***Rolling mills:*** One 1-stand 2-high reversing, 6-stand 2-high, and 4-stand 2-high blooming and 2-high billet (1,270,000 tonnes); One fully-continuous 3-strand 26-stand bar and wire rod (804,000 tonnes).

■ Hirohata Works, Himeji City, Hyogo Pref. (Tel: 0792 36 1111). Director: Masahisia Kozaki. ***Coke ovens:*** No 1 (annual capacity 361,000 tonnes) and No 4 (442,000 tonnes). ***Sinter plant:*** Dwight Lloyd 113.1 sq m (1,241,000 tonnes) and 150 sq m

Nippon Steel Corp (Shin Nippon Seitetsu) (continued)

stand (1,643,000 tonnes). ***Iron ore pelletizing plant:*** One (not in operation) (2,500,000 tonnes). ***Blast furnaces:*** No 4 11.7m hearth dia, 2,950 cu m (2,449,000 tonnes). ***Steelmaking plant:*** Three 100-tonne (No 1 shop) basic oxygen converters (2,400,000 tonnes). ***Continuous casting machines:*** One 2-strand bending slab (1,620,000 tonnes). ***Rolling mills:*** One 1-vertical stand, 1-stand 4-high reversing and 4-stand 6-high finishing hot strip and hoop (3,000,000 tonnes); single stand Sendzimir for silicon steel (84,000 tonnes); cold reduction one FIPL (fully integrated processing line) – pickling, tandem cold mill, continuous annealing and processing line. ***Coil coating lines:*** Two electrolytic tinning (276,000 tonnes) and one electro-galvanizing (215,00 tonnes).

■ Hikari Works, Hikari City, Yamaguchi Pref. (Tel: 0833 72 1300). Director: Takashi Furuya. ***Steelmaking plant:*** One 40-tonne and one 60-tonne electric arc furnaces (annual capacity 289,000 tonnes). ***Continuous casting machines:*** One 2-strand vertical bloom (50,000 tonnes) and one 1-strand vertical slab (230,000 tonnes). ***Rolling mills:*** One wire rod – (No 2) fully-continuous 2-strand 28-stand (515,000 tonnes); cold reduction three 1-stand 20-high Sendzimir for stainless steel (No 1 48,000 tonnes, No 2 48,000 tonnes and No 3 144,000 tonnes). ***Tube and pipe mills:*** seamless One Ugine Séjournet extrusion mill (55,000 tonnes). Welded Three ERW (573,000 tonnes). ***Other plant:*** Stainless steel tube mill (3,000 tonnes).

■ Nagoya Works, Tokai City, Aichi Pref. (Tel: 056 604 2111). Managing director: Motoi Yasue. ***Coke ovens:*** No 1 (475,000 tonnes), No 2 (697,000 tonnes), No 3 (810,000 tonnes) and No 4 (898,000 tonnes). ***Sinter plant:*** Three Dwight Lloyd – 182 sq m (2,008,000 tonnes), 196 sq m (2,154,000 tonnes) and 280 sq m (3,066,000 tonnes). ***Blast furnaces:*** No 1 13 m hearth dia, 3,890 cu m (3,073,000 tonnes), No 2 (not in operation) 11 m hearth dia, 2,520 cu m (1,719,000 tonnes) and No 3 12.5m hearth dia, 3,240 cu m (2,778,000 tonnes). ***Steelmaking plant:*** Five basic oxygen converters (No 1 shop) two 160-tonne and one 170-tonne (3,901,000 tonnes) and (No 2 shop) two 250-tonne (4,504,000 tonnes). ***Continuous casting machines:*** Two 2-strand bending slab (1,920,000 and 2,400,000 tonnes). ***Rolling mills:*** One (No 2 primary mill – universal) slabbing (1,630,000 tonnes); one 1-stand 4-high roughing and 1-stand 4-high finishing heavy plate (1,967,000 tonnes); one fully-continuous wide hot strip 1-stand 2-high, 4-stand 4-high and 7-stand 4-high finishing (4,440,000 tonnes); three cold reduction No 1 5-stand 4-high (888,000 tonnes), No 2 5-stand 4-high (1,960,000 tonnes) and No 3 6-stand 4-high (824,000 tonnes). ***Tube and pipe mills:*** welded Four ERW (528,000 tonnes) and one continuous butt-weld (300,000 tonnes). ***Coil coating lines:*** One electrolytic tinning (216,000 tonnes); one tin-free steel (ECCS) (246,000 tonnes); four hot-dip galvanizing (786,000 tonnes) and one electro-galvanizing (360,000 tonnes).

■ Sakai Works, Sakai City, Osaka-fu. (Tel: 0722 33 1131). Director: Shigeru Omori. ***Sinter plant:*** Dwight Lloyd 183 sq m (annual capacity 2,081,000 tonnes). ***Blast furnaces:*** No 2 11.6m hearth dia, 2,797 cu m (2,245,000 tonnes). ***Steelmaking plant:*** Three 170-tonne basic oxygen converters (4,465,000 tonnes). ***Continuous casting machines:*** One 2-strand bending slab (1,920,000 tonnes). ***Rolling mills:*** One 1-stand 2-high reversing, 2 universal stands, one universal finishing stand heavy section (1,254,000 tonnes).

■ Kimitsu Works, Kimitsu City, Chiba Pref. (Tel: 0439 52 4111). Managing director: Akira Shinoda. ***Coke ovens:*** No 1 (annual capacity 730,000 tonnes), No 2 (770,000 tonnes), No 3 (810,000 tonnes), No 4 (905,000 tonnes) and No 5 (905,000 tonnes). ***Sinter plant:*** Dwight Lloyd 183 sq m (2,446,000 tonnes), 245 sq m (2,555,000 tonnes) and 500 sq m (5,475,000 tonnes). ***Blast furnaces:*** No 2 12m hearth dia, 2,884 cu m (1,697,000 tonnes) and No 4 14m hearth dia, 4,930 cu m (4,139,000 tonnes). ***Steelmaking plant:*** Three 250-tonne (No 1 shop) (6,700,000 tonnes) and two 300-tonne (No 2 shop) basic oxygen converters (3,822,000 tonnes). ***Continuous casting machines:*** One 2-strand bending bloom (1,200,000 tonnes); two 2-strand vertical/bending slab (2,520,000 and 2,760,000 tonnes). ***Rolling mills:*** One blooming 2 high-lift stands and billet mill (2,880,000 tonnes) and; one heavy section (1-stand 2-high roughing, 1-stand 2-high breakdown two 2-high roughing and 7-stand universal finishing train) (1,116,000 tonnes); one 4-strand fully-continuous 25-stand

Japan

Nippon Steel Corp (Shin Nippon Seitetsu) (continued)

wire rod (972,000 tonnes); one 1-stand 4-high roughing and 1-stand 4-high finishing heavy plate (2,160,000 tonnes); one 2-stand 2-high, 3-stand 4-high, 7-stand 4-high wide hot strip (6,000,000,tonnes); two cold reduction No 1 5-stand 4-high and No 2 6-stand 4-high (combined capacity 4,320,000 tonnes). ***Tube and pipe mills:*** welded - one ERW (84,000 tonnes), one continuous butt-weld (276,000), one UO pipe mill. ***Coil coating lines:*** Three hot-dip galvanizing (924,000 tonnes) and " t electro-galvanizing (744,000 tonnes).

■ Oita Works, Oita City, Oita Pref. (Tel: 0975 58 1313). Director: Seiro Hiwatashi. ***Coke ovens:*** No 1 (annual capacity 719,000 tonnes), No 2 (719,000 tonnes), No 3 (785,000 tonnes) and No 4 (785,000 tonnes). ***Sinter plant:*** Two Dwight Lloyd 400 sq m (4,928,000 tonnes) and 600 sq m (6,570,000 tonnes). ***Blast furnaces:*** No 1 14m hearth dia, 4,158 cu m (3,650,000 tonnes) and No 2 14.8m hearth dia, 5,070 cu m (4,526,000 tonnes). ***Steelmaking plant:*** Three 340-tonne basic oxygen converters (8,434,000 tonnes). ***Continuous casting machines:*** Three slab No 1 2-strand bending (2,280,000 tonnes), No 2 2-strand bending (1,046,000 tonnes), and No 4 2-strand bending (2,724,000 tonnes). ***Rolling mills:*** One 1-stand 4-high reversing heavy plate (1,258,000 tonnes) and one ¾ continuous 1-stand 2-high, 1-stand 4-high, 2-stand 4-high reversing, 7-stand 4-high wide hot strip (5,031,000 tonnes).

■ Tokyo Works, Itabashi-ku, Tokyo. (Tel: 03 968 6801). Executive counsellor: Teruo Goda. ***Tube and pipe mills:*** seamless One Mannesmann plug mill (annual capacity 183,000 tonnes).

Products: Steelmaking pig iron. Foundry pig iron. Carbon steel – slabs; blooms; billets; wire rod (including low-carbon, high carbon, wire rod for electrodes, cold-heading quality, steel tyre cord wire rod, piano wire rod); reinforcing bars; round bars; medium angles; medium tees; medium joists; medium channels; wide-flange beams (medium); universals (medium); heavy angles (equal, unequal and invert); heavy tees; heavy joists; heavy channels; wide-flange beams (heavy); universals (heavy); heavy rails (standard, hardened and special); sheet piling (u-type, z-type, flat type, H-type, box-type); pipe piling; medium plates; heavy plates; floor plates; hot rolled sheet/coil (uncoated); cold rolled sheet/coil (uncoated); hot-dip galvanized sheet/coil; electro-galvanized sheet/coil, aluminized sheet/coil; terne-plate (lead coated) sheet/coil; electrical grain oriented silicon sheet/coil; electrical non-oriented silicon sheet/coil; electrical non-oriented non-silicon sheet/coil; electrolytic tinplate (single-reduced); electrolytic tinplate (double-reduced); seamless tubes and pipes; longitudinal weld tubes and pipes (ERW and continuous butt-weld); spiral-weld tubes and pipes; large-diameter tubes and pipes; oil country tubular goods. Stainless steel – hot rolled band (for re-rolling); wire rod; medium plates; heavy plates; universal plates; clad plates; hot rolled sheet/coil (uncoated); cold rolled sheet/coil (uncoated); bright annealed sheet/coil; seamless tubes and pipes; longitudinal weld tubes and pipes. Carbon steel beam blanks, sheet bars, bearing piles, vibration-damping steels, heavy column sections, bulb flats, universal box (box-shaped structural sections), mining joists, mining channels, extra-heavy section plates, Zn-Fe alloy-coated steel sheets, one-side galvanized steel sheets, submerged arc weld pipes, UOE pipe, coated pipe. Output in Japanese fiscal year, 1986 – Pig iron 26,528,000 tonnes, raw steel 25,567,000 tonnes and finished products 242733,000 tonnes.

Brands: Durgrip (galvanized coil/sheets, Zn-Fe alloy coated steel sheets) Zincote, Durzinklite, Durpaint (electro-galvanized coil/sheets). Alsheet (aluminized sheet). Super Zinc (hot-dip Al-Zn alloy coated steel sheets). Orientcore, Orientcore Hi-B (grain-oriented electrical sheets). Hilitecore, Homecore, Thin-Gauge Hilitecore (non-oriented electrical sheets). Semicore (non-oriented semi-processed electrical sheets). Wel-Ten (weldable high-strength steels). Cor-Ten, S-Ten, Naw (atmospheric corrosion-resistant steels). N-Tuf (low-temperature steels). ML (seawater resistant steels). Fabricated steels – Panzermast sectional pole, H-beam bridge-composite, frame H, frame Z, honeycomb frame, steel segments, ductile segments, press concrete segments, grating. Metrodeck road cover deck.

Sales offices/agents: USA – Nippon Steel USA Inc, NY (Tel: 212 486 7150. Telex: 12366 nipponst nyk); Nippon Steel USA Inc, CA (Tel: 213 624 4101. Telex: 0673511 nipponst lsa); Nippon Steel USA Inc, TX (Tel: 713 652 0922. Telex: 774116 nsusa); West Germany – Nippon Steel Corp , Düsseldorf (Tel: 320791,

Nippon Steel Corp (Shin Nippon Seitetsu) (continued)

329366. Telex: 8587850 npst d); Italy – Nittetsu Italiana SpA, Rome (Tel: 4746015, 4746017, 4746033, 6799955. Telex: 621211 npstri); UK – Nippon Steel Corp, London (Tel: 248 8646. Telex: 8811551 nsc In); Australia – Nippon Steel Australia Pty Ltd, Sydney, NSW (Tel: 233 3988. Telex: 22603); Singapore – Nippon Steel Corp, Singapore (Tel: 2236777. Telex: 21042 nipponst rs); Brazil – Praia de Botafogo, Rio, RJ (Tel: 552 7997. Telex: 2122360 nses br), Nippon Steel Siderurgia Ltda, São Paulo, SP (Tel: 251 4533, 287 1582, 287 1638. Telex: 1125053 nses br).
Additional information: The company is also engaging in the following businesses: Manufacture and sales of special metals such as titanium products, and special steel products such as shape memory alloys and magnetic materials; manufacture and sales of non-ferrous metals, such as silicon products, semi-conductor bonding wire, fine ceramics (eg ceramic capacitors), and other chemicals and composite materials such as coal tar pitch-based carbon fibers, structural fine ceramics, and functional fine ceramics; manufacture, sale and construction of industrial machinery and plant such as blast furnaces, basic oxygen converters, rolling mills and other steelworks equipment, industrial machinery, moulds and rolls, industrial furnaces, chemical plants and pollution control devices; technical co-operation, providing technical consultancy services for steelworks construction and operation, supply of know-how and patents; civil engineering and marine construction, such as overland and submarine pipelines, offshore structures, suspension and other types of bridges, piling foundation work; manufacture and sale of standard steel stuctures, for factories, housing, offices and warehouses (from planning to fabrication and construction), businesses related to urban development, transaction and leasing of residential land and buildings; manufacture and sales of systems related to education, medical care, sports, etc, such as planning and implementation of various educational programmes and development of teaching techniques, education on non-Japanese residents in Japn of the Japanese language and business customs, and oranisation of international exhibitions and conventions; sales of technology related to the above businesses, such as material analysis services, engineering and patent information research services, and commercialisation of new sol and gel manufacturing methods for fine ceramics.

Nippon Steel Metal Products Co Ltd (Nittetsu Kenzai Kogyo KK)

Head office: 16-3 Ginza 7-chome, Chuo-ku, Tokyo 104. **Tel:** 03 542 8111. **Telex:** 02524428.
Ownership: Nippon Steel Corp.
Products: Carbon steel – hollow sections. Light-gauge steel sections, corrugated pipe, guard rails and other fabricated products.

Nippon Yakin Kogyo Co Ltd

Head office: 15-1 1-chome, Chuo-ku, Kyobashi, Tokyo. **Tel:** 03 561 8911. **Telex:** 24359 yakinko. **Cables:** Yakinko Tokyo. **Telecopier (Fax):** 03 561 5492.
Management: President – Yoshito Ishiguro; executive managing directors – Kozo Yokota, Susumu Ikeda; managing directors – Haruo Arai, Masaakira Kurimoto, Hiroshi Tanimoto, Sadao Ono, Kanji Ohama; directors – Haruo Suzuki, Nobuyuki Yasunaga, Kazuo Ebato, Koichiro Osozawa, Tetsuo Yanagida, Shohei Shimada, Yuhnosuke Uda, Zenjiro Hatsushiba, Hiroshi Matsuzaki. **Established:** 1940. **Capital:** Y 11,582,845,000. **No. of employees:** Works – 1,477. Office – 410. **Subsidiaries:** Toa Seiki Co Ltd (100%) (tube and pipe – eleven mills); Nippon Seisen Co Ltd (44.4%) (wire); NAS Stainless Co Ltd (73.9%) (sinks, bath-tubs); NAS Trading Co Ltd (90.4%) (trading company).
Works:
■ Kawasaki Plant, 4-2 Kojima-cho, Kawasaki-ku, Kawasaki City, Kanagawa. (Tel: 044 271 3111). Works director: Sadao Ono. ***Steelmaking plant:*** Four electric arc furnaces – one 10-tonne Heroult, one 30-tonne Lectromelt and two 60-tonne Lectromelt (combined annual capacity 330,000 tonnes); one 1.5-tonne HF induction furnace, one 1-tonne vacuum induction melting furnace and one 3-tonne consumable

Japan

Nippon Yakin Kogyo Co Ltd (continued)

electrode arc furnace (combined annual capacity 2,900 tonnes). ***Refining plant:*** One 60-tonne AOD and one 60-tonne VOD (330,000 tonnes). ***Continuous casting machines:*** Two 1-strand Rossi vertical (Concast) slab (max 154mm x 5') (330,000 tonnes). ***Rolling mills:*** One Sendzir PL-64-57 wide hot strip, 5' wide (coil weight 19 tonnes) (336,000 tonnes). Four cold reduction – two Sendzimir ZR 22-50, (coil weight 9 tonnes) and one Sendzimir ZR-21b-62 (coil weight 19 tonnes) 0.25-8.0mm x 5' (combined annual capacity 153,000 tonnes); one Sendzimir ZS-07-60, (0.6-25mm x 1,600 x 10,150mm). two temper/skin pass – one reversing (0.3-3mm x 4') (coil weight 7.5 tonnes) and one non-reversing (0.3-3mm x 5') (coil weight 19 tonnes) (combined annual capacity 153,000 tonnes). ***Tube and pipe mills:*** welded – one 4,000-tonne hydraulic forming press (max 1,016mm dia x 12,500mm long (5,760 tonnes). ***Other plant:*** Forge: One 1,500-tonne hydraulic press and one 1,000-tonne hydraulic press (combined annual capacity 21,000 tonnes).

■ Kanazawa Plant, 1-475 Miike-cho, Kanazawa City, Ishikawa. (Tel: 0762 52 3251). Works director: Seiji Shibata. ***Steelmaking plant:*** Four high frequency induction furnaces – one 0.6-tonne, one 0.3-tonne, one 0.5-tonne and one 2.0-tonne; one 0.7-tonne LF vacuum induction melting furnace; one 2.0-tonne LF induction furnace. ***Other plant:*** Foundry (annual capacity 2,400 tonnes). Metal fibre equipment.

■ Oheyama Plant, 413 Suzu, Miyazu City, Kyoto. (Tel: 077246 3121). Works director: Haruo Arai. Ferro-nickel plant: Four rotary kilns, 3.6m dia x 70m long (annual capacity 12,600 tonnes of nickel).

Products: Stainless steel – slabs 600mm dia and 600mm sq (forging); blooms 600mm dia and 600mm sq (forging); round bars 410mm dia x 10m long (forging); square bars 410mm dia x 10m long (forging); flats 2-12mm thick x 15-120mm wide x 4-6m long; light angles 2-12mm thick x 15-120mm wide x 6m long; medium plates up to 60mm thick x 3.7m wide x 12m long; hot rolled sheet/coil (uncoated) 2.5-8mm thick x 5' wide; cold rolled sheet/coil (uncoated) 0.25-5mm thick x 5' wide; longitudinal weld tubes and pipes $\frac{1}{4}$-66" dia x 12m long. Total output 232,000 tonnes. Castings and forgings.

Brands: NAS.

Sales offices/agents: Sales offices in major cities throughout Japan. European office – Kasernenstrasse 59, D 4000 Düsseldorf, West Germany (Tel: 001 49 0211 327805. Telex: 8587911 yakn d). Singapore office – Room 1608, 16th Floor, Robina House, 1 Shenton Way, Singapore 0106 (Tel: 001 65 2239233. Telex: 3347 yakins).

Nisshin Kokan Co Ltd—see Nisshin Steel Co Ltd

Nisshin Steel Co Ltd

Head office: 4-1 Marunouchi 3-chome, Chiyoda-ku, Tokyo 100. **Tel:** 03 216 5511. **Telex:** 03-2222788 nissin j. **Telecopier (Fax):** 03 415 1895.

Management: Chairman and chief executive officer – Yuzuru Abe; president – Tsuyoshi Kai; executive vice presidents – Akio Shigekuni, Saiji Hayahshi, Yoshihiro Hisamatsu; senior managing director – Koichi Doi, Tamotsu Takeda, Seiji Mori; managing directors – Sohei Nabeshima, Hisao Tanaka, Kazuyoshi Nishimura, Shigeho Tanaka, Seio Hachiya, Sanemi Yamamoto, Michihiko Kusuda, Noboru Hirota; directors – Teruo Kitamura, Shigeaki Maruhashi, Kazuo Nakamara, Yukihiko Kumagai, Michio Kubota, Kaneshige Yamamoto, Yasunori Takashima, Takamasa Yamamura, Tamotsu Omoto. **Established:** 1928. **Capital:** Y 41,762,695,000. **No. of employees:** 7,940. **Subsidiaries:** Tsukiboshi Kogyo Co Ltd (100%) (manufacture of construction materials and light gauge steel shapes; processing and sale of steel sheet and cutting of strip); Nisshin Kokan Co Ltd (100%) (manufacture, processing, and sale of steel pipes and tubes); Tsukiboshi Kaiun Co Ltd (67.9%) (sea and land transportation of iron and steel products); Nisshin Koki Co Ltd (100%) (installation, repair, and maintenance of Nisshin's steelmaking facilities); Tsukiboshi Art Co Ltd (100%) (manufacture, processing, and sale of colour-developed stainless steel using Nisshin's stainless steel); NSK Finance Co Ltd (100%) (financing and investment in domestic market); Shinwa Kigyo Co Ltd (95.3%) (maintenance and management of Nisshin's company housing and welfare facilities; property and life insurance agency, real estate, advertising

Nisshin Steel Co Ltd (continued)

agency, and management of golf courses); Shinsei Kogyo Co Ltd (94.7%) (manufacture, processing, and sale of prepainted galvanized steel sheet); Nisshin Ferrite Co Ltd (75%) (manufacture and sale of calcined hard ferrite); RHD Inc (70%) (research and development of advanced technology); Shunan Shigyo Co Ltd (66.7%) (recycling and sale of kraft paper); Osaka Stainless Center Co Ltd (52%) (shearing, slitting, processing, and sale of stainless steel sheet and strip); The Thinsheet Metals Co (100%) (rerolling of copper and copper alloys, nickel alloys, and stainless steel); Nisshin Steel Australia Pty Ltd (100%) (manufacture and sale of iron ore pellets); Nisshin USA Inc (100%) (taking over of the former New York office); Nisshin Steel (Canada) Ltd (100%) (coal mine development for manufacture and sale of coking coal); Nisshin International Finance (Netherlands) BV (100%) (financing and investment in overseas markets); Wheeling-Nisshin Inc (67%) (manufacture and sale of hot-dip aluminzned steel sheets and hot-dip galvanized steel sheets – facility under construction). **Associates:** Nikken Stainless Fittings Co Ltd (50%) (manufacture and sale of stainless steel pipe fittings); Tsukiboshi Trading Co Ltd (41.2%) (processing and sale of steel products; reerhousing and transportation business); Sunwave Kogyo Co Ltd (38.3%) (manufacture and sale of kitchen sinks, bath-tubs, sanitary devices, and air conditioners using stainless steel); Nihon Teppan Co Ltd (33.3%) (domestic agent for sale of galvanized steel sheet and coated steel sheet and strip); Mizukami Metal Industry (33.3%) (manufacture and sale of steel products, construction materials, and kitchen apparatus); Shimbishi Kaiun Co Ltd (32%) (domestic sea transportation service, and agency for insurance against loss); Kano Tekko Co Ltd (20.8%) (shearing, processing, and sale of steel); Shunan Denko Co Ltd (20%) (manufacture of ferrochromium and silicon chromium); Kure Oxygen Center Co Ltd (20%) (manufacture and supply of oxygen); Ube Nisshin Lime Co Ltd (20%) (processing and supply of lime for steelmaking). **Annual capacity:** Pig iron: 3,185,000 tonnes. Raw steel: 3,784,000 tonnes. Hot-rolled finished steel: 6,241,000 tonnes.

Works:

■ Kure Works, 11-1 Showacho, Kure City, Hiroshima Pref (Tel: 0823 24 111. Fax: 0823 22 7770). Works general manager: Seio Hachiya. ***Sinter plant:*** Two Dwight-Lloyd – one 330 sq m strand (annual capacity 2,190,000 tonnes); one 150 sq m strand (1,460,000 tonnes). ***Blast furnaces:*** No 1 10.2m hearth dia, 2,150 cu m (1,960,000 tonnes); No 2 9.25m hearth dia, 1,650 cu m (1,225,000 tonnes). ***Steelmaking plant:*** Four LD basic oxygen converters – three 90-tonne (2,730,000 tonnes); one 180-tonne (two vessels, one blowing station) (2,381,000 tonnes). ***Refining plant:*** One 180-tonne RH vacuum degassing unit (1,445,000 tonnes); one 80-tonne VAD unit (192,000 tonnes). ***Continuous casting machines:*** One 2-strand USSR straight-type for slab (190 x 610mm to 190 x 1,350mm) (418,000 tonnes); one 2-strand Hitachi-Zosen bending-type for slabs (190 x 800mm to 250 x 1,325mm) (1,992,000 tonnes). ***Rolling mills:*** One 1-stand universal-type slabbing (3,816,000 tonnes); two wide hot strip – No 1 1.15 to 12.7mm with three roughing and six finishing stands (2,565,000 tonnes); No 2 1.20 to 12.7mm thickness with two roughing and six finishing stands (3,676,000 tonnes).

■ Shunan Works, 4976 Tonda, Shinnanyo City, Yamaguchi Pref (Tel: 0834 63 0111. Fax: 0834 63 1995). Works general manager: Kazuyoshi Nishimura. ***Steelmaking plant:*** Two 40-tonne LD basic oxygen converters (annual capacity 362,000 tonnes); five Heroult electric arc furnaces – one 6.5-tonne (14,000 tonnes), four 30-tonne (362,000 tonnes – melting only). ***Refining plant:*** One 75-tonne VAD unit (362,000 tonnes). ***Continuous casting machines:*** One 1-strand Demag curved mould for stainless steel slab (200mm thick x 600 to 1,300mm wide) (236,000 tonnes). ***Rolling mills:*** Four Sendzimir cold reduction – No 1 1-stand 0.1 to 4mm thick; No 2 1-stand 0.1 to 4mm thick; No 3 4-stand 0.20 to 3.2mm thick; No 4 1-stand 0.70 to 6.0mm thick. The annual production capacity combined amounts to 208,000 tonnes.

■ Amagasaki Works, 1 Tsurumachi, Amagasaki City, Hyogo Pref (Tel: 06 416 1031. Fax: 06 416 0141). Works general manager: Kazuo Mitani. ***Tube and pipe mills:*** welded – One HF welding unit 12.7 to 76.3mm dia x 0.5 to 3mm thick (12,000 tonnes); ten TIG units 10 to 219.2mm dia x 0.4 to 10mm thick (19,000 tonnes); one PFO unit 267.4 to 508mm dia x 3 to 10mm thick (2,000 tonnes).

■ Sakai Works, 5 Ishizunishimachi, Sakai City, Osaka Pref (Tel: 0722 41 1131. Fax: 0722 44 0067). Works general manager: Koichi Doi. ***Rolling mills:*** Three cold

Japan

Nisshin Steel Co Ltd (continued)

reduction comprising one 5-stand 0.2 to 2.3mm (annual capacity 1,179,000 tonnes); three 2-stand combination 0.2 to 2.3mm (198,000 tonnes); two 3-stand 0.1 to 2.3mm (735,000 tonnes). ***Coil coating lines:*** One continuous hot-dip galvanizing 0.25 to 1.0mm thick and 600 to 1,250mm wide (183,000 tonnes); one continuous electro-galvanizing 0.3 to 1.6mm thick and 600 to 1,610mm wide (123,600 tonnes). – one continuous aluminizing line 0.27 to 1.6mm thick and 630 to 1,240mm wide (164,000 tonnes); one continuous galvanizing and aluminizing line 0.27 to 2.3mm thick and 630 to 1,250mm wide (279,000 tonnes); one continuous electrolytic copper-plating line 0.25 to 2.3mm thick and 508 to 1,150mm wide (43,000 tonnes); one continuous coating line 0.27 to 1.0mm thick and 600 to 1,250mm wide (136,000 tonnes).

■ Osaka Works, 1-171, Sakurajima 2-chome, Konohana-ku, Osaka City, Osaka Pref (Tel: 06 468 1231. Fax: 06 468 0861). Works general manager: Yutaka Okuno. ***Rolling mills:*** Two cold reduction – No 1 1-stand 0.1 to 2.3mm (157,000 tonnes); No 2 1-stand 0.2 to 4.0mm (227,000 tonnes).

■ Ichikawa Works, 7-1 Koyashinmachi, Ichikawa City, Chiba Pref (Tel: 0473 28 1111. Fax: 0473 27 5544). Works general manager: Michio Kubota. ***Coil coating lines:*** Two continuous hot-dip galvanizing – one 0.27 to 1.6mm thick and 630 to 1,250mm wide (274,000 tonnes); one 1.0 to 4.5mm thick and 710 to 1,219mm wide (212,200 tonnes); – two continuous coating lines – one 0.27 to 1mm thick and 610 to 1,060mm wide (52,800 tonnes); one 0.27 to 1.6mm thick and 610 to 1,250mm wide (72,200 tonnes).

Products: Steelmaking pig iron (1986 output, 3,007,000 tonnes). Carbon steel – ingots (total incl semis (all grades) output, 3,218,000 tonnes); slabs; hot rolled coil (for re-rolling); cold roll-formed sections (from strip); hot rolled hoop and strip (uncoated); skelp (tube strip); hot-dip galvanized hoop and strip; medium plates; heavy plates; clad plates; hot rolled sheet/coil (uncoated) (614,000 tonnes); cold rolled sheet/coil (uncoated) (468,000 tonnes); hot-dip galvanized sheet/coil; electro-galvanized sheet/coil; galvanized corrugated sheet/coil; aluminized sheet/coil; colour coated sheet/coil (coated products total output, 1,325,000 tonnes); longitudinal weld tubes and pipes; galvanized tubes and pipes; cold drawn tubes and pipes. Stainless steel – ingots; slabs; hot rolled band (for re-rolling); hot rolled hoop and strip (uncoated); cold rolled hoop and strip (uncoated); medium plates; heavy plates; hot rolled sheet/coil (uncoated); cold rolled sheet/coil (uncoated); longitudinal weld tubes and pipes; cold drawn tubes and pipes. Alloy steel (excl stainless) – ingots; slabs. Spring steels; bearing steels.

Brands: Moonstar Color (prepainted galvanized and galvannealed sheet and strip); Moonstar Vinytite (polyvinyl chloride-metal laminated sheet and strip); Moonstar Polyfoam (prepainted galvanized sheet and strip laminated with polyethylene foam); Lamitite (vinyl-coated galvanized sheet laminated with acrylic film); Moonstar Coppertite (electrolytic copper-plated sheet and strip); Alstar (aluminized sheet and shapes); White LG (galvanized light gauge sections); White Deck Plate (flooring and roofing); White Keystone Plate (flooring and roofing); Moonstar Bainite (special steel strip); Tecstar (prepainted galvanized and galvannealed sheet and strip); Weather-resistant Alstar (sheet and strip with extra-thick aluminium coating); Flowmate (indoor stainless pipes); Fine Color (coloured stainless sheet); Coppersoftain (electrolytic copper-plated stainless sheet and strip); Colorsoftain (prepainted stainless sheet); Moonstar Zinc (electrogalvanized sheet and strip); Fine Color N (colour stainless sheet and strip); Moonstar Color F (prepainted galvanized and galvannealed sheet and strip); Finesoftain (coloured stainless sheet).

Sales offices/agents: Singapore Representative Office, 16 Raffles Quay 20-01, Hong Leong bldg, Singapore 0104 (Tel: 221 1277-8. Telex: RS 24496. Fax: 65 224 2357); Nisshin USA Inc, 375 Park Avenue, New York, NY 10152, USA (Tel: 212 980 0580. Telex: 6801384. Fax: 212 421 0496); Düsseldorf Office, Immermannstrasse 45, D 4000 Düsseldorf 1, West Germany (Tel: Düsseldorf 353921/3. Telex: 8587316. Fax: 49 211 365460); Beijing Office, Room No 1701, Beijing Hotel, Beijing, China (Tel: 500 7766 Ext 1701. Telex: 2106).

NKK – Nippon Kokan KK

Head office: 1-2, 1-chome, Marunouchi, Chiyoda-ku, Tokyo 100. **Tel:** 03 212 7111. **Telex:** 222 2811 nkkj, 22 2816 nkship j. **Cables:** Kokannk Tokyo; Kokan Jukoo Tokyo; Kokanship Tokyo.
Management: Chairman of the board – Minoru Kanao; president – Yoshinari Yamashiro; executive vice presidents – Tsuneo Sekigawa, Toshio Isago, Akira Dote, Sachio Hatori; senior managing director – Hisanobu Nishiyama, Shigeyoshi Horie, Hiroshi Ono, Tetsuo Kumashiro, Shoichi Ide, Hiroshi Otaka, Hiroji Hirano; managing directors – Yasuyo Ishihara, Shunkichi Miyoshi, Noboru Hirayama, Tadashi Nishikawa, Hiroyoshi Matsubara, Yukio Tani, Tatsuo Era, Hachiro Tomoda, Osamu Sawaragi, Akifusa Sekizawa, Hisaya Nakata; director and executive adviser – Yukio Miyake; directors – Takashi Kawai, Sosuke Doi, Yukio Fujimoto, Eiichi Sakamoto, Seiji Funjii, Suemasa Imaizumi, Shigeyoshi Fukazawa, Haruo Kubotera, Minoru Hashimoto, Shigemasa Ogura, Soji Omura, Shunichi Tanaka, Tatsumi Osuka, Haruki Kamiya, Toshio Sugisaki. **Established:** 1912. **Capital:** Y 159,247,000,000 (at Mar 31st 1987). **No. of employees:** 29,150. **Subsidiaries:** Toshin Steel Co Ltd (42.0%) (ordinary steel ingots, sections, deformed, bars, fence materials and other fabricated products); Azuma Steel Co Ltd (41.4%) (wire rods, bar in coil, small bars, spiral-weld pipe); Tokyo Shearing Co Ltd (46.6%) (shearing, cutting, fabrication and sales of steel materials; manufacture and sale of press-formed products and automotive parts; construction); Nippon Kokan Light Steel Co Ltd (100%) (light sections quard rails and other road-related products; sheet piling and other civil engineering products; standard bridge members and other steel structures; related construction services); Kokan Kenzai Co Ltd (67.5%) (manufacture and sale of ERW pipe); Kawasaki Kokan KK (53.6%) (small dia, heavy wall seam welded pipe); Fuji Trading Co Ltd (53.6%) (sale and purchase of various types of steel pipe, cast iron pipe, and pipe fittings); Nippon Kokan Steel Warehouse Co (67.7%) (port transport, warehousing, automobile transport handling, and steel fabrication); Fuyo Kaiun KK (43.6%) (coastal shipping harbour transport service, customs clearance service, etc); Nissan Senpaku Ltd (45.0%) (marine transport, marine transport agency, and shipping agency); Nichiei Unyu Sohko Co Ltd (50.0%) (freight and automobile transport and warehousing); Kokan Mining Co Ltd (70.9%) (mining, quarrying, mine products, steelmaking by-products, limestone and other materials for steel production, mixed concrete production and sale); Nippon Kokan Pipe Fitting Mfg Co Ltd (40.3%) (production and sale of pipe fittings and attachments, malleable cast iron products, steel castings); Kokan Kikai Kogyo KK (100%) (design and manufacture of industrial machinery and steel structures of various types, construction contracting business, maintenance and consulting services for machinery and equipment); Fukuyama Kyodo Kiko KK (50%) (repairs, installation and building of machinery, electrical equipment and instrumentation for Fukuyama Works and associated companies); Nippon Kokan Koji KK (100%) (civil engineering and road pavement projects; gas, water and other piping installation projects; bridges and other structures, petrochemical and other construction projects); Nippon Rotary Nozzle Co Ltd (50%) (rotary nozzles and marketing of nozzle manufacturing technology); Japan Casting Co Ltd (34.7%) (design, manufacture and sale of steel castings, iron castings, rolls, moulds, valves of various types, machinery and plants); Nippon Cast Iron Pipe Co Ltd (33.1%) (cast iron pipe, valves, water supply equipment, civil engineering works, warehousing, trucking); NK Lease Co Ltd (50%) (leasing and sale of installment plans); NK Home Co Ltd (100%) (design, building and sale of homes; sale of interior and exterior products, sale of real estate). **Associates:** International Light Metal Corp (40%) (production and sale of bars, extrusions and forgings of aluminium and titanium alloys). **Annual capacity:** Pig iron: 20,270,000 tonnes. Raw steel: 21,030,000 tonnes.
Works:
■ Keihin works, 1-1 Minamiwatarida, Kawasaki City, Kanagawa Pref. (Tel: 044 355 1111. Telex: 2222811 nkk j). Works director: Tetsuo Kumashiro. ***Coke ovens:*** Two Carl Still batteries (198 ovens) (annual capacity 2,480,000 tonnes). ***Sinter plant:*** One Dwight Lloyd 450 sq m strand (5,260,000 tonnes). ***Blast furnaces:*** Two 13.5 m hearth dia, 4,050 cu m and 4,052 cu m (combined capacity 5,260,000 tonnes). ***Steelmaking plant:*** Three 250-tonne (top and bottom blowing) basic oxygen converters; one 50-tonne electric arc furnace (combined capacity 6,130,000 tonnes); hot metal dephosphorising plant. ***Refining plant:*** One 250-tonne NKK (arc process) ladle furnace; one 250-tonne RH vacuum degassing unit; one 50-tonne VAD; one

Japan

NKK – Nippon Kokan KK (continued)

50-tonne VOD; one 250-tonne and one 50-tonne powder injection units. ***Continuous casting machines:*** One 4-strand USSR vertical bloom (400x520mm) (1,050,000 tonnes); slab – one 2-strand Concast vertical bending (254 x 2,300mm) (1,420,000 tonnes), one 2-strand USSR bending (254 x 1,850mm) (1,300,000 tonnes), one 2-strand Demag bending (254 x 1,950mm) (1,440,000 tonnes); one 6-strand Demag bending round (170-230mm dia) (840,000 tonnes); one 2-strand NKK-Davy horizontal (170-330mm dia). ***Rolling mills:*** 1-stand 2-high (1,300 x 2,800mm) high-lift universal slabbing /blooming (annual capacity 2,550,000 tonnes); 1-stand 2-high (960 x 2,800mm) blooming; 4-stand vertical/horizontal billet; 1-stand 4-high 216" heavy plate (1,230 x 5,500 and 2,400 x 5,400mm) (1,445,000 tonnes); 94" semi-continuous wide hot strip with two roughing and six finishing stands (820 x 2,400mm, 1,624 x 2,350mm) (3,060,000 tonnes); 56" continuous tandem cold reduction with 5-stand 4-high (615 x 1,422mm, 1,400 x 1,422mm) (1,272,000 tonnes); one 2-stand 4-high tandem temper/skin pass (560 x 1,422mm, 1,346 x 1,422mm) (785,000 tonnes). ***Tube and pipe mills:*** seamless – Mandrel mill (Mannesman), 25.4 to 141.3mm od, 2.5 to 40mm wall thickness (245,000 tonnes); Mannesmann retrained mandrel mill (114.3 to 244.4mm od, 4.0 to 50mm wall thickness (600,000 tonnes); Mannesmann plug mill 152.4 to 426mm od, 5.5 to 80mm wall thickness (509,000 tonnes); Ugine-Séjournet extrusion mill, 40 to 230mm od, 3 to 40mm wall thickness (24,000 tonnes); Ehrhardt push bench, 241.8 to 1,066.8mm od, 30 to 200mm wall thickness (2,000 tonnes); Cold drawing plant, 6 to 290mm od, 1 to 30mm wall thickness; stainless cold drawing plant, 14 to 219mm od, 1 to 22mm wall thickness. welded – 24" Thermatool ERW mill, 267.4 to 609.6mm od, 3.2 to 16.1mm wall thickness (244,000 tonnes); 11¾" Thermatool ERW mill, 88.9 to 298.4mm od, 2.1 to 15.9mm wall thickness (223,000 tonnes); 4" ERW mill, 34.0 to 114.3mm od, 2.3 to 8.6mm wall thickness (35,000 tonnes); 4" Butt-weld mill, 21.7 to 114.3mm od, 1.9 to 6.02mm wall thickness (265,000 tonnes). ***Coil coating lines:*** One electrolytic tinning line, thickness 0.15 to 0.6mm, width up to 965mm (111,000 tonnes); continuous galvanizing lines – No 2 thickness 0.2-0.6mm, width up to 1,219mm (78,000 tonnes) No 3 thickness 0.27-2.3mm, width up to 1,300mm wide (216,000 tonnes); colour coating – No 1 continuous line (one coat one bake), thickness 0.2 to 1.0mm, width up to 1,219mm (12,000 tonnes), No 2 continuous line (two coat two bake), thickness 0.2 to 1.6mm, width up to 1,219mm (84,000 tonnes).

■ Fukuyama Works, 1 Kokan-cho, Fukuyama City, Hiroshima, Pref. (Tel: 0849 41 2111. Telex: 2222811 nkk j). Works director: Shunkichi Miyoshi. ***Coke ovens:*** Two Wilputte batteries (214 ovens) (annual capacity 1,660,000 tonnes); three Wilputte-Otto batteries (444 ovens) (4,460,000 tonnes). ***Blast furnaces:*** No 1 10.5m hearth dia, 2,323 cu m; No 2 11.5m hearth dia, 2,828 cu m; No 3 12.4m hearth dia, 3,223 cu m; No 4 14.0m hearth dia, 4,288 cu m; No 5 14.4m hearth dia, 4,617 cu m (combined capacity 14,510,000 tonnes). ***Steelmaking plant:*** basic oxygen – No 1 shop three 180-tonne LD converters, No 2 shop three 250-tonne (top and bottom-blown) converters, No 3 shop two 300-tonne (top and bottom-blown) converters (combined capacity 14,900,000 tonnes); hot metal dephosphorising plant. ***Refining plant:*** One 250-tonne NKK-arc process ladle furnace; two 250-tonne and one 300-tonne RH vacuum degassing units; one 250-tonne powder injection unit; ***Continuous casting machines:*** One 4-strand Concast bending bloom (250 x 490, 400 x 480) (1,290,000 tonnes); slab – One 2-strand Concast bending (300 x 1,600) (1,510,000 tonnes); one 2-strand Demag vertical bending (250 x 1,950) and one 2-strand Demag bending (300 x 2,100) (combined capacity 3,400,000 tonnes); one 2-strand Concast vertical bending (250 x 1,650) (2,050,000 tonnes). ***Rolling mills:*** 1-stand 2-high 1,168 x 2,432mm universal slabbing (3,960,000 tonnes); 1-stand 2-high 1,350 x 3,400mm high lift slabbing/blooming (2,410,000 tonnes). 1-stand 2-high 1,150 x 2,250mm blooming. 4-stand vertical/horizontal billet. Heavy section / rail – No 1 universal section mill comprising one 1,100 x 2,800mm breakdown stand, two universal (or 2-high) roughing stands, horizontal 1,575 x 1,530mm, vertical 1,070 x 560mm, (1,250 x 2,500mm) and one universal (or 2-high) finishing stand, horizontal 1,575 x 1,400mm, vertical 1,070 x 560mm (1,250 x 2,500mm) (1,132,000 tonnes); No 2 universal section mill comprising 1,050 x 2,500mm breakdown stand, one 2-high 1,050 x 2,500mm roughing stand, two universal (or 2-high) roughing stands, horizontal 1,130 x 800mm, vertical 780 x 260mm, (950 x 2,200mm), one universal

NKK - Nippon Kokan KK (continued)

(or 2-high) finishing stand, horizontal 1,060 x 770mm, vertical 780 x 260mm (950 x 1,600mm) (762,000 tonnes). 2-stand 4-high 185" heavy plate 1,050 x 4,700mm, 1,980 x 4,600mm, rougher and finisher (2,169,000 tonnes). Wide hot strip - No 1 80" continuous with five roughing and seven finishing stands, 762 x 2,057mm, 1,524 x 2,032mm (4,300,000 tonnes); No 2 70" continuous with four roughing and seven finishing stands 750 x 1,830mm, 1,570 x 1,780mm (3,690,000 tonnes). Cold reduction - No 1 with 5-stand 4-high 80" continuous tandem 603 x 2,032mm, 1,530 x 1,981mm (1,811,000 tonnes); No 2 with 5-stand 4-high 56" continuous tandem 610 x 1,425mm, 1,455 x 1,370mm (1,280,000 tonnes). Temper/skin pass - No 1 with 1-stand 4-high 610 x 2,032mm, 1,524 x 1,981mm (1,248,000 tonnes); No 2 tandem 2-stand 4-high 620 x 1,425mm, 1,435 x 1,400mm (900,000 tonnes); No 3 tandem 2-stand 4-high 520 x 1,420mm, 1,435 x 1,400mm (600,000 tonnes). ***Tube and pipe mills:*** welded - UOE submerged arc weld mill, 406.4 to 1,424mm od, 6.4 to 44.5mm wall thickness (581,000 tonnes); Spiral-weld pipe mill, 600 to 2,540mm od, 6.0 to 30.0mm wall thickness (76,000 tonnes). ***Coil coating lines:*** electrolytic tinning - Ferrostan, thickness 0.1 to 0.5mm, width up to 1,067mm (167,000 tonnes); tin-free steel (ECCS) - thickness 0.10 to 0.5mm, width up to 1,061mm (180,000 tonnes); hot-dip galvanizing - thickness 0.35 to 4.5mm, width up to 1,829mm (187,000 tonnes); electro-galvanizing - No 1 thickness 0.31 to 2.3mm, width up to 1,835mm (170,000 tonnes); No 2 thickness 0.31 to 1.2mm, width up to 1,270mm (98,000 tonnes); No 3 thickness 0.4 to 1.6mm, width up to 1,880mm (240,000 tonnes); No 1 universal (two coat two bake) thickness 0.5-2.3mm, width up to 1,880mm (60,000 tonnes).

■ Toyama Works, 11-1 2-chome, Shosai-cho, Shimminato City, Toyama Pref. (Tel: 0849 41 2111). Works director: Yasuo Harada.Toyama Works (established 1917) : Seven electric furnaces with an aggregate capacity of 119,000 KVA (annual capactiy 149,000 tonnes). Main products-ferro-chromium, ferro-molybdenum, ferro-vanadium, nitrogen-bearing alloys, electro-fused products (calcium silicate, magnesia, magnesia chromia, zirconia), ferro-boron, Silicon nitride, high-purity zirconia and master alloys (V-Al, Mo-Al etc).

■ Niigata Works, 2-3 Kamiose-cho, Niigata City, Niigata Pref. Works director: Hiroshi Iwabuchi. Niligata Works (established 1935) : Five electric furnaces with an aggregate transformer capacity of 109,500 kVA (annual capacity 240,000 tonnes) and precision metal products plants. Main products-ferro-manganese, silico-manganese, cold forged products, powder metallurgy products, investment casting and high-purity fused silica.

Products: Steelmaking pig iron (12,065,000 tonnes). Carbon steel - (output covers carbon, stainless and alloys steel for Apr 86 - March 87): Ingots (235,000 tonnes); slabs (9,435,000 tonnes); blooms (903,000 tonnes); billets (350,000 tonnes); semis for seamless tubemaking; hot rolled coil (for re-rolling); reinforcing bars (3,000 tonnes); heavy angles 65sq mm, 250mm sq, 100 x 75-550 x 150mm; heavy tees 250 x 6-300 x 32mm and heavy joists 250 x 125-300 x 150mm (combined capacity 291,000 tonnes); heavy channels 75 x 40-380 x 100; wide-flange beams (heavy) 4 x 4-27 x 10, 100 x 50-900 x 300mm, 100mm sq, 500mm sq; heavy rails JIS, JRS 37-60 kg, 115-136 lb; sheet piling U type (400 x 85-500 x 225) Z type (400 x 305-400 x 367) and pipe piling 318.5-2,500mm od, 6.9-250mm thickness (combined capacity 156,000 tonnes); hot rolled hoop and strip (uncoated) 2.3-4.5 x 914-1,524mm; cold rolled hoop and strip (uncoated) 0.7-1.2 x 914-1,524mm; hot-dip galvanized hoop and strip 1.0-1,000mm; medium plates and heavy plates 6-250 x 1,200-5,300 x 3,000 x 25,000mm, universal plates (combined capacity 1,436,000 tonnes); floor plates; clad plates; hot rolled sheet/coil (uncoated) 1.2-25.4 x 610-2,300 (2,614,000 tonnes); cold rolled sheet/coil (uncoated) 0.2-3.2 x 508-1,829 (2,216,000 tonnes); hot-dip galvanized sheet/coil 0.27-4.5 x 610-1,829 (630,000 tonnes); electro-galvanized sheet/coil 0.3-2.3 x 762-1,829 (554,000 tonnes); colour coated sheet/coil 0.27-1.6 x 610-1,219 (63,000 tonnes); electrolytic tinplate (single-reduced) 0.18-0.6 x 600-1,067 (382,000 tonnes); tin-free steel (ECCS) 0.2-0.5 x 600 x 1,067 (197,000 tonnes); seamless tubes and pipes 25.4-426mm od, 2.5-80mm wall thickness (564,000 tonnes); longitudinal weld tubes and pipes ERW 31.8-609mm od 2.1-16.1 wall thickness, UOE 406.4-1,424mm od 6.4-44.5mm wall thickness (588,000 tonnes ERW/buttweld, 429,000 tonnes UOE); spiral-weld tubes and pipes

Japan

NKK – Nippon Kokan KK (continued)

406.4-2,540mm od 6-30mm wall thickness (72,000 tonnes); large-diameter tubes and pipes 241.8-1,066.8 od, 30-200mm wall thickness; oil country tubular goods; galvanized tubes and pipes; plastic coated tubes and pipes; cold drawn tubes and pipes. Stainless steel – ingots; slabs; blooms : Billets; semis for seamless tubemaking; medium plates; heavy plates; clad plates; seamless tubes and pipes; oil country tubular goods and cold drawn tubes and pipes 14-406.4mm od, 1-35mm wall thickness.
Brands: NK-LTC (low-temperature steel); NK-HITEN (high tensile strength steel); CUPLOY, CUPTEN (atmospheric corrosion-resistant steel); NK-MARINE (seawater-resistant steel); NK-EH (abrasion-resistant steel); NAC (anti-corrosion steel); Palm-Zinc, Palm-Zinc Alloy (hot-dip galvanized coil/sheets), Uni-Zinc Alloy, Excel-Zinc (electrolytically galvanized coil/sheets); Light-well (electrolytic tinplate); Britecote (tin-free steel); NKK-Colour, NKK-high-bake (colour-coated coil/sheets); TEMPALOY (high-temperatured high tensile strength steel); NK-PLP (plastic lining pipe).
Sales offices/agents: Nippon Kokan KK, Bangkok Liaison Office, Thaniya Building 6th Floor, 62 Silom Road, Bangkok, Thailand (Tel: 2369124, 2360126, 2363514. Telex: 87002 nkk th); Nippon Kokan KK, Beijing Office, Room 3, Floor 12, CITIC Building, Jianguomenwai Street, Beijing, People's Republic of China (Tel: 500 4375. Telex: 22630 nnk cn); Nippon Kokan KK, Hong Kong Office, Room 1502-3, 15th Floor, Bank of America Tower, 12 Harcourt Road, Central, Hong Kong (Tel: Hong Kong 5-215120, 215129. Telex: 80283003 kokan hx); Jakarta Representative of Nippon Kokan KK, 30th Floor, Ratu Plaza, Kav 9 Jl Jend, Sudirman, Jakarta, Indonesia (Tel: 711874, 711875. Telex: 47346 nkk jkt); Nippon Kokan KK, Malaysia Office, 2nd Floor Mui Plaza, Jalan Parry, Kuala Lumpur 01-04, Malaysia (Tel: 03 2484369, 2431208. Telex: 32528); Nippon Kokan KK, Singapore Office, Unit 3008, UIC Building, 5 Shenton Way, 0106 Singapore (Tel: 2217277. Telex: 21308 nkk sp rs. Cable: Steeltube Singapore); Nippon Kokan Arabia Ltd, Al-Bustan Compound, Al-Khobar-Damman Highway, Saudi Arabia (Tel: 03 894 8011 Ext 1257. Telex: 670028 wateka sj); Nippon Kokan KK, Düsseldorf Office, Immermannstrasse 43, D 4000 Düsseldorf 1, West Germany (Tel: 0211 353481-3. Telex: 8587839 nkk d. Cable: Kokannk Dusseldorf); NKK (UK) Ltd, 4th Floor, West Block, 11 Moorsfields High Walk, London EC2Y 9DE (Tel: 01 628 2161-5. Telex: 886310 nkk ln g, 887602 nkk ln g. Cable: Kokannk London); NKK Netherlands BV, 9th Floor, Room 913 Calandstraat 66, 3016CD, Rotterdam, Netherlands (Tel: Rotterdam 010 436 13 44. Telex: 26700 nkkrt nl); NKK America Inc, Houston Office, Four Allen Center, 1400 Smith Street, Suite 750, Houston, TX 77002 USA (Tel: 713 658 0611-4. Telex: 775691 nkk hou. Cable: Kokannk Houston); NKK America Inc, Los Angeles Office, 444 South Flower Street, Suite 2430, Los Angeles, CA 90017, USA (Tel: 213 624 6651. Telex: 9103212456 nkk lsa. Cable: Kokannk Los Angeles California); NKK America Inc, New York Office, 450 Park Avenue, New York, NY 10022, USA (Tel: 212 826 6250. Telex: 233495 nkk ur. Cable: Kokannk New York); Nippon Kokan KK, Vancouver Office, PO Box 49168, Four Bentall Ce ntre, Suite 3394-1055, Dunsmuir Street, Vancouver, BC V7X-1J1, Canada (Tel: 604 687 0091. Telex: 4507539 nkk vcr. Cable: Kokannk Vcr); Nippon Kokan KK, Oficina Tecnica Informativa en Mexico, Paseo de la Reforma, 300-1905 Col. Juarez, CP 06600 Mexico, DF Me xico (Tel: 5333439, 5333451, 5330238. Telex: 1771781 nkk me); Nippon Kokan-Indústria Siderúrgica Ltda, Rua da Assembleia 10 S/3501, Edificio Centro Candido Mendes, 20011 Centro Rio de Janeiro, RJ Brazil (Tel: 021 224 4570, 6477. Telex: 2122843 nm mr br).
Expansion programme: Keihin Works: One new ERW mill, 31.8 to 114.3mm od, 2.1 to 13.5mm thickness – August 1987 (capacity 42,000 tonnes/year); one new continuous colour line (2-coat 2-bake) – December 1988 (capacity 42,000 tonnes/year); Fukuyama Works: One electrolytic galvanizing line – September 1987 (capacity 300,000 tonnes/year); one uiversal coating line – Spetember 1987 (capacity 150,000 tonnes/year); one tandem cold strip mill – July 1987 (capacity 1,560,000 tonnes/year).
Additional information: NKK's, aim is to launch ventures in new fields and to bring them to fruition as fully fledged, successful businesses. Among the fields that it considers as having especially strong growth potential are new materials, machinery, community development, health and sports, food, and aerospace. The company has established a New Materials Division – a fourth business sector alongside its steel, shipbuilding, and engineering and construction divisions.

Osaka Steel (Osaka Seitetsu)

Head office: 9-3, 1-chome Minamiokajima, Taisho-ku, Osaka 551. **Tel:** 06 552 1441.
Management: President – Hitoshi Shitara; managing directors – Masanori Umehara, Toranosuke Kawazoe, Tsukasa Nagase; managers – Tetsuo Matsuoka, Osamu Yashiki, Koji Nishino. **Established:** 1978. **Capital:** Y 960,000,000.
Works:
■ Okajima (Osaka) and Tsumori (Osaka). ***Rolling mills:*** light section and bar.
Products: Carbon steel – light angles; light channels; special light sections (elevator rails, car wheel trims); light rails; rail accessories (fishplates). Expanded metal, deck plate.
Additional information: Tokyo office: Nihon Building, 6-2, 2-chome, Otemachi, Chiyoda-ku, Tokyo 100 (Tel: 03 279 0576).

Pacific Metals Co Ltd (Taiheiyo Kinzoku)

Head office: Ohtemachi Bldg, 6-1 Ohtemachi 1-chome, Chiyoda-ku, Tokyo 100. **Tel:** 03 201 6681. **Telex:** 2222248 pamco j. **Cables:** Pacificpamco Tokyo. **Telecopier (Fax):** 03 212 7876.
Management: President – Yoshishige Nagoya; senior managing directors – Hideo Jingu, Shozo Tsunoda, Shinkichi Koike; managing directors – Kotaro Ishii, Tohru Shiroi, Noboru Tsumita. **Established:** 1949. **Capital:** Y 7,362,000,000. **No. of employees:** Works – 869. Office – 116. **Ownership:** Nippon Steel Corp. **Associates:** Sowa Kinzoku Co Ltd (sales company for iron and steel, etc); Pacific Rundum Co Ltd (Nisso Rundum); Pacific Special Alloy Castings Co Ltd (special alloy castings); Pacific Steel Mfg Co Ltd (ordinary and special forged steels, NT rolls); Pacific Machinery & Engineering Co Ltd (pumps, construction machinery, etc). **Annual capacity:** Finished steel: 150,000 tonnes.
Works:
■ Hachinohe Works, Toyama Shinden, Hachinohe City, Aomori Pref. (Tel: 0178 43 7111. Telex: 0178 45 8118). Works director: Shinkichi Koike. ***Steelmaking plant:*** One 30-tonne LD basic oxygen converter (annual capacity 75,000 tonnes); one 25-tonne electric arc furnace (75,000 tonnes). ***Refining plant:*** Two 30-tonne AOD (150,000 tonnes); one 30-tonne Asea-SKF. ***Continuous casting machines:*** One 2-strand Mannesmann straight mould billet and bloom (105-205mm sq) (80,000 tonnes); one 2-strand curved mould slab (130-150mm thick, 500-1,300mm wide) (70,000 tonnes). ***Other plant:*** Ferro-nickel plant (four rotary kilns, three closed electric furnaces); ferro-alloy plant (three rotary kilns, four closed electric furnaces, one shaking converter); stainless steel wire rod plant (pickling and annealing line); stainless steel round bar plant (annealing, bar straightner, turning, polishing machine); stainless steel powder plant; rock wool plant; blast scavenger plant.
■ Niigata Works, Tarohdai Sambre, Niigata City, Niigata Pref. (Tel: 0252 55 3161. Telex: 3122 447). Works director: Kotaro Ishii. ***Rolling mills:*** 2-high Sendzimir cold strip (450mm max width) (coil weight 3.5 tonne) (annual capacity 4,800 tonnes, stainless steel). ***Other plant:*** Plant for magnetic material; cold strip plant; activated carbon plant.
Products: Stainless steel – ingots, 1-45 tonnes; slabs 130-150mm thick, 500-1,300mm wide; blooms 165, 175 and 205mm sq; billets 105, 115 130 and 140mm sq; wire rod 5.5-13.5mm dia; round bars 9-100mm dia; cold rolled sheet/coil (uncoated) 0.1-4.0mm thick x 200-450mm wide; powder, under 100 mesh. Grades: Austenitic stainless steel – AISI 302, 303, 304, 305, 308, 309, 310, 316, 317, 321, 347 and low-carbon grade; SUS 630 (17-4PH), 305 JI, 329 JI; AWS 300 series; high-nickel alloy. Ferro-nickel; ferro-manganese; rock wool; blast scavenger.
Sales offices/agents: Tokyo office – Business Dept II, Ohtemachi Bldg, 6-1 Ohtemachi 1-chome, Chiyoda-ku, Tokyo (Tel: 03 201 6672).

Japan

Rinko Steel Works Co Ltd

Head office: 18-5-2, Itachibori, Nishi-ku, Osaka 550. **Tel:** 6 532 2771. **Telecopier (Fax):** 06 532 2766.
Management: President – Tamotsu Sato; chairman – Nobuhisa Hakaridani; vice president – Shuko Nakano; general managers – Isamu Tanaka (executive), Kazuyuki Kamei, Hidenori Anami. **Established:** 1933. **Capital:** Y 360,000,000. **No. of employees:** Works – 350. Office – 55. **Annual capacity:** Raw steel: 275,000 tonnes. Finished steel: 320,000 tonnes.
Works:
■ 4700, Hoshida, Katano City, Osaka. (Tel: 0720 91 0621). Works director: Kamei Kazuyuki. ***Steelmaking plant:*** 50-tonne NK electric arc (annual capacity 275,000 tonnes). ***Refining plant:*** 40-tonne Daido steel ladle furnace. ***Continuous casting machines:*** 3-strand Mitsubishi-Olsson billet (100-120mm) (270,000 tonnes). ***Rolling mills:*** light section (channels etc) (40,000 tonnes). Bar – One 3-high, (12-80mm) (160,000 tonnes and one 3-high for flats (3 x 25-25 x 150mm) (120,000 tonnes).
Products: Carbon steel – billets 100-210mm (1986 output 265,000 tonnes); round bars 12-80mm (105,000 tonnes); square bars 12-75mm (40,000 tonnes); flats 6 x 25-25 x 105mm (110,000 tonnes) and octagans 16-32mm (1,000 tonnes); light channels 5 x 25 x 50mm (2,000 tonnes); special light sections (2,000 tonnes).

Sanwa Metal Industries Ltd

Head office: 15-27 Shinmachi 2-chome, Nishi-ku, Osaka 550. **Tel:** 06 543 3037. **Telex:** 5253057. **Cables:** Sanmetal Osaka. **Telecopier (Fax):** 06 541 8519.
Management: President – Harushige Den; directors – Shuzo Adachi (executive), Yasuo Tsuda (managing, sales), Torao Kamel (production), Hideaki Kumazawa (finance), Toshihro Adachi, Akinobu Taguchi. **Established:** 1946. **Capital:** Y 300,000,000. **No. of employees:** Works – 70. Office – 160. **Subsidiaries:** Kasadera Kogyo (cold drawn steel tube and pipe); Houkoku Kinzoku Kogyo KK (cold drawn steel tube and pipe); Shinshin Kogyo KK (steel pipes and tubes); Osaka Kokusai Kikai KK; Sanwa Nougei Shisetsu KK (agricultural equipment); Sanwa Kougaku KK; Sanwa Kensetsu Kigyo KK (building construction); Toua Jitsugyo KK (agricultural equipment); Sanwa Kinzoku Kogyo Tokyo Kojo KK; Touzai Kogyo KK (cold toll-formed products, etc); Sanwa Suido Kigyo KK (water supply and drainage works); Sanwa Metal America Inc (import and export). **Annual capacity:** Finished steel: 12,000 tonnes.
Works:
■ 39 Osadano-cho, 1-chome, Fukuchiyama, Kyoto Prefecture. (Tel: 0773 27 6255). Works director: Torao Kamel. Cold drawing plant for tubes (draw-benches, automatic pickling, straighten, bright annealing.
Products: Carbon steel – cold drawn tubes and pipes 3 x 0.3 to 70 x 10mm. Stainless steel – cold drawn tubes and pipes. Alloy steel (excl stainless) – cold drawn tubes and pipes 3 x 0.3 to 70 x 10mm.
Brands: SK.
Sales offices/agents: Overseas branch office: Los Angeles, USA.
Additional information: Cold drawn seamless and welded tubes are produced for hydraulic, refrigeration and condenser applications in industrial equipment, petrochemical, shipbuilding, the motor industry and nuclear power. The company also represents other Japanese producers of seamless and welded (including spiral) steel tubes, alloy steel tubes, gas and water pipes, wide-flange beams, etc.

Sanyo Special Steel Co Ltd

Head office: 3007 Nakashima, Shikama-ku, Himeji-shi, Hyogo-ken. **Tel:** 0792 35 6111. **Telex:** 5772110. **Telecopier (Fax):** 0792 34 8571.
Management: Chairman – Shunji Ohuchi; president – Toshikazu Uesugi; executive vice presidents – Tomoyasu Genma, Hiroyuki Inoue; senior managing directors – Kunihiro Nishio, Koichi Asano, Masao Kimura; managing directors – Takeo Tomita, Nobuaki Sugiyama, Kozo Seto, Yoshinori Fujihana, Hiroshi Tateoka, Takehiro Hara. **Established:** 1933. **Capital:** Y 8,025,730,394. **No. of employees:** Works – 2,314.

Sanyo Special Steel Co Ltd (continued)

Office – 752. **Subsidiaries:** Santoku Kogyo Co Ltd (100%) (subcontract general work); Yoko Bussan Co Ltd (100%) (trading); Santoku Fudosan Co Ltd (50%) (administration of property); Santoku Kako Co Ltd (100%) (subcontract bearing ring machining from tubes). **Annual capacity:** Raw steel: 1,512,000 tonnes. Finished steel: 960,000 tonnes.
Works:
■ At head office. ***Steelmaking plant:*** Three electric arc furnaces – one 30-tonne (annual capacity 108,000 tonnes), one 60-tonne (1973) (540,000 tonnes), one 90-tonne (1982) (864,000 tonnes). ***Refining plant:*** One 2-tonne ESR (2,400 tonnes); one ladle furnace (864,000 tonnes); three RH vacuum degassing units (1,512,000 tonnes); one 1-tonne VAR (1,200 tonnes); one 1-tonne VIM (1,200 tonnes). ***Continuous casting machines:*** One 3-strand vertical bloom with electro-magnetic stirring (370 x 470mm) (864,000 tonnes). ***Rolling mills:*** Kawasaki/SMS billet – 3-roll planetary mill (828,000 tonnes). Two bar – Kobelco large-diameter (660,000 tonnes) and No 1 medium and small-diameter (540,000 tonnes); No. 2 medium and small-diameter bar and wire rod (624,000 tonnes). ***Tube and pipe mills:*** seamless – one 2,000-tonne Ugine-Séjournet hot extrusion press (1959) (30-148mm od, 3-25mm wall thickness (37,200 tonnes); one 1,250-tonne hot extrusion press (25-70mm od, 3-20mm wall thickness) (28,800 tonnes); one Assel-type (1970) (50-180mm od, 5-30mm wall thickness) (138,000 tonnes); fourteen Meer cold Pilger mills (81,600 tonnes). ***Other plant:*** 3,000-tonne Mitsubishi/Pahnke hydraulic forging press (36,000 tonnes). Bar turning shop – seven 15-80mm dia peeling machines. Cold bar drawing shop – one 10-30mm dia combination machine; one 8-14mm dia coil-to-bar straightener. Wire shop – six 4-23mm dia block (drum) drawing machines.
Products: Stainless steel – wire rod 5.5-40mm dia; round bars 12-550mm dia; seamless tubes and pipes 16-148mm od, 1.6-30mm wall thickness; hollow bars 16-148mm od, 5-30mm wall thickness. Alloy steel (excl stainless) – (including tool, free cutting and bearing steels and superalloy) round bars 12-550mm, wire 4-23mm dia, seamless tubes and pipes 16-180mm od and 1.6-30mm wall thickness. Round shapes with contoured internal diameter, made in the cold Pilger mill; shapes and die blocks machined on the plano-miller.
Brands: QD4, QD61, QD62, QD10, QDR1, QT41 (hot work tool steels); QKS3, QKS95, QC1, QC11, QCP1, QCM8, QJ2, QJ2L (cold work tool steels); QH51, QH50, QHS (QH-Super) (high-speed steels).
Sales offices/agents: Export department – Nittetsu Nihonbashi Bldg, 1-13-1 Nihonbashi, Chuo-ku (Tel: 03 278 8323). New York office – Rm 2104, 445 Park Avenue, New York, NY 10022, USA (Tel: 212 935 9033. Telefax: 212 980 8838).
Additional information: Steel made in a combination of the 30-tonne electric arc furnace and R-H vacuum degassing vessel or the 60-tonne electric arc furnace and the R-H vacuum degassing vessel is bottom-poured to ingots. Steel made in a combination of the 90-tonne electric arc furnace and the ladle furnace and the R-H vacuum degasser is continuously cast to blooms and immediately, upon reheating, rolled in the 3-roll planetary mill to billets. 2.6-tonne ingots are rolled in the large-diameter bar mill to 80-250mm dia bars and billets. 5.4-tonne and 8-tonne ingots are forged in the 3,000-tonne hydraulic press to rounds up to 550mm dia. Bars 12-80mm dia are re-rolled in No 1 medium and small-diameter bar mill. Bars 15-80mm dia and wire rods 5.5-40mm are re-rolled in No 2 medium and small-diameter bar mill.

Shinsei Kogyo Co Ltd—see Nisshin Steel Co Ltd

Sumikin Stainless Steel Tube Co Ltd

Head office: Oorza Okasato 3-2, Souwa-cho, Sashima-gun, Ibaraki Pref 306.
Ownership: Sumitomo Metal Industries. Stainless steel tubes.

Japan

Sumikin Weld Pipe Co Ltd

Head office: 2 Dejima Nishi-cho, Sakai, Osaka.
Ownership: Sumitomo Metal Industries.
Products: Carbon steel – large-diameter tubes and pipes (welded).

Sumitomo Electric Industries Ltd (Sumitomo Denki Kogyo)

Head office: 1-1 Koyatita, Itami 664. **Tel:** 0727 81 5151. **Telex:** 5326430.
Established: 1897.
Works:
■ At head office. ***Steelmaking plant:*** electric arc. ***Continuous casting machines:*** One 2-strand progressive bending billet. ***Rolling mills:*** Two wire rod (1961 and 1975). Wire drawing plant.
Products: Carbon steel – wire rod (high-carbon); bright wire; black annealed wire. Stainless steel – wire rod; wire. Alloy steel (excl stainless) – wire rod; wire.

Sumitomo Metal Industries Ltd

Head office: 15 Kitahama 5-chome, Higashi-ku, Osaka. **Tel:** 06 220 5111. **Telex:** 63490. **Cables:** Sumitomometal Osaka. **Telecopier (Fax):** 06 223 0305.
Management: Chairman – H. Hyuga (honorary), Y. Kumagai (board); president – Y. Shingu; executive vice presidents – H. Kojima, K. Kinoshita; senior managing directors – T. Tanaka, S. Tamamoto, Y. Masuko, R. Fukase, K. Kawakami, S. Matsuoka; managing directors – R. Mori, T. Nakamura, Y. Kasahara, K. Tamaki, H. Takeuchi, Y. Ito, S. Fujii, S. Tatemoto, Y. Yoshitomi; directors – S. Naruse, T. Nakamura, H. Fukui, H. Hara, T. Nishizeki, D. Kanmuri, H. Nashiwa, T. Yukitoshi, T. Kurashige, K. Maruyama, I. Takai, H. Kugimiya, T. Tanaka. **Established:** 1897. **Capital:** Y 132,172,000,000. **No. of employees:** Works – 21,780. Office – 2,460. **Subsidiaries:** Nippon Pipe Mfg Co Ltd (conduit, welded pipe, seamless tubes, metal tube and pipe); Sumikin Weld Pipe Co Ltd (large diameter welded pipe); Sumikin Piping & Construction Co Ltd (engineering of pipelines, pipe structures); Sumikin Metal Products Co Ltd (manufacture and production sale of cold formed sections, pipe fabricating); Sumikin Kiko Co Ltd (processing of gas containers, pipe specialities); Sumikin Welding Industries Ltd (welding materials and equipment); Nippon Stainless Steel Co Ltd (stainless steel products); Igeta Steel Sheet Co Ltd (colour-coated sheets and plates); Sumikura Kogyo Co Ltd (industrial machinery); Umebachi Kogyo Co Ltd (production and sale of wire products); Sumikin Precision Forge Inc (production and sale of automobile parts); Kizugawa Shearing Steel Center Ltd (shearing and processing of steel sheets, plates and coils); Wako Steel Co Ltd (shearing and processing of steel sheets, plates and coils); Sumikin Stainless Steel Tube Co Ltd (stainless steel tubes); Sumikin Coke & Chemicals Co Ltd (coke, chemicals); Kyodo Oxygen Co Ltd (oxygen, nitrogen, argon); Chuo Denki Kogyo Co Ltd (ferro-alloys, electrolytic manganese metal); Sumimetal Mining Co Ltd (limestone); Wakayama Kyodo Power Co Inc (supply electricity); Kashima Kyodo Power Co Ltd (supply electricity); Sumikin Bussan Kaisha Ltd (sale of Sumitomo Metals products, etc); Sumikin Shipping Co Ltd (marine transport, marine insurance agency service); Daiichi Chuo Kisen Kaisha (marine transport, shipping agency); Sumitomo Special Metals Co Ltd (permanent magnets, powder alloys, electro-magnetics); Osaka Titanium Co Ltd (titanium sponge, high-purity silicon); Sumitomo Precision Products Co Ltd (aircraft equipment, heat exchangers, hydraulic controls, environmental equipment); Kanto Special Steel Works Ltd (rolls); Daikin Industries Ltd (freezers, compressors, freon gas, industrial robots); Narumi China Corp (pottery and ceramics); Kashiwara Machine Mfg Co (processing of pipe couplings, industrial machines); Western Tube & Conduit Corp, USA (steel conduit, steel tubes for structural purposes); Quality Metals Inc, USA (sale of flat products and others); Tube Turns Inc, USA (elbows and fittings, upset forgings, engineered products); Thai Steel Pipe Industry Co Ltd, Thailand (electric welded pipe, galvanized pipe); National Pipe Co Ltd, Saudi Arabia (large diameter welded pipe). **Annual capacity:** Pig iron: 22,390,000 tonnes. Raw steel: 21,512,000 tonnes. Rolled steel: 15,807,000 tonnes.
Works:
■ Wakayama Steelworks, 1850 Minato, Wakayama (Tel: 0734 51 2345). Works director: Y. Masuko. ***Sinter plant:*** Five 757 sq m strand Dwight Lloyd (annual capacity

Sumitomo Metal Industries Ltd (continued)

9,053,000 tonnes). ***Blast furnaces:*** Five – 8.9m hearth dia, 1,633 cu m (1,060,000 tonnes); 10.0m hearth dia, 2,100 cu m (1,680,000 tonnes);10.0m hearth dia, 2,150 cu m (1,680,000 tonnes); 11.1m hearth dia, 2,700 cu m (2,170,000 tonnes); 11.1m hearth dia, 2,700 cu m (2,170,000 tonnes). ***Direct reduction plant:*** One 80m x 3.9m SDR rotary kiln (156,000 tonnes). ***Steelmaking plant:*** Seven basic oxygen converters – two 160-tonne LD (2,428,000 tonnes), four 160-ton STB (top and bottom blown) (5,814,000 tonnes), and one 70-tonne LD (810,000 tonnes). ***Refining plant:*** Three vacuum degassing units – one 100-tonne DH, and two 160-tonne RH. ***Continuous casting machines:*** two Concast S for bloom – one 1-strand (450 dia, 370 to 400 x 600mm) (450,000 tonnes), and one 4-strand (370 x 600mm (1,170,000 tonnes); one 2-strand Concast S for slab and rounds (213 to 282mm o.d., 190 to 300mm x 960 to 1,840mm) (941,000 tonnes); and two 1-strand Concast S for slabs and blooms (210 to 400 x 600 to 1,900mm (2,160,000 tonnes). ***Rolling mills:*** 2-stand 2-high reversing slabbing /blomming 115mm sq to 420 x 480mm, 147-400mm o.d. (1,840,000 tonnes); one universal slabbing 110 x 680-380 x 1,195mm (4,590,000 tonnes); blooming / billet comprising 2-stand 2-high reversing blooming and 6-stand continuous billet mill (90mm sq-350 x 450mm, 147-335mm od (2,310,000 tonnes); one 4-high 170" reversin8 heavy plate – 6-150mm x 1,000 to 4,280mm (1.020,000 tonnes); wide hot strip – 80" semi-continuous 1-stand 2-high roughing scale breaker, 1-stand 4-high reversing roughing and 7-stand 4-high finishing (coil weight 25-tonnes) 1.2-19.0mm x 560-1950mm (3,000,000 tonnes); two cold reduction – 1-stand 4-high 80" reversing 0.4 to 3.2mm x 610 to 880mm (coil weight 33 tonnes) and 5-stand 56" tandem cold strip 0.15 to 2.3mm x 610 to 1,270 mm (coil weight 33 tonnes) (combined annual capacity 960,000 tonnes); 1-stand 4-high 56" temper/skin pass 0.15 to 2.3mm x 610 to 1,270mm (coil weight 33 tonnes). ***Tube and pipe mills:*** one Mannesmann plug mill seamless 219.1 to 426mm o.d. (570,000 tonnes). Welded – five ERW 19 to 609.6mm o.d. (687,000 tonnes), one UOE 406.4 to 1,066.8mm o.d. (240,000 tonnes). ***Coil coating lines:*** Two Sendzimir hot-dip galvanizing 0.198 to 3.2mm thick, 1,230mm wide (228,000 tonnes); one horizontal electro-galvanizing 0.25 to 3,2mm thick, 1,880mm wide (204,000 tonnes).

■ Kokura Steelworks, 1 Konomi-cho, Kokurakita-k, Kitakyushu, Fukuoka. (Tel: 093 561 4121). Works director: T. Nakamura. ***Sinter plant:*** One 222 sq m strand Dwight Lloyd (annual capacity 2,555,000 tonnes). ***Blast furnaces:*** Two – 8.4m hearth dia, 1,400 cu m (1,030,000 tonnes); 9.6m hearth dia, 1,850 cu m (1,460,000 tonnes). ***Steelmaking plant:*** Three 70-ton LD basic oxygen converters (1,945,000 tonnes). ***Refining plant:*** One 70-tonne VAD unit. ***Continuous casting machines:*** One 3-strand Concast vertical with bending for bloom (180mm sq, 300 x 400mm (640,000 tonnes). ***Rolling mills:*** One 1-stand 2-high reversing blooming mill and 6-stand continuous billet (72-500mm sq) (1,560,000 tonnes); One 1-strand 21-stand continuous bar (13-105mm dia) (coil weight 3-tonnes) (800,000 tonnes); 4-strand 27-stand continuous wire rod (5.5-18mm dia) (coil weight 2-tonnes) (900,000 tonnes).

■ Kashima Steelworks, 3 Hikari, Kashima-cho, Kashima-gun, Ibaraki. (Tel: 0299 82 2111). Works director: S. Tamamoto. ***Sinter plant:*** One 223 sq m strand Dwight Lloyd (annual capacity 2,522,000 tonnes); one 500 sq m strand Dwight Lloyd (5,658,000 tonnes); one 600 sq m strand Dwight Lloyd (6,789,000 tonnes). ***Blast furnaces:*** Three 13.4m hearth dia, 3,680 cu m (3,150,000 tonnes); 13.8m hearth dia, 4,080 cu m (3,570,000 tonnes); 15.0m hearth dia, 5,050 cu m (4,420,000 tonnes). ***Direct reduction plant:*** One 71m long x 4.5m SPM rotary kiln (150,000 tonnes). ***Steelmaking plant:*** Five 250-tonnes STB (top and bottom blown) basic oxygen converters (10,150,000 tonnes). ***Refining plant:*** One 270-tonne ladle furnace; two 270-tonne RH vacuum degassing units. ***Continuous casting machines:*** Two 2-strand Concast S for slab (220 to 300mm x 700 to 2,300mm) (2,469,000 tonnes), one 1-strand Concast SH for slabs and blooms (300 x 500 to 1,700mm) (430,000 tonnes), and one 2-strand Concast vertical with bending for slabs (270 x 700 to 1,600mm (2,880,000 tonnes); ***Rolling mills:*** One universal slabbing (80 x 610-500 x 2,300mm) (5,780,000 tonnes); one heavy section – 1-stand 2-high reversing breakdown roughing, 1-stand universal roughing, 1-stand 2-high reversing edging and 1-stand universal finishing producing wide flange beams 100 x 50 to 900 x 300mm (1,200,000 tonnes); light section – Thermatool welded wide flage beam mill,

Japan

Sumitomo Metal Industries Ltd (continued)

producing wide flange beams 100 x 50mm to 450 x 300mm (180,000 tonnes); heavy plate – 4-high 210″ reversing roughing and 4-high 186″ finishing (4.5 to 300mm x 1,000 to 5,200mm) (1,920,000 tonnes); wide hot strip – 70″ continuous 3-stand 2-high roughing, 3-stand 4-high roughing and 7-stand 4-high finishing (1.0 to 25.0mm x 610 to 1,673mm) (coil weight 31-tonnes) (4,320,000 tonnes); cold strip – 5-stand 68″ tandem (0.25 to 3.2mm x 600 to 1,625mm) (coil weight 45 tonnes) (2,202,000 tonnes); temper/skin pass – 1-stand 4-high 68″ (0.25 to 3.2mm x 600 to 1,625mm) (coil weight 45 tonnes) (2,202,000 tonnes). ***Tube and pipe mills:*** Two welded – one hot ERW and butt welded pipe mill (21.7 to 114.3mm od) (294,000 tonnes); one UOE mill (609.6 to 1,625.6mm od) (576,000 tonnes). ***Coil coating lines:*** One Sendzimir hot-dip galvanizing (0.35 to 4.5mm thick, 1,230mm wide) (360,000 tonnes); one vertical electro-galvanizing (0.3 to 1.6mm thick, 1,600mm wide) (360,000 tonnes); one colour coating (2-coat 2-bake) (0.25 to 1.6mm thick, 1,600mm wide) (144,000 tonnes).

■ Osaka Steelworks, 1-109 Shimaya 5-chome, Konohana-ku, Osaka. (Tel: 06 461 2331). Works director: S. Naruse. ***Steelmaking plant:*** Five electric arc furnaces – one 3-tonne, one 8-tonne, one 15-tonne, one 25-tonne, and one 80-tonne (combined annual capacity 263,000 tonnes). ***Refining plant:*** one 60-tonne ladle furnace ; one 80-tonne DH vacuum degassing unit. ***Other plant:*** Three hot mills – two 280 to 3,000mm od tyre (96,000 tonnes), and one 860 to 1,385mm od wheel (180,000 tonnes); two tyre mills (280-3,000mm) (96,000 tonnes), one wheel tolling mill (860-1,385mm od) (180,000 tonnes); 16,000-tonne, 11,000-tonne and three 6,000-tonne high-speed forging presses; two 35-metre-tonne counter-blow hammers; high-speed precision forging machine; 3,000-tonne vertical hydraulic press; 3,000-tonne horizontal hydraulic press.

■ Kainan Production Centrek, 260-100 Funoo, Kainan, Wakayama. (Tel: 07348 2 5111). Works director: I. Takai. ***Tube and pipe mills:*** seamless – Two 25.4 to 177mm o.d. Mandrel mills (annual capacity 1,224,000 tonnes).

■ Steel Tube Works, 1 Higashimukojima Nishino-cho, Amagasaki, Hyogo. (Tel: 06 411 1221). Works director: I. Takai. ***Steelmaking plant:*** Two electric arc furnaces – one 8-tonne, and one 50-tonne (combined annual capacity 102,000 tonnes). ***Refining plant:*** One 10-tonne AOD unit; one 3-tonne ESR unit; one 50-tonne VOD unit. ***Rolling mills:*** One 3-stand 2-high reversing blooming and billet 150 to 400mm sq, 130 to 253mm dia (420,000 tonnes). ***Tube and pipe mills:*** seamless – One Mannesmann plug (127 to 273mm o.d.) (433,000 tonnes); one Enrhardt (260 to 711.2mm o.d) (25,000 tonnes); two Ugine Sejournet (32 to 273mm o.d.) (119,000 tonnes).

■ Kainan Production Centre, 260-100 Funoo, Kainan, Wakayaue (Tel: 07348 2-5111). ***Tube and pipe mills:*** Two Mandrell seamless (25.4-177mm od) (annual capacity 1,224,000 tonnes).

Products: Steelmaking pig iron; foundry pig iron (combined 1986 output for pig iron, 11,238,000 tonnes); Carbon steel – ingots; slabs; blooms; billets; hot rolled coil (for re-rolling); wire rod rounds 5.5 to 50.8mm o.d. and hexagonals 17.5 to 33.3mm reinforcing bars; round bars 12.7 to 330.0mm o.d.; square bars 45.0 to 550.0mm sq; flats 7.0 x 60.0 to 25.0 x 160.0mm; hexagons 17.5 to 41.0mm; heavy angles 200 x 95mm to 400 x 100mm; wide-flange beams (heavy) 100 x 50mm to 900 x 300mm; welded sections (heavy) 150 x 75mm to 450 x 250mm; sheet piling 400 x 85mm to 500 x 225mm; pipe piling 318.5 to 2,650.0mm o.d.; hot rolled hoop and strip (uncoated); skelp (tube strip); cold rolled hoop and strip (uncoated); hot-dip galvanized hoop and strip; medium plates; heavy plates 4.3 to 270mm x 900 to 5,100mm; universal plates; floor plates; clad plates 6.0 to 100mm x 900 to 4,500mm; hot rolled sheet/coil (uncoated) 1.2 to 19.0mm x 680 to 1,880mm; cold rolled sheet/coil (uncoated) 0.15 to 3.2mm x 600 to 1,880mm; hot-dip galvanized sheet/coil 0.23 to 4.5mm x 40 to 1,230mm; electro-galvanized sheet/coil 0.4 to 3.2mm x 40 to 1,829mm; galvanized corrugated sheet/coil 0.25 to 0.80mm x 762 to 1,000mm; enamelling grade sheet/coil 0.5 to 2.3mm x 610 to 1,829mm (not coated); colour coated sheet/coil; coated sheet/coil 0.5 to 1.6mm x 600 to 1,270mm Zincrometal; electrical non-oriented silicon sheet/coil 0.50 to 0.80mm x 40 to 1,200mm; blackplate; seamless tubes and pipes 25.4 to 711.2mm o.d.; longitudinal weld tubes and pipes ERW 19.0 to 609.6mm o.d.; spiral-weld tubes and pipes 406.4 to 2,650mm o.d.; large-diameter tubes and pipes 406.4 to 1,625mm o.d.; oil country

Sumitomo Metal Industries Ltd (continued)

tubular goods; galvanized tubes and pipes 21.7 to 318.5mm o.d.; plastic coated tubes and pipes 21.7 to 1,625.6mm o.d.; cold drawn tubes and pipes 3 to 508mm o.d.; hollow bars; precision tubes and pipes 15.88 to 482.60mm o.d.; hollow sections bimetallic 25.4 to 1,219.2mm o.d.. Stainless steel – ingots; slabs; blooms; seamless tubes and pipes; longitudinal weld tubes and pipes; oil country tubular goods; cold drawn tubes and pipes; precision tubes and pipes. Steel castings and forgings; rolling stock parts; fabricated steel products; titanium. Output in 1986 – Pig iron 11,238,000 tonnes; Raw steel 10,114,000 tonnes, wire rods 746,000 tonnes; bars 535,000 tonnes; sections 362,000 tonnes. Plates 1,428,000 tonnes; hot rolled steel 5,041,000 tonnes cold rolled steel 2,275,000 tonnes, coated steel 1,002,000 tonnes, seamless tube and pipe 1,142,000 tonnes, ERW tube and pipe 425,000 tonnes, butt welded tube and pipe 19,000 tonnes, large dia tube and pipe 504,000 tonnes and spiral welded tube and pipe 122,000 tonnes.
Sales offices/agents: USA: Sumitomo Metal America Inc – 420 Lexington Avenue, New York, NY 10017 (Tel: 212 949 4760. Telex: 224328. Fax: 212 490 3949, 212 370 1390); 700 South Flower Street, Suite 2106, Los Angeles, CA 90017 (Tel: 213 485 0433, 213 485 1713. Telex: 674626. Fax: 213 485 1781); 230 WFax: 312 372 4431); 1635 One Allen Center, Houston, TX 77002 (Tel: 713 654 7111. Telex: 762362. Fax: 713 654 1261). Australia: Sumitomo Metal Australia Pty Ltd 30th Floor, Westpac Plaza, 60 Margaret Street, Sydney, NSW 2000 (Tel: 231 4287, 27 2171-3. Telex: 27439. Fax: 233 7996). UK – Sumitomo Metal Industries Ltd, 33 Cavendish Square, London W1M 9HF (Tel: 01-493 6472. Telex: 24948. Fax: 01-499 5736). West Germany: Königsalle 48, Düsseldorf (Tel: 211 80641-4. Telex: 8589627. Fax: 211 328302). Austria: Landstrasser Hauptstrasse 2/17st, A 1030 Vienna (Tel: 222 723571, 222 724319. Telex: 136038. Fax: 222 714423). Singapore: 6 Shenton Way 39-03, DBS Building, Singapore 0106 (Tel: 220 9193. Telex: 21194. Fax: 2240386). Mexico: Darwin 32, piso 6, Col. Anzures, Delegacion Miguel Hidalgo, 11590 Mexico 5, DF (Tel: 254 3958, 545 7865. Telex: 1772997. Fax: 254 3948). China: 2162 Minzu Hotel, 51 Fuxinmen-nei, Beijing (Tel: 65 5161. Telex: 22990, 22991).
Additional information: The company also has a Tokyo office at 1-3 Ohtemachi 1-chome, Chiyoda-ku, Tokyo, Japan.

Taiyo Steel Co Ltd (Taiyo Seiko)

Head office: 6-1 chome Nihonbashi-honcho, Chuo-ku, Tokyo 103. **Tel:** 241 4131. **Cables:** Taiyosteel Tokyo.
Established: 1961.
Works:
■ Funabashi. ***Rolling mills:*** Two cold reduction. ***Coil coating lines:*** Two Sendzimir hot-dip galvanizing; colour coating.
Products: Carbon steel – cold rolled sheet/coil (uncoated), hot-dip galvanized sheet/coil, colour coated sheet/coil. Sungrip (galvanized sheet and coil); Sungrip color (colour-coated sheet and coil).

Takasago Tekko KK

Head office: 1-1, 1-chome, Shin Gashi, Itabashi-ku, Tokyo. **Tel:** 03 934 1151. **Telex:** 222 2835. **Telecopier (Fax):** 293 5520.
Management: Tatsuo Yamamoto (president), Akira Mori (managing director, general sales manager). **Established:** 1923. **Capital:** Y 1,504,000,000. **No. of employees:** Works – 350. Office – 200. **Ownership:** Nippon Steel Corp (19.3%).
Works:
■ At head office. ***Rolling mills:*** single-stand cold reduction (0.08 – 0.012" thick, 40" wide) (annual capacity 55,000 tonnes); single-stand cold strip (0.08" – 0.008" thick, 40" wide).
Products: Stainless steel – cold rolled hoop and strip (uncoated), cold rolled sheet/coil (uncoated) 0.1-2.0mm thick; bright annealed sheet/coil 0.2-1.5mm thick.
Sales offices/agents: 12 2-chome, Kanda, Ogawamachi, Chiyoda-ku, Tokyo.

Japan

Toa Steel Co Ltd

Head office: Homat Horizon Building, 6-2 Gobancho, Chiyoda-ku, Tokyo. **Tel:** 03 221 7172. **Telecopier (Fax):** 03 221 9320.
Management: President – Mr Kamiya; managing directors – Yoshio Shibutani (general manager, Hemeji works), Ichiro Oyama (general manager, Sendai works), Akiji Cho (general manager, Adachi works, Tokyo), Yasushi Ono (sales); directors – Shibronori Aibe (personnel and general affairs), Akira Kijima (accounts and purchasing), Koji Furutachi (Osaka sales office manager), Yoshisuke Ikegami (technical services and claims), Minoru Tanaka (plant). **Established:** 1987. **Capital:** Y 5,718,000,000. **No. of employees:** Works – 1,650. Office – 140. **Ownership:** NKK (41.70%).
Works:
■ 1-6-1 Minato, Sendai City, Miyagi Pref. ***Steelmaking plant:*** Two 100-tonne electric arc (annual capacity 840,000 tonnes). ***Refining plant:*** One 100-tonne ladle furnace with vacuum degassing unit (840,000 tonnes). ***Continuous casting machines:*** One 4-strand curved mould bloom (280 x 350mm) (840,000 tonnes). ***Rolling mills:*** One vertical/horizontal billet (114mm sq); one Hitachi Zozen bar (installed 1986); one wire rod – Morgan with Stelmor for rods (5.5-13mm dia) (coil weight 2 tonnes) (900,000 tonnes).
■ Shinden 3-chome, Adachi-ku, Tokyo 123. ***Steelmaking plant:*** Two electric arc furnaces – one 70-tonne Daido-Lectromelt (annual capacity 300,000 tonnes), and one 70-tonne American Bridge (300, 000 tonnes). ***Continuous casting machines:*** Two 4-strand Mitsubishi low-head billet (300,000 tonnes each). ***Rolling mills:*** Three heavy section, light section and bar – 7-stand 3-high (240,000 tonnes), 8-stand 3-high (150,000 tonnes), 11-stand 2-high (270,000 tonnes).
■ Hosoe 1280-Banchi, Shikama-ku, Himeji 672. ***Steelmaking plant:*** Two electric arc furnaces – one 50-tonne IHI-American Bridge (annual capacity 280,000 tonnes), and one 150-tonne NKK-Swindell (700,000 tonnes). ***Continuous casting machines:*** Two billet – one 4-strand Concast S (300,000 tonnes), one 8-strand Concast S (600,000 tonnes); one 4-strand Mitsubishi bloom, beam blank and billet (600,000 tonnes). ***Rolling mills:*** Four heavy section, light section and bar – 5-stand 3-high (310,000 tonnes), 16-stand 2-high (430,000 tonnes), 15-stand 2-high (180,000 tonnes), 16-stand 2-high (600,000 tonnes).
■ 7-1 Anesaki Kaigan, Ichihara City, Chiba Pref (Tel: 0436 662252). ***Tube and pipe mills:*** welded – Two Hoesch spiral (annual capacity 150,000 tonnes).
Products: Carbon steel – blooms; billets; wire rod including high-carbon; reinforcing bars; light angles; light channels; special light sections (vees); heavy angles; heavy tees; heavy channels; wide-flange beams (heavy); spiral-weld tubes and pipes. Alloy steel (excl stainless) – blooms; billets; wire rod; round bars.
Additional information: The company was formed in October 1987 through the merger of Toshin Steel Co Ltd and Azuma Steel Co Ltd.

Toho Sheet & Frame Co Ltd

Head office: 12-2, 3-chome, Nihonbashi, Chuo-ku, Tokyo 103. **Tel:** 03 274 6213. **Telex:** 2222413 tohosf j.
Established: 1937.
Works:
■ 1812 Kamikoya, Yachiyo City, Chiba Pref 276. ***Coil coating lines:*** hot-dip galvanizing; two colour coating. Corrugating line.
Products: Carbon steel – hot-dip galvanized sheet/coil, galvanized corrugated sheet/coil, colour coated sheet/coil.

Tohoku Special Steel Works Ltd (Tohoku Tokushuko)

Head office: 1-20 7-chome Nagamachi, Sendai City, Miyagi 982. **Tel:** 48 3151.
Established: 1937.
Works:
■ ***Steelmaking plant:*** electric arc. ***Refining plant:*** ESR. ***Rolling mills:*** bar.
Products: Stainless steel – round bars. Alloy steel (excl stainless) – round bars (including tool steel).

Tokai Steel Works Ltd (Tokai Kogyo)

Head office: 7-8 Kyobashi 3-chome, Chuo-ku, Tokyo. **Tel:** 03 561 0191/4. **Telex:** 2524137 tokaia.
Management: Osamu Araki (president). **Established:** 1916. **Capital:** Y 2,000,000,000.
Works:
■ Wakamatsu. ***Rolling mills:*** bar. ***Coil coating lines:*** colour coating.
Products: Carbon steel – reinforcing bars, round bars, colour coated sheet/coil (galvanized).

Tokushu Kinzoku Kogyo Co Ltd (Tokkin)

Head office: 3-41 Maeno-cho 3-chome, Itabashi-ku, Tokyo. **Tel:** 03 960 6151. **Cables:** Specialmetal Tokyo. **Telecopier (Fax):** 03 960 6159.
Management: Eiichi Taniguchi (president), Yoshita Taniguchi (managing director), Tsukasa Hasegawa (executive vice president, sales). **Established:** 1940. **Capital:** Y 320,000,000. **No. of employees:** Works – 165. Office – 65. **Annual capacity:** Finished steel: 13,000 tonnes.
Works:
■ Saitama. Works director: Eizou Yoshioka. ***Rolling mills:*** cold strip – Z-high mill-4 (55/115/460 x 450mm), and 4-high mill-10 (150/460 x 450mm, etc) (coil weight 1,600kg) (annual capacity 11,000 tonnes).
Products: Carbon steel – cold rolled hoop and strip (uncoated) 0.03-2.00mm thick, 2-300mm wide. Stainless steel – cold rolled hoop and strip (uncoated) 0.03-2.00mm thick, 2-300mm wide.

Tokyo Kohtetsu Co Ltd

Head office: 1-6-11 Yaesu, Chuo-ku, Tokyo. **Tel:** 03 278 1436. **Telex:** 3552453.
Established: 1918.
Products: Carbon steel – medium angles.

Tokyo Steel Manufacturing Co Ltd

Head office: Fukokuselmel Building 22F, 2-2-2 Uchisaiwaicho, Chiyoda-ku, Tokyo. **Tel:** 03 501 7721. **Telex:** 2222568 tkysty. **Telecopier (Fax):** 03 580 8859.
Management: Chairman – T. Iketani; president – M. Iketani; executive vice-president – T. Iketani; executive directors – H. Ozeki, K. Takahashi; directors – Y. Miyao, T. Kimura, Y. Ishil, H. Soda, N. Takao, H. Yasuda. **Established:** 1934. **Capital:** Y 4,584,896,345. **No. of employees:** 1,587. **Subsidiaries:** Amazing Inc (100%) (leisure). **Associates:** Tamco (25%) (production of steel bar). **Annual capacity:** Raw steel: 3,200,000 tonnes. Finished steel: 3,450,000 tonnes.
Works:
■ Okayama. (Tel: 0864 55 7151). Works director: K. Takahashi. ***Steelmaking plant:*** Two 140-tonne electric arc furnace (annual capacity 1,600,000 tonnes). ***Continuous casting machines:*** Two S-type 4 and 5-strand bloom (170 x 170, 250 x 310; 170 x 235, 250 x 355; 220 x 250mm) (1,800,000 tonnes). ***Rolling mills:*** One 4-stand heavy section – one high, two universal, one edger (200-400mm),and one 4-stand medium section – one high, two universal, one edger (100-250mm) (combined capacity 800,000 tonnes); one 1-stand horizontal light section (25-50mm) (200,000 tonnes); one 22-stand bar – three vertical, nineteen horizontal (10-16mm) (600,000 tonnes).
■ Kyushu. (Tel: 093 791 2635). Works director: N. Takao. ***Steelmaking plant:*** Two 60-tonne electric arc furnace (annual capacity 800,000 tonnes). ***Refining plant:*** One 60-tonne ladle furnace (900,000 tonnes). ***Continuous casting machines:*** One 1-strand S-type (curved mould) slab (1,030 × 200mm) and one 3-strand dog-bone (480 x 420mm) (combined annual capacity 1,000,000 tonnes). ***Rolling mills:*** One Universal heavy section (250-700mm) and one 15-stand horizontal medium section

Japan

Tokyo Steel Manufacturing Co Ltd (continued)

(50-150mm) (combined capacity 250,000 tonnes). One universal heavy plate (1,025mm) (800,000 tonnes).
■ Takamatsu. (Tel: 0878 22 3111). Works director: H. Yasuda. ***Steelmaking plant:*** Two 60-tonne electric arc furnaces (annual capacity 800,000 tonnes). ***Continuous casting machines:*** Two 3-strand S-type billet (170mm sq) (1,000,000 tonnes). ***Rolling mills:*** One 16-stand H-V bar (16-40mm) (800,000 tonnes).
Products: Carbon steel – slabs, blooms, billets, reinforcing bars, round bars, light angles, light channels, special light sections, medium angles, medium channels, wide-flange beams (medium), heavy channels, wide-flange beams (heavy), medium plates, heavy plates, universal plates. Blue Bar (deformed bar).

Topy Industries Ltd

Head office: 5-9 Yonbancho, Chiyoda-ku, Tokyo. **Tel:** 03 265 0111. **Telex:** 22639 topywise. **Cables:** Topywise. **Telecopier (Fax):** 03 230 1840.
Management: President – A. Adachibara; executive vice president – T. Ohishi; managing director – A. Eguchi; general managers, overseas operations centre – T. Matsunaga, S. Tohyama. **Established:** 1921. **Capital:** Y 7,897,674,614. **No. of employees:** Works – 2,800. Ofice – 500. **Ownership:** Nippon Steel Corp (10.9%). **Subsidiaries:** Topy Enterprises Ltd, Topy Marine Transport Ltd, Topy Fasteners Ltd, Topy Buidling & Construction Co Ltd, Topy Metalli Ltd, Topy Corp. **Associates:** Hokuetsu Metal Co Ltd, NHK Steel Ltd. **Annual capacity:** Raw steel: 740,000 tonnes. Finished steel: 740,000 tonnes.
Works:
■ Toyohashi Works, Akemicho-1, Toyohashi City, Aichi Pref. (Tel: 0532 25 1111). Works director: Y. Ishikawa. ***Steelmaking plant:*** electric arc furnaces – one 120-tonne (annual capacity 650,000 tonnes) and two 30-tonne (90,000 tonnes). ***Refining plant:*** ladle furnace – one 120-tonne (650,000 tonnes) and two 30-tonne (90,000 tonnes). One Ladle vacuum degassing unit. ***Continuous casting machines:*** Two bloom – one 3-strand straight bending type (160 x 185 – 200 x 560) (160,000 tonnes) and one 3-strand curved type (200 x 300 – 280 x 550) (580,000 tonnes). ***Rolling mills:*** including one heavy section and one medium section.
Products: Carbon steel – flats; special light sections (rim and side ring sections, tractor shoes and cutting edges – 4-8" (output 200,000 tonnes, incl. rail accessories, etc); medium angles equal – 65-90mm (2½ -3½" (300 tonnes); unequal – 75 x 50-100 x 90mm (3 x 2"-4 x 3½" (5,000 tonnes)); heavy angles equal 100-200" (4-8") (33,000 tonnes), unequal 125 x 75-200 x 100mm (5 x 3"-8 x 6") (28,000 tonnes), inverted 200 x 90mm 350-100mm (8,000 tonnes); heavy joists 100 x 75 to 250 x 200mm (4-12") (2,000 tonnes); heavy channels 100 x 50-380 x 100mm (4-15") (36,000 tonnes); wide-flange beams (heavy) 100 x 50-400 x 200mm (4-14") (350,000 tonnes); rail accessories (tie plate sections, fish plate sections). Universal plates 100 x 9-550 x 125mm (36,000 tonnes).
Sales offices/agents: Topy Enterprises Ltd, Topy International Inc, Topy Enterprises USA Inc.

Toshin Steel Co Ltd—see Toa Steel Co Ltd

Toyo Kohan Co Ltd

Head office: Kohan Building, 4-3 Kasumigaseki 1-chome, Chiyoda-ku, Tokyo 100. **Tel:** 03 502 6611. **Telex:** 33818 yokohan j. **Cables:** Toyokohan Tokyo. **Telecopier (Fax):** 03 502 6637.
Management: Kozo Yoshizaki (chairman), Ichiro Kuno (president). **Established:** 1934. **Capital:** Y 5,040,000,000.
Works:
■ 1302 Higashitoyoi, Kudamatsu City Yamaguchi Pref. ***Rolling mills:*** cold reduction. ***Coil coating lines:*** Ferrostan electrolytic tinning; vertical tin-free steel (ECCS); electro-

Toyo Kohan Co Ltd (continued)

galvanizing; colour coating.
Products: Carbon steel – cold rolled sheet/coil (uncoated), electro-galvanized sheet/coil, colour coated sheet/coil, electrolytic tinplate (single-reduced). Top (cold rolled sheet and coil); Hi-Top (tin-free steel); Vinytop (vinyl coated steel); Silver-Top (elctro-galvanized steel); SS-Top (steel strapping).

Tsukiboshi Art Co Ltd—see Nisshin Steel Co Ltd

Tsukiboshi Kogyo Co Ltd—see Nisshin Steel Co Ltd

Wheeling-Nisshin Inc—see Nisshin Steel Co Ltd

Yahagi Iron Co Ltd (Yahagi Seitetsu KK)

Head office: 18 Showa-cho, Minato-ku, Nagoya 455. **Tel:** 052 611 1511.
Established: 1937.
Works:
■ At head office. ***Sinter plant***. Electric smelting furnaces producing ferro-silicon.
Products: Foundry pig iron.

Yamato Kogyo Co Ltd

Head office: 380 Kibi, Otsu-ku, Himeji City, Hyogo Pref 671-11. **Telex:** 5772129.
Established: 1937.
Works:
■ Himeji. ***Steelmaking plant:*** electric arc. ***Continuous casting machines:*** Two bloom – one Mitsubishi and one Concast; also casts beam blanks. ***Rolling mills:*** One heavy section.
Products: Carbon steel – heavy angles, heavy joists, heavy channels, wide-flange beams (heavy), heavy rails, rail accessories.

Yodogawa Steel Works Ltd (Yodogawa Seikosho)

Head office: 36, 4-chome, Minami Honmachi, Higashi-ku, Osaka. **Tel:** 06 245 1111. **Telex:** 5228611 ydk oj. **Cables:** Yodoseko Osaka.
Management: President – Masao Omori; senior executive directors – Masazumi Umebara, Isamu Mikoshiba, Kazuhiro Chikamori, Tosuke Shibata, Hideaki Sato; directors – Shoichiro Yoshimura, Kazuo Toyoda, Noboru Okumura, Masaru Tanaka, Kiyofumi Nonaka, Ryogiro Ageta, Osamu Tokunaga. **Established:** 1935. **Capital:** Y10,292,000,000. **No. of employees:** 1,730. **Subsidiaries:** Takada Steel Materials Industries Ltd; Yodoliving Industries Ltd, Hakuyo Sangyo Co Ltd; Sadoshima Co Ltd; Keiyo Steel Quay Co Ltd. **Annual capacity:** Finished steel: 770,000 tonnes (rolled).
Works:
■ Kure Works, 9-1, Showa-cho, Kure, Hiroshima. (Tel: 0823 25 1111). Works director: Kiyofumi Nonaka. ***Rolling mills:*** Two cold reduction No 1 440 x 1,372mm (annual capacity 190,000 tonnes) and No 2 508 x 1,372mm (260,000 tonnes) (coil weight 18 tonnes). ***Coil coating lines:*** Two continuous hot-dip galvanizing (240,000 tonnes); one colour coating (72,000 Tonnes).
■ Ichikawa Works, 5 Kooya Sinmachi, Ichikawa, Chiba. (Tel: 0473 28 1231). Works director: Masazumi Umebara. ***Rolling mills:*** One 4-high reversing cold reduction (493 x 1,370mm) (coil weight 25 tonnes) (annual capacity 260,000 tonnes). ***Coil coating lines:*** One continuous hot-dip galvanizing (140,000 tonnes); one continuous zinc-aluminium (190,000 tonnes); one continuous colour coating (120,000 tonnes).
■ Osaka Works, 1-21, 2-chome, Hyakushima, Nishi Yodogawa-ku, Osaka. (Tel: 06 472 1251). Works director: Tosuke Shibata. ***Coil coating lines:*** Three hot-dip galvanizing

Yodogawa Steel Works Ltd (Yodogawa Seikosho) (continued)

sheet (annual capacity 90,000 tonnes) and three continuous colour coating (100,000 tonnes). ***Other plant:*** Four induction furnaces (70,000 tonnes); one reverberatory furnace (7,200 tonnes); two centrifugal casting machine; roll foundry; forming lines.
■ Izumi-Otsu Works, 18-14 Nishi Minato-cho, Izumi-Otsu, Osaka. (Tel: 0725 32 5681). Works director: Noboru Okumura. ***Steelmaking plant:*** Two electric arc furnace (annual capacity 90,000 tonnes). ***Refining plant:*** One 25-tonne ladle furnace. ***Continuous casting machines:*** One 2-strand billet (120,000 tonnes). ***Other plant:*** Roll forming lines; grating plant.
Products: Carbon steel – billets (1986 output 190,000 tonnes – including alloy steel billets); cold rolled hoop and strip (uncoated) 9"-500mm wide, 0.4-2.5mm thick (110,000 tonnes); cold rolled sheet/coil (uncoated) 28-48" wide, 6-10ft long, 0.11-2.3mm thick (130,000 tonnes); hot-dip galvanized sheet/coil 28-48" wide, 6-16ft long, 0.11-1.2mm thick (310,000 tonnes); colour coated sheet/coil (292,000 tonnes). Alloy steel (excl stainless) – billets. Cast iron and cast steel rolls (10,000 tonnes). Also buildings, warehouses, storage sheds, garages, corrugated roofing, shutters, exterior wall materials, siding, steel sheet spandrels, gratings and treads. Coat-hangers, cupboards, kitchen cabinets, racks. Projects undertaken for cold strip mills, galvanizing lines, colour-coating lines, slitting lines, forming lines, roll foundries and continuous casting plants.
Brands: Cherry Yodogawa (with symbol) and Yodoko.
Sales offices/agents: At head office and Tokyo (Tel: 03 551 1171).

Jordan

Jordan Iron & Steel Industry Co Ltd

Head office: Jabal Amman 1st Circle, Amman. **Telex:** 21279 jisico jo.
Established: 1965.
Works:
■ Zarqa-Awajan. ***Steelmaking plant:*** electric arc. ***Continuous casting machines:*** One 2-strand Danieli billet (80-140mm sq).
Products: Carbon steel – reinforcing bars, round bars.

The Jordan Pipes Manufacturing Co Ltd (JPMC)

Head office: Arab Bank Building, Station Road, Amman. **Postal address:** PO Box 6899. **Tel:** 651451-2. **Telex:** 21517 anabeb. **Cables:** Anabeeb Jordan. **Telecopier (Fax):** 651451.
Management: Moh'd Ali Budeir (chairman), Zakariya N. Stetieh (vice chairman), Khaled M. Kanaan (general manager), Suleiman Kh. Halasa (assistant general manager and administration manager). **Established:** 1974. **Capital:** JD 2,500,00. **No. of employees:** Works – 122. Office – 21. **Ownership:** Public company.
Works:
■ At head office. ***Tube and pipe mills:*** welded – One Yoder ½-3" (annual capacity 12,000 tonnes (per one shift)). ***Other plant:*** Tube galvanizing equipment.
Products: Carbon steel – longitudinal weld tubes and pipes (class A & B) (1986 output 12,628 tonnes); galvanized tubes and pipes, hollow sections. Alloy steel (excl stainless) – longitudinal weld tubes and pipes, hollow sections.

National Steel Industry Co Ltd

Head office: Insurance Building, 1st Circle, Jabel Amman, Amman. **Postal address:** PO Box 2870. **Tel:** 066 38191. **Telex:** 21092 steel jo.
Management: H.E. Riad Al-Mufleh (president and board chairman), Hasan Haj Hasan (deputy chairman), Hassan R. Al-Mufleh (vice president). **Established:** 1979.

National Steel Industry Co Ltd (continued)

Capital: JD 4,000,000. **Ownership:** Public company. **Annual capacity:** Finished steel: 120,000 tonnes.
Works:
■ Awajan/Zarqa. Works director: Hassan R. Al-Mufleh. ***Rolling mills:*** One light section / bar comprising 3-high 450mm roughing stand, 4-stand 350mm continuous train and 6-stand 300mm continuous train (annual capacity 120,000 tonnes rolled from billets 120mm sq, 8,000mm long). ***Other plant:*** Straightening line.
Products: Carbon steel – reinforcing bars and round bars 10-30mm; square bars 10-25mm; flats 25-50mm; light angles; light tees; light channels 25-50mm.

Kenya

Kenya United Steel Co Ltd

Head office: PO Box 90550, Miritini, Mombasa. **Tel:** 433601-8. **Telex:** 21197. **Cables:** Kusco.
Management: Directors – J.G. Rajani (managing), M.P. Chandaria (British), S.J. Rajani, S.M. Chandaria (British). **Established:** 1966. **Capital:** KShs 50,000,000. **No. of employees:** Works – 395. Office – 45. **Subsidiaries:** Associated Nail Manufacturers Ltd (100%). **Annual capacity:** Raw steel: 18,000 tonnes. Finished steel: 30,000 tonnes.
Works:
■ At head office. Works director: U.K. Ghoshi. ***Steelmaking plant:*** Two 4/5-tonne electric arc furnaces (annual capacity 24,000 tonnes). ***Continuous casting machines:*** 1-strand vertical billet (80 x 80mm) (25,000 tonnes). ***Rolling mills:*** bar, rod and light section – 11-stand x 3-high 350mm (30,000 tonnes).
Products: Carbon steel – wire rod, reinforcing bars, round bars, square bars, bright (cold finished) bars, light angles, bright wire, black annealed wire, galvanized wire (plain), galvanized wire (barbed).
Expansion programme: Section mill – under construction (1987); revamp of existing wire rod facilities (1988/89); ship-breaking yard (1987/88) and sponge iron plant (1990/1992).

Steel Africa Ltd

Head office: Pamba Road, Mombasa. **Postal address:** PO Box 81827. **Telex:** 21036 zincot.
Established: 1964.
Works:
■ ***Coil coating lines:*** One sheet hot-dip galvanizing; colour coating. Cold roll-forming line.
Products: Carbon steel – cold roll-formed sections (from strip), hot-dip galvanized sheet/coil, colour coated sheet/coil.

Steel Rolling Mills Ltd

Head office: Nairobi. **Postal address:** Po Box 41733. **Telex:** 45037.
Works:
■ ***Rolling mills:*** light section and bar.
Products: Carbon steel – reinforcing bars, round bars, flats, light angles, light tees.

North Korea

Chongjin Works

Head office: North Kankyo, Hamgyong.
Works:
■ ***Direct reduction plant:*** Krupp-Renn rotary kilns. **Steelmaking plant:**. ***Continuous casting machines:*** Three 2-strand USSR slab (170 x 650mm-250 x 1,550mm). **Rolling mills:**. ***Other plant:*** Foundry.

Hwanghai Iron Works

Head office: Songnim, North Hwanghai Province.
Works:
■ **Iron ore pelletizing plant.** ***Blast furnaces:*** three. ***Steelmaking plant:*** electric arc furnaces and open hearth furnaces. ***Rolling mills:*** blooming, bar, wire rod, heavy plate and sheet.
Products: Steelmaking pig iron. Carbon steel – ingots, blooms, wire rod, reinforcing bars, medium plates. Heavy plates, hot rolled sheet/coil (uncoated). Cast iron – pipes.

Kangson Works

Head office: Kangson, North Hwanghai Province.
Works:
■ ***Steelmaking plant:*** Eight electric arc furnaces (annual capacity over 960,000 tonnes). Bessemer converters. ***Rolling mills:*** blooming (annual capacity 900,000 tonnes), wire rod , strip mill.
Products: Carbon steel – wire rod; round bars; round bars; sections; medium plates; heavy plates; bright wire; wire rope; seamless tubes and pipes. Stainless steel – medium plates; heavy plates.

Kimchaek Works

Head office: Kimchaek.
Works:
■ ***Coke ovens:*** Four batteries. ***Sinter plant:*** including 3,000,000 tonnes/year continuous. ***Iron ore pelletizing plant:*** including one 1,500 cu m. ***Blast furnaces:*** three. ***Steelmaking plant:*** basic oxygen Bessemer converters, electric arc furnaces, open hearth. ***Rolling mills:*** wire rod, medium plate, hot strip and hoop, cold strip. ***Tube and pipe mills:*** seamless, welded. ***Coil coating lines:*** hot-dip galvanizing.
Products: Carbon steel – wire rod, medium plates, hot-dip galvanized sheet/coil, seamless tubes and pipes.

Songjin Works

Head office: Songjin.
Works:
■ ***Steelmaking plant:*** electric arc. ***Rolling mills:*** slabbing, bar, sheet including one 500mm and one 600mm, medium plate.
Products: Carbon steel – medium plates. Alloy steel (excl stainless) – round bars molybdenium high-speed steels (20-70mm dia); medium plates.

South Korea

Boo-Kuk Steel Industrial Co Ltd

Head office: 7-ga Bongrae-dong, Jung-gu, Seoul. **Telex:** 28489.
Works:
■ ***Rolling mills:*** light section; wire rod.
Products: Carbon steel – wire rod, light angles, light channels.

Daehan Sangsa Co Ltd

Head office: 210 Anrak-dong, Dongnae-ku, Pusan 607. **Telex:** 53416.
Works:
■ ***Rolling mills:*** wire rod.
Products: Carbon steel – wire rod, reinforcing bars sections.

Dong Bu Steel Co Ltd

Head office: 21-9 Cho-dong, Jung-ku, Seoul 100. **Telex:** 27643.
Products: Carbon steel – cold rolled sheet/coil (uncoated), hot-dip galvanized sheet/coil.

Dong Yang Tinplate Industrial Co Ltd

Head office: 93 4-Ka Dangsan-dong, Youngdeungpo-ku, Seoul. **Telex:** 27486.
Established: 1959.
Works:
■ ***Coil coating lines:*** electrolytic tinning; tin-free steel (ECCS).
Products: Carbon steel – electrolytic tinplate (single-reduced), tin-free steel (ECCS).

Dongjin Steel Co Ltd

Head office: 180-28 Gajwa-Dong, Buk-Gu, Inchon 160-05. **Postal address:** KPO Box 919. **Tel:** 83 0141. **Telex:** 27643.

Dongkuk Steel Mill Co Ltd

Head office: 50 Suha-Dong, Chung-Ku, Seoul. **Tel:** 757 3001 20. **Telex:** 23681 dongkuk k. **Telecopier (Fax):** 02 756 0814.
Management: Chairman – Sang-Tai Chang; president – Sang-Don Chang; vice president – Chan Kye; executive directors – Byeong-Joo Han, Byung-Yong Kwun, Chul-Woo Lee; directors – Kyung-Du Chun, Hak-Su Lee, Dal-Seung Lee, Hong-Jae Park. **Established:** 1954. **Capital:** US$27,000,000. **No. of employees:** Works – 1,934. Office – 465. **Associates:** Chosun Wire Industries Co Ltd; Korea Iron & Steel Co Ltd; Pusan Gas Co Ltd; Changwon Ceramics Co Ltd; Central Investment & Finance Corp; Dongkuk Industries Co Ltd; Dong-Il Steel Mfg Co Ltd; Chunyang Transportation Co Ltd; Pusan Cast Iron Co Ltd; Kukje Machinery Co Ltd. **Annual capacity:** Raw steel: 1,710,000 tonnes. Finished steel: 1,730,000 tonnes.
Works:
■ Pusan Works, 177 Yongho-Dong, Nam-Ku, Pasan. (Tel: 622 4191-8. Telex: 53331 dongkuk k). Works director: Hak-Su Lee. ***Steelmaking plant:*** Six electric arc furnaces – three 15-tonne (annual capacity 180,000 tonnes), two 40-tonne (600,000 tonnes) and one 50-tonne (429,000 tonnes). ***Continuous casting machines:*** billet – one 6-strand Concast S-type (100-160mm sq) (600,000 tonnes), and one 5-strand MHI S-type (100-160mm sq) (500,000 tonnes); one 1-strand Concast S-type slab (100 x 400mm) (150,000 tonnes). ***Rolling mills:*** bar – one 20-stand 2-high 480mm continuous (300,000 tonnes), two 13-stand 2-high 320mm continuous (142,000 tonnes), one 7-stand 3-high 500mm (60,000 tonnes) and one 6-stand 3-high 300mm

Dongkuk Steel Mill Co Ltd (continued)

(60,000 tonnes); One 25-stand 2-high 360mm wire rod (coil weight 400kg) (250,000 tonnes); one heavy plate 910/560mm x 2,550mm (300,000 tonnes).
■ Ichon Works, 1 Songhyun-Dong, Dong-Ku, Inchon. (Tel: 73 0021-5). Works director: Kwan-Ho Suh. ***Steelmaking plant:*** Two 30-tonne electric arc (annual capacity 500,000 tonnes). ***Continuous casting machines:*** One 6-strand Concast S-type billet (115mm sq) (600,000 tonnes). ***Rolling mills:*** Three bar – one 20-stand 2-high 480mm (340,000 tonnes), one 19-stand 2-high 450mm (150,000 tonnes) and one 13-stand 2-high 320mm (120,000 tonnes).
Products: Carbon steel – billets 100-160mm sq; wire rod 5.5-12mm dia; reinforcing bars 6-42mm dia; round bars 6-42mm dia; flats 30-95mm; light angles 20 x 20-100 x 100mm; light channels 75-150mm; heavy plates 6 x 910 x 1,830-50 x 1,830 x 6,100mm.
Sales offices/agents: Tokyo branch (Ocean Trading Co Ltd), 3rd Floor, Shimizu Building 2-6, Toranomon-chome, Minato-ku, Tokyo, Japan. (Tel: 03 503 6841-3. Telex: 28357 dongkuk); Los Angeles branch, 21515 Hawthorne Boulevard, Suite 1007, Torrance, CA 90503, USA. (Tel: 213 540 8511. Telex: 664857 dongkukam trnc); New York branch (Deawonsa International Inc), One World Trade Center, Suite 8469, New York, NY 10048, USA. (Tel: 212 466 1424. Telex: 661788 daewonsa nyk).

Hyundai Pipe Co Ltd

Head office: No 1 2-Ka Shinmoon-Ro, Jongro-ku, Seoul. **Telex:** 24656.
Established: 1975.
Works:
■ Ulsam ***Tube and pipe mills:*** welded.
Products: Carbon steel – longitudinal weld tubes and pipes, galvanized tubes and pipes, plastic coated tubes and pipes, hollow sections.

Inchon Iron & Steel Co Ltd

Head office: 1 Songhyun-dong, Dong-ku, Inchon. **Tel:** 032 763 5000. **Telex:** 27460 westeel. **Cables:** Ironsteel Seoul. **Telecopier (Fax):** 032 763 5046.
Management: Mong-Gu Jung (chairman), Soo-Il Choi (president), Bon-Soong Koo (vice president), Soo-Ung Kwon (executive managing director, Steel Product & Engineering Division). **Established:** 1953. **Capital:** Won 55,000,000,000. **No. of employees:** Works – 2,345. Office – 362. **Annual capacity:** Raw steel: 1,650,000 tonnes. Finished steel: 1,530,000 tonnes.
Works:
■ Inchon. ***Steelmaking plant:*** Four electric arc furnaces – 70-tonne (42,000kVA) (annual capacity 520,000 tonnes), 60-tonne (37,000kVA) (460,000 tonnes), 50-tonne (36,000kVA) (420,000 tonnes), 40-tonne (15,000kVA) (210,000 tonnes). ***Refining plant:*** 40-tonne VAD (100,000 tonnes). ***Continuous casting machines:*** One 4-strand curved mould billet (120 x 120mm) (450,000 tonnes) (fed by 50-tonne arc furnace); one 5-strand curved mould billet (118 x 118mm) (500,000 tonnes) (fed by 60-tonne arc furnace); one 4-strand curved mould bloom (180 x 230mm, 250 x 300mm, 270 x 400mm, 350 x 480mm, beam blanks 400 x 490 x 120mm) (600,000 tonnes) (fed by 70-tonne arc furnace); one 2-strand curved mould bloom (160 x 180mm, 160 x 240mm, 200 x 275mm, 200 x 310mm) (260,000 tonnes) (fed by 40-tonne arc furnace). ***Rolling mills:*** heavy section comprising one 2-high breakdown, one universal roughing, one 2-high edging and one universal finishing stand; and rail comprising one 2-high breakdown, one 2-high roughing, and one 2-high finishing stand (combined annual capacity 500,000 tonnes). Medium section comprising one 3-high blooming, two 3-high intermediate, one 2-high pre-finishing, and one universal combination finishing stand (200,000 tonnes), light section comprising one 3-high roughing, two 2-high intermediate tandem, and six 3-high finishing stands (200,000 tonnes); bar comprising one 3-high roughing, four 2-high intermediate, and eight 2-high tandem finishing stands (400,000 tonnes); wire rod comprising eight 2-high

Inchon Iron & Steel Co Ltd (continued)

roughing, ten 2-high intermediate and finishing stands, and eight-stand horizontal/ vertical finishing block (coil weight 370kg) (250,000 tonnes). ***Other plant:*** One 25,200kVA ferro-silicon furnace (16,000 tonnes). Steel foundries – 10-tonne electric furnace (5,000kVA) producing steel for castings (12,000 tonnes); 40-tonne electric furnace (15,000kVA) producing forging ingots (50,000 tonnes).
Products: Carbon steel – billets; wire rod 5.5-10mm dia; reinforcing bars D10-D35; light angles 45 x 45mm to 75 x 75mm; light channels 75 x 40mm; medium angles 80 x 80mm to 150 x 150mm; medium channels 100 x 50mm to 200 x 90mm; heavy angles 175 x 175mm to 250 x 250mm; heavy channels 250 x 90mm to 300 x 90mm; wide-flange beams (heavy) 100 x 100mm to 600 x 300mm; heavy rails 50 and 60kg/metre; light rails 15, 20, 22 and 30kg/metre. Ferro-silicon; forging ingots; steel castings.

Kang Won Industries Ltd

Head office: 6, 2-Ka, Shinmoon-ro, Chongro-ku, Seoul. **Tel:** 730 7111, 730 7211. **Telex:** 27323 kwonind. **Cables:** Kangwoninco Seoul. **Telecopier (Fax):** 739 1104.
Management: Chairman – In Wook Chung; vice chairman – In Yub Chung; president – Moon Won Chung; vice president – Do Won Chung; managing directors – Tong Won Part (steel products and engineering division), Byung Jun Park (overseas operation division), Chang Kuk Bae (casting and rolled products and sales division); executive director – Ki Shik Hwang (steel sales division). **Established:** 1952. **Capital:** US$27,500,000. **No. of employees:** Works – 5,400. Office – 1,220. **Annual capacity:** Raw steel: 940,000 tonnes. Finished steel: 970,000 tonnes.
Works:
■ 444 Songnae-dong, Pohang City, Kyungsang Buk-do (Tel: 72 1701, 72 1801). Works director: Do Won Chung. ***Steelmaking plant:*** Three electric arc furnaces – two 70-tonne (annual capacity 740,000 tonnes), one 30-tonne (200,000 tonnes). ***Refining plant:*** One 30-tonne ladle furnace (140,000 tonnes). ***Continuous casting machines:*** Two 4-strand Concast low-head curved mould billet (110mm sq, 120mm sq, 135mm sq, 140 x 180mm) (800,000 tonnes). ***Rolling mills:*** blooming and heavy section comprising 1-stand 3-high 970mm (400,000 tonnes) and 3-stand 2-high 850mm (300,000 tonnes); light section (160-310mm) (70,000 tonnes); bar – No 1 2-stand 3-high 470mm and seven 330mm stands; No 2 one 3-high 490mm, eight tandem420-460mm stands and four 330-370mm stands (600,000 tonnes) (these mills are being modified); ***Other plant:*** Foundries – two 3-tonne induction furnaces; two 5-tone Cupola; one ladle car 100-tonne; and two Arc furnaces 2-tonne (90,000 tonnes).
Products: Carbon steel – (Output 1,853 tonnes); ingots; billets 120 x 120 x 1,500-5,000mm (650,166 tonnes); reinforcing bars D10-D35 (546,558 tonnes); round bars 9-250mm dia (13,128 tonnes); light angles (equal) 50 x 4-100 x 13 (25,684 tonnes); light channels 100 x 50, 125 x 65 (3,221 tonnes); medium angles (equal) 130 x 9-150 x 19 (10,801 tonnes); medium channels 150 x 75 (8,900 tonnes); heavy angles (equal) 175 x 12-250 x 35 (7,030 tonnes); heavy joists 200 x 100-300 x 150 (13,659 tonnes); heavy channels 200 x 80-380 x 100 (19,466 tonnes); wide-flange beams (heavy); heavy rails 37kg/m-70kg/m (47,278 tonnes); light rails 22kg/m (3,690 tonnes); rail accessories fishplates, base plates (2,827 tonnes); sheet piling U type 400 x 100 x 10.5-400 x 170 x 16 (20,029 tonnes).

Korea Heavy Machinery Industrial Ltd

Head office: 446 Shindorim-dong, Kuro-ku, Seoul 150. **Telex:** 284533.
Products: Carbon steel – round bars, square bars, light rails.

South Korea

Korea Steel Pipe Co Ltd

Head office: 27-1 Supyo-dong, Jung-ku, Seoul 100. **Telex:** 28301 kspipe k.
Works:
■ ***Tube and pipe mills:*** welded.
Products: Carbon steel – longitudinal weld tubes and pipes, hollow sections.

Pohang Iron & Steel Co Ltd

Head office: 680, 1 Goedong-dong, Pohang. **Tel:** 0562 70 0114. **Telex:** 54474, 54352 posdo. **Cables:** Pohangsteel Pohang. **Telecopier (Fax):** 0562 70 0599, 0562 72 7590.
Management: Chairman – Tae Joon Park; president – Myung Sik Chung; executive vice presidents – Duk Hyon Baik (general superintendent, Pohang works), Tuk Pyo Park (general administration, budget and finance, investment control), Il Yong Oh (Kwangyang works, constructio n, production and plant engineering), Mal Soo Cho (foreign procurement, raw materials, transport, sales); standing auditor – Yun Jhong Kang; managing directors – Yong Sun Chough (production engineering division), Ki Hwan Song (budget and finance, investme nt control dept), Young Soo Han (foreign procurement, raw materials central terminal system dept), Chong Chin Kim (general superintendent, Kwangyang works), Dai Kong Lee (public information, computer system planning dept), Jong Whan Oh (plant engineering d ivision), Geun Suk Son (corporate strategic planning, management information research dept); vice presidents – Chol Woo Kim (general superintendent of technical research laboratories), Keon Yu Hong (resident director, Tokyo, for Japan and South-East Asia) ; directors – Sang Boo Yoo (deputy director, plant engineering division), Taek Joong Kim (construction division), Sun Koo Lee (overseas sales), Hak Bong Ko (secretary and domestic sales), Sang Dal Suh (general manager, Seoul office and sports team managin g group), Jeong Joon Ahn (deputy superintendent, Pohang works), Jeong Soo Kim (personnel management, education and training, general affairs, housing control, landscaping purchasing dept). **Established:** 1968. **Capital:** W432,275,000,000. **No. of employees:** Works – 14,319. Office – 4,401. **Subsidiaries:** Pohang Steel America Corp, USA (100%) (steel trading); USS-Posco Industries, USA (50%) (production of cold rolled steel sheets in coil); Tanoma Coal Co Inc, USA (100%) (coal mining); Pohang Steel Canada Ltd, Canada (100%) (coal mining); Pohang Steel Australia Pty Ltd, Australia (100%) (coal mining); Puya Co Ltd, Hong Kong (100%) (steel trading); Posco Engineering Co Ltd, South Korea (100%) (plant engineering), Sema Architects Co Ltd, Korea (100%) (architectural design); Samhwa Chemical Co Ltd, Korea (50%) (production of refractory bricks); Posco Electric Control Co Ltd, Korea (51.1%) (manufacture and maintenance of industrial electric control gauges). **Associates:** Feralloy West Co, USA (45%) (steel service center); Woo-Jin Osk Corp, South Korea (20%) (manufacture of industrial sensors for steelmaking and shipbuilding); Pohang Steel Industries Co Ltd, South Korea (49%) (production of colour-coated steel sheet). **Annual capacity:** Raw steel: 11,800,000 tonnes.
Works:
■ Pohang Steelworks. ***Coke ovens:*** No 1 Otto (174 ovens) (annual capacity 1,122,000 tonnes); No 2 Otto (146 ovens) (1,302,000 tonnes); No 3 Otto (150 ovens) (1,338,000 tonnes); No 4 (75 ovens) (700,000 tonnes). Total 4,462,000 tonnes (lump coke). ***Sinter plant:*** Five Dwight-Lloyd including one 130 sq m strand (1,309,000 tonnes), one 200 sq m strand (2,264,000 tonnes), one 400 sq m strand (3,601,000 tonnes), one 400 sq m strand (3,751,000 tonnes), one 40 sq m strand (392,000 tonnes).Total 11,317,000 tonnes (pure sinter). ***Blast furnaces:*** No 1 8.9m hearth dia, 1,660 cu m (1,097,000 tonnes); No 2 11m hearth dia, 2,550 cu m (1,862,000 tonnes); No 3 and No 4 13.2m hearth dia, 3,795 cu m (3,020,000 tonnes each); furnace producing foundry iron – 5m hearth dia, 330 cu m (200,000 tonnes). ***Steelmaking plant:*** Two LD basic oxygen converters – No 1 shop comprising three 100-tonne (2,300,000 tonnes); No 2 shop comprising of three 300-tonne (6,800,000 tonnes). ***Refining plant:*** One 100-tonne ladle furnace (600,000 tonnes); four vacuum degassing units – two 300-tonne (4,130,000 tonnes), two 100-tonne (1,360,000 tonnes); ine PI (660,000 tonnes). ***Continuous casting machines:*** 14-strand billet (1,000,000 tonnes); 4-strand bloom (500,000 tonnes); 7-strand slab (5,850,000 tonnes). ***Rolling mills:*** Two wire rod – Ashlow 25-stand and Mannesmann Demag

Pohang Iron & Steel Co Ltd (continued)

Sack (max coil weight 2 tonnes) total 5-strands (combined annual capacity 860,000 tonnes); two heavy plate (1,860,000 tonnes); two wide hot strip (3,200,000 tonnes); two wide hot strip (1,950,000 tonnes); one cold mill for electrical sheet/coil (80,000 tonnes).
■ Kwang Yang Steelworks. ***Coke ovens:*** One Otto battery (132 ovens) (annual capacity 1,227,000 tonnes). ***Sinter plant:*** One 400 sq m Dwight-Lloyd (3,983,000 tonnes). ***Blast furnaces:*** One 13.2m hearth dia, 3,800 cu m (2,840,000 tonnes). ***Steelmaking plant:*** Two x 250-tonne LD basic oxygen converters (2,784,000 tonnes). ***Refining plant:*** Two 250-tonne vacuum degassing units (1,830,000 tonnes). ***Continuous casting machines:*** 2-strand slab (2,700,000 tonnes). ***Rolling mills:*** one wide hot strip (2,638,000 tonnes).
Products: Steelmaking pig iron (1986 output, 8,812,400 tonnes); foundry pig iron 4 to 6kg/each (152,800 tonnes). Carbon steel – slabs 200 to 250mm x 870 to 2,200mm x 3,000 to 10,000mm (7,881,300 tonnes); blooms 250 x 300 x 6,690mm (1,164,100 tonnes); billets 120 x 120 x 6,000 to 18,000mm, 135 x 135 x 9,600 to 14,200mm (76,900 tonnes); hot rolled coil (for re-rolling) 1.2 to 22.0mm x 600 to 1,880mm (coil weight 35 tonnes max) (5,295,900 tonnes); wire rod 5.5 to 42.0mm (coil weight 2 tonnes max) (877,600 tonnes); skelp (tube strip) 1.2 to 12.7mm x 65 to 1,860mm (coil weight 35 tonnes max) (60,200 tonnes); medium plates 1.2 to 12.7mm x 600 to 1,860mm x 1,500 to 6,124mm (58,700 tonnes); heavy plates 6.0 to 100mm x 1,200 to 4,500mm x 1,200 to 25,000mm (1,655,600 tonnes); hot rolled sheet/coil (uncoated) 1.2 to 22.0mm x 600 to 1,880mm (coil weight 35 tonnes max) (4,231,300 tonnes); cold rolled sheet/coil (uncoated) 0.18 to 2.0mm x 340 to 1,650mm (522,300 tonnes); hot-dip galvanized sheet/coil 0.18 to 2.3mm x 700 to 1,255mm (111,400 tonnes); electro-galvanized sheet/coil 0.4 to 2.0mm x 800 to 1,650mm (4,400 tonnes); galvanized corrugated sheet/coil 0.2 to 0.8mm x 762 to 1,000mm (1,600 tonnes); electrical grain oriented silicon sheet/coil 0.27 to 0.35mm x 75 to 914mm (24,600 tonnes); electrical non-oriented silicon sheet/coil 0.35 to 0.50mm x 75 to 1,000mm (71,200 tonnes); blackplate 0.18 to 0.60mm x 685 to 980mm (144,700 tonnes).
Brands: Posco (all products); Posten (atmospheric steel plate); Posgal (galvanealled steel).
Expansion programme: Kwangyang – increase in raw steel capacity by 2,700,000 tonnes (October 1988); new cold rolling mill (1,225,000 tonnes) (March 1989). Pohang – No 2 billet caster (500,000 tonnes) (December 1988); continuous annealing line No 2 (300,000 tonnes) (December 1987); No 3 wire rod mill (540,000 tonnes) (January 1989); stainless steel plant (250,000 tonnes) (June 1989).

Pusan Steel Pipe Corp

Head office: 40-153 3GA, Hangang-Ro, Youngsan-Ku, Seoul. **Tel:** 02 797 1881. **Telex:** 28287 pspsel. **Cables:** Haiduck Seoul. **Telecopier (Fax):** 02 797 2462.
Management: Woon Hyung Lee (president); Chul Hee Lee (Pohang plant manager) and Kwang Suk Kim (planning and project div) (senior executive managing directors); Si Won Kim (managing director, sales); Kil Sin Koo (managing director, production and plant maintenance); Hyo Il Lee (managing director, Seoul plant manager); Jae Chul Cho (director, purchasing and general management). **Established:** 1960. **Capital:** W 9,400,000,000. **No. of employees:** Works – 600. Office – 250. **Subsidiaries:** Pusan Pipe America Inc (40%) (importer and distributor); Haiduk Machinery CoLtd (52%) (machinery and plant manufacture); Pusan Bundy Corp (51%) (producer of copper-brazed double-wall steel tubes); Haiduck Express Ltd (100%) (haulage); Busan Steel Pipe Co Ltd (94%) (distributor of steel pipe); ARKC (Alloy Rods Korea Corp) (50%) (producer of welding rods). **Associates:** Korea Cast Iron Pipe Industrial Co Ltd (25.5%). **Annual capacity:** Finished steel: 700,000 tonnes.
Works:
■ 14-1 Jang-Heung Dong Kyung Sang Buk Do. (Tel: 0562 72 1401. Telex: 54351 pspph). Works director: Chul Hee Lee. ***Tube and pipe mills:*** welded – five ½-2½" ERW (annual capacity 130,000 tonnes); two 2-8" ERW (210,000 tonnes); one 6-20" ERW

South Korea

Pusan Steel Pipe Corp (continued)

(240,000 tonnes) and one SRM ½-4" (120,000 tonnes).
■ 180-15 Gaebong-Dong, Kuro-Ku, Seoul. (Tel: 02 613 0171. Telex: 28287 pspsel). Works director – Hyo Il Lee. ***Tube and pipe mills:*** welded – one 12-88" spiral (24,000 tons).
Products: Carbon steel – longitudinal weld tubes and pipes ½-20" (anticipated 1987 output 600,000 tonnes); spiral-weld tubes and pipes 12-88" (24,000 tonnes); oil country tubular goods 2-⅜-20" (100,000 tonnes); galvanized tubes and pipes ½-20" (160,000 tonnes); hollow sections ⅜ x ⅜"-14 x 14" (50,000 tonnes). Stainless steel – longitudinal weld tubes and pipes ⅜-2" (3,000 tonnes); Alloy steel (excl stainless) – longitudinal weld tubes and pipes ½-8" (5,000 tonnes). Pre-insulated pipes (½-20") (annual capacity 150km); stepped electrical poles (0.4kv-11.9kv) (annual capacity 20,000 tonnes).
Sales offices/agents: At head office. Pusan Pipe American Inc, CA. (Tel: 213 324 5544. Telex: 181903).

Sammi Steel Co Ltd

Head office: 935 – 34 Bangbae-Dong, Kangman-ku, Seoul. **Postal address:** CPO Box 1434. **Tel:** 02 583 6911, 02 584 6911. **Telex:** 26273. **Cables:** Stainless Seoul. **Telecopier (Fax):** 02 584 4944.
Management: Chairman – Hyun Chul Kim; president – Zixang Yun; vice president – In Soo Park; directors – Yong Kyun Lee, Wook Jin Lee, Jae Wan Shim, Huyn Zong Kim, Young Kook Kim, Kwan Seop Kim, Jung Ho Lee, Hong Jo Yun, Hyng Ki Kim. **Established:** 1966. **Capital:** US$ 82,143,000. **No. of employees:** Works – 2,100. Office – 1,100. **Ownership:** Sammi Corp (stainless steel, logs and lumber, mining, general trade). **Subsidiaries:** Sammi Metal Products Co Ltd (forgings, stainless steel welded pipe, industrial fasteners, engineers' cutting tools, welding electrodes, stainless steel wire). **Associates:** Sammi Youna Department Store Co Ltd, Sammi Cultural Foundation. **Annual capacity:** Raw steel: 250,000 tonnes. Finished steel: 385,000 tonnes.
Works:
■ 25 Chuk Hyun-Dong, Chang Won. (Tel: 0551 86 2611. Telex: 53746). Works director: Young Kook-Kim. ***Steelmaking plant:*** One 30-tonne and two 15-tonne electric arc furnace and one 1-tonne and one ¾-tonne high frequency furnaces (combined annual capacity 250,000 tonnes). ***Refining plant:*** One 2-tonne ESR, one 15-tonne ladle furnace, one 30-tonne and one 15-tonne vacuum degassing units, one 2-tonne VIM and one 30-tonne VOD (total annual capacity 250,000 tonnes). ***Continuous casting machines:*** billet and bloom with vertical combination – 2-strand (130 x 130mm sq) and 1-strand (250 x 410mm) (250,000 tonnes). ***Rolling mills:*** Two blooming – one 2-high reversing and one 3-high 5-stand; one 3-high 4-stand billet, four bar and wire rod – continuous mill comprising nine 3-high and nine 2-high stands (total capacity 180,000 tonnes). two 20-roll Sendzimir cold reduction 52" width (150,000 tonnes) (75,000 tonnes each mill); one 2-high reversing temper/skin pass. ***Tube and pipe mills:*** seamless – one 2,000-ton horizontal extrusion press (25,000 tonnes).
■ 250 Yochum-Dong, Ulsan. (Tel: 0522 72 7181-9). Works director: Kwan Seop Kim. ***Rolling mills:*** One 42" width 20-roll Sendzimir cold reduction (35,000 tonnes); one 2-high non-reversing temper/skin pass.
Products: Carbon steel – ingots 1-8 tonnes; billets 53-170mm RCS; wire rod 8-25mm dia; round bars 8-400mm dia; square bars 20-50mm a/f; flats 4 x 4mm-30 x 105mm, 30 x 405mm-200 x 405mm; special light sections hinge, mast rail (combined output 60,000 tonnes). Seamless tubes and pipes 10-170mm od, 1-40mm wall; cold drawn tubes and pipes 10-120mm od, 1-14mm wall (combined output 10,000 tonnes). Stainless steel – ingots (1-8 tonne); billets 100, 130, 150, 170mm RCS; wire rod 8-25mm dia; round bars 6-400mm dia; square bars 40-90mm sq; flats 4 x 30mm-30 x 150mm, 30 x 200-180mm-405mm; hexagons 18-52g; light angles 3 x 30mm-6 x 60mm (combined output 40,000 tonnes). Cold rolled sheet/coil (uncoated) 0.3-6mm x 1,250mm max (180,000 tonnes); bright annealed sheet/coil 0.3-2mm x 1,250mm max (36,000 tonnes); seamless tubes and pipes 10-170mm-1-40mm wall and cold drawn tubes and pipes 10-120mm-1-14mm wall (combined 5,000 tonnes). Alloy steel (excl stainless) – ingots 1-8 tonnes; billets 100, 130, 150, 170mm RCS; round bars

Sammi Steel Co Ltd (continued)

(hot rolled and finished) 6-400mm dia, square bars 40-90mm and flats 4 x 30mm-30 x 105mm, 30 x 200mm, 180mm and 405mm (combined 65,000 tonnes); seamless tubes and pipes 10-170mm od, 1-40mm wall and cold drawn tubes and pipes 10-120mm, 1-14mm wall (combined 5,000 tonnes).
Sales offices/agents: Sammisa America Corp, CA, USA (Tel: 714 895 5535, Telex: 6836043 sammi uw); Sammi Steel Co Ltd, Dusselforf, W. Germany (Tel: 0211 354041/2, Telex: 8586904 sam d); Sammisa (Japan) Corp, Tokyo, Japan (Tel: 574 1623/5, Telex: 26331 sammisa j); Samstar Trading Pte Ltd, Jurong, Singapore (Tel: 8624788, 8624789, Telex: 25882 samsin rs).
Expansion programme: Specialty steel bar and wire rod mill project will be completed by 1989 comprising 330,000 tonnes/year bar and wire rod mill and steel making facilities (electric arc furnace, VD/VOD and continuous casting). Revamping of profile rolling mill will be completed by 1987. Installation of two Pilger mills will be completed by 1988.

Union Steel Mfg Co Ltd

Head office: 13th Floor, Kyobo Building, 1-1 1-ka Chong-ro, Chongro-ku, Seoul. **Telex:** 53361 unistil k.
Works:
■ ***Rolling mills:*** Three cold reduction; one temper/skin pass. ***Tube and pipe mills:*** welded. ***Coil coating lines:*** Two hot-dip galvanizing; two colour coating.
Products: Carbon steel -, cold rolled sheet/coil (uncoated), hot-dip galvanized sheet/coil, colour coated sheet/coil, longitudinal weld tubes and pipes, oil country tubular goods, galvanized tubes and pipes.

Kuwait

Kuwait Metal Pipe Industries KSC

Head office: Opposite Central Prison, Sulaibiah Industrial Area, Kuwait. **Postal address:** PO Box 3416, 13035 Safat. **Tel:** 4875622, 4873697, 4874843. **Telex:** 30065, 23064. **Cables:** Pipes Kuwait. **Telecopier (Fax):** 4875897.
Management: Vice chairman and managing director – Abdull Essa Al Saleh; general manager – Salem A. Al-Musallam; managers – Emile R. Khair (marketing and projects), Khalid Al Ali (materials and purchase). **Established:** 1967. **Capital:** KD 15,206,986. **No. of employees:** Works – 250. Office – 50. **Subsidiaries:** Kuwait Engineering Operations & Management Co (Kenomac) (engineering supervision and consultants). **Annual capacity:** Finished steel: 200,000 tonnes.
Works:
■ Shuwaikh Industrial Area. ***Tube and pipe mills:*** welded – large dia (36-120" dia) (annual capacity 65,000 tonnes).
■ Sulaibiyah. ***Tube and pipe mills:*** welded – spiral-weld to API-5L, etc (6-48") (annual capacity 120,000 tonnes); ERW ½-2" dia (15,000 tonnes).
Products: Carbon steel – spiral-weld tubes and pipes 6-48" dia (output 20,000 tonnes); large-diameter tubes and pipes 36-120" dia (2,500 tonnes); galvanized tubes and pipes and black ERW to BS1387, ½-2" (3,500 tonnes). Carbon steel tanks and pressure vessels; corrosion protection coating and lining by enamel, epoxy, cement, rubber, etc; hot dip galvanization.

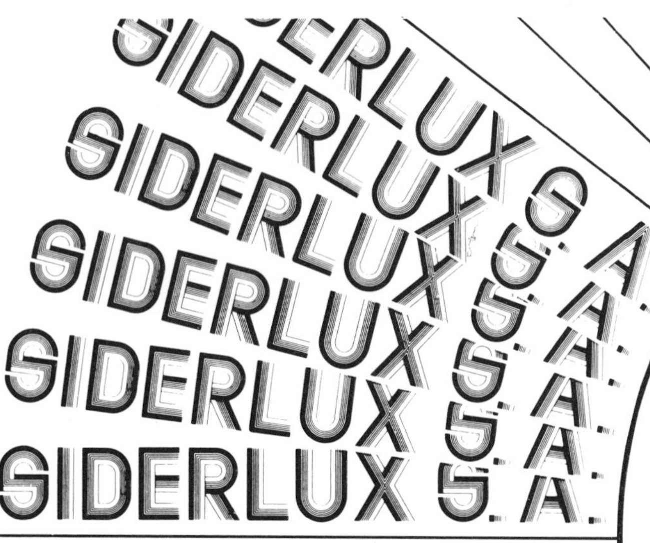
SIDERLUX S.A.

Lebanon

Consolidated Steel Lebanon SAL (CSL)

Head office: Abounayan Building, Jal El Dib Blvd, Beirut. **Postal address:** PO Box 5198. **Tel:** 417468, 417431, 403016, 940191. **Telex:** 43068, 45182 le steel. **Cables:** Steeliban.
Management: Board chairman – Miguel Najib Abizaid (general manager); board member – Jose Najib Abizaid; managers – Hatem Hatem (general), Emile Cristo (financial), Antoine Kiwan (sales), Abdo Younis (personnel), Michael Saleeby (plant), Emile Badawy (purchasing). **Established:** 1959. **Capital:** L£30,000,000. **No. of employees:** Works – 200. Office – 50. **Annual capacity:** Rolled steel: 220,000 tonnes.
Works:
■ Byblos-Amchit. ***Rolling mills:*** One light section; one bar.
Products: Carbon steel – reinforcing bars; round bars 6-32mm; square bars 10 x 10mm-25 x 25mm; flats 12-40mm. Costeel (reinforcing bars).

Luxembourg

Arbed SA

Head office: Avenue de la Liberté 19, L 2930 Luxembourg City. **Tel:** 47 921. **Telex:** 3407 arbe lu. **Cables:** Centralarbed Luxembourg. **Telecopier (Fax):** 4792 2675.
Management: President of board – Georges Faber; managers – René Brück (general), Pierre Everard (marketing), Joseph Kinsch (finance), François Schleimer (operations), Pierre Thein (planning). **Established:** 1882. **Capital:** Subscribed and paid – LuxFr 17,610,110,000; authorised – LuxFr 25,750,025,000. **No. of employees:** Works – 9,400. Office – 2,800. **Ownership:** State of Luxembourg and SNCI (42.9%); Sté Générale des Belgique (20.4%); Schneider Group (5.6%); others (31.1%). **Subsidiaries:** Raw materials and supplies – Eschweiler Bergwerks-Verein AG, West Germany (97%); CoalArbed Inc, USA (100%); Samitri SA, Brazil (55%); Samarco SA, Brazil (51%); Refralux Sarl, Luxembourg (50%). Iron and steelmaking – Sidmar SA, Belgium (52%); Métallurgique et Minière de Rodange-Athus SA, Luxembourg (41%); Saarstahl Völklingen GmbH, West Germany (24%); Cia Siderúrgica Belgo-Mineira SA, Brazil (30%). Steel processing – TechnoArbed Luxembourg Sarl, Luxembourg (100%). Wireworks – TrefilArbed Bissen Sarl, Luxembourg (100%); TrefilArbed Bettembourg Sarl, Luxembourg (100%); TrefilArbed Grembergen SA, Luxembourg (100%); TrefilArbed Bouwstaal Gent NV, Belgium (100%); TrefilArbed Welding NV, Belgium (100%); TrefilArbed Bouwstaal Roermond BV, Netherlands (100%); TrefilArbed Korea Co Ltd, South Korea (50%). Steel construction and mechanical engineering – SA Paul Wurth, Luxembourg (41%); MecanArbed Dommeldange Sarl, Luxembourg (100%). Other processing – Cogifer SA, France (27%); SA Mecavoie, Belgium (100%); Ets Durieux-Rails SA, Belgium (100%). Miscellaneous – Continental Alloys SA, Luxembourg (82%); Intermoselle Sarl, Luxembourg (33%); Ciments Luxembourgeois SA, Luxembourg (50%); Galvalange Sarl, Luxembourg (50%); Ewald Giebel Luxemburg GmbH, Luxembourg (33%); Laborlux SA, Luxembourg (75%); InfoArbed Sarl, Luxembourg (100%); Sotel – Sté de Transport de l'Energie Électrique du Grand-Duché de Luxembourg Soc. Coop., Luxembourg (75%). Marketing and sales – TradeArbed SA, Luxembourg (100%); TradeArbed Rails SA, Luxembourg (100%); Brasilux SA, Luxembourg (100%); TrefilArbed Luxembourg/Saarbrücken Sarl, Luxembourg (100%); CoalArbed International Trading Co, USA (100%). Management and finance – SA Luxembourgeoise d'Exploitations Minières, Luxembourg (98%); Arbed Finance SA, Luxembourg (100%); Electro Holding Co SA, Luxembourg (75%); TradeArbed Participations Sarl, Luxembourg (100%); TrefilArbed Participations SA, Belgium (100%); Ciments de Buda SA, Belgium (61%); NV Zolder, Belgium (25%); Sidarfin SA, Belgium (100%); Arbed-Finanz Deutschland GmbH, West Germany (100%); Sibral

Luxembourg

Arbed SA (continued)

Participaçôes Ltda, Brazil (100%); Arbed International Insurance Consultants SA, Luxembourg (100%). **Annual capacity:** Raw steel: 4,500,000 tonnes. Finished steel: 4,200,000 tonnes.

Works:

■ Esch-Belval Works, L 4002 Esch-sur-Alzette. (Tel: 55501. Telex: 3487 arbel lu). Works director: Léon Helbach. ***Sinter plant:*** One 320 sq m strand Dwight-Lloyd (annual capacity 2,300,000 tonnes); one 400 sq m strand Dwight-Lloyd (2,900,000 tonnes). ***Blast furnaces:*** (A) 8m hearth dia, 1,376 cu m (600,000 tonnes); (B) 9.2m hearth dia, 1,781 cu m (780,000 tonnes); (C) 11.2m hearth dia, 2,465 cu m (1,400,000 tonnes). ***Steelmaking plant:*** Two 180-tonne LD-AC basic oxygen converters (LBE process) (2,000,000 tonnes). ***Rolling mills:*** Two 2-high 1,150mm reversing blooming; one 6-stand 2-high 500mm continuous billet. One heavy section – No II with three 2-high 900mm reversing stands, two 2-high 1,200mm reversing stands, two 1,200mm universal stands and one edging stand. three medium section – No III with three 3-high 750mm stands; No IV with two 3-high 750mm reversing stands, one 3-high 1,000mm reversing stand and one edging stand, No V with four 3-high 500mm reversing stands. Two hot strip and hoop – No VIII strip mill with four horizontal 2-high continuous stands, two vertical and four horizontal 2-high continuous stands 480mm wide, one vertical and six horizontal 2-high continuous stands and two continuous groups comprising four vertical and one horizontal 2-high and four horizontal 3-high continuous stands; No IX strip mill with two vertical and four horizontal 2-high continuous stands 600mm wide, and three vertical and six horizontal 4-high continuous stands. Combined annual capacity of rolling mills – 1,600,000 tonnes. ***Other plant:*** One oxygen plant (640 tonnes/day); one MW thermal power station; one dry slag granulation plant.

■ Differdange Works, POB 60, L 4620 Differdange. (Tel: 584021-1. Telex: 3415 and 3436 ardif lu). Works director: Fernand Wagner. ***Steelmaking plant:*** One 165-tonne LD-AC basic oxygen converter (LBE process) (annual capacity 1,800,000 tonnes). ***Refining plant:*** One vacuum degassing unit (6,000 tonnes/day). ***Rolling mills:*** One 2-high 1,200mm reversing slabbing blooming; one 2-high 1,620mm reversing grey blooming; one grey universal heavy section. Combined annual capacity 1,000,000 tonnes. ***Tube and pipe mills:*** welded – one 300mm (annual capacity 150,000 tonnes).

■ Esch-Schifflange Works, POB 141, L 4002 Esch-sur-Alzette. (Tel: 5551-1. Telex: 2520 archi lu). Works director: Arthur Schummer. ***Steelmaking plant:*** Two 95-tonne LD-AC basic oxygen converters (LBE process) (annual capacity 1,450,000 tonnes). ***Refining plant:*** One ladle furnace. ***Continuous casting machines:*** One 6-strand MecanArbed/KSW billet (125mm sq) and one 6-strand MecanArbed/KSW bloom (220mm sq) (combined annual capacity 1,300,000 tonnes). ***Rolling mills:*** One 5-stand 600mm continuous billet; one 21-stand 330mm continuous light section; one 2-strand 28-stand continuous wire rod (90 metres/second). Combined annual capacity 1,000,000 tonnes.

■ Dudelange Works, POB 74, L 3401 Dudelange. (Tel: 515151. Telex: 3439 ardud lu). Works director: Charles Hosch. ***Rolling mills:*** One 4-high reversing CVC4-HS cold reduction (1,630/600mm, 3.0/0.20mm) (coil weight 20 tonnes) (annual capacity 360,000 tonnes). ***Coil coating lines:*** One Sendzimir hot-dip galvanizing (610/ 1,060mm) (75,000 tonnes). ***Other plant:*** Iron foundry; basic slag granulation plant.

Products: Carbon steel – wire rod; reinforcing bars (plain and deformed); round bars; square bars; flats; light angles; light tees; light channels; special light sections; medium angles; medium tees; medium joists (I-beams, IPEs, columns, bearing piles, dissymetric European light parallel-flange beams); medium channels (Continental standard sections and American channels); wide-flange beams (medium) (grey beams and American parallel-flange shapes); universals (medium); heavy angles; heavy tees; heavy joists (I-beams, IPEs, columns, bearing piles, dissymetric European light parallel-flange beams); heavy channels (Continental standard sections and American channels); wide-flange beams (heavy) (grey beams and American parallel-flange shapes); universals (heavy); rail accessories; sheet piling (BZ and BU sections, PBP and ABP flat

Arbed SA (continued)

piles, HZ steel wall; hot rolled hoop and strip (uncoated); skelp (tube strip); cold rolled hoop and strip (uncoated); hot-dip galvanized hoop and strip; cold rolled sheet/coil (uncoated); hot-dip galvanized sheet/coil; galvanized corrugated sheet/coil; longitudinal weld tubes and pipes (2.3-114.2mm); galvanized tubes and pipes. Box piles, anchorage material and pipe-driving accessories.
Sales offices/agents: Luxembourg – TradeArbed, Luxembourg. Belgium – TradeArbed Belgium, Brussels. France – TradeArbed France, Paris. UK – TradeArbed UK Ltd, London and Birmingham. West Germany – TradeArbed Deutschland GmbH, Cologne. Netherlands – TradeArbed SA, Rotterdam. Switzerland – TradeArbed (Suisse) SA, Basle. Italy – TradeArbed Italia Srl, Milan. Austria – Roechling Austria, Roechling GmbH & Co KG, Vienna. Spain – Blitz & Parser, Madrid. Portugal – Alberto Maria Bravo & Filhos, Lisbon; A. Spratley da Silva & Filhos, Porto. Sweden – G. & J. Beijer, Import och Export Ab i Stockholm, Stockholm. Norway – TradeArbed Norge A/S, Oslo. Denmark – S.N. Simonsen & Co A/S, Copenhagen. Finland – Oy Ferro-Tekno AB, Helsinki. Greece – Angelos A. Perrakis, Kifissia-Athens. Cyprus – Theocharis S. Mitzis, Larnaca. Turkey – Korimpeks Mumeselik ve Ticaret Ltd Sirketi, Istanbul. Lebanon – Georges Daoud, Beirut. Jordan – Ayoub Trading Agency Co, Amman. Syria – Mas Import-Export Co, Alep. Kuwait – Alabdeen Construction Co WLL, Safat. United Arab Emirates – Obaid Humaid Al-Tayer, Dubai. Egypt – Mohamed Ghanem & Co, Trade & Investment Consultant Ltd, Cairo. South Africa – Arcona (Pty) Ltd, Cape Town and Randburg. USA – TradeArbed Inc, New York, NY; Larkspur, CA; Houston, TX. Canada – TradeArbed Canada Inc, Burlington, Ont. Mexico – TradeArbed de Mexico SA, Mexico, DF. Uruguay – Madeco, Eduardo Salomon, Montevideo. Brazil – Cia Siderúrgica Belgo-Mineira, Sabara, Belo Horizonte, Rio de Janeiro and São Paulo. Argentina – Estal, Buenos Aires. Chile – Jorge Moya Raggio, Santiago. Peru – Umetal – Union Metalurgica SA, Lima. Ecuador – Comercial Eurotex C. Ltda, Quito and Guayaquil. Colombia – Remapril Ltda, Bogotá. Singapore – TradeArbed Pvt Ltd, Singapore. Thailand – Meridien-Marketing (Thailand) Ltd, Bangkok. Hong Kong - TradeArbed Pvt Ltd, Central Hong Kong. Philippines – McJufelen International, Metro Manila. Malaysia – TradeArbed Pvt Ltd, Selangor. South Korea – Top Corp, Seoul. India – Nathani Steel Pvt Ltd, Bombay, New Delhi and Calcutta. Pakistan – Globe Trade Corp Ltd, Karachi and Lahore. China – TradeArbed (China) Ltd, Central Hong Kong. Japan – Takeshi Yano, Tokyo.

Galvalange Sarl

Head office: Zone Industrielle Wolser, Dudelange. **Postal address:** POB 92, L 3401 Dudelange. **Tel:** 518686-1. **Telex:** 3240. **Telecopier (Fax):** 519223.
Management: Managers – André Laux (general), Hans Klubert (sales), Jean-Paul Schuler (product). **Established:** 1981. **Capital:** LuxFr 650,000,000. **No. of employees:** Works – 77. Office – 28. **Ownership:** Arbed SA, Luxembourg (50%); Phenix Works SA, Belgium (50%).
Works:
■ At head office. Works director: André Laux. **Coil coating lines** – one Aluzinc aluminizing line (annual capacity 250,000 tonnes).
Products: Carbon steel – aluminized sheet/coil (700-1,500mm wide, 0.28-2.0mm thick) (1986 output 2,095 tonnes); zinc-aluminium alloy coated sheet/coil (Aluzinc) (700-1,500mm wide, 0.28-2.0mm thick) (143,458 tonnes).
Brands: Aluzinc (zinc-aluminium alloy coated sheet/coil); Alugal (aluminized sheet/coil).
Sales offices/agents: TradeArbed SA, L 2930 Luxembourg. Eurinter SA, 10 Quai du Halage, B 4110 Flémalle, Belgium.

Ewald Giebel-Luxembourg

Head office: Zone Industrielle Wolser, L 3401 Dudelange. **Postal address:** BP 96. **Tel:** 51 87 21-1. **Telex:** 2919. **Telecopier (Fax):** 51 98 68.
Management: André Feiereisen (director). **Established:** 1980. **Capital:** LuxFr

Ewald Giebel-Luxembourg (continued)

201,000,000. **No. of employees:** Works – 66. Office – 22. **Ownership:** Ewald Giebel KG, West Germany (66%); Arbed (33%).
Works:
■ At head office. ***Coil coating lines:*** One electro-galvanizing, vertical cells (acid electrolyte) (annual capacity 100,000 tonnes). ***Other plant:*** One cut-to-length line; two slitting lines.
Products: Carbon steel – electro-galvanized sheet/coil – single-side and two-side coated (output 100,000 tonnes).

Métallurgique & Minière de Rodange-Athus (MMRA)

Head office: Case Postale 24, Rodange. **Tel:** 50 19 1. **Telex:** 3416 mmra lu. **Cables:** Usine Rodange.
Management: Jean-Marc Wagener (president), Raymond Kirsch (vice president), Jacques Bouchoms (director). **Established:** 1872. **Ownership:** Arbed (41%) other shareholders include Groupe Bruxelles Lambert, Cockerill Sambre and the Luxembourg Government.
Works:
■ At head office. ***Rolling mills:*** One heavy section / rail and light section / bar.
Products: Carbon steel – reinforcing bars (1986 output 243,235 tonnes); special light sections (38,424 tonnes); heavy joists, heavy channels and wide-flange beams (heavy) (combined output 102,502 tonnes); heavy rails (134,115 tonnes).

Malaysia

Amalgamated Industrial Steel Bhd

Head office: Jalan Utas, Shah Alam, Selangor. **Postal address:** PO Box 22, 40700 Shah Alam. **Tel:** 5591616-9, 5593411. **Telex:** 38663 pipe ma. **Cables:** Amalindus Shah Alam.
Management: Managers – Eng Meng Swee (general), Noriaki Inomata (export), Teo Choon Chye (technical sales), Adrian Tan (marketing), Fong Tiang Gee (sales). **Established:** 1969. **Capital:** Authorised M$ 100,000,000. Paid-up M$ 51,289,000. **No. of employees:** Works – 122. Office – 55. **Annual capacity:** Finished steel: 45,000 tonnes/year approx.
Works:
■ At head office. Factory manager: Chong Tuck On. ***Tube and pipe mills:*** welded – ERW for round ($\frac{1}{2}$-6″), square ($\frac{1}{2}$ x $\frac{1}{2}$″-6 x 6″), rectangular ($\frac{3}{4}$ x 1$\frac{1}{2}$″-4 x 8″) tubes (annual capacity 45,000 tonnes approx).
Products: Carbon steel – longitudinal weld tubes and pipes and hollow sections (output 45,000 tonnes).
Expansion programme: Mill for stainless steel pipes expected completion date mid-1988.

Amalgamated Steel Mills Bhd

Head office: 4th & 5th Floor, Wisma SPS, 32, Jln Imbi, Kuala Lumpur, 55100. **Tel:** 03 2412155, 03 2413166. **Telex:** 33195 asm ma; 33194 lion ma. **Telecopier (Fax):** 03 2411036.
Management: Director – William Cheng Heng Jem (managing); managers – Lim Ee Pai (general), Anthony Chin (commercial). **Established:** 1976. **Capital:** Authorised: M$ 500,000,000; paid-up M$ 170,000,000. **No. of employees:** 798. **Ownership:** Main shareholders – Lion Corp Bhd; Lembaga Tabung Angkatan Tentera, Titra Enterprise Sdn Bhd; Aseam Malaysia Nominees Sdn Bhd; Lion Enterprise (Kuala Lumpur) Sdn Bhd. **Subsidiaries:** Angkasa Marketing Sdn Bhd (100%) (sales and distribution of iron and steel products); Antkasa Marketing (Singapore) Pte Ltd (100%) (sales and distribution of iron and steel products); Umatrac Enterprises Sdn Bhd (100%) (insurance brokers and agents); Simen Dagangan Sdn Bhd (51%) (sales and distribution

Amalgamated Steel Mills Bhd (continued)

of cement); Lion Suzuki Marketing Sdn Bhd (51%) (sales and distribution of Suzuki motorcycles); Lion Metal Industries Sdn Bhd (100%) (manufacture, sales and distribution of nuts and bolts); Akurjaya Sdn Bhd, Bungawang Sdn Bhd, Aquabio (Sabah) Sdn Bhd and Seritawan Sdn Bhd (70%) (aquaculture and integrated farming); Natvest Sdn Bhd (70%) (supermarket). **Associates:** Asia Commercial Finance (M) Bhd (20%) (licensed finance company); Suzuki Assemblers Malaysia Sdn Bhd (49%) (assembly and sales of Suzuki motorcycles). **Annual capacity:** Raw steel: 350,000 tonnes. Finished steel: 400,000 tonnes.
Works:
■ Lot 6, Solok Waja 2, Bukit Raja Industrial Estate, 41050, Klang, Selangor. (Tel: 3412322/422. Telex: 39543/39534. Telefax: 03 3412354). Works director: Senior production manager – Lam Kok Kee. ***Steelmaking plant:*** One 70-tonne 55MVA electric arc (annual capacity 350,000 tonnes). ***Refining plant:*** One 70-tonne 33MVA ladle furnace (350,000 tonnes). ***Continuous casting machines:*** 6-strand Danieli billet (100-160 mm sq) (350,000 tonnes). ***Rolling mills:*** bar with 3-high reversing roughing stand, three cross-country intermediate stands, 5-stand tandem train and 6-stand, horizontal/vertical finishing block (9-40mm dia) (150,000 tonnes); continuous wire rod with 10-stand finishing block (5.5mm-32mm dia) (coil weight 1,500 kg) (250,00 tonnes). ***Other plant:*** Scrap treatment plant (baling press and scrap shear).
Products: Carbon steel – wire rod – low-carbon 5.5mm-32mm; reinforcing bars – high-tensile deformed 9.0mm-40mm; round bars – mild steel 9.0mm-40mm; flats – thickness 4.5/6/9/12mm, width 25/32/38/50/65/75mm.
Sales offices/agents: Angkasa Marketing Sdn Bhd, Block A, 14, Jln Vivekanada, Brickfields, 50470, Kuala Lumpur. (Tel: 03 2746911/22. Telex: 31480 simen ma. Telefax: 03 2747060). Angkasa Marketing (Singapore) Pte Ltd, Teck Chiang Realty Building, 14 Arumugam Road, 08 01 Singapore 1440. (Tel: 02 7459577/78. Telex: 24513 lion. Telefax: 02 7479493).

Antara Steel Mills Sdn Bhd

Head office: PLO 277, Jalan Gangsa Satu, 81700 Pasir Gudang, Johore. **Postal address:** PO Box 79. **Tel:** 07 512021-3. **Telex:** 60673.
Management: Directors – Wan Abdul Ghani Bin Wan Ahmad (managing), Horst D. Tiegelkamp (technical); manager – Madam Azizah Binti Abdul Rahman. **Established:** 1979. **Capital:** M$25,000,000. **No. of employees:** Works – 167. Office – 49. **Annual capacity:** Finished steel: 120,000 tonnes.
Works:
■ At head office. ***Rolling mills:*** light section 8-stand (angles 25 x 3mm, 30 x 3mm, 38 x 3mm, 50 x 4mm) (annual capacity 20,000 tonnes); 2-stand (65 x 6mm); bar 12-stand (8, 10, 11 and 12mm dia) (10,000 tonnes), 8-stand (16, 20, 25mm dia) (100,000 tonnes).
Products: Carbon steel – reinforcing bars 8, 10, 12, 16, 20, 25, 28, 32, 40mm (1986 output 30,000 tonnes); round bars 8, 10, 12, 16, 20, 25, 28, 32, 40mm (5,000 tonnes); flats 32 x 9mm, 38 x 9mm, 50 x 6mm (200 tonnes); light angles 25 x 3mm, 38 x 3mm, 50 x 6mm, 65 x 6mm (13,000 tonnes).

Choo Bee Metal Industries Sdn Bhd

Head office: 42-44 Jalan Bendahara, 31650 Ipoh, Perak. **Tel:** 513111. **Telex:** 44323. **Cables:** Choobeeco.
Management: Directors – Soon Ah Khun (managing), Soon Lian Huat (managing), Soon Lian Lim, Soon Cheng Hai, Soon Cheng Boon, Soon Chuan Hock (sales manager); manager – Beh Tock Lum (factory). **Established:** 1971. **Capital:** M$7,312,000,000. **No. of employees:** Works – 45. Office – 12. **Ownership:** Choo Bee Hardwares Sdn Bhd (Ultimate holding company – Soon Lian Huat Holdings Sdn Bhd). **Subsidiaries:** Taik Bee Hardware Sdn Bhd; Kepong Warehouse Sdn Bhd.
Works:
■ ***Tube and pipe mills:*** welded Three – TFM-65 ($\frac{1}{2}$-2") (annual capacity 4,800 tonnes), TFM-55SP ($\frac{1}{2}$-1") (5,400 tonnes), TFM-45SP ($\frac{1}{2}$-$\frac{3}{4}$") (3,600 tonnes). ***Other plant:*** Fully

Choo Bee Metal Industries Sdn Bhd (continued)

automatic lip-channel forming machine – model SF-460K (200 x 75 x 20 x 4.5mm max) (coil weight 3.5 tonnes) (3,600 tonnes); Loopco 20-tonne x 13mm x 7,620mm cut-to-length line (1-13mm) (max coil weight 20 tonnes) (20,400 tonnes); Iowa Precision Line SLR6012-2 (0.38-2.74mm) (coil weight 9 tonnes) (10,200 tonnes); Kent cut-to-length line: 10-tonnes x 7.5mm (1.0-7.5mm) (coil weight 10 tonnes) (12,000 tonnes); slitting line (1.0-4.5mm) (coil weight 10 tonnes) (12,000 tonnes), slitting line (0.3-1.2mm) (coil weight 10 tonnes) (9,600 tonnes).
Products: Carbon steel – cold roll-formed sections (from strip), longitudinal weld tubes and pipes.

Dah Yung Steel (M) Sdn Bhd

Head office: 19 Jalan Empat, Off Jalan Chan Sow Lin, Kuala Lumpur. **Telex:** 30672.
Works:
■ ***Steelmaking plant:*** electric arc. ***Rolling mills:*** light section and bar.
Products: Carbon steel – reinforcing bars, round bars, square bars, flats, light angles.

Federal Iron Works Sdn Bhd (FIW)

Head office: 14 Jalan Tandang, Petaling Jaya, Selangor. **Tel:** 7571177. **Telex:** 37631 fedion ma. **Cables:** Fediron. **Telecopier (Fax):** 03 7916721.
Management: Directors – Goh Geok Khim (chairman), Lim Chin Guan (managing), Goh Geok Chuan (executive), Takashi Yamasaki (executive), Takaaki Kikuyama, Shogoro Toyoda, Goh Geok Loo. **Established:** 1959. **Capital:** M$39,752,679. **No. of employees:** Works – 320. **Ownership:** Private limited company. **Subsidiaries:** None.
Works:
■ At head office. Assistant works director: Philip Ding Chin Kwok. ***Coil coating lines:*** Two hot-dip galvanizing (annual capacity 100,000 tonnes).
Products: Carbon steel – hot-dip galvanized sheet/coil. LPG cylinders.
Brands: Tiger.

Hume Industries Sdn Bhd

Head office: Petaling Jaya, Jalan 219, Off Federal Highway, Petaling Jaya. **Postal address:** PO Box 21. **Tel:** 760544, 304650, 304633. **Telex:** 37654.
Products: Carbon steel – longitudinal weld tubes and pipes, large-diameter tubes and pipes.

Malayawata Steel Bhd

Head office: 1st Floor, Komplek Antarabangsa, Jalan Sultan, Sulaiman, Kuala Lumpur. **Tel:** 03 2935155. **Telex:** 31688 wata ma. **Cables:** Besiwaja Kuala Lumpur.
Management: Y.M. Tunku Dato' Shahriman Bin Tunku Sulaiman (chairman), Choo Kean Hin (general manager), Mohd Shah Bin Mohd Ali (secretary), Kok Kee Boo (business manager), Mohd Rejab Hashim (business development manager). **Established:** 1961. **Capital:** M$67,200,000 (paid-up) **No. of employees:** Works – 1,106. Office – 169. **Subsidiaries:** Malayawata Charcoal Sdn Bhd (charcoal production). Malaysian Steel Corp Sdn Bhd. **Associates:** Pernas Realty Development Sdn Bhd. **Annual capacity:** Pig iron: 157,000 tonnes. Raw steel: 186,000 tonnes. Finished steel: 412,000 tonnes.
Works:
■ Prai, Province Wellesley (Tel: 307144. Telex: 40052 mawata ma. Fax: 04 308863). Works director: Choo Kean Hin. One Dwight Lloyd ***Sinter plant:*** (1.2m) (annual capacity 132,000 tonnes). Two ***Blast furnaces:*** – one 3.7m hearth dia (204 cu m)

Malayawata Steel Bhd (continued)

(96,000 tonnes), and one 3.1m hearth dia (148 cu m) (62,000 tonnes). Steelmaking plant: Two 15-ton basic oxygen converters (160,000 tonnes); one 10-ton electric arc furnace (24,000 tonnes). ***Continuous casting machines:*** One 2-strand billet (72,000 tonnes). ***Rolling mills:*** One continuous cross country bar, incorporating wire rod train (180,000 tonnes); one continuous tandem wire rod (232,000 tonnes).
Products: Carbon steel – wire rod 10, 12, 16, 20, 22, 25, 28, 32mm; reinforcing bars 10, 12, 16, 20, 22, 25, 28, 32mm; round bars 10, 12, 16, 20, 22, 25, 32mm.

Maruichi Malaysia Steel Tube Bhd

Head office: Lot 49, Jalan Utas, Shah Alam, Selangor. **Postal address:** PO Box 18, 40700 Shah Alam. **Tel:** 03 5592455. **Telex:** 38655 mmst ma. **Cables:** Auroratube Shah Alam. **Telecopier (Fax):** 03 5592033.
Management: Chairman – Pan Chin Ho (works director); vice chairman – Zain Azahari; directors – Yang Yen Fang (managing), Tham Wai Mun (executive), T. Horikawa, Zain Azlan. **Established:** 1969. **Capital:** Authorised – M$100,000,000. Paid-up – M$39,000,000. **No. of employees:** Works – 131. Office – 37. **Subsidiaries:** Tokyo Steel Wire Sdn Bhd (100%) (steel wire drawing). **Associates:** Sumiputeh Steel Centre Sdn Bhd (19%) (steel service centre). **Annual capacity:** Finished steel: 120,000 tonnes.
Works:
■ At head office. ***Tube and pipe mills:*** welded – nine 9-400mm ERW (annual capacity 108,000 tonnes). ***Other plant:*** Seven cold forming lines for sections up to 250mm (27,000 tonnes); two slitting lines for thicknesses up to 6mm (144,000 tonnes); one steel sheet shearing line for thicknesses up to 6mm (60,000 tonnes); one hot-dip galvanizing plant (24,000 tonnes).
■ PLO 17, Jalan Gangsa, Pasir Gudang Industrial Estate, Johore. (Tel: 07511861. Telex: 60254). ***Tube and pipe mills:*** welded – one 12-65mm ERW (annual capacity 12,000 tonnes). ***Other plant:*** One slitting line for thicknesses up to 6mm (60,000 tonnes).
Products: Carbon steel – longitudinal weld tubes and pipes 9-400mm, galvanized tubes and pipes 12-400mm and hollow sections 12-200mm (combined 1986 output 72,000 tonnes). Angles (lip, plain, hat-type, U gate channels), tubes, triangular tubes (conduit).
Brands: Aurora (electrical conduits); Maruichi (pipes).

Oriental Metal (M) Sdn Bhd

Head office: Jalan Gudang 16/9, Shan Alam, Selangor. **Postal address:** PO Box 24. **Tel:** 03 362706-7, 362992. **Telex:** 38613 omi ma.
Established: 1970.
Works:
■ Cold roll-forming plant.
Products: Carbon steel – cold roll-formed sections (from strip).

Perstima – Perusahaan Sadur Timah Malaysia (Perstima) Sdn Bhd

Head office: Plo 255, Jalan Timah Tiga, Kawasan Perindasterian Pasir Gudang, 81707 Pasir Gudang, Johor. **Tel:** 07 512001. **Telex:** 60554 sadur ma.
Management: Chairman – YB Dato' Haji Mohd Noor Bin Ismail; directors – Abdul Rahman Omar (managing), Jaafar Yusof, Tuan Haji Faisal Bin Haji Siraj, Ahmad Zukni Johari, Lee Seng Toh, Rin Kei Mei, Motoi Inokuchi, Kazumasa Suzuki, Toshihiro Murai, Naoichi Abe, Tadakatsu Nishikawa, Shoshichi Fujisawa, Lim Kay Seng, Yoshinari Higo (deputy managing); managers – Yoshihisa Shimamura (works); Fujihiro Uchiyama (sales), Kamarudin Awang (finance and accounting). **Established:** 1979. **Capital:**

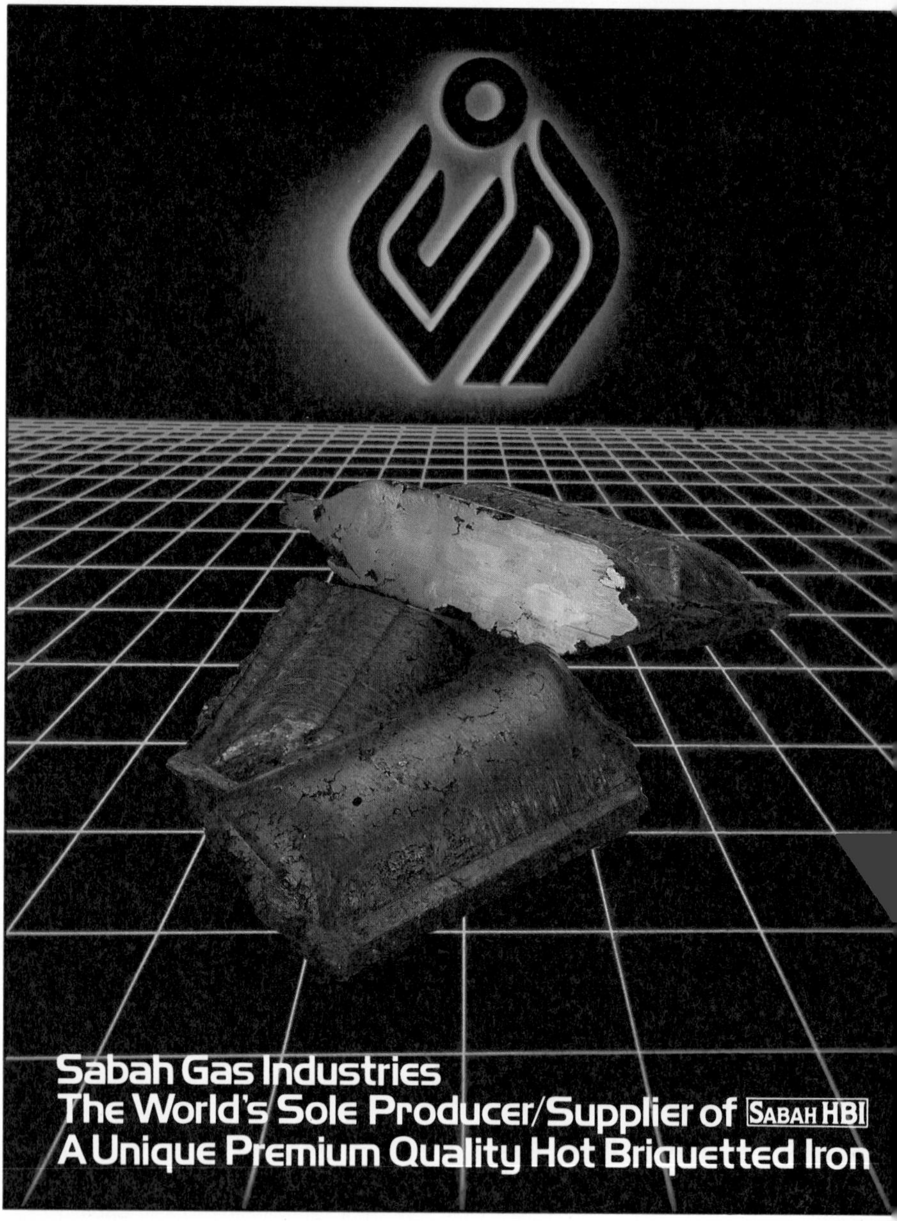

Perstima – Perusahaan Sadur Timah Malaysia (Perstima) Sdn Bhd (continued)

Paid-up – M$ 30,000,000. Authorised – M$ 30,000,000. **No. of employees:** Works – 201. Office – 44. **Ownership:** Shareholders – Kumpulan Fima Bhd, Kawasho Corp, Mitsui & Co Ltd, Malaysia Mining Corp, Johore State Economic Development Corp, Rin Seiko & Sons Co Pte Ltd, Lee Pineapple Co Pte Ltd, United Malayan Pineapple Growers & Canners Co Pte Ltd.
Works:
■ At head office. Works director: Yoshihisa Shimamura. ***Coil coating lines:*** One Halogen KSC electrolytic tinning (annual capacity 90,000 tonnes).
Products: Carbon steel – electrolytic tinplate (single-reduced) (electrolytic tinplate to JIS G 3303), thickness 0.19-0.40mm, width 457-965mm, length 457-1,038mm.

Perwaja Trengganu Sdn Bhd

Head office: Kawasan Perindustrian Telok Kalong, 24007 Kemaman, Terengganu. **Postal address:** PO Box 61. **Tel:** 09 531435. **Telex:** 51351 pewaja ma. **Management:** Director – Hj Razali Hj Ismail (general manager); managers – Woo Khye Thye (sales and marketing), Daud Wahid (purchasing and contract). **Established:** 1982. **Capital:** M$250,000,000. **No. of employees:** Works – 746. Office – 136. **Ownership:** Heavy Industries of Malaysia Bhd. **Subsidiaries:** None. **Associates:** None. **Annual capacity:** Finished steel: 559,000 tonnes.
Works:
■ At head office. ***Steelmaking plant:*** Three 70-tonne electric arc furnaces (combined annual capacity 559,000 tonnes). ***Continuous casting machines:*** Two 4-strand MC-5 billet (100 and 120mm sq) (559,000 tonnes).
Products: Carbon steel – billets 100 and 120mm sq (output 124,321 tonnes).
Sales offices/agents: Marketing office – 17th Floor, Menara Dato Onn, Kompleks UMNO Malaysia, Jl Tun Ismail, 50480 Kuala Lumpur (Tel: 03 2934577, 03 2934401-2. Telex: 33703 wajama ma).

Sabah Gas Industries Sdn Bhd

Head office: PO Box 10123, Kota Kinabalu, Sabah 88801. **Tel:** 88 219826. **Telex:** 80827 sabgas. **Cables:** Sabgas. **Telecopier (Fax):** 88 216577.
Management: Tuan Haji G.R. Ismail (managing director), Victor Sutian Lourdes (marketing manager). **Established:** 1980. **Capital:** M$ 371,900,000. **No. of employees:** Works – 71. Office – 15. **Ownership:** Government of Malaysia. **Annual capacity:** 715,000 tonnes (hot briquetted iron).
Works:
■ At head office. ***Direct reduction plant:*** – one 600 series Midrex (annual capacity 715,000 tonnes).
Products: Direct-reduced iron (DRI) hot briquetted iron (output 715,000 tonnes).
Brands: Sabah HBI.

Southern Iron & Steel Works Sdn Bhd

Head office: 2435 Lorong Perusahaan 12, Prai Industrial Complex, 13600 Perai, Pulau Pinang. **Postal address:** PO Box 138, 10710 Pulua Pinang. **Tel:** 04 306540. **Telex:** 40334 sisw ma. **Cables:** Southmetal. **Telecopier (Fax):** 04 308060. **Management:** Tan Tat Wai (director and chief executive), Lim Hong Sun (general manager), Saw Teow Hong (commercial manager). **Established:** 1964. **Capital:** M$32,000,000. **No. of employees:** Works – 140. Office – 114. **Subsidiaries:** Southern Iron & Steel Trading Sdn Bhd (100%) (trading in iron and steel products).
Works:
■ At head office. ***Rolling mills:*** bar – 15-stand (3-high rougher with horizontal tandem mill) and wire rod – 8-stand twist-free block 5.5-14mm dia (coil weight 5,404 kg) (combined annual capacity 200,000 tonnes).
Products: Carbon steel – wire rod, reinforcing bars, round bars.

Steel Pipe Industry of Malaysia Sdn Bhd

Head office: 4457 Mk 15, Chain Ferry Road, Butterworth. **Tel:** 348822, 348759, 348931. **Telex:** 40765.
Established: 1967.
Products: Carbon steel – longitudinal weld tubes and pipes, galvanized tubes and pipes.

United Malaysian Steel Mills Bhd

Head office: 1-5 Jalan 13/6, Petaling Jaya, Selangor.
Works:
■ ***Steelmaking plant:*** electric arc. ***Rolling mills:*** bar.
Products: Carbon steel – reinforcing bars.

Malta

Hompesch Steel Co Ltd

Head office: Dejma Road, Fgura. **Postal address:** PO Box 62, Valletta. **Tel:** 821325-27. **Telex:** 1337 mw. **Cables:** Wintab Malta.
Management: Edwin Tabona (chairman), Anthony Tabona (managing director). **Established:** 1960. **Capital:** M£98,000. **No. of employees:** Works – 21. Office – 3. **Ownership:** Privately owned. **Annual capacity:** Finished steel: 20,000 tonnes.
Works:
■ At head office. ***Rolling mills:*** medium section – one 14″ cogging and six 12″ finishing (annual capacity 20,000 tonnes).
Products: Carbon steel – reinforcing bars, round bars, flats (not produced regularly).

Mauritania

Sté Arabe du Fer et d'Acier (Safa)

Head office: PO Box 114, Nouadhibou. **Tel:** 45603, 45389. **Telex:** 444 mtn.
Management: Mohamed Saleck ould Heyine (president) and Ahmedov ould Jiddov (director general). **Established:** 1985. **Capital:** US$7,500,000. **No. of employees:** Works – 176. Office – 24. **Ownership:** Sté Nationale Industrielle et Minière (Snim) (33.33%); Arab Iron & Steel Co (Bahrain) (33.33%); Arab Mining C (Jordan) – Armico (33.33%). **Annual capacity:** Raw steel: 15,000 tonnes. Finished steel: 36,000 tonnes.
Works:
■ At head office. ***Steelmaking plant:*** One 7-tonne electric arc furnace (annual capacity 15,000 tonnes). ***Rolling mills:*** One 14-stand 3-high and 2-high bar (36,000 tonnes).
Products: Carbon steel – ingots 110mm sq (not for sale); reinforcing bars 8-32mm dia (output 5,000 tonnes).

Mauritius

Desbro International Ltd

Head office: Plaine Lauzun, Port Louis. **Postal address:** PO Box 60. **Tel:** 084043, 083047. **Telex:** 4260 steel iw. **Cables:** Finance.
Management: Dennis A. Taylor (chairman), Francis Masson (managing director), Phillip Rae (company secretary). **Established:** 1966. **Capital:** MRs 25,000,000. **Ownership:** Rogers & Co. **Subsidiaries:** None. **Annual capacity:** Finished steel: 30,000 tonnes.
Works:
■ At head office. ***Rolling mills:*** Two – bar one 7-stand 12" (annual capacity 24,000 tonnes), and one 6-stand 8" (6,000 tonnes).
Products: Carbon steel – reinforcing bars 8-40mm.

Mexico

Ahmsa – Altos Hornos de Mexico SA de CV

Head office: Av Yucatán 15, Mexico City, Distrito Federal. **Postal address:** Apartado Postal 1077. **Tel:** 2 11 00 85. **Telex:** 1774307, 1773229. **Cables:** Ahmsa Mexico. **Telecopier (Fax):** 2 11 00 85 Ext 101.
Management: Directors – J. Guillermo Becker Arreola (director general, Sidermex), David Galván Haro (corporativo de adminstración, difusión y relaciones públicas), Juan de Dios Román Pineda (corporativo de operaciones), José Arango Rojas (corporativo de finanzas), Juan Cederborg Almeida (corporativo de asuntos jurídicos y relaciones industriales), Javier Castillo Ayala (corporativo de planeación, presupuesto e informática), Rodolfo Tapia Abarca (corporativo de política de comercialización); controlor interno corpora tivo – Rodolfo Delgado Duhart. **Established:** 1944. **Capital:** Pesos 713,330,000,000(stock). **No. of employees:** Works – 18,983. Office – 5,770. **Ownership:** Government supported. **Subsidiaries:** Rassini SA de CV (vehicle springs); Sidermex International Inc (steel export sales); Fund de Hierro y Acero Sa (castings); Asmsa Ingeniería SA (tubes); Torres Mexicanas SA (electricity transmission towers); Avíos de Acero SA (sales of Ahmsa products); Cia Mexicana de Tubos SA (tubes); Cabezas Acero Kikapoo (dished ends). **Associates:** Consorcio Minero BJP Colorada SA (iron ore pellets); Tubacero SA (welded steel tubes); Serv y Suministros Siderúrgicos SA (research and development); Hotel Chulavista de Monclova SA (company hotel); Inmobiliaria Guardiana SA (leasing); Eléctrica Monclova SA (electricity distribution); Inmobiliaria Hierro y Acero SA (leasing).
Works:
■ Prol B. Juárez S/N, Apartado Postal 121, Monclova, Coahuila, Mexico. (Tel: 3 70 11, 3-11-11, 3-81-11. Telex: 039904, 039909. Fax: 3-29-16, 3-66-09. Cables: Ahmsa Monclova). Works director: Pedro García Altamirano. ***Coke ovens:*** (No 1) Two Heinrich Koppers compound circular batteries (218 ovens) 21.5 cu m and (No 2) two Krupp-Koppers compound circular (95 oven ***Sinter plant:*** (No 2) 140 sq m IES Delattre-Levivier. ***Iron ore pelletizing plant:*** One Dravo de Mexico 464 sq m (606 tonnes/hour). ***Blast furnaces:*** No I 5.18m hearth dia, 482.3 cu m, No II 7.16m hearth dia 932.3 cu m, No III 7.54m hearth dia 1,035 cu m, No IV 7.50m hearth dia 1,033.2 cu m, No V 11.5m hearth dia 2,163 cu m. ***Steelmaking plant:*** basic oxygen converters – No 1 shop three 78-tonne, No 2 shop two 125-tonne. one 145-tonne, one 225-tonne, four 235-tonne open hearth furnaces (Nos 1 and 3 out of use since 1980). ***Continuous casting machines:*** slab One Demag 2-strand curved mould (7-8" x 33-42" x 180-245") and one Demag-Hitachi 2-strand curved mould (7-8" x 33-62" x 180-265"). ***Rolling mills:*** United Engineering and Foundry 41 x 67" 2-high reversing roughing slabbing; United 41 x 86" 1-stand 2-high reversing roughing blooming; 8-stand 2-high continuous Birdsboro billet with two 32 x 54", three 26 x 48" and three 24 x 30" stands; 1-stand 2-high reversing United with 44 x 86" rougher, 50 x 38 x 15" universal stand and 17" edger heavy section. Light section and bar with 22" 1-stand 2-high reversing rugher, 2-stand 3-high reversing 14", 3-stand 2-high

Ahmsa – Altos Hornos de Mexico SA de CV (continued)

reversing 14", 5-stand 2-high continuous 12", 3-stand 2-high continuous 10" and 1-stand 2-high continuous 10" finishing mill; Morgan with 13-stand 2-high continuous roughing and 2-strand 10-stand 2-high continuous wire rod; Wean United 4-high reversing 36.5 x 130" heavy plate; wide hot strip United 2-high reversing with 2 horizontal and 2 vertical rolls, 6-stand 62" tandem mill with 2 roughing and 4 continuous stands; Hitachi 4-stand 18.1 x 56" tandem cold reduction; three temper/ skin pass (44, 42 and 56"); five reversing cold mills (four 1-stand and one 5-stand). ***Coil coating lines:*** Three electrolytic tinning (width 18-36", thickness 0.007-0.01495", 0.0075-0.01495", 0.0077-0.01495".
Products: Steelmaking pig iron (1986 output 2,422,800 tonnes). Monclova pellets (1,274,600 tonnes), sinter (1,218,700 tonnes), coke (1,545,800 tonnes), raw steel (2,868,200). Carbon steel – ingots (1,835,200 tonnes); slabs 3½-8" x 42-62" (1,745,800 tonnes produced, 46,600 tonnes for sale); blooms 7-7"-13 x 17" (735,800 tonnes produced). Billets 100 × 100mm, 2¾ × 2¾", 39/16 × 3: 9/16", 5 x 5" (461,600 tonnes produced, 149,000 tonnes for sale); wire rod (105,000 tonnes); reinforcing bars (62,800 tonnes); flats and light sections (110,900 tonnes); heavy angles 6 x 4 – 6 x 6", heavy joists and wide-flange beams (heavy) (combined 70,100 tonnes); medium plates, heavy plates thickness ⅜-8", width 60-120", floor plates and clad plates (combined output 357,000 tonnes), hot rolled sheet/coil (uncoated) thickness 0.075-0.375", width 23-60" (384,000 tonnes); pickled – thickness 0.091-0.250", width 23-38"; cold rolled sheet/coil (uncoated) thickness 0.012-0.035", width 24.5-48.5" (386,900 tonnes); electrolytic tinplate (single-reduced) thickness – sheet 0.0083-0.0149", width 30-36", length 18-42"; coil – 0.0083-0.0149", width 30-36" (184,500 tonnes);
Brands: Ahmsa.
Sales offices/agents: Monclova and Sidermex International Inc, TX, USA (Tel: 512 736 4111).

Ansa – Aceros Nacionales SA

Head office: Avenida Hidalgo 132, 54000 Tlalnepantla, Edo de Mexico. **Postal address:** Apartado Postal 22. **Telex:** 0172266.
Established: 1945.
Works:
■ ***Rolling mills:*** One wire rod. ***Other plant:*** Wire plant.

Laminadora Atzcapotzalco SA – Lasa—see Industrias Nylbo SA, Division Acero

Laminados Barniedo SA de CV

Head office: Insurgentes 70, San Juan Ixhuatepec, Tlalnepantla, Edo de Mexico 54180. **Tel:** 5 77 46 36, 5 77 00 58.
Management: José Calixto Perez del Blanco (presidente), Nemesio Perez del Blanco (gerente general), Vicente Carlos Velasco Perez (gerente administrativo), C.P. Jorge Javier Medrano Ortega (contador). **Established:** 1985. **Capital:** Pesos 10,000,000. **No. of employees:** Works – 16. Office – 2. **Annual capacity:** Finished steel: 3,000 tonnes.
Products: Alloy steel (excl stainless) – square bars 10mm and ⅜"; light angles 3 x 19mm and ¾".

Camas y Tubos SA de CV

Head office: Av Electricidad y Carlos B Zetina, Fracc Ind Xalostoc, Ecatepec Edo de México 55540. **Postal address:** PO Box, Apartado Postal 11, Sta Clara Coatitla, Edo de México. **Tel:** 569 33 55.
Management: President – Jacques Porteny; vice-presidents – Eduardo Porteny

Camas y Tubos SA de CV (continued)

(engineering dept), Luis Porteny (financial and purchasing dept), Carlos Porteny (commercial dept). **Established:** 1929. **Capital:** US$1,500,000. **No. of employees:** Works – 300. Office – 60. **Ownership:** Grupo Catusa SA de CV (85%). **Annual capacity:** Finished steel: 150,000 tonnes.
Works:
■ At head office. ***Tube and pipe mills:*** Two Yoder high frequency welded (annual capacity 25,000,000 metres).
Products: Carbon steel – longitudinal weld tubes and pipes gauges 9-16 dia, ½-4" (output 6,000,000 metres); galvanized tubes and pipes gauges 9-16 dia, ½-4" (4,000,000 metres).

Campos Hermanos—see Industrias CH SA

Cold Rolled de México SA

Head office: Calz. Javier Rojo Gomez 186, Iztapalapa, Mexico 13, DF. **Postal address:** Apartado Postal 55 441. **Telex:** 1772693 crolme.
Established: 1966.
Works:
■ Bright bar plant.
Products: Carbon steel – bright (cold finished) bars. Alloy steel (excl stainless) – bright (cold finished) bars.

Deacero – Productos de Acero SA

Head office: Hidalgo 540 Pte, Monterrey, NL. **Telex:** 382435.
Works:
■ ***Rolling mills:*** One Danieli bar and wire rod (annual capacity 140,000 tonnes).
Products: Carbon steel – wire rod, reinforcing bars.

Aceros Ecatepec SA (Aesa)

Head office: Km 19.5 Carretera México-Laredo, Tulpetlac, Edo de Mexico. **Telex:** 01771326.
Established: 1951.
Works:
■ At head office. ***Steelmaking plant:*** Three electric arc furnaces. ***Continuous casting machines:*** Three Concast for billet – two 1-strand and one 2-strand. ***Rolling mills:*** One Danieli billet; three bar comprising one Danieli, two Lewis.
Products: Carbon steel – billets, reinforcing bars, round bars, light angles, light channels.

Fundiciones de Hierro y Acero SA (Fhasa)

Head office: Prolg Fl Madero, Zona Industrial, Monclova, Coah. **Telex:** 3923 fhsame.
Established: 1943.
Works:
■ At head office. ***Steelmaking plant:*** Four electric arc furnaces. ***Rolling mills:*** three bar. ***Other plant:*** Foundry.
Products: Carbon steel – reinforcing bars. Alloy steel (excl stainless) – ingots.

Galvak SA

Head office: Ave de la Juventud 340 Nte, San Nicolas de Los Garza, Nuevo Leon. **Postal address:** PO Box 140, Monterey, NL. **Tel:** 83 53 01 00. **Telex:** 382881 lafame.
Management: Luis Garza T. Fernandez (general director), Abel Moreno P. (sales manager), Raul Franco Morones (export manager). **Established:** 1980. **Capital:** US$26,223,064. **No. of employees:** Works – 213. Office – 297. **Ownership:** Alfa Industries SA CV (Ultimate holding company – Hylsa SA CV). **Subsidiaries:** Commercial Zincacero SA CV (98%) (export, import, distribution and sales agency; products handled – carbon steel, hot-dip galvanized coils, sheets and corrugated sheets for roofing, longitudinal weld tube). **Annual capacity:** Finished steel: 135,000 tonnes.
Works:
■ At head office. ***Tube and pipe mills:*** welded one Yoder $\frac{3}{4}$-2$\frac{1}{2}$" (annual capacity 7,500 tonnes), and one Wean $\frac{3}{4}$-2" (7,500 tonnes). ***Coil coating lines:*** One Sendzimir Wean United hot-dip galvanizing (12,000 tonnes).
Products: Carbon steel – pipe piling 0.75-2.5" thin wall type; hot-dip galvanized sheet/coil 0.010-0.062" thick 24"-48" width; galvanized corrugated sheet/coil ASTM-A-361, JIS-G-3302. Alloy steel (excl stainless) – longitudinal weld tubes and pipes $\frac{3}{4}$ and 2$\frac{1}{2}$" (thin wall).
Brands: Galvak.

Galvanizadora Nacional SA de CV

Head office: Suarez Peredo Esq Independencia Nte s/n, Veracruz. **Postal address:** Apartado Postal 124. **Tel:** 34 61 20/34 62 90. **Cables:** Galvacional.
Management: Juan Carlos Cobo-Losey Alvarez (president); Jose Luis Cobo Martinez (director). **Established:** 1934. **Capital:** Pesos 40,000,000. **No. of employees:** Works – 28. Office – 10.
Works:
■ At head office. ***Coil coating lines:*** One zinc-lead alloy hot-dip galvanizing (annual capacity 10,000 tonnes).
Products: Carbon steel – hot-dip galvanized sheet/coil, galvanized corrugated sheet/coil.
Brands: Cobos – hot-dip galvanized and corrugated sheets.

Hylsa SA de CV

Head office: Av Munich y Guerrero, 64000 Monterrey, NL. **Postal address:** PO Box 996. **Tel:** 5283 51 8020. **Telex:** 382866 hlhome. **Telecopier (Fax):** 5283 51 2120.
Management: President – Felipe Cortes; corporate vice presidents – Manuel G. Gutierrez (planning), Ernesto Ortiz (finance), Alberto Valdes (human resources); executive vice presidents – Mateo Quiroga (raw materials and services), Fernando Maltos (Flat Rolled Division), Alejandro M. Elizondo (Bar & Rod Division), Joaquin Guzman (Tubular Division), Raul Quintero (Technology Division); vice presidents – Luis Padilla (government relations), Ricardo Ortiz (legal counsel), Roberto X. Margain (treasurer), Alejandro Frias (procurement); assistant vice presidents – Roberto F. Cavazos (external affairs), Regulo Salinas (strategic planning), Ricardo J. Rendon (information systems), Luis G. Villarreal (comptroller), Roberto Garza (finance administration), Roberto M. Quinones (auditing), José A. Ramirez (labour affaris), Gilberto J. Lozano (organisational planning), Martin Espino (total quality project). **Established:** 1942. **No. of employees:** 7,350. **Ownership:** Grupo Industrial Alfa (100%). **Subsidiaries:** Las Encinas SA (100%) (mining and pelletizing); Marcacero SA (100%) (steel trader); Materiales y Aceros SA (100%) (steel distributor); HYL (100%) (direct reduction technology and services). **Annual capacity:** Raw steel: 1,700,000 tonnes.
Works:
■ Pelletizing Plant, PO Box 130, 28000 Colima, Col (Tel: 52331 2 0116. Telex:

Hylsa SA de CV (continued)

62213 hlcome). Works director: José Maciel (vice president, mining and pelletizing). ***Iron ore pelletizing plant:*** One straight grate (annual capacity 1,800,000 tonnes).
■ Monterrey Works, PO Box 996, 64000 Monterrey, NL (Tel: 5283 51 2020. Telex: 382866 hlmome). Works director: Hector Sanchez (vice president, production Flat Rolled Division). ***Direct reduction plant:*** – one HYL (batch) (annual capacity 105,00 tonnes), two HYL-III (continuous) (250,000 and 500,000 tonnes). ***Steelmaking plant:*** electric arc furnaces – one 120-tonne and three 100-tonne (1,100,000 tonnes). ***Rolling mills:*** Two reversing roughing slabbing (45¼" max) (1,000,000 tonnes); one 6-stand continuous wide hot strip (23-43") (1,000,000 tonnes); three cold reduction (22" min to 42" max) (500,000 tonnes); one reversing cold strip temper/skin pass (300,000 tonnes); one temper mill (300,000 tonnes). ***Coil coating lines:*** electrolytic tinning (100,000 tonnes). ***Other plant:*** One continuous pickling and oiling line; two continuous cleaning lines. Iron and steel foundry.
■ Puebla Works, PO Box 842, 72000 Puebla, Pue (Tel: 5222 46 6000. Telex: 178254 hlpume). Works director: Ernesto Sanchez (vice president, production, Bar & Rod Division). One HYL ***Direct reduction plant:*** (batch) (annual capacity 630,000 tonnes). ***Steelmaking plant:*** Three 65-tonne electric arc furnaces (600,000 tonnes). ***Continuous casting machines:*** Two 5-strand curved type billet (4 x 4") (600,000 tonnes). ***Rolling mills:*** One 2-strand continuous combination merchant bar (5/16-1½") and wire rod (¼-0.587") mill equipped with controlled cooling (500,000 tonnes).
■ Apodaca Plant, PO Box 996, 64000 Monterrey, NL (Tel: 5283 53 8020. Telex: 382866 hlmome). Works director: Hectro Cuellar (assistant vice president, Production, Bar & Rod Division). ***Rolling mills:*** One continuous bar (⅜-¾") (annual capacity 150,000 tonnes).
■ Tube & Pipe Plant, PO Box 996, 64000 Monterrey, NL (Tel: 5283 51 2020. Telex: 382866 hlmome). Works director: Ricardo Cantu (vice president, production, Tubular Division). ***Tube and pipe mills:*** welded – one forming and welding and one reducing (4-½") (annual capacity 140,000 tonnes), one galvanizing and conduit line, one square and rectangular tubing line.
Products: Direct-reduced iron (DRI) Carbon steel – billets; wire rod (low-carbon and high-carbon); reinforcing bars; medium plates; hot rolled sheet/coil (uncoated); cold rolled sheet/coil (uncoated); electro-galvanized sheet/coil; electrolytic tinplate (single-reduced); longitudinal weld tubes and pipes (standard A-53 grade A ERW); oil country tubular goods; galvanized tubes and pipes; hollow sections. Direct reduction technology (HYL process); technical assistance (direct reduction and melting shop); engineering consulting. Special steels. Iron and steel castings.
Brands: Hylsa.
Sales offices/agents: At head office (Fernando Cavazos, vice president, international marketing).

Industrias CH SA (formerly Campos Hermanos SA)

Head office: Agustín Melgar 23, Col Niños Heroes, Tlalnepantla, 54030 Edo de Mexico. **Postal address:** PO Box M-29 bis, Mexico 1, DF. **Tel:** 5 65 36 00. **Telex:** 172240 chasame. **Telecopier (Fax):** 3 90 50 42.
Management: President – Carlos O. Murguia; directors – Jorge Aldrete (commercial), Florentino Fernandez (technical). **Established:** 1934. **Capital:** Pesos 1,550,000,000. **No. of employees:** Works – 940. Office – 74. **Ownership:** Grupo Industrial Hermes SA. **Associates:** Cerrey; Aralmex; Famsa; Purina. **Annual capacity:** Raw steel: 120,000 tonnes. Finished steel: 80,000 tonnes.
Works:
■ ***Steelmaking plant:*** One 40-tonne, 13ft dia electric arc furnaces (120,000 tonnes). ***Refining plant:*** One 5MVA ladle furnace; two vacuum degassing units. ***Rolling mills:*** One 34" blooming (150,000 tonnes) and billet (110,000 tonnes); One 7-stand 14/18" bar (70,000 tonnes). ***Other plant:*** Bright bar plant.
Products: Carbon steel – ingots 15¾" x 20: 5/16" x 72½" sq (output 38,203 tonnes); blooms 6¼" to 14" in ¼" increments; billets 4" to 6" in ¼" increments; round bars 1 to 4". Bright (cold finished) bars (6,225 tonnes); Stainless steel – ingots 15¾" x 20: 5/16" x 72½"; blooms 6¼ to 14" increments; billets 4" to 6" in 14"; round

Industrias CH SA (formerly Campos Hermanos SA) (continued)

bars from 1: to 4". Alloy steel (excl stainless) – ingots 15¾" x 20: 5/16" x 72½" (30,958 tonnes); blooms 6¼" up to 14" in ¼" increments (4,200 tonnes); billets 4" up to 6" in 1/4" increments (22,509 tonnes); round bars 1" to 4" (20,592 tonnes); square bars. Bright (cold finished) bars (3,913 tonnes); Forged bars (output 4,781 tonnes).

Tubería Laguna SA de CV (Tulsa)

Head office: Valle del Guadiana y Piedras Negras, Parque Industrial Lagunero, Gómez Palacio, Dgo. **Telex:** 032562.
Established: 1968.
Works:
■ At head office.
Products: Carbon steel – longitudinal weld tubes and pipes.

Mexinox SA de CV

Head office: Paseo de la Reforma 116, Desp 140, Mexico City, DF. **Tel:** 5 92 10 88. **Telex:** 01776210 mexime. **Telecopier (Fax):** 546 77 90.
Management: Directors – Juan Autrique Gómez (general), Luis Almeida, Francisco Guillermo Jáuregui (promotion and services), Federico Reims (administrative and financial, José Luis Rodríguez (operations), Homero Garza (projects), Rafael Lujan (sales), Juan Casanueva (exports). **Established:** 1973. **Capital:** Pesos 91,886,000,000. **No. of employees:** 615. **Subsidiaries:** Inssco, USA; Mexinox Far East Ltd, Japan; Comercinox, Mexico. **Annual capacity:** Finished steel 45,000 tonnes.
Works:
■ Carr. Querétaro, Piedras Negras, Km 192 San Luis Potosí (Tel: 8 10 12. Telex: 013864 mexime). Works director: José Luis Rodríguez Gan. ***Rolling mills:*** Sendzimir cold reduction (annual capacity 45,000 tonnes).
Products: Stainless steel – cold rolled sheet/coil (uncoated) AISI 301-2D, 301-2B, 3013, 304-2D, 304-2B, 3043, 316-2D, 316-2B, 3163, 430-2B, 4303, 3 x 8', 3 x 10', 4 x 8', 4 x 10' in thicknesses from 10 to 30 USG (output 33,400 tonnes).

Industrials Nylbo SA, Division Acero (formerly Laminadora Atzcapotzalco SA – Lasa)

Head office: Pte 128 No 665, Col Industrial Vallejo, 02300 Mexico, DF. **Tel:** 567 95 11. **Telex:** 1777292 protme.
Management: José Jorge Carrillo Schuldes (general manager). **Established:** 1945. **Capital:** US$4,000,000. **No. of employees:** Works – 350. Office – 80. **Ownership:** Grupo Camyna (100%). **Annual capacity:** Raw Steel: 60,000 tonnes. Finished steel: 80,000 tonnes.
Works:
■ At head office. Works director: Benjamin Nuñez Fernandez. ***Steelmaking plant:*** One 7.5 MVA electric arc furnace (annual capacity 60,000 tonnes). ***Continuous casting machines:*** One 2-strand Mitsubishi vertical billet (77 sq mm) (120,000 tonnes). ***Rolling mills:*** Three light section / bar.
Products: Carbon steel – billets 7mm sq (output 60,000 tonnes); reinforcing bars 5/16, ⅜, ½, ¾, 1 (44,400 tonnes): round bars ⅞, ¾, 1⅛ (6,000 tonnes); square bars ⅝, 9/16 (4,000 tonnes); flats ½ × 1½ (2,000 tonnes); hexagons ¾ (1,000 tonnes).

Precitubo SA de CV

Head office: Km 11.4 Carretera al Castillo, El Salto, Jalisco 44940 /Guadalajara. **Postal address:** Apdo Postal 1515. **Tel:** 39 64 73, 39 64 92. **Telex:** 682881 swecome.
Management: Gonzalo Guerra (director), Dionisio Tostado (general manager), Manuel

Precitubo SA de CV (continued)

Godoy (commercial manager). **Established:** 1979. **Capital:** Pesos 955,236,000. **No. of employees:** Works – 135. Office – 77. **Ownership:** Indelta SA de CV, Mexico (51%); Transmesa, Spain (49%) (Ultimate holding company – Industrias Nacobre SA de CV). **Annual capacity:** Finished steel: 10,000 tonnes.
Works:
■ At head office. Works director: Klaus Hellemann. **Draw benches:** (annual capacity 10,000 tonnes).
Products: Carbon steel – cold drawn tubes and pipes for use in boilers, heat exchangers, high pressure boilers; alloy steel tubes, including molybdenum and chrome-molydbenum, and tubes for low and high temperatures; mechanical tubes; welded and seamless.

Productora Mexicana de Tubería SA de CV (PMT)

Head office: Insurgentes Sur 664, 11° piso, Deleg. Benito Juarez, 03100 Mexico, DF. **Tel:** 356 12 10, 536 23 20. **Telex:** 1176510 pmtme.
Established: 1980.
Products: Carbon steel – longitudinal weld tubes and pipes, large-diameter tubes and pipes 16 to 48″.

Protumsa – Productos Tubulares Monclova SA

Head office: Cd. Frontera, Coah. **Postal address:** PO Box 32. **Telex:** 3931.
Established: 1963.
Works:
■ ***Tube and pipe mills:*** welded – Three ERW.
Products: Carbon steel – longitudinal weld tubes and pipes.

Aceros San Luis SA

Head office: Eje 114 s/n, Zona Industrial, San Luis Potosi, SLP. **Tel:** 481 8 11 88. **Telex:** 13685 aslme.
Management: Josefina Garcia de Valladares (president), Rafael del Blanco Garrido (director general), Miguel R. Valladares Garcia (vice president), Miguel F. Valladares Garcia (sub director), Juan Carlos Valladares Garcia (sub director), Jose Marco Campos (director de ventas), Roberto Esparza Carreño (director de production), Julio Fernandez Ferrer (administrador). **Established:** 1968. **Capital:** Pesos 400,000,000. **No. of employees:** Works – 380. Office – 120. **Ownership:** Josefina Garcia Valladares; Miguel Valladares Garcia; Rafael del Blanco Garrido; Luis Arenas Garcia. **Annual capacity:** Raw steel: 120,000 tonnes. Finished steel: 105,000 tonnes.
Works:
■ At head office. Works director: Roberto Esparza Carreño. ***Steelmaking plant:*** Two electric arc furnaces – one 12.5 MVA (11″ dia), and one 7.5 MVA (11″ dia) (combined annual capacity 120,000 tonnes). ***Continuous casting machines:*** Two 2-strand Continua International billet (90mm sq) (120,000 tonnes). ***Rolling mills:*** One semi-continuous bar / wire rod with two 3-high 16″ roughing stands, five 2-high 13″ intermediate, four 2-high 12″ finishing, one 9″ dia monoblock finishing (coil weight 240kg) (105,000 tonnes).
Products: Carbon steel – wire rod 6mm (output 25,000 tonnes); reinforcing bars 8mm, 10mm, 13mm (80,000 tonnes).
Expansion programme: Power of 7.5 MVA electric arc furnace being increased to 12.5 MVA.

pmt
PROMOTES COOPERATIO
WITH UOE-DSAW QUAL
Strict QUALITY CONTROLS have been established throughout the manufacturing process, from the selection of raw materials to the approval of the finished pipe.
INTEGRATED INSPECTION procedures are carri out on each pipe in all phases of manufacturin compliance with domestic and international standards, creating "State of the Art" technol using the UOE process to produce 16" to 48" diameter pipe with 1/4" to 1 1/4" wall thickne
PMT is pleased to contribute in widening tr channels between Mexico and the develop countries within the framework of South-S cooperation.
Productora Mexicana de Tubería, S.A. de C.V.
Insurgentes Sur No. 664-10th and 11th floors
01300 Mexico, D.F. Mexico
Telephones (5) 6-60-41-68 and (5) 5-43-33-07
Telex 1776510 PMTME.
QUALITY
pmt

Sicartsa – Siderurgica Lazaro Cardenas "Las Truchas SA"

Head office: Avenida Yucatan 15, Mexico 7, DF. **Postal address:** Apartado Postal 46-A, Sicartsa, Lazaro Cardenas, Mich. **Tel:** 211 00 85. **Telex:** 017 71113 silcme, 017 74307 sicame.
Established: 1969. **Annual capacity:** Sicartsa I: Pig iron – 1,100,000 tonnes; raw steel – 1,166,000 tonnes; finished steel – 1,000,000 tonnes. Sicartsa II: DRI – 2,000,000 tonnes; raw steel – 2,000,000 tonnes; finished steel – 1,500,000 tonnes.
Works:
■ Sicartsa I, Apartado Postal 45, Co Lazaro Cardenas, Mich (Tel: 2 0333 37. Telex: 017 74242 silcme). ***Coke ovens:*** Two NKK by-product type batteries (30 ovens each) (annual capacity 560,000 tonnes). ***Iron ore pelletizing plant:*** Parrilla Viajera (traveling grate) (1,850,000 tonnes). ***Blast furnaces:*** One 9m, 1,728 cu m (1,100,000 tonnes). ***Steelmaking plant:*** Two 120-tonne basic oxygen converters (1,166,000 tonnes).
Products: Carbon steel – wire rod 5.5 to 12.7mm (1986 output, 194,172 tonnes); reinforcing bars 12.0 to 40.0mm (547,877 tonnes); round bars 12.0 to 40.0mm (144,196 tonnes).

Acero Solar SA de CV

Head office: Via Dr Gustavo Baz 3995, Tlalnepantla, Estado de Mexico. **Postal address:** Apartado Postal 64. **Tel:** 5 65 37 00. **Telex:** 01771218.
Management: Directors – Charles G. Seifert III (administration), Sven Klindtberg (financial), Fernando W. Fest (operations). **Established:** 1958. **Capital:** Pesos 856,000,000. **No. of employees:** Works – 448. Office – 200. **Annual capacity:** Finished steel: 8,000 tonnes.
Works:
■ At head office. Works director: Charles G. Seifert III. ***Steelmaking plant:*** electric arc furnace – one 5-tonne and one 8-tonne (annual capacity 12,000 tonnes). ***Refining plant:*** One 8-tonne AOD (8,000 tonnes). ***Rolling mills:*** One 1-stand 20" billet (8,000 tonnes); two bar – one 3-stand 16" (4,000 tonnes), and one 4-stand 12" (6,000 tonnes); one 8-stand 10" and 8" wire rod (coil weight 35 kg) (2,000 tonnes). ***Other plant:*** Green sand foundry for stainless and alloy steels with 1,000 tonne annual capacity. Cold drawing facilities with annual capacity of 1,800 tonnes.
Products: Stainless steel – wire rod ¼-⅜" (in increments of $\frac{1}{32}$") (1986 output 200 tonnes); round bars 5/16"-5$\frac{1}{2}$" (in increments of $\frac{1}{16}$) (1,800 tonnes); square bars 5/16-4" (in increments of $\frac{1}{8}$ ths) (50 tonnes); flats $\frac{1}{8}$-3" thick x $\frac{1}{2}$ to 6" wide (200 tonnes); hexagons $\frac{3}{8}$ to 1" (in increments of $\frac{1}{16}$") (20 tonnes); wire $\frac{3}{8}$-0.010" (1,000 tonnes). Alloy steel (excl stainless) – round bars 5/16 to 5$\frac{1}{2}$: (in increments of $\frac{1}{16}$") (2,400 tonnes); square bars 5/16 to 4" (in increments of $\frac{1}{8}$") (500 tonnes); flats from $\frac{1}{8}$-3" thick x $\frac{1}{2}$ to 6" wide (1,500 tonnes); wire $\frac{3}{8}$-0.010" (500 tonnes).

Tamsa – Tubos de Acero de Mexico SA

Head office: Campos Eliseos No 400, CP 11000, Mexico City, DF. **Postal address:** Apartado Postal 32139, Mexico, DF. **Tel:** 5 202 00 03. **Telex:** 017 71307, 017 71819. **Telecopier (Fax):** 5592 6059.
Management: Directors – Luis Bossi (general), Guillermo Vogel (general, financial), Gunter Zwingmann (sales, export), G. Vatta Bianchi (sales, national), M. Garcia Ramos (legal), A. Roji Uribe (operational), J. Marentes Manzanilla (purchasing), J. Brogeras Oliva (personnel), J. Peña Flores (technical). **Established:** 1952. **No. of employees:** Works – 3,500. Office – 500. **Annual capacity:** Raw steel: 940,000 tonnes. Finished steel: 900,000 tonnes.
Works:
■ Veracruz Plant, Km 433.5 Carretera Mexico-Veracruz, Apartado Postal 402, Veracruz (Tel: 81 01 44. Telex: 151843). Works director: Adrian Roji Uribe. ***Rolling mills:*** One bar (1$\frac{1}{2}$"-10") (70,000 tonnes).
Products: Carbon steel – round bars, square bars, seamless tubes and pipes, oil

Mexico

Tamsa – Tubos de Acero de Mexico SA (continued)

country tubular goods, cold drawn tubes and pipes. Alloy steel (excl stainless) – round bars, square bars, seamless tubes and pipes, oil country tubular goods, cold drawn tubes and pipes.

Tubacero SA

Head office: Avenida Los Angeles, Monterrey, NL. **Telex:** 0382829.
Established: 1943.
Works:
■ At head office. ***Tube and pipe mills:*** welded – ERW and SAW.
Products: Carbon steel – longitudinal weld tubes and pipes, large-diameter tubes and pipes, oil country tubular goods.

Tubería Nacional SA

Head office: Monterrey, NL. **Postal address:** Apartado Postal 1547.
Established: 1956.
Works:
■ One 22-stand Mannesmann-Meer stretch reducing mill (annual capacity 60,000 tonnes). Hot-dip pipe galvanizing plant (30,000 tonnes).
Products: Carbon steel – longitudinal weld tubes and pipes $\frac{3}{8}$ to 4"; galvanized tubes and pipes.
Brands: Tuna.

Siderúrgica de Yucatán SA de CV

Head office: Km 8.5 Carretera Merida-Progreso, Merida, Yucatan 97110. **Postal address:** PO Box 17, Unidad Cordemex. **Tel:** 99 27 6544. **Telex:** 753745 siyume.
Management: Vicente Erosa Sr (board chairman), Victor M. Erosa (president and chief executive), Jorge H. González (comptroller), Alfonso Paez (plant director), José A. Ledezma (purchasing manager), Rafael Gama (sales manager), Manuel A. Rosado (traffic manager). **Established:** 1974. **Capital:** Pesos 1,027,000,000. **No. of employees:** Works – 400. Office – 60. **Annual capacity:** Raw steel: 120,000 tonnes per year.
Works:
■ At head office. Works director: Alfonso Paez. ***Steelmaking plant:*** Two electric arc furnaces – one 10' 6.5 MVA and one 11' 8.5 MVA (annual capacity 120,000 tonnes). ***Continuous casting machines:*** One 1-strand Koppers billet (4"/sq) (78,000 tonnes). ***Rolling mills:*** One 19-strand cross-country bar.
Products: Carbon steel – billets 4 x 4" (1986 output 40,000 tonnes); reinforcing bars 5/16, $\frac{3}{8}$, $\frac{1}{2}$ $\frac{5}{8}$, $\frac{3}{4}$", 1", 1$\frac{1}{4}$", 1$\frac{1}{2}$" (36,000 tonnes); flats; light angles $\frac{1}{8}$ x $\frac{3}{4}$, $\frac{1}{8}$ x 1$\frac{1}{8}$ x 1$\frac{1}{4}$, $\frac{3}{16}$ x 1" (5,000 tonnes); special light sections Zeta $\frac{1}{8}$ x $\frac{5}{8}$ x 1" (3,000 tonnes); Zeds $\frac{1}{8}$ x $\frac{5}{8}$ x 1".
Expansion programme: Possible installation of a new 3-stand casting machine.

Morocco

Sonasid – Sté Nationale de Siderurgie

Head office: Avenue Youssef Ibn Tachfine, Rue No 2, Nador. **Postal address:** BP 151, Nador. **Tel:** 060 40 28, 060 49 54. **Telex:** 65787, 65778 sonasid.
Management: Abdellah Souibri (director general), Nacer Bouimadaghene (works director), Mohamed Benamar (commercial division chief), Mostapha Rabhi (financial division chief). **Established:** 1974. **Capital:** Dh 500,000,000. **No. of employees:** Works – 460. Office – 114. **Ownership:** State owned.
Works:

Sonasid – Sté Nationale de Siderurgie (continued)

■ At head office. ***Rolling mills:*** One 25-stand Morgan wire rod (5.5-25mm) (coil weight 1.5 tonnes) (annual capacity 420,000 tonnes)
Products: Carbon steel – wire rod 5.5, 6, 7, 7.5, 8, 10, 11, 12, 12.5, 14, 16, 20, 25 (1986 output 252,000 tonnes); reinforcing bars (smooth and ribbed) 06-25mm.
Expansion programme: The mill is currently supplied with imported billets. Expansion plans are under examination to produce billets locally and also to produce merchant sections.

Mozambique

Ima – Indústria Moçambicana do Aço Sarl

Head office: Av 24 de Julho 2373, 12° Esq, Maputo. **Postal address:** PO Box 2566.
Tel: 23446, 23515, 21141. **Telex:** 6-323 tubos mo.
Established: 1970.
Works:
■ At head office. ***Tube and pipe mills:*** High frequency welded. ***Coil coating lines:*** One hot-dip galvanizing (sheet).
Products: Carbon steel – galvanized corrugated sheet/coil, longitudinal weld tubes and pipes.

Nepal

Himal Iron & Steel Co

Head office: Kathmandu, Kingsway. **Telex:** 2264.
Works:
■ **Mini steelworks:** (annual capacity 40,000 tonnes).

Netherlands

Avesta Sandvik Tube BV—see Avesta Sandvik Tube AB, Sweden

Iron and Steel Works of the World

The entry for Hoogovens Groep BV, the Netherlands (page 319) is incorrect and should read as follows:

Hoogovens Groep BV

Head office: Vondellaan 10, Beverwijk. **Postal address**: PO Box 10000, 1970 CA IJmuiden. **Tel**: 02510-99111. **Telex**: 35203. **Telecopier (Fax)**: 02510-10542.
Management: From May 1, 1988: Olivier H. A. van Royen (chairman), Max Albrecht (finance, control, trading division), Maarten C. van Veen (steel division), J. Fokko van Duyne (other divisions). **Hoogovens IJmuiden**: Maarten C. van Veen (chairman), Henk van Appeldoorn (purchasing), Maarten Frese (controlling), Wim J. A. Hoenselaar (technology), Kees Koetzier (sales), Freek J. Scholten (personnel), Wim Singer (production). **Established**: 1918. **No. of employees**: Steel Division — 17,400 (end-1986). **Ownership**: Koninklijke Nederlandsche Hoogovens en Staalfabrieken NV (Ultimate ownership — private investors (80.5%), Netherlands government (13.6%), municipality of Amsterdam (5.9%)). **Subsidiaries**: Steel division subsidiaries — Hoogovens IJmuiden (100%), Hoogovens IJmuiden Verkoopkantoor (100%), Neban (100%), Holco (100%), Segal (33.33%), Cindu Chemicals (50%). Aluminium division subsidiaries — Alumined Beheer (100%), Aldel (100%), Alu/Premetaal (100%), Sidal (100%). Technical Services Division — Hoogovens Technische Dienstverlening (HTD) (100%), ESTS (100%), Esmil Water Systems (100%), HAS (100%), Nucon (33.33%). Industrial Supply Division — Hoogovens Industriele Toeleveringsbedrijven (HIT) (100%), Nederlands Draadindustrie (NDI) (100%), Van Schothorst (100%), Staalcenter Roermond (100%), Cirex (100%), Thibodraad (100%), De Globe (50%). Trading Division — Hoogovens Handel (100%), Hoogovens Handel Constructiestaal (100%), Hoogovens Handel Metals (100%), Hoogovens Handel Wapeningsstaal (100%), Sternotte (99.98%). Other subsidiaries — Cemij (100%), Hoogovens Delfstoffen (100%), Hoogovens UK (100%), Tac-Groep (100%), VBF-Buizen (100%).
Annual capacity: Pig iron: 6,580,000 tonnes. Raw steel: 7,730,000 tonnes. Finished steel: 4,735,000 tonnes.
Works: Hoogovens IJmuiden (Tel: 02510 99111, 02510 99222. Telex: 35211. Fax: 02510 97475). **Coke ovens**: Two plants with 10 batteries altogether, totalling 346 ovens (3 Otto batteries — 102 ovens; 1 Didier battery — 64 ovens; 2 Coppee batteries — 72 ovens; 4 Still batteries — 108 ovens) total annual capacity 2,400,000 Dwight Lloyd type, strand size [illegible] 4,000,000 tonnes). **Iron ore pelletizing plant**: Grate type 428sq m (annual capacity 3,800,000 tonnes). **Blast furnaces**: No 4 — hearth dia 8.5m, working volume 1,490cu m, total volume 1,670cu m; No 5 — hearth dia 9.0m, working volume 1,420cu m, total volume 1,610cu m; No 6 — hearth dia 11m, working volume 2,330cu m, total volume 2,680cu m; No 7 — hearth dia 13m, working volume 3,520cu m, total volume 4,200cu m (combined annual capacity 6,580,000 tonnes). **Steelmaking plant**: No 1 — three 100-tonne combined blowing BOF converters (1,200,000 tonnes); No 2 — three 315-tonne combined blowing BOF converters (4,900,000 tonnes). **Refining plant**: (steelmaking plant No 2) one 315-tonne RH-OB (2,000,000 tonnes). **Continuous casting machines**: (steelmaking plant No 1) two 6-strand Concast for billets (1,200,000 tonnes), (steelmaking plant No 2) one 2-strand and one 4-strand Demag for slabs (4,900,000 tonnes). **Rolling mills**: One universal slabbing (3,500,000 tonnes), one rod/merchant (400,000 tonnes), one heavy and medium plate (595,000 tonnes), one 88in hot strip (4,290,000 tonnes). Cold mill dept 1 — two pickle lines, one 4-stand 56in cold strip mill, one 5-stand 48in cold strip mill, two continuous annealing and processing lines, three temper mills. Cold mill dept 2 — one pickle line, one 5-stand 87in cold strip mill, one temper mill. **Coil coating lines**: Three electrolytic tinning lines, one ECCS line, one hot-dip galvanizing and processing line (total capacity 3,000,000 tonnes, of which 600,000 tonnes tinplate and ECCS).
Products: Carbon steel — slabs, blooms and billets (rolled 70-350mm sq, continuously cast 100-200mm sq) (1986 output 1,356,000 tonnes); reinforcing bars 8-40mm (258,000 tonnes); hot rolled hoop and strip (uncoated) (257,000 tonnes); medium and heavy plates (237,000 tonnes); hot rolled sheet/coil (uncoated) (592,000 tonnes); cold rolled sheet/coil (uncoated) (1,427,000 tonnes, including further processing); hot-dip galvanized sheet/coil; electro-galvanized sheet/coil; zinc-aluminium coated sheet/coil; blackplate (11,000 tonnes); electrolytic tinplate (single reduced) and electrolytic tinplate (double reduced) (484,000 tonnes); tin-free steel (ECCS). Alloy steel (excl stainless) — slabs, blooms, billets, hot rolled hoop and strip (uncoated), medium plates, heavy plates, hot rolled sheet/coil (uncoated), cold rolled sheet/coil (uncoated).

InterStahl

The Netherlands
Hoogbrugstraat 64
6221 CS Maastricht
Tel. 31 (43) 25 41 44
Telefax 31 (43) 25 28 12
Telex 56 888 inter nl

W-Germany
In der Beckuhl 4
D4224 Hünxe-1
Tel. 49 (2858) 70 71
Telefax 49 (2858) 70 76
Telex 81 29 42 insta d

UK
4 Yarm Road
Stockton on Tees
Cleveland TS 18 3NA
Tel. 44 (642) 60 34 43
Telefax 44 (642) 678 678
Telex 58 71 19 g

Telemarker Revolving Head RH
Fully automatic absolutely correct deep marking
on rough uneven surfaces, for slabs, blooms and billets

36584

Telemarker Hot Spray HS
Fully automatic absolutely correct marking on uneven
surfaces for slabs and blooms. Flexible heights of characters

Telesis Pinstamp/I-Dent PI/ID
Fully automatic, stamping and printing head, within a 5x7 dot matrix
system for alphanumeric characters and with variable heights

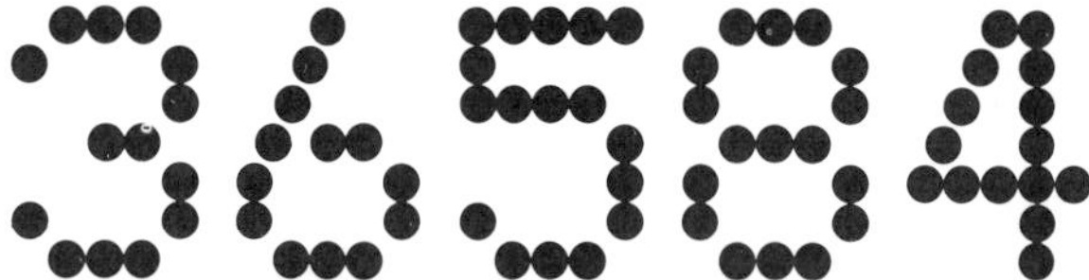

Rotel Deburring Machine DB
Complete flexible integrated in all existing lines,
for head and tail deburring of slabs and blooms

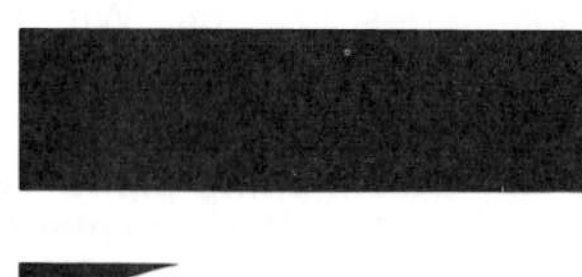

Nedstaal BV (continued)

dia) (coil weight 1,200kg) (257,000 tonnes); one Schloemann rod (11-27mm dia) (coil weight 700kg) (248,000 tonnes). ***Other plant:*** Wire drawing plant (35,000 tonnes finished products). Annealing furnaces for bright annealed wire rod and wire (55,000 tonnes). Pickling plant.
Products: Carbon steel – wire rod 5.50-27mm dia (1986 output 140,000 tonnes, non-alloyed); bright wire 1.50-12mm dia (9,500 tonnes). Alloy steel (excl stainless) – wire rod 5.50-27mm dia (105,000 tonnes); wire (incl 0.60-1.60mm dia Co2 welding wire) (1.50-12.00mm dia) (12,000 tonnes).
Sales offices/agents: P. Klincke, Altena, West Germany; J. Scalini, Nancy, France; Harlow & Jones Ltd, London, UK; Masider Materiali Siderurgici e Affini Srl, Milan, Italy; Sutinsa, Barcelona, Spain; Thyssen Steel Co, New York, USA; Th. Smith-Christensen, Oslo, Norway; Scansider AB, Gothenburg, Sweden; Schubarth & Co AG, Basle, Switzerland; A. Felthaus, c/o Axel Felthaus & Co, Lynbgy, Denmark.

Prins NV

Head office: Rondweg 35, 9101 BE Dokkum. **Postal address:** PO Box 4, 9100 AA Dokkum. **Tel:** 05190 3715. **Telex:** 46110 prima. **Telecopier (Fax):** 05190 8137.
Management: Managers – T.G. de Gier (general), A. Hooysma (works), H.P. Herwig (sales), **Established:** 1940. **No. of employees:** 250.
Works:
■ At head office. Cold forming mills, presses.
Products: Carbon steel – cold roll-formed sections (from strip). Frames for glass houses, wheels, guard rails.

Robur Buizenfabriek BV

Head office: 5700 AA Helmond. **Postal address:** Postbus 27. **Tel:** 04920 22345. **Telex:** 51202.
Established: 1841.
Works:
■ At head office. **Plant:** Six tube drawbenches.
Products: Carbon steel – seamless tubes and pipes, precision tubes and pipes 10-50mm.

P. van Leeuwen Jr's Buizenhandel BV

Head office: Lindtsedijk 20, 3330 AA Zwijndrecht. **Postal address:** PO Box 1. **Tel:** 078 11 59 11. **Telex:** 29176. **Cables:** Buisleeuw. **Telecopier (Fax):** 078 19 44 44.
Management: Board directors – C.C. Th. Rietberg (chairman), S. Duvekot, P. van der Woude, Jan van Breugel. **Established:** 1934. **Subsidiaries:** Australia – Van Leeuwen Pipe & Tube, West Perth. Belgium – Van Leeuwen Buizen, Vilvoorde. Canada – Van Leeuwen Pipe & Tube (Canada) Inc, Calgary, Alta. France – Liontube SA, Chalette-sur-Loing. Malaysia – Van Leeuwen Pipe & Tube (Malaysia) Sdn Bhd, Kuala Lumpur. Netherlands Antilles – Van Leeuwen Pipe & Tube Corp, NV, Curaçao. Singapore – Van Leeuwen Pipe & Tube (Far East) Pte Ltd, Jurong Town. Spain – Leon Tubos y Aceros Iberica SAE, Alcala de Henares. UK – The Lion Tube & Steel Co Ltd, London and Middlesbrough. USA – Van Leeuwen Pipe & Tube Corp, Houston, TX.
Works:
■ At head office.
Products: Carbon steel – longitudinal weld tubes and pipes 13-127mm dia; hollow sections. Alloy steel (excl stainless) – longitudinal weld tubes and pipes 13-127mm dia; hollow sections 100mm sq, 120 x 80mm.
Additional information: The company's main activity is stockholding and international trading in tubes and pipes, fittings and flanges. Its own production accounts for about a quarter of its turnover.

Netherlands

VBF Buizen BV

Head office: Wilhelminakanaal Zuid 150, 4903 RA Oosterhout. **Postal address:** Postbus 39, 4900 BB Oosterhout. **Tel:** 01620 82000. **Telex:** 54285 vbf o. **Telecopier (Fax):** 01620 82345.
Management: A. de Ruyter, N. Schutte. **Established:** 1969. **No. of employees:** 900. **Ownership:** Hoogovens Groep. **Subsidiaries:** Rijnstaal BV (100%) (manufacture of cord wire); Kleinekoort Poot & Baade (100%) (transport). **Annual capacity:** Finished steel: 210,000 tonnes.
Works:
■ At head office. ***Tube and pipe mills:*** welded – one SRW $\frac{3}{8}$-4" dia (annual capacity 80,000 tonnes); seven HF-LON 9-76mm dia (170,000 tonnes).
Products: Carbon steel – longitudinal weld tubes and pipes $\frac{3}{8}$-4"; galvanized tubes and pipes $\frac{3}{8}$-4"; plastic coated tubes and pipes $\frac{3}{8}$-4" PE; precision tubes and pipes 9-76mm dia, 0.8-4mm; hollow sections $\frac{3}{8}$-4", 40 x 40mm, 100 x 100 x 6mm. Stainless steel – longitudinal weld tubes and pipes 30-76mm dia; precision tubes and pipes 30-76mm dia.
Sales offices/agents: Intertube BV, Brussels, Belgium; Salzgitter France SA, Sante-Mandé, France; Intertube (UK) Ltd, London, UK; Svend Møller & Co, Copenhagen, Denmark; AB Vedeve Bolaget, Gothenburg, Sweden; Kanito Oy, Helsinki, Finland; Norsk Stal AS, Oslo, Norway; Flemming Birk, Broendby Strand, Dmaamrk; VBF Verkaufsbureau Sud, Schwaebisch Hall, West Germany; VBF Verkaufsbureau Nord, Solingen, West Germany; Fa Stahlhandel Mag Hubert Isopp, Hargelsberg, Austria; VBF France J.P. Sineux, Buc, France.
Expansion programme: G 60,000,000 modernisation plan includes installation of three new welding lines for precision and stainless tubing, and a new galvanizing plant.

New Zealand

Centralloy Industries – Division of Stainless Castings Ltd

Head office: 53 Curries Road, Woolston, Christchurch. **Postal address:** PO Box 19-579. **Telex:** 4601.
Established: 1960.
Works:
■ At head office. Centrifugal casting plant for tubes.
Products: Carbon steel – hollow bars. Stainless steel – seamless tubes and pipes.

Email Industries Ltd

Head office: Penrose, Auckland. **Postal address:** PO Box 12-078. **Telex:** 21214.
Works:
■
Products: Stainless steel – longitudinal weld tubes and pipes.

Humes SWP

Head office: 224 Neilson Street, Onehunga, Auckland. **Postal address:** PO Box 3550. **Tel:** 665 129. **Telex:** 21647. **Telecopier (Fax):** 09 593 639.
Management: M.S. Johnson (sales manager). **Established:** 1904. **Capital:** NZ$ 6,000,000. **No. of employees:** Works – 60. Office – 20. **Ownership:** Humes Ltd, Melbourne, Australia.
Works:
■ At head office and Wangammi. ***Tube and pipe mills:*** welded – 6 spiral (100-2,000mm dia) (10,000 tonnes).
Products: Carbon steel – spiral-weld tubes and pipes, plastic coated tubes and pipes.

New Zealand Steel Ltd

Head office: Woolf Fisher Works, Glenbrook, South Auckland. **Postal address:** Private Bag, Auckland. **Tel:** 09 758 999. **Telex:** 2493. **Telecopier (Fax):** 09 758 959.
Management: Chairman – S.D. Pasley, N.M.T. Geary (deputy); directors – A.L. Fergusson (managing), J.A. Anderson, D.A. Clark, A.M. McConnell, D.A.R. Williams Q.C.; general managers – C.P. Bates (production), D.A. Bold (deputy, production), J.S. Brewer (general, personnel), D.R. Buist (mining), N.E. Clark (technology & systems), G.M. Jones (secretarial and legal), P.W. King (planning and marketing), G.S. Peterson (finance and accounting), W.H. Vernon (sales and trading), R.B. Wylie (acting, development); chief engineer – R.G. Williams. **Established:** 1965. **No. of employees:** Works – 1,564. Office – 130. **Ownership:** New Zealand Government (90%). **Subsidiaries:** New Zealand Steel Mining Ltd (ironsand mining); New Zealand Steel Development Ltd. **Associates:** McDonald's Lime Ltd (34%); The Slag Reduction Co (New Zealand) Ltd (50%). **Annual capacity:** Pig iron: 700,000 tonnes. Raw steel: 830,000 tonnes. Finished steel: 690,000 tonnes.
Works:
■ Woolf Fisher Works, Private Bag, Glenbrook, South Auckland (Tel: 09 758 990. Telex: 2493). ***Direct reduction plant:*** One 4 x 75m SLRN (annual capacity 150,000 tonnes); four 5 x 60m SLRN (225,000 tonnes each). ***Other ironmaking plant:*** Two 26 x 7.5m Elkem six-in-line melters (675,000 tonnes). ***Steelmaking plant:*** One 70-tonne K-OBM basic oxygen converter (570,000 tonnes); two 50-tonne electric arc furnaces (1 operating) (160,000 tonnes). ***Refining plant:*** Ladle treatment, stirring argon and nitrogen, alloy trim, temperature control. ***Continuous casting machines:*** One 6-strand Concast billet (90 to 180mm) (500,000 tonnes); one 1-strand Sack slab (185 to 250mm, 650 to 1,550mm) (550,000 tonnes). ***Rolling mills:*** IHI rougher, coilbox 4-finishing 1,550mm wide wide hot strip (coil weight 20 tonnes) (700,000 tonnes); UBE skinpass 1,530mm (coil weight 20 tonnes) (500,000 tonnes); cold

New Zealand

New Zealand Steel Ltd (continued)

reduction comprising 6-high Hitachi 1,350mm reversing (coil weight 20 tonnes) (230,000 tonnes), and 4-high IHI 1,250mm reversing (combination mill) (coil weight 20 tonnes) (250,000 tonnes); 4-high IHI (combination mill) temper/skin pass (coil weight 20 tonnes) (250,000 tonnes). ***Tube and pipe mills:*** welded – One 15-100mm nb linear forming (30,000 tonnes); one 15-50mm nb linear forming (30,000 tonnes). ***Coil coating lines:*** Sendzimir Sumitomo hot-dip galvanizing (150,000 tonnes); floater oven Nippon Herr colour coating (60,000 tonnes).
Products: Carbon steel – slabs 210mm thick, 650 to 1,550mm (available 1988); billets 90 x 90mm, 100 x 100mm, 120 x 120mm; (output 149,275 tonnes); hot rolled coil (for re-rolling) (available 1988); hot rolled hoop and strip (uncoated) (available 1988); cold rolled hoop and strip (uncoated); hot-dip galvanized hoop and strip gauge 0.32 to 1.85mm, width 100 to 500mm (7,653 tonnes); coated hoop and strip plastic laminate, gauge 0.32 to 1.15mm, width 100 to 500mm (included in colour coated); medium plates 3.0 to 12.0mm, 650 to 1,550mm; heavy plates 16.0 to 50.0mm, 650 to 1,550mm; floor plates 3.0 to 12.0mm, 650 to 1,550mm; hot rolled sheet/coil (uncoated) 2.0 to 12.0mm, 650 to 1,550mm (available 1988); cold rolled sheet/coil (uncoated) 0.30 to 2.5mm, 610 to 1,550mm (available 1988); hot-dip galvanized sheet/coil width 500 to 1,225mm, gauge 0.32 to 1.85mm (88,012 tonnes); colour coated sheet/coil gauge 0.32 to 1.15mm, width 100 to 1,225mm (43,552 tonnes); longitudinal weld tubes and pipes internal diameter 15 to 120mm (4,639 tonnes); galvanized tubes and pipes internal diameter 15 to 120mm (9,939 tonnes); hollow sections RHS exterior dimensions 25 x 25mm to 102 x 102mm black and galvanized (5,635 tonnes).
Sales offices/agents: Northern Region Sales Office, 407 Great South Road, Penrose,(Tel: 09 593 552. Telex: 21718. Fax: 09 279 8448); Export Sales Office, Cnr Ryan Place and Great South Road, Wiri (Tel: 09 278 5112. Telex: 21718. Cables: Glenzinc. Fax: 09 379 8448); Central Region Sales Office, 134 Qucon's Drive, PO Box 30-256, Lower Hutt (Tel: 04 692 850. Telex: 31006. Fax: 04 692 047); Southern Region Sales Office, 6 Vulcan Lane, PO Box 6032, Upper Riccarton, Christchurch (Tel: 03 385634. Fax: 03 380 111); New Zealand Steel – Ferrostaal Pte Ltd, 510 Thomson Road, 16-01 SLF Complex (Block A), Singapore 1129 (Tel: 65 258 9266. Telex: 8722124. Fax: 0065 0258 1933).

Pacific Steel Ltd

Head office: Favona Road, Otahuhu, Auckland. **Postal address:** PO Box 22201. **Tel:** 09 276 1849. **Telex:** 2334. **Cables:** Pastel. **Telecopier (Fax):** 09 276 3155.
Management: Managing director – J.B. Williams; managers – J. Pitt-Stanley (general, operations), J.E. Hilliard (commercial), D.B. McNeill (project), D. McNabb (employee relations); company secretary – J.F. Ovens. **Established:** 1962. **Capital:** NZ$7,000,000. **No. of employees:** Works – 434. Office – 128. **Ownership:** Fletcher Challenge Ltd (88%), Steel & Tube Holdings (9%), Cable Price Downer (3%). **Subsidiaries:** Pacific Metal Industries Ltd. **Annual capacity:** Raw steel: 190,000 tonnes. Finished steel: 240,000 tonnes.
Works:
■ At head office. ***Steelmaking plant:*** One 45-tonne UHP, 36 MVA electric arc furnace (190,000 tonnes). ***Refining plant:*** One Krupp 8 MVA ladle furnace. ***Continuous casting machines:*** One 3-strand Rokop billet (100-150mm sq) (200,000 tonnes). ***Rolling mills:*** One 3-high 22" cogging and 11-stand 14 x 12" merchant bar (110,000 tonnes); single-strand 27-stand wire rod (controlled cooling equipment) (coil weight 0.65 tonnes) (130,000 tonnes).
Products: Carbon steel – blooms 150mm sq and billets 100mm sq (1986/87 output 160,000 tonnes); wire rod 5.5-12mm (64,000 tonnes); reinforcing bars 6-40mm (114,000 tonnes); flats 25-100mm (8,000 tonnes); light angles 25-80mm (11,000 tonnes); light channels 3" and 4" (3,000 tonnes).
Expansion programme: New bar and section mill under construction (completion August 1988).

Spiral Welded Pipes Ltd—see Humes SWP

Nigeria

Ajaokuta Steel Co Ltd

Head office: Ajaokuta City, Kwara State. **Postal address:** PMB 100. **Tel:** 00234 058220931. **Telex:** 36390 NG.
Established: 1979. **Capital:** Naira 2,065,379,000. **Ownership:** Nigerian Federal Government.
Works:
■ At head office. ***Rolling mills:*** One light section (annual capacity 400,000 tonnes); one wire rod (130,000 tonnes).
Products: Carbon steel – wire rod, reinforcing bars, round bars, square bars, light angles, light channels, medium joists.
Sales offices/agents: At head office; Wire Industry Ltd, Lagos; Sanusi Steel Co Ltd, Lagos; Pamak Ltd, Benin; A.C. Aso & Bros, Onitsha; and Emi-Edo Steel, Benin City.
Expansion programme: First phase to produce 1,300,000 tpy of finished steel products. Expansion to 2,6000,000 tpy including 1,300,000 tpy flat products. Ultimate production capacity of 5,200,000 tpy envisaged.
Additional information: The works is initially operating as a re-roller using imported billets. It will eventually be integrated using the blast furnace/LD converter route. Continuous casting facilities will consisit of three 4-strand casters for blooms. Three will also be a billet mill. Plate and strip mills are envisaged at a later stage. Considerable delay has been experienced with the project.

Allied Steel Industry (Nigeria) Ltd

Head office: Atani Road, Onitsha, Anambra State. **Postal address:** PO Box 1711. **Tel:** 00234046 210963.
Established: 1980. **Capital:** Naira 2,000,000. **Ownership:** Chief J.E. Muoghalu.
Annual capacity: Finished steel: 20,000 tonnes.
Works:
■ At head office. ***Rolling mills:*** One bar.
Products: Carbon steel – round bars, light angles, light channels.

Delta Steel Co Ltd

Head office: Ovwian-Aladja, Bendel. **Postal address:** PMB 1220, Warri. **Tel:** 00234 053232622, 00234 053232064, 00234 053231900, 00234 053621002-11. **Telex:** 43326 dsteel ng.
Management: Managers – Tachia Jooji (general, chief executive), Steve Nwachukwu Ojobor (deputy general, technical services), Donatus K. Uko (deputy general, commercial), P.N. Atanmo (deputy general, ironmaking), Abiodun Titus Abe (deputy general, steel production), Suleiman Umaru (deputy general, administration), F.E. Okorefe (deputy general, finance), E.T. Obodo (deputy general, estate services).
Established: 1979. **Capital:** Naira 1,261,365,791. **No. of employees:** Works – 3,965. Office – 2,614. **Ownership:** Federal Government of Nigeria (100%).
Associates: Cross River Limestone Ltd (42%) (quarrying and supply of limestone); Marine Recovery Service (10%) (supply of scrap); Metallurgical & Engineering Consultancy (Nigeria) Ltd (30%) (engineering consultancy); Medical Technical Services Ltd (15%) (medica services and medical equipment supply). **Annual capacity:** Raw steel: 1,000,000 tonnes. Finished steel: 300,000 tonnes.
Works:
■ At head office. Works directors: A.T. Abe (steel production), P.N. Atanmo (ironmaking). ***Iron ore pelletizing plant:*** Lurgi travelling grate (annual capacity 1,500,000 tonnes). ***Direct reduction plant:*** Two Midrex 600-Series (1,020,000 tonnes). ***Steelmaking plant:*** Four 110-tonne, 6.4m internal dia electric arc furnaces (1,000,000 tonnes). ***Continuous casting machines:*** Three 6-strand Concast/ Schloemann Siemag, billet (120mm sq) (960,000 tonnes). ***Rolling mills:*** One light

Nigeria

Delta Steel Co Ltd (continued)

section (320,000 tonnes). ***Other plant:*** Foundry with two 2,300kg medium frequency induction furnaces (1,200 tonnes).
Products: Direct-reduced iron (DRI) 3-15mm dia (80% 6-12mm) (1986 output, 108,638 tonnes). Carbon steel – billets 120mm x 120mm x 12 metres (125,622 tonnes); reinforcing bars 12-28mm dia (25,396 tonnes); round bars 12-50mm dia (23,261 tonnes); flats 20 x 5mm-125 x 12mm (3,341 tonnes); light angles equal – 20 x 20 x 3mm-65 x 65 x 11mm, unequal – 40 x 20 x 3mm-60 x 50 x 9mm (13,757 tonnes); light tees 20-60mm; light channels 40 x 20mm-100 x 50mm (1,188 tonnes). Gases (oxygen, argon and nitrogen); lime products (burnt and hydrated limes).
Sales offices/agents: 1 Ozumba Mbadiwe Street, PMB 12786, Victoria Island, Lagos (Tel: 0023 401 617769).
Expansion programme: Expansion of foundry pattern and mould shops; provision of an industrial estate to attract supporting services to the steelplant.

General Steel Mills Ltd

Head office: Km 1, Asaba Ibusa Road, Asaba, Bendel State. **Postal address:** PO Box 1053.
Established: 1981. **Capital:** Naira 8,000,000.
Works:
■ At head office. ***Steelmaking plant:*** Two electric arc. ***Rolling mills:*** One bar.
Products: Carbon steel – reinforcing bars 12 and 18mm.

Hoesch Pipe Mills (Nigeria) Ltd

Head office: Henry Carr Street, Ikeja, Lagos. **Postal address:** PMB 21149. **Tel:** 900090. **Telex:** 26870 ng.
Management: Chairman – Cheif H.T.O. Coker; managing director – A.A. Raman; managers – P.I. Vaidya (general), G.R. Moorthy (factory), S.M. Patel (training), F.I. Mormah (finance), D.O. Ofuasia (personnel). **Established:** 1975. **Capital:** Naira 2,500,000. **No. of employees:** Works – 206. Office – 61. **Ownership:** Private. **Subsidiaries:** Industrial Mineral Products, Division of Hoesch Pipe Mills (manufacturers of asbestos roofing sheets, ceiling sheets). **Annual capacity:** Finished steel: 100,000 tonnes.
Works:
■ At head office. ***Tube and pipe mills:*** welded – RS76 1-2½" (annual capacity 41,000 tonnes), RS40 ½-1¼" (17,000 tonnes), T55R ½-2" (42,000 tonnes). ***Other plant:*** Plants producing galvanized buckets (3,600,000 pieces/year), expanded metal mesh (10,000 tonnes/year), and allied steel products (300,000 pieces/year).
Products: Carbon steel – longitudinal weld tubes and pipes; galvanized tubes and pipes ½", ¾", 1", 1¼", 1½", 2", 2¼" (1986 output 3,000 tonnes (maximum output achieved 12,000 tonnes in 1982)); hollow sections ¾ x ¾", 1 x 1", 2 x 1", 2 x 2", ¾" dia, 1" dia, 2" dia (2,000 tonnes (maximum output achieved 12,000 tonnes in 1982)).

The Jos Steel Rolling Co

Head office: Industrial Area, Old Airport Road, Jos, Plateau State. **Tel:** 00234073 050060-64. **Telex:** 81387 josrc.
Established: 1979. **Capital:** Naira 120,000,000. **Ownership:** Nigerian Federal Government. **Annual capacity:** Finished steel: 210,000 tonnes.
Works:
■ At head office. ***Rolling mills:*** 17-stand bar and wire rod with 10-stand Demag finishing block and Stelmor cooling system (annual capacity 210,000 tonnes).
Products: Carbon steel – wire rod 6-12mm; reinforcing bars 12-40mm (plain and ribbed).
Sales offices/agents: At head office.
Expansion programme: First phase of expansion to 420,000 tpy and second phase

The Jos Steel Rolling Co (continued)

to 720,000 tpy.
Additional information: Billets are supplied by Delta Steel Co, Aladja, with the shortfall being imported. It is envisaged that Ajaokuta Steel Co will also supply billets when it starts production.

Katsina Steel Rolling Co Ltd (KSRC)

Head office: Kano Road, Katsina, Kaduna State. **Postal address:** PMB 2056. **Tel:** 00234065 30815-953.
Established: 1981. **Ownership:** Federal Government of Nigeria.
Works:
■ ***Rolling mills:*** bar and wire rod (annual capcity 210,000 tonnes).
Products: Carbon steel – wire rod plain and deformed 5.5-12mm; reinforcing bars plain and deformed 6-40mm.
Sales offices/agents: At head office and 8A Ribadu Road, Ikoyi, Lagos.
Expansion programme: Wire drawing plant, double strand rolling and straightening facilities.
Additional information: Billets are supplied by Delta Steel Co Ltd with shortfall being imported.

Kolorkote Nigeria Ltd

Head office: Industrial Estate, Ota-Idiroko Road, Ota, Ogun State. **Postal address:** PMB 1043. **Tel:** 900490-3. **Telex:** 26408 tower ng. **Cables:** Tower Lagos.
Management: Chairman – A.M. Ferguson; managing director – K. Venkatramana; non-executive directors – A.A. Raman, J.O. Emanuel, Alhaji Y. A. Salami, Chief N.A. Mene-Afejuku. **Established:** 1982. **Capital:** Naira 4,500,000. **No. of employees:** Works – 27. Office – 6. **Associates:** Tower Aluminium (Nigeria) Ltd (60%) (manufacturers of aluminium coils, sheets, circles, hollow ware, long span roofing sheets, extruded profiles).
Works:
■ At head office. ***Coil coating lines:*** One Hunter USA colour coating, two coat/two bake (annual capacity 18,000 tonnes).
Products: Carbon steel – colour coated sheet/coil 600-1,250mm wide (1986 output 1,600 tonnes).

Kwara Commercial, Metal & Chemical Industries Ltd

Head office: Foge Road, Industrial Estate, Ilorin, Kwara State. **Postal address:** PO Box 709. **Telex:** 20333.
Established: 1974. **Capital:** Naira 2,500,000. **Ownership:** Partnership. **Annual capacity:** Finished steel: 54,000 tonnes.
Works:
■ ***Rolling mills:*** bar.
Products: Carbon steel – round bars 8-25mm; square bars 8-12mm; flats.
Expansion programme: Icreasing production capacity to 108,000 tpy, billet production and establishment of a foundry.

Major Engineering Co Ltd

Head office: Plot 68, Ikorodu Industrial Estate, Ikorodu, Lagos. **Postal address:** PO Box 252.
Established: 1979. **Capital:** Naira 4,000,000. **Annual capacity:** Finished steel: 310,000 tonnes.
Works:
■ ***Rolling mills:*** bar (annual capacity 310,000 tonnes).
Products: Carbon steel – round bars, flats, light angles.
Expansion programme: Melting shop planned.

Nigeria

Mandarin Industries Co Ltd

Head office: Plot B, Block XII Industrial Estate, Alhaji Adejumo Avenue, Ikeja, Lagos State. **Postal address:** PO Box 21033.
Established: 1968. **Subsidiaries:** Asiatic Industries Ltd. **Annual capacity:** Finished steel: 75,000 tonnes.
Works:
■ Ikeja. ***Rolling mills:*** bar (annual capacity 75,000 tonnes).
Products: Carbon steel – reinforcing bars.

Nigersteel Co Ltd

Head office: Emene, Enugu, Anambra State. **Postal address:** PMB 1229. **Telex:** 51106.
Established: 1961.
Works:
■ At head office. ***Steelmaking plant:*** electric arc. ***Rolling mills:*** bar.
Products: Carbon steel – reinforcing bars, round bars.

Osogbo Steel Rolling Co Ltd

Head office: Dagbolu Area, New Ikirun Road, Osogbo, Oyo. **Postal address:** PMB 4309.
Management: Managers – A.B. Kolade (acting general), A. Babalola-Adubi (acting technical services), A.O. Fadare (deputy, administration), S.Y. Lawal (assistant, projects and estates), I.O. Fakolujo (assistant, production), G.O. Adelakun (assistant, commercial), H.O. Oshinbanjo (deputy, finance). **Established:** 1979. **Capital:** Naira 150,000,000 (1979). **No. of employees:** Works – 285. Office – 221. **Ownership:** Federal goernment of Nigeria. **Annual capacity:** Finished steel 210,000 tonnes (first phase).
Works:
■ At head office. Works director: B.A. Kolade. ***Rolling mills:*** bar / wire rod – with 17 2-high stands and single-strand finishing block with bar line. Sizes and annual capacity – bar 12-40mm dia (140,000 tonnes), rod 6-12mm dia (coil weight 1,300kg) (70,000 tonnes).
Products: Carbon steel – wire rod 6, 8, 10, and 12mm dia (1986 output, 26,246 tonnes); reinforcing bars 12, 14, 16, 18, 22, and 25mm dia (1,573 tonnes); round bars 12, 14, 16, 18, 22, 25, 32 and 40mm dia (4,307 tonnes).
Expansion programme: Expansion to separate bar line and 2-strand wire rod line to increase output from 210,000 to 720,000 tonnes/year (1990).

Qua Steel Products Ltd

Head office: Ibeno Road, PMB 164, Eket, Cross River State. **Postal address:** PMB 1164.
Established: 1978. **Capital:** Niara 10,000,000. **No. of employees:** Works – 95. Office – 45. **Ownership:** State-owned. Danieli & Co SpA, Italy. **Annual capacity:** Finished steel 60,000 tonnes.
Works:
■ At head office. Works director: E.O. Udoanya. ***Rolling mills:*** bar and wire rod – single-strand with eight 300mm and sixteen 200mm horizontal (annual capacity 60,000 tonnes).
Products: Carbon steel – wire rod (5.5mm, 8mm, 10mm and 12mm dia), reinforcing bars (10mm, 12mm, 16mm, 18mm, 20mm and 32mm), flats (36 x 6mm a/f) and light angles (30 x 5mm) (total output 60,000 tonnes).
Expansion programme: Long-term plan to install furnace to recycle scrap generated during rolling.

Universal Steels Ltd

Head office: Universal Steels Crescent, 77 Surulere Industrial Road, OGBA Scheme, Lagos State. **Postal address:** PO Box 144, Ikeja. **Tel:** 961028, 963145, 964185, 961074. **Telex:** 26378 usteel. **Cables:** Usteels Ikeja. **Telecopier (Fax):** 962489, 964631.
Management: Chairman – Sunday Dankaro; managing director – H.H.T. Wong; managers – Johnson W.H. Chiang, W.Y. Lian (production), James Ajayi (sales); directors – T.Y. Danjuma, C.C. Lee, Msutapha Aliyu (Emir of Bui); Alhaji Abba Kyari, Mrs Olisa Chukwura, Joshua O. Aina. **Established:** 1970. **Capital:** Naira 4,200,000. **No. of employees:** Works – 690. Office – 39. **Ownership:** Private (60% Nigerian, 40% foreign). **Associates:** Sellmetals Ltd (rolling mill). **Annual capacity:** Raw steel: 49,000 tonnes. Finished steel: 72,000 tonnes.
Works:
■ At head office.
Products: Carbon steel – round bars 9-25mm (output 10,000 tonnes); light angles 25 x 25 x 3mm, 40 x 40 x 3mm, 50 x 50 x 6mm (12,200 tonnes).
Expansion programme: It is planned to scrap the 10-tonne electric furnace and replace with a new 12-tonne electric furnace, upgrade the light section mill to produce 65mm and 75mm equal angles, and to install a continuous casting machine in the next four years.

Norway

Christiania Spigerverk AS

Head office: Torshov, N 0401 Oslo 4. **Postal address:** PB 4224. **Tel:** 02 234090. **Telex:** 71131. **Telecopier (Fax):** 02 238205.
Management: Per Havdal (chairmam), Knut Misvær (managing director). **Established:** 1853. **Capital:** NKr 130,000,000. **No. of employees:** 926 (Dec 1986).
Works:
■ Nydalen. Works director: Knut Misvær. ***Steelmaking plant:*** One 60-tonne electric arc furnace (annual capacity 175,000 tonnes). ***Continuous casting machines:*** One 4-strand Demag for billet (175,000 tonnes). ***Rolling mills:*** bar and wire rod – comprising one 3-high jumping roughing stand, four 3-high intermediate stands and ten 2-high finishing stands (160,000 tonnes); and one 6-stand rod finishing block and controlled cooling system. ***Other plant:*** Wire drawing and galvanizing plant. Reinforcement mesh plant. Nail factory.
■ Mandal Stål, PO Box 23, N 4501 Mandal. (Tel: 043 62211. Telex: 21352. Fax: 04 363122). Works director: Øistein Thorkildsen. **Wire drawing and fencing plant**.
■ Stavanger-Hinnaverket, PO Box 100, N 4031 Hinna. (Tel: 04 575644. Telex: 73757. Fax: 04 576251). Works director: Paul B. Mauland. **Nail factory**.
Products: Carbon steel – wire rod and reinforcing bars (plain and reformed) (1986 output 161,000 tonnes). Raw steel (184,000 tonnes). Mesh and other reinforcement products. Drawn wire and wire products (63,000 tonnes).

Norsk Jernverk AS

Head office: N 8600 Mo i Rana. **Tel:** 087 50000. **Telex:** 55025 jverk n. **Cables:** Jernverket Morana. **Telecopier (Fax):** 087 53033.
Management: Board of directors – Per Ditlev-Simonsen (chairman), Gyrd Skråning, Jan Balstad, Anne Cathrine Høeg Rasmussen, Erik Lundgaard, Rolf Nordheim, Hallstein Sjøvoll, John Monsen; administration directors (Jan 1st 1987) – Per Bakken (managing), Thore Michaelsen (production), Per Havdal (production), Ørnulf Valla (corporate development), Gorm Rosnæss (corporate staff), Bjørn Brudvik (financial), Walther Nielsen (accounts), Kåre Saxvik (legal services), Jomar Melen (corporate marketing). **Established:** 1946. **Capital:** NKr 485,000,000. **No. of employees:** 3,565 (end 1986). **Ownership:** Norwegian State (80%); Elkem A/S (20%). **Subsidiaries:** Christiania Spigerverk AS (100%); Norsk Jernverk Mülheim GmbH,

Norway

Norsk Jernverk AS (continued)

Mülheim, West Germany (100%) (steel stockholder); Stålfisk AS (70%) (fishfarming); NB-Kvalitetssikring (51%) (quality assurance, documentation control and maintenance control for offshore oil industry); Polarstål IS (50%) (developing a new steelmaking process). **Associates:** Norsk Stål AS (49%) (steel stockholder); ScanRope AS (37%) (steel and fibrerope); Norwegian Petroleum Consultants (10%) (offshore consultants). **Annual capacity:** Pig iron: 600,000 tonnes. Raw steel: 940,000 tonnes. Rolled products: 600,000 tonnes.

Works:

■ At head office. Works director: Thore Michaelsen. ***Sinter plant:*** One 5-pan Greenawalt (annual capacity 750,000 tonnes). ***Iron ore pelletizing plant:*** Dwight Lloyd (600,000 tonnes). **Ironmaking plant:** Six electric furnaces – four 23 MW (350,000 tonnes), two 35 MW (250,000 tonnes). ***Steelmaking plant:*** Two 70-tonne LD basic oxygen converters (650,000 tonnes); one 75-tonne, 55 MVA electric arc furnace (350,000 tonnes). ***Continuous casting machines:*** One 5-strand Concast billet (350,000 tonnes). ***Rolling mills:*** One 1-stand 2-high 108 x 43" blooming (800,000 tonnes); one 3-stand 2-high heavy section (300,000 tonnes); one 18-stand 2-high light section and bar (3000,000 tonnes). ***Other plant:*** Welding and priming line for sections (40,000 tonnes).

■ Bergen, N 5034 Ytre Laksevåg. (Tel: 05 34 04 00. Telex: 42105 bverk n. Fax: 05 34 04 00/155).

■ Storforshei, N 8630 Storforshei. (Tel: 087 60111. Telex: 55025 jverk n). Fax: 087 53033. Works director: Egil Nordvik. Open pit. Ore dressing plant.

Products: Steelmaking pig iron and foundry pig iron (external sales in 1986, 42,000 tonnes); Carbon steel – slabs (20,000 tonnes); billets 115-160mm (23,000 tonnes), round billets 110-140mm (3,000 tonnes); reinforcing bars 8-40mm (232,000 tonnes); flats 150-200mm (3,000 tonnes); light angles 120 x 120-180 x 90mm (3,000 tonnes); heavy channels VNP 100 x 300mm (15,000 tonnes); wide-flange beams (heavy) columns HE/WF 100-300mm (200,000 tonnes); universals (heavy) 150-700mm (5,000 tonnes); welded sections (heavy) 400-2,500mm (30,000 tonnes); electrical non-oriented silicon sheet/coil 600-1,250mm x 0.05-0.65mm (12,200 tonnes); pole steel (HR electro mechanical sheet for electric generators) 150-1,250mm x 1.6-4.0mm (10,000 tonnes); electrolytic tinplate (single-reduced) 600-1,000mm x 0.15-0.50mm (88,000 tonnes). Iron ore concentrate (external sales in 1986, 556,000 tonnes).

Brands: Noriron (pig iron); Nortin (tinplates); Norcor (electrical sheet); Norpol (pole steel).

Sales offices/agents: Mining Division – Drammensveien 40, Oslo (Tel: 02 603890. Telex: 76068 jverk n. Fax: 02 552629). Iron and steelmaking division – N 8600 Moi Rana (Tel: 087 50000. Telex: 55025 jverk n. Fax: 087 53033). Structural steel division – Drammensveien 40, Oslo (Tel: 02 603890. Telex: 76068 jverk n. Fax: 02 552629). Light products division – Christiania Spigerverk AS, Nydalen, Oslo (Tel: 02 234090. Telex: 02 71131. Fax: 02 238205). Strip mill division – Ytre Laksevåg, Bergen (Tel: 05 340400. Telex: 42105. Fax: 05 340400/155). Norsk Jernverk (UK) Ltd, Nicholson House, Maidenhead, Berks, UK (Tel: 0628 74248. Telex: 848801. Fax: 628 35028). Norsk Jernverk Mülheim GmbH, Mülheim a.d. Ruhr, West Germany (Tel: 208 5860. Telex: 0856551. Fax: 208 586153).

A/S Sønnichsen Rørvalseverket

Head office: Sandakerveien 116, Postbox 4210 T, Oslo 4. **Tel:** 02 18 31 80. **Telex:** 71253 sotub n. **Cables:** Sontube.

Management: Chairman – Arne Sønnichsen; directors – Sven Erik Jansby (managing), Hans Breder (marketing). **Annual capacity:** Finished steel: 35,000 tonnes/shift.

Works:

■ Forus (Stavanger). ***Tube and pipe mills:*** welded – one Sønnichsen $\frac{1}{4}$-$\frac{1}{2}$" (capacity 1,500 tonnes/shift); one HF $\frac{3}{4}$-2" and one HF 1$\frac{1}{4}$-5" (combined capacity 20,000 tonnes/shift).

■ Lisleby (Fredrikstad). ***Tube and pipe mills:*** welded – one ERW 10-15mm (capacity

1,000 tonnes/shift); one HF 12-25mm (6,000 tonnes/shift); one HF 25-90mm (8,000 tonnes/shift).
Products: Carbon steel – longitudinal weld tubes and pipes 10-140mm dia; galvanized tubes and pipes ¼-5"; hollow sections 40 x 40 x 2.5mm-100 x 100 x 5mm. Steel conduit rigid and flexible, PVC conduit rigid and flexible, PVC pressure and sewage pipe.

Pakistan

Ahmed Investments (Pvt) Ltd

Head office: Plot 22, Sector 19, Korangi Industrial Area, Karachi 3113. **Tel:** 310312, 310344. **Telex:** 24650 ingot pk. **Cables:** Steel Ingot.
Established: 1972. **No. of employees:** Works – 150. Office – 12.
Works:
■ At head office. Works director: Asghar Mahmood Asmad. ***Steelmaking plant:*** Two 12-tonne electric arc furnaces (annual capacity 26,000 tonnes each) ***Continuous casting machines:*** 2-strand billet (80-130sq mm).
Products: Carbon steel – ingots; billets 80 x 80, 100 x 100, 130 x 130mm;

Hashoo Steel Industries Ltd

Head office: 1st Floor, Overseas Investors Chamber of Commerce Building, Talpur Road, Karachi. **Tel:** 222604, 222693. **Telex:** 2706 hashu pk.
Management: Chairman – Sadruddin Hashwani; chief executive – M. Ateequllah; directors – S.H. Tehsin, Arif Hashwani, Rajab Ali Panjwani, Mrs Zaver Hashwani, Parvez Ajmal. **Established:** 1982. **Capital:** Rs 21,000,000. **No. of employees:** Works – 140. Office – 25. **Associates:** Hashwani Hotels Ltd (owners of Hashwani Hotels in Pakistan); Pakistan Services Ltd (owners of Pearl Continental Hotels in Pakistan); Premier Tobacco Industries Ltd (cigarette manufacturer); New Jubilee Insurance Co Ltd (insurance). **Annual capacity:** Finished steel: 39,000 tonnes.
Works:
■ Plot DSU-2, Bin Qasim, Karachi. (Tel: 418911-20). ***Rolling mills:*** wire rod comprising one 3-high 350 x 1,200mm and one 2-high 350 x 600mm roughing, three 2-high 250 x 500mm and three 3-high 250 x 500mm intermediate, two 2-high 250 x 500mm finishing stands (coil weight 140kg) (39,000 tonnes).
Products: Carbon steel – wire rod 5.5-12mm dia (1986 output 10,000 tonnes, first year of operation).

Indus Steel Pipes Ltd

Head office: 6th Floor, PIDC House, Dr Ziauddin Ahmed Road, Karachi 4, Sind. **Tel:** 520578, 514837. **Telex:** 23552 indus pk. **Cables:** Gaspipe.
Management: Chairman – Sharbat Ali Changazi; directors – Latif E. Jamal, M.B. Farooqui,Z.I. Bokhari, Aziz E. Jamal, Rashid A. Latif; S.Z. Shami (managing); chief engineer – Syed Husbanallah. **Established:** 1969. **Capital:** Rs 13,000,000. **No. of employees:** Works – 129. Office – 29. **Ownership:** PIDC (Pvt) Ltd (50%), Hussain Industries Ltd (50%). **Annual capacity:** Finished steel: 33,000 tonnes (spiral-weld pipe).
Works:
■ Nr WAPDA thermal power plant, Site Kotri. (Tel: 51169). Factory manager: Chowdhri Saleem Umer ***Tube and pipe mills:*** Four welded – 16-72" dia SR 1500 (annual capacity 15,000 tonnes); 6-18" dia SR 600 (6,000 tonnes); 4-12" dia SR 450 (4,000 tonnes) and 8-24" dia SR 800 (8,000 tonnes).
Products: Carbon steel – pipe piling, spiral-weld tubes and pipes 6-24" dia (output 1985-86 8,556 tonnes). Alloy steel (excl stainless) – spiral-weld tubes and pipes.
Brands: API specification.

Pakistan

Ittefaq Ltd

Head office: Lahore
Works:
■ ***Steelmaking plant:*** Eight electric arc furnaces – one 40-tonne, two 12-tonne, and five 7.5-tonne. ***Continuous casting machines:*** Two billet – one 3-strand Concast and one 5-strand. ***Rolling mills:*** One 19-stand Schloemann bar and wire rod with 8-stand wire rod train (max finishing speed 35m/second).
Products: Carbon steel – round bars, square bars, flats, light angles, light channels.
Additional information: The bar/rod mill was transferred from Klöckner-Werke's Haspe works in West Germany. The three-strand Concast billet caster was originally installed as a two-strand machine.

Metropolitan Steel Corp Ltd (MSC)

Head office: National Press Trust House, 11 Chundrigar Road, Karachi. **Postal address:** PO Box 470. **Telex:** 25521 msc pk.
Works:
■ Landhi Industrial Area, Karachi. ***Rolling mills:*** light section; bar; wire rod; hot strip and hoop.
Products: Carbon steel – wire rod, reinforcing bars, round bars, light angles, light channels, cold roll-formed sections (from strip), cold rolled hoop and strip (uncoated), bright wire, black annealed wire, galvanized wire (plain), galvanized wire (barbed).

New Era Steels

Head office: 3-B Al-Haroon, 10 Garden Road, Karachi-3. **Tel:** 726231-32. **Telex:** 25393 rsl pk. **Cables:** Steeltor.
Management: Sultan Mowjee (proprietor), Ahmed Latif (general manager). **No. of employees:** Works – 80. Office – 12. **Ownership:** Razaque Steels (Pvt) Ltd. **Annual capacity:** Finished steels: 12,000 tonnes.
Works:
■ ***Rolling mills:*** One 5-stand 10" bar (annual capcity 12,000 tonnes).
Products: Carbon steel – reinforcing bars, round bars, light angles, medium angles.

Nowshera Engineering Co Ltd

Head office: Sarhad Chamber House, G.T. Road, Peshawar. **Tel:** 0521 62281, 65257, 0521 64195. **Telex:** 52316 sears pak. **Cables:** Badshahco PSH, Neco NSR.
Management: Syed Phool Badshah (managing director), Pir Jamaat Ali Shah (works director), Khawaja Muhammad Tahir (secretary). **Established:** 1958. **Capital:** Rs 10,000,000. **No. of employees:** Works – 410. Office – 22. **Ownership:** Public limited company. **Subsidiaries:** None. **Annual capacity:** Finished steel: 18,000 tonnes.
Works:
■ PO Ferozsons Laboratories, G.T. Road, Amangarh (Nowshera), North West Frontier Piovina. (Tel: 05231 2630). ***Steelmaking plant:*** Four electric arc furnaces, two 5-tonne, two 2.5-tonne (annual capacity 30,000 tonnes). ***Rolling mills:*** Three light section and bar (6", 8" and 12") (18,000 tonnes) (single-shift). ***Other plant:*** Iron foundry with two cupoles. Two 1-tonne forging hammers. Galvanising plant. Structural fabrication shop.
Products: Carbon steel – reinforcing bars (defromed); round bars; flats; light angles; light tees. Forged shafts and axles. Iron and steel castings.
Sales offices/agents: 29/30 Grain Marhet, Main Beco Road, Badami Bagh, Lahore. (Tel: 200883, 202640).

Pakistan Steel Mills Corp Ltd

Head office: Bin Qasim, Karachi 50. **Tel:** 418911/2622/2230. **Telex:** 2380 psmc pk. **Cables:** Pak Steel.
Management: Chairman – Shujat Ali Bokharee; directors – M. Akram Shaikh (managing), Margbub Ahmad (finance), M. Salim (accounts and personnel), Parvez Rahim (commercial), Z.A. Khan (acting director, general services); principal executives – Anis A.K. Lodhi (planning & investment), Shahdad Sial (development & services), Ali Hashim (systems & training), Habibullah Siddiqui (PSFCL). **Established:** 1968. **Capital:** Authorised Rs 500,000,000. Paid up Rs 50,000,000. **No. of employees:** Works – 14,806. Office – 7,203. **Ownership:** Private. **Subsidiaries:** Pakistan Steel Fabricating Co Ltd (PSFCL). **Annual capacity:** Pig iron: 1,230,000 tonnes. Raw steel: 1,100,000 tonnes. Finished steel: 1,015,000 tonnes.
Works:
■ At head office. ***Coke ovens:*** two batteries (49 ovens) (annual capacity 970,000 tonnes). ***Sinter plant:*** two 75 sq m strand (1,500,000 tonnes of self-fluxed sinter). ***Blast furnaces:*** two 7.2m hearth dia, 1,033 cu m (1,230,000 tonnes). ***Steelmaking plant:*** Two 130-tonne LLD basic oxygen converter (1,100,000 tonnes). ***Continuous casting machines:*** One 4-strand bow-type bloom (260 x 260mm) (275,000 tonnes); two 2-strand bow-type slab (150-200mm x 700-1,550mm) (825,000 tonnes). ***Rolling mills:*** 2-high single-stand 800mm billet (50 x 50 – 125 x 125mm) (260,000 tonnes); wide hot strip – 1,700mm with 4-high universal reversing roughing stand and continuous finishing train (1.6mm-10.00mm thickness, 1,000-1,500mm width) (coil weight 14.5 tonnes, max) (790,000 tonnes); cold reduction – 1,700mm universal four-high reversing stand (0.30-2.5mm thickness 60-1,500mm width) (coil weight 14.5 tonnes, max) (200,000 tonnes). ***Coil coating lines:*** hot-dip galvanizing – continuous (100,000 tonnes). ***Other plant:*** Foundry producing iron, steel and non-ferrous castings up to 10-tonnes, 5-tonnes and 0.8-tonne respectively, equipped with a 5-tonne/hour cupola, a 6-tonne electric arc furnace and an induction furnace with two 1-tonne pots.
Products: Steelmaking pig iron (1985/86 output 793,268 tonnes); foundry pig iron 18, 23 and 45 kg pigs (55,183 tonnes); slabs 150-200mm x 700-1,500mm (363,342 tonnes); blooms 260 x 260mm (239 013 tonnes); billets 50 x 50-125 x 125mm (248 289 tonnes); cold roll-formed sections (from strip) 80 x 80-150 x 150mm (angles – up to 12 metres length; channels – 140 x 60-250 x 80mm, 2-8mm thickness; other shapes – 100-600mm wide up to 12 metres length) (2,977 tonnes); medium plates; heavy plates; hot rolled sheet/coil (uncoated) (1.6-10mm thick) 630-1,500mm wide, 2.5-6 metre long/coils; weight 14.5 tonnes max (301,272 tonnes); cold rolled sheet/coil (uncoated) (0.3-2.5mm thick, 700-1,500mm wide, 1-4 metre long/coil, weight 14.5 tonnes max) (63,556 tonnes); hot-dip galvanized sheet/coil 0.35-1.5mm thick, 700-1,500mm wide, 1.5-4 metres long, 6.5 tonnes max weight) (11,345 tonnes).
Sales offices/agents: General manager (marketing), State Life Building 2, 5th Floor, Wallace Road, Karachi 2.
Expansion programme: The works has been designed for eventual expansion to 2,000,000 tonnes of raw steel per annum.

Ramna Pipe & General Mills Ltd

Head office: Lahore. **Tel:** 851413. **Telex:** 44867.
Works:
■ ***Tube and pipe mills:*** welded One spiral (annual capacity 24,000 tonnes) (single-shift).
Products: Carbon steel – spiral-weld tubes and pipes $\frac{1}{2}$-6" dia; galvanized tubes and pipes.

Razaque Steels (Pvt) Ltd

Head office: 3B Al-Haroon, 10 Garden Road, Karachi 3. **Tel:** 726231-2. **Telex:** 25393 rsl pk. **Cables:** Steeltor Karachi.
Management: Directors – Sultan Mowjee (managing), Irfan Mowjee, Irshad Mowjee, Ahmed Latif. **Established:** 1971. **Capital:** Rs 3,000,000. **No. of employees:** Works – 185. Office – 16. **Annual capacity:** Finished steel: 18,000 tonnes.
Works:
■ Mills: B-30 (B), Estate Avenue, S.I.T.E., Karachi-16. (Tel: 291825, 293275).
Products: Carbon steel – reinforcing bars, round bars, light angles, medium angles, heavy angles.

Sind Steel Corp Ltd

Head office: Steel House, West Wharf Road, Karachi 2. **Tel:** 200416, 201625. **Telex:** 2848 milwa. **Cables:** Millwala.
Management: Fakhruddin Millwala (chairman), Noorbhai Millwala (managing director), Qutubuddin Millwala (director). **Established:** 1950. **Capital:** Rs 4,000,000 (paid-up). **No. of employees:** Works – 110. Office – 20. **Ownership:** Millwala Sons Pvt Ltd (majority). **Subsidiaries:** Kemari Docks Pvt Ltd (wire and galvanizing); Crescent Star Insurance Co Ltd (insurance); Pak Arab Shipping & Trading Agencies Ltd. **Associates:** Undok Millsons Foundries Pvt Ltd (malleable pipe fittings). **Annual capacity:** Raw steel: 20,000 tonnes.
Works:
■ F-169 S.I.T.E, Karachi. (Tel: 294034. Telex: 2848 milwa). Works director: Mr. Mohammadi. ***Steelmaking plant:*** One 5-tonne electric arc furnace (annual capacity 20,000 tonnes). ***Continuous casting machines:*** 2-strand 75mm for billet (planned) (20,000 tonnes). ***Rolling mills:*** 6-stand 10" bar ($\frac{1}{4}$-2" dia) (5,000 tonnes). ***Other plant:*** 1$\frac{1}{2}$-tonnes/hour Cupola being installed.
Products: Carbon steel – ingots; reinforcing bars, round bars, square bars and flats (combined annual output, 5,000 tonnes).
Brands: Millwala/SSC.
Expansion programme: Continuous casting (1987/88); special steel casting (1987/88); foundry (1987).

Panama

Aceros Panamá SA

Head office: Carretera, Tocumen, Panama City. **Telex:** 3114.
Established: 1961.
Works:
■ At head office. ***Rolling mills:*** Two bar.
Products: Carbon steel – reinforcing bars, galvanized wire (barbed).

Paraguay

Acepar – Acero del Paraguay SA

Head office: Azara 197, 8° piso, Asunci6n. **Telex:** 287 py.
Established: 1976.
Works:
■ Villa Hayes. ***Blast furnaces:*** Two (charcoal) (annual capacity 175,000 tonnes). ***Steelmaking plant:*** Two 15-tonne basic oxygen (180,000 tonnes). ***Continuous casting machines:*** Two 2-strand billet. ***Rolling mills:*** One light section, bar and wire rod (combined capacity 150,000 tonnes).
Products: Carbon steel – wire rod, reinforcing bars, round bars, square bars, flats, hexagons, light angles.

Peru

Aceros Arequipa SA (Acersa)

Head office: Avenida Enrique Meiggs 297, Parque Industrial s/n, Lima. **Tel:** 232430. **Telex:** 51041 pe acersa. **Cables:** Acersa.
Established: 1967.
Works:
■ Oficina en Arequipa, Calle 11, Parque Industrial, Arequipa (232430. Telex: 51041 pe acersa). ***Rolling mills:*** One light section / bar.
Products: Carbon steel – round bars, square bars, flats, light angles, light tees, light channels.

Siderperú – Empresa Siderúrgica del Perú

Head office: Av. Tacna 543, piso 11, Lima. **Tel:** 283450. **Telex:** 25465 pe. **Cables:** Siderperu.
Management: Cesar Garay Ghilardi (general manager). **Established:** 1958. **Capital:** Intis 1,615,204,608. **No. of employees:** Works – 3,179. Office – 1,134. **Ownership:** State-owned. **Annual capacity:** Pig iron: 260,000 tonnes. Raw steel: 445,000 tonnes. Finished steel: 320,000 tonnes.
Works:
■ Av. Santiago Antúnez de Mayolo s/n, Chimbote, Ancash. ***Blast furnaces:*** One 5.5m hearth dia, 531 cu m (annual capacity 260,000 tonnes). ***Direct reduction plant:*** Three rotary kilns (90,000 tonnes). ***Steelmaking plant:*** Two LD basic oxygen (245,000 tonnes); four electric arc furnaces (200,000 tonnes). ***Continuous casting machines:*** One Concast 4-strand billet (100mm sq) (120,000 tonnes), one Fives-Cail 3-strand bloom (200 x 250mm) (85,000 tonnes). ***Rolling mills:*** One 650mm billet (110,000 tonnes). Bar merchant (450mm and 500mm) (150,000 tonnes and 70,000 tonnes). One 2-high/4-high reversing heavy plate (2,750mm) (250,000 tonnes); one Steckel wide hot strip (1,425mm) (100,000 tonnes); one 4-high cold reduction (1425mm) (120,000 tonnes). ***Coil coating lines:*** One electrolytic tinning (70,000 tonnes); one hot-dip galvanizing (29,000 tonnes).
Products: Carbon steel – billets (60, 80 and 100mm sq), wire rod (¼:), reinforcing bars (½-⅝"), round bars (½, ⅝, 1, 1½"), heavy plates (6.4-32mm), hot rolled sheet/coil (uncoated) (2-9.5mm), cold rolled sheet/coil (uncoated) (0.3-2mm), hot-dip galvanized sheet/coil (0.3-1.2mm), galvanized corrugated sheet/coil (0.3mm), electrolytic tinplate (single-reduced) (0.17-0.28mm). Alloy steel (excl stainless) – round bars (grinding – 2, 2½, 3, 3½").
Expansion programme: It is to regain previous installed capacity level.

Philippines

Apollo Enterprises Inc (Apollo Steel Mill)

Head office: 0165 Quirino Avenue, Cheng Tsai Jun Bldg, Parañaque, Metro Manila. **Tel:** 833 41 26-9. **Telex:** 22389 philvic ph.
Management: Han Ching Chung (vice president and general manager). **Established:** 1970. **No. of employees:** Works – 164. Office – 43.
Works:
■ 818 E Pantaleon Street, Mandaluyong, Metro Manila. Tel: 78 24 51, 78 24 25, 78 24 44. ***Steelmaking plant:*** 15-tonne, 7,500 kVA electric arc furnace (annual capacity 40-45,000 tonnes). ***Continuous casting machines:*** 3-strand billet (80 and 100mm sq) (42,000 tonnes). ***Rolling mills:*** bar (10-36mm) (20,000 tonnes).
Products: Carbon steel – reinforcing bars.

Armco-Marsteel Alloy Corp

Head office: 2nd Floor, Alpap I Bldg, 140 Alfaro Street, Salcedo Village, Makati Metro, Manila. **Telex:** 64065.
Works:
■ Taguig Steelworks, Napindan, Taguig. ***Continuous casting machines:*** One 2-strand Concast billet and bloom (100mm sq, 160 x 200mm). ***Rolling mills:*** bar – 3-high breakdown and 8-stand continuous.
■ 555 T. Sora Avenue, Baesa, Novaliches, Quezon City. Grinding ball plant.
■ Dr Nativdad Avenue, Napindan, Taguig, Metro Manila. Cut-to-length line for sheet; curving rolls; slitter; forming press.
Products: Carbon steel – blooms, billets, round bars. Alloy steel (excl stainless) – blooms, billets, round bars. Grinding rods and balls; large-dia galvanized corrugated culvert.

Armstrong Industries Inc

Head office: 2804 Cabreta Street, Metro Manila. **Telex:** 63024 arms.
Established: 1966.
Works:
■ ***Steelmaking plant:*** One electric arc furnace. ***Continuous casting machines:*** One 2-strand Concast billet (80, 100mm sq). ***Rolling mills:*** One 280/230/255mm bar.
Products: Carbon steel – reinforcing bars.

Marcelo Steel Corp (MSC)

Head office: Punta Sta Ana, Manila. **Telex:** 3392.
Established: 1949.
Works:
■ At head office. ***Steelmaking plant:*** Two electric arc furnaces. ***Rolling mills:*** One 22" billet. Bar – one 3-high 20" roughing, 3-stand 2-high intermediate and 4-stand 2-high finishing; one 17-stand wire rod. ***Other plant:*** Grinding ball plant.
Products: Carbon steel – wire rod, round bars. Grinding balls.

National Steel Corp (NSC)

Head office: NSC Building, 337 Sen Gil J Puyai Ave, Makati, Metro Manila. **Postal address:** PO Box 631, MCC. **Tel:** 816 20 36-55. **Telex:** 22524 RCA ism ph. **Cables:** Natsteel. **Telecopier (Fax):** 7420173 ITT issmi.
Management: Jose Ben R. Laraya (president); Rolando S. Narciso (executive vice president). **Established:** 1974. **Capital:** Pesos 4,600,000,000. **No. of employees:** Works 4,420. Office – 409. (3,770 regular, 1,059 contract). **Ownership:** National Development Co (100%). **Associates:** Semirara Coal Corp (30%) (mining); Refractories Corp of the Philippines (71%) (refratroy manufacturing); Integrated Air Corp (100%) (air transport). **Annual capacity:** Semi-finished steel: 2,040,000 tonnes.
Works:
■ Iligan Plant (Cold Mill, Hot Mill, Bar Mill and Billet Shop), Camp Overton, Iligan City, Lanao del Norte, Mindanao. (Tel: 20615/20942). Works director: Benito M. Mauricio. ***Steelmaking plant:*** Two 45-tonne electric arc furnaces (annual capacity 300,000 tonnes). ***Continuous casting machines:*** One 4-strand continuous billet (80-150mm) (300,000 tonnes). ***Rolling mills:*** billet and bar – 6-stand 3-high for billets 50mm sq and 3-stand 3-high for bars 10-36mm dia (output 60,000 tonnes); heavy plate and one medium plate – 4-high combination (4.76-50.0mm); one wide hot strip – reversing (1.50-12.70mm) (coil weight 13.6 tonnes max) (700,000 tonnes); one 5-stand cold reduction (0.18-1.52mm) (10 tonnes max) (800,000 tonnes/year) and one 4-stand temper/skin pass.
■ Pasig Plant (Tinning Line), Elisco Road, Bo Kalawaan Sur Paisig, Metro Manila. (Tel: 693 22 80/84). Works director: Panfilo M. Tejada. ***Coil coating lines:*** Two electrolytic tinning – one Ferrostan and one Halogen (annual capacity 130,000

Philippines

National Steel Corp (NSC) (continued)

tonnes).
Products: Carbon steel – billets (output 210,000 tonnes); hot rolled coil (for re-rolling) (128,000 tonnes); reinforcing bars and round bars (39,000 tonnes); medium plates, heavy plates and hot rolled sheet/coil (uncoated) (51,000 tonnes); cold rolled sheet/coil (uncoated) (186,000 tonnes); blackplate; electrolytic tinplate (single-reduced).
Sales offices/agents: At head office and RSA Building R. Palma Street, Cebu.
Expansion programme: Cold mill and hot mill upgrading, tinning line expansion, billet shop expansion and ironmaking and steelmaking.

Pag-Asa Steel Works Inc (PSWI)

Head office: Amang Rodriguez Avenue, Manggahan, Pasig, Metro Manila. **Postal address:** PO Box 7553, Mia, Pasay City, Metro Manila. **Tel:** 60 95 66, 947 9185-7. **Telex:** RCA 27677, 22031 (IMC PH (Attn. Pagasa Steel)). **Cables:** Steelworks Manila.
Management: Francisco Tong (board chairman), Wellington Y. Tong (president), Gabriel Y. Tong (director, general manager). **Established:** 1964. **Capital:** Pesos 20,000,000. **No. of employees:** Works – 100. Office – 30. **Annual capacity:** Finished steel: 45,000-60,000 tonnes (reinforcing bars and wire rods).
Works:
■ At head office. ***Rolling mills:*** bar / wire rod – fully automatic cross-country (annual capacity 45,000-60,000 tonnes, depending on size mix).
Products: Carbon steel – wire rod 7.8, 10, 12, and 16mm dia; reinforcing bars 10, 12, 16, 20, and 25mm; round bars 28, 32, and 36mm dia.

Philippine Sinter Corp

Head office: 11th Floor, Allied Bank Center, Ayala Avenue, Makati, Metro Manila. **Postal address:** PO Box 645. **Tel:** 818 72 21. **Telex:** 22134 RCA sin ph. **Cables:** Sintermla.
Management: Board Chairman – Osamu Endo (president); vice presidents – Gabriel Evangelista (senior, resident manager), Harutoshi Ito (senior, treasurer), Yuji Tsuruoka (assistant resident manager); management staff – Harumi Kondo (chief); manager – Keita Hiraki (import-export dept); director – Hiroshi Sato. **Established:** 1974. **Capital:** Pesos 510,000,000. **No. of employees:** Works – 430. Office – 197. **Ownership:** Kawasaki Steel Corp.
Works:
■ Phividec Industrial Estate, Misamis Oriental (Tel: 33 28, 34-21. Telex: ITT 48406 psc). Works director Gabriel B. Evangelista. ***Sinter plant:*** One 90 x 5m strand Dwight-Lloyd (annual capacity 5,000,000 tonnes). Sintered iron ore and limestone.

Philippine Steel Coating Corp

Head office: Buendia Ave Ext, New Solid Bldg, Makati, Metro Manila. **Telex:** 4404.
Products: Carbon steel – cold roll-formed sections (from strip), galvanized corrugated sheet/coil, colour coated sheet/coil.

Tubemakers Philippines Inc

Head office: 2nd Floor, CMS Bldg, 2293 Pasong Tamo Extension, Makati, Metro Manila. **Telex:** 22049 tpi ph.
Works:
■ Km 40, Sta. Rosa, Laguna. ***Tube and pipe mills:*** welded One TIG.
Products: Stainless steel – longitudinal weld tubes and pipes.

Poland

Export sales of Polish iron and steel products are handled by Stalexport, Mickiewicza 29, 40055 Katowice (Tel: 512 211 19, 513 221 24. Telex: 0315751. Cables: Stalex Katowice) (for rolled steel products). Centrozap Foreign Trade Co Ltd, Mickiewicza 40085 Katowice (Tel: 513401. Telex: 0315771. Cables: Centrozap Katowice) (for foundry products). Centrostal, Centrala Zbytu Stali, Ul. Wita Stwosza 7, Katowice.

Huta Baildon

Head office: Zelazna 9, 40 952 Katowice. **Tel:** 587 221, 588 221. **Telex:** 0315582.
Established: 1823.
Works:
■ ***Steelmaking plant:*** electric arc furnaces including electric melting shop (annual capacity 75,000 tonnes of high-alloy steel), induction furnace; open hearth furnaces. ***Refining plant:*** three ESR – 1, 2, and 10-tonne; vacuum degassing unit; VAD; VOD. ***Continuous casting machines:*** one 1-strand billet (100 sq m). ***Rolling mills:*** rail, light section, heavy plate; 4-high. Cold strip. Wire drawing. ***Other plant:*** Plant producing welding electrodes, sintered carbide, forged bars, drills, permanent magnets.
Products: Carbon steel – heavy rails. Stainless steel – ingots, billets, cold rolled hoop and strip (uncoated), medium plates, heavy plates. Alloy steel (excl stainless) – ingots, billets, cold rolled hoop and strip (uncoated), medium plates, heavy plates. Welding electrodes; sintered carbide products; indexable inserts; forged bars; discs; rings and die blocks.

Huta Batory

Works:
■ Chorz/ow-Batory. ***Steelmaking plant:*** electric arc furnace and open hearth furnace. ***Rolling mills:*** heavy plate (annual capacity 200,000 tonnes).
Products: Carbon steel – heavy plates; tubes and forgings.

Huta Bierut

Annual capacity: Raw steel: 1,000,000 tonnes.
Works:
■ Raków and Mirów, Czestochowa. **Coke ovens and blast furnaces. *Steelmaking plant:*** electric arc furnace and open hearth furnace. ***Rolling mills:*** heavy plate (annual capacity 1,400,000 tonnes); wide hot strip. ***Tube and pipe mills:*** seamless.
Products: Carbon steel – medium plates; heavy plates 4-40mm thick and 2,800-3,300mm wide; hot rolled sheet/coil (uncoated); seamless tubes and pipes.

Huta Bobrek

Works:
■ Bobrek, near Bytom. **Sinter plant and blast furnaces. *Steelmaking plant:*** open hearth furnace. ***Rolling mills:*** slabbing (annual capacity over 1,300,000 tonnes).

Huta Cedlera

Works:
■ Sosnowiec. ***Rolling mills:*** 2-strand 23-stand Morgan No-Twist wire rod with Stelmor controlled cooling.
Products: Carbon steel – wire rod from 5mm dia.

Poland

Huta Centrum—see Huta Katowice

Huta Dzierzyński

Works:
■ Dabrowa Górnicza. **Blast furnaces.** ***Steelmaking plant:*** open hearth furnace. **Rolling mill. Foundry.**

Huta Ferrum

Works:
■ Katowice. ***Tube and pipe mills:*** Spiral welded (annual capacity over 120,000 tonnes).
Products: Carbon steel – longitudinal weld tubes and pipes 150-1,016mm dia.

Huta Florian

Head office: 41-600 Swietochlowice. **Tel:** 452431-39. **Telex:** 0315580.
Works:
■ Swietochlowice. **Blast furnaces.** ***Steelmaking plant:*** open hearth furnace. **Rolling mills.** ***Coil coating lines:*** hot-dip galvanizing (annual capacity 200,000 tonnes). Colour coating.
Products: Carbon steel – hot rolled sheet/coil (uncoated), cold rolled sheet/coil (uncoated), hot-dip galvanized sheet/coil, colour coated sheet/coil, bars and sections.

Huta Jednośc

Works:
■ Siemianowice. ***Steelmaking plant:*** 70-tonne electric arc furnace; open hearth furnace. ***Continuous casting machines:*** 4-strand billet (tube billets 150mm sq). ***Tube and pipe mills:*** seamless (annual capacity 400,00 tonnes).

Huta Katowice

Head office: 41-303 Dabrowa Gornicza. **Tel:** 6622569, 645333, 641611. **Telex:** 0315562, 0315365, 0312741-11.
Works:
■ Dabrowa Gornicza, Katowice. ***Coke ovens:*** Two batteries (total annual capacity 1,600,000 tonnes). ***Blast furnaces:*** Three 3,200 cu m (annual capacity 4,400,000 tonnes). ***Steelmaking plant:*** Two 350-tonne LD basic oxygen converters (4,500,000 tonnes). About twenty open hearth furnaces. ***Refining plant:*** Vacuum refining plant, 300-tonne Argon unit. ***Rolling mills:*** slabbing, blooming and billet (4,500,000 tonnes); heavy section; medium section (850,000 tonnes); wide hot strip (1,200,000 tonnes).
Products: Steelmaking pig iron. Carbon steel – slabs; blooms; billets; heavy rails; medium and heavy sections; hot rolled sheet/coil (uncoated);
Expansion programme: Capacity due to rise to about 7,000,000 tonnes, with billets and blooms being produced for other Polish mills. Third converter (top and bottom blowing) – late 1980s. Two additional coke batteries will raise capacity to 3,200,000 tonnes; one will be operational by 1989.

Huta Kościuszko

Works:
■ Chorzów. **Blast furnaces.** ***Steelmaking plant:*** open hearth furnace. ***Refining plant:*** 70-tonne DH vacuum degassing unit. ***Rolling mills:*** blooming, heavy section, medium section, light section, Morgan wire rod.
Products: Carbon steel – blooms; wire rod; heavy, medium and light sections; heavy rails.

Huta Labedy

Works:
■ ***Steelmaking plant:*** open hearth furnaces.

Huta im Lenina

Head office: Skrytka 44, Cracow 28. **Telex:** 0322441 hil pl.
Annual capacity: Raw steel: over 5,000,000 tonnes.
Works:
■ ***Coke ovens:***. Eight ***Blast furnaces:*** including two 1,033 cu m, two 1,1719 cu m and one 2,002 cu m. ***Steelmaking plant:*** Three 140-tonne basic oxygen (annual capacity 3,000,000 tonnes); one 140-tonne electric arc furnace. Open hearth furnaces – seven 370-140-tonne, one 185-210-tonne, and two tandem 200-tonne. ***Rolling mills:*** One 2-high 1,150mm reversing slabbing and blooming; one 2-strand continuous wire rod; one 10-stand continuous wide hot strip; hot strip and hoop; three cold reduction – one 4-high 1,700mm, one 5-stand 1,200mm tandem and one Sendzimir; two temper/skin pass. ***Tube and pipe mills:*** two high frequency induction welded – 3/8-3'' dia, 2.35-4.05mm wall thickness and 2-6⅝'' dia, 3.25-8mm wall thickness. ***Coil coating lines:*** One continuous electrolytic tinning; two hot-dip tinning and Sendzimir continuous hot-dip galvanizing. ***Other plant:*** Hot-dip tube galvanizing plant; cold-formed section line; ingot mould foundry.
Products: Carbon steel – slabs 100-230 x 760-1,580mm; blooms 100-230 x 550-1,550mm; billets 60 and 80mm sq and 140 x 125mm; reinforcing bars (plain and deformed, 10-22mm dia); light angles 25 x 25 x 3-45 x 45 x 5mm; cold roll-formed sections (from strip) (open and closed); hot rolled hoop and strip (uncoated) 2-6mm thick and 100-500mm wide; cold rolled hoop and strip (uncoated); hot-dip galvanized hoop and strip; medium plates and heavy plates 5-12.7mm thick and 1,000-1,500 wide (ship-plates), 2.25-12.7mm thick and 900-1,500mm wide (carbon and low-alloy commercial); floor plates 3.5-8mm thick and 1,000-1,500mm wide; hot rolled sheet/coil (uncoated) 2.5-5mm thick and 700-1,524mm wide; cold rolled sheet/coil (uncoated) 0.2-0.8mm thick, including autobody sheets; hot-dip galvanized sheet/coil; galvanized corrugated sheet/coil; electrical non-oriented silicon sheet/coil; electrolytic tinplate (single-reduced); hot-dip tinplate; longitudinal weld tubes and pipes $\frac{3}{8}$-ø$\frac{5}{8}$" dia. Cable tape; ingot moulds.

Mostostal

Works:
■ Bochnia, near Cracow. **Rolling mill:** Mill for electrical sheets (annual capacity 50,000 tonnes).
Products: Carbon steel – electrical non-oriented non-silicon sheet/coil.

Huta Nowotko

Works:
■ Oştrowiec. ***Steelmaking plant:*** Three 140-tonne electric arc furnaces (annual capacity 1,000,000 tonnes); ***Refining plant:*** One VOD. ***Continuous casting machines:*** Two 4-strand Concast bloom (220 x 220mm-280 x 320mm). ***Rolling mills:*** light section (800,000 tonnes). Carbon and low-alloy steel. Light sections.

Huta Pokój

Works:
■ Nowy Bytom. ***Blast furnaces:*** Two. ***Steelmaking plant:*** electric arc furnace and six open hearth furnaces. ***Rolling mills:*** blooming (sheet); rail; tube mill.
Products: Carbon steel – blooms, heavy rails tubes, sheet, sections.

Poland

Poldi Steelworks

Works:
■ Kladno. ***Steelmaking plant:*** Two 100-tonne electric arc furnaces and open hearth furnace (combined capacity 1,750,000 tonnes). ***Refining plant:*** One 40-tonne and one 115-tonne vacuum degassing units. ***Continuous casting machines:*** One 3-strand Olsson (220 x 300mm) and one 4-strand USSR (280 x 320mm) bloom. ***Rolling mills:*** billet; medium section and light section (450,000 tonnes).
Products: Carbon steel – blooms; billets; semis for seamless tubemaking. Stainless steel – blooms; billets. Alloy steel (excl stainless) – blooms; billets; round bars 16-74mm; square bars; flats; hexagons 50-120mm x 5-50mm.

Stawola Wola

Works:
■ Stawola Wola. **Steelmaking plant:** Prototype plasma furnace for special steel (annual capacity 500,000 tonnes). Die-casting plant (52,000 tonnes).

Huta Swierczewski

Works:
■ Opole.
Products: Carbon steel – precision tubes and pipes and rolled steel.

Szczecin Works

Works:
■ Szczecin. ***Steelmaking plant:*** Two (annual capacity approx 400,000 tonnes).

Huta Warszawa

Works:
■ Mlociny, near Warsaw. ***Steelmaking plant:*** electric arc furnace and open hearth furnace. ***Rolling mills:*** blooming, roughing mill, bar and wire rod and 4-high reversing cold reduction (annual rolled steel capacity about 450,000 tonnes).
Products: Carbon steel – blooms, wire rod, bright (cold finished) bars, cold rolled sheet/coil (uncoated), electrical non-oriented non-silicon sheet/coil; stainless steel.
Sales offices/agents: This works is Poland's major special steel producer.

Huta Zabrze

Works:
■ Zabrze. ***Steelmaking plant:*** electric arc furnace.

Huta Zawiercie

Works:
■ Zawiercie. **Blast furnaces.** ***Steelmaking plant:*** electric arc furnace (including four 140-tonne) and open hearth furnace. ***Continuous casting machines:*** One 2-strand billet (140mm and 200mm sq). **Rolling and tube mills**.
Products: Carbon steel – billets.
Additional information: This works is Poland's main producer of semis, supplying a number of works including Huta Cedlera, Huta Swierczewski and Huta Jednosć.

Portugal

Oliva – Indústrias Metalúrgicas SA

Head office: Rua da Fundição, 3701 S. João da Madeira. **Tel:** 056 24041. **Telex:** 22683. **Telecopier (Fax):** 32 2 377521.
Management: General manager and president of board – João Miguel Duarte Cebola; board members – Rogério Elias Costa Fernandes (comptroller), Manuel Vaz Sousa (materials director), Eduardo José Felizardo (manufacturing manager), José Bernardo Costa e Silva (technical manager); personnel director – Manual Mendes de Oliveira; export manager – Manuel Soares. **Established:** 1925. **Capital:** Esc 1,904,000,000. **No. of employees:** Works – 798. Office – 258. **Associates:** Oliva Trade (70%) (import and export). **Annual capacity:** Finished steel: 25,000 tonnes.
Works:
■ At head office. Works director: Eduardo Felizardo. ***Tube and pipe mills:*** welded – $\frac{3}{8}$-4″ (annual capacity 25,000 tonnes).
Products: Carbon steel – longitudinal weld tubes and pipes $\frac{3}{8}$-4″ (output 25,000 tonnes); galvanized tubes and pipes (hot dip) $\frac{3}{8}$-4″ (20,000 tonnes).

Cia Portuguesa de Fornos Eléctricos Sarl

Head office: Largo de São Carlos 4-4°, 1200 Lisbon. **Tel:** 36 55 55. **Telex:** 16616 forno p, 53528. **Cables:** Electrofornos Lisbon.
Products: Foundry pig iron, silicon metal, ferro-silicon, calcium carbide, calcium cynamide.
Additional information: This company's works was closed down as we went to press.

F. Ramada, Aços e Indústrias Sarl

Head office: Rua Luís de Camões, P 3881 Ovar Codex, Apartado 10. **Tel:** 54111. **Telex:** 22581 and 22653 ramada p. **Cables:** Framada.
Management: Chairman – Manuel Ramada. Managing director – Francisco Correia de Almeida. Managers – Sousa Uva (works), Manuel Reis Costa (sales), Zeferino Almeida (purchasing). **Established:** 1935. **Capital:** Esc 400,000,000.
Works:
■ At head office. ***Rolling mills:*** Three cold strip – one 4-high reversing, 508mm max wide, 0.27mm min thick; one 4-high reversing, 280mm max wide, 0.12mm min thick; and one 2-high (combined capacity 25,000 tonnes). ***Other plant:*** Draw bench; two continuous drawing machines; bar turning machine.
Products: Carbon steel – bright (cold finished) bars turned and drawn (hexagons 2-62mm, rounds 2-80mm, squares 2-60mm, and flats (annual capacity 6,000 tonnes); cold rolled hoop and strip (uncoated) max 508mm wide, min 0.12mm thick.
Sales offices/agents: Rua Manuel Pinto de Azevedo 171, Porto; Av Infante D. Henrique, lote 575 A, Lisbon; Rua Serpa Pinto 33-A/33-B, Luanda, Angola.

Portugal

Siderurgia Nacional EP (SN)

Head office: Rua Braamcamp 7, P 1297 Lisbon Codex. **Tel:** 533151. **Telex:** 13771, 12229. **Telecopier (Fax):** 535051.
Management: Board – José Almeida Serra (president), Eduardo Santa Marta (vice president), Carlos Prieto Traguelho, Joaquim Pinto Leal, António Silva Carnei; managers – José Tavares Domingues (purchasing), José Manuel Forjó (marketing and sales). **Established:** 1961. **Capital:** Esc 13,170,000,000. **No. of employees:** Works – 5,202. Office – 393. (At Dec 1986). **Ownership:** State-owned. **Subsidiaries:** Portosider, Actividades Portuárias Lda (40%) (port operator). **Annual capacity:** Pig iron: 450,000 tonnes. Raw steel: 840,000 tonnes. Finished steel: 980,000 tonnes.
Works:
■ Seixal Plant, Paio Pires, P 2840 Seixal. (Tel: 2212021. Telex: 18141). Works director: Lemos Pereira. ***Coke ovens:*** One Still battery (28 ovens) (annual capacity 300,000 tonnes); one CEC battery (14 ovens) (not in operation). ***Sinter plant:*** One 42 sq m strand Dwight Lloyd (480,000 tonnes). ***Blast furnaces:*** One 6.5m hearth dia, 670 cu m (450,000 tonnes). ***Steelmaking plant:*** One 48-tonne LD/LBE basic oxygen converter (500,000 tonnes); one 42-tonne 24 MVA electric arc furnace (120,000 tonnes). ***Continuous casting machines:*** One 4-strand Danieli billet (80-100mm) (120,000 tonnes). ***Rolling mills:*** One 3-stand 2-high reversing 800mm heavy section (380,000 tonnes); one medium section – 4-stand 3-high, single-stand 2-high and edger (476mm) (85,000 tonnes); one light section – 3-stand 3-high and 2-stand 3-high (350mm) (45,000 tonnes); one bar – fifteen continuous stands (293 x 450mm dia); one 2-strand wire rod with eight continuous stands (263 x 293mm dia) (300,000 tonnes); two 4-high reversing cold reduction (1,100 and 1,500mm wide) (coil weight 20 tonnes) (285,000 tonnes); one 4-high reversing temper/skin pass (1,500mm wide) (coil weight 20 tonnes) (180,000 tonnes). ***Coil coating lines:*** electrolytic tinning – one Davy-Wean Ferrostan (82,000 tonnes); hot-dip galvanizing – one Heurtey Sendzimir (92,000 tonnes).
■ Maia Plant, S. Pedro de Fins, P 4445 Ermesinde. (Tel: 9670531. Telex: 25129). Works director: Américo Azevedo. ***Steelmaking plant:*** One 75-tonne 60 MVA electric arc furnace (annual capacity 220,000 tonnes). ***Continuous casting machines:*** One 4-strand Demag billet (105-125mm) (220,000 tonnes). ***Rolling mills:*** One light section with nineteen continuous stands (300-500mm dia) (161,000 tonnes); one 2-strand Morgan wire rod with eight continuous stands (288.3mm dia) (148,000 tonnes).
Products: Carbon steel – billets; wire rod 6-32mm (output 159,600 tonnes); reinforcing bars 6-32mm (314,300 tonnes); round bars 6-70mm, square bars 10-80mm, flats 5-25mm x 16-150mm, light angles 20 x 3mm, light tees 25 x 25 x 3.5mm, light channels 4 x 15 x 30mm, medium angles 150 x 15mm, medium tees 70 x 70 x 8mm, medium joists 3.5 x 30 x 50mm-6.3 x 74 x 160mm and medium channels 8 x 70 x 180mm (combined output 107,900 tonnes); heavy rails 54kg/m (3,000 tonnes); rail accessories; cold rolled sheet/coil (uncoated) 0.25-2.99mm x 500-1,550mm x 900-6,000mm (20,100 tonnes); hot-dip galvanized sheet/coil 0.32-3.00mm x 600-1,550mm x 900-6,000mm and galvanized corrugated sheet/coil 0.32-1.25mm x 874mm x 2,000-6,000mm (67,700 tonnes); blackplate 0.19-0.44mm x 680-1,000mm x 460-1,005mm and electrolytic tinplate (single-reduced) 0.19 x 0.44mm x 680/1,000mm x 460/1,005mm (47,000 tonnes).
Expansion programme: Restructuring plan: Seixal Plant – new continuous casting machine and modernisation of Danieli continuous casting machine, new wire rod rolling mill, modernisation of cold rolled plant. Maia Plant – modernisation of light section rolling mill. Expected date of completion – 1991.

Fábrica de Aços Tomé Fèteira SA

Head office: Tv. da Industria, 2425 Monte Real, Vieira de Leiria. **Postal address:** Apartado 2. **Tel:** 44 65153. **Telex:** 43798 aomegap.
Management: Carlos Jorge Lage de Almeida (director), António Pimenta Feijão (sales), Armando José do Mar Alves Coimbra (purchasing). **Established:** 1950.

Fábrica de Aços Tomé Fèteira SA (continued)

Capital: Esc 100,000,000. **No. of employees:** Works – 190. Office – 10. **Annual capacity:** Finished steel: 30,000 tonnes.
Works:
■ At head office. Works director: Armando Alves Coimbra. ***Steelmaking plant:*** Two electric arc furnaces – 3.5-tonne and 10-tonne (annual capacity 30,000 tonnes). ***Continuous casting machines:*** One square billet (140 x 140 x 1,350mm) (30,000 tonnes). ***Rolling mills:*** Sack/Danieli bar (425/325/250mm) (30,000 tonnes).
Products: Carbon steel – round bars 5.5-75mm; square bars 5-75mm; flats 11 x 3mm-120 x 20mm; hexagons 8-40mm. Alloy steel (excl stainless) – round bars 5.5-75mm; square bars 5-75mm; flats 12 x 3mm-120 x 20mm; hexagons 8-40mm.
Brands: Omega.

Qatar

Qatar Steel Co Ltd (Qasco)

Head office: Umm Said. **Postal address:** PO Box 10090. **Telex:** 4606.
Established: 1974. **Ownership:** Kobe Steel, Japan (20%).
Works:
■ Doha. ***Direct reduction plant:*** One Midrex (annual capacity 400,000 tonnes). ***Steelmaking plant:*** Two 70-tonne (37 MVA) electric arc furnaces (400,000 tonnes). ***Continuous casting machines:*** Two 4-strand Kobe straight mould bending billet (110 x 150mm sq) (1,000,000 tonnes). ***Rolling mills:*** One 20-stand 2-high bar (330,000 tonnes).
Products: Carbon steel – reinforcing bars (mild steel and high yield, 10-32mm dia).

Rumania

Export sales of Rumanian iron and steel products are handled by Metalimportexport, 21-25 Mendeleev Street, Bucharest (Tel: 620; 621. Telex: 11515. Cables: Metalimportexport).

Beclean Works

Works:
■ Beclean. ***Rolling mills:*** Including a wire rod (annual capacity 140,000 tonnes).

Brǎila Works

Works:
■ Brǎlia. ***Rolling mills:*** medium section and wire rod.
Products: Carbon steel – wire rod and sections.

Bucharest Welded Pipe Rolling Mill

Works:
■ Bucharest. ***Tube and pipe mills:*** Spiral welded (annual capacity 70,000 tonnes).
Products: Carbon steel – spiral-weld tubes and pipes.

Rumania

Buzau Works

Works:
■ ***Rolling mills:*** Demag wire rod.
Products: Carbon steel – wire rod 6-12mm.

Calarasi Iron & Steel Works

Works:
■ Calarasi. Coke oven plant and sinter plant. Steelmaking plant – two 150-tonne basic oxygen converters (annual capacity 3,600,000 tonnes), electric arc furnace and open hearth were due to start-up mid-1980's; three 250-tonne basic oxygen converters (6,000,000 tonnes) have been projected for the late 1980's. Rolling mills – including a medium and heavy section.

Cîmpia Turzii Works

Works:
■ Cîmpia Turzii. ***Steelmaking plant:*** electric arc and open hearth furnaces. ***Rolling mills:*** light section; wire rod (annual capacity 280,000 tonnes).
Products: Carbon steel – wire rod including 5-14mm dia; light sections.

Cristea Nicolas Works

Works:
■ Galaţi. **Rolling mill:** Sheet mill (annual capacity 98,000 tonnes).

Focsani Metallurgical Enterprise

Works:
■ Focsani. ***Rolling mills:*** slabbing and wire rod.
Products: Carbon steel – wire rod, medium plates.

Combinat Siderurgic Galaţi (CSG)

Annual capacity: Rolled steel: over 5,000,000 tonnes.
Works:
■ Galaţi. ***Coke ovens:*** seven batteries including one oven (annual capacity 850,000 tonnes) and two ovens (600,000 tonnes). ***Sinter plant:*** Dwight Lloyd – one 2-strand 156 sq (3,120,000 tonnes) and No 3 (2,000,000 tonnes). ***Blast furnaces:*** one 3,500 cu m (2,500,000 tonnes); one 2,700 cu m; four 1,700 cu m. ***Steelmaking plant:*** Nine 150-tonne basic oxygen converters (10,000,000 tonnes). ***Continuous casting machines:*** Three 5-strand Concast bloom (260 x 350mm) (600,000 tonnes each); four 2-strand Concast slab (150 x 700 – 300 x 1,900mm) (850,000 tonnes). ***Rolling mills:*** One slabbing (2,500,000 tonnes); one blooming and section mill (2,500,000 tonnes); two heavy plate – one 3,000mm (1,300,000 tonnes) and one other (1,500,000 tonnes); one semi-continuous wide hot strip (2,000,000 tonnes); one 5-stand cold reduction (450,000 tonnes). ***Coil coating lines:*** hot-dip galvanizing. ***Other plant:*** Cold-formed section plant (42,000 tonnes); ingot mould foundry.
Products: Carbon steel – heavy rails; heavy plates 4-150mm thick; hot rolled sheet/coil (uncoated) 1.5-12mm x 700-1,500mm; cold rolled sheet/coil (uncoated) 0.10-3mm thick; blackplate. Stainless steel – hot rolled sheet/coil (uncoated) and cold rolled sheet/coil (uncoated) (low alloy). Sections.

Combinat Siderurgic Hunedoara

Works:
■ Hunedoara. ***Coke ovens:*** four batteries (annual capacity 1,000,000 tonnes). ***Sinter plant:*** Dwight Lloyd (3,000,000 tonnes). ***Blast furnaces:*** two 1,000 cu m and six 450-750cu m. ***Steelmaking plant:*** Two 50-tonne and four 25-tonne electric arc furnaces; open hearth furnaces – five 100-tonne and eight 400-tonne (400,000

Combinat Siderurgic Hunedoara (continued)

tonnes and 2,800,000 tonnes). ***Rolling mills:*** two blooming – one 1,200mm (3,000,000 tonnes) and one 1,100mm (2,000,000 tonnes); one billet (1,480,000 tonnes); two heavy section – one 800mm (320,000 tonnes) and one 650mm (550,000 tonnes); one hot strip and hoop and medium section (540,000 tonnes); one light section (300,000 tonnes); one 3-strand Schloemann wire rod (200,000 tonnes). ***Other plant:*** Wire drawing plant.
Products: Carbon steel, alloy and stainless steel – heavy, medium and light sections, hot rolled hoop and strip and wire rods.

Oţelul Roşu Works

Works:
■ ***Steelmaking plant:*** One 60-tonne and one 110-tonne electric arc furnaces and open hearth furnace. ***Continuous casting machines:*** bloom one 4-strand Concast billet and bloom (100mm sq and 180 x 200mm sq) and one 4-strand Concast bloom (160-260mm sq). ***Rolling mills:*** One hot strip and hoop and light section (annual capacity 60,000 tonnes); one bar (flats) (240,000 tonnes); one sheet (45,000 tonnes). ***Other plant:*** Bar drawing plant.
Products: Carbon steel – flats; bright (cold finished) bars; light sections; sheets; hot rolled hoop and strip (uncoated).

Reşita Steel Works

Works:
■ Reşita. ***Coke ovens:*** (annual capacity 120,000 tonnes). ***Sinter plant:*** Dwight Lloyd (1,500,000 tonnes). ***Blast furnaces:*** including one of 2,700 cu m. ***Steelmaking plant:*** open hearth furnace including one 250-tonne (750 tonnes/day). ***Rolling mills:*** One 800mm heavy section (200,000 tonnes); one medium section (85,000 tonnes); one 280mm light section (65,000 tonnes) and one heavy plate (130,000 tonnes). ***Other plant:*** Wheel and tyre plant.
Products: Carbon steel – medium plates, heavy plates. Heavy, medium and light sections; wheels and tyres.

Uzina de Tevi Roman

Works:
■ Roman. Tube mills, including 406 and 152mm units (total tubemaking capacity 400,000 tonnes/year).

Uzina de Tevi Republica (UTR)

Works:
■ Bucharest. ***Rolling mills:*** light section and bar. One cold reduction. ***Tube and pipe mills:*** seamless (annual capacity 40,000 tonnes). Welded 3 spiral – 3″ and 6″ (300,000 tonnes). Hot reduction plant. Cold drawing plant.
Products: Carbon steel – reinforcing bars, cold rolled hoop and strip (uncoated) 20-95 x 1-3.5mm; hot-dip tinplate; bright wire; seamless tubes and pipes hot rolled, hot reduced, cold rolled 6-168.3mm od; spiral-weld tubes and pipes 419-1,020mm od; oil country tubular goods casings with couplings, 114.3-139.7mm dia; internal upset drill pipe 60.3-139.7mm dia; cold drawn tubes and pipes 6-80mm dia. Nails.

Tirgovişte Works

Works:
■ Tirgovişte. ***Steelmaking plant:*** Six electric arc furnaces including 100-tonne and 50-tonne. ***Refining plant:*** One 125-tonne DH vacuum degassing unit. ***Rolling mills:*** One Moeller and Neumann bar (annual capacity 120,000 tonnes); one Sendzimir

Rumania

Tirgovişte Works (continued)

reversing cold reduction (100,000 tonnes); one 4-high Voest-Alpine temper/skin pass (100,000 tonnes). ***Other plant:*** Strip processing lines.
■ Otelinox Works, Tirgovişte. ***Rolling mills:*** cold reduction – Nisshin Sendzimir for stainless.
Products: Carbon steel – electrical non-oriented silicon sheet/coil. Stainless steel – cold rolled sheet/coil (uncoated) 1,320mm x 0.18-1mm thick. Alloy steel (excl stainless) – (including bearing steels) cold rolled sheet/coil (uncoated) 1,320 x 0.18-1mm thick.
Additional information: Alloy steel annual capacity is about 1,000.000 tonnes and electrical sheet about 160,000 tonnes.

Uzina Metalurgică Iaşi

Works:
■ Iaşi. ***Steelmaking plant:*** electric arc furnaces including 50-tonne. ***Tube and pipe mills:*** welded Mannesmann-Meer. ***Other plant:*** Stretch-reduction plant; cold-formed section plant.
Products: Carbon steel – longitudinal weld tubes and pipes $\frac{3}{8}$-4"; galvanized tubes and pipes; cold drawn tubes and pipes 8-60mm od, 1-3mm wall thickness.

Victoria Works

Works:
■ Calan. **Blast furnaces.**

Saudi Arabia

Jeddah Rolling Mill Co

Head office: Jeddah. **Postal address:** PO Box 1826. **Telex:** 603749 sulb sj.
Established: 1967. **Ownership:** Sabic (Saudi Arabian Basic Industries Corp).
Works:
■ ***Rolling mills:*** One bar for 14-28mm dia (annual capacity 90,000 tonnes). ***Other plant:*** Coil staightening and cut-up line for bar 6-12mm dia (50,000 tonnes).
Products: Carbon steel – reinforcing bars 6-28mm dia.
Additional information: Billets are supplied from the Al-Jubail steelworks of Hadeed.

Saudi Iron & Steel Co (Hadeed)

Head office: Tareeg 230, Madinat Al-Jubail Al-Sinaiyah. **Postal address:** PO Box 10053, Madinat Al-Jubail Al-Sinaiyah 31961. **Tel:** 357 1100. **Telex:** 832022 hade sj. **Telecopier (Fax):** 341 7385.
Management: President – Abdullah A. Al-Sulaim; vice president – Mubarak M. Al-Kahtany (commercial), Abdullah A. Al-Gahtani (finance and administration),Heinz Jung (technical). **Established:** 1979. **Capital:** Srl 3,000,000,000. **No. of employees:** 1,830 (Dec. 1986). **Ownership:** Saudi Basic Industries Corp (Sabic) (95%); DEG, West Germany (5%). **Subsidiaries: Associates:** Steel Rolling Co (Sulb), Jeddah (100%) (re-rolling steel billets to reinforcing bars). **Annual capacity:** Direct reduced iron: 1,050,000 tonnes. Raw steel: 1,400,000 tonnes. Finished steel: 1,200,000 tonnes (rebar and wire rod).
Works:
■ At head office. ***Direct reduction plant:*** Two 400 series Midrex (annual capacity 1,050,000 tonnes). ***Steelmaking plant:*** 135-tonne tap weight electric arc furnace (1,400,000 tonnes, liquid steel). ***Continuous casting machines:*** 6-strand Voest-Alpine billet (130 x 130mm and 100 x 100mm) (1,350,000 tonnes). ***Rolling mills:***

Saudi Iron & Steel Co (Hadeed) (continued)

One Demag bar (12-32mm) (720,000 tonnes); one Schloemann-Siemag wire rod (5.5-14mm) (approx coil weight 1,800kg) (480,000 tonnes). ***Other plant:*** Straightening shop 6-14mm (196,000 tonnes).
Products: Carbon steel – wire rod 5.5-14mm (output 290,000 tonnes); reinforcing bars 6-32mm (800,000 tonnes).

Singapore

Bee Huat Industries Pte Ltd

Head office: 10 Gul Drive, Jurong 2262. **Telex:** 36111.
Works:
■ At head office. ***Tube and pipe mills:*** welded – one 2" and one 1½".
■ Lot A5803, Benoi Crescent, Singapore (Tel: 2682371. Telex: 36111 bhpipe rs). ***Tube and pipe mills:*** One 4" welded.
Products: Carbon steel – longitudinal weld tubes and pipes, galvanized tubes and pipes, hollow sections.

First Rolling Mills Singapore Pte Ltd

Head office: 26 Pioneer Sector 2, Jurong 2262. **Tel:** 8611713-4. **Telex:** 26445 rs. **Cables:** Sichewstel.
Management: Chairman – Kwong Chik Yung; directors – Shee Kok Chong (managing), Margaret Tung Yu Lien, Kwong Lap Kee, Tan Chew Kim. **Established:** 1977. **Capital:** S$5,000,000. **No. of employees:** Works – 50. Office – 12. **Annual capacity:** Finished steel: 15,000 tonnes.
Products: Carbon steel – square bars 9-12mm (output 2,000 tonnes); flats 3-12mm thickness (3,000 tonnes); light angles 25 x 25 x 3mm (2,000 tonnes); special light sections (window and door sections) (8,000 tonnes).

Leong Huat Industries Ltd

Head office: 132 Hillview Avenue, Sinapore 2366. **Telex:** 23323.
Works:
■ Cold roll-forming line.
Products: Carbon steel – cold roll-formed sections (from strip).

National Iron & Steel Mills Ltd

Head office: 22 Tanjong Kling Road, Singapore 2262. **Tel:** 2651233. **Telex:** 21569 nism rs. **Cables:** Singasteel. **Telecopier (Fax):** 2658317.
Management: Board of directors: Chairman – Yeo Seng Teck; managing director – Ang Kong Hua; directors – Goh Eng Chew, Lim Chee Onn, Tan I. Tong, Tan Keong Choon (Alternate – Lim Nai Tian), Oliver Tan Kok Kheng, Wee Cho Yaw, Michael Wee Soon Lock. **Established:** 1961. **Capital:** S$63,847,900. **No. of employees:** 1,245. **Ownership:** DBS Bank (24.19%); Temasek Holdings Pte Ltd (21.47%). **Subsidiaries:** Eastern Industries Pte Ltd (100%) (investment holding, indent trading, rebar fabrication and wire processing); Jurong Industries Ltd (100%) (production of lime products and detinning of scrap); National Shipbreakers Pte Ltd (100%) (shipbreaking); NISM Trading Pte Ltd (100%) (trading in steel and steel-related products, securities investment and trading); NISM Properties Pte Ltd (100%) (property investment); PL Properties Pte Ltd (100%) (property investment); Singacon Investments Pte Ltd (100%) (investment holding); Taoson Engineering Pte Ltd (100%) (trailer, crane and steel fabrication, mechanical and electrical engineering works); Eastern Wire Manufacturing Co (Pte) Ltd (70.7%) (manufacturing of wire and welded wire mesh); Eastern Partek Pte Ltd (55%) (manufacturing and trading in prestressed

Singapore

National Iron & Steel Mills Ltd (continued)

concrete components); Eastern Erico Industries Pte Ltd (55%) (manufacturing of splices and couplers for steel reinforcement); Easteel Construction Services Pte Ltd (100%) (construction servicing); Eastern Vetonit Industries Pte Ltd (55%) (manufacturing and marketing of wall and ceiling plastering compounds). **Associates:** National Oxygen Pte Ltd (36.8%) (production and trading industrial gases); Southern Iron & Steel Works Sdn Bhd, Malaysia (24%) (manufacture of steel and steel-related products); Keenblast Manufacturing Pte Ltd (50%) (abrasives); Cybernex Advanced Storage Technology Pte Ltd (39%) (disk drives); Tye Li Quick Lime Manufacturing Co Sdn Bhd, Malaysia (50%) (lime products); Transtech Venture Management Pte Ltd (40%) (venture capital funds management); Intan Murni Sdn Bhd (50%) (lime products).
Works:
■ At head office. Works director: Ang Cheok Sai. ***Steelmaking plant:*** Three 50-tonne 30MVA electric arc furnaces fitted with water-cooled wall and roof panels and oxy-fuel burners (annual capacity 750,000 tonnes). ***Refining plant:*** One 50-tonne 10MVA ladle furnace with automatic alloy charging system (500,000 tonnes). ***Continuous casting machines:*** Two 4-strand MHI-Olsson billet (100-140mm sq) (450,000 tonnes). ***Rolling mills:*** One light section and bar – 2-strand 18-stand (max rolling speed 12 metres/sec) (annual capacity 250,000 tonnes). One bar and wire rod – single-strand 18-stand with slit rolling and 8-stand No-Twist finishing block and controlled cooling. Max finishing speed 60 metres/sec for rod and 12 metres/sec for bar. Fed by oil-fired 60-tonnes/hour walking beam reheating furnace. Coil weight 1 tonne. Annual capacity 250,000 tonnes (bar) or 270,000 tonnes (rod). ***Other plant:*** Scrap preheating system.
Products: Carbon steel – billets 100 and 120mm sq (1986 output 390,000 tonnes); wire rod 5.5-14mm dia (150,000 tonnes); reinforcing bars 9-50mm dia (350,000 tonnes); round bars 6-50mm dia (80,000 tonnes); square bars 9-30mm dia; flats 3 x 15mm to 25 x 75mm; light angles 3.0 x 25 x 25mm to 6 x 75 x 75mm.

Steel Tubes of Singapore (Pte) Ltd (STS)

Head office: 11 Gul Crescent, Singapore 2262. **Telex:** 39515.
Established: 1981.
Works:
■ ***Tube and pipe mills:*** welded – ERW.
Products: Carbon steel – longitudinal weld tubes and pipes, hollow sections.

South Africa

Cape Town Iron & Steel Works (Pty) Ltd

Head office: Fabriek Street, Kuils River 7580, Cape Province. **Postal address:** PO Box 121. **Tel:** 903 2141. **Telex:** Teletex – 555071. **Cables:** Ciscosteel.
Management: Directors – K.D. Kynoch (managing), I.J.M. de Vries (financial); managers – F. Schlagbauer (production), B. Laing (engineering), A.R. Deenik (sales).
Established: 1964. **No. of employees:** Works – 355. Office – 12. **Ownership:** Iscor Ltd. **Annual capacity:** Finished steel: 130,000 tonnes.
Works:
■ At head office. ***Steelmaking plant:*** One 30-tonne 15/18 MVA electric arc furnace (annual capacity 140,000 tonnes). ***Continuous casting machines:*** One 2-strand Danieli billet (120 x 120mm) (140,000 tonnes). ***Rolling mills:*** bar with two 3-high and eleven 2-high stands (500/350/290/270mm) (130,000 tonnes).
Products: Carbon steel – billets 100 and 120mm sq (output 80,000 tonnes); reinforcing bars 8-40mm dia (full range) (60,000 tonnes).

Davsteel (Pty) Ltd

Head office: 33 De Beer Street, Braamfontein, Johannesburg 2017. **Postal address:** PO Box 31136. **Tel:** 011 339 6001. **Telex:** 4 22012. **Cables:** Gateworks. **Telecopier (Fax):** 011 339 3935.
Management: Directors – M.I. Kaplan (chairman), B.H.H.A.J. Eras (managing), I.W. Joffe, R.H. Kaplan; alternate directors – I. Myers, I. Joffe, P.J. Fairhead, D.C. Viljoen. **Established:** 1974. **Ownership:** Cape Gate Holdings (Pty) Ltd.
Works:
■ Nobel Boulevard, Van Der Byl Park. (Tel: 016 31 6091. Telex: 4 22012). Works director: B.H.H.A.J. Eras. ***Steelmaking plant:*** 60-tonne 5,200mm dia electric arc furnace (annual capacity 280,000 tonnes). ***Refining plant:*** 60-tonne Asea-SKF (280,000 tonne). ***Continuous casting machines:*** 4-strand Danieli billet (115mm sq) (280,000 tonnes). ***Rolling mills:*** bar and light section – 14-stand Danieli (2-stand 450, 4-stand 330, 4-stand 380 and 4-stand 260) (150,000 tonnes); 24-stand single-strand Danieli wire rod (4-stand 450, 4-stand 300, 6-stand 260 and 8-stand 200/160) (coil weight 1,000kg) (170,000). ***Other plant:*** Wire works.
■ Sonderwater, Transvaal. ***Direct reduction plant:*** one kiln (annual capacity 30,000 tonnes).
Products: Carbon steel – wire rod 5.5-14.5mm; reinforcing bars 8-40mm; round bars 8-40mm; square bars 8-16mm; flats 25 x 3-80 x 12mm; light angles 25 x 25 x 3-50 x 50 x 8mm.

Dunswart Iron & Steel Works Ltd—see Iscor Ltd

Flather Bright Steels (Pty) Ltd

Head office: Springs, Tvl 1560. **Telex:** 420757 sq.
Established: 1950.
Works:
■ Jansen Road, Nuffield, Springs. Bright bar plant.
Products: Carbon steel – bright (cold finished) bars. Stainless steel – bright (cold finished) bars.

Highveld Steel & Vanadium Corp Ltd

Head office: Old Pretoria Main Road, Witbank, District Witbank, Transvaal. **Postal address:** PO Box 111, Witbank 1035. **Tel:** 27 0 1351 9011. **Telex:** 420088. **Cables:** Hiveldsteel. **Telecopier (Fax):** 01351 97561.
Management: Chiarman – L. Boyd; managing director – J. Hall; general managers – D. Bruckmann (engineering), R.R. Callanan (finance and administration), R.A. Herbertson (marketing), T.E. Jones (steel and vanadium operations), A.L. Melvill (ferro alloy operations). **Established:** 1960. **Capital:** R 75,000,000 (authorised). **No. of employees:** 7,430. **Ownership:** Anglo American Industrial Corp Ltd (Ultimate holding company – Anglo Amercian Corp of South Africa Ltd). **Subsidiaries:** Iron Stone Minerals (Pty) Ltd (100%); Multisport Witbank (Pty) Ltd (100%); Rand Carbide (Pty) Ltd (100%); Rheem South Africa (Pty) Ltd (100%); Transalloys Pty (Ltd) (100%); Transvaal Vanadium Co (Pty) Ltd (100%); Vansteel Estates (Pty) Ltd (100%); Steelbus Services (Pty) Ltd (100%). **Associates:** Ferroveld (Pty) Ltd (50%). **Annual capacity:** Hot metal (1985): 776,103 tonnes. Continuously cast blocks (1985): 865,497 tonnes. Finished steel (1985): 669,793 tonnes.
Works:
■ At head office. Works director: T.E. Jones (general manager, steel and vanadium operations).
Pre-reduction plant: No 1 Iron plant – 10 and No 2 Iron pant – 3 rotary pre-reduction kilns (4m dia x 61m long). **Ironmaking plant:** No 1 plant submerged arc smelting furnaces (two 33 MVA and four 45 MVA); No 2 plant submerged arc smelting furnace (one 63 MVA) (combined annual capacity 900,000 tonnes). ***Steelmaking plant:*** Three 75-tonne LD basic oxygen (1,000,000 tonnes); four shaking ladles for vanadium extraction. ***Continuous casting machines:*** One 4-strand billet (98 x 98 to 160 x 160mm); one 4-strand and two 2-strand bloom (180 x 230mm, 570 x 310mm, 310 x 260mm); one 1-strand slab (180 x 1,200-1,600mm, 250 x 1,200-2,000mm). ***Rolling mills:*** One medium section / light section with 2-strand and two universal stands (350,000 tonnes); one 4-high reversing medium plate (barrel width 2,620mm) and one 4-high reversing wide hot strip (barrel width 2,620mm) (coil weight 27 tonnes) (combined annual capacity 500,000 tonnes); one 2-high temper/skin pass (roll width 2,000mm). ***Other plant:*** Ferro-alloy Works (Vantra – vanaduim pentoxide and ferro-vanaduim production). Rand Carbide: Three submerged arc furnaces (16 MVA, 20 MVA and 46.5 MVA) (55,000 tonnes). Transalloys: Four submerged arc furnaces (one 48 MVA and three 22 MVA) (manganese alloys) (118,000 tonnes); two 7 MVA ore-lime melting furnaces (manganese alloys) (35,000 tonnes).
Products: Carbon steel – slabs 150 x 1,100-1,400mm, 180 x 1,300-1,600mm, 250 x 1,700-2,000mm (1985 output 355,409 tonnes); blooms and billets 63.5-190.5mm blooms – 328,696 tonnes, billets – 181,392 tonnes continuously cast and 40,344 tonnes rolled); semis for seamless tubemaking; hot rolled coil (for re-rolling); round bars 80-160mm; square bars; flats 150-550mm wide, 16-65mm thick; medium angles 120 x 120-200 x 200mm and 150 x 75-152 102mm, medium joists 152 x 89-203 x 152mm, medium channels 127 x 64-381 x 102mm, wide-flange beams (medium), universals (medium) (beams and columns) 203 x 133-533 x 210mm and 152 x 152-305 x 305mm, heavy angles, heavy joists, heavy channels, wide-flange beams (heavy), universals (heavy), heavy rails 23 kg/m-57 kg/m and light rails (combined output of sections 258,092 tonnes); skelp (tube strip); medium plates and heavy plates 5-60mm thickness, 2,540mm (max) width, 29m (max) length and universal plates (combined out of plate 201,787 tonnes); hot rolled sheet/coil (uncoated) 1.6-16mm guage, 1,000-2,450 width (169,570 tonnes); blackplate. Containers and closures (mainley for local market); ferro-alloys (190,937 tonnes); vanadium spinel (57,340 tonnes).

South Africa

Iscor Ltd

Head office: Roger Dyason Road, Pretoria, Transvaal. **Postal address:** PO Box 450. **Tel:** 012 298 1111. **Telex:** 322007. **Telecopier (Fax):** 264721.
Management: Board of directors – M.T. de Waal (chairman), W. van Wyk (managing), R.J. Ironside, J.P. Kearney, J.J. Kitshaff, L.B. Knall, P.R. Morkel, G.P. Morum, P.F. Theron. **Established:** 1928. **Capital:** R 865,200,000. **No. of employees:** 60,000. **Ownership:** Government. **Annual capacity:** Liquid steel: 7,185,000 tonnes. Finished steel: 5,689,000 tonnes (products sold).
Works:
■ Pretoria Works, PO Box 19, Pretoria 0001. ***Coke ovens:*** Three Woodall Duckham batteries (120 ovens) (annual capacity 657,000 tonnes). ***Sinter plant:*** Two Dwight Lloyd (area 100 m sq, width 2.39m) (828,550 tonnes). ***Blast furnaces:*** Two – P2 5.130m hearth dia, 503 cu m (292,000 tonnes), P4 7.8m hearth dia, 1,210 cu m (693,500 tonnes). ***Steelmaking plant:*** Two 125-tonne Birlec 80 MVA electric arc furnaces (900,000 tonnes). ***Refining plant:*** One 125-tonne 15MVA VAD (480,000 tonnes). ***Rolling mills:*** One 2-high reversing blooming (1,050 x 2,540mm) (1,200,000 tonnes); One heavy section / rail comprising one breakdown (2,500 x 1,250mm), two roughing (2,000 x 950mm), one finishing (1,700 x 950mm), and two universal (1,195mm) (combined capacity 800,000 tonnes); one 15-stand 2-high Demag light section (350mm) (coil weight 370kg) (120,000 tonnes); one 2-strand 24-stand Schloemann Morgan wire rod (300,000 tonnes).
■ Iscor Works, Vanderbijlpark, PO Box 2, Vanderbijlpark 1900. (Tel: 016 889 9111. Telex: 421227). Works director: B.H.P. Havenga. ***Coke ovens:*** Five Woodall Duckham (Becker) and three Didier batteries (total – 443 ovens) (combined annual capacity 3,120,800 tonnes). ***Sinter plant:*** One Lurgi Dwight Lloyd, 100m sq strand (860,900 tonnes); one Lurgi Dwight Lloyd, 100m sq strand (926,500 tonnes). ***Blast furnaces:*** Four – 7.01m hearth dia, 890 cu m (557,700 tonnes); 7.01m hearth dia, 890 cu m (598,600 tonnes); 9.1m hearth dia, 1,500 cu m (1,111,200 tonnes); and 10.1m hearth dia, 2,007 cu m (1,141,400 tonnes). ***Direct reduction plant:*** SL/RN – 4 kilns (80 x 4.8m in dia) (720,000 tonnes). ***Steelmaking plant:*** Three 150-tonne (nominal) LD-KGC basic oxygen (3,500,000 tonnes); three 155-tonne, 80/96 MVA electric arc furnaces (1,600,000 tonnes). ***Refining plant:*** One 155-tonne RH vacuum degassing unit (1986) (500,000 tonnes); one 155-tonne VAD (1976) (400,000 tonnes); one 155-tonne TN (400,000 tonnes). ***Continuous casting machines:*** Three slab – two 2-strand Concast Distington (750-2,000mm wide, 210-240mm thick) (2,300,000 tonnes); one 2-strand Voest-Alpine (800-2,000 wide, 210-240mm thick) (1,680,000 tonnes). ***Rolling mills:*** Two slabbing – one 1-stand IHI (3,050mm) (1,500,000 tonnes) and one 1-stand Mesta (1,420mm) (1,250,000 tonnes). one 1-stand 4-high heavy plate (3,650mm) (600,000 tonnes); two wide hot strip – one roughing 4-high, six finishing 4-high (1,420mm) (1,400,000) and one roughing 4-high, 6 finishing 4-high (2,050mm) (coil weight 36 tonnes) (2,400,000 tonnes); Three cold reduction – one 5-stand 4-high (1,050mm) (coil weight 27.5 tonnes) (500,000 tonnes), one 4-stand 4-high (1,225mm) (coil weight 27.5 tonnes) (500,000 tonnes) and one 5-stand 4-high (0.45-2.0mm thick, 1,850mm wide) (coil weight 27 tonnes) (1,000,000 tonnes); three temper/skin pass – 2-stand 4-high (1,050mm) (coil weight 27.5 tonnes) (361,200 tonnes), one 1-stand 4-high (1,225mm) (coil weight 27.5 tonnes) (350,000 tonnes) and one 1-stand 4-high (0.45-2.0mm thick, 1,850mm wide) (coil weight 27.5 tonnes) (660,000 tonnes). ***Coil coating lines:*** One Halogen Wean electrolytic tinning (318,000 tonnes); Four hot-dip galvanizing – one Selas non-retractable Furnace-Wean (285,600 tonnes), one annealing hot-dip Selas-Head Wrightson (200,000 tonnes), one Sendzimer Mesta (1951) (95,000 tonnes) and one Sendzimer Mesta (1952) (95,000 tonnes); one colour coating double-deck catenary oven B & K (70,000 tonnes).
■ Newcastle Works, PO Box 2, Newcastle 2940. ***Coke ovens:*** – Four DR. C. Otto batteries (200 ovens) (annual capacity 2,400,000 tonnes dry coal). ***Sinter plant:*** – One Delattre-Levivier on strand C, 4.5m wide (2,100,000 tonnes unscreened). ***Blast furnaces:*** – One 10.14m hearth dia, 2,367 cu m (1,560,000 tonnes). ***Steelmaking plant:*** Three 165-tonne LD-CB basic oxygen (2,800,000 tonnes). ***Continuous casting machines:*** Five Concast bloom with a total of eighteen strands (315mm sq and 315 x 205mm) (1,800,000 tonnes). ***Rolling mills:*** billet – one 820mm reversing stand, four 710mm continuous mill stands (1,600,000 tonnes); medium section – two

Iscor Ltd (continued)

800mm reversing stands, 9-stands in continuous mill consisting of three edging stands and six universal conventional/convertible stands (700mm dia conventional, 1,000mm dia universal) (500,000 tonnes); bar – 26-stand multi-line mill of which 4-stands are convertible horizontal/vertical (550,000 tonnes); wire rod – 15 twin-strand and 10-stand No-Twist mill (330,000 tonnes per year).
■ Dunswart. ***Direct reduction plant:*** One Codir 76m (annual capacity 120,000 tonnes). ***Steelmaking plant:*** Two electric arc furnaces – 50-tonne and 40-tonne (200,000 tonnes). ***Continuous casting machines:*** One 4-strand Concast curved mould billet (100-160mm sq) (200,000 tonnes). ***Rolling mills:*** light section – 12 vertical/horizontal, 2-high and one 3-high breakdown (10-40mm rounds also flats and squares) (120,000 tonnes).
Products: Steelmaking pig iron. Carbon steel – slabs, blooms, billets, semis for seamless tubemaking, wire rod, round bars, square bars, flats, hexagons, light angles (equal and unequal), light tees, special light sections, medium angles (equal and unequal), medium tees, medium joists, wide-flange beams (medium), universals (medium), heavy angles (equal and unequal), heavy tees, heavy joists, heavy channels, wide-flange beams (heavy), universals (heavy), heavy rails, light rails, rail accessories, medium plates, heavy plates, universal plates, floor plates, cold rolled sheet/coil (uncoated), hot-dip galvanized sheet/coil, colour coated sheet/coil, electrolytic tinplate (single-reduced). Fencing materials, forgings, castings. Total despatches of rolled steel in 1986 were 5,304,050 tonnes.
Additional information: Dunswart works due to be closed from October 1987 except for the direct reduction plant.

McWillaw Steel (Pty) Ltd

Head office: 4110 Isipingo, Natal. **Postal address:** PO Box 23036. **Tel:** 92 2313. **Telex:** 6-23068 sa. **Telecopier (Fax):** 902 8530.
Ownership: Raphaely International Holdings Pty Ltd
Works:
■ ***Steelmaking plant:*** electric arc furnace. ***Continuous casting machines:*** One 2-strand Concast for billet 57-115mm sq. ***Rolling mills:*** bar mills.
Products: Carbon steel – reinforcing bars, light sections.

Middelburg Steel & Alloys (Pty) Ltd – Steel Division

Head office: 3rd Floor, Zenex House, Sandton City Office Park, 5th Street, Sandton, Transvaal. **Postal address:** PO Box 781815, Sandton 2146. **Tel:** 011 783 2060. **Telex:** 4 – 21601. **Cables:** Lucarbon Sandton. **Telecopier (Fax):** 011 883 1416.
Management: Group executives – J.C. Hall (chairman and chief executive), J.E. G omersall (managing director), J.S. Tennant (finance), M.L. Melvill (marketing), A.D. Tonkin (human resources), L.A. Kok; steel division executives – J.E. Gomersall (managing director), M.G. Gomersall (finance), R.J. Linnell (marketing), A.W. Bagnall, D. Luyt, R.H. Southey, J.R.J. Hoffman. **Established:** 1965. **No. of employees:** Works – 1,391. Office – 125. **Ownership:** Barlow Rand Ltd (100%). **Annual capacity:** Raw steel: 150,000 tonnes. Finished steel: 100,000 tonnes.
Works:
■ PO Box 133, Middelburg 1050, Hendrina Road, Middelburg, Transvaal . (Tel: 0132 79111. Telex: 4 23554). ***Steelmaking plant:*** One 50-tonne electric arc furnace (annual capacity 150,000 tonnes). ***Refining plant:*** One 60-tonne AOD (150,000 tonnes). ***Continuous casting machines:*** 1-strand Voest-Alpine slab (180mm thick, 950-1,600mm wide) (150,000 tonnes). ***Rolling mills:*** One single stand Steckel medium plate (slab thickness 180mm, final gauge 3.0mm) (950-1,600mm wide) (max slab weight 15 tonnes); cold reduction and cold strip Sendzimir ZR22-B52; 2-high temper/skin pass; 4-high cold mill (total cold rolled capacity 45,000 tonnes).
Products: Stainless steel – slabs 180mm thick, 950-1,600mm wide (110,000 tonnes); medium plates (20,000 tonnes); hot rolled sheet/coil (uncoated) (25,000 tonnes); cold rolled sheet/coil (uncoated) (45,000 tonnes).
Brands: Grades Aisi 301, 304, 304L, 309, 309S, 310, 310S, 316, 316L, 3167S, 3174, 321, 409, 430.

South Africa

Mono-Die Engineers (Pty) Ltd

Head office: Paul Smit Street, Boksburg North. **Postal address:** PO Box 6105, Dunswart 1508. **Tel:** 27 011 894 4274. **Telex:** 27 9 451614. **Cables:** Mono-Die Benoni.
Management: J.B. Annegaru (managing director). **Established:** 1951. **No. of employees:** Works – 70. Office – 10. **Annual capacity:** Finished steel: 12,000 tonnes.
Products: Carbon steel – bright (cold finished) bars 3-105mm dia.
Brands: Mono-Bright Steels.

Quadrisection (Pty) Ltd

Head office: Main Reef Road, Cleveland. **Postal address:** PO Box 25670, Denver 2027. **Telex:** 421635.
Works:
■ Tube draw-benches.
Products: Stainless steel – cold drawn tubes and pipes.

Rand Bright Steels (Pty) Ltd

Head office: Liverpool Road, Industrial Sites, Benoni South 1502, Tvl. **Postal address:** PO Box 5035.
Works:
■ At head office. Bar drawing and peeling plant.
Products: Carbon steel – bright (cold finished) bars. Stainless steel – bright (cold finished) bars. Alloy steel (excl stainless) – bright (cold finished) bars.

Salmac Stainless Tube (Pty) Ltd

Head office: 21 Chenik Street, Chamdor, Krugersdorp.
Works:
■ At head office. ***Tube and pipe mills:*** welded – Demag, Abbey Etna, Driam.
Products: Stainless steel – longitudinal weld tubes and pipes.

Sandvik (Pty) Ltd

Head office: Benoni, Tvl 1500. **Postal address:** PO Box 206. **Telex:** 29196.
Products: Stainless steel – longitudinal weld tubes and pipes.

Scaw Metals Ltd

Head office: 2nd Floor, 45 Main Street, Johannesburg, Transvaal. **Postal address:** PO Box 61721, Marshalltown, Tvl 2107. **Tel:** 011 902 1001. **Telex:** 4-29203. **Cables:** Scaworks.
Management: Chairman – W.G. Boustred; managing directors – R.A. Boustred, E.A. Barry (Flather Bright Steels); directors – J. Parkin (mill manager), H.M.S. Ferreira (engineering manager), W.B. Ballantyne (melting manager), A. Harris (technical manager) B.C. Stevens (electrical engineering manager); alternate directors – F.E. Ferreira (financial manager), A.P.S. Jansen (mill sales manager), B.R. King (group financial manager), K. Mannhardt (development engineering), W.M. Wedderburn (metalurgist); managers – J.M. Bird (general, RandScrap Iron), D. Bennett (administration), R.A. Kastell (foundry sales), A.G. Murray (industrial relations). **Established:** 1937. **Capital:** Issued 3,500,000 shares at 40 cents each – R 1,400,000. **No. of employees:** 3,800. **Ownership:** Anglo American Industrial Corp Ltd. **Subsidiaries:** Flather Bright Steels (100%) (bright drawers); Rand Scrap Iron & Metal Co (100%) (scrap dealers); Scaw Metals Housing Co (100%); Scaw Alloys Ltd (100%) (property owning); Scaw Industrial Property (Pty Ltd (100%) (property owning); Birkdale Holdings (51%) (investment). **Associates:** Haggie Rand (35.6%) (rope manufacturers). **Annual capacity:** Finished steel: 400,000 tonnes).
Works:
■ Black Reef Road, Dinwiddie, Germiston. (Tel: 902 1001. Telex: 4-29203). Works manager: R.A. Boustred. ***Direct reduction plant:*** one 60 x 4.5ft DRC kiln (annual capacity 90,000 tonnes). ***Steelmaking plant:*** Four electric arc furnaces – 10, 12, 20, and 60-tonne (400,000 tonnes). ***Refining plant:*** One 60-tonne Daido ladle furnace (300,000 tonnes). ***Continuous casting machines:*** Three billet – one Concast and two Koppers (total 7-strands) (100mm and 140mm sq) (combined annual capacity 400,000 tonnes). ***Rolling mills:*** One billet – 3-stand (forms part of Morgan rod and bar mill). Bar mill: one Brightside merchant, 3-high 18" roughing stand, 5-stand 12" open train with 17-stand continuous bar mill and 6-stand Morgan rod block (40,000 tonnes rounds and hexagons and 300,000 tonnes straight bars and coil); one Hille bar and section – 3-high breakdown combined with 2-high sizing stand feeding (a) 3-high stand with further nine stands, and (b) 3-high shaping stand and 4-stand finishing train (120,000 tonnes angles, channels, flats and rounds). ***Other plant:*** Iron and steel foundries with induction furnaces.
Products: Direct-reduced iron (DRI). Carbon steel – ingots; billets; wire rod; reinforcing bars 6 to 42mm; round bars special purposes up to 53mm, spring steel up to 40mm; square bars special purposes up to 53mm, up to 100 x 16mm; flats special purposes up to 53mm, up to 100 x 16mm; hexagons special purposes up to 53mm, up to 100 x 16mm; light angles equal up to 100 x 100mm; light tees; light channels 96 x 38mm to 120 x 55mm. Alloy steel (excl stainless) – billets; round bars special purposes up to 53mm; square bars special purposes up to 53mm; flats special purposes up to 53mm; hexagons special purposes up to 53mm. Alloy iron castings; steel and alloy steel castings up to 25-tonnes; cast and forged grinding media. The company is licensed by Abex Corp, USA, to manufacture the Southern railroad wheel and Amsco products; GSI Inc, USA to manufacture Commonwealth cast steel devices for railway rolling stock; and by Fonderies Magotteaux, Belgium for the manufacture of high chromium alloy iron grinding media.
Expansion programme: Morgan mill modernisation to increase finishing speeds; 60-tonne arc furnace to EBT; modernisation of scrap processing plant.

South Africa

Steel Pipe Industries (Pty) Ltd (SPI)

Head office: North Reef Road, Elandsfontein 1406. **Postal address:** PO Box 8013. **Telex:** 428993.
Established: 1970.
Works:
■ ***Tube and pipe mills:*** welded – Spiral and Longitudinal.
Products: Carbon steel – longitudinal weld tubes and pipes, spiral-weld tubes and pipes.

Strip Steel (A Division of GIC Engineering (Pty) Ltd)

Head office: 4 Nywerheid Street, Manufacta, Roodepoort. **Tel:** 763 4261. **Telex:** 425869. **Cables:** Cold Roll Johannesburg.
Management: Directors – T.G. Brooks (managing), J.D. Mirk (works), D.C.J. Uys (marketing), M.N. van Niekerk (technical), R.J.S. Venter (financial). **Established:** 1953. **Capital:** R 353,000. **No. of employees:** Works – 143. Office – 29. **Ownership:** Holding company – Goldfields Industrial Corp Ltd (Ultimate holding company – B. Elliott & Co PLC, UK). **Annual capacity:** Finished steel: 40,000 tonnes.
Works:
■ At head office. Works director: J.D. Mirk. ***Rolling mills:*** Seven cold strip mills – one Sendzimir (450 x 2mm max input) (annual capacity 2,400 tonnes), one 4-high Boeker & Volkenborn (380 x 6.35mm max input) (3,000 tonnes), one Robertson 4-high (200 x 2mm max input) (2,000 tonnes), two Victor Bauer 2-high (380 x 6mm max input) (1,500 tonnes), one B & V 2-high (150 x 6mm max input) (1,000 tonnes), one B & V 2-high (130 x 6mm max input) (1,000 tonnes); and one cold strip/temper skin pass Loewy-Robertson (450 x 7mm max input), 4-high, with automatic gauge control (7,000 tonnes).
■ 2130 Main Reef Road, Manufacta, Roodepoort. (Tel: 763 4261. Telex: 425869). Works director: J.D. Mirk. **Slitting facilities**.
Products: Carbon steel – hot rolled hoop and strip (uncoated) (including spring steel) 0.30-7mm thick; cold rolled hoop and strip (uncoated) (precision 0.14-4.50mm thick, annealed spring steel 0.50-4.50mm); hot-dip galvanized hoop and strip 0.30-2mm; coated hoop and strip (electro-galvanized, nickel and brass-coated from 0.30-1.20mm. Widths depend on gauge: mild and spring steels 8 to 600mm, precision 8-450mm).

Titan Industrial Corp (Pty) Ltd

Head office: 40 Potgieter Street, Alrode, Alberton. **Telex:** 422521.
Established: 1973.
Works:
■ Bar drawing, centreless grinding, peeling and polishing plant.
Products: Carbon steel – bright (cold finished) bars, bright wire.

Tosa (Division of Dorbyl Ltd)

Head office: Jack Pienaar Street, Germiston South, Tvl. **Postal address:** PO Box 9222, Elsburg 1407. **Tel:** 011 825 7970. **Telex:** 4 28955. **Telecopier (Fax):** 011 825 6560.
Management: Directors – P. Waetzel (managing), T.S. McCurdie (sales). **Established:** 1929. **No. of employees:** Works – 3,000. Office – 500. **Ownership:** Dorbyl Ltd. **Subsidiaries:** Salmac Stainless Tube (100%) (stainless steel tubes); Steel Pipe Industries (100%) (large bore pipe to 4,000mm); Afgate (100%) (steel profiles and heavy-wall hollow sections). **Annual capacity:** Finished steel: 300,000 tonnes approx (tubing).
Works:
■ Vereeniging. ***Tube and pipe mills:*** seamless – two 5-170mm od push benches (annual capacity 25,000 tonnes). Welded – ten ERW 15-4,000mm od (270,000 tonnes).
Products: Carbon steel – seamless tubes and pipes 5-170mm od; longitudinal weld

Tosa (Division of Dorbyl Ltd) (continued)

tubes and pipes 15-450mm nb; spiral-weld tubes and pipes up to 1,800mm nb; large-diameter tubes and pipes up to 4,000mm od; oil country tubular goods up to 170mm; galvanized tubes and pipes up to 1,800mm; cold drawn tubes and pipes up to 250mm; hollow bars; precision tubes and pipes up to 63.5mm; hollow sections up to 400 x 400 x 16mm. Stainless steel – longitudinal weld tubes and pipes up to 101.6mm od; oil country tubular goods up to 3½", schedule 40 only; spiral-weld up to 508mm od.

Usco – The Union Steel Corp (of South Africa) Ltd

Head office: Gen. Hertzog Road, Peacehaven, Vereeniging. **Postal address:** PO Box 48, Vereeniging 1930. **Tel:** 4 5122. **Telex:** 421230 sa. **Cables:** Jumbola.
Management: Chairman – K.P. Kotzee; directors – J.H. Kaltwasser (managing), K.N. Jenkins, G.F. u.d. Merwe, J. Vermooten, P.A. Olivier, G. Steinmetz, W.V. Wyk.
Established: 1911. **Capital:** R 97,700,000. **No. of employees:** Works – 3,270. Office – 832. **Ownership:** Iscor, Metcor and members of the public. By government legislation Iscor has the authority to appoint the majority of directors and chairman. **Annual capacity:** Raw steel: 300,000 tonnes. Rolled steel: 250,000 tonnes.
Works:
■ Vaal Works. ***Steelmaking plant:*** Three 45-tonne electric arc furnaces (annual capacity 250,000 tonnes). ***Refining plant:*** ladle furnace (tank). ***Continuous casting machines:*** one 4-strand Concast for billet (100mm sq) (100,000 tonnes); one 4-strand Danieli for billet (100 and 130mm sq) (100,000 tonnes). ***Rolling mills:*** 2-stand 3-high cogging billet using 11" ingots (65,000 tonnes); 3-stand 3-high in-line light section and wire rod with one 3-high roughing train, 12-stand intermediate train and 6-stand finishing block (coil weight 1-tonne, 10-30.6mm dia) (220,000 tonnes).
■ Klip Works. ***Steelmaking plant:*** Two 15-tonne electric arc furnaces. ***Refining plant:*** One ESR unit, max ingot size 700mm dia. ***Rolling mills:*** 2-high 24" cogging billet (3⅝ to 5¼" sq product) (annual capacity 60,000 tonnes). ***Other plant:*** One GFM forging machine for material 80-200mm dia (20,000 tonnes); one 1,600-tonne press.
Products: Carbon steel – billets – continuously cast 100 and 130mm sq; round bars 12-100mm; square bars 10.5mm and 12-75mm, round-cornered 25.4mm and 92.1mm; flats 25 x 6mm-150 x 25mm, 80 x 40mm-90 x 35mm; special light sections, Alloy steel (excl stainless) – billets – continuously cast 100 and 130mm sq; round bars 12-100mm; flats 25 x 8mm-150 x 25mm. Hollow drill steel hexagons (alloy – stainless steel and homogeneous lined; carbon – stainless and homogeneous lined) – 19mm, 25.4mm, 31.8mm, 22.2mm and 28.6mm. Spring steels – rounds 12-100mm; flats 25 x 6mm-150 x 25mm; squares 10.5 x 12mm-50mm; round-cornered squares 25.4-76.2mm. Tool steel – rounds 12-199mm; hexagons 19mm, 22.2mm, 25.4mm, 28.6mm, 31.8mm; octagons 12.5mm, 16mm, 19mm, 22.2mm, 25.4mm, 28.6mm, 31.8mm, 38.1mm. Forgings: Carbon steel – rounds 150-400mm; squares 150-370mm; flats on application. Alloy steel – rounds 108-400mm; squares 108-370mm; flats 50 x 25mm to 45mm and 356 x 65mm-150mm. Tool steel – rounds 40-320mm; squares 108-285mm; flats 50 x 25mm-45mm and 356 x 75mm-152mm.

Spain

Acenor SA

Head office: Plaza Museo 1, Bilbao. **Tel:** 4444100. **Telex:** 34197.
Management: Francisco Antepara (president), Vicente Carretero (vice president).
Additional information: Acenor is a special steels group whose associate companies include: Olarra SA, SA Echevarría, Pedro Orbegozo, Aceros de Llodio and Forjas Alavesas. Acenor has largely absorbed the Aceriales group, which was established as a consortium to promote rationalisation in the speical steel sector.

Spain

Aceralava—see Tubacex - C.E. de Tubos por Extrusion SA

Acerinox SA

Head office: Dr Fleming 51, 28036 Madrid. **Tel:** 91 4 57 86 50. **Telex:** 23271 acnox e, 45156 aca e. **Telecopier (Fax):** 259 4290.
Management: President - José María Aguirre Gonzalo; chief executive officer - Victoriano Muñoz Cava (managing director); directors - José Luis Lejeune Castrillo (works), Rafael Naranjo Olmedo (commercial), Manuel López de la Parte (financial).
Established: 1971. **Capital:** Ptas 6,050,000,000. **No. of employees:** Works - 1,535. Office - 192. **Annual capacity:** Raw steel: 350,000 tonnes. Finished steel: 216,000 tonnes.
Works:
■ Palmones (Los Barrios), Cadiz. (Tel: 956 66 08 54. Telex: 23271 acnox e, 45156 acxae e). Works director: José Luis Lejeune. ***Steelmaking plant:*** One 90-tonne and one 100-tonne electric arc furnaces (annual capacities 200,000 tonnes and 300,000 tonnes). ***Refining plant:*** One 80-tonne AOD (350,000 tonnes). ***Continuous casting machines:*** 3-strand billet (180 x 200mm, 150 x 150mm, 175 x 175mm) and 1-strand slab (130/200mm x 900/1,600mm) (350,000 tonnes). ***Rolling mills:*** cold reduction - No 1 ZR-22 B-50 (45,000 tonnes), No 2 ZR-22 B-50 (45,000 tonnes), No 3 ZR-21 B-63 (126,000 tonnes), No 4 ZS -07-62 (24,000 tonnes); 2-stand Steckel hot mill - 4-high roughing and 6-high finishing (coil weight 18 kg/mm) (759,000 tonnes); 2-high 1,580mm temper/skin pass.
Products: Stainless steel - ingots; slabs 130 x 200mm, 900 x 1,600mm; billets, 180 x 200mm, 150 x 150mm, 175 x 175mm; hot rolled band (for re-rolling) 2.50-60mm x 100-3,000mm; flats 0.8-25m x 12.5 x 250mm; hot rolled hoop and strip (uncoated); skelp (tube strip); cold rolled hoop and strip (uncoated); medium plates 2.5-15mm x 3,000mm; heavy plates 16-60mm x 1,580mm; hot rolled sheet/coil (uncoated), 2.5-15mm x up to 3,000mm; cold rolled sheet/coil (uncoated) 0.25-6.0mm x up to 1,524mm; bright annealed sheet/coil 0.25-1.5mm x up to 1,250mm.
Sales offices/agents: Acerinox Duetschland GmbH, Langenfeld, West Germany; Acerinox France Sarl, Le Blanc Mesnil, France; Siderinox SA, Buenos Aires, Argentina; Acerinox Chile SA, Santiago, Chile; Inoxlee Int Ltd, Tsuen Wan NT, Hong Kong; Acerol Corp, Hasbrouck Heights, NJ, USA; Acerol Scandinavia A/S, Oslo, Norway.

Aforasa - Acerias y Forjas de Azcoitia SA

Head office: Carretera de Zumárraga s/n, 20720 Azkoitia, Guipuzcoa. **Postal address:** Apartado 20. **Tel:** 43 81 02 00. **Telex:** 36810 ayfa e. **Cables:** Afora. **Telecopier (Fax):** 43 81 56 69.
Management: Directors - Cesáreo Garcia (general), Pedro Echave (sub), Sabin Aguirre (administration), Javier Otaño (personnel), J.I. Urreta-Vizcaya (técnico); Jefe de Producción - Javier Gurruchaga. **Established:** 1939. **Capital:** Ptas 900,000,000. **No. of employees:** Works - 454. **Ownership:** Bilboil SA. **Annual capacity:** Raw steel: 100,000 tonnes.
Products: Carbon steel - (Engineering steel) round bars 16-200mm; square bars 20-35mm; hexagons 16-67mm. Alloy steel (excl stainless) - round bars 16-200mm; square bars 20-35mm; hexagons 16-67.

AHM - Altos Hornos del Mediterráneo SA—see Sidmed - Siderurgica del Mediterráneo SA

AHV - Altos Hornos de Vizcaya SA

Head office: Carmen 2, Baracaldo (Vizcaya). **Tel:** 250 000, 960011. **Telex:** 32777. **Cables:** Altos Hornos.
Established: 1902. **Ownership:** US Steel Corp (25%). **Subsidiaries:** Laminaciones de Lesaca SA; Agruminsa SA.
Works:
■ At head office. ***Coke ovens:*** Two Otto batteries (29 ovens each) (combined annual capacity 304,000 tonnes); one Didier battery (40 ovens) (240,000 tonnes) and one

AHV - Altos Hornes de Vizcaya SA (continued)

Didier battery (50 ovens) (345,000 tonnes). ***Sinter plant:*** Three Dwight Lloyd – two 74.5 sq m strand (739,000 tonnes), and one 102 sq m strand (1,022,000 tonnes). ***Blast furnaces:*** Two 6.5m hearth dia (500,000 tonnes); one 9m hearth dia (800,000/1,100,000 tonnes). ***Steelmaking plant:*** Three 110-tonne LD basic oxygen (2,000,000 tonnes); one 50-tonne electric arc furnace (120,000 tonnes). ***Continuous casting machines:*** Three Mannesmann Demag slab (2,000,000 tonnes). ***Rolling mills:*** One semi-continuous wide hot strip (1,525mm) (1,700,000 tonnes).

■ Echévarri. ***Rolling mills:*** Two single stand 4-high cold reduction (1,219mm) (annual capacity 324,000 tonnes); one single stand 4-high cold strip (863mm) (71,000 tonnes); one 2-stand temper/skin pass (1,219mm) (272,000 tonnes). ***Coil coating lines:*** One electrolytic tinning (100,370 tonnes); hot-dip tinning – two Wean (22,060 tonnes), three Aetna-Abercarn (19,070 tonnes); one Armco-Sendzimir continuous hot-dip galvanizing (106,000 tonnes).

Products: Steelmaking pig iron, foundry pig iron. Carbon steel – slabs, blooms, hot rolled coil (for re-rolling), cold rolled hoop and strip (uncoated), cold rolled sheet/coil (uncoated), hot-dip galvanized sheet/coil, electrolytic tinplate (single-reduced). Metallurgical coke, sinter.

Forjas Alavesas SA (Fasa)

Head office: Portal de Gamarra 22, Vitoria (Alava). **Postal address:** PO Box 130. **Tel:** 261900. **Telex:** 35245 foral. **Cables:** Fasa. **Telecopier (Fax):** 256879.
Products: Carbon steel – ingots, blooms, billets, wire rod (incl. high carbon), round bars, flats, bright (cold finished) bars, special light sections. Stainless steel – ingots, blooms, billets, wire rod, round bars, flats, bright (cold finished) bars, special light sections. Alloy steel (excl stainless) – ingots, blooms, billets, wire rod, round bars, flats, bright (cold finished) bars, special light sections.

Aprovechamientos Siderurgicos SA

Head office: La Vega 29, Guernica, Vizcaya. **Postal address:** Apartado 19. **Tel:** 4 685 03 71, 685 07 08. Light sections and bright bars.

Perfiles Aragón SA

Head office: c/ San Juan de la Peña 262, Zaragoza 15. **Tel:** 29 33 86.
Products: Carbon steel – longitudinal weld tubes and pipes.

Arania SA

Head office: Bo San Antonio, Euba-Amorebieta, Vizcaya. **Postal address:** PO Box 1486. **Tel:** 344 6730250. **Telex:** 34548 arni e.
Management: A. Segovia (export manager).
Products: Carbon steel – cold roll-formed sections (from strip), cold rolled hoop and strip (uncoated), cold rolled sheet/coil (uncoated), longitudinal weld tubes and pipes.

Grupo José María Aristrain SA

Head office: Apartado 008, Olaberria, Guipúzcoa. **Tel:** 943 880300. **Telex:** 36253. **Cables:** Aristrain.
Management: José María Aristrain Noain (presidente de consejo), Javier Imaz Buenechea (director general), Angel Zabalegui Vitoria (director, comercial). **Established:** 1955. **Capital:** Ptas 5,575,000,000. **No. of employees:** 1,600. **Subsidiaries:** José María Aristrain SA, Olaberria (see below); José María Aristrain Madrid SA (see below); José María Aristrain SA, Artierro, Valencia (see below); Nucleos

Grupo José María Aristrain SA (continued)

Magneticos Aristrain SA, Barcelona (stampings); Aristrain International AG (sales).
Annual capacity: Raw steel: 600,000 tonnes. Rolled steel: 500,000 tonnes.
Works:
■ José María Aristrain SA, Olaberria. (At head office). ***Steelmaking plant:*** Two 70-tonne electric arc furnaces (annual capacity 15,000 tonnes each); One induction furnace. ***Continuous casting machines:*** One 7-strand billet. ***Rolling mills:*** Two 3-high medium section and 5-stand mill for electrical sheet and strip. ***Other plant:*** Insulating and coil forming lines for electrical steel.
■ José María Aristrain Madrid SA, Carretera Toledo Km 9, Madrid. (Tel: 91 7972300. Telex: 22120). ***Steelmaking plant:*** Three electric arc furnaces – two 15-tonne (50,000 tonnes each) and one 100-tonne (200,000 tonnes). ***Continuous casting machines:*** One 1-strand bloom. ***Rolling mills:*** One 3-high heavy section; one Steckel hot strip and hoop up to 450mm wide. ***Tube and pipe mills:*** welded – one Abbey Etna 2-6", and one Yoder $\frac{3}{8}$-1$\frac{1}{2}$". ***Other plant:*** Tube galvanizing line.
■ José María Aristrain SA, Artierro, Camino Viejo de Silla a Beniparell s/n, Artierro, Valencia. (Tel: 96 1200800. Telex: 26266). ***Steelmaking plant:*** One electric arc furnace. ***Rolling mills:*** Two 3-high medium section and 5-stand mill for electrical sheet and strip. One wire rod (annual capacity 180,000 tonnes).
Products: Carbon steel – ingots, wire rod, medium joists, heavy angles, heavy joists, heavy channels, hot rolled hoop and strip (uncoated), skelp (tube strip), electrical non-oriented silicon sheet/coil, longitudinal weld tubes and pipes, galvanized tubes and pipes. Alloy steel (excl stainless) – ingots.

Armco SA, Spain

Head office: Muntaner 374-6, Barcelona 6. **Tel:** 209 41 77. **Telex:** 52337, 53087. **Cables:** Armcosun. **Telecopier (Fax):** 93 2090844.
Management: Juan A. Saldaña (general manager). **Established:** 1963. **No. of employees:** Works – 600. Office – 100.
Products: Stainless steel – longitudinal weld tubes and pipes 4-104mm dia.

Arregui SA

Head office: Portal de Gamarra 38, Vitoria (Alava). **Postal address:** Apartado 177. **Tel:** 26 00 00. **Telex:** 35554 arrg e. **Cables:** Arreguisa.
Products: Carbon steel – light angles, light tees, light channels, hot rolled hoop and strip (uncoated), longitudinal weld tubes and pipes.

Artierro SA

Head office: Cno. Viejo de Silla a Beniparrell s/n, Silla, 46460 Valencia. **Tel:** 1 20 08 00. **Telex:** 62216. **Telecopier (Fax):** 1200231.
Products: Carbon steel – billets 120 x 120mm; reinforcing bars 6-25mm.

Austinox SA

Head office: Ctra Calafell Km 9.3, Sant Boi de Llobregat (Barcelona). **Tel:** 6610450, 6611808, 6611862. **Telex:** 57418 itea e.
Products: Stainless steel – longitudinal weld tubes and pipes austenitic and ferritic $\frac{1}{8}$-42"; cold drawn tubes and pipes up to 3". Welded pipe fittings.

Azma SA

Head office: Paseo de la Habana 16, 28026 Madrid. **Tel:** 695 28 00. **Telex:** 23626 azma e.
Established: 1942. **Capital:** Ptas 530,000,000. **No. of employees:** Works – 324. Office – 32. **Subsidiaries:** Tecnicas del Hierro SA (Tecrosa). **Associates:** Aceros y

Azma SA (continued)

Tecnologia de la Construccion SA (Atecsa).
Works:
■ C/Fundidores 4, Poligono Industrial Los Angeles, Getafe (Madrid). (Tel: 695 28 00. Telex: 23626 azma e). Works director: José Ramón Gayol. ***Steelmaking plant:*** Two electric arc furnaces – 50-tonne and 35-tonne (annual capacity 360,000 tonnes). ***Continuous casting machines:*** billet – one 3-strand Demag (Sidernaval) and one 3-strand Danieli (120 x 120mm, 160 x 160mm) (560,000 tonnes). ***Rolling mills:*** One 500mm light section for special sections, tees and channels (100-160mm); rebars 2mm-57.3mm dia (combined annual capacity 220,000 tonnes). One 350mm bar and wire rod for coils 6-12mm dia and rebars 6-25mm dia (180,000 tonnes).
Products: Carbon steel – billets; wire rod 6-10mm; reinforcing bars 6-57.3mm; light tees; light channels 100mm and 120mm; special light sections (including edge rails); medium tees 100mm, 120mm, 140mm and 160mm; medium joists 100mm, 120mm, 140mm and 160mm; medium channels 120-160mm.
Brands: Altres (rebars); GEWI systems.

Babcock & Wilcox Española SA Tubular Products Division

Head office: Alameda de Recalde 27, Bilbao 9. **Tel:** 94 4415700. **Telex:** 32776, 32235 bwbil e. **Cables:** Babcock.
Management: President – Manual Fernández García la Rubia; directors – José Antonio Vallejo, Ernesto Gastelu-Iturri (production), Pedro Basurco (technical and commercial), José Fouassier (sales); manager – Eduardo Gorostiza (export).
Established: 1918.
Products: Carbon steel – blooms, billets, semis for seamless tubemaking, round bars, square bars, flats, bright (cold finished) bars, pipe piling, seamless tubes and pipes, oil country tubular goods, hollow bars. Alloy steel (excl stainless) – blooms, billets, semis for seamless tubemaking, round bars, square bars, flats, bright (cold finished) bars, seamless tubes and pipes, oil country tubular goods, hollow bars.

Talleres Balcells SA (Tabalsa)

Head office: Paseo Zona Franca 45-51, Barcelona 4. **Tel:** 3 31 73 00. **Cables:** Tabalsa.
Established: 1952.
Products: Carbon steel – cold rolled hoop and strip (uncoated), longitudinal weld tubes and pipes, galvanized tubes and pipes.

Industrias Bérriz

Head office: Calle Estación 3, Bérriz (Vizcaya).
Products: Carbon steel – bright wire. Bright bars.

Industrias del Besós SA (Inbesa)

Head office: Calle Angel Guimerá 27, San Adrián de Besós (Barcelona). **Tel:** 381 06 00. **Telex:** 53105, 51857 besos e. **Cables:** Inbesa S. Adrian Besos.
Management: Jaime Vila Mariné (majority shareholder), Jaime Carol Bernaus (executive president), José Perramón Palmada (technical director). **Established:** 1950.
Products: Carbon steel – square bars 10-50mm; flats 20 x 3-220 x 16mm.

Bornay SA

Head office: Av San Fernando 39, Ibi (Alicante). **Postal address:** Apartado 110. **Tel:** 55 05 12. **Telex:** 68421 born e.
Products: Carbon steel – longitudinal weld tubes and pipes.

La Calibradora Mecánica SA

Head office: Camino de Solicrup s/n, Zona Industrial, Villanueva y Geltrú (Barcelona). **Postal address:** Apartado 40. **Telex:** 52109 cali e.
Products: Carbon steel – bright (cold finished) bars. Stainless steel – bright (cold finished) bars.

Tubos y Calibrados de Cataluña SA

Head office: Sicilia 95, 5th Floor, Barcelona. **Tel:** 3 246 2486. **Telex:** 52818 aadtt e.
Products: Stainless steel – longitudinal weld tubes and pipes, tube fittings.

Celsa – Cia Española de Laminacion SA

Head office: Poligono Industrial Can Pelegrí, 08740 San Andrés de la Barca (Barcelona). **Postal address:** Apartado 4. **Tel:** 3 653 05 11. **Telex:** 52915 celsa e, 51885 tder e, 50576 rbar e. **Cables:** Celsa Martorell. **Telecopier (Fax):** 3 653 07 03.
Management: President – Francisco Rubiralta; director – Juan Puiggali (managing); sub-director – Luis Tejada (managing); managers – Ricardo Hugas (sales), Antonio Falco (purchasing), Carmen Muñoz (financial). **Subsidiaries:** Siderurgica de Galicia SA (Sidegasa); Nueva Montaña Quijano SA.
Works:
■ At head office. ***Steelmaking plant:*** One 110-tonne electric arc furnace (annual capacity 600,000 tonnes). ***Continuous casting machines:*** One 6-strand Concast billet (100-160mm sq) (700,000 tonnes). ***Rolling mills:*** One 18-stand bar 8-32mm dia (475,000 tonnes).
Products: Carbon steel – reinforcing bars 8-32mm (output 380,000 tonnes).

Centracero SA

Head office: Ctra de Estella s/n, Murieta (Navarra). **Tel:** 48 53 40 92. **Telex:** 37761 cana e.
Products: Carbon steel – Centrifugal cast products including seamless tubes and pipes; Stainless steel – seamless tubes and pipes; and Alloy steel (excl stainless) – seamless tubes and pipes.

Chinchurreta SL

Head office: B. San Pedro s/n, Oñate (Guipúzcoa). **Postal address:** Apartado 6. **Tel:** 78 13 12, 78 13 90, 78 14 11. **Telex:** 36612 chta e. **Cables:** Chinchurreta.
Products: Carbon steel – longitudinal weld tubes and pipes.

Condesa – Conducciones y Derivados SA

Head office: Ctra. de Vergara Km 14, Villarreal de Alava (Alava). **Tel:** 945 45 50 60. **Telex:** 35459.
Products: Carbon steel – cold roll-formed sections (from strip), longitudinal weld tubes and pipes.

Cufesa SA

Head office: Apartado 52, La Felguera (Oviedo). **Tel:** 69 49 11, 69 48 51. **Cables:** Cufesa.
Products: Carbon steel – reinforcing bars and

Industrias Duero SA

Head office: Carretera de Villaverde 50, Madrid 31. **Tel:** 203 09 40, 203 14 31-34.
Products: Carbon steel – longitudinal weld tubes and pipes.

Teodoro Dura SA

Head office: Ctra Barcelona 34, Puzol (Valencia). **Tel:** 142 06 00. **Telex:** 62083 hlla e. **Cables:** Dura. Merchant bars, light sections, heavy sections, bright bars.

Fundiciones Echevarría SA

Head office: Apartado 14, Beasain (Guipúzcoa). **Tel:** 88 15 00. **Telex:** 36614 aefe e. **Cables:** Fesa. Carbon and alloy engineering steels.

SA Echevarría (SAE)

Head office: Simón Bolívar 7, 2° Bilbao (Vizcaya). **Tel:** 4444100. **Telex:** 32146. **Cables:** Echevarria.
Products: Carbon steel – ingots, billets, wire rod high-carbon, cold rolled hoop and strip (uncoated). Stainless steel – ingots, billets, wire rod, round bars, square bars, flats, wire. Alloy steel (excl stainless) – ingots, billets, wire rod, round bars, square bars, flats, cold rolled hoop and strip (uncoated), wire. Special steels – Carbon and alloy tool and stainless steels, cold-heading and spring steels. Bright bars, forged axles, crankshafts, mill rods etc.
Additional information: The company is associated with the Acenor group.

Patricio Echeverría SA

Head office: Urola s/n, Legazpia (Guipúzcoa). **Tel:** 43 730000. **Telex:** 36685. **Cables:** Echeverria.
Products: Carbon steel – billets, medium plates, hot rolled sheet/coil (uncoated). Stainless steel – billets, round bars, hot rolled sheet/coil (uncoated). Alloy steel (excl stainless) – billets, round bars. Special steels, carbon and alloy, tool steel, engineering steel, stainless steel; forgings; hand tools.

Ensidesa – Empresa Nacional Siderurgica SA

Head office: Plaza de América 10, Oviedo, Asturias. **Postal address:** Apartado 93, Avilés, Asturias. **Tel:** 571000. **Telex:** 88215 enssa e. **Cables:** Ensidesa. **Telecopier (Fax):** 570477.
Management: Presidente – Fernando Lozano Cuervo. Directors – Rogelio Bodelón López (general), Tomás Gutiérrez Montes (económico-financiero), Arsenio Saldaña Albillos (desarrollo industrial), Ricardo Díez Serrano (adjunto a la presidencia), Antonio Arranz Espino (comercial), Carlos Avello Pérez (factoría da Avilés), Angel Préstamo Díaz (factoría de Gijón), José María Herrero Cuesta (aprovisionamientos), Humberto Hurtado Sánchez (tecnología), Pedro Fanego Valle (organización), Manuel Martín Martín (relaciones industriales), Manuel Gorostidi Sustaeta (planificación). **Established:** 1950. **Capital:** (incl reserves) Ptas 110,331,000,000 (31.12.86.). **No. of employees:** 18,200 (31.12.86.). **Ownership:** INI – Instituto Nacional de Industria. **Subsidiaries:** Sidmed (cold rolling mill and electrogalvanizing); Perfrisa (cold formed sections, panels and welded pipe); Metalsa (rolling stock repair and other

Spain

Ensidesa – Empresa Nacional Siderurgica SA (continued)

manufacturing) (see below); Parque Aboño SA (blending and homogenization of coal). **Associates:** Hispanobrás (iron ore pellet plant). **Annual capacity:** Raw steel: 4,700,000 tonnes.

Works:

■ Avilés, Asturias. (Tel: 571000. Telex: 88215 enssa e). Works director: Carlos Avello. ***Coke ovens:*** Ten batteries (300 ovens) (annual capacity 1,850,000 tonnes). ***Sinter plant:*** Two 67 sq m strand (each) (annual capacity 1,375,000 tonnes); one 212 sq m strand (2,350,000 tonnes). ***Blast furnaces:*** Four – No 1 8,686mm hearth dia, 1,712 cu m; No 2 8,992mm hearth dia, 1,790 cu m; No 3 8,992mm hearth dia, 1,794 cu m; No 4 8,992mm hearth dia, 1,794 cu m (combined annual capacity 3,725,000 tonnes). ***Steelmaking plant:*** basic oxygen converters – three 70-tonne and two 100-tonne (2,500,000 tonnes); two 250-tonne (under construction to replace existing steelplant). **Refining plant:** RH and Argon, etc. ***Refining plant:*** RH and Argon, etc. ***Continuous casting machines:*** Three 6-strand vertical with bending bloom (100 x 100mm-200 x 200mm) (1,000,000 tonnes); two 2-strand curved mould slab (210-280mm x 1,600mm) (under construction) (2,500,000 tonnes). ***Rolling mills:*** 2-high universal slabbing (two 1,219 x 2,286mm and two 965 x 2,134mm) (2,000,000 tonnes – scheduled for closure); 2-high reversing blooming /slabbing (two 1,050 x 2,743mm) (850,000 tonnes – scheduled for closure); three 2-high reversing heavy section and rail (three 850 x 2,286mm) (300,000 tonnes – scheduled for closure); one 2-high and one 4-high heavy plate and medium plate (3,454mm) (500,000 tonnes – scheduled for closure); two wide hot strip – one roughing and seven finishing stands (1,727mm) (2,300,000 tonnes) and one 4-high Steckel (1,206mm) (300,000 tonnes – scheduled for closure); two cold reduction – four 4-high (1,676mm) (800,000 tonnes) and five 4-high (1,422mm) (450,000 tonnes); three temper/skin pass – one 4-high, two 4-high and two 4-high (1,676/1,219/1,219mm). ***Coil coating lines:*** Two Ferrostan vertical electrolytic tinning (300,000 tonnes); one Sendzimir hot-dip galvanizing (150,000 tonnes).

■ Veriña, Asturias. (Tel: 320700. Telex: 87348 onuve e). Works director: Angel Prestamo. ***Coke ovens:*** Two batteries (90 ovens) (annual capacity 1,050,000 tonnes). ***Sinter plant:*** One 71.4 sq m strand (725,000 tonnes), one 200.0 sq m strand (2,200,000 tonnes). ***Blast furnaces:*** Two – No 6 9,500mm hearth dia, 1,981 cu m (1,200,000 tonnes); No 7 9,500mm hearth dia, 1,981 cu m (1,200,000 tonnes). ***Steelmaking plant:*** Three 115-tonne basic oxygen furnaces (2,200,000 tonnes). ***Refining plant:*** One 120-tonne vacuum degassing unit (180,000 tonnes); one TN desulphurising unit (200,000 tonnes). ***Continuous casting machines:*** One 3-strand curved mould bloom (380 x 505-600mm) (850,000 tonnes); one 2-strand curved mould slab (250 x 2,000mm max and 180-250 x 2,000mm max) (800,000 tonnes). ***Rolling mills:*** blooming /slabbing – one 2-high (1,330 x 3,000mm) (1,525,000 tonnes); billet – one 2-high reversing and four 2-high (650mm) (800,000 tonnes); heavy section – one roughing, three intermediate and two finishing stands (400,000/450,000 tonnes); light section and bar – seven roughing, four intermediate and four finishing stands (260,000 tonnes); wire rod – 2-strand 7 roughing, 8 intermediate and 2 Morgan stands (400,000/480,000 tonnes); 4-high reversing heavy plate (660,000 tonnes).

■ Metalsa, La Felguera. ***Rolling mills:*** One light rail and light section – two 3-high and one 2-high stands (annual capacity 85,000 tonnes).

Products: Steelmaking pig iron (1986 output 3,847,030 tonnes). Carbon steel – ingots (3,933,094 tonnes); slabs (2,195,768 tonnes); blooms (962,772 tonnes); billets (960,563 tonnes); hot rolled coil (for re-rolling) (893,655 tonnes); wire rod (280,667 tonnes); reinforcing bars (145,479 tonnes); square bars (751 tonnes); flats (7,663 tonnes); light angles less than 80mm (10,518 tonnes); heavy angles 80-150mm (5,576 tonnes); heavy joists (IPN) 80-600mm (46,562 tonnes); heavy channels (UPN) 80-300mm (43,476 tonnes); wide-flange beams (heavy) (IPE) 100-600mm (65,307 tonnes); universals (heavy) (HE) 100-600mm (168,099 tonnes); heavy rails 45-60kg/metre (34,433 tonnes); light rails 20kg/metre (1,975 tonnes); hot rolled sheet/coil (uncoated) 45-16 x 500-1,500mm (308,528 tonnes); cold rolled sheet/coil (uncoated) 0.513 x 500-1,500mm (580,688 tonnes); hot-dip galvanized

Ensidesa – Empresa Nacional Siderurgica SA (continued)

sheet/coil 0.513 x 500-1,500mm (135,934 tonnes); zinc-aluminium alloy coated sheet/coil (11,438 tonnes); electrolytic tinplate (single-reduced) (143,431 tonnes). Mine sections (36,818 tonnes); bulb flats (5,155 tonnes); rail accessories (661 tonnes).
Brands: Ensacor; Galvafort; Supergalvafort; Algafort (zinc-aluminium alloy coated sheet); Redur; Tetracero.

Flejes Industriales SA (Flinsa)

Head office: Ctra Ibi-Alicante Km. 1.5, Ibi, Alicante. **Postal address:** Apartado 51. **Tel:** 55 29 36. **Telex:** 68432 fli e.
Products: Carbon steel – cold rolled hoop and strip (uncoated).

Funditubo SA

Head office: P/O Pereda 223/0, Santander. **Telex:** 35615.
Ownership: Pont-à-Mousson SA, France.
Products: Cast iron – pipes, pipe fittings.

Industrias Garaeta

Head office: Calle de la Rioja, 1 y 3 Polígono Industrial de Coslada, Coslada, Madrid. **Postal address:** Apartado 67. **Telex:** 43511 inga e.
Products: Carbon steel – cold roll-formed sections (from strip), longitudinal weld tubes and pipes.

Hijos de Juan de Garay SA

Head office: Obispo Otaduy 7, 9 & 11, Oñati, Guipuzcoa. **Postal address:** PO Box 9. **Tel:** 943 78 00 11. **Telex:** 36638. **Cables:** Garay.
Established: 1921. **Capital:** Ptas 250,000,000. **No. of employees:** Works – 250. Office – 50. **Ownership:** Garay family.
Works:
■ At head office. ***Tube and pipe mills:*** welded – three 8-108mm dia Yoder (annual capacity 54,000 tonnes).
Products: Carbon steel – longitudinal weld tubes and pipes, cold drawn tubes and pipes, precision tubes and pipes, hollow sections.

Sdad General de Electro-Metalurgia SA

Head office: Plaza de Urquinaona 5, Barcelona 10. **Tel:** 317 59 58.
Products: Carbon steel – cold rolled hoop and strip (uncoated); wire, wire products.

Helisold Iberica SA

Head office: Carretera de Ilárraza a Salvatierra s/n, Alegria de Alava, Alava. **Postal address:** PO Box 589, Vitoria. **Tel:** 45 42 00 50. **Telex:** 35456 heal e. **Cables:** Helisold.
Management: Board – Pedro Basurco (president), Eduardo Sancho (executive director), José M. Jacob (secretary); managers – Juan R. Irurzun (export), Javier Pérez de Albeniz (technical), Julian Asurmendi (personnel); secretary of the board – José M. Jacob. **Established:** 1974. **Capital:** Ptas 400,000,000. **No. of employees:** Works – 217. Office – 29. **Ownership:** Babcock & Wilcox Española SA (42.4%). **Subsidiaries:** Helisold SA (100%); Helisold de Venezuela SA. **Annual capacity:** Finished steel: 84,150 tonnes (spiral welded steel pipe).
Works:
■ At head office. Works director: Javier Pérez de Albeniz. ***Tube and pipe mills:***

Spain

Helisold Iberica SA (continued)

welded – spiral, 168-2,000mm dia, 3-18mm thick (annual capacity 84,150 tonnes).
Other plant: Pipe coating plant.
Products: Carbon steel – spiral-weld tubes and pipes 327.8-812.8mm dia (output 30,000 tonnes).
Expansion programme: Polyethelene pipe and coating plant.

Indusla SA

Head office: c/Alvarez Garaya 1, 4°, Gijon, Asturias.

Inoxca – Inoxidables de Catalunya SA

Head office: Caracas 35, 08030 Barcelona. **Tel:** 345 42 12. **Telex:** 98162 xca e. **Telecopier (Fax):** 345 42 54.
Management: J. Diaz Albacete (managing director), M. Adua Malet (export manager). **Established:** 1982. **No. of employees:** 103. **Ownership:** Private. **Annual capacity:** Finished steel: 3,500 tonnes.
Works:
■ At head office.
Products: Stainless steel – wire rod, round bars, square bars, flats, light angles, wire.

Inoxservi SA (formerly Aceros Completos Españoles Recocidos y Aleados SA)

Head office: c/Juan de la Cierva s/n, San Just Desvern (Barcelona). **Postal address:** Apartao 28. **Tel:** 371 79 08. **Telex:** 52450 finox e.
Works:
■ At head office.
Products: Stainless steel – cold rolled hoop and strip (uncoated).

Aceros de Irura SA

Head office: Ctra Madrid-Irún Km 444, Irura, Guipúzcoa. **Postal address:** Apartado 94, Tolosa. **Telex:** 38796 saai e.
Products: Carbon steel – (engineering steels) billets; round bars and flats (carbon). Stainless steel – billets, round bars and flats. Alloy steel (excl stainless) – (spring and tool steels) billets, round bars and flats.

Itinsa – Iberica de Tubos Inoxidables SA

Head office: Paseo de la Riera s/n, Pol. Ind Cova Solera, 08191 Rubi, Barcelona. **Postal address:** Apartado de Correos 41. **Tel:** 93 699 15 00. **Telex:** 56130 mitub e. **Cables:** Itinsa. **Telecopier (Fax):** 93 697 13 07.
Management: Directors – Enrique Manubens (managing), Antonio Serramia (finance), Jose Mascaro (works), Angel dal Maschio (export), J. Lopez del Castillo (sales). **Established:** 1961. **Capital:** Ptas 56,000,000. **No. of employees:** Works – 60. Office – 17. **Annual capacity:** Welded stainless steel tubes: 1,400-1,500 tonnes.
Works:
■ At head office. ***Tube and pipe mills:*** welded (annual capacity 1,400-1,500 tonnes).
Products: Stainless steel – bright (cold finished) bars; longitudinal weld tubes and pipes 4 x 0.5-63.5 x 2.5mm – AISI grades 304, 304L, 321, 316, 316L, 316 and TI.

Laminación y Derivados SA

Head office: c/ San Fausto 6, Durango, Vizcaya. **Postal address:** Apartado 22. **Tel:** 94 6814900. **Telex:** 34413 layde e.
Products: Carbon steel – cold rolled hoop and strip (uncoated), cold rolled sheet/coil (uncoated).

Laminados Especiales SA

Head office: Avda. Generalísimo 3, 3°, Eibar, Guipúzcoa. **Postal address:** Apartado 312. **Tel:** 681 35 00, 681 35 04, 681 35 08. **Telex:** 34455 irbe e.
Products: Carbon steel – cold rolled hoop and strip (uncoated).

Laminor SA

Head office: Bo. Larrea, Amorebieta, Vizcaya. **Postal address:** Apartado 11. **Tel:** 4 673 07 50. **Telex:** 34536 ianor e, 91150. Light and heavy sections.

Lecanda SA

Head office: Polígono Industrial Bakiola, Arrancudiaga, Vizcaya. **Postal address:** Apartado 568, Bilbao. **Tel:** 4 671 1711. Bright bars.

Laminaciones de Lesaca SA (LL)

Head office: Barrio Arratzubi s/n, Lesaca (Navarra). **Tel:** 948 63 70 00. **Telex:** 36149 lesac e.
Management: Managers – Carlos Sancho Coscolluela (factory), José Fernández D'Arlás (strip and tube sales), José Ignacio Laffitte Mesa (profiles and roof sales), José María Gorostiza (foreign trade); purchase department chief – Alejandro Hernández Asensio. **Capital:** Ptas 1,100,000,000. **No. of employees:** Works – 1,962. Office – 490. **Ownership:** AHV – Altos Hornos de Vizcaya SA. **Subsidiaries:** Industrias Cusí SA (cold roll forming – see below); Industrias Metálicas Castellón (Imcasa) (greenhouses – see below). **Annual capacity:** Rolled steel: 900,000 tonnes.
Works:
■ At head office. Works director: Carlos Sancho Coscolluela. ***Rolling mills:*** Four cold strip – one 4-high 640/400mm reversing RWF-Still (annual capacity 90,000 tonnes), one 4-high 425/200mm reversing Loewy Robertson (80,000 tonnes), one 4-high 1,300/600mm reversing secim (350,000 tonnes), one 4-high 1,300/600mm reversing Davy (350,000 tonnes); one temper/skin pass. ***Coil coating lines:*** One Sendzimir hot-dip galvanizing (180,000 tonnes); one 30" Ruthner electro-plating line for zinc, nickel, tin, copper and brass plating. ***Other plant:*** Levelling, slitting, annealing, pickling, cleaning and stretch levelling lines.
■ Barrio Zalain Zoko, Lesaca (Navarra) (Tel: 948 63 07 11. Telex: 36149 lesac e). ***Tube and pipe mills:*** welded – one Abbey Etna and one Wean Damiron for casing tubes and hollow sections 2-8$\frac{5}{8}$" od and 2-9mm thick (annual capacity 180,000 tonnes); two triple draw benches $\frac{1}{2}$-4" od (10,000 tonnes). ***Other plant:*** Twelve lines for precision tube and metallic coating 8-75mm od and 0.6-4mm thick (165,000 tonnes).
■ Industrias Cusí SA, San Justo Desvern, Barcelona (Tel: 93 371 1350). Works director: Miguel Valencia Muñoz. Twelve cold roll formers for sections (annual capacity 22,000 tonnes). Three slitting lines (21,000 tonnes).
■ Industrias Metálicas Castellon (Imcasa), Ctra Valencia, Km 63.5, Castellón de la Plana (Tel: 964 21 14 00, 964 21 14 11). Works director: Juan Luis Llonis Barrio. Two cold roll formers for sections (annual capacity 6,000 tonnes). The works' main activity is manufacture of greenhouses.
■ Lesaca (Navarra) (Tel: 948 59 42 00. Telex: 36149 lesac e). ***Coil coating lines:*** colour coating – one B & K paint line 48" wide (annual capacity 70,000 tonnes). ***Other plant:*** Three roll forming lines.
Products: Carbon steel – cold roll-formed sections (from strip), skelp (tube strip), cold

Spain

Laminaciones de Lesaca SA (LL) (continued)

rolled hoop and strip (uncoated), hot-dip galvanized hoop and strip, coated hoop and strip, cold rolled sheet/coil (uncoated), hot-dip galvanized sheet/coil, electrical non-oriented silicon sheet/coil, longitudinal weld tubes and pipes, galvanized tubes and pipes, cold drawn tubes and pipes, hollow sections, line pipe.
Sales offices/agents: Home market – head office and seventeen warehouses in Spain. Export sales – AHV, Carmen 2, 48901 Baracaldo, Vizcaya.

Aceros de Llodio SA (Allosa)

Head office: Barrio Gardea 12, Llodio, Alava. **Tel:** 672 00 50, 672 07 00, 433 79 00. **Telex:** 34318, 34311 llosa e. **Telecopier (Fax):** 672 0008.
Products: Carbon steel – (including engineering steels) blooms; billets; round bars; square bars; flats; hexagons. Bright (cold finished) bars; Stainless steel – blooms; billets; round bars; square bars; flats; hexagons. Bright (cold finished) bars; Alloy steel (excl stainless) – (including engineering, hot and cold worked tool, high-speed, nitriding, free-cutting, spring and bearing steels) blooms; billets; round bars; square bars; flats; hexagons. Bright (cold finished) bars. Forged bars, forgings and castings.

Macosa – Material y Construcciones SA

Head office: Plaza de la Independencia 8-4, 28001 Madrid. **Tel:** 522 47 87, 522 47 86. **Telex:** 22168 mayco e. **Telecopier (Fax):** 522 71 89.
Management: Board chairman – Eduardo Santos; chief executive officer – Juan Ignacio Muñiz; directors – Emilio Daroca (managing), José Sanz Roca (sales), Juan Jose Sanchis (Barcelona works'). **Established:** 1881. **Capital:** Ptas 725,000,000. **No. of employees:** 2,800. **Annual capacity:** Finished steel: 150,000 tonnes.
Works:
■ Herreros 2, 08019 Barcelona. (Tel: 307 05 00. Telex: 52286 mayco e). Works director: Juan Jose Sanchis. ***Rolling mills:*** Two medium section – 3-stand (600mm) and 3-stand (500mm) (combined annual capacity 150,000 tonnes).
Products: Carbon steel – medium joists IPN 80 up to 260mm; medium channels UPN 80 up to 240mm.

Megasa – Metalurgica Galaica SA

Head office: Ctra. Castilla 820, Naron (La Coruña). **Tel:** 38 05 00. **Telex:** 85536.
Management: Board member – Enrique Freire Artera; managers – José Antonio Coira Castro (sales), Juan Ignacio Montalvo (financial). **Established:** 1952. **Capital:** Ptas 550,000,000. **No. of employees:** Works – 230. Office – 20. **Ownership:** Private.
Works:
■ At head office. Works director: José Manuel Malde Varela. ***Steelmaking plant:*** One 5.50-tonne 46 MW AEG electric arc furnace (annual capacity 350,000 tonnes). ***Refining plant:*** One 15 MW AEG ladle furnace (350,000 tonnes). ***Continuous casting machines:*** 4-strand Concast billet (120 x 120mm) (350,000 tonnes). ***Rolling mills:*** 20-stand bar (600,000 tonnes).
Products: Carbon steel – billets 80 x 80mm, 100 x 100mm and 120 x 120mm; reinforcing bars 6-40mm (output 180,000 tonnes); round bars 6-40mm (70,000 tonnes).

Metalsa—see Ensidesa

Nervacero SA

Head office: Plaza Sagrado Corazón 5, 48011 Bilbao. **Postal address:** PO Box 201. **Tel:** 4410250. **Telex:** 31141 nerv e. **Telecopier (Fax):** 4420766.
Works:
■ At head office. ***Steelmaking plant:*** One 100-tonne electric arc furnace.

Nervacero SA (continued)

Continuous casting machines: One 6-strand Concast billet. ***Rolling mills:*** One Danielli bar.
Products: Carbon steel – reinforcing bars.

Nueva Montaño Quijano SA

Head office: Paseo de Pereda 32, Santander, Cantabria. **Postal address:** Apartado 36, 39080 Santander. **Tel:** 942 336000. **Telex:** 35839 nmqsa e. **Cables:** Nuquisa. **Telecopier (Fax):** 942 339029.
Management: Eduardo Santos Andres (presidente), Francisco Rubiralta Vilaseca (vice presidente), Rafael Villaseca Marco (consejero delegado), Fernando Garcia Perez (director general), Jose M. Hormaechea Cazon (director, commercial). **Established:** 1899. **Capital:** Ptas 3,426,459,000. **No. of employees:** Works – 1,102. Office – 572. **Ownership:** Cia Española de Laminacion SA (Celsa). **Subsidiaries:** Aceria de Santander (see below) (68.73%); Cia Española de Electrodos, SA (Cedesa) (99.9%) (welding equipment); SA de Trefileria y Derivados (29.49%); Sdad de Empresas de Trefileria SA (Setresa) (25%); Mallacero SA (39.29%) (weld mesh); Hilos Para Soldadurs SA (Hisosa) (55%); Marismas del Astillero SA (40%); Trefilerias Quijano SA (51%); Electrodos Sideros SA (welding elctrodes) (34.73%); Tetracero SA (11.93%) (reinforcing bars); Soldadura Actarc Quijano SA (80%) (welding electrodes); Hidrotecar SA (15.58%) (hydraulic pumps). **Annual capacity:** Raw steel: 700,000 tonnes. Finished steel: 450,000 tonnes.
Works:
■ Aceria de Santander, SA, Nueva Montaña, Santander, Cantabria. (Tel: 942 338855). Works director: Javier Hierro Sierra. ***Steelmaking plant:*** One 110-tonne AEG-Demag electric arc (annual capacity 700,000 tonnes). ***Refining plant:*** One 110-tonne Daido ladle furnace (700,000 tonnes). ***Continuous casting machines:*** Two 6-strand Demag-Sidernaval billet (100/200mm square) (500,000 tonnes).
■ Rolling Mills Division. (At head office). Works director: Felix Lopez Diaz. ***Rolling mills:*** One 23-stand wire rod (5.5-13mm) (coil weight 1,250kg) (annual capacity 450,000 tonnes).
■ Wire Division, Los Corrales de Buelna. (Tel: 942 836000). Works director: Faustino Puente Gainza. Wire drawing and wire product plant for – (annual capacity 106,000 tonnes), rope (20,000 tonnes), nails (10,000 tonnes), other wire products (4,000 tonnes).
Products: Carbon steel – wire rod 5.5-13mm dia (output 294,000 tonnes); bright wire (8,900 tonnes); black annealed wire (2,800 tonnes); galvanized wire (plain) (7,200 tonnes); galvanized wire (barbed) (1,800 tonnes).
Expansion programme: Installation of ladle furnace and controlled cooling on wire rod mill.

Olarra SA (Olsa)

Head office: Avenida del Ejército 29, 4º, Bilbao 14. **Tel:** 435 63 00. **Telex:** 32376, 31361 olsa e. **Cables:** Olarrasa.
Established: 1957.
Works:
■ Larrondo, Lujua, Bilbao. ***Steelmaking plant:*** Induction and electric arc furnace. ***Refining plant:*** One 15-tonne and two 10-tonne AOD. ***Rolling mills:*** bar (roughing and finishing).
Products: Carbon steel – ingots, billets, round bars, square bars, flats. Bright (cold finished) bars, Stainless steel – ingots, billets, round bars, square bars, flats. Bright (cold finished) bars, Alloy steel (excl stainless) – ingots, billets, round bars, square bars, flats. Bright (cold finished) bars.

Spain

Esteban Orbegozo SA

Head office: Barrio Artiz 20, Zumarraga, Guipuzcoa. **Postal address:** PO Box 3. **Tel:** 943 720011. **Telex:** 38835, 38836. **Cables:** Orbegozo.
Established: 1944. **Capital:** Ptas 1,968,000,000. **No. of employees:** Works – 1,200. Office – 200. **Ownership:** Private (Orbegozo family). **Subsidiaries:** Aceros Corrugados SA (100%) (steel manufacturing, rebar production). **Associates:** Cebusa, Pasajes (ship's agent, stevedoring, port agents, shipbrokers). **Annual capacity:** Raw steel: 800,000 tonnes. Finished steel: 710,000 tonnes.
Works:
■ At head office. ***Steelmaking plant:*** Three 80-tonne electric arc furnaces (annual capacity 800,000 tonnes). ***Refining plant:*** One 80-tonne ladle furnace (240,000 tonnes). ***Continuous casting machines:*** Three 6-strand Concast billet (120 x 120mm) (750,000 tonnes). ***Rolling mills:*** 20-stand Danieli bar (250,000 tonnes); one 4-strand Pomini/Demag wire rod – 25 Horizontal/Vertical stands (coil weight 1,150 kg) (310,000 tonnes), 22-stand Morgan bar/rod (coil weight 1,150 kg) (150,000 tonnes). ***Other plant:*** Wire works (35,000 tonnes).
Products: Carbon steel – billets 120 x 120mm (1986 output 710,000 tonnes); wire rod 5.5-32mm (300,000 tonnes); reinforcing bars 6-40mm (240,000 tonnes); round bars 14-60mm (160,000 tonnes); bright wire 1-9mm (8,500 tonnes); black annealed wire 0.8-6mm (7,500 tonnes); galvanized wire (plain) 1.8-6mm (10,000 tonnes); galvanized wire (barbed) 1.8-4mm (2,500 tonnes). Alloy steel (excl stainless) – round bars 5.5-60mm (120,000 tonnes). Wire – prestressed concrete 3-7mm (7,200 tonnes); welded mesh 4-10mm (10,000 tonnes); concrete reinforcements 4-10mm (10,000 tonnes).

Pedro Orbegozo y Cía SA (Acenor)

Head office: Apartado 21. 20120 Hernani, Guipuzcoa. **Tel:** 43 55 01 00. **Telex:** 36279 posa e. **Telecopier (Fax):** 43 55 17 79.
Management: President – Pedro Orbegozo Eizaguirre; directors – Javier de Isusi del Villar (managing), Juan I. Barcelona Abaitua (works); manager – Andres Redondo Rodriguez (sales). **Established:** 1953. **Capital:** Ptas 1,501,000,000. **No. of employees:** Works – 510. Office – 90. **Associates:** Acenor Group. **Annual capacity:** Raw steel: 266,000 tonnes. Rolled steel: 280,000 tonnes.
Works:
■ At head office. ***Steelmaking plant:*** One 70-tonne electric arc furnace (annual capacity 266,000 tonnes). ***Refining plant:*** 70-tonne Daido ladle furnace and vacuum degassing (266,000 tonnes). ***Continuous casting machines:*** One 4-strand Concast billet (145 and 200 x 250 mm sq) (266,000 tonnes). ***Rolling mills:*** One 2-high 750mm blooming (163,000 tonnes); medium section – comprising 1-stand 3-high 600mm, 5-stand 3-high 450mm, 3-stand 2-high 450mm, 2-stand 2-high 400mm, 6-stand 2-high 290mm (117,000 tonnes). ***Other plant:*** Bright steel plant – four turning machines, four grinding machines, two polishing machines, automatic inspection. Heat treatment plant – two roller hearth type continuous furnaces for annealing, one roller hearth type continuous furnace for hardening and tempering.
Products: Carbon steel – billets 80-200 x 250mm; round bars 15-140mm; square bars 80-160mm; flats 40 x 4-140 x 20mm. Bright (cold finished) bars 10 by 125mm. Alloy steel (excl stainless) – billets 80-200 x 250mm; round bars 15-140mm; square bars 80-160mm; flats 40 x 4-140 x 20mm. Bright (cold finished) bars 10 by 125mm;
Sales offices/agents: Acenor, Dirección Comerial, Simón Bolívar 7, 48010 Bilbao; Cavexsa, Agente Overseas, Edificio Albia, I-Planta 11, 48001 Bilbao.

Marcos Ormaechea Múgica

Head office: Calle la Vega 13, Guernica, Vizcaya. **Postal address:** Apartado 17. **Tel:** 685 1400, 685 06 48. **Telecopier (Fax):** 6854976.
Products: Carbon steel – bright (cold finished) bars. Special light sections Bright bars.

Pecofrisa – Perfiles Conformados en Frio SA

Head office: c/Albarracin 13 y 15, Zaragoza 15. **Postal address:** Apartado 546. **Tel:** 76 51 17 33.
Products: Carbon steel – cold roll-formed sections (from strip), longitudinal weld tubes and pipes.

Perfil en Frio SA (Perfrisa)

Head office: Sta Engracia 2, Pamplona, Navarra. **Postal address:** Apartado 141. **Tel:** 48 121300. **Telex:** 37710.
Products: Carbon steel – longitudinal weld tubes and pipes, precision tubes and pipes.

Manufacturas Permet SA

Head office: Barrio de Aperriaby 8, Galdácano, Vizcaya. **Tel:** 4 44 91 500.
Products: Carbon steel – cold roll-formed sections (from strip), cold rolled hoop and strip (uncoated).

Forjas y Aceros de Reinosa SA

Head office: Paseo Alejandro Calonje 1, Reinosa, Cantabria. **Tel:** 942 750100. **Telex:** 35671. **Telecopier (Fax):** 942 753062.
Management: Presidente – Jose Luis Alvarez; gerente linea de piezas – Angel Andérez Obeso; director técnico – Antonio Caballero Moreno, director de relaciones industriales – Jose Antono García Calvo. **Established:** 1919. **Capital:** Ptas 1,000,000,000. **No. of employees:** Works – 1,734. **Ownership:** I.N.I. (Instituto Nacional de Industria).
Works:
■ At head office. ***Steelmaking plant:*** Two electric arc furnaces – 65-tonne and 30-tonne (commbined annual capacity 150,000 tonnes). ***Refining plant:*** One 24-tonne ESR (1,781 tonnes); one VAD (39,310 tonnes). ***Rolling mills:*** One bar (68,752 tonnes).
Products: Carbon steel – ingots 1,000, 1,850, 2,500, 30,000 and 60,000mm; round bars 60-230mm, square bars 50-130mm and flats 60-230mm (combined output 57,311 tonnes). Stainless steel – round bars 25-140mm, square bars 25-130mm (combined output 791 tonnes). Alloy steel (excl stainless) – (38,512 tonnes).

Rico y Echeverria SA

Head office: PO Box 546, 50080 Zaragoza. **Tel:** 76 51 51 11. **Telex:** 58060.
Management: Antonio Rico Gambarte, Fernando Echeverría Polo, Javier Tambo, J. Manuel Iglesias. **Established:** 1933.
Works:
■ At head office. ***Steelmaking plant:*** One 50-tonne electric arc furnace (annual capcity 225,000 tonnes). ***Refining plant:*** One 50-tonne ladle furnace furnace. ***Continuous casting machines:*** 5-strand 4 metre radius billet (100 x 100-180 x 120) (500,000 tonnes). ***Rolling mills:*** medium section comprising two 3-high horizontal stands (550mm dia), five 2-high horizontal stands (480mm dia) and one 2-high vertical stand (650mm dia) (300,000 tonnes); light section comprising two 3-high stands (450mm) and eight 2-high stands (380mm) (250,000 tonnes).
Products: Carbon steel – light angles 25 x 25-50 x 50mm; medium angles 60 x 60-120 x 120mm; medium joists 80, 100, 120mm; medium channels 80, 100, 120.

Perfiles Rioja SA

Head office: Ctra de Zaragoza km 4,400, Logroño. **Postal address:** Apartado 10. **Tel:** 41 23 15 84, 41 23 38 96.
Products: Carbon steel – longitudinal weld tubes and pipes.

Industrias Rocandio

Head office: c/Chonta 28, 20600 Eiber, Guipúzcoa. **Postal address:** Apartado 52. **Tel:** 43 12 10 08.
Products: Carbon steel – cold rolled hoop and strip (uncoated).

Roldan SA

Head office: Félix Boix 3, 28036 Madrid. **Tel:** 1 259 15 86. **Telex:** 47429.
Management: Jorge Guarner (general manager). **Established:** 1957. **Capital:** Ptas 794,420,000. **No. of employees:** Works – 404. Office – 15.
Works:
■ Santo Tomas de las Ollas, Ponferrada (Leon). (Tel: 87 41 05 19. Telex: 47429). Works director: Jorge Rodriguez. **Refining Plant:** Two induction furnaces – 6.7-tonne and 3-tonne (annual capacity 38,500 tonnes). ***Refining plant:*** Two induction furnaces – one 6.7-tonne and one 3-tonne (annual capacity 38,500 tonnes). ***Continuous casting machines:*** One 1-strand Technica Guss billet (75 x 75mm to 150 x 150mm) (55,800 tonnes). ***Rolling mills:*** Pomini billet (65,600 tonnes); Pomini medium section (45,900 tonnes); Morgårdshammar bar and wire rod (coil weight 450kg) (38,600 tonnes).
Products: Stainless steel – ingots; billets 75 x 75mm-150 x 150mm; wire rod 5.5-23.5mm; round bars 2-83.5mm; flats 15 x 3mm-60 x 20mm; wire 0.8-21mm. Alloy steel (excl stainless) – ingots; billets 75 x 75mm-150 x 150mm; round bars 2-83.5mm; flats 15 x 3mm-60 x 20mm; wire 0.8-21mm.

Aceria de Santander—see Nueva Montana Quijano SA

Siderúrgica Sevillana SA

Head office: Carretera de Málaga Km 10, Alcalá de Guadaira, 41080 Seville. **Postal address:** Apartado 882. **Telex:** 72166.
Products: Carbon steel – reinforcing bars, round bars.
Additional information: This company's associated with Sidersur (see seperate entry).

Sidegasa – Siderúrgica de Galicia SA

Head office: Crta C-544 km 7 Teixeiro, Curtis La Coruña. **Tel:** 981 789025. **Telex:** 82350 siga e.
Products: Carbon steel – wire rod, reinforcing bars.

Sidersur SA

Head office: Carretera de Málaga Km 10, Alcalá de Guadaira, 41080 Seville. **Postal address:** Apartado 304. **Tel:** 02000. **Telex:** 72166.
Works:
■ Dos Hermanos (Seville). ***Rolling mills:*** Merchant bar.

Sidmed – Siderurgica del Mediterráneo SA (formerly AHM – Altos Hornes del Mediterráneo SA)

Head office: Ctra Acceso IV Planta pk 3-9, 46520 Puerto de Sagunto (Valencia), Comunidad Valenciana. **Postal address:** Apartado 112. **Tel:** 2472000. **Telex:** 62336. **Telecopier (Fax):** 96 2471433.
Management: Ignacio Lopez Aranguren Marcos (presidente del consejo administración); Jose Manuel Riera Fernandez Raigoso (director gral operaciones); Francisco Muñoz de Morales (director financiero); Jose Ma Ruiz Oyaga (director de relaciones laborales); Antonio Ortola Almeda (director industrial); Jose Antonio Millan Echevarria (director commercial). **Established:** 1985. **Capital:** Ptas 17,824,000,000. **No. of employees:** 1,861 (31.12.86). **Ownership:** Ensidesa (100%) (Ultimate holding company – INI – Instituto Nacional de Industria). **Annual capacity:** Finished steel: 1,200,000 tonnes.
Works:
■ At head office. Works director: Mr Ortola. ***Rolling mills:*** One 6-stand 4-high cold reduction (1,520 x 610 x 2,032) (coil weight 45 tonnes/max) and two 56 and 80" temper/skin pass (coil weight 45 tonnes/max (combined total annual capacity 1,200,000 tonnes). ***Coil coating lines:*** One horizontal NKK electro-galvanizing (90,000 tonnes).
Products: Carbon steel – cold rolled hoop and strip (uncoated), cold rolled sheet/coil (uncoated), electro-galvanized sheet/coil.
Brands: Zincmed – electro-zinc coated and oiled; Zincfos – electro-zinc coated and phosphated; Zincfilm – elector-zinc coated and chromium and organic.
Sales offices/agents: At head office. Paseo Castellana 83-85, Madrid. Overseas – Ensidesa.
Expansion programme: Revamp of the pickling and tandem mill.

Laminaciones Sondica SA

Head office: Goronda Bekoa 9, Sondica, Vizcaya. **Tel:** 94 4531600 48. **Telex:** 31823.
Management: Managers – Gonzalo Basterra (general), Juan Ignacio Elorza (purchasing). **Established:** 1958. **No. of employees:** Works – 27. Office – 5. **Annual capacity:** Finished steel: 15,000 tonnes.
Products: Carbon steel – light tees, light channels, medium tees, medium channels.

Tamesa – Talleres Meleiro SA

Head office: El Barco de Valdeorras, Orense. **Postal address:** Apartado 18. **Tel:** 88 32 02 00, 88 32 02 25, 88 32 07 57. Light sections.

Torras, Herrería y Construcciones SA

Head office: c/ Bailén 71 bis 6°, Barcelona. **Postal address:** Apartado 148. **Tel:** 93 225 60 17. **Telex:** 51651 thyc e.
Products: Carbon steel – wire rod; reinforcing bars plain and deformed; round bars.

Laminaciones Torrente SA

Head office: c/Industria s/n, Puebla de Farnals, Valencia. **Tel:** 6 144 18 26.
Products: Carbon steel – reinforcing bars.

Transmesa – Transformaciones Metalurgicas SA

Head office: c/ Industria 10, 08330 Premia de Mar Barcelona. **Postal address:** Apartado Postal 7. **Tel:** 3 751 05 00. **Telex:** 51362 draw.
Management: Directors – Ricardo Santoma (gerente), Miguel Porta (gerente), José

Transmesa - Transformaciones Metalurgicas SA (continued)

Amela (comercial), Pedro Hernandez (producción), Javier Alonso (administrativo), Fernando Otero (proyectos e inversiones). **Established:** 1944. **Capital:** Ptas 450,000,000. **No. of employees:** Works – 125. Office – 50. **Ownership:** Tubacex – C.E. de Tubos por Extrusión SA (66%). **Subsidiaries:** Precitubo SA de CV (49%), Guadalajara, Mexico. **Annual capacity:** Finished steel: 6,000 tonnes (cold drawn tubes).
Products: Carbon steel – seamless tubes and pipes, longitudinal weld tubes and pipes, cold drawn tubes and pipes, precision tubes and pipes. Stainless steel – cold drawn tubes and pipes, precision tubes and pipes. Alloy steel (excl stainless) – cold drawn tubes and pipes, precision tubes and pipes. Cold drawn precision tube 3-100mm o.d. and 0.5-10mm wall thickness; diesel injection fuel lines; double-wall copper brazed steel tubing; rifled tubes for barrels of firearms; special heat exchangers and boilers.
Brands: Drawn; Conexion-35; PHP; Turbulent; TM/WCD; Resistent.
Sales offices/agents: Sr D. Angel Soler Frances, Sancho Dávila 22 1 C, 28028 Madrid.

Tubacex - C.E. de Tubos por Extrusion SA

Head office: B° Gardea s/n, Llodio, Alava. **Postal address:** PO Box 22. **Tel:** 4 6721408. **Telex:** 34325 tuco e. **Cables:** Tubacex. **Telecopier (Fax):** 4 4334507.
Management: Managers – J.I. Garin (general), O. Negro (vice-general), D. Amigo (vice-general), L.M. Villar (quality), A. Claver (commercial), A. Videgain (export).
Established: 1963. **Capital:** Ptas 3,585,098,000. **No. of employees:** Works – 239. Office – 691. **Subsidiaries:** T.T.A. (100)%) (manufacturer of fittings); Aceralava (100%) (melting shop); Tubacex Commercial (100%) (commercial activities).
Associates: Transmesa (manufacturer of special tubing (66%). **Annual capacity:** Raw steel: 200,000 tonnes. Finished steel: 130,000 tonnes.
Works:
■ At head office. Works director: J.M. Bilbao. ***Tube and pipe mills:*** seamless one extrusion press – (annual capacity 80,000 tonnes).
■ Aceralava (melting shop), Políg Ind De Saracho s/n, Amurrio (Alava). Works director: Mr Guerrero. ***Steelmaking plant:*** One electric furnace (annual capacity 200,000 tonnes).
■ Tubacex, Amurrio, Políg Ind de Saracho s/n. Works director: Mr Lasuen. ***Tube and pipe mills:*** seamless One Assel-mill, and cold drawing shop (annual capacity 50,000 tonnes).
Products: Carbon steel – ingots, billets, semis for seamless tubemaking, seamless tubes and pipes, hollow bars, precision tubes and pipes. Stainless steel – seamless tubes and pipes, cold drawn tubes and pipes, hollow bars. Alloy steel (excl stainless) – ingots, billets, seamless tubes and pipes, cold drawn tubes and pipes, hollow bars, precision tubes and pipes. Steamless steel tubes & pipes for high temperature service (in carbon, alloy & stainless steel), general pipes, boiler tubes, heat exchanger tubes.
Sales offices/agents: Tubacex SA (head office); Tubacex Commercial SA, Dtor Fleming 3-10, 28036 Madrid, Spian; Tubacex Europe BV, Burg de Bruinelaan 142, 3331 AJ Zwijndrecht, Netherlands; Tubacex Italia Srl, Paizza, IV Novembre 4, I 20124 Milan, Italy; Tubacex Inc, 1800 West Loop South, Suite 1790, Houston, TX 77027, USA; Tubacex Canada Inc, 1351 Matheson Blvd East Suite 24, Mississauga, Ont, Canada L4W 2A1; Tubacex Latinoamericana SA de CV, Bosque de Ciruelos 140-1001, Bosques de las Lomas, 11700 Mexico, D.F.

Tubos de Precisión SA (Tupre)

Head office: Calle Otalora 31, Arechavaleta, Guipúzcoa. **Postal address:** Apartado 8. **Tel:** 43 79 05 98-99, 79 07 98. **Telex:** 36473 ucem e.
Products: Carbon steel – longitudinal weld tubes and pipes.

Tubos Reunidos SA

Head office: Barrio Sagarribay, Amurrio, Alava. **Postal address:** Apartado 108, Bilbao. **Tel:** 94 441 9562. **Telex:** 32209 tr bi e; 34097 tr bi e.
Management: President of the board – Emilio de Ybarra; general manager – F. Javier Déniz; directors – F. Javier M. Múgica (plants), José A. Erice (sales), Fernando Noguera (financial), Jesús Villanueva (development); secretary of the board – Juan J. López de Maturana. **Established:** 1968. **Capital:** Ptas 1,000,000,000. **No. of employees:** Works – 1,220. Office – 198. **Subsidiaries:** Almacenes Metalurgicos SA (Almesa) (75%) (tube and pipe stockist); TR Aplicaciones Tubulares de Andalucia (Trandsa) (100%) (piping); Industria Auxiliar Alavesa SA (Inauxa) (62.5%) (car industry); Aceros y Calibrado SA (Acecsa) (100%) (precision cold drawn tubes); Tubos Reunidos America Inc, USA (100%) (sales); Tubos Reunidos France (100%) (sales); Proel SA de CV, Mexico (24%) (sales). **Annual capacity:** Raw steel: 330,000 tonnes. Finished steel: 200,000 tonnes (seamless tubes).
Works:
■ Amurrio Works, Barrio de Sagarribay, Amurrio, Alava. (Tel: 945 891511. Telex: 31932 trbra e). Steel plant director: Carlos Ortuondo. Tube plant director: Jesus Abascol. ***Steelmaking plant:*** One 90-tonne electric arc furnace (annual capacity 330,000 tonnes). ***Refining plant:*** One 90-tonne ladle furnace (330,000 tonnes). ***Continuous casting machines:*** One 4-strand Demag round (160mm and 200mm) (330,000 tonnes). ***Tube and pipe mills:*** seamless – one 7" max CPE (200,000 tonnes).
■ Galindo Works, Galindo, San Salvador del Valle, Vizcaya. (Tel: 94 4966811. Telex: 34215 tr gl). Works director: Jesús Aguirrebengoa. ***Tube and pipe mills:*** seamless – two 8" Pilger (annual capacity 40,000 tonnes). ***Other plant:*** Cold drawing facilities.
Products: Carbon steel – seamless tubes and pipes (line pipe) $\frac{1}{2}$-7" od (1986 output 100,400 tonnes); oil country tubular goods $\frac{3}{4}$-7" od (32,600 tonnes); hollow bars up to 7" od (3,400 tonnes); precision tubes and pipes (seamless cold drawn) $\frac{1}{2}$-3$\frac{1}{2}$" od (13,600 tonnes). Alloy steel (excl stainless) – seamless tubes and pipes $\frac{1}{2}$-7" od (1,500 tonnes); oil country tubular goods $\frac{3}{4}$-7" od (5,500 tonnes); precision tubes and pipes (seamless cold drawn) $\frac{3}{4}$-3$\frac{1}{2}$" od (1,400 tonnes). Ancillary parts for cars, heat exchangers, finned tubes, tank heaters, tubular coils, elbows, etc.
Sales offices/agents: Tubos Reunidos America Inc, 7700 San Felipe, Suite 540, Houston, TX 77063, USA (Tel: 1 713 975 70 08. Telex: 11 66 159 tra inc hou. Fax: 713 975 82 28); Tubos Reunidos France, 103 Rue de Pont à Mousson, F 57158 Montigny Les Metz, France (Tel: 33 87 62 46 15. Telex: 930301); Bin Malik Oilfield & Industrial Est, PO Box 3242, Sharjah (Tel: 971 6 59 10 33. Telex: 68707); DVA Peking Office, Xiyuan Hotel, Bldg 1, Room 105, Erligou, Xijiao, Peking, China (Tel: 89 81 52); Proel SA de CV, Paseo de la Reforma 76, Mexico 6, DF (Tel: 52 5 592 56 11. Telex: 1771918).

Marcial Ucin SA

Head office: Paseo de los Fueros, s/n, 20.730 Azpeitia, Guipuzcoa. **Postal address:** PO Box 54. **Tel:** 943 810100. **Telex:** 36259 ucin e. **Telecopier (Fax):** 943 815468.
No. of employees: Works – 295. Office – 22. **Annual capacity:** Raw steel: 610,000 tonnes/year. Finished steel: 800,000 tonnes/year.
Works:
■ At head office. ***Steelmaking plant:*** One 100-tonne electric arc furnace (annual capacity 727,000 tonnes). ***Refining plant:*** One 100-tonne Heats ladle furnace (727,000 tonnes). ***Continuous casting machines:*** One 6-strand Concast billet (120 x 120-140 x 140mm) (610,000 tonnes). ***Rolling mills:*** One 20-stand Danieli bar (coil weight 1,000kg) (375,000 tonnes).
Products: Carbon steel – wire rod 6-14mm (1986 output 75,000 tonnes); reinforcing bars 6-40mm (325,000 tonnes); flats 30 x 4-75 x 122mm (35,000 tonnes).

Spain

Cooperativa Industrial Talleres Ulma

Head office: Oñate, Guipúzcoa. **Postal address:** Apartado 13. **Tel:** 43 78 00 51.
Products: Carbon steel – longitudinal weld tubes and pipes.

Unión Cerrajera SA

Head office: Avda de Viteri 16, Mondragón, Guipúzcoa. **Postal address:** Apartado 1. **Tel:** 79 00 55. **Telex:** 38828 uceme.
Works:
■ At head office and Barrio de Altos Hornos, Apartado 5, Vergara, Guipúzcoa.
Products: Carbon steel – (engineering steels) billets for re-rolling and forging; reinforcing bars; round bars; flats; Light sections including light channels; heavy sections including heavy angles and heavy channels; hot rolled hoop and strip (uncoated); cold rolled hoop and strip (uncoated). Bright bars, tyres and wheels.

Aceros Valdemoro SA

Head office: Carretera de Andalucia Km 30, 28340 Valdemoro, Madrid. **Postal address:** Apartado de Correos 40. **Tel:** (34) (1) 895 01 09, 895 01 96. **Telex:** 48050.
Management: Prudencio Villamor (director). **Established:** 1981. **Capital:** Ptas 150,000,000. **No. of employees:** Works – 30. Office – 3. **Annual capacity:** Finished steel: 12,000 tonnes.
Works:
■ At head office. ***Rolling mills:*** Five 3-high and two 2-high bar (annual capacity 12,000 tonnes).
Products: Carbon steel – flats 12 x 4, 14 x 3, 14 x 4, 16 x 3, 16 x 4, 18 x 3, 18 x 4, 20 x 3, 20 x 4, 20 x 5, 20 x 6, 25 x 3, 25 x 4, 25 x 5, 25 x 6, 30 x 3, 30 x 4, 30 x 5, 30 x 6. (output 10,000 tonnes).

Metalúrgicas Vallisoletana SA (Meva)

Head office: Carretera de Circunvalación 1123, Valladolid. **Tel:** 83 291 067.
Products: Carbon steel – longitudinal weld tubes and pipes.

Alvarez Vázquez SA

Head office: Bilbao. **Postal address:** Apartado 290. **Tel:** 4 449 18 00. **Telex:** 32452 avsa e.
Products: Carbon steel – cold rolled hoop and strip (uncoated).

Aceros de Vizcaya SA (Acevisa)

Head office: Bilbao 48080. **Postal address:** Apartado 1267. **Tel:** 94 671 13 04, 94 671 13 51, 94 671 13 64, 94 671 14 01. **Telex:** 44409. Merchant bars, light sections, heavy sections.

Laminación Vizcaya SA

Head office: c/Iturrigorri 10, 48970 San Miguel de Basauri, Vizcaya. **Tel:** 4 449 02 00, 4 449 95 00. **Telex:** 33153 lauiz.
Products: Carbon steel – cold rolled hoop and strip (uncoated).

Industrias Zarra SA

Head office: Barrio Bekea 3, 48960 Galdacano, Vizcaya. **Tel:** 4560563. **Telex:** 31664 arza e.
Management: Juan Bta Eguia Mandaluniz (consejero delegado). **Established:**

Industrias Zarra SA (continued)

1957. **Capital:** Ptas 22,176,000. **No. of employees:** Works – 24. Office – 3. **Associates:** Laminados Velasco SA.
Works:
■ At head office.
Products: Carbon steel – heavy tees 80mm and over (output 6,000 tonnes), heavy channels 80mm and over (6,000 tonnes).

Sweden

AB Alvenius Industrier

Head office: S 63107 Eskilstuna. **Postal address:** PO Box 522. **Tel:** 016 11 01 60. **Telex:** 46076 alve s. **Cables:** Alvenius.
Products: Carbon steel – longitudinal weld tubes and pipes, spiral-weld tubes and pipes. Quick-coupling pipe system.

Ansab AB

Head office: Box 203, S 68800 Storfors. **Tel:** 0550 62000. **Telex:** 66190 seamles s. **Telecopier (Fax):** 0550 60867.
Established: 1985. **Capital:** Skr 1,000,000. **No. of employees:** Works – 45. Office – 10. **Ownership:** Avesta (50%); Sandvik (50%).
Works:

Ansab AB (continued)

■ At head office. ***Tube and pipe mills:*** seamless Four cold pilger – one 2½", two 3½", and one 4½" (operated at a rate of about 1,000 tonnes/year).
Products: Stainless steel – seamless tubes and pipes ¾-4½" (1986 output 1,200).

Asea Powdermet AB

Head office: S 73500 Surahammar. **Tel:** 0220 348 00.
Management: Björn Ericsson (managing director). **Established:** 1983.
Ownership: Asea AB (100%).
Works:
■ At head office. **Metal powder plant:** (annual capacity 3,000 tonnes). Hot Isostatic press (workpiece dimensions – dia 1,400mm, height 3,600mm).
Products: Alloy steel (excl stainless) – billets. Alloy and special steel powder and components.

Avesta AB

Head office: Birger Jarlsgatan 25, Stockholm. **Postal address:** Box 7782, S 10396 Stockholm. **Tel:** 46 8 7885050. **Telex:** 12378. **Telecopier (Fax):** 46 8 208481.
Management: Chairman – G. Engman; president – P. Molin; vice presidents – H.J. Waern (marketing), A. Wedmalm (personnel/information), B. Lindstrom (finance).
Established: 1984 (Avesta Jernverk 1883). **Capital:** Skr 3,600,000,000. **No. of employees:** 5,800 (group total). **Ownership:** Nordstjernan AB (Johnson Group).
Subsidiaries: ABE (100%) (fabricated products); Nords (100%) (fabricated products); Calamo (100%) (fabricated products); Nordform (100%) (fabricated products); Billing Metal Trading (100%); Avesta Nyby Powder (60%); Avesta Welding (100%); Avesta Sandvik Tube (AST) (75%); Avesta Uddeholm Stainless Bar (65%). **Associates:** Fagersta Stainless (wire rod, drawn wire). **Annual capacity:** Raw steel: 500,000 tonnes. Finished steel: 350,000 tonnes.
Works:
■ S 77401 Avesta. (Tel: 46 226 81800. Telex: 40976). Works director: G. Sjöberg. ***Steelmaking plant:*** Two 50-tonne electric arc furnaces. ***Refining plant:*** One 55-tonne AOD; one 9-tonne ESR. ***Continuous casting machines:*** One singe-strand Voest slab (1,600 x 200mm). ***Rolling mills:*** 4-high Schloemann heavy plate (max width 2.5m); 2-high cold reduction (max width 2m). ***Other plant:*** Forge; sheet polishing; wire drawing; welding products plant; plate prefabrication plant.
■ Box 102, S 69301 Degerfors. (Tel: 46 586 47000. Telex: 73215). Works director: S. Malm. ***Steelmaking plant:*** One 70-tonne electric arc furnace. ***Refining plant:*** One 70-tonne CLU. ***Continuous casting machines:*** 1-strand Voest bloom and slab. ***Rolling mills:*** 3-high MH-type billet; 2-stand 2-high Moeller & Neumann bar; 4-high MH-type heavy plate.
■ Nyby, S 64480 Torshälla. (Tel: 46 16 349000. Telex: 46013). Works director: L. Pajkull. ***Rolling mills:*** One Sendzimir cold reduction (max width 1.35m); one Sendzimir cold strip (max width 1.0m). ***Other plant:*** Brushing and polishing plant; metal powder production; pressed powder products.
■ Kloster, Box 100, S 77070 Långshyttan. (Tel: 46 225 61500. Telex: 74142). Works director: L. Lundstrom. ***Rolling mills:*** Two Sendzimir cold strip (max width 0.8m). ***Other plant:*** Bright annealing.
Products: Stainless steel – (including subsidiaries) slabs; billets; semis for seamless tubemaking; round bars; square bars; flats (cut from plate); light angles and light tees (welded or threaded fittings); light channels; special light sections; hot rolled hoop and strip (uncoated); skelp (tube strip); cold rolled hoop and strip (uncoated) up to 0.8m, 0.20-6.35mm thick; medium plates and heavy plates (5-250mm thick, up to 3.2 x 14m, max weight 12 tonnes); hot rolled sheet/coil (uncoated); cold rolled sheet/coil (uncoated) up to 2m wide, 0.20-6.35mm thick; bright annealed sheet/coil; wire; seamless tubes and pipes; longitudinal weld tubes and pipes; hollow bars.
Sales offices/agents: Avesta and A. Johnson & Co sales companies worldwide.

Sweden

Avesta Nyby Powder AB

Head office: PO Box 45, S 64400 Torshälla. **Tel:** 4616 349800. **Telex:** 46011 anps. **Telecopier (Fax):** 4616 357620.
Management: President – Christer Åslund; managers – Lars Brokvist (marketing), Claes Tornberg (production), Valdis Andersson (technical). **Established:** 1983. **Capital:** Skr 30,000,000. **No. of employees:** Works – 100. Office – 30. **Ownership:** Avesta AB, Sweden (60%), Vallourec SA, France (40%). **Annual capacity:** Finished steel: 8,000 tonnes.
Works:
■ At head office. ***Tube and pipe mills:*** seamless – one 1,750-tonne extrusion press (annual capacity 2,000 tonnes). ***Other plant:*** Metal powder plant with 5-tonne induction furnace.
Products: Stainless steel – semis for seamless tubemaking, round bars, seamless tubes and pipes, hollow bars. Alloy steel (excl stainless) – seamless tubes and pipes, hollow bars, hollow sections. Stainless and high alloy powder.

Avesta Sandvik Tube AB (AST)

Head office: S 77301 Fasersta. **Postal address:** PO Box 510. **Tel:** 46 223 45100. **Telex:** 7525. **Telecopier (Fax):** 46 223 10094.
Management: Dan Johansson (president), Björn Torssell (finance), Sven Wamer (marketing), Mats Jansson (logistics). **Established:** 1984. **Capital:** Skr 60,000,000. **No. of employees:** Works – 559. Office – 268. **Ownership:** Avesta AB, Sweden (75%). **Subsidiaries:** AST Stainless ApS, Denmark (100%) (sales); Avesta Sandvik Tube BV, Netherlands (100%) (production and sales). **Associates:** AST Rostfria VA-System AB, Sweden (50%). **Annual capacity:** Finished steel: 35,000 tonnes.
Works:
■ At head office. Works director: Åke Kölerud. ***Tube and pipe mills:*** welded – stainless (6-206mm dia) (annual capacity 14,000 tonnes).
■ Nyby. (Tel: 46 16 349500. Telex: 46010). Works director: Jonny Eriksson. ***Tube and pipe mills:*** welded – stainless (16-1,600mm dia) (annual capacity 13,000 tonnes).
■ Storfors. (Tel: 46 550 60900. Telex: 66074). Works director: Göran Persson. ***Tube and pipe mills:*** welded – stainless (114-610mm dia) (annual capacity 3,000 tonnes).
■ Avesta Sandvik Tube BV, Helmond, Netherlands. (Tel: 31 49 20 43455. Telex: 51622). Works director: Ben Custers. ***Tube and pipe mills:*** welded – stainless (15-88.9mm dia) (annual capacity 4,000 tonnes).
Products: Stainless steel – longitudinal weld tubes and pipes, oil country tubular goods, cold drawn tubes and pipes, precision tubes and pipes.

Avesta Uddeholm Stainless Bar AB

Head office: S 65105 Karlstad. **Postal address:** PO Box 324. **Tel:** 054 190300. **Ownership:** Avesta AB (65%).
Works:
■ Hagfors.
Products: Stainless steel – round bars, square bars, flats, light angles.

Björneborgs Jernverks AB

Head office: S 68071 Björneborg. **Tel:** 46 550 27100. **Telex:** 66087 bjoern s. **Telecopier (Fax):** 46 550 27420.
Management: President – Lars Almhed; managers – Thomas Karlsson (production), John Roslund (sales), Lars Schütt (finance). **Established:** 1656. **Capital:** Skr 6,000,000. **No. of employees:** 287. **Ownership:** Avesta AB (member of the Axel Johnson Group).
Works:
■ At head office. ***Steelmaking plant:*** Two electric arc furnaces – one 30-tonne and

Björneborgs Jernverks AB (continued)

50-tonne. ***Other plant:*** Forge with one oil hydraulic press of max 4500-tonne; heat treatment; machine shop.
Products: Carbon steel – blooms, round bars, square bars, flats. Alloy steel (excl stainless) – blooms, round bars, square bars, flats, heavy plates. Heavy open-die forgings (black, rough or finish machined).

Dobel AB

Head office: S 78184 Borlänge. **Tel:** 46 243 82000. **Telex:** 74053 dobeld s. **Telecopier (Fax):** 46 243 12274.
Management: Board chairman – Owe Brandes; directors – Lenmart Mix (managing), Carl-Henrik Ahlström (technical), John-Arne Neldemo (marketing). **Established:** 1968 (1981 limited company). **Capital:** Skr 60,100,000. **No. of employees:** Works – 240. Office – 210. **Ownership:** Svenskt Stål AB. **Subsidiaries:** Dobel Coated Steel Ltd, Lye, Birmingham, UK (100%) (marketing and distribution); Dobel Coated Steel AS, Copehagen, Denmark (100%) (marketing); Dobel Fönsterprofiler AB, Borlänge, Sweden (100%) (production and marketing of window frames for restoring windows); Ortic AB, Borlänge, Sweden (100%) (consultants, roll-forming technology); Verform AS, Gudå, Norway (100%) (profiling building sheet). **Annual capacity:** Finished steel: 80,000 tonnes.
Works:
■ At head office. ***Coil coating lines:*** One colour coating (80,000 tonnes); sheet laminating (15,000 tonnes).
Products: Carbon steel – galvanized corrugated sheet/coil, colour coated sheet/coil.

Structo DOM Europe AB

Head office: 68800 Storfors. **Postal address:** Box 204. **Tel:** 0550 62010. **Telex:** 66169 struc. **Telecopier (Fax):** 0550 61681.
Ownership: Avesta AB, Copperweld Corp, Rautaruukki Oy.
Products: Carbon steel – seamless tubes and pipes (hot finished). Cold drawn tubes and pipes.

Fagersta Stainless AB

Head office: Box 508, S 77301 Fagersta. **Tel:** 46 223 45500. **Telex:** 7525. **Cables:** Fagsta. **Telecopier (Fax):** 46 223 10094.
Management: Directors – Bjorn Fredriksson (managing), Bo G. Persson (marketing and sales), Per-Olov Andersson (finance); managers – Bertil Nordin (wire rod division), Åke Haglund (strip division), Stig Tätnblom (drawn wire division), Claes Plamborg (personnel), Barry Solly (research and development). **Established:** 1984 (original company 1873). **Capital:** Skr 80,000,000. **No. of employees:** Works – 660. Office – 240. **Ownership:** Sandvik Steel AB (50%). Avesta AB (50%). **Annual capacity:** Raw steel: 4,000 tonnes. Finished steel: 100,000 tonnes.
Works:
■ At head office. ***Steelmaking plant:*** Three 9-tonne high freguency-furnaces (annual capacity 4,000 tonne, (one shift). ***Rolling mills:*** blooming – two 2-high reversing stands (60,000 tonnes); wire rod – 1-strand, two 3-high and twenty 2-high stands (coil weight 330kg) (50,000 tonnes); 4-high Steckel hot strip and hoop (max product width 840mm, thickness 2.5-6.5mm) (coil weight 7.8kg/mm) (100,000 tonnes, three shifts). ***Other plant:*** Wire drawing plant for stainless.
Products: Stainless steel – ingots (1986 output 4,000 tonnes); billets (40,000 tonnes); hot rolled band (for re-rolling) up to 84.0mm wide, 2.5-6.5mm thick (50,000 tonnes); wire rod 5.6-13.5mm dia (50,000 tonnes); wire (6,000 tonnes).

Forsbacka Jernverks AB

Head office: Bruleskonforet, S 81041 Forsbacka. **Postal address:** Box 100. **Tel:** 026 35180. **Telex:** 47136 farsba.
Ownership: Industri Förvaltnings AB, Kinnevik. **Associates:** Halmstads, Jernverks AB. **Annual capacity:** Finished steel: 130,000 tonnes.
Works:
■ At head office. ***Rolling mills:*** One bar. Carbon and low-alloy bars, grinding balls – sales in 1985 were 80,000 tonnes.

Gavle Verken AB

Head office: Lötängsg 10, S 80133 Gävle. **Postal address:** Box 967. **Tel:** 026 17 2500. **Telex:** 47159. **Telecopier (Fax):** 026 172599.
Ownership: Skåne-Gripen AB.
Works:
■ At head office. ***Coil coating lines:*** One colour coating (annual capacity 50-70,000 tonnes). ***Other plant:*** Profiling line, sheet metal fabricating plant.
Products: Carbon steel – cold roll-formed sections (from strip), colour coated sheet/coil. Profiled sheets for roofing, decking, wall panels and other construction uses.

Halmstads Järnverks AB

Head office: Stationsgatan 55, S 30103 Halmstad. **Postal address:** Box 119. **Tel:** 035 10 95 00. **Telex:** 38081 welbond. **Telecopier (Fax):** 035 1295 13.
Ownership: Industriförvaltniugs AB, Kinnevik. **Associates:** Forsbacka Jernverks AB. **Annual capacity:** Finished steel: 760,000 tonnes.
Works:
■ At head office. ***Refining plant:*** CLU. ***Continuous casting machines:*** billet. ***Rolling mills:*** bar and wire rod. ***Other plant:*** Reinforcement fabrication plant.
Products: Carbon steel – reinforcing bars. Fabricated reinforcements – sales in 1985 were 255,000 tonnes.

Hedpipe AB (formerly Gränges Hedlund)

Head office: Jordbro, S 13650 Handen. **Tel:** 990 46 75021120. **Telex:** 11432 hedpipe s. **Telecopier (Fax):** 990 46 27559.
Ownership: Rautaruukki Oy. Welded pipe.

Höganäs AB

Head office: S 26301 Höganäs. **Postal address:** Box 501. **Tel:** 042 38000. **Telex:** 72651, 72368.
Ownership: Kanthal Höganäs AB. **Associates:** Kanthal AB. **Annual capacity:** Iron: 500,000 tonnes.
Products: Direct-reduced iron (DRI), iron powder.

Kanthal AB

Head office: S 73401 Hallstahammar. **Telex:** 40401, 40741, 40690.
Ownership: Kanthal Höganäs AB. **Associates:** Höganäs AB. **Annual capacity:** Finished steel: 600,000 tonnes. Electrical resistance strip and wire.

Kloster Speedsteel AB

Head office: S 81060 Söderfors. **Tel:** 0293 17000. **Telex:** 76198 speeds s. **Cables:** Speedsteelab Söderfors.
Management: Hans Harvig (president), Kjell Hammerin (bar division), Jan Sondell (metallurgy), Göran Nybratt (group marketing), Tord Sjöström (controller).

Sweden

Kloster Speedsteel AB (continued)

Established: 1982 (by merger of Uddeholm's Söderfors works and the high speed steel works of Fagersta). **Capital:** Skr 100,000,000. **No. of employees:** Works – 740. Office – 300. **Ownership:** Fagersta AB (49%), Uddeholms AB (45%). **Subsidiaries:** Speedsteel of New Jersey Inc, Fairfield, USA; Aciers de Champagnole, Champagnole, France; Aciers de Champagnole, Le Blanc Mesnil, France; Speedsteel (UK), Alcester, UK; Skinner Steels Ltd, Sheffield; KSS Speedsteel GmbH, Düsseldorf, W. Germany; Stà Italiana Acciaierie Champagnole (SIAC), Nichelino-Turin, Italy. **Annual capacity:** Raw steel: 44,000 tonnes.

Works:

■ At head office. ***Steelmaking plant:*** One 30-tonne electric arc furnace. ***Refining plant:*** One 30-tonne Asea-SKF. ***Rolling mills:*** One Morgårdshammer bar (6-90mm rounds and corresponding sizes of square and flats); 2-high 700mm reversing cogging mill; ***Other plant:*** Powder metallurgy plant for production of ASP high-speed steel; plant for manufacture of work rolls.

■ Långshyttan/Vikmannshyttan. (Tel: 0225 60380. Telex: 74518). ***Rolling mills:*** One wire rod (5.5-22mm dia); cold strip (12-60mm wide). ***Other plant:*** Drawing plant – bars and flattened wire.

Products: Alloy steel (excl stainless) – (High-speed steels) round bars 1-300mm and corresponding sizes of square bars and flats.

Brands: WKE; K; ASP.

Ovako Steel AB

Head office: Svärdvägen 3, S 18212 Danderyd. **Postal address:** PO Box 133. **Tel:** 08753 3220. **Telex:** 15341. **Telecopier (Fax):** 0875 31565.

Management: Matti Sundberg (chief executive), Ingeman Eliassan (finance), Gösta Bergman (legal), Olof Karling (marketing), Sven Nilsson (personnel and information), Erkki Ström (strategic planning), Jan Åkessan (technology). **Established:** 1986 (merging SKF Steel with Ovako Oy). **No. of employees:** 7,149 (Dec 31 1986). **Ownership:** AB SKF (50%), Oy Wärtsilä Ab (25%), Fiskars Oy Ab (20%), Union Bank of Finland (5%). **Subsidiaries:** Ovako Steel Oy Ab (Finland).

Works:

■ Ovako Steel Hofors AB, Box 202, 81300 Hofors. (Tel: 0290 250 00. Telex: 47300). Works director: Christen Olson. ***Steelmaking plant:*** One electric arc furnace (annual capacity 400,000 tonnes, planned). ***Refining plant:*** One Asea-SKF. ***Rolling mills:*** One billet, one wire rod. ***Tube and pipe mills:*** seamless – One Assel, cold rolling. ***Other plant:*** Forging press.

■ Ovako Steel Hellefors AB, Box 77, 71200 Hällefors. (Tel: 0591 60000. Telex: 5648). Works director: Johnny Tidström. ***Steelmaking plant:*** One electric arc (scheduled for closure). ***Rolling mills:*** One billet, one bar. ***Other plant:*** Bright drawing mill.

■ Ovaka Steel Ring AB, Box 202, 81300 Hofors. (Tel: 0290 25000. Telex: 47300). **Ring mills.**

■ Ovako Steel Mora AB, Örjasvägen 21, 79200 Mora. (Tel: 0250 15950. Telex: 74242). Works director: Åke Paulsson. Plating plant for hard-chromed bars, especially for hydraulic cylinders (annual capacity 7,000 tonnes).

■ Ovaka Steel Kilsta AB, Box 411, 69127 Karlsekoga. (Tel: 0568 36075. Telex: 73494). Work director: Johnny Tidström. Heat treatment and finishing plant for bars (annual capacity 30,000 tonnes).

Products: Carbon steel – billets, wire rod, round bars, bright (cold finished) bars, seamless tubes and pipes. Alloy steel (excl stainless) – billets, wire rod, round bars, bright (cold finished) bars, seamless tubes and pipes. Rolled wire for valve springs, rolled and forged rings, special steel including bearing steel, case-hardening steel and structural steel. Hard-chromed bars.

Sales offices/agents: At head office; von Utfallsgatan 9, Box 13094, 402 52 Gothenberg, Sweden; Stålkulegan 7, Box 1004, 21228 Malmö, Sweden; Björnnäsvägen 2, 11347 Stockholm, Sweden; Förrådsgatan 5, Box 836, 85123 Sundsvall, Sweden; Ovako Stål A/S, Horskaetten 5, DK 263 0 Tåstrup, Denmark;

Ovako Steel AB (continued)

Ovako Teräsmarkkinointi Oy, SF 02430 Masaby, Finland; SKF Stål A/S, Vestbyveien 50, Kaldbakken, PO Box 27, N 0901 Oslo 9, Norway; Ovako Steel BeNeLux BV, De Smalle Zdidje 3, Postbus 81, NL 3900 Veenendaal, Netherlands; SKF Steel Ltd, 60 North Wind Place, Agincourt, Ont, Canada M1S 3R5; Ovako Steel Ltd, Neachells Lane, Wednesfield, Wolverhampton, West Midlands WV11 3QF, UK; Ovako Acier SA, 1 Avenue Jean-Lolive, F 93177 Bagnolet, France; Ovako Stahl AG Industriestrasse 27, Volketswil, Postfach, CH 8603 Schwerzenbach, Switzerland; SKF Acciai SpA, Via Donizetti 18, I 20020 Lainate (Milan), Italy; Ovako Stahl GmbH, Am Tonisberg 6, Postfach, D 4006 Erkrath 1, West Germany; Ovako Steel Inc, 22 Waterville Road, Avon, CT 06001, USA; SKF Österreich GmbH, Industriezentrum NöSüd, Strasse 1, Objekt 50, Postfach 87, A 2351 Wr Neudorf, Austria; SKF Steel Asia Pacific Area Office, Holian House 3/F, 4-6 On Lan Street, Hong Kong.

Plannja AB

Head office: Fack, S 95188 Luleå. **Tel:** 0920 65440. **Telex:** 80402.
Management: Harold Bernhardsson (president). **Ownership:** SSAB.
Products: Carbon steel – cold roll-formed sections (from strip); cold rolled sheet/coil (uncoated) including boron-alloy hardenable sheet; colour coated sheet/coil. Painted building profiles and sheet for roofing and cladding. Engineering products.

AB Sandvik Steel

Head office: S 81181 Sandviken. **Tel:** 26 26 30 00. **Telex:** 47000 sandvik s. **Cables:** Sandvikenswed.
Management: Gunnar Björklund (president), Rolf Gullberg (vice president, tube), Björn Sköld (vice president, strip and wire), Gunnar Folke (vice president, semi-finished products). **Established:** 1862. **Capital:** About Skr 1,110,000,000. **No. of employees:** 5,700. **Ownership:** Sandvik AB.
Subsidiaries: Sandvik Stål Försäljnings AB, Stockholm; Sandvik Steel UK, Halesowen; Sandvik Steel Germany; Düsseldorf, Sandvik Aciers, Orleans; Sandvik Steel Co, Scranton, USA; Sandvik Special Metals Corp, Kennwick, USA and several other Sandvik group subsidiaries as well as Guldsmedshytte Bruks AB; Edmeston AB, Sweden; Osprey Ltd, UK and Nor-Sand Metals Inc, Canada. **Associates:** Fagersta Stainless AB (50%), Avesta Sandvik Tube AB (25%), ANSAB AB (50%) and Uddeholm Strip Steel AB (50%). **Annual capacity:** Finished steel: 15,000 tonnes approx.
Works:
■ At head office. ***Steelmaking plant:*** One 75-tonne electric arc furnace. One 10-tonne and one 7-tonne high frequency furnaces. ***Refining plant:*** One 75-tonne AOD; two VAR. ***Continuous casting machines:*** One 3-strand Concast bloom (265mm sq-265 x 365mm) (140,000 tonnes). ***Rolling mills:*** One 400mm hot strip and hoop. ***Tube and pipe mills:*** seamless – Three extrusion presses (1,250, 1,700 and 3,000-tonnes); cold pilger mills; drawbenches. ***Other plant:*** Heat treatment furnace, wire drawing plant.
Products: Carbon steel – cold rolled hoop and strip (uncoated) (hardened and tempered); bright wire. Stainless steel – blooms, billets, round bars, hot rolled hoop and strip (uncoated), cold rolled hoop and strip (uncoated), wire, seamless tubes and pipes. Alloy steel (excl stainless) – blooms, billets, hot rolled hoop and strip (uncoated), cold rolled hoop and strip (uncoated), wire. Hollow drill steel, flat rolled wire welding materials, zirconium and titanium tubes. Other Sandvik group companies produce cemented carbide products, tools, conveyer systems and electronic instruments.

Smedjebacken-Boxholm Stål AB

Head office: S 77701 Smedjebacken. **Tel:** 240 71120. **Telex:** 74514. **Telecopier (Fax):** 240 74362.
Management: Leif Gustafsson (managing director), Rune Johansson (metallurgical division), Claes Häger (steel bar division), Hans Andersson (building products division),

Smedjebacken-Boxholm Stål AB (continued)

Bengt Stenberg (purchasing manager). **Established:** 1856. **Capital:** SwKr 100,000,000. **No. of employees:** Works – 1,050. Office – 325. **Ownership:** Förvaltningsaktiebolaget Ratos, AB Iggesunds Bruk. **Annual capacity:** Raw steel: 450,000 tonnes. Finished steel: 400,000 tonnes.
Works:
■ At head office. ***Steelmaking plant:*** One 120 tonne electric arc furnace (annual capacity 400,000 tonnes). ***Continuous casting machines:*** One 6-strand billet (115, 150, 260 x 150mm) (400,000 tonnes). ***Rolling mills:*** medium section (125,000 tonnes), light section (125,000 tonnes).
■ Boxholm. ***Rolling mills:*** medium section (110,000 tonnes). Light section (140,000 tonnes).
Products: Carbon steel – reinforcing bars, round bars, square bars, flats, bright (cold finished) bars, light angles, light channels, special light sections, medium angles, medium channels, universals (medium). Alloy steel (excl stainless) – round bars, square bars, flats, bright (cold finished) bars (incl. spring steel), light angles, light channels, universals. Welded wire mesh; welded reinforcement products; SWL steel joists.
Sales offices/agents: Smedjebacken-Boxholm Stahl GmbH, Hohenstaufenstrasse 4, D 4000 Düsseldorf 11, West Germany; Smedjebacken-Boxholm Nederland BV, Postbus 124, 2501 CC, Het Kleine Loo 378 2592 CK Den Haag, Netherlands; Smedjebacken-Boxholm (UK) Ltd, Rotherham Road, Parkgate, Rotherham S62 6EZ, UK.

SSAB – Svenskt Stål AB

Head office: Birger Jarlsgatan 58, S 10326 Stockholm. **Postal address:** PO Box 16344. **Tel:** 08 242310. **Telex:** 12670. **Telecopier (Fax):** 0810 7974.
Management: Board chairman – Björn Wahlström; president – Leif Guntafsson; vice-presidents – Orvar Nyquist, Åke Sander, Torsten Sandia, Inge Selinder, Anden Ullberg. **Established:** 1977. **No. of employees:** 15,354 (1986). **Ownership:** Majority state-owned. **Subsidiaries:** These include Surahammars Bruks AB, Dobel AB, Plannja AB, Primdata AB, Tibner AB, Dickson Stål Metall AB, Plåtluna AB. **Annual capacity:** Raw steel: 3,100,000.
Works:
■ SSAB Luleå, S 95188 Luleå. (Tel: 0920 92000. Telex: 68321 ssablu). **Coke ovens**. ***Blast furnaces:*** Two. ***Steelmaking plant:*** Two 102-tonne LD basic oxygen. ***Continuous casting machines:*** One billet; two bloom; one slab. ***Rolling mills:*** Two heavy section; two medium section. ***Other plant:*** Fabrication plant for welded sections.
■ SSAB Domnarvet, S 78184 Borlänge. (Tel: 0243 700 00. Telex: 74050 ssabdos). ***Steelmaking plant:*** Two electric arc furnaces (to be closed). ***Continuous casting machines:*** Two slab. ***Rolling mills:*** One rail; one bar and wire rod; one wide hot strip; one cold reduction. ***Coil coating lines:*** One hot-dip galvanizing; one zinc-aluminium; one plastic coating line (Dobel AB division). ***Other plant:*** Fabricated reinforcement plant.
■ SSAB Oxelösund, S 61301 Oxelösund. (Tel: 0155 5400. Telex: 64160). **Coke ovens and Sinter plant**. ***Blast furnaces:*** Two. ***Steelmaking plant:*** One 180-tonne LD basic oxygen. ***Continuous casting machines:*** Two slab. ***Rolling mills:*** One 4-high heavy plate (3,600mm wide).
Products: Carbon steel – ingots, slabs, blooms, billets, wire rod, reinforcing bars, round bars, medium angles, medium tees, medium joists, medium channels, wide-flange beams (medium), universals (medium), welded sections (medium), heavy angles, heavy tees, heavy joists, heavy channels, wide-flange beams (heavy), universals (heavy), welded sections (heavy), heavy rails, light rails, cold roll-formed sections (from strip), medium plates, heavy plates, universal plates, hot rolled sheet/coil (uncoated), cold rolled sheet/coil (uncoated), hot-dip galvanized sheet/coil, zinc-aluminium alloy coated sheet/coil.
Sales offices/agents: SSAB Swedish Steel SA/NV, 6 Rue du Charnoy, Postbox 9, B 1348 Louvain La Neuve, Belgium; SSAB Svensk Stål A/S, Mitchellsgade 9, DK 1568 Copenhagen V, Denmark; Dobel Coated Steel A/S, Mitchellsgade 9, DK 1568

SSAB - Svenskt Stål AB (continued)

Copenhagen V, Denmark; Oy Svenskt Stål AB, Glogatan 8, SF 00100 Helsinki 10, Finland; Swedish Steel SA, 114 Avenue Charles de Gaulle, F 92522 Neuilly Sur Seine Cedex, France; SSAB Swedish Steel BV, Postbus 589, NL 6500 An Nijmegen, Netherlands; Svensk Stål A/S, Postboks 340, N 1301 Sandvika, Norway; Plåtluna A/S, Postboks 340, N 1301 Sandvika, Norway; Plannja A/S, Fridtjof Nansens Plass 6, N 0160 Oslo 1, Norway; Swedish Steel Ltd, 11th Floor Berkshire House, Queen Street, Maidenhead SL6 1NF, UK; Dobel Coated Steel Ltd, PO Box 16, Stourbridge, West Midlands, UK; Plannja Ltd, 69 High Street, Maidenhead, Berkshire SL6 1JX, UK; SSAB Swedish Steel, Mörsenbroicher Weg 200, D 4000 Düsseldorf 30, West Germany.

Surahammars Bruks AB

Head office: Box 201, S 735 00 Surahammar. **Tel:** 0220 34500. **Telex:** 40644 surav s. **Telecopier (Fax):** 0220 30372.
Management: Aake Sander (board chairman), Jorbjoern Henningson (managing director). **Established:** 16th century. **Capital:** Skr 60,000,000. **No. of employees:** 520. **Ownership:** SSAB – Svenskt Stål AB (100%). **Subsidiaries:** NorMag Inc, USA (100%) (steel service centre and transformer core manufacturing). Cor-Magnetics Inc, Canada (100%) (steel service centre and transformer core manufacturing). **Annual capacity:** Finished steel: 200,000 tonnes.
Works:
■ At head office. ***Rolling mills:*** One Steckel wide hot strip 60", (1,524mm) (coil weight 10 tonnes) (annual capacity 200,000 tonnes); one 4-high reversing cold reduction 56", (1,422mm) (200,000 tonnes). ***Other plant:*** Electrical steel annealing and coating lines.
Products: Carbon steel – hot rolled sheet/coil (uncoated) thickness 2-8mm, width 800-1300mm; electrical grain oriented silicon sheet/coil max 920mm width, electrical non-oriented silicon sheet/coil max 1250 width, electrical non-oriented non-silicon sheet/coil max 1250mm width, Stainless steel – hot rolled band (for re-rolling) (hire rolling only).
Sales offices/agents: USA – NorMag Inc, Bridgeport, CT (Tel 203 335 9921, Telex 703969); Canada – Cor-Magnetics Inc, Burlington, Ont (Tel 416 335 3490, Telex 361316); Finland – Valtameri Oy, Helsinki, (Tel 90 7420, Telex 124449); UK – Ernst B. Westman Ltd, London (Tel 01 692 1933, Telex 888228); Switzerland, Austria, Yugoslavia and Turkey – Estamag AG, Zurich, Switzerland (Tel 553433, Telex 55601).

Uddeholm Strip Steel AB

Head office: S 68301 Munkfors. **Postal address:** Box 503. **Tel:** 0563 16000.
Ownership: Uddeholm Tooling AB (50%), AB Sandvik Steel (50%). **Annual capacity:** Finished steel: 700,000 tonnes.
Products: Carbon steel – cold rolled hoop and strip (uncoated), precision strip. Stainless steel – cold rolled hoop and strip (uncoated).

Uddeholm Tooling AB

Head office: Uddeholm, S 68305 Hagfors. **Tel:** 46 563 17000. **Telex:** 5644. **Cables:** Uddeholms, Uddeholm. **Telecopier (Fax):** 23535.
Management: Bo Jonsson (president), Bo Mellin (vice president). **Established:** 1981 (1668). **Capital:** Skr 90,000,000. **No. of employees:** 2,915. **Ownership:** Uddeholms AB (Ultimate holding company – AGA, Sweden). **Subsidiaries:** AB Alfix (100%); Brukens Härdverkstäder AB (91.7%); AB Brukens Härdtjanster (91.7%); Ramnäs Värmebehandling AB (91.7%); Risto Ingenjörsfirma AB (91.7%); Brukshotellet i Uddeholm AB (100%); Hagfors Malsegment AB (91%); Hagfors Mekaniska Verkstads AB (100%); Hagfors Tooling Center AB (95.2%); Nordmarks-Klarälvens Järnvags AB (100%); Oy Uddeholm Ab, Finland (100%); Sintool AB (100%); Uddamelt AB (100%); Uddeholm Castings AB (100%); Uddeholm Vägverktyg AB (100%); AB

Uddeholm Tooling AB (continued)

Uddeholmsagenturen (100%); Uddeholms Forsknings AB (100%). Sale companies – Udde holjm Tooling Svenska AB (100%); Uddeholm A/S, Norway (100%); Uddeholm A/S, Denmark (100%); Uddeholm Tooling Europe GmbH, West Germany (100%); Uddeholm Sustan GmbH, West Germany (100%); Sunorm Werkzeugnormalien GmbH, West Germany (100%); Uddeholm Tooling BV, Netherlands (100%); Uddeholm Tooling NV, Belgium (100%); Uddeholm Ltd, UK (100%); Uddeholm Pension Trustees Ltd, UK (100%); Uddeholm SA, France (100%); Uddeholm SA, Spain (100%); Ud deholm SpA, Italy (100%); Uddeholm Tooling Ges mbH, Austria (100%); Uddeholm KK, Japan (100%); Uddeholm Ltd, Canada (100%); Uddeholm Corp, USA (100%). **Associates:** Avesta Uddeholm Stainless Bar AB (35.0%); Bruksinvest AB (25.0%); Ekshärads Sömnadsindustri AB (50.0%); Ellwood Uddeholm Steel Co, USA (20.0%); Föreningen Hagforshälsan (41.1%); Hagfors Flyg AB (50.0%); AB Klarinvest (44.7%); Kloster Speedsteel AB (45.0%); Kohlswa Uddeholm Gjuteri AB (50.0%); Lundgrens Formverktyg AB (41.0%); Storfors Invest AB (21.7%); Söderfors Polymerteknik AB (30.0%); Trienta AB (47.6%); Uddeholm Alpha Tooling Inc, USA (50.0%); Uddeholm Strip Steel AB (50.0%); Uddenor SA, Spain (50.0%); Värmlands Vakt AB (40.0%); AB Järnbruksförnödenheter (12.0%); TIAB Transportköparnas Intresse AB (7.4%). **Annual capacity:** Raw steel: 160,000 tonnes. Finished steel: 100,000 tonnes.
Works:
■ At head office. ***Steelmaking plant:*** One 65-tonne electric arc furnace (155,000 tonnes); one 6-tonne induction furnace (5,000 tonnes). ***Refining plant:*** One 65-tonne Asea-SKF (155,000 tonnes); four 300-1,000mm ESR (140,000 tonnes) (two located in Kistaverken, Karlskoga). ***Rolling mills:*** bar – 6-stand 3-high and 6-stand 2-high continuous train (550 x 1,500/1,800mm) (50,000 tonnes);3-stand 2-high cogging mill (850/800 x 2,150/2,000mm) (100,000 tonnes). ***Other plant:*** One 3,000-tonne forging press.
Products: Alloy steel (excl stainless) – round bars, square bars, flats. Hot and cold work tool steel, steel for plastic moulding.

Wirsbo Bruks AB

Head office: S 73061 Virsbo, Västmanland. **Tel:** 0223 34200. **Telex:** 7517 s. **Telecopier (Fax):** 0223 34012.
Management: Board chairman – Ossi Virolainen; president – Åke Forssell; vice president – Karl-Gunnar Boman (marketing); managers – Sune Bodén (sales, steel tubes), Birger Kronberg (production, steel tubes). **Established:** 1620. **Capital:** Skr 20,000,000. **No. of employees:** Works – 1,000. Office – 400. **Ownership:** Outokumpu Oy, Finland (100%). **Subsidiaries:** Wholly-owned – Wirsbo Aquawarm AB, Sweden; Wirsbo GmbH, Wirsbo Pex and Velta GmbH, West Germany; Wirsbo UK Ltd, UK; Wirsbo SA, Spain; Wirsbo AG, Switzerland, Wirsbo Co, USA. **Annual capacity:** Finished steel: 60,000 tonnes.

Sweden

Wirsbo Bruks AB (continued)

Works:
■ At head office. ***Tube and pipe mills:*** Four high frequency welded – One stretch reducer ⅜-4″ nb (annual capacity 40,000 tonnes); one ERW 20-76mm od (15,000 tonnes); one ERW 12-40mm od (5,000 tonnes) and one ERW 8-20mm od (2,000 tonnes). ***Other plant:*** Forging and machine plant.
Products: Carbon steel – longitudinal weld tubes and pipes 8-114.3mm od (output 55,000 tonnes); oil country tubular goods 2-4″ od (5,000 tonnes); plastic coated tubes and pipes 10-28mm od (3,000 tonnes). Steel forgings, copper tubes, cross-linked polyethylene tubes, heating and sanitary systems.
Sales offices/agents: See under subsidiaries.

Switzerland

Arfa Röhrenwerke AG

Head office: Postfach, 4002 Basle. **Tel:** 061 505555. **Telex:** 962239. **Telecopier (Fax):** 061 502263.
Established: 1917. **Subsidiaries:** Microtube SA, Lausanne (manufacture of small-diameter stainless tubes) (see below).
Works:
■ At head office. ***Tube and pipe mills:*** Two high frequency longitudinal welded.
■ Arfa Röhrenwerke AG, Moehlin AG (Tel: 061 884311. Telex: 964998. Fax: 061 881823). ***Tube and pipe mills:*** Eight TIG longitudinal welded.
■ Microtube SA, Lausanne (Tel: 021 372222. Telex: 25322. Fax: 021 365370). ***Tube and pipe mills:*** Four longitudinal welded.
Products: Carbon steel – longitudinal weld tubes and pipes 38.8-168.3mm od, 1.6-7.1mm wall thickness. Stainless steel – longitudinal weld tubes and pipes 7.5-76.1mm od, 0.40-3.75mm wall thickness, and 6.0-19.0mm od, 0.3-1.5mm wall thickness (Microtube SA); cold drawn tubes and pipes (or bright annealed).

Ferrowohlen AG

Head office: Industriestrasse 21, CH 5610 Wohlen. **Tel:** 057 22 25 77. **Telex:** 59420 ferwo ch.
Management: M. Stratighiou (board member, director). **Established:** 1955. **Capital:** SwFr 3,000,000. **No. of employees:** 200. **Annual capacity:** Raw Steel: 150,000 tonnes. Rolled Steel: 150,000 tonnes.
Works:
■ At head office. Works director: M. Stratighiou. ***Steelmaking plant:*** One electric arc furnace. ***Continuous casting machines:*** One billet. ***Rolling mills:*** One bar and wire rod.
Products: Carbon steel – wire rod, reinforcing bars.

Forming AG

Head office: Steinligasse Neubau, CH 4313 Möhlin. **Tel:** 061 882244. **Telex:** 62572.
Products: Carbon steel – cold roll-formed sections (from strip), sheet piling. Stamped metal parts, slit strip.

Hermann Forster AG, Steel Tube Division

Head office: Romanshornerstrasse 4, CH 9320 Arbon. **Tel:** 071 469191. **Telex:** 77244.
Management: Jakob Züllig (board president). **Established:** 1874.
Products: Carbon steel – longitudinal weld tubes and pipes, cold drawn tubes and pipes, profiled tubes.

Jansen AG Stahlröhrenwerke

Head office: CH 9463 Oberreit SG. **Tel:** 071 780111. **Telex:** 77159 jano.
Management: J. Biedermann (director).
Works:
■ At head office.
Products: Carbon steel – longitudinal weld tubes and pipes 5-108mm; cold drawn tubes and pipes 5-160mm; precision tubes and pipes; plastic tube.

Kaltband AG

Head office: Unterwerkstrasse, CH 5734 Reinach. **Tel:** 064 71 44 77. **Telex:** 981538.
Established: 1965. **Ownership:** Private.
Works:
■ At head office. ***Rolling mills:*** cold strip. ***Other plant:*** Controlled atmosphere annealing furnaces. Flat wire mills.
Products: Carbon steel – cold rolled hoop and strip (uncoated) – mild steel (heat treatable), high carbon steel (5-330mm wide, 0.25-6mm thick); black annealed wire – flat 5-13mm wide, 0.8-3mm thick. Alloy steel (excl stainless) – cold rolled hoop and strip (uncoated) 5-330mm wide, 0.25-6mm thick; wire – flat 5-13mm wide, 0.8-3mm thick.

LN Industries SA

Head office: 2 Rue des Falaises, CH 1205 Geneva. **Tel:** 22 21 5133. **Telex:** 422831.
Products: Carbon steel – cold roll-formed sections (from strip). Stainless steel – precision tubes and pipes 0.3-40mm od. Cold drawn special sections.

Merz & Cie Drahtwerke AG

Head office: Haupstrasse 150, CH 5733 Leimbach AG. **Tel:** 064 712466. **Telex:** 981560. **Telecopier (Fax):** 064 718488.
Products: Carbon steel – cold rolled sheet/coil (uncoated); hot-dip galvanized sheet/coil and tinned steels; colour coated sheet/coil; wire products. Stainless steel – cold rolled sheet/coil (uncoated).

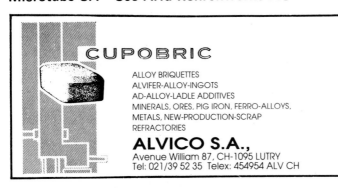

Microtube SA – See Arfa Röhrenwerke AG

Montanstahl AG

Head office: CH 6855 Stabio. **Tel:** 0041 91471141. **Telex:** 842867. **Telecopier (Fax):** 0041 91473718.
Products: Carbon steel – special light sections and bars – min 0.15kg/m (0.1 lb/ft), max 4.5kg/m (3 lb/ft).

Monteforno Acciaierie e Laminatoi SA

Head office: CH 6743 Bodio. **Tel:** 092 741841. **Telex:** 846447. **Telecopier (Fax):** 092 741777.
Management: Markus Voser (president); Luigi P. Pesce (technical); René Trauffer (sales). **Established:** 1946. **Capital:** Sfr 40,000,000. **Ownership:** Von Roll AG (93.6%). **Subsidiaries:** (Comfer SA (100%) (steel trading); Icsa SA (100%), (wire products); Elin Sa (100%), (hydroelectric power plants).
Works:
■ At head office. ***Steelmaking plant:*** One EBT 85-tonne electric arc furnace (annual capacity 380,000 tonnes). ***Refining plant:*** One 85-tonne ladle furnace (380,000 tonnes). ***Continuous casting machines:*** One 5-strand Demag billet and bloom (billets 130mm sq, blooms 180 x 220mm) (380,000 tonnes). ***Rolling mills:*** One 1-stand Siemag blooming (750mm); one 18-stnad Danieli/Demag bar (450-280mm) (300,000 tonnes).
Products: Carbon steel – blooms, billets, reinforcing bars, round bars.
Brands: Box Ultra (rebar).

Rheintub Ltd

Head office: CH 4338 Rheinsulz. **Postal address:** PO Box 100. **Tel:** 064 642485. **Telex:** 982271.
Products: Carbon steel – longitudinal weld tubes and pipes; spiral-weld tubes and pipes; galvanized tubes and pipes; sewer pipes, Rapid coupling pipes, irrigation pipes.

Georges Robert, Usines de Laminage de la Jaluse

Head office: CH 2400 La Locle. **Tel:** 039 313519. **Telex:** 972317.
Products: Carbon steel – cold rolled hoop and strip (uncoated), non-ferrous strip.

Romag – Röhren und Maschinen AG

Head office: CH 3186 Düdingen. **Tel:** 037 439131. **Telex:** 942087.
Products: Carbon steel – spiral-weld tubes and pipes, galvanized tubes and pipes. Pressure vessels.

Rothrist Tube Ltd

Head office: CH 4852 Rothrist. **Postal address:** Box 136. **Tel:** 062 45 62 45. **Telex:** 981911 rorch. **Telecopier (Fax):** 062 443320.
Products: Carbon steel – high frequency longitudinal weld tubes and pipes; cold drawn tubes and pipes; precision tubes and pipes; precision DOM cylinder bore and precision profile low carbon and alloy steel tubes. Alloy steel (excl stainless) – cold drawn tubes and pipes; precision tubes and pipes.

Tubofer SA

Head office: Steel Tube Mill, CH 6849 Mezzovico, TI. **Tel:** 091 95 17 41. **Telex:** 79398 tubo ch.
Products: Carbon steel – longitudinal weld tubes and pipes and precision tubes and pipes 21-63.5mm od. Round and shaped radiator pipes for heating industries.

Vereinigte Drahtwerke AG

Head office: Neumarkstrasse 33, CH 2500 Biel. **Tel:** 32 229911. **Telex:** 934230. **Telecopier (Fax):** 233647.
Products: Stainless steel – bright (cold finished) bars, rounds, square, flats and special sections.

Von Moos Stahl AG

Head office: Kasernenplatz 1, CH 6002 Lucerne. **Postal address:** PO Box 176. **Tel:** 041 595111. **Telex:** 78131.
Established: 1842.
Works:
■ Emmenbrücke. ***Steelmaking plant:*** One electric arc furnace. ***Continuous casting machines:*** One billet. ***Rolling mills:*** Hot rolling mills including light section, bar and wire rod with 10-stand Morgan finishing block and Stelmor cooling (coil weight 1,450kg) (annual capacity 300,000 tonnes). ***Other plant:*** Drawing plant.
Products: Carbon steel – wire rod, reinforcing bars, round bars, square bars, flats, hexagons, bright (cold finished) bars, special light sections, bright wire, black annealed wire, galvanized wire (plain). Alloy steel (excl stainless) – round bars, square bars, flats, hexagons. Bright (cold finished) bars. Free cutting steel.

Von Roll Group

Head office: CH 4563 Gerlafingen. **Tel:** 065 34 11 51. **Telex:** 934 240. **Telecopier (Fax):** 065 35 14 84.
Management: Hans Rüegg (chairman), Heinz W. Frech (chief executive officer/president), D. Bedenig (corporate planning), M. Reinhardt (technical operation). **Established:** 1823. **Capital:** Sfr 90,000,000. **No. of employees:** 1,700 (steel business). **Subsidiaries:** New Jersey Steel Corp, Sayreville, NJ, USA (100%) (reinforcing bars); Robert Aebi AG, Zürich (100%) (machines and equipment for building and civil construction, road construction and mining). **Associates:** Monteforno SA, Giornico (93.6%).
Works:
■ Von Roll AG, Steel Division (At head office). Works director: E. Schlatter. ***Steelmaking plant:*** One 60-70 tonne 50 MVA 5.2m dia electric arc furnace (annual capacity 300,000 tonnes). ***Refining plant:*** One 60-70-tonne 6 MVA ladle furnace (300,000 tonnes); 60-tonne VAD; 60-tonne VOD. ***Continuous casting machines:*** 7-strand 4m radius curved mould billet (130mm, 130 x 250mm, 130 x 300mm) (300,000 tonnes). ***Rolling mills:*** One combination light section, bar and wire rod – 2-strand, seven horizontal/vertical, nine horizontal, three vertical and eight horizontal stands (560-280mm) (coil weight 700kg) (170,000 tonnes); light section and bar –

Switzerland

Von Roll Group (continued)

3-high stand plus six horizontal stands (650-450mm) (70,000 tonnes). ***Other plant:*** Reinforcing mesh plant (30,000 tonnes). Forging plant (open-die forgings up to 15-tonnes and closed-die forgings up to 800 kg) (10,000 tonnes).
Products: Carbon steel – Von Roll AG (steel division) – reinforcing bars hot rolled 8-30mm (output 80,000 tonnes); cold finished 6-12mm (12,000 tonnes); flats 30-300mm (125,000 tonnes); light angles 80-120mm (3,000 tonnes); light channels 50-65mm (2,000 tonnes).
Sales offices/agents: Von Roll AG, Department Stahlprodulcte, CH 4563 Gerafingen; Monteforno Acciaierie haminatoi SA, CH 6743 Bodio; New Jersey Steel Corp, PO Box 11, Sayreville, NJ 08872, USA.

Zwahlen & Mayr

Head office: Zone Industrielle 2, CH 1860 Aigle.
Ownership: Sitindustrie SpA. **Associates:** Stainless steel tube and pipe – Sitai SpA and Nuova CMC SpA (Italy), BSL tubes & Raccords SA (France). **Annual capacity:** Finished steel: 7,000 tonnes (stainless steel tubes).
Products: Stainless steel – longitudinal weld tubes and pipes for condeners and heat exchangers.
Sales offices/agents: Sitai International BV, Rotterdam.

Taiwan

Chia Hsin Metal Industrial Co Ltd

Head office: 2 Lane, 473 Sui Yuan Road, Taichung Hsien, Fung Yuen. **Telex:** 51255 chiahsin.
Established: 1970.
Works:
■ ***Steelmaking plant:*** One electric arc and induction furnace. ***Rolling mills:*** One 360mm bar. ***Other plant:*** Foundry.
Products: Carbon steel – ingots, round bars. Mill rolls.

Chin Ho Fa Steel & Iron Co Ltd

Head office: 104 Cheng-Tu Road, Taipei. **Tel:** 02 3118161-3. **Telex:** 21699 tp. **Telecopier (Fax):** 02 3817408.
Management: H.C. Chen (president). **Established:** 1961. **Capital:** NT$ 25,200,000. **No. of employees:** Works – 40. Office – 10. **Associates:** Shiong Yek Steel Corp (shipbreaking); Tien Cheng Steel Mfg Co Ltd. **Annual capacity:** Finished steel: 5,000 tonnes.
Works:
■ 3 Li-Chiang Street, Lin Hai 2nd Road, Chiao Kang, Kaohsiung. (Tel: 07 8020881-2. Telex: 21699 tp). Works director: Hou Weng Yean. ***Rolling mills:*** bar for flat and squares.
Products: Carbon steel – square bars, flats 3-4mm x 12, 15, 19, 25mm (output 5,000 tonnes).

Taiwan

China Steel Corp

Head office: 1 Chung Kung Road, Kaohsiung. **Postal address:** PO Box 47-29, Kaohsiung 81233. **Tel:** 07 802 1111. **Telex:** 71108 stlmill. **Cables:** Steelmill Kaohsiung. **Telecopier (Fax):** 07 802 2511.
Management: Chairman – M.H. King; president – C.C. Hsiang; vice presidents – Y.C. Chang (executive), C.S. Huang (administration), C.Y. Wang (commercial), L.M. Chung (finance), T-Lin Cheng (technology), Y.T. Kuo (steel production), F.C. Shih (aluminium production). **Established:** 1971. **Capital:** NT$55,220,000,000. **No. of employees:** Works – 8,320. Office – 1,126. **Subsidiaries:** China Steel Structure Co Ltd (manufacture of steel structures) (23.6%). **Annual capacity:** Pig iron: 150,000 tonnes. Raw steel: 3,250,000 tonnes. Finished steel: 2,600,000 tonnes.
Works:
■ At head office. ***Coke ovens:*** Two Carl Still batteries (39 ovens each) (annual capacity 821,925 tonnes); two Carl Still batteries (49 ovens each) (1,029,600 tonnes). ***Sinter plant:*** One 3 x 50m strand Dwight Lloyd (1,700,000 tonnes); one 4 x 70m strand Dwight Lloyd (3,011,000 tonnes). ***Blast furnaces:*** No 1 10.3m hearth dia, 2,101 cu m (1,400,000 tonnes); No 2 11.3m hearth dia, 2,360 cu m (1,750,000 tonnes). ***Steelmaking plant:*** Three 160-tonne LBE type basic oxygen furnaces (3,250,000 tonnes). ***Refining plant:*** 160-tonne vacuum degassing unit (1,500,000 tonnes). ***Continuous casting machines:*** 8-strand Mitsubishi-Olsson low-head bloom (214 x 258mm) (896,000 tonnes); 6-strand Circular arc slab (155 x 1,140-1,400mm, 210 x 950-1,680mm, 270 x 950-1,950mm) (2,104,000 tonnes). ***Rolling mills:*** 3-stand billet (100mm, 115mm, 120mm) (842,000 tonnes); 16-stand bar (14-36mm) (coil weight 1.4 tonnes) (300,000 tonnes); 2-strand 25-stand wire rod (5.5-13mm) (ciol weight 1.4 tonnes) (300,000 tonnes); 1-stand 4-high heavy plate and medium plate (6-125mm thick, 1,524-3,800mm wide, 2,438-18,000mm long) (400,000 tonnes);

China Steel Corp (continued)

wide hot strip – 3/4-continuous with seven 4-high 1,730mm finishing stands (coil weight 28.5 tonnes max) (3,451,000 tonnes); 5-stand 4-high tandem cold reduction carbon steel – (0.3-3.2mm, 610-1,270mm),(coil weight 24.2 tonnes max) and blackplate – (0.2-0.3mm, 610-1,015mm), (coil weight 19.5 tonnes max) (705,000 tonnes); 2-stand 4-high temper/skin pass (687,000 tonnes).
Products: Foundry pig iron 8kg pigs (output 34,650 tonnes). Carbon steel – blooms 214 x 258mm (859,306 tonnes); billets 100mm, 115mm, 120mm (820,474 tonnes); hot rolled coil (for re-rolling) 1.5-6.0mm thick and 900-1,295mm wide (782,679 tonnes); wire rod 5.5-13mm (476,717 tonnes); reinforcing bars 13-40mm (40,457 tonnes); round bars 14-40mm (228,091 tonnes); medium plates and heavy plates 12-25mm (435,021 tonnes); hot rolled sheet/coil (uncoated) 1.5-12.7mm thick, 900-1,575mm wide (1,293,252 tonnes); cold rolled sheet/coil (uncoated) 0.4-2.5mm (732,279 tonnes); blackplate. Alloy steel (excl stainless) – medium plates and heavy plates 12-25mm (165,374 tonnes). Aluminium products – alloy ingot (10kg/pcs and 20kg/pcs) and wire rod, plate 6-75mm, sheet 6-0.18mm, coil 0.15-3.2mm, foil and laminated foil 0.12mm-7.
Sales offices/agents: Representative offices – 4th Floor, Nantai Bldg, 40-8 Minamihonmachi 4-chome, Higashi-ku, Osaka 541, Japan (Tel: 06 281 1021); Suite 2273, One World Trade Center, New York, NY 10048, USA (Tel: 212 755 1088); B1-09 UIC Bldg, 5 Shenton Way, Singapore 0106 (Tel: 223 8777).
Expansion programme: The company's Phase III expansion project, scheduled to be completed by April 30, 1988, will result in annual capacities of: Pig iron 150,000 tonnes, raw steel 5,652,000 tonnes, finishing steel 4,661,000 tonnes.

Feng Hsin Iron & Steel Co Ltd

Head office: 702 Chia Hou Road, Hou Li Hsiang, Taichung Hsien. **Tel:** 045 565101-9. **Telex:** 51243 fenghsin. **Cables:** Fhis Houli, Taiwan. **Telecopier (Fax):** 045 565110.
Management: W.K. Lin (president), Mark Lin (manager, business and management department). **Established:** 1969. **Capital:** US$5,000,000. **No. of employees:** Works – 350. Office – 100. **Annual capacity:** Finished steel: 300,000 tonnes.
Works:
■ At head office. ***Steelmaking plant:*** Two 25-tonne electric arc furnaces (annual capacity 336,000 tonnes). ***Continuous casting machines:*** 3-strand vertical mould, with bending, billet (100/130/160mm sq) (336,000 tonnes). ***Rolling mills:*** billet (130-160mm sq); 5-stand 3-high and 1-stand 2-high horizontal medium section (50, 65, 75, 90 and 100mm) (108,000 tonnes); 1-stand 3-high and 10-stand 2-high horizontal light section (25, 30, 38, 40 and 50mm) (192,000 tonnes).
Products: Carbon steel – light angles 25, 30, 38, 40 and 50mm; medium angles 50, 65, 75, 90 and 100mm.

Feng Lung Steel Factory Co Ltd

Head office: 5 Lane 15, Ta Guan Road, Sec 2 Pan Chiao, Taipei Hsien. **Telex:** 24369 flsteel.
Works:
■ ***Rolling mills:*** light section and bar.
Products: Carbon steel – reinforcing bars, round bars, square bars, flats, light angles, light channels.

Hai Kwang Enterprise Corp

Head office: 12 Yen-Hai 2nd Road, Hsiao Kang, Kaohsiung. **Tel:** 07 802 1011. **Telex:** 81955 Haikwang. **Cables:** Haikwang Kaohsiung. **Telecopier (Fax):** 07 802 2030.
Management: President – T.H. Huang; vice president – Stanley Y.T. Wu; managers – Tony W.C. Lin (sales and purchase) T.M. Huang (research and development).
Established: 1969. **Capital:** US$ 300,000. **No. of employees:** Works – 250. Office

Taiwan

Hai Kwang Enterprise Corp (continued)

– 30. **Annual capacity:** Finished steel: 450,000 tonnes.
Works:
■ At head office. ***Steelmaking plant:*** One 30-tonne electric arc furnace (annual capacity 250,000 tonnes). ***Refining plant:*** One 40-tonne ladle furnace ***Continuous casting machines:*** 4-strand billet (115 and 120mm sq) (250,000 tonnes). ***Rolling mills:*** wire rod (5.5, 6, 7, 8, 9, 10mm) (coil weight 350kg) (200,000 tonnes).
Products: Carbon steel – billets 115 and 120mm sq x 3 metres long; wire rod 5.5, 6, 7, 8, 9, 10mm;

Kao Hsing Chang Iron & Steel Corp

Head office: 1600 Chung Hwa 1st Road, Kaohsiung. **Postal address:** PO Box 274. **Tel:** 07 3122111. **Telex:** 71540. **Cables:** Khpipe Kaohsiung. **Telecopier (Fax):** 07 3211203.
Management: Chairman – Tzer-Shaang Leu; president – Teh-Hsing Leu; executive director – Tzer-Fang Leu; vice president – Chiu-Yueh Yang. **Established:** 1952. **Capital:** NT$ 750,000,000. **No. of employees:** Works – 605. Office – 56. **Annual capacity:** Finished steel: 372,000 tonnes.
Works:
■ At head office. ***Rolling mills:*** cold reduction Two 4-high and one Sendzimir (0.4-2.0mm) (coil weight 7-11 tonnes) (annual capacity 240,000 tonnes); one cold strip (0.4-2.0mm) (50-100kg) (12,000 tonnes); two temper/skin pass (0.4-2.0mm) (7-11 tonnes) (240,000 tonnes). ***Tube and pipe mills:*** welded – one API 5L/5LX 2-16" (24,000 tonnes); one ASTM A53A/B ½-16" (24,000 tonnes); one BS 1387/ ASTM A120 ½-16" (72,000 tonnes).
Products: Carbon steel – cold rolled sheet/coil (uncoated) (sheets) 0.3-1.2mm thick; longitudinal weld tubes and pipes 1-14"; galvanized tubes and pipes.

Li-Chong Steel & Iron Works Co Ltd

Head office: 7 Chiang Kong Road, Chia-Tai Industrial Estate, Chia-Yi Hsien. **Tel:** 05 2372193. **Cables:** Chia-Yi.
Capital: NT$ 180,000,000. **No. of employees:** Works – 125. Office – 20. **Annual capacity:** Finished steel: 100,000 tonnes.
Works:
■ At head office. ***Steelmaking plant:*** One 15-tonne, 4,000mm dia electric arc (annual capacity 70,000 tonnes). ***Continuous casting machines:*** One 2-strand Japanese billet (117-150mm sq) (70,000 tonnes). ***Rolling mills:*** bar and wire rod (9-28mm dia) (coil weight 400kg) (80,000 tonnes); high tensile bar mill (½-1¼" dia) (20,000 tonnes).
Products: Carbon steel – billets, wire rod (output 80,000 tonnes); reinforcing bars (6,000 tonnes).
Brands: LC (within circle device).

Mayer Steel Pipe Corp

Head office: 8th Floor, 71 Nanking E. Road, Sec 2, Taipei. **Tel:** 02 551 7155. **Telex:** 21253 mayerspc. **Cables:** Mayerpipe. **Telecopier (Fax):** 02 563 4623.
Management: Board chairman – T.C. Lo; presidents – K.L. Wu; executive directors – B.C. Lee, H.C. Cheng, S.L. Meng; vice presidents – T.L. Lee, Y.K. Chang; sales manager – K.C. Hsu. **Established:** 1959. **Capital:** NT$ 200,000,000. **No. of employees:** Works – 120. Office – 40. **Associates:** Mason Steel Co Ltd (wholesaler). **Annual capacity:** Finished steel: 50,000 tonnes (steel pipe) and 30,000 tonnes (sheet cutting).
Works:
■ No 2 You-Shih Road, Yang-Mei, Taoyuan, Taiwan. (Tel: 03 4782821-3). Works director: C.T. Peng. ***Tube and pipe mills:*** welded – $\frac{1}{2}$-8" HF (annual capacity 50,000 tonnes). ***Other plant:*** Hot-dip galvanizing plant (8 metres/hour) (24,000 tonnes).
Products: Carbon steel – hot rolled hoop and strip (uncoated), skelp (tube strip), cold rolled hoop and strip (uncoated), hot-dip galvanized hoop and strip, hot rolled sheet/coil (uncoated), cold rolled sheet/coil (uncoated), hot-dip galvanized sheet/coil, electro-galvanized sheet/coil, aluminized sheet/coil, longitudinal weld tubes and pipes, galvanized tubes and pipes. Stainless steel – hot rolled sheet/coil (uncoated), cold rolled sheet/coil (uncoated), longitudinal weld tubes and pipes.

Shang Tai Steel Co Ltd

Head office: 112 Tai Yuan Road, Taipei. **Tel:** 02 – 5623773, 5317300, 5114908, 5410502. **Telex:** 11829 wuluntra. **Telecopier (Fax):** 02 – 5212014.
Management: President – Mrs A.O. Chen; purchasing managers – Wen Hong Chen and Wen Chuen Chen. **Established:** 1967. **Capital:** NT$ 1,500,000,000. **No. of employees:** Works – 45. Office – 13. **Associates:** Wu Lung International (USA) Inc, CA 90638, USA.
Works:
■ At head office. Corrugating facilities.
Products: Carbon steel – hot-dip galvanized sheet/coil, galvanized corrugated sheet/coil.

Taiwan Machinery Manufacturing Corp

Head office: 3 Tai Chi Road, Hsiaokang, Kaohsiung. **Postal address:** PO Box 29-87. **Telex:** 73119 taimaco.
Established: 1946.
Works:
■ ***Steelmaking plant:*** One electric arc. ***Refining plant:*** One ESR; one ladle furnace. ***Coil coating lines:*** electrolytic tinning; hot-dip tinning.
Products: Alloy steel (excl stainless) – ingots.

Tang Eng Iron Works Co Ltd

Head office: 109 San-To 4th Road, Kaohsiung. **Telex:** 71554.
Established: 1940.
Works:
■ Chung Hsin Steel Plant, Kaohsiung. ***Steelmaking plant:*** Two 30-tonne electric arc furnaces. ***Continuous casting machines:*** One 3-strand Concast billet (100, 120mm sq). ***Rolling mills:*** One 26" heavy section; two bar one 17" and one semi-tandem; one wire rod. ***Other plant:*** Foundry, machine shop.
■ Stainless Steel Plant, Kaohsiung. ***Steelmaking plant:*** One 40-tonne electric arc furnace (annual capacity 130,000 tonnes). ***Refining plant:*** One 45-tonne AOD (130,000 tonnes). ***Continuous casting machines:*** One 1-strand slab (250,000 tonnes). ***Rolling mills:*** Two Sendzimir ZR-22WF-B-52" cold reduction and one reversing temper/skin pass (800 x 1,420) (combined capacity 50,000 tonnes).
Products: Carbon steel – reinforcing bars, heavy angles, heavy channels, wide-flange beams (heavy), light rails. Stainless steel – slabs. Alloy steel (excl stainless) – medium plates, heavy plates, hot rolled sheet/coil (uncoated), cold rolled sheet/coil (uncoated).

Ton Yi Industrial Corp

Head office: 146 Niaosung Village, Yungkang, Kachsiung. **Telex:** 72303.
Works:
■ ***Coil coating lines:*** One Halogen electrolytic tinning (annual capacity 60,000 tonnes).
Products: Carbon steel – electrolytic tinplate (single-reduced).

Tanzania

Aluminium Africa Ltd (Alaf)

Head office: Plot 18, Pugu Road, Dar es Salaam. **Postal address:** PO Box 2070. **Tel:** 64011-9. **Telex:** 41265 alaf tz. **Cables:** Aluminium.
Established: 1960. **Capital:** TShs 83,000,000. **No. of employees:** Works – 1,200. Office – 100. **Annual capacity:** Finished steel: 48,000 tons.
Works:
■ Sheet and pipe galvanizing (annual capacity sheet – 34,000 tons, pipe – 10,000 tons).
Products: Carbon steel – light angles, cold rolled hoop and strip (uncoated), galvanized corrugated sheet/coil, galvanized tubes and pipes.
Sales offices/agents: Galco Division of Alaf, PO Box 2641, Dar es Salaam; Aluco Division of Alaf, PO Box 9293, Dar es Salaam; Pipeco Division of Alaf, PO Box 9550, Dar es Salaam; Steelco Division of Alaf, PO Box 9464, Dar es Salaam; Steelcast Division of Alaf, PO Box 9032, Dar es Salaam; Asbesco Division of Alaf, PO Box 3241, Dar es Salaam.
Additional information: The company also produces aluminium circles and sheets (annual capacity 9,000 tons) and asbestos cement sheets.

Steel Rolling Mills Ltd (Tanzania)

Head office: PO Box 5034, Tanga. **Telex:** 45037.
Works:
■ ***Rolling mills:*** light section, bar and wire rod.
Products: Carbon steel – wire rod, reinforcing bars, round bars, flats, light angles.

Thailand

The Bangkok Iron & Steel Works Co Ltd

Head office: 627 Klongyom Lane, Yaowaraj Road, Bangkok. **Telex:** 84617 bisteel th.
Established: 1964.
Works:
■ 42 Suksawad Road, Phra Pradaeng District. ***Steelmaking plant:*** electric arc furnace. ***Continuous casting machines:*** One 3-strand billet (75 x 130mm sq). ***Rolling mills:*** light section and bar.
Products: Carbon steel – reinforcing bars, round bars, light angles.

Bangkok Steel Industry Co Ltd

Head office: 205 Floor 5th UFM Building, Rajawongse, Sampantawongse, Bonkok. **Tel:** 2230114 20. **Telex:** 72362 metbsi th. **Telecopier (Fax):** 66 02 3842316.
Management: Pichit Nithivasin (managing director), Srinakorn Poonipat (deputy managing director), Jaray Bhoommijitr (director), Hin Navawongse (plant manager).
Established: 1965. **Capital:** Baht 300,000,000. **No. of employees:** Works – 710.

Bangkok Steel Industry Co Ltd (continued)

Office – 30. **Ownership:** Metro Co (27.5%). **Subsidiaries:** Bangkok Metal Works Co Ltd (99%) (marketing arm of Bangkok Steel Industry Co Ltd). **Annual capacity:** Finished steel: 120,000 tonnes.
Works:
■ 27 Poochao Samingprai Road, Prapradaeng District, Samuthprakarn Province (Tel: 3943241-5. Telex 72121 fabsi th). Works director: Hin Navawongse. ***Steelmaking plant:*** Two 25-tonne electric arc furnaces (annual capacity 120,000 tonnes). ***Continuous casting machines:*** One 2-strand Concast billet (100mm sq) (117,000 tonnes). ***Rolling mills:*** One 11-stand, two and three-high stands bar (120,000 tonnes). ***Coil coating lines:*** One continuous wet flux process hot-dip galvanizing (50,000 tonnes).
Products: Carbon steel – billets 100mm dia (1986 output, 72,000 tonnes); reinforcing bars 9, 10, 12, 15, 16, 19, 20, 25, 28, 32mm (68,000 tonnes); hot-dip galvanized sheet/coil 0.19-1.60mm thick (18,000 tonnes); galvanized corrugated sheet/coil USG 35 x 2.5 ft.width (25,000 tonnes).
Brands: BSI (reinforcing bar), Cow Head (corrugated sheet); Singha (galvanized sheet and coil).

The Siam Iron & Steel Co Ltd

Head office: 1 Siam Cement Road, Bangsue, Bangkok. **Postal address:** PO Box 1474. **Tel:** 02 587 2197. **Telex:** 73426 siscotl th. **Telecopier (Fax):** 066 02 587 2199.
Management: Chaisak Saeng-Xuto (managing director), Krit Kulnate (marketing manager), Somlak Jiumteranat (purchasing manager). **Established:** 1966. **Capital:** US$14,000,000. **No. of employees:** Works – 920. Office – 100. **Ownership:** The Siam Cement Co Ltd (99.8%). **Annual capacity:** Finished steel: 190,000 tonnes).
Works:
■ Ta Luang Factory, Ban Moh, Saraburi (Tel: 035 341040. Telex 73426 siscotl th). Works director: Tavisak Serirak. ***Steelmaking plant:*** Two 30-tonne electric arc furnaces (annual capacity 230,000 tonnes); ***Continuous casting machines:*** Two 3-strand Concast billet (100 x 100 x 4,000mm) (230,000 tonnes). ***Rolling mills:*** One combined multi-purpose Schloemann bar and wire rod – 18-stand repeater type (coil weight 300kg) (180,000 tonnes).
Products: Carbon steel – wire rod 5.5-9.0mm dia (1986 output, 50,000 tonnes); reinforcing bars 10-32mm dia (80,000 tonnes); round bars 6-25mm dia (50,000 tonnes).

Thai Steel Pipe Industry Co Ltd

Head office: 36 Thanon Poochao, Samingphrai Phra, Pradaeng, Samudprakarn 10130. **Telex:** 84170 thosia th.
Works:
■ ***Tube and pipe mills:*** welded Longitudinal. ***Other plant:*** Tube galvanizing plant.
Products: Carbon steel – longitudinal weld tubes and pipes, galvanized tubes and pipes.

Thai Tinplate Manufacturing Co Ltd

Head office: 33 Mu 13, Soi Salakbandh, Poochao Samingprai Road, Prapradaeng, Samutprakarn. **Tel:** 3941671-4. **Telex:** 20890 ttp th. **Cables:** Tinplateco.
Management: Chamni Visvapolboon (chairman and acting managing director); Taizo Mikami (deputy managing director, administration); Yoshihiro Okawa (deputy managing director, operation); Hideki Namigai (executive director); Kazuhiko Shimizu (finance director); Toru Kishida (sales director), Attapong Kateratorn (director of the office of managing director and secretary of board of director meeting). **Established:** 1958. **Capital:** Baht 140,000,000. **No. of employees:** Works – 341. Office – 32. **Ownership:** Mitsui & Co Ltd (14%); Kawaksaki Steel Corp (8%); C. Itoh

Thai Tinplate Manufacturing Co Ltd (continued)

Co (6%),: Kawasho Corp (12%). **Annual capacity:** Tinplate: 120,000 tonnes. TFS: 30,000 tonnes.
Works:
■ At head office. ***Coil coating lines:*** One halogen electrolytic tinning – line No 1 speed 153 m/min; line No 2 dual (183m/min) electrolytic tinning/tin-free steel – 2-stage process. Total tinplate capacity 120,000 tonnes, tin-free steel 30,000 tonnes.
Products: Carbon steel – electrolytic tinplate (single-reduced), tin-free steel (ECCS).

Thai-Asia Steel Pipe Co Ltd

Head office: 627/633 Soi Kongtom Thanon, Yaoswarat, 10100 Bangkok. **Telex:** 82011 udomtai th.
Established: 1963.
Works:
■ Samudprakarn. ***Tube and pipe mills:*** welded Longitudinal. ***Other plant:*** Tube galvanizing plant.
Products: Carbon steel – longitudinal weld tubes and pipes, galvanized tubes and pipes, hollow sections.

Togo

SNS – Sté Nationale de Sidérurgie

Head office: Lomé. **Telex:** 5317.
Management: John Moore (president).
Works:
■ Lomé. ***Rolling mills:*** bar.
Products: Carbon steel – reinforcing bars.
Additional information: Bars are rolled from used rails.

Sté Togolaise de Galvanisation de Toles – Sototoles

Head office: Lomé. **Postal address:** BP 9103. **Tel:** Lome 21 42 75, 21 44 36. **Telex:** 5375 toles to.
Works:
■ ***Coil coating lines:*** hot-dip galvanizing (sheet).
Products: Carbon steel – hot-dip galvanized hoop and strip.

Trinidad & Tobago

Central Trinidad Steel Ltd – Centrin

Head office: Point Lisas Industrial Estate, Point Lisas. **Telex:** 31356 centrin wg.
Established: 1981.
Works:
■ ***Rolling mills:*** light section and bar.
Products: Carbon steel – reinforcing bars, round bars, square bars, flats, light angles, light channels.

Iscott – Iron & Steel Co of Trinidad & Tobago Ltd

Head office: PO Box 183, Point Lisas, Couva. **Tel:** 636 2211-5. **Telex:** 31369, 31254 iscott wg. **Telecopier (Fax):** 809 636 5696.
Management: Chairman – Ian Dasent; president – Samuel A. Martin. **Established:**

Iscott – Iron & Steel Co of Trinidad & Tobago Ltd (continued)

1979. **Ownership:** Government of Trinidad & Tobago.
Works:
■ At head office. ***Direct reduction plant:*** Two Midrex 400 series (annual capacity 420,000 tonnes each). ***Steelmaking plant:*** Two 90-tonne Mannesmann-Demag electric arc furnaces (600,000 tonnes). ***Continuous casting machines:*** Two 4-strand Concast billet (100-150mm sq) (350,000 tonnes). ***Rolling mills:*** 2-strand Morgan wire rod comprising 7-stand roughing, 6-stand intermediate and two lines of Morgan No-Twist finishing blocks with Stelmor cooling (coil weight 2,000kg max) (500,000 tonnes).
Products: Direct-reduced iron (DRI). Carbon steel – billets 100-150mm sq; wire rod 5.5-16mm for extra fine wire, drawing qualities, cold heading qualities; hollow sections (CO^2 welding wire).

Tunisia

El Fouladh – Sté Tunisienne de Sidérurgie

Head office: Route de Tunis Km3, 7050 Menzel Bourguiba, Bizerte. **Postal address:** BP 23-24. **Tel:** 02 60 522. **Telex:** 21 036, 21 055, 21 060. **Cables:** Tunsider.
Management: President directeur general/chairman – Mohamed Fadhel Zerelli, directeur general adjoint/assistant manager – Mokhtar Robei. **Established:** 1962. **Capital:** Dinars 5,909,920. **No. of employees:** Works – 2,165. Office – 335. **Ownership:** 93% state owned. **Annual capacity:** Pig iron: 160,000 tonnes. Raw steel: 200,000 tonnes (billets). Finished steel: 200,000 tonnes.
Works:
■ At head office. ***Sinter plant:*** One 1 x 20m strand (annual capacity 200,000 tonnes). ***Blast furnaces:*** One 4m hearth dia, 308 cu m (160,000 tonnes). ***Steelmaking plant:*** Two 22-tonne basic oxygen converter (155,000 tonnes); one 15-tonne, 8.4 MVA electric arc furnace (45,000 tonnes). ***Continuous casting machines:*** Three 2-strand Concast billet (110 x 110 x 3,500mm) (200,000 tonnes). ***Rolling mills:*** One 10-stand bar (10-40mm) (130,000 tonnes); wire rod one 19-stand (6-12mm dia) (coil weight 320kg) (70,000 tonnes).
Products: Steelmaking pig iron (1986 output 150,000 tonnes). Carbon steel – billets 110 x 110 x 3,500mm dia (181,000 tonnes); wire rod 6-12mm dia (75,000 tonnes); reinforcing bars 10-32mm dia, round bars 10-40mm dia and square bars 10-18mm dia (133,000 tonnes); bright wire, black annealed wire and galvanized wire (plain) (low and high carbon) (25,000 tonnes). Steel structures made from round galvanized bars for building and electricity pylons (10,000 tonnes/year).

Tunisacier

Head office: 5 Rue d'Angola, 1002 Tunis.
Works:
■ ***Rolling mills:*** One temper/skin pass. ***Coil coating lines:*** One hot-dip galvanizing (annual capacity 40,000 tonnes). ***Other plant:*** Annealing line.
Products: Carbon steel – hot-dip galvanized sheet/coil.
Additional information: Provision for expansion to 70,000 tonnes a year. Feedstock is 0.2-3mm thick and up to to 1,500mm wide in coils up to 20 tonnes. Max line speed 60 metres/min.

Turkey

Asil Çelik Sanayi ve Ticaret AŞ (Asil Çelik)

Head office: Büyükdere cad. 121/6-7, Gayrettepe, Istanbul. **Tel:** 167 55 60, 167 41 74. **Telex:** 26101 acel tr. **Cables:** Asil Celik Istanbul.
Management: Chairman of board of directors – Yahya Keskin; general manager – Yahya Keskın; assistant general managers – Ali Zengil (technical), Cüneyt Alten (finance), Lalif Anbarli (sales). **Established:** 1974. **Capital:** TL 50,000,000,000.
Works:
■ Gemic Köyü, Orhangazi Bursa. Works manager: Ali Zengil. ***Steelmaking plant:*** Two electric arc furnaces – 15-tonne (annual capacity 30,900 tonnes) one 45-tonne (150,000 tonnes). ***Refining plant:*** One RH vacuum degassing unit. ***Rolling mills:*** One 2-high reversing blooming (820mm) (227,000 tonnes); one 2-high reversing billet (750mm) (215,000 tonnes); one medium section – four 3-high (550mm) and one 2-high (420mm) finishing stand (132,000 tonnes).
Products: Carbon steel – blooms 160-350mm sq; billets 46-160mm sq; round bars 22-180mm; flats 50 x 6-120 x 40mm; hexagons 23.5-47.5mm. Alloy steel (excl stainless) – blooms 160-350mm sq; billets 46-160mm sq; round bars 22-180mm; flats 50 x 6-120 x 40mm; hexagons 23.5-47.5mm.

Assan Demir ve Sac Sanayi AŞ

Head office: Tersane Caddesi, Aslan Han, Karaköy, Istanbul. **Tel:** 901 155 58 62, 901 155 58 12. **Telex:** 24180 kbar, 25310 asac tr. **Telecopier (Fax):** 901 155 57 79.
Management: President and chairman – Asim Kibar; vice chairman – Ali Kibar; managers – Mertol Soydas (import), Atilay Sanag (export); deputy manager – Mustafa Atasagun (export). **Established:** 1976. **Capital:** TL 5,000,000,000. **No. of employees:** Works – 250. Office – 60. **Ownership:** Kibar Holding AŞ. **Subsidiaries:** Assan Noral Aluminium AŞ; Kibar Holding AŞ. **Associates:** Assac AŞ, Kibarlar AŞ, Kobar AŞ. **Annual capacity:** Finished steel: 200,000 tonnes.
Works:
■ E-5 Karayolu Ustü Paşabahçe/Sungurlar Fabrikasi Yani, Tuzla, Istanbul. (Tel: 901 395 29 68. Telefax: 901 395 29 71). Works director: Memis Ali Usta. ***Coil coating lines:*** H.H. Robertson (UK) hot-dip galvanizing (annual capacity 120,000 tonnes); Assan (local and imported) hot dip continuous (80,000 tonnes).
Products: Carbon steel – cold roll-formed sections (from strip); hot-dip galvanized hoop and strip, coated hoop and strip, hot-dip galvanized sheet/coil and galvanized corrugated sheet/coil 0.30-2.0mm (combined output 60,000 tonnes); colour coated sheet/coil 0.30-2.0mm x 1,250mm. Aluminium sheets, coils, tubes, etc.

Birlik Galvaniz Sac Sanayii ve Ticaret AŞ – See Borusan Holding AŞ

Borusan Boru Sanayii AŞ – See Borusan Holding AŞ

Borusan Gemlik Boru Tesisleri AŞ – See Borusan Holding AŞ

Turkey

Borusan Holding AŞ

Head office: Borusan Building, 523/533 Salipazari, Istanbul. **Postal address:** PO Box 502 Karaköy. **Tel:** 144 87 50. **Telex:** 24190 brs tr, 24586 brsn te. **Cables:** Borusan. **Telecopier (Fax):** 149 37 22.
Management: Chairman – Asim Kocabiyik; deputy chairmen – Necip Yüksel (production), Hulusi Çetinoğlu; board members – A. Ahmet Kocabiyik (general co-ordinator, foreign trade), Bülent Demircioğlu (planning, sales), Sait Köseoğlu (chief financial officer); secretary general – Mohammad Hamedi. **Established:** 1972. **Subsidiaries:** Borusan Ihracat Ithalat ve Dağitim AŞ (62%); Borusan Boru Sanayii AŞ (93%); Borusan Gemlik Boru Tesisleri AŞ (29%); Borusan Elektronik Sanayii ve Ticaret AŞ (50%); Istikbal Ticaret TAŞ (82%;I Borusan Oto Servis ve Ticaret AŞ (49%); Tubex (Jersey) Ltd (30%).

Kerim Çelik Mamulleri Imalat ve Ticaret AŞ Meclisi Mebusan Cad, Borusan Buildijg,Salipazari,Istanbul. ***Tel:*** 151 34 10. ***Telex:*** 24586 brs tr, 24586 brsn tr. ***Telecopier (Fax):*** 149 37 22.
Management: Suat Sarisoy (general manager). **Established:** 1973. **Ownership:** Borusan Group.
Works:
■ Inönü Cad, 162 Sefaköy-Istanbul. Tel: 579 62 50. Telex: 28797 bspr tr. **Plant details:** Metal-plated steel strip plant (nickel, brass copper and zinc electrolytic plating). Hot-dip sheet galvanising plants wire galvanizing lines, steel curtain rails and accessories, metallic furniture accessories, slitters lines.

Borusan Boru Sanayii AŞ Meclisi Mebusan Cad, Borusan Building,Salipazari,Istanbul. ***Tel:*** 1 151 34 10. ***Telex:*** 24 190 brs tr, 24 586 brsn tr. ***Telecopier (Fax):*** 149 37 22.
Management: Chairmen – Asim Kocabiyik, Fahrettin Ermutlu (deputy); managin director – Necip Yüksel; board member – A. Ahmet Kocabiyik, Mehmet L. Kurşuncu; general manager – Orhan Kesimgil. **Established:** 1958. **Capital:** TL 3.1,000,000,000. **No. of employees:** Works – 270. Office – 40. **Ownership:** Borusan Holding AŞ (93.73%). **Subsidiaries:** Borusan Ihracat Ithalat ve Dağtim AŞ (8%); Borusan Gemlik Boru Tes. AŞ (6.50%); Borusan Amortisör Imalat ve Tic AŞ (88%); Borusan Manisa Metal San AŞ (94%); Süpsan Motor Supaplari Sanayii ve Tic AŞ (9.9%); Istikbal Ticaret TAŞ (15.8%).
Works:
■ Halkali Asfalti, Sefaköy, Istanbul. Tel: 579 07 14. Telex: 28797 bspr tr. ***Rolling mills:*** One 4-high reversing cold reduction (annual capacity 24,000 tonnes). ***Tube and pipe mills:*** welded (80,000 tonnes); stainless steel (2,000 tonnes) and drawbenches (5,000 tonnes).

Birlik Galvaniz Sac Sanayii ve Ticaret AŞ Meclisi Mebusan Cad, Borusan Building, Salipazari,Istanbul.
Established: 1968. **Ownership:** Borusan Group.
Works:
■ Küçükçekmece, Avcilar Firuzköy Yolu, Istanbul. Tel: 573 58 61. ***Coil coating lines:*** Continuous hot-dip galvanizing for flat, corrugated, trapezoid and roller-shutter sheets (annual capacity 40,000 tonnes).

Borusan Gemlik Boru Tesisleri AŞ, Meclisi Mebusan Cad, Borusan Building,Salipazari,Istanbul. ***Tel:*** 1 151 34 10. ***Telex:*** 24190 brs tr, 24586 brs tr. ***Telecopier (Fax):*** 149 37 22.
Management: Chairmen – Asim Kocbiyik, Necip Yüksel (deputy); board member – A. Ahmet Kocabiyik, Basri Tüfekçioğlu, Teoman Ekim, Halil Sözer; general manager – Mehmet L. Kurşuncu. **Established:** 1973. **Capital:** TL 5,800,000,000. **No. of employees:** Works – 785. Office – 70. **Ownership:** Borusan Holding AŞ (29.3%); Istikbal Ticaret TAŞ (17.6%); Brown Shipley Ltd (31.7%); Borusan Boru Sanayii AŞ (6.54%); other individual shareholders. **Subsidiaries:** Borusan Makina Imalat AŞ (86.9%).
Works:
■ Hisar Mah Gemsas Mevkii-Gemlik. Tel: 4601. Telex: 32433 brmi tr. ***Tube and pipe***

Borusan Holding AŞ (continued)

mills: welded including stretch reducing mill and 12¾" API pipe mill (annual capacity 300,000 tonnes). ***Other plant:*** Three tube galvanizing plants; finishing lines.

Products: Carbon steel – hot-dip galvanized hoop and strip nickel, brass and copper coated; coated hoop and strip nickel, brass and copper coated; hot-dip galvanized sheet/coil up to 1,300mm wide, 0.20-3mm thick; electro-galvanized sheet/coil; galvanized corrugated sheet/coil, trapezoid and roller-slitter sheets; galvanized wire (plain); round, square and rectangular longitudinal weld tubes and pipes 4-76-101.4mm od, 0.6-3mm wall thickness; galvanized tubes and pipes cooler and brake tubes 4.70-6.35mm od and 0.70mm wall thickness, specially cleaned, annealed and pressure tested; API pipes up to 12¾"; square and rectangular hollow sections up to 250 x 250mm; black galvanised, threaded sleeve and varnished water, gas, petroleum and boiler pipes and tubes 17.2-323.8mm and 2-12.7mm od; steel strapping and coated strip to DIN 1624, DIN 1,544 and ASTM A-109. Stainless steel – longitudinal weld tubes and pipes 10-300mm od 0.80-4.5mm wall thickness. Steel furniture accessories.
Sales offices/agents: Local – Borusan Export-Import and Distribution Inc; England – Ferco Intertrade Ltd, Kent; Tubex (Jersey) Ltd, London.

Cemtas AŞ

Head office: PO Box 228, Bursa. **Tel:** 024 431318. **Telex:** 32172.
Products: Carbon steel – round bars, square bars, light angles, light channels.

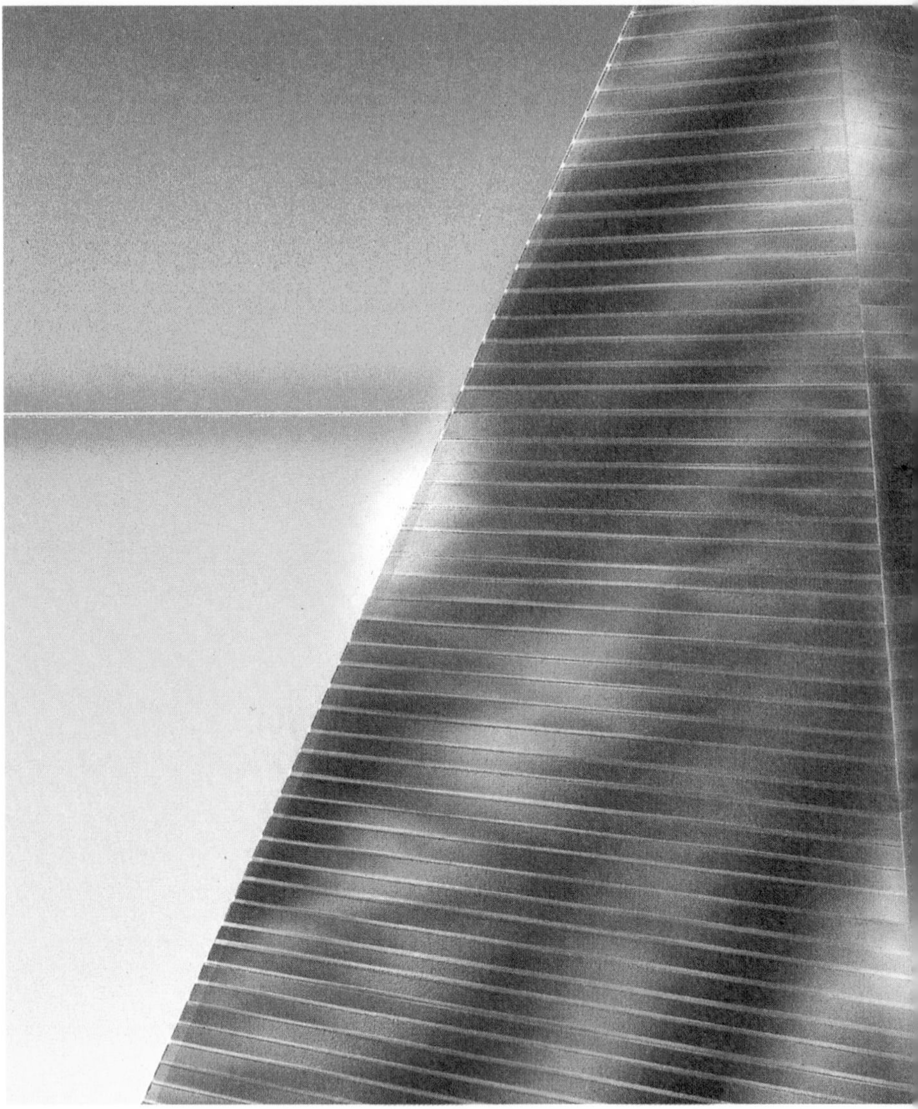

Çolakoğlu Metalurji AŞ

Head office: Kemeralti Cad 24/6, Karaköy 80030, Istanbul. **Tel:** 1 1434630. **Telex:** 25739 cmc tr, 24709 cmas tr. **Cables:** Colmetalurji Istanbul. **Telecopier (Fax):** 1 1495588.
Management: Chairman – Mehmet Çolakoğlu; managers – Ergin Erel (general), Nezihi Bayrak (purchasing), Tuncay Sergen (sales). **Established:** 1969. **Capital:** TL 4,300,000,000. **No. of employees:** Works – 900. Office – 50. **Annual capacity:** Raw steel: 600,000 tonnes. Finished steel: 450,000 tonnes.
Works:
■ Diliskelesi Mevkii, Gebze, Izmit. (Tel: 1991 5030. Telex: 34125 cmg tr, 34194 cfg tr). Works director: Cahit Askin. ***Steelmaking plant:*** Four 35-tonne electric arc furnaces (annual capacity 600,000 tonnes). ***Refining plant:*** Two 40-tonne ladle furnace (600,000 tonnes). ***Continuous casting machines:*** Two 4-strand Concast billet (100-130mm sq) (600,000 tonnes). ***Rolling mills:*** 26-stand bar and wire rod (5-10.5mm coil, 8-20mm rebar) (coil weight 1,100kg) (450,000 tonnes).
Products: Carbon steel – billets 100-130mm sq; wire rod 5-10.5mm dia; reinforcing bars 8-20mm dia; round bars 8-20mm dia.
Sales offices/agents: Export sales – At head office. EEC – Fercom Trading Ltd, 40 Basinghall Street, London EC2V 5DE (Tel: 01 588 5256. Telex: 263363 fercom g).

Çukurova Çelik Endüstrisi AS

Head office: Büyükdere Cad 14, Sişli, Istanbul. **Tel:** 146 47 06. **Telex:** 23345 cce tr.
Established: 1978. **Ownership:** The Çukurova group. **Annual capacity:** Raw steel: 1,500,000 tonnes (from 1988).
Works:
■ Horozgediği Köyü Kospet Mevki, Aliağa, Izmir (Tel: Aliağa 1135-36). Works director: Oktay Sunata. ***Steelmaking plant:*** Four electric arc furnaces – two BBC 75-tonne, 30 MVA, two Krupp 75-tonne, 60 MVA (annual capacity 1,500,000 tonnes). ***Continuous casting machines:*** Four billet – two 6-strand Concast S type, two 4-strand (being expanded to 6) Concast S type (100-130mm sq).
Products: Carbon steel – billets.

Demir Sanayi Demir Çelik Ticaret ve Sanayi AŞ

Head office: Silahtar Cad 36, Halicioğlu, Istanbul. **Tel:** 1461915. **Telex:** 28735.
Management: Board director – Yusuf Aslan; managers – Şinasi Mansur (general), Fehamettin Acarer (assistant general), Yilmaz Erdogu (trade), Selahattin Toksoy (factory). **Established:** 1976. **Capital:** TL 1,500,000,000. **No. of employees:** Works – 123. Office – 19. **Associates:** Kardemir Demir Çekme ve Makina Sanayi AŞ.
Works:
■ At head office. Works director: Selahattin Toksoy. ***Rolling mills:*** 18-stand bar (400/330mm) (annual capacity 100,000 tonnes).
Products: Carbon steel – reinforcing bars 10mm, 12mm, 14mm and 16mm dia (output 100,000 tonnes).

Demirsaç Galvaniz Ticaret ve Sanayi AŞ

Head office: Demirsaç Building, Tersane cad 10, Perşembepazari, Karaköy, Istanbul. **Tel:** 155 52 15. **Telex:** 24536 des tr.
Management: Chairman – Fikret Gazioğlu; vice chairman – Metin Gazioğlu; board members – Mustafa Gazioğlu, Mehmet Murat Gazioğlu, Ahmet Gazioğlu, Mustafa Gazioğlu. **No. of employees:** 100. **Ownership:** Joint stock company.
Products: Carbon steel – hot-dip galvanized sheet/coil 0.30-2.00mm and galvanized corrugated sheet/coil (output 70,000 tonnes).

Ekinciler Demir ve Çelik San AŞ

Head office: Mebusan Yokusu SSK Han Kat 1, Findikli, Istanbul. **Tel:** 143 25 10. **Telex:** 24330 ekge tr. **Cables:** Ekinciler/Istanbul. **Telecopier (Fax):** 1513528.
Management: President – Orhan Ekinci; board members – Visaleddin Unsal, Ahmet Eker, Fikret Ince (general manager), Recep Ekinci, Dogen Kotanoglij. **Established:** 1962. **Capital:** TL 10,000,000,000. **No. of employees:** Works – 650. Office – 50. **Ownership:** Ekinciler Holding Inc. **Subsidiaries:** Ekinciler International Trade Inc; Ektrans International Transport & Trade Inc; Ektel Wire & Nail Inc; Ekçelik Marketing Inc; Ekmak Steel Industry Manufacturing Inc; Ekintaş Construction Industry Inc; Ekdata Elektronic Trade & Industry Inc; Ekfa International Trade; Ekinciler Insurance Services Inc; Ekha Handelsgesellschaft GmbH (Düsseldorf); Ekco International Trade Inc; Ektrade America Inc (New York). **Annual capacity:** Finished steel: 450,000 tonnes.
Works:
■ Iskenderun Rolling Mill. Works director: Selahattin Gürer. ***Rolling mills:*** 16-stand bar and wire rod – one 3-high 420mm stand and fifteen 2-high 400-260mm stands (8-36mm rod) (annual capacity 250,000 tonnes).
■ Karabük. ***Rolling mills:*** One bar (annual capacity 100,000 tonnes).
■ Izmir. ***Rolling mills:*** One bar (annual capacity 100,000 tonnes).
■ Ekinciler Çukurora Iron & Steel Industry Inc, Adana. ***Rolling mills:*** One bar / wire rod (annual capacity 132,000 tonnes).
Products: Carbon steel – wire rod; reinforcing bars (8-32mm dia); round bars (8-32mm dia); square bars (12-30mm); flats (30 x 3 – 100 x 10mm); hexagons (12-36mm); light angles (30 x 30 x 3mm – 70 x 70 x 7mm). Combined output 400,000 tonnes.
Expansion programme: Electric furnace to be installed at Iskenderun works with (400,000 tonnes capacity) and a continuous caster for billets (100-120mm sq).

Elektrofer Çelik Sanayi AŞ

Head office: Yalikosku Cad 23, Sirkece, Istanbul. **Tel:** 5281544. **Telex:**
Products: Carbon steel – billets.

Erbosan – Erciyas Boru Sanayii ve Ticaret AŞ

Head office: Ambar Köyü Bogazköprü Mevkii, Kayseri. **Tel:** 368 62. **Telex:** 49581 erbs.
Management: Board management – Mhuarrem Altunbağ (chairman), Mehemt Tahtasakal, Sami Yangin, Yalcin Besçeli, Bekir Özbiyik, Mustafa Gürkan, Hüsamettin Çetinbulut; managers – Ibrahim Yardimci (general), Mahmut Özkan (purchasing), Mahmut Özkan (export). **Established:** 1977. **Capital:** TL 3,520,000,000. **No. of employees:** Works – 150. Office – 30. **Annual capacity:** Finished steel: 150,000 tonnes.
Works:
■ At head office. ***Tube and pipe mills:*** welded – No 1 ½-4" (annual capacity 40,905 tonnes); No 2 10 x 20 x 1 and 120 x 60 x 4 profile (28,936 tonnes); No 3 13-114mm industrial (10,720 tonnes); No 4 44.5-114mm boiler and steam pipes (5,200 tonnes).
Products: Carbon steel – longitudinal weld tubes and pipes ½-4", 10 x 20 x 1-120 x 60 x 4mm (1986 output 85,761 tonnes).

Ereğli Demir ve Çelik Fabrikalari TAS (Erdemir)

Head office: Atatürk Bulvari 127, Kat 6, Ankara. **Tel:** 41 342730. **Telex:** 42428 eran tr. **Telecopier (Fax):** 170387.
Management: Chairman – Celal Kurtuluş; president – Tümer Özenç; vice presidents – Kerim Dervişoğlu (operations), Şevket Önder (finance), Mehmet Kip (purchasing), Ahmet Arslan (personnel and social services), Erol Onar (sales), Mete Gökbayrak (engineering and investments). **Established:** 1960. **Capital:** TL38,400,000,000. **No. of employees:** Works – 5,675. Office – 2,616. **Ownership:** T. Demir ve Çelik

Turkey

Ereğli Demir ve Çelik Fabrikalari TAS (Erdemir) (continued)

Işletmeleri Genel Müdürlüğü, Sümerbank Genel Müdürlüğü, Türkiye Iş Bankasi A.Ş., Ankara Ticaret Odasi and other shareholders. **Annual capacity:** Pig iron: 2,230,000 tonnes. Raw steel: 1,800,000. Finished steel: 1,650,000.
Works:
■ Uzunkum 7, KDZ.Ereğli ***Coke ovens:*** – two under jet type batteries (37 ovens each) (annual capacity 512,000 tonnes), one low pressure battery (85 ovens) (588,000 tonnes). ***Sinter plant:*** One 3.00 x 44.67m strand Lurgi (1,440,000 tonnes). ***Blast furnaces:*** Two – 8.99m hearth dia, 1,582 cu m (980,000 tonnes) and 9.70m hearth dia, 1,707 cu m (1,250,000 tonnes). ***Steelmaking plant:*** Three basic oxygen converters – 47.4 cu m (1,800,000 tonnes). ***Refining plant:*** Automatic Al-wire feeding, Al measurement, temperature measurement sampler (1,800,000 ton/yil – steel can be processed in these stations. ***Continuous casting machines:*** Two single-strand slab – one 750-1,300 x 200mm and one 1,000-1,600 x 200 (combined 1,100,000 tonnes). ***Rolling mills:*** 1-stand 2-high slabbing (47.5 x 114") (coil weight 13,000kg) (800,000 tonnes); 1-stand 4-high heavy plate (38 x 53 x 114") (100,000 tonnes); 6-stands 4-high hot strip and hoop (27.5 x 53 x 66") (1,600,000 tonnes); Steckel mill (26 x 49 x 66") (500,000 tonnes); 5-stand 4-high cold strip (21.5 x 53 x 66:") (750,000 tonnes); 1-stand 4-high temper/skin pass (21.5 x 66") (300,000 tonnes); 2-stand 4-high cold mill (53 x 66") (500,000 tonnes). ***Coil coating lines:*** One Ferrostan electrolytic tinning (100,000 tonnes).
Products: Steelmaking pig iron (1986 output 1,432,841 tonnes); foundry pig iron (27,693 tonnes). Carbon steel – ingots (518,645 tonnes); slabs continuously cast 1,200 x 200mm (957,880 tonnes); skelp (tube strip) 1,000 x 2mm (28,657 tonnes); medium plates up to 12mm thick; heavy plates 1,500 x 25mm (109,511 tonnes); universal plates 1,500 x 6mm (169,547 tonnes); hot rolled sheet/coil (uncoated) 1,200 x 3.0mm (870,652 tonnes); cold rolled sheet/coil (uncoated) 1,200 x 1.25mm (367,964 tonnes); electrolytic tinplate (double-reduced) 762 x 0.38mm (95,470 tonnes).
Sales offices/agents: Uzunkum 7, Kdz. Ereğli (Tel: 388 19500. Telex: 45523 erec tr and 48575 edc tr); Atatürk Bulvari 127 Kat 5, Ankara (Tel: 41 342730. Telex: 42428 eran tr); Gazipaşa Bulvari 1/1, Adana (Tel: 711 41541. Telex: 62651 erdc); Istanbul Cad.Kasabali iş Hani Kat 3 105/301, Konya Istanbul (Tel: 331 22114); Hükümet Cad.Sim Iş Hani Kat 4, Kayseri-Hükümet (Tel: 351 24007); Inönü Cad.90, Dersan Iş Hani Kat 4, Istanbul. (Tel: 1 1451400. Telex: 24297 ecis tr); Cumhuriyet Bulvari 139 Kat 2, Izmir (Tel: 51 21971. Telex: 53362 erdi tr); Çakmak Cad.Beyhan Iş Hani 69/4, Bursa Fevzi (Tel: 241 21971); Kahramanmaraş Cad. Yavuz Han 29, Kat 3, Trabzon (Tel: 031 17431).
Expansion programme: Steel producing facilities – development plan 1988-1992 includes increasing the oxygen steelmaking capacity; two additional continuous casters; modernisation of hot steel mill No 1; third slab reheating furnace; third pickling line and coil preparation line. The plan aims to raise production to 2,392,000 tonnes/year of finished products.

Erkboru Profil Sanayi ve Ticaret AŞ

Head office: Yakacik Asfalti, Istanbul. **Telex:** 22063.

Habas Sinai ve Tibbi Gazlar Istihsal Endustrisi AŞ

Head office: Bahriye Cad 261, Kasimpasa, Istanbul. **Tel:** 150 48 65/150 13 90. **Telex:** 24024 sgaz tr.
Management: Board president – Mehmet Başaran. **Established:** 1964. **Capital:** TL 1,150,000,000. **No. of employees:** Works – 300. Office – 30. **Ownership:** Başaran Holding AŞ. **Associates:** Habas Endustri Tesisleri AŞ, Asgaz Anadolu Sinai Gazlar AŞ. **Annual capacity:** Raw steel: 600,000 tonnes.
Works:
■ Habas Steel & Iron Plant, Yeni Foça Yolu Üzeri, Aliağa/Izmir. (Tel 5436 2097/ 1571. Telex: 53845 htgz tr). ***Steelmaking plant:*** Two 80-tonne electric arc furnaces

Habas Sinai ve Tibbi Gazlar Istihsal Endustrisi AŞ (continued)

(annual capacity 600,000 tonnes). ***Continuous casting machines:*** Two 5-strand Concast billet (80-140mm sq) (600,000 tonnes).
Products: Carbon steel – billets.

Içdaş Istanbul Çelik ve Demir Izabe Sanayii AŞ

Head office: Mahmutbey Cad.Halkali Yolu Üzeri, Şirinevler, Istanbul. **Postal address:** PK 14. **Tel:** 557 15 27. **Telex:** 28735 icds tr.
Management: Board director – Yusuf Aslan; managers – Şinasi Mansur (general), Fehamettin Acarer (assistant general), Yilmaz Erdoğu (trade), Fahrettin Pamuk (production). **Established:** 1970. **Capital:** TL 2,050,000,000. **No. of employees:** Works – 300. Office – 36. **Annual capacity:** Finished steel: 150,000 tonnes.
Works:
■ At head office. Works director: Fahrettin Pamuk. ***Steelmaking plant:*** One electric arc furnace (annual capacity 150,000 tonnes). ***Continuous casting machines:*** One 4-strand Concast billet (100mm sq) (360,000 tonnes).
Products: Carbon steel – billets 100mm sq (1986 output 96,000 tonnes).

Izmir Demir Çelik Sanayi AŞ

Head office: Gazi Bulvari 68/1, 35210 Izmir. **Tel:** 51 25 67 10. **Telex:** 18 53015 idc tr. **Cables:** Izcelik Izmir. **Telecopier (Fax):** 51 25 70 23.
Management: Chairman – Atilla Yurtçu; vice chairman – Samim Sivri; board members – Ali Sait Özcivril, Mete Yurtçu, Saim Sivri, Erdem Karaismail, Saleh Mohd. Abdullaziz Al Rajhi; managers – Esat Özalp (general), Emin Tüfekciler (vice general, works), A. Oğuz Dağli (finance and accounting), M. Cengiz Torun (marketing and sales), Mehmet Gültekingil (purchasing), Sevinc Fişek (personnel and administration), Süleyman Dogan (investment and research). **Established:** 1959. **Capital:** TL 15,000,000,000. **No. of employees:** Works – 566. Office – 57. **Ownership:** Izdas Holding (34.86%). Public shares (65.14%). **Associates:** Izdas – Izmir Dis Ticaret AS (export/import); ADC – Antalya Demir Celik Sanayi ve Ticaret AS (bar and profile rolling mill); Ege Celik Sanayi ve Ticaret AS (steel trading); Cemas Celik ve Metal Ticaret AS (steel sales); Asmas Agir Sanayi Makinalari AS (manufacturing heavy industrial machinery); Izdas Dagitim ve Pazarlama AS (steel marketing and sales); Poyraz Denizcilik AS (shipping and transport); Nemtas – Nemrut Liman Isletmeleri AS (Nemrut Port Loading & Unloading Corp); Santes Sanayi Tesisleri AS (construction and consultancy). **Annual capacity:** Raw steel: 400,000 tonnes. Finished steel: 350,000 tonnes.
Works:
■ Foca Celik Fabrikasi, Horozgedigi Köyü, Aliağa, Izmir. (Tel: 5436 2783 1801. Telex: 53919 idcf tr). Works director: Emin Tufekciler. ***Steelmaking plant:*** 65-tonne NKK EBT UHP electric arc furnace (annual capacity 400,000 tonnes). ***Refining plant:*** 65-tonne NKK 10 MWA ladle furnace (400,000 tonnes). ***Continuous casting machines:*** Two 4-strand Concast model S14-6 billet (80-120mm and 120-140mm) (400,000 tonnes). ***Rolling mills:*** medium section comprising one 3-high 450 x 1,400mm roughing stand and six 2-high 450 x 800mm continuous finishing stands (65-140mm joists (IPN), 65-160mm channels (UPN)); bar comprising one 3-high 450 x 1,400mm roughing stand, four 2-high 450 x 800mm stands, four 2-high 350 x 700mm stands and four 2-high 320 x 600mm continuous finishing stands (8-32mm deformed and plain rebars; sections under 200mm sq are rolled using Slitting Method). Combined annual capacity of medium section and bar mills 350,000 tonnes.
Products: Carbon steel – blooms 120-140mm; billets 80-120mm; reinforcing bars 8-32mm; round bars 8-32mm; square bars 10-60mm; flats 10-100mm; light angles 25-100mm; light tees 25-100mm; medium joists 65-140mm; medium channels 65-160mm.
Brands: IDÇ TSE NO (profiles); IDÇ BÇ IIIA (deformed reinforcing bars).
Sales offices/agents: Exclusive exporter – Izdas – Izmir Dis Ticaret AS, Buyuk Dere Cad 112, Zincirlikiuyu, Istanbul, Turkey (Tel: 1 1361200. Telex: 28314 izda tr). Overseas – 53 Blvd Mohamed V., Algiers, Algeria (Tel: 213 618683. Telex: 66336

Izmir Demir Çelik Sanayi AŞ (continued)

izdas dz); Feldstrasse 54, D 4000 Dusseldorf, West Germany (Tel: 211 492004. Telex: 8588744 izda d); Izdas Trading Co Ltd, 350 5th Avenue, Empire State Bldg, Suite 7906, New York, NY 10118, USA. Sales offices – Ege Celik Sanayi ve Ticaret AS, Gazi Bulvari 68/7, 35210 Izmir (Tel: 51 147393-4. Telex: 53460 idc tr. Fax: 51 25 70 23. Teletex: 18 53015 idc tr); Cemas Celik ve Metal Tic AS, Sair Esref Bul. 25, 35210 Izmir (Tel: 51 254364-6).

Kerim Çelik Mamulleri Imalat ve Ticaret AŞ – See Borusan Holding AŞ

Kroman Steel Industries Co (Kroman Celik)

Head office: PK 24, Cayirova-Gebze. **Tel:** 1991 2076-7. **Telex:** 34136, 24391. **Management:** Y. Yücel (owner), S. Oktar (general director), S. Sarikaeeilor (works director), Suhad Oktar (project manager). **Established:** 1966. **Ownership:** Y. Yücel.
Works:
■ At head office. ***Steelmaking plant:*** One 30-tonne and one 17-tonne electric arc furnaces. ***Continuous casting machines:*** One 4-strand Continua billet. ***Rolling mills:*** One light section and bar.
Products: Carbon steel – billets; round bars and light sections.
Expansion programme: Following an ownership change in 1986, capacity is being expanded to over 100,000 tonnes per year.

Mannesmann-Sümerbank Boru Endüstrisi TAS (MSBE)

Head office: Izmit, Kocaeli. **Postal address:** PK 5. **Tel:** 11626-7, 11808. **Telex:** 33128 26336, 33128 boru tr. **Cables:** Boru Endustrisi Izmit.
Established: 1955. **Ownership:** Mannesmann group (W. Germany).
Works:
■ At head office. ***Tube and pipe mills:*** welded – Three HF $\frac{1}{4}$-6$\frac{5}{8}$", two submerged arc spiral weld 8-48". ***Other plant:*** Two drawbenches for precision tube, machining centre for OCTG 2$\frac{7}{8}$-13$\frac{3}{8}$".
Products: Carbon steel – longitudinal weld tubes and pipes, spiral-weld tubes and pipes, oil country tubular goods, precision tubes and pipes.

Metaş Izmir Metalurji Fabrikasi Türk AŞ

Head office: Kemalpaşa Caddesi, Işikkent Girişi, Izmir. **Postal address:** PO Box 458 35213 Izmir. **Tel:** 51 162200. **Telex:** 53384 mtas tr. **Cables:** Metas Izmir. **Telecopier (Fax):** 51 163844.
Management: Board of directors: Honorary chairman – Raşit F. Özsaruhan; chairman – M. Haluk Özsaruhan; deputy chairman – Osman Türkmenler; members – Şahap Kocatopçu, Enis Özsaruhan, Tezcan Yaramanci, A. Tahir Türetken, Göral Okta; secretary general – Selçuk A. Inan. Executives: General manager – Işin Çelebi; advisor to general manager – Feridun Bilginer; technical advisor to general manager – Erdoğan Reisoğlu; managers – Gürcan Heper (financial), Mustafa D. Somersan (purchasing), Gürol Beğen (sales), Ünal Kamali (personnel), Ataman Ilgaz (research and development), Aydin Telseren (works). **Established:** 1956. **Capital:** TL 15,000,000,000 (registered); TL 6,000,000,000 (paid). **No. of employees:** Works – 786. Office – 250. **Ownership:** Ege Yatirim AŞ group. **Subsidiaries:** Satmaş – Sanayi ve Tarim Mallari Diş Tic AŞ (99.76%) (export, marketing); Limaş – Liman Işletmeleri AŞ (99.51%) (ports and port administration); Izmir Çelik Ticaret AŞ (75%) (trade and marketing); Edaş – Ege Demir Sanayii AŞ (93.35%) (sponge iron production); Entaş – Ege Inşaat Nakliyat ve Ticaret AŞ (66%) (contracting and transportation); Aysan – Anadolu Yay Sanayi ve Ticaret AŞ (97.45%) (automotive spring production and marketing); Metar Denizcilik AŞ (51%) (shipping and ship leasing); Ebim – Elektronik Bilgi Işlem Sanayi ve Ticaret AŞ (39%) (data processing services and marketing); Tüstaş Sinai Tesisler AŞ (17.5%) (engineering, consultancy and project planning). **Associates:** Betontaş TAŞ, Demaş AŞ, Endüstri Yatirim AŞ, Meptaş AŞ, Yapi Metal Ticaret AŞ, Binta AŞ, Demta AŞ, Celik Nakliyat ve Ticaret AŞ, Ege Yatirim AŞ. **Annual capacity:** Raw steel: 450,000 tonnes. Finished steel: 350,000 tonnes.
Works:
■ At head office. Works director: Aydin Telseren. ***Steelmaking plant:*** Two 45-tonne electric arc furnaces (annual capacity 450,000 tonnes). ***Refining plant:*** Two 45-tonne ladle furnace (450,000 tonnes). ***Continuous casting machines:*** 3/4-strand Demag 7.5m dia billet and bloom (100 x 100mm to 220 x 220mm) (annual capacity 450,000 tonnes). ***Rolling mills:*** bar comprising two 3-high stands for rebars and plain bars, nine 2-high stands for rounds 8-26mm dia (120,000 tonnes); wire rod comprising one swing forging machine 160mm sq and 80mm sq, four 45° 2-high stands 80 x 80mm (40-55mm dia), ten 370mm three roll stands (12-32mm dia) and ten 290mm dia three roll stands (5.5-12.5mm dia) (coil weight 800kg) (230,000 tonnes).
Products: Carbon steel – blooms 160 x 160mm (1986 output 261,709 tonnes); billets 100 x 100mm; hot rolled coil (for re-rolling) 5.5-12mm (169,168 tonnes); wire rod 5.5-12mm (21,830 tonnes); reinforcing bars 8-32mm (148,046 tonnes); round bars 8-32mm (236 tonnes).

MKEK Çelik Çekme Boru Fabrikasi

Head office: Tandoğan Meydani, Kirikkale, Ankara. **Postal address:** PK 21, Kirikkale. **Tel:** 4 222 61 20. **Telex:** 42223 mkga tr. **Cables:** Makkim Ankara. **Telecopier (Fax):** 2230140 ankara.
Established: 1974. **No. of employees:** Works – 420. Office – 110. **Ownership:** Makina ve Kimya Endüstrisi Kurumu (state owned). **Annual capacity:** Finished steel:

SINCE THE BEGINNING OF CIVILIZATIONS
SINCE 1937
TURKISH IRON & STEEL WORKS
PRODUCTION PROGRAMME • ALL GRADES OF PIG IRON • BLOOMS, BILLETS • RE-BARS, ANGLES, BEAMS • COKE OVEN BY-PRODUCTS • METALLURGICAL COKE AND COKE BREEZE • ELECTROLYTIC GRADE PITCH • FERTILIZERS • CASTINGS • STEEL CASTINGS UP TO 50 TONS WT. • IRON CASTINGS UP TO 20 TONS WT. • NON FERROUS CASTINGS • DESİGN & MANUFACTURE & ERECTION OF • STEEL STRUCTURES • RADIO & TV ANTENNA TOWERS • RAILWAY BRIDGES, CRANES, ETC.
GENERAL DIRECTORATE OF TURKISH IRON & STEEL WORKS
TLX: 42 506 DEIS TR/43 174 SDÇ TR

MKEK Çelik Çekme Boru Fabrikasi (continued)

15,000 tonnes.**Works:**
■ At head office. Works director: M. Yavuz Yücel. ***Tube and pipe mills:*** seamless – one piercing press 1-10" (annual capacity 15,000 tonnes), one extrusion press, and one strech reducing mill.
Products: Carbon steel – seamless tubes and pipes 25.4-273mm dia (output 6,120 tonnes); oil country tubular goods $2\frac{7}{8}$"-$9\frac{5}{8}$" dia (930 tonnes).

MKEK Celik Fabrikasi

Head office: Tandoğan, Ankara. **Tel:** 45915150/3501. **Telex:** 42172 kmke tr. **Established:** 1930. **No. of employees:** Works – 950. Office – 150. **Ownership:** MKEK – Makina Kimya Endüstrisi Kurumu (Ultimate holding company – MKEK (stated owned)). **Annual capacity:** Raw steel: 55,000 tonnes. Finished steel: 40,000 tonnes.
Works:
■ At head office. Work director: Salim Çelik. ***Steelmaking plant:*** Three electric arc furnaces – two 25-tonne and one 7-tonne (annual capacity 55,000 tonnes). ***Refining plant:*** ESR – 450mm, 750mm, 1,050mm dia (2,000 tonnes). ***Rolling mills:*** One 3-high reversing blooming (700mm) (30,000 tonnes); one 3-high reversing billet (700mm) (30,000 tonnes); one 7-stand medium section (550-440mm), one 7-stand light section (440mm) and one 7-stand bar (290mm) (coil weight 40) (combined capacity 15,000 tonnes). ***Other plant:*** Forging shop; cold drawing and peeling shop.
Products: Carbon steel – ingots, blooms, billets, round bars, square bars, flats, hexagons, bright (cold finished) bars, light channels, special light sections. Stainless steel – ingots, slabs, blooms, billets, round bars, square bars, flats, hexagons. Alloy steel (excl stainless) – ingots, slabs, blooms, billets, round bars, square bars, flats, hexagons, bright (cold finished) bars.

Orpaş Metal Sanayi ve Ticaret AŞ

Head office: Mahmutbey Köy Yolu 40, Kirazli Köyu, Bakirköy-Istanbul. **Tel:** 901 575 88 87. **Telex:** 28819 orpm.
Management: Board members – Ahmet Vefa Kükük (managing director), Mehmet Sefa Küçük, Köksal Öncel (sales manager), Mustafa Aldikaçti (purchasing manager), Feyyaz Aldikaçti. **Capital:** TL 500,000,000. **Ownership:** Aker Holdings AŞ (50%). **Associates:** Aymetal Pazarlama ve Yatirim AŞ (marketing firm), Cenk Demir Sanayi ve Ticaret AŞ. **Annual capacity:** Raw steel: 161,000 tonnes. Finished steel: 90,000 tonnes.
Works:
■ At head office and Davutpaşa Cad 97, Topkapi-Istanbul. (Tel: 575 14 25. Telex: 28819 orpm). Works director: Yaşar Güven. ***Steelmaking plant:*** Two electric arc furnaces – one 15-tonne KGV (annual capcity 45,000 tonnes) and one 45-tonne Man GHH electromelt (120,000 tonnes). ***Continuous casting machines:*** One 4-strand Concast billet (110 x 110 x 3,000) (250,000 tonnes). ***Rolling mills:*** bar – (coil weight 250kg) (80,000 tonnes).
■ Kartal Works, Plant No 2, Yakacik Asfalti, Hayrat Kuyu Mevki, 30/2 Kartal-Istanbul. (Tel: (901) 353 46 85). Works director: Metin Ejder. ***Steelmaking plant:*** One 15-tonne KGV electric arc furnace (35,000 tonnes). ***Rolling mills:*** bar (35,000 tonnes).
Products: Carbon steel – ingots (output 32,000 tonnes); billets (107,000 tonnes); wire rod; reinforcing bars.

TDCI – Türkiye Demir ve Çelik Işletmeleri Genel Müdürlügü

Head office: Ziya Gökalp Cad 80, Kurtuluş, Ankara. **Tel:** 41 33 99 85. **Telex:** 42506 deis tr, 43174 sdc tr. **Telecopier (Fax):** 41 34 4705-6.
Works:
■ Karabük. ***Coke ovens:*** Two batteries (44 ovens) (annual capacity 285,000 tonnes); two batteries (56 ovens) (408,000 tonnes); three batteries (63 ovens) (407,000

Turkey

TDCI – Türkiye Demir ve Çelik Işletmeleri Genel Müdürlügü (continued)

tonnes). ***Sinter plant:*** One 60 sq m strand (450,000 tonnes); one 90 sq m strand (890,000 tonnes). ***Blast furnaces:*** two 4,572mm hearth dia, 416 cu m (150,000 tonnes each); one 7,400mm hearth dia, 925 cu m (300,000 tonnes). ***Steelmaking plant:*** Six 150-tonne open hearth furnaces (600,000 tonnes). ***Rolling mills:*** 1-stand blooming (515,000 tonnes); 3-stand billet (450,000 tonnes); 4-stand heavy section; 4-stand rail (100,000 tonnes); 4-stand medium section; 3-stand light section; 2-stand bar (45,000 tonnes); 25/19-stand wire rod (coil weight 400kg) (300,000 tonnes).
■ Iskenderun Iron & Steel Works. ***Coke ovens:*** Four batteries (69 ovens each) (annual capacity 2,000,000 tonnes). ***Sinter plant:*** Four 75 sq m strand (3,200,000 tonnes). ***Blast furnaces:*** Two 8,200mm hearth dia, 1,386 cu m (550,000 tonnes each); one 9,750mm hearth dia, 2,000 cu m (1,350,000 tonnes). ***Steelmaking plant:*** Three 130-tonne basic oxygen converters (2,200,000 tonnes). ***Continuous casting machines:*** Two bloom – vertical type and bow type (265 x 340mm, 260 x 260mm, 200 x 200mm, 260 x 335mm) (2,200,000 tonnes). ***Rolling mills:*** 10-stand billet (1,510,000 tonnes); 13-stand medium section (700,000 tonnes); 23-stand light section (500,000 tonnes); 37-stand (4 finishing stands) wire rod (coil weight 600kg) (500,000 tonnes).
Products: Steelmaking pig iron (1986 output 1,977,286 tonnes). Foundry pig iron (277,337 tonnes). Carbon steel – ingots 500 x 500mm (608,328 tonnes); blooms 260 x 260mm-260 x 340mm (1,416,059 tonnes); billets 75 x 75mm-112 x 112mm (1,862,685 tonnes); wire rod 8-32mm dia (937,414 tonnes); flats 10-20mm and 70-150mm wide; light angles 40 x 40mm-60 x 60mm (50,688 tonnes); light channels 65mm (7,515 tonnes); medium angles 80 x 80mm-120 x 120mm (12,317 tonnes); medium channels 80-120mm (4,645 tonnes); wide-flange beams (medium) 80-120mm (7,920 tonnes); heavy angles 150 x 150mm (1,727 tonnes); heavy channels 160-300mm (19,945 tonnes); wide-flange beams (heavy) 160-380mm (38,028 tonnes); heavy rails 46.3-47/49/59kg/m (7,323 tonnes); light rails 10-19kg/m (2,043 tonnes).
Expansion programme: Karabük: Blast furnace capacity to be raised to 900,000 tpy. Sinter plant, 35 sq m to be extended to 90 sq m. Power plant capacity to be raised to 25MW (now 8MW).

Ümran Spiral Kaynakli Boru Sanayii AŞ

Head office: Meclisi Mebusan Caddesi 323/5, Salipazari 80040, Istanbul. **Tel:** 1430372. **Telex:** 24340 orya tr. **Cables:** Oryahan-Istanbul. **Telecopier (Fax):** 1494349.
Management: Orhan Yavuz (president); Murat Yavuz (board member, foreign affairs); Fehmi Başkut (foreign relations and marketing executive); Hamit Gözaydin (general co-ordinator). **Established:** 1968. **Capital:** TL2,500,000,000. **No. of employees:** Works – 480. Office – 47. **Subsidiaries:** Ümran Taşimacilik AŞ (transport); Oryataş Inşaat ve Ticaret AŞ (construction). **Annual capacity:** Finished steel: 147,000 tonnes (spirally welded pipes).
Works:
■ Alemdağ Cad 153, Ümraniye-Istanbul. (Tel: 3353504. Telex: 29462 orya tr). Works director: Nazim Miras. ***Tube and pipe mills:*** welded – one spiral 8″-80″ (annual capacity 147,000 tonnes). ***Other plant:*** Pipe coating plant – polyethylene, coal tar, epoxy, bitumen. Pipe lining plant – cement mortar, epoxy, bitumen, coal tar.
Products: Carbon steel – pipe piling 8″-48″ (output 10,000 tonnes); spiral-weld tubes and pipes 8″-80″ (48,000 tonnes).
Expansion programme: A new ERW longitudinally welded pipe plant 1/2″-16″ under construction; estimated time for starting production end 1987, estimated capacity over 200,000 tonnes.

Yücel Boru ve Profil Endustrisi AŞ

Head office: Rihtim Caddesi 16, Kadikoy, Istanbul. **Tel:** 1 337 8680, 337 8681. **Telex:** 29157, 29360.
Management: S.O. Oktar (general manager). **Annual capacity:** Pipe and Tube: 600,000 tonnes.
Works:
■ Gebze.
Products: Carbon steel – longitudinal weld tubes and pipes.

United Kingdom

Accles & Pollock

Head office: Rounds Green Road, Oldbury, Warley, W Midlands. **Tel:** 021 552 1500. **Telex:** 338141.
Established: 1897. **No. of employees:** 250. **Ownership:** TI Group PLC.
Works:
■ At head office. Tube drawing, bright annealing and finishing plant.
Products: Stainless steel – seamless tubes and pipes, cold drawn tubes and pipes, precision tubes and pipes.

Airedale Iron & Steel Co Ltd

Head office: Unit 20, Butterfields Industrial Estate, Otley Road, Shipley, W Yorks. **Tel:** 0274 596392.
Management: Directors – Bryan Lunn (managing), Martin Karl Lunn (works). Andrew Marshall(technical). **Established:** 1985. **Capital:** £400,000. **No. of employees:** Works – 4. Office – 2. **Subsidiaries:** Biywell Products Ltd (ferrous and non-ferrous castings). **Annual capacity:** Pig iron 3,000 tonnes.
Works:
■ At head office. **Iron making plant**. Two 3-tonne induction furnaces (annual capacity 3,000 tonnes).
Products: Steelmaking pig iron, foundry pig iron (1986 output 1,500 tonnes).

Allied Steel & Wire Ltd

Head office: PO Box 83, Castle Works, Cardiff CF1 5XQ. **Tel:** 0222 471333. **Telex:** 498295. **Telecopier (Fax):** 0222 488256.
Management: Chief executive – Alan Cox, directors – Malcolm Wallace (managing), Paul Rich (executive, corporate development), Chris Lyddon (finance); company secretary – George Barlett. **Established:** 1981. **Capital:** Share capital: ordinary – £117,000,000. Preference – £51,000,000. **No. of employees:** Works – 2,124. Office – 1,011. **Subsidiaries:** Barfab Reinforcements, Smethwick; General Steel Services, Belfast; Castle Wire, Cardiff; Reinforcement Steel Services, Sheffield; McCalls Special Products, Sheffield; Power Steels & John Smith Wire, Birmingham; Somerset Wire, Cardiff; Castle Nails, Cardiff; Cardiff Carbides, Cardiff. **Annual capacity:** Raw steel: 780,000 tonnes. Finished steel: 1,570,000 tonnes.
Works:
■ Tremorfa Steelworks, PO Box 37, Tremorfa Works, Cardiff. CF2 2YX. (Tel: 0222 471333). Telex: 498295. Works director: G. Sheehan. ***Steelmaking plant:*** Two 100-tonne EBT electric arc (annual capacity 780,000 tonnes). ***Continuous casting machines:*** Two 6-strand Concast/Rokop billet (product dimensions – 90,100, 110, 115, and 127mm sq), (780,000 tonnes).
■ Cardiff Rod Mill, PO Box 15, Castle Works, Cardiff CF1 5XN. (Tel: 0222 471333. Telex: 498295). Works director: R.E. Judd. ***Rolling mills:*** One 2-strand wire rod

United Kingdom

Allied Steel & Wire Ltd (continued)

comprising five 18", five 16", four 14", two 10" stands; and two 8" and eight 16" non-twist stands (coil weight 1.5 tonnes) (annual capacity 430,000 tonnes).
■ Cardiff Bar and Section Mills, PO Box 16, Tremorfa Works, Cardiff CF2 2YW. (Tel: 0222 482501. Telex: 498295). Works director: G.W. Richards. ***Rolling mills:*** One medium section comprising one vertical stand (430mm dia) and seven horizontal stands (460-565 mm dia (annual capacity 240,000 tonnes); One bar 15 horizontal stands (350-550mm dia) (300,000 tonnes).
■ Scunthorpe Rod Mill, PO Box 7, Scunthorpe, South Humberside DN16 1BQ. (Tel: 0724 282222. Telex: 527477). Works director: W.E. Bagnall. ***Rolling mills:*** One 4-strand wire rod comprising fifteen traditional and ten non-twist stands (coil weights 1,360kg, 1,500kg, 1,820kg (annual capacity 600,000 tonnes).
Products: Carbon steel – billets 90, 100, 110, 115 and 127mm sq; wire rod 5.0-13.0mm (in 0.5mm increments); reinforcing bars 16-50mm; round bars 16.50mm; flats 40 x 5-150 x 25mm; light angles 25 x 25-80 x 80mm; light channels 76 x 38mm; medium channels 102 x 51-127 x 64mm; bright wire; black annealed wire; galvanized wire (plain). Welded mesh and reinforcement; coppered wire; nails.

Alloy Steel Rods Ltd—see BSC – British Steel Corp, BSC Stainless

Alphasteel Ltd

Head office: 77 South Audley Steet, London W1Y 5TA. **Tel:** 01 493 8791. **Telex:** 23151 alpha g. **Telecopier (Fax):** 01 499 2208.
Management: Werner Meyerhans, Theobald Skoda, Maurice Webb. **Established:** 1974. **Annual capacity:** Raw steel: 1,500,000 tonnes.
Works:
■ Corporation Road, Newport, Gwent NP9 0XE (Tel: 0633 290288. Telex: 498502 alpha g. Fax: 0633 290240). ***Steelmaking plant:*** Four 100-tonne electric arc (60 MVA) (annual capacity 1,500,000 tonnes). ***Continuous casting machines:*** Three 4-strand billet; one 1-strand slab.
Products: Carbon steel – slabs up to 1,550mm wide and 180mm thick; billets 120-140mm sq.

Thomas Andrews & Co Ltd

Head office: St Thomas Steelworks, Aizlewood Road, Sheffield S8 0YW. **Tel:** 0742 58 9511. **Telex:** 54174.
Associates: F.M. Parkin (Sheffield) Ltd. Alloy engineering and tool steels.

Armco Ltd

Head office: Jubilee Road, Lethworth, Herts SG6 1NQ. **Tel:** 046 26 6588. **Telex:** 825040. **Cables:** Armco Letchworth.
Products: Carbon steel – longitudinal weld tubes and pipes – double wall brazed pipe.
Additional information: In January 1987, TI Group acquired most of Armco's steel operations in Europe.

Armstrong Tube Co Ltd

Head office: Manor Lane, Shipton Road, York YO3 6UA. **Tel:** 0904 59833. **Telex:** 57826.
Products: Carbon steel – longitudinal weld tubes and pipes and cold drawn tubes and pipes 31-70mm.

A.S. Rolling Mills Ltd

Head office: Lower High Street, Cradley Heath, Warley, West Midlands B64 5AQ. **Tel:** 0384 67961.
Products: Carbon steel – flats 10-35mm wide.

Bar Bright Ltd

Head office: Lichfield Road, Brownhills, West Midlands WS8 6LD. **Tel:** 0543 371071. **Telex:** 338561.
Products: Carbon steel – bright (cold finished) bars. Alloy steel (excl stainless) – bright (cold finished) bars.

Barlow Bright Steels, Glynwed Steels Ltd

Head office: Mounts Road, Wednesbury, West Midlands WS10 0DU. **Tel:** 021 502 2411. **Telex:** 337446.
Management: Directors – B.R. Davies (managing), G. Walker (commercial), L. Clark (engineering). **Established:** 1980. **Capital:** Works – 31. Office – 18. **Ownership:** Glynwed Steels Ltd (Ultimate holding company – Glynwed International PLC). **Annual capacity:** Finished steel: 17,000 tonnes.
Products: Carbon steel – bright (cold finished) bars – rounds, squares and hexagons – (Output in January-June 1987 7,551 tonnes). Alloy steel (excl stainless) – bright (cold finished) bars – rounds, squares and hexagons – (January-June 1987 515 tonnes).

United Kingdom

Barton Engineering Ltd

Head office: Birchills, Walsall, W Midlands WS2 8QE. **Tel:** 0922 26581. **Telex:** 338282 barwal g. **Cables:** Ingenuity, Walsall. **Telecopier (Fax):** 0922 646675.
Management: Directors – W.R. Pickering (managing), J.W. Parkinson (assistant managing), M.E. Jones (manufacturing), M.C. Evans (engineering), J.I. Clark (financial); Precision Tube Div sales manager – R.J. Sweet. **Established:** 1936. **Capital:** £3,600,000. **No. of employees:** Works – 430. Office – 130. **Ownership:** Carparo Industries PLC (100%) (Ultimate holding company – Caparo Group Ltd).
Works:
■ At head office. ***Tube and pipe mills:*** welded – five ERW precision (annual capacity 40,000 tonnes).
Products: Carbon steel – longitudinal weld tubes and pipes (1.30-17mm) precision tubes and pipes.

Barworth Flockton Ltd

Head office: Ecclesfield, Sheffield, South Yorks S30 3X. **Tel:** 0742 468291. **Telex:** 54657. **Telecopier (Fax):** 0742 454230.
Management: Directors – W.N. Edwards (managing), A.J. Briggs (financial); managers – D. Wragg (works), B. Jackson (production), A. Colley (sales), A. Robertson (technical). **Established:** 1950. **Capital:** £500,000. **No. of employees:** Works – 140. Office – 67. **Ownership:** Barworths Holding Ltd, Sheffield. **Associates:** Moss & Gamble Bros Ltd (steel forgers); G. & J. Hall Ltd (engineers tool manufacturers) – Wholly owned by Barworth Holdings Ltd. **Annual capacity:** Finished steel: approx 5,700 tonnes.
Works:
■ At head office. One medium frequency induction furnace (1.5 tonnes) (annual capacity 2,000 tonnes). **hot mills:** Two four-stand 2-high (250m/m) (1,200 tonnes). 800-tonnes push-down forging press (4,500 tonnes). GFM 5 × 16 forging machine (4,500 tonnes).
Products: Alloy steel (excl stainless) – billets; forgings; round bars; square bars; flats – full range of tool steels and high-speed steels. Bright (cold finished) bars; Full range of tool steels & high speed steels to BS4659, AISI and Werkstoff specifications.
Sales offices/agents: At head office and warehouses in London and Birmingham.

Beam Tubes Ltd

Head office: Moulton Park Industrial Estate, Northampton NN3 1QQ. **Tel:** 0604 48246.
Products: Carbon steel – longitudinal weld tubes and pipes $\frac{3}{8}$-$1\frac{1}{4}$".

Bedford Steels

Head office: 151 Effingham Road, Sheffield S4 7YS. **Tel:** 0742 769643. **Telex:** 54172. **Cables:** Bedsteels Sheffield.
Management: D.L. Oldale (chairman), P. Barlow (general manager). **Established:** 1972. **Capital:** £100. **No. of employees:** Works – 125. Office – 25. **Ownership:** Gardner Denver Holdings (UK) Ltd, Dronfield, Sheffield. (Ultimate holding company – Cooper Industries Inc, Houston, TX, USA). **Annual capacity:** Finished steel: 25,000 tonnes.
Works:
■ At head office. ***Rolling mills:*** Two bar one 305mm double duo (annual capacity 20,000 tonnes) and one 228mm 3-high (5,000 tonnes). ***Other plant:*** Facilities for centreless bar peeling/turning and precision grinding, billet boring, core extraction, reeling and cold sawing.
Products: Carbon steel – round bars, square bars, flats, hexagons, bright (cold finished) bars, hollow bars, hollow sections. Stainless steel – round bars, square bars, flats, hexagons, bright (cold finished) bars, hollow bars. Alloy steel (excl stainless) – round bars, square bars, flats, hexagons, bright (cold finished) bars, hollow bars. Bolt, nut and rivet steels; bright drawing quality black bars; heat-treated bars, octagons, half-rounds, three-squares – mostly file steel; valve steel for motor vehicles; centreless ground alloy steel bright bars; mining drill steels, hollow or solid; stainless steel lined hollow bar; leaded steels and spring steels.
Brands: Bedrock (material specified for rock-drilling steel).
Sales offices/agents: At head office. Agents – DES Inc, Glen Afton Drive, Burlington, Ont, Canada (for USA/Canada); T.F. Schultz KG, Ross Strasse 13, D 4000 Düsseldorf 30, West Germany.

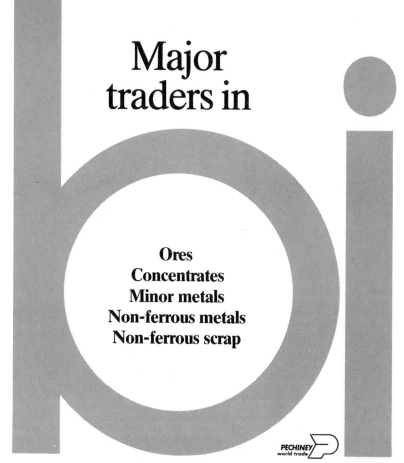

Ben Bennett Jr Ltd

Head office: Lisle Road, Rotherham, S Yorks S60 2RL. **Tel:** 0709 382251. **Telex:** 922488 bureau g, ref bbj. **Cables:** Silent Rotherham.
Management: Ben Bennett Sr (joint managing director), Ben Bennett Jr (chairman and joint managing director),Ben Bennett (director) Philipp Udell (director, secretary). **Established:** 1885. **Capital:** £25,000. **No. of employees:** Works – 61. Office – 20. **Subsidiaries:** None. **Associates:** None.
Works:
■ Eastwood Rolling Mills, Fitzwilliam Road, Rotherham S65 1SH. (Tel: 0709 382006. Telex: 922488 bureau g ref: bbj). Works director: Ben Bennett. ***Rolling mills:*** Three cold strip – two single-stand 4-high 10" (coil weight 1,000kg), and one single-stand 4-high 4" (coil weight 75kg). ***Other plant:*** Annealing and slitting plant, hardening and tempering furnace lines , including primary strip polishing and edge-dressing plant.
Products: Carbon steel – cold rolled hoop and strip (uncoated) 0.004-0.050" thick, 10/0.187" wide; cold rolled strip, hardened and tempered, 0.004-0.050" thick, 10/0.187" wide.

Bentham International Ltd

Head office: Bentham Works, Witcombe, Gloucester GL3 4UB. **Tel:** 0452 863301. **Telex:** 43189. **Telecopier (Fax):** 0452 864718.
Management: M.J. Smith (managing director), A.J. Eden (sales director). C. Sherwood (sales manager). **Established:** 1980. **No. of employees:** Works – 100. Office – 20. **Ownership:** Independent. **Annual capacity:** Finished steel: 4,000 tonnes.
Works:
■ At head office. ***Tube and pipe mills:*** welded – two 6-72" od Brake press (annual capacity 4,000 tonnes).
Products: Stainless steel – light angles from approx 3 x 3" (output 1986 300 tonnes); light channels, special light sections, longitudinal weld tubes and pipes 6-72" od (4,000 tonnes). Alloy steel (excl stainless) – longitudinal weld tubes and pipes 6-72" od (300 tonnes); large-diameter tubes and pipes 6-72" od (300 tonnes).

Leon M. Berner & Co Ltd

Head office: Ledra Works, Reservoir Place, Walsall, West Midlands WS2 9SN. **Tel:** 0922 612545. **Telex:** 337546. **Cables:** Berner Walsall Telex.
Management: J.P. McGowan (chairman), J.H. Hussey (managing director), J.G. Oebel (director and general manager), L. Payne (works director), R. Savage. **Established:** 1952. **Capital:** £150,000. **No. of employees:** Works – 18. Office – 8. **Ownership:** Leon Berner Group Ltd (100%) (Ultimate holding company – Wheway PLC (100%)).
Works:
■ At head office. Works director: Mr L. Payne. ***Rolling mills:*** Two cold strip – one 4-high Robertson (10½/5½") (coil weight 1.50 tonnes) and one 2-high Robertson (15½/12") (coil weight 1.75 tonnes) (combined annual capacity 5,000 tonnes).
Products: Carbon steel – cold rolled hoop and strip (uncoated).

Brasway PLC, Bright Bar Division

Head office: All Saints Road, Wednesbury, West Midlands WS10 9LN. **Tel:** 021 526 3089. **Telex:** 339594.
Works:
■ At head office. Bright drawing equipment, heat treatment, pickling and phosphating plant.
Products: Carbon steel – bright (cold finished) bars. Stainless steel – bright (cold finished) bars, Alloy steel (excl stainless) – bright (cold finished) bars, rounds, hexagons, squares.

Brasway PLC, Tubes Division

Head office: All Saints Road, Wednesbury, West Midlands. **Tel:** 021 526 3089.
Works:
■ At head office. ***Tube and pipe mills:*** welded Three ERW.
Products: Carbon steel – longitudinal weld tubes and pipes, round, oval, square and rectangular 12-76mm od (up to 38,000 tonnes).

The Bright Steel Co (Tipton) Ltd

Head office: Castle Street Works, Tipton, West Midlands. **Tel:** 021 557 1831-3. **Telex:** 336830.
Products: Carbon steel – cold rolled hoop and strip (uncoated).

Bright Steels Ltd

Head office: Norton Works, Malton, N Yorks YO17 9BD. **Tel:** 0653 694961. **Telex:** 57925. **Telecopier (Fax):** 0653 695856.
Management: Directors – P.N. Chouler (managing, chairman), L.P. Chouler, B. Shepherd (company secretary), R.W. Bates (commercial), K. Gardner (sales), P.J. Knaggs (works). **Established:** 1919. **No. of employees:** Works – 100. Office – 40. **Annual capacity:** Finished steel: 40,000 tonnes.
Works:
■ At head office.
Products: Carbon steel – bright (cold finished) bars – flats 12-300mm wide and 6-100mm thick; squares 12-150mm, hexagons 12-95mm; special shapes. Alloy steel (excl stainless) – (including free cutting, case hardening, key steels) bright (cold finished) bars – Flats 12-300mm wide and 6-100mm thick; squares 12-150mm, hexagons 12-95mm, special sections.

British Bright Bar Ltd

Head office: Bloomfield Road, Tipton, W Midlands DY4 9EP. **Tel:** 021 520 8141. **Telex:** 336214 bbbtpt g. **Telecopier (Fax):** 021 520 8929.
Management: Chairman – B.D. Insch; deputy chairman – N. Parker; directors – J.B. Atherton (managing, chief executive), R. Peake (financial, company secretary), D. Jackson (commercial), I.T. Dalloway (production); non-executive directors – A.D.P. Milne, D.C. Canham; managers – P. Bramall (technical), D.C. Thomas (works), M. Parker (export sales), R. Norton (commercial). **Established:** 1983 (merger of bright bar activities of Exors of James Mills Ltd, Flather Bright Steels Ltd and British Rolling Mills Ltd). **Capital:** Authorised – £5,000,000. Issued – £1,500,000. **No. of employees:** Works – 170. Office – 75. **Ownership:** British Steel Corp (40%); GKN PLC (40%); Brymill Ltd (20%). **Subsidiaries:** Trading division – Nationwide Steelstock (100%) (bright steel bar stockholding warehouses at Oadby, Leicester; Hetton-Le-Hole, Tyne and Wear; Ossett, W Yorks; Thornbury, Avon). **Annual capacity:** Finished steel: 60,000 tonnes.
Works:
■ At head office. Five Schumags – one 3B, two 2B, one 1B and one 1S; Etibar combined drawing machine; three bar drawbenches – 80-tonne, 60-tonne and 40-tonne; three Farmer Norton centreless turning machines – one HS50, one MS104 and one 7"; three Lidkoping centreless grinding machines – two 5B and one 4B;three Grunewald hardening and tempering furnaces for heat treating bars (up to 9.5m length and coils to 2,000 kg weight); one Wellman controlled atmosphere annealing/normislling furnace; precision sawing facilities; bar and coishot blast equipment.
Products: Carbon steel – bright (cold finished) bars – rounds 6-180mm, hexagons 6-75mm, squares 6-75mm, flats; and special sections to customers requirements. Alloy steel (excl stainless) – (heat treated steels) bright (cold finished) bars rounds 6-180mm, hexagons 6-75mm, squares 6-75mm, flats and special sections to customers requirements.
Brands: Hitenspeed 45A and 55; Hitest; Brytest; Disco 21.
Sales offices/agents: At head office. Agents – Belgium, Irish Republic, France, West Germany, Holland, USA.

United Kingdom

British Rolling Mills Ltd

Head office: Brymill Steel Works, Tipton, West Midlands DY4 9EF. **Postal address:** PO Box 10. **Tel:** 021 557 3939. **Telex:** 337724 brymil g.
Products: Carbon steel – cold rolled hoop and strip (uncoated) upto 14½" wide.

British Uralite PLC

Head office: Higham Works, Higham, Rochester, Kent ME3 7JA. **Tel:** 0474 82 3451. **Telex:** 27837 . **Telecopier (Fax):** 0474 82 3961.
Management: R.W. Powell (chairman), N.O. Lance (managing director), D.M. Phillips (marketing director), W.D. Ritchie (company secretary(. **Established:** 1928. **No. of employees:** 260.
Works:
■ At head office. ***Other plant:*** Cold roll-forming line for sections.
Products: Carbon steel – cold roll-formed sections (from strip). Pipes and fittings, heat shields.

Bromford Iron & Steel Co Ltd

Head office: Bromford Lane, West Bromwich, W Midlands B70 7JJ. **Tel:** 021 553 6121. **Telex:** 338380. **Telecopier (Fax):** 021 553 2019.
Management: Directors – R.W.F. Yates (managing), G.J. Godbert (hot mill); F.G. Butler, C.H. Payne, C.M. Crew. **Established:** 1780. **Capital:** £1,750,000. **No. of employees:** Works – 100. Office – 42. **Ownership:** CI Group PLC.
Works:
■ At head office. ***Rolling mills:*** Two light section – 200 and 250mm. ***Other plant:*** Pickling plant for sheets, plates, wire, rod, components, etc.
Products: Carbon steel – flats 80-12mm wide, 12-20mm thick; light angles 25 x 25 x 3-6 to 30 x 30 x 3-6. Reinforcing mesh.
Sales offices/agents: Klaus J. Ragoss GmbH, Hamburg, West Germany; Nillesen BV, Rotterdam, Netherlands; British Steel Corp, Copenhagen, Denmark.

Bruntons (Musselburgh) Ltd

Head office: Musselburgh, Lothian EH21 7UG. **Tel:** 031 665 3888. **Telex:** 72212. **Cables:** Wiremill, Musselburgh. **Telecopier (Fax):** 031 665 0486.
Management: Directors – B. Mycock (executive), D.R. Frazer (general manager, wire rope division), P.C. Adams (general manager, steel products division). **Established:** 1876. **No. of employees:** Works – 336. Office – 100. **Ownership:** Carclo Engineering PLC.
Products: Carbon steel – cold rolled hoop and strip (uncoated). Steel wire – rope wire, piano wire, high-duty spring wire, armature binding wire, stainless steel wire, wire ropes; cables; tie rods; aerial rods; tool steel rods; bright drawn steel sections in alloy, stainless and heat-resisting steels.
Brands: Kilindo (non-rotating wire ropes); Beacon (stainless steel wire and prelay control cables); Jupiter (valve spring wire); Atlas (armature binding wire).

Brymbo Steel Works—see UES – United Engineering Steels Ltd

BSC – British Steel Corp

Head office: 9 Albert Embankment, London SE1 7SN. **Tel:** 01 735 7654. **Telex:** 916061. **Cables:** Bristeelon SE1. **Telecopier (Fax):** 01 587 1142.
Management: Members of the Corporation at 31 March 1987 – Sir R. Scholey (chairman); Sir Ronald Halstead (deputy chairman, non-executive); M.E. Llowarch (chief executive); F. Fitzgerald (managing director, technical); D. Grieves (managing director, personnel and social policy); G.H. Sambrook (chairman and chief executive, BSC

BSC - British Steel Corp (continued)

General Steels); J.G. Stewart (chairman and chief executive, BSC Strip Products); J.D. Birkin (non-executive); J.F. Eccles (non-executive); Lord Gregson (non-executive); S.J. Gross (non-executive); H.L.I Runciman (non-executive); A.E. Wheatley (non-executive). Group executives and managing directors, operating groups and businesses – G.H. Sambrook (chairman and chief executive, BSC General Steels); J.G. Stewart (chairman and chief executive, BSC Strip Products); P.D. Allen (managing director, operations, BSC Strip Mill Products); H. Ford (managing director, BSC Tubes); G.D. Saul (managing director, operations, BSC General Steels); A.P. Pedder (director and chief executive, BSC Stainless). Head office – F. Fitzgerald (managing director, technical and chairman, BSC (Overseas Services) Ltd); D. Grieves (managing director, personnel and social policy and chairman, BSC Diversified Activities); B.S. Moffat (managing director, finance); T.J. MacDonald (corporation secretary). **Established:** 1967. **Capital:** £3,046,000,000. **No. of employees:** 52,000. **Ownership:** State-owned.
Subsidiaries: These include associated companies as at 28th March, 1987 – UK: Holding companies (100%) – British Steel Corp (International) Ltd; British Steel Corp (Investments) Ltd; British Steel Corp Liaison Services Ltd; Brierley Hill Investments Ltd; St Andrews Shipping Co Ltd ; Stanton & Staveley (Overseas) Ltd; Stewarts & Lloyds (Overseas) Ltd.
Steel producing, further steel processing and allied activities – Afon Tinplate Co Ltd (25%); Air Products Llanwern Ltd (50%); Allied Steel & Wire (Holdings) Ltd (48%); Alloy Steel Rods Ltd (100%); Bridon PLC (10%); British Bright Bar Ltd (40%) ; British Steel Service Centres Ltd (100%); British Tubes Stockholding Ltd (100%); Cold Drawn Tubes Ltd (25%); Clyde Shaw Ltd (50%); Darlington & Simpson Rolling Mills PLC (50%); Fox Wire Ltd (100%); GR-Stein Refractories Ltd (22%) ; Hadfields Holdings Ltd (38%); London Works Steel Co Ltd (100%); Midland Rollmakers Ltd (50%); Railway and Ring Rolled Products Ltd (100%); Wheelset Manufacturers Ltd (100%); Ring Rolled Products Ltd (100%); Seamless Tubes Ltd (75%); Sheffield Forgemasters Holdings PLC (50%); The Templeborough Rolling Mills Ltd (50%); The Victaulic Co PLC (30%) Tinsley Bridge Ltd (100%); TWIL Ltd (20%); United Engineering Steels Ltd (50%); United Merchant Bar PLC (25%); Whitehead (Narrow Strip) Ltd (100%); Vacmetal (UK) Ltd (45%).
Metal recovery and slag processing – Appleby Slag Co Ltd (33%); Appleby Slag Reduction Ltd (33%); Cambrian Stone Ltd (49%); Colvilles Clugston Shanks (Holdings) Ltd (33%); Margam Slag Co Ltd (49%); Scunthorpe Slag Ltd (40%); Teesside Slag Ltd (50%).
Chemicals and chemical by-products – Benzole Producers Ltd (26%); Staveley Chemicals Ltd (45%).
Selling and other companies – British Steel Corp (Exports) Ltd (100%); British Steel Corp (Tubes) Exports Ltd (100%); British Steel Corp (Overseas Services) Ltd (100%); British Steel Corp (Industry) Ltd (100%); British Steel Corp (Property) Ltd (100%); British Steel Corp Liaison Services (India) Ltd (100%); BSC Brokers Ltd (100%); BSC Overseas Management Services Ltd (100%); BSC Pension Fund Trustee Ltd (100%); Flixborough Wharf Ltd (25%); Workington Welfare Hall Ltd (50%).
Overseas companies: Argentina – Ostrimet SA (40%); Belgium – Fischer Profielen GVC (100%), Fischer en Co-Handelsvennootschap NV (100%); Canada – British Steel Canada Inc (100%), Firth Brown Stainless Ltd (100%), H.M. Long Inc (100%), Metal West, (Alberta), Normines Inc (42%); France – British Steel Corp (France) SA (100%), BSC Stainless SA (100%), Profilacier Sarl (100%), United Channel Commerce SNC (100%), Huet et Lanoe SA (34%); West Germany – Walter Blume GmbH (100%), British Steel Corp (Deutschland) GmbH (100%), BSC Stainless GmbH (100%), Fischer Profil GmbH (100%); Hong Kong – British Steel Corp (China) Ltd (50%); India – Stewarts & Lloyds of India Ltd (40%); Irish Republic – British Steel Corp (Ireland) Ltd (100%); Netherlands Antilles – KDMP Co NV (100%). Netherlands – Feyen Steelservice BV (100%), Van der Vliet & de Jouge BV; New Zealand - British Steel Corp (NZ) Ltd (100%); Norway – BSC Norge AS (100%); South Africa – British Steel Corp (South African Sales) (Pty) Ltd (100%); Sweden – British Steel Corp Svenska AB (100%); Switzerland – Concast Holding AG (22%); USA – BSC Alloys Inc (100%), British Steel Corp Inc (100%), British Steel (Sheet & Tube) Corp Inc (100%), Energy

United Kingdom

BSC – British Steel Corp (continued)

Steel Products Corp (100%), Fox Metals Inc (100%), Oremco Inc (30%); Turkey – British Steel Ticaret (50%). **Annual capacity:** Iron: 14,600,000 tonnes. Steel: 17,100,000 tonnes.

BSC General Steels
Works:
■ Scunthorpe Works, Scunthorpe, South Humberside DN16 1BP. (Tel: 0724 280280. Telex: 52601). Works director: H. Homer. ***Coke ovens:*** Three WD Koppers batteries (75 ovens) (annual capacity 606,000 tonnes), three Koppers batteries (99 ovens) (467,000 tonnes) and one WD Koppers battery (43 ovens) (287,000 tonnes). ***Sinter plant:*** One 2-strand 155 sq m and one 125 sq m (combined capacity 1,150,000 tonnes) and one 2-strand 291 sq m (under construction). ***Blast furnaces:*** One 8.28m hearth dia, 1,290 cu m (1,040,000 tonnes), one 8.30m hearth dia, 1,315 cu m (936,000 tonnes), one 9.51m hearth dia, 1,488 cu m (1,092,200 tonnes) and one 9.36m hearth dia, 1,437 cu m (1,092,200 tonnes) (normal operation 3 out of 4). ***Steelmaking plant:*** Three 300-tonne LD basic oxygen converters (3,500,000 tonnes). ***Refining plant:*** One 300-tonne RH vacuum degassing ladle unit; Ca-Si injection/Argon stirring/Pb facilities. ***Continuous casting machines:*** 8-strand Rokop billet (100, 115 and 140mm sq) (500,000 tonnes); 4-strand Demag bloom (750 x 355mm) (900,000 tonnes); 2-strand Demag slab (1,020-2,000 x 230-305mm) (9,000 tonnes). ***Rolling mills:*** Two 2-high reversing blooming (1,400 x 2,600/1,060 x 2,200) and 10 continuous stand billet (1,015 x 1,370/670 x 1,220) (combined capacity 2,000,000 tonnes); 2-high reversing blooming (1,092 x 2,670) and heavy section comprising five 2-high reversing and one continuous stand (combined capacity 333,000 tonnes); 2-stand 2-high reversing (incl universal) and 10-stand continuous medium section (550,000 tonnes); one 4-high reversing (990 x 3,700) and one 2-high reversing (1,180 x 3,700) heavy plate (550,000 tonnes). ***Other plant:*** Two pit arch machines (1,000 tonnes and 2,500 tonnes per week).
■ Teesside Works, Steel House, Redcar, Cleveland TS10 5QW. (Tel: 0642 474111. Telex: 587401). Works director: D. Ward. ***Coke ovens:*** Two OSC batteries (132 ovens) (annual capacity 1,260,000 tonnes) and two OSC batteries (88 ovens) (640,000 tonnes). ***Sinter plant:*** One 1-strand 3,364 sq m (4,000,000 tonnes). ***Blast furnaces:*** one 14m hearth dia, 3,628 cu m (3,380,000 tonnes), one 5.64m hearth dia, 567 cu m (300,000 tonnes) and one 5.49m hearth dia, 568 cu m (280,000 tonnes). ***Steelmaking plant:*** Three 260-tonne LD basic oxygen converters (3,600,000 tonnes). ***Refining plant:*** One 260-tonne RH Ladle vacuum degassing unit; Argon stirring/powder injection. ***Continuous casting machines:*** 8-strand Concast bloom (254-483 x 254 x 305mm) (900,000 tonnes); Two 2-strand Concast slab (1,024-2,030 x 195-254mm) (2,100,000 tonnes). ***Rolling mills:*** one 2-high reversing 1,240mm slabbing and one 2-high 1,340mm blooming (combined capacity 880,000 tonnes); heavy section – two 2-high edgers 1,170mm and two 2-high universal 1,400mm (612,000 tonnes); wide hot strip – one 4-high roughing 997mm and six 4-high continuous 745mm (coil weight 23 tonnes) (1,400,000 tonnes). ***Tube and pipe mills:*** Hartlepool – one submerged arc welded 20-44" (250,000 tonnes).
■ Skinningrove Works, Carlin How, Saltburn-by-the-Sea, Cleveland TS13 4ET. ***Rolling mills:*** heavy section – one 2-high breakdown 1,067mm and two 2-high reversing 914mm stands (annual capacity 152,000 tonnes); medium section – one 2-high breakdown 813mm and three 3-high reversing stands 457mm (92,000 tonnes).
■ Shelton Works, Stoke-on-Trent ST1 5JH. ***Rolling mills:*** heavy section – one 2-high breakdown 1,067mm, one 2-high edger 840mm and two 2-high universal 1,245mm stands (annual capacity 335,000 tonnes).
■ Dalzell Works, Park Street, Motherwell, Lanarks MLP 1PU. ***Rolling mills:*** One 4-high reversing heavy plate (4.13m) (annual capacity 370,000 tonnes).
■ Brinsworth Strip Mills, PO Box 69, Sheffield Road, Rotherham S60 1SZ. ***Rolling mills:*** Continuous hot strip and hoop – three vertical, four three-high roughers and six 4-high finishers (coil weight 2 tonne) (annual capacity 351,000 tonnes); Two cold strip – four 4-high tandem (65,000 tonnes) and one 4-high reversing (20,000 tonnes); one 2-high non-reversing temper/skin pass (20,000 tonnes). ***Other plant:*** Annealing, pickling, slitting, cutting-to-length.
Products: Steelmaking pig iron, foundry pig iron, blast furnace ferro-manganese.

BSC – British Steel Corp (continued)

Carbon steel – ingots, slabs, blooms, billets, hot rolled coil (for re-rolling), medium angles, medium tees, medium joists, medium channels, wide-flange beams (medium), universals (medium), heavy angles, heavy tees, heavy joists, heavy channels, wide-flange beams (heavy), universals (heavy), sheet piling, hot rolled hoop and strip (uncoated), cold rolled hoop and strip (uncoated), heavy plates, floor plates, hot rolled sheet/coil (uncoated), large-diameter tubes and pipes. Stainless steel – hot rolled hoop and strip (uncoated). Alloy steel (excl stainless) – hot rolled hoop and strip (uncoated), cold rolled hoop and strip (uncoated).

Strip Products Group
Works:

■ Llanwern Works, Newport, Gwent NP9 OXN. (Tel: 0633 290011. Telex: 497601, 50102). Works director: W.R. Harrison. ***Coke ovens:*** Four Simon-Carves batteries (158 ovens) (annual capacity 933,000 tonnes). ***Sinter plant:*** Two Lurgi 180 sq m strand and one Lurgi 120 sq m strand (3,019,000 tonnes). ***Blast furnaces:*** Two 9.15m hearth dia, 1,488 cu m and one 11.2m hearth dia, 2,289 cu m (combined capacity 2,825,000 tonnes). ***Steelmaking plant:*** Three 180-tonne LD basic oxygen converter (3,304,000 tonnes). ***Rolling mills:*** One 2-high universal slabbing (1,575mm max wide); wide hot strip – five 4-high roughing, six 4-high finishing (1,575-12.7mm x 720-1,550mm) (coil weight 27 tonnes) (2,750,000 tonnes); 4-stand cold reduction (710-1,524 x 0.5-3.0mm) (821,000 tonnes); temper/skin pass – one hot rolled skin pass (1,524 mm wide) (400,000 tonnes) and two cold temper (1,524mm wide) (850,000 tonnes).

■ Port Talbot Works, Port Talbot, West Glamorgan, SA13 2NG. Tel: 0639 87111. Telex: 497601, 60101. Works director: J. Marden. ***Coke ovens:*** One OSC battery (80 ovens) (annual capacity 295,000 tonnes) and two OSC batteries (84 ovens) (938,000 tonnes). ***Sinter plant:*** One Lurgi 336 cu m strand (3,440,000 tonnes). ***Blast furnaces:*** one 8.31m hearth dia, 1,364 cu m; one 8.31m hearth dia, 1,374 cu m; one 9.17 m hearth dia, 1,637 cu m and one 9.45m hearth dia, 1,789 cu m (annual capacity on three furnace operation 2,525,000 tonnes). ***Steelmaking plant:*** Two 330-tonne LD basic oxygen converters (2,978,000 tonnes). ***Refining plant:*** One DH 330-tonne vacuum degassing unit. ***Continuous casting machines:*** 2-strand Concast slab (196-254 x 675-1,875mm) (2,000,000 tonnes). ***Rolling mills:*** 2-high universal slabbing (not in use); wide hot strip – one reversing rougher and seven finishing (1.2-17.5 x 685-1,880mm) (coil weight 34 tonnes) (2,490,000 tonnes); cold reduction 14-stand (686-1,829 x 0.63-2.0mm) (coil weight 22.7 tonnes) and 15-stand (686-1,270 x 0.305-1.22mm) (coil weight 25.0 tonnes) (combined capacity 1,130,000 tonnes); temper/skin pass – one hot rolled skin pass (roll frame – 813mm, table width – 2.032mm) (coil weight 34 tonnes) (580,000 tonnes) and two cold temper mills (roll frame – 546mm, table width – 2,032mm) (coil weight 25 tonnes) and one cold temper mill (roll frame – 559mm, table width – 1,422mm) (coil weight 18 tonnes) (combined capacity 942,000 tonnes). ***Coil coating lines:*** Continuous Armco-Sendzimir hot-dip galvanizing (165,000 tonnes).

■ Ravenscraig Works, Motherwell, Lanarks ML1 1SW. (Tel: 0698 66211. Telex: 779282). Works director: R.E. Mercer. ***Coke ovens:*** Six Woodhall-Duckham batteries (210 ovens) (annual capacity 833,000 tonnes). ***Sinter plant:*** One 252 sq m strand Head-Wrightson (2,250,000 tonnes). ***Blast furnaces:*** one 8.69m hearth dia, 1,340 cu m; one 8.69m hearth dia, 1,360 cu m and one 8.69m hearth dia, 1,315 cu m (combined capacity 2,685,000 tonnes). ***Steelmaking plant:*** Three 130-tonnes LD basic oxygen converters (2,870,000 tonnes). ***Refining plant:*** 130-tonne DH vacuum degassing unit. 130-tonne VAD; SL ladle injection. ***Continuous casting machines:*** slab – two 1-strand (770-1,580 x 229mm) and one 2-strand (770-1,830 x 200-305mm) (1, 610,000 tonnes). ***Rolling mills:*** wide hot strip – one roughing and six finishing stands (1.575-13 thick x 610-1,560mm wide) (coil weight 25 tonnes) (1,686,000 tonnes) and temper/skin pass – one hot rolled skin pass (roll dia 1,168mm, table width 1,574mm) (coil weight 23 tonnes) (500,000 tonnes).

■ Hunterston Terminal (Ore and coal discharge and reloading), Hunterston, Fairlie, Largs, Aryshire KA29 OAZ. (Tel: 04758 8181. Telex: 777677). Works director: R.E.

United Kingdom

BSC – British Steel Corp (continued)

Mercer. ***Direct reduction plant:*** Two Midrex (annual capacity 800,000 tonnes – not commissioned).

■ Velindre Works, Swansea, Glamorgan SA5 7LP. (Tel: 0792 310011. Telex: 497601). Director-in-charge: J.A. Sowerbutts. Works manager: D. Veck. ***Rolling mills:*** One 5-stand 4-high tandem cold reduction (406-952 x 0.17-1.2mm) (coil weight 16 tonnes) (annual capacity 505,000 tonnes); two 2-stand temper/skin pass (1,067mm) (coil weight 16 tonnes) (530,000 tonnes). ***Coil coating lines:*** Four Ferrostan electrolytic tinning No 1 Mills, No 2 Wean, No 3 Head Wrightson and No 4 Wean (510,000 tonnes). ***Other plant:*** One sulphuric acid continuous pickling line, two electrolytic cleaning lines, batch annealing (single stock and multi stock), one continuous annealing line, two coil preparation lines, one blackplate shearline and two tinplate shear-lines.

■ Trostre Works, Llanelli, Dyfed SA14 9SD. (Tel: 0554 741111. Telex: 497601). Director-in-charge: J.A. Sowerbutts. Works manager: D. Veck. ***Rolling mills:*** One 5-stand 4-high tandem cold reduction (406-1,110 x 0.16-0.6mm) (coil weight 16.3 tonnes) (annual capacity 500,000 tonnes); two 2-stand temper/skin pass (1,220mm) (coil weight 13.6 tonnes) (total capacity 480,000 tonnes); one 3-stand double reduction (1,422mm) (coil weight 18.0 tonnes) (240,000 tonnes). one 3-stand double reduction (1,422mm) (coil weight 18 tonnes) (240,000 tonnes). ***Coil coating lines:*** Two Ferrostan Wean and Wean/Mills electrolytic tinning (203,000 tonnes); one Wean (BSC Modified) tin-free steel (ECCS) (60,000 tonnes); one (dual ET/TFS) Ferrostan/ ECCS Head Wrightson/Wean (162,000 tonnes). ***Other plant:*** One sulphuric acid continuous pickling line, two electrolytic cleaning lines, batch annealing (single stock and multi stock), one continuous annealing line, four coil preparation lines, one blackplate shear-line, two tinplate shear-lines and one coil inspection line.

■ Ebbw Vale Works, Ebbw Vale, Gwent NP3 6XD. (Tel: 0495 350011. Telex: 497601). Director-in-charge: J.A. Sowerbutts. Works manager: J.H. Bell. ***Rolling mills:*** One 5-stand 4-high tandem cold strip (533-990 x 0.17-0.8mm) (coil weight 18.1 tonnes) (annual capacity 530,000 tonnes); two 2-stand temper/skin pass (1,120 and 1,067mm) (coil weight 16.2 tonnes) (235,000 and 349,000 tonnes) and one 2-stand double reduction (1,065mm) (coil weight 20 tonnes) (240,000 tonnes). one double reduction 2-stand (1,065mm) (coil weight 20 tonnes) (240,000 tonnes). ***Coil coating lines:*** Three Ferrostan Head Wrightson electrolytic tinning (476,000 tonnes) and one non-oxidising Wean/ElFurno hot-dip galvanizing (210,000 tonnes). ***Other plant:*** One hydrochloric acid continuous pickling line, two electrolytic cleaning lines, batch annealing (single stock), one continuous annealing line, three coil preparation lines, five shear-lines and one coil inspection line.

■ Shotton Works, Deeside, Clwyd CH5 2NH. (Tel: 0244 812345. Telex: 61241). Works director: E.A. Cotterill. ***Rolling mills:*** One 5-stand tandem cold reduction (610-1,350 x 0.35-3.175mm) (annual capacity 864,000 tonnes); two single stand temper/skin pass (2,032 and 1,422mm wide). ***Coil coating lines:*** hot-dip galvanizing (two lines, one final purpose); electro-galvanizing (two lines); colour coating (one line) and one line for plastic laminate.

■ Tafarnaubach Works, Tredegar, Gwent. (Tel: 0495 254321. Telex: 497864). Works director: E.A. Cotterill. ***Coil coating lines:*** colour coating One organic paint (max strip width 1,380mm) (42,000 tonnes). ***Other plant:*** Cut-up, blanking units.

■ Bryngwyn Works, Lime Street, Gorseinon, West Glamorgan SA4 2JA. (Tel: 0792 893055. Telex: 48114). Works director: E.A. Cotterill. ***Coil coating lines:*** colour coating – two organic paint (max strip width 1,370mm) (annual capacity 110,000 tonnes). ***Other plant:*** Cut-up, blanking, slitter units.

■ Whiteheads Narrow Strip, Mendalgief Road, Newport. (Tel: 0633 65401. Telex: 497601 50600). Works director: P.J.K. Ferguson. ***Rolling mills:*** One 4-stand 2- high reversing cold strip (up to 622mm wide) (coil weight 5 tonnes) (annual capacity 90,000 tonnes); seven single stand temper/skin pass (8-610mm) (coil weight 5 tonnes) (100,000 tonnes). ***Coil coating lines:*** Continuous electro-galvanizing (50-450mm) (7,500 tonnes) and bitumen varnish line (20-60mm) (7,500 tonnes). ***Other plant:*** Cold formed section lines – one 30-stand (1.2mm thick) (20,000 tonnes); one 16-stand (5.0mm thick) and one 12-stand (5.0mm thick) (combined capacity 40,000 tonnes).

■ Orb Works, Corporation Road, Newport, Gwent NPT OX. (Tel: 0633 290033. Telex:

BSC – British Steel Corp (continued)

497601 50401). Works director: A.J. Edwards. ***Rolling mills:*** Two 4-high cold reduction (600-1,118 x 0.20-2.30mm) (coil weight 20 tonnes) (annual capacity 132,000 tonnes). ***Coil coating lines:*** One thermal flattening/coating, three decarburisation/coating and one varnishing for electrical sheet (200,000 tonnes).
■ Newton Aycliffe Works, Aycliffe Industrial Estate, Newton Aycliffe, Co Durham DL5 6AY. (Tel: 0325 312343. Telex: 58465). Works director: E.H. Cotterill . Plant for fabricating profiled cladding sheets for roofing and side walling. One roll form line for coil up to 1,524mm wide. Press brakes up to 10.93m.
■ Cookley Works, PO Box 13, Brierley Hill, W Midlands DY5 3UT. (Tel: 0384 480041. Telex: 338091). Works director: E.A. Cotterill. ***Coil coating lines:*** One continuous, terne/lead coating incorporating nickel flash facility (annual capacity 18,000 tonnes). ***Other plant:*** Cut-up, side trim, slitter units, shears.
Products: Carbon steel – ingots, slabs, hot rolled coil (for re-rolling), hot rolled hoop and strip (uncoated), cold rolled hoop and strip (uncoated), hot rolled sheet/coil (uncoated), cold rolled sheet/coil (uncoated), hot-dip galvanized sheet/coil, electro-galvanized sheet/coil, galvanized corrugated sheet/coil, aluminized sheet/coil, enamelling grade sheet/coil, terne-plate (lead coated) sheet/coil, colour coated sheet/coil, zinc-aluminium alloy coated sheet/coil, coated sheet/coil, electrical grain oriented silicon sheet/coil, electrical non-oriented silicon sheet/coil, electrical non-oriented non-silicon sheet/coil, blackplate, electrolytic tinplate (single-reduced), electrolytic tinplate (double-reduced), tin-free steel (ECCS).

BSC Tubes
Works:
■ Clydesdale Works, Bellshill, Lanarkshire, Scotland ML4 2RR. (Tel: 0698 749233. Telex: 779179). Works director: J. Hollis. ***Steelmaking plant:*** One 70-tonne 60 MVA and one 70 tonne 35 MVA electric arc furnaces (combined annual capacity 375,000 tonnes). ***Refining plant:*** One 70-tonne Asea-SKF. ***Continuous casting machines:*** One 3-strand DEC Concast bloom (310-430mm od) (330,000 tonnes). ***Tube and pipe mills:*** seamless Two rotary forge (273-457 and 139-273mm od) (190,000 tonnes on fifteen shifts).
■ Imperial Works, Martyn Street, Airdrie, Lanarks, ML6 9BJ. (Tel: 02364 54561. Telex: 779179). Works director: J. Hollis. Finishing plant for threading casing and couplings.
■ Corby Works, PO Box 101, Weldon Road, Corby, Northants NN17 1UA. Tel: 0536 202121. Telex: 341561. Works director: K.W.A. Wood. ***Tube and pipe mills:*** welded One continuous weld 13.5-42.4mm od (annual capacity 89,000 tonnes); two electric weld stretch reducing 42.4-139.7mm od (including squares and equivalent rectangles) (380,000 tonnes); one 6″ electric weld 139.7-193.7mm (100,000 tonnes) and one electric resistance weld 19-139.7 (16,000 tonnes).
■ Bromford Works, Wheelwright Road, Erdington, Birmingham, W Midlands B24 8DS. (Tel: 021 327 1311. Telex: 341561). Works director: J. Hollis. ***Tube and pipe mills:*** seamless One 203-508mm od Pilger (annual capacity 72,000 tonnes on 15 shifts).
■ Coombs Wood Works, Halesowen, W Midlands B62 8AD. (Tel: 021 559 1511. Telex: 341561). Works director: K.W.A. Wood. Tube finishing, plastic cladding, fittings and lighting columns manufacture.
■ Brenda Road, Hartlepool, Cleveland TS25 2EG. (Tel: 0429 266611. Telex: 341561). Works director: K.W.A. Wood. ***Tube and pipe mills:*** welded – one 20″ ERW 1 (178-508mm od) (annual capacity 175,000 tonnes).
■ Stockton Works, PO Box 5, Stockton-on-Tees, Clevelands TS18 2NF. (Tel: 0642 672188. Telex: 341561). Works director: K.W.A. Wood. ***Tube and pipe mills:*** welded – one 84″ submerged arc (508-2,134mm od) (annual capacity 26,000 tonnes).
Products: Carbon steel – seamless tubes and pipes, longitudinal weld tubes and pipes, large-diameter tubes and pipes, oil country tubular goods, hollow sections.

United Kingdom

BSC – British Steel Corp (continued)

BSC Stainless

Works:

■ Alloy Steel Rods Ltd, Stevenson Road, Sheffield S9 3XG (Tel: 441046-8). ***Rolling mills:*** One 3-high roughing, twenty finishing stands wire rod (round, flat and square) (coil weight 500kg) (annual capacity 30,000 tonnes).

■ Shepcote Lane, Sheffield S9 1TR. Tel: 0742 443311. Telex: 547025. Works director: A.P. Pedder. ***Steelmaking plant:*** One 130-tonne 60 MVA electric arc furnace (annual capacity 315,000 tonnes). ***Refining plant:*** One 130-tonne AOD. ***Continuous casting machines:*** 1-strand slab (140-200 x 800-1,650mm) (275,000 tonnes0. ***Rolling mills:*** One 4-high reversing medium plate (2m max) (40,000 tonnes); cold reduction – two single stand Sendzimir (1,600mm max) (coil weight 25 tonnes) (86,000 tonnes); one single stand Sendzimir (1,320mm max) (coil weight 25 tonnes) (45,500 tonnes) and two single stand (330mm max) (coil weight 2.7 tonnes) (5,000 tonnes). ***Other plant:*** Continuous bright annealing, pickling.

■ Panteg Works, Griffithstown, Pontypool, Gwent NP4 5YN. (Tel: 04955 3261. Telex: 498494). Works director: A.P. Pedder. ***Steelmaking plant:*** One 42-tonne 15 MVA electric arc furnace (annual capacity 72,000 tonnes). ***Refining plant:*** One 42-tonne AOD (72,000 tonnes). ***Rolling mills:*** Two cold reduction – one 4-high (1,220mm wide) (coil weight 11 tonnes) (40,500 tonnes) and one Sendzimir (1,270mm wide) (coil weight 11 tonnes) (18,900 tonnes); one temper/skin pass (1,270mm) (coil weight 10 tonne) (43,200 tonnes). ***Other plant:*** Continuous bright annealing, pickling.

Products: Stainless steel – ingots, slabs, hot rolled band (for re-rolling), wire rod, flats, medium plates, hot rolled sheet/coil (uncoated), cold rolled sheet/coil (uncoated), bright annealed sheet/coil. Alloy steel (excl stainless) – wire rod, flats.

BSC Diversified Activities

Works:

■ BSC Cumbria – Track Products, Moss Bay, Workington, Cumbria CA14 5AE. (Tel: 0900 4321. Telex: 64147). Works director: A.V.L. Williams. ***Rolling mills:*** Six-stand horizontal rail (40, 34, 32, 18, 18, 18" dia) (annual capacity 310,000 tonnes). ***Other plant:*** Off-line induction rail head hardening; in-line mill rail hardening; steel sleeper rolling and forming.

■ BSC Precision Strip, Stocksbridge, Sheffield S30 5JA. (Tel: 0742 888021. Telex: 54224). Manager: D. Hall. ***Rolling mills:*** cold strip – two Sendzimir (400mm max); two 4-high (400mm max); three single-stand (230mm max) and four 4-high (150mm max). ***Other plant:*** Shot blast and pickle; batch and continuous annealing; continuous hardening and tempering; precision slitting.

Products: Carbon steel – (Cumbria unless otherwise marked) – heavy rails 25-167kg/m (1985/86 output 182,000 tonnes); light rails 9-25kg/m (7,000 tonnes); rail accessories (sleepers, baseplates, fishplates) (13,000 tonnes); cold rolled hoop and strip (uncoated) (Stocksbridge). Stainless steel – cold rolled hoop and strip (uncoated) (Stocksbridge) (including razor strip, surgical steel, hardened and tempered, bandsaw strip etc).

Sales offices/agents: Europe – British Steel Corp, Copenhagen, Denmark (Tel: 01121758. Telex: 22696 steel dk); BSC (France) SA, Nanterre, France (Tel: 4729 1600. Telex: 611443); BSC Stainless SA, Paris, France (Tel: 01 766 0173. Telex: 642425); British Steel Corp (Deutschland) GmbH, Düsseldorf, W. Germany (Tel: 0211 498090. Telex: 8585533 bscd); BSC Stainless GmbH, Ratingen, W. Germany (Tel: 499041. Telex: 8585311 bscr d); British Steel Corp (Spain) Ltd, Madrid, Spain (Tel:2486207. Telex: 22886 bscs e); British Steel Corp (Norge) AS, Oslo, Norway (Tel: 02 37 3160. Telex: 72373 bscn); British Steel Corp Svenska AB, Gothenburg, Sweden (Tel: 031 800770. Telex: 27350 bscgbg s); British Steel Corp (Ireland), Dublin, Irish Republic (Tel: 685499. Telex: 31591). India – British Steel Corp Liaison Services (India) Ltd, New Delhi (Tel: 674240. Telex: 31 62790 bsclj). New Zealand – British Steel Corp (NZ) Ltd, Auckland (Tel: 641 179. Telex: 21794). Middle East – British Steel Corp Middle East, Dubai (Tel: 375 233/377 297. Telex: 48896 bscmeem). North America – British Steel Canada Inc, Montreal (Tel: 514481 8145. Telex: 05566336); BSC Alloys Division, British Steel Canada Inc, BC (Tel: 604278 3151. Telex: 04355708); British Steel Corp Inc, Houston, TX (Tel: 713 820 2828. Telex: 775 298). South Africa – British Steel Corp (South African Sales) (Pty) Ltd,

BSC – British Steel Corp (continued)

Randburg (Tel: 789 2597. Telex: 4 28528). South East Asia – British Steel Corp Singapore, Singapore (Tel: 733 3677. Telex: 39078 bsc sea). Turkey – BS Celik Ticaret AS, Istanbul (Tel: 131 1585/141 0244. Telex 26925 bsct tr). China – BSC (China) Ltd, Hong Kong (Tel: 5 8070196. Telex: 82007 steel hx).

Burn Tubes Ltd

Head office: Burcol Works, Radway Road, Shirley, Solihull, West Midlands B90 4NR. **Tel:** 021 704 2211. **Telex:** 339758.
Products: Carbon steel – longitudinal weld tubes and pipes, galvanized tubes and pipes, plastic coated tubes and pipes, precision tubes and pipes. Chrome-plated and brass-plated tubes.

Byard Kenwest Eng Ltd

Head office: Muir Road, Livingston, W Lothian. **Tel:** 0506 32882. **Telex:** 72685.
Management: R.G. Alexander (chairman), A. Jack (director and general manager), R. Dobson (accountant), B. Molland (sales). **Established:** 1969. **Capital:** £82,500 (issued). **No. of employees:** Works – 32. Office – 11. **Ownership:** Richard Costain UK Ltd (Ultimate holding company – Costain Group PLC).
Works:
■ At head office. Works manager: J. Swarbrick. ***Tube and pipe mills:*** welded – two 16/80" spiral (annual capacity 30,000 tonnes).
Products: Carbon steel – pipe piling 16-80" dia (1986 output 6,000 tonnes); spiral-

United Kingdom

Byard Kenwest Eng Ltd (continued)

weld tubes and pipes 16-80" dia (5,000 tonnes).
Brands: BACUA (drainage pipe systems).
Sales offices/agents: 145 High Street, Sevenoaks, Kent.

Cameron Iron Works Ltd, Forged Products Division

Head office: Houstoun Road, Livinston, W Lothian EH54 5BZ. **Tel:** 0506 31122. **Telex:** 72434, 727331 camliv g. **Telecopier (Fax):** 0506 31122 ext 3489.
Management: Managing director – W.B. Campbell; managers – C.I. Leviston (sales), F.J. Logan (technical), C.G. Webster (engineering), D.G. Pattle (plant). **Ownership:** Cameron Iron Works Inc (100%).
Works:
■ At head office. ***Refining plant:*** Three 36" dia VAR and one 10-tonne VIM (annual capacity 3,000 tonnes). ***Tube and pipe mills:*** seamless – 30,000-tonne extrusion press/9,000-tonne forging press. ***Other plant:*** 9,000-tonne forging press.
Products: Carbon steel – seamless tubes and pipes 200-1,200mm od. Stainless steel – seamless tubes and pipes 200-1,200mm od. Alloy steel (excl stainless) – seamless tubes and pipes 200-1,200mm od. Closed-die forgings (15-tonnes max) in steel, nickel, titanium and aluminium alloys. Extruded seamless pipe and sections in nickel, titanium and aluminium alloys.
Sales offices/agents: At head office (main sales office). 53 Grosvenor Street, London W1X 9FH (Tel: 01 493 7921. Telex: 24868. Fax: 01 629 1666).

Caparo Industries PLC

Head office: Caparo House, 103 Baker Street, London W1M 1FD. **Tel:** 01 486 1417. **Telex:** 8811343. **Telecopier (Fax):** 01 935 3242.
Management: Chairman – Swraj Paul; chief executive – James A. Leek; directors – Steven G. Mills (finance), Robert L. Lewis, Richard G.C. Hanna, John M. Wardle (non-executive). **Established:** 1980. **Ownership:** Caparo Group Ltd. **Subsidiaries:** Barton Conduits Ltd (Barton Engineering) (100%) (see separate entry); United Merchant Bar PLC (75%) (see separate entry); Wrexham Wire Co Ltd (100%) (wire works); Clydesdale Stamping Co Ltd (100%) (press and drop forge).

C.G. Carlisle & Co Ltd

Head office: 629 Penistone Road, Sheffield, Yorks. **Tel:** 0742 346161. **Telex:** 54411.
Management: Directors – K.F. Booth (chairman), D.H. Stagg (chief executive), G.E. Atkinson (finance), L.P. Race (production); technical general manager – E.K. Wise.
Ownership: C.F. Booth Ltd, Rotherham. **Associates:** Rotherham Stainless & Nickel Alloys Ltd. **Annual capacity:** Raw steel: 13,000 tonnes. Finished steel: 10,000 tonnes.
Works:
■ Northfield Road, Rotherham, S Yorks. (Tel: 0709 820358. Telex: 547693). ***Steelmaking plant:*** One 6.5-tonne electric arc (annual capacity 13,000 tonnes). ***Refining plant:*** One 6.5-tonne AOD (13,000 tonnes).
■ 629 Penistone Road, Sheffield. (Tel: 0742 346161. Telex: 54411). ***Rolling mills:*** One billet (annual capacity 6,000 tonnes); two bar – 14" (6,000 tonnes) and 11" (4,000 tonnes).
■ Grange Lane, Stairfoot, Barnsley. (Tel: 0226 205631. Telex: 54403). ***Rolling mills:*** One 26" billet (annual capacity 13,000 tonnes).
Products: Stainless steel – ingots, slabs, blooms, billets, semis for seamless tubemaking, round bars, square bars, flats, hexagons. Bright (cold finished) bars. Alloy steel (excl stainless) – ingots, slabs, blooms, billets, semis for seamless tubemaking, bright (cold finished) bars.

Richd W. Carr & Co Ltd

Head office: Pluto Works, Wadsley Bridge, Sheffield S6 1LL. **Tel:** 0742 349451. **Telex:** 54416.
Subsidiaries: These include associated companies: G.P. Wall Ltd, F.M. Parkin (Sheffield) Ltd and Thomas Andrews & Co Ltd.
Products: Alloy steel (excl stainless) – bright (cold finished) bars. Alloy tool and high-speed steel, bars and forgings. Stainless steel bars.

George Clark (Sheffield) Ltd

Head office: Warren Street, Sheffield S4 7WR. **Tel:** 0742 701882. **Telex:** 54512 gclark g. **Telecopier (Fax):** 0742 701379.
Products: Stainless steel – cold roll-formed sections (from strip) for the constuction industry.

The Clay Cross Co Ltd

Head office: Clay Cross, Chesterfield, Derbys S45 9NG. **Tel:** 0246 862151. **Telex:** 54301. **Telecopier (Fax):** 0246 886 352.
Management: Directors – B.W.S. Nuttall (managing), T.W.A. Barker (operations), Peter Smith (technical), T.A. Dexter (production), J.W. McKenna (company secretary and financial), A.M. Bull (sales), A.J. Wood (export sales). **Established:** 1837. **No. of employees:** 570. **Ownership:** Biwater Pipes and Castings (Ultimate holding company – Biwater Ltd).
Works:
■ At head office. ***ironmaking plant:*** Two hot blast cupolas. ***Other plant:*** Plant for the production of SG iron spun pipe and fittings, and general engineering castings in SG iron.
Products: Cast iron – pipes (80-600mm) and pipe fittings (SG iron). Spun SG iron pipe 80-600mm and fittings. General engineering SG iron castings.

CM Steel Mills Ltd

Head office: Elwell Street, Great Bridge., West Bromwich, West Midlands. **Tel:** 021 557 3811.
Products: Carbon steel – cold rolled sheet/coil (uncoated).

Coated Metals Ltd

Head office: Glamorgan Works, Pontardulais, Swansea, W Glam SA4 1SB. **Tel:** 0792 882548. **Telex:** 48454.
Management: Chairman – E. Palethorpe (non-executive director); directors – D. Kedward (joint managing), T.M. Molossi (joint managing), E.A. Thomas (works). **Established:** 1954. **Capital:** £200,000. **No. of employees:** Works – 60. Office – 22. **Ownership:** Coated Metals (Holdings) Ltd (100%) (Ultimate holding company – C. Walker & Sons Ltd). **Annual capacity:** Finished steel: 50,000 tonnes.
Works:
■ At head office. **Coil coating lines** One continuous hot-dip (annual capacity 50,000 tonnes).
Products: Carbon steel – aluminized sheet/coil 0.4-2.0mm thick, up to 1,250mm wide. (1986 output 46,000 tonnes).
Brands: Aludip Type 1, Aludip BQ.
Sales offices/agents: At head offices address. Export Sales – Henley Road, Slough Industrial Estate, Slough, Berks SL1 4JF. (Tel: 0753 37934, Telex: 848141).

Coated Strip Ltd—see Firsteel Ltd

United Kingdom

Coilcolor

Head office: Whiteheads Estate, Docks Way, Newport, Gwent NP1 2NS. **Tel:** 0633 52191.
Products: Carbon steel – colour coated sheet/coil.

Cold Drawn Tubes Ltd

Head office: Broadwell Works, Birmingham Road, Oldbury, Warley, W Midlands B69 4BZ. **Tel:** 021 552 1585. **Telex:** 339024. **Telecopier (Fax):** 021 544 5931.
Management: Directors – F. Laverick (managing), J.N. Tweedale (operations), R.R. Stand (financial), R.R.T. Gullick (sales and marketing), D.C. Crawford (personnel).
Established: 1984. **Ownership:** TI (75%), BSC (25%).
Works:
■ At head office.
■ Weldon Road, Corby, Northants NN17 1UA. (Tel: 0536 205000).
Products: Carbon steel – seamless tubes and pipes and DOM tubes 50-230mm od; cold drawn tubes and pipes up to 50mm od, wall thickness up to 12mm, thin wall tubing up to 100mm od; tube honed, skived and burnished, and centreless ground for hydraulic applications.
Sales offices/agents: At works addresses.

Cooper Coated Coil Ltd

Head office: Great Bridge Street, West Bromwich, West Midlands. **Tel:** 021 557 8161. **Telex:** 337855.
Products: Carbon steel – coated hoop and strip colour coated strip up to 780mm wide.

Coventry Tubes Permatube (Glynwed Tubes & Fittings Ltd)

Head office: Paynes Lane, Coventry CV1 5LG. **Tel:** 0203 29291. **Telex:** 311506.
Management: J. Jones (managing director).
Products: Stainless steel – longitudinal weld tubes and pipes and nickel alloy tubes.

Darlington & Simpson Rolling Mills Ltd (DSRM)

Head office: PO Box 9, Rise Carr House, Darlington DL3 0RG. **Tel:** 0325 382382. **Telex:** 58610 dsrm g. **Telecopier (Fax):** 0325 380038.
Management: Chairman – J.T. Carter; directors – D.B. Hale, R.A. Barras; company secretary – J. Murphy. **Established:** 1935. **Capital:** £5,000,000. **No. of employees:** Works – 500. Office – 100. **Ownership:** DSRM Group PLC (100%) (Ultimate holding company – British Steel Corp; Norcros PLC). **Associates:** DSRM International Ltd, London (international marketing); DSRM Inc, Houston and San Francisco, USA and Canada (American marketing). **Annual capacity:** Finished steel: 142,000 tonnes.
Works:
■ At head office. ***Rolling mills:*** North works – three-stand 2-high medium section (annual capacity 48,000 tonnes). South works – Four-stand 3-high and three-stand 2-high light section (24,000 tonnes). West works – one-stand 3-high and eight-stand 2-high bar (70,000 tonnes).
Products: Carbon steel – light angles, light tees, special light sections, medium joists.

Ductile Cold Mill, Glynwed Steels Ltd

Head office: Jubilee Works, Charles Street, Willenhall, West Midlands WV13 1HQ. **Tel:** 0902 366941. **Telex:** 339752. **Telecopier (Fax):** 0902 633274.
Management: Directors – C.J.H. Smith (managing), D.F.P. Hockey (sales); managers – T. Harris (commercial), D.F. Jones (general works), K. Taylor (sales), M.R. Andrews (technical). **Established:** 1936. **No. of employees:** Works – 76. Office – 23. **Ownership:** Glynwed Steels Ltd (Ultimate holding company – Glynwed International PLC). **Subsidiaries:** None. **Associates:** None. **Annual capacity:** Finished steel: 42,300 tonnes.
Works:
■ At head office. ***Rolling mills:*** Four cold strip – one 4-high Farmer Norton Hydraulic (508mm) (coil weight 5 tonnes) (annual capacity 21,500 tonnes), one 4-high R.W.F. (530mm) (coil weight 3 tonnes) (15,950 tonnes), one 4-high Robertson (15"/380mm) (coil weight 1 tonne) (1,650 tonnes), and one 4-high Farmer Norton (508mm) (1,650 tonnes). Two temper/skin pass – one 2-high (508mm) (coil weight 3 tonne) (21,500 tonnes), and one 2-high (305mm) (coil weight 1 tonne) (7,000 tonnes); **Other cold mills:** One 2-high Tormanco Edge Roll (102mm) (coil weight 1 tonne) (1,550 tonnes).
Products: Carbon steel – cold rolled hoop and strip (uncoated) 6.35mm/470mm x 0.25mm/6.35mm (1986 output 25,500 tonnes).

Ductile Hot Mill, Glynwed Steels Ltd

Head office: Jubilee Works, Charles Street, Willenhall, West Midlands WV13 1HQ. **Tel:** 0902 366941. **Telex:** 339752. **Telecopier (Fax):** 0902 633274.
Management: Directors – C.J.H. Smith (managing), D.F.P. Hockey (sales); managers – J.J. Byers (commercial), R. Broomhall (works), M.A. Andrews (technical). **Established:** 1936. **No. of employees:** Works – 25. Office – 19. **Ownership:** Glynwed Steels Ltd (Ultimate holding company – Glynwed International PLC). **Subsidiaries:** None. **Associates:** None. **Annual capacity:** Finished steel: 32,000 tonnes.
Works:
■ At head office. ***Rolling mills:*** One 5-stand hot strip and hoop with 4 edgers

The specialists.

DSRM: creators of special profiles in hot rolled steel

·DARLINGTON & SIMPSON ROLLINC MILLS LTD
PO BOX 9, RISE CARR HOUSE, DARLINGTON, ENGLAND DL3 0RG TELEPHONE: (0325) 382382 TELEX: 58610 FAX: (0325) 380038

DSRM INTERNATIONAL LIMITED
40 GEORGE STREET, LONDON, ENGLAND W1H 5RE TELEPHONE: 01-487 2737 TELEX: 264378 FAX: 01-487 4069

DSRM INC
363 NORTH BELT, HOUSTON, TEXAS 77060-2409 TELEPHONE: (713) 847-9320 TELEX: 372-5728 FAX: (713) 847-2152

Ductile Hot Mill, Glynwed Steels Ltd (continued)

(457mm) (annual capacity 32,000 tonnes).**Products:** Carbon steel – flats 80-150 x 8-52mm (1986 output 3,266 tonnes); universals (medium) 151-350 x 15-52mm (2,256 tonnes); hot rolled hoop and strip (uncoated) 80-350 x 6-15mm (11,875 tonnes). Alloy steel (excl stainless) – flats 80-150 x 8-52mm (297 tonnes); hot rolled hoop and strip (uncoated) 80-350 x 6-15mm (1,188 tonnes); universal plates 151-350 x 15-52mm (119 tonnes).

Ductile Sections, Glynwed Steels Ltd

Head office: Planetary Road, Willenhall, West Midlands WV13 3SW. **Tel:** 0902 739739. **Telex:** 338111.
Management: T.D. Morris (director and general manager), G.J. Nicklin (technical director). **Established:** 1948. **No. of employees:** 109. **Ownership:** Glynwed Steels Ltd (Ultimate holding company – Glynwed International PLC).
Products: Carbon steel – cold roll-formed sections (from strip) including channels and special sections.

Ductile Steel Processors, Glynwed Steels Ltd

Head office: Planetary Road, Willenhall, West Midlands WV13 3SP. **Tel:** 0902 305200. **Telex:** 337884. **Telecopier (Fax):** 0902 861427.
Management: Director – C.J.H. Smith; managers – P.M. Haselock (general), S. Adey (sales). **Established:** 1957. **No. of employees:** Works – 43. Office – 18. **Ownership:** Glynwed Steels Ltd (Ultimate holding company – Glynwed International PLC).
Works:
■ At head office. ***Coil coating lines:*** One Multi-cell strand electro-galvanizing (annual capacity 9,500 tonnes). ***Other plant:*** Wide coil pickling upto 1,470mm x 8.2mm; Wide coil slitting upto 1,500mm x 8.2mm.
Products: Carbon steel – coated hoop and strip electro-galvanized (6.35-508mm x 0.13-2.4mm).

Dudley Port Rolling Mills, Glynwed Steels Ltd

Head office: Lower Church Lane, Tipton, W Midlands. **Tel:** 021 520 6211. **Telex:** 339124. **Telecopier (Fax):** 021 520 2451.
Established: 1919. **No. of employees:** Works – 110. Office – 30. **Ownership:** Glynwed Steels Ltd (100%) (Ultimate holding company – Glynwed International PLC). **Annual capacity:** Finished steel: 50,000 tonnes.
Works:
■ At head office. ***Rolling mills:*** One 9-stand 14" medium section (annual capacity 35,000 tonnes); one 7-stand 9" bar (10,000 tonnes); one 5-stand 7" hand mill (5,000 tonnes).
Products: Carbon steel – round bars 12-71.5mm, (output 7,000-7,500 tonnes); square bars 12-66mm (6,500/7,000 tonnes); flats 12 x 6-100 x 65 mm (22,500/24,300 tonnes); hexagons 20-58.5mm, (700/750 tonnes); special light sections (8,500/9,000 tonnes). Stainless steel – round bars 12-71.5mm square bars 12-6mm, flats 12 x 6 – 100 x 65mm and hexagons 20-58.5mm (combined output 5 tonnes). Special light sections. Alloy steel (excl stainless) – round bars 12-71.5mm (375 tonnes); square bars 12-66mm (120 tonnes); flats 12 X 6 – 100 x 65mm (825 tonnes); hexagons 20-50.8mm (25 tonnes); special light sections 20-50.8mm (100 tonnes).

Durham Tube Ltd

Head office: Brookside Works, Middleton St. George, Darlington, Co Durham DL2 1JX **Tel:** 0325 332521-3. **Telex:** 58340.
Products: Stainless steel – cold drawn tubes and pipes.

United Kingdom

Elmtube Ltd

Head office: Northern Road, Aylesbury, Bucks HP19 3RA. **Tel:** 0296 415541. **Telex:** 83620.
Management: Directors – E.T. Jones (managing), R. Mather (sales), F. Pickles, R. Stanaway (works). **Established:** 1955. **Capital:** £1,000. **No. of employees:** 120.
Ownership: Wolseley PLC. **Annual capacity:** Finished steel: 6,000 tonnes.
Works:
■ At head office. ***Tube and pipe mills:*** welded – one 12-38mm and one 12-30mm Mannesman-Meer, and one 19-75mm Perkins (combined annual capacity 6,000 tonnes).
Products: Carbon steel – longitudinal weld tubes and pipes (output 630 tonnes). Longitudinally welded aluminium tube and section.

Energy Tubes Ltd

Head office: Britannia Works, Brandon Road, Binley, Coventry CV3 2AH. **Tel:** 0203 447777. **Telex:** 312226. **Telecopier (Fax):** 0203 451774.
Management: Jim McKinlay (director). **Established:** 1973. **No. of employees:** 35.
Ownership: Metsec PLC.
Works:
■ At head office. ***Tube and pipe mills:*** Eight longitudinal weld mills.
Products: Stainless steel – longitudinal weld tubes and pipes up to 2½" dia. Element sheaths.

English Seamless Tube Co

Head office: Westminster Works, Alcester Street, Birmingham B12 OPZ. **Tel:** 021 772 03952. **Telex:** 336518.
Management: Chairman and company secretary – A.G. Cottam; directors – R.P. King (managing), R.A. Gilbert (production), R. Kenyon (engineering). **Established:** 1932. **No. of employees:** 70. **Ownership:** Universal Steel Tube Co Ltd.
Works:
■ At head office. ***Tube and pipe mills:*** seamless – one cone roll piercer with Diescher discs elongating mill and 3-roll Kocks stretch reducing mill (annual capacity 10/12,000 tonnes).
Products: Carbon steel – seamless tubes and pipes.

Eurocast Bar Ltd

Head office: Empress Road, Loughborough, Leics LE11 1RQ. **Tel:** 0509 230621. **Telex:** 342349.
Management: Board members – A. Hastie (managing director), M. Rawlins (sales director), N.J. Marriott (company secretary), R. Springthorpe (works director), P.D. Walker (sales manager). **Established:** 1979. **Capital:** £28,400. **No. of employees:** Works – 52. Office – 14. **Ownership:** Warwick Engineering Investments (Ultimate holding company – Warwick Industries Ltd). **Associates:** Mestra GmbH. **Annual capacity:** Cast iron bar: 11,000 tonnes. Cast and SG iron rounds, squares, rectangles and tubes. Meehanite continuously cast iron bar.
Sales offices/agents: Bernareggi Srl, Milan, Italy; Ademetal, Drancy, France; AB Castings, Malmo, Sweden; Maatmetaal, Rotterdam, Netherlands; Kilsby Roberts, Brea, CA, USA; Rico y Echeverria SA, Zaragoza, Spain.
Expansion programme: 6-tonne electric melting furnace was due to be installed 1987.

European Profiles Ltd

Head office: Llandybie, Ammanford, Dyfed SA18 3JG. **Tel:** 0269 850691. **Telex:** 48171 epclad. **Telecopier (Fax):** 0269 851081.
Management: M. Davis (managing director). **Ownership:** RTZ Group.
Works:

European Profiles Ltd (continued)

■ At head office. Cold-forming lines.
Products: Carbon steel – cold roll-formed sections (from strip) (galvanised and PVC-coated roofing and cladding sheets).

Fine Tubes Ltd

Head office: Estover Works, Plymouth, Devon PL6 7LG. **Tel:** 0752 775851. **Telex:** 45252.
Products: Stainless steel – seamless tubes and pipes; longitudinal weld tubes and pipes; precision tubes and pipes and nickel alloy tube upto 1⅜".
Expansion programme: New weld mill for diameters up to 2" – operational February-March 1988.

Firsteel Ltd

Head office: Brockhurst Crescent, Bescot, Walsall WS5 4AX. **Tel:** 0922 38238. **Telex:** 338569. **Cables:** Firsteel Walsall. **Telecopier (Fax):** 0922 640601.
Management: Chairman – H.J. Nash; directors – J.V. Palmer (managing), A.A. Vurlan (financial), K.T. Jones (works), W.D. Davies (sales). **Established:** 1937. **Capital:** £285,175. **No. of employees:** Works – 62. Office – 35. **Ownership:** Firsteel Manufacturing Ltd (100%) (a subsidiary of Lonrho PLC). **Associates:** Coated Strip Ltd (see below). **Annual capacity:** Finished steel: 25,000 tonnes.
Works:
■ At head office. Works director: K.T. Jones. ***Rolling mills:*** Four Farmer Norton cold strip – 4-high 150 x 100 x 350mm reversing (annual capacity 14,000 tonnes), 4-high 210 x 500 x 600mm reversing (17,000 tonnes), 4-high 160 x 350 x 300mm reversing (11,000 tonnes), 2-high 400 x 600mm non-reversing (21,000 tonnes). ***Coil coating lines:*** One electro-galvanizing. ***Other plant:*** One pickle line for material 500mm wide, 1.2-6mm thick; one wide slitting line for material 1,200mm wide, 4.5mm thick; three finishing slitting lines for material, 300mm, 450mm and 600mm wide; one fifteen head precision slitter; two flattening and cut-to-length lines (total annual capacity 29,000 tonnes).
■ Coated Strip Ltd. At head office address. ***Coil coating lines:*** colour coating One 510mm wide.
Products: Carbon steel – cold rolled hoop and strip (uncoated) up to 530mm wide and 0.25-4mm thick; coated hoop and strip electro-galvanized; organic coated strip – paint, lacquer, plastisol or organisol, and PVC film coatings, up to 500mm wide, 0.15-1.2mm thick, in coils, straight lengths or blanks.
Brands: Firsteel (cold rolled bright, matt or electro-galvanized steel).

Firth Cleveland Steel Strip, Glynwed Steels Ltd

Head office: Locarno Road, Tipton, W Midlands DY4 9SE. **Tel:** 021 557 6871. **Telex:** 339011. **Cables:** Hardentem Tipton.
Established: 1949. **Capital:** £2,679,000. **No. of employees:** Works – 140. Office – 70. **Ownership:** Glynwed Steels ltd (Ultimate holding company – Glynwed International Ltd). Glynwed Steels Ltd. **Associates:** Firth Cleveland Steels, NY (North American sales). **Annual capacity:** Finished steel: 14,000 tonnes.
Works:
■ At head office.
Products: Carbon steel – cold rolled hoop and strip (uncoated) ¾ – 13" x 0.010 – 0.100, (1986 output 4,573 tonnes). Hardened and tempered steel strip.
Sales offices/agents: Firth Cleveland Steels Inc, New York, USA; BA-Stahl GmbH, Dusseldorf, West Germany; Meko Bandstaal AB, Sweden, Scandinavia.

United Kingdom

Firth Steels Ltd

Head office: Brighton Mills, Little Green Lane, Heckmondwike, West Yorks. **Tel:** 0924 406106. **Telex:** 557963. **Telecopier (Fax):** 0924 403090.
Management: Chairman – E.G.T. Firth; directors – R.T. Firth (managing), D.K. MacDonald (sales). **Established:** 1983. **Capital:** £100,000 (authorised) (£75,000 paid up). **No. of employees:** Works – 6. Office – 12. **Annual capacity:** Finished steel; 4,000 tonnes.
Works:
■ At head office. Cold roll forming plant for cladding and roofing profiles, 914mm cover width, plastic coated and galvanized.
Products: Carbon steel – cold roll-formed sections (from strip) – for cladding and roofing. Norclad (roofing and cladding sheets).
Additional information: The company is sales agent for Cardiff Bar & Section Mills (Allied Steel & Wire Ltd) covering Denmark and Norway.

Firth Vickers Special Steels Ltd—see Sheffield Forgemasters Ltd

Forgemasters Steels Ltd – Sheffield Forgemasters Ltd

Fulton (TI) Ltd

Head office: Halesfield Industrial Estate, Telford, Salop TF7 4ET. **Tel:** 0952 586301. **Telex:** 35344.
Established: 1963. **Ownership:** TI Group PLC.
Products: Carbon steel – Small-dia longitudinal weld tubes and pipes for the motor industry and refrigeration systems.

George Gadd & Co, Glynwed Steels Ltd

Head office: Blackbrook Road, Woodside, Dudley, W Midlands DY2 0QS. **Tel:** 0384 77111. **Telex:** 339667. **Cables:** Gadds Dudley West Mid. **Telecopier (Fax):** 0384 265243.
Management: Directors – W. Devney (managing), R. Gresswell (commercial); sales manager – C. Burns. **Established:** 1865. **No. of employees:** Works – 107. Office – 15. **Ownership:** Glynwed Steels Ltd (Ultimate holding company – Glynwed International PLC). **Annual capacity:** Finished steel: 85,000 tonnes.
Works:
■ At head office. **Rolling mills** Two light bar and section mills – 11 stand horizontal 10" (annual capacity 22,000 tonnes) and 8-stand horizontal 14" (63,000 tonnes).
Products: Carbon steel – round bars 19-66mm dia (1986 output incl alloy steel 28,000 tonnes); square bars 19-63mm (5,000 tonnes); flats 27 x 6-91 x 46mm (7,000); hexagons 19-60mm a/f (7,000); light angles 32 x 32 x 5-50 x 50 x 6, light tees 44 x 44 x 5-50 x 50 x 10, light channels 50 x 25 x 5-6mm and special light sections (3,000 tonnes). Stainless steel – round bars 19-66mm dia square bars 19-63mm, flats 27 x 6-91 x 46mm, hexagons 19-60mm a/f. Alloy steel (excl stainless) – round bars 19-66mm dia, square bars 19-63mm, flats 27 x 6-91 x 46mm, hexagons 19-60mm a/f.

Joseph Gillott & Sons, Glynwed Steels Ltd

Head office: Wharf Road, Kilnhurst, Nr Rotherham, South Yorks S62 5ST. **Tel:** 0709 582986. **Telex:** 547387. **Telecopier (Fax):** 0709 587771.
Management: Managers – R. Mullis (general), P. Bunting (commercial), R. Lodge (mill), D. Wilson (quality/warehouse). **Established:** Circa 1869. **No. of employees:** Works – 60. Office – 12. **Ownership:** Glynwed Steels Ltd (Ultimate holding company –

Joseph Gillott & Sons, Glynwed Steels Ltd (continued)

Glynwed International PLC). **Annual capacity:** Finished steel: 20,000 tonnes.
Works:
■ At head office. ***Rolling mills:*** bar – one 8-stand cross country (10") (annual capacity 13,500 tonnes), and one 4-stand cross country (18") (6,500 tonnes).
Products: Carbon steel – round bars 16-40mm and 50-105mm; square bars 16-103mm; flats 25-100mm x 6-45mm thick; bright (cold finished) bars 15-39.5mm dia; special light sections; rail accessories fishplates. Stainless steel – round bars 50-92mm. Alloy steel (excl stainless) – round bars 16-40mm and 50-105mm; square bars 16-103mm; flats 25-100mm wide 6-45mm thick; bright (cold finished) bars centreless ground 15-39.5mm dia; special light sections.

Glynwed International PLC

Head office: Headland House, New Coventry Road, Sheldon, Birmingham B26 3AZ. **Tel:** 021 742 2366. **Telex:** 336608. **Telecopier (Fax):** 021 742 0403.
Management: Executive directors – G. Davies (group chairman and chief executive), D.L. Milne (group finance), D. Gripton, T. O'Neill, D.W. Richardson; non-executive directors – J.D. Eccles, Sir Eric Pountain. J.C. Blakeley (group secretary and legal adviser). **Subsidiaries:** Glynwed Consumer & Building Products Ltd (domestic appliances, catering equipment, iron castings, drainage systems, building and construction products, steel fabrications). Glynwed Steel & Engineering Ltd (hot rolled steel bars and sections, cold drawn, precision ground and turned steel bars, hot rolled and cold rolled steel strip, bright machined edge steel flats, high-carbon cold rolled and hardened and tempered steel strip, cold rolled formed sections, perforated strip, general presswork, electrical cable tray and support systems, coil slitting, pickling, electro-galvanizing, steel conduit and steel precision tubes, steel and aluminium stockholders, stainless steel stockholders. Distributors of non-standard ball bearings

United Kingdom

Glynwed International PLC (continued)

and fastenings. Lifting and mooring equipment, automotive trim manufacturers, fastening manufacturers, steel flooring, stairways and fencing. Rivets, cable clips, masonry nails, wall plugs). Glynwed Tubes & Fittings Ltd (manufacturers of copper and stainless steel tubes, pipe fitting, thermoplastic pipework products, plant and tooling designers, ERW steel tube, Flo-coat and Spectra-coat steel tube, precision rubber mouldings, polyethylene and polypropylene drainage systems, injection mouldings, vacuum forming, UPVC windows). Other trading companies – Apollo Medical Supplies, Glynwed Wholesale Chemists. Other manufacturing plants worldwide.
Additional information: Activities: Holding company. London office – First floor, Pollen House, 10/12 Cork Street, London W1X 1PD (Tel: 01 734 0693-6, 734 1542-3).

A.E. Godrich & Son Ltd

Head office: Wharf Street, Aston, Birmingham B6 5SB. **Tel:** 021 327 1306. **Telex:** 338755.
Management: I.P. Mobley (managing director). **No. of employees:** 40. **Ownership:** GEI International PLC. **Associates:** Midland Bright Drawn Steel Ltd.
Works:
■ At head office. Continuous bar and wire drawing machines; drawbenches; high-speed reeling machines; wire drawing blocks; heat treatment furnaces.
Products: Carbon steel – bright (cold finished) bars rounds, squares, flats and hexagons, bright wire. Alloy steel (excl stainless) – bright (cold finished) bars – rounds, squares, flats and hexagons, wire.

Hayes Tubes Ltd

Head office: Balds Lane, Lye, Nr Stourbridge, W Midlands DYP 8NN. **Tel:** 038482 2373. **Telex:** 335494 mt.
Management: Chairman – T.O. Penn; joint managing directors – J.O. Penn, A.S. Penn (company secretary). **Established:** 1936. **Capital:** £6,000. **No. of employees:** Works – 23. Office – 7. **Annual capacity:** Finished steel: 5,000 tonnes.
Works:
■ At head office. Works director: J.O. Penn. ***Tube and pipe mills:*** welded – one 2" (annual capacity 5,000 tonnes).
Products: Carbon steel – longitudinal weld tubes and pipes 16-45mm od. (circular only) (5,000 tonnes).
Expansion programme: Conversion of existing weld mill to quick change by duplication of forming and finishing stands. Due to be commissioned September 1987.

K.W. Haywood & Son Ltd

Head office: Abrasteel Works, Little London Road, Sheffield, S Yorks S8 0UH. **Tel:** 0742 557371. **Telex:** 547927 kwhay g.
Management: Chairman – K.W. Haywood; directors – C.J. Haywood (joint managing), D. Geoffrey Smith (joint managing), P. Gordon Smith (commercial), J.L. Musson, K.I. Haywood (secretary). **Established:** 1950's. **Capital:** £50,000. **No. of employees:** Under 100. **Subsidiaries:** K.W. Haywood & Son (Abrasives) Ltd.
Products: Carbon steel – bright (cold finished) bars. Stainless steel – bright (cold finished) bars. Alloy steel (excl stainless) – bright (cold finished) bars.

Hemmings Ltd

Head office: Valley Works, Grange Lane, Sheffield, S Yorks S5 0DQ. **Tel:** 0742 467225. **Telex:** 54535.
Management: S.F. Telford (managing director). **Established:** 1900. **Ownership:** GEI

Hemmings Ltd (continued)

International PLC **Annual capacity:** Finished steel: 2,500 tonnes.
Works:
■ At head office. Wire drawing machines, Schumag bright drawing machines, Cincinatti and Schumag grinding machines.
Products: Stainless steel – round bars 1.6-12.7mm dia; square bars 5.0-9.5mm square; hexagons 5.0-11.0mm a/f; wire 1.2-12.7mm dia; wire shapes.

Hobson, Houghton & Co Ltd—see Spencer Clark Metal Industries PLC.

Howmet Turbine (UK) Inc, Howmet Alloys International

Head office: Heron Road, Exeter, South Devon. **Tel:** 0392 32751. **Telex:** 42524. **Telecopier (Fax):** 0392 211774.
Management: Managers – B.H. Elver (general), D.H. Shaw (quality), J.W. Day (commerical), G.L.R. Durber (technical), R. Dodd (production); financial controller – R.A. Bates. **Established:** 1975. **No. of employees:** 100. **Ownership:** Howmet Turbine Components Corp (Ultimate holding company – Pechiney Group). **Associates:** Howmet Alloy Operations – Dover Alloy Division, NJ, USA; Metal Products Division, MI, USA.
Works:
■ At head office. **Steelmaking plant** Three air melt induction melt furnaces – two 1-tonne and one 2-tonne. ***Refining plant:*** Two 5/2½-tonne VIM.
Products: Carbon steel – ingots 5kg (50 x 50 x 250mm) and 1.5kg (33 x 250mm). Stainless steel – ingots 5kg (50 x 50 x 250mm) and 1.5kg (33 x 250mm); round bars 1¼-6" (32-150mm) dia, 39" (1,000mm) length. Alloy steel (excl stainless) – ingots 5kg (50 x 50 x 250mm), 1.5kg (33 x 250mm). Electrodes 5 tonnes, 22" (560mm) dia x 13" (4.0mm) length, superalloy bars 32-150mm dia.
Sales offices/agents: Howmet Alloys International (at head office address); Jesco Ltd, Great River Enterprises.

Kiveton Park Steel & Wire Works Ltd

Head office: Kiveton Park, Sheffield, S Yorks S31 8NQ. **Tel:** 0909 770252. **Telex:** 54179 kpsww g. **Telecopier (Fax):** 0909 772949.
Management: Joint managing directors – A. Maycock (chairman), G. Blatherwick (deputy chairman); directors – G.B. Parkinson (works), E. Mozley (commercial), G.L. Wharton, C. Firth (sales), P.W. Wright (technical); company secretary – K. Reynolds. **Established:** 1923. **Capital:** £400,000. **No. of employees:** Works – 174. Office – 24. **Ownership:** Kiveton Park (Holdings) Ltd. **Annual capacity:** Finished steel: 25,000 tonnes.
Products: Carbon steel – round bars 6-75mm (1986 output 5,000 tonnes); hexagons 6-50mm (1,500 tonnes); bright wire (6-18mm) (4,000 tonnes). Stainless steel – round bars 10-18mm (50 tonnes). Alloy steel (excl stainless) – round bars 6-75mm (6,000 tonnes); square bars 6-50mm (100 tonnes); hexagons 6-50mm (2,000 tonnes); wire 6-18mm (1,500 tonnes).

Arthur Lee & Sons PLC

Head office: PO Box 54, Meadow Hall, Sheffield S9 1HU. **Tel:** 0742 437272. **Telex:** 54165 crown g. **Cables:** Crown Sheffield. **Telecopier (Fax):** 0742 439782.
Management: Chairman and managing director – Peter W. Lee; directors – D.S.W. Lee, N.H.W. Lee, C.G. Holland, H.G. Mutkin, G.A. Ashton, J.V. Henderson. **Established:** 1874. **Subsidiaries:** Steel Division – Lee Steel Wire Ltd, Ecclesfield, Sheffield (wire); Lee Steel Strip Ltd, Sheffield (cold rolled stainless and speciality carbon strip); Lee Bright Bars Ltd, Sheffield and Warrington (bright steel bar). Wire and Rope Division – Smith Wires Ltd, Halifax (fine carbon wires); John Shaw Ltd, Worksop, Notts

Arthur Lee & Sons PLC (continued)

(wire ropes) and its New Zealand subsidiary John Shaw (NZ) Ltd. Steel Stockholding Division – Bell & Harwood Ltd, West Bromwich; C. Roberts & Co (Steel) Ltd, Rotherham and Manchester; Roberts Magnetics Ltd, Rotherham; Lindley Steels Ltd, Slough and Sittingbourne; Stainless Strip Supplies Ltd, Rotherham. Plastics Division – Barrington Products (Leicester) Ltd; Leicester. Plasro Plastics Ltd, Mitcham, Surrey. Group Export Sales Company – Lee of Sheffield Ltd.
Works:
■ At head office. ***Rolling mills:*** cold strip. ***Other plant:*** Bright bar plant.
■ Ecclesfield **Wire drawing plant**.
■ Warrington **Bright bar plant**.
■ Halifax **Wire drawing plant**.
Products: Carbon steel – bright (cold finished) bars, cold rolled hoop and strip (uncoated) (special carbon). Stainless steel – bright (cold finished) bars, cold rolled hoop and strip (uncoated) (precision). Alloy steel (excl stainless) – bright (cold finished) bars, cold rolled hoop and strip (uncoated). Precision rolled flat wire, carbon steel wire, steel wire rope; also injection moulded plastic products.
Brands: Steelex (stainless roofing steel).
Sales offices/agents: See under subsidiaries. Agents in most industrialised countries.

J.B. & S. Lees Ltd

Head office: Trident Steel Works, Albion Road, West Bromwich, W Midlands B70 8BH. **Tel:** 021 553 3031. **Telex:** 336501. **Telecopier (Fax):** 021 553 7680.
Management: Chairman – M.E. Doherty; directors – D.N. Smith (managing), T.A. Cooper (production), M.E. Horton (finance), D.G. Riley (sales and marketing), I.A. Warner (engineering). **Established:** 1852. **Capital:** £220,000. **No. of employees:** Works – 278. Office – 34. **Ownership:** Cope Allman International PLC (100%) (Ultimate holding company – Henlys Group Ltd, Canada). **Associates:** J.B. & S. Lees Inc, USA; J.B. & S. GmbH, West Germany. **Annual capacity:** Finished steel: 20,000 tonnes.
Works:
■ At head office. ***Rolling mills:*** Three 4-high cold strip (450mm max) (max coil weight 2,700kg) (annual capacity 20,000 tonnes); one 2-high/4-high temper/skin pass
Products: Carbon steel – cold rolled hoop and strip (uncoated) 6-450mm wide, 0.1-2mm thick (1986 output 17,000 tonnes); coated hoop and strip (lead/antimony) up to 150mm wide and 0.25mm thick (500 tonnes). Hardened and tempered strip up to 250mm wide and 2.5mm thick. Metal cutting bandsaw strip in all standard sizes. Bi-metal strip for manufacture of metal cutting saws.
Brands: Trident.
Sales offices/agents: At head office. J.B. & S. Lees Inc, 4 Edison Place, Fairfield, NJ 07006 3590, USA; J.B. & S. Lees GmbH, Hochstrasse, Lindenhof, D 8903 Bobingen bei Augsburg, West Germany (covering Europe).

F.H. Lloyd & Co Ltd

Head office: James Bridge Steelworks, Wednesbury, West Midlands WS10 9SD. **Tel:** 021 526 3121. **Telex:** 339502.
Management: W.M. Tillotson (managing director), A.E. Johnson (works director), G.P.D. Goodreds (sales manager), M.W. Day (secretary). **Established:** 1980. **Ownership:** Triplex Lloyd Group. **Annual capacity:** Raw steel: 100,000 tonnes.
Works:
■ At head office. Works director: A.E. Johnson. ***Steelmaking plant:*** One 35-tonne electric arc furnace. Scandinavian Lancers ladle injection station (annual capacity 100,000 tonnes). ***Continuous casting machines:*** One 3-strand Danieli billet (160mm square, 160mm round) (100,000 tonnnes). ***Rolling mills:*** One 26" billet.
Products: Carbon steel – blooms 160 x 160mm; billets 40 x 75mm; semis for seamless tubemaking 160mm dia. Alloy steel (excl stainless) – blooms 160 x 160mm; billets 40 x 75mm; semis for seamless tubemaking 160mm dia.

United Kingdom

London Tinning Co Ltd

Head office: Unit 4, Burwell Industrial Estate, Burwell Road, Leyton, London E10 7QG. **Tel:** 01 558 5225. **Telex:** 897665. **Telecopier (Fax):** 01 558 8348.
Management: Chairman – R.C. Buxton; directors – F.H. Sach (financial and company secretary), D.H. Webb (general manager), P. Hurrell, J. Hertel, P.H. Buxton.
Established: 1983. **Capital:** £100,000. **No. of employees:** 12. **Ownership:** H. Young Co Ltd. **Annual capacity:** Finished steel: 2,000 tonnes (tinned sheet).
Works:
■ At head office. One hot-dip sheet tinning line.
Products: Carbon steel – hot-dip tinplate (2,000 x 1,000mm and 2,500 x 1,250mm, 0.4-2.0mm sheets (1986 output 1,200 tonnes).
Sales offices/agents: At head office. Billiton Witmetaal Belgium, Antwerp, Beligum; Metaalmaatschappij Van Houten BV, Rotterdam, Netherlands; Ulf Schele AB, Gothenburg, Sweden; Tritin Ltd, Newdigate, Surrey (Eastern Europe); K. Grutter & Co AG, Wallisellen, Switzerland; E. Fischer Co, Chelles, France.

Longmore Brothers, Glynwed Steels Ltd

Head office: Springfield Steel Works, Mill Street, Darlaston, Wednesbury, West Midlands WS10 8TJ. **Tel:** 021 526 4122. **Telex:** 338826 lonbs g. **Telecopier (Fax):** 021 526 5876.
Management: K. Owen (managing director). **Established:** 1921. **No. of employees:** 127. **Ownership:** Glynwed Steels Ltd (Ultimate holding company – Glynwed International PLC). **Annual capacity:** Finished steel: 50,000 tonnes.
Works:
■ At head office. ***Tube and pipe mills:*** welded – One 2" ARM, and one 2½" Perkins (combined capacity 6,560 tonnes). ***Other plant:*** Bright bar drawing plant.
Products: Carbon steel – bright (cold finished) bars – rounds, squares and hexagons; longitudinal weld tubes and pipes – round, oval, square and rectangular (⅝-1½"); galvanized tubes and pipes, precision tubes and pipes. Alloy steel (excl stainless) – bright (cold finished) bars – rounds, squares and hexagons.

Manchester Cold Rollers Ltd

Head office: Broadway Industrial Area, Trafford Road, Salford M5 2UQ. **Tel:** 061 848 0835, 061 872 8081. **Telex:** 669755 mcold. **Telecopier (Fax):** 061 834 1558.
Management: Derek Flaherty (sales director) and Stephen Cadney (managing and technical director). **Established:** 1981. **Capital:** £49,000. **No. of employees:** Works – 8. Office – 2. **Annual capacity:** Finished steel: 1,000 tonnes.
Works:
■ At head office. ***Tube and pipe mills:*** welded – one 10-stand roll-forming and tube welding line (own manufacture), 4mm wallthickness (annual capacity 300 tonnes).
Products: Carbon steel – cold roll-formed sections (from strip) (450 tonnes). Stainless steel – cold roll-formed sections (from strip) (100 tonnes). Alloy steel (excl stainless) – longitudinal weld tubes and pipes (5 tonnes).

Martins (Dundyvan) Ltd

Head office: Dundyvan Steel Works, Coatbridge, Strathclyde ML5 1EB. **Tel:** 0236 33601. **Telex:** 779606. **Cables:** Martin Coatbridge.
Management: Directors – H.H. Martin (chairman and managing), K.R. Martin (non-executive), R. Scougall (finance); managers – G. Cameron (general sales), P. Miles (sales). **Established:** 1884. **Capital:** £150,000. **No. of employees:** Works – 100. Office – 20. **Ownership:** Private. **Annual capacity:** Finished steel: 120,000 tonnes.
Works:
■ At head office. ***Rolling mills:*** One 9-stand 490/440mm medium section (annual capacity 90,000 tonnes); one 7-stand 375/300mm bar (30,000 tonnes).
Products: Carbon steel – flats 25-100mm wide; light channels 3 x 1½, 3 x 2; special

Martins (Dundyvan) Ltd (continued)

light sections (to customers requirements). Medium angles 50 x 50mm, 60 x 60mm; medium channels 4 x 2"; heavy channels 5 x 2½, 6 x 3".
Brands: Dundyvan.
Expansion programme: Modernisation of 300mm bar mill consisting of new cold shearing line (1987), and continuous finishing train (1988).

MET Steel Ltd

Head office: Hydepark, Mallusk, Co Antrim BT36 8LU. **Tel:** 02313 7311. **Telex:** 747295.
Management: Directors – R. McMahon (managing), N. Kennedy (financial); managers – J. Caughey (sales), T. Marshall (administration), H. Marron (purchasing).
Established: 1977. **Capital:** £620,000. **No. of employees:** Works – 11. Office – 13. **Ownership:** Hill Street Trustees Ltd. **Annual capacity:** Finished steel: 10,000 tonnes.
Works:
■ Newtonabbey (at head office). Works director: R. McMahon.
Products: Carbon steel – galvanized corrugated sheet/coil.

Metalon Steels Ltd—see Ductile Steel Processors, Glynwed Steels Ltd

United Kingdom

Metsec PLC

Head office: Broadwell Works, Birmingham Road, Oldbury, West Midlands. **Tel:** 021 552 1541. **Telex:** 338338. **Telecopier (Fax):** 021 544 5520.
Management: Directors – K.R. Hirst (chairman and managing), J. McKinlay, D.G. Jeavons, K.D. Dodd, S.R. Tilsley, D. Lum; non-executive director – D.S. Haggett.
Established: 1981. **Capital:** £1,250,000 (issued). **No. of employees:** 215.
Subsidiaries: Energy Tubes Ltd (see separate entry), Metsec Business Systems Ltd (computer hardware and software); Air Tube Conveyors Ltd (document transfer systems); Metframe Ltd (lightweight steel frame building systems).
Works:
■ At head office. Cold roll forming mills, longitudinal welding machines, section piercing machines.
Products: Carbon steel – cold roll-formed sections (from strip). Stainless steel – cold roll-formed sections (from strip). Alloy steel cold roll-formed sections. Aluminium alloy cold roll-formed sections.

Midland Bright Drawn Steel Ltd

Head office: Richmond Works, Brickhouse Lane, West Bromwich B70 0DX. **Tel:** 021 557 4971. **Telex:** 336911.
No. of employees: 45. **Ownership:** GEI International PLC. **Associates:** A.E. Godrich & Son Ltd.
Works:
■ At head office. Draw benches; high speed reeling machines.
Products: Carbon steel – bright (cold finished) bars. Alloy steel (excl stainless) – bright (cold finished) bars.

Monmore Tubes

Head office: Dixon Street, Wolverhampton, West Midlands WV2 2BT. **Tel:** 0902 351135. **Telex:** 339751. **Telecopier (Fax):** 0902 351098.
Management: Directors – M. Morgan (managing), D.H. Harris (sales), R. Beardsmore (works), A.J. Lakin (financial). **No. of employees:** Works – 104. Office – 51.
Ownership: Glynwed Tubes & Fittings Ltd (Ultimate holding company – Glynwed International PLC).
Works:
■ At head office. ***Tubing and pipe mills:*** one semi-bright mill for electric-weld tube (annual capacity 22,000 tonnes). One SpectraCoat line for nylon-coated tube, size range $\frac{1}{2}$-2$\frac{5}{8}$"; Two FloCoat galvanizing lines.
Products: Carbon steel – longitudinal weld tubes and pipes, galvanized tubes and pipes, plastic coated tubes and pipes. Conduit and Precison mechanical tubes.

Motherwell Bridge Pipe

Head office: Logans Road, Motherwell, Lanarks. **Tel:** 0698 66111. **Telex:** 77198.
Products: Carbon steel – spiral-weld tubes and pipes 900-3,000mm dia, 7-19mm wall thickness.

Natural Gas Tubes Ltd

Head office: Caparo House, 103 Baker Street, London W1M 1FD. **Tel:** 01 486 1417. **Telex:** 8811343 caparo g.
Management: S. Paul (chairman), C.N. Shaw (managing director), T. Wilkinson (financial controller), P. Waterland (sales director). **Established:** 1966. **Ownership:** Caparo group (100%). **Annual capacity:** Finished steel: 50,000 tonnes.
Works:
■ 3 Tafarnaubach Industrial Estate, Tredegar, Gwent NP2 3AA. (Tel: 049525 3881. Telex: 497303). Works director: C.N. Shaw. ***Tube and pipe mills:*** Three welded –

Natural Gas Tubes Ltd (continued)

PRD Oceanics 1,000mm spiral (annual capacity 5,000 tonnes), Abbey Etna 7XU ERW, max 7.5″ od, Abbey-Etna ERW. $\frac{3}{4}$-$2\frac{3}{8}$″ od. ***Other plant:*** Tube straightener, Abbey-Etna slitting line (50,000 tonnes).
■ Brampton Road, Huntingdon PE1 6BQ. (Tel: 0480 52815. Telex: 32682). ***Tube and pipe mills:*** welded – Driam 457mm spiral-weld.
Products: Carbon steel – longitudinal weld tubes and pipes and spiral-weld tubes and pipes 26-1,500mm dia.
Expansion programme: Third ERW mill at Tredegar – similar in design to present two and about the size of the smaller mill.

Newman-Tipper Tubes Ltd

Head office: Holyhead Road, Wednesbury, West Midlands WS10 7BJ. **Tel:** 021 556 1200. **Telex:** 336996 newtbe g. **Telecopier (Fax):** 021 503 4719.
Management: W.M. Thomas (managing director). **Ownership:** Tubes and Fittings division, Glynwed International PLC.
Works:
■ At head office. ***Tube and pipe mills:*** ERW. ***Other plant:*** Draw-bench for seamless and cold-drawn welded tubes. Presswork for production of slotted tubes for shopfitting.
Products: Carbon steel – longitudinal weld tubes and pipes, galvanized tubes and pipes, plastic coated tubes and pipes, precision tubes and pipes.

North Eastern Iron Refining Co Ltd

Head office: Stillington, Stockton-on-Tees, Cleveland TS21 1LE. **Tel:** 0740 30212. **Telex:** 58158 neiron g.
Works:
■ At head office. Three induction furnaces.
Products: Foundry pig iron (low phosphorus, high phosphorus, malleable and alloyed.

Osborn Steel Extrusions Ltd

Head office: Brighouse Road, Low Moor, Bradford BD12 0QL. **Tel:** 0274 677331-5. **Telex:** 51333. **Telecopier (Fax):** 0274 607858.
Management: R. Cox (managing director). **Established:** 1952. **Ownership:** Aurora PLC.
Works:
■ At head office. 1,265-tonne hydraulic extrusion press.
Products: Carbon steel – special light sections. Stainless steel – special light sections.

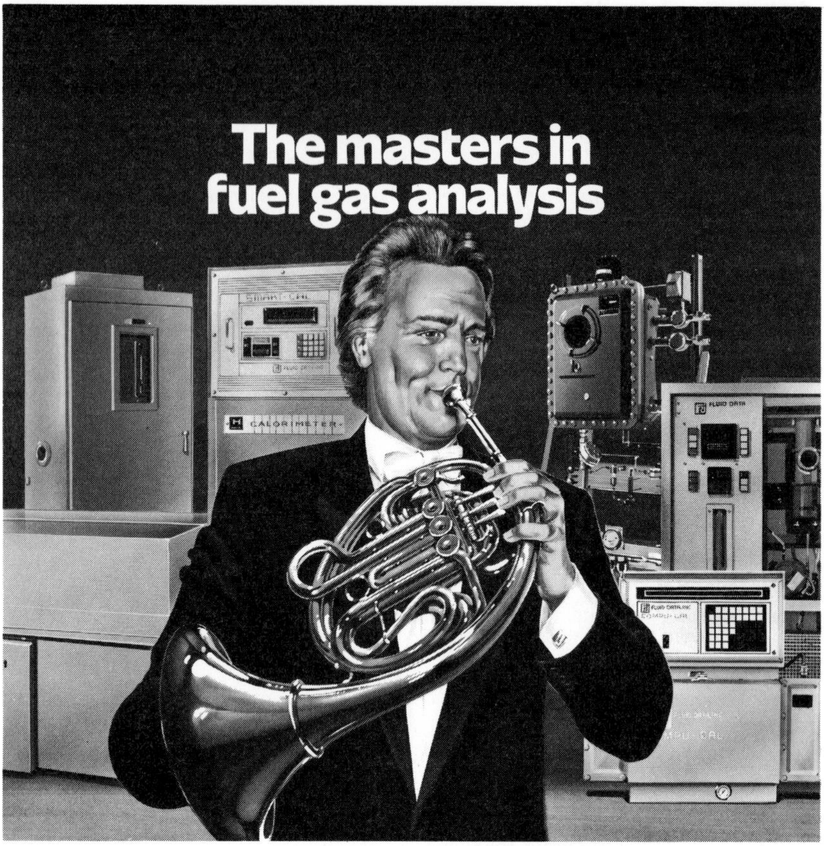
The masters in
fuel gas analysis
CALORIMETER

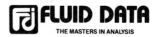

Osborn Steel Extrusions Ltd (continued)

Alloy steel (excl stainless) – special light sections. Hot extruded solid and hollow sections up to 130mm circumscribed circle in structural, carbon and stainless steels.

F.M. Parkin (Sheffield) Ltd

Head office: St Thomas Steelworks, Aizlewood Road, Sheffield S8 0YW. **Tel:** 0742 550083-7.
Associates: Thomas Andrews & Co Ltd. High speed steel, tool steel, ground flat stock.

T.W. Pearson Ltd

Head office: Wyvern Steel Works, Matilda Lane, Sheffield 1. **Tel:** 0742 25151. **Telex:** 547003.
Management: Chairman – P.P. Hague; directors – J. Hague; R.P. Hague (managing); M.G. Baker (financial). **Established:** 1896. **No. of employees:** 25. **Ownership:** Private. **Associates:** T.W. Pearson Forge Ltd, Sheffield (products include stainless, alloy, carbon, high speed forgings and forged bar).
Products: Alloy steel (excl stainless) – (high speed steels in all qualities: 10% cobalt, 5% cobalt, 22% tungsten, 18% tungsten-Ti; 14% tungsten; 6/5/2-M2; 9/2/1-M1; 6/5/4/4-M4) round bars, square bars, flats, hexagons.

The Phoenix Steel Tube Co Ltd

Head office: Phoenix Street, West Bromwich, W Midlands B70 0AS. **Tel:** 021 553 1811. **Telex:** 337301.
Management: Chairman – J. Bell; directors – R.A. Baker (managing), R. Cartwright (financial), L. Rawles (sales), T. Wood (production). **Established:** 1930. **Capital:** £150,000. **No. of employees:** Works – 163. Office – 62. **Ownership:** Senior Engineering Group Ltd.
Works:
■ At head office. ***Tube and pipe mills:*** Four high frequency welded (16-82.55mm dia). ***Other plant:*** Controlled atmosphere furnaces, pickling and wet processing for subsequent cold drawing (tube size 6-76mm dia), also seamless redraw mill within same size range.
Products: Carbon steel – seamless tubes and pipes 16-76mm; longitudinal weld tubes and pipes 16-82.55mm; cold drawn tubes and pipes 6-76mm; precision tubes and pipes 6-76mm.

Frank Pickering & Co Ltd

Head office: Admiral Steel Works, Sedgley Road, Sheffield S6 2DN. **Tel:** 0742 348721. **Telex:** 54667.
Management: Directors – W. Burkinshaw (managing), M. Flounders, D.J. Woodhouse. **Established:** 1923. **Capital:** £100. **No. of employees:** Works – 12. Office – 11. **Ownership:** Wilfred Burkinshaw Ltd. **Associates:** Jackstead Engineering Co Ltd; Burkinshaws (Alloys & Metals) Ltd.
Works:
■ At head office. Five forging hammers ranging from 0.35-1.5 tonnes; cut off machines; annealing furnaces; bump straightener.
Products: Carbon steel – round bars $\frac{1}{2}$-10″; square bars $\frac{1}{2}$-6″; flats 1 x $\frac{1}{2}$-10 x 2″. Stainless steel – round bars $\frac{1}{2}$-8″. Bright (cold finished) bars; bright (cold finished) bars $\frac{1}{2}$-3″; high-speed steels (rounds, squares, flats and forgings); spring steels (round and forgings); bearing steels (rounds and forgings); hot and cold work tool steels; shock-resisting steels; stainless, alloy and carbon steel forgings. Bars and forgings may be rough or finish-machined. Admiral (stainless steel); Tenax (tool steels); Dreadnought (high-speed steels).

United Kingdom

Plated Strip (International) Ltd

Head office: Wharfdale Road, Tyseley, Birmingham, W Midlands B11 2PL. **Tel:** 021 708 2244. **Telex:** 337875.
Management: Directors – B.K. Johnson (managing), G.W. Cooper (works); manager – K.A. Stokes (commercial). **Established:** 1933. **Capital:** £175,000. **No. of employees:** Works – 60. Office – 10. **Ownership:** Ultimate holding company – Banro Industries PLC. **Subsidiaries:** Concept Coated Steels Ltd (100%) (colour coated steels); Birmingham Steel Processors Ltd (100%). **Associates:** Edward Rose (Birmingham) Ltd. **Annual capacity:** Finished steel: 7,000 tonnes.
Works:
■ At works address. ***Coil coating lines:*** One hot-dip tinning (annual capacity 1,000 tonnes); electro nickel, chrome, copper and brass plating lines (7,000 tonnes). Hot-dip galvanizing
Products: Carbon steel – hot-dip tinplate 4-450mm x 0.1-2.00mm (1986 output 800 tonnes); chrome/nickel/brass/copper electro plated (4,000 tonnes). Stainless steel – bright annealed sheet/coil 6-457mm x 0.20-2.0mm (1,000 tonnes).
Brands: Presbrite (electro plated mild and stainless steel).
Sales offices/agents: At head office. Meko Bandstal, Halmstad, Sweden (Scandinavia); Solac Steel Ltd, Montreal, Quebec, Canada; Umcos Trading Pty Ltd, Victoria, Australia; Fletcher Steel, Auckland, New Zealand; Gabriel Arghyriadis, Thessaloniki, Greece; Morrell & Lee, Kowloon, Hong Kong.
Expansion programme: New finishes.

Premier Tubes Ltd

Head office: Walsall Road, Norton Canes, Nr Cannock WS11 3PU. **Tel:** 0543 75642.
Products: Carbon steel – longitudinal weld tubes and pipes – mild steel.

Pressfab Sections Ltd

Head office: Fountain House, 50 Fountain Lane, Oldbury, Warley, W Midlands. **Tel:** 021 541 1991. **Telex:** 334345.
Management: Managing director – Stuart Hobbs; managers – Peter Bennett (works), Mark Dicken (production), Nick Leek (administration). **Established:** 1976. **Capital:** £20,000. **No. of employees:** Works – 15. Office – 7. **Associates:** Pressfab Steels (metal stockholders and processors). **Annual capacity:** Finished steel 4,000 tonnes.
Works:
■ At head office.
Products: Carbon steel – light angles, light tees, light channels, special light sections, medium angles, medium tees, medium channels, cold roll-formed sections (from strip). Stainless steel – light angles, light tees, light channels, special light sections, cold roll-formed sections (from strip).

Queenborough Rolling Mill Co Ltd

Head office: Rushenden Road, Queenborough, Kent. **Tel:** 0795 668080. **Telex:** 965150. **Telecopier (Fax):** 0795 661165.
Management: Joint managing directors – M.G. Gandini, I.R. Watson; non-executive director – P. Wood. **Established:** 1972. **Capital:** £520,000. **No. of employees:** Works – 70. Office – 10. **Ownership:** Parconsult, Lugano, Switzerland. **Annual capacity:** Finished steel 100,000 tonnes.
Works:
■ At head office. ***Rolling mills:*** Two 17-stand Pomini bar – 480/360/220/110mm (annual capacity 100,000 tonnes).
Products: Carbon steel – reinforcing bars 8-16mm (1986 output 60,000 tonnes).

Raine & Co Ltd

Head office: Delta Works, Derwenthaugh, Swalwell, Newcastle Upon Tyne NE16 3BB. **Tel:** 091 488 2911. **Telex:** 53249 raine g.
Management: Chairman – A.N. Rudd; directors – N. Mc.Gillivray (managing), D. Janes (sales), G. Hudson (production), J.A. Futers (company secretary). **Established:** 1911. **Capital:** £400,000. **No. of employees:** Works – 175. Office – 25.
Ownership: Raine Industries PLC, Ashbourne Road, Mackworth, Derby DE3 4NB.
Subsidiaries: None.
Works:
■ At head office. ***Rolling mills:*** Two 17" medium section, light section and rail – 2-stand 3-high and 3-stand 3-high (annual capacity 30,000 tonnes).
Products: Carbon steel – round bars, square bars, light channels, special light sections, medium joists, medium channels, heavy joists, heavy channels, heavy rails, light rails, rail accessories.
Sales offices/agents: Railway Mine and Plantation Equipment Ltd, PO Box 41, Royal London House, 22/25 Finsbury Square, London EC2P 2AS.

H.H. Robertson (UK) Ltd

Head office: Cromwell Road, Ellesmere Port, South Wirral, Cheshire L65 4DS. **Tel:** 051 355 3622. **Telex:** 629639.
Products: Carbon steel – cold roll-formed sections (from strip), hot-dip galvanized sheet/coil, galvanized corrugated sheet/coil, colour coated sheet/coil.

Rotherham Engineering Steels—see UES – United Engineering Steels Ltd

United Kingdom

Sanderson Kayser Ltd

Head office: Attercliffe Steelworks, Newhall Road, Sheffield, S Yorks S9 2SD. **Tel:** 0742 449994. **Telex:** 54194. **Telecopier (Fax):** 0742 430171.
Management: Directors – M.J. Heeley, T. Worrall, S.A. Rawlinson; company secretary – P.R. Andrews. **Established:** 1776. **Capital:** £1,695,426. **No. of employees:** Works – 310. Office – 80. **Ownership:** SK Holdings Ltd (Ultimate holding company – GEI International PLC). **Subsidiaries:** Sanderson Kayser (Lancashire) Ltd; Sanderson Newbould (Australia) Pty Ltd; Sanderson Newbould SA (Pty) Ltd, South Africa; Sanderson Technitool (Pty) Ltd, South Africa; Sanderson Newbould Ltd, Canada.
Annual capacity: Raw steel: 19,000 tonnes. Finished steel: 12,000 tonnes.
Works:
■ Darnall Works, PO Box 6, Newhall Road, Sheffield S9 2SD. (Telex: 54194. Fax: 0742 430171). Works director: T. Worrall. ***Steelmaking plant:*** One 15-tonne electric arc furnace (annual capacity 16,000 tonnes). ***Refining plant:*** One ESR (3 x 1-tonne or 1 x 3-tonne) (1,100 tonnes). ***Rolling mills:*** One 2-high reversing 760mm billet (4,000 tonnes); one 5-stand 3-high 355mm bar (1,000 tonnes); one 15-stand 1-strand cross country 255mm wire rod (coil weight 100kg) (5,000 tonnes).
■ Newhall Works, PO Box 6, Newhall Road, Sheffield S9 2SD. (Tel: 0742 449994. Telex: 54194. Fax: 0742 430171). Works director: T. Worrall. **Steelmaking plant**: One 1-tonne high frequency furnace (annual capacity 3,000 tonnes). ***Rolling mills:*** Two bar – one 5-stand 3-high 355mm (3,000 tonnes) and one 6-stand double-duo 266mm (2,000 tonnes); one 3-stand 2-high 560mm medium plate (2,000 tonnes).
Products: Stainless steel – wire rod 6.0-15.0mm dia (1986 output 100 tonnes); round bars 3.0-75.0mm dia (300 tonnes); hexagons 4.0-10.0mm dia (20 tonnes); wire 6.0-15.0mm dia (output 300 tonnes); Alloy steel (excl stainless) – round bars 8.0-305.0mm (1,400 tonnes); square bars 6.0-205.0mm (300 tonnes); flats 20 x 10-400 x 120mm (900 tonnes); hexagons 4.0-40.0mm (20 tonnes); medium plates 2 x 1 metre x 2.0mm-40.0mm thick (1,500 tonnes); wire 3.0-12.0mm dia (300 tonnes). Tool steels; free cutting stainless steels; silver steel; forgings; armour plate; machine knives; circular cutters; segmental circular saws; circular wood and hot saw; hacksaw blades; woodworking saws; ground flat stock.
Brands: All black or bright steel products – KE.A162 (A2); KE.A180 (D2); KE.970 (D3); KE.396 (L6); KE.672 (01); KE.839 (L1); KE.960 (S1); KE.1006 (W2); Silver Steel (BS1407); KE.A145 (H13); KE.896 (L2); KE.805 (4340); KE.40A (416SE); KE.A507 (303SE); KE.A505 (431SE); KE.A521 (316SE).
Sales offices/agents: At head office. London – Shakespeare House, Newport Road, Hayes, Middx UB4 8JU; Midlands – Station Road, Coleshill, Birmingham B46 1JB; Glasgow – 19 Siemens Place, Blochairn Industrial Estate, Glasgow G21 2BN; Rainford – M & S Steels, Unit 19, Rainford Industrial Estate, Mill Lane, Rainford, St Helens, Merseyside WA11 8LS.
Expansion programme: Modernisation of forge and mills re-heating furnace linings (1988); rolling mills (1988); machine shop to CNC Centres (1987).

Seamless Tubes Ltd

Head office: Waddens Brook Lane, Wednesfield, Wolverhampton, West Midlands WV12 3SQ. **Tel:** 0902 305000. **Telex:** 338661.
Ownership: British Steel Corp. TI Group.
Works:
■ At head office. ***Tube and pipe mills:*** seamless and cold drawing plant.
■ Weldon Road, Corby, Northants (Tel: 0563 205005). Finishing plant for seamless tube.
Products: Carbon steel – seamless tubes and pipes, hot rolled and cold drawn tubes and pipes 1-5" nominal bore. Alloy steel (excl stainless) – seamless tubes and pipes, hot rolled and cold drawn tubes and pipes 1-5" nominal bore.

Sections & Tubes Ltd

Head office: Hall Street, West Bromwich, W Midlands. **Tel:** 021 553 2721. **Telex:** 336454.
Management: Chairman – Bernard Carlyle Hackett; director – Bernard Henry Turvey. **Established:** 1945. **Capital:** £20,000. **No. of employees:** Works – 50. Office – 10. **Annual capacity:** Finished steel: 10,000 tonnes approx.
Works:
■ At head office. Cold-forming lines (annual capacity 10,000 tonnes).
Products: Carbon steel – cold roll-formed sections (from strip) including light and medium sections – angles, tees, channels, special sections – up to 10″ wide (output year ended October 31st, 1986, 10,000 tonnes). Stainless steel – (incl alloy steel) – cold roll-formed sections (from strip) including light sections – angles, tees, channels, special sections;

Sheerness Steel Co PLC

Head office: Brielle Way, Sheerness, Kent ME12 1TH. **Tel:** 0795 663333. **Telex:** 965022, 965764. **Telecopier (Fax):** 0795 668051.
Management: Director and chief executive – C.E.H. Morris; deputy chairman – P.A. Learmond; directors – M.J. Shirley (finance), M.G. Smith (works), D.L. Williams (commercial), J.S. Moses (Car Fragmentation Ltd), H.E. Billot (personnel), J.B. Price (engineering); general superintendent – J.W. Clayton (steelmaking); superintendents – W.J. Hewitt (bar mill), S.M. Maxwell (rod mill), K.A. Plowman (quality control); controller – G. Bell. **Established:** 1971. **Capital:** £20,000,000. **No. of employees:** Works – 566. Office – 74. **Ownership:** Co-Steel Inc (Cananda) (65%). **Subsidiaries:** Car Fragmentation Ltd (raw materials division and scrap processing). **Associates:** Raritan River Steel Co, New Jersey, USA; Lake Ontario Steel Co, Ontario, Canada; Chaparral Steel Co, Texas, USA. **Annual capacity:** Raw steel: 600,000 tonnes. Finished steel: 640,000 tonnes.
Works:
■ At head office. ***Steelmaking plant:*** Two electric arc furnaces – 85-tonne 37 MVA (annual capacity 600,000 tonnes). ***Continuous casting machines:*** Two 4-strand Concast billet (120/130/140mm) (600,000 tonnes). ***Rolling mills:*** One single-strand 15-stand bar (400,000 tonnes); one single-strand 25-stand wire rod (coil weight 1.4 tonne) (240,000 tonnes).
Products: Carbon steel – billets 120,130 and 140mm sq (1986 output 500,000 tonnes); wire rod 5.5-16.5mm (180,000 tonnes); reinforcing bars 8-50mm (200,000 tonnes); round bars 18-52mm (36,000 tonnes); square bars 50-63mm (20,000 tonnes); flats 30-150mm wide (48,000 tonnes); light angles 30 x 30-70 x 70mm (30,000 tonnes).

Sheffield Forgemasters Ltd

Head office: Don Valley House, Brightside Lane, Sheffield, S Yorks, **Tel:** 0742 449071. **Telex:** 54185. **Telecopier (Fax):** 449071 3411.
Management: Chairman – Thomas Kenny; deputy chairman – E. Thompson; directors – P.M. Wright (managing), M.A. Brand (financial), R.G. Hardie (non-executive), W.D. Morton (non-executive). **Established:** 1982. **Capital:** £29,196,000. **No. of employees:** 2,721. **Ownership:** Sheffield Forgemasters Holdings PLC (Ultimate holding companies – BSC (50%), Johnson & Firth Brown PLC (50%)). **Subsidiaries:** Forgemasters Steels Ltd (100%) (ingot and forged black bar); Forgemasters Engineering Ltd (100%) (offshore and nuclear engineering products); River Don Castings Ltd (100%) (heavy castings); Special Melted Products Ltd (100%) (special steel products); Firth Vickers Special Steels Ltd (100%) (re-rolled bars); River Don Stampings Ltd (100%) (drop, press and upset forgings); Forged Rolls (UK) Ltd (100%) (forged rolls/back-up rolls); Midland Rollmakers Ltd (70%); R.B. Tennent Ltd (100%) (manufacture of rolls/ingots); Miller & Co Ltd (100%) (chilled iron rollmaking); Sheffield Forgemasters (Canada) Ltd (100%) (selling agents); Staballoy Inc (100%) (selling agents). **Annual capacity:** Raw steel: 100,000 tonnes. Finished steel: 77,500

Durmech Engineering Limited

High Street, Pensnett, Kingswinford,
West Midlands DY6 8XH, England.

Tel: 0384 298541 Telex: 337159 Fax: 0384 271117

This wide, heavy gauge slitting line with triple arbor turret head slitter and twin banding line facility gives an indication of our experience in the design and manufacture of a wide variety of process lines.

From a modest start in 1969 we have quickly grown to become one of the leading suppliers of coil processing equipment. This position has been gained by our policy of combining the finest engineering skills with modern manufacturing methods.

Whilst many standard designs are available we maintain a high degree of flexibility, enabling us to provide specially designed plant to meet the exacting individual requirements of our customers.

We engineer, manufacture, install and commission, all to the highest standards.

These services are at your disposal.

- Slitting Lines
- Cut to Length Lines
- Coating Lines
- Annealing Lines
- Pickling Lines
- Press Feed Lines
- Banding Lines
- Tension Level Lines
- Roll Forming Lines
- Combined Lines
- Individual Machines

Refendage de tole large et de grosse epaisseur a tourelle 3-mandrins de refendage et appareil de cerclage double, illustrant notre experience dans le domaine de l'etude et de la construction d'une vaste gamme de chaines de traitement.

Après un modeste démarrage en 1969, nous nous sommes développés rapidement pour devenir un des principaux fournisseurs d'équipements de parachèvement des tôles en rouleaux. Cette position a été obtenue grâce à notre politique combinant un bureau d'études de tout premier plan avec les méthodes les plus modernes de fabrication.

Tandis que nous disposons de nombreuses conceptions standard, nous conservons un niveau élevé de flexibilité qui nous permet de fournir des ensembles spécialement conçus suivant les besoins exacts individuels de nos clients.

Nous concevons, construisons, installons et mettons en route suivant les standards les plus élevés. Ces services sont à votre disposition.

- Lignes de refente
- Lignes de coupe à longueur
- Lignes de revêtement
- Lignes de recuit
- Lignes de décapage
- Lignes d'alimentation de presses
- Lignes de cerclage
- Lignes de bobinage sous tension
- Lignes de formage à galets
- Lignes combinées
- Machines individuelles

Diese breite Grobblech-Langsteil-Scherenstrasse mit Dreistationen-Revolverekopf für Scherendorne und doppe Banderolierstrasse lasst unsere Erfahrungen der Konstruktion und Herstellung eines umfangreichen Spektrums von Verarbeitung strassen erkennen.

Wir haben 1969 mit der Produktion Von Coi Verarbeitungs-Maschinen begonner. Wir sir heute einer der fuhrenden Hersteller auf dies Sektor.

Diese Position wurde durch gute Firmenpoli die ausgezeichneten Fachkenntnisse unsere Ingenieure, als auch durch die moderne Herstellungsmethode erreicht.

Neben unserer Standard-modellreihe haben unsere Produktion so flexibel gehalten, daß im Bedartsfalle auch Sonder ausfuhrungen, den Wunschen des Kunden entsprechen, herstellen konnen.

Wir entwickeln, bauen und installieren unse Maschinen. Auch die Inbetriebnahme erfolg durch qualifizierte Fachkrafte.

Dieser Service steht Ihnen zur Verfugung.

- Streifenscheren
- Streifenschneideinrichtuungen
- Beschichtungsanlagen
- Gluhanlagen
- Galvano-Einrichtungen
- Preßstraßen Einrichtungen
- Haspelautomaten
- Spannungsausgleicheinrichtungen
- Walzenprofilierstraßen
- Kombinierte Anlagen
- Einzelne Maschinen

Sheffield Forgemasters Ltd (continued)

tonnes.
Works:
■ Forgemasters Steels Ltd, River Don Works, Brightside Lane, Sheffield S9 2RU. (Tel: 0742 449071. Telex: 54185). Works director: P. Marsh. ***Steelmaking plant:*** 90-tonne electric arc furnace (annual capacity 100,000 tonnes). ***Refining plant:*** 35/100-tonne VAD (100,000 tonnes); 35/55-tonne VOD (10,000 tonnes); stream degassing tanks for ingots up to 250-tonnes. ***Other plant:*** Open die forging plant – 10,000-tonne, 1,750-tonne and 1,500-tonne presses.
■ Special Melted Products Ltd, Atlas House, Attercliffe Road, Sheffield S4 7UY. (Tel: 0742 20081. Telex: 547448 smp g). Works director: T. Marrison. ***Refining plant:*** 14-tonne ESR (up to 1,000mm dia ingots) (annual capacity 3,500 tonnes); 7-tonne VAR (up to 650mm dia ingots) (7,500 tonnes); 6-tonne VIM (up to 580mm dia ingots) (2,200 tonnes).
■ Firth Vickers Special Steels Ltd, Staybrite Works, Weedon Street, Sheffield S9 2FU. (Tel: 0742 449955. Telex: 54491). Works director: C.J. Stevens. ***Rolling mills:*** Two-stand 18" cogging billet (annual capacity 2,200 tonnes). 6-stand 10½" double duo bar (1,850 tonnes) 3-stand 8" cold mill (200 tonnes).
Products: Carbon steel – ingots 305mm and over (1986 output 15,408 tonnes); slabs 75 x 100mm-254 x 406mm (5 tonnes); blooms 100mm and over (2,705 tonnes); billets 100-450mm; round bars 9.5mm and over (5,597 tonnes); square bars 15mm and over; flats 3 x 9mm-254 x 406mm; hexagons 14-60mm Stainless steel – ingots 305mm and over (200 tonnes); slabs 75 x 100mm-254 x 406mm; blooms 100mm and over (5 tonnes); billets 100-450mm (868 tonnes); round bars 9.5mm and over (1,572 tonnes); square bars 15mm and over (105 tonnes); flats 3 x 70mm-254 x 406mm (1,032 tonnes); hexagons 14-60mm (51 tonnes). Alloy steel (excl stainless) – ingots 305mm and over (9,686 tonnes); slabs 75 x 100mm to 254 x 406mm; blooms 100mm and over (3,300 tonnes); billets 100-450mm (874 tonnes); round bars 9.5mm and over (9,763 tonnes); square bars 15-350mm (49 tonnes); flats 3 x 70mm-254 x 406mm (336 tonnes); hexagons 14-60mm (30 tonnes). Open die forgings; rolled rings; tool steel; nitrided steels.
Brands: Staybrite (stainless grades); FV 520 (B).
Sales offices/agents: At works addresses. France – Ets Goldring & Cie, Levallois-Perret. Israel – Pladot Yemini Ltd, Tel-Aviv. Italy – Metallco, Turin. Sweden and Norway – G. Kjellstrom & Co AB, Örebro, Sweden. USA – Erickson Aerospace, Redmond, WA.Canada – Sheffield Forgemasters (Canada) Ltd.
Expansion programme: 10/12-tonne VIM melting furnace.

Spartan Redheugh Ltd

Head office: Teams, Gateshead, Tyne & Wear NE8 2RD. **Tel:** 091 460 4245. **Telex:** 53259.
Management: Chairman – J.A. Nutt; chief executive – D.M. Bejarano; directors – M.A. Charlton (non-executive), B. Scott (deputy managing), P. Beckwith (commercial), D. Wilkin (financial). **Established:** 1957. **Capital:** £1,313,000. **No. of employees:** Works – 146. Office – 31. **Ownership:** Spartan Redheugh Holdings Ltd (100%).
Annual capacity: Finished steel: 50,000 tonnes.
Works:
■ At head office. Works director: B. Scott. ***Rolling mills:*** One 4-high reversing heavy plate 90 x 23" (annual capacity 50,000 tonnes).
Products: Carbon steel – heavy plates max width 2m, 6-150mm thick (1986 output 36,000 tonnes). Stainless steel – heavy plates max width 2m, 6-40mm thick (1,700 tonnes). Alloy steel (excl stainless) – heavy plates max width 2m, 6-75mm thick (5,000 tonnes).
Brands: Red Diamond (wear-resistant plates).

United Kingdom

Spartan Sheffield Ltd

Head office: Attercliffe Works, Attercliffe Road, Sheffield, S Yorks S9 3QQ. **Tel:** 0742 447551, 0742 449551. **Telex:** 54431.
Management: Chairman and managing director – D.B. Moody; directors – P. Flaherty (works), A.J. Street (sales); non-executive – M.L. Carr. **Established:** 1981. **No. of employees:** Works – 38. Office – 17. **Annual capacity:** Finished steel: 5,000 tonnes.
Works:
■ At head office. ***Rolling mills:*** Two bar – 6-stand 10" (two 3-high and four 2-high); 3-stand 2-high 10" (2,000 tonnes). Single-stand 2-high 12" cold mill (500 tonnes) (annual capacity 3,000 tonnes);
Products: Stainless steel – round bars (1986 approx output 1,200 tonnes); square bars (50 tonnes); flats (400 tonnes); hexagons (400 tonnes).

Special Melted Products Ltd—see Sheffield Forgemasters Ltd

Spencer Clark Metal Industries PLC

Head office: Greasbrough Street, Rotherham, S Yorks S60 1RG. **Tel:** 0709 363281. **Telex:** 54131 scmi g. **Cables:** Spenclark. **Telecopier (Fax):** 0709 361507.
Management: Directors – M.T. Davies (managing), D.R. Thorpe (general manager), K.L. Riley (production), I. Harding (financial), G. Parkin (technical). **Established:** 1972. **No. of employees:** Workss – 225. Office – 45. **Ownership:** Williams Holdings PLC, Derby. **Annual capacity:** Finished steel: 15,000 tonnes.
Works:
■ At head office. Works director: K.L. Riley. ***Rolling mills:*** bar – one 8", one 13", one 14", one 16", one Schloemann single pass (annual capacity 13,000 tonnes).
■ Savile Street, Sheffield S4 7UH. (Tel: 0742 700869. Telex: 54131 scmi g). Works director: K.L. Riley. **Rolling mills:** Two sheet mills, one 26" and one 24" (hand) (annual capacity 1,000 tonnes).
■ Warren Street, Sheffield S4 7WR. (Tel: 0742 768914. Telex: 54131 scmi g). Works director: K.L. Riley. Six forging hammers – one-250kg, two 500-kg, two 750-kg; heat treatment and machining facilities.
Products: Carbon steel – round bars, square bars, flats, hexagons, special light sections, hot rolled sheet/coil (uncoated). Stainless steel – round bars, square bars, flats, hexagons, special light sections, hot rolled sheet/coil (uncoated). Alloy steel (excl stainless) – round bars, square bars, flats, hexagons, special light sections, hot rolled sheet/coil (uncoated). Open die-hammer forgings in stainless, alloy and superalloys; bars and sections in superalloys; composite steels in bar and sheet form.
Sales offices/agents: At head office (rolled products); Warren Street, Sheffield S4 7WR (forged products). Spencer Clark Metal Industries Inc, 17066A South Park Avenue, Box 551, Chicago, IL 60473, USA (Tel: 0101 3123311070. Telex: 0230 910 257 0839 spenclark shnd. Fax: 3123315064).

Stanton PLC

Head office: Stanton By Dale, Ilkeston, Derbys NG10 5AA. **Postal address:** PO Box 72. **Tel:** 0602 322121. **Telex:** 37671 sands. **Telecopier (Fax):** 0602 329513.
Management: G.J. Nicholls (board member and chief executive), D. Rolland (deputy chief executive); associate directors – G.E. Else (technical), V.G. Harper (commercial), P.J. Webb (operations). **Established:** 1898. **Capital:** Authorised – £50,000,000. Issued – £12,604,000. **No. of employees:** Works – 1,850. Office – 650. **Ownership:** Pont à Mousson SA, France (100%). (Ultimate holding company – Cie de Saint Gobain (100%). **Subsidiaries:** Stanton (Exports) Ltd, 100%, (exports Stanton Plc products), Nutbrook Canal Navigation, 100%, (Dormant), Stanton Trustees Ltd, 100% (pension fund administration), Stanton Technicast Ltd, 55% (continuous cast s.g. (ductile) iron bar).
Works:
■ At head office; Holwell, Melton Mowbray (Tel: 0664 812812) and Staveley, Nr

Stanton PLC (continued)

Chesterfield (Tel: 0246 475331).
Products: Cast iron – pipes, pipe fittings.

Staystrip Ltd

Head office: Eyre Street, Spring Hill, Birmingham, W Midlands B18 7AA. **Tel:** 021 455 0111. **Telex:** 336236 stayst g. **Cables:** Staystrip.
Management: Directors – D. Myers, L.W. Myers, S.J. Myers, K.R. Simpson, W.F. Aherne, A.R. Taylor, K.H. Hill. **Established:** 1966. **No. of employees:** 100 approx. **Ownership:** Staystrip Group Ltd. **Subsidiaries:** Welded Stainless Steel Tubes Ltd (see separate entry). **Annual capacity:** Finished steel: 4,000 tonnes approx.
Works:
■ At head office. Works director: W.F. Aherne. ***Rolling mills:*** Three 4-high Robertson cold strip (coil weight 5 tonnes max) (annual capacity 1,500 tonnes). ***Other plant:*** Decoiling lines; rotary slitters 1,250mm max wide.
Products: Carbon steel – cold rolled sheet/coil (uncoated) Temper rolled strip in coil/length; annealed strip in coil/length; perforated strip.

Steel Bars & Sections Ltd

Head office: Castle Street Works, Tipton, West Midlands. **Tel:** 021 520 2433. **Telex:** 336830 steels g.
Products: Carbon steel – bright (cold finished) bars rounds and flats.

Steel of Staffs

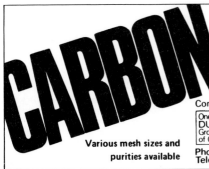

Head office: Link 51 Ltd, Haldane House, Halesfield, Telford, Salop TF7 4LN. **Tel:** 0952 680051, 0384 296151 (sales office). **Telex:** 35404. **Telecopier (Fax):** 0952 680444.
Management: Chairman – J.L. Hudson; directors – D.E. Poyner (managing), M. Bailey, A.G. Jones, E. Billington. **Established:** 1958. **Capital:** £500,000. **No. of employees:** 100. **Ownership:** Link 51 Ltd.
Works:
■ Kingswinford, W Midlands. (Tel: 0384 296151. Telex: 336216). Divisional managing director: M. Bailey. Divisional sales director: G. Perry. ***Rolling mills:*** cold strip – 4-high 13 x 5½" Robertson non-reversing (annual capacity 10,000 tonnes), 2-high 12 x 10" Robertson non-reversing (20,000 tonnes), 2-high 8 x 8" ARM non-reversing (5,000 tonnes); temper/skin pass – one 20 x 16" Robertson non-reversing (15,000 tonnes). ***Other plant:*** Two Grunewald bright annealing furnaces (20,000 tonnes); One 54 x 0.250" DAC wide slitter (100,000 tonnes).
Products: Carbon steel – cold rolled hoop and strip (uncoated) in grades from CS2 to

United Kingdom

Steel of Staffs (continued)

CS 80, size range 0.010-0.125" thick, 0.375-10.5" wide – supplied in a variety of tempers from fully annealed to hard bright, in coils or straight flat lengths. Hot rolled strip and hot rolled pickled and oiled, in coils and straight flat lengths, sheared from wide sheet coils, in thicknesses up to 0.250" is also supplied as for cold rolled strip but in cold reduced finish and gauges.

Steel Parts, Glynwed Steels Ltd

Head office: Spartan Works, Brickhouse Lane, West Bormwich, West Midlands B70 0DS. **Tel:** 021 557 2861. **Telex:** 337667.
Established: 1929. **No. of employees:** Works – 52. Office – 24. **Ownership:** Glynwed Steels Ltd (Ultimate holding company – Glynwed International PLC). **Annual capacity:** Finished steel: 35,000 tonnes.
Products: Carbon steel – bright (cold finished) bars. Stainless steel – bright (cold finished) bars. Alloy steel (excl stainless) – bright (cold finished) bars. Sparcut 45 (carbon steel). Spartcut 55 (alloy steel).

Stelco Hardy Ltd

Head office: Blaenrhondda, Treorchy, Glam CF42 5BY. **Tel:** 0443 771774. **Telex:** 498272 tubmac g. **Telecopier (Fax):** 0443 776009.
Management: Ken Hall (managing director). **Ownership:** McKechnie PLC. **Annual capacity:** Finished steel: 9,000 tonnes.
Works:
■ At head office. ***Tube and pipe mills:*** welded.
Products: Stainless steel – longitudinal weld tubes and pipes and hollow sections $\frac{3}{8}$-4", austenitic and ferritic, also in special alloys.

Stocksbridge Engineering Steels—see UES – United Engineering Steels Ltd

Stourbridge Rolling Mills, Glynwed Steels Ltd

Head office: Bradley Road, Stourbridge, West Midlands DY8 1UY. **Postal address:** PO Box 13. **Tel:** 0384 392411-7. **Telex:** 338487.
Management: Directors – I.M. Jones (general manager), J.D. Mobberley (finance), F.G. Lee (production). **Established:** 1905. **Capital:** £3,000,000. **No. of employees:** Works – 80. Office – 25. **Ownership:** Glynwed Steels Ltd (100%) (Ultimate holding company – Glynwed International PLC (100%)). **Annual capacity:** Finished steel: 23,000 tonnes.
Works:
■ At head office. ***Rolling mills:*** Three cold strip – one 2-stand 4-high (300mm), one 1-stand 2-high (400mm) and one 3-stand 2-high (450mm) (combined annual capacity 25,000 tonnes).
Products: Carbon steel – cold rolled hoop and strip (uncoated) 0.5-5.5mm (1986 output 23,000 tonnes).

The Templeborough Rolling Mills Ltd (TRM)

Head office: Ekim Street, Rotherham, S Yorks S60 1DT. **Postal address:** PO Box 19. **Tel:** 0709 378153. **Telex:** 54638. **Cables:** Rods Rotherham.
Management: F. Holloway (chairman), J. Churchfield (managing director), C.E. Morgan (director and general manager), A.C. Budge (commercial manager), J.B. Stickland (chief engineer), D.T. Parkinson (technical manager), R.F. Graham (company secretary). **Established:** 1916. **Capital:** £216,004. **No. of employees:** Works – 284. Office – 77. **Ownership:** British Steel Corp (50%). Bridon PLC (50%).

The Templeborough Rolling Mills Ltd (TRM) (continued)

Subsidiaries: None. **Annual capacity:** Finished steel 465,000 tonnes.
Works:
■ At head office. ***Rolling mills:*** 2-strand 27-stand wire rod (with single stand 10-strand No-Twist finishing trains) (coil weight 1,800kg) (annual capacity 465,000 tonnes).
Products: Carbon steel – wire rod 5-13.5mm (1986 output 180,564 tonnes). Stainless steel – wire rod. Alloy steel (excl stainless) – wire rod 5-13.5mm.
Expansion programme: Microprocessor to be installed.

TI Desford Tubes Ltd

Head office: Kirby Muxloe, Leicester, Leics LE9 9BJ. **Tel:** 04557 2321. **Telex:** 341026 (sales); 342801. **Telecopier (Fax):** 04557 3441.
Management: Directors – G. Jones (managing), D.C. Beevers (personnel), D. Clark (financial), F.A. Cope (production), J.G.H. Morris (sales and marketing), T.J.W. Bostock (production services), J.E. Meredith (technical services), B. Thompson (engineering development). **Established:** 1972. **Capital:** £100,000. **No. of employees:** Works – 524. Office – 241. **Ownership:** TI Group PLC.
Works:
■ At head office. ***Tube and pipe mills:*** seamless – three 45-200mm od Assel (annual capacity 150,000 tonnes). ***Other plant:*** Twelve cold reducing machines (18-127mm od) (23,500 tonnes). Upset forging and swaging plant (25-500mm od) (10,000 tonnes). Heat treatment plant including quench and temper.
Products: Carbon steel – seamless tubes and pipes 18-200mm od; oil country tubular goods 18-200mm od; hollow bars 18-200mm od; cold reduced tube 18-127mm od.

United Kingdom

TI Desford Tubes Ltd (continued)

Alloy steel (excl stainless) – seamless tubes and pipes 18-200mm od; oil country tubular goods 18-200mm od; hollow bars 18-200mm od; cold reduced tubes 18-127mm od. Hot and cold reduced tubes in bearing steels; solid and tubular forgings.
Brands: TI52 and TI6U (hollow bar).
Sales offices/agents: Official stockists and agents worldwide.
Expansion programme: Increase in capacity of cold reducing machines to include 127mm od finished product.

TI Group PLC

Head office: 50 Curzon Street, London W1Y 7P. **Tel:** 01 499 9131. **Telex:** 263740.
Management: Board of Directors: R.E. Utiger (chairman), C. Lewinton (deputy chairman and chief executive), Sir John Cuckney (deputy chairman), J. Sawkill (research), M.F. Garner (finance), A.M. Davies (non-executive), D. Edwards (non-executive), S. Taylor (automotive), G.R. Mackenzie (specialised tube), H.J. Atkins (domestic applicances) **Subsidiaries:** TI Group Specialised Tube Business Area: Box 356, Popes Lane, Oldbury, Warley, West Midlands B69 4UH (Management: G.R. Mackenzie – chairman). The companies in the TI Specialised Tube business area are: TI Desford Tubes, TI Chesterfield, TI Hollow Extrusions, TI Accles & Pollock, TI Apollo, TI Reynolds 531, TI Stainless Tubes, Cold Drawn Tubes (in which TI has a 75% interest) (see separate entries). Other TI Subsidiary and associate companies include: Standard Tube Canada Inc; Seamless Tubes Ltd, UK; TI Reynolds Rings Ltd, UK; Armco Ltd, UK; Fulton (TI) Ltd, UK; Armco SA, Belgium; Armco SA, France; Valti SA, France; TI Metal Sections, India; Tube Products of India.

TI Hollow Extrusions Ltd

Head office: Redfern Road, Tyseley, Birmingham B11 2BW. **Tel:** 021 706 3333. **Telex:** 339999 he man g. **Telecopier (Fax):** 021 707 6141.
Management: Directors – J.A. Winchester (general manager), R.J. Edwards (manufacturing), M.W. Fisher (sales and marketing), J.C. Hindmarsh (technical). **Established:** 1898/1986. **No. of employees:** Works – 130. Office – 40. **Ownership:** TI Group PLC. **Associates:** Other companies in TI Group.
Works:
■ At head office. ***Tube and pipe mills:*** seamless One extrusion press.
Products: Carbon steel – seamless tubes and pipes. Stainless steel – seamless tubes and pipes. Alloy steel (excl stainless) – seamless tubes and pipes. Aluminium and copper alloy tubes.

TI Reynolds Rings Ltd

Head office: Hay Hall, Redfern Road, Tyseley, Birmingham B11 2BG. **Tel:** 021 706 3333. **Telex:** 338181. **Telecopier (Fax):** 021 707 0127.
Management: Manager – S.M. Bates (general); directors: P.A. Bullivant (sales), M.H. Burden (technical). **Established:** 1917. **No. of employees:** Works – 90. Office – 80.
Works:
■ Plant for flash welding, bending, cold rolling, truform rolling and machining. Heat treatment. Rings and annular components in alloy and stainless steels, nickel alloys, and titanium alloys for aero-engines.
Brands: Truform.

TI Stainless Tubes Ltd

Head office: Green Lane, Walsall, W Midlands WS2 7BW. **Postal address:** PO Box 21. **Tel:** 0922 21222. **Telex:** 337587. **Cables:** Tistan G. **Telecopier (Fax):** 0922 721812.
Management: Directors – F.J. England (managing), P.G. Christensen (commercial), J.R. McGregor (manufacturing, hot products), M.E. Giles (manufacturing, cold products), W.M. Good (financial), R.P. De'Ath (personnel). **Established:** 1961. **Capital:** £10,000,000. **No. of employees:** Works – 379. Office – 133. **Ownership:** TI Group PLC (100%). **Annual capacity:** Finished steel: 10,000 tonnes.
Works:
■ At head office. Manufacturing director: M.E. Giles. ***Tube and pipe mills:*** seamless – six cold Pilger machines and five drawbenches (annual capacity 5,000 tonnes). ***Other plant:*** One hydrogen atmosphere annealing furnace.
■ Derby Road, Chesterfield, Derbys S40 2EA. (Tel: 0246 211994. Telex: 547028). Manufacturing director: J.R. McGregor. ***Tube and pipe mills:*** seamless – one 3,000-tonne hot extrusion press (annual capacity 10,000 tonnes).
Products: Carbon steel – special light sections within 180mm circumscribing circle. Stainless steel – light angles, light tees, light channels and special light sections – within 180mm circumscribing circle; seamless tubes and pipes, oil country tubular goods and cold drawn tubes and pipes – hot and cold finished in austenitic and martensitic qualities, high-nickel alloys 600, 800 and 825, and duplex alloys (hot finished sizes 25-171mm id and 194mm max od, cold finished sizes 12-180mm); hollow bars – round and hexagon up to 190mm od. Alloy steel (excl stainless) – special light sections within 180mm circumscribing circle. Co-extruded bi-metallic boiler, superheater and heat exchanger tube; mechanically bonded bi-metallic heat exchanger tube. Combined 1986 output 9,000 tonnes. Co-extruded bi-metallic boiler, superheater and heat exchanger tube; mechanically bonded bi-metallic heat exchanger tube.
Sales offices/agents: At head office. TI International Ltd, PO Box 356, Popes Lane, Oldbury, Warley, W Midlands B69 4UH (Tel: 021 552 3345. Telex: 338031. Fax: 021 552 4077).

TI Tube Products Ltd

Head office: PO Box 13, Popes Lane, Oldbury, Warley, W Midlands B69 4PF. **Tel:** 021 552 1511. **Telex:** 337307.
Management: Directors – Maurice Andrews (managing), Nick Burton-Carter (financial), Malcolm Armson (manufacturing), Mike Hurst (sales and marketing), Roy Baverstock general manager, drawn products), Peter Groom (manipulated products); Nigel Watkins abd David Haggett. **Established:** 1926. **No. of employees:** Works – 420. Office – 180.
Works:
■ At head office. Works director: M. Armson. ***Rolling mills:*** Robertson cold strip equalising mill up to 5mm thick. ***Tube and pipe mills:*** welded – 10 high frequency (12.70-111.12mm od x 0.70-5.00mm thick) (annual capacity 85,000 tonnes). ***Other plant:*** Slitting line for wide coil, annealing furnaces, drawbenches cutting and straightening machines. Tube forming and manipulation facilities.
Products: Carbon steel – longitudinal weld tubes and pipes 12.70-111.12mm od (inc non circular sections); galvanized tubes and pipes; cold drawn tubes and pipes; precision tubes and pipes, hollow sections; conduit, refrigeration and pressure tube. Alloy steel (excl stainless) – precision tubes and pipes (chrome-molybdenum alloy tube).
Additional information: Until October 1987 the company was a subsidiary of TI Group PLC, and was known as TI Tube Products Ltd. It wa then the subject of a management buy-out, and the name was expected to revert to Tube Products Ltd.

UES – United Engineering Steels Ltd

Head office: Sheffield Road, Rotherham, S Yorks S60 1DQ. **Postal address:** PO Box 29. **Tel:** 0709 371234. **Telex:** 54209. **Telecopier (Fax):** 0709 371234 Ext 5273.
Management: Board – Ian Donald (chairman), Gordon H. Sambrook (deputy chairman BSC), John Pennington (chief executive director), Norman H. Broadhurst (executive director), Alistair Brown (executive director), Brian Insch, David B. Lees, Alistair R. Love (secretary). Executive committee John S. Pennington, Norman Broadhurst (financial director), Alistair J. Brown (chief executive, United Engineering & Forging), Alistair R. Love (secretary), Ken Knaggs (managing director, Rotherham Engineering Steels), David R. Stone (managing director, Stocksbridge Engineering Steels). **Established:** 1986. **Capital:** £427,000,000. **No. of employees:** 7,300 (at 31 Dec 1986, steel divisions only). **Ownership:** British Steel Corp; GKN. **Annual capacity:** Raw steel: 2,420,000 tonnes. Finished steel: 1,920,000 tonnes.

Brymbo Steel Works. Brymbo, Wrexham,Clwyd LL11 5BT. ***Tel:*** 0978 756 333. ***Telex:*** 61883.
Management: J.M. Marsh (commercial director)
Works:
■ Brymbo ***Steelmaking plant:*** One 92-tonne UHP electric arc furnace (annual capacity 370,000 tonnes). ***Refining plant:*** Two ladle furnace, one VAD and one leading plant (combined capacity 370,000 tonnes). ***Rolling mills:*** 2-high cogging slabbing / blooming and close tolerance billet (annual capacity 320,000 tonnes). ***Other plant:*** Billet finishing – controlled cooling, annealing, normalising.
■ Cable Street Bar Mill, Wolverhampton. ***Rolling mills:*** 8-stand bar (annual capacity 130,000 tonnes).
Products: Carbon steel – ingots (forging) 2.9-42-tonne; blooms and billets 50-180mm, round billets 76-180mm; round bars 22-76mm; square bars 30-51mm; flats 50-127 x 12-40mm; hexagons 32-71.5mm; bright (cold finished) bars 22-175mm. Alloy steel (excl stainless) – ingots 2.49-42-tonne; blooms and billets 50-180mm, round billets 76-180mm; round bars 22-76mm; square bars 30-51mm; flats 50-27 x 12-40mm; hexagons 32-71.5mm; bright (cold finished) bars 22-175mm. Engineering steels including carbon and carbon-manganese steels; micro-alloyed steels; boron, spring and bearing steels; case-hardening and nitriding steels.

Rotherham Engineering Steels, PO Box 50, Aldwarke Lane, Rotherham S60 1DW ***Tel:*** 0709 382141. ***Telex:*** 54209.
Management: K. Knaggs (managing director), G. Forster (commercial director).
Works:
■ Templeborough. ***Steelmaking plant:*** One 180-tonne electric arc furnace (annual capacity 400,000 tonnes). ***Refining plant:*** One ladle furnace, arc re-heating and one vacuum degassing unit. ***Continuous casting machines:*** One 6-strand curved billet (100, 115, 140mm) (400,000 tonnes).
■ Aldwarke. ***Steelmaking plant:*** Three 175-tonne electric arc furnaces (1,150,000 tonnes). ***Refining plant:*** one vacuum degassing unit; ladle leading facilities; under construction ladle steelmaking, arc re-heating. ***Continuous casting machines:*** twin 2-strand bloom (560 x 400mm) (under construction). ***Rolling mills:*** 2-high cogging/ blooming and 6-stand continuous billet (combined annual capacity 800,000 tonnes).
■ Roundwood. ***Rolling mills:*** 18-stand continuous bar for rounds, squares, hexagons (coil weight 1,800kg) (annual capacity 300,000 tonnes). ***Other plant:*** Bar processing plant – six computer controlled heat-treatment furnaces, seven reeling machines, six grinding machines, two turning machines and cold sawing, abrasive cutting.
■ Thrybergh. ***Rolling mills:*** 18-stand continuous bar for rounds, squares and hexagons (in straight lengths) (annual capacity 500,000 tonnes).
■ Tinsley Park. ***Rolling mills:*** 13-stand cross country bar for rounds and flats (annual capacity 75,000 tonnes).
■ Wolverhampton. ***Rolling mills:*** 6-stand bar for rounds, squares, hexagons and special sections (annual capacity 18,000 tonnes).
Products: Carbon steel – slabs 76 x 152-355 x 3,000-18,000mm; billets 50-140mm; wire rod round – 5.5-24mm and square – 9.0, 10.5 and 11.0mm. Round bars 13-115; square bars 31-51mm; flats 15-140 x 5-54mm; hexagons 16-71.5mm; bright (cold

United Kingdom

UES – United Engineering Steels Ltd (continued)

finished) bars 13-100mm. Alloy steel (excl stainless) – slabs 152-355 x 76 x 3,000-18,000mm; billets 50-140mm; wire rod round – 5.5-24mm and square 9.0, 10.5 and 11.0mm; round bars 13-115mm; square bars 13-51mm; flats 15-140 x 5-54mm; hexagons 16-71.5mm. Engineering steels including: carbon and carbon-manganese steels; free-cutting steels; micro-alloyed steels; boron, spring and bearing steels; case-hardening and nitriding steels; improved machining steel.

Stocksbridge Engineering Steels, Stocksbridge, Sheffield S30 5JA ***Tel:*** 0742 882361. ***Telex:*** 54224.
Management: D.R. Stone (managing director), P.K. Fletcher (commercial director).
Works:
■ Stocksbridge. ***Steelmaking plant:*** One 95-tonne and one 150-tonne electric arc furnaces (annual capacity 500,000 tonnes). ***Refining plant:*** One ESR, one ladle furnace, one VAD, one VAR, one VOD and Stokes degasser. ***Continuous casting machines:*** One 4-strand vertical billet (rounds – 115, 140, 155, 160, 175 and 200mm; squares – 100, 108 and 140mm) (150,000 tonnes). ***Rolling mills:*** slabbing and blooming; intermediate and finishing close tolerance billet (combined annual capacity 400,000 tonnes). ***Other plant:*** Peeling and precision turning of rounds; four heat treatment furnaces; dedicated inspection and finishing facilites for aerospace industry.
Products: Carbon steel – slabs 76-1,054 x 44-317mm; blooms 150-457mm; billets 50-184mm (square), 76-381mm (round). Stainless steel – slabs 76-1,054 x 44-317mm; blooms 150-457mm; billets 50-184mm (square), 76-381mm (round). Alloy steel (excl stainless) – slabs 76-1,054 x 44-317mm; blooms 150-457mm; billets 50-184mm (square), 76-381mm (round). Turned rounds 70-350mm; engineering steels including: carbon and carbon-manganese; micro-alloyed, boron, spring, tool and die steels, case hardening and nitriding steels; bearing steels, improved machining steels; high-temperature bolting steels; stainless and corrosion resistant steels; heat-treatment steels; aircraft and maraging steels.

Brands: Brymbo – Brymach (engineering steels with improved machinability and properties); Brysav (micro-alloyed precipitation air-hardening steels). Rotherham – XL and XL Cut (low and medium carbon freecutting steels); Tentuff (micro-alloyed steels for spanners). Rotherham and Stocksbridge – IM (standard engineering steels with improved machinability); Maxim (re-sulphurised engineering steels with superior machinability); Vanard (micro-alloyed precipatation air-hardeni steels); Durehete (high strength high temperature bolting steels, for use in power generating units and process plant). Stocksbridge – Esshete (alloy and stainless steels with a combination of high strength, corrosion resistance and high temperature properties); Jethete (ultra high strength hig h temperature steels particularly designed for use in aircraft engines); Red Fox – (austenitic steels for use in furnace equipment); Silver Fox (stainless steels); Nimar (three grades of 18% nickel maraging steels); Foxalloy 101 (high strength fully austenitic stainless alloy suitable for non-magnetic drill collars).
Sales offices/agents: Steel divisions: Belgium – Roger Hec Sprl, Braine-L'Alleud (Tel: 02 3846037. Telex: 65575); Canada – BSC Canada Inc, Montreal (Tel: 514 481 8145. Telex: 05 267572); China – BSC (China) Ltd, Hong Kong (Tel: 5-716213. Telex: 60004); Denmark – BSC Dansk, Copenhagen (Tel: 01 121758. Telex: 22696); Finland – Oy Mercantile Ab, Helsinki (Tel: 358 03450 239. Telex: 124416); France – JC Gay, Cluses (Tel: 503 4776. Telex: 385275) (forging and automotive steels) and F.D. Metals France S Charenton-Le-Pont (Tel: 43 96 30 57. Telex: 262549) (other steels), West Germany – United Engineering Steels (Deutschland) GmbH, Düsseldorf (Tel: 0211 499041. Telex: 8585533); India – BSC Liaison Services (India) Ltd, New Delhi (Tel: 674240. Telex: 031 62790); Italy – Metalbi Srl, Milan (Tel: 02 436087. Telex: 335267) and Brymbo International (Italy) Srl, Milan (Tel: 02 2151589. Telex: 334169); Netherlands – Ang lo Staal Holland BV, Rotterdam (Tel: 10 128840. Telex: 27403); Norway – BSC Norge A/S, Oslo (Tel: 02 373160. Telex: 72373); Spain – BSC (Spain) Ltd, Madrid (Tel: 01 248 6 207. Telex: 22886); Sweden – BSC Svenska AB, Gothenburg (Tel: 031 800770. Telex: 27350) and Cevece Montan AB, Gothenburg

BEDA OXYGENTECHNIK

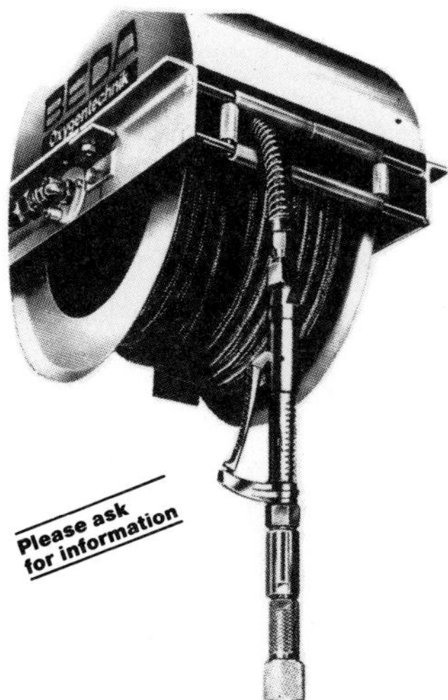

Safety-fittings for oxygen lances

- Lance-holders (pat.)
- Slag return safety devices (pat.)
- Quick shut-valves (pat.)
- Oxygen hoses (fiberglas, asbestos, metal braiding)
- Hose drums
- Automatic lancing advance

"World leaders in the development of quality oxygen fittings"

Armaturen GmbH & Co KG

P.O. Box 22 46 · D-4030 Ratingen 1
Tel. (0 21 02) 49 04 61 · Tx. 8 585 240 beda d
Fax. (02102) 499134

BEDA OXYGENTECHNIK

UK SOLE AGENT FOSTER GARTON
A DIVISION OF FURMANITE ENGINEERING LTD.
Owens Road, Skippers Lane Industrial Estate
South Bank, Middlesbrough, Cleveland, TS6 6HE

Tel: (0642) 455111 After Hours: (0642) 478312
Telex: 65262 Furmte G, Telefax: (0642) 465692

UES – United Engineering Steels Ltd (continued)

(Tel: 031 80 36 80. Telex: 2241); Switzerland – K. Grütter Handels AG, Wallisellen (Tel: 01 8300578. Telex: 54425); USA – BSC Inc, Houston (Tel: 713 820 2828. Telex: 775298), Brymbo Steel Inc, New Jersey (Tel: 201 431 5708. Telex: 132137), Interstel Inc, Chicago (Tel: 312 332 3923), British Steel Co, San Pedro, CA (Tel: 213 547 4471. Telex: 68 6267).
Additional information: The company's forging division, United Engineering & Forging, is at PO Box 4, Bromsgrove, Worcestershire B60 3D2. (Tel: 0527 579200. Telex: 337469).

Unbrako Steel Co Ltd

Head office: Burnaby Road, Coventry CV6 4AE. **Tel:** 0909 770431. **Telex:** 54176. **Telecopier (Fax):** 772848.
Management: Directors – D. Wales (managing), M. Broomhead (marketing), J. Eyre (manufacturing), A. Whitehead (engineering), H.J. Wilkinson (USA), R.L. Sproat (USA). **Established:** 1959. **Capital:** £250,000. **No. of employees:** Works – 286. Office – 44. **Ownership:** SPS Technologies Inc, Newtown, PA, USA. **Annual capacity:** Raw steel: 35,000 tonnes. Finished steel: 30,000 tonnes.
Works:
■ Manor Road, Kiveton Park, Sheffield S31 8PB. ***Steelmaking plant:*** One 25-tonne electric arc furnace (annual capacity 35,000 tonnes). ***Rolling mills:*** billet – single-stand 3-high non-reversing 28" cogging (30,000 tonnes); 14-stand 11" bar and wire rod (coil weights 400kg and 800kg) (28,000 tonnes). ***Other plant:*** Controlled atmosphere annealing furnaces and heat treatment, wire drawing blocks, bar drawing machine, two Schumag centreless grinders and peeling machines.
Products: Carbon steel – billets 3-5$\frac{1}{8}$" square; wire rod 0.406-0.739" dia; round bars (hot rolled) 0.740-1.032" dia; bright (cold finished) bars 0.703-1.625" dia; bright wire 0.035-0.702". Alloy steel (excl stainless) – billets 3-5$\frac{1}{8}$" square; wire rod 0.406-0.739" dia; round bars (hot rolled) 0.740-1.032" dia; bright (cold finished) bars 0.703-1.625" dia; wire 0.035-0.702" dia.
Sales offices/agents: At works address. Agents – G. Härtel GmbH, Düsseldorf, West Germany; Meko Bandstål, Halmstad, Sweden (Scandinavia).

United Merchant Bar PLC

Head office: UMB House, Brigg Road, Scunthorpe, S Humbers DN16 1XL. **Postal address:** PO Box 15. **Tel:** 0724 853333. **Telex:** 527419. **Telecopier (Fax):** 0724 867698.
Management: J. Crossman (chief executive and director). **Established:** 1985. **Capital:** £13,000,000. **No. of employees:** Works – 100. Office – 21. **Ownership:** Caparo (75%), British Steel Corp (25%). (Ultimate holding company – Caparo Group). **Annual capacity:** Finished steel: 204,000 tonnes.
Works:
■ At head office. Works director: R.F. Lormor. ***Rolling mills:*** light section – fifteen horizontal/vertical stands (annual capacity 200,000 tonnes).
Products: Carbon steel – flats 40-150mm wide, 5-50mm thick (1987 expected output, 60,000 tonnes); light angles – equal 30-70mm, 5-10mm thick (35,000 tonnes) and unequal 50 x 40mm, 65 x 50mm, 75 x 50mm (7,000 tonnes); light channels 76 x 51mm and 76 x 38mm (6,000 tonnes); medium angles 80 x 80mm and 80 x 60mm, 5-10mm thick (2,000 tonnes); medium channels 102 x 51mm (5,000 tonnes).
Sales offices/agents: British Steel Corp agents in France, Denmark, West Germany, Netherlands, Belgium, Norway, Spain, Sweden and Irish Republic.
Expansion programme: Additional roller straightening capacity being installed.

Universal Steel Tube Co Ltd—see English Seamless Tube Co.

United Kingdom

G.P. Wall Ltd

Head office: Penistone Road North, Sheffield S6 1LL. **Tel:** 0702 348676.
Associates: Richard W. Carr & Co Ltd. Spring steel, piano wire.

Warwick-Finspa Ltd, Warwick Sections Division

Head office: Bagnall Street, Golds Green, West Bromwich, W Midlands B27 0TN. **Tel:** 021 556 9881. **Telex:** 337064 browar g. **Telecopier (Fax):** 021 556 0606.
Management: Directors – David Oxtahey (managing), Terry McEllroy (plant), Andrew Coles (financial); sales manager – Alan Briley. **Established:** 1929. **No. of employees:** Works – 72. Office – 31. **Ownership:** Evered PLC.
Products: Carbon steel – cold roll-formed sections (from strip) (output 15,000 tonnes). Stainless steel – cold roll-formed sections (from strip) (3,000 tonnes).

Welded Stainless Steel Tubes Ltd

Head office: 11-16 Eyre Street Springhill, Birmingham B18 2AA. **Tel:** 021 455 0111. **Telex:** 336236 stayst. **Cables:** Staystrip Ltd. **Telecopier (Fax):** 021 454 5524.
Management: Directors – D. Myers (chairman and managing), K.V. Hill (sales), A.R. Taylor (production). **Established:** 1974. **No. of employees:** Works – 35. Office – 10. **Ownership:** Staystrip Group Ltd. **Subsidiaries:** Staystrip Ltd (manufacturers of stainless steel strip in coil, temper rolled stainless steel strip, stainless steel flat bar).
Annual capacity: Finished steel: 3,500 tonnes.
Works:
■ At head office. **Cold rolling mills**. ***Tube and pipe mills:*** welded – five $\frac{1}{2}$-4" od (own design) (annual capacity 2,000 tonnes).
Products: Stainless steel – round bars (cold finished) (output 100 tonnes); flats (cold finished) (1000 tonnes); cold roll-formed sections (from strip) (1,500 tonnes); skelp (tube strip) (cold rolled) (1,500 tonnes); cold rolled sheet/coil (uncoated) up to 54in wide (1,500 tonnes); bright annealed sheet/coil (1,500 tonnes); wire (300 tonnes); seamless tubes and pipes (200 tonnes); longitudinal weld tubes and pipes (2,000 tonnes). Perforated stainless steel tubing and strip.

W. Wesson, Glynwed Steels Ltd

Head office: Bull Lane, Moxley, Wednesbury, West Midlands WS10 8RS. **Tel:** 021 556 1331. **Telex:** 338240. **Cables:** Iron Wednesbury. **Telecopier (Fax):** 021 502 5325.
Management: Directors – B. Doyle (managing), P.D. Holder (commercial), M. Pye (sales); managers – C. Williams – (sales, hot rolled products), P. Jackson (sales, cold finished – bright products). **Established:** 1898. **No. of employees:** Works – 139.

W. Wesson, Glynwed Steels Ltd (continued)

Office – 35. **Ownership:** Glynwed Steels Ltd (Ultimate holding company – Glynwed International PLC). **Subsidiaries:** Leabrook Steel Stock (1985) Ltd (turned and ground bright bars). **Associates:** See Glynwed International PLC. **Annual capacity:** Finished steel: 80,000 tonnes.
Works:
■ At head office. ***Rolling mills:*** bar; hot strip and hoop – Five 2-high stands (90-405 x 25mm max), four edging stands (90-150 x 50mm max) (annual capacity 50,000 tonnes). One 4-high cold strip (5,000 tonnes); two 2-high cold mills (5,000 tonnes). ***Other plant:*** Cold finished/bright drawing plant – four benches, one continuous flat from coil, heat treatment plant, machining facilities for flats and squares.
Products: Carbon steel – flats (hot rolled universal), bright (cold finished) bars, hot rolled hoop and strip (uncoated) (Output 42,000 tonnes); cold rolled hoop and strip (uncoated) (4,000 tonnes). Stainless steel – bright (cold finished) bars. Alloy steel (excl stainless) – flats, bright (cold finished) bars, hot rolled hoop and strip (uncoated).
Sales offices/agents: Bright Bar Steel Co Inc, 103 South Center, Northville, Mich 48167, USA; Metal Mill Specialties Inc, 2 Park Avenue, Manhasset, NY 11030.
Expansion programme: Stainless steel bars (End 1987).

George Whitehouse Engineering Ltd

Head office: Whitehouse Road, off Worcester Road, Kidderminster, Worcs DY10 1HU. **Tel:** 0562 820820. **Telex:** 335246.
Management: A.J. Cross (chairman), D.A. Baldwin (managing director), A. Snelgrove (sompany secretary), T. Jenks (sales manager). **Established:** About 1880. **Capital:** £98,000. **No. of employees:** Works – 67. Office – 20. **Annual capacity:** Finished steel: 5,000 tonnes.
Works:
■ At head office. Cold roll-forming line for sections.
Products: Carbon steel – cold roll-formed sections (from strip).

Wiggin Steel & Alloys

Head office: St Stephens Street, Aston, Birmingham B6 4RE. **Tel:** 021 3593115. **Telex:** 338525.
Products: Stainless steel – ingots.

Woodstone Rolling Mills Ltd

Head office: Bannerley Road, Garretts Green, Birmingham B33 OSL. **Postal address:** Old Park Road, Wednesbury, W Midlands WS10 9TB. **Tel:** 021 556 7385. **Telex:** 337733 wsrm g.
Established: 1977. **Capital:** £200,000. **No. of employees:** Works – 134. Office – 18. **Ownership:** Woodstone Metals Ltd (50%). James Wood (Merchants) Ltd (50%). **Annual capacity:** Finished steel: 75,000 tonnes.
Works:
■ At head office. ***Rolling mills:*** 4-stand 18″ cross country bar (hand mill) (annual capacity 32,000 tonnes); 17″ roughing stand and 5-stand 12″ bar (hand mill) (43,000 tonnes).
Products: Carbon steel – round bars 112-24mm (1986 (nine months) output 21,800 tonnes); square bars 103-25.5mm square and round cornered. (11,384 tonnes); hexagons 78-33.5mm a/f (743 tonnes); special light sections fluted squares 39-32mm (60 tonnes). Alloy steel (excl stainless) – round bars 112-24mm (1,536 tonnes); square bars 103-25.5mm square cornered (439 tonnes); hexagons 78-33.5mm a/f (11 tonnes).

United States

AB Steel Mill Inc

Head office: 204 W. North Bend Road, Cincinnati, OH 45216. **Tel:** 513 761 6625.
Established: 1976.
Works:
■ At head office. ***Rolling mills:*** One continuous bar comprising four 16″ stands, three 12″ stands and eight 8″ stands (annual capacity 30,000 short tons).
Products: Carbon steel – reinforcing bars, round bars.

Acme Roll Forming Co

Head office: 812 North Beck Street, Sebewaing, MI. **Postal address:** PO Box 706. **Tel:** 517 883 2050. **Telecopier (Fax):** 517 883 2052.
Management: James Wineman (president), Charles Wineman (vice president), Marjorie Finkbeiner (secretary), David Wineman (treasurer). **Established:** 1960. **Capital:** $2,500,000. **No. of employees:** Works – 42. Office – 8. **Annual capacity:** Finished steel: 45,000 short tons.
Works:
■ At head office. Tel: 517 883 2050. Works director: Charles Wineman. ***Tube and pipe mills:*** welded – two Yoder (annual capacity 25,000 short tons) and one Yoder (20,000 tons).
Products: Carbon steel – longitudinal weld tubes and pipes 1.75-6″sq (0.083-0.250″ wall) (output 21,000 tons).

Acme Steel Co

Head office: 13500 South Perry Avenue, Riverdale, IL 60627. **Tel:** 312 841 2500. **Telex:** 253506. **Telecopier (Fax):** 312 849 2500 (ext 2567).
Management: Chairman – Reynold C. MacDonald; president and chief executive officer – Brian W.H. Marsden; vice presidents – J.F. Williams (finance), L.D. Camino (operations), S.D. Oker (marketing of steel products), R.J. Stefan (employee relations), R.P. Zenere (marketing of steel strapping), E.P. Weber, jr (general counsel and secretary); treasurer – D.M. McMahon; controller – J.P. Shaughnessy; director – R.J. Sayre (public relations). **Established:** 1986. **Associates:** None. **Annual capacity:** Pig iron: 1,600,000 short tons. Raw steel: 1,350,000 tons.
Works:
■ 135th Street and Perry Avenue, Riverdale (Tel: 312 849 2500. Telex: 25 3819) Plant manager: L.D. Camino. ***Steelmaking plant:*** Two 75-ton basic oxygen (annual capacity 900,000 short tons). ***Rolling mills:*** One 35″ blooming (737,000 tons); one 3-stand billet (24″/31″/24″) (580,000 tons); two hot strip and hoop – one 36″ reversing/continuous (574,000 tons) and one 16″ continuous (170,000 tons); four cold strip – one 4-stand 16″ (35,500 tons), one 4-stand (14½″) (76,700 tons) and two 5-stand 14″ (96,600 tons). ***Other plant:*** Equipment for producing steel strapping.
■ 10730 Burley Avenue, Chicago, IL 60617. (Tel: 312 221 3131). Superintendent: Jack Garzella. ***Coke ovens:*** Two Wilputte underjet batteries (100 ovens) (annual capacity 436,000 tons). ***Sinter plant:*** One Lurgi 8ft 3″ strand (1,000,000 tons). ***Blast furnaces:*** Two (A) 25′ 3″ hearth dia, 41,448 cu ft (750,000 tons), (B) 19′ 8″ hearth dia, 27,027 cu ft (438,000 tons).
Products: Foundry pig iron. Carbon steel – ingots, blooms, billets, hot rolled coil (for re-rolling) (under 1m), hot rolled sheet/coil (uncoated) (under 1m), cold rolled sheet/coil (uncoated) (under 1m). Alloy steel (excl stainless) – ingots, blooms, billets.
Additional information: The company was formerly known as Interlake Inc.

United States

Al Tech Specialty Steel Corp

Head office: Willowbrook Avenue, Dunkirk, NY. **Tel:** 716 366 1000. **Telex:** 5102467719. **Telecopier (Fax):** 716 366 1000 (Ext 347).
Management: Vice presidents – J.H. Mintun Jr (general manager), Clark K. Riley (sales and marketing), J. Ron Hansen (finance and controller), K. Jack Shuster (administration), Fred H. Ruff (operations), George F. Taylor (general manager, extrusion). **Ownership:** Rio Algom Ltd. **Annual capacity:** Finished steel: 108,000 short tons.
Works:
■ At head office; and Spring Street, Watervliet, NY (Tel: 518 273 4110). Works director: F.H. Ruff. ***Steelmaking plant:*** Two 30-ton electric arc furnaces (annual capacity 117,000 short tons). ***Refining plant:*** Two 32-tonne AOD (125,000 short tonnes). ***Continuous casting machines:*** One 2-strand RoKop billet ($4\frac{1}{2}$" sq) (46,000 short tons). ***Rolling mills:*** One 1-reversing stand (22") blooming (43,000 short tons) and three fixed cross country stands (22" and 21") (13,000 short tons). Three bar – 5-stand 3-high cross country (14") (17,000 short tons), 5-stand 3-high cross country (14") (5,000 short tons), 6-stand 3-high corss country (10") (1,600 short tons). One wire rod comprising – two 3-high (20") (31,200 short tons), one 2-high (18"), three 2-high (14"), three 2-high (12"), four 2-high (10"), six 2-high horizontal/vertical (8"). ***Tube and pipe mills:*** One extrusion and piercing press seamless (2,200 ton) (7,650 short tons). ***Other plant:*** 2,000-ton extrusion press; 2,000-ton forging press; wire drawing equipment.
Products: Stainless steel – wire rod 0.217-8.78"; round bars 1-$7\frac{7}{8}$"; square bars 0.75-2.75"; flats 0.625 x 0.50"-4 x 2"; hexagons 0.656-1.5"; wire 0.035-0.750"; seamless tubes and pipes 1-5.50"; cold drawn tubes and pipes 1-4.50"; hollow bars 1.25-5.50". Alloy steel (excl stainless) – round bars 1-7.75"; square bars 0.75-2.75"; flats 0.625 x 0.50"-4 x 2"; hexagons 0.656-1.50"; seamless tubes and pipes 1-5.50"; cold drawn tubes and pipes 1-4.50; hollow bars 1.25-5.50".

Allegheny Ludlum Corp

Head office: 1000 Six PPG Place, Pittsburgh, PA 15222. **Tel:** 412 394 2800. **Telex:** 866444. **Telecopier (Fax):** 412 394 2805.
Management: Richard P. Simmons (chief executive officer and board chairman), Robert P. Bozzone (president and chief operating officer), Richard C. O'Sullivan (senior vice president finance and chief financial officer). **Established:** 1936. **No. of employees:** Works – 3,200. Office – 2,300. **Annual capacity:** Raw steel: 1,000,000 short tons.
Works:
■ Brackenridge Works, River Road, Brackenridge, PA 15014. (Tel: 412 224 1000. Telex: 866444). Works director: Victor P. Ardito. **Ironmaking plant:**. Three 70-ton coreless induction furnace (annual capacity 450,000 tons). ***Steelmaking plant:*** Two 80-ton basic oxygen converters (600,000 tons); four electric arc furnaces (two 17ft

Allegheny Ludlum Corp (continued)

and two 18ft dia) (400,000 tons). ***Refining plant:*** One 100-ton AOD (450,000 tons). ***Continuous casting machines:*** One 1-strand Concast slab (8″ thick x 26-52.5″ wide) (430,000 tons). ***Rolling mills:*** Single-stand 2-high 40″ blooming (can roll 25 x 55″ to 51″ max width and narrower); semi-continuous 56″ wide hot strip comprising one 2-high reversing roughing and six 4-high finishing stands (coil weight 550lb PIW up to 51″ wide – 28,050lb max) (over 600,000 tons); cold reduction comprising 4-stand 4-high 56″ finishing (coil weight 40,000lb – 49″ wide) (185,000 tons), 1-stand 4-high 56″ reversing (coil weight 37,500lb – 51″ wide) (122,000 tons) and 56″ Sendzimir cluster (coil weight 34,000lb – 51″ wide) (121,600 tons); 2-high temper/skin pass.

■ West Leechburg Works, Leechburg, PA 15656. (Tel: 412 845 0600. Telex: 866444). Works director: E.A. Westwood. ***Rolling mills:*** Four 4-high cold strip – one 27″ max (annual capacity 36,000 tons), one 38″ max (40,000 tons), and two 28″ max (72,000 tons); temper/skin pass comprising one 2-high 38″ max (50,000 tons), one 4-stand 28″ max (150,000 tons), and one 3-stand 38″ max (180,000 tons). **Coating lines:** Five Alsco – special electrical steel (both sides dip) (120,000 tons).

■ New Castle Plant, State Route 38, PO Box 309, New Castle, IN 47362. (Tel: 317 529 9570. Telex: 866444). Works director: J.B. Bell. ***Rolling mills:*** One Sendzimir ZR 2250 cold reduction (coil weight 50,000lb) (annual capacity 120,000 tons); one 2-high 50″ temper/skin pass (coil weight 50,000lb) (150,000 tons).

■ Wallingford Plant, 80 Valley Street, PO Box 249, Wallingford, CT 06492. (Tel: 203 269 3361. Telex: 866444). Works director: L.A. Greco. ***Rolling mills:*** Five cold strip – one 2-stand TM, one 4-high and three Sendzimir (25″ wide) (coil weight 12,500lb) (total annual capacity 126,672 tons); two 2-high temper/skin pass (13/28″ wide) (coil weight 5,500/12,500lb) (38,438 tons).

■ Tubular Products Division, Highway 66 and Lowry Road, Claremore, OK. (Tel: 918 341 8711. Telex: 866444). Works director: W.G. Bieber. ***Tube and pipe mills:*** welded – three WU-20H $\frac{5}{8}$-1$\frac{1}{2}$″, three IKU $\frac{5}{8}$-1$\frac{1}{4}$″, three RS-625 $\frac{5}{8}$-3$\frac{1}{2}$″ and one WU-1710 $\frac{5}{8}$-1$\frac{1}{4}$″ (total annual capacity 6,600 tons).

■ Special Materials Division, 695 Ohio Street, Lockport, NY 14094. (Tel: 716 433 4411. Telex: 866444). Works director: R.W. Fountain. ***Steelmaking plant:*** Three 15-ton Lectromelt electric arc furnaces. ***Refining plant:*** One 180-ton AOD (annual capacity 36,000 tons); 4-6-ton/10-ton ESR (max 30″ dia ingots or 12 x 42″ slab size) (6,000 tons); VAR – one station (two heads) 33″ dia max (1,200 tons of 20″ dia size); one 9-ton VIM (6,000 tons).

Products: Stainless steel – hot rolled band (for re-rolling), hot rolled hoop and strip (uncoated), cold rolled hoop and strip (uncoated), medium plates, heavy plates, clad plates, hot rolled sheet/coil (uncoated), cold rolled sheet/coil (uncoated), bright annealed sheet/coil, longitudinal weld tubes and pipes. Alloy steel (excl stainless) – hot rolled sheet/coil (uncoated), cold rolled sheet/coil (uncoated). High strength alloys, nickel alloys and high temperature alloys – in sheet, strip and plate.

Sales offices/agents: Western region – Woodfield Office Plaza III, 999 Plaza Drive, Suite 610, Schaumburg, IL 60173-4985. Eastern region – PO Box 1409, 80 Valley Street,Wallingford, CT 06492.

Allied Tube & Conduit Corp, an ATCOR Company

Head office: 16100 South Lathrop Avenue, Harvey, IL 60426. **Tel:** 312 339 1610. **Telex:** 254352.

Management: Chairman, chief executive officer and president – Theodore H. Krengel; vice presidents – Lawrence Vollmuth (engineering), David Shotts (operations), Anthony M. Santone (distribution), R.J. Conklin (human resources), John Ranelli (finance), J. Randal (treasurer), James J. Strenk (electrical sales), Stanley W. Nochowitz (logistics), Elliott L. Avery (fire protection), L. James LeBla Jr c (MIS); executive vice president – Michael Lowenthal; division managers – Joshua H. Krengel (mechanical), Robert R. Carpenter (fence). **Established:** 1960. **Capital:** $113,000,000. **No. of employees:** 2,400.

Works:

■ At head office. Works director: Vijay Patel. ***Tube and pipe mills:*** welded One Wean

United States

Allied Tube & Conduit Corp, an ATCOR Company (continued)

United ($3\frac{1}{2}$") (annual capacity 55,000 short tons), five Abbey-Etna (2") (110,000 tons).
■ 11350 Norcom Road, Philadelphia, PA (Tel: 215 676 6464. Telex: 834671). Works director: Pasquale Domanico. ***Tube and pipe mills:*** welded Two Abbey-Etna (3") (annual capacity 65,000 short tons).
Products: Carbon steel – longitudinal weld tubes and pipes fence tubing 1.315", 1.660", 1.900", 2.375"; fence pipe (SS-20 and SS-40) 1.315", 1.660", 1.900", 2.375, 2.875", 3.500"; sprinkler pipe (same sizes as fence pipe); raw ERW tube $\frac{1}{2}$-$3\frac{1}{2}$ od 22-8g; high strength tube 1-$2\frac{3}{8}$" od, 15-8g; galvanized tubes and pipes (zinc Flo-Coat) $\frac{1}{2}$- $3\frac{1}{2}$" od, 22-8g;/ hollow sections (square) $\frac{1}{2}$-$1\frac{3}{4}$", 20-16g and $1\frac{1}{2}$-$2\frac{1}{2}$", 16-10g.
Sales offices/agents: Regional distribution centers: 14489 Industry Circle, La Mirada, CA, 90637 (Tel: 714/521-4511); 6349 Old Peachtree Road, Norcross, GA, 30071 (Tel: 404/448-0540); 151st & Halsted, Harvey, IL, 60426 (Tel: 312/339-1610); 11350 Norcom Road, Philadelphia, PA, 19154 (Tel: 215/676-6464); 11539 N. Houston Rosslyn Road, Houston, TX, 77088 (Tel: 713/445-2900). Allied foreign licensees: Eluma SA – Indústria e Comércio, Division Bundy, Caixa Postal 30509, 01000, São Paulo, SP, Brazil; Cintac SA – Cia Industrial de Tubos de Acero SA, Casilla 14294, Correo 21, Santiago, Chile; Monmore Tubes Ltd, Dixon Street, Wolverhampton, West Midlands WV2 2BT, UK; Nagarjuna Steel Ltd, Nargarjuna Hills, Panjagutta, Hyderabad 500004, Andra Pradesh, India; Matshushita Electric Works Ltd, 10458 Kadoma, Kadoma-shi, Osaka, Japan; Peasa – Productos Especializados de Acero SA, Poniente 134 No 854, Colonia Industrial Vallejo, Delegación Azcapotzalco, 02300 Mexico, DF; VBF Buizen BV, Postbus 39, 4900 BB Oosterhout, Netherlands; CA Coduven, Aptdo 60299, Caracas 1060A, Venezuela; Metaliku, Stan-Banesa, S. Stavileci 66, Djakovica, Yugoslavia.

American Tube Co

Head office: 2525 N 27th Avenue, Phoenix, AZ 85009. **Postal address:** PO Box 6633, Phoenix, AZ 85005. **Tel:** 602 272 6606. **Telex:** 187129 fence ut. **Telecopier (Fax):** 602 269 1324.
Management: Board chairman – J.P. Van Denbergh; president – D.S. Van Denburgh; vice presidents – Will Boggs (executive), Steve Brock (purchasing), Mike Bertola (marketing and sales). **Established:** 1974. **Capital:** $3,700,000. **No. of employees:** 124. **Ownership:** American Fence Corp. **Subsidiaries:** Kokomo Works, Division of American Tube (100%) (pipe and tube manufacturer). **Associates:** American Fence Co, American Security Fence Corp, Studco, Metco.
Works:
■ At head office. ***Tube and pipe mills:*** welded – one Yoder (annual capacity 20,000 short tons); one Turek & Heller (25,000 tons) and one Turek and Heller (25,000 tons). ***Other plant:*** Rollformer for galvanized and colour-coated sheets.
Products: Carbon steel – galvanized corrugated sheet/coil 24" and 36" wide (1986

American Tube Co (continued)

output 6,000 tons); colour coated sheet/coil 24" and 36" wide (6,000 tons); galvanized wire (barbed) 2 point and 4 point (1,500 tons); longitudinal weld tubes and pipes 1-4" od (70,000 tons); galvanized tubes and pipes 1-4" od (70,000 tons); plastic coated tubes and pipes (PVC) 1-4" od (8,000 tons). Barbed tape concertina – security fencing.
Brands: Tube – TUF-20, TUF-40, TUF-15, TUF-Tube. Barbed tape – Razor Ribbon, Instabarrier.
Expansion programme: A barbed tape plant is being opened in Wexford, Ireland in the first quarter of 1988.

Armco Inc

Head office: 300 Interpace Parkway, Parsippany, NJ 07054-0324.
Management: Officers – Robert E. Boni (chairman and chief executive), Robert L. Purdum (president and chief operating), Wallace B. Askins (executive vice president, and chief financial); vice presidents – Robert W. Kent (corporate – law, general counsel and secretary), John M. Bilich (corporate – corporate human resources), Carl R. Fiora (president – Eastern Steel Division); Lawrence W. Hicks (president – Midwestern Steel Division), Kempton B. Jenkins (corporate – international and government affairs), David E. Todd (president – Specialty Steel Division), William H. Schaar (corporate – controller), William L. Trubeck (corporate – treasurer). **Established:** 1900.
Subsidiaries: Including Associates: Divisions and affiliated companies – Eastern Steel Division; Specialty Steel Division; Midwestern Steel Division; Latin American Division; National-Oilwell (50/50 joint venture with USX Corp); Cumberland Group; Armco Assets Management; Armco Financial Services Group.
Works:
■ Ashland Works, PO Box 191, Ashland, KY 41105 (Tel: 606 329 7111). Works manager: C.T. Gorder. ***Coke ovens:*** Two Wilputte underjet batteries – No 3 (76 ovens) (capacity 1,042 short tons per day); No 4 (70 ovens) (1,674 tons per day). ***Sinter plant:*** 96" wide Dravo/Lurgi 807sq ft grate area (annual capacity 620,000 tons). ***Blast furnaces:*** Amanda – 33' 5" hearth dia, 72,000 cu ft (1,533,000 tons); Bellefonte – 28' 9" hearth dia, 64,729 cu ft (1,007,000 tons). ***Steelmaking plant:*** Two 180-ton basic oxygen converters (2,000,000 tons). ***Refining plant:*** Electro magnetic stir ladle furnace (1,656,000 tons). ***Continuous casting machines:*** One 6-strand Concast curved mould for bloom (8½", 10½", and 13½" square) (720,000 tons). ***Rolling mills:*** One 1-stand 2-high 45 x 100" slabbing (1,800,000 short tons); one wide hot strip comprising 4-stand 4-high 27" and 54 x 80" roughing, and 6-stand 4-high 27" and 54 x 80" finishing (coil weight 35,000lb) (l,800,000 short tons); one 5-stand 4-high 66" cold reduction (coil weight 60,000lb) (900,000 tons); one 1-stand 4-high 80" continuous temper/skin pass (400,000 tons). ***Coil coating lines:*** Two hot-dip galvanizing comprising one 60" Sendzimir (128,000 tons), and one 60" Selas non-oxidising (319,000 tons).
■ Kansas City Works, 7000 Roberts Street, Kansas City, MO 64125 (Tel: 816 242 5100). President: L.W. Hicks. ***Steelmaking plant:*** Six electric arc furnaces – two 100-ton, four 150-ton. ***Continuous casting machines:*** One 6-strand curved mould bloom (7¾" square). ***Rolling mills:*** One 1-stand 2-high 32" reversing blooming (annual capacity 830,000 tons); One billet comprising 2-stand 2-high 21", 4-stand 2-high continuous (830,000 tons); breakdown comprising 2-stand 2-high 21", 4-stand 2-high 19" (900,000 tons); one 12" bar comprising 3-stand 2-high 18", one stand vertical 12", 2-stand 2-high 14", 1-stand vertical 12", 3-stand 2-high 14", one stand vertical 12", 4-stand 2-high 12" (360,000 tons); one continuous wire rod comprising 4-stand 2-high 18", 5-stand 2-high 16", 4-stand 2-high 14", four single strand No-Twist finishing blocks, 2-stand 2-high 8" and 8-stand 2-high 6" (670,000 tons). ***Other plant:*** Grinding ball forging facilities.
■ Middletown Works, PO Box 600, Middletown, OH 45043. ***Coke ovens:*** Three Slot By-product batteries – one underjet (76 ovens) (annual capacity 396,000 tons); two multistage combination (57 ovens each) (630,000 tons each). ***Sinter plant:*** One 96" wide, 768 ft sq McKee (2,600,000 tons). ***Blast furnaces:*** No 1 (Hamilton, Ohio) 18'

Armco Inc (continued)

6" hearth dia, 22,839 cu ft (420,000 short tons);No 3 (Middletown, Ohio) 29' 6" hearth dia, 55,324 cu ft (1,460,000 short tons). ***Steelmaking plant:*** Two 220-ton top blown, bottom stir basic oxygen converter (2,400,000 short tons). ***Refining plant:*** 220-ton RH vacuum degassing unit (1,020,000 short tons); CAS/CAS OB (1,620,000 short tons). ***Continuous casting machines:*** One 2-strand Demag-curved mould slab (38 to 84" wide x 9" thick) (1,500,000 tons). ***Rolling mills:*** 1-stand 2-high 48 x 96" slabbing (2,600,000 short tons); one wide hot strip comprising 3-stand 4-high 48 x 86" roughing, 3-stand 4-high 38 x 86" roughing, 7-stand 4-high 29 x 86" finishing (coil weight 11,000 to 84,000lb) (3,400,000 tons); three cold reduction comprising one 3-stand 4-high 74" wide (coil weight 50,000lb) (260,000 tons), one 4-stand 4-high 49" wide (coil weight 50,000lb) (140,000 tons), and one 5-stand 4-high 86" wide (coil weight 88,000lb) (1,700,000 tons); three temper/skin pass – all 1-stand 4-high $15\frac{1}{4}$ & 36 x 80", 27 & 60 x 80", $22\frac{1}{2}$ x 53 x 80" (coil weight 80,000lb) (2,244,000 tons). ***Tube and pipe mills:*** welded – six $\frac{3}{8}$ x 3" Armco/Yoder ERW, and two $\frac{3}{8}$ x 3" Armco/Yoder HF (42,000 tons). ***Coil coating lines:*** One continuous horizontal/anneal hot-dip galvanizing (iron pot) (352,000 tons); one 72" wide Gravitel continuous electro-galvanizing (254,000 tons); one 60" wide continuous horizontal terne/lead coating (199,000 tonnes); one ceramic pot vertical anneal aluminizing (326,000 tonnes).

■ Baltimore Works, 3501 E. Biddle Street, PO Box 1697, Baltimore, MD 21203 (Tel: 301 563 5500). General manager; Ray Heln. ***Steelmaking plant:*** One 50-ton electric arc furnace. ***Refining plant:*** One 25/50-ton AOD unit; three consumable electrode furnaces – one 8-ton and two 15-ton. ***Continuous casting machines:*** One 2-strand horizontal billet. ***Rolling mills:*** One bar comprising 1-stand 2-high 20", 1-stand 3-high 18", 1-stand 3-high 16", 3-stand 2-high 14", one stand 2-high 12". ***Other plant:*** 2,000-ton forging press; one GFM continuous precision forging machine; wire processing facility; cold finished bar facility with Schumag and other shape mills.

■ Butler Works, Butler Plant, PO Box 832, Butler, PA 16003 (Tel: 412 284 2000). Manager: G.L. Puff. ***Steelmaking plant:*** Three 165-ton top charged electric arc furnaces. ***Refining plant:*** One 175-ton AOD unit; one 175-ton max DH vacuum degassing unit. ***Continuous casting machines:*** Two Demag curved mould slab (annual capacity 250,000 tons). ***Rolling mills:*** One 1-stand 2-high 34" and 56" slabbing (for hot strip mill) (946,000 tons); one 1-stand 2-high 36" x 58" scale breaker (915,000 tons); one wide hot strip comprising one 1-stand 4-high 36" and 54 x 58" reversing rougher, one 5-stand 4-high 25" and 54 x 58" continuous finisher (915,000 tons); five cold reduction comprising one 4-stand 4-high $15\frac{1}{2}$" and 49 x 56" continuous (315,000 short tons), one 1-stand cluster $2\frac{3}{16}$ x 42" reversing (21,800 tons), one 1-stand cluster $2\frac{1}{8}$ x 54" reversing (33,300 short tons), one 1-stand 4-high $16\frac{1}{2}$" and 45 x 42" reversing (133,000 tons), one 1-stand 4-high $16\frac{1}{2}$" and $35\frac{1}{2}$ x 54" continuous (83,000 tons); two temper/skin pass – one 1-stand 2-high 34 x 56" continuous (31,200 tons), one 1-stand 2-high $19\frac{1}{2}$" x 42" continuous (91,500 short tons). ***Other plant:*** Bright annealing line.

■ Butler Works, Zanesville Plant, 1730 Linden Avenue, Zanesville, OH 43702 (Tel: 614 452 6321). ***Rolling mills:*** One 1-stand cluster $3\frac{1}{2}$ x 49" reversing cold reduction (annual capacity 136,000 tons).

Products: Foundry pig iron. Carbon steel – ingots, wire rod, reinforcing bars, round bars, bright (cold finished) bars, hot rolled hoop and strip (uncoated), cold rolled hoop and strip (uncoated), medium plates, heavy plates (including sheared), universal plates, hot rolled sheet/coil (uncoated), cold rolled sheet/coil (uncoated), hot-dip galvanized sheet/coil, electro-galvanized sheet/coil, aluminized sheet/coil, enamelling grade sheet/coil, terne-plate (lead coated) sheet/coil, electrical grain oriented silicon sheet/coil, longitudinal weld tubes and pipes, spiral-weld tubes and pipes, oil country tubular goods. Stainless steel – ingots, wire rod. Alloy steel (excl stainless) – ingots; wire rod. Cold finished bars; grinding media; wire rope.

Brands: Univit; Zincquip; Electro-smooth (electro galvanized steel).

Atlantic Steel Co

Head office: 1300 Mecaslin St NW, Atlanta, GA 30318. **Postal address:** PO Box 1714. **Tel:** 404 897 4500. **Telex:** 542504. **Telecopier (Fax):** 404 897 4623. **Management:** President and chief executive officer – J.J. Webb; vice presidents – E.M. Vannerson (operations), C.W. Mynard (finance), R.B. Webb (sales). **Established:** 1901. **Ownership:** Ivaco Inc, Montreal, Que, Canada (100%). **Annual capacity:** Raw steel: 750,000 short tons.
Works:
■ Atlanta, GA. ***Steelmaking plant:*** Two 90-ton electric arc furnaces. ***Continuous casting machines:*** One 6-strand billet ($4\frac{1}{2}$ x $4\frac{1}{2}$"-$8\frac{1}{2}$ x 4"). ***Rolling mills:*** One bar; one single-strand wire rod (coil weight 3,000lb). ***Other plant:*** Wire mill includes drawing, galvanizing, annealing, nail cutting and galvanizing, straighten and cutting.
■ Cartersville, GA. ***Steelmaking plant:*** One 100-ton electric arc furnace. ***Continuous casting machines:*** One 4-strand billet (4 x 4"-6 x 4"). ***Rolling mills:*** One bar (coil weight 2,000 lb).
Products: Carbon steel – billets, wire rod, reinforcing bars, round bars, square bars, flats, bright wire, black annealed wire, galvanized wire (plain). Alloy steel (excl stainless) – billets, round bars, square bars, flats.

Auburn Steel Co Inc (Austeel)

Head office: Quarry Road, Auburn, NY 13021. **Postal address:** PO Box 2008. **Tel:** 315 253 4561.
Management: President – Hideshi Numaguchi; executive vice president – William J. Humes Jr; vice presidents – Yoshihiro Yasuda (secretary/treasurer), John Rossi (operations), Martin E. Fanning (industrial relations), Mack White (sales). **Established:** 1973. **Capital:** $8,000,000. **No. of employees:** Works – 250. Office – 50. **Ownership:** Sumitomo Corp, Tokyo, Japan; Kyoei Steel, Osaka, Japan. **Annual capacity:** Raw steel: 400,000 tons. Finished steel: 320,000 tons.
Works:
■ At head office. Superintendents: Gunter Altermann (melting and casting), Ronald Carter (rolling mills). ***Steelmaking plant:*** 16 ft Lectromelt electric arc furnace (split shell design) (annual capacity 400,000 tons). ***Continuous casting machines:*** One 3-strand Concast billet (400,000 tons). ***Rolling mills:*** One 14-stand Birdsboro bar (320,000 tons).
Products: Carbon steel – billets $4\frac{1}{2}$ x $4\frac{1}{2}$", 5 x 5" (10-30ft) (1986 output 335,000 tons); reinforcing bars No 3 to No 11; round bars $\frac{1}{2}$-$1\frac{1}{2}$"; square bars :FR:7/16-1"; flats $\frac{1}{4}$ x 1"-1 x 4"; light angles $1\frac{1}{4}$ x $1\frac{1}{4}$ x $\frac{1}{8}$" to 3 x 3 x $\frac{1}{4}$"; light channels 3" and 4".
Brands: Austeel (all products).

Avesta Inc

Head office: State Road 38 W, New Castle, IN 47362. **Postal address:** PO Box 370. **Tel:** 317 529 0120. **Telex:** WUT 501578 avesta ud.
Management: E. Wayne Pokorney (president). **Established:** Originally 1898 (formerly Ingersoll Johnson Steel Co).
Works:
■ At head office. ***Steelmaking plant:*** One electric arc furnace and pressure slab casting facility. ***Rolling mills:*** One heavy plate and medium plate.
Products: Stainless steel – medium plates, heavy plates. Alloy steel (excl stainless) – medium plates, heavy plates.

The Babcock & Wilcox Co, Tubular Products Division

Head office: Beaver Fall, PA 15010. **Postal address:** PO Box 401. **Tel:** 412 846 0100.
Ownership: The Babcock & Wilcox Co, New Orleans, LA (Ultimate holding company – McDermott International).
Works:
■ At head office. ***Steelmaking plant:*** Six electric arc furnaces – one 50-ton, one 75-ton, one 95-ton UHP and three 100-ton. ***Refining plant:*** One 3-ton ESR; one

The Babcock & Wilcox Co, Tubular Products Division (continued)

95-ton ladle furnace; two (Stokes ladle) vacuum degassing units with induction stirring. ***Continuous casting machines:*** One 4-strand round and square. ***Rolling mills:*** 1-stand 2-high 40" blooming (annual capacity 500,000 short tons). two bar – one 3-stand 3-high 28" (180,000 tons) and one 5-stand 2-high 22" (240,000 tons). ***Tube and pipe mills:*** four seamless and one extrusion mill ½-10".
■ Alliance, OH. ***Tube and pipe mills:*** Four high frequency electric welded ⅝-12¾".
Products: Carbon steel – ingots, semis for seamless tubemaking, round bars. Alloy steel (excl stainless) – ingots including high strength low-alloy; semis for seamless tubemaking; round bars; seamless tubes and pipes; longitudinal weld tubes and pipes; oil country tubular goods; mechanical and pressure tubing.
Additional information: In autumn 1987 the company proposed closing Beaver Falls, apart from a small specialty tube operation, by the end of the year.

Baron Drawn Steel Corp

Head office: 1400 Hastings Street, Toledo, OH 43607. **Tel:** 419 531 5525.
Established: 1964. **Ownership:** Champion Spark Plug Co.
Works:
■ At head office. Two drawbenches, three draw blocks, 2 Schumag continuous bar drawers, etc.
Products: Carbon steel – bright (cold finished) bars. Alloy steel (excl stainless) – bright (cold finished) bars.

Bayou Steel Corp

Head office: River Road, LaPlace, LA 70069-1156. **Postal address:** PO Box 5000. **Tel:** 504 652 4900. **Telex:** 78 4958. **Telecopier (Fax):** 504 525 9810.
Management: President – Howard M. Meyers; general managers – Jerry M. Pitts (Bayou Steel Corp), Rodger A. Malehorn (commercial operations), Richard J. Gonzalez (finance and data processing), Timothy R. Postlewait (plant operations). **Established:** 1975. **No. of employees:** Works – 635. Office – 183. **Ownership:** Private. **Annual capacity:** Raw steel: 700,000 short tons. Finished steel: 500,000 tons.
Works:
■ At head office. ***Steelmaking plant:*** Two 65-ton Voest-Alpine/Krupp electric arc furnaces (annual capacity 700,000 tons). ***Continuous casting machines:*** Two 4-strand Voest-Alpine straight mould billet (100, 120, 125, 160 and 200mm sq) (700,000 tons). ***Rolling mills:*** One 15-stand 21" Daniel medium section (500,000 tons).
Products: Carbon steel – billets 100-200mm sq; flats 4-6" wide; light angles from 2" x 2"; medium angles to 7" x 4"; medium channels 3-6"; wide-flange beams (medium) to 4.13".
Sales offices/agents: At head office. The company also has offices at – 301 W. New Avenue, Lemont, IL 60439 (Tel: 312 257 7171); 900 Ebenezer Road, Knoxville, TN 37923 (Tel: 615 691 9865); 5140 Bird Creek, Catoosa, OK 74015 (Tel: 918 266 3145); Leetsdale Industrial Pk (Tel: 412 266 0610).

Berg Steel Pipe Corp

Head office: 1415 C. Avenue, Panama City, FL. **Postal address:** PO Box 2039. **Tel:** 904 769 2273. **Telex:** 755961 bspcoff ud.
Management: President – S.E. Koehle; directors – Carl Zingler (administrative services), John Heindemann (manufacturing services). **No. of employees:** Works – 164. Office – 28. **Associates:** Bergrohr GmbH, Herne, West Germany
Products: Carbon steel – longitudinal weld tubes and pipes, large-diameter tubes and pipes.

United States

Bethlehem Steel Corp

Head office: Bethlehem, PA 18016. **Tel:** 215 694 2424. **Telex:** 847417. **Cables:** Mesbeth Bethlehem PA.
Management: Officers: Chairman, president and chief executive – Walter F. Williams; vice presidents – Larry L. Adams (union relations), Lonnie A. Arnett (controller), D. Sheldon Arnot (executive), Curtis H. Barnette (senior, general counsel and secretary), Benjamin C. Boylston (human resources), Robert W. Cooney (planning), George T. Fugere (information services), John A. Jordan Jr (senior), James F. Kegg (materials management), Gary L. Millenbruch (senior and chief executive), C. Adams Moore (steel sales), Roger P. Penny (senior), James C. Van Vliet (steel marketing); treasurer – Andrew M. Weller. **Established:** Incorporated 1919. **Capital:** Stockholders' equity: $1,236,800,000. **No. of employees:** 31,500. **Annual capacity:** Raw steel: 16,000,000 short tons.

Steel Group
Works:
■ Sparrows Point, MD. ***Coke ovens:*** 249 Koppers gun-flue ovens, 130 Koppers underjet ovens and 80 Otto hairpin flue underjet ovens. ***Sinter plant:*** 1-strand Dravo Lurgi, 3,800 sq ft grate area. ***Blast furnaces:*** Four – (H) 30' hearth dia, 54,792 cu ft; (J) 30' hearth dia, 54,505 cu ft; (K) 30' hearth dia, 54,610 cu ft; (L) 44'6" hearth dia, 130,399 cu ft. ***Steelmaking plant:*** Two 220-ton basic oxygen converters and four 420-ton open hearth. ***Continuous casting machines:*** one slab (7" and 10" x 24-86") (2,900,000 tons). ***Rolling mills:*** Two slabbing – one 1-stand 2-high horizontal and 1-stand vertical 45" (3,540,000 tons); one 1-stand 2-high horizontal and 1-stand vertical 40" (2,320,000 tons); one 1-stand 2-high 54" blooming (1,730,000 tons, based on twenty shifts/week); one billet comprising 1-stand 2-high horizontal and 1-stand vertical 30", 6-stand 2-high 24" and 6-stand 2-high 18" (1,040,000 tons); one wire rod comprising 5-stand 2-high 18", 4-stand 2-high 14", 4-stand 2-high 12", 2-stand 2-high 8" and 8-stand 2-high 6" (467,000 tons); two heavy plate – one sheared comprising 1-stand 3-high 44" and 27 x 160", 1-stand 4-high 59" and 38 x 160" (890,000 tons); one universal comprising 1-stand 2-high 60" (36 x 84") (270,000 tons); two wide hot strip – one 1-stand 2-high 36 x 70", 2-stand 4-high 38" and 53 x 69", 2-stand 4-high 34½" and 53 x 69", 6-stand 4-high 27½" and 53 x 69" (2,730,000 tons); and one 1-stand 2-high 36 x 56", 1-stand 2-high 45 x 58", 3-stand 4-high 25½" and 49½ x 58" 6-stand 4-high 25½: and 49½ x 58" (2,480,000 tons); seven cold reduction – two 5-stand 4-high 18¾" and 49½ x 42" continuous (for tinplate) (660,000 tons), one 5-stand 4-high 21" and 53 x 45" continuous (tinplate) (720,000 tons), one 5-stand 4-high 21" and 56 x 56" continuous (660,000 tons), one 4-stand 4-high 21" and 56 x 66" continuous (1,270,000 tons), two 2-stand 4-high 22" and 56 x 49" and 24" and 56 x 49" continuous (tinplate) (660,000 tons); six temper/skin pass – one 2-stand 4-high 18¾" and 49½" x 42" continuous (tinplate) (300,000 tons), two 2-stand 4-high 19" and 53 x 48" continuous (tinplate) (750,000 tons); one 2-stand 4-high 16½" x 56 x 57" and 21" and 56 x 56" continuous sheet (350,000 tons), one 1-stand 4-high 21" and 56 x 66" continuous sheet (375,000 tons), one 1-stand 4-high 21" and 56 x 56" continuous and levelling sheet (500,000 tons). ***Coil coating lines:*** One 38" alkaline and two 45" Halogen electrolytic tinning; two 38" tin-free steel (ECCS); 49" continuous hot-dip galvanizing; two 49" continuous zinc-aluminium (Galvalume) for sheets and strip. ***Other plant:*** 27 single spindle wire drawing blocks and 50 multiple spindle continuous wire drawing machines; 43 take-up blocks for galvanizing wire; 159 take-up blocks for patenting wire; 4-strand Galvalume wire line; 42 wire nail machines, two wire staple machines; wire stand operation; brass foundry.
■ Burns Harbor, IN. ***Coke ovens:*** 82 Mckee-Otto underjet ovens and 82 Koppers underjet ovens. ***Sinter plant:*** 1-strand McKee, 2,020 sq ft grate area. ***Blast furnaces:*** Two – (C) 38'3" hearth dia, 89,204 cu ft; (D) 35' hearth dia, 86,646 cu ft. ***Steelmaking plant:*** Three 300-ton basic oxygen converters. ***Continuous casting machines:*** Two slab – one 2-strand 10 x 32"-76" (annual capacity 1,800,000 tons, based on twenty shift/week) and one other (2,200,000 tons). ***Rolling mills:*** One 1-stand 2-high universal 50" slabbing (3,400,000 tons); Two heavy plate – one 160" sheared comprising 1-stand 2-high and 1-stand 4-high (1,140,000 tons), one 110"

Bethlehem Steel Corp (continued)

sheared comprising 1-stand 2-high and 1-stand 4-high (700,000 tons); one wide hot strip comprising 2-stand 2-high 48 x 80", 3-stand 4-high 44" and 60 x 80", 7-stand 4-high 28½" and 61 x 80" (3,500,000 tons); one 5-stand 4-high 24"/26", 60 x 80" cold reduction (1,700,000 tons); two temper/skin pass – one 2-stand 4-high 24" and 56 x 53" (500,000 tons), one 1-stand 4-high 26" and 60 x 80" (1,200,000 tons).
■ Bethlehem, PA. ***Coke ovens:*** 284 Koppers gun-flue ovens; 80 McKee-Otto underjet ovens. ***Sinter plant:*** 4-strand Dwight Lloyd, 496 sq ft grate area. ***Blast furnaces:*** Four – (B) 30' hearth dia, 54,657 cu ft; (C) 27'11" hearth dia, 49,818 cu ft; (D) 30' hearth dia, 54,834 cu ft; (E) 24' hearth dia, 41,254 cu ft. ***Steelmaking plant:*** Two 270-ton basic oxygen converters; one 28-ton and four 50-ton electric arc furnaces; ***Refining plant:*** Two ESR (annual capacity approx 12,000 ingot tons). Six vacuum degassing units – four 250-ton, one 350-ton, and one 12-ton for ingot production and ladle to ladle degassing. ***Rolling mills:*** Two blooming – one 1-stand 2-high 40" (1,760,000 tons); one 1-stand 2-high 32" (1,400,000 tons); Four heavy section – one 6-stand 2-high 44" combination (800,000 tons), one one 1-stand 2-high 46" bloomer and two 2-stand 2-high 48" (1,080,000 tons), one 1-stand 2-high 40" bloomer and two 2-stand 2-high 42" (450,000 tons), one 2-stand 3-high 18" and one 1-stand 2-high 18" (200,000 tons). ***Other plant:*** Facilities for bar processing; iron, brass, ingot mould and roll foundries; press and hammer, drop and speciality forge departments with five steam forging hammers (1,500 to 8,000lb), four board forging hammers (2,000 to 4,000lb), five hydraulic forging presses (500 to 10,000 tons); two draw benches.
■ Lackawanna, NY. ***Coke ovens:*** 76 Wilputte underjet ovens, 152 Wilputte gun-flue ovens. ***Rolling mills:*** One bar comprising 1-stand 2-high 24", 3-stand 2-high 22", 6-stand 2-high 18", 6-stand 2-high 16" and 6-stand 2-high 14" (annual capacity 832,000 tons). ***Coil coating lines:*** One 72" continuous hot-dip galvanizing.
■ Johnstown, PA. ***Steelmaking plant:*** Two 185-ton electric arc furnaces. ***Rolling mills:*** One 1-stand 2-high 46" blooming (annual capacity 2,640,000 tons based on twenty shifts/week). Three billet – one 1-stand 2-high 34" (665,000 tons), one 2-high 30" with two horizontal and two vertical stands and one 9-stand 2-high 21" (1,248,000 tons). Three bar – one 3-stand 3-high 12" and 2-stand 2-high 12" (38,500 tons); one two horizontal and two vertical stands 2-high 16", six horizontal and four vertical stands 2-high 13", and two horizontal and two vertical stands 2-high 11" (390,000 tons); one 6-stand 2-high 12", 3-stand 2-high 9", and 2-stand 3-high 9" (93,000 tons). One wire rod – 4-stand 2-high 16", 5-stand 2-high 14", 4-stand 2-high 13", 2-stand 2-high 12¼:, 6-stand 2-high 10½", 2-stand 2-high 11" (300,000 tons). One wheel plant – (No 1) Slick forging and rolling (50,000 tons). ***Other plant:*** Wire drawing facilities; one hot dip and two electrogalvanizing lines for wire; iron and brass foundries; steel freight car shop; mine tie and rail anchor shop; speciality products shop; axle forge shop, and a circular forgings shop; two mechanical axle forging hammers.
■ Steelton, PA. ***Steelmaking plant:*** One 5-ton, one 15-ton and three 150-ton electric arc furnaces. ***Continuous casting machines:*** One bloom (14.6 x 23.6") (annual capacity 1,200,000 tons on twenty shifts/week). ***Rolling mills:*** one 1-stand 2-high 44" blooming (1,325,000 tons, based on twenty shifts/week); one rail – 1-stand 2-high 35", 2-stand 3-high 28" and 1-stand 2-high 28" (1,160,000 tons); two bar – one 1-stand 2-high 28", 3-stand 3-high 20" and 2-stand 2-high 20" (280,000 tons); one 3-stand 2-high 16", 6-stand 2-high 14" and 8-stand 2-high 11" (680,000 tons). ***Tube and pipe mills:*** welded – one 'U' press (rated capacity 2,400 tons) and one 'O' press (rated capacity 18,000 tons), one hydraulic pipe expander (5,400 psi max), for electric fusion weld pipe, 18 to 42" inclusive. ***Other plant:*** Steel foundry; splice bar and tie plate shops; frog and switch shops.

Manufactured Products Group
Works:
■ Wire rope division – Williamsport, PA. Wire mill: Two single spindle wire drawing blocks, 16 cone type wet continuous wire drawing machines, 39 multiple spindle continuous wire drawing machines, 51 take-up blocks for patenting wire. Rope mill:

United States

Bethlehem Steel Corp (continued)

65 tubular stranders, seven planetary stranders, ten tubular closers and five planetary closers. Splice shop: For attaching end fittings to wire rope or strand. Fourteen wire rope service centres.
■ Supply Division (Headquarters Tulsa, OK). Twenty-four supply stores serving the oil country market.
Other Facilities: Marine construction: Three yards for drill rig and ship construction and repair in the USA, and a majority interior in a drill rig construction and repair yard in Singapore. Mineral wool: Two plants producing wool insulation for the building trade and wool for manufactured building products. Natural resources: Coal – fourteen mines and eleven cleaning plants in Pennsylvania, West Virginia and Kentucky producing metallurgical and steam coal and three operations in which Bethlehem has a partial interest: one in Minnesota, one in Canada, one in Brazil. Limestone – one wholly-owned quarry and one partly-owned quarry. Others – interests in one operation producing silver. Bethlehem also has interests in oil and gas wells. Water transport – three owned vessels and one long-term chartered vessel for the transport of ore on the Great Lakes. Rail transport – five short-line railroads in the vicinities of the Sparrows Point, Lackawanna, Bethelehem, Johnstown and Steelton plants and one railroad serving coal mines near Johnston, PA.

Products: Carbon steel – slabs, blooms, billets, hot rolled coil (for re-rolling), wire rod (including high carbon), round bars, square bars, flats, light angles, light tees, light channels, heavy angles, heavy tees, heavy joists, heavy channels, wide-flange beams (heavy), heavy rails, light rails, rail accessories, cold roll-formed sections (from strip), sheet piling, hot rolled hoop and strip (uncoated), medium plates, heavy plates, universal plates, hot rolled sheet/coil (uncoated), cold rolled sheet/coil (uncoated), electro-galvanized sheet/coil, galvanized corrugated sheet/coil, aluminized sheet/coil, electrical grain oriented silicon sheet/coil, electrolytic tinplate (single-reduced), electrolytic tinplate (double-reduced), tin-free steel (ECCS), bright wire, black annealed wire, galvanized wire (plain), large-diameter tubes and pipes. Alloy steel (excl stainless) – slabs, blooms, billets, round bars, square bars, flats, wire. Alloy steel wire rods; aluminized wire; axles, crane track wheels, stapler, bridge flooring, flanged and dished heads. Wire rope and strand steel forgings (including rolls), iron castings (including rolls and ingot moulds). Steel castings, coke and by-products.
Sales offices/agents: Domestic offices are located in more than 30 US cities and a Bethlehem Steel Export Corp office in New York and Bethlehem, Pennsylvania.
Expansion programme: Modernisation of 48″ structural mill at Bethlehem, PA.
Additional information: Bethlehem Steel jointly owns with Inland Steel and Pre Finish Metals Inc an electrogalvanizing line which started up in 1986 at Walbridge, OH.

Birmingham Steel Corp

Head office: 1 Independence Plaza, Suite 710, Birmingham, AL 35201. **Postal address:** PO Box 1208. **Tel:** 205 871 9290. **Telex:** TWX 810 733 5548.
Management: President, chief executive officer and director – James A. Todd; vice president and chief financial officer – Phillip E. Casey; executive vice president – Jim C. Nuckels. **Established:** 1984. **No. of employees:** 2000. **Annual capacity:** Raw steel: 1,020,000 short tons. Finished steel: 930,000 short tons.
Works:
■ Southern United Steel Division, PO Box 2764, Birmingham, AL 35202. (Tel: 205 252 8777). Works director: Ray H. Rickey. ***Steelmaking plant:*** One 50-ton electric arc furnace (annual capacity 240,000 tonnes). ***Continuous casting machines:*** One 3-strand Koppers billet (4 x 6″, 4 x 4″) (240,000 tonnes). ***Rolling mills:*** bar – Two 16″ 3-high and seven 12″ 2-high (220,000 tonnes).
■ Illinois Steel Division, PO Box 628, Kankakee, IL 60901. (Tel: 815 937 3131). Works director: Joe Bercik. ***Steelmaking plant:*** Two 25-ton electric arc furnace (160,000 tonnes). ***Continuous casting machines:*** One 3-strand Rokop billet (4½ x 4½″) (160,000 tonnes). ***Rolling mills:*** bar – One 3-high 16″, two 2-high 14″ and seven 2-high 12″ (150,000 tonnes).
■ Mississippi Steel Division, PO Box 5780, Jackson, MS 39208. (Tel: 601 939 1623). Works director: J. D. Stutz. ***Steelmaking plant:*** One 50-ton electric arc furnace

Birmingham Steel Corp (continued)

(annual capacity 220,000 tonnes). ***Continuous casting machines:*** One 3-strand Rokop billet (4 x 4", 4½ x 4½") (220,000 tonnes). ***Rolling mills:*** Four continuous bar – 2-high 11-stand (two 22", five 16", two 14", four 12" and two 10") (200,000 tonnes).
■ Salmon Bay Steel Division, 4315 Ninth Ave NW, Seattle, WA 98107. (Tel: 206 783 4000). Works director: Ed Katai. ***Steelmaking plant:*** Two 35-ton electric arc furnace (annual capacity 250,000 tonnes). ***Continuous casting machines:*** One 3-strand Rokop billet (4 x 4", 5½ x 5½") (250,000 tonnes). ***Rolling mills:*** bar – Two 30-high 16", four 2-high 14" and four 2-high 12" (220,000 tonnes).
■ Norfolk Steel Division, 1500 Steel Street, Chesapeake, VA 23323. (Tel: 804 485 0600). Works director: Gary Baker. ***Steelmaking plant:*** Two 25-ton electric arc furnace (annual capacity 150,000 tonnes). ***Continuous casting machines:*** One 2-strand Danieli billet (4½ x 4½") (150,000 tonnes). ***Rolling mills:*** bar – One comprising 3-high 16", five stand 2-high 14" and 2-stand 2-high 12" (140,000 tonnes).
Products: Carbon steel – billets, reinforcing bars, round bars square bars, flats, light angles, light channels, medium angles, medium channels.
Additional information: In August local government approval was given for the re-opening of the Judson Steel plant at Emeryville, CA, which Birmingham Steel was negotiating to acquire. It was hoped to re-start steelmaking in November 1987 and build up production to the full 200,000 tons per year capacity. The plant will mostly produce rebar from its continuous combination mill. Other equipment is a 60-ton Whiting electric furnace and a Concast billet caster.

Blair Strip Steel Co

Head office: 1209 Butler Avenue, New Castle, PA 16107. **Postal address:** PO Box 7159, New Castle, PA 16107. **Tel:** 412 658 2611.
Established: 1923. **Annual capacity:** Finished steel: 68,000 short tons.
Products: Carbon steel – cold rolled hoop and strip (uncoated). Spring steel.

Bliss & Laughlin Steel Co

Head office: 281 East 155th Street, Harvey, IL 60426. **Tel:** 312 264 1800. **Telecopier (Fax):** 312 333 3786.
Management: President – Gregory H. Parker (chief executive officer and chairman); vice presidents – Anthony J. Romanovich (commercial), William P. Daugherty (operations), Michael A. de Bias (purchasing).
Works:
■ At head office. Drawbenches, turning mills, centreless grinders, annealing furnaces.
Products: Carbon steel – bright (cold finished) bars (rounds, hexagons, squares, flats). Stainless steel – bright (cold finished) bars (rounds, hexagons, squares, flats). Alloy steel (excl stainless) – bright (cold finished) bars (rounds, hexagons, squares, flats).

Bock Ind Inc

Head office: 57540 SR 19 S, Elkhart, IN. **Postal address:** PO Box B 1027. **Tel:** 219 295 8070. **Telecopier (Fax):** 219 293 6593.
Management: Augie Bock (board chairman), Calvin Bock (president), James Bock (executive vice president), Dale Hoogenboom (secretary/treasurer), F. Sailor (purchasing director), Joe McMahon (sales manager). **Established:** 1960. **No. of employees:** Works – 210. Office – 20.
Works:
■ At head office. ***Tube and pipe mills:*** welded – five for welded hollow steel sections 1½-12" square, 0.083-⅝" wall thickness.
Products: Carbon steel – hollow sections square and rectangular.

United States

Border Steel Mills Inc

Head office: PO Box 12843, El Paso, TX 79912. **Tel:** 915 545 2802.
Management: President and chief executive officer – A.W. Lupia; vice presidents – M.J. Barber (operations), F.P. Boyd (administration and finance), J.A. Perko (sales).
Established: 1962. **No. of employees:** Works – 324. Office – 120. **Ownership:** Private. **Subsidiaries:** Metal Processing Inc (ferrous and non-ferrous scrap processing). **Annual capacity:** Raw steel: 206,000 short tons. Finished steel: 195,700 short tons.
Works:
■ Vinton Road, El Paso, TX. (Tel: 915 545 2802). Vice president, operations: M.J. Barber. ***Steelmaking plant:*** Two 20-ton 12ft dia NOT Lectromelt electric arc furnaces (annual capacity 206,000 tons). ***Continuous casting machines:*** 3-strand curved mould Concast billet 4", 4¾ and 5¼" sq (206,000 tons). ***Rolling mills:*** Two bar – one 8-stand 16-12" Bliss and one 3-stand 14" Krupp (195,700 tons). ***Other plant:*** Forging plant with three Hill Acme ball forging machines.
Products: Carbon steel – reinforcing bars No 3 to No 18; round bars ⅜-4"; square bars 7/16-3½". Forged grinding balls 1½", 2", 2½", 3", 4", 4½" and 5".
Expansion programme: Ball heat treating line – hardening furnace, quench system, tempering furnace (completed June 1986); new baghouse (July 1987); induction hardening (bar) system (September 1987).

Braeburn Alloy Steel Division, CCX Inc

Head office: River Road, Lower Burrell, PA 15068. **Tel:** 412 224 6900. **Telex:** TWX 510 467 4649.
Management: Thomas R. Wiseman (president). **Established:** 1898. **Ownership:** CCX Inc. **Annual capacity:** Finished steel: 12,000 tonnes.
Works:
■ At head office. ***Steelmaking plant:*** Two 10-ton electric arc furnaces. ***Refining plant:*** One 11-ton VAR. ***Rolling mills:*** Two bar – one 5-stand 14" and one 6-stand 10". ***Other plant:*** One 2,000-ton forging press; three hammers, etc. High speed steels, hot work steels, die and alloy steels in rolled and forged bars; high speed tool bits.

Bristol Metals Inc

Head office: Weaver Pike, Bristol, TN 37621. **Postal address:** PO Box 1589.
Management: C.L. Reid (president), J.N. Avento (vice president). **Established:** 1940. **No. of employees:** 250 approx. **Ownership:** Synalloy Corporation.
Works:
■ At head office. ***Tube and pipe mills:*** welded – six (up to 12") (annual capacity 18,000 short tons).
Products: Stainless steel – longitudinal weld tubes and pipes.

California Steel Industries (CSI)

Head office: 14000 San Bernardino Avenue, Fontana, CA 92335. **Tel:** 714 350 6300.
Established: 1941 (as Kaiser Steel Corp). **Ownership:** CVRD and Kawasaki Steel (50% each).
Works:
■ At head office. ***Rolling mills:*** One 148" heavy plate; one 86" wide hot strip; one 5-stand 4-high cold reduction 21" and 53" x 44" continuous; one temper/skin pass.
Products: Carbon steel – medium plates, heavy plates, hot rolled sheet/coil (uncoated), cold rolled sheet/coil (uncoated), hot-dip galvanized sheet/coil.

Calumet Steel Co

Head office: 317 East 11th Street, Chicago Heights, IL 60411. **Tel:** 312 757 7300.
Management: President – James Mertz; vice presidents – Richard E. Salisbury (marketing), John R. Fazzini (administration), Joe A. Belk (operations), John P. Mullin (sales). **No. of employees:** 240. **Ownership:** Private.
Works:
■ At head office. Works director: James Mertz. ***Steelmaking plant:*** Two electric arc furnaces. ***Continuous casting machines:*** One 3-strand and one 2-strand billet (4 x 4" to 6 x 6"). ***Rolling mills:*** One 13" bar.
Products: Carbon steel – reinforcing bars, round bars, flats, light angles, light tees, special light sections.

Cameron Iron Works Inc

Head office: 13013 NW Freeway Houston, TX 77040. **Postal address:** PO Box 1212, Houston, TX 77251. **Tel:** 713 939 2211. **Telex:** ITT 4620021 cmrnoi, WUT 775422 camiron hou a. **Telecopier (Fax):** 713 939 2620 (93).
Established: 1920.
Works:
■ At head office. ***Refining plant:*** Five consumable electrode remelt furnaces. ***Other plant:*** Presses including 10,000-ton, 20,000-ton and 35,000-ton; 23,000-ton forging press; 8,000-ton isothermal press.
Products: Carbon steel – ingots. Alloy steel (excl stainless) – ingots. Forgings and extrusions for aerospace, nuclear and similar high performance applications.

Carpenter Technology Corp

Head office: 101 West Bern Street, Reading, PA 19603. **Tel:** 215 371 2000.
Established: 1968 (as successor to company incorporated in 1904). **Annual capacity:** Finished steel 150,000-200,000 short tons.
Works:
■ Steel Division – at head office. ***Steelmaking plant:*** Six electric arc furnaces. ***Refining plant:*** Two AOD; four ESR; ten VAR; three VIM. ***Continuous casting machines:*** One 2-strand billet. ***Other plant:*** Numerous hot and cold rolling mills; one GFM rotary forge; two hydraulic forging preses.
■ Tube Division, Springfield Road, Union NJ 07083 (Tel: 201 686 7230). ***Tube and pipe mills:*** welded for stainless steel.
■ Bridgeport Plant, 837 Seaview Avenue, Bridgeport CT 06607 (Tel: 203 382 5400). ***Steelmaking plant:*** Two electric arc furnaces. ***Refining plant:*** One AOD. ***Rolling mills:*** One blooming / billet. One light section. ***Other plant:*** Cold drawing plant.
■ Braeburn Alloy Steel Division, River Road, Lower Burrell, PA 15068 (Tel: 412 224 6900). ***Steelmaking plant:*** Two 10-ton electric arc furnaces. ***Rolling mills:*** Two bar – one 3-high 6-stand and one 3-high 14". ***Other plant:*** Three forging hammers and a 2,000-ton press.
Products: Stainless steel – slabs, blooms, billets, bright (cold finished) bars, longitudinal weld tubes and pipes, hollow bars. Alloy steel (excl stainless) – slabs, blooms, billets, round bars, bright (cold finished) bars, cold rolled sheet/coil (uncoated), wire, longitudinal weld tubes and pipes, hollow bars. High-speed steel; alloy tool and die steels; tool bits; forgings; forged bars; fine wire; thin wall alloy and stainless tubing; electronic, magnetic and special purpose alloys.
Additional information: There are also works (Special Products Division) at El Cajon, CA (producing thin wall alloy and stainless tubing); Orangeburg, SC (producing fine wire); and Fryeburg, Maine (Eagle Precision Metals Corp) (producing stainless and alloy hollow forms).

Cascade Steel Rolling Mills Inc

Head office: 3200 N. Highway 99W, McMinnville, OR. **Postal address:** PO Box 687, McMinnville, OR 97128. **Tel:** 503 472 4181. **Telex:** 910 452 7011. **Telecopier (Fax):** 503 434 5739.
Management: L. Schnitzer (president Schnitzer Steel Products); G. Schnitzer (board chairman Schnitzer Steel Products); M. Michelson (vice president, general manager Cascade); A. Robare (manager, operations), G. Peterson (manager, general sales); C. Burnett (controller); B. Crume (superintendent, melt shop), C. Cyr (superitendent, rolling mill), J. Stevens (superitendent, maintenance and engineering). **Established:** 1969. **No. of employees:** Works – 440. **Ownership:** Schnitzer Steel Products. **Annual capacity:** Finished steel: 360,000 short tons.
Works:
■ At head office. ***Steelmaking plant:*** Two 30-ton 12 ft dia electric arc furnaces (annual capacity 280,000 tons). ***Continuous casting machines:*** One 3-strand Rokop billet (5¼ sq) (280,000 tons). ***Rolling mills:*** 12" bar comprising 17 horizontal/vertical stands (11 are cantilever type) (360,000 tons).
Products: Carbon steel – reinforcing bars No 3 to No 18 (1987 fiscal year output 253,000 tons); round bars ½-2½" (11,000 tons); square bars 7/16", ½" (2,000 tons); flats 3/16 x 1"-1 x 6" (70,000 tons); light tees with studs for fence posts (16,000 tons).

Central Nebraska Tubing

Head office: 14631 US Highway 6, Waverly, NE 68462. **Tel:** 402 786 5005.
Management: President – Scott Brown Jr; vice president operations – Bill Schneider. **Established:** 1981. **Capital:** $5,000,000. **No. of employees:** Works – 30. Office – 5. **Ownership:** Brownie Manufacturing Co. **Annual capacity:** Finished steel: 17,000 short tons.
Works:
■ At head office. ***Tube and pipe mills:*** welded – one 3" Ardcor longitudinal weld (annual capacity 8,000 tons) and one 3" Turek & Heller (9,000 tons).
Products: Carbon steel – longitudinal weld tubes and pipes ½"-2", 11-18g (square, rectangular and round), (output 17,000 tons).

CF&I Steel Corp

Head office: Pueblo, CO 81002. **Postal address:** PO Box 1830. **Tel:** 303 561 6000.
Management: F.J. Yaklich, Jnr (president). **Established:** Incorporated in 1936. **Annual capacity:** Raw steel: 800,000 short tons.
Works:
■ At head office. ***Steelmaking plant:*** Two 150-ton electric arc furnaces (annual capacity 800,000 short tons). ***Continuous casting machines:*** One 6-strand billet; one 6-strand round. ***Rolling mills:*** One 45" rail; one 11" continuous bar; one wire rod comprising 16" breakdown and 10" continuous finishing with 8 non-twist stands. ***Tube and pipe mills:*** seamless One with piercing, plug mills and stretch reducing stands. ***Other plant:*** Plant for wire drawing, galvanizing, etc. Rail end and rail head hardening facilites (30,000 short tons); iron foundry.
Products: Carbon steel – ingots; billets; wire rod; reinforcing bars; round bars (incl SBQ); light angles; heavy rails; bright wire; galvanized wire (barbed); seamless tubes and pipes; oil country tubular goods. Alloy steel (excl stainless) – ingots; billets; wire rod; round bars. Baling wire; rail anchors; fencing and fence posts; nails; welded fabric.

The Champion Steel Co

Head office: 8247 Penniman Road, Orwell, OH 44076. **Tel:** 216 437 5161.
Annual capacity: Raw steel: 8,000 short tons.
Works:
■ At head office. Two hydraulic forging presses (600 and 1,500-ton) and four open

United States

The Champion Steel Co (continued)

die hammers (1,200lb, 2,000lb, 2,500lb and 5,000lb).
Products: Carbon steel – ingots, billets. Alloy steel (excl stainless) – ingots, billets. High speed and tool steels; forgings and castings.

Chaparral Steel Co

Head office: 300 Ward Road, Midlothian, TX 76065. **Tel:** 214 775 8241. **Telex:** 75 2406. **Telecopier (Fax):** 214 775 6120.
Management: Directors – Robert D. Rogers (chairman of the board); Gordon E. Forward (president and chief executive officer); Ronald P. Fournier (executive vice president, Co-Steel Inc, Ont); Richard I. Galland; Alex J. Hamilton (manager, Winbam Investments Ltd, Alberta); Gerald R. Heffernan (president and chief executive officer, Co-Steel Inc, Ont); Michael M. Koerner (president, Canada Overseas Investments Ltd, Ont); George A. Meihaus (Texas Industries Inc); Ralph B. Rogers (Texas Industries Inc); Lionel H. Schipper (president, Schipper Enterprises Inc, Ont); Christopher G. Schmelzle (director, Co-Steel Inc, Ont); Virgil L. Sewell (Texas Industries Inc); Fergus J. Walker Jr (Texas Industries Inc). Officers – Gordon E. Forward (president and chief executive officer); Dennis E. Beach (vice president, administration); Larry L. Clark (controller and assistant treasurer); Richard M. Fowler (vice president, finance); Richard T. Jaffre (vice president, raw materials); Ronald E. Lincoln (vice president, steel production); Libor F. Rostik (vice president, engineering); Lloyd M. Schmelzle (vice president, rolling mills); Jeffry A. Werner (senior vice president, commercial); Peter H. Wright (vice president, quality control). **Established:** 1973. **Capital:** $82,000,000. **No. of employees:** Works – 676. Office – 292. **Ownership:** Texas Industries Inc (100%). **Annual capacity:** Finished steel: 1,300,000 short tons.
Works:
■ At head office. ***Steelmaking plant:*** One 150-ton 22ft dia and one 135-ton 18ft dia electric arc furnaces (total annual capacity 1,300,000 tons). ***Refining plant:*** 150-ton ladle furnace (750,000 tons). ***Continuous casting machines:*** One 4-strand Concast billet (5 x 7") (550,000 tons); one 5-strand Concast bloom (6 x 10") (750,000 tons); one 2-strand SCE horizontal (5 x 7"-6 x 10") (200,000 tons). ***Rolling mills:*** medium section – Four vertical stands, twelve horizontal stands, 23" (900,000 tons); bar – Three vertical and thirteen horizontal stands (seven 18", six 14" and three 12" (600,000 tons).
Products: Carbon steel – reinforcing bars, round bars, square bars, flats, light angles, medium angles, medium channels, wide-flange beams (medium). Alloy steel (excl stainless) – round bars.

Charter Electric Melting Inc

Head office: 3151 S California Av, Cook, Chicago, IL. **Tel:** 312 376 9109. **Established:** 1981. **No. of employees:** Works – 126. Office – 7. **Ownership:** Charter Manufacturing Co Inc. **Annual capacity:** Raw steel 100,000 short ton.
Works:
■ At head office. ***Steelmaking plant:*** One 50-ton electric arc (annual capacity 150,000 tons). ***Refining plant:*** One 50-ton ladle furnace (100,000 tons). ***Continuous casting machines:*** 3-strand Rokop billet (4 x 4") (100,000 tons).
Products: Carbon steel – billets 4 x 4 and $4\frac{1}{2}$ x $4\frac{1}{2}$" (output 100,000 tons).

Charter Rolling, Division of Charter Manufacturing Inc

Head office: 1658 Gold Spring Drive, Saukville, WI 53080. **Tel:** 414 284 9486. **Annual capacity:** Finished steel: 200,000 short tons.
Works:
■ At head office. ***Steelmaking plant:*** One 45-ton electric arc. ***Continuous casting machines:*** One 3-strand billet. ***Rolling mills:*** One single-strand wire rod with Morgan No-Twist and Stelmor controlled cooling.
Products: Carbon steel – wire rod 11/64", $\frac{3}{16}$", 7/32", 9/16" and 5mm. Cold heading wire.

Chicago Heights Steel Co

Head office: 211 East Main Street, Chicago Heights, IL 60411. **Annual capacity:** Finished steel: 140,000 short tons.
Works:
■ ***Rolling mills:*** light section (using rails and billets).
Products: Carbon steel – flats, light angles, light tees, light channels, special light sections. Fence posts and subpurlins.

Cold Metal Products Co Inc

Head office: 45 South Montgomery Avenue, Youngstown, OH. **Postal address:** PO Box 6078. **Tel:** 216 746 6311. **Telex:** 8104352925. **Telecopier (Fax):** 216 742 1200.
Management: James R. Harpster (president and chief executive officer), Gordon A. Wilber (executive vice president), Robert R. Albert (vice president, sales), Allen R. Morrow (chief financial officer). **Established:** 1980. **No. of employees:** Works – 650. Office – 330. **Ownership:** Aarque Management Corp. **Annual capacity:** Finished steel; 225,000 short tons.
Works:
■ At head office. Works director: C.N. Lee. ***Rolling mills:*** Six cold strip – 4-high (27") Steckel reversing (annual capacity 55,000 short tons), 4-high (26") Steckel reversing (29,000 tons), (28") reversing (27,000 tons), 4-high (10") Steckel reversing (15,000 tons), 4-high (10") Steckel reversing (3,800 tons), 4-high (10") Steckel reversing (14,000 tons); two temper/skin pass – one 2-high (14") Temper (10,000 tons) and one 2-high (28") Temper (30,000 tons). **Coil coating lines:** One electro plating for copper and zinc (16,000 tons).
■ 65 Burritt Street, New Britain, CT 06050 (Tel: 203 225 3050). Works director: T.C. Kausel, Jr. ***Rolling mills:*** Two cold strip – one 4-high (12" x 24" x 26") United reversing (annual capacity 75,000 short tons) and one 20-high (26") Sendzimir reversing (40,000 tons).
■ 65 Imperial Street, Hamilton, Ont, Canada (Tel: 416 544 2803). Works director: R. Weir. ***Rolling mills:*** Two cold strip – one 6-high Hitachi (12 x 15 x 39 x 39") reversing (annual capacity 90,000 short tons) and one 4-high Davy United (12 x 24 x 24") reversing (15,000 tons).
■ 2301 South Holt Road, Indianapolis, IN 46241 (Tel: 317 248 6620). Works director: D.W. Hughes. ***Rolling mills:*** Two cold strip – one 4-high (10") Steckel reversing (annual capacity 13,000 short tons) and one 4-high (28") Steckel reversing

United States

Cold Metal Products Co Inc (continued)

(70,000 tons).
Products: Carbon steel – cold rolled hoop and strip (uncoated) 0.004-0.160" x 0.250-36" (1986 output 180,000 short tons).

Columbia Tool Steel Co

Head office: 400 Lincoln Highway, Chicago Heights, IL 60411. **Tel:** 312 757 5253.
Established: 1904.
Products: Alloy steel (excl stainless) – ingots, round bars, bright (cold finished) bars. High-speed; water hardening tool and die steels; forgings.

Commercial Steel Corp

Head office: Allegheny Avenue and West 7th Street, Glassport, PA 15045. **Tel:** 412 664 6330.
Annual capacity: Finished steel; 45,000 short tons.
Works:
■ Rail slitting and re-rolling.
Products: Carbon steel – reinforcing bars (from rails) No 3, 4, and 5.

Connecticut Steel Corp

Head office: 35 Toelles Road, Wallingford, CT 06492. **Postal address:** PO Box 928. **Tel:** 203 265 0615.
Management: Chairman of the board – Willy Korf; president – D.W. Schlett; vice presidents – Charles E. Volpert (finance), R.D. Jones (steel operations), Monty L. Pastorick (wire operations); general sales manager – Brenda D. Pastorick. **Established:** 1984. **Ownership:** Private. **Annual capacity:** Raw steel: 250,000 tonnes. Finished steel: 238,000 tonnes.
Works:
■ At head office. ***Steelmaking plant:*** One 30-tonne energy optimizing furnace (EOF) (annual capacity 250,000 tonnes – 1988). ***Refining plant:*** One 30-tonne ladle furnace (250,000 tonnes – 1988). ***Continuous casting machines:*** One 2-strand curved mould billet (120mm sq) (250,000 tonnes). ***Rolling mills:*** One 2-high 17 stand bar (two 19", four 16", nine 14" and two 11") and one 10-stand no-twist wire rod (two 8" and eight 6") (coil weight 1 tonne) (combined capacity 238,000 tonnes). ***Other plant:*** Wire mill, welded wire fabric plant, nail mill.
Products: Carbon steel – billets 4¾sq x 30'; wire rod 7/32-½"; reinforcing bars No 3 to No 8 inclusive; round bars ⅜-1"; bright wire 10 gauge, 6 gauge. Welded wire fabric, nails.
Expansion programme: Bar/rod rolling mill commissioned in 1987, EOF steelmaking and billet casting facilities commissioning 1988.

Continuous Coating Corp

Head office: 520 West Grove Avenue, Orange, CA 92665. **Tel:** 714 637 4643.
Established: 1965. **Ownership:** Robert Larrabee is largest holder of shares.
Works:
■ ***Coil coating lines:*** One 48" electro-galvanizing.
Products: Carbon steel – electro-galvanized sheet/coil (coil only).

Copperweld Corp

Head office: Four Gateway Center, Suite 2000, Pittsburgh, PA 15222. **Tel:** 412 263 2000.
Established: 1915.
Works:

Copperweld Corp (continued)

■ Ohio Steel Tube Division, 132 West Main Street, Shelby, OH 44875 (Tel: 347 2424). ***Tube and pipe mills:*** seamless Assel rotary piercer (annual capacity 55,000 short tons). Welded One ERW (1-4" od) (50,000 short tons), and one ERW ($3\frac{1}{4}$-$7\frac{1}{2}$" od (30,000 tons). ***Other plant:*** Cold drawing facilities; tube forging equipment.
■ Regal Tube Division, 7401 South Linder Avenue, Chicago IL 60638 (Tel: 312 458 4820). ***Tube and pipe mills:*** Three ERW, own 2½5'' sq and two 10'' max sq (190,000 short tons).
■ Copperweld Southern Division, PO Box Elk Cotton Mill Road, Fayetteville, Tenn 37334 (Tel: 615 433 7177). Equipment for cladding, wire drawing, etc.
Products: Carbon steel – seamless tubes and pipes, longitudinal weld tubes and pipes, cold drawn tubes and pipes. Mechanical tubing; DOM tubing; Copper clad aluminium.
Brands: Copperweld.

Copperweld Steel Co

Head office: 4000 Mahoning Avenue, Warren, OH 44483. **Tel:** 216 841 6011.
Ownership: CSC Industries Inc.
Works:
■ ***Steelmaking plant:*** Five 90-ton electric arc furnaces. ***Refining plant:*** Two DH vacuum degassing units. ***Rolling mills:*** Several bar. ***Other plant:*** Draw benches, etc.
Products: Carbon steel – bright (cold finished) bars. Alloy steel (excl stainless) – blooms, billets, round bars, bright (cold finished) bars. High grade carbon and alloy steels; aircraft steels; corrosion and heat resistant steels; carbon and alloy tool steels; valve steels; leaded steels.

Crucible Specialty Metals Division, Crucible Materials Corp

Head office: Syracuse, NY 13201. **Postal address:** PO Box 977. **Tel:** 315 487 4111.
Ownership: Crucible Materials Corp, Syracuse, NY 13201. **Associates:** Trent Tube Division (see separate entry); Crucible Magnetics Division, Elizabethstown, KY (magnets).
Works:
■ At head office. ***Steelmaking plant:*** Three electric arc furnaces; one electric induction. ***Refining plant:*** One AOD; one VAR. ***Rolling mills:*** One billet. Two bar / wire rod. ***Other plant:*** Cold drawing benches, etc; forging hammers.
Products: Carbon steel – bright (cold finished) bars. Alloy steel (excl stainless) – round bars incl ground; bright (cold finished) bars. High speed and carbon and alloy tool steels; high temperature steels; corrosion resistant steels; titanium.

Cumberland Steel Co

Head office: 101-115 Winston Street, Cumberland, MD 21502. **Tel:** 301 724 1370.
Established: 1892. **Annual capacity:** Finished steel: 12,000 short tons.
Products: Carbon steel – bright (cold finished) bars.

Cuyahoga Steel & Wire Inc

Head office: 31000 Solon Road, Solon, OH 44139. **Tel:** 216 248 0290.
Established: 1913.
Works:
■ At head office. One drawbench, six wire drawing blocks, controlled atmospheric annealing facilities.
Products: Carbon steel – bright (cold finished) bars up to 3" dia in rounds, squares and

United States

Cuyahoga Steel & Wire Inc (continued)

hexagons; bright wire, black annealed wire. Alloy steel (excl stainless) – bright (cold finished) bars up to 3″ dia in rounds, squares and hexagons; wire.

Cyclops Corp

Head office: 650 Washington Road, Pittsburgh, PA 15228. **Tel:** 412 343 4000. **Management:** William H. Knoell (chairman); James F. Will (president); William D. Dickey (senior vice president); Robert A. Kushner (vice president, secretary and general counsel). **Ownership:** Cyclops Industries.

Coshocton Stainless Division
Works:
■ PO Box 548, Coshocton, OH 43812. (Tel: 614 829 2341). ***Rolling mills:*** One 40″ and one 52″ Sendzimir reversing cold reduction; one 30″ Z-high and one 18″ Sendzimir reversing and one 18″ and one 30″ 4-high reversing cold strip; one 50″, one 30″ and one 18″ temper/skin pass. ***Other plant:*** Three bright annealing lines: two 36″ and one 18″.

Cytemp Specialty Steel Division 701 East Spring Street, Titusville, PA 16354. ***Postal address:*** PO Box 247. ***Tel:*** 814/827 3641. ***Telex:*** 5106070718. ***Telecopier (Fax):*** 814/827 3641-421.
Management: Vice presidents – M.P. Sullivan (operations), C.A. Burhoe (sales), L.W. Lherbier (technology); directors – D.K. Carmichael (industrial relations), W.B. Kent (marketing); controller – P.W. Land. **Established:** 1887. **No. of employees:** Works – 840. Office – 200. **Ownership:** Cyclops Corp.
Works:
■ Bridgeville Plant, Mayer Street, Bridgeville, PA 15017. (Tel: 412/221-8000. Telex: 5106970718). Plant manager: R.D. Williams. ***Steelmaking plant:*** Three electric arc furnaces. ***Refining plant:*** 25/50-ton AOD (annual capacity 100,000 tons); three 5-ton ESR; two 5-ton VAR. ***Rolling mills:*** One 38″ reversing Tippins blooming (100,000 tons). ***Other plant:*** Advanced powder facility.
■ Pittsburgh Plant, 31st and AVRR, Pittsburgh, PA 15201. (Tel: 412/261 4388. Telex: 5106979718). Plant manager: D.E. Daube. ***Rolling mills:*** Hand sheet mill; 4-high continuous cold reduction.
■ Titusville Plant, PO Box 247, Titusville, PA 16354. (Tel: 814/827 3641. Telex: 5106970718). Vice president, operations – M.D. Sullivan. ***Refining plant:*** Five 5-ton VAR and 30-ton VIM. ***Rolling mills:*** Five 3-high bar. ***Other plant:*** Air hammers, forging press.

Detroit Strip Division 913 Bowman Street, Mansfield, OH 44905. ***Postal address:*** PO Box 247. ***Tel:*** 419/755 3011. ***Telex:*** 510 601 5758. ***Telecopier (Fax):*** 419/755 3430.
Management: President – John A. Frecka; vice presidents – Charles G. Kennedy (sales), Raymond H. Enderle (operations); controller – James R. Watt. **No. of employees:** Works – 178. Office – 74. **Ownership:** Cyclops Corp.
Works:
■ 1025 South Oakwood Avenue, PO Box 09200, Detroit, MI 48217. (Tel: 313 297 9800). Plant manager: Robert J. Sisneros. ***Rolling mills:*** Two 26″ 4-high reversing cold reduction (annual capacity 85,000 tons); two 26″ 2-high tandem temper/skin pass (100,000 tons) and two 6″ 2-high non-reversing cold mills (10,000 tons).
■ 2061 State Street, New Haven, CT 06507. (Tel: 203 773 5200). Plant manager: Dennis M. Kovach. ***Rolling mills:*** Two 26″ 4-high reversing cold reduction (annual capacity 150,000 tons); one 26″ 2-high tandem temper/skin pass (75,000 tons) and one 6″ 2-high non-reversing cold mill (2,000 tons).

Empire-Detroit Steel Division 913 Bowman Street, Mansfield, OH 44905. ***Postal address:*** PO Box 247. ***Tel:*** 419/755 3011. ***Telex:*** 510 601 5758. ***Telecopier (Fax):*** 419/755 3430.
Management: President – John A. Frecka; vice presidents – Charles G. Kennedy

Cyclops Corp (continued)

(sales), Raymond H. Enderle (operations); controller – James R. Watt. **Established:** 1900. **No. of employees:** Works – 1,236. Office – 426. **Ownership:** Cyclops Corp.
Works:
■ 913 Bowman Street, PO Box 247, Mansfield, HO 44901. (Tel: 419 755 3011). General superintendent: Clendon N. Parr. ***Steelmaking plant:*** Two 100-ton Herult and American Bridge electric arc furnaces (annual capacity 550,000 tons). ***Refining plant:*** 100-ton Salem AOD (100,800 tons). ***Rolling mills:*** 52" primary mill; 6-stand 52" wide hot strip primary mill (912,000 tons); 5-stand 52" cold reduction (462,000 tons); three 2-high 48" temper/skin pass (588,000 ton). ***Coil coating lines:*** One Wean and one Blaw Knox hot-dip tinning (318,000 tons).

Sawhill Tubular Division
Works:
■ PO Box 11, Sharon, PA 16146. (Tel: 412 347 7771). ***Tube and pipe mills:*** seamless One 3" od Pilger. Welded One continuous weld and three electric weld. ***Other plant:*** Plug type draw benches and mandrel draw benches.

Tex Tube Division, Cyclops Corp North Post Oak, Houston, TX. ***Postal address:*** PO Box 7707. ***Tel:*** 713 686 4351. ***Telex:*** 77270.
Management: D.W. Biggers (president); Carl Farnsworth (general manager). **Established:** 1949. **No. of employees:** Works – 55. Office – 35. **Ownership:** Cyclops Corp.
Works:
■ ***Tube and pipe mills:*** welded – two Mckay ERW (2-4", 4-8") (annual capacity 120,000 tons).

Products: Carbon steel – ingots, hot rolled coil (for re-rolling), sheet piling, pipe piling, hot rolled hoop and strip (uncoated), cold rolled hoop and strip (uncoated), hot-dip galvanized hoop and strip, hot rolled sheet/coil (uncoated), cold rolled sheet/coil (uncoated), hot-dip galvanized sheet/coil, electrical non-oriented silicon sheet/coil, longitudinal weld tubes and pipes, oil country tubular goods, galvanized tubes and pipes, cold drawn tubes and pipes. Stainless steel – ingots, slabs, blooms, billets, hot rolled band (for re-rolling), wire rod, round bars, square bars, flats, hexagons, cold roll-formed sections (from strip), cold rolled hoop and strip (uncoated), hot rolled sheet/coil (uncoated), cold rolled sheet/coil (uncoated), cold drawn tubes and pipes, precision tubes and pipes. Alloy steel (excl stainless) – ingots, slabs, blooms, billets, hot rolled coil (for re-rolling), round bars, square bars, flats, hexagons, hot rolled sheet/coil (uncoated), cold rolled sheet/coil (uncoated), cold drawn tubes and pipes. High temperature alloys in ingots, semis, bars, plates and sheet; electrical conduit; powder products.
Sales offices/agents: Detroit, MI (Tel: 313/297 9877); Chicago, IL (Tel: 312/355 2929); Dayton, OH (Tel: 513/898 6446); Cleveland, OH (Tel: 216/464 8910); Charlotte, NC (Tel: 704/436 9704); Indianapolis, IN (Tel: 317/873 6811).

Dart Rollformers Inc

Head office: 3651 Sausalito St, Los Alamitos, CA 90720. **Postal address:** PO Box 497. **Tel:** 213 430 3541.
Management: President – Dennis de Zonia; vice president – Sharron de Zonia; managers – Cyndi de Zonia (office), Dave de Zonia (production), Dennis Moore (sales). **Established:** 1963. **No. of employees:** Works – 7. Office – 5. **Subsidiaries:** Dudley Buildings (100%) (prefab steel buildings). **Annual capacity:** Finished steel: 600 short tons.
Works:
■ At head office. Four Yoder rollformers – cold forming of purlins, etc.
Products: Carbon steel – light angles, light channels, medium angles, medium channels, heavy channels, cold roll-formed sections (from strip) 3 to 18" (openbox and zed section) (1986 output 600 tons).

Eastern Stainless Steel Co

Head office: Baltimore, MD 21203. **Postal address:** PO Box 1975. **Tel:** 301 522 6200. **Telex:** 87826, 87840.
Established: 1919. **Ownership:** Eastmet Corp. **Annual capacity:** Raw steel: 100,000 short tons.
Works:
■ At head office. ***Steelmaking plant:*** One 50-ton electric arc furnace. ***Refining plant:*** One 50-ton AOD. ***Continuous casting machines:*** One 1-strand Concast slab. Hot and cold mills for plate and plate-sheet. ***Other plant:*** Bar straightening machines.
Products: Stainless steel – ingots; slabs; flats produced from plates and coil; hot rolled hoop and strip (uncoated); skelp (tube strip); medium plates; heavy plates; universal plates; floor plates; hot rolled sheet/coil (uncoated); cold rolled sheet/coil (uncoated).

Edgewater Steel Co

Head office: College and Allegheny Avenues, Oakmont, PA 15139. **Tel:** 412 826 7200.
Established: 1916. **Ownership:** Edgewater Corp.
Works:
■ At head ofice. **Rolling mills**. Wheel and ring. Railway wagon and locomotive wheels; seamless rolled rings for bearings, gears, aero engines, in carbon, alloy and stainless steels; superalloys; titanium

Electralloy Corp

Head office: 475 Park Avenue South, New York, NY 10016. **Tel:** 212 481 0770. **Telex:** 126564 electrokral nyk. **Cables:** Chromore New York. **Telecopier (Fax):** 710 581 5630.
Management: Chairman – Michael Kral; vice presidents – R.L. Pflueger (senior), W.H. Brewster (sales wrought products), W.W. Weaver (sales master alloy and ingots); Comptroller – L. Zander; managers – R.V. Nordquist (sales tool steels), J. Bruzek (export sales), D. Montgomery (rolling operations), J. Simmons (forging operations), J. Toy (plant operations). **Established:** 1967. **No. of employees:** Works – 100. Office – 35. **Ownership:** Michael Kral Industries Inc. **Subsidiaries:** Ven-Met (permanent mould and sand static castings for heat, wear and corrosion-resistant applications); Kokomo Tube Co (corrosion and heat resistant centrifugally cast petrochemical tubing, retorts, radiant tubing, furnace rolls, refermer and catalyst tube assemblies, cast valve seat rings, pump liners, gear and flange blanks, crushing rollers, forming dies and inserts in tool steels and copper base, nickel base, and cobalt base alloys).
Works:
■ 175 Main Street, Oil City, PA 16301 (Tel: 814 676 1894). Works director: R.L. Pflueger. ***Steelmaking plant:*** One 35-ton electric arc furnace; one 3-ton vacuum induction and twelve induction furnaces – one 4,000 lb, three 2,000 lb, five 1,500 lb and three 780 lb (combined annual capacity 60,000 tons). ***Refining plant:*** Two AOD – one 30-ton and one 20-ton (60,000 tons). ***Other plant:*** Seven horizontal and one vertical casting machines.
Products: Stainless steel – ingots, billets, round bars. Cast iron – pipes (centrifugally cast). Tool steel and nickel base alloy ingots, billets and bars.

Elliott Bros Steel Co

Head office: 1038 N. Cedar Street, New Castle, PA 16103. **Postal address:** PO Box 551. **Tel:** 412 658 5561. **Telecopier (Fax):** 412 658 7661.
Management: President – Thomas C. Elliott (treasurer); vice president – John Peluso Jr (secretary), Thomas C. Elliott III (sales), John Peluso III (manufacturing). **Established:** 1893. **No. of employees:** Works – 23. Office – 10. **Annual capacity:** Finished steel: 10,950 short tons.
Works:

United States

Elliott Bros Steel Co (continued)

■ At head office. ***Rolling mills:*** cold strip – One 6-high 6½" x 12" x 14" reversing cluster mill (annual capacity 4,000 tons), one 2-high 12¾" x 14½" Bliss one-way reduction and skin pass mill (5,000 tons), 4-high 17 x 24 Bliss mill (6,000 tons).
Products: Carbon steel – cold rolled hoop and strip (uncoated).

Ellwood Uddeholm Steel Corp

Head office: Park Industrial Centre, 700 Moravia, New Castle, PA 16101. **Tel:** 412 658 6504. **Telex:** 866140 elcyforge ewcy.
Annual capacity: Finished steel: 100,000 short tons.
Works:
■ At head office. ***Steelmaking plant:*** One 35-tonne electric arc furnace. ***Refining plant:*** One Asea-SKF and one VOD. ***Other plant:*** 1,500-ton and 2,500-ton forging presses.
Products: Carbon steel – ingots, blooms (forged). Stainless steel – ingots, blooms (forged). Alloy steel (excl stainless) – ingots, blooms (forged). Forged bars in carbon, alloy and stainless steel.

A. Finkl & Sons Co

Head office: 2011 N. Southport, Chicago, IL 60614. **Tel:** 312 975 2500. **Telex:** 25254.
Management: C.W. Finkl (chairman and chief executive officer), A.L. Lehman (president, operations), E.E. Byrne (vice president, finance, and secretary), C.P. Laurenson (vice president, sales). **Established:** 1879. **Capital:** $10,000,000. **No. of employees:** Works – 374. Office – 125. **Ownership:** Wholly-owned by officers.
Annual capacity: Raw steel: 100,000 short tons.
Works:
■ At head office. ***Steelmaking plant:*** Two 65-ton electric arc furnaces (annual capacity 100,000 tons). ***Other plant:*** Presses and hammers, etc. Heavy forgings and hot work die steels.
Brands: FX, Durodi, Shell Die, Cuprodie, Press-X, Mold Die.
Sales offices/agents: At head office; 2122 Morrissey Street, Detroit, MI 48030; 10735 Sessler Street, South Gate, Los Angeles, CA 90280.

Florida Steel Corp

Head office: 1715 Cleveland Street, Tampa, FL 33630. **Postal address:** PO Box 23328. **Tel:** 813 251 8811. **Telex:** WU 52731.
Management: Chairman of the board and chief executive officer – Edward L. Flom; president and chief operating officer – Thomas G. Creed; vice presidents – Alfred D. Gres (chief financial officer), Ralph R. Boswell (steel mills), J. Donald Haney (fabricated products), James C. Hogue (employee relations), Donald S. Ballard (mill product sales). **Established:** Incorporated 1956. **Subsidiaries:** Wholly-owned – Stafford Rail Product Inc, Lancaster, SC; Atlas Steel & Wire Corp, New Orleans, LA. Reinforcing steel fabricating group has plants at Fort Lauderdale, Fort Myers, Jacksonville, Orlando and Tampa (Florida); Atlanta (Georgia); New Orleans (Lousiana); Charlotte and Raleigh (N. Carolina) and Aiken (S. Carolina). **Annual capacity:** Raw steel 1,540,000 short tons.
Works:
■ Tampa Steel Mill, PO Box 23328, 7105 East 6th Avenue, Tampa, FL 33630. (Tel: 813 621 3511). Division manager: Earl Hendry. ***Steelmaking plant:*** Two electric arc furnaces 12.5 and 13.5 ft dia (annual capacity 270,000 tons). ***Rolling mills:*** One 9-stand bar (one 18" and eight 12").
■ Jacksonville Steel Mill, Highway 217 Yellow Water Road, PO Box 518, Baldwin, FL 32234. (Tel: 904 266 4261). Division manager: Alton W. Davis. ***Steelmaking plant:*** One 18 ft dia electric arc furnace (annual capacity 400,000 tons). ***Rolling mills:*** One

Florida Steel Corp (continued)

16-stand (four 18", six 14", four 12" and two 10") bar one 10-stand rod block (coil weight 2,500lb).
■ Charlotte Steel Mill, PO Box 31067, Lakeview Road, Charlotte, NC 28231. (Tel: 704 596 0361). Division manager: F.C. Pilkinton. ***Steelmaking plant:*** Two 13.5 ft dia electric arc furnaces (annual capacity 270,000 tons). ***Rolling mills:*** One 15-stand (three 19", four 14" and eight 12") bar.
■ West Tennessee Steel Mill, US 45 North, PO Box 3855, Jackson, TN 38303. (Tel: 901 424 5600). Division manager: W.G. Manuel. ***Steelmaking plant:*** One 20 ft dia electric arc furnace (annual capacity 400,000 tons). ***Rolling mills:*** One 16-stand (two 600mm, three 495mm, seven 420mm, four 360mm) bar.
■ Knoxville Steel Mill, 1919 Tennessee Avenue, PO Box 4165, Knoxville, TN 37921. (Tel: 615 546 0102). Division manager: Kurt C. Zetzsche. ***Steelmaking plant:*** Two 12 ft dia electric arc furnaces (annual capacity 200,000 short tons). ***Rolling mills:*** One 16-stand (four 530mm, four 370mm and eight 320mm) bar.
Products: Carbon steel – billets; reinforcing bars; round bars 5/16"-2"; square bars $\frac{1}{2}$"; flats; light angles; light channels; heavy angles.

Fort Worth Pipe Co

Head office: 3128 West Bolt, Forth Worth, TX 76113. **Postal address:** PO Box 2108. **Tel:** 817 921 6261. **Telex:** 758224.
Management: President – C.V. Hester; vice presidents – Jerry McDonald (executive, operations), Wesley Dalley (marketing). **Established:** 1900. **No. of employees:** Works – 190. **Ownership:** Whittaker Corp, Los Angeles, CA. **Annual capacity:** Finished steel: 120,000 short tons.
Works:
■ PO Box 659, Conroe, TX 77301. (Tel: 409 539 2136). Works director: Dave Appel. ***Tube and pipe mills:*** welded – one $2\frac{3}{8}$-$8\frac{5}{8}$" Wean-United (annual capacity 120,000 tons).
Products: Carbon steel – longitudinal weld tubes and pipes $2\frac{3}{8}$, $2\frac{7}{8}$, $4\frac{1}{2}$", $6\frac{5}{8}$, $8\frac{5}{8}$" (1986 output 18,000 short tons); oil country tubular goods $2\frac{3}{8}$, $2\frac{7}{8}$, $4\frac{1}{2}$, $5\frac{1}{2}$, 7,$8\frac{5}{8}$" (80,000 tons); Alloy steel (excl stainless) – oil country tubular goods $2\frac{3}{8}$, $2\frac{7}{8}$, $4\frac{1}{2}$, $5\frac{1}{2}$, 7, $8\frac{5}{8}$" (15,000 tons).

L.B. Foster Co

Head office: 415 Holiday Drive, Pittsburgh, PA 15220. **Tel:** 419 928 3400. **Telex:** 166072166122 ibfco ut, 794548 foster co hou.
Established: 1902.
Works:
■ PO Box 1461, Parkersburg WV 26101. ***Tube and pipe mills:*** Spiral weld.
■ Route 3, Box 100, Tampa, FL 33619. ***Tube and pipe mills:*** Spiral weld.
■ PO Box 816, Savannah, GA 31402. ***Tube and pipe mills:*** Spiral weld.
■ PO Box 356 Windsor, NJ 08561. ***Tube and pipe mills:*** Spiral weld.
Products: Carbon steel – pipe piling, spiral-weld tubes and pipes 12-144" od; large-diameter tubes and pipes. Fusion bond coated tube and pipe.
Brands: Fosterweld (spiral weld pipe).

Franklin Steel Co

Head office: 600 Atlantic Av, Franklin, PA 16323. **Postal address:** PO Box 671. **Tel:** 814 676 8511.
Subsidiaries: Missouri Rolling Mill, MI, USA (100%).
Works:
■ At head office. ***Rolling mills:*** One 12-stand bar (annual capacity 75,000 short tons).
Products: Carbon steel – reinforcing bars 4-No 11; round bars $\frac{1}{2}$-$1\frac{1}{2}$"; square bars $\frac{3}{4}$-$1\frac{1}{2}$"; flats $\frac{1}{4}$ x 1-$\frac{1}{2}$ x 4"; hexagons $\frac{5}{8}$-$1\frac{1}{8}$"; light angles $1\frac{1}{4}$ x $1\frac{1}{4}$ x $\frac{1}{8}$-2 x $1\frac{1}{2}$ x $\frac{1}{4}$; light tees $1\frac{1}{2}$ x $1\frac{1}{2}$ x $\frac{1}{8}$; light channels 2lb/ft-4lb/ft; special light sections various.

United States

Geneva Steel

The former Geneva works of USS at Provo, Utah was acquired during 1987 by Basic Manufacturing & Technologies of Utah Inc (BMT). The company's plan is to operate one of the two blast furnaces and other equipment to give an annual output of 750,000 short tons of steel products, including 200,000 tons of plate and 60,000 tons of pipe.

Georgetown Steel Corp

Head office: 420 South Hazard Street, Georgetown, SC 29440. **Postal address:** PO Box 619, Georgetown PA 29442. **Tel:** 803 546 2525. **Telex:** 803053. **Telecopier (Fax):** 803 546 2525 Ext 187.
Management: Chairman of the board and president – Roger R. Regelbrugge; vice presidents – Donald B. Daily (operations), Richard C. Holzworth (sales and marketing); controller – E. Michael McMahan; director of purchasing – Kent Baumgardner. **Established:** 1968. **No. of employees:** Works – 584. Office – 153. **Ownership:** Georgetown Industries Inc. **Annual capacity:** Raw steel: 700,000 tons. Finished steel: 650,000 tons.
Works:
■ At head office. Vice president, operations: Don B. Daily. ***Direct reduction plant:*** One 400 Midrex (annual capacity 420,000 tons). ***Steelmaking plant:*** Two 70-ton electric arc furnace (700,000 tons). ***Refining plant:*** One 75-ton ladle furnace (500,000 tons). ***Continuous casting machines:*** billet – one 4-strand Concast and one 4-strand Pecor (230mm sq) (700,000 tons). ***Rolling mills:*** One 2-strand 24-stand wire rod (coil weight 2 tonnes) (650,000 tons).
Products: Direct-reduced iron (DRI) (output 161,000 tonnes); Carbon steel – billets (658,000 short tons); wire rod (620,000 tons); reinforcing bars (5,900 tons).

Gibraltar Steel Corp

Head office: 2555 Walden Avenue, Buffalo, NY 14225. **Tel:** 716 684 1020. **Telex:** 646215.
Management: Brian J. Lipke (chief executive officer). **Established:** 1954. **No. of employees:** About 150. **Annual capacity:** Finished steel: 80,000 short tons.
Works:
■ At head office. ***Rolling mills:*** Two cold strip – one 4-high 3-stand 16″ tandem, and one 4-high reversing 16″.
Products: Carbon steel – cold rolled hoop and strip (uncoated). Cold rolled high tensile steel strapping.

Greer Steel Co

Head office: 624 Boulevard, Dover, OH 44622. **Postal address:** PO Box 280. **Tel:** 216 343 8811.
Management: President, steel div – Russell E. Geisinger; vice presidents – H. Philip Seibel (sales), J. Lynn Korns (operations); controller – Ronald L. Patterson. **Established:** 1917. **No. of employees:** Works – 250. Office – 25. **Annual capacity:** Finished steel: 60,000 short tons.
Works:
■ Dover, OH and Detriot, MI. ***Rolling mills:*** Five cold strip – two reversing and three 3-stand Tandem (28" max width); temper/skin pass Six C.R. continuous (28" max width) (combined coil weight 10,000 lb max) (combined annual capacity 60,000 short tons).
Products: Carbon steel – bright (cold finished) bars; cold rolled hoop and strip (uncoated); coated hoop and strip electro galvanized. Alloy steel (excl stainless) – bright (cold finished) bars; cold rolled hoop and strip (uncoated).

Gulf States Steel Inc

Head office: 174 South 26th Street, Gadsden, AL 35901-1035. **Tel:** 205 543 6100.
Ownership: Brenlin Group. **Annual capacity:** Finished steel: 1,400,000 short tons.
Works:
■ ***Coke ovens:***. ***Blast furnaces:*** One. ***Steelmaking plant:*** Two 150-ton basic oxygen converters. ***Rolling mills:*** One 2-high 48" blooming. One 2-stand 134" heavy plate; one continuous 58" wide hot strip. One 4-stand 54" cold reduction; two temper/skin pass. ***Coil coating lines:*** One 48" hot-dip galvanizing.
Products: Carbon steel – medium plates, heavy plates, hot rolled sheet/coil (uncoated), cold rolled sheet/coil (uncoated), galvanized corrugated sheet/coil.
Expansion programme: A Stelco coil box is to be installed.
Additional information: The Gadsden works, formerly belonged to Republic Steel Corp.

Hawaiian Western Steel Ltd

Head office: Campbell Industrial Park, Ewa, HI 96707.
Management: John M. Willson (president), Thomas B. Holden (vice president and treasurer), W.M. Swope (secretary). **Established:** 1958. **Ownership:** Western Canada Steel Ltd, Canada (Ultimate holding company – Cominco).
Works:
■ At head office. ***Steelmaking plant:*** One 15/20-ton electric arc furnaces (annual capacity 60,000 short tons). ***Rolling mills:*** One bar comprising 2-stand 18" roughing and 5-stand 3-high 12" cross-country (104,000 tons).
Products: Carbon steel – ingots, reinforcing bars.

Heidtman Steel Products Inc

Head office: 5601 T Enterprise, Toledo, OH 43612. **Tel:** 419 729 4646.
Management: John C. Bates (president and chief executive officer).
Works:
■ Toledo (projected). ***Rolling mills:*** One 48" wide hot strip with Stelco coil box (coil weight 48,000 lb) (annual capacity 1,400,000 tonnes) (start-up scheduled for early 1989).

I/N Tek

Head office: c/o Inland Steel Co, 30 W. Monroe Street, Chicago, IL 60603.
Established: 1987. **Capital:** $150,000,000 (initial). **Ownership:** A joint venture between Inland Steel (60%) and Nippon Steel (40%).
Works:

United States

I/N Tek (continued)

■ Near South Bend, IN (under construction). ***Rolling mills:*** One cold reduction (annual capacity 1,000,000 short tons). ***Other plant:*** CAPL (continuous annealing and processing line).
Additional information: Hot coil for South Bend will be supplied from Inland Steel's Indiana Harbor works. Nippon will take up to 200,000 tons from I/N Tek's output when the mill starts up in early 1990.

Independence Tube Corp

Head office: 6226 West 74th Street, Chicago, IL. **Tel:** 312 496 0380. **Telecopier (Fax):** 563 1950.
Management: Jack M. Woodcock (regional sales manager). **Established:** 1972. **No. of employees:** Works – 40. Office – 10.
Works:
■ At head office. ***Tube and pipe mills:*** welded -one (6"sq, 8 x 4", $\frac{3}{8}$" wall thickness).
Products: Carbon steel – hollow sections square and rectangular structural steel tubing.

Indiana Tube Corp

Head office: 2100 Lexington, Evansville, Vanderburgh, IN. **Postal address:** PO Box 3005. **Tel:** 812 424 9028. **Telecopier (Fax):** 812 424 0340.
Management: Vice president sales/marketing – Jerry D. Stohler; purchasing – Jerry O. Blake; department heads – Charlie Oran and Leon Axsom. **No. of employees:** Works – 88. Office – 33. **Ownership:** Handy & Harman, NY.
Works:
■ At head office. ***Tube and pipe mills:*** Four ERW welded ($\frac{3}{16}$-$\frac{3}{4}$" od, (annual capacity 400,000,000 ft)
Products: Carbon steel – longitudinal weld tubes and pipes 4.75-30mm od; precision tubes and pipes. Stainless steel – longitudinal weld tubes and pipes 4.75-20mm od; precision tubes and pipes. Alloy steel (excl stainless) – longitudinal weld tubes and pipes 4.76-19.6mm od.

Ingersoll-Rand Oilfield Products Co, Specialty Steel Division —see IRI International Corp

Inland Steel Co

Head office: 30 West Monroe Street, Chicago, IL 60603. **Tel:** 312 346 0300. **Telex:** 25 3363. **Telecopier (Fax):** 312 899 3665.
Management: Chairman and chief executive officer – Frank W. Luerssen; president – O. Robert Nottelmann; vice presidents – Joseph D. Corso (materials and energy), Ian F. Hughes (research and quality), Robert E. Powell (sales), Delmar R. Rediger (manufacturing), Norman A. Robins (technological assessment and strategic planning), Walter C. Wingenroth (human resources); treasurer – Jay E. Dittus; controller and assistant treasurer – Ludwig A. Streck; secretary – Clark L. Wagner; assistant treasurer – Stephen R. Wilson; assistant secretary – Charles B. Salowitz. **Established:** 1893. **Capital:** Assets – $2,500,000,000 (parent company). **No. of employees:** 17,323. **Ownership:** Inland Steel Industries Inc. **Associates:** Joseph T. Ryerson & Son Inc and J.M. Tull Metals Co Inc (materials service centre organisations) – owned by Inland Steel Industries through Inland Steel Services Holding Inc. **Annual capacity:** Raw steel: 5,700,000 short tons. Finished steel: (shipped) 4,900,000 tons.
Works:
■ Indiana Harbor Works, East Chicago, IN. ***Coke ovens:*** Three Koppers underjet batteries (203 ovens) (annual capacity 225,000 short tons), two Koppers gun-flue batteries (174 ovens) (450,000 tons), one Koppers preheated underjet battery (69 ovens) (725,000 tons). ***Sinter plant:*** One 8 x 168ft Dwight Lloyd (1,000,000 tons).

Inland Steel Co (continued)

Blast furnaces: One 21ft 6" hearth dia, 32,179 cu ft (550,000 tons); one 19ft 10" hearth dia, 25,689 cu ft (450,000 tons); one 21ft 6" hearth dia, 30,793 cu ft (inactive); one 20ft 10" hearth dia, 29,585 cu ft (inactive); one 26ft 6" hearth dia, 48,218 cu ft (900,000 tons); one 26ft 6" hearth dia, 48,097 cu ft (900,000 tons); one 26ft 6" hearth dia, 48,182 cu ft (900,000 tons); one 26ft 6" hearth dia, 48,042 cu ft (900,000 tons); one 45ft hearth dia, 123,897 cu ft (3,285,000 tons). ***Steelmaking plant:*** Two 212-ton basic oxygen furnaces (3,400,000 tons), two 256-ton W/LBE basic oxygen furnaces (2,200,000 tons); two 120-ton electric arc furnaces (500,000 tons); open hearth facilities down permanently. ***Refining plant:*** Two 218-ton 31.5 MVA Asea-SKF (2,200,000 tons); one 256-ton RH-OB vacuum degassing unit (2,400,000 tons). ***Continuous casting machines:*** One 4-strand curved mould billet (5⅝ x 5⅝" to 7 x 7") (500,000 tons); two 2-strand Demag-Hitachi Zosen bloom (15 x 20" and 15 x 24") (1,100,000 tons); two slab – one 1-strand Demag-Hitachi Zosen (9¼ x 35- 76") (1,400,000 tons) and one 2-strand Demag (9¼ x 30-66") (1,900,000 tons). ***Rolling mills:*** One 2-stand 2-high 46" universal slabbing (3,200,000 tons); two blooming – one 1-stand 2-high 46" (1,500,000 tons) and one 1-stand 2-high 40" (1,100,000 tons); one 8-stand 2-high 21" billet (900,000 tons); two bar – one 18-stand 2-high 12" (600,000 tons) and one 15-stand 2-high 10" (300,000 tons); one 1-stand 3-high 100" heavy plate (325,000 tons); one 10-stand 4-high 76" medium plate (1,300,000 tons); one 12-stand 4-high 80" continuous wide hot strip (4,000,000 tons); three cold reduction continuous strip – one 5-stand 4-high 19½" and 49 x 42" (450,000 tons), one 4-stand 4-high 21½" and 56 x 56" (1,200,000 tons), and one 5-stand 4-high 23.8" and 60 x 80" (1,800,000 tons); temper/skin pass – one 1-stand 2-high 27½ x 56" continuous coil (338,000 tons), one 1-stand 2-high 28½ x 62" continuous sheet (178,000 tons), one 1-stand 2-high 30.5 x 62" continuous coil (225,000 tons), one 1-stand 4-high 20½ x 74½" continuous coil (300,000 tons), one 1-stand 4-high 20½ x 74½" continuous sheet (135,000 tons), one 3-stand 4-high 23" and 56 x 54" continuous coil (300,000 tons), one 1-stand 4-high 22½ and 56 x 56" continuous coil (670,000 tons), one 2-stand 4-high 24" and 60 x 80" continuous coil (800,000 tons), one 1-stand 4-high 23" and 60 x 80" continuous coil (1,020,000 tons), one 1-stand 2-high 40 x 80" HR processing (450,000 tons). ***Coil coating lines:*** Four Sendzimir and modified Sendzimir horizontal hot-dip galvanizing (one 48", two 60" and one 72" wide) (650,000 tons); one 60" wide Sendzimir horizontal aluminizing (200,000 tons).
Products: Carbon steel – hot rolled coil (for re-rolling) 0.076-0.50" gauge, 28-73" wide; round bars 2 3/64-6"; square bars ⅜-1½"; hexagons ⅜-2 5/16"; wide-flange beams (medium) 8-18"; wide-flange beams (heavy) 21-24"; pipe piling (bearing pile) 8-12"; medium plates and heavy plates 3/16-3", 30-96" wide, up to 720" long; floor plates (4-way plate) 16 gauge-1", 24-96" wide, up to 720" long; hot rolled sheet/coil (uncoated) 0.076-0.50" gauge, 28-73" wide; cold rolled sheet/coil (uncoated) 0.119-0.0014" gauge, 10-72" wide; hot-dip galvanized sheet/coil; electro-galvanized sheet/coil; aluminized sheet/coil; enamelling grade sheet/coil 0.071-0.020 gauge, 10-72" wide; colour coated sheet/coil; zinc-aluminium alloy coated sheet/coil; coated sheet/coil; electrical non-oriented silicon sheet/coil 0.031-0.017" gauge, 10-39" wide; electrical non-oriented non-silicon sheet/coil 0.063-0.014" gauge, 10-48" wide. Alloy steel (excl stainless) – hot rolled coil (for re-rolling) 0.076-0.375" gauge, 24-72" wide; round bars 2 3/64-6"; square bars ⅜-1½"; hexagons ⅜-2 5/16"; medium plates and heavy plates 3/16-3", 30-96", up to 720" long; floor plates (4-way) 16 gauge-1", 24-96", up to 720" long; hot rolled sheet/coil (uncoated) 0.076-0.357" gauge, 12-73" wide; cold rolled sheet/coil (uncoated).

Inmetco

Head office: Route 488, Ellwood City, PA 16117. **Postal address:** PO Box 720. **Tel:** 412 758 5515. **Telex:** 62888018. **Telecopier (Fax):** 412 758 9311.
Management: Dr Frank W. Schaller (president), R.H. Hanewald (vice president, operations). **Established:** 1976. **Capital:** US$35,000,000. **No. of employees:** Works – 50. Office – 26. **Ownership:** Inco Ltd (100%).
Works:

United States

Inmetco (continued)

■ ***Steelmaking plant:*** One 6.3 MVA electric arc furnace (21′ dia) (annual capacity 25,000 short tons). **Products:** Stainless steel pigs (25,000 short tons).

IRI International Corp

Head office: PO Box 1101, Pampa, TX 79065. **Tel:** 806 665 3701.
Management: President – V.P. Raymond; senior vice presidents – W.J. Orr (finance), W.P. McFatridge (operations), C.M. White (marketing), R.F. Hupp (administration); vice presidents – P.J. Trgovac (sales), W.L. Hallerberg (metals). **No. of employees:** 285.
Works:
■ At head office. ***Steelmaking plant:*** One 22-ton electric arc furnace (annual capacity 25,000 short tons, including vacuum stream degassing plant). ***Refining plant:*** Two ESR (10,000 tons). ***Other plant:*** 2,000-ton hydraulic press.
Products: Alloy steel (excl stainless) – billets, forgings.
Additional information: Company formerly known as Ingersoll-Rand Oilfield Products Co, Specialty Steel Division.

J & L Specialty Products Corp

Head office: Pittsburgh, PA. **Postal address:** PO Box 3373. **Tel:** 412 338 1600. **Telex:** 9102502954. **Telecopier (Fax):** 412 338 1676-77.
Management: President – Claude F. Kronk (chief executive officer); chairman of the board – James J. Paulos; vice presidents – Richard A. Ferrari (operations), Donald A. Zackman (commercial), G.L. Kosko (purchasing and traffic), K.F. Vincent (secretary and general counsel), Joseph J. Bellas (ieeif financial officer). **Ownership:** Private company. **Subsidiaries:** None. **Associates:** None.
Works:
■ Midland Plant, 12th and Midland Avenue, Midland, PA 15059. (Tel: 412 773 2700). Works director: John A. Wallace (plant manager). ***Steelmaking plant:*** Two electric arc furnaces – one (175 short tons) and one utilized at 115 short tons. ***Refining plant:*** One (115 short tons) AOD (annual capacity 390,000 short tons). ***Continuous casting machines:*** One 1-strand Mannesman-Demag curved mould slab (6 and 7½ x 26-52″) (390,000 short tons). ***Rolling mills:*** cold reduction – three 4-high reversing (two 52″, one 42″) (coil weight 30,000 lb), three Z mills (52″ width) (coil weight 30,000 lb) (combined capacity 415,000 short tons). ***Other plant:*** Detroit Plant, 330 South Livernois, Detroit, MI 48209; Louisville Plant, 1500 West Main Street, Louisville, OH 44641.
Products: Stainless steel – slabs 6 and 7½″ x 26-52″; hot rolled band (for re-rolling) 0.100-0.313 x 26-52″; medium plates (continuous hot mill – rolled); hot rolled sheet/coil (uncoated) 0.100-0.313″ x 26-52″; cold rolled sheet/coil (uncoated) 0.015-0.1,874 by up to 48″; bright annealed sheet/coil 0.015-0.090 by up to 48″.
Sales offices/agents: 1829 Independence Square, Atlanta, GA 30338; 303 East Army Trail Road, Suite 120, Bloomingdale (Chicago) IL 60108; PO Box 393, Chester, New Jersey 07930 (New York City); PO Box 3373, Pittsburgh, PA 15230-3373.
Additional information: Slabs are converted to hot band on toll basis.

Jackson Tube Service Inc

Head office: 8210 Industry Park Drive, Piqua, OH 45356. **Tel:** 513 773 8550. **Telecopier (Fax):** 778 8689 888 596.
Management: Chairman of the board – Sam Jackson; president – Don Massa; vice president, sales – T. Gordon Ralph. **Established:** 1974. **No. of employees:** Works – 93. Office – 25. **Ownership:** Private.
Works:
■ At head office. ***Tube and pipe mills:*** welded – 5 ERW (⅜-2⅜) (annual capacity 60,000,000 ft).
Products: Carbon steel – longitudinal weld tubes and pipes (output 23-25,000 short tons).

James Steel & Tube Co

Head office: 29774 Stephenson Highway, Madison Heights, MI 48095. **Tel:** 313 547 4200. **Telecopier (Fax):** 313 547 0430.
Management: Chairman – Leland Boren; president – Robert Becker; sales manager – James Sutton. **Established:** 1955. **No. of employees:** Works – 35. Office – 13. **Ownership:** Avis Industrial Corp.
Works:
■ At head office. ***Tube and pipe mills:*** welded – two Abbey Etna up to 6$\frac{5}{8}$ x 0.25" wall (annual capacity 50,000 short tons).
Products: Carbon steel – longitudinal weld tubes and pipes, hollow sections.

Jersey Shore Steel Co Inc

Head office: Jersey Shore, PA 17740. **Postal address:** PO Box 5055. **Tel:** 717 753 3000.
Established: 1938. **No. of employees:** 350. **Annual capacity:** Finished steel: 80,000 short tons.
Works:
■ At head office. **Rolling mills:**. Rerolling from rails – 3-high 2-stand 16" slitting, 3-high 5-stand 12" rolling.
Products: Carbon steel – flats, light angles.

Jessop Steel Co

Head office: Jessop Place, Washington, PA 15301. **Tel:** 412 222 4000. **Telex:** 81 2451.
Management: President – Guy McCracken; executive vice president – Ronald T. Orie (operations), Carl R. Moulton; vice presidents – E.J. Houser (purchasing and traffic), Ronald E. Bailey (technical services), Albert J. Morack (treasurer); controller – Richard L. Ross. **Established:** 1901. **No. of employees:** 700. **Ownership:** Athlone Industries Inc (100%).
Works:
■ At head office. ***Steelmaking plant:*** Three 20-ton electric arc furnaces (annual capacity 60,000 short tons). ***Rolling mills:*** One combination 110" slabbing / blooming plate (95,000 tons).
Products: Carbon steel – ingots, slabs, billets, medium plates (sheared), heavy plates (sheared), hot rolled sheet/coil (uncoated). Stainless steel – ingots, slabs, billets, medium plates, heavy plates. Alloy steel (excl stainless) – ingots, slabs, billets. High speed steels; non-magnetic, saw, die and tool steels; super-alloy grades; armour plate; precision ground flat stock, tool bits.
Brands: Econ-O-Miser (decarb-free die steel flats).

Earle M. Jorgensen Co, Forge Division

Head office: 8531 East Marginal Way S, Seattle, Wash 98108. **Tel:** 206 762 1100. **Telex:** 320080.
Management: Earle M. Jorgensen (chairman and chairman, executive committee, Earle M. Jorgensen Co). Forge Division: vice president and general manager – Jack Bunt (forge operations); manager forge sales – Jan Hansen. **Established:** Inc 1923. **No. of employees:** 270.
Works:
■ At head office. General superintendent: Jess Farmer. ***Steelmaking plant:*** Two 40-ton electric arc furnaces. ***Refining plant:*** One 42-ton AOD; one vacuum degassing unit; one Electro-slag hot topping system. ***Other plant:*** Forging hammers, forging presses, ring rolling mill, expander, upsetters, heat treatment and machining plant for steel and aluminium.
Products: Carbon steel – billets. Stainless steel – billets, aircraft steel billets. Alloy steel (excl stainless) – billets (high strength low alloy) Machined forgings in carbon, high strength low-alloy and stainless steels.

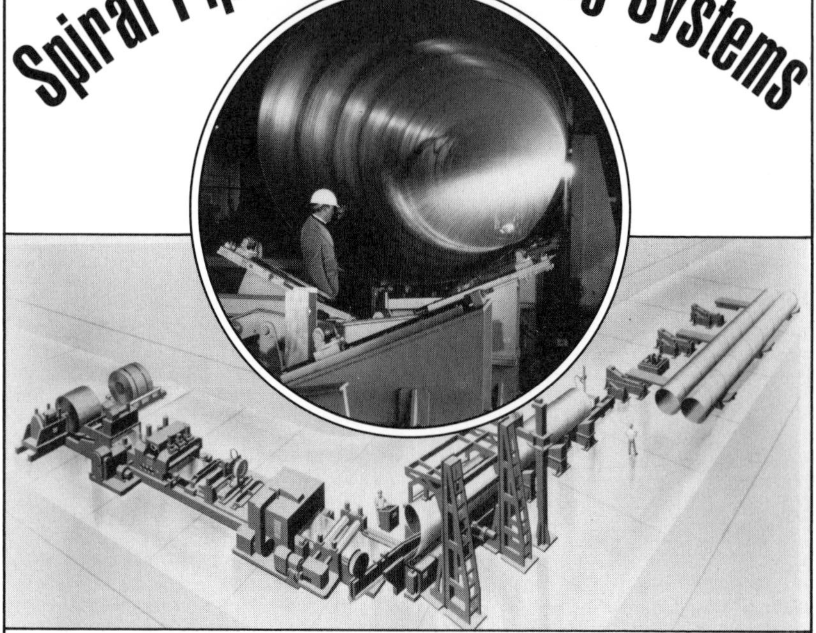

Kentucky Electric Steel Corp

Head office: Old U.S. Route 60 West, Coalton, Boyd Co, KY. **Postal address:** PO Box 3500, Ashland, KY 41101. **Tel:** 606 928 6441. **Telecopier (Fax):** 606 928 2305.
Management: Managers – Jack W. Mehalko (vice president and general), Walter S. Taylor (plant), William D. Skeens (sales), Warren R. Bocard (purchasing). **Established:** 1964. **No. of employees:** Works – 300. Office – 75. **Ownership:** NS Group Inc, Newport, KY, USA. **Annual capacity:** Raw Steel: 300,000 short tons. Finished Steel: 250,000 tons.
Works:
■ At head office. Works director – Walter S. Taylor. ***Steelmaking plant:*** Two 50-ton electric arc furnace (annual capacity 300,000 tons). ***Continuous casting machines:*** One 3-strand Concast 14" radius billet (4¾ x 8") (300,000 tons). ***Rolling mills:*** One 13-stand in line 14" bar (250,000 short tons).
Products: Carbon steel – round bars (1985 output, 20,000 tons); flats (80,000 tons). Alloy steel (excl stainless) – round bars (30,000 tons); flats (75,000 tons).

Keystone Steel & Wire Co

Head office: 7000 S. Adams Street, Peoria, IL 61641. **Tel:** 309 697 7020. **Telex:** 910 652 0135. **Telecopier (Fax):** 309 697 7422.
Management: President – Nick Owens; vice presidents – Larry Whitler (sales and finance), Les Phillips (manufacturing), Lyle Pfeffinger (administration services); General superinteedants – Tom Subbert (steel works), Barry Howard (wire mill); controller – Ed Campbell; general sales managers – John Lenart (wire production), Pete Smith (industrial and construction products); directors – Rick Cloyd (marketing), Jim Ring (engineering). **Established:** 1889. **No. of employees:** Works – 1,200. Office – 300. **Ownership:** Keystone Consolidated Industries Inc, Dallas, TX. **Annual capacity:** Raw steel: 650,000 short tons.
Works:
■ At head office. ***Steelmaking plant:*** Two 175-tonne (22') electric arc furnace (annual capacity 650,000 tons). ***Continuous casting machines:*** One 6-strand Koppers billet. ***Rolling mills:*** One 16½"-10" stand 6-line wire rod (coil weight 970lb).
Products: Carbon steel – billets, wire rod, bright wire, black annealed wire, galvanized wire (plain), galvanized wire (barbed), fabricated wire products and nails.

Laclede Steel Co

Head office: Equitable Building, 10 Broadway, St. Louis, MO 63102. **Tel:** 314 425 1400.
Management: John McKinney (president and chief executive officer), J.W. Hebenstreit (vice president, operations). **Established:** 1911 (incorporated 1969). **Annual capacity:** Raw steel: 840,000 short tons. Finished steel: 717,000 tons.
Works:
■ Alton, IL 62002 (Tel: 618 474 2100). ***Steelmaking plant:*** Two 225-ton electric arc furnace (annual capacity 840,000 tons) ***Continuous casting machines:*** One 6-strand Koppers/Demag bloom (7" sq). ***Rolling mills:*** One 36/19" blooming (550,000 tons); one 14-stand continuous bar (216,000), one 10-stand continuous bar/strip/skelp (181,000 tons), one 7-stand continuous bar/strip (43,000 tons); one 20-stand 12/6" continuous wire rod with No-Twist Stelmor (180,000); ***Tube and pipe mills:*** one Aetna Standard butt-weld (97,000 tons). ***Other plant:*** Pipe galvanizing plant; wire drawing equipment.
Products: Carbon steel – round bars (special quality), hot rolled hoop and strip (uncoated), universal plates, longitudinal weld tubes and pipes ½" to 4" (continuous-weld), wire, galvanized tubes and pipes, Alloy steel (excl stainless) – round bars (special quality).

LaSalle Steel Corp—see Quanex Corp

United States

Latrobe Steel Co

Head office: 2626 Ligonier Street, Latrobe, PA 15650.
Established: 1913. **Ownership:** The Timken Co.
Works:
■ At head office. ***Steelmaking plant:*** 30-ton electric arc furnaces. ***Refining plant:*** ESR furnaces and VAR unit up to 42" dia, VIM. ***Rolling mills:*** One 32" blooming; three bar; one continuous wire rod (down to 5mm). ***Other plant:*** Forging presses (3,200-ton and 2,000-ton).
Products: Alloy steel (excl stainless) – wire rod, round bars. High speed steels, high alloy hot and cold work die steels, mould steels, shock-resistant steels; VIM high strength steels and bearing steels; corrosion-resistant high-fatique alloys.

Lone Star Steel Co

Head office: 2200 W Mockingbird Lane, Dallas, TX 75235 0888. **Postal address:** PO Box 35888. **Tel:** 214 352 3981. **Telex:** 214 353 6490.
Management: Wm. Howard Beasley III (chairman of the board and chief executive officer); James E. Chenault Jr (president and director); W. Preston Holsinger (executive vice president, chief financial officer and treasurer); Dean G. Wilson (executive vice president, oper ations and engineering); James R. Davisson (vice president, purchasing); W. Byron Dunn (vice president, sales); Bedford F. Foster Jr (vice president and general manager, Specialty Tubing); Warren P. Schneider (vice president, sales, Specialty Tubing); Wil liam A. Osborn (assistant secretary); M. Don Hairston (assistant treasurer). **Established:** 1966. **Capital:** $301,600,000. **No. of employees:** Works – 700. Office – 300. **Ownership:** Lone Star Technologies Inc. **Subsidiaries:** Texas Northern Industries Inc; Hydro-Sonic Systems Inc.
Works:
■ At head office. (Telex: 214 656 6215). Works director: Earl A. Tankesley. ***Coke ovens:*** Koppers underjet battery (78 ovens). ***Sinter plant:*** One Dwight Lloyd 5ft single strand, grate area 550 sq ft. ***Blast furnaces:*** One 27' hearth dia, 44,403 cu ft. ***Steelmaking plant:*** Two 60-ton electric arc furnaces; five 250-ton basic open hearth furnaces. ***Continuous casting machines:*** One 2-strand billet / bloom. ***Rolling mills:*** One 1-stand 2-high slabbing (46 x 80"); one 4-stand 2-high bar (10 x 16"); one 1-stand 4-high reversing wide hot strip (27½ x 72"). ***Tube and pipe mills:*** seamless – two extrusion presses (2⅜-10¾), five drawbenches for mechanical tubing and one Mannesmann-Meer stretch reducing pipe mill. Welded – Two ⅞-16" Yoder resistance and one 24-80" Intercontinental Enterprises submerged arc spiral weld.
■ Ft Collins Pipe Company, 1925 Timberline Road, Ft Collins, CO 80525. (Tel: 303 484 1110). ***Tube and pipe mills:*** one McKay ERW type welded – (2⅜-8⅝).
Products: Carbon steel – ingots, slabs, hot rolled coil (for re-rolling), hot rolled hoop and strip (uncoated), skelp (tube strip), hot rolled sheet/coil (uncoated), seamless tubes and pipes, longitudinal weld tubes and pipes, spiral-weld tubes and pipes, large-diameter tubes and pipes, oil country tubular goods, cold drawn tubes and pipes, hollow bars, precision tubes and pipes, hollow sections. Stainless steel – seamless tubes and pipes, cold drawn tubes and pipes, precision tubes and pipes. Alloy steel (excl stainless) – seamless tubes and pipes, longitudinal weld tubes and pipes, spiral-weld tubes and pipes, large-diameter tubes and pipes, oil country tubular goods, cold drawn tubes and pipes, hollow bars, precision tubes and pipes, hollow sections.
Brands: OCTG – AO Smith, Star Seal Connector.

LTV Steel Co Inc

Head office: 25 West Prospect Avenue, Cleveland, OH 44115. **Postal address:** Box 6778, Cleveland, OH 44101. **Tel:** 216 622 5000.
Management: David H. Hoag (president and chief executive officer, LTV Steel Co Inc); Russell W. Maier (president, LTV Steel Flat Rolled and Bar Co); Robert T. Buck (senior vice president, commercial, LTV Steel Co); Michael J. Hiemstra (senior vice president, finance, LTV Steel Co); Paul N. Wigton (president, LTV Steel Tubular Products Co); J.T. Wood, (vice president – bar, commercial, LTV Steel Flat Rolled and Bar Co).

LTV Steel Co Inc (continued)

Established: 1984. **Ownership:** The LTV Corp, Dallas, TX, USA. **Associates:** LTV Steel Flat Rolled and Bar Co, LTV Steel Tubular Products Co.

Works:

■ Cleveland Works, OH. General manager operations: J. Bush. ***Coke ovens:*** Three Koppers batteries (153 ovens) (annual capacity 777,500 short tons) and two Koppers batteries (126 ovens) (728,000 tons). ***Iron ore pelletizing plant:*** Vertical Shaft furnace (24 furnaces) (8,600,000 tons of pellets). Three ***Blast furnaces:*** C2 27′ hearth dia, 44,870 cu ft (1,000,000 tons), C5 29′6″ hearth dia, 56,143 tons (1,300,000 tons), C6 29′6″ hearth dia, 56,159 tons (1,000,000 tons). ***Steelmaking plant:*** Four basic oxygen – No 1 shop two 200-ton converters (2,700,000 tons), No 2 shop two 250-ton converters (3,500,000 tons); two 185-ton electric arc (22′ shell dia) (840,000 tons). ***Continuous casting machines:*** One 2-strand slab (9 x 73″) (2,700,000 tons). ***Rolling mills:*** Two 2-stand 45 x 90″ universal horizontal/vertical blooming (combined capacity 3,200,000 tons); two wide hot strip – one 80″ with one reversing rougher, one continuous rougher and six finishing stands (coil weight 60,000 lb) (2,000,000 tons) and one 84″ with 5-stand roughing and seven finishing stands (coil weight 90,000 lb) (3,700,000 tons); two cold reduction – one 4-stand 77″ (74″ wide coil) (coil weight 60,000 lb) (1,200,000 tons); three temper/skin pass 84″ 2-stand 78″ wide (coil weight 75,000 lb) (1,200,000 tons), 84″ 2-stand 78″ wide (coil weight 40,000lb) (1,200,000 tons) and 77″ 1-stand 74″ wide (coil weight 60,000lb) (6,000,000 tons). ***Coil coating lines:*** Two electro-galvanizing – one 17-cell 85,000 amp 60″ electroplate (144,000 tons), one 21-cell vertical anode counterflow (43,000 tons).

■ Indiana Harbor Works, IN. General manager, operations – R.A. Veitch. ***Coke ovens:*** Two batteries (60 ovens each) (annual capacity 648,000 tons each). ***Sinter plant:*** One McKee 8 x 168′ (1,200,000 tons). ***Blast furnaces:*** H3 29′-6″ hearth dia, 56,112 cu ft (1,600,000 tons), H4 32′-9″ hearth dia, 69,775 cu ft (1,700,000 tons). ***Steelmaking plant:*** Two 285-ton basic oxygen converters (4,200,000 tons). ***Continuous casting machines:*** 2-strand slab (10 x 88″) (3,500,000 tons). ***Rolling mills:*** 6 roughers, 7 finishing, 84″ max width wide hot strip (coil weight 100,000lb) (4,100,000 tons); 4-high 5-stand 75″ max width cold reduction (coil weight 100,000lb) (1,400,000 tons); 4-high 2-stand 75″ max width temper/skin pass (coil weight 100,000lb) (1,100,000 tons); 6-stand tin cold mill 46″ (coil weight 62,500 tons) (520,000 tons) and 2-stand tin temper 46″ (coil weight 62,500 tons) (520,000 tons). ***Coil coating lines:*** One 48″ wide, 0.006″ min Halogen electrolytic tinning (320,000 tons); one 46″ wide electrolytic tin-free steel (ECCS) (220,000 tons); two hot-dip galvanizing No 1 Sendzimir Wean 62″, horizontal (also for zinc aluminium) (20,000 tonnes) and No 2 Sendzimir Aetna 72″ horizontal (43,000 tonnes). ***Other plant:*** Ingot mould and stool foundry (iron).

■ Hennepin Works, IL. Plant manager: T.T. Ames. ***Rolling mills:*** 84″ 5-stand tandem cold reduction 76″ wide (coil weight 90,000lb) (annual capacity 1,440,000 tons); 84″ 2-stand temper/skin pass 76″ wide (coil weight 90,000lb) (900,000 tons). ***Coil coating lines:*** Sendzimir horizontal Blaw Knox hot-dip galvanizing 60″ wide (430,000 tons).

■ Warren Works, OH. General manager, operations: J.Stack. ***Coke ovens:*** One Koppers battery (85 ovens) (annual capacity 511,400 tons). ***Blast furnaces:*** W1 28′-1″ hearth dia, 53,164 cu ft (1,400,000 tons). ***Steelmaking plant:*** Two 180-ton basic oxygen converters (2,300,000 tons). ***Rolling mills:*** 1-stand 56″ slabbing (1,400,000 tons); 56″ reversing rougher, 6 finishing stands wide hot strip (coil weight 40,000lb) (1,900,000 tons); 54″ 4-stand cold reduction (500,000 tons); 54″ 4-high 1-stand temper/skin pass (400,000 tons). ***Coil coating lines:*** Two Flux type hot-dip galvanizing (40,000lb coils) (230,000 tons); 20-48″ width, hot dip Flux type terne/lead coating (40,000lb coil) (120,000 tons).

■ Flat rolled products – Aliquippa Works, PA. Superintendent, Tin mill: W.E. Stephens. ***Rolling mills:*** One 3-stand 4-high 48″ double cold reduction (coil weight 40,000lb) (annual capacity 340,000 short tons) and one 5-stand 42″ 4-high tandem mill (coil weight 40,000lb) (720,000 tons); No 4 2-stand 48″ temper/skin pass (coil weight 40,000lb) (430,000 tons). ***Coil coating lines:*** electrolytic tinning No 3 Halogen 24

LTV Steel Co Inc (continued)

cells, 180,000 amp (180,000 tons) and No 4 Halogen 12 cells, 120,000 amp (40,000 tons).
■ Bar Division – Canton and Massillon (including Chicago, IL), PO Box 700, Canton, OH 44701. (Tel: 216 438 5201. Telex: 985281). Vice president, operations: J.T. Anderson. ***Steelmaking plant:*** electric arc furnaces – three 200-ton (75,000 KVA Transformers) (annual capacity 1,344,000 short tons) and three 90-ton (20,000 KVA Transformers) (402,000 tons). ***Refining plant:*** One 100-ton (Salem design) AOD (402,000 tons); one 240-ton (own design) ladle furnace (1,200,000 tons); one 240-ton (Stokes design) vacuum degassing (300,000 tons). ***Continuous casting machines:*** One 4-strand Babcock and Wilcox bloom (10 x 8¾" x 14' 6"-27' 6") (468,000 tons). ***Rolling mills:*** Canton – one slabbing (1-stand 2-high 35") (37¾ x 5"-16 x 4 x 28"), blooming and billet – 1-stand 2-high 34" (24 x 15" to 4¾" x 4¾" x 30') combined capacity 1,020,000 tons); Massillon – 2-stand 3-hig billet (6¼-8" dia x 34' and 7¼"-8¼") (456,000 tons); three bar Canton – one 10-stand 2-high 12" (1⅜-3" dia x 45', 1 1/6"-2 9/16") (360,000 tons); 18-stand 2-high 8" (23/64"-1 13/64" dia x 42') (coil weight 2,400lb) (168,000 tons); Massillon one 18" comprising 1-stand 3-high rougher, 3-stand 2-high intermediate and 1-stand 2-high finishing (2⅝-6¼" dia x 45', 2⅝"-7" sq) (480,000 tons); Chicago – 16-stand 2-high 11" (35/64"-2⅓2" dia) (coil weight 2,400lb) (348,000 tons).
■ Youngstown Works. Works director: D.G. Kundich. ***Tube and pipe mills:*** One ERW 4½-8⅝" (annual capacity 150,000 short tons) and one ERW 6-16" (360,000 tons).
Products: Carbon steel – ingots, slabs, blooms, billets, semis for seamless tubemaking, hot rolled coil (for re-rolling), wire rod, round bars, square bars, flats, hexagons, special light sections, medium plates, heavy plates, universal plates, hot rolled sheet/coil (uncoated), cold rolled sheet/coil (uncoated), hot-dip galvanized sheet/coil, electro-galvanized sheet/coil, galvanized corrugated sheet/coil, aluminized sheet/coil, enamelling grade sheet/coil, terne-plate (lead coated) sheet/coil, zinc-aluminium alloy coated sheet/coil, coated sheet/coil electrical non-oriented silicon sheet/coil, blackplate, electrolytic tinplate (single-reduced), electrolytic tinplate (double-reduced), tin-free steel (ECCS), bright wire, black annealed wire, longitudinal weld tubes and pipes, large-diameter tubes and pipes, oil country tubular goods, galvanized tubes and pipes, cold drawn tubes and pipes, hollow bars. Stainless steel – billets, round bars, square bars, flats, hexagons, special light sections, wire, longitudinal weld tubes and pipes, cold drawn tubes and pipes, hollow bars. Alloy steel (excl stainless) – ingots, slabs, blooms, billets, round bars, square bars, flats, hexagons, special light sections, longitudinal weld tubes and pipes, large-diameter tubes and pipes, oil country tubular goods, cold drawn tubes and pipes. Drainage products, culvert; special metals products produced from consumable electrode vacuum remelt (CEVM) and electroslag remelt (ESR) steels – alloy, stainless, special alloys; rolled products – hot rolled or cold finished, rounds, squares, flats; forged products – rounds, squares, rectangles; ingots – CEVM (rounds); ESR (rounds).
Sales offices/agents: District sales offices: Atlanta, Chicago, Cincinnati, Cleveland, Dallas, Detroit, Kansas City, Memphis, Philadelphia, Pittsburgh, Indianapolis; other sales offices – Indianapolis, Los Angeles, Milwaukee, New York, St Louis, International (Cleveland).

Lukens Steel Co

Head office: 50 S. First Avenue, Coatesville, PA 19320. **Tel:** 215 383 2000.
Established: 1810. **Ownership:** Division of Lukens Inc.
Works:
■ At head office. ***Steelmaking plant:*** Two 150-ton electric arc furnaces. ***Refining plant:*** One 150-ton DH vacuum degassing; one Lectrefine consumable electrode unit. ***Continuous casting machines:*** One 1-strand slab. ***Rolling mills:*** Three heavy plate – one-stand (120"), one-stand (140"), and one-stand (206").
■ Conshohocken, PA 19420 (Tel: 215 825 6020). ***Rolling mills:*** One 4-high 2-stand (110") heavy plate.
Products: Carbon steel – ingots; slabs; medium plates; heavy plates in thickness up to

Lukens Steel Co (continued)

25", width up to 195"; clad plates. Stainless steel – clad plates. Alloy steel (excl stainless) – ingots; slabs. Clad plates in nickel, Monel, Inconel and copper; pressed and spun heads and fittings; sheared and flame cut components.
Brands: Lectrefine; Frostline; Fineline; Surefoot.
Expansion programme: In the late summer of 1987 Lukens was contemplating the purchase of the arc furnace and slab caster from Phoenix Steel's closed Claymont works.

Marathon LeTourneau Co, Longview Division

Head office: Longview, TX 75601. **Postal address:** PO Box 2307. **Tel:** 214 753 4411. **Telex:** 730371 marlet gv.
Ownership: Marathon Manufacturing Co, Houston TX. **Annual capacity:** Finished steel: 75,000 short tons.
Works:
■ At head office. ***Steelmaking plant:*** Two 25-tonne electric arc furnaces. ***Rolling mills:*** One slabbing and heavy plate 1-stand 4-high (154 x 36") (annual capacity 135,000 tons).
Products: Carbon steel – ingots, slabs, heavy plates. Alloy steel (excl stainless) – ingots, slabs, heavy plates.

Marion Steel Co

Head office: 912 Cheney Avenue, Marion, OH 43302. **Tel:** 612 383 4011.
Works:
■ At head office. ***Steelmaking plant:*** Two 40-50-ton electric arc furnaces (annual capacity 350,000 short tons). ***Continuous casting machines:*** One 4-strand Olsson billet. ***Rolling mills:*** One 12-stand bar (300,000 short tons).
Products: Carbon steel – reinforcing bars No 4-11; round bars $\frac{7}{8}$-1$\frac{1}{2}$"; flats $\frac{1}{4}$" x 2"-$\frac{1}{2}$" x 4"; light angles $\frac{1}{2}$-3". Sign posts.
Additional information: This works formerly belonged to Armco Steel.

Maruichi American Corp

Head office: 11529 S Greenstone Avenue, Sante Fe Springs, Los Angeles, CA 90670. **Tel:** 213 946 1881. **Telex:** 194759. **Telecopier (Fax):** 213 941 0047.
Management: President – K. Shiota; managers – K. Marumoto (general), S. Nakajima (plant). **Established:** 1975. **Capital:** $7,500,000. **No. of employees:** Works – 42. Office – 14. **Annual capacity:** Finished steel: 108,000 tonnes.
Works:
■ At head office. ***Tube and pipe mills:*** welded – one 6" Abbey Etna (annual capacity 40,000 tonnes); two 2" Abbey Etna (50,000 tonnes) and one 2" Maruichi (18,000 tonnes).
Products: Carbon steel – longitudinal weld tubes and pipes $\frac{1}{2}$ to 6" (1986 output 50,000 tonnes).

McDonald Steel Corp

Head office: 100 Ohio Avenue, McDonald, OH 44437. **Tel:** 216 530 9118.
Management: David S. Houck (president). **Established:** 1981. **Capital:** $3,000,000. **No. of employees:** Works – 130. Office – 10. **Annual capacity:** Finished steel: 150,000 short tons.
Works:
■ At head office. ***Rolling mills:*** One 11-stand 8" Belgian Looping light section (coil weight 200lb) (annual capacity 30,000 short tons) and 11-stand 14" cross country (120,000 tons).
Products: Carbon steel – flats, light angles, light channels, special light sections. Stainless steel – flats, special light sections. Alloy steel (excl stainless) – flats, special light sections.

United States

McLouth Steel Products Corp

Head office: 1650 W Jefferson, Trenton, MI 48183. **Tel:** 313 285 1200. **Telex:** 235581. **Telecopier (Fax):** 313 675 4491.
Management: Cyrus Tang (chairman of the board); Frank D. Sove (executive vice president, finance and treasurer); Thomas G. Whittingham (vice president, commercial); Michael Tang (secretary); David H. Clark (president, chief executive officer); Edward W. Stotz (vice president, quality control) and Gerald C. Jacobson (controller). **No. of employees:** 2,000. **Ownership:** Tang Industries Inc. **Annual capacity:** Raw steel: 2,000,000 short tons.
Works:
■ 1491 West Jefferson Avenue, Trenton, MI 48183. Tel: 313 285 1200. Works director: H. Herin. ***Blast furnaces:*** Two 30' hearth dia, 57,238 cu ft (annual capacity 2,000,000 tons). One 150-ton iron desulphurisation station (1,000,000 tons). ***Steelmaking plant:*** Five 120-ton basic oxygen converters (2,000,000 tons) and two 200-ton electric arc furnaces (350,000 tons). ***Refining plant:*** 100-ton ladle furnace (1,000,000 tons). ***Continuous casting machines:*** One 4-strand Concast slab (12 x 36, 12 x 44 and 12 x 57") (2,400,000 tons). ***Rolling mills:*** One 60" wide hot strip (2,400,000 tons).
■ Gibraltar Works. ***Rolling mills:*** One 60" tandem cold strip.
Products: Carbon steel – cold rolled hoop and strip (uncoated), hot rolled sheet/coil (uncoated), cold rolled sheet/coil (uncoated).

Mercury Stainless Corp

Head office: 475 West Allendale Drive, Wheeling, IL 60090. **Tel:** 312 459 6060.
Works:
■ Massillon, OH (Tel: 216 837 6900). ***Rolling mills:*** cold reduction; temper/skin pass.
Products: Stainless steel – cold rolled hoop and strip (uncoated) up to 24"; medium plates; cold rolled sheet/coil (uncoated) up to 60"; bright annealed sheet/coil up to 52".
Additional information: Mercury stainless is a service centre company which had been contemplating building a CR stainless mill in Wisconsin, but instead purchased the Republic Stels Enduro mill in Massillon, OH.

Metalsource-Triumph

Head office: 8687 S 77th Avenue, Bridgeview, Cook, IL **Postal address:** PO Box 96331. **Tel:** 312 598 5100. **Telecopier (Fax):** 312 430 5732.
Management: President – Thomas J. Fournier; vice president, manufacturing – Richard Carney; managers – Barbara Reisse (sales), Robert Hess (purchasing). **Established:** 1960. **No. of employees:** Works – 48. Office- 24. **Ownership:** Alco Standard Corp. **Subsidiaries:** Metalsource-Great Western; Metalsource-Wheelock Lovejoy; Metalsource-Good Steel; Metalsource-Commerical Heat Treating.
Works:
■ At head office. ***Coil coating lines:*** Two 48" and 13" horizontal electro-galvanizing (annual capacity 60,000 short tons).
Products: Carbon steel – electro-galvanized sheet/coil.
Brands: Trizin, Trizin Tribrite and Trizin Trichrome – electro-galvanized.

Metaltech

Head office: 2400 Second Avenue, Pittsburgh, PA. **Tel:** 412 391 0483. **Telecopier (Fax):** 412 391 0402.
Management: Chairman and chief executive officer – G. Watts Humphrey Jr; president and chief operating officer – Edwin H. Gott Jr; managers – Wilson J. Farmerie (plant), Robert W. Riordan (administration), James H. Donahue (sales), Mark M. Sherwin (material control), Armand A. Chick (operations and maintenance); controller – Gerald

Metaltech (continued)

L. Gilbert. **Established:** 1984. **No. of employees:** Works – 56. Office – 21. **Annual capacity:** Finished steel: 180,000 short ton.
Works:
■ At head office. ***Coil coating lines:*** One Sendzimir Aetna hot-dip galvanizing (electric furnace) (annual capacity 180,000 tons).
Products: Carbon steel – hot-dip galvanized sheet/coil 24-48″ coils; 30,000 lb max; 60-168″ sheets (1987 output 163,000 tons).

Milton Manufacturing Co

Head office: 230 Lower Market Street, Milton, PA 17847 0337. **Postal address:** PO Box 337. **Tel:** 717 742 7661.
Management: Chairman – Fred W. Cormell; president – Peter H. Davidson; vice presidents – Ralph J. Webb (executive, operations), Edward K. Boynton (administration), Charles A. Thomas (finance), Anthony T. Boldwrian (sales). **Established:** 1893 (originally), 1982 (present owners). **No. of employees:** Works – 375. Office – 25. **Annual capacity:** Raw steel: 200,000 short tons. Finished steel: 180,000 tons.
Works:
■ At head office. Works manager: John V. Benoit. ***Steelmaking plant:*** Three 22-ton 12′ dia electric arc (annual capacity 200,000 tons). ***Continuous casting machines:*** One 3-strand Danieli billet (4 x 4″) (200,000 tons). ***Rolling mills:*** bar – six roughing and five finishing stands, 18″ and 12″ (180,000 tons).
Products: Carbon steel – ingots (discontinued 1986) (output 6,829 ton); billets 4 x 4″ (in-house use) (126,597 ton); reinforcing bars Nos 3,4,5,6,7,8,9,10,14, and 18 (127,622 ton); round bars 0.468 to 1.0″ (16,036 ton).
Expansion programme: Electric furnace upgrade (1988), reheat furnace modernisation (1990), roughing mill improvement (1992).

Missouri Rolling Mill Corp

Head office: 6800 Manchester Avenue, St Louis, MO 63143. **Tel:** 314 645 3500.
Established: 1921. **No. of employees:** 250. **Ownership:** Leggett & Platt Inc, Carthage, MO.
Works:
■ At head office. ***Rolling mills:*** Two bar – one 3-high 5-stand 9″ and one (rail slitting) 3-high 4-stand 14″.
Products: Carbon steel – reinforcing bars, round bars, light angles. Fence posts, sign posts, delineator posts.

Moltrup Steel Products Co

Head office: 14th Street and Second Avenue, Beaver Falls, PA 15010. **Tel:** 412 846 3100.
Established: 1914. **No. of employees:** About 170.
Products: Carbon steel – bright (cold finished) bars in rounds, squares, hexagons and flats. Alloy steel (excl stainless) – bright (cold finished) bars in rounds, squares, hexagons and flats.

MP Metal Products Inc

Head office: W 1250 Elmwood Road, Ixonia, Jefferson, WI 53036. **Postal address:** PO Box 170. **Tel:** 414 261 9650.
Management: Donnell H. Geib (vice president). **Established:** 1968. **No. of employees:** Works – 20. Office -5.
Works:
■ At head office. Plant for the production of cold formed sections, including channels,

United States

MP Metal Products Inc (continued)

angles, hollow sections and special sections.
Products: Carbon steel – cold roll-formed sections (from strip), hollow sections. Stainless steel – cold roll-formed sections (from strip).

National Forge Co

Head office: 100 Front Street, Irvine, PA 16329. **Tel:** 814 563 7522. **Telex:** 091 4446.
Management: R.O. Wilder (chairman and chief executive officer); G.H. Wells (president and chief operating officer) and R.J. Fulton (executive vice president, materials and components group). **Established:** 1915. **No. of employees:** Works – 900. Office – 250. **Annual capacity:** Raw steel: 125,000 short tons. Finished steel: 70,000 tons.
Works:
■ 1341 W 16th Street, Erie, PA 16502. (Tel: 814 452 2300). Works director: R.J. Pike. ***Steelmaking plant:*** One 75-ton and one 35-ton electric arc furnace (annual capacity 125,000 tons). ***Refining plant:*** One 30-40" ESR (10,000 tons); one 125-ton ladle furnace furnace (125,000 tons), three 14-60" VAR units (10,000 tons). One 2,500-ton and one 4,000-ton open die forge press.
■ Irvine PA 16329. Works director: W.E. Erickson. One 4,000-ton multiple ram closed die forge press.
Products: Stainless steel – ingots, slabs, blooms, billets, forgings. Alloy steel (excl stainless) – ingots, slabs, blooms, billets (combined output 20,000 tons); round bars; square bars; forgings (output 30,000 tons).

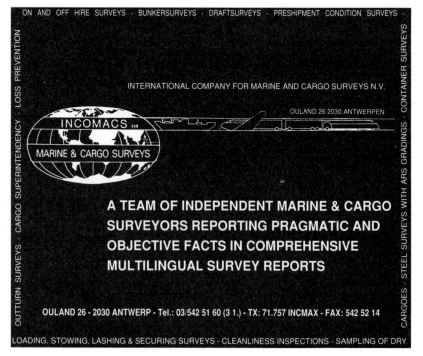

National Steel Corp

Head office: National Steel Center, 20 Stanwix Street, Pittsburgh, PA 15222. **Tel:** 412 394 4100.
Management: Board chairman and chief executive officer – H.M. Love; president and chief operating officer – K. Hagiwara; executive vice presidents – M. Ito, R.H. Doerr; president, National Mines – V.R. Knapp; vice presidents – Y. Tokumitsu (distribution and services, assistant to the president), E. Krivoshia, Jr (general counsel and secretary), L.L. Symons (finance), R.P. Coffee (human resources), R.S. Brzenk (information systems), R.E. Westergren (marketing and sales); treasurer – J.W. Hoekwater, J. Marentes Manzanilla (purchasing), J. Brogeras Oliva (personnel), J. Peña Flores (techn
Subsidiaries: Great Lakes Div, Detroit, MI; Midwest Division, Portage, IN; Granite City Division, Granite City, IL; National Mines Corp, Lexington, KY.

Great Lakes Division 1 Quality Drive, Ecorse, MI 48229.
Management: Vice president – J.N. Howell (general manager); directors – L.L. Judd (human resources), J. Householder (plant engineering and maintenance), D. Simmons (quality assurance), S. Sczytco (financial services), M. Christy (production planning and scheduling), J. oadman (operations support purchasing and traffic).
Works:
■ Blast Furnace Division, River Rouge, MI. ***Coke ovens:*** 78 Wilputte underjet type ovens. ***Sinter plant:*** One single strand 12′ x 199′6″, grate area 2,400 sq ft. ***Blast furnaces:*** one 30′6″ hearth dia, 60,780 cu ft; one 29′ hearth dia, 55,468 cu ft; one 28′ hearth dia, 53,252 cu ft.
■ At head office. ***Steelmaking plant:*** Two 240-ton basic oxygen converters; two 180-ton electric arc furnaces. ***Refining plant:*** One metallurgy ladle furnace facility (24 heats/day). ***Continuous casting machines:*** slab one 1-strand Concast (104 x

United States

National Steel Corp (continued)

9.5") (annual capacity 2,184,000 short tons) and one 2-strand Concast variable width (24-74 x 9.5") (2,280,000 tons). ***Rolling mills:*** One 1-stand 2-high 45 x 90" slabbing (3,300,000 tons); one wide hot strip comprising 1-stand 2-high 47 x 81", 4-stand 4-high 44 and 60 x 80", 7-stand 4-high 29 and 60 x 80" (combined capacity 4,400,000 tons); one 5-stand 4-high 24 and 60 x 80" continuous cold reduction (1,360,000 tons); three temper/skin pass - one 1-stand 2-high 42 x 80" continuous hot rolled (242,000 tons) and two 1-stand 4-high 20½ and 53 x 93" continuous (1,000,000 tons). ***Coil coating lines:*** One 74" vertical type electro-galvanizing (220,000 tons).

Granite City Division 20th and State Streets, Granite City, IL 62040.
Tel: 618 451 3456.
Management: Vice president – Bill Swanson (general manager); assistant general manager – Hank Sangster; directors – John Wandling (ironmaking), Jim Squires (steelmaking), Harry Wright (hot rolled products), Larry Siebenberger (cold roll and coated), Fred Steinkuehler (facility services), Tom Huyer (financial services), Roy Paulsen (human resources), Jim Little (customer services), Jim Eisenbeis (material control), Jeff Jolly (information services), Bill Wilson (internal co-ordination).
Works:
■ At head office. ***Coke ovens:*** Two McKee Otto underjet batteries (90 ovens) (annual capacity 580,000 short tons). ***Sinter plant:*** One 8", 1,024 ft² grate area (not in service since 1982). ***Blast furnaces:*** One 27'3" hearth dia, 50,652 cu ft and one 27'3" hearth dia, 50,490 cu ft (combined capacity 1,769,000 tons). ***Steelmaking plant:*** Two 235-ton (one LBE, one LD) basic oxygen converters (2,162,000 tons). ***Refining plant:*** One 245-ton NK-AP ladle furnace (to match caster). ***Continuous casting machines:*** One 1-strand SMS Concast slab (8-¾ x 49" to 86") (1,387,000 tons). ***Rolling mills:*** One 1-stand 2-high slabbing (45 x 115") (1,770,000 tons); one 80" continuous wide hot strip comprising 45 x 17" vertical edger, 2-stand 2-high 50 x 83", 2-stand 4-high 44 x 83" work roll and 60 x 80" back-up roll, and 7-stand 4-high 28½ x 83" and 60 x 80" (2,134,000 tons); one 4-stand 4-high cold reduction (19 x 56" work roll, 53 x 51" backup roll (549,000 tons); one 1-stand 4-high temper/skin pass 17" work roll, 49½ x 56" backup roll (525,000 tons). ***Coil coating lines:*** Two continuous hot-dip galvanizing one 42" wide (No 7) (197,000 tons) and one 48" wide (No 6) (225,000 tons).

National Mines Corp 333 West Vine Street, Lexington, KY 40507.
Management: President – V.R. Knapp; vice presidents – D.J. Patton (operations and engineering), H.L. Mays (sales), N. Willard (human resources); controller – L.F. Tonkel. **Established:** 1952 (incorporated).
Works:
■ Coal mines, properties and preparation plants in Pennsylvania, West Virginia and Kentucky. Production capability of 5,000,000 short tons.

Midwest Division US Route 12, Portage, IN 46368.
Management: Vice president – D.A. Chatfield (general manager); division controller – L.A. Menor; directors – G. Orr (human resources), J.J. Miller (materials control), K.E. Beall (technology), P.J. Ferrari (cold roll product operation), J.T. Welsh (tin mill production operation), R.A. Strikwerda (galvanize product operation), W.D. Schuetz (maintenance and service), R.C. Earl (organizational effect); sales manager – J.T. Strick (containers controller);
Works:
■ At head office. ***Rolling mills:*** Four continuous cold reduction one 5-stand 4-high 23 x 56 x 52" (annual capacity 600,000 short tons), one 5-stand 4-high 23 x 60 x 80" (1,400,000 tons), one 2-stand 4-high 21 x 56 x 52" (300,000 tons), one 2-stand 4-high 21 x 56 x 54" (600,000 tons); one temper/skin pass – 1-stand 4-high 21 x 56 x 80" continuous cold finishing sheet (750,000 tonnes. ***Coil coating lines:*** One electrolytic tinning; one tin-free steel (ECCS) and two hot-dip galvanizing.

Products: Carbon steel – medium plates, hot rolled sheet/coil (uncoated), cold rolled sheet/coil (uncoated), hot-dip galvanized sheet/coil, electro-galvanized sheet/coil,

National Steel Corp (continued)

enamelling grade sheet/coil, colour coated sheet/coil, electrical grain oriented silicon sheet/coil, blackplate, electrolytic tinplate (single-reduced), tin-free steel (ECCS). Low carbon electrical sheets; Galvalume and Zincrometal; Culvert sheets; Siding; NSQ, GLX, NAPAC and NAX high strength steels; Nax 9100 alloy steels; NACOR lamination steels; Coal chemicals; metallurgical, steam and industrial coal.
Brands: Strongbarn, Strongpanel.

National-Standard Co

Head office: 1618 Terminal Road, Niles, MI 49120. **Tel:** 616 8100. **Telex:** 258453.
Works:
■ Athenia Steel Division, Clifton, NJ. Plant for the production of tempered and polished flat spring steel motor, typewriter and clock springs, piston rings, hand saws etc.
Products: Stainless steel – wire. Tempered and polished flat spring steel.
Additional information: Other plants are at Mount Joy, PA (spring and specialty wire); Niles, MI (stainless steel and alloy wire, music wire, copper and nickel plated wire etc) Rubber Industry Division; and Welding Products Division, Stillwater, OK and Columbiana, AL (tyre bead and welding wire); and Woven Products Division, Corbin, KY (industrial wire cloth).

Nelsen Steel & Wire Co Inc

Head office: 9400 West Belmont Av, Franklin Park, Cook, IL 60131. **Tel:** 312 671 9700. **Telex:** TWX 910 227 0063. **Telecopier (Fax):** 312 671 9700.
Management: President – C.D. Nelsen; vice president/treasurer – L.A. Nelsen; vice president/secretary – C.D. Nelsen; vice president/general manager – J.R. McVicker. **Established:** 1939. **No. of employees:** Works – 200. Office – 35. **Subsidiaries:** Nelsen Steel Co (100%) (producer of cold finished bars); Nelsen Wire Co (100%) (producer of cold heading wire) (see below). **Annual capacity:** Finished steel: 150,000 short tons. Finished wire: 90,000 tons.
Works:
■ Nelson Steel Co (at head office address). Plant for the production of cold finished steel bars (rounds, sqaures, hexagons, flats): drawn, annealed, turned, high strength, chamfered, etc (0.060-4½") (annual capacity 150,000 tons).
■ Nelsen Wire Co, 1060 East Irving Park Road, Bensenville, IL 60106. (Tel: 312 595 4960. Telex: 910 227 0063). Vice president and general manager: T. Rodhouse. Plant for the production of spherodized annealed cold heading and scrapless nut quality carbon and alloy wire in coil (0.030-2") and custom processing of rods and wire (annual capacity 90,000 tons).
Products: Carbon steel – round bars, square bars, flats, hexagons, bright wire, black annealed wire. Stainless steel – round bars, square bars, flats, hexagons, wire. Alloy steel (excl stainless) – round bars, square bars, flats, hexagons, wire.
Brands: Viking "100" (100,000 psi minimum yield strength stress relieved bars).
Expansion programme: Nelsen Steel – modernisation of draw bench lines for increased speeds (October 1, 1987) and installation of a continuous coil to bar line (January 1, 1988).

New Jersey Steel Corp

Head office: North Crossman Road, Sayreville, NJ 08872. **Postal address:** PO Box 11. **Tel:** 201 721 8800. **Telex:** 138959.
Management: President – R.J. Pasquarelli (chief executive officer); vice presidents- W.P. Mullen (operations), G.L. Hudson (sales and marketing), J. Sullivan (industrial relations), W.J. Maier (executive), P. Roik (finance and tresurer). **Established:** 1972. **No. of employees:** Works – 375. Office – 35. **Ownership:** Von Roll Ltd, Switzerland (through majority ownership of Monteforno SA, Switzerland). **Subsidiaries:** New

New Jersey Steel Corp (continued)

Jersey Fab Div (100%) (rebar fabrications). **Annual capacity:** Raw steel: 440,000 short tons. Finished steel: 420,000 tons.
Works:
■ At head office. Works director W.P. Mullen. ***Steelmaking plant:*** One 75-tonne electric arc furnace (annual capacity 440,000 tons). ***Continuous casting machines:*** One 5-strand Danieli billet (440,000 short tons). ***Rolling mills:*** One 14-stand 12" Danieli bar (420,000 tons).
Products: Carbon steel – billets 115 x 115mm (1986 output 440,000 tons); reinforcing bars No 3 to No 11 (440,000 tons); hand rail (2,000 tons).

Newman Crosby Steel Inc

Head office: 10 Dean Street, Pawtucket, RI 02861. **Tel:** 401 725 3750.
Established: 1910. **No. of employees:** 170.
Works:
■ At head office. ***Rolling mills:*** Four cold strip – one reversing Bliss 13", two one-way 10" and one cluster 6" (combined annual capacity 10,000 short tons); one 4-stand Waterbury Farrel wire flattening mill (2,000 tons). One continuous copper coating coil line (approx 2,000 tons). ***Coil coating lines:*** One continuous copper coating (about 2,000 short tons).
Products: Carbon steel – cold rolled hoop and strip (uncoated) 0.005-0.180" up to 13" wide. Flattened wire $\frac{1}{8}$-$\frac{1}{2}$" wide.

Newport Steel Corp

Head office: Ninth and Lowell Streets, Newport, KY 41072. **Postal address:** PO Box 1670. **Tel:** 606 292 6000. **Telex:** TWX 810 541 8512.
Management: D.L. Eggleston (president). **Established:** 1981. **Capital:** $35,000,000. **No. of employees:** Works – 450. Office – 150. **Subsidiaries:** Imperial Adhesives, Blue Ash, OH (100%) (adhesives); Erlanger Tubular, Tulsa, OK (100%) (pipe finishing and treating). **Annual capacity:** Raw steel: 500,000 short tons. Finished steel: 450,000 tons.
Works:
■ At head office. ***Steelmaking plant:*** Three 100-ton electric arc furnaces (19' hearth dia) (500,000 tons). ***Rolling mills:*** One wide hot strip – one roughing, one intermediate and one reversing finishing train (50") (coil weight 30,000lb) (456,000 tons). ***Tube and pipe mills:*** welded – one 8" and one 16" ERW (450,000 tons).
Products: Carbon steel – longitudinal weld tubes and pipes $4\frac{1}{2}$, $6\frac{5}{8}$, $8\frac{5}{8}$, $10\frac{3}{4}$, $12\frac{3}{4}$, $13\frac{3}{8}$" (Output 70,826 tons); spiral-weld tubes and pipes $10\frac{3}{4}$, $12\frac{3}{4}$, 14" (10,481 tons); oil country tubular goods $4\frac{1}{2}$, $5\frac{1}{2}$, 7, $8\frac{5}{8}$, $9\frac{5}{8}$, $10\frac{3}{4}$, $12\frac{3}{4}$" (62,973 tons).

North Star Steel Co (NSS)

Head office: Dept 51, Minneapolis, MN 55440. **Postal address:** PO Box 9300. **Tel:** 612 475 5471. **Telex:** 290625 cargill mps a nss.
Management: Chairman of the board – C.H. Humphries; president – Robert A. Garvey; executive vice presidents – G.H. Geiger (technology), B.F. Hill (sales and marketing); vice presidents – J.F. Smythe (industrial relations), H.L. Anderson (controller), R.A. Jakse, J.C. Gay (general manager, St Paul and Duluth, MN), E.J. Fox (general manager, Monroe, MI), Steve Filips (general manager, Beaumont, TX), G.A. Giovannetti (general manager, Youngstown, OH); general managers – Pete Anderson (Wilton , IA), Harold Olden (Calvert City, KY). **Established:** 1966. **Ownership:** Wholly-owned subsidiary of Cargill, Inc. **Subsidiaries:** Magnimet, PO Box 868, Monroe, MI (scrap processing and brokerage; shredding and scrap production and scrap sales). **Annual capacity:** Raw steel: 2,400,000 short tons. Finished steel: 2,500,000 tons.
Works:
■ PO Box 64189, St Paul, MN 55164 (Tel: 612 735 2100. Telex: 290625 cargill mps a stp). General manager: James C. Gay. ***Steelmaking plant:*** Two 75-ton electric

North Star Steel Co (NSS) (continued)

arc furnaces (annual capacity 450,000 short tons). ***Continuous casting machines:*** Two 3-strand curved mould billet (450,000 short tons). ***Rolling mills:*** 12" and 16" continuous bar (450,000 short tons).
■ PO Box 468, Calvert City, KY 42029 (Tel: 502 395 3100). General manager: Harold Olden. ***Rolling mills:*** One 5-stand heavy section / medium section (250,000 short tons).
■ PO Box 749, Wilton, IA 52778 (Tel: 319 732 3231. Telex: 290625 cargill mps a is). General manager: Pete Anderson. ***Steelmaking plant:*** One 80-ton electric arc (annual capacity 300,000 short tons). ***Continuous casting machines:*** One 3-strand curved mould billet (300,000 short tons). ***Rolling mills:*** One 12" continuous bar (300,000 short tons).
■ 3000 East Front Street, Monroe, MI 48161 (Tel: 313 243 2446. Telex: 290625 cargill mps a ms). General manager: Edward J. Fox. ***Steelmaking plant:*** One 120-ton electric arc furnace (500,000 short tons). ***Continuous casting machines:*** One 4-strand curved mould billet (500,000 short tons). ***Rolling mills:*** One 12" continuous bar.
■ PO Box 2390, Beaumont, TX 77701 (Tel: 409 768 1211. Telex: 290625 cargill mps a ts). General manager: Steve Filips. ***Steelmaking plant:*** Two 120-ton electric arc furnaces (750,000 short tons). ***Continuous casting machines:*** Two 4-strand curved mould billet. ***Rolling mills:*** One 2-strand. Wire rod (700,000 short tons).
■ 2669 West Federal Street, Youngstown, OH 44510 (Tel: 216 742 3000. Telex: 290625 cargill mps a os). General manager: Gary A. Giovannetti. ***Steelmaking plant:*** Two 85-ton electric arc furnaces (400,000 short tons). ***Continuous casting machines:*** One 3-strand curved mould billet (400,000 short tons). ***Tube and pipe mills:*** seamless One retained mandrel (350,000 short tons).
■ Universal Tubular Services, PO Box 24659, Houston, TX 77229 (Tel: 713 456 9716). Manager: Jim Thompson. Tubing and casing finishing facilities.
Products: Carbon steel – wire rod :FR:7/32-$\frac{1}{2}$" in coil; reinforcing bars No 3-18, 4-11" (incl MBQ); round bars :FR:9/16-3"; square bars RCSQ 1-2$\frac{1}{4}$"; flats $\frac{3}{4}$ x $\frac{1}{4}$"-4 x 1", 5-12"; light angles 2 x 2 x $\frac{3}{16}$-5 x 3$\frac{1}{2}$" x $\frac{1}{2}$"; medium angles 6 x 4"-8 x 4"; medium joists 4-10"; medium channels 3-12" (including special sections); seamless tubes and pipes 4$\frac{1}{2}$-9$\frac{5}{8}$" od casing; oil country tubular goods. Alloy steel (excl stainless) – round bars. Nails; forged grinding balls.
Sales offices/agents: These is a forge shop at Duluth, MN (Tel: 218 722 5614) (producing grinding balls and a rail plant at New Orleans, LA (Tel: 504 733 2244).
Expansion programme: A pipe sizing mill (5$\frac{1}{2}$-9$\frac{5}{8}$") is planned for Youngstown.

Northwest Pipe & Casing Co

Head office: 12005 N. Burgard, Portland, OR 97203. **Postal address:** PO Box 03149. **Telecopier (Fax):** 503 285 2913.
Management: William R. Tagmyer (president), Virgil C. Vandenburg (sales manager). **Established:** 1966. **No. of employees:** Works – 150. Office – 30. **Annual capacity:** Finished steel: 90,000 short tons.
Works:
■ At head office. ***Tube and pipe mills:*** welded – two 3-10" ERW, two 6-16" ERW, one 30-90" submerged arc, one 17-48" spiral, one 16-146" spiral and one 18-146" spiral (combined annual capacity 90,000 tons).
Products: Carbon steel – pipe piling, longitudinal weld tubes and pipes, spiral-weld tubes and pipes, large-diameter tubes and pipes, galvanized tubes and pipes. Stainless steel – longitudinal weld tubes and pipes. Pipeline linings and coatings – cement mortar, polyethylene tape, coal tar enamel, specialty paints.

Northwestern Steel & Wire Co

Head office: 121 Wallace Street, Sterling, IL 61081. **Tel:** 815 625 2500. **Telex:** 9106423894. **Telecopier (Fax):** 815 625 2500 (Ext 369).
Management: Chairman of board – Peter W. Dillon; president and chief executive –

United States

Northwestern Steel & Wire Co (continued)

Robert M. Wilthew; vice presidents – Edward G. Maris (finance, secretary and treasurer), Charles H. Biermann (sales, steel division), Tom L. Galanis (operations, steel division), David C. Oberbillig (sales, wire products division), Michael J. Mullen (operations, wire products division), Robert W. Martin (purchasing). **Established:** 1879. **No. of employees:** Works – 2,400. Office – 400. **Subsidiaries:** None. **Associates:** None. **Annual capacity:** Raw steel: 2,400,000 short tons.
Works:
■ ***Steelmaking plant:*** Three 400-ton electric arc furnaces (annual capacity 2,400,000 short tons). ***Continuous casting machines:*** One 8-strand billet (1,000,000 short tons); one 6-strand bloom (1,400,000 short tons). ***Rolling mills:*** One 46" blooming (1,500,000 short tons). One 24" medium section (440,000 short tons); one 14" light section (400,000 short tons); one 12" wire rod (400,000 short tons); ***Other plant:*** Wire mill.
Products: Carbon steel – ingots, slabs, blooms, billets, wire rod, flats, light channels, medium angles, medium channels, wide-flange beams (medium), bright wire, black annealed wire, galvanized wire (plain), galvanized wire (barbed).
Additional information: In August 1987 the company was negotiating to take over, with an unnamed partner, the Armco wide flange beam mill in Texas, which has a design capacity of 600,000 short tons a year and which operated only from 1972-1977.

Nucor Corp

Head office: 4425 Randolph Road, Charlotte, NC. **Tel:** 704 366 7000.
Management: Directors – Kenneth Iverson (chief executive officer), Dave Aycock (chief operating officer). **Established:** 1958. **No. of employees:** 4,200. **Subsidiaries:** Divisions – Nucor Steel Divisions (see below); Vulcraft Divisions, producing steel joists, joist girders, steel deck and standing seam roof, at Florence (SC), Norfolk (NE), Fort Payne (AL), Grapeland (TX), Saint Joe (IN) and Brigham City (UT). Nucor Cold Finish Divisions at Norfolk (NE), Darlington (SC) and Brigham City (UT). Nucor Grinding Balls Division at Brigham City (UT). Nucor Fastener Division at Saint Joe (IN). Nucor Machined Products Inc (NC). Research Chemicals Division; producing rare earth oxides, metals and salts. **Annual capacity:** Raw steel: 2,400,000 short tons. Finished steel: 2,200,000 tons.
Works:
■ Box 525, Darlington, SC. (Tel: 803 393 5841). Works director: Bill Dauksch. ***Steelmaking plant:*** Three 11' 6" and two 13' electric arc furnaces (annual capacity 550,000 tons). ***Continuous casting machines:*** billet One 2-strand curved mould – (4¾" square) and one 2-strand curved mould (7½ x 5") (550,000 tons); one Hazelett for thin slab (on trial at time of going to press). **Rolling mills:** one 1-stand rougher, 9-stand in-line finishing and one 5-stand rougher, 9-stand in-line finishing (550,000 tons).
■ Box 309, Norfolk, NE. (Tel: 402 644 0200). Works director: John Doherty. ***Steelmaking plant:*** Four 13-0" electric arc furnace (annual capacity 550,000 tons). ***Refining plant:*** One 70-ton ladle furnace. ***Continuous casting machines:*** One 2-strand curved mould billet (5¼ sq) and one 2-strand curved mould billet 4¼ sq) (550,000 tons). **Rolling mills:** One 1-stand rougher, 10-stand in-line finishing, one 4-stand rougher and 9-stand in-line finishing (500,000 tons).
■ Box 126, Jewett, TX. (Tel: 214 626 4461). Works director: Bill Kontor. ***Steelmaking plant:*** Four 13-0" electric arc furnace (annual capacity 600,000 short tons). ***Refining plant:*** One 75-ton ladle furnace. ***Continuous casting machines:*** One 2-strand curved mould billet (8½ x 5") and one 2-strand curved mould billet (5¼ sq) (600,000 tons). **Rolling mills:** One 1-stand rougher and 9-stand finishing; and one 4-stand rougher, 7-stand finishing (500,000 tons).
■ Box 448, Plymouth, UT. (Tel: 811 458 3691). Works director: John Correnti. ***Steelmaking plant:*** Two 15' electric arc (annual capacity 700,000 tonnes approx). ***Continuous casting machines:*** One 4-strand billet (6¼" sq). **Rolling mills:** Two including one 1-stand rougher 9-stand in-line finishing (650,000 tons).
Products: Carbon steel – wire rod, reinforcing bars, round bars, square bars, flats,

Nucor Corp (continued)

bright (cold finished) bars (rounds, hexagons, flats, squares). Light angles, light channels, medium angles, Alloy steel (excl stainless) – round bars, bright (cold finished) bars (rounds, hexagons, flats, squares). Steel grinding balls; fasteners. Sales were 108,000 tons. Total raw steel output in 1986 was 1,706,000 tons and total rolled steel sales 1,140,000 tons.
Expansion programme: Nucor and Yamato Kogyo are to construct a 650,000 tons/year mill to produce wide-flange beams and other structural products at Blytheville, Arkanasas to be in operation by late 1988. A mill to produce hot and cold rolled sheet products based on an SMS Schloemann – Siemag thin slab caster will be built at Crawfordsville, Indiana, with operations due to start in the second quarter of 1989; production will be 400,000 tons/year each of hot and cold rolled sheets.

Oklahoma Steel Mill Inc

Head office: 5000 N.W. 31st Street, Oklahoma City, OK 73122.
Established: 1951. **Annual capacity:** Finished steel: 18,000 short tons.
Works:
■ At head office. ***Rolling mills:*** bar – Single-stand 3-high 12″, 4-stand 3-high 10″, 4-stand 8″ (annual capacity 18,000 short tons).
Products: Carbon steel – reinforcing bars.

Opelika Welding, Machine & Supply Inc

Head office: 1200 Steel Street, Opelika, AL. **Postal address:** POB 2209. **Tel:** 205 745 3524.
Management: G.P. Mann (president and chairman of the board), W.S. Harris (vice president and chief engineer). **Established:** In the 1940s. **No. of employees:** Works – 100. Office – 7. **Annual capacity:** Finished steel over 15,000 short tons.
Works:
■ At head office. ***Tube and pipe mills:*** welded – one W20D Yoder (annual capacity over 6,000 short tons); one W25 Yoder (over 8,000 tons).
Products: Carbon steel – longitudinal weld tubes and pipes $\frac{3}{4}$-3″ (output 10,000 tons).

Oregon Steel Mills, Division of Gilmore Steel Corp

Head office: 14000 M. Rivergate Blvd, Portland, OR 97203. **Postal address:** PO Box 2760. **Tel:** 503 286 9651. **Telecopier (Fax):** 503 286 9651.
Management: C. Lee Emerson (chairman and board director); Thomas B. Boklund (president, chief executive officer and director); V. Neil Fulton (vice President, secretary-treasurer and director. Edward C. Gendron (director, non-executive); Robert R. Mausshardt (vice president, marketing); Robert J. Sikora (vice president, operations); Malcolm G. Putnam (vice president, employee relations); David J. Tikks (assistant treasurer). **Established:** 1928 (Gilmore Steel Corp). **Capital:** $40,000,000. **No. of employees:** Works – 290. Office – 170. **Annual capacity:** Finished steel: 300,000 short tons.
Works:
■ At head office. ***Steelmaking plant:*** One 100-ton 18′ dia electric arc furnace (annual capacity 400,000 tons). ***Refining plant:*** Ladle injection (calcium silicide and argon) (400,000 tons). ***Rolling mills:*** One 4-high 108″ wide medium plate (335,000 tons).
Products: Carbon steel – slabs 6-7″ thick, 100 x 400″ wide (output 400,000 ton); medium plates $\frac{3}{16}$-3″ thick, 48-102″ wide (335,000 tons). Alloy steel (excl stainless) – slabs 6-7″ thick, 100 x 400″ (30,000 tons); medium plates $\frac{3}{16}$-2″ thick (25,000 ton). Heat treated alloy plates, steel armour plates, normalized plates.
Expansion programme: Heat treat expansion – from 2,500 short tons per month to 3,750 short ton. Completion date August 1, 1987.

Owen Electric Steel Co of South Carolina

Head office: Cayce, SC 29171. **Postal address:** PO Box 2005.
Management: Lawrence P. Flood (vice president and general manager).
Works:
■ At head office. ***Continuous casting machines:*** One 3-strand billet. ***Rolling mills:*** bar – One 10" breakdown and 5-stand cross-country finishing.
Products: Carbon steel – reinforcing bars, round bars, light angles.

Pacific Tube Co

Head office: 5710 Smithway Street, Los Angeles, CA. **Postal address:** PO Box 91-1222. **Tel:** 213 728 2611. **Telex:** 910 580 3147, 67 3166. **Telecopier (Fax):** 213 724 4453.
Management: G.C. McEvoy (president); W.J. Kepp (vice president sales), D.O Hukkannen (vice president operations); Mrs M.J. Johnson (treasurer); A.H. Johnson (chief engineer); Mrs B.S. Shuptrine (director of purchasing). **Established:** 1943. **No. of employees:** Works – 150. Office – 80. **Ownership:** Cawsl Corp.
Works:
■ At head office. **Redraw mill:**
Products: Carbon steel – bright (cold finished) bars round, square and hexagonal; seamless tubes and pipes 4.25 x 0.50" od wall max; longitudinal weld tubes and pipes 4.50 x 0.18" od max wall; cold drawn tubes and pipes 4.25 x 0.50" od wall max; precision tubes and pipes 4.25 x 0.50" od wall max. Alloy steel (excl stainless) – seamless tubes and pipes 4.25 x 0.50" od wall max; longitudinal weld tubes and pipes 4.25 x 0.50" od wall max.

Pinole Point Steel Co (PPSC)

Head office: 5000 Giant Road, PO Box 4050, Richmond, CA 94804. **Tel:** 415 223 8883. **Telex:** 171298. **Telecopier (Fax):** 415 223 2580.
Management: Marshall I. Wais (president); Peter E. Wais (vice president and general manger); E.R. Smith (general manager operations). **Established:** 1978. **Ownership:** Marwais Steel Co.
Works:
■ At head office. ***Coil coating lines:*** One hot-dip galvanizing. ***Other plant:*** Pre painting line.
Products: Carbon steel – hot-dip galvanized sheet/coil, galvanized corrugated sheet/coil.

Pittsburgh Tube Co

Head office: 2060 Pennsylvania Avenue, Monaca, PA 15061. **Tel:** 412 774 6830. **Telex:** 866639. **Telecopier (Fax):** 412 774 6830 (Ext 210).
Management: R.C. Huemme (president and cheif executive officer), W.J. Allison (Sr vice president, sales and marketing), J.R. Meneely (secretary-treasurer), L.H. Whitver (general manager, operations). **Established:** 1924. **No. of employees:** Works – 375. Office – 100. **Subsidiaries:** General Tube Company, Chicago, IL 60627 (Tel: 312389-1925). General manager: Ben Boyd. Four drawbenches (annual capacity 8,000 short tons).
Works:
■ Fairbury, IL (Tel: 815 692 2311). Works director: W.J. Allison. ***Tube and pipe mills:*** welded – 5 Electric weld (½-5" od) (annual capacity 40,000 short tons) (with two drawbenches).
■ Monaca, IL (Tel: 412 774 6830). Works director: L.H. Whitver. Five drawbenches (annual capacity 50,000 short tons).
■ Darlington, PA (Tel: 412 774 680). Plant manager: J.V. Presutti. ***Tube and pipe mills:*** welded – 1 Electric weld with stretch/reduction mill (2-7" od) (annual capacity 100,000 short tons).
■ Jane Lew, WV (Tel: 304 884 7801). Plant manager: G.D. Evans. Two drawbenches

United States

Pittsburgh Tube Co (continued)

(annual capacity 10,000 short tons).
Products: Carbon steel – longitudinal weld tubes and pipes (cold drawn mechanical $\frac{1}{2}$-4$\frac{1}{2}$" od, 0.083-0.44" wall thickness; hot rolled and cold rolled electric weld $\frac{1}{2}$-4" od, 0.035-0.20" wall thickness); oil country tubular goods; cold drawn tubes and pipes (electric weld, DOM $\frac{3}{8}$-3$\frac{1}{2}$" od, 0.035-0.20" wall thickness).

Pixley Tube Corp

Head office: Highway 81 South, Marlow, OK. **Postal address:** PO Box 31. **Tel:** 654 1663/522 1681.
Management: Carlis Pixley (owner and president); Bill Hambrick (manager sales and purchasing). **Established:** 1970. **Capital:** $3,000,000. **No. of employees:** Works – 10. Office – 5. **Annual capacity:** Finished steel: 9,000 short ton.
Works:
■ At head office. ***Tube and pipe mills:*** welded – one 2$\frac{1}{2}$ Yoder (annual capacity 6,000 tons) and one 1$\frac{1}{2}$ Yoder (3,000 tons).
Products: Carbon steel – longitudinal weld tubes and pipes. Alloy steel (excl stainless) – longitudinal weld tubes and pipes.

Plymouth Tube Co

Head office: 29 W 150 Warrenville Road, Warrenville, IL. **Postal address:** PO Box 45. **Tel:** 312 393 3550. **Telex:** 910 252 2126. **Telecopier (Fax):** 312 393 3551.
Management: D.C. Van Pelt Sr (president), J.W. Fikejs (treasurer), H.F. Hugh (general manager, stainless sales), D.A. Fergusan (manager, extrusion sales), C.M. Foster Jr (general manager, carbon and alloy sales), G. Lloyd (divisional manufacturing manager), F.H. Foglesong (manager, corporate development). **Established:** 1924. **No. of employees:** Works – 400. Office – 75.% **Ownership:** D.C. Van Pelts Sr (majority owner).
Works:
■ PO Box 278, Winamac, IN 46966 (Tel: 219 946 3125). ***Tube and pipe mills:*** welded – three $\frac{3}{4}$-2" od. ***Other plant:*** Redraw mill facilities for carbon and alloy tubing, welded/seamless, $\frac{1}{4}$-1$\frac{1}{2}$" od.
■ PO Box 600, Streator, IL 61364 (Tel: 815 673 1515). Redraw mill facilities for carbon and alloy tubing, welded/seamless, $\frac{1}{8}$-1$\frac{1}{8}$" od.
■ PO Box 1678, W. Monroe, LA 71291 (Tel: 318 388 3360). ***Tube and pipe mills:*** welded – six $\frac{5}{8}$-1$\frac{1}{2}$" od. ***Other plant:*** Redraw mill facilities for stainless steel tubing, welded/seamless, $\frac{1}{4}$-1$\frac{1}{2}$" od.
■ 201 Commerce Court, Hopkinsville, KY 42240 (Tel: 502 886 6631). Extrusion and redraw facilities for carbon, alloy and stainless steel tubing, $\frac{3}{4}$-3$\frac{5}{8}$" od.
■ Horsham Road, Horsham, PA 19044 (Te!: 215 625 9000). Redraw facilities for stainless steel tubing, welded/stainless, $\frac{1}{8}$-1" od.
Products: Carbon steel – special light sections, seamless tubes and pipes, longitudinal weld tubes and pipes, large-diameter tubes and pipes, cold drawn tubes and pipes, precision tubes and pipes. Stainless steel – light angles, light tees, light channels, special light sections, seamless tubes and pipes, longitudinal weld tubes and pipes, cold drawn tubes and pipes, hollow bars, precision tubes and pipes. Alloy steel (excl stainless) – seamless tubes and pipes, longitudinal weld tubes and pipes, large-diameter tubes and pipes, cold drawn tubes and pipes, precision tubes and pipes.
Expansion programme: New hot mill for larger diameter carbon steel and alloy steel tubular products – under construction at Winamac, Indiana, expected completion mid-January 1988.

Quanex Corp

Head office: 1900 West Loop South, Suite 1500, Houston, TX 77027. **Tel:** 713 961 4600. **Telex:** TWX 910 881 5000.
Management: Directors – Gerald B. Haeckel, Wilfred D. MacDonnell, Donald J. Morfee, John D. O'Connell, Carl E. Pfeiffer (president and chief executive officer), James D. Tracy, Robert L. Walker, Robert C. Snyder (executive vice president); vice presidents – Robert V. Kelly, Jr, James C. Hill, Wayne M. Rose (finance); controller – James D. Parrish. **Established:** 1927. **Capital:** $90,800,000. **No. of employees:** 1,925 (at 31st October 1986). **Subsidiaries:** Viking Metallurgical, LaSalle Steel Co. **Associates:** MacSteel Division, MI; MacSteel Division, AR; Bellville Tube Division, TX; Gulf States Tube Division, TX; Heat Treating Division, IN; Michigan Seamless Tube Division, MI.
Works:
■ Bellville, TX (Tel: 713 463 3441). ***Tube and pipe mills:*** High frequency ERW (max size 4½ od x 0.40" wall) (annual capacity 100,000 short tons).
■ MacSteel – Michigan, PO Box 1101, 3100 Brooklyn Road, Jackson, MI 49203. (Tel: 517 764 0311). General manager: Dennis Beck. ***Steelmaking plant:*** Two 40-ton electric arc furnaces. ***Refining plant:*** One DH vacuum degassing unit. ***Continuous casting machines:*** two 2-strand round. ***Rolling mills:*** One 6-stand Voest-Alpine bar (2-5").
■ MacSteel – Arkansas, 4700 Planter Road, Fort Smith, AR 72902. (Tel: 501 646 0223). General manager: Ronald Miller. ***Steelmaking plant:*** Two 50-ton electric arc furnaces. ***Refining plant:*** One VAD. ***Continuous casting machines:*** One 3-strand Rotary billet. ***Rolling mills:*** One 10-stand Birdsboro bar for 1½-6" rounds and round cornered squares.
■ Michigan Seamless Tube Division, 400 William N. McMunn Street, South Lyon, MI 48178. (Tel: 313 437 8117). General manager – John J. Yetso. ***Tube and pipe mills:*** One Mannesmann piercer 2-4'' and seven drawbenches ½-3½''.
■ Gulf States Tube Divsion, PO Box 952, Rosenberg, TX 77471. (Tel: 713 342 5401. Telex: 910 880 4091). General manager: Lynn Branan. ***Tube and pipe mills:*** seamless Extrusion press 1-2½'' and four drawbenches ¼-2''. ERW ¾-2½''.
■ Heat Treat Division, 25 Commercial Road, Huntington, IN 46750. (Tel: 219 356 9520). General manager: Robert Anderson. Quench and temper line for tubing 4½:" max, bar 2½" max.
■ LaSalle Steel Corp, 1412 150th Street, Hammond, IN 46327. (Tel: 219 853 6000. Telex: 725451, 910 696 2025). General manager: Richard Treder. Ten drawing benches, max 4½" (bar); four Schumag wire drawers, max 1½".
■ Viking Metallurgical, 1 Erik Circle, Verdi, NV 89439. (Tel: 702 345 0345). General manager: John J. Eikleberry. Plant for the production of titanium and super alloy seamless rings; Wagner seamless ring rolling mills, hydraulic open die forging presses, production electron beam melting furnaces.
Products: Carbon steel – round bars (2-5"); square bars (2-5"); flats; bright (cold finished) bars; seamless tubes and pipes; longitudinal weld tubes and pipes; oil country tubular goods; cold drawn tubes and pipes. Alloy steel (excl stainless) – round bars; square bars; flats; bright (cold finished) bars; seamless tubes and pipes; longitudinal weld tubes and pipes; oil country tubular goods; cold drawn tubes and pipes. Leaded steels and free-cutting steels; titanium and super alloy contoured rings, forgings, ingots and slabs.
Brands: Mac +, Stressproof, Fatigueproof and ETD 150 (all cold finished bars).
Sales offices/agents: Bar Group, MI 49201; Tube Group, MI 48152; LaSalle Steel Co, IN 46327; Viking Metallurgical, NV 89439.

Raritan River Steel Co

Head office: 225 Elm Street, Perth Amboy, NJ 08862. **Tel:** 201 442 1600. **Telex:** 844683 raritan nbk.
Established: 1977. **No. of employees:** 580. **Ownership:** Co-Steel Inc, Whitby, Ont. **Annual capacity:** Finished steel: 800,000 short tons.
Works:
■ At head office. ***Steelmaking plant:*** One 150-ton electric arc furnace (575,000

United States

Raritan River Steel Co (continued)

short tons). ***Continuous casting machines:*** One 5-strand low head curved mould billet ($5\frac{1}{4}$ x $5\frac{1}{4}$"). ***Rolling mills:*** One 25-stand twin strand Morgan/Stelmor wire rod (coil weight 2-ton) (740,000 short tons).
Products: Carbon steel – wire rod.

Razorback Steel Corp (RSC)

Head office: Van Dyke Road, Newport, AR 72112. **Postal address:** PO Box 310. **Tel:** 501 523 3693.
Management: Joseph E. Knight (president), J.L. Leadbetter (general manager), H.J. Moorheart (controller). **Established:** 1981.
Works:
■ At head office. ***Steelmaking plant:*** Two 30-ton Whiting electric arc furnaces. ***Continuous casting machines:*** One 2-strand Pecor (modified) billet / slab. ***Rolling mills:*** Two 18" 3-high light section with four 14" intermediate and finishing mills. ***Other plant:*** Punch presses.
Products: Carbon steel – billets, tie plates $\frac{1}{4}$" x 4" and up. Alloy steel (excl stainless) – billets, semis for seamless tubemaking.

Roanoke Electric Steel Corp (Resco)

Head office: Roanoke, VA 24034. **Postal address:** PO Box 13948.
Management: William M. Meador (chairman of the board), Donald G. Smith (president, treasurer and chief executive officer), T.J. Crawford (secretary), J.E. Morris (assistant treasurer), D.R. Higgins (vice president, sales), W.B. King (rolling mill superintendent), W.J. Akers (melting superintendent), R.A. Tekamp (superintendent, engeenering and maintenance), P.D. Hatcher (quality control supervisor). **Established:** 1955. **Subsidiaries:** John W. Hancock Jr Inc, Salem, VA; Shredded Products Corp, Montvale, VA.
Works:
■ At head office. ***Steelmaking plant:*** Three electromelt and Whiting electric arc furnaces, 18', 14' and 11' dia. ***Continuous casting machines:*** Two billet one 2-strand vertical curve high head Babcock & Willcox, and one 5-strand vertical curve low head Danieli ($3\frac{1}{2}$", 4" and $4\frac{1}{2}$"). ***Rolling mills:*** One 15-stand Danieli and Birdsboro 12" bar. ***Other plant:*** Cold drawing, turning and centreless grinding; galvanizing; wire drawing; rebar fabrication; there is also an auto-shredder.
Products: Carbon steel – wire rod, reinforcing bars No 3 to No 8; round bars $\frac{1}{2}$-1"; square bars; flats $\frac{1}{4}$" x 1"-$\frac{1}{2}$" x 5"; light angles $\frac{3}{4}$" x $\frac{3}{4}$" x $\frac{1}{8}$"-3" x 3" x $\frac{3}{8}$"; light channels 2" x 1" x $\frac{1}{8}$", 2" x 1" x $\frac{3}{16}$", 3" x 4.1" and 3" x 5".

Roll Forming Corp

Head office: Industrial Park, Shelbyville, KY 40065. **Postal address:** PO Box 369. **Tel:** 502 633 4435. **Telex:** 510 543 2591.
Management: President and chief executive officer – B.W. Brooks; vice presidents – J.W. Brooks, J.A. Hagan (treasurer), J.D. Tolliver (marketing). **Established:** 1947. **No. of employees:** 155.
Works:
■ At head office.
Products: Carbon steel – cold formed sections (angles, channels and special sections). Stainless steel – cold roll-formed sections (from strip) (angle, channels and special sections). Alloy steel (excl stainless) – cold formed sections (angles, channels, special section and hollow sections). Custom roll former.

Rome Strip Steel Co Inc

Head office: 530 Henry Street, Rome, NY 13440. **Postal address:** PO Box 189. **Tel:** 315 336 5500. **Telecopier (Fax):** 315 336 5510.
Management: President – A. Buol Hinman; vice presidents – Jeffrey H. Hinman (sales), David N. Hinman (operations), Mark Hinman (purchasing), Kirk Hinman (treasurer).
Established: 1926. **No. of employees:** Works – 200. Office – 20. **Annual capacity:** Finished steel: 65,000 tons.
Works:
■ At head office. ***Rolling mills:*** 3-stand 4-high 18" wide cold strip (coil weight 440 p/w); one single-stand 6-high reversing 18" wide temper/skin pass (coil weight 440 p/w); 18" Sendzimir; and 2-high and 4-high mills.
Products: Carbon steel – cold rolled hoop and strip (uncoated) 0.005" to 0.20" thick 0.375 to 17" wide, high carbon, alloy, low carbon, high strength-low alloy. Alloy steel (excl stainless) – cold rolled hoop and strip (uncoated) 0.010" to 0.1875" thick, 0.375" to 17" wide.

Rouge Steel Co

Head office: 3001 Miller Road, Dearborn, MI 48121. **Tel:** 313 323 9001. **Telex:** 231239. **Telecopier (Fax):** 313 845 8581.
Management: Chairman – Edson Williams; president – David E. Blackwell; vice presidents – M.P. Wojtowicz (operations), H.I. Weinberg (engineering), D.L. Armstrong (sales), W.L. Petek (supply), J.R. Gilson (labour relations), B.G. Spengler (finance).
Established: 1920. **No. of employees:** Works – 4000. Office – 900. **Ownership:** Ford Motor Co. **Subsidiaries:** Eveleth Taconite Co (85%) (iron ore). **Associates:** Double Eagle Steel Coating Co, MI, USA (50%) (electrolytic galvanizing). **Annual capacity:** Pig iron: 1,690,000 short tons. Raw steel: 2,980,000 tons. Finished steel: 2,400,000 tons.
Works:
■ At head office. ***Iron ore pelletizing plant:*** Rotary kiln (annual capacity 3,600,000 tons). ***Blast furnaces:*** Two – one 20' hearth dia, 28,058 cu ft (585,000 tons); one 29' hearth dia, 54,987 cu ft (1,105,000 tons). ***Steelmaking plant:*** Two 250-ton basic oxygen converters (3,000,000 tons); two 225-ton electric arc furnaces (900,000 tons). ***Continuous casting machines:*** One 2-strand Mannesman Demag/Hitachi Zosen slab (8" x variable x 29') (1,800,000 tons). ***Rolling mills:*** One reversing stand slabbing 8" x variable x variable (3,500,000 tons); one 5 roughing and 7 finishing stand wide hot strip (6.8"-8.0" x 28"-63" x up to 384") (coil weight 49,050lb) (3,000,000 tons); one 4-stand 4-high cold reduction (0.187" to 0.025") (60,000 lb) (832,000 tons); two 1-stand 4-high temper/skin pass (0.025") (60,000lb) (1,000,000 tons). ***Coil coating lines:*** One US Steel radial cell electro-galvanizing (350,000 tons (RSC share)).
Products: Foundry pig iron. Carbon steel – ingots, slabs, hot rolled sheet/coil (uncoated), cold rolled sheet/coil (uncoated), hot-dip galvanized sheet/coil, electro-galvanized sheet/coil. Alloy steel (excl stainless) – ingots, slabs, hot rolled coil (for re-rolling), hot rolled sheet/coil (uncoated), cold rolled sheet/coil (uncoated).
Expansion programme: Vacuum degassing (1989).

Scientific Tube Inc

Head office: 410 Interstate Road, Addison, IL 60101. **Tel:** 312 543 0858.
Management: Gerald Brunken (president), Don Vandevoorde (secretary).
Established: No. of employees: Works – 22. Office – 3. **Annual capacity:** Finished steel: 350 short tons.
Works:
■ At head office. Works director: D. Vandevoorde. ***Tube and pipe mills:*** Two AME welded – $\frac{1}{4}$-1$\frac{3}{4}$ (annual capacity 350 tons).
Products: Stainless steel – longitudinal weld tubes and pipes, oil country tubular goods and precision tubes and pipes $\frac{1}{4}$-1$\frac{3}{4}$" (output 3,500,000 ft – 350 tons).

United States

Seattle Steel Inc

Head office: 4001 28th Avenue SW, Seattle, WA 98126. **Postal address:** Box C 3827, Seattle, WA 98124. **Tel:** 206 938 6800. **Telecopier (Fax):** 206 938 7095.
Management: C.E. Meitzen (chairman and chief executive officer); J.R. Powers (president and chief operating officer); J.B. Parrish (executive vice president and chief financial officer) and J.A. Agnello (senior vice president, commercial). **Established:** 1985. **No. of employees:** Works – 525. Office – 130. **Ownership:** CEM Associates Inc. **Annual capacity:** Raw steel: 500,000 short tons. Finished steel: 350,000 tons.
Works:
■ At head office. ***Steelmaking plant:*** Two 120-ton 20′ dia electric arc furnace (annual capacity 500,000 tons). ***Continuous casting machines:*** One 4-strand Concast billet (5x5 and $5\frac{1}{2}$x$5\frac{1}{2}$) (500,000 tons). ***Rolling mills:*** One 1-stand 2-high reversing 32″ blooming (350,000 tons); one bar comprising 1-stand 2-high 18″, 3-stand 2-high 16″, 2-stand 2-high 12″, 3-stand 2-high 14″, 3-stand 2-high 12″, 1-stand vertical 10″ and 1-stand 2-high 10″ (225,000 tons); bar/structural/universal plate comprising 3-stand 3-high 22″ and 1-stand 3-high 22″ (125,000 tons).
Products: Carbon steel – ingots, slabs, blooms, billets, reinforcing bars No 3 to No 18 (up to No 8 in coils); round bars $\frac{3}{8}$ to 6″ (up to $1\frac{1}{4}$″ in coils); flats, medium angles, medium channels, heavy angles, universal plates $\frac{1}{4}$ to $3\frac{1}{2}$″ thick. Alloy steel (excl stainless) – ingots, slabs, blooms, billets, round bars, flats, universal plates. Railroad tie plates and spikes.
Additional information: The works was formerly the Seattle Division of Bethlehem Steel Corp.

Seneca Steel Corp

Head office: 1050 Military Road, Buffalo, NY 14217. **Tel:** 716 875 7920.
No. of employees: 150.
Works:
■ At head office. ***Rolling mills:*** One 4-high reversing 27″ cold strip. ***Other plant:*** Slitters.
Products: Carbon steel – cold rolled hoop and strip (uncoated).

Sharon Steel Corp

Head office: Sharon, PA 16146. **Postal address:** PO Box 290, Farrell, PA 16121. **Tel:** 412 981 1375.
Management: Victor Posner (chairman, chief executive officer and president). **Established:** 1900. **Subsidiaries:** Union Steel Corp, Union, NJ (see separate entry).
Works:
■ At head office. Two ***Blast furnaces:*** (23′ 1″ hearth dia) (annual capacity 1,400

Sharon Steel Corp (continued)

short tons per day each). ***Steelmaking plant:*** One 150-ton basic oxygen converter. Two 120-ton electric arc furnaces. ***Refining plant:*** One DH vacuum degassing unit; two Stokes stream degassers. ***Rolling mills:*** One 60″ semi-continuous wide hot strip. Two cold reduction – one 5-stand continuous (60″), and one Sendzimir (60″); cold strip – One Sendzimir (25″), one 4-stand (26″) tandem, and one 3-stand (18″) tandem. ***Coil coating lines:*** One (36″) hot-dip galvanizing.
■ Damascus Tube Division, PO Box 71, Greenville, PA 16125 (Tel: 412 646 1500). ***Tube and pipe mills:*** Sixteen welded; drawbenches.
■ Brainard Strapping Division, Larchmont Aveune, Ex, N.E., Warren, OH 44482 (Tel: 216 372 4040). ***Coil coating lines:*** One electro-galvanizing; one colour coating.
■ Dearborn Division, 7951 Maple Street, Box 4438, Detroit, MI 48228 (Tel: 313 846 1500). ***Rolling mills:*** cold strip.
Products: Carbon steel – ingots, slabs, blooms, billets, hot rolled hoop and strip (uncoated), cold rolled hoop and strip (uncoated), medium plates, hot rolled sheet/coil (uncoated), cold rolled sheet/coil (uncoated), hot-dip galvanized sheet/coil, electro-galvanized sheet/coil, coated sheet/coil. Alloy steel (excl stainless) – hot rolled hoop and strip (uncoated), cold rolled hoop and strip (uncoated).
Brands: Sharon Galvanite (galvanized sheets); Sharonart (textured surface sheets).

Sharon Tube Co

Head office: 134 Mill Street, Sharon, PA 16146 0492. **Tel:** 412 981 5200.
Management: Irwin M. Yanowitz (president and chief executive officer).
Established: 1929. **Annual capacity:** Finished steel: 40,000 short tons.
Works:
■ At head office. ***Tube and pipe mills:*** welded – continuous butt weld. Five drawbenches.
Products: Carbon steel – longitudinal weld tubes and pipes, galvanized tubes and pipes up to 1″; cold drawn tubes and pipes up to 3″ (welded) and up to 2″ (seamless).

Sheffield Steel Corp

Head office: 2300 South Highway 97, Sand Springs, OK 74063. **Tel:** 918 245 1335. **Telex:**
Established: 1981. **Annual capacity:** Finished steel: 350,000 short tons.
Works:
■ At head office. ***Steelmaking plant:*** Two 80-ton Lectromelt electric arc furnaces. ***Continuous casting machines:*** One 6-strand low level billet (4¼″) (annual capacity 475,000 short tons). ***Rolling mills:*** One 12-stand bar (12″) (300,000 short tons).
Products: Carbon steel – reinforcing bars. Studded fence posts.

Shenango Inc

Head office: 200 Neville Road, Pittsburgh, PA 15225. **Tel:** 412 777 6600.
Management: Chairman – M.A. Aloe; vice chairman and chief executive officer – J.S. Spatz; president and chief operating officer – P.M. Reese; vice president – C.A. Coffey (customer services), D.C. Kay (purchasing and transportation), G.R. Muhl (finace and treasurer); managers – F.G. Christie (customer services), G.L. Owens (coke and iron division), D.W. Demoise (foundry division), D.L. Carpenter (coke and pig iron sales).
Established: 1962.
Works:
■ Neville Island, PA. ***Coke ovens:*** Koppers underjet battery (35 ovens) and Koppers gunflue battery (56 ovens). ***Blast furnaces:*** Two – (A&B) both 22′ hearth dia, 29,182 cu ft. ***Other plant:*** Ingot mould and stool foundries.
Products: Foundry pig iron to all specifications; Blast furnace coke; ingot moulds and stools.

United States

Slater Steels, Fort Wayne Specialty Alloys Division

Head office: 2400 Taylor Street W, Fort Wayne, IN. **Postal address:** PO Box 630. **Tel:** 219 432 2561. **Telex:** 023 2432. **Telecopier (Fax):** 219 436 2072. **Management:** President – Douglas K. Pinner (member of board); vice presidents – Thomas F. Frappier (sales and marketing), Joseph F. O'Hara (marketing development); directors – Gary Naber (finance), James E. Rinehart (purchasing), Joseph M. Fallon (engineering services), Raymond Teders (manufacturing) and Timothy Bower (industrial relations). **Established:** 1932. **No. of employees:** Works – 390. Office – 122. **Ownership:** Slater Steels Holding BV. **Associates:** Slacan Division, Hamilton Specialty Bar, Sorel Forge. **Annual capacity:** Finished steel: 60,000 short tons.
Works:
■ At head office. ***Steelmaking plant:*** One 20-ton and one 17-ton electric arc furnace (annual capacity 69,000 tons). ***Rolling mills:*** One 1-stand 38" Treadwell blooming (90,000 tons); one 3-stand 24-22" Treadwell billet (90,000 tonnes); one 6-stand Birdsboro bar (one 14" and five 12") (16,320 tons).
Products: Carbon steel – billets. Stainless steel – ingots, billets, round bars, square bars, flats, hexagons, light angles. Alloy steel (excl stainless) – round bars.

SMI Steel Inc

Head office: 101 South 50th St, Birmingham, AL 35212. **Postal address:** PO Box 2875-A. **Tel:** 205 592 8981. **Telecopier (Fax):** 205 591 7429.
Management: Dolphus Morrison (vice president, general manager), Dan Thompson (vice president, sales), Jack Wheeler (vice president, works manager); Phil Seidenberger (assistant works manager, production), Willard Johnson (assistant works manager, services); controller – Robert Unfried; Ed Shaw (director of purchasing). **Established:** 1983. **No. of employees:** Works – 400. **Ownership:** Commercial Metals Co (100%). **Annual capacity:** Raw steel: 280,000 short tons. Finished steel: 250,000 tons.
Works:
■ At head office. ***Steelmaking plant:*** Two 45-ton electric arc furnace (annual capacity 280,000 tons). ***Continuous casting machines:*** One 4-strand Rokop billet (5$\frac{1}{4}$", 5" x 7$\frac{3}{4}$", 7" x 7") (280,000 tons). ***Rolling mills:*** One 15-stand light section (110,000 tons of 2"-6" angles, 40,000 tons of 3"-8" channels, 65,000 tons of 3"-8" flats and 35,000 tons of 3"-8" beams).
Products: Carbon steel – flats 3-8" (1986 output 69,000 tons); light angles 2-6" (104,000 tons); light channels 3-8" (41,000 tons).
Expansion programme: Rolling mill was due to be replaced in October 1987.

Somers Thin Strip, Olin Corp

Head office: 215 Piedmont Street, Waterbury, CT 06706. **Tel:** 203 597 5000.
Works:
■ ***Rolling mills:*** Three Sendzimir cold strip 37", 27" and 19"; various strip finishing cold mills. ***Other plant:*** Plant for hot tinning and tin-lead coating.
Products: Carbon steel – coated hoop and strip (tinned and tin-lead coated). Stainless steel – cold rolled hoop and strip (uncoated); hot rolled sheet/coil (uncoated) (coil only). CR strip in copper and copper alloys and some nickel alloys.

Standard Steel

Head office: 500 N. Walnut Street, Burnham, PA 17009. **Tel:** 717 248 4911.
Management: D.D. Borland (executive vice president and general manager). **Established:** 1811. **Ownership:** Freedom Forge Corp.
Works:
■ At head office. Steam hammers; hydraulic presses; GFM axle forging plant; ring rolling mill.
■ Latrobe, PA. Hydraulic presses; spring plant. Steel production is used for fabricating railroad car wheels and axles, and industrial rolled rings and industrial forging in carbon and alloy steels and high temperature alloys.

The Stanley Works—see Cold Metal Products Co Inc

Steel of West Virginia Inc

Head office: Huntington, WV 25726-2547. **Tel:** 304 529 7171.
Established: Reorganised 1982 (formerly part of Connors Steel Co). **No. of employees:** 360. **Annual capacity:** Finished steel: 250,000 short tons.
Works:
■ 17th Street and Second Avenue, Huntington, WV. ***Steelmaking plant:*** Two 70-ton electric arc furnaces. ***Continuous casting machines:*** One 3-strand billet. ***Rolling mills:*** One 21" blooming; two light section 12" and 18".
Products: Carbon steel – special light sections and mini beams; light rails.

Structural Metals Inc

Head office: 8111 Bill Dwyer Road, Seguin, TX 78156. **Postal address:** PO Box 911. **Tel:** 512 379 7520. **Telex:** 767356.
Management: President – Marvin Selig; vice presidents – Clyde Selig (operations), Byron Henderson (works manager), Marvin Rinn (sales). **Established:** 1947. **No. of employees:** Works – 650. **Ownership:** Commercial Metals Co (100%). **Annual capacity:** Raw steel 475,000 short tons. Finished steel 425,000 short tons.
Works:
■ At head office. ***Steelmaking plant:*** One 92-ton electric arc furnace (annual capacity 475,000 tons). ***Continuous casting machines:*** One 4-strand Rokop billet (4", 4½:, 5" and 6¼") (475,000 tons). ***Rolling mills:*** bar and section – eight-stand 12" Sakai, four-stand 14" Sakai, seven-stand 18" Morgan and two-stand 24" (425,000 tons).
Products: Carbon steel – billets 4", 4½", 5" and 6¼; reinforcing bars No 3 to No 18 (1986 output 309,676 tons); round bars ½'' to 1 9/16'' (1986 output 21,730 tons); square bars ½''to 1 9/16'' (1986 output 3,397 tons); light angles 1¼ to 4'' (1986 output 31,580 tons); light tees 1.20 and 1.33 (1986 output 24,479 tons); special T's and Flats (1986 output 8,685 tons).
Expansion programme: In-line rolling mill – August 1988.

Stupp Corp

Head office: Baton Rouge, LA 70821. **Postal address:** PO Box 3558. **Tel:** 504 775 8800. **Telex:** WU 586422.
Management: Chairman of the board – John P. Stupp; president – Ralph V. Schmidt; vice presidents – W.H. Bassett (executive), D.L. Efferson (sales), N.J. LeBlanc (plant manager); comptroller – W.L. Irvin. **Established:** 1952. **No. of employees:** 200/300 (80% works, 20% office). **Ownership:** Wholly-owned sudsidiary of Stupp Bros. Bridge & Iron Co, St Louis, MO. **Subsidiaries:** None.
Works:
■ Thomas Road, Baton Rouge. Works director: Numa J. LeBlanc. ***Tube and pipe mills:*** welded One HF ERW 8⅝-24" od (annual capacity 200,000-300,000 short tons).
Products: Carbon steel – pipe piling; longitudinal weld tubes and pipes for oil and gas pipelines; water pipe; pipe for slurry, irrigation, water well casing etc. Normal lengths 64', wall thickness up to about ½".

Superior Drawn Steel Co, Division of Standard Steel Specialty Co

Head office: 1585 Beaver Avenue, Monaca, PA 15061. **Tel:** 412 774 9110.
Established: 1925.
Works:
■ At head office. Cold drawbenches and coil blocks; atmosphere controlled annealing line.
Products: Carbon steel – bright (cold finished) bars. Alloy steel (excl stainless) – bright (cold finished) bars (including high strength low alloy).

United States

Techalloy Co Inc

Head office: Rte 113, Rahns, PA 19426. **Tel:** 215 489 1110. **Telex:** TWX 5106606918 tech vahns ud.
Established: 1954.
Works:
■ At head office. ***Rolling mills:*** One Sendzimir 24" cold strip mills. ***Other plant:*** Vaughn wire drawing mills; Turk's head shaped wire drawing machine; wire metal coating lines.
Products: Stainless steel – wire rod, round bars, square bars, hexagons, cold rolled hoop and strip (uncoated), wire. Alloy steel (excl stainless) – round bars, square bars, wire. Nickel alloys in strip, rod, bar and wire; welding wire; coated electrodes.

Teledyne Columbia-Summerill

Head office: Foot of Woodkirk Street, Carnegie, PA 15106. **Postal address:** PO Box 1557. **Tel:** 412 923 2040. **Telecopier (Fax):** 412 887 9686.
Management: President – William E. Manby; vice presidents – M.D. Ottaviani (general manager), Gerald L. Cordis (sales); marketing manager – Joseph P. Stockhausen; controller – J. Michael Woods. **Established:** 1898. **No. of employees:** Works – 125. Office – 35. **Ownership:** Teledyne Inc. **Annual capacity:** Finished steel: 90,000 short tons.
Works:
■ Bright drawing (cold finishing) plant.
Products: Carbon steel – bright (cold finished) bars. Alloy steel (excl stainless) – bright (cold finished) bars.

Teledyne Pittsburgh Tool Steel

Head office: 1535 Beaver Avenue, Monaca, PA 15061. **Tel:** 412 774 8330. **Telex:** 866447 tdyptsw moca.
Established: 1902. **Ownership:** Teledyne Inc, Los Angeles, CA.
Works:
■ At head office. **Rolling mills:** One cold mill for rod. ***Other plant:*** Drawing and cold finishing equipment.
Products: Alloy steel (excl stainless) – bright (cold finished) bars – tool steel drill rod; precision ground flat stock in tool steels and low carbon steel; cold drawn steel shapes; wire.

Teledyne Rodney Metals

Head office: 1357 East Rodney French Blvd, New Bedford, MA 02742. **Tel:** 617 996 5691. **Telex:** TWX 5106656710 rodney ivy l, WUI 6817344 td yrodn uw.
Established: 1941. **Ownership:** Teledyne Inc, Los Angeles, CA.
Works:
■ At head office. ***Rolling mills:*** One 42" Sendzimir cold reduction; three Sendzimir cold strip – two 25" and one $12\frac{1}{2}$". ***Coil coating lines:*** Strip coating (paints and adhesives).
Products: Stainless steel – cold rolled hoop and strip (uncoated) down to very narrow widths and gauges; cold rolled sheet/coil (uncoated) (precision, up to 42" wide, down to very fine gauges). Strip and sheet in nickel and titanium and their alloys; coated products (paints and adhesives on many metal sub-strates).

Teledyne Vasco

Head office: Latrobe, PA 15650. **Postal address:** PO Box 151. **Tel:** 412 537 5551. **Telex:** 866653.
Management: Frederick A. Kaufman (chairman), Robert C. Rubino (president). **Ownership:** Teledyne Inc, Los Angeles, CA.
Works:
■ At head office. ***Steelmaking plant:*** electric arc. ***Refining plant:*** One 20-tonne

Teledyne Vasco (continued)

AOD. ***Rolling mills:*** Four bar. ***Other plant:*** Forging hammers.
■ Colonial Plant, Monaca, PA 15061. ***Rolling mills:*** One bar. One medium plate; one hot sheet mill.
Products: Alloy steel (excl stainless) – round bars, medium plates, hot rolled sheet/coil (uncoated), wire. High speed steels; carbon and alloy tool and die steels; ultra high strength steels; high temperature steels; hot and cold drawn products; drill rod; forgings; circles.

The Thinsheet Metals Co

Head office: 271 Railroad Hill Street, Waterbury, CT 06708. **Tel:** 203 756 7414. **Telecopier (Fax):** 6714039.
Management: Chairman of the board – T. Omoto; president – R. Scott; vice president – H. Furusawa; secretary treasurer – Richard J. Lavallee. **Established:** 1918. **No. of employees:** Works – 70. Office – 10. **Ownership:** Nisshin Steel Co Ltd. **Annual capacity:** Finished steel: 8,000 short tons.
Works:
■ At head office. ***Rolling mills:*** Five cold reduction – two 25" Sendzimir and three 4-high (annual capacity 20,000 short tons).
Products: Stainless steel – cold rolled hoop and strip (uncoated).

Thomas Steel Corp

Management: Robert C. Thomas (president and chief executive officer). **Established:** 1983 (works formerly owned by Ceco Corp).
Works:
■ Lemont, IL 60439. ***Steelmaking plant:*** Three 45-ton electric arc furnaces. ***Continuous casting machines:*** One 3-strand billet. ***Rolling mills:*** One 14-stand 12" merchant bar (annual capacity 250,000 short tons).
Products: Carbon steel – reinforcing bars, round bars, flats, light angles.

Thomas Steel Strip Corp

Head office: Delaware Avenue NW, Warren, OH 44485. **Tel:** 216 841 6111.
Management: J.R. Thumann (chiarman), M.T. Schmader (president). **Annual capacity:** Finished steel: 100,000 short tons.
Products: Carbon steel – cold rolled hoop and strip (uncoated), coated hoop and strip (tin, electro-zinc, copper, brass and nickel coatings); patterned strip.
Brands: NIZN-COTE (cold rolled coated strip).

Thompson Steel Co Inc

Head office: 120 Royall Street, Canton, MA 02021. **Tel:** 617 828 8800.
Established: 1974.
Works:
■ 9470 King Street, Franklin Park, IL 60131. (Tel: 312 678 0400). ***Rolling mills:*** One sizing/finishing cold strip (annual capacity 80,000 short tons).
■ North Point Road, PO Box 6610, Sparrows Point, MD 21219. (Tel: 301 477 0400). ***Rolling mills:*** One sizing/finishing cold strip (annual capacity 70,000 tons).
Products: Carbon steel – cold rolled hoop and strip (uncoated) (high and low carbon). Alloy steel (excl stainless) – cold rolled hoop and strip (uncoated). Flat wire in cut lengths and coil; bright and polished strip.
Additional information: There is also a works for heat treatment of strip at Route 2, Country Road, Paulding, OH 45879.

United States

The Timken Co

Head office: 1835 Dueber Avenue SW, Canton, OH. **Tel:** 216 438 3000. **Telecopier (Fax):** 216 438 3452.
Management: W.R. Timken Jr (chairman, board of directors), J.F. Toot Jr (president), C.H. West (executive vice president, steel). **Established:** 1899. **No. of employees:** 16,500 (worldwide). **Subsidiaries:** Latrobe Steel Co (100%). **Annual capacity:** Melting capacity over 1,000,000 ingot tons.
Works:
■ Harrison Steel Plant (at head office). Works director: R.C. Caldas. ***Steelmaking plant:*** Six electric arc furnaces – three 100-ton and three 150-ton (annual capacity 800,000 short tons). ***Refining plant:*** Two D-H vacuum degassing units (400,000 short tons); one 130-ton ladle refining system (300,000 short tons). ***Continuous casting machines:*** One 4-strand Concast bloom (9½ X 12") (230,000 short tons). ***Rolling mills:*** One 1-stand 3-high 35" reversing blooming; two 1-stand 2-high 28" reversing billet; one 4-stand 3½-9"dia cross-country bar; one 14-stand 1¼-4" dia cross country bar. ***Tube and pipe mills:*** seamless – one 2¼-5⅜" dia (150,000 short tons).
■ Faircrest Steel Plant. Works director: J.K. Preston. ***Steelmaking plant:*** One 160-ton electric arc furnace (annual capacity 500,000 short tons). ***Refining plant:*** One 150-ton ladle refining system (500,000 short tons). ***Rolling mills:*** One 1-stand 2-high 46" reversing blooming; one 1-stand 2-high 36" reversing billet.
■ Gambrinus Steel Plant. Works director: N.P. Luchitz. ***Tube and pipe mills:*** seamless – four 1¾-11¼" (annual capacity 500,000 short tons).
Products: Carbon steel – ingots, blooms, billets, semis for seamless tubemaking, wire rod, round bars, square bars, seamless tubes and pipes, oil country tubular goods, cold drawn tubes and pipes, hollow bars, precision tubes and pipes. Stainless steel – ingots, blooms, billets, semis for seamless tubemaking, wire rod, round bars, square bars, seamless tubes and pipes, oil country tubular goods, cold drawn tubes and pipes, hollow bars, precision tubes and pipes. Alloy steel (excl stainless) – ingots, blooms, billets, semis for seamless tubemaking, hot rolled coil (for re-rolling), round bars, square bars, flats, wire, seamless tubes and pipes, oil country tubular goods, cold drawn tubes and pipes, hollow bars, precision tubes and pipes.
Sales offices/agents: Atlanta District, 5335 Fulton Industrial Blvd, PO Box 43224, Atlanta, GA 30336; Chicago District, 16 West 221 Shore Court, Hinsdale, IL 60521; Cincinnati District, 4010 Executive Park Drive, Suite 420, Cincinnati, OH 45241; Cleveland District, 30000 Aurora Road, Solon, OH 44139; Dallas District, 1610 Walnut Hill Lane, PO Box 152610, Irving, TX 75015-2610; Detroit District, 31100 Telegraph Road, PO Box 3073, Birmingham, MI 48012-3073; Hartford District, 415 Silas Deane Highway, Wethersfield, CT 06109; Houston District, 3535 Briarpark, Suite 250, Houston, TX 77042; Los Angeles District, 85 Brookhollow Drive, Santa Ana, CA 92705; Milwaukee District, 9800 W. Bluemound Road, Wauwatosa, WI 53226; Philadelphia District, 575 Pinetown Road, Ft. Washington, PA 19034; Pittsburgh District, 1370 Washington Pike, Bridgeville, PA 15017.

Trent Tube

Head office: Box 77, 2188 South Church Street, East Troy, Walworth County, WI 53120. **Tel:** 414 642 7321. **Telex:** TWX 910 260 1460. **Telecopier (Fax):** 414 642 9571.
Management: President – William Grant; vice presidents – Michael Douglas (sales, marketing), John Tverberg (engineering), David Coppock (manufacturing); director – John Thackray (marketing); manager – Paul Carpenter (marketing). **No. of employees:** Works – 300. Office – 100. **Ownership:** Crucible Materials Corp.
Works:
■ Trentweld plant, Young Street, East Troy, WI 53120. Works director – Jeff Stam. ***Tube and pipe mills:*** Twelve Turek & Heller welded (two M1, six M2, four M3); ***Other plant:*** Forming/forging, od and id polishing, bright anneal and oxide furnaces and five draw benches (4-⅛" od).
■ C.W.A. Plant, Young Street, East Troy, WI 53120. Works director: Ian Greeley. ***Tube***

Trent Tube (continued)

and pipe mills: Eleven Turek & Heller welded (eight M2, two M1, one spoke mill); ***Other plant:*** Forming/forging to 150″ lengths, bright anneal and oxide furnaces, U-bending and integral finning.
■ Carrollton Plant, 141 Hammond Street, Carrollton, GA 30117. (Tel: 404 832 9047). Works director: William Brossmann. ***Tube and pipe mills:*** welded – two M6 continuous, three spindle 2⅞-12¾″ od, one spindel 6⅝-24″ od, and four bedwelders 2⅞-72″ od. ***Other plant:*** 1,500-ton press brake (up to 45′ lengths), Bersch roll (width 24-72″ od) and large od pipe expander 14-60″ od.
Products: Carbon steel – longitudinal weld tubes and pipes, cold drawn tubes and pipes (welded). Alloy steel (excl stainless) – longitudinal weld tubes and pipes, large-diameter tubes and pipes, cold drawn tubes and pipes, precision tubes and pipes.
Brands: Sea-Cure (power and chemical industry tubing).

Tuscaloosa Steel Corp

Head office: 1700 Holt Road, Tuscaloosa, AL 35404. **Postal address:** PO Box 3369. **Tel:** 205 556 1310. **Telex:** 593114.
Management: George W. Tippins (chairman), John E. Thomas (chief executive officer). **Established:** 1985 (operations began).
Works:
■ ***Rolling mills:*** One Steckel combination (112″) slabbing / heavy plate.
Products: Carbon steel – hot rolled coil (for re-rolling); medium plates (incl coiled up to ¾″); heavy plates.

Ulbrich Stainless Steels and Special Metals Inc

Head office: 57 Dodge Avenue, North Haven, CT 06473. **Tel:** 203 239 4481. **Telex:** 160110. **Telecopier (Fax):** 203 239 7479.
Management: Fred Ulbrich (chairman), Dana Stanziale (president). **Established:** 1924. **No. of employees:** 350. **Ownership:** Privately owned.
Works:
■ 1 Dudley Avenue, Wallingford, CT. ***Rolling mills:*** Eight cold strip – five Sendzimir and three 4-high (max width 14").
■ Ulbrich Wire Inc, 55 Defco Park Road, North Haven, CT 06473. ***Other plant:*** Turks heads; multi-head 2-high wire mills, etc.
Products: Stainless steel – cold rolled hoop and strip (uncoated), wire. Wide range of stainless steels and special metals including nickel base alloys; titanium; controlled expansion metals; high temperature alloys; cobalt base alloys; flat and shaped wire.
Brands: Ulbra-Dul (dull grey matt finish). Ultra-Brite (bright, very highly reflective finish).

Union Steel Corp

Head office: Union, NJ 07083. **Tel:** 201 687 2000.
Ownership: Sharon Steel Corp.
Works:
■ At head office. ***Rolling mills:*** One 4-high reversing cold strip.
Products: Carbon steel – cold rolled hoop and strip (uncoated).

Unistrut Corp

Head office: 35005 Michigan Ave W, Wayne, MI. **Tel:** 313 721 4040. **Telex:** 4320813 strutui.
Management: President – C.J. Malfese; senior vice presidents – E. Cady (finance), R. Snyder (sales and distribution), T. Condit (manufacturing products). **Established:** 1924. **No. of employees:** Works – 342. **Annual capacity:** Finished steel: 20,000 short tons/year.
Works:
■ At head office. ***Tube and pipe mills:*** Three Yoder welded – 13-stand (annual capacity 27,500,000ft), 13-stand (32,000,000ft) and 12-stand (32,000,000ft).
Products: Carbon steel – cold roll-formed sections (from strip) (channels). Perforated tube.
Brands: Unistrut – metal framing system. Telespar – tubing.

Universal Molding Co

Head office: 10807 Stanford Avenue, Lynwood, CA 90262. **Tel:** 213 636 0721.
Management: Syd Sossin (executive vice president and chief executive officer); Mike Ingram (purchasing manager); Bert Smith (vice president sales); Jim Johnson (vice president manufacturing). **Established:** 1948. **No. of employees:** Works – 220. Office – 20. **Subsidiaries:** Anaheim Extrusion (100%).
Works:
■
Products: Stainless steel – cold rolled hoop and strip (uncoated).

UNR-Leavitt, Div of UNR Inc

Head office: 1717 West 115th Street, Chicago, IL 60643. **Tel:** 312 239 7700. **Telex:** 25 4297. **Telecopier (Fax):** 312 239 1023.
Management: Roy Herman (president); Robert Hunt (senior vice president, operations); Tony Ventura (director purchasing); William A. Spamos (director sales administration); Parry Katsafana (regional sales manager); Craig Iammarimo (regional sales manager). **Established:** 1956. **No. of employees:** Works – 350. Office –

United States

UNR-Leavitt, Div of UNR Inc (continued)

100.
Works:
■ At head office. ***Tube and pipe mills:*** Fifteen welded – ½"to 12¾" round, ½" to 10" square; 22g-0.5" walls.
Products: Carbon steel – pipe piling, longitudinal weld tubes and pipes, oil country tubular goods, galvanized tubes and pipes, hollow sections structural A500B, mechanical A513.

USS, A Division of USX Corp

Head office: 600 Grant Street, Pittsburgh, PA 15230. **Tel:** 412 433 1121. **Telex:** 866425. **Cables:** Ussteel Pittsburgh.
Management: David M. Roderick (board chairman, chief executive officer, USX Corp); Thomas C. Graham (president, USS). **Established:** 1901 (as United States Steel Corp).
Works:
■ Fairless Works, Fairless Hills, PA. ***Coke ovens:*** Wilputte underjet battery (174 ovens). ***Sinter plant:*** Two 8' strand McKee. ***Blast furnaces:*** No 1 29' 6" hearth dia, 58,507 cu ft; No 2 and No 3 30' 10" hearth dia, 58,835 cu ft. ***Steelmaking plant:*** Two 200-ton electric arc furnaces; nine 400-ton open hearth furnaces. ***Continuous casting machines:*** 2-strand curved rack bloom (10 x 23"). ***Rolling mills:*** 1-stand 2-high 45 x 90" universal slabbing (annual capacity 3,280,000 short tons); one 1-stand 2-high 40 x 90" blooming, and two billet – one 6-stand 2-high 30", one 4-stand 2-high 21" (combined annual capacity 1,725,000 short tons); one 18-stand 2-high 10" bar (342,000 short tons); 11-stand 4-high continuous wide hot strip comprising 40 x 130" and 54 x 124", 38 x 81" and 53½ x 76", 27 x 81" and 53½ x 76" (2,650,000 tons); one 18" Skelp comprising 10-stand horizontal and 5-stand 2-high vertical (837,00 tons); three cold reduction – one 5-stand 4-high 21" and 53 x 48" continuous (581,000 tons), one 4-stand 4-high 21" and 56 x 80" continuous (1,017,000 tons), one 2-stand 4-high 21" and 53 x 48" continuous (281,000 tons); three temper/skin pass – one 2-stand 4-high 19" and 53 x 48" continuous (436,000 tons), one 1-stand 4-high 21" and 53 x 80" continuous (800,000 tons), one 1-stand 2-high 32 x 81½" continuous (500,000 tons). ***Tube and pipe mills:*** welded – Two ½ to 4" Aetna Standard continuous buttweld. ***Coil coating lines:*** Two 38" electrolytic tinning; one 45" tin-free steel (ECCS); one 62" hot-dip galvanizing. ***Other plant:*** Pipe galvanizing plant.
■ Fairfield Works, Fairfield, AL. ***Blast furnaces:*** No 8 32" hearth dia, 77,520 cu ft. ***Steelmaking plant:*** Three 200-ton basic oxygen converters. ***Rolling mills:*** One continuous wide hot strip comprising 1-stand 2-high 36 x 56", 1-stand 4-high 25" and 49 x 54" and 6-stand 4-high 27⅜" and 65 x 60" (1,710,000 tons); two cold reduction – one 4-stand 4-high 21½" and 53 x 54" continuous tandem (753,000 tons), and one 6-stand 4-high 21" and 23 x 56 x 52" continuous tandem (581,000 tons); two temper/skin pass – one continuous temper comprising 2-stand 4-high 18" and 42.5 x 42", 2-stand 18" and 2-stand 4-high 22" and 53 x 48" (475,000 tons) and one continuous skin pass comprising 1-stand 2-high 16½", 24" and 30 x 55", and 1-stand 4-high 21" and 53 x 54" (630,000 tons); other mills – one 2-stand 4-high 21" and 53 x 48" DCR (194,000 tons). ***Coil coating lines:*** Three electrolytic tinning lines – one 34⅞" wide and two 38" wide; one continuous coil painting line. three continuous hot-dip galvanizing lines – two 48" wide and one 61" wide. ***Other plant:*** Four steam hammers – 1,500-lb to 2,500-lb; one 1,000-lb air hammer; one 1,000-ton steam and hydraulic press. Axle producing facilites including one 4-stage axle finishing machine; one GFM precision forging machine; also plant for tie plates and track spikes.
■ Lorain Works, Lorain, OH. ***Blast furnaces:*** No 1 23' hearth dia, 28,628 cu ft; No 2 23' 3" hearth dia, 28,973 cu ft; No 3 28'6", 48,505 cu ft; No 4 29', 49,196 cu ft. ***Steelmaking plant:*** Two 225-ton basic oxygen converters. ***Continuous casting machines:*** One 6-strand round (6", 9¼", 10½", and 12¼" rounds; 6 x 6", and 10½ x 10½" squares. ***Rolling mills:*** One 18-stand 2-high 46" blooming (annual capacity 2,295,000 tons); one 1-stand 2-high 38" bar (590,000 short tons); one 6-stand 2-high

USS, A Division of USX Corp (continued)

continuous vertical/horizontal 1,215,000 tons), one 4-stan000 tons) and 2-stand tin temper 46" (coil weight 62,500 tons) (520,000 tons). ***Tube and pipe mills:*** seamless – Three 2 to 26" od. Welded – One 1 to 4" Aetna Standard continuous buttweld; one $2\frac{3}{8}$ to $6\frac{5}{8}$" od. ***Other plant:*** Pipe galvanizing.

■ Texas Works, Baytown, TX. ***Steelmaking plant:*** Two 200-ton and two 220-ton electric arc furnaces. ***Refining plant:*** One 200-ton and one 220-ton Lectromelt DH vacuum degassing units. ***Continuous casting machines:*** Two 8 x 76" for slab. ***Rolling mills:*** Two heavy plate – one 1-stand 2-high 53 x 160", and one 1-stand 4-high 39 x 160" and 72 x 155" (annual capacity 1,233,000 short tons). ***Tube and pipe mills:*** welded – One double submerged arc weld 24 to 48" od, up to 1.0" wall thickness.

■ Mon Valley Works. ***Coke ovens:*** Clairton, PA: Four batteries – Koppers underjet (174 ovens); Koppers gun-flue (192 ovens); Wilputte gun-flue (192 ovens); Still gun-flue (258 ovens). ***Blast furnaces:*** Edgar Thomson Plant, Braddock PA: Three furnaces – No 1 28' 10" hearth dia, 57,218 cu ft; No 2 28' 10" hearth dia, 48,094 cu ft; No 3 26' hearth dia, 38,837 cu ft. ***Steelmaking plant:*** Two 220-ton basic oxygen converters. ***Rolling mills:*** One 1-stand 2-high 44" slabbing (Edgar Thomson Plant, Braddock, PA) (annual capacity 1,907,000 short tons). one continuous wide hot strip comprising 1-stand 2-high 36 x 70" scale-breaker, 1-stand 4-high 38" and 51 x 130" and 124" broadside, 1-stand 4-high 38" and 51 x 81" and 76" roughing, 2-stand 4-high 34" and $53\frac{1}{2}$ x 81" and 76" roughing, 6-stand 4-high $27\frac{1}{4}$" and $53\frac{1}{2}$ x 81" and 76" finishing (Irvin Plant, Dravosburg, PA) (annual capacity 2,520,000 short tons); one 5-stand 4-high 21"/24" and 60 x 84" continuous cold reduction (Irvin Plant, Dravosburg, PA) (1,357,000 tons); one 1-stand 4-high $21\frac{1}{2}$" and 53 x 54" reversing cold reduction (Vandergrift Plant, Vandergrift, PA) (73,000 tons); one 1-stand 4-high 26" and 61 x 84" continuous temper/skin pass (Irvin Plant, Dravosburg, PA) (959,000 tons); one 1-stand 4-high $20\frac{1}{2}$" and 48 x 66" continuous temper/skin pass (Vandergrift Plant, Vandergrift, PA) (119,900 tons). ***Coil coating lines:*** Irvin Plant, Dravosburg, PA – two continuous hot-dip tinning lines (54" and 48" wide); one continuous hot-dip long nickel-terne line.

■ Gary Works, Gary Plant, Gary, IN. ***Coke ovens:*** Two batteries – Wilputte underjet (308 ovens); Still single divided gun fired (114 ovens). ***Sinter plant:*** Three 8' 3" Dravo-Lurgi. ***Blast furnaces:*** Five furnaces – No 4 28' 3", 53,929 cu ft; No 6 28', 53,198", No 7 28', 44,478 cu ft; No 8 26' 6", 45,668 cu ft; No 13 40', 101,987 cu ft. ***Steelmaking plant:*** Six basic oxygen converters – three 210-ton and three 246-ton. ***Continuous casting machines:*** One slab straight with bend (9.3" x 76 x 480" and 9.3 x 64 x 480"); one 2-strand and one 1-strand slab 9' 3" x 90" x 480" (3,300,000 tons and 2,000,000 tons). ***Rolling mills:*** one Wean United 1-stand 2-high 46" reversing slabbing; wide hot strip – continuous comprising 1-stand 2-high 48 x 24", 2-stand 2-high 48 x 84", 3-stand 4-high 42 x 48", 2-stand 2-high 48 x 84" and 7-stand 4-high $28\frac{1}{2}$ x 84" (4,670,000 short tons). four cold reduction – one 4-stand 4-high 21" and 53" x 56" continuous (698,000 short tons), one 2-stand 4-high $21\frac{1}{4}$" and 53 x 48" continuous tandem for tinplate (242,000 short tons), one 5-stand 4-high 23" and 60 x 80" continuous tandem (2,157,000 short tons); one 6-stand 4-high 23" and 56 x 52" continuous tandem for tinplate (680,000 tons); six temper/skin pass – one 1-stand 4-high $20\frac{1}{2}$" and 48 x 84" (509,000 tons); one 2-stand 4-high 23" and 53 x 48" for tinplate (649,000 tons); one 2-stand 4-high 18" and 42 x 42" for tinplate (194,000 tons); one 2-stand 4-high 24" and 48 x 84" (727,000 tons); one 1-stand 4-high 24" and 60" x 80" (1,211,000 tons); one 1-stand 2-high 44 x 84" (563,000 tons). ***Coil coating lines:*** Two electrolytic tinning – 46" and 37"; one 36" tin-free steel (ECCS); two hot-dip galvanizing – width: light gauge 48", intermediate gauge 60"; one electro-galvanizing for one-side coating, width 62". ***Other plant:*** Steel foundry with one 25-ton and one 10-ton open hearth furnaces.

■ South Works, South Chicago, IL. ***Steelmaking plant:*** Three electric arc furnaces – two 200-ton and one 100-ton. ***Refining plant:*** One 90-ton AOD converter. ***Rolling mills:*** One 1-stand 2-high 54" blooming (annual capacity 803,000 short tons); one 52" heavy section with 3-stand 2-high, two horizontal edgers (720,000 tons). one 1-stand 4-high 160/210" reversing heavy plate (927,000 tons);

Products: Carbon steel – ingots; slabs; blooms; billets; round bars; wide-flange beams (heavy) (and H-piles); hot rolled hoop and strip (uncoated); skelp (tube strip); cold rolled

United States

USS, A Division of USX Corp (continued)

hoop and strip (uncoated); heavy plates; hot rolled sheet/coil (uncoated); cold rolled sheet/coil (uncoated); electro-galvanized sheet/coil; enamelling grade sheet/coil; terne-plate (lead coated) sheet/coil (and nickel-terne plate); zinc-aluminium alloy coated sheet/coil (Galvalume); electrical non-oriented silicon sheet/coil; blackplate; electrolytic tinplate (single-reduced); electrolytic tinplate (double-reduced); tin-free steel (ECCS); seamless tubes and pipes; longitudinal weld tubes and pipes; oil country tubular goods. Stainless steel – heavy plates (high-strength low-alloy); hot rolled sheet/coil (uncoated) (high-strength low-alloy). Alloy steel (excl stainless) – ingots; slabs; blooms; billets. Raw steel ouput in 1986 was 9,579,000 short tons. Pressure tubing, zinc-iron-alloy coated coil/sheets.

USS-Posco Industries Inc

Head office: Loveridge Road, Pittsburg, CA.
Management: E. Roskowesky (president and chief executive officer). **No. of employees:** 1,150. **Ownership:** USX Corp (50%) and Pohang Iron & Steel Co (50%).
Works:
■ ***Rolling mills:*** cold reduction; temper/skin pass. ***Coil coating lines:*** electrolytic tinning; hot-dip galvanizing.
Products: Carbon steel – cold rolled sheet/coil (uncoated), hot-dip galvanized sheet/coil, electrolytic tinplate (single-reduced).
Expansion programme: Modernisation programme, due for completion in 1989, will bring annual capacity to plant's original 1,350,000 short tons.

Wallace Barnes Steel

Head office: 300 Broad Street, Bristol, CT 06010. **Tel:** 203 589 5511.
Ownership: Theis of America Inc, Bristol, CT.
Products: Carbon steel – cold rolled hoop and strip (uncoated) including spring.

Washington Steel, Division of Blount Inc

Head office: PO Box 494, Washington, PA 15241.
Management: President and chief executive officer – Robert E. Heaton; vice presidents – Thomas S. Foust (sales), James J. Pierotti (executive), Robert E. Grote (administration). **Established:** 1945. **No. of employees:** 670. **Ownership:** Blount Inc, Montgomery, AL. **Annual capacity:** Raw steel: 150,000 short tons.
Works:
■ At head office. ***Steelmaking plant:*** Two 50-ton electric arc (annual capacity 150,000 tons). ***Refining plant:*** One 50-ton AOD (150,000 tons). ***Continuous casting machines:*** 1-strand Concast slab (27-52" width) (170,000 tons). ***Rolling mills:*** Steckel mill 50" wide (coil weight 20,000 lb) (170,000 tons), three Sendzimir (36", 42" and 52") (coil weight 20,000 lb) and one 2-high temper/skin pass (50" wide) (coil weight 20,000 lb).
Products: Stainless steel – slabs 5 x 27-52"; hot rolled band (for re-rolling) 27-52" wide; universal plates (continuous plate coils); cold rolled sheet/coil (uncoated) 36-52" wide.
Expansion programme: 52" Sendzimir was to be rebuilt during 1987.

Weirton Steel Corp

Head office: 400 Three Springs Drive, Weirton, WV. **Tel:** 304 797 2000.
Management: Robert Loughhead (chairman, president and chief executive) was due to retire during 1987. **Established:** 1984. **Capital:** $7,000,000 (shares authorised). $303,073,000. **No. of employees:** 8,400. **Ownership:** Employee-owned. **Annual capacity:** Raw steel: 3,700,000 short tons.
Works:
■ Weirton, WV. ***Coke ovens:*** (idle). ***Sinter plant:*** One Koppers (12' wide and 18"

Weirton Steel Corp (continued)

deep strand) (annual capacity 1,768,900 short tons). ***Blast furnaces:*** (No 1) 27'0" hearth dia, 54,048 cu m (1,095,000 tons); (No 2) 27'0" hearth dia, 44,000 cu m (949,000 tons); (No 3) 26'3" hearth dia, 46,615 cu m (985,500 tonnes); (No 4) 26'3" hearth dia, 46,670 cu m (985,500 tons) (total annual hot metal capacity limited to 3,416,400 tons). ***Steelmaking plant:*** Two 365-ton basic oxygen converters (3,648,500 tons). ***Refining plant:*** vacuum degassing – Dravo-RH 2 vessel system – 365 ton heats (3,648,500 tons). ***Continuous casting machines:*** One 4-strand Concast vertical slab with bending (30-41" x 9") (1,800,000 tons). ***Rolling mills:*** wide hot strip comprising three 2-high roughing, two 4-high roughing and six 4-high finishing stands (width 23/49", gauge 0.067/0.250) (coil weight 30,000lb) (3,148,000 tons); three cold reduction – (No 7 tandem) 5-stand 4-high (width 22/48", gauge 0.0065/0.045) (coil weight 60,000lb) (679,000 tons), (No 8 tandem) 4-stand 4-high (width 21/48:FR:5/16", gauge 0.019/0.135) (coil weight 60,000lb) (730,900 tons), (No 9 tandem) 5-stand 4-high continuous (width 24/50", gauge 0.007/0.06) (coil weight 70,000lb) (847,800 tons); temper/skin pass – (No 7 SS) 1-stand 4-high (width 18/48", gauge 0.010/0.097) (coil weight 22,000lb) (199,500 tons), (No 8 SS) 1-stand 4-high (width 17-48", gauge 0.025/0.135) (coil weight 60,000lb) (390,200 tons), (No 1 SM) 1-stand 2-high (width 20/48", gauge 0.013/0.074) (coil weight 25,000lb) (240,800 tons), (No 4 TM) (one 1-stand 4-high and one 1-stand 2-high) (width 22/45", gauge 0.006/0.065) (coil weight 60,000lb) (482,900 tons), (No 5 TM) 2-stand 4-high (width 20/45", gauge 0.004/0.065) (coil weight 60,000lb) (527,500 tons). Other cold mills No 1 Weirlite – 2-stand 4-high (width 22/45" (coil weight 30,000lb) (335,800 tons); No 2 Weirlite – 2-stand 4-high (width 24/45", gauge 0.032) (coil weight 30,000lb) (400,400 tons). ***Coil coating lines:*** Two Halogen horizontal electrolytic tinning (844,800 tons) (including tinplate capability of one TFS/tinning line); two Hexavalent chrome bath tin-free steel (ECCS) (2 stage vertical) (one is dual-purpose tinning/TFS) (549,000 tons); three continuous horizontal hot-dip galvanizing (658,600 tons); one zinc sulphate bath horizontal electro-galvanizing (156,800 tons).
Products: Carbon steel – hot rolled sheet/coil (uncoated), cold rolled sheet/coil (uncoated), hot-dip galvanized sheet/coil, electro-galvanized sheet/coil, zinc-aluminium alloy coated sheet/coil, blackplate, electrolytic tinplate (single-reduced), electrolytic tinplate (double-reduced), tin-free steel (ECCS).

Welded Tube Co of America

Head office: 1855 East 122nd Street, Chicago, IL 60633. **Tel:** 312 646 4500. **Telex:** 9102211347.
Management: Bill Nostrand **Established:** 1944.
Works:
■ ***Tube and pipe mills:*** welded One Wean (51 x 51-127 x 127mm) RHS, one Abbey Etna (100 x 100-200 x 200mm) RHS, and one McKay (127 x 127-400 x 400mm) RHS (combined annual capacity 280,000 short tons).
Products: Carbon steel – longitudinal weld tubes and pipes, hollow sections.

Western Tube & Conduit Corp

Head office: 2001 E. Dominguez St, PO Box 2720, Long Beach, CA 90801-2720. **Tel:** 213 537 6300. **Telex:** 656 396. **Telecopier (Fax):** 213 604 9785.
Management: President – Arthur C. Geldner; vice presidents – T. Tokoyoda (executive), Peter Chifo (senior); managers – Merle Bassett Jr (electrical division), Don Finn (mechanical tube and fence post divisions); controller – William Yoshimoto. **Established:** 1964. **No. of employees:** Works – 200. Office – 50. **Ownership:** Sumitomo Metals Ltd and affiliated companies. **Subsidiaries:** Omega Tube & Conduit Corp, Little Rock, AR (100%).
Works:
■ At head office. ***Tube and pipe mills:*** welded – seven high-frequency ($\frac{1}{2}$-3$\frac{1}{2}$" round, $\frac{1}{2}$-2$\frac{1}{2}$" square, $\frac{1}{2}$-1$\frac{1}{2}$" rectangular, 0.024-0.160" wall) (annual capacity 150,000 short tons).
Products: Carbon steel – longitudinal weld tubes and pipes; mechanical tubing, galvanized fence posts, electrical metallic tubing, intermediate metallic conduit.

United States

Wheatland Tube Co

Head office: 900 Haddon Avenue, Collingswood, NJ. **Tel:** 412 342 6857. **Telex:** 812 561. **Telecopier (Fax):** 412 342 6851 263.
Management: James O'Donnel (president and chairman of board); James Feeney (senior vice president of operations). **Established:** 1931. **No. of employees:** Works – 450. Office – 50.
Works:
■ At head office. ***Tube and pipe mills:*** welded – one continuous butt weld (½ to 4 NPS) (annual capacity 250,000 short tons). ***Other plant:*** Hot dip galvanizing, steel pipe and conduit couplings.
Products: Carbon steel – longitudinal weld tubes and pipes (output 250,000 tons of which 125,000 tons galvanized); galvanized tubes and pipes.

Wheeling Pittsburgh Steel Corp

Head office: Pittsburgh, PA 15230. **Postal address:** PO Box 18. **Tel:** 412 288 3600.
Established: 1968 through merger of Pittsburgh Steel Corp into Wheeling Steel Corp. **Subsidiaries:** Include Wheeling – Nisshin Coaters Inc (Nisshin 67%) – see separate entry.
Works:
■ Steubenville Plant, Steubenville, OH 43952 (Tel: 614 283 5000). ***Coke ovens:***. ***Sinter plant:***. ***Blast furnaces:*** Five. ***Steelmaking plant:*** Two 280-ton basic oxygen converters. ***Continuous casting machines:*** One 2-strand slab. ***Rolling mills:*** One universal slabbing; one 80" wide hot strip. cold reduction 4-stand tandem; three temper/skin pass. ***Coil coating lines:*** One semi-continuous terne/lead coating.
■ Monnessen Plant, Monesse, PA 15062 (Tel: 412 684 4360). ***Rolling mills:*** One blooming; two continuous billet.
■ Allenport Plant, Allenport, PA 15412 (Tel: 412 326 5601). ***Rolling mills:*** cold reduction.
■ Yorkville Plant, Yorkville, OH 43971 (Tel: 614 859 6000). ***Rolling mills:*** cold reduction; temper/skin pass. ***Coil coating lines:*** Two electrolytic tinning.
■ Martin's Ferry Plant, Martin's Ferry, OH 43935 (Tel: 304 234 0000). ***Coil coating lines:*** Three hot-dip galvanizing 36", 48" and 60".
■ Benwood Plant, Benwood, WV 26031. ***Tube and pipe mills:*** welded No 1 continuous butt weld (½-5"). ***Other plant:*** Tube galvanizing pot.
Products: Carbon steel – hot rolled sheet/coil (uncoated), cold rolled sheet/coil (uncoated), galvanized corrugated sheet/coil, terne-plate (lead coated) sheet/coil, blackplate, electrolytic tinplate (single-reduced), electrolytic tinplate (double-reduced), longitudinal weld tubes and pipes.

Wheeling-Nisshin Inc

Head office: Penn & Main Streets, Follansbee, WV 26037. **Postal address:** PO Box 635. **Tel:** 304 527 2800. **Telex:** 304 232 3296.
Management: T. Yamamura (chairman and chief executive officer); J.E. Wright III (chief operating officer); Y. Hamada (executive vice president); sales dept – M. Manatsu (vice president and general manager), K. Yamamoto (general manager). **Established:** 1986, (commercial operations to begin April 1, 1988). **Capital:** $10,000,000. **No. of employees:** Works – 75. Office – 25. **Ownership:** Nisshin Steel Co Ltd (67%), Wheeling Pittsburgh Steel Corp (33%). **Annual capacity:** Finished steel: 270,000 short tons.
Works:
■ At head office. ***Coil coating lines:*** Non-oxygenated furnace, one continuous aluminizing and galvanizing line (annual capacity 270,000 tons).
Products: Carbon steel – hot-dip galvanized sheet/coil and aluminized sheet/coil 0.014"-0.10" x 24"-60" in coil, galvanised sheets.

Whittar Steel Strip

Head office: 20001 Sherwood Avenue, Detroit, MI 48234. **Tel:** 313 893 5000.
Ownership: Dofasco, Canada.
Works:
■ At head office. ***Rolling mills:*** Five cold strip – one Bliss 26", two Lewis 26", one Ruesch 9½", and one Schloemann 26".
Products: Carbon steel – cold rolled hoop and strip (uncoated) high and low carbon.
Additional information: Formerly known as Whittaker Steel Strip Division, Whittaker Industries Inc.

Whittymore Tube Co, Div Whittymore Enterprises Inc

Head office: First & D Avenue, Seymour, IN 47274. **Postal address:** PO Box 756 **Tel:** 812 522 6039. **Telecopier (Fax):** 812 522 9932.
Management: President – Dillard Whittymore, Jr; personal assistant – Dillard Whittymore, III; sales – Leon K. Toms, Bill Bracker, Rick Pennington; chief financial officer – Douglass Booth. **Established:** 1984. **No. of employees:** 60-65.
Ownership: Whittymore Enterprises Inc.
Works:
■ At head office. ***Tube and pipe mills:*** welded – two ⅝ – 2" (8,000 tonnes) and one 1½ – 4" (1,200 tonnes).
Products: Carbon steel – longitudinal weld tubes and pipes ⅝ – 3" dia; galvanized tubes and pipes ⅝ – 3" dia; cold drawn tubes and pipes ⅝ – 3" dia; precision tubes and pipes ⅝ – 3" dia (combined total output 5,000 tonnes). Alloy steel (excl stainless) – longitudinal weld tubes and pipes ⅝ – 3" dia; cold drawn tubes and pipes ⅝ – 3" dia; precision tubes and pipes ⅝ – 3" dia (combined total output 1,000 tonnes).

Wismarq Corp

Head office: 930 E Armour Road, Oconomowoc, WI 53066. **Postal address:** PO Box 47. **Tel:** 414 567 1112. **Telecopier (Fax):** 414 567 8289.
Management: President – Charles G. Terrizzi; vice presidents – Gregory G. Wisniewski (executive), Jerome D. Stanke (manufacturing), Michael E. Hartenstein (finance), Jan M. Terrizzi (administration). **Established:** 1982. **No. of employees:** Works – 40. Office – 10.
Works:
■ At head office. ***Coil coating lines:*** Two roll (Gasway) colour coating (70,000 short tons).
Products: Carbon steel – colour coated sheet/coil (output 20,000 tons).

Wyckoff Steel Division, Ampco-Pittsburgh Corp

Head office: 700 Porter Building, Pittsburgh, PA 15219. **Tel:** 412 456 4400.
Works:
■ Chicago, IL; Putnam, CT and Plymouth, MI. Equipment for cold drawing, centreless grinding, turning and polishing.
Products: Carbon steel – bright (cold finished) bars. Alloy steel (excl stainless) – bright (cold finished) bars.

USSR

Sales are handled by V/O Promsyrioimport, 13 Chaikovsky Street, Moscow 121834 (Tel: 203 05 77. Telex: 7151, 7152. Cables: Primsi Moscow G99).

Met Zavod Amurstal (Amur Steel Works)

Annual capacity: Raw steel: over 1,000,000 tonnes.
Works:
■ Komsomolsk-on-Amur. ***Blast furnaces:***. ***Steelmaking plant:*** electric arc furnaces; open hearth. ***Continuous casting machines:*** Two 2-strand slab (160 x 750-200 x 1,550mm). ***Rolling mills:*** medium section. Cold reduction; one 1,700mm mill. ***Tube and pipe mills:*** ERW.
Products: Carbon steel – Alloy steels, sheet, medium sections, tinplate, longitudinal weld tubes and pipes.

Ashinskiy Met Zavod (Asha Iron & Steel Works)

Works:
■ Asha. ***Blast furnaces:*** two. ***Steelmaking plant:*** Four open hearth.
Products: Carbon steel – hot rolled hoop and strip (uncoated), cold rolled hoop and strip (uncoated).

Met Zavod Azovstal (Azovstal Iron & Steel Works)

Annual capacity: Raw steel: over 6,500,000 tonnes.
Works:
■ Zhdanov ***Blast furnaces:*** Six (No 3 has a capacity of 2,800-2,900 tonnes/day). ***Steelmaking plant:*** Four 400-tonne basic oxygen converters; electric arc furnaces (700,000 tonnes); open hearth. ***Refining plant:*** ESR. Vacuum degassing. ***Continuous casting machines:*** Five 2-strand slab (200 x 900-315 x 1,900mm). ***Rolling mills:*** rail; medium section; bar. Heavy plate; medium plate (3,600mm) (1,750,000 tonnes).
Products: Carbon steel – reinforcing bars, heavy rails, medium plates, heavy plates.

Beloretskiy Met Zavod (Beloretsk Iron & Steel Works)

Works:
■ Beloretsk. ***Blast furnaces:*** two. ***Other ironmaking plant:*** One shaft furnace. ***Steelmaking plant:*** One electric arc furnace; four open hearth furnaces. ***Rolling mills:*** wire rod including a 150mm mill (annual capacity 400,000 tonnes); one rod and bar (220,000 tonnes); 150mm section mill; and plate mill. Fine wire drawing plant ferro-alloy plant.
Products: Alloy steel (excl stainless) – wire rod, wire plates and sections. Fine wire drawing plant. Ferro-alloy plant.

Chelyabinskiy Met Zavod (Chelyabinsk Iron & Steel Works)

Works:
■ Chelyabinsk. ***Coke ovens:*** No 3 battery (annual capacity 440,000 tonnes). **Sinter plant**. ***Sinter plant:***. ***Blast furnaces:*** Six. ***Steelmaking plant:*** Three 130-tonne basic oxygen converters (3,000,000 tonnes); twelve electric arc furnaces; fifteen open hearth furnaces. ***Refining plant:*** ESR; VAR. ***Rolling mills:*** Two blooming – one (1,300mm); one 2-high Demag reversing; one 1-stand 3-high billet (790mm). One (250mm) light section; five bar – 15-stand continuous; 11-stand Demag looping; 3-stand 3-high (760mm) for heavy rounds and squares; intermediate hand mill; looping mill for light bar; two (250mm) wire rod. Heavy plate; one semi-continuous wide hot strip; 2,300mm sheet mill. One Sendzimir ZRZ1 BB 60 cold reduction for stainless steel up to 1,550mm wide; 20-roll cluster mill for stainless steel. ***Other plant:*** Forging presses.
Products: Carbon steel – heavy angles, heavy joists, heavy channels, wide-flange beams (heavy). Stainless steel – hot rolled sheet/coil (uncoated), cold rolled sheet/coil (uncoated), seamless tubes and pipes. Alloy steel (excl stainless) – round bars, square bars, medium plates, heavy plates, hot rolled sheet/coil (uncoated), cold rolled sheet/coil (uncoated), seamless tubes and pipes. Tool steel, bearing steels, electrical sheets.
Additional information: Tube mills include 140mm, 250mm and 1,120mm mills.

Cherepovetskiy Met Zavod (Cherepovets Iron & Steel Works)

Annual capacity: Finished steel: over 10,000,000 tonnes.
Works:
■ Cherepovets. ***Coke ovens:*** (9 batteries). ***Sinter plant:*** including a plant with a capacity of 3,000,000 tonnes. ***Blast furnaces:*** One (5,580 cu m) (capacity 12,600 tonnes/day, 4,500,000 tonnes/year, one (2,700 cu m), one (1,580 cu m), one (730 cu m), and one other. ***Steelmaking plant:*** Two 400-tonne basic oxygen converters (5,000,000 tonnes); twelve open hearth, of which one is a tandem furnace. ***Continuous casting machines:*** Two 1-strand slab (150 x 650-250 x 1,650mm). ***Rolling mills:*** One slabbing and blooming; one continuous billet. One medium section (250mm); light section; wire rod (9280mm). Rod and bar (220,000 tonnes). One heavy plate (2,800mm); two wide hot strip – one (1,700mm) semi-continuous; and one (2,000mm) continuous (6,000,000 tonnes). One (1,700mm) continuous cold reduction. ***Coil coating lines:*** One continuous hot-dip galvanizing. ***Other plant:*** Slab reheating furnace (420 tonnes per hour).
Products: Carbon steel – wire rod, round bars, light angles, hot rolled hoop and strip (uncoated), cold rolled hoop and strip (uncoated), medium plates, heavy plates, hot rolled sheet/coil (uncoated), cold rolled sheet/coil (uncoated); Electrical sheets.

Dneprodzerzhinskiy Zavod Imeni Dzerzhinskogo (Dzerzhinsky Works)

Annual capacity: Rolled steel capacity about 5,000,000 tonnes.
Works:
■ Dneprodzerzhinsk. ***Blast furnaces:***. ***Steelmaking plant:*** Two 200-tonne basic oxygen converters (2,200,000 tonnes). ***Rolling mills:*** blooming ; mill for seamless tube semis. Heavy section; rail; medium section – automated semi-continuous 350mm; light section. Heavy plate; medium plate.
Products: Carbon steel – blooms, billets, semis for seamless tubemaking, round bars, light angles, light channels, heavy angles, heavy channels, wide-flange beams (heavy), heavy rails, medium plates, heavy plates, strip, axles.

Dnepropetrovskiy Met i Trubnoprokatny Savod Imeni Lenina (Lenin Iron & Steel Tube Rolling Works)

Works:
■ Dnepropetrovsk. ***Steelmaking plant:*** Three 50-tonne basic oxygen converters (annual capacity 1,500,000 tonnes); six open hearth furnaces (700,000 tonnes). ***Rolling mills:*** One slabbing; one billet. Steckel mill. ***Tube and pipe mills:*** seamless Stiefel mill for tapered tubes in engineering and heat-resistant steels. Welded.
Products: Carbon steel – seamless tubes and pipes, longitudinal weld tubes and pipes. Alloy steel (excl stainless) – seamless tubes and pipes. Automobile axle castings.

Met Zavod Dneprospetstal (Dnieper Special Steel Works)

Works:
■ Zaporozhye. ***Steelmaking plant:*** Twenty electric arc furnaces. ***Refining plant:*** Six ESR; VIM. ***Rolling mills:*** One blooming. Three bar. ***Other plant:*** Forge. Alloy and special steels, including tool steels.

Donetskiy Metallurgicheskiy Zavod (Donetsk Iron & Steel Works)

Works:
■ Donetsk. ***Blast furnaces:***. ***Steelmaking plant:*** electric arc furnaces including one 100-tonne; open hearth. ***Refining plant:*** One 130-tonne DH vacuum degassing unit.

USSR

Donetskiy Metallurgicheskiy Zavod (Donetsk Iron & Steel Works) (continued)

Continuous casting machines: Two slab including 1 4-strand (150 x 600-200 x 1,200mm). ***Rolling mills:*** One billet (950/990mm continuous). Section mill.
Products: Carbon steel – slabs, billets. Alloy steel (excl stainless) – slabs, billets, sections.

Elektrostal Mashinostroitelny Zavod (Elektrostal (Novokramatorsk) Machine Building Works)

Works:
■ Kramatorsk. ***Steelmaking plant:*** electric arc furnaces; four open hearth furnaces. ***Refining plant:*** VAR. **Rolling mills:** ***Other plant:*** The steel plant supplies the engineering works, which makes plant for the steel industry.

Elektrostal Metallurgechesky Zavod Imeni Tevosyan

Annual capacity: 400,000-600,000 tonnes.
Works:
■ Elektrostal (formerly Noginsk), near Moscow. ***Steelmaking plant:*** electric arc. ***Refining plant:*** ESR; VAR. ***Continuous casting machines:*** One 2-strand billet (82mm sq). ***Tube and pipe mills:*** Continuous weld (25-83mm dia, 4-8m long). ***Other plant:*** 1,020mm tube mill. Vacuum casting plant.
Products: Alloy steel (excl stainless) – longitudinal weld tubes and pipes.

Izhevskiy Met Zavod (Izhevsk Iron & Steel Works)

Works:
■ Izhevsk. ***Steelmaking plant:*** electric arc furnaces; open hearth furnaces. ***Rolling mills:*** One blooming (annual capacity 1,000,000 tonnes).

Karagandinskiy Met Zavod (Kazakhstanskaya Magnitka) (Karaganda Iron & Steel Works)

Works:
■ Karaganda. ***Coke ovens:*** (8 batteries) (annual capacity 7,000,000 tonnes). ***Blast furnaces:*** Two (1,700 cu m) (1,500,000 tonnes), one (2,700 cu m), and one (3,200 cu m) (2,000,000 tonnes). ***Steelmaking plant:*** basic oxygen converters – three 250-tonne, three 300-tonne, and a third shop with 250-tonne vessels (4,000,000-5,000,000 tonnes); four open hearth furnaces, two 500-tonnes. ***Rolling mills:*** One slabbing (1,150mm). One 12-stand wide hot strip (1,700mm) (1.2-12mm). Two cold reduction – one (1,700mm) (0.4-2mm), and one 6-stand (1,400mm) continuous. ***Coil coating lines:*** Four electrolytic tinning (1,125,000 tonnes).
Products: Carbon steel – slabs, hot rolled sheet/coil (uncoated), cold rolled sheet/coil (uncoated), electrolytic tinplate (single-reduced).

Kazakhskiy Met Zavod (Kazakh Iron & Steel Works)

Works:
■ Temir-Tau (near Karaganda). ***Blast furnaces:*** Four including one (2,700 cu m) and one (3,200 cu m). ***Steelmaking plant:*** Three 310-tonne LD basic oxygen converters (annual capacity 5,000,000 tonnes); four open hearth furnaces, one 150-tonne, two 100-tonne and one 65-tonne. **Continuous casting machines:**. ***Rolling mills:*** rail; wire rod. One wide hot strip. One cold reduction (1,700mm). ***Coil coating lines:*** one electrolytic tinning, tin-free steel (ECCS).
Products: Carbon steel – wire rod, heavy rails, hot rolled sheet/coil (uncoated), cold rolled sheet/coil (uncoated), blackplate, electrolytic tinplate (single-reduced), tin-free steel (ECCS).

Makeyevskiy Met Zavod Imeni Kirova (Kirov Iron & Steel Works)

Works:
■ Makeyevka. ***Coke ovens:*** including one battery with an annual capacity of (417,000 tonnes). ***Blast furnaces:*** Five (3,000,000 tonnes). ***Steelmaking plant:*** One electric arc furnace; eighteen open hearth furnaces (3,500,000 tonnes); 2 converters. ***Rolling mills:*** One heavy section; one continuous wire rod. ***Other plant:*** Forge. Tube and pipe mill (65-100mm dia). Centrifugal casting plant.
Products: Carbon steel – wire rod, reinforcing bars, heavy joists, heavy channels, heavy rails, rail accessories. Strip, bars, forgings. Frost-resistant steels.

Kirovskiy Zavod (Kirov Works)

Works:
■ Leningrad. ***Steelmaking plant:*** Three electric arc furnaces; Six open hearth furnaces. ***Rolling mills:*** One billet. Section mills including one 350mm and one 900/680mm. ***Tube and pipe mills:*** seamless.
Products: Carbon steel – billets, sections and tubes.
Additional information: The steel plant forms part of a large engineering works.

Kommunarskiy Met Zavod (Kommunarsk Iron & Steel Works)

Works:
■ Alchevsk (formerly Voroshilovsk). **Sinter plant**. ***Blast furnaces:*** Five including one of (3,000 cu m). ***Steelmaking plant:*** Ten open hearth furnaces including 250-tonne and 500-tonne. ***Rolling mills:*** blooming. Heavy section. heavy plate (2,800mm).
Products: Carbon steel – medium plates, heavy plates, stuctural sections.

Konstantinovskiy Met Zavod Imeni Frunze (Konstantinovka Frunze Iron & Steel Works)

Works:
■ Konstantinovka. ***Blast furnaces:*** Two. ***Steelmaking plant:*** Five open hearth furnaces. ***Rolling mills:*** bar; wire rod; Sheet, section.
Products: Carbon steel – wire rod, reinforcing bars, sheets, sections, bars.

Met Zavod Krasny Oktyabr (Red October Steel Works)

Works:
■ Volgograd. ***Steelmaking plant:*** electric arc furnaces including one 200-tonnes; sixteen open hearth furnaces. ***Refining plant:*** One ESR. ***Rolling mills:*** One Demag blooming (1,150mm); two billet and bar – Schloemann (750mm) and Schloemann (450mm); one Klein Demag light bar wire rod. Two heavy plate – one Uralmash 3-high universal, and one (96"); one mill for helical rolling of drill steel.
Products: Carbon steel – wire rod, medium plates, heavy plates.

Met Zavod Krivorizhstal (Krivoy Rog Iron & Steel Works)

Works:
■ Krivoi Rog. Coke ovens. Sinter plant. ***Iron ore pelletizing plant:*** 2-strand (annual capacity 8,000,000 tonnes). ***Blast furnaces:*** One 5,000 cu m (13,000 tonnes/day), one 2,700 cu m, three 2,000 cu m, one 1,719 cu m, and three others. ***Steelmaking plant:*** Ten basic oxygen converters – four 50-tonne (2,000,000 tonnes), six 150-tonne (6,500,000 tonnes); five 600-tonne open hearth. ***Rolling mills:*** blooming – several including one 1,300mm with an annual capacity of 3,200,000 tonnes; two continuous billet. Light section (250mm) for alloy and low alloy steels; bar; one 4-strand

USSR

Met Zavod Krivorizhstal (Krivoy Rog Iron & Steel Works) (continued)

continuous Sket wire rod (250mm) (500,000 tonnes); one rod and bar (220,000 tonnes). ***Other plant:*** Tinning line.
Products: Carbon steel – blooms, wire rod, round bars, square bars, flats, light angles, hot rolled hoop and strip (uncoated), skelp (tube strip), bright wire. Alloy steel (excl stainless) – blooms, wire rod, round bars, square bars, light angles. Alloy steels.

Kursk Works—see Oskolskiy Elektrometallurgicheskiy Kombinat (OEMK)

Met Zavod Kuybyshev (Kuibyshev Iron & Steel Works)

Works:
■ Kramatorsk. ***Blast furnaces:*** Four (950,000 tonnes). ***Steelmaking plant:*** electric arc furnaces; Five open hearth furnaces. ***Rolling mills:*** blooming; billet; light section; bar. Hot strip and hoop; cold strip. ***Other plant:*** Foundry equipped with 50-tonne arc furnace.
Products: Carbon steel – reinforcing bars; medium and light sections. Stainless steel – round bars 6-230mm dia; tool steels; heavy machinery castings.

Kuznetskiy Met Kombinat

Works:
■ Novokuznetsk (formerly Stalinsk), Kuzbas region. ***Coke ovens:*** (5 batteries). ***Blast furnaces:*** including one (3,000 cu m) (annual capacity 2,000,000 tonnes). Total annual capacity (4,000,000 tonnes). ***Steelmaking plant:*** 250/300-tonne basic oxygen converters; Five electric arc furnaces (250,000 tonnes each); fifteen open hearth furnaces. ***Rolling mills:*** One (1,300mm) slabbing; two blooming; one continuous billet. Heavy section and rail; one medium section; three bar. One heavy plate; (450mm) section mill (1,500,000 tonnes).
Products: Carbon steel – blooms, billets, reinforcing bars, round bars, heavy rails, medium plates, heavy plates.

Met Zavod Imeni Karl Liebknecht (Karl Liebknecht Steel Works)

Works:
■ Dnepropetrovsk. ***Steelmaking plant:*** Six open hearth furnaces (annual capacity 700,000 tonnes). ***Tube and pipe mills:*** seamless Pilger mill for bearing tube. ***Other plant:*** 140mm tube mill (200,000-215,000 tonnes).
Products: Carbon steel – seamless tubes and pipes; railway wheels.

Liepajas Rupnica Sarkanais Metallurgs (Red Metal Worker Plant)

Works:
■ Liepaja (formerly Libau, Latvia). ***Steelmaking plant:*** electric arc furnaces; five open hearth furnaces (600,000 tonnes). ***Continuous casting machines:*** Two 8-strand billet (120mm and 150mm sq). **Rolling mills:** Section and sheet.
Products: Carbon steel – billets; Sections and sheets.

Magnitogorskiy Metallurgischeskiy Kombinat (Magnitogorsk Iron & Steel Works Combine)

Annual capacity: Raw steel: about 14,000,000 tonnes.
Works:
■ Magnitogorsk. ***Blast furnaces:*** Twelve including one of (1,370 cu m). ***Steelmaking plant:*** Thirty-five open hearth furnaces including two 900-tonne. ***Refining plant:*** One

Magnitogorskiy Metallurgischeskiy Kombinat (Magnitogorsk Iron & Steel Works Combine) (continued)

300-tonne DH vacuum degassing unit. ***Rolling mills:*** One (49") slabbing; two blooming one (45") Demag, one 2-stand (1,150mm); one continuous billet; one Demag 6-stand cross-country billet and medium section; one Siemag light section; one bar and light section – Siemag; two wire rod – one 8-stand 2-strand, one Siemag 4-stand. One (98") wide hot strip; one hot strip and hoop and rod – Sack; one light plate. One (98") cold reduction; one 20-roll cold strip. ***Coil coating lines:*** One Head Wrightson electrolytic tinning; one Sendzimir hot-dip galvanizing.
Products: Carbon steel – wire rod, round bars, light angles, light channels, heavy angles, heavy joists, heavy channels, wide-flange beams (heavy), heavy rails, hot rolled hoop and strip (uncoated), skelp (tube strip), cold rolled hoop and strip (uncoated), hot rolled sheet/coil (uncoated), cold rolled sheet/coil (uncoated), hot-dip galvanized sheet/coil, electrolytic tinplate (single-reduced). Castings.

Novolipetskiy Met Zavod (Novolipetsk Iron & Steel Works)

Works:
■ Lipetsk. ***Coke ovens:*** including one of (annual capacity 900,000 tonnes). ***Sinter plant:***. ***Blast furnaces:*** Six – two (1,260 cu m), two (2,000 cu m), one (3,200 cu m) (2,200,000 tonnes), and one (5,580 cu m) (14,000 tonnes per day). ***Steelmaking plant:*** Six basic oxygen converters – three 160-tonne (4,200,000 tonnes), three 300-tonne (6,000,000 tonnes); two electric arc furnaces. ***Continuous casting machines:*** Two 2-strand slab (200 x 1,000-300 x 2,050mm (2,000,000 tonnes). ***Rolling mills:*** Two wide hot strip – one 12-stand 2,000mm (6,000,000 tonnes), one 2,400mm (8,000,000 tonnes); Steckel mill; five cold reduction – one Sendzimir (1,200mm), one (2,000mm), one 5-stand continuous tandem producing strip up to 1,850mm wide and 0.35-2mm thick (2,500,000 tonnes), one continuous mill for sheet 0.5-1.5mm thick and up to 2,000mm wide (500,000 tonne and one (1,300,000 tonnes)e Electrical sheet mill (480,000 tonnes). ***Coil coating lines:*** One hot-dip galvanizing (500,000 tonnes).
Products: Carbon steel – hot rolled sheet/coil (uncoated), cold rolled sheet/coil (uncoated), hot-dip galvanized sheet/coil, electrical non-oriented silicon sheet/coil.

Novomoskovskiy Zavod (Novomoskovsk Works)

Works:
■ Novomoskovsk. ***Tube and pipe mills:*** welded ERW Spiral.
Products: Carbon steel – longitudinal weld tubes and pipes upto 2,500mm dia for oil and gas pipelines; spiral-weld tubes and pipes (double-layer, for gas); oil country tubular goods.
Additional information: Skelp from Karaganda works or Ilyich works, Zhdanov, is used.

Novosibirskiy Met Zavod Kuzmin

Works:
■ Novosibirsk. ***Blast furnaces:***. ***Steelmaking plant:*** Six open hearth furnaces. ***Tube and pipe mills:*** welded Longitudinal. ***Other plant:*** Hot and cold rolling including a sheet mill and a narrow strip mill.
Products: Carbon steel – hot rolled hoop and strip (uncoated), skelp (tube strip), cold rolled hoop and strip (uncoated), hot rolled sheet/coil (uncoated), cold rolled sheet/coil (uncoated), longitudinal weld tubes and pipes upto 114mm dia.

Novo-Tagilskiy Met Zavod (Novo-Tagil Iron & Steel Works)

Works:
■ Nizhniy-Tagil. ***Blast furnaces:*** Six (including one 2,700 cu m). ***Steelmaking plant:*** Three 160-tonne basic oxygen converters (3,500,000 tonnes); eighteen open hearth furnaces. ***Continuous casting machines:*** One 1-strand slab (200 x 1,500mm-250

Novo-Tagilskiy Met Zavod (Novo-Tagil Iron & Steel Works) (continued)

x 1,800mm). ***Rolling mills:*** One blooming (1,500mm). Two heavy section – one universal beam, one wide flange beam (1,000,000 tonnes); one rail (500,000 tonnes); One light section and bar One heavy plate; one wide hot strip. One (1,300mm) cold reduction.
Products: Carbon steel – semis for seamless tubemaking, reinforcing bars, heavy sections including wide-flange beams (heavy), heavy rails, medium plates, heavy plates, cold rolled sheet/coil (uncoated), tyres.

Orsko-Khalilovskiy Met Kombinat (Orsk-Khalilovo Iron & Steel Combine)

Works:
■ Novo-Troitsk. ***Blast furnaces:*** Four (including one of 2,000 cu m). ***Steelmaking plant:*** Two electric arc furnaces; seven open hearth furnaces; three Bessemer converters. ***Refining plant:*** Two 130-tonne DH vacuum degassing units. **Continuous casting**. ***Rolling mills:*** One (1,120mm) blooming. One heavy section; (800mm universal). One (2,800mm) heavy plate; Strip mill.
Products: Steelmaking pig iron Carbon steel – medium plates; heavy plates 9-10mm and 2,530mm wide; low-alloy steels including chromium steels; ferro-nickel.

Oskolskiy Elektrometallurgicheskiy Kombinat (OEMK) (Kursk Works)

Works:
■ Stary Oskol. ***Iron ore pelletizing plant:*** (annual capacity 2,400,000 tonnes). Four Midrex ***Direct reduction plant:*** (400 series) (1,670,000 tonnes). ***Steelmaking plant:*** Four 150-tonne Krupp electric arc furnaces. ***Refining plant:*** Two 150-tonne DH vacuum degassing units; two Argon refining plants, two Desulphurising plants. ***Continuous casting machines:*** Four 4-strand bloom (300 x 360mm). ***Rolling mills:*** SMS heavy section with reversing rougher and 8-stand continuous billet train (1,200,000 tonnes).
Expansion programme: A further eight DR modules are planned, bringing annual capacity to 5,000,000 tonnes, together with additional arc furnace and continuous casting capacity.

Pervouralskiy Novotrubny Zavod (Pervouralsk New Tube Works)

Works:
■ Pervouralsk. ***Tube and pipe mills:*** seamless (102mm) (650,000 tonnes%. ***Other plant:*** 160mm tube mill. Mill making boiler tube 10-219mm.
Products: Carbon steel – seamless tubes and pipes for the oil, gas and chemical industries, thick-wall high-alloy steel tube and pipe.

Met Zavod Imeni Petrovskogo (Petrovsky Iron & Steel Works)

Works:
■ Dnepropetrovsk. ***Blast furnaces:*** six. ***Steelmaking plant:*** Three basic oxygen converters; eight open hearth furnaces. ***Rolling mills:*** blooming; wire rod. Heavy plate; Sheet, section.
Products: Carbon steel – blooms, wire rod, medium plates, heavy plates, sheets, sections.

Serovskiy Metallurgicheskiy Zavod (Serov Iron & Steel Works)

Works:
■ Serov. ***Blast furnaces:*** Seven. ***Steelmaking plant:*** electric arc furnaces; Nine open hearth furnaces. ***Rolling mills:*** medium section; light section; plate and sheet. Plate, sheet, light and medium sections, including special steels.

Met Zavod Serp i Molot (Sickle & Hammer Works)

Works:
■ Moscow. ***Steelmaking plant:*** Four 10-tonne electric arc furnaces. ***Continuous casting machines:*** Two 1-strand (100-160mm sq) billet and slab (130 x 400-160 x 500mm). ***Rolling mills:*** light section and wire rod; bar (220,000 tonnes); 20-roll cold reduction Mill for stainless sheets. ***Other plant:*** Wire drawing plant.

Severskiy Met Zavod (Seversk Iron & Steel Works)

Works:
■ Polevskoy. ***Steelmaking plant:*** Four open hearth furnaces. ***Tube and pipe mills:*** seamless including (10-500mm dia). Welded including (220mm dia). ***Other plant:*** Strip mill (320,000 tonnes). Tinning lines.

Zavod Svobodniy Sokol (Free Falcon Works)

Works:
■ Lipetsk. ***Blast furnaces:*** Being increased to about 1,500,000 tonnes. ***Other plant:*** Tube plant (320,000 tonnes).

Taganrogskiy Metallurgicheskiy Zavod (Taganrog Iron & Steel Works, formerly Andreyevskiy Works)

Works:
■ Taganrog. ***Steelmaking plant:*** Two electric arc furnaces; eight open hearth furnaces. **Continuous casting machines:**. ***Tube and pipe mills:*** welded.
Products: Carbon steel – oil country tubular goods.

Uzbekskiy Metallurgicheskiy Zavod (Uzbek Iron & Steel Works)

Works:
■ Bekabad (formerly Begovat) near Leninabad, Uzbekistan. ***Steelmaking plant:*** Five electric arc furnaces (annual capacity 1,000,000 tonnes). ***Continuous casting machines:*** One 4-strand bloom (200mm sq). ***Rolling mills:*** medium section 33-stand (300mm) (1,000,000 tonnes). Bar.
Products: Carbon steel – semis for seamless tubemaking. Stainless steel – blooms, semis for seamless tubemaking, round bars.

Verkh-Isetskiy Met Zavod (Verkh-Isetsk Iron & Steel Works)

Works:
■ Sverdlovsk. ***Blast furnaces:*** One (3,000 cu m) (annual capacity 2,000,000 tonnes). ***Steelmaking plant:*** Three electric arc furnaces; open hearth. ***Rolling mills:*** One Sendzimir cold strip; mills for electrical sheet (200,000 tonnes).

USSR

Volzhskiy Truboprokatny Zavod (Volga Pipe Works)

Works:
■ Volzskiy, near Volgograd. ***Tube and pipe mills:*** seamless including two extrusion presses and plant producing 170,000 tonnes/year of pipes 42-245mm dia in carbon, low alloy and special steels . Welded Ten.
Products: Carbon steel – seamless tubes and pipes, longitudinal weld tubes and pipes, large-diameter tubes and pipes.

West Siberian Steel Works—see Zapsib Met Zavod

Yenakiyevskiy Met Zavod (Yenakiyevo Iron & Steel Works)

Works:
■ Yenakiyevo. ***Blast furnaces:*** Six. ***Steelmaking plant:*** Three 135-tonne basic oxygen converters (annual capacity 3,500,000 tonnes). ***Rolling mills:*** rail. Heavy plate.
Products: Carbon steel – heavy rails, medium plates, heavy plates.

Yuzhnotrubny Zavod (Southern Tube Works)

Works:
■ Nikopol. ***Tube and pipe mills:*** seamless Continuous extrusion plant. Mannesmann piercers and plug mills. Welded Continuous butt-weld (3-102mm). ERW.
Products: Carbon steel – seamless tubes and pipes (small-diameter); longitudinal weld tubes and pipes; oil country tubular goods. Thin-wall high-alloy tubes; low-alloy pipes.

Zakavkazskiy Met Zavod (Transcaucasian Metallurgical Works)

Works:
■ Rustavi, Georgian SSR. ***Sinter plant:*** (annual capacity 2,200,000 tonnes). ***Blast furnaces:*** Two (700 cu m) (800,000 tonnes), one (1,030 cu m) (700,000 tonnes). ***Steelmaking plant:*** open hearth. ***Continuous casting machines:*** One 4-strand bloom (200 x 280mm). ***Rolling mills:*** billet. Rail. Sheet mills. ***Tube and pipe mills:*** seamless American-built Mannesmann-type piercers. Plug mills. ***Other plant:*** Forge. Tube galvanizing/aluminizing plant.
Products: Carbon steel – round bars, heavy rails, seamless tubes and pipes upto 400mm dia. Sections, engineering steel.

Zaporozhstal Zavod (Zaporozhye Steel Works)

Works:
■ Zaporozhye. **Coke ovens. Sinter plant.** ***Blast furnaces:*** Five.
Steelmaking plant: *Eighteen open hearth including one tandem with* an annual capacity of 1,000,000 tonnes. ***Refining plant:*** VAR. ***Rolling mills:*** slabbing. Wide hot strip (1,700mm). Cold reduction (1,700mm and others). ***Coil coating lines:*** One electrolytic tinning; five continuous hot-dip tinning; Sendzimir hot-dip galvanizing. ***Other plant:*** 2,800mm mill.
Products: Carbon steel – cold roll-formed sections (from strip), hot rolled hoop and strip (uncoated), cold rolled hoop and strip (uncoated), hot rolled sheet/coil (uncoated), cold rolled sheet/coil (uncoated), hot-dip galvanized sheet/coil, colour coated sheet/coil, electrical non-oriented silicon sheet/coil, electrolytic tinplate (single-reduced). Stainless steel – hot rolled sheet/coil (uncoated), cold rolled sheet/coil (uncoated). Special steels. Ferro-alloys, including ferro manganese and ferro-silicon.

Zapsib Met Zavod (West Siberian Steel Works)

Works:
■ Antonovskaya, near Novokuznetsk (Kuzbas). ***Coke ovens:*** (at least 7 batteries). ***Blast furnaces:*** – One (2,000 cu m), one (2,700 cu m) and one (3,000 cu m) (combined annual capacity 2,000,000 tonnes). ***Steelmaking plant:*** Three 130/250-tonne basic oxygen converters (4,000,000 tonnes); electric arc furnaces for special steel. ***Rolling mills:*** blooming; one continuous billet. Wire rod. One wide hot strip (2,500,000 tonnes).
Products: Carbon steel – blooms, billets, wire rod, light angles, hot rolled sheet/coil (uncoated), alloy steels.

Zhdanovskiy Met Zavod Ilyich (Ilyich Works)

Works:
■ Zhdanov. ***Blast furnaces:*** Five (annual capacity 3,500,000 tonnes). ***Steelmaking plant:*** Three 150-tonne basic oxygen converters; eighteen open hearth furnaces (2,600,000 tonnes). ***Refining plant:*** One 30/60-tonne Asea-SKF with induction stirring vacuum decarburising and powder injection facilities. ***Continuous casting machines:*** One 2-strand slab (250 x 1,150-350 x 2,200mm). ***Rolling mills:*** 4-high medium plate (3,100mm) (2,500,000 tonnes). ***Tube and pipe mills:*** seamless upto (355mm dia). Welded Three Spiral – one (1,020mm), and two (1,220mm). ***Other plant:*** 1,700mm continuous strip mill.
Products: Carbon steel – slabs, medium plates, hot-dip galvanized sheet/coil, seamless tubes and pipes, spiral-weld tubes and pipes.

Zhlobin Metallurgical Works

Works:
■ Zhlobin, Belorussia. ***Steelmaking plant:*** Two 100-tonne UHP electric arc (750,000 tonnes). ***Refining plant:*** ladle furnace. ***Continuous casting machines:*** Two 6-strand Danieli billet (100-300mm sq); one 4-strand Voest-Alpine bloom (200mm sq – 300 x 400mm). ***Rolling mills:*** One light section, bar and wire rod (annual capacity 500,000 tonnes). ***Other plant:*** Tyre cord plant.

Zlatoustovskiy Met Zavod (Zlatoust Iron & Steel Works)

Works:
■ Zlatoust. ***Steelmaking plant:*** electric arc furnaces; nine open hearth furnaces; two vacuum melting shops with seven furnaces. **Continuous casting machines:**. ***Rolling mills:*** blooming. Heavy section; medium section including a 560/400mm mill; light section; bar; 350/500mm mill (annual capacity over 500,000 tonnes). Alloy, stainless, high-speed and bearing steels.

Venezuela

Acerex CA

Head office: Edif Centro-Ven, Gonzalez Rincones, La Trinidad, Caracas 1010A, DF. **Postal address:** PO Box 1392. **Tel:** 02 93 97 11. **Telex:** 21516 telen vc. **Cables:** Acerexcar. **Telecopier (Fax):** 58 02 936379.
Management: President – Ricardo Rojas; managers – Boris Goldstein (general), Jose A. Gomez (plant), Pilar Menendez (sales); comptroller – Zully Vargas. **Established:** 1976. **Capital:** Bs 13,000,000. **No. of employees:** Works – 85. Office – 20. **Annual capacity:** Raw steel: 6,000 tonnes. Finished steel: 5,000 tonnes.
Works:
■ C Ave, Zona Industrial Los Montones, Barcelona, E. Anzoategui. (Tel: 081 770459. Telex: 82108 acrxb vc). Works director: Jose A. Gomez. ***Steelmaking plant:*** One

Acerex CA (continued)

4-tonne electric arc furnace (annual capacity 6,000 tonnes). ***Other plant:*** Open die hydraulic press (1,000-tonnes) (5,000 tonnes).
Products: Carbon steel – round bars 6-20" dia; square bars 5-16"; flats 2-8" thick x 6-26" wide. Stainless steel – round bars 6-20" dia. Alloy steel (excl stainless) – round bars 6-20" dia (including tool steels); square bars 5-16" (including tool steels); flats 2-8" thick x 6-26" wide (including tool steels).
Additional information: All bars produced by the company are forged.

CA Conduven

Head office: Avda Beethoven, Torre Financiera, piso 9, Colinas de Bello Monte, Caracas. **Postal address:** Apartado Postal 60299. **Tel:** 752 41 11, 752 79 99. **Telex:** 27885, 21301.
Management: Dezider Weisz (president); Ricardo Rojas (executive vice president); Evangelina Sánchez (chief financial officer); Eliyahu Weisz (chief executive officer); Pedro Contreras (general plant manager); Laudelino Martin (operations manager).
Established: 1959. **Capital:** Bs 160,000,000. **No. of employees:** Works – 960. Office – 140. **Ownership:** Century CA. **Subsidiaries:** Intermodal La Flecha (55%) (storage for export, services). **Associates:** Tuboauto CA (30%) (pipe producer), Metaltubos CA (25%) (galvanized pipe distributor). **Annual capacity:** Finished steel: 360,000 tonnes.
Works:
■ Urb Industrial La Chapa, Av Gran Colombia, La Victoria, Edo Aragua (Tel: (044) 23311, 21211. Telex: 44136). Works director: Pedro Contreras. ***Tube and pipe mills:*** Eight ERW high frequency welded – one $12\frac{3}{4}$" od (annual capacity 120,000 tonnes); four $2\frac{3}{8}$"od (60,000 tonnes); two $4\frac{1}{2}$" od (72,000 tonnes) and one 3" od ERW-HF (18,000 tonnes).
Products: Carbon steel – pipe piling $2\frac{3}{8}$-$12\frac{3}{4}$"od; longitudinal weld tubes and pipes $\frac{3}{8}$"-12" nominal; oil country tubular goods 2 $\frac{3}{8}$"-12 $\frac{3}{4}$"od; galvanized tubes and pipes $\frac{3}{8}$", 6" nominal; cold drawn tubes and pipes $\frac{1}{8}$"-8 $\frac{1}{2}$" od; hollow sections – rounds $\frac{1}{2}$" to 12" nominal, squares 12 x 12mm to 220mm, rectangles 25 x 12mm to 320 x 120mm.

Fior de Venezuela

Head office: Torre America, piso 13, Avda Venezuela, Urb. Bello Monte, Caracas. **Postal address:** PO Box 4022, Carmelitas CCS 1010. **Tel:** 708 6300. **Telex:** 28390 fior vc. **Cables:** Fiovensa Caracas. **Telecopier (Fax):** 71 30 17.
Management: President – Oscar Augusto Machado; directors – Francisco J. Layrisse R. (steel division), Alberto Hassan, Eduardo Gisbert (works), Hans Petersen (sales).
Established: 1973. **Capital:** Bs 256,000,000. **No. of employees:** Works – 340. Office – 70. **Ownership:** Siderurgica Venezolana SA (Sivensa) (60%); Corporacion Venezolana de Guayana (CVG) (40%).
Works:
■ ***Direct reduction plant:*** One fluidised bed iron ore reduction (Fior process) (annual capacity 400,000 tonnes).
Products: Direct-reduced iron (DRI) briquettes (output 383,656 tonnes).

Imosa – Industria Mecanica Orion SA

Head office: Av Romulo Gallegos, Edificio Zulia, piso 2, Boleita Norte, Caracas 1070. **Tel:** 02 35 34 71. **Telex:** 29564 imosa vc. **Telecopier (Fax):** 239 69 76.
Management: Paul Otamendi B. (president), Giuseppe R. Lupo (vice president), Natale Clemenza (sales and purchasing manager). **Established:** 1960. **Capital:** Bs 78,000,000. **No. of employees:** Works – 420. Office – 50.
Works:
■ At head office. ***Tube and pipe mills:*** welded One SAW 20-120" (annual capacity 100,000 tonnes).
Products: Carbon steel – longitudinal weld tubes and pipes and accessories for water

Imosa - Industria Mecanica Orion SA (continued)

supply mains, oil transfer pipelines and hydro-electric plants; pressure vessels for thermal plants, refineries and petrochemical plants; heavy structural steelwork, gates and valves for - hydraulic and hydro-electric projects.
Sales offices/agents: USA - Speri Associates, 2600 South Gessner, Suite 310, Houston, TX 77063 (Telex: 166295). Colombia - Ospa SA, Calle 26 No 44/45, Edf KLM, Bogota (Telex: 44417).

Lamigal - Láminas Galvanizadas CA

Head office: Edif. Centro Plaza Torre A, piso 13, Oficina A. Ave Francisco de Miranda, Los Palos Grandes, Caracas.
Established: 1964.
Works:
■ Zona Industrial Sur, Valencia, Edo Carabobo. ***Coil coating lines:*** hot-dip galvanizing.
Products: Carbon steel - hot-dip galvanized sheet/coil.

Metalanca - Metalúrgica de Laminados Nacionales CA

Head office: Av. Los Jabillos, Edif Centro Bancor, piso 8, Sabana Grande, Caracas.
Telex: 21974 met.
Works:
■ ***Rolling mills:*** light section (360/260mm) and bar.
Products: Carbon steel - round bars, flats, light angles, light tees.

Siderzulia - Siderurgica del Zulia CA

Head office: Edif Corpozulia, Avda 4 Con Calle 83, Maracaibo Edo, Lulia. **Postal address:** Apartado 1153. **Tel:** 70425-28. **Telex:** 61101.
Expansion programme: This ambitious project for an integrated steelworks has been considerably delayed and there are doubts as to whether construction will proceed.

Sidetur - Siderúrgica del Turbio SA

Head office: Avda Venezuela Torre America piso 11, Caracas. **Telex:** 21911.
Management: Directors - Francisco Layrisse (general), Malcolm Caldwell (manufacturing), Hans Petersen (sales), Miguel Grosso (purchasing). **Established:** 1972. **Capital:** BS 500,000,000. **No. of employees:** Works - 1,159. Office - 1,439. **Ownership:** Siderúrgica Venezolana SA (Sivensa). **Annual capacity:** Raw steel: 477,000 tonnes. Finsihed steel: 370,000 tonnes.
Works:
■ Barquisimeto. ***Steelmaking plant:*** Five electric arc furnaces - three 20-tonne and one 40-tonne (annual capacity 500,000 tonnes); one 75-tonne (under construction) (300,000 tonnes). ***Continuous casting machines:*** Two 4-strand billet - one Concast-Rokop 100-130mm (300,000 tonnes) and one Concast (under construction) (300,000 tonnes). ***Rolling mills:*** One cross country medium section 320mm (140,000 tonnes); one cross country light section 260mm (80,000 tonnes); two bar - one continuous 500-260mm (300,000 tonnes) and one cross country 480-260mm (150,000 tonnes).
Products: Carbon steel - ingots (Output 190,000 tonnes); billets (167,000 tonnes); reinforcing bars (175,000 tonnes); square bars (9,281 tonnes); flats (3,891 tonnes); light angles (25,612 tonnes); light tees (54,388 tonnes).

Sidor – CVG Siderúrgica del Orinoco CA

Head office: Zona Industrial Matanzas, Puerto Ordaz, Estado Bolivar. **Tel:** 86229. **Telex:** 86370.
Management: Chairman – Cesar Mendoza Osio; vice presidents – Angel Barreto (operations), Rafael A. Carrasquel (sales), Oswaldo Herrera Malpica (personnel), Ramon Azpurua (financial). **Established:** 1961. **Capital:** Bs 18,172,800,000. **No. of employees:** Works – 13,920. Office – 3,961. **Ownership:** Fondo de Inversiones de Venezuela (71%); National Executive (21%); CVG – Corporación Venezolana de Guyana (8%). **Associates:** CVG-International (15%) (trading company); Fesilven (5%) (ferro-alloy plant). **Annual capacity:** Pig iron: 680,000 tonnes. Raw steel: 4,800,000 tonnes. Finished steel: 3,700,000 tonnes.
Works:
■ Edificio General, Avenida La Estancia, Chuao (Tel: 02 2080900. Telex: 23409 sidor vc). ***Sinter plant:*** (not in operation). ***Iron ore pelletizing plant:*** Continuous – dry (annual capacity 6,600,000 tonnes). ***Direct reduction plant:*** One HYL (363,000 tonnes), three HYL II (2,100,000 tonnes), four Midrex (1,600,000 tonnes). ***Other ironmaking plant:*** Nine Elkem electric pig iron furnaces (40 tonnes/hour) (700,000 tonnes). ***Steelmaking plant:*** Ten electric arc furnaces – six 200-tonne and four 150-tonne (3,600,000 tonnes); 300-tonne open hearth furnace (1,200,000 tonnes). ***Continuous casting machines:*** Three 3-strand Hitachi billet (1,500,000 tonnes); three 2-strand Demag slab (2,400,000 tonnes). ***Rolling mills:*** One 1-stand Mesta slabbing and blooming (890,000 tonnes). one 3-stand Innocenti billet (500,000 tonnes). one 3-stand Innocenti medium section (500,000 tonnes); one 3-stand Innocenti-Schloemann light section (95,000 tonnes); two 26-stand Schloemann-Siemag bar (750,000 tonnes); two 2-strand 25-stand Schloemann-Siemag wire rod (450,000 tonnes); one 2-stand heavy plate comprising one scale-breaker and one 4-high reversing (90,000 tonnes); one 6-stand 4-high wide hot strip (coil weight 20 tonnes max) (2,100,000 tonnes); two 10-stand 4-high cold reduction (coil weight 20 tonnes max) (1,000,000 tonnes and 450,000 tonnes); three 5-stand 4-high temper/skin pass (coil weight 15 tonnes max) (1,000,000 tonnes). ***Tube and pipe mills:*** seamless – two Pilger (120,000 tonnes); one Mannesmann push bench (30,000 tonnes). ***Coil coating lines:*** One Wean United sulphuric acid electrolytic tinning (120,000 tonnes); one Wean United 1-stand 2-high tin-free steel (ECCS) (140,000 tonnes). ***Other plant:*** Ferro-alloy furnaces. Forging, cold forging and stamping plant. Foundry.
Products: Steelmaking pig iron (1986 output 493,300 tonnes). Direct-reduced iron (DRI) (2,555,300 tonnes). Carbon steel – ingots (603,380 tonnes); slabs (continuous cast – 170-200mm x up to 840mm x up to 2,040mm; rolled – 790 x 185 x 7,730mm) (1,638,165 tonnes); blooms 240-279mm sq x 6,900-7,620mm (289,900 tonnes); billets (continuously cast – 131mm sq x 15m; rolled – 80-178mm sq x 4.2-15m) (983,600 tonnes); semis for seamless tubemaking (104,556 tonnes); hot rolled coil (for re-rolling) (881,083 tonnes); wire rod 5.5-12mm (coil weight 1,800kg) (394,530 tonnes); reinforcing bars (375,100 tonnes); round bars $\frac{1}{4}$-2.257" x 6-12m long; light angles (9,691 tonnes); light tees (15,365 tonnes); light channels (17,094 tonnes); medium angles 20 x 20 x 3mm-120 x 120 x 12mm x 6-12m long; medium tees (9,056 tonnes); medium joists 80-300mm x 10-12m long; medium channels 80-300mm x 10-12m long (5,646 tonnes); cold roll-formed sections (from strip) (775,020 tonnes); sheet piling; pipe piling; skelp (tube strip) (401,000 tonnes); medium plates (43,749 tonnes); heavy plates 8-76mm thick, 1,500-3,000mm wide, 2-12m long (45,321 tonnes); floor plates (11,202 tonnes); hot rolled sheet/coil (uncoated) 2-12mm x 600-1,250mm; cold rolled sheet/coil (uncoated) 0.27-2mm x 600-1,200mm; blackplate (162,701 tonnes); electrolytic tinplate (single-reduced) and tin-free steel (ECCS) (189,900 tonnes); seamless tubes and pipes (linepipe – $2\frac{3}{8}$-24" dia x 2.11-39.67mm wall thickness; API casing – $4\frac{1}{2}$-$7\frac{5}{8}$" dia x 5.21-12.7mm wall thickness; API tubing – 1.050-$4\frac{1}{2}$" dia x 2.87-6.88mm thick); large-diameter tubes and pipes; oil country tubular goods.

Venezuela

Sivensa – Siderúrgica Venezolana SA

Head office: Caracas 101. **Postal address:** Apartado 1488. **Telex:** 21682 sivensa.
Established: 1948. **Subsidiaries:** Sidetur – Siderúrgica del Turbio SA (mini steelworks); Fior de Venezuela (direct reduction plant); Sidecar (direct reduction plant).
Works:
■ Matanzas (closed since 1977). ***Direct reduction plant:*** One Midrex (400 series) (Sidecar – due to start up 1990) (annual capacity 400,000 tonnes). ***Steelmaking plant:*** Two electric arc furnaces – one 85-tonne (closed since 1977), one 75-tonne (proposed). ***Continuous casting machines:*** One slab (proposed). ***Rolling mills:*** One bar (closed since 1977) (350,000 tonnes).
■ Antímano, Caracas. ***Steelmaking plant:*** Two 23-tonne electric arc furnaces. ***Continuous casting machines:*** Two 2-strand Concast billet. ***Rolling mills:*** Two bar and wire rod.

Univensa – Unión Industrial Venezolana SA

Head office: Barquisimeto, Edo. Lara. **Postal address:** Apartado 113. **Telex:** 52118 univensa.
Works:
■ ***Tube and pipe mills:*** welded Lonitudinal. ***Other plant:*** Tube galvanizing plant.
Products: Carbon steel – longitudinal weld tubes and pipes, galvanized tubes and pipes.

Vietnam

Steel sales are handled by Vietnam National Minerals Export/Import Corp, 35 Hai-Ba-Trung, Hanoi (Tel: 55264, 53336, 52661. Telex: 4515. Cables: Minexport Hanoi).

Bien Hoa Works

Works:
■ ***Rolling mills:*** bar (annual capacity 25,000 tonnes).
Products: Carbon steel – reinforcing bars.

Gia Sang Works

Products: Carbon steel – wire rod.

Thai Nguyen Iron & Steel Works

Works:
■ Located about 70km north of Hanoi. Blast furnaces. Steelmaking plant. Rolling mill.

Yugoslavia

Zeljezara Iljaš

Head office: Iljaš, Sarajevo.
Established: 1956. **Ownership:** RMK-Zenica.
Works:

Zeljezara Iljaš (continued)

■ Iljaš. Electric pig iron furnaces (annual capacity 264,000 tonnes).
Products: foundry pig iron.

IMK – Organizacija Za Proizvodnju Cévićno Zavarenih cevi i Profila Uroševac

Head office: Provog Maja 7, Ferizaj Uroševac. **Tel:** 038 70 411. **Telex:** 18160 yuimk. **Telecopier (Fax):** 24.
Management: Ilaz Ilazi (generalni direktor), Evic Bošidar (zamenik-kotoro), Jahja Idrizi (teghnioki direktor), Ahmet Hysenaj (komercialni direktor), Njazi Aslani (financiski direktor). **Established:** 1972. **No. of employees:** Works – 850. Office – 450. **Annual capacity:** Finished steel: 120,000 tonnes.
Works:
■ ***Tube and pipe mills:*** welded Longitudinal-weld (114-406mm dia) (annual capacity 40,000 tonnes), Spiral weld (355-2,020mm dia) (40,000 tonnes).
Products: Carbon steel – longitudinal weld tubes and pipes 114-406mm; spiral-weld tubes and pipes and large-diameter tubes and pipes upto 2,020mm; oil country tubular goods.

Boris Kidrič Steelworks

Head office: Vuka Karadžića BB, 81400 Nikšić. **Tel:** 083 31 422 31 622. **Telex:** 61120. **Cables:** Željezara Nikšić.
Management: Radoslav Bulajić (president). **Established:** 1950. **No. of employees:** 6,530. **Annual capacity:** Raw steel: 360,000 tonnes. Finished steel: 310,000 tonnes.
Products: Carbon steel – wire rod 3-8mm dia; round bars 8-150mm dia; square bars 50 x 50-200 x 200mm; flats 5-40 x 30-60mm; hexagons 10-60mm; medium plates 3-5mm. Stainless steel – wire rod 2-8mm dia (cold rolled); round bars 8-135mm dia; square bars 50-160mm; flats 5-40 x 30-140mm; hexagons 10-60mm. Alloy steel (excl stainless) – billets 60 x 60-200 x 200mm; round bars 8-150mm dia; square bars 50 x 50-200 x 200mm; flats 5-40 x 30-140mm; hexagons 10-60mm; medium plates 3-5mm; heavy plates 5-6mm; wire 2-8mm (colled rolled).

Kosjeric Works

Head office: Kosjeric, Serbia.
Works:
■ Kosjeric. Strip mill (annual capacity 20,000 tonnes).

11 Oktomvri Tube Works, Kumanovo

Head office: 91300 Kumanovo.
Works:
■ ***Tube and pipe mills:*** welded Longitudinal. ***Other plant:*** Coating plants for pipes.
Products: Carbon steel – longitudinal weld tubes and pipes, hollow sections.

RMK-Zenica – Rudarsko-Metalurski Kombinat Zenica

Head office: 72000 Zenica, Dzemala Bijedica 15. **Tel:** 072 33322. **Telex:** 43129 rmkze yu.
Works:
■ At head office. ***Coke ovens:*** Four batteries (39 ovens) (annual capacity 720,000 tonnes), one battery (65 ovens) (720,000 tonnes). ***Sinter plant:*** One 4-strand (2,300,000 tonnes), one sinter pan (800,000 tonnes). ***Blast furnaces:*** One (2,000 cu m), three (1,000 cu m) and one (236 cu m). ***Other ironmaking plant:*** Three

RMK-Zenica – Rudarsko-Metalurski Kombinat Zenica (continued)

Tysland-Hole electric pig iron furnaces (100,000 tonnes). ***Steelmaking plant:*** Two 100-tonne basic oxygen converters; two electric arc furnaces; open hearth. ***Continuous casting machines:*** Three 4-strand USSR vertical bloom (265 x 340mm). ***Rolling mills:*** One blooming (915/2,143mm) (850,000 tonnes). One heavy section (800mm) (400,000 tonnes); one medium section (550mm) (230,000 tonnes); three light section – two (320mm) and one (280mm); one 2-strand 25-stand wire rod (230mm) (coil weight 2,700 lb). ***Tube and pipe mills:*** welded Longitudinal. ***Other plant:*** Cast iron pipe plant. Wire plant. Forge. Foundry. Cold roll-formed section plant.

Zeljezara Sisak Integrated Metallurgical Works

Head office: Božidara Adžije 2, 44103 Sisak. **Tel:** 23617; 23618.
Works:
■ Sisak. ***Coke ovens:*** One battery (65 ovens) (annual capacity 850,000 tonnes). ***Sinter plant:*** One (4 x 8sq m) Greenawalt (160,000 tonnes). ***Blast furnaces:*** Two 3.8m hearth dia (202 cu m0 (225,000 tonnes). ***Steelmaking plant:*** One 30-tonne electric arc furnace (70,000 tonnes); two 150-tonne open hearth (326,000 tonnes). ***Continuous casting machines:*** One 6-strand Demag bloom and slab (240,000 tonnes). ***Rolling mills:*** One blooming and hot strip and hoop (400,000 tonnes). ***Tube and pipe mills:*** seamless (154,000 tonnes). Welded (136,000 tonnes), precision (70,000 tonnes). ***Other plant:*** Foundry.
Products: Steelmaking pig iron. Carbon steel – ingots, slabs, blooms, billets, semis for seamless tubemaking, hot rolled hoop and strip (uncoated), seamless tubes and pipes, longitudinal weld tubes and pipes, oil country tubular goods, galvanized tubes and pipes, cold drawn tubes and pipes, precision tubes and pipes. Castings.
Expansion programme: New seamless pipe mill (annual capacity 200,000 tonnes) – 1989.

Rudnici i Zelezara Skopje

Head office: 91000 Skopje.
Established: 1967.
Works:
■ ***Sinter plant:*** One Delattre Levivier (annual capacity 260,000 tonnes). ***ironmaking plant:*** Five rotary kilns for ore pre-reduction, five Tysland-Hole electric pig iron furnaces (430,000 tonnes). ***Steelmaking plant:*** Two 110-tonne LD/AC basic oxygen converters; one 130-tonne Birlec electric arc furnace (combined capacity 980,000 tonnes). ***Continuous casting machines:*** Three 1-strand USSR curved mould slab (175 x 900mm-250 x 1,500mm). ***Rolling mills:*** One 1-stand 4-high universal reversing slabbing (1,584 x 3,350mm) and heavy plate; one semi-continuous wide hot strip (1,700mm). Two cold reduction – one reversing (1,700mm), one 5-stand 4-high tandem; temper/skin pass. ***Coil coating lines:*** One hot-dip galvanizing; and one colour coating. ***Other plant:*** Welded beam plant.
Products: Carbon steel – hot rolled hoop and strip (uncoated), cold rolled hoop and strip (uncoated), medium plates, heavy plates, hot rolled sheet/coil (uncoated), cold rolled sheet/coil (uncoated), hot-dip galvanized sheet/coil, colour coated sheet/coil. Welded beams.

Slovenske Železarne – Iron & Steel Works of Slovenia

Head office: M. Pijadejeva 5, Ljubljana. **Telex:** 31372 yu slez.
Ownership: State-owned.

Zelezarna Jesenice *Head office:* Cesta Železarjev 8,Jesenice,SRS.
Established: 1869.
Works:
■ Ljubljana. ***Sinter plant:*** One 2 metre Dwight Lloyd (annual capacity 220,000 tonnes). ***Blast furnaces:*** two (200,000 tonnes). ***Steelmaking plant:*** Two electric arc furnaces (180,000 tonnes); six open hearth furnaces (320,000 tonnes). ***Rolling mills:***

Slovenske Železarne - Iron & Steel Works of Slovenia (continued)

One semi-continuous wire rod (135,000 tonnes); one 3-high heavy plate (103,000 tonnes); one 2-stand 4-high universal roughing and 4-high reversing finishing wide hot strip (540,000 tonnes); two cold strip one Sendzimir (115,000 tonnes) and 4-high Siemag (30,000 tonnes). ***Other plant:*** Foundry; cold forming plant for sections; bar drawing plant; wire plant; nail plant; electrode plant.

Zelezarna Ravne
Established: 1620.
Works:
■ Ravne na Koroškem. ***Steelmaking plant:*** Four electric arc furnaces: two 40-tonne (annual capacity 110,000 tonnes), one 30-tonne (50,000 tonnes) and one 10-tonne (2,000 tonnes). ***Refining plant:*** One 2-tonne ESR (2,000 tonnes); one 30-tonne Stokes vacuum degassing unit (10,000 tonnes). ***Rolling mills:*** One 750mm blooming; one 550mm medium section and one 300mm light section (combined capacity 95,000 tonnes). ***Other plant:*** Steel foundry; bright bar plant; machine shop.

Zelezarna Store
Works:
■ Store Pri Celju. ***Sinter plant:*** One 8sq m Greenawalt (annual capacity 50,000 tonnes). ***ironmaking plant:*** One Demag 18,000kVA (50,000 tonnes). ***Steelmaking plant:*** Two electric arc furnaces: one 40-tonne (60,000 tonnes) and one 60-tonne (80,000 tonnes). ***Continuous casting machines:*** One 4-strand Concast billet (200,000 tonnes). ***Rolling mills:*** One light section and bar – 3-high 550mm and 2-high 300mm and 250mm (120,000 tonnes); one section mill – two 425mm trains and one 300mm train (35,000 tonnes). ***Other plant:*** Foundry for mill rolls, ingot moulds and heavy castings; foundry for engineering castings.

Products: Carbon steel – billets; wire rod; round bars; flats; bright (cold finished) bars; special light sections; cold roll-formed sections (from strip); cold rolled hoop and strip (uncoated); medium plates; heavy plates; hot rolled sheet/coil (uncoated); electrical non-oriented silicon sheet/coil; bright wire. Stainless steel – round bars; bright (cold finished) bars; wire. Alloy steel (excl stainless) – round bars; flats (spring steel); bright (cold finished) bars; cold rolled hoop and strip (uncoated); hot rolled sheet/coil (uncoated); wire. Carbon and alloy steel forgings; nails; electrodes; castings; cast mill rolls, ingot moulds and heavy iron castings – up to 20 tonnes; repetition engineering castings.
Expansion programme: A new special steel works built by Mannesmann Demag was due to start up in mid-1986, with an annual capacity of 210,000 tonnes of special, stainless, silicon and low-alloy steels.

Metalurski Kombinat Smederevo

Head office: Goranska 12, 11300 Smederevo. **Tel:** 026 23413. **Telex:** 11621 zel sm yu. **Cables:** Metkos. **Telecopier (Fax):** 026 21696.
Management: President – Svetislav Radivojevic; vice presidents – Zdeslav Luketic (sales and metallurgy), Zarko Stankovic (research and development). **Established:** 1913. **No. of employees:** 10,500. **Annual capacity:** Pig iron: 1,300,000 tonnes. Raw steel: 1,400,000 tonnes. Hot rolled coils and sheets: 1,300,000 tonnes. Cold rolled coils and sheets: 800,000 tonnes.
Works:
■ At head office. ***Sinter plant:*** Two 75 sq m strand (annual capacity 1,200,000 tonnes) and two 75 sq m (under construction) (1,200,000 tonnes). ***Blast furnaces:*** One 7.2m hearth dia, 1,033 cu m (600,000 tonnes) and one 8.2m hearth dia, 1,386 cu m (700,000 tonnes). ***Steelmaking plant:*** Three LD basic oxygen – two 80-tonne (to be uprated to 100-tonne) and one 100-tonne (1,300,000 tonnes); One 4-tonne electric arc (8,000 tonnes); Two 50-tonne open hearth (100,000 tonnes). ***Refining plant:*** One 100-tonne RH vacuum degassing unit (500,000 tonnes); three 100-tonne ladle Argon stirring units. ***Continuous casting machines:*** slab – three 2-strand vertical (200 x 1,000-1,500 x 6,000mm) and two 1-strand curved mould (under construction) (200 x 1,200-2,000 x 1,200mm) (combined capacity 1,300,000

Yugoslavia

Metalurski Kombinat Smederevo (continued)

tonnes). ***Rolling mills:*** One 88" SMS semi-continuous wide hot strip with 5-finishing stands (2,240mm) (coil weight 15 tonnes – to be 35 tonnes) (1,200,000 tonnes); two cold reduction – one Davy 5-stand tandem (1,700mm) (coil weight 15 tonnes – to be 35 tonnes) (60,000 tonnes) and one single-reversing 4-high (1,700mm) (coil weight 15 tonnes) (200,000 tonnes); two temper/skin pass – one 2-stand DCR (1,450mm) (coil weight 15 tonnes – to be 35 tonnes) (600,000 tonnes) and one single-stand (1,700mm) (coil weight 15 tonnes) (200,000 tonnes). ***Other plant:*** Forge, steel foundry.
Products: Steelmaking pig iron. Carbon steel – ingots 2 tonnes; slabs; hot rolled coil (for re-rolling) 3-15mm x 1,000-1,500mm x (15 tonnes); sheet piling 3-5mm x 1,000-1,500mm x 2,000-6,000mm; hot rolled hoop and strip (uncoated) 3-15mm x 50-1,500mm; cold rolled hoop and strip (uncoated) 0.15-2.50mm x 25-1,500mm; medium plates 3.0-15.0mm x 1,000-1,500 x 4,000-12,000mm; floor plates 3.0-15.0mm x 1,000-1,500mm; hot rolled sheet/coil (uncoated) 3-15 x 1,000-1,500mm; cold rolled sheet/coil (uncoated) 0.15-2.50mm x 1,000-1,500mm; enamelling grade sheet/coil 0.50-2.0, 1,000-1,500mm; blackplate 0.15-0.35mm x 700-1,200mm. Steel castings, forgings.

Jadranska Zelejezara Split (Adriatic Steelworks, Split)

Head office: 58212 Split. **Postal address:** PO Box 259. **Telex:** 26169 adsid yu.
Works:
■ Split. ***Steelmaking plant:*** Two electric arc furnaces – one 20/23-tonne Brown Boveri (annual capacity 55,000 tonnes), one 25-tonne Brown Boveri UHP (65,000 tonnes). ***Continuous casting machines:*** Two billet – one 2-strand Concast (75-125mm sq), one 2-strand Concast (80-130mm sq) (combined capacity 120,000 tonnes). ***Rolling mills:*** One Pomini Farrel wire rod with (450mm) roughing train, 5-stand (290/340mm) intermediate, (310mm) tandem and 6-stand (260mm) monobloc finishing train (85,000 tonnes). ***Other plant:*** Wire drawing plant. Bi-steel light beam plant.
Products: Carbon steel – billets, wire rod, reinforcing bars,. Bright wire. Bi-steel light beams.

Trepca Rudarsko-Metalurski-Hemijski Kombinat

Head office: Olova i cinta-Zvecan Fabrika Pocinkovanoglima, Vucitrn. **Telex:** 18146 koslim.
Works:
■ Vucitrn. ***Coil coating lines:*** One hot-dip galvanizing.
Products: Carbon steel – hot-dip galvanized hoop and strip.

Unis (Associated Metal Industry in Sarajevo)

Head office: Bulevar Borisa Kidrica bb, Sarajevo.
Works:
■ Borisa Kidrica 15, Banja Luka. ***Rolling mills:*** One cold strip (annual capacity 115,000 tonnes).
■ Derventa. ***Tube and pipe mills:*** welded.
Products: Carbon steel – cold rolled hoop and strip (uncoated), bright wire, longitudinal weld tubes and pipes, cold drawn tubes and pipes.

Rudnik i Zelezara Vareš

Works:
■ Vareš. ***Blast furnaces:***.
Products: Foundry pig iron. Cast iron – pipes.

Zorka Chemical Works – Hemijska Industrija Zorka

Head office: Obilicev Venac 4, Belgrade. **Telex:** 12049.
Works:
■ Narodni heroja 1, Sabac. ***Coil coating lines:*** One Head Wrightson electrolytic tinning (annual capacity 150,000 tonnes).
Products: Carbon steel – electrolytic tinplate (single-reduced).

Zaire

Sté Nationale de Sidérurgie

Head office: Kinshasa 1. **Postal address:** BP 13098. **Telex:** 21784.
Works:
■ ***Steelmaking plant:*** One 55-tonne electric arc furnace (annual capacity 120,000 tonnes). ***Continuous casting machines:*** One 4-strand Demag billet (100-136mm sq). ***Rolling mills:*** wire rod and merchant bar. One cold strip. ***Other plant:*** Galvanizing plant.
Products: Carbon steel – billets, wire rod, reinforcing bars, cold rolled hoop and strip (uncoated).

Zimbabwe

Lancashire Steel (Pvt) Ltd

Head office: Bessemer Street Industrial Sites, Kwekwe. **Postal address:** PO Box 315. **Telex:** 3327 lawire zw.
Ownership: Ziscosteel.
Works:
■ At head office. ***Rolling mills:*** wire rod and wire plant.
Products: Carbon steel – wire rod, bright wire, black annealed wire, galvanized wire (plain), galvanized wire (barbed).

Zimbabwe Iron & Steel Co Ltd – Ziscosteel

Head office: Private Bag 2, Redcliff. **Tel:** 62400. **Telex:** 7042, 7312 zisco zw.
Management: Directors – D. Divaris (chairman), D.J. Fry, B.V. Mancama, W.P. Furusa, N.A.F. Williams, S. Biyam, F. Grobler, M.J. Harris, X.M. Kadhani, H. Mashanyare, N.T. Mawema, E. Mutowo; alternate – I.M. Mathieson. **Established:** 1957. **Capital:** Z$350,000,000. **No. of employees:** 5,800. **Ownership:** Government of Zimbabwe (90%). **Subsidiaries:** Lancashire Steel (Pvt) Ltd (wire and rod production) (100%); Buchwa Iron Mining Co (Pvt) Ltd (iron ore mining) (100%). **Annual capacity:** Raw steel: 1,000,000 tonnes.
Works:
■ At head office. ***Coke ovens:*** Two CRC batteries (110 ovens) (annual capacity 545,000 tonnes). ***Sinter plant:*** One Dwight Lloyd (37m sq) (200,000 tonnes). ***Blast furnaces:*** One 5.5m (562 cu m) (210,000 tonnes), one 8.75m (1,348 cu m) (525,000 tonnes). ***Steelmaking plant:*** Two x 50 tonne LD basic oxygen converters (840,000 tonnes). ***Continuous casting machines:*** One 2-strand Voest-Alpine billet (150 x 160mm) (150,000 tonnes). ***Rolling mills:*** One 1 x 2-high reversing blooming (1,055 x 2,440mm) (720,000 tonnes); one 8-stand continuous billet (24" roughing, 20/18" finishing) (650,000 tonnes). One 3 x 3-high cross country medium section (530mm (21")) (100,000 tonnes); one 4 x 3-high and 2 x 2-high cross country light section (12" x 10") (40,000 tonnes); one 1-strand bar / wire rod 28-stand 2-high (450 – 260mm dia) (coil weight 600kg) (200,000 tonnes). ***Other plant:*** Foundry – for iron, steel and non-ferrous castings (100,000 tonnes per annum).
Products: Steelmaking pig iron 8.4kg per half pig (1986 output 3,000 tonnes). Foundry pig iron 8.4kg per half pig (6,000 tonnes). Carbon steel – blooms 150/160/

Zimbabwe

Zimbabwe Iron & Steel Co Ltd – Ziscosteel (continued)

165mm sq (74,000 tonnes); billets 55/60/63.5/70/80/92/100/110mm sq (265,000 tonnes); wire rod (47,000 tonnes); reinforcing bars 10-30mm (37,600 tonnes); round bars 10-90mm (900 tonnes); square bars 8-40mm (9,300 tonnes); flats 20 x 4mm – 230 x 25mm; light angles 25 x 4mm – 50 x 8mm (12,900 tonnes); special light sections grader blade 1,100mm/plough share 1,700mm/w section 6,000mm/ fen standard 2,300mm (combined output 12,800 tonnes); medium angles 60 x 6mm – 100 x 12mm (9,200 tonnes); medium channels 3 x 1½, 4" x 2", 5 x 2½, 6" x 3" (6,200 tonnes); light rails 20lb, 30lb, 45lb (2,000 tonnes).

Buyers' Guide

Index to Buyers' Guide

MILL AGENTS
& EXPORTERS
Ferofin S.A.
4, Rue de l'Hôpital
2000 Neuchâtel, Switzerland
Phone: (38) 25.01.02
Telex: 952680 fero ch
Fax: (38) 25.01.05
Your Partner for
Prime & Secondary Steel
Blooms Billets Slabs
H.R. Coils/Plates
C.R. Coils/Sheets
Blackplate Tinplate TFS
Pipes & Tubes
Wire Rods
Carbon & Alloy Steel
Stainless Steel
Forgings

Index to Buyers' Guide – cont.
Stainless Steel – cont.

Buyers' Guide

Steelmaking pig iron

ALGERIA
SNS – Sté Nationale de Sidérurgie

ARGENTINA
Establecimiento Altos Hornos Zapla

AUSTRALIA
The Broken Hill Pty Co Ltd

BELGIUM
Forges de Clabecq SA

BRAZIL
Siderurgica Amaral SA
Aços Anhanguera SA (Acoansa)
Cimetal Siderurgia SA
Siderúrgica São Cristóvão Ltda
Usinas – Usinas Siderúrgicas de Minas Gerais SA
Usipa – Usina Siderúrgica Paraense SA

CANADA
The Algoma Steel Corp Ltd
Stelco Inc

CHINA
Anshan Iron & Steel Co
Baoshan Iron & Steel Works
Fushun Xinfu Steel Plant
Guangzhou Iron & Steel Works
Shoudu Iron & Steel Co
Taiyuan Iron & Steel Co
Wuhan Iron & Steel (Wisco)

CZECHOSLOVAKIA
Vitkovice Steelworks

FRANCE
Usinor Sacilor

WEST GERMANY
Duisburger Kupferhütte GmbH
Klöckner Stahl GmbH
Maxhütte – Eisenwerke-Gesellschaft Maximilianshütte mbH

WEST GERMANY —*continued*
Stahlwerke Peine-Salzgitter AG (P&S)
Rogesa Roheisengesellschaft Saar mbH

HUNGARY
Danube Works
Diosgyör Works (Lenin Metallurgical Works)

INDIA
Sail – Steel Authority of India Ltd
The Tata Iron & Steel Co Ltd (Tisco)
Visvesvaraya Iron & Steel Ltd (formerly The Mysore Iron & Steel Ltd)

IRAN
National Iranian Steel Corp (Nisc)

ITALY
Nuova Italsider SpA

JAPAN
Godo Steel Ltd (Godo Seitetsu)
Kawasaki Steel Corp
Kobe Steel Ltd (Kobe Seikosho)
Nakayama Steel Works Ltd (Nakayama Seikosho)
Nippon Steel Corp (Shin Nippon Seitetsu)
Nisshin Steel Co Ltd
NKK – Nippon Kokan KK
Sumitomo Metal Industries Ltd

NORTH KOREA
Hwanghai Iron Works

SOUTH KOREA
Pohang Iron & Steel Co Ltd

MEXICO
Ahmsa – Altos Hornos de Mexico SA de CV

NORWAY
Norsk Jernverk AS

Steelmaking pig iron
—continued

PAKISTAN
Pakistan Steel Mills Corp Ltd

POLAND
Huta Katowice

SOUTH AFRICA
Iscor Ltd

SPAIN
AHV – Altos Hornes de Vizcaya SA
Ensidesa – Empresa Nacional Siderurgica SA

TUNISIA
El Fouladh – Sté Tunisienne de Sidérurgie

TURKEY
Ereğli Demir ve Çelik Fabrikalari TAS (Erdemir)
TDCI – Türkiye Demir ve Çelik Işletmeleri Genel Müdürlügü

UNITED KINGDOM
Airedale Iron & Steel Co Ltd
BSC – British Steel Corp

USSR
Orsko-Khalilovskiy Met Kombinat (Orsk-Khalilovo Iron & Steel Combine)

VENEZUELA
Sidor – CVG Siderúrgica del Orinoco CA

YUGOSLAVIA
Zeljezara Sisak Integrated Metallurgical Works
Metalurski Kombinat Smederevo

ZIMBABWE
Zimbabwe Iron & Steel Co Ltd – Ziscosteel

Foundry pig iron

ALBANIA
Steel of the Party Metallurgical Combine

ARGENTINA
Tamet – SA Talleres Metal/urgicos San Martin

AUSTRALIA
The Broken Hill Pty Co Ltd

BRAZIL
Acesita – Cia Aços Especiais Itabira
Siderúrgica Alterosa Ltda
Siderurgica Amaral SA
Aços Anhanguera SA (Acoansa)
Cia Metalúrgica Barbará
Cimetal Siderurgia SA
Usina Queiroz Júnior SA Indústria Siderúrgica
Siderúrgica São Cristóvão Ltda
Sidersete – Siderúrgica Sete Lagoas Ltda
Usinas – Usinas Siderúrgicas de Minas Gerais SA
Usipa – Usina Siderúrgica Paraense SA

CANADA
QIT – Fer et Titane Inc

CHILE
Cia Siderúrgica Huachipato SA

CHINA
Fushun Xinfu Steel Plant

CZECHOSLOVAKIA
Vitkovice Steelworks

FINLAND
Dalsbruk Oy Ab

FRANCE
Usinor Sacilor

WEST GERMANY
Duisburger Kupferhütte GmbH
Stahlwerke Peine-Salzgitter AG (P&S)
Thyssen Stahl AG

INDIA
Sail – Steel Authority of India Ltd
The Sandur Manganese & Iron Ores Ltd (Smiore)

ITALY
Athirta Industriale Triestine SpA

JAPAN
Kobe Steel Ltd (Kobe Seikosho)
Nippon Steel Corp (Shin Nippon Seitetsu)
Sumitomo Metal Industries Ltd
Yahagi Iron Co Ltd (Yahagi Seitetsu KK)

Foundry pig iron *—continued*

SOUTH KOREA
Pohang Iron & Steel Co Ltd

NORWAY
Norsk Jernverk AS

PAKISTAN
Pakistan Steel Mills Corp Ltd

PORTUGAL
Cia Protuguesa de Fornos Eléctricos Sarl

SPAIN
AHV – Altos Hornes de Vizcaya SA

TAIWAN
China Steel Corp

TURKEY
Ereğli Demir ve Çelik Fabrikalari TAS (Erdemir)
TDCI – Türkiye Demir ve Çelik Işletmeleri Genel Müdürlügü

UNITED KINGDOM
Airedale Iron & Steel Co Ltd
BSC – British Steel Corp
North Eastern Iron Refining Co Ltd

UNITED STATES
Acme Steel Co
Armco Inc
Rouge Steel Co
Shenango Inc

YUGOSLAVIA
Zeljezara Iljaš
Rudnik i Zelezara Vareš

ZIMBABWE
Zimbabwe Iron & Steel Co Ltd – Ziscosteel

Blast furnace ferro-manganese

WEST GERMANY
Thyssen Stahl AG

INDIA
The Tata Iron & Steel Co Ltd (Tisco)

UNITED KINGDOM
BSC – British Steel Corp

Direct-reduced iron (DRI)

ARGENTINA
Acindar Industria Argentina de Aceros SA

BRAZIL
Usiba – Usina Siderúrgica da Bahia SA

WEST GERMANY
Hamburger Stahlwerke GmbH

INDIA
Sponge Iron India Ltd

INDONESIA
PT Kratkatau Steel

IRAN
National Iranian Steel Corp (Nisc)

MALAYSIA
Sabah Gas Industries Sdn Bhd

MEXICO
Hylsa SA de CV

NIGERIA
Delta Steel Co Ltd

SOUTH AFRICA
Scaw Metals Ltd

SWEDEN
Höganäs AB

TRINIDAD & TOBAGO
Iscott – Iron & Steel Co of Trinidad & Tobago Ltd

UNITED STATES
Georgetown Steel Corp

VENEZUELA
Fior de Venezuela
Sidor – CVG Siderúrgica del Orinoco CA

Iron powder

CANADA
Domfer Metal Powders Ltd
QIT – Fer et Titane Inc

EAST GERMANY
VEB Eisen- und Hüttenwerke Thale

JAPAN
Kawasaki Steel Corp

Iron powder *—continued*

JAPAN *—continued*
Kobe Steel Ltd (Kobe Seikosho)

SWEDEN
Höganäs AB

Carbon steel – ingots

ARGENTINA
Acindar Industria Argentina de Aceros SA
Establecimiento Altos Hornos Zapla

AUSTRALIA
The Broken Hill Pty Co Ltd
Commonwealth Steel Co Ltd

AUSTRIA
Eisenwerk Breitenfeld Ges.mbH Nachf KG

BELGIUM
SA Fabrique de Fer de Charleroi

BRAZIL
Acesita – Cia Aços Especiais Itabira
Aços Anhanguera SA (Acoansa)
Cobrasma SA
Cofavi – Cia Ferro e Aço de Vitória
Cosipa – Cia Siderúrgica Paulista
Mannesmann SA

BULGARIA
Kremikovtsi Iron & Steel Works
Lenin Iron & Steel Works

CANADA
The Algoma Steel Corp Ltd
Atlas Specialty Steels Division of Rio Algom Ltd
Canadian Steel Wheel, Division of Hawker Siddeley Inc
Ipsco Inc (formerly Interprovincial Steel & Pipe Corp Ltd)
Slater Steels, Sorel Forge Division
Stelco Inc
Sydney Steel Corp (Sysco)
Western Canada Steel Ltd

CHINA
Anshan Iron & Steel Co
Baotau Iron & Steel Co
Jiangxi Steel Works
Kunming Iron & Steel Works
Shoudu Iron & Steel Co
Taiyuan Iron & Steel Co

CHINA *—continued*
Wuhan Iron & Steel (Wisco)

CZECHOSLOVAKIA
Vitkovice Steelworks

EGYPT
National Metal Industries Co

EL SALVADOR
Corinca SA – Corporación Industrial Centroamericana SA

FRANCE
Usinor Sacilor

WEST GERMANY
Böhler AG, Edelstahlwerke
Klöckner Stahl GmbH
Krupp Stahl AG (KS)
Lemmerz-Werke KGaA
Saarstahl Völklingen GmbH
Thyssen Stahl AG

HUNGARY
Csepel Steel & Tube Works

INDIA
Agarwal Foundry & Engineering Works
BD Steel Castings Ltd
Bhartia Electric Steel Co Ltd
Bihar Alloy Steels Ltd
Ellora Steels Ltd (ESL)
Indore Steel & Iron Mills Pvt Ltd (Isim)
Kalyani Steels Ltd
Kap Steel Ltd
Man Industrial Corp Ltd
Modi Industries Ltd, Steel Division
Mohan Steels Ltd
Punjab Con-Cast Steels Ltd
Raipur Wires & Steel Ltd (RWSL)
Sail – Steel Authority of India Ltd
Sanghvi Steels Ltd
Steel Ingots Pvt Ltd
The Tata Iron & Steel Co Ltd (Tisco)
Upper India Steel Manufacturing & Engineering Co Ltd

ITALY
ASO SpA – Acciai Speciali Ospitaletto
Fonderie Acciaierie Giovanni Mandelli
Metallurgica Marcora SpA
Acciaieria di Rubiera SpA

JAPAN
The Japan Steel Works Ltd
Kobe Steel Ltd (Kobe Seikosho)

Carbon steel - ingots
—continued

JAPAN *—continued*
Nisshin Steel Co Ltd
NKK – Nippon Kokan KK
Sumitomo Metal Industries Ltd

NORTH KOREA
Hwanghai Iron Works

SOUTH KOREA
Kang Won Industries Ltd
Sammi Steel Co Ltd

MAURITANIA
Sté Arabe du Fer et d'Acier (Safa)

MEXICO
Ahmsa – Altos Hornos de Mexico SA de CV
Industrias CH SA (formerly Campos Hermanos SA)

PAKISTAN
Ahmed Investments (Pvt) Ltd
Sind Steel Corp Ltd

SOUTH AFRICA
Scaw Metals Ltd

SPAIN
Forjas Alavesas SA (Fasa)
Grupo José María Aristrain SA
SA Echevarría (SAE)
Ensidesa – Empresa Nacional Siderurgica SA
Olarra SA (Olsa)
Forjas y Aceros de Reinosa SA
Tubacex – C.E. de Tubos por Extrusion SA

SWEDEN
SSAB – Svenskt Stål AB

TAIWAN
Chia Hsin Metal Industrial Co Ltd

TURKEY
Ereğli Demir ve Çelik Fabrikalari TAS (Erdemir)
MKEK Celik Fabrikasi
Orpaş Metal Sanayi ve Ticaret AŞ
TDCI – Türkiye Demir ve Çelik Işletmeleri Genel Müdürlügü

UNITED KINGDOM
BSC – British Steel Corp
Howmet Turbine (UK) Inc, Howmet Alloys International

UNITED KINGDOM *—continued*
Sheffield Forgemasters Ltd
UES – United Engineering Steels Ltd

UNITED STATES
Acme Steel Co
Armco Inc
The Babcock & Wilcox Co, Tubular Products Division
Cameron Iron Works Inc
CF&I Steel Corp
The Champion Steel Co
Cyclops Corp
Ellwood Uddeholm Steel Corp
Hawaiian Western Steel Ltd
Jessop Steel Co
Lone Star Steel Co
LTV Steel Co Inc
Lukens Steel Co
Marathon LeTourneau Co, Longview Division
Milton Manufacturing Co
Northwestern Steel & Wire Co
Rouge Steel Co
Seattle Steel Inc
Sharon Steel Corp
The Timken Co
USS, A Division of USX Corp

VENEZUELA
Sidetur – Siderúrgica del Turbio SA
Sidor – CVG Siderúrgica del Orinoco CA

YUGOSLAVIA
Zeljezara Sisak Integrated Metallurgical Works
Metalurski Kombinat Smederevo

Carbon steel - slabs

ALGERIA
SNS – Sté Nationale de Sidérurgie

ARGENTINA
Acindar Industria Argentina de Aceros SA
Somisa – Sdad Mixta Siderurgia Argentina

AUSTRALIA
The Broken Hill Pty Co Ltd

BELGIUM
SA Fabrique de Fer de Charleroi
Forges de Clabecq SA
Sidmar NV

Carbon steel – slabs
—continued

BRAZIL
Acesita – Cia Aços Especiais Itabira
Açominas – Aço Minas Gerais SA
Cofavi – Cia Ferro e Aço de Vitória
Cosipa – Cia Siderúrgica Paulista
Cia Siderúrgica de Tubarão
Usinas – Usinas Siderúrgicas de Minas Gerais SA

BULGARIA
Kremikovtsi Iron & Steel Works

CANADA
The Algoma Steel Corp Ltd
Atlas Specialty Steels Division of Rio Algom Ltd
Ipsco Inc (formerly Interprovincial Steel & Pipe Corp Ltd)
Sidbec-Dosco Inc
Slater Steels, Sorel Forge Division
Stelco Inc
Sydney Steel Corp (Sysco)

CHILE
Cia Siderúrgica Huachipato SA

CHINA
Jiangxi Steel Works
Shoudu Iron & Steel Co
Wuhan Iron & Steel (Wisco)

CZECHOSLOVAKIA
NHKG – Nová Hut Klementa Gottwalda
Vitkovice Steelworks

EGYPT
The Egyptian Iron & Steel Co (Hadisolb)

FRANCE
Usinor Sacilor

EAST GERMANY
VEB Stahl- und Walzwerk Brandenburg

WEST GERMANY
Benteler AG
Hoesch AG
Klöckner Stahl GmbH
Krupp Stahl AG (KS)
Stahlwerke Peine-Salzgitter AG (P&S)
Thyssen Edelstahlwerke AG
Thyssen Stahl AG

GREECE
Halyvourgiki Inc

HUNGARY
Danube Works

INDIA
Agarwal Foundry & Engineering Works
Mahindra Ugine Steel Co Ltd (Muscosteel)
Sail – Steel Authority of India Ltd
The Tata Iron & Steel Co Ltd (Tisco)

INDONESIA
PT Kratkatau Steel

IRAN
National Iranian Steel Corp (Nisc)

ITALY
Acciaierie e Ferriere Lombarde Falck SpA
Nuova Italsider SpA

JAPAN
Kobe Steel Ltd (Kobe Seikosho)
Nakayama Steel Works Ltd (Nakayama Seikosho)
Nippon Steel Corp (Shin Nippon Seitetsu)
Nisshin Steel Co Ltd
NKK – Nippon Kokan KK
Sumitomo Metal Industries Ltd
Tokyo Steel Manufacturing Co Ltd

SOUTH KOREA
Pohang Iron & Steel Co Ltd

MEXICO
Ahmsa – Altos Hornos de Mexico SA de CV

NETHERLANDS
Hoogovens Groep BV

NEW ZEALAND
New Zealand Steel Ltd

NORWAY
Norsk Jernverk AS

PAKISTAN
Pakistan Steel Mills Corp Ltd

POLAND
Huta Katowice
Huta im Lenina

Carbon steel – slabs

—continued

SOUTH AFRICA
Highveld Steel & Vanadium Corp Ltd
Iscor Ltd

SPAIN
AHV – Altos Hornes de Vizcaya SA
Ensidesa – Empresa Nacional Siderurgica SA

SWEDEN
SSAB – Svenskt Stål AB

TURKEY
Ereğli Demir ve Çelik Fabrikalari TAS (Erdemir)

UNITED KINGDOM
Alphasteel Ltd
BSC – British Steel Corp
Sheffield Forgemasters Ltd
UES – United Engineering Steels Ltd

UNITED STATES
Bethlehem Steel Corp
Jessop Steel Co
Lone Star Steel Co
LTV Steel Co Inc
Lukens Steel Co
Marathon LeTourneau Co, Longview Division
Northwestern Steel & Wire Co
Oregon Steel Mills, Division of Gilmore Steel Corp
Rouge Steel Co
Seattle Steel Inc
Sharon Steel Corp
USS, A Division of USX Corp

USSR
Donetskiy Metallurgicheskiy Zavod (Donetsk Iron & Steel Works)
Karagandinskiy Met Zavod (Kazakhstanskaya Magnitka) (Karaganda Iron & Steel Works)
Zhdanovskiy Met Zavod Ilyich (Ilyich Works)

VENEZUELA
Sidor – CVG Siderúrgica del Orinoco CA

YUGOSLAVIA
Zeljezara Sisak Integrated Metallurgical Works
Metalurski Kombinat Smederevo

Carbon steel – blooms

ARGENTINA
Acindar Industria Argentina de Aceros SA
Somisa – Sdad Mixta Siderurgia Argentina

AUSTRALIA
The Broken Hill Pty Co Ltd

AUSTRIA
Voest-Alpine AG

BRAZIL
Acesita – Cia Aços Especiais Itabira
Açominas – Aço Minas Gerais SA
Aços Anhanguera SA (Acoansa)
Cofavi – Cia Ferro e Aço de Vitória
Cosim – Cia Siderúrgica de Mogi das Cruzes
Mannesmann SA
Villares Group

BULGARIA
Kremikovtsi Iron & Steel Works

CANADA
The Algoma Steel Corp Ltd
Atlas Specialty Steels Division of Rio Algom Ltd
Slater Steels, Sorel Forge Division
Stelco Inc
Sydney Steel Corp (Sysco)

CHILE
Cia Siderúrgica Huachipato SA

CHINA
Anshan Iron & Steel Co
Jiangxi Steel Works
Shoudu Iron & Steel Co

COLOMBIA
Sidelpa – Siderúrgica del Pacifico SA

CZECHOSLOVAKIA
NHKG – Nová Hut Klementa Gottwalda
Vitkovice Steelworks

FRANCE
Usinor Sacilor

EAST GERMANY
VEB Stahl- und Walzwerk Brandenburg

WEST GERMANY
Benteler AG

Carbon steel - blooms

—continued

WEST GERMANY *—continued*
Hoesch AG
Klöckner Stahl GmbH
Krupp Stahl AG (KS)
Maxhütte - Eisenwerke-Gesellschaft Maximilianshütte mbH
Stahlwerke Peine-Salzgitter AG (P&S)
Saarstahl Völklingen GmbH
Thyssen Edelstahlwerke AG
Thyssen Stahl AG

INDIA
Agarwal Foundry & Engineering Works
Bhoruka Steel Ltd
Bihar Alloy Steels Ltd
Ferro Alloys Corp Ltd (Facor) (Steel division), (Vidarbha Iron & Steel Corp Ltd)
Mahindra Ugine Steel Co Ltd (Muscosteel)
Panchmahal Steel Ltd
The Tata Iron & Steel Co Ltd (Tisco)

IRAN
National Iranian Steel Corp (Nisc)

ITALY
Nuova Deltasider SpA
Acciaierie e Ferriere Luigi Leali SpA
Metallurgica Marcora SpA
Acciaieria e Ferriere Stefana F.lli fu Girolamo SpA

JAPAN
Godo Steel Ltd (Godo Seitetsu)
Kobe Steel Ltd (Kobe Seikosho)
Nakayama Steel Works Ltd (Nakayama Seikosho)
Nippon Steel Corp (Shin Nippon Seitetsu)
NKK - Nippon Kokan KK
Sumitomo Metal Industries Ltd
Toa Steel Co Ltd
Tokyo Steel Manufacturing Co Ltd

NORTH KOREA
Hwanghai Iron Works

SOUTH KOREA
Pohang Iron & Steel Co Ltd

MEXICO
Ahmsa - Altos Hornos de Mexico SA de CV

MEXICO *—continued*
Industrias CH SA (formerly Campos Hermanos SA)

NETHERLANDS
Hoogovens Groep BV

NEW ZEALAND
Pacific Steel Ltd

PAKISTAN
Pakistan Steel Mills Corp Ltd

PHILIPPINES
Armco-Marsteel Alloy Corp

POLAND
Huta Katowice
Huta Kościuszko
Huta im Lenina
Huta Pokój
Poldi Steelworks
Huta Warszawa

SOUTH AFRICA
Highveld Steel & Vanadium Corp Ltd
Iscor Ltd

SPAIN
AHV - Altos Hornes de Vizcaya SA
Forjas Alavesas SA (Fasa)
Babcock & Wilcox Española SA Tubular Products Division
Ensidesa - Empresa Nacional Siderurgica SA
Aceros de Llodio SA (Allosa)

SWEDEN
Björneborgs Jernverks AB
SSAB - Svenskt Stål AB

SWITZERLAND
Monteforno Acciaierie e Laminatoi SA

TAIWAN
China Steel Corp

TURKEY
Asil Çelik Sanayi ve Ticaret AŞ (Asil Çelik)
Izmir Demir Çelik Sanayi AŞ
Metaş Izmir Metalurji Fabrikasi Türk AŞ
MKEK Celik Fabrikasi
TDCI - Türkiye Demir ve Çelik Işletmeleri Genel Müdürlügü

UNITED KINGDOM
BSC - British Steel Corp

Carbon steel - blooms

—continued

UNITED KINGDOM *—continued*
F.H. Lloyd & Co Ltd
Sheffield Forgemasters Ltd
UES - United Engineering Steels Ltd

UNITED STATES
Acme Steel Co
Bethlehem Steel Corp
Ellwood Uddeholm Steel Corp
LTV Steel Co Inc
Northwestern Steel & Wire Co
Seattle Steel Inc
Sharon Steel Corp
The Timken Co
USS, A Division of USX Corp

USSR
Dneprodzerzhinskiy Zavod Imeni Dzerzhinskogo (Dzerzhinsky Works)
Met Zavod Krivorizhstal (Krivoy Rog Iron & Steel Works)
Kuznetskiy Met Kombinat
Met Zavod Imeni Petrovskogo (Petrovsky Iron & Steel Works)
Zapsib Met Zavod (West Siberian Steel Works)

VENEZUELA
Sidor - CVG Siderúrgica del Orinoco CA

YUGOSLAVIA
Zeljezara Sisak Integrated Metallurgical Works

ZIMBABWE
Zimbabwe Iron & Steel Co Ltd - Ziscosteel

Carbon steel - billets

ARGENTINA
Acindar Industria Argentina de Aceros SA
Somisa - Sdad Mixta Siderurgia Argentina
Establecimiento Altos Hornos Zapla

AUSTRALIA
The Broken Hill Pty Co Ltd
Commonwealth Steel Co Ltd
Smorgon Steel

AUSTRIA
Eisenwerk Breitenfeld Ges.mbH Nachf KG
Stahl- und Walzwerk Marienhütte GmbH
Voest-Alpine AG

BANGLADESH
Chittagong Steel Mills Ltd

BELGIUM
Usines Gustave Boël

BRAZIL
Acesita - Cia Aços Especiais Itabira
Açominas - Aço Minas Gerais SA
Aços Anhanguera SA (Acoansa)
Siderúrgica Nossa Senhora Aparecida SA
Cofavi - Cia Ferro e Aço de Vitória
Cosigua - Cia Siderurgia da Guanabara
Cosim - Cia Siderúrgica de Mogi das Cruzes
Dedini SA Sideúrgica
Siderúrgica Fi-El SA
Siderúrgica Hime SA
Mannesmann SA
Cia Siderúrgica Pains
Aços Finos Piratini SA
Siderúrgica Riograndense SA
Usiba - Usina Siderúrgica da Bahia SA
Villares Group

BULGARIA
Kremikovtsi Iron & Steel Works

CANADA
The Algoma Steel Corp Ltd
Atlas Specialty Steels Division of Rio Algom Ltd
Courtice Steel Ltd
Ivaco Inc
Manitoba Rolling Mills, Amca International
QIT - Fer et Titane Inc
Sidbec-Dosco Inc
Slater Steels, Sorel Forge Division
Stelco Inc
Sydney Steel Corp (Sysco)

CHINA
Anshan Iron & Steel Co
Changzhi Iron & Steel Works
Fushun Xinfu Steel Plant
Shoudu Iron & Steel Co
Wuhan Iron & Steel (Wisco)

Carbon steel - billets

—continued

COLOMBIA
Acerías Paz del Rio SA
Simesa - Siderúrgica de Medellín

CZECHOSLOVAKIA
Vitkovice Steelworks

ECUADOR
Funasa - Fundiciones Nacionales SA

EGYPT
The Egyptian Iron & Steel Co (Hadisolb)

EL SALVADOR
Siderúrgica Centroamericana del Pacífico SA (Sicepasa)

FINLAND
Dalsbruk Oy Ab

FRANCE
Iton Seine SA
Aciéries et Laminoirs de Rives
Usinor Sacilor

EAST GERMANY
VEB Stahl- und Walzwerk Brandenburg
VEB Stahl- und Walzwerk Wilhelm Florin, Hennigsdorf

WEST GERMANY
Badische Stahlwerke AG
Walzwerk Becker-Berlin Verwaltungsges.mbH & Co Strangguss KG
Benteler AG
Edelstahlwerke Buderus AG
Hamburger Stahlwerke GmbH
Klöckner Stahl GmbH
Krupp Stahl AG (KS)
Lech-Stahlwerke GmbH
Maxhütte - Eisenwerke-Gesellschaft Maximilianshütte mbH
Moselstahlwerk GmbH & Co KG
Stahlwerke Peine-Salzgitter AG (P&S)
Saarstahl Völklingen GmbH
Thyssen Edelstahlwerke AG
Thyssen Stahl AG

GREECE
Halyvourgiki Inc
Helliniki Halyvourgia SA

HONDURAS
Industria Nacional Del Acero SA

HUNGARY
Diosgyör Works (Lenin Metallurgical Works)
Ozdi Kohászati Uzemek (Okü)

INDIA
Agarwal Foundry & Engineering Works
Bhoruka Steel Ltd
Bhushan Industrial Corp
Bihar Alloy Steels Ltd
Indore Steel & Iron Mills Pvt Ltd (Isim)
Kalyani Steels Ltd
Kap Steel Ltd
Maharashtra Elektrosmelt Ltd
Mahindra Ugine Steel Co Ltd (Muscosteel)
Modi Industries Ltd, Steel Division
Mukand Iron & Steel Works Ltd
Panchmahal Steel Ltd
Partap Steel Rolling Mills (1935) Ltd
Raipur Wires & Steel Ltd (RWSL)
Sail - Steel Authority of India Ltd
Sanghvi Steels Ltd
Singh Electrosteel Ltd (SES)
Star Steel Pvt Ltd (Starsteel)
The Tata Iron & Steel Co Ltd (Tisco)
Upper India Steel Manufacturing & Engineering Co Ltd
Usha Alloys & Steels Ltd (UASL)

INDONESIA
PT Ispat Indo
PT Kratkatau Steel
PT Master Steel Mfg Co

IRAN
National Iranian Steel Corp (Nisc)

ISRAEL
United Steel Mills Ltd (USM)

ITALY
AFP - Acciaierie Ferriere del Po SpA
A.F.V. Beltrame SpA
Ferriere Bertoli SpA
Acciaieria e Ferriera del Caleotto SpA (AFC)
Acciaierie di Calvisano SpA
Acciaierie e Ferriere di Caronno SpA
Acciaieria di Cividate al Piano SpA
Nuova Deltasider SpA
Acciaierie e Ferriere Lombarde Falck SpA
Acciaierie e Ferriere Feralpi SpA

Carbon steel – billets

—continued

ITALY *—continued*

Indumetal – Industrie Metallurgiche SpA
Acciaierie e Ferriere Luigi Leali SpA
Acciaierie di Lonato SpA
Lucchini Siderurgica SpA
Metallurgica Marcora SpA
Ferriere Nord SpA
Pasini Siderurgica SpA
Pietra SpA – Acciaierie Ferriere e Tubifici
Acciaieria di Porto Nogaro Srl
Profilati Nave SpA
Prolafer
Raimondi SpA
Acciaierie e Ferriere Riva SpA
Acciaieria di Rubiera SpA
Siderurgica Commerciale Santo Stefano SpA
Seta Acciai SpA
Ferriera Siderscal SpA
Sisva Srl
Acciaieria e Ferriere Stefana F.lli fu Girolamo SpA
Acciaierie Venete SpA

JAPAN

Funabashi Steel Works Ltd (Funabashi Seito)
Godo Steel Ltd (Godo Seitetsu)
Kobe Steel Ltd (Kobe Seikosho)
Kyoei Steel Ltd
Nakayama Steel Works Ltd (Nakayama Seikosho)
Nippon Steel Corp (Shin Nippon Seitetsu)
NKK – Nippon Kokan KK
Rinko Steel Works Co Ltd
Sumitomo Metal Industries Ltd
Toa Steel Co Ltd
Tokyo Steel Manufacturing Co Ltd
Yodogawa Steel Works Ltd (Yodogawa Seikosho)

SOUTH KOREA

Dongkuk Steel Mill Co Ltd
Inchon Iron & Steel Co Ltd
Kang Won Industries Ltd
Pohang Iron & Steel Co Ltd
Sammi Steel Co Ltd

MALAYSIA

Perwaja Trengganu Sdn Bhd

MEXICO

Ahmsa – Altos Hornos de Mexico SA de CV
Aceros Ecatepec SA (Aesa)
Hylsa SA de CV
Industrias CH SA (formerly Campos Hermanos SA)
Industrials Nylbo SA, Division Acero (formerly Laminadora Atzcapotzalco SA – Lasa)
Siderúrgica de Yucatán SA de CV

NETHERLANDS

Hoogovens Groep BV

NEW ZEALAND

New Zealand Steel Ltd
Pacific Steel Ltd

NIGERIA

Delta Steel Co Ltd

NORWAY

Norsk Jernverk AS

PAKISTAN

Ahmed Investments (Pvt) Ltd
Pakistan Steel Mills Corp Ltd

PERU

Siderperú – Empresa Siderúrgica del Perú

PHILIPPINES

Armco-Marsteel Alloy Corp
National Steel Corp (NSC)

POLAND

Huta Katowice
Huta im Lenina
Poldi Steelworks
Huta Zawiercie

PORTUGAL

Siderurgia Nacional EP (SN)

SINGAPORE

National Iron & Steel Mills Ltd

SOUTH AFRICA

Cape Town Iron & Steel Works (Pty) Ltd
Highveld Steel & Vanadium Corp Ltd
Iscor Ltd
Scaw Metals Ltd
Usco – The Union Steel Corp (of South Africa) Ltd

Carbon steel - billets

—continued

SPAIN
Forjas Alavesas SA (Fasa)
Artierro SA
Azma SA
Babcock & Wilcox Española SA Tubular Products Division
SA Echevarría (SAE)
Patricio Echeverría SA
Ensidesa - Empresa Nacional Siderurgica SA
Aceros de Irura SA
Aceros de Llodio SA (Allosa)
Megasa - Metalurgica Galaica SA
Olarra SA (Olsa)
Esteban Orbegozo SA
Pedro Orbegozo y Cía SA (Acenor)
Tubacex - C.E. de Tubos por Extrusion SA
Unión Cerrajera SA

SWEDEN
Ovako Steel AB
SSAB - Svenskt Stål AB

SWITZERLAND
Monteforno Acciaierie e Laminatoi SA

TAIWAN
China Steel Corp
Hai Kwang Enterprise Corp
Li-Chong Steel & Iron Works Co Ltd

THAILAND
Bangkok Steel Industry Co Ltd

TRINIDAD & TOBAGO
Iscott - Iron & Steel Co of Trinidad & Tobago Ltd

TUNISIA
El Fouladh - Sté Tunisienne de Sidérurgie

TURKEY
Asil Çelik Sanayi ve Ticaret AŞ (Asil Çelik)
Çolakoğlu Metalurji AŞ
Çukurova Çelik Endüstrisi AS
Elektrofer Çelik Sanayi AŞ
Habas Sinai ve Tibbi Gazlar Istihsal Endustrisi AŞ
Içdaş Istanbul Çelik ve Demir Izabe Sanayii AŞ
Izmir Demir Çelik Sanayi AŞ
Kroman Steel Industries Co (Kroman Celik)

TURKEY *—continued*
Metaş Izmir Metalurji Fabrikasi Türk AŞ
MKEK Celik Fabrikasi
Orpaş Metal Sanayi ve Ticaret AŞ
TDCI - Türkiye Demir ve Çelik Işletmeleri Genel Müdürlügü

UNITED KINGDOM
Allied Steel & Wire Ltd
Alphasteel Ltd
BSC - British Steel Corp
F.H. Lloyd & Co Ltd
Sheerness Steel Co PLC
Sheffield Forgemasters Ltd
UES - United Engineering Steels Ltd
Unbrako Steel Co Ltd

UNITED STATES
Acme Steel Co
Atlantic Steel Co
Auburn Steel Co Inc (Austeel)
Bayou Steel Corp
Bethlehem Steel Corp
Birmingham Steel Corp
CF&I Steel Corp
The Champion Steel Co
Charter Electric Melting Inc
Connecticut Steel Corp
Florida Steel Corp
Georgetown Steel Corp
Jessop Steel Co
Earle M. Jorgensen Co, Forge Division
Keystone Steel & Wire Co
LTV Steel Co Inc
Milton Manufacturing Co
New Jersey Steel Corp
Northwestern Steel & Wire Co
Razorback Steel Corp (RSC)
Seattle Steel Inc
Sharon Steel Corp
Slater Steels, Fort Wayne Specialty Alloys Division
Structural Metals Inc
The Timken Co
USS, A Division of USX Corp

USSR
Dneprodzerzhinskiy Zavod Imeni Dzerzhinskogo (Dzerzhinsky Works)
Donetskiy Metallurgicheskiy Zavod (Donetsk Iron & Steel Works)
Kirovskiy Zavod (Kirov Works)
Kuznetskiy Met Kombinat
Liepajas Rupnica Sarkanais Metallurgs (Red Metal Worker Plant)

Carbon steel – billets
—continued

USSR *—continued*
Zapsib Met Zavod (West Siberian Steel Works)

VENEZUELA
Sidetur – Siderúrgica del Turbio SA
Sidor – CVG Siderúrgica del Orinoco CA

YUGOSLAVIA
Zeljezara Sisak Integrated Metallurgical Works
Slovenske Železarne – Iron & Steel Works of Slovenia
Jadranska Zelejezara Split (Adriatic Steelworks, Split)

ZAIRE
Sté Nationale de Sidérurgie

ZIMBABWE
Zimbabwe Iron & Steel Co Ltd – Ziscosteel

Carbon steel – semis for seamless tubemaking

BRAZIL
Cofavi – Cia Ferro e Aço de Vitória
Cosim – Cia Siderúrgica de Mogi das Cruzes
Mannesmann SA

BULGARIA
Kremikovtsi Iron & Steel Works

CHINA
Fushun Xinfu Steel Plant

CZECHOSLOVAKIA
NHKG – Nová Hut Klementa Gottwalda
Sverma Steel Works
Vitkovice Steelworks

FRANCE
Usinor Sacilor

WEST GERMANY
Benteler AG
Maxhütte – Eisenwerke-Gesellschaft Maximilianshütte mbH

HUNGARY
Csepel Steel & Tube Works

INDIA
Sail – Steel Authority of India Ltd
The Tata Iron & Steel Co Ltd (Tisco)

ITALY
Acciaierie di Bolzano SpA
Acciaierie e Ferriere Lombarde Falck SpA
Acciaierie e Ferriere Luigi Leali SpA
Metallurgica Marcora SpA
Pietra SpA – Acciaierie Ferriere e Tubifici
Acciaieria di Rubiera SpA
Seta Acciai SpA

JAPAN
NKK – Nippon Kokan KK

POLAND
Poldi Steelworks

SOUTH AFRICA
Highveld Steel & Vanadium Corp Ltd
Iscor Ltd

SPAIN
Babcock & Wilcox Española SA Tubular Products Division
Tubacex – C.E. de Tubos por Extrusion SA

UNITED KINGDOM
F.H. Lloyd & Co Ltd

UNITED STATES
The Babcock & Wilcox Co, Tubular Products Division
LTV Steel Co Inc
The Timken Co

USSR
Dneprodzerzhinskiy Zavod Imeni Dzerzhinskogo (Dzerzhinsky Works)
Novo-Tagilskiy Met Zavod (Novo-Tagil Iron & Steel Works)
Uzbekskiy Metallurgicheskiy Zavod (Uzbek Iron & Steel Works)

VENEZUELA
Sidor – CVG Siderúrgica del Orinoco CA

YUGOSLAVIA
Zeljezara Sisak Integrated Metallurgical Works

Carbon steel – hot rolled coil (for re-rolling)

AUSTRALIA
The Broken Hill Pty Co Ltd
Smorgon Steel

AUSTRIA
Voest-Alpine AG

BELGIUM
Usines Gustave Boël
Cockerill Sambre SA
Sidmar NV

BRAZIL
Cofavi – Cia Ferro e Aço de Vitória
CSN – Cia Siderúrgica Nacional
Usinas – Usinas Siderúrgicas de Minas Gerais SA

COLOMBIA
Acerías Paz del Rio SA

EGYPT
The Egyptian Iron & Steel Co (Hadisolb)

FRANCE
Usinor Sacilor

WEST GERMANY
Klöckner Stahl GmbH
Thyssen Stahl AG

GREECE
Halyvourgiki Inc

INDONESIA
PT Kratkatau Steel

ITALY
Nuova Italsider SpA

JAPAN
Nisshin Steel Co Ltd
NKK – Nippon Kokan KK
Sumitomo Metal Industries Ltd

SOUTH KOREA
Pohang Iron & Steel Co Ltd

NEW ZEALAND
New Zealand Steel Ltd

PHILIPPINES
National Steel Corp (NSC)

SOUTH AFRICA
Highveld Steel & Vanadium Corp Ltd

SPAIN
AHV – Altos Hornes de Vizcaya SA
Ensidesa – Empresa Nacional Siderurgica SA

TAIWAN
China Steel Corp

TURKEY
Metaş Izmir Metalurji Fabrikasi Türk AŞ

UNITED KINGDOM
BSC – British Steel Corp

UNITED STATES
Acme Steel Co
Bethlehem Steel Corp
Cyclops Corp
Inland Steel Co
Lone Star Steel Co
LTV Steel Co Inc
Tuscaloosa Steel Corp

VENEZUELA
Sidor – CVG Siderúrgica del Orinoco CA

YUGOSLAVIA
Metalurski Kombinat Smederevo

Carbon steel – wire rod

ALGERIA
SNS – Sté Nationale de Sidérurgie

ARGENTINA
Acindar Industria Argentina de Aceros SA
Aceros Bragado Sacif
Establecimiento Altos Hornos Zapla

AUSTRALIA
The Broken Hill Pty Co Ltd
Smorgon Steel

AUSTRIA
Stahl- und Walzwerk Marienhütte GmbH
Joh. Pengg, Draht- & Walzwerke
Voest-Alpine AG

BANGLADESH
Bengal Steel Works Ltd

BELGIUM
Usines Gustave Boël
Cockerill Sambre SA

Carbon steel - wire rod
—continued

BRAZIL
Siderúrgica Aconorte SA
Siderúrgica J.L. Aliperti SA
Siderúrgica Nossa Senhora Aparecida SA
Siderúrgica Barra Mansa SA
Cia Siderúrgica Belgo-Mineira
Cofavi – Cia Ferro e Aço de Vitória
Cosigua – Cia Siderurgia da Guanabara
Siderúrgica Fi-El SA
Lafersa – Laminação de Ferro SA
Mannesmann SA
Siderúrgica Mendes Junior SA (SMJ)
Cia Siderúrgica Pains
Aços Finos Piratini SA
Siderúrgica Riograndense SA
Laminação Santa Maria SA – Indústria e Comércio
Siderama – Cia Siderúrgica da Amazônia
Usiba – Usina Siderúrgica da Bahia SA
Villares Group

BULGARIA
Kremikovtsi Iron & Steel Works
Lenin Iron & Steel Works

BURMA
Ywama Steel Mill

CANADA
Ivaco Inc
Sidbec-Dosco Inc
Stelco Inc

CHILE
Cia Siderúrgica Huachipato SA
Cia de Acero Rengo SA

CHINA
Changcheng Steel Works
Changzhi Iron & Steel Works
Dalian Steel Rolling Mill
Echeng Iron & Steel Works
Fushun Xinfu Steel Plant
Guiyang Steel Mill
Handan Iron & Steel Works
Jiangxi Steel Works
Sanming Iron & Steel Works
Taiyuan Iron & Steel Co
Tianjin Steel Plant

COLOMBIA
Acerías Paz del Rio SA
Sidemuña – Siderúrgica de Muña SA
Simesa – Siderúrgica de Medellín

CUBA
Empressa Metalúrgica José Marti

CZECHOSLOVAKIA
Hlohovec-Galgoc Works
NHKG – Nová Hut Klementa Gottwalda

DOMINICAN REPUBLIC
Metaldom – Complejo Metalúrgico Dominicano C por A

ECUADOR
Andec – Acerías Nacionales del Ecuador SA
Funasa – Fundiciones Nacionales SA

EGYPT
Alexandria National Iron & Steel Co

EL SALVADOR
Siderúrgica Centroamericana del Pacífico SA (Sicepasa)

FINLAND
Dalsbruk Oy Ab
Ovako Steel Oy Ab

FRANCE
Usinor Sacilor

EAST GERMANY
VEB Stahl- und Walzwerk Brandenburg

WEST GERMANY
Badische Stahlwerke AG
Hamburger Stahlwerke GmbH
Krupp Stahl AG (KS)
Lech-Stahlwerke GmbH
Maxhütte – Eisenwerke-Gesellschaft Maximilianshütte mbH
Moselstahlwerk GmbH & Co KG
Saarstahl Völklingen GmbH
Thyssen Edelstahlwerke AG
Thyssen Stahl AG

GREECE
Metallurgiki Halyps SA
Halyvourgia Thessalias SA
Halyvourgiki Inc
Helliniki Halyvourgia SA
Sidenor SA (formerly Steel Works of Northern Greece SA)

Carbon steel – wire rod
—continued

GUATEMALA
Hornos SA

INDIA
Agarwal Foundry & Engineering Works
Bhoruka Steel Ltd
Bihar Alloy Steels Ltd
Calcutta Steel Co Ltd
Indore Steel & Iron Mills Pvt Ltd (Isim)
Man Industrial Corp Ltd
Modi Industries Ltd, Steel Division
Mohan Steels Ltd
Mukand Iron & Steel Works Ltd
Partap Steel Rolling Mills (1935) Ltd
The Punjab Steel Rolling Mills
Rathi Udyog Ltd
Sail – Steel Authority of India Ltd
The Tata Iron & Steel Co Ltd (Tisco)

INDONESIA
PT Gunung Gahapi Steel
PT Ispat Indo
PT Jakarta Kyoei Steel Works Ltd
PT Kratkatau Steel

IRAN
National Iranian Steel Corp (Nisc)

ISRAEL
United Steel Mills Ltd (USM)

ITALY
AFP – Acciaierie Ferriere del Po SpA
Acciaierie di Bolzano SpA
Acciaieria e Ferriera del Caleotto SpA (AFC)
Acciaierie e Ferriere di Caronno SpA
Nuova Deltasider SpA
Acciaierie e Ferriere Feralpi SpA
Ferdofin SpA
Officine e Fonderie Galtarossa SpA
Lucchini Siderurgica SpA
Ferriere Nord SpA
Mini Acciaieria Odolese Srl
Ori Martin SpA
Ferriera Padana SpA
Acciaierie e Ferriere Riva SpA
Acciaieria e Ferriere Stefana F.lli fu Girolamo SpA

JAPAN
Daido Steel Co Ltd
Godo Steel Ltd (Godo Seitetsu)
Kawasaki Steel Corp

JAPAN *—continued*
Kobe Steel Ltd (Kobe Seikosho)
Nakayama Steel Works Ltd (Nakayama Seikosho)
Nippon Steel Corp (Shin Nippon Seitetsu)
Sumitomo Electric Industries Ltd (Sumitomo Denki Kogyo)
Sumitomo Metal Industries Ltd
Toa Steel Co Ltd

KENYA
Kenya United Steel Co Ltd

NORTH KOREA
Hwanghai Iron Works
Kangson Works
Kimchaek Works

SOUTH KOREA
Boo-Kuk Steel Industrial Co Ltd
Daehan Sangsa Co Ltd
Dongkuk Steel Mill Co Ltd
Inchon Iron & Steel Co Ltd
Pohang Iron & Steel Co Ltd
Sammi Steel Co Ltd

LUXEMBOURG
Arbed SA

MALAYSIA
Amalgamated Steel Mills Bhd
Malayawata Steel Bhd
Southern Iron & Steel Works Sdn Bhd

MEXICO
Ahmsa – Altos Hornos de Mexico SA de CV
Deacero – Productos de Acero SA
Hylsa SA de CV
Aceros San Luis SA
Sicartsa – Siderurgica Lazaro Cardenas "Las Truchas SA"

MOROCCO
Sonasid – Sté Nationale de Siderurgie

NETHERLANDS
Nedstaal BV

NEW ZEALAND
Pacific Steel Ltd

NIGERIA
Ajaokuta Steel Co Ltd
The Jos Steel Rolling Co
Katsina Steel Rolling Co Ltd (KSRC)
Osogbo Steel Rolling Co Ltd
Qua Steel Products Ltd

Carbon steel - wire rod

—continued

NORWAY
Christiania Spigerverk AS

PAKISTAN
Hashoo Steel Industries Ltd
Metropolitan Steel Corp Ltd (MSC)

PARAGUAY
Acepar - Acero del Paraguay SA

PERU
Siderperú - Empresa Siderúrgica del Perú

PHILIPPINES
Marcelo Steel Corp (MSC)
Pag-Asa Steel Works Inc (PSWI)

POLAND
Huta Cedlera
Huta Kościuszko
Huta Warszawa

PORTUGAL
Siderurgia Nacional EP (SN)

RUMANIA
Brăila Works
Buzau Works
Cîmpia Turzii Works
Focsani Metallurgical Enterprise

SAUDI ARABIA
Saudi Iron & Steel Co (Hadeed)

SINGAPORE
National Iron & Steel Mills Ltd

SOUTH AFRICA
Davsteel (Pty) Ltd
Iscor Ltd
Scaw Metals Ltd

SPAIN
Forjas Alavesas SA (Fasa)
Grupo José María Aristrain SA
Azma SA
SA Echevarría (SAE)
Ensidesa - Empresa Nacional Siderurgica SA
Nueva Montaño Quijano SA
Esteban Orbegozo SA
Sidegasa - Siderúrgica de Galicia SA
Torras, Herrería y Construcciones SA
Marcial Ucin SA

SWEDEN
Ovako Steel AB
SSAB - Svenskt Stål AB

SWITZERLAND
Ferrowohlen AG
Von Moos Stahl AG

TAIWAN
China Steel Corp
Hai Kwang Enterprise Corp
Li-Chong Steel & Iron Works Co Ltd

TANZANIA
Steel Rolling Mills Ltd (Tanzania)

THAILAND
The Siam Iron & Steel Co Ltd

TRINIDAD & TOBAGO
Iscott - Iron & Steel Co of Trinidad & Tobago Ltd

TUNISIA
El Fouladh - Sté Tunisienne de Sidérurgie

TURKEY
Çolakoğlu Metalurji AŞ
Ekinciler Demir ve Çelik San AŞ
Metaş Izmir Metalurji Fabrikasi Türk AŞ
Orpaş Metal Sanayi ve Ticaret AŞ
TDCI - Türkiye Demir ve Çelik Işletmeleri Genel Müdürlügü

UNITED KINGDOM
Allied Steel & Wire Ltd
Sheerness Steel Co PLC
The Templeborough Rolling Mills Ltd (TRM)
UES - United Engineering Steels Ltd
Unbrako Steel Co Ltd

UNITED STATES
Armco Inc
Atlantic Steel Co
Bethlehem Steel Corp
CF&I Steel Corp
Charter Rolling, Division of Charter Manufacturing Inc
Connecticut Steel Corp
Georgetown Steel Corp
Keystone Steel & Wire Co
LTV Steel Co Inc
North Star Steel Co (NSS)
Northwestern Steel & Wire Co
Nucor Corp
Raritan River Steel Co

Carbon steel – wire rod

—continued

UNITED STATES *—continued*
Roanoke Electric Steel Corp (Resco)
The Timken Co

USSR
Cherepovetskiy Met Zavod (Cherepovets Iron & Steel Works)
Kazakhskiy Met Zavod (Kazakh Iron & Steel Works)
Makeyevskiy Met Zavod Imeni Kirova (Kirov Iron & Steel Works)
Konstantinovskiy Met Zavod Imeni Frunze (Konstantinovka Frunze Iron & Steel Works)
Met Zavod Krasny Oktyabr (Red October Steel Works)
Met Zavod Krivorizhstal (Krivoy Rog Iron & Steel Works)
Magnitogorskiy Metallurgischeskiy Kombinat (Magnitogorsk Iron & Steel Works Combine)
Met Zavod Imeni Petrovskogo (Petrovsky Iron & Steel Works)
Zapsib Met Zavod (West Siberian Steel Works)

VENEZUELA
Sidor – CVG Siderúrgica del Orinoco CA

VIETNAM
Gia Sang Works

YUGOSLAVIA
Boris Kidrič Steelworks
Slovenske Železarne – Iron & Steel Works of Slovenia
Jadranska Zelejezara Split (Adriatic Steelworks, Split)

ZAIRE
Sté Nationale de Sidérurgie

ZIMBABWE
Lancashire Steel (Pvt) Ltd
Zimbabwe Iron & Steel Co Ltd – Ziscosteel

Carbon steel – reinforcing bars

ALGERIA
SNS – Sté Nationale de Sidérurgie

ANGOLA
Siderurgia Nacional (Sina UEE)

ARGENTINA
Acindar Industria Argentina de Aceros SA
Aceros Bragado Sacif
Rosati y Cristófaro Saic
Establecimiento Altos Hornos Zapla

AUSTRALIA
The Broken Hill Pty Co Ltd
Smorgon Steel

AUSTRIA
Stahl- und Walzwerk Marienhütte GmbH
Voest-Alpine AG

BANGLADESH
Bengal Steel Works Ltd
Dhaka Steel Group of Industries

BELGIUM
Cockerill Sambre SA

BRAZIL
Siderúrgica Aconorte SA
Siderúrgica J.L. Aliperti SA
Siderúrgica Barra Mansa SA
Cia Siderúrgica Belgo-Mineira
Siderúrgica Cearense SA
Cimetal Siderurgia SA
Comesa – Cia Siderúrgica de Alagoas
Cosigua – Cia Siderurgia da Guanabara
Cosinor – Cia Siderúrgica do Nordeste
Dedini SA Sideúrgica
Siderúrgica Guaíra SA
Cia Industrial Itaunense
Lafersa – Laminação de Ferro SA
Siderúrgica Mendes Junior SA (SMJ)
Cia Siderúrgica Pains
Siderúrgica Riograndense SA
Siderama – Cia Siderúrgica da Amazônia
Usiba – Usina Siderúrgica da Bahia SA

BULGARIA
Burgas Steelworks
Kremikovtsi Iron & Steel Works
Lenin Iron & Steel Works

CANADA
Courtice Steel Ltd
Lake Ontario Steel Co (Lasco Steel)

Carbon steel - reinforcing bars *—continued*

CANADA *—continued*
Manitoba Rolling Mills, Amca International
Sidbec-Dosco Inc
Slater Steels – Hamilton Speciality Bar Division
Stelco Inc
Western Canada Steel Ltd

CHILE
Famae – Fabrica y Maestranzas del Ejercito
Cia Siderúrgica Huachipato SA
Cia de Acero Rengo SA

CHINA
Chengdu Metallurgical Experimental Plant
Fushun Xinfu Steel Plant
Guangzhou Alloy Steel Plant
Guangzhou Iron & Steel Works
Jiangxi Steel Works
Shanghai Iron & Steel Works
Shanghai Xihu Iron & Steel Works
Shoudu Iron & Steel Co
Tangshan Iron & Steel Co

COLOMBIA
Futec – Fundiciones Technicas SA
Acerías Paz del Rio SA
Sidelpa – Siderúrgica del Pacifico SA
Sidemuña – Siderúrgica de Muña SA
Simesa – Siderúrgica de Medellín

CUBA
Empressa Metalúrgica José Marti

CZECHOSLOVAKIA
NHKG – Nová Hut Klementa Gottwalda

DENMARK
Danish Steel Works Ltd (Det Danske Stålvalseværk A/S)

DOMINICAN REPUBLIC
Metaldom – Complejo Metalúrgico Dominicano C por A

DUBAI
Ahli Steel Co

ECUADOR
Adelca – Acería del Ecuador CA
Andec – Acerías Nacionales del Ecuador SA

EGYPT
Alexandria National Iron & Steel Co
Delta Steel Mill SAE
The Egyptian Copper Works
The Egyptian Iron & Steel Co (Hadisolb)
National Metal Industries Co

EL SALVADOR
Corinca SA – Corporación Industrial Centroamericana SA
Siderúrgica Centroamericana del Pacífico SA (Sicepasa)

FINLAND
Dalsbruk Oy Ab

FRANCE
Iton Seine SA
Manufer SA
Aciéries et Laminoirs de Rives
Usinor Sacilor

EAST GERMANY
VEB Stahl- und Walzwerk Wilhelm Florin, Hennigsdorf

WEST GERMANY
Max Aicher KG Stahlwerk Annahütte
Badische Stahlwerke AG
Walzwerk Becker (Wabec) GmbH & Co
Hamburger Stahlwerke GmbH
Lech-Stahlwerke GmbH
Maxhütte – Eisenwerke-Gesellschaft Maximilianshütte mbH
Saarstahl Völklingen GmbH
Thyssen Stahl AG

GREECE
Halyvourgia Thessalias SA
Halyvourgiki Inc
Helliniki Halyvourgia SA
Sidenor SA (formerly Steel Works of Northern Greece SA)

HONDURAS
Industria Nacional Del Acero SA

HONG KONG
Shiu Wing Steel Ltd
Shun Fung Iron Works Ltd

HUNGARY
Ozdi Kohászati Uzemek (Okü)

INDIA
Agarwal Foundry & Engineering Works

Carbon steel – reinforcing bars —*continued*

INDIA —*continued*
Andhra Steel Corp Ltd
BD Steel Castings Ltd
Calcutta Steel Co Ltd
Indore Steel & Iron Mills Pvt Ltd (Isim)
Kap Steel Ltd
Mukand Iron & Steel Works Ltd
Punjab Con-Cast Steels Ltd
Rathi Udyog Ltd
Sail – Steel Authority of India Ltd
Sanghvi Steels Ltd
Singh Electrosteel Ltd (SES)
Special Steel Co
Star Steel Pvt Ltd (Starsteel)
Sunflag Iron & Steel Co
Taloja Rolling Mills
The Tata Iron & Steel Co Ltd (Tisco)
Trichy Steel Rolling Mills Ltd
Usha Alloys & Steels Ltd (UASL)

INDONESIA
PT Budidharma Jakarta
PT Gunung Gahapi Steel
PT Inti General Jaya Steel
PT Jakarta Kyoei Steel Works Ltd
PT Kratkatau Steel
PT Master Steel Mfg Co
PT National Union Steel Ltd

IRAQ
State Company for Steel Industries

IRISH REPUBLIC
Irish Steel Ltd

ISRAEL
United Steel Mills Ltd (USM)

ITALY
Afim – Acciaierie Ferriere Industrie Metallurgiche
AFP – Acciaierie Ferriere del Po SpA
Alfer – Azienda Laminazione Ferro SpA
Ferriere di Barghe SpA
Ferriere Bertoli SpA
Bredina Srl
Ferriera Acciaieria Casilina SpA
Diano SpA
Acciaierie e Ferriere Feralpi SpA
Ferdofin SpA
Acciaierie Ferrero SpA
Officine e Fonderie Galtarossa SpA

ITALY —*continued*
Ilfo – Industria Laminati Ferrosi Odolesi Srl
Acciaierie e Ferriere Luigi Leali SpA
Ori Martin SpA
Ferriera Padana SpA
Pasini Siderurgica SpA
Predalva Acciaieria e Ferriera Srl
Presider
Prolafer
Acciaierie e Ferriere Riva SpA
Ferriere San Carlo SpA
Nuova Sidercamuna SpA
Siderpotenza SpA
Simet – Stà Industriale Metallurgica di Napoli SpA
Ferriera Valchiese SpA
Ferriera Valsabbia SpA

JAPAN
Asahi Industries Co Ltd
Funabashi Steel Works Ltd (Funabashi Seito)
Godo Steel Ltd (Godo Seitetsu)
Kobe Steel Ltd (Kobe Seikosho)
Kokko Steel Works Ltd (Kokko Seiko)
Kyoei Steel Ltd
Nakayama Steel Products Co Ltd (Nakayama Kogyo)
Nippon Steel Corp (Shin Nippon Seitetsu)
NKK – Nippon Kokan KK
Sumitomo Metal Industries Ltd
Toa Steel Co Ltd
Tokai Steel Works Ltd (Tokai Kogyo)
Tokyo Steel Manufacturing Co Ltd

JORDAN
Jordan Iron & Steel Industry Co Ltd
National Steel Industry Co Ltd

KENYA
Kenya United Steel Co Ltd
Steel Rolling Mills Ltd

NORTH KOREA
Hwanghai Iron Works

SOUTH KOREA
Daehan Sangsa Co Ltd
Dongkuk Steel Mill Co Ltd
Inchon Iron & Steel Co Ltd
Kang Won Industries Ltd

LEBANON
Consolidated Steel Lebanon SAL (CSL)

Carbon steel – reinforcing bars —*continued*

LUXEMBOURG
Arbed SA
Métallurgique & Minière de Rodange-Athus (MMRA)

MALAYSIA
Amalgamated Steel Mills Bhd
Antara Steel Mills Sdn Bhd
Dah Yung Steel (M) Sdn Bhd
Malayawata Steel Bhd
Southern Iron & Steel Works Sdn Bhd
United Malaysian Steel Mills Bhd

MALTA
Hompesch Steel Co Ltd

MAURITANIA
Sté Arabe du Fer et d'Acier (Safa)

MAURITIUS
Desbro International Ltd

MEXICO
Ahmsa – Altos Hornos de Mexico SA de CV
Deacero – Productos de Acero SA
Aceros Ecatepec SA (Aesa)
Fundiciones de Hierro y Acero SA (Fhasa)
Hylsa SA de CV
Industrials Nylbo SA, Division Acero (formerly Laminadora Atzcapotzalco SA – Lasa)
Aceros San Luis SA
Sicartsa – Siderurgica Lazaro Cardenas "Las Truchas SA"
Siderúrgica de Yucatán SA de CV

MOROCCO
Sonasid – Sté Nationale de Siderurgie

NETHERLANDS
Hoogovens Groep BV

NEW ZEALAND
Pacific Steel Ltd

NIGERIA
Ajaokuta Steel Co Ltd
Delta Steel Co Ltd
General Steel Mills Ltd
The Jos Steel Rolling Co
Katsina Steel Rolling Co Ltd (KSRC)
Mandarin Industries Co Ltd
Nigersteel Co Ltd
Osogbo Steel Rolling Co Ltd

NIGERIA —*continued*
Qua Steel Products Ltd

NORWAY
Christiania Spigerverk AS
Norsk Jernverk AS

PAKISTAN
Metropolitan Steel Corp Ltd (MSC)
New Era Steels
Nowshera Engineering Co Ltd
Razaque Steels (Pvt) Ltd
Sind Steel Corp Ltd

PANAMA
Aceros Panamá SA

PARAGUAY
Acepar – Acero del Paraguay SA

PERU
Siderperú – Empresa Siderúrgica del Perú

PHILIPPINES
Apollo Enterprises Inc (Apollo Steel Mill)
Armstrong Industries Inc
National Steel Corp (NSC)
Pag-Asa Steel Works Inc (PSWI)

POLAND
Huta im Lenina
Třinec Iron & Steel Works

PORTUGAL
Siderurgia Nacional EP (SN)

QATAR
Qatar Steel Co Ltd (Qasco)

RUMANIA
Uzina de Tevi Republica (UTR)

SAUDI ARABIA
Jeddah Rolling Mill Co
Saudi Iron & Steel Co (Hadeed)

SINGAPORE
National Iron & Steel Mills Ltd

SOUTH AFRICA
Cape Town Iron & Steel Works (Pty) Ltd
Davsteel (Pty) Ltd
McWillaw Steel (Pty) Ltd
Scaw Metals Ltd

SPAIN
Artierro SA

Carbon steel – reinforcing bars —*continued*

SPAIN —*continued*
Azma SA
Celsa – Cia Española de Laminacion SA
Cufesa SA
Ensidesa – Empresa Nacional Siderurgica SA
Megasa – Metalurgica Galaica SA
Nervacero SA
Esteban Orbegozo SA
Siderúrgica Sevillana SA
Sidegasa – Siderúrgica de Galicia SA
Torras, Herrería y Construcciones SA
Laminaciones Torrente SA
Marcial Ucin SA
Unión Cerrajera SA

SWEDEN
Halmstads Järnverks AB
Smedjebacken-Boxholm Stål AB
SSAB – Svenskt Stål AB

SWITZERLAND
Ferrowohlen AG
Monteforno Acciaierie e Laminatoi SA
Von Moos Stahl AG
Von Roll Group

TAIWAN
China Steel Corp
Feng Lung Steel Factory Co Ltd
Li-Chong Steel & Iron Works Co Ltd
Tang Eng Iron Works Co Ltd

TANZANIA
Steel Rolling Mills Ltd (Tanzania)

THAILAND
The Bangkok Iron & Steel Works Co Ltd
Bangkok Steel Industry Co Ltd
The Siam Iron & Steel Co Ltd

TOGO
SNS – Sté National de Sidérurgie

TRINIDAD & TOBAGO
Central Trinidad Steel Ltd – Centrin

TUNISIA
El Fouladh – Sté Tunisienne de Sidérurgie

TURKEY
Çolakoğlu Metalurji AŞ

TURKEY —*continued*
Demir Sanayi Demir Çelik Ticaret ve Sanayi AŞ
Ekinciler Demir ve Çelik San AŞ
Izmir Demir Çelik Sanayi AŞ
Metaş Izmir Metalurji Fabrikasi Türk AŞ
Orpaş Metal Sanayi ve Ticaret AŞ

UNITED KINGDOM
Allied Steel & Wire Ltd
Queenborough Rolling Mill Co Ltd
Sheerness Steel Co PLC

UNITED STATES
AB Steel Mill Inc
Armco Inc
Atlantic Steel Co
Auburn Steel Co Inc (Austeel)
Birmingham Steel Corp
Border Steel Mills Inc
Calumet Steel Co
Cascade Steel Rolling Mills Inc
CF&I Steel Corp
Chaparral Steel Co
Commercial Steel Corp
Connecticut Steel Corp
Florida Steel Corp
Franklin Steel Co
Georgetown Steel Corp
Hawaiian Western Steel Ltd
Marion Steel Co
Milton Manufacturing Co
Missouri Rolling Mill Corp
New Jersey Steel Corp
North Star Steel Co (NSS)
Nucor Corp
Oklahoma Steel Mill Inc
Owen Electric Steel Co of South Carolina
Roanoke Electric Steel Corp (Resco)
Seattle Steel Inc
Sheffield Steel Corp
Structural Metals Inc
Thomas Steel Corp

USSR
Met Zavod Azovstal (Azovstal Iron & Steel Works)
Makeyevskiy Met Zavod Imeni Kirova (Kirov Iron & Steel Works)
Konstantinovskiy Met Zavod Imeni Frunze (Konstantinovka Frunze Iron & Steel Works)
Met Zavod Kuybyshev (Kuibyshev Iron & Steel Works)
Kuznetskiy Met Kombinat

Carbon steel - reinforcing bars *—continued*

USSR *—continued*
Novo-Tagilskiy Met Zavod (Novo-Tagil Iron & Steel Works)

VENEZUELA
Sidetur – Siderúrgica del Turbio SA
Sidor – CVG Siderúrgica del Orinoco CA

VIETNAM
Bien Hoa Works

YUGOSLAVIA
Jadranska Zelejezara Split (Adriatic Steelworks, Split)

ZAIRE
Sté Nationale de Sidérurgie

ZIMBABWE
Zimbabwe Iron & Steel Co Ltd – Ziscosteel

Carbon steel - round bars

ALBANIA
Steel of the Party Metallurgical Combine

ARGENTINA
Acindar Industria Argentina de Aceros SA
La Cantábrica SA Metalúrgica, Industrial y Comercial
Rosati y Cristófaro Saic
Establecimiento Altos Hornos Zapla

AUSTRALIA
The Broken Hill Pty Co Ltd
Commonwealth Steel Co Ltd
Rydal Steel – Division of LNC Industrial Products Pty Ltd
Smorgon Steel

AUSTRIA
Stahl- und Walzwerk Marienhütte GmbH
Voest-Alpine AG

BANGLADESH
Chittagong Steel Mills Ltd

BELGIUM
Laminoirs du Ruau SA

BRAZIL
Acesita – Cia Aços Especiais Itabira
Aços Anhanguera SA (Acoansa)
Siderúrgica Nossa Senhora Aparecida SA
Siderúrgica Barra Mansa SA
Siderúrgica Cearense SA
Cimetal Siderurgia SA
Cofavi – Cia Ferro e Aço de Vitória
Comesa – Cia Siderúrgica de Alagoas
Cosigua – Cia Siderurgia da Guanabara
CSN – Cia Siderúrgica Nacional
Dedini SA Sideúrgica
Siderúrgica Hime SA
Cia Industrial Itaunense
Lafersa – Laminação de Ferro SA
Mannesmann SA
Siderúrgica Mendes Junior SA (SMJ)
Cia Siderúrgica Pains
Aços Finos Piratini SA
Siderúrgica Riograndense SA
Laminação Santa Maria SA – Indústria e Comércio
Villares Group

BULGARIA
Burgas Steelworks
Lenin Iron & Steel Works

CANADA
Atlas Specialty Steels Division of Rio Algom Ltd
Courtice Steel Ltd
Lake Ontario Steel Co (Lasco Steel)
Manitoba Rolling Mills, Amca International
Sidbec-Dosco Inc
Slater Steels – Hamilton Speciality Bar Division
Slater Steels, Sorel Forge Division
Stelco Inc
Western Canada Steel Ltd

CHILE
Cia Siderúrgica Huachipato SA

CHINA
Changcheng Steel Works
Chengdu Metallurgical Experimental Plant
Chongqing Iron & Steel Co
Dalian Steel Rolling Mill
Daye Steel Works
Echeng Iron & Steel Works
Fushun Xinfu Steel Plant
Guangzhou Alloy Steel Plant

Carbon steel - round bars
—continued

CHINA *—continued*
Guangzhou Iron & Steel Works
Guiyang Steel Mill
Hefei Iron & Steel Co
Lingyuan Iron & Steel Works
Shanghai Iron & Steel Works
Shoudu Iron & Steel Co

COLOMBIA
Sidelpa – Siderúrgica del Pacifico SA

CZECHOSLOVAKIA
NHKG – Nová Hut Klementa Gottwalda
Vitkovice Steelworks

DENMARK
Danish Steel Works Ltd (Det Danske Stålvalseværk A/S)

DOMINICAN REPUBLIC
Metaldom – Complejo Metalúrgico Dominicano C por A

DUBAI
Ahli Steel Co

ECUADOR
Adelca – Acería del Ecuador CA

FINLAND
Ovako Steel Oy Ab

FRANCE
Aciéries de Bonpertuis
Grousset
Iton Seine SA
Aciéries et Laminoirs de Rives
Aciéries et Laminoirs du Saut du Tarn (ALST)
Usinor Sacilor

EAST GERMANY
VEB Stahl- und Walzwerk Wilhelm Florin, Hennigsdorf

WEST GERMANY
Böhler AG, Edelstahlwerke
Remscheider Walz- und Hammerwerke, Böllinghaus & Co
Edelstahlwerke Buderus AG
Klöckner Stahl GmbH
Krupp Stahl AG (KS)
Lech-Stahlwerke GmbH
Maxhütte – Eisenwerke-Gesellschaft Maximilianshütte mbH

WEST GERMANY *—continued*
Saarstahl Völklingen GmbH
Thyssen Edelstahlwerke AG
Thyssen Stahl AG

GREECE
Halyvourgia Thessalias SA

HUNGARY
Csepel Steel & Tube Works
Ozdi Kohászati Uzemek (Okü)

INDIA
Agarwal Foundry & Engineering Works
Ahmedabad Steelcraft & Rolling Mills Pvt Ltd
Andhra Steel Corp Ltd
Bihar Alloy Steels Ltd
Calcutta Steel Co Ltd
Ellora Steels Ltd (ESL)
Ferro Alloys Corp Ltd (Facor) (Steel division), (Vidarbha Iron & Steel Corp Ltd)
Guest Keen Williams Ltd (GKW)
Indore Steel & Iron Mills Pvt Ltd (Isim)
Kalyani Steels Ltd
Kap Steel Ltd
Mahindra Ugine Steel Co Ltd (Muscosteel)
Modi Industries Ltd, Steel Division
Mohan Steels Ltd
Mukand Iron & Steel Works Ltd
Panchmahal Steel Ltd
Parikh Steel Pvt Ltd
Partap Steel Rolling Mills (1935) Ltd
Punjab Con-Cast Steels Ltd
The Punjab Steel Rolling Mills
Rathi Udyog Ltd
Sail – Steel Authority of India Ltd
Sanghvi Steels Ltd
Singh Electrosteel Ltd (SES)
Special Steel Co
Star Steel Pvt Ltd (Starsteel)
Steel Ingots Pvt Ltd
Sunflag Iron & Steel Co
Taloja Rolling Mills
The Tata Iron & Steel Co Ltd (Tisco)
Trichy Steel Rolling Mills Ltd
Usha Alloys & Steels Ltd (UASL)
Visvesvaraya Iron & Steel Ltd (formerly The Mysore Iron & Steel Ltd)

INDONESIA
PT Budidharma Jakarta
PT Kratkatau Steel

Carbon steel - round bars
—continued

IRAN
National Iranian Steel Corp (Nisc)

IRISH REPUBLIC
Irish Steel Ltd

ISRAEL
United Steel Mills Ltd (USM)

ITALY
AFP – Acciaierie Ferriere del Po SpA
SpA Officine Fratelli Bertoli fu Rodolfo
Acciaierie di Bolzano SpA
Nuova Deltasider SpA
DeltaValdarno
Acciaierie e Ferriere Lombarde Falck SpA
Ferrosider SpA
Las – Laminazione Acciai Speciali SpA
Acciaierie e Ferriere Luigi Leali SpA
Lucchini Siderurgica SpA
Metallurgica Marcora SpA
Pasini Siderurgica SpA
Stilma SpA
Acciaierie Valbruna SpA

JAPAN
Aichi Steel Works Ltd
Asahi Industries Co Ltd
Daido Steel Co Ltd
Godo Steel Ltd (Godo Seitetsu)
Kawasaki Steel Corp
Kobe Steel Ltd (Kobe Seikosho)
Kyoei Steel Ltd
Nakayama Steel Works Ltd (Nakayama Seikosho)
Nippon Steel Corp (Shin Nippon Seitetsu)
Rinko Steel Works Co Ltd
Sumitomo Metal Industries Ltd
Tokai Steel Works Ltd (Tokai Kogyo)
Tokyo Steel Manufacturing Co Ltd

JORDAN
Jordan Iron & Steel Industry Co Ltd
National Steel Industry Co Ltd

KENYA
Kenya United Steel Co Ltd
Steel Rolling Mills Ltd

NORTH KOREA
Kangson Works

SOUTH KOREA
Dongkuk Steel Mill Co Ltd
Kang Won Industries Ltd
Korea Heavy Machinery Industrial Ltd
Sammi Steel Co Ltd

LEBANON
Consolidated Steel Lebanon SAL (CSL)

LUXEMBOURG
Arbed SA

MALAYSIA
Amalgamated Steel Mills Bhd
Antara Steel Mills Sdn Bhd
Dah Yung Steel (M) Sdn Bhd
Malayawata Steel Bhd
Southern Iron & Steel Works Sdn Bhd

MALTA
Hompesch Steel Co Ltd

MEXICO
Aceros Ecatepec SA (Aesa)
Industrias CH SA (formerly Campos Hermanos SA)
Industrials Nylbo SA, Division Acero (formerly Laminadora Atzcapotzalco SA – Lasa)
Sicartsa – Siderurgica Lazaro Cardenas "Las Truchas SA"
Tamsa – Tubos de Acero de Mexico SA

NIGERIA
Ajaokuta Steel Co Ltd
Allied Steel Industry (Nigeria) Ltd
Delta Steel Co Ltd
Kwara Commercial, Metal & Chemical Industries Ltd
Major Engineering Co Ltd
Nigersteel Co Ltd
Osogbo Steel Rolling Co Ltd
Universal Steels Ltd

PAKISTAN
Ittefaq Ltd
Metropolitan Steel Corp Ltd (MSC)
New Era Steels
Nowshera Engineering Co Ltd
Razaque Steels (Pvt) Ltd
Sind Steel Corp Ltd

PARAGUAY
Acepar – Acero del Paraguay SA

PERU
Aceros Arequipa SA (Acersa)

Carbon steel – round bars
—continued

PERU *—continued*
Siderperú – Empresa Siderúrgica del Perú

PHILIPPINES
Armco-Marsteel Alloy Corp
Marcelo Steel Corp (MSC)
National Steel Corp (NSC)
Pag-Asa Steel Works Inc (PSWI)

PORTUGAL
Siderurgia Nacional EP (SN)
Fábrica de Aços Tomé Fèteira SA

SINGAPORE
National Iron & Steel Mills Ltd

SOUTH AFRICA
Davsteel (Pty) Ltd
Highveld Steel & Vanadium Corp Ltd
Iscor Ltd
Scaw Metals Ltd
Usco – The Union Steel Corp (of South Africa) Ltd

SPAIN
Aforasa – Acerias y Forjas de Azcoitia SA
Forjas Alavesas SA (Fasa)
Babcock & Wilcox Española SA Tubular Products Division
Aceros de Irura SA
Aceros de Llodio SA (Allosa)
Megasa – Metalurgica Galaica SA
Olarra SA (Olsa)
Esteban Orbegozo SA
Pedro Orbegozo y Cía SA (Acenor)
Forjas y Aceros de Reinosa SA
Siderúrgica Sevillana SA
Torras, Herrería y Construcciones SA
Unión Cerrajera SA

SWEDEN
Björneborgs Jernverks AB
Ovako Steel AB
Smedjebacken-Boxholm Stål AB
SSAB – Svenskt Stål AB

SWITZERLAND
Monteforno Acciaierie e Laminatoi SA
Von Moos Stahl AG

TAIWAN
Chia Hsin Metal Industrial Co Ltd
China Steel Corp
Feng Lung Steel Factory Co Ltd

TANZANIA
Steel Rolling Mills Ltd (Tanzania)

THAILAND
The Bangkok Iron & Steel Works Co Ltd
The Siam Iron & Steel Co Ltd

TRINIDAD & TOBAGO
Central Trinidad Steel Ltd – Centrin

TUNISIA
El Fouladh – Sté Tunisienne de Sidérurgie

TURKEY
Asil Çelik Sanayi ve Ticaret AŞ (Asil Çelik)
Cemtas AŞ
Çolakoğlu Metalurji AŞ
Ekinciler Demir ve Çelik San AŞ
Izmir Demir Çelik Sanayi AŞ
Kroman Steel Industries Co (Kroman Celik)
Metaş Izmir Metalurji Fabrikasi Türk AŞ
MKEK Celik Fabrikasi

UNITED KINGDOM
Allied Steel & Wire Ltd
Bedford Steels
Dudley Port Rolling Mills, Glynwed Steels Ltd
George Gadd & Co, Glynwed Steels Ltd
Joseph Gillott & Sons, Glynwed Steels Ltd
Kiveton Park Steel & Wire Works Ltd
Frank Pickering & Co Ltd
Raine & Co Ltd
Sheerness Steel Co PLC
Sheffield Forgemasters Ltd
Spencer Clark Metal Industries PLC
UES – United Engineering Steels Ltd
Unbrako Steel Co Ltd
Woodstone Rolling Mills Ltd

UNITED STATES
AB Steel Mill Inc
Armco Inc
Atlantic Steel Co
Auburn Steel Co Inc (Austeel)

Carbon steel – round bars
—continued

UNITED STATES *—continued*
The Babcock & Wilcox Co, Tubular Products Division
Bethlehem Steel Corp
Birmingham Steel Corp
Border Steel Mills Inc
Calumet Steel Co
Cascade Steel Rolling Mills Inc
CF&I Steel Corp
Chaparral Steel Co
Connecticut Steel Corp
Florida Steel Corp
Franklin Steel Co
Inland Steel Co
Kentucky Electric Steel Corp
Laclede Steel Co
LTV Steel Co Inc
Marion Steel Co
Milton Manufacturing Co
Missouri Rolling Mill Corp
Nelsen Steel & Wire Co Inc
North Star Steel Co (NSS)
Nucor Corp
Owen Electric Steel Co of South Carolina
Quanex Corp
Roanoke Electric Steel Corp (Resco)
Seattle Steel Inc
Structural Metals Inc
Thomas Steel Corp
The Timken Co
USS, A Division of USX Corp

USSR
Cherepovetskiy Met Zavod (Cherepovets Iron & Steel Works)
Dneprodzerzhinskiy Zavod Imeni Dzerzhinskogo (Dzerzhinsky Works)
Met Zavod Krivorizhstal (Krivoy Rog Iron & Steel Works)
Kuznetskiy Met Kombinat
Magnitogorskiy Metallurgischeskiy Kombinat (Magnitogorsk Iron & Steel Works Combine)
Zakavkazskiy Met Zavod (Transcaucasian Metallurgical Works)

VENEZUELA
Acerex CA
Metalanca – Metalúrgica de Laminados Nacionales CA
Sidor – CVG Siderúrgica del Orinoco CA

YUGOSLAVIA
Boris Kidrič Steelworks
Slovenske Železarne – Iron & Steel Works of Slovenia

ZIMBABWE
Zimbabwe Iron & Steel Co Ltd – Ziscosteel

Carbon steel – square bars

ALBANIA
Steel of the Party Metallurgical Combine

ARGENTINA
Acindar Industria Argentina de Aceros SA
La Cantábrica SA Metalúrgica, Industrial y Comercial
Rosati y Cristófaro Saic
Establecimiento Altos Hornos Zapla

AUSTRALIA
The Broken Hill Pty Co Ltd
Rydal Steel – Division of LNC Industrial Products Pty Ltd

AUSTRIA
Stahl- und Walzwerk Marienhütte GmbH
Voest-Alpine AG

BANGLADESH
Bengal Steel Works Ltd

BELGIUM
Laminoirs du Ruau SA

BRAZIL
Acesita – Cia Aços Especiais Itabira
Siderúrgica J.L. Aliperti SA
Aços Anhanguera SA (Acoansa)
Siderúrgica Nossa Senhora Aparecida SA
Siderúrgica Barra Mansa SA
Siderúrgica Cearense SA
Cofavi – Cia Ferro e Aço de Vitória
Cosigua – Cia Siderurgia da Guanabara
Cosim – Cia Siderúrgica de Mogi das Cruzes
CSN – Cia Siderúrgica Nacional
Dedini SA Sideúrgica
Siderúrgica Hime SA
Mannesmann SA
Siderúrgica Mendes Junior SA (SMJ)
Aços Finos Piratini SA

Carbon steel – square bars —*continued*

BRAZIL —*continued*
Laminação Santa Maria SA – Indústria e Comércio
Villares Group

CANADA
Courtice Steel Ltd
Lake Ontario Steel Co (Lasco Steel)
Manitoba Rolling Mills, Amca International
Sidbec-Dosco Inc
Slater Steels – Hamilton Speciality Bar Division
Slater Steels, Sorel Forge Division
Western Canada Steel Ltd

CHINA
Guangzhou Alloy Steel Plant
Guiyang Steel Mill
Hefei Iron & Steel Co

CZECHOSLOVAKIA
Vitkovice Steelworks

DENMARK
Danish Steel Works Ltd (Det Danske Stålvalseværk A/S)

DOMINICAN REPUBLIC
Metaldom – Complejo Metalúrgico Dominicano C por A

FRANCE
Aciéries de Bonpertuis
Iton Seine SA
Aciéries et Laminoirs du Saut du Tarn (ALST)
Usinor Sacilor

EAST GERMANY
VEB Stahl- und Walzwerk Brandenburg
VEB Stahl- und Walzwerk Wilhelm Florin, Hennigsdorf

WEST GERMANY
Böhler AG, Edelstahlwerke
Remscheider Walz- und Hammerwerke, Böllinghaus & Co
Edelstahlwerke Buderus AG
Klöckner Stahl GmbH
Krupp Stahl AG (KS)
Lech-Stahlwerke GmbH
Maxhütte – Eisenwerke-Gesellschaft Maximilianshütte mbH
Saarstahl Völklingen GmbH

WEST GERMANY —*continued*
Thyssen Edelstahlwerke AG
Thyssen Stahl AG

INDIA
Agarwal Foundry & Engineering Works
Ahmedabad Steelcraft & Rolling Mills Pvt Ltd
Andhra Steel Corp Ltd
Bihar Alloy Steels Ltd
Calcutta Steel Co Ltd
Ellora Steels Ltd (ESL)
Ferro Alloys Corp Ltd (Facor) (Steel division), (Vidarbha Iron & Steel Corp Ltd)
Guest Keen Williams Ltd (GKW)
Indore Steel & Iron Mills Pvt Ltd (Isim)
Kalyani Steels Ltd
Kap Steel Ltd
Mahindra Ugine Steel Co Ltd (Muscosteel)
Modi Industries Ltd, Steel Division
Panchmahal Steel Ltd
Punjab Con-Cast Steels Ltd
Rathi Udyog Ltd
Sail – Steel Authority of India Ltd
Special Steel Co
Star Steel Pvt Ltd (Starsteel)
Taloja Rolling Mills
Trichy Steel Rolling Mills Ltd

ISRAEL
United Steel Mills Ltd (USM)

ITALY
AFP – Acciaierie Ferriere del Po SpA
A.F.V. Beltrame SpA
Nuova Deltasider SpA
DeltaValdarno
Falci SpA, Reparto Laminazione
Ferdofin SpA
Ferdofer SpA
Acciaierie Ferrero SpA
Ferrosider SpA
Las – Laminazione Acciai Speciali SpA
Acciaierie e Ferriere Luigi Leali SpA
Lucchini Siderurgica SpA
Metallurgica Marcora SpA
Acciaierie e Ferriere Riva SpA
Stilma SpA
Acciaierie Valbruna SpA

JAPAN
Daido Steel Co Ltd
Kobe Steel Ltd (Kobe Seikosho)
Kyoei Steel Ltd

Carbon steel - square bars
—continued

JAPAN *—continued*
Rinko Steel Works Co Ltd
Sumitomo Metal Industries Ltd

JORDAN
National Steel Industry Co Ltd

KENYA
Kenya United Steel Co Ltd

SOUTH KOREA
Korea Heavy Machinery Industrial Ltd
Sammi Steel Co Ltd

LEBANON
Consolidated Steel Lebanon SAL (CSL)

LUXEMBOURG
Arbed SA

MALAYSIA
Dah Yung Steel (M) Sdn Bhd

MEXICO
Industrials Nylbo SA, Division Acero (formerly Laminadora Atzcapotzalco SA – Lasa)
Tamsa – Tubos de Acero de Mexico SA

NIGERIA
Ajaokuta Steel Co Ltd
Kwara Commercial, Metal & Chemical Industries Ltd

PAKISTAN
Ittefaq Ltd
Sind Steel Corp Ltd

PARAGUAY
Acepar – Acero del Paraguay SA

PERU
Aceros Arequipa SA (Acersa)

PORTUGAL
Siderurgia Nacional EP (SN)
Fábrica de Aços Tomé Fèteira SA

SINGAPORE
First Rolling Mills Singapore Pte Ltd
National Iron & Steel Mills Ltd

SOUTH AFRICA
Davsteel (Pty) Ltd
Highveld Steel & Vanadium Corp Ltd
Iscor Ltd
Scaw Metals Ltd
Usco – The Union Steel Corp (of South Africa) Ltd

SPAIN
Aforasa – Acerias y Forjas de Azcoitia SA
Babcock & Wilcox Española SA Tubular Products Division
Industrias del Besós SA (Inbesa)
Ensidesa – Empresa Nacional Siderurgica SA
Aceros de Llodio SA (Allosa)
Olarra SA (Olsa)
Pedro Orbegozo y Cía SA (Acenor)
Forjas y Aceros de Reinosa SA

SWEDEN
Björneborgs Jernverks AB
Smedjebacken-Boxholm Stål AB

SWITZERLAND
Von Moos Stahl AG

TAIWAN
Chin Ho Fa Steel & Iron Co Ltd
Feng Lung Steel Factory Co Ltd

TRINIDAD & TOBAGO
Central Trinidad Steel Ltd – Centrin

TUNISIA
El Fouladh – Sté Tunisienne de Sidérurgie

TURKEY
Cemtas AŞ
Ekinciler Demir ve Çelik San AŞ
Izmir Demir Çelik Sanayi AŞ
MKEK Celik Fabrikasi

UNITED KINGDOM
Bedford Steels
Dudley Port Rolling Mills, Glynwed Steels Ltd
George Gadd & Co, Glynwed Steels Ltd
Joseph Gillott & Sons, Glynwed Steels Ltd
Frank Pickering & Co Ltd
Raine & Co Ltd
Sheerness Steel Co PLC
Sheffield Forgemasters Ltd
Spencer Clark Metal Industries PLC
UES – United Engineering Steels Ltd

Carbon steel - square bars —*continued*

UNITED KINGDOM —*continued*
Woodstone Rolling Mills Ltd

UNITED STATES
Atlantic Steel Co
Auburn Steel Co Inc (Austeel)
Bethlehem Steel Corp
Birmingham Steel Corp
Border Steel Mills Inc
Cascade Steel Rolling Mills Inc
Chaparral Steel Co
Florida Steel Corp
Franklin Steel Co
Inland Steel Co
LTV Steel Co Inc
Nelsen Steel & Wire Co Inc
North Star Steel Co (NSS)
Nucor Corp
Quanex Corp
Roanoke Electric Steel Corp (Resco)
Structural Metals Inc
The Timken Co

USSR
Met Zavod Krivorizhstal (Krivoy Rog Iron & Steel Works)

VENEZUELA
Acerex CA
Sidetur - Siderúrgica del Turbio SA

YUGOSLAVIA
Boris Kidrič Steelworks

ZIMBABWE
Zimbabwe Iron & Steel Co Ltd - Ziscosteel

Carbon steel - flats

ALBANIA
Steel of the Party Metallurgical Combine

ARGENTINA
Acindar Industria Argentina de Aceros SA
La Cantábrica SA Metalúrgica, Industrial y Comercial
Rosati y Cristófaro Saic
Establecimiento Altos Hornos Zapla

AUSTRALIA
The Broken Hill Pty Co Ltd

AUSTRALIA —*continued*
Rydal Steel - Division of LNC Industrial Products Pty Ltd
Smorgon Steel

AUSTRIA
Voest-Alpine AG

BANGLADESH
Bengal Steel Works Ltd

BELGIUM
Cockerill Sambre SA
Laminoirs du Ruau SA

BRAZIL
Acesita - Cia Aços Especiais Itabira
Siderúrgica J.L. Aliperti SA
Aços Anhanguera SA (Acoansa)
Siderúrgica Nossa Senhora Aparecida SA
Siderúrgica Cearense SA
Cimetal Siderurgia SA
Cofavi - Cia Ferro e Aço de Vitória
Cosigua - Cia Siderurgia da Guanabara
Siderúrgica Hime SA
Mannesmann SA
Cia Siderúrgica Pains
Aços Finos Piratini SA
Laminação Santa Maria SA - Indústria e Comércio
Villares Group

BULGARIA
Burgas Steelworks

BURMA
Ywama Steel Mill

CANADA
Courtice Steel Ltd
Manitoba Rolling Mills, Amca International
Sidbec-Dosco Inc
Slater Steels - Hamilton Speciality Bar Division
Slater Steels, Sorel Forge Division

CHINA
Chengdu Metallurgical Experimental Plant
Dalian Steel Rolling Mill
Fushun Xinfu Steel Plant
Guangzhou Alloy Steel Plant
Hefei Iron & Steel Co
Jiangxi Steel Works
Shanghai Iron & Steel Works

Carbon steel - flats
—continued

CHINA *—continued*
Shoudu Iron & Steel Co

COLOMBIA
Sidelpa - Siderúrgica del Pacifico SA
Simesa - Siderúrgica de Medellín

CZECHOSLOVAKIA
Vitkovice Steelworks

DENMARK
Danish Steel Works Ltd (Det Danske Stålvalseværk A/S)

ECUADOR
Adelca - Acería del Ecuador CA

FINLAND
Ovako Steel Oy Ab

FRANCE
Aciéries de Bonpertuis
Aciéries et Laminoirs de Rives
Aciéries et Laminoirs du Saut du Tarn (ALST)
Forges de Syam
Usinor Sacilor

EAST GERMANY
VEB Stahl- und Walzwerk Wilhelm Florin, Hennigsdorf
VEB Maxhütte Unterwellenborn

WEST GERMANY
Böhler AG, Edelstahlwerke
Remscheider Walz- und Hammerwerke, Böllinghaus & Co
Edelstahlwerke Buderus AG
Klöckner Stahl GmbH
Krupp Stahl AG (KS)
Lech-Stahlwerke GmbH
Maxhütte - Eisenwerke-Gesellschaft Maximilianshütte mbH
Saarstahl Völklingen GmbH
Thyssen Edelstahlwerke AG
Thyssen Stahl AG

INDIA
Agarwal Foundry & Engineering Works
Ahmedabad Steelcraft & Rolling Mills Pvt Ltd
Andhra Steel Corp Ltd
Bihar Alloy Steels Ltd
Calcutta Steel Co Ltd
Ellora Steels Ltd (ESL)

INDIA *—continued*
Ferro Alloys Corp Ltd (Facor) (Steel division), (Vidarbha Iron & Steel Corp Ltd)
Guest Keen Williams Ltd (GKW)
Indore Steel & Iron Mills Pvt Ltd (Isim)
Mahindra Ugine Steel Co Ltd (Muscosteel)
Mukand Iron & Steel Works Ltd
Parikh Steel Pvt Ltd
Partap Steel Rolling Mills (1935) Ltd
Punjab Con-Cast Steels Ltd
The Punjab Steel Rolling Mills
Sail - Steel Authority of India Ltd
Sanghvi Steels Ltd
Special Steel Co
Star Steel Pvt Ltd (Starsteel)
Sunflag Iron & Steel Co
Taloja Rolling Mills
Trichy Steel Rolling Mills Ltd
Usha Alloys & Steels Ltd (UASL)
Visvesvaraya Iron & Steel Ltd (formerly The Mysore Iron & Steel Ltd)

INDONESIA
PT Gunung Gahapi Steel
PT National Union Steel Ltd

ISRAEL
United Steel Mills Ltd (USM)

ITALY
AFP - Acciaierie Ferriere del Po SpA
A.F.V. Beltrame SpA
Ferriera Castellana SpA
Nuova Deltasider SpA
DeltaValdarno
Dora Acciaierie e Ferriere (formerly Acciaierie Ferriere Alpine SpA (AFB))
Falci SpA, Reparto Laminazione
Ferdofin SpA
Ferdofer SpA
Acciaierie Ferrero SpA
Ferrosider SpA
Las - Laminazione Acciai Speciali SpA
Acciaierie e Ferriere Luigi Leali SpA
Metallurgica Marcora SpA
Acciaierie e Ferriere Riva SpA
Stilma SpA
Acciaierie Valbruna SpA

JAPAN
Daido Steel Co Ltd
Funabashi Steel Works Ltd (Funabashi Seito)

Carbon steel - flats

—continued

JAPAN *—continued*
Kobe Steel Ltd (Kobe Seikosho)
Kyoei Steel Ltd
Nakayama Steel Works Ltd (Nakayama Seikosho)
Rinko Steel Works Co Ltd
Sumitomo Metal Industries Ltd
Topy Industries Ltd

JORDAN
National Steel Industry Co Ltd

KENYA
Steel Rolling Mills Ltd

SOUTH KOREA
Dongkuk Steel Mill Co Ltd
Sammi Steel Co Ltd

LEBANON
Consolidated Steel Lebanon SAL (CSL)

LUXEMBOURG
Arbed SA

MALAYSIA
Amalgamated Steel Mills Bhd
Antara Steel Mills Sdn Bhd
Dah Yung Steel (M) Sdn Bhd

MALTA
Hompesch Steel Co Ltd

MEXICO
Ahmsa – Altos Hornos de Mexico SA de CV
Industrials Nylbo SA, Division Acero (formerly Laminadora Atzcapotzalco SA – Lasa)
Siderúrgica de Yucatán SA de CV

NEW ZEALAND
Pacific Steel Ltd

NIGERIA
Delta Steel Co Ltd
Kwara Commercial, Metal & Chemical Industries Ltd
Major Engineering Co Ltd
Qua Steel Products Ltd

NORWAY
Norsk Jernverk AS

PAKISTAN
Ittefaq Ltd

PAKISTAN *—continued*
Nowshera Engineering Co Ltd
Sind Steel Corp Ltd

PARAGUAY
Acepar – Acero del Paraguay SA

PERU
Aceros Arequipa SA (Acersa)

PORTUGAL
Siderurgia Nacional EP (SN)
Fábrica de Aços Tomé Fèteira SA

RUMANIA
Oţelul Roşu Works

SINGAPORE
First Rolling Mills Singapore Pte Ltd
National Iron & Steel Mills Ltd

SOUTH AFRICA
Davsteel (Pty) Ltd
Highveld Steel & Vanadium Corp Ltd
Iscor Ltd
Scaw Metals Ltd
Usco – The Union Steel Corp (of South Africa) Ltd

SPAIN
Forjas Alavesas SA (Fasa)
Babcock & Wilcox Española SA Tubular Products Division
Industrias del Besós SA (Inbesa)
Ensidesa – Empresa Nacional Siderurgica SA
Aceros de Irura SA
Aceros de Llodio SA (Allosa)
Olarra SA (Olsa)
Pedro Orbegozo y Cía SA (Acenor)
Forjas y Aceros de Reinosa SA
Marcial Ucin SA
Unión Cerrajera SA
Aceros Valdemoro SA

SWEDEN
Björneborgs Jernverks AB
Smedjebacken-Boxholm Stål AB

SWITZERLAND
Von Moos Stahl AG
Von Roll Group

TAIWAN
Chin Ho Fa Steel & Iron Co Ltd
Feng Lung Steel Factory Co Ltd

Carbon steel - flats
—continued

TANZANIA
Steel Rolling Mills Ltd (Tanzania)

TRINIDAD & TOBAGO
Central Trinidad Steel Ltd - Centrin

TURKEY
Asil Çelik Sanayi ve Ticaret AŞ (Asil Çelik)
Ekinciler Demir ve Çelik San AŞ
Izmir Demir Çelik Sanayi AŞ
MKEK Celik Fabrikasi
TDCI - Türkiye Demir ve Çelik Işletmeleri Genel Müdürlügü

UNITED KINGDOM
Allied Steel & Wire Ltd
A.S. Rolling Mills Ltd
Bedford Steels
Bromford Iron & Steel Co Ltd
Ductile Hot Mill, Glynwed Steels Ltd
Dudley Port Rolling Mills, Glynwed Steels Ltd
George Gadd & Co, Glynwed Steels Ltd
Joseph Gillott & Sons, Glynwed Steels Ltd
Martins (Dundyvan) Ltd
Frank Pickering & Co Ltd
Sheerness Steel Co PLC
Sheffield Forgemasters Ltd
Spencer Clark Metal Industries PLC
UES - United Engineering Steels Ltd
United Merchant Bar PLC
W. Wesson, Glynwed Steels Ltd

UNITED STATES
Atlantic Steel Co
Auburn Steel Co Inc (Austeel)
Bayou Steel Corp
Bethlehem Steel Corp
Birmingham Steel Corp
Calumet Steel Co
Cascade Steel Rolling Mills Inc
Chaparral Steel Co
Chicago Heights Steel Co
Florida Steel Corp
Franklin Steel Co
Jersey Shore Steel Co Inc
Kentucky Electric Steel Corp
LTV Steel Co Inc
Marion Steel Co
McDonald Steel Corp
Nelsen Steel & Wire Co Inc

UNITED STATES *—continued*
North Star Steel Co (NSS)
Northwestern Steel & Wire Co
Nucor Corp
Quanex Corp
Roanoke Electric Steel Corp (Resco)
Seattle Steel Inc
SMI Steel Inc
Thomas Steel Corp

USSR
Met Zavod Krivorizhstal (Krivoy Rog Iron & Steel Works)

VENEZUELA
Acerex CA
Metalanca - Metalúrgica de Laminados Nacionales CA
Sidetur - Siderúrgica del Turbio SA

YUGOSLAVIA
Boris Kidrič Steelworks
Slovenske Železarne - Iron & Steel Works of Slovenia

ZIMBABWE
Zimbabwe Iron & Steel Co Ltd - Ziscosteel

Carbon steel - hexagons

ALBANIA
Steel of the Party Metallurgical Combine

ARGENTINA
Acindar Industria Argentina de Aceros SA

AUSTRALIA
The Broken Hill Pty Co Ltd
Rydal Steel - Division of LNC Industrial Products Pty Ltd

BRAZIL
Mannesmann SA
Laminação Santa Maria SA - Indústria e Comércio

CANADA
Slater Steels, Sorel Forge Division

CHINA
Fushun Xinfu Steel Plant
Hefei Iron & Steel Co

COLOMBIA
Sidelpa - Siderúrgica del Pacifico SA

Carbon steel – hexagons

—continued

FRANCE
Grousset
Aciéries et Laminoirs du Saut du Tarn (ALST)
Usinor Sacilor

EAST GERMANY
VEB Stahl- und Walzwerk Brandenburg
VEB Stahl- und Walzwerk Wilhelm Florin, Hennigsdorf

WEST GERMANY
Remscheider Walz- und Hammerwerke, Böllinghaus & Co
Edelstahlwerke Buderus AG
Krupp Stahl AG (KS)
Saarstahl Völklingen GmbH
Thyssen Edelstahlwerke AG

INDIA
Agarwal Foundry & Engineering Works
Guest Keen Williams Ltd (GKW)
Mahindra Ugine Steel Co Ltd (Muscosteel)
Punjab Con-Cast Steels Ltd

ISRAEL
United Steel Mills Ltd (USM)

ITALY
Nuova Deltasider SpA
Falci SpA, Reparto Laminazione
Acciaierie e Ferriere Riva SpA
Stilma SpA
Acciaierie Valbruna SpA

JAPAN
Kobe Steel Ltd (Kobe Seikosho)
Sumitomo Metal Industries Ltd

MEXICO
Industrials Nylbo SA, Division Acero (formerly Laminadora Atzcapotzalco SA – Lasa)

PARAGUAY
Acepar – Acero del Paraguay SA

PORTUGAL
Fábrica de Aços Tomé Fèteira SA

SOUTH AFRICA
Iscor Ltd
Scaw Metals Ltd

SPAIN
Aforasa – Acerias y Forjas de Azcoitia SA
Aceros de Llodio SA (Allosa)

SWITZERLAND
Von Moos Stahl AG

TURKEY
Asil Çelik Sanayi ve Ticaret AŞ (Asil Çelik)
Ekinciler Demir ve Çelik San AŞ
MKEK Celik Fabrikasi

UNITED KINGDOM
Bedford Steels
Dudley Port Rolling Mills, Glynwed Steels Ltd
George Gadd & Co, Glynwed Steels Ltd
Kiveton Park Steel & Wire Works Ltd
Sheffield Forgemasters Ltd
Spencer Clark Metal Industries PLC
UES – United Engineering Steels Ltd
Woodstone Rolling Mills Ltd

UNITED STATES
Franklin Steel Co
Inland Steel Co
LTV Steel Co Inc
Nelsen Steel & Wire Co Inc

YUGOSLAVIA
Boris Kidrič Steelworks

Carbon steel – bright (cold finished) bars

AUSTRALIA
Martin Bright Steels Ltd

BRAZIL
Acesita – Cia Aços Especiais Itabira
Aços Anhanguera SA (Acoansa)
Siderúrgica Nossa Senhora Aparecida SA
Mannesmann SA
Siderúrgica Riograndense SA

CANADA
Stelco Inc
Union Drawn Steel Co Ltd

CHINA
Changcheng Steel Works
Changzhi Iron & Steel Works
Chengdu Metallurgical Experimental Plant

Carbon steel - bright
—continued

CHINA *—continued*
Hefei Iron & Steel Co
Laiwu Iron & Steel Plant
Shanghai Iron & Steel Works

FINLAND
Ovako Steel Oy Ab

FRANCE
Grousset

WEST GERMANY
Saarstahl Völklingen GmbH
Thyssen Edelstahlwerke AG
Westig GmbH

INDIA
BP Steel Industries Pvt Ltd
Chase Bright Steel Ltd
Damehra Steels & Forgings Pvt Ltd
Guest Keen Williams Ltd (GKW)
Mukand Iron & Steel Works Ltd

ITALY
SpA Officine Fratelli Bertoli fu Rodolfo
Nuova Deltasider SpA
Rodacciai Industria Trafilati SpA
Acciaierie Valbruna SpA

JAPAN
Daido Steel Co Ltd

KENYA
Kenya United Steel Co Ltd

MEXICO
Cold Rolled de México SA
Industrias CH SA (formerly Campos Hermanos SA)

POLAND
Huta Warszawa

PORTUGAL
F. Ramada, Aços e Indústrias Sarl

RUMANIA
Oţelul Roşu Works

SOUTH AFRICA
Flather Bright Steels (Pty) Ltd
Mono-Die Engineers (Pty) Ltd
Rand Bright Steels (Pty) Ltd
Titan Industrial Corp (Pty) Ltd

SPAIN
Forjas Alavesas SA (Fasa)
Babcock & Wilcox Española SA Tubular Products Division
La Calibradora Mecánica SA
Aceros de Llodio SA (Allosa)
Olarra SA (Olsa)
Pedro Orbegozo y Cía SA (Acenor)
Marcos Ormaechea Múgica

SWEDEN
Ovako Steel AB
Smedjebacken-Boxholm Stål AB

SWITZERLAND
Von Moos Stahl AG

TURKEY
MKEK Celik Fabrikasi

UNITED KINGDOM
Bar Bright Ltd
Barlow Bright Steels, Glynwed Steels Ltd
Bedford Steels
Brasway PLC, Bright Bar Division
Bright Steels Ltd
British Bright Bar Ltd
Joseph Gillott & Sons, Glynwed Steels Ltd
A.E. Godrich & Son Ltd
K.W. Haywood & Son Ltd
Arthur Lee & Sons PLC
Longmore Brothers, Glynwed Steels Ltd
Midland Bright Drawn Steel Ltd
Steel Bars & Sections Ltd
Steel Parts, Glynwed Steels Ltd
UES – United Engineering Steels Ltd
Unbrako Steel Co Ltd
W. Wesson, Glynwed Steels Ltd

UNITED STATES
Armco Inc
Baron Drawn Steel Corp
Bliss & Laughlin Steel Co
Copperweld Steel Co
Crucible Specialty Metals Division, Crucible Materials Corp
Cumberland Steel Co
Cuyahoga Steel & Wire Inc
Greer Steel Co
Moltrup Steel Products Co
Nucor Corp
Pacific Tube Co
Quanex Corp

Carbon steel - bright
—continued

UNITED STATES *—continued*
Superior Drawn Steel Co, Division of Standard Steel Specialty Co
Teledyne Columbia-Summerill
Wyckoff Steel Division, Ampco-Pittsburgh Corp

YUGOSLAVIA
Slovenske Železarne – Iron & Steel Works of Slovenia

Carbon steel - light angles

ALBANIA
Steel of the Party Metallurgical Combine

ARGENTINA
La Cantábrica SA Metalúrgica, Industrial y Comercial
Establecimiento Altos Hornos Zapla

AUSTRALIA
The Broken Hill Pty Co Ltd
Smorgon Steel

AUSTRIA
Voest-Alpine AG

BELGIUM
Cockerill Sambre SA
Laminoirs du Ruau SA

BRAZIL
Siderúrgica J.L. Aliperti SA
Siderúrgica Barra Mansa SA
Cia Brasileira de Aço
Cimetal Siderurgia SA
Cofavi – Cia Ferro e Aço de Vitória
Comesa – Cia Siderúrgica de Alagoas
Copalam – Cia Paulista de Laminação
Cosigua – Cia Siderurgia da Guanabara
CSN – Cia Siderúrgica Nacional
Siderúrgica Hime SA
Siderúrgica Riograndense SA

BULGARIA
Burgas Steelworks
Kremikovtsi Iron & Steel Works
Lenin Iron & Steel Works

CANADA
Courtice Steel Ltd
Lake Ontario Steel Co (Lasco Steel)

CANADA *—continued*
Manitoba Rolling Mills, Amca International
Sidbec-Dosco Inc
Stelco Inc

CHINA
Chongqing Iron & Steel Co
Echeng Iron & Steel Works
Fushun Xinfu Steel Plant
Jiangxi Steel Works
Laiwu Iron & Steel Plant
Sanming Iron & Steel Works
Shanghai Iron & Steel Works
Shoudu Iron & Steel Co
Taiyuan Iron & Steel Co
Tianjin Steel Plant

CUBA
Empressa Metalúrgica José Marti

CZECHOSLOVAKIA
Vitkovice Steelworks

DOMINICAN REPUBLIC
Metaldom – Complejo Metalúrgico Dominicano C por A

ECUADOR
Adelca – Acería del Ecuador CA

EGYPT
The Egyptian Iron & Steel Co (Hadisolb)

EL SALVADOR
Siderúrgica Centroamericana del Pacífico SA (Sicepasa)

FRANCE
Aciéries et Laminoirs de Rives
Usinor Sacilor

WEST GERMANY
Remscheider Walz- und Hammerwerke, Böllinghaus & Co
Lech-Stahlwerke GmbH
Maxhütte – Eisenwerke-Gesellschaft Maximilianshütte mbH
Saarstahl Völklingen GmbH
Thyssen Stahl AG

INDIA
Agarwal Foundry & Engineering Works
Ahmedabad Steelcraft & Rolling Mills Pvt Ltd
BD Steel Castings Ltd
Ellora Steels Ltd (ESL)

Carbon steel - light angles
—continued

INDIA *—continued*
Indore Steel & Iron Mills Pvt Ltd (Isim)
Kap Steel Ltd
Man Industrial Corp Ltd
Partap Steel Rolling Mills (1935) Ltd
The Punjab Steel Rolling Mills
Sail – Steel Authority of India Ltd
The South India Steel & Starch Industries
Special Steel Co
Steel Ingots Pvt Ltd
Sunflag Iron & Steel Co
Taloja Rolling Mills
The Tata Iron & Steel Co Ltd (Tisco)
Trichy Steel Rolling Mills Ltd

INDONESIA
PT Inti General Jaya Steel

IRAN
National Iranian Steel Corp (Nisc)

IRISH REPUBLIC
Irish Steel Ltd

ISRAEL
United Steel Mills Ltd (USM)

ITALY
A.F.V. Beltrame SpA
DeltaValdarno
Dora Acciaierie e Ferriere (formerly Acciaierie Ferriere Alpine SpA (AFB))
Falci SpA, Reparto Laminazione
Ferdofin SpA
Ferdofer SpA
Lucchini Siderurgica SpA
Metalgoi SpA
Ferriera Olifer Srl
Profilati Nave SpA
Acciaierie Valbruna SpA

JAPAN
Daido Steel Co Ltd
Nakayama Steel Works Ltd (Nakayama Seikosho)
Osaka Steel (Osaka Seitetsu)
Toa Steel Co Ltd
Tokyo Steel Manufacturing Co Ltd

JORDAN
National Steel Industry Co Ltd

KENYA
Kenya United Steel Co Ltd
Steel Rolling Mills Ltd

SOUTH KOREA
Boo-Kuk Steel Industrial Co Ltd
Dongkuk Steel Mill Co Ltd
Inchon Iron & Steel Co Ltd
Kang Won Industries Ltd

LUXEMBOURG
Arbed SA

MALAYSIA
Antara Steel Mills Sdn Bhd
Dah Yung Steel (M) Sdn Bhd

MEXICO
Aceros Ecatepec SA (Aesa)
Siderúrgica de Yucatán SA de CV

NEW ZEALAND
Pacific Steel Ltd

NIGERIA
Ajaokuta Steel Co Ltd
Allied Steel Industry (Nigeria) Ltd
Delta Steel Co Ltd
Major Engineering Co Ltd
Qua Steel Products Ltd
Universal Steels Ltd

NORWAY
Norsk Jernverk AS

PAKISTAN
Ittefaq Ltd
Metropolitan Steel Corp Ltd (MSC)
New Era Steels
Nowshera Engineering Co Ltd
Razaque Steels (Pvt) Ltd

PARAGUAY
Acepar – Acero del Paraguay SA

PERU
Aceros Arequipa SA (Acersa)

POLAND
Huta im Lenina

PORTUGAL
Siderurgia Nacional EP (SN)

SINGAPORE
First Rolling Mills Singapore Pte Ltd
National Iron & Steel Mills Ltd

Carbon steel – light angles
—continued

SOUTH AFRICA
Davsteel (Pty) Ltd
Iscor Ltd
Scaw Metals Ltd

SPAIN
Arregui SA
Ensidesa – Empresa Nacional Siderurgica SA
Rico y Echeverria SA

SWEDEN
Smedjebacken-Boxholm Stål AB

SWITZERLAND
Von Roll Group

TAIWAN
Feng Hsin Iron & Steel Co Ltd
Feng Lung Steel Factory Co Ltd

TANZANIA
Aluminium Africa Ltd (Alaf)
Steel Rolling Mills Ltd (Tanzania)

THAILAND
The Bangkok Iron & Steel Works Co Ltd

TRINIDAD & TOBAGO
Central Trinidad Steel Ltd – Centrin

TURKEY
Cemtas AŞ
Ekinciler Demir ve Çelik San AŞ
Izmir Demir Çelik Sanayi AŞ
TDCI – Türkiye Demir ve Çelik Işletmeleri Genel Müdürlügü

UNITED KINGDOM
Allied Steel & Wire Ltd
Bromford Iron & Steel Co Ltd
Darlington & Simpson Rolling Mills Ltd (DSRM)
George Gadd & Co, Glynwed Steels Ltd
Pressfab Sections Ltd
Sheerness Steel Co PLC
United Merchant Bar PLC

UNITED STATES
Auburn Steel Co Inc (Austeel)
Bayou Steel Corp
Bethlehem Steel Corp
Birmingham Steel Corp
Calumet Steel Co

UNITED STATES *—continued*
CF&I Steel Corp
Chaparral Steel Co
Chicago Heights Steel Co
Dart Rollformers Inc
Florida Steel Corp
Franklin Steel Co
Jersey Shore Steel Co Inc
Marion Steel Co
McDonald Steel Corp
Missouri Rolling Mill Corp
North Star Steel Co (NSS)
Nucor Corp
Owen Electric Steel Co of South Carolina
Roanoke Electric Steel Corp (Resco)
SMI Steel Inc
Structural Metals Inc
Thomas Steel Corp

USSR
Cherepovetskiy Met Zavod (Cherepovets Iron & Steel Works)
Dneprodzerzhinskiy Zavod Imeni Dzerzhinskogo (Dzerzhinsky Works)
Met Zavod Krivorizhstal (Krivoy Rog Iron & Steel Works)
Magnitogorskiy Metallurgischeskiy Kombinat (Magnitogorsk Iron & Steel Works Combine)
Zapsib Met Zavod (West Siberian Steel Works)

VENEZUELA
Metalanca – Metalúrgica de Laminados Nacionales CA
Sidetur – Siderúrgica del Turbio SA
Sidor – CVG Siderúrgica del Orinoco CA

ZIMBABWE
Zimbabwe Iron & Steel Co Ltd – Ziscosteel

Carbon steel – light tees

ALBANIA
Steel of the Party Metallurgical Combine

BELGIUM
Cockerill Sambre SA
Laminoirs du Ruau SA

BRAZIL
Cofavi – Cia Ferro e Aço de Vitória
Copalam – Cia Paulista de Laminação

Carbon steel – light tees
—continued

BRAZIL *—continued*
Cosigua – Cia Siderurgia da Guanabara

BULGARIA
Kremikovtsi Iron & Steel Works

CANADA
Stelco Inc

CHINA
Laiwu Iron & Steel Plant
Taiyuan Iron & Steel Co

CZECHOSLOVAKIA
Vitkovice Steelworks

FRANCE
Aciéries et Laminoirs de Rives
Usinor Sacilor

WEST GERMANY
Remscheider Walz- und Hammerwerke, Böllinghaus & Co
Lech-Stahlwerke GmbH
Saarstahl Völklingen GmbH
Thyssen Stahl AG

INDIA
Agarwal Foundry & Engineering Works
Man Industrial Corp Ltd
Partap Steel Rolling Mills (1935) Ltd
Special Steel Co
Taloja Rolling Mills
The Tata Iron & Steel Co Ltd (Tisco)

ITALY
A.F.V. Beltrame SpA
DeltaValdarno
Ferdofin SpA
Ferdofer SpA
Lucchini Siderurgica SpA
Metalgoi SpA
Profilati Nave SpA

JORDAN
National Steel Industry Co Ltd

KENYA
Steel Rolling Mills Ltd

LUXEMBOURG
Arbed SA

NIGERIA
Delta Steel Co Ltd

PAKISTAN
Nowshera Engineering Co Ltd

PERU
Aceros Arequipa SA (Acersa)

PORTUGAL
Siderurgia Nacional EP (SN)

SOUTH AFRICA
Iscor Ltd
Scaw Metals Ltd

SPAIN
Arregui SA
Azma SA
Laminaciones Sondica SA

TURKEY
Izmir Demir Çelik Sanayi AŞ

UNITED KINGDOM
Darlington & Simpson Rolling Mills Ltd (DSRM)
George Gadd & Co, Glynwed Steels Ltd
Pressfab Sections Ltd

UNITED STATES
Bethlehem Steel Corp
Calumet Steel Co
Cascade Steel Rolling Mills Inc
Chicago Heights Steel Co
Franklin Steel Co
Structural Metals Inc

VENEZUELA
Metalanca – Metalúrgica de Laminados Nacionales CA
Sidetur – Siderúrgica del Turbio SA
Sidor – CVG Siderúrgica del Orinoco CA

Carbon steel – light channels

ALBANIA
Steel of the Party Metallurgical Combine

AUSTRALIA
The Broken Hill Pty Co Ltd

AUSTRIA
Voest-Alpine AG

BELGIUM
Cockerill Sambre SA

Carbon steel – light channels *—continued*

BELGIUM *—continued*
Laminoirs du Ruau SA

BRAZIL
Cia Brasileira de Aço
Cofavi – Cia Ferro e Aço de Vitória
Cosigua – Cia Siderurgia da Guanabara
Siderúrgica Riograndense SA

BULGARIA
Kremikovtsi Iron & Steel Works

BURMA
Ywama Steel Mill

CANADA
Courtice Steel Ltd
Lake Ontario Steel Co (Lasco Steel)
Manitoba Rolling Mills, Amca International
Slater Steels – Hamilton Speciality Bar Division
Stelco Inc

CHINA
Sanming Iron & Steel Works
Shanghai Iron & Steel Works
Taiyuan Iron & Steel Co
Tangshan Iron & Steel Co

CZECHOSLOVAKIA
Vitkovice Steelworks

FRANCE
Usinor Sacilor

WEST GERMANY
Lech-Stahlwerke GmbH
Saarstahl Völklingen GmbH
Thyssen Stahl AG

INDIA
Agarwal Foundry & Engineering Works
Ellora Steels Ltd (ESL)
Indore Steel & Iron Mills Pvt Ltd (Isim)
Sail – Steel Authority of India Ltd
Sunflag Iron & Steel Co
Taloja Rolling Mills
The Tata Iron & Steel Co Ltd (Tisco)
Trichy Steel Rolling Mills Ltd

IRAN
National Iranian Steel Corp (Nisc)

IRISH REPUBLIC
Irish Steel Ltd

ITALY
A.F.V. Beltrame SpA
DeltaValdarno
Ferdofin SpA
Ferdofer SpA
Lucchini Siderurgica SpA
Metalgoi SpA
Profilati Nave SpA
Nuova Sidercamuna SpA

JAPAN
Osaka Steel (Osaka Seitetsu)
Rinko Steel Works Co Ltd
Toa Steel Co Ltd
Tokyo Steel Manufacturing Co Ltd

JORDAN
National Steel Industry Co Ltd

SOUTH KOREA
Boo-Kuk Steel Industrial Co Ltd
Dongkuk Steel Mill Co Ltd
Inchon Iron & Steel Co Ltd
Kang Won Industries Ltd

LUXEMBOURG
Arbed SA

MEXICO
Aceros Ecatepec SA (Aesa)

NEW ZEALAND
Pacific Steel Ltd

NIGERIA
Ajaokuta Steel Co Ltd
Allied Steel Industry (Nigeria) Ltd
Delta Steel Co Ltd

PAKISTAN
Ittefaq Ltd
Metropolitan Steel Corp Ltd (MSC)

PERU
Aceros Arequipa SA (Acersa)

PORTUGAL
Siderurgia Nacional EP (SN)

SOUTH AFRICA
Scaw Metals Ltd

SPAIN
Arregui SA
Azma SA
Laminaciones Sondica SA

Carbon steel – light channels —*continued*

SPAIN —*continued*
Unión Cerrajera SA

SWEDEN
Smedjebacken-Boxholm Stål AB

SWITZERLAND
Von Roll Group

TAIWAN
Feng Lung Steel Factory Co Ltd

TRINIDAD & TOBAGO
Central Trinidad Steel Ltd – Centrin

TURKEY
Cemtas AŞ
MKEK Celik Fabrikasi
TDCI – Türkiye Demir ve Çelik İşletmeleri Genel Müdürlügü

UNITED KINGDOM
Allied Steel & Wire Ltd
George Gadd & Co, Glynwed Steels Ltd
Martins (Dundyvan) Ltd
Pressfab Sections Ltd
Raine & Co Ltd
United Merchant Bar PLC

UNITED STATES
Auburn Steel Co Inc (Austeel)
Bethlehem Steel Corp
Birmingham Steel Corp
Chicago Heights Steel Co
Dart Rollformers Inc
Florida Steel Corp
Franklin Steel Co
McDonald Steel Corp
Northwestern Steel & Wire Co
Nucor Corp
Roanoke Electric Steel Corp (Resco)
SMI Steel Inc

USSR
Dneprodzerzhinskiy Zavod Imeni Dzerzhinskogo (Dzerzhinsky Works)
Magnitogorskiy Metallurgischeskiy Kombinat (Magnitogorsk Iron & Steel Works Combine)

VENEZUELA
Sidor – CVG Siderúrgica del Orinoco CA

Carbon steel – special light sections

ARGENTINA
La Cantábrica SA Metalúrgica, Industrial y Comercial
Rosati y Cristófaro Saic

AUSTRALIA
The Broken Hill Pty Co Ltd
Rydal Steel – Division of LNC Industrial Products Pty Ltd
Siddons Steel Mills

BELGIUM
Laminoirs du Ruau SA

BRAZIL
Cofavi – Cia Ferro e Aço de Vitória
Copalam – Cia Paulista de Laminação
Laminação Santa Maria SA – Indústria e Comércio

CANADA
Manitoba Rolling Mills, Amca International
Slater Steels – Hamilton Speciality Bar Division

CHINA
Jiangxi Steel Works
Shanghai Xihu Iron & Steel Works

FRANCE
Forges de Clairvaux
Aciéries et Laminoirs de Rives
Forges de Syam
Usinor Sacilor

EAST GERMANY
VEB Stahl- und Walzwerk Wilhelm Florin, Hennigsdorf

WEST GERMANY
Remscheider Walz- und Hammerwerke, Böllinghaus & Co
Hoesch AG
Klöckner Stahl GmbH
Stahlwerke Peine-Salzgitter AG (P&S)
Thyssen Edelstahlwerke AG
Thyssen Stahl AG

INDIA
Agarwal Foundry & Engineering Works
Ahmedabad Steelcraft & Rolling Mills Pvt Ltd
Ellora Steels Ltd (ESL)
Guest Keen Williams Ltd (GKW)

Carbon steel – special light sections —*continued*

INDIA —*continued*
Man Industrial Corp Ltd
Partap Steel Rolling Mills (1935) Ltd
Sail – Steel Authority of India Ltd
Taloja Rolling Mills

ITALY
Metallurgica Calvi SpA
Falci SpA, Reparto Laminazione
Las – Laminazione Acciai Speciali SpA
Metalgoi SpA
Profilati Nave SpA

JAPAN
Osaka Steel (Osaka Seitetsu)
Rinko Steel Works Co Ltd
Toa Steel Co Ltd
Tokyo Steel Manufacturing Co Ltd
Topy Industries Ltd

SOUTH KOREA
Sammi Steel Co Ltd

LUXEMBOURG
Arbed SA
Métallurgique & Minière de Rodange-Athus (MMRA)

MEXICO
Siderúrgica de Yucatán SA de CV

SINGAPORE
First Rolling Mills Singapore Pte Ltd

SOUTH AFRICA
Iscor Ltd
Usco – The Union Steel Corp (of South Africa) Ltd

SPAIN
Forjas Alavesas SA (Fasa)
Azma SA
Marcos Ormaechea Múgica

SWEDEN
Smedjebacken-Boxholm Stål AB

SWITZERLAND
Montanstahl AG
Von Moos Stahl AG

TURKEY
MKEK Celik Fabrikasi

UNITED KINGDOM
Darlington & Simpson Rolling Mills Ltd (DSRM)

UNITED KINGDOM —*continued*
Dudley Port Rolling Mills, Glynwed Steels Ltd
George Gadd & Co, Glynwed Steels Ltd
Joseph Gillott & Sons, Glynwed Steels Ltd
Martins (Dundyvan) Ltd
Osborn Steel Extrusions Ltd
Pressfab Sections Ltd
Raine & Co Ltd
Spencer Clark Metal Industries PLC
TI Stainless Tubes Ltd
Woodstone Rolling Mills Ltd

UNITED STATES
Calumet Steel Co
Chicago Heights Steel Co
Franklin Steel Co
LTV Steel Co Inc
McDonald Steel Corp
Plymouth Tube Co
Steel of West Virginia Inc

YUGOSLAVIA
Slovenske Železarne – Iron & Steel Works of Slovenia

ZIMBABWE
Zimbabwe Iron & Steel Co Ltd – Ziscosteel

Carbon steel – medium angles

ARGENTINA
La Cantábrica SA Metalúrgica, Industrial y Comercial
Establecimiento Altos Hornos Zapla

AUSTRALIA
The Broken Hill Pty Co Ltd

BANGLADESH
Chittagong Steel Mills Ltd

BELGIUM
Cockerill Sambre SA

BRAZIL
Cofavi – Cia Ferro e Aço de Vitória
CSN – Cia Siderúrgica Nacional

CANADA
Manitoba Rolling Mills, Amca International
Sidbec-Dosco Inc

Carbon steel - medium angles —*continued*

CHINA
Laiwu Iron & Steel Plant
Sanming Iron & Steel Works
Taiyuan Iron & Steel Co
Wuhan Iron & Steel (Wisco)

COLOMBIA
Acerías Paz del Rio SA
Sidelpa - Siderúrgica del Pacifico SA

CUBA
Empressa Metalúrgica José Marti

CZECHOSLOVAKIA
Vitkovice Steelworks

EGYPT
The Egyptian Iron & Steel Co (Hadisolb)

FRANCE
Usinor Sacilor

EAST GERMANY
VEB Maxhütte Unterwellenborn

WEST GERMANY
Saarstahl Völklingen GmbH

INDIA
Agarwal Foundry & Engineering Works
Kap Steel Ltd
The South India Steel & Starch Industries
The Tata Iron & Steel Co Ltd (Tisco)

INDONESIA
PT Kratkatau Steel
PT National Union Steel Ltd

IRISH REPUBLIC
Irish Steel Ltd

ITALY
A.F.V. Beltrame SpA
DeltaValdarno
Falci SpA, Reparto Laminazione

JAPAN
Nakayama Steel Works Ltd (Nakayama Seikosho)
Nippon Steel Corp (Shin Nippon Seitetsu)
Tokyo Kohtetsu Co Ltd
Tokyo Steel Manufacturing Co Ltd
Topy Industries Ltd

SOUTH KOREA
Inchon Iron & Steel Co Ltd
Kang Won Industries Ltd

LUXEMBOURG
Arbed SA

PAKISTAN
New Era Steels
Razaque Steels (Pvt) Ltd

PORTUGAL
Siderurgia Nacional EP (SN)

SOUTH AFRICA
Highveld Steel & Vanadium Corp Ltd
Iscor Ltd

SPAIN
Rico y Echeverria SA

SWEDEN
Smedjebacken-Boxholm Stål AB
SSAB - Svenskt Stål AB

TAIWAN
Feng Hsin Iron & Steel Co Ltd

TURKEY
TDCI - Türkiye Demir ve Çelik Işletmeleri Genel Müdürlügü

UNITED KINGDOM
BSC - British Steel Corp
Martins (Dundyvan) Ltd
Pressfab Sections Ltd
United Merchant Bar PLC

UNITED STATES
Bayou Steel Corp
Birmingham Steel Corp
Chaparral Steel Co
Dart Rollformers Inc
North Star Steel Co (NSS)
Northwestern Steel & Wire Co
Nucor Corp
Seattle Steel Inc

VENEZUELA
Sidor - CVG Siderúrgica del Orinoco CA

ZIMBABWE
Zimbabwe Iron & Steel Co Ltd - Ziscosteel

Carbon steel – medium tees

ARGENTINA
Establecimiento Altos Hornos Zapla

BELGIUM
Cockerill Sambre SA

BRAZIL
Cofavi – Cia Ferro e Aço de Vitória

CHINA
Taiyuan Iron & Steel Co
Wuhan Iron & Steel (Wisco)

CZECHOSLOVAKIA
Vitkovice Steelworks

FRANCE
Usinor Sacilor

WEST GERMANY
Saarstahl Völklingen GmbH

INDIA
Agarwal Foundry & Engineering Works
Ellora Steels Ltd (ESL)
Partap Steel Rolling Mills (1935) Ltd
Sail – Steel Authority of India Ltd
The Tata Iron & Steel Co Ltd (Tisco)

ITALY
A.F.V. Beltrame SpA
DeltaValdarno

JAPAN
Nippon Steel Corp (Shin Nippon Seitetsu)

LUXEMBOURG
Arbed SA

PORTUGAL
Siderurgia Nacional EP (SN)

SOUTH AFRICA
Iscor Ltd

SPAIN
Azma SA
Laminaciones Sondica SA

SWEDEN
SSAB – Svenskt Stål AB

UNITED KINGDOM
BSC – British Steel Corp
Pressfab Sections Ltd

VENEZUELA
Sidor – CVG Siderúrgica del Orinoco CA

Carbon steel – medium joists

ARGENTINA
La Cantábrica SA Metalúrgica, Industrial y Comercial
Establecimiento Altos Hornos Zapla

AUSTRALIA
The Broken Hill Pty Co Ltd

AUSTRIA
Voest-Alpine AG

BELGIUM
Cockerill Sambre SA

BULGARIA
Lenin Iron & Steel Works

CANADA
Manitoba Rolling Mills, Amca International

CHINA
Laiwu Iron & Steel Plant
Sanming Iron & Steel Works
Taiyuan Iron & Steel Co
Wuhan Iron & Steel (Wisco)

EGYPT
The Egyptian Iron & Steel Co (Hadisolb)

FRANCE
Usinor Sacilor

EAST GERMANY
VEB Maxhütte Unterwellenborn

WEST GERMANY
Maxhütte – Eisenwerke-Gesellschaft Maximilianshütte mbH
Stahlwerke Peine-Salzgitter AG (P&S)
Saarstahl Völklingen GmbH

INDIA
Agarwal Foundry & Engineering Works
Ellora Steels Ltd (ESL)
Partap Steel Rolling Mills (1935) Ltd
Sail – Steel Authority of India Ltd
The Tata Iron & Steel Co Ltd (Tisco)

Carbon steel – medium joists *—continued*

INDONESIA
PT Kratkatau Steel

ITALY
A.F.V. Beltrame SpA
DeltaValdarno
Eurocolfer Acciai SpA
Lucchini Siderurgica SpA
Nuova Sidercamuna SpA

JAPAN
Nippon Steel Corp (Shin Nippon Seitetsu)

LUXEMBOURG
Arbed SA

NIGERIA
Ajaokuta Steel Co Ltd

PORTUGAL
Siderurgia Nacional EP (SN)

SOUTH AFRICA
Highveld Steel & Vanadium Corp Ltd
Iscor Ltd

SPAIN
Grupo José María Aristrain SA
Azma SA
Macosa – Material y Construcciones SA
Rico y Echeverria SA

SWEDEN
SSAB – Svenskt Stål AB

TURKEY
Izmir Demir Çelik Sanayi AŞ

UNITED KINGDOM
BSC – British Steel Corp
Darlington & Simpson Rolling Mills Ltd (DSRM)
Raine & Co Ltd

UNITED STATES
North Star Steel Co (NSS)

VENEZUELA
Sidor – CVG Siderúrgica del Orinoco CA

Carbon steel – medium channels

ARGENTINA
Somisa – Sdad Mixta Siderurgia Argentina
Establecimiento Altos Hornos Zapla

AUSTRALIA
The Broken Hill Pty Co Ltd

AUSTRIA
Voest-Alpine AG

BELGIUM
Cockerill Sambre SA

BRAZIL
Cofavi – Cia Ferro e Aço de Vitória
CSN – Cia Siderúrgica Nacional

BULGARIA
Lenin Iron & Steel Works

CANADA
Manitoba Rolling Mills, Amca International
Sidbec-Dosco Inc

CHINA
Wuhan Iron & Steel (Wisco)

COLOMBIA
Acerías Paz del Rio SA

CZECHOSLOVAKIA
Vitkovice Steelworks

EGYPT
The Egyptian Iron & Steel Co (Hadisolb)

FRANCE
Usinor Sacilor

EAST GERMANY
VEB Maxhütte Unterwellenborn

WEST GERMANY
Maxhütte – Eisenwerke-Gesellschaft Maximilianshütte mbH
Stahlwerke Peine-Salzgitter AG (P&S)
Saarstahl Völklingen GmbH

INDIA
Sail – Steel Authority of India Ltd
The Tata Iron & Steel Co Ltd (Tisco)

INDONESIA
PT Kratkatau Steel

Carbon steel – medium channels —*continued*

IRISH REPUBLIC
Irish Steel Ltd

ITALY
A.F.V. Beltrame SpA
DeltaValdarno
Nuova Sidercamuna SpA

JAPAN
Nakayama Steel Works Ltd (Nakayama Seikosho)
Nippon Steel Corp (Shin Nippon Seitetsu)
Tokyo Steel Manufacturing Co Ltd

SOUTH KOREA
Inchon Iron & Steel Co Ltd
Kang Won Industries Ltd

LUXEMBOURG
Arbed SA

PORTUGAL
Siderurgia Nacional EP (SN)

SOUTH AFRICA
Highveld Steel & Vanadium Corp Ltd

SPAIN
Azma SA
Macosa – Material y Construcciones SA
Rico y Echeverria SA
Laminaciones Sondica SA

SWEDEN
Smedjebacken-Boxholm Stål AB
SSAB – Svenskt Stål AB

TURKEY
Izmir Demir Çelik Sanayi AŞ
TDCI – Türkiye Demir ve Çelik Işletmeleri Genel Müdürlügü

UNITED KINGDOM
Allied Steel & Wire Ltd
BSC – British Steel Corp
Martins (Dundyvan) Ltd
Pressfab Sections Ltd
Raine & Co Ltd
United Merchant Bar PLC

UNITED STATES
Bayou Steel Corp
Birmingham Steel Corp
Chaparral Steel Co

UNITED STATES —*continued*
Dart Rollformers Inc
North Star Steel Co (NSS)
Northwestern Steel & Wire Co
Seattle Steel Inc

VENEZUELA
Sidor – CVG Siderúrgica del Orinoco CA

ZIMBABWE
Zimbabwe Iron & Steel Co Ltd – Ziscosteel

Carbon steel – wide-flange beams (medium)

ARGENTINA
Somisa – Sdad Mixta Siderurgia Argentina

AUSTRALIA
The Broken Hill Pty Co Ltd

AUSTRIA
Voest-Alpine AG

BELGIUM
Cockerill Sambre SA

BRAZIL
Cofavi – Cia Ferro e Aço de Vitória

CANADA
The Algoma Steel Corp Ltd

CHINA
Wuhan Iron & Steel (Wisco)

FRANCE
Usinor Sacilor

WEST GERMANY
Stahlwerke Peine-Salzgitter AG (P&S)
Saarstahl Völklingen GmbH

INDIA
Partap Steel Rolling Mills (1935) Ltd

IRISH REPUBLIC
Irish Steel Ltd

ITALY
Eurocolfer Acciai SpA

JAPAN
Nakayama Steel Works Ltd (Nakayama Seikosho)

Carbon steel – wide-flange beams —*continued*

JAPAN —*continued*
Nippon Steel Corp (Shin Nippon Seitetsu)
Tokyo Steel Manufacturing Co Ltd

LUXEMBOURG
Arbed SA

SOUTH AFRICA
Highveld Steel & Vanadium Corp Ltd
Iscor Ltd

SWEDEN
SSAB – Svenskt Stål AB

TURKEY
TDCI – Türkiye Demir ve Çelik Işletmeleri Genel Müdürlügü

UNITED KINGDOM
BSC – British Steel Corp

UNITED STATES
Bayou Steel Corp
Chaparral Steel Co
Inland Steel Co
Northwestern Steel & Wire Co

Carbon steel – universals (medium)

AUSTRALIA
The Broken Hill Pty Co Ltd

BELGIUM
Cockerill Sambre SA

FRANCE
Usinor Sacilor

WEST GERMANY
Stahlwerke Peine-Salzgitter AG (P&S)
Saarstahl Völklingen GmbH

IRISH REPUBLIC
Irish Steel Ltd

ITALY
Lucchini Siderurgica SpA

JAPAN
Nippon Steel Corp (Shin Nippon Seitetsu)

LUXEMBOURG
Arbed SA

SOUTH AFRICA
Highveld Steel & Vanadium Corp Ltd
Iscor Ltd

SWEDEN
Smedjebacken-Boxholm Stål AB
SSAB – Svenskt Stål AB

UNITED KINGDOM
BSC – British Steel Corp
Ductile Hot Mill, Glynwed Steels Ltd

Carbon steel – welded sections (medium)

FRANCE
Usinor Sacilor

WEST GERMANY
Saarstahl Völklingen GmbH

SWEDEN
SSAB – Svenskt Stål AB

Carbon steel – heavy angles

AUSTRALIA
The Broken Hill Pty Co Ltd

AUSTRIA
Voest-Alpine AG

BRAZIL
CSN – Cia Siderúrgica Nacional

CANADA
The Algoma Steel Corp Ltd
Lake Ontario Steel Co (Lasco Steel)
Manitoba Rolling Mills, Amca International
Stelco Inc

CHINA
Anshan Iron & Steel Co
Laiwu Iron & Steel Plant
Panzhihua Works
Shoudu Iron & Steel Co
Taiyuan Iron & Steel Co
Wuhan Iron & Steel (Wisco)

CZECHOSLOVAKIA
NHKG – Nová Hut Klementa Gottwalda
Vitkovice Steelworks

Carbon steel – heavy angles
—continued

EGYPT
The Egyptian Iron & Steel Co (Hadisolb)

FRANCE
Usinor Sacilor

WEST GERMANY
Hoesch AG
Saarstahl Völklingen GmbH
Thyssen Stahl AG

INDIA
The Tata Iron & Steel Co Ltd (Tisco)

IRAN
National Iranian Steel Corp (Nisc)

ITALY
Acciaieria e Ferriere Stefana F.lli fu Girolamo SpA

JAPAN
Godo Steel Ltd (Godo Seitetsu)
Nippon Steel Corp (Shin Nippon Seitetsu)
NKK – Nippon Kokan KK
Sumitomo Metal Industries Ltd
Toa Steel Co Ltd
Topy Industries Ltd
Yamato Kogyo Co Ltd

SOUTH KOREA
Inchon Iron & Steel Co Ltd
Kang Won Industries Ltd

LUXEMBOURG
Arbed SA

MEXICO
Ahmsa – Altos Hornos de Mexico SA de CV

PAKISTAN
Razaque Steels (Pvt) Ltd

SOUTH AFRICA
Highveld Steel & Vanadium Corp Ltd
Iscor Ltd

SPAIN
Grupo José María Aristrain SA
Ensidesa – Empresa Nacional Siderurgica SA
Unión Cerrajera SA

SWEDEN
SSAB – Svenskt Stål AB

TAIWAN
Tang Eng Iron Works Co Ltd

TURKEY
TDCI – Türkiye Demir ve Çelik Işletmeleri Genel Müdürlügü

UNITED KINGDOM
BSC – British Steel Corp

UNITED STATES
Bethlehem Steel Corp
Florida Steel Corp
Seattle Steel Inc

USSR
Chelyabinskiy Met Zavod (Chelyabinsk Iron & Steel Works)
Dneprodzerzhinskiy Zavod Imeni Dzerzhinskogo (Dzerzhinsky Works)
Magnitogorskiy Metallurgischeskiy Kombinat (Magnitogorsk Iron & Steel Works Combine)

Carbon steel – heavy tees

CHINA
Taiyuan Iron & Steel Co
Wuhan Iron & Steel (Wisco)

FRANCE
Usinor Sacilor

WEST GERMANY
Saarstahl Völklingen GmbH
Thyssen Stahl AG

INDIA
Sail – Steel Authority of India Ltd
The Tata Iron & Steel Co Ltd (Tisco)

JAPAN
Nippon Steel Corp (Shin Nippon Seitetsu)
NKK – Nippon Kokan KK
Toa Steel Co Ltd

LUXEMBOURG
Arbed SA

SOUTH AFRICA
Iscor Ltd

SPAIN
Industrias Zarra SA

Carbon steel – heavy tees
—continued

SWEDEN
SSAB – Svenskt Stål AB

UNITED KINGDOM
BSC – British Steel Corp

UNITED STATES
Bethlehem Steel Corp

Carbon steel – heavy joists

AUSTRALIA
The Broken Hill Pty Co Ltd

AUSTRIA
Voest-Alpine AG

CANADA
The Algoma Steel Corp Ltd

CHINA
Anshan Iron & Steel Co
Baotau Iron & Steel Co
Laiwu Iron & Steel Plant
Panzhihua Works
Taiyuan Iron & Steel Co
Wuhan Iron & Steel (Wisco)

CZECHOSLOVAKIA
NHKG – Nová Hut Klementa Gottwalda
Vitkovice Steelworks

EGYPT
The Egyptian Iron & Steel Co (Hadisolb)

FRANCE
Usinor Sacilor

WEST GERMANY
Krupp Stahl AG (KS)
Stahlwerke Peine-Salzgitter AG (P&S)
Saarstahl Völklingen GmbH
Thyssen Stahl AG

INDIA
Sail – Steel Authority of India Ltd
The Tata Iron & Steel Co Ltd (Tisco)

ITALY
Lucchini Siderurgica SpA
Acciaieria e Ferriere Stefana F.lli fu Girolamo SpA
Acciaierie del Tirreno SpA

JAPAN
Godo Steel Ltd (Godo Seitetsu)
Kawasaki Steel Corp
Nippon Steel Corp (Shin Nippon Seitetsu)
NKK – Nippon Kokan KK
Topy Industries Ltd
Yamato Kogyo Co Ltd

SOUTH KOREA
Kang Won Industries Ltd

LUXEMBOURG
Arbed SA
Métallurgique & Minière de Rodange-Athus (MMRA)

MEXICO
Ahmsa – Altos Hornos de Mexico SA de CV

SOUTH AFRICA
Highveld Steel & Vanadium Corp Ltd
Iscor Ltd

SPAIN
Grupo José María Aristrain SA
Ensidesa – Empresa Nacional Siderurgica SA

SWEDEN
SSAB – Svenskt Stål AB

UNITED KINGDOM
BSC – British Steel Corp
Raine & Co Ltd

UNITED STATES
Bethlehem Steel Corp

USSR
Chelyabinskiy Met Zavod (Chelyabinsk Iron & Steel Works)
Makeyevskiy Met Zavod Imeni Kirova (Kirov Iron & Steel Works)
Magnitogorskiy Metallurgischeskiy Kombinat (Magnitogorsk Iron & Steel Works Combine)

Carbon steel – heavy channels

AUSTRALIA
The Broken Hill Pty Co Ltd

AUSTRIA
Voest-Alpine AG

Carbon steel – heavy channels —*continued*

BELGIUM
Cockerill Sambre SA

BRAZIL
CSN – Cia Siderúrgica Nacional

CANADA
The Algoma Steel Corp Ltd
Lake Ontario Steel Co (Lasco Steel)
Manitoba Rolling Mills, Amca International

CHINA
Panzhihua Works
Shoudu Iron & Steel Co
Wuhan Iron & Steel (Wisco)

CZECHOSLOVAKIA
NHKG – Nová Hut Klementa Gottwalda
Vitkovice Steelworks

EGYPT
The Egyptian Iron & Steel Co (Hadisolb)

FRANCE
Usinor Sacilor

WEST GERMANY
Hoesch AG
Krupp Stahl AG (KS)
Stahlwerke Peine-Salzgitter AG (P&S)
Saarstahl Völklingen GmbH
Thyssen Stahl AG

INDIA
Sail – Steel Authority of India Ltd
The Tata Iron & Steel Co Ltd (Tisco)

IRAN
National Iranian Steel Corp (Nisc)

ITALY
Acciaieria e Ferriere Stefana F.lli fu Girolamo SpA

JAPAN
Godo Steel Ltd (Godo Seitetsu)
Kawasaki Steel Corp
Nippon Steel Corp (Shin Nippon Seitetsu)
NKK – Nippon Kokan KK
Toa Steel Co Ltd
Tokyo Steel Manufacturing Co Ltd
Topy Industries Ltd

JAPAN —*continued*
Yamato Kogyo Co Ltd

SOUTH KOREA
Inchon Iron & Steel Co Ltd
Kang Won Industries Ltd

LUXEMBOURG
Arbed SA
Métallurgique & Minière de Rodange-Athus (MMRA)

NORWAY
Norsk Jernverk AS

SOUTH AFRICA
Highveld Steel & Vanadium Corp Ltd
Iscor Ltd

SPAIN
Grupo José María Aristrain SA
Ensidesa – Empresa Nacional Siderurgica SA
Unión Cerrajera SA
Industrias Zarra SA

SWEDEN
SSAB – Svenskt Stål AB

TAIWAN
Tang Eng Iron Works Co Ltd

TURKEY
TDCI – Türkiye Demir ve Çelik Işletmeleri Genel Müdürlügü

UNITED KINGDOM
BSC – British Steel Corp
Martins (Dundyvan) Ltd
Raine & Co Ltd

UNITED STATES
Bethlehem Steel Corp
Dart Rollformers Inc

USSR
Chelyabinskiy Met Zavod (Chelyabinsk Iron & Steel Works)
Dneprodzerzhinskiy Zavod Imeni Dzerzhinskogo (Dzerzhinsky Works)
Makeyevskiy Met Zavod Imeni Kirova (Kirov Iron & Steel Works)
Magnitogorskiy Metallurgischeskiy Kombinat (Magnitogorsk Iron & Steel Works Combine)

Carbon steel – wide-flange beams (heavy)

AUSTRALIA
The Broken Hill Pty Co Ltd

BELGIUM
Cockerill Sambre SA

CHINA
Baotau Iron & Steel Co
Wuhan Iron & Steel (Wisco)

FRANCE
Usinor Sacilor

WEST GERMANY
Stahlwerke Peine-Salzgitter AG (P&S)
Saarstahl Völklingen GmbH
Thyssen Stahl AG

INDIA
Sail – Steel Authority of India Ltd

IRAN
National Iranian Steel Corp (Nisc)

ITALY
Acciaieria e Ferriere Stefana F.lli fu Girolamo SpA
Acciaierie del Tirreno SpA

JAPAN
Godo Steel Ltd (Godo Seitetsu)
Kawasaki Steel Corp
Nakayama Steel Works Ltd (Nakayama Seikosho)
Nippon Steel Corp (Shin Nippon Seitetsu)
NKK – Nippon Kokan KK
Sumitomo Metal Industries Ltd
Toa Steel Co Ltd
Tokyo Steel Manufacturing Co Ltd
Topy Industries Ltd
Yamato Kogyo Co Ltd

SOUTH KOREA
Inchon Iron & Steel Co Ltd
Kang Won Industries Ltd

LUXEMBOURG
Arbed SA
Métallurgique & Minière de Rodange-Athus (MMRA)

MEXICO
Ahmsa – Altos Hornos de Mexico SA de CV

NORWAY
Norsk Jernverk AS

SOUTH AFRICA
Highveld Steel & Vanadium Corp Ltd
Iscor Ltd

SPAIN
Ensidesa – Empresa Nacional Siderurgica SA

SWEDEN
SSAB – Svenskt Stål AB

TAIWAN
Tang Eng Iron Works Co Ltd

TURKEY
TDCI – Türkiye Demir ve Çelik Işletmeleri Genel Müdürlügü

UNITED KINGDOM
BSC – British Steel Corp

UNITED STATES
Bethlehem Steel Corp
Inland Steel Co
USS, A Division of USX Corp

USSR
Chelyabinskiy Met Zavod (Chelyabinsk Iron & Steel Works)
Dneprodzerzhinskiy Zavod Imeni Dzerzhinskogo (Dzerzhinsky Works)
Magnitogorskiy Metallurgischeskiy Kombinat (Magnitogorsk Iron & Steel Works Combine)
Novo-Tagilskiy Met Zavod (Novo-Tagil Iron & Steel Works)

Carbon steel – universals (heavy)

AUSTRALIA
The Broken Hill Pty Co Ltd

AUSTRIA
Voest-Alpine AG

FRANCE
Usinor Sacilor

WEST GERMANY
Stahlwerke Peine-Salzgitter AG (P&S)
Saarstahl Völklingen GmbH

JAPAN
Nippon Steel Corp (Shin Nippon Seitetsu)

Carbon steel – universals

—continued

LUXEMBOURG
Arbed SA

NORWAY
Norsk Jernverk AS

SOUTH AFRICA
Highveld Steel & Vanadium Corp Ltd
Iscor Ltd

SPAIN
Ensidesa – Empresa Nacional Siderurgica SA

SWEDEN
SSAB – Svenskt Stål AB

UNITED KINGDOM
BSC – British Steel Corp

Carbon steel – welded sections (heavy)

FINLAND
Rautaruukki Oy

FRANCE
Usinor Sacilor

ITALY
DeltaValdarno

JAPAN
Sumitomo Metal Industries Ltd

NORWAY
Norsk Jernverk AS

SWEDEN
SSAB – Svenskt Stål AB

Carbon steel – heavy rails

ARGENTINA
Somisa – Sdad Mixta Siderurgia Argentina

AUSTRALIA
The Broken Hill Pty Co Ltd

AUSTRIA
Voest-Alpine AG

BRAZIL
CSN – Cia Siderúrgica Nacional

CANADA
The Algoma Steel Corp Ltd
Sydney Steel Corp (Sysco)

CHINA
Anshan Iron & Steel Co
Baotau Iron & Steel Co
Panzhihua Works
Taiyuan Iron & Steel Co
Wuhan Iron & Steel (Wisco)

CZECHOSLOVAKIA
Vitkovice Steelworks

EGYPT
The Egyptian Iron & Steel Co (Hadisolb)

FRANCE
Usinor Sacilor

WEST GERMANY
Krupp Stahl AG (KS)
Maxhütte – Eisenwerke-Gesellschaft Maximilianshütte mbH
Thyssen Stahl AG

INDIA
Sail – Steel Authority of India Ltd

ITALY
Nuova Deltasider SpA

JAPAN
Godo Steel Ltd (Godo Seitetsu)
Nippon Steel Corp (Shin Nippon Seitetsu)
NKK – Nippon Kokan KK
Yamato Kogyo Co Ltd

SOUTH KOREA
Inchon Iron & Steel Co Ltd
Kang Won Industries Ltd

LUXEMBOURG
Métallurgique & Minière de Rodange-Athus (MMRA)

POLAND
Huta Baildon
Huta Katowice
Huta Kościuszko
Huta Pokój

PORTUGAL
Siderurgia Nacional EP (SN)

RUMANIA
Combinat Siderurgic Galaţi (CSG)

Carbon steel – heavy rails

—continued

SOUTH AFRICA
Highveld Steel & Vanadium Corp Ltd
Iscor Ltd

SPAIN
Ensidesa – Empresa Nacional Siderurgica SA

SWEDEN
SSAB – Svenskt Stål AB

TURKEY
TDCI – Türkiye Demir ve Çelik Işletmeleri Genel Müdürlügü

UNITED KINGDOM
BSC – British Steel Corp
Raine & Co Ltd

UNITED STATES
Bethlehem Steel Corp
CF&I Steel Corp

USSR
Met Zavod Azovstal (Azovstal Iron & Steel Works)
Dneprodzerzhinskiy Zavod Imeni Dzerzhinskogo (Dzerzhinsky Works)
Kazakhskiy Met Zavod (Kazakh Iron & Steel Works)
Makeyevskiy Met Zavod Imeni Kirova (Kirov Iron & Steel Works)
Kuznetskiy Met Kombinat
Magnitogorskiy Metallurgischeskiy Kombinat (Magnitogorsk Iron & Steel Works Combine)
Novo-Tagilskiy Met Zavod (Novo-Tagil Iron & Steel Works)
Yenakiyevskiy Met Zavod (Yenakiyevo Iron & Steel Works)
Zakavkazskiy Met Zavod (Transcaucasian Metallurgical Works)

Carbon steel – light rails

ALBANIA
Steel of the Party Metallurgical Combine

ARGENTINA
La Cantábrica SA Metalúrgica, Industrial y Comercial
Rosati y Cristófaro Saic

AUSTRALIA
The Broken Hill Pty Co Ltd
Siddons Steel Mills

BRAZIL
CSN – Cia Siderúrgica Nacional

CANADA
The Algoma Steel Corp Ltd

CHINA
Nanjing Iron & Steel Works

CZECHOSLOVAKIA
NHKG – Nová Hut Klementa Gottwalda

EGYPT
The Egyptian Iron & Steel Co (Hadisolb)

FINLAND
Ovako Steel Oy Ab

FRANCE
Usinor Sacilor

EAST GERMANY
VEB Maxhütte Unterwellenborn

WEST GERMANY
Krupp Stahl AG (KS)
Thyssen Stahl AG

INDIA
Sail – Steel Authority of India Ltd

IRAN
National Iranian Steel Corp (Nisc)

ITALY
Nuova Deltasider SpA

JAPAN
Godo Steel Ltd (Godo Seitetsu)
Osaka Steel (Osaka Seitetsu)

SOUTH KOREA
Inchon Iron & Steel Co Ltd
Kang Won Industries Ltd
Korea Heavy Machinery Industrial Ltd

SOUTH AFRICA
Highveld Steel & Vanadium Corp Ltd
Iscor Ltd

SPAIN
Ensidesa – Empresa Nacional Siderurgica SA

Carbon steel – light rails
—continued

SWEDEN
SSAB – Svenskt Stål AB

TAIWAN
Tang Eng Iron Works Co Ltd

TURKEY
TDCI – Türkiye Demir ve Çelik İşletmeleri Genel Müdürlügü

UNITED KINGDOM
BSC – British Steel Corp
Raine & Co Ltd

UNITED STATES
Bethlehem Steel Corp
Steel of West Virginia Inc

ZIMBABWE
Zimbabwe Iron & Steel Co Ltd – Ziscosteel

Carbon steel – rail accessories

AUSTRALIA
The Broken Hill Pty Co Ltd
Siddons Steel Mills

AUSTRIA
Voest-Alpine AG

BRAZIL
CSN – Cia Siderúrgica Nacional
Siderúrgica Hime SA

CANADA
The Algoma Steel Corp Ltd
Sydney Steel Corp (Sysco)
Western Canada Steel Ltd

FINLAND
Ovako Steel Oy Ab

WEST GERMANY
Krupp Stahl AG (KS)
Thyssen Stahl AG

INDIA
The Punjab Steel Rolling Mills
Sail – Steel Authority of India Ltd
Star Steel Pvt Ltd (Starsteel)

ITALY
Nuova Deltasider SpA
DeltaValdarno

JAPAN
Daido Steel Co Ltd
Osaka Steel (Osaka Seitetsu)
Topy Industries Ltd
Yamato Kogyo Co Ltd

SOUTH KOREA
Kang Won Industries Ltd

LUXEMBOURG
Arbed SA

PORTUGAL
Siderurgia Nacional EP (SN)

SOUTH AFRICA
Iscor Ltd

UNITED KINGDOM
BSC – British Steel Corp
Joseph Gillott & Sons, Glynwed Steels Ltd
Raine & Co Ltd

UNITED STATES
Bethlehem Steel Corp

USSR
Makeyevskiy Met Zavod Imeni Kirova (Kirov Iron & Steel Works)

Carbon steel – cold roll-formed sections (from strip)

ALGERIA
SNS – Sté Nationale de Sidérurgie

AUSTRALIA
The Broken Hill Pty Co Ltd

BELGIUM
Métal Profil Belgium SA

BULGARIA
Kremikovtsi Iron & Steel Works

CANADA
Brockhouse Canada Ltd

FINLAND
Rautaruukki Oy

FRANCE
Usinor Sacilor

Carbon steel – cold roll-formed sections —*continued*

EAST GERMANY
VEB Eisenhüttenkombinat Ost

WEST GERMANY
Hoesch AG
Kronenberg GmbH & Co KG
Tillmann Kaltprofilwalzwerk GmbH & Co KG
Wickeder Eisen- und Stahlwerk GmbH

HUNGARY
Danube Works

ITALY
Giuseppe & Fratelli Bonaiti SpA
Profilati a Freddo Brollo SpA
Profilati Leggeri Cogoleto Srl
Profilerie Trentine SpA

JAPAN
Nisshin Steel Co Ltd

KENYA
Steel Africa Ltd

MALAYSIA
Choo Bee Metal Industries Sdn Bhd
Oriental Metal (M) Sdn Bhd

NETHERLANDS
Prins NV

PAKISTAN
Metropolitan Steel Corp Ltd (MSC)
Pakistan Steel Mills Corp Ltd

PHILIPPINES
Philippine Steel Coating Corp

POLAND
Huta im Lenina

SINGAPORE
Leong Huat Industries Ltd

SPAIN
Arania SA
Condesa – Conducciones y Derivados SA
Industrias Garaeta
Laminaciones de Lesaca SA (LL)
Pecofrisa – Perfiles Conformados en Frio SA
Manufacturas Permet SA

SWEDEN
Gavle Verken AB

SWEDEN —*continued*
Plannja AB
SSAB – Svenskt Stål AB

SWITZERLAND
Forming AG
LN Industries SA

TURKEY
Assan Demir ve Sac Sanayi AŞ

UNITED KINGDOM
British Uralite PLC
Ductile Sections, Glynwed Steels Ltd
European Profiles Ltd
Firth Steels Ltd
Manchester Cold Rollers Ltd
Metsec PLC
Pressfab Sections Ltd
H.H. Robertson (UK) Ltd
Sections & Tubes Ltd
Warwick-Finspa Ltd, Warwick Sections Division
George Whitehouse Engineering Ltd

UNITED STATES
Bethlehem Steel Corp
Dart Rollformers Inc
MP Metal Products Inc
Unistrut Corp

USSR
Zaporozhstal Zavod (Zaporozhye Steel Works)

VENEZUELA
Sidor – CVG Siderúrgica del Orinoco CA

YUGOSLAVIA
Slovenske Železarne – Iron & Steel Works of Slovenia

Carbon steel – sheet piling

AUSTRALIA
The Broken Hill Pty Co Ltd

CZECHOSLOVAKIA
Vitkovice Steelworks

FRANCE
Usinor Sacilor

EAST GERMANY
VEB Maxhütte Unterwellenborn

WEST GERMANY
Hoesch AG

Carbon steel – sheet piling
—continued

WEST GERMANY *—continued*
Stahlwerke Peine-Salzgitter AG (P&S)

INDIA
Sail – Steel Authority of India Ltd

JAPAN
Kawasaki Steel Corp
Nippon Steel Corp (Shin Nippon Seitetsu)
NKK – Nippon Kokan KK
Sumitomo Metal Industries Ltd

SOUTH KOREA
Kang Won Industries Ltd

LUXEMBOURG
Arbed SA

SWITZERLAND
Forming AG

UNITED KINGDOM
BSC – British Steel Corp

UNITED STATES
Bethlehem Steel Corp
Cyclops Corp

VENEZUELA
Sidor – CVG Siderúrgica del Orinoco CA

YUGOSLAVIA
Metalurski Kombinat Smederevo

Carbon steel – pipe piling

AUSTRALIA
The Broken Hill Pty Co Ltd

BRAZIL
Persico Pizzamiglio SA

CANADA
Sonco Steel Tube, Division of Ferrum Inc

FINLAND
Rautaruukki Oy

WEST GERMANY
Benteler AG
Bergrohr GmbH Herne
Mannesmannröhren-Werke AG

JAPAN
Nippon Steel Corp (Shin Nippon Seitetsu)
NKK – Nippon Kokan KK
Sumitomo Metal Industries Ltd

MEXICO
Galvak SA

PAKISTAN
Indus Steel Pipes Ltd

SPAIN
Babcock & Wilcox Española SA Tubular Products Division

TURKEY
Ümran Spiral Kaynakli Boru Sanayii AŞ

UNITED KINGDOM
Byard Kenwest Eng Ltd

UNITED STATES
Cyclops Corp
L.B. Foster Co
Inland Steel Co
Northwest Pipe & Casing Co
Stupp Corp
UNR-Leavitt, Div of UNR Inc

VENEZUELA
CA Conduven
Sidor – CVG Siderúrgica del Orinoco CA

Carbon steel – hot rolled hoop and strip (uncoated)

AUSTRALIA
The Broken Hill Pty Co Ltd

BELGIUM
Cockerill Sambre SA

BRAZIL
Cosipa – Cia Siderúrgica Paulista

BULGARIA
Kremikovtsi Iron & Steel Works

CHINA
Changcheng Steel Works
Changzhi Iron & Steel Works
Echeng Iron & Steel Works

CZECHOSLOVAKIA
Vitkovice Steelworks

Carbon steel – hot rolled hoop and strip *—continued*

FRANCE
Usinor Sacilor

EAST GERMANY
VEB Walzwerk Finow

WEST GERMANY
Edelstahlwerke Buderus AG
Hoesch AG
Klöckner Stahl GmbH
Krupp Stahl AG (KS)
Eisen- und Stahlwalzwerke Rötzel GmbH
Thyssen Bandstahl Berlin GmbH

GREECE
Metallurgiki Halyps SA
Halyvourgiki Inc
Sidenor SA (formerly Steel Works of Northern Greece SA)

HUNGARY
Csepel Steel & Tube Works
Ozdi Kohászati Uzemek (Okü)

INDIA
Calcutta Steel Co Ltd
The Tata Iron & Steel Co Ltd (Tisco)

IRAN
Navard Va Luleh Ahwaz

ITALY
Albasider SpA
Acciaierie e Ferriere Lombarde Falck SpA
Nuova Italsider SpA

JAPAN
Daido Steel Co Ltd
Kobe Steel Ltd (Kobe Seikosho)
Nakayama Steel Works Ltd (Nakayama Seikosho)
Nisshin Steel Co Ltd
NKK – Nippon Kokan KK
Sumitomo Metal Industries Ltd

LUXEMBOURG
Arbed SA

NETHERLANDS
Hoogovens Groep BV

NEW ZEALAND
New Zealand Steel Ltd

POLAND
Huta im Lenina

RUMANIA
Oţelul Roşu Works

SOUTH AFRICA
Strip Steel (A Division of GIC Engineering (Pty) Ltd)

SPAIN
Grupo José María Aristrain SA
Arregui SA
Unión Cerrajera SA

TAIWAN
Mayer Steel Pipe Corp

UNITED KINGDOM
BSC – British Steel Corp
Ductile Hot Mill, Glynwed Steels Ltd
W. Wesson, Glynwed Steels Ltd

UNITED STATES
Armco Inc
Bethlehem Steel Corp
Cyclops Corp
Laclede Steel Co
Lone Star Steel Co
Sharon Steel Corp
USS, A Division of USX Corp

USSR
Ashinskiy Met Zavod (Asha Iron & Steel Works)
Cherepovetskiy Met Zavod (Cherepovets Iron & Steel Works)
Met Zavod Krivorizhstal (Krivoy Rog Iron & Steel Works)
Magnitogorskiy Metallurgischeskiy Kombinat (Magnitogorsk Iron & Steel Works Combine)
Novosibirskiy Met Zavod Kuzmin
Zaporozhstal Zavod (Zaporozhye Steel Works)

YUGOSLAVIA
Zeljezara Sisak Integrated Metallurgical Works
Rudnici i Zelezara Skopje
Metalurski Kombinat Smederevo

Carbon steel – skelp (tube strip)

AUSTRALIA
The Broken Hill Pty Co Ltd

CANADA
Dofasco Inc
Sidbec-Dosco Inc

CZECHOSLOVAKIA
Vitkovice Steelworks

FRANCE
Usinor Sacilor

WEST GERMANY
Benteler AG
Klöckner Stahl GmbH

ITALY
Acciaierie e Ferriere Lombarde Falck SpA
Nuova Italsider SpA

JAPAN
Nisshin Steel Co Ltd
Sumitomo Metal Industries Ltd

SOUTH KOREA
Pohang Iron & Steel Co Ltd

LUXEMBOURG
Arbed SA

SOUTH AFRICA
Highveld Steel & Vanadium Corp Ltd

SPAIN
Grupo José María Aristrain SA
Laminaciones de Lesaca SA (LL)

TAIWAN
Mayer Steel Pipe Corp

TURKEY
Ereğli Demir ve Çelik Fabrikalari TAS (Erdemir)

UNITED STATES
Lone Star Steel Co
USS, A Division of USX Corp

USSR
Met Zavod Krivorizhstal (Krivoy Rog Iron & Steel Works)
Magnitogorskiy Metallurgischeskiy Kombinat (Magnitogorsk Iron & Steel Works Combine)
Novosibirskiy Met Zavod Kuzmin

VENEZUELA
Sidor – CVG Siderúrgica del Orinoco CA

Carbon steel – cold rolled hoop and strip (uncoated)

ARGENTINA
Laminfer SA

AUSTRALIA
The Broken Hill Pty Co Ltd

AUSTRIA
Martin Miller KG

BELGIUM
Armco SA, Belgium
Usines Gustave Boël
Cockerill Sambre SA

BRAZIL
Armco do Brasil SA Indústria e Comércio
Cosipa – Cia Siderúrgica Paulista
Mangels São Bernardo SA

CANADA
Dofasco Inc
Stanley Steel Co Ltd

CHINA
Changcheng Steel Works
Changzhi Iron & Steel Works
Echeng Iron & Steel Works
Jiangxi Steel Works
Laiwu Iron & Steel Plant
Lingyuan Iron & Steel Works

CZECHOSLOVAKIA
Vitkovice Steelworks

FRANCE
Usinor Sacilor

EAST GERMANY
VEB Kaltwalzwerk Oranienburg
VEB Kaltwalzwerk Salzungen

WEST GERMANY
Andernech & Bleck KG
Walzwerk Becker (Wabec) GmbH & Co
Benteler AG
Kaltwalzwerk Brockhaus GmbH (KWB)
Edelstahlwerke Buderus AG
J.N. Eberle & Cie GmbH
Ewald Giebel KG

Carbon steel - cold rolled hoop and strip *—continued*

WEST GERMANY *—continued*
Hille & Müller Kaltwalzwerk
Hoesch AG
J.P. Hüsecken & Co GmbH
Klöckner Stahl GmbH
Krupp Stahl AG (KS)
Rasselstein AG
Risse & Wilke GmbH & Co
Eisen- und Stahlwalzwerke Rötzel GmbH
Friedrich Gustav Theis Kaltwalzwerke GmbH
Thyssen Edelstahlwerke AG
C.D. Wälzhölz KG
Westig GmbH
Wickeder Eisen- und Stahlwerk GmbH

GREECE
Metallurgiki Athinon SA
Halyvourgiki Inc

HUNGARY
Salgótarján Works

INDIA
Graham Firth Steel Products (India) Ltd
KMA Ltd
Mohta Electro Steel Ltd
Nagarjuna Steels Ltd (NSL)
Tube Products of India (TPI)

ITALY
Albasider SpA
Giuseppe & Fratelli Bonaiti SpA
Acciaierie e Ferriere Lombarde Falck SpA
Forsidera SpA
Nuova Italsider SpA
Ocsa – Officine di Crocetta SpA
Olan – Officina Laminazione Nastri Srl
ProReNa – Produttori Reggetta Nastri SpA
Redaelli Tecna SpA
Cesare Rizzato & C SpA
Metallurgica Tognetti SpA

JAPAN
Daido Steel Co Ltd
Kobe Steel Ltd (Kobe Seikosho)
Nakayama Steel Products Co Ltd (Nakayama Kogyo)
Nippon Kinzoku Co Ltd
NKK – Nippon Kokan KK
Sumitomo Metal Industries Ltd

JAPAN *—continued*
Tokushu Kinzoku Kogyo Co Ltd (Tokkin)
Yodogawa Steel Works Ltd (Yodogawa Seikosho)

LUXEMBOURG
Arbed SA

NEW ZEALAND
New Zealand Steel Ltd

PAKISTAN
Metropolitan Steel Corp Ltd (MSC)

POLAND
Huta im Lenina

PORTUGAL
F. Ramada, Aços e Indústrias Sarl

RUMANIA
Uzina de Tevi Republica (UTR)

SOUTH AFRICA
Strip Steel (A Division of GIC Engineering (Pty) Ltd)

SPAIN
AHV – Altos Hornes de Vizcaya SA
Arania SA
Talleres Balcells SA (Tabalsa)
SA Echevarría (SAE)
Flejes Industriales SA (Flinsa)
Sdad General de Electro-Metalurgia SA
Laminación y Derivados SA
Laminados Especiales SA
Laminaciones de Lesaca SA (LL)
Manufacturas Permet SA
Industrias Rocandio
Sidmed – Siderurgica del Mediterráneo SA (formerly AHM – Altos Hornes del Mediterráneo SA)
Unión Cerrajera SA
Alvarez Vázquez SA
Laminación Vizcaya SA

SWEDEN
AB Sandvik Steel
Uddeholm Strip Steel AB

SWITZERLAND
Kaltband AG
Georges Robert, Usines de Laminage de la Jaluse

TAIWAN
Mayer Steel Pipe Corp

Carbon steel – cold rolled hoop and strip *—continued*

TANZANIA
Aluminium Africa Ltd (Alaf)

UNITED KINGDOM
Ben Bennett Jr Ltd
Leon M. Berner & Co Ltd
The Bright Steel Co (Tipton) Ltd
British Rolling Mills Ltd
Bruntons (Mussellburgh) Ltd
BSC – British Steel Corp
Ductile Cold Mill, Glynwed Steels Ltd
Firsteel Ltd
Firth Cleveland Steel Strip, Glynwed Steels Ltd
Arthur Lee & Sons PLC
J.B. & S. Lees Ltd
Steel of Staffs
Stourbridge Rolling Mills, Glynwed Steels Ltd
W. Wesson, Glynwed Steels Ltd

UNITED STATES
Armco Inc
Blair Strip Steel Co
Cold Metal Products Co Inc
Cyclops Corp
Elliott Bros Steel Co
Gibraltar Steel Corp
Greer Steel Co
McLouth Steel Products Corp
Newman Crosby Steel Inc
Rome Strip Steel Co Inc
Seneca Steel Corp
Sharon Steel Corp
Thomas Steel Strip Corp
Thompson Steel Co Inc
Union Steel Corp
USS, A Division of USX Corp
Wallace Barnes Steel
Whittar Steel Strip

USSR
Ashinskiy Met Zavod (Asha Iron & Steel Works)
Cherepovetskiy Met Zavod (Cherepovets Iron & Steel Works)
Magnitogorskiy Metallurgischeskiy Kombinat (Magnitogorsk Iron & Steel Works Combine)
Novosibirskiy Met Zavod Kuzmin
Zaporozhstal Zavod (Zaporozhye Steel Works)

YUGOSLAVIA
Rudnici i Zelezara Skopje
Slovenske Železarne – Iron & Steel Works of Slovenia
Metalurski Kombinat Smederevo
Unis (Associated Metal Industry in Sarajevo)

ZAIRE
Sté Nationale de Sidérurgie

Carbon steel – hot-dip galvanized hoop and strip

ARGENTINA
Armco Argentina SA

CHILE
Cintac – Cĭa Industrial de Tubos de Acero SA

FRANCE
Usinor Sacilor

WEST GERMANY
Ewald Giebel KG

INDONESIA
PT Witikco

ITALY
Nuova Italsider SpA

JAPAN
Kobe Steel Ltd (Kobe Seikosho)
Nisshin Steel Co Ltd
NKK – Nippon Kokan KK
Sumitomo Metal Industries Ltd

LUXEMBOURG
Arbed SA

NEW ZEALAND
New Zealand Steel Ltd

POLAND
Huta im Lenina

SOUTH AFRICA
Strip Steel (A Division of GIC Engineering (Pty) Ltd)

SPAIN
Laminaciones de Lesaca SA (LL)

TAIWAN
Mayer Steel Pipe Corp

Carbon steel – hot-dip galvanized hoop and strip —*continued*

TOGO
Sté Togolaise de Galvanisation de Toles – Sototoles

TURKEY
Assan Demir ve Sac Sanayi AŞ
Borusan Holding AŞ

UNITED STATES
Cyclops Corp

YUGOSLAVIA
Trepca Rudarsko-Metalurski-Hemijski Kombinat

Carbon steel – coated hoop and strip

BELGIUM
Cockerill Sambre SA

FRANCE
Usinor Sacilor

WEST GERMANY
Ewald Giebel KG
Hille & Müller Kaltwalzwerk
Wickeder Eisen- und Stahlwerk GmbH

ITALY
Olan – Officina Laminazione Nastri Srl

NEW ZEALAND
New Zealand Steel Ltd

SOUTH AFRICA
Strip Steel (A Division of GIC Engineering (Pty) Ltd)

SPAIN
Laminaciones de Lesaca SA (LL)

TURKEY
Assan Demir ve Sac Sanayi AŞ
Borusan Holding AŞ

UNITED KINGDOM
Cooper Coated Coil Ltd
Ductile Steel Processors, Glynwed Steels Ltd
Firsteel Ltd
J.B. & S. Lees Ltd

UNITED STATES
Greer Steel Co

UNITED STATES —*continued*
Somers Thin Strip, Olin Corp
Thomas Steel Strip Corp

Carbon steel – medium plates

ALBANIA
Steel of the Party Metallurgical Combine

ALGERIA
SNS – Sté Nationale de Sidérurgie

ARGENTINA
Somisa – Sdad Mixta Siderurgia Argentina

AUSTRALIA
The Broken Hill Pty Co Ltd
Bunge Industrial Steels Pty Ltd

AUSTRIA
Voest-Alpine AG

BANGLADESH
Chittagong Steel Mills Ltd

BELGIUM
Forges de Clabecq SA
Cockerill Sambre SA
Sidmar NV

BRAZIL
Cosipa – Cia Siderúrgica Paulista
Usinas – Usinas Siderúrgicas de Minas Gerais SA

BULGARIA
Kremikovtsi Iron & Steel Works
Lenin Iron & Steel Works

CANADA
The Algoma Steel Corp Ltd
Atlas Specialty Steels Division of Rio Algom Ltd
Dofasco Inc
Ipsco Inc (formerly Interprovincial Steel & Pipe Corp Ltd)
Stelco Inc

CHILE
Cia Siderúrgica Huachipato SA

CHINA
Anshan Iron & Steel Co
Changcheng Steel Works
Kunming Iron & Steel Works

Carbon steel – medium plates —*continued*

CHINA —*continued*
Shanghai Iron & Steel Works
Taiyuan Iron & Steel Co
Wuhan Iron & Steel (Wisco)

CZECHOSLOVAKIA
East Slovak Iron & Steel Works
NHKG – Nová Hut Klementa Gottwalda

EGYPT
The Egyptian Iron & Steel Co (Hadisolb)

FINLAND
Rautaruukki Oy

FRANCE
Usinor Sacilor

EAST GERMANY
VEB Walzwerk Hermann Matern, Burg
VEB Walzwerk Michael Niederkirchner, Ilsenburg
VEB Blechwalzwerk Olbernhau

WEST GERMANY
Edelstahlwerke Buderus AG
AG der Dillinger Hüttenwerke
Klöckner Stahl GmbH
Krupp Stahl AG (KS)
Stahlwerke Peine-Salzgitter AG (P&S)
Thyssen Edelstahlwerke AG

GREECE
Halyvourgiki Inc

HUNGARY
Danube Works
Ozdi Kohászati Uzemek (Okü)

INDIA
Sail – Steel Authority of India Ltd
The Tata Iron & Steel Co Ltd (Tisco)

INDONESIA
PT Kratkatau Steel

ITALY
Acciaierie e Ferriere Lombarde Falck SpA
Nuova Italsider SpA
Siderurgica Villalvernia SpA

JAPAN
The Japan Steel Works Ltd
Kawasaki Steel Corp

JAPAN —*continued*
Kobe Steel Ltd (Kobe Seikosho)
Nakayama Steel Works Ltd (Nakayama Seikosho)
Nippon Steel Corp (Shin Nippon Seitetsu)
Nisshin Steel Co Ltd
NKK – Nippon Kokan KK
Sumitomo Metal Industries Ltd
Tokyo Steel Manufacturing Co Ltd

NORTH KOREA
Hwanghai Iron Works
Kangson Works
Kimchaek Works
Songjin Works

SOUTH KOREA
Pohang Iron & Steel Co Ltd

MEXICO
Ahmsa – Altos Hornos de Mexico SA de CV
Hylsa SA de CV

NETHERLANDS
Hoogovens Groep BV
Laura Metaal BV

NEW ZEALAND
New Zealand Steel Ltd

PAKISTAN
Pakistan Steel Mills Corp Ltd

PHILIPPINES
National Steel Corp (NSC)

POLAND
Huta Bierut
Huta im Lenina

RUMANIA
Focsani Metallurgical Enterprise
Reşita Steel Works

SOUTH AFRICA
Highveld Steel & Vanadium Corp Ltd
Iscor Ltd

SPAIN
Patricio Echeverría SA

SWEDEN
SSAB – Svenskt Stål AB

TAIWAN
China Steel Corp

Carbon steel – medium plates —*continued*

TURKEY
Ereğli Demir ve Çelik Fabrikalari TAS (Erdemir)

UNITED STATES
Armco Inc
Bethlehem Steel Corp
California Steel Industries (CSI)
Gulf States Steel Inc
Inland Steel Co
Jessop Steel Co
LTV Steel Co Inc
Lukens Steel Co
National Steel Corp
Oregon Steel Mills, Division of Gilmore Steel Corp
Sharon Steel Corp
Tuscaloosa Steel Corp

USSR
Met Zavod Azovstal (Azovstal Iron & Steel Works)
Cherepovetskiy Met Zavod (Cherepovets Iron & Steel Works)
Dneprodzerzhinskiy Zavod Imeni Dzerzhinskogo (Dzerzhinsky Works)
Kommunarskiy Met Zavod (Kommunarsk Iron & Steel Works)
Met Zavod Krasny Oktyabr (Red October Steel Works)
Kuznetskiy Met Kombinat
Novo-Tagilskiy Met Zavod (Novo-Tagil Iron & Steel Works)
Orsko-Khalilovskiy Met Kombinat (Orsk-Khalilovo Iron & Steel Combine)
Met Zavod Imeni Petrovskogo (Petrovsky Iron & Steel Works)
Yenakiyevskiy Met Zavod (Yenakiyevo Iron & Steel Works)
Zhdanovskiy Met Zavod Ilyich (Ilyich Works)

VENEZUELA
Sidor – CVG Siderúrgica del Orinoco CA

YUGOSLAVIA
Boris Kidrič Steelworks
Rudnici i Zelezara Skopje
Slovenske Železarne – Iron & Steel Works of Slovenia
Metalurski Kombinat Smederevo

Carbon steel – heavy plates

ALBANIA
Steel of the Party Metallurgical Combine

ARGENTINA
Somisa – Sdad Mixta Siderurgia Argentina

AUSTRALIA
The Broken Hill Pty Co Ltd
Bunge Industrial Steels Pty Ltd

AUSTRIA
Voest-Alpine AG

BANGLADESH
Chittagong Steel Mills Ltd

BELGIUM
SA Fabrique de Fer de Charleroi
Forges de Clabecq SA
Sidmar NV

BRAZIL
Cosipa – Cia Siderúrgica Paulista
Usinas – Usinas Siderúrgicas de Minas Gerais SA

BULGARIA
Kremikovtsi Iron & Steel Works

CANADA
The Algoma Steel Corp Ltd
Dofasco Inc
Stelco Inc

CHILE
Cia Siderúrgica Huachipato SA

CHINA
Changcheng Steel Works
Chongqing Iron & Steel Co
Kunming Iron & Steel Works
Shanghai Iron & Steel Works
Taiyuan Iron & Steel Co
Wuhan Iron & Steel (Wisco)

CZECHOSLOVAKIA
East Slovak Iron & Steel Works
NHKG – Nová Hut Klementa Gottwalda
Vitkovice Steelworks

DENMARK
Danish Steel Works Ltd (Det Danske Stålvalseværk A/S)

Carbon steel – heavy plates
—continued

FINLAND
Rautaruukki Oy

FRANCE
Usinor Sacilor

EAST GERMANY
VEB Stahl- und Walzwerk Brandenburg
VEB Walzwerk Michael Niederkirchner, Ilsenburg

WEST GERMANY
AG der Dillinger Hüttenwerke
Klöckner Stahl GmbH
Krupp Stahl AG (KS)
Mannesmannröhren-Werke AG
Stahlwerke Peine-Salzgitter AG (P&S)
Thyssen Edelstahlwerke AG
Thyssen Stahl AG

GREECE
Halyvourgiki Inc

HUNGARY
Csepel Steel & Tube Works
Diosgyör Works (Lenin Metallurgical Works)

INDIA
Sail – Steel Authority of India Ltd
The Tata Iron & Steel Co Ltd (Tisco)

INDONESIA
PT Kratkatau Steel

ITALY
Acciaierie e Ferriere Lombarde Falck SpA
Nuova Italsider SpA
Siderurgica Villalvernia SpA

JAPAN
The Japan Steel Works Ltd
Kawasaki Steel Corp
Kobe Steel Ltd (Kobe Seikosho)
Nakayama Steel Works Ltd (Nakayama Seikosho)
Nippon Steel Corp (Shin Nippon Seitetsu)
Nisshin Steel Co Ltd
NKK – Nippon Kokan KK
Sumitomo Metal Industries Ltd
Tokyo Steel Manufacturing Co Ltd

NORTH KOREA
Hwanghai Iron Works
Kangson Works

SOUTH KOREA
Dongkuk Steel Mill Co Ltd
Pohang Iron & Steel Co Ltd

MEXICO
Ahmsa – Altos Hornos de Mexico SA de CV

NETHERLANDS
Hoogovens Groep BV

NEW ZEALAND
New Zealand Steel Ltd

PAKISTAN
Pakistan Steel Mills Corp Ltd

PERU
Siderperú – Empresa Siderúrgica del Perú

PHILIPPINES
National Steel Corp (NSC)

POLAND
Huta Batory
Huta Bierut
Huta im Lenina

RUMANIA
Combinat Siderurgic Galaţi (CSG)
Reşita Steel Works

SOUTH AFRICA
Highveld Steel & Vanadium Corp Ltd
Iscor Ltd

SWEDEN
SSAB – Svenskt Stål AB

TAIWAN
China Steel Corp

TURKEY
Ereğli Demir ve Çelik Fabrikalari TAS (Erdemir)

UNITED KINGDOM
BSC – British Steel Corp
Spartan Redheugh Ltd

UNITED STATES
Armco Inc
Bethlehem Steel Corp
California Steel Industries (CSI)
Gulf States Steel Inc

Carbon steel – heavy plates —*continued*

UNITED STATES —*continued*
Inland Steel Co
Jessop Steel Co
LTV Steel Co Inc
Lukens Steel Co
Marathon LeTourneau Co, Longview Division
Tuscaloosa Steel Corp
USS, A Division of USX Corp

USSR
Met Zavod Azovstal (Azovstal Iron & Steel Works)
Cherepovetskiy Met Zavod (Cherepovets Iron & Steel Works)
Dneprodzerzhinskiy Zavod Imeni Dzerzhinskogo (Dzerzhinsky Works)
Kommunarskiy Met Zavod (Kommunarsk Iron & Steel Works)
Met Zavod Krasny Oktyabr (Red October Steel Works)
Kuznetskiy Met Kombinat
Novo-Tagilskiy Met Zavod (Novo-Tagil Iron & Steel Works)
Orsko-Khalilovskiy Met Kombinat (Orsk-Khalilovo Iron & Steel Combine)
Met Zavod Imeni Petrovskogo (Petrovsky Iron & Steel Works)
Yenakiyevskiy Met Zavod (Yenakiyevo Iron & Steel Works)

VENEZUELA
Sidor – CVG Siderúrgica del Orinoco CA

YUGOSLAVIA
Rudnici i Zelezara Skopje
Slovenske Železarne – Iron & Steel Works of Slovenia

Carbon steel – universal plates

AUSTRALIA
The Broken Hill Pty Co Ltd

BRAZIL
Cosipa – Cia Siderúrgica Paulista

WEST GERMANY
Hoesch AG
Klöckner Stahl GmbH
Lemmerz-Werke KGaA
Saarstahl Völklingen GmbH

WEST GERMANY —*continued*
Thyssen Stahl AG

JAPAN
NKK – Nippon Kokan KK
Sumitomo Metal Industries Ltd
Tokyo Steel Manufacturing Co Ltd
Topy Industries Ltd

SOUTH AFRICA
Highveld Steel & Vanadium Corp Ltd
Iscor Ltd

SWEDEN
SSAB – Svenskt Stål AB

TURKEY
Ereğli Demir ve Çelik Fabrikalari TAS (Erdemir)

UNITED STATES
Armco Inc
Bethlehem Steel Corp
Laclede Steel Co
LTV Steel Co Inc
Seattle Steel Inc

Carbon steel – floor plates

AUSTRALIA
The Broken Hill Pty Co Ltd

AUSTRIA
Voest-Alpine AG

BELGIUM
Forges de Clabecq SA

BRAZIL
Cosipa – Cia Siderúrgica Paulista

CANADA
The Algoma Steel Corp Ltd
Dofasco Inc

FINLAND
Rautaruukki Oy

EAST GERMANY
VEB Blechwalzwerk Olbernhau

WEST GERMANY
Klöckner Stahl GmbH

JAPAN
Kobe Steel Ltd (Kobe Seikosho)
Nippon Steel Corp (Shin Nippon Seitetsu)
NKK – Nippon Kokan KK

Carbon steel – floor plates
—continued

JAPAN *—continued*
Sumitomo Metal Industries Ltd

MEXICO
Ahmsa – Altos Hornos de Mexico SA de CV

NEW ZEALAND
New Zealand Steel Ltd

POLAND
Huta im Lenina

SOUTH AFRICA
Iscor Ltd

UNITED KINGDOM
BSC – British Steel Corp

UNITED STATES
Inland Steel Co

VENEZUELA
Sidor – CVG Siderúrgica del Orinoco CA

YUGOSLAVIA
Metalurski Kombinat Smederevo

Carbon steel – clad plates

AUSTRIA
Voest-Alpine AG

CHINA
Chongqing Iron & Steel Co

WEST GERMANY
Klöckner Stahl GmbH

ITALY
Nuova Italsider SpA

JAPAN
Kobe Steel Ltd (Kobe Seikosho)
Nisshin Steel Co Ltd
NKK – Nippon Kokan KK
Sumitomo Metal Industries Ltd

MEXICO
Ahmsa – Altos Hornos de Mexico SA de CV

UNITED STATES
Lukens Steel Co

Carbon steel – hot rolled sheet/coil (uncoated)

ALGERIA
SNS – Sté Nationale de Sidérurgie

ARGENTINA
Somisa – Sdad Mixta Siderurgia Argentina

AUSTRALIA
The Broken Hill Pty Co Ltd

AUSTRIA
Voest-Alpine AG

BELGIUM
Usines Gustave Boël
Cockerill Sambre SA
Sidmar NV

BRAZIL
Acesita – Cia Aços Especiais Itabira
Cosipa – Cia Siderúrgica Paulista
CSN – Cia Siderúrgica Nacional
Usinas – Usinas Siderúrgicas de Minas Gerais SA

BULGARIA
Kremikovtsi Iron & Steel Works

CANADA
The Algoma Steel Corp Ltd
Dofasco Inc
Ipsco Inc (formerly Interprovincial Steel & Pipe Corp Ltd)
Sidbec-Dosco Inc
Stelco Inc

CHILE
Cia Siderúrgica Huachipato SA

CHINA
Anshan Iron & Steel Co
Jiangxi Steel Works
Nanjing Iron & Steel Works
Shanghai Iron & Steel Works
Wuhan Iron & Steel (Wisco)

COLOMBIA
Acerías Paz del Rio SA

CZECHOSLOVAKIA
East Slovak Iron & Steel Works
NHKG – Nová Hut Klementa Gottwalda

EGYPT
The Egyptian Iron & Steel Co (Hadisolb)

Carbon steel – hot rolled sheet/coil *—continued*

FINLAND
Rautaruukki Oy

FRANCE
Usinor Sacilor

WEST GERMANY
Hoesch AG
Klöckner Stahl GmbH
Krupp Stahl AG (KS)
Stahlwerke Peine-Salzgitter AG (P&S)
Thyssen Edelstahlwerke AG
Thyssen Stahl AG

GREECE
Halyvourgiki Inc

HUNGARY
Danube Works

INDIA
Sail – Steel Authority of India Ltd
The Tata Iron & Steel Co Ltd (Tisco)
The Tinplate Co of India Ltd

INDONESIA
PT Kratkatau Steel

ITALY
Acciaierie e Ferriere Lombarde Falck SpA
Nuova Italsider SpA

JAPAN
Kawasaki Steel Corp
Kobe Steel Ltd (Kobe Seikosho)
Nippon Steel Corp (Shin Nippon Seitetsu)
Nisshin Steel Co Ltd
NKK – Nippon Kokan KK
Sumitomo Metal Industries Ltd

NORTH KOREA
Hwanghai Iron Works

SOUTH KOREA
Pohang Iron & Steel Co Ltd

MEXICO
Ahmsa – Altos Hornos de Mexico SA de CV
Hylsa SA de CV

NETHERLANDS
Hoogovens Groep BV

NEW ZEALAND
New Zealand Steel Ltd

PAKISTAN
Pakistan Steel Mills Corp Ltd

PERU
Siderperú – Empresa Siderúrgica del Perú

PHILIPPINES
National Steel Corp (NSC)

POLAND
Huta Bierut
Huta Florian
Huta Katowice
Huta im Lenina

RUMANIA
Combinat Siderurgic Galaţi (CSG)

SOUTH AFRICA
Highveld Steel & Vanadium Corp Ltd

SPAIN
Patricio Echeverría SA
Ensidesa – Empresa Nacional Siderurgica SA

SWEDEN
SSAB – Svenskt Stål AB
Surahammars Bruks AB

TAIWAN
China Steel Corp
Mayer Steel Pipe Corp

TURKEY
Ereğli Demir ve Çelik Fabrikalari TAS (Erdemir)

UNITED KINGDOM
BSC – British Steel Corp
Spencer Clark Metal Industries PLC

UNITED STATES
Acme Steel Co
Armco Inc
Bethlehem Steel Corp
California Steel Industries (CSI)
Cyclops Corp
Gulf States Steel Inc
Inland Steel Co
Jessop Steel Co
Lone Star Steel Co
LTV Steel Co Inc
McLouth Steel Products Corp
National Steel Corp

Carbon steel – hot rolled sheet/coil *—continued*

UNITED STATES *—continued*
Rouge Steel Co
Sharon Steel Corp
USS, A Division of USX Corp
Weirton Steel Corp
Wheeling Pittsburgh Steel Corp

USSR
Cherepovetskiy Met Zavod (Cherepovets Iron & Steel Works)
Karagandinskiy Met Zavod (Kazakhstanskaya Magnitka) (Karaganda Iron & Steel Works)
Kazakhskiy Met Zavod (Kazakh Iron & Steel Works)
Magnitogorskiy Metallurgischeskiy Kombinat (Magnitogorsk Iron & Steel Works Combine)
Novolipetskiy Met Zavod (Novolipetsk Iron & Steel Works)
Novosibirskiy Met Zavod Kuzmin
Zaporozhstal Zavod (Zaporozhye Steel Works)
Zapsib Met Zavod (West Siberian Steel Works)

VENEZUELA
Sidor – CVG Siderúrgica del Orinoco CA

YUGOSLAVIA
Rudnici i Zelezara Skopje
Slovenske Železarne – Iron & Steel Works of Slovenia
Metalurski Kombinat Smederevo

Carbon steel – cold rolled sheet/coil (uncoated)

ALGERIA
SNS – Sté Nationale de Sidérurgie

ARGENTINA
Propulsora Siderurgica Saic
Somisa – Sdad Mixta Siderurgia Argentina

AUSTRALIA
The Broken Hill Pty Co Ltd

AUSTRIA
Voest-Alpine AG

BELGIUM
Usines Gustave Boël

BELGIUM *—continued*
Cockerill Sambre SA
Sidmar NV

BRAZIL
Acesita – Cia Aços Especiais Itabira
Cosipa – Cia Siderúrgica Paulista
CSN – Cia Siderúrgica Nacional
Usinas – Usinas Siderúrgicas de Minas Gerais SA

BULGARIA
Kremikovtsi Iron & Steel Works

CANADA
The Algoma Steel Corp Ltd
Dofasco Inc
Sidbec-Dosco Inc
Stelco Inc

CHILE
Cia Siderúrgica Huachipato SA

CHINA
Wuhan Iron & Steel (Wisco)

CZECHOSLOVAKIA
East Slovak Iron & Steel Works

EGYPT
The Egyptian Iron & Steel Co (Hadisolb)

FINLAND
Rautaruukki Oy

FRANCE
Laminoirs de Beautor SA
Forges de Froncles
Galvameuse SA – Sté Meusienne de Produits d'Usines Métallurgiques
Usinor Sacilor

EAST GERMANY
VEB Eisenhüttenkombinat Ost

WEST GERMANY
Stahlwerke Bochum AG (SWB)
Ewald Giebel KG
Hoesch AG
Klöckner Stahl GmbH
Krupp Stahl AG (KS)
Stahlwerke Peine-Salzgitter AG (P&S)
Rasselstein AG
Thyssen Edelstahlwerke AG
Thyssen Stahl AG

GREECE
Halyvourgiki Inc

Carbon steel – cold rolled sheet/coil *—continued*

GREECE *—continued*
Hellenic Steel Co

HUNGARY
Danube Works

INDIA
Sail – Steel Authority of India Ltd

ITALY
Acciaierie e Ferriere Lombarde Falck SpA
Nuova Italsider SpA
Cesare Rizzato & C SpA

JAPAN
Igeta Steel Sheet Co Ltd
Kawasaki Steel Corp
Kobe Steel Ltd (Kobe Seikosho)
Nippon Steel Corp (Shin Nippon Seitetsu)
Nisshin Steel Co Ltd
NKK – Nippon Kokan KK
Sumitomo Metal Industries Ltd
Taiyo Steel Co Ltd (Taiyo Seiko)
Toyo Kohan Co Ltd
Yodogawa Steel Works Ltd (Yodogawa Seikosho)

SOUTH KOREA
Dong Bu Steel Co Ltd
Pohang Iron & Steel Co Ltd
Union Steel Mfg Co Ltd

LUXEMBOURG
Arbed SA

MEXICO
Ahmsa – Altos Hornos de Mexico SA de CV
Hylsa SA de CV

NETHERLANDS
Hoogovens Groep BV

NEW ZEALAND
New Zealand Steel Ltd

PAKISTAN
Pakistan Steel Mills Corp Ltd

PERU
Siderperú – Empresa Siderúrgica del Perú

PHILIPPINES
National Steel Corp (NSC)

POLAND
Huta Florian
Huta im Lenina
Huta Warszawa

PORTUGAL
Siderurgia Nacional EP (SN)

RUMANIA
Combinat Siderurgic Galaţi (CSG)

SOUTH AFRICA
Iscor Ltd

SPAIN
AHV – Altos Hornes de Vizcaya SA
Arania SA
Ensidesa – Empresa Nacional Siderurgica SA
Laminación y Derivados SA
Laminaciones de Lesaca SA (LL)
Sidmed – Siderurgica del Mediterráneo SA (formerly AHM – Altos Hornes del Mediterráneo SA)

SWEDEN
Plannja AB
SSAB – Svenskt Stål AB

SWITZERLAND
Merz & Cie Drahtwerke AG

TAIWAN
China Steel Corp
Kao Hsing Chang Iron & Steel Corp
Mayer Steel Pipe Corp

TURKEY
Ereğli Demir ve Çelik Fabrikalari TAS (Erdemir)

UNITED KINGDOM
BSC – British Steel Corp
CM Steel Mills Ltd
Staystrip Ltd

UNITED STATES
Acme Steel Co
Armco Inc
Bethlehem Steel Corp
California Steel Industries (CSI)
Cyclops Corp
Gulf States Steel Inc
Inland Steel Co
LTV Steel Co Inc
McLouth Steel Products Corp
National Steel Corp
Rouge Steel Co
Sharon Steel Corp

Carbon steel – cold rolled sheet/coil —*continued*

UNITED STATES —*continued*
USS, A Division of USX Corp
USS-Posco Industries Inc
Weirton Steel Corp
Wheeling Pittsburgh Steel Corp

USSR
Cherepovetskiy Met Zavod (Cherepovets Iron & Steel Works)
Karagandinskiy Met Zavod (Kazakhstanskaya Magnitka) (Karaganda Iron & Steel Works)
Kazakhskiy Met Zavod (Kazakh Iron & Steel Works)
Magnitogorskiy Metallurgischeskiy Kombinat (Magnitogorsk Iron & Steel Works Combine)
Novolipetskiy Met Zavod (Novolipetsk Iron & Steel Works)
Novosibirskiy Met Zavod Kuzmin
Novo-Tagilskiy Met Zavod (Novo-Tagil Iron & Steel Works)
Zaporozhstal Zavod (Zaporozhye Steel Works)

VENEZUELA
Sidor – CVG Siderúrgica del Orinoco CA

YUGOSLAVIA
Rudnici i Zelezara Skopje
Metalurski Kombinat Smederevo

Carbon steel – hot-dip galvanized sheet/coil

ALGERIA
SNS – Sté Nationale de Sidérurgie

ARGENTINA
Comesi Saic
Ostrilion SA Comercial e Industrial

AUSTRALIA
The Broken Hill Pty Co Ltd

AUSTRIA
Voest-Alpine AG

BANGLADESH
Chittagong Steel Mills Ltd

BELGIUM
Phenix Works SA

BELGIUM —*continued*
Segal SC – Sté Européenne de Galvanisation

BRAZIL
CSN – Cia Siderúrgica Nacional

BULGARIA
Kremikovtsi Iron & Steel Works
Lenin Iron & Steel Works

CANADA
Dofasco Inc
Stelco Inc

CHILE
Cia Siderúrgica Huachipato SA

CHINA
Anshan Iron & Steel Co
Shanghai Iron & Steel Works

COLOMBIA
Acesco – Acerias de Colombia SA

COSTA RICA
Metalco SA

EGYPT
The Egyptian Iron & Steel Co (Hadisolb)

FINLAND
Rautaruukki Oy

FRANCE
Galvameuse SA – Sté Meusienne de Produits d'Usines Métallurgiques
Fabrique de Fer de Maubeuge
Usinor Sacilor

EAST GERMANY
VEB Eisenhüttenkombinat Ost

WEST GERMANY
Ewald Giebel KG
Hoesch AG
Krupp Stahl AG (KS)
Stahlwerke Peine-Salzgitter AG (P&S)
Thyssen Stahl AG

GREECE
Hellenic Steel Co

INDIA
Sail – Steel Authority of India Ltd
Sipta Coated Steels Ltd
The Tinplate Co of India Ltd

Carbon steel - hot-dip galvanized sheet/coil
—continued

INDONESIA
PT Fumira, Galvanized Sheet Works
PT Industri Badja Garuda (IBG)
PT Sermani Steel Corp
PT Witikco

ITALY
Nuova Italsider SpA
La Magona d'Italia SpA

JAPAN
Daido Steel Sheet Corp (Daido Kohan)
Kawasaki Steel Corp
Kobe Steel Ltd (Kobe Seikosho)
Nippon Steel Corp (Shin Nippon Seitetsu)
Nisshin Steel Co Ltd
NKK - Nippon Kokan KK
Sumitomo Metal Industries Ltd
Taiyo Steel Co Ltd (Taiyo Seiko)
Toho Sheet & Frame Co Ltd
Yodogawa Steel Works Ltd (Yodogawa Seikosho)

KENYA
Steel Africa Ltd

NORTH KOREA
Kimchaek Works

SOUTH KOREA
Dong Bu Steel Co Ltd
Pohang Iron & Steel Co Ltd
Union Steel Mfg Co Ltd

LUXEMBOURG
Arbed SA

MALAYSIA
Federal Iron Works Sdn Bhd (FIW)

MEXICO
Galvak SA
Galvanizadora Nacional SA de CV

NETHERLANDS
Hoogovens Groep BV

NEW ZEALAND
New Zealand Steel Ltd

PAKISTAN
Pakistan Steel Mills Corp Ltd

PERU
Siderperú - Empresa Siderúrgica del Perú

POLAND
Huta Florian
Huta im Lenina

PORTUGAL
Siderurgia Nacional EP (SN)

SOUTH AFRICA
Iscor Ltd

SPAIN
AHV - Altos Hornes de Vizcaya SA
Ensidesa - Empresa Nacional Siderurgica SA
Laminaciones de Lesaca SA (LL)

SWEDEN
SSAB - Svenskt Stål AB

SWITZERLAND
Merz & Cie Drahtwerke AG

TAIWAN
Mayer Steel Pipe Corp
Shang Tai Steel Co Ltd

THAILAND
Bangkok Steel Industry Co Ltd

TUNISIA
Tunisacier

TURKEY
Assan Demir ve Sac Sanayi AŞ
Borusan Holding AŞ
Demirsaç Galvaniz Ticaret ve Sanayi AŞ

UNITED KINGDOM
BSC - British Steel Corp
H.H. Robertson (UK) Ltd

UNITED STATES
Armco Inc
California Steel Industries (CSI)
Cyclops Corp
Inland Steel Co
LTV Steel Co Inc
Metaltech
National Steel Corp
Pinole Point Steel Co (PPSC)
Rouge Steel Co
Sharon Steel Corp
USS-Posco Industries Inc
Weirton Steel Corp

Carbon steel – hot-dip galvanized sheet/coil
—continued

UNITED STATES *—continued*
Wheeling-Nisshin Inc

USSR
Magnitogorskiy Metallurgischeskiy Kombinat (Magnitogorsk Iron & Steel Works Combine)
Novolipetskiy Met Zavod (Novolipetsk Iron & Steel Works)
Zaporozhstal Zavod (Zaporozhye Steel Works)
Zhdanovskiy Met Zavod Ilyich (Ilyich Works)

VENEZUELA
Lamigal – Láminas Galvanizadas CA

YUGOSLAVIA
Rudnici i Zelezara Skopje

Carbon steel – electro-galvanized sheet/coil

ARGENTINA
Aceros Revestidos SA

AUSTRALIA
The Broken Hill Pty Co Ltd

AUSTRIA
Voest-Alpine AG

BELGIUM
Cockerill Sambre SA
Tôleries Delloye-Matthieu SA

CZECHOSLOVAKIA
East Slovak Iron & Steel Works

FRANCE
Usinor Sacilor

WEST GERMANY
Ewald Giebel KG
Hoesch AG
Klöckner Stahl GmbH
Krupp Stahl AG (KS)
Stahlwerke Peine-Salzgitter AG (P&S)
Thyssen Stahl AG

ITALY
Figli di E. Cavalli SpA
Zincor Italia SpA

JAPAN
Kawasaki Steel Corp
Kobe Steel Ltd (Kobe Seikosho)
Nippon Steel Corp (Shin Nippon Seitetsu)
Nisshin Steel Co Ltd
NKK – Nippon Kokan KK
Sumitomo Metal Industries Ltd
Toyo Kohan Co Ltd

SOUTH KOREA
Pohang Iron & Steel Co Ltd

LUXEMBOURG
Ewald Giebel-Luxembourg

MEXICO
Hylsa SA de CV

NETHERLANDS
Hoogovens Groep BV

SPAIN
Sidmed – Siderurgica del Mediterráneo SA (formerly AHM – Altos Hornes del Mediterráneo SA)

TAIWAN
Mayer Steel Pipe Corp

TURKEY
Borusan Holding AŞ

UNITED KINGDOM
BSC – British Steel Corp

UNITED STATES
Armco Inc
Bethlehem Steel Corp
Continuous Coating Corp
Inland Steel Co
LTV Steel Co Inc
Metalsource-Triumph
National Steel Corp
Rouge Steel Co
Sharon Steel Corp
USS, A Division of USX Corp
Weirton Steel Corp

Carbon steel – galvanized corrugated sheet/coil

ARGENTINA
Comesi Saic

BANGLADESH
Chittagong Steel Mills Ltd

Carbon steel – galvanized corrugated sheet/coil

—continued

COSTA RICA
Metalco SA

FRANCE
Usinor Sacilor

GREECE
Hellenic Steel Co

INDIA
The Tinplate Co of India Ltd

INDONESIA
PT Sermani Steel Corp
PT Witikco

JAPAN
Kobe Steel Ltd (Kobe Seikosho)
Nisshin Steel Co Ltd
Sumitomo Metal Industries Ltd
Toho Sheet & Frame Co Ltd

SOUTH KOREA
Pohang Iron & Steel Co Ltd

LUXEMBOURG
Arbed SA

MEXICO
Galvak SA
Galvanizadora Nacional SA de CV

MOZAMBIQUE
Ima – Indústria Moçambicana do Aço Sarl

PERU
Siderperú – Empresa Siderúrgica del Perú

PHILIPPINES
Philippine Steel Coating Corp

POLAND
Huta im Lenina

PORTUGAL
Siderurgia Nacional EP (SN)

SWEDEN
Dobel AB

TAIWAN
Shang Tai Steel Co Ltd

TANZANIA
Aluminium Africa Ltd (Alaf)

THAILAND
Bangkok Steel Industry Co Ltd

TURKEY
Assan Demir ve Sac Sanayi AŞ
Borusan Holding AŞ
Demirsaç Galvaniz Ticaret ve Sanayi AŞ

UNITED KINGDOM
BSC – British Steel Corp
MET Steel Ltd
H.H. Robertson (UK) Ltd

UNITED STATES
American Tube Co
Bethlehem Steel Corp
Gulf States Steel Inc
LTV Steel Co Inc
Pinole Point Steel Co (PPSC)
Wheeling Pittsburgh Steel Corp

Carbon steel – aluminized sheet/coil

BULGARIA
Kremikovtsi Iron & Steel Works

CZECHOSLOVAKIA
East Slovak Iron & Steel Works

FRANCE
Usinor Sacilor

WEST GERMANY
Thyssen Stahl AG

JAPAN
Nippon Steel Corp (Shin Nippon Seitetsu)
Nisshin Steel Co Ltd

LUXEMBOURG
Galvalange Sarl

TAIWAN
Mayer Steel Pipe Corp

UNITED KINGDOM
BSC – British Steel Corp
Coated Metals Ltd

UNITED STATES
Armco Inc
Bethlehem Steel Corp
Inland Steel Co
LTV Steel Co Inc
Wheeling-Nisshin Inc

Carbon steel – enamelling grade sheet/coil

AUSTRALIA
The Broken Hill Pty Co Ltd

BELGIUM
Cockerill Sambre SA

BRAZIL
Cosipa – Cia Siderúrgica Paulista

CANADA
Dofasco Inc

FRANCE
Usinor Sacilor

JAPAN
Sumitomo Metal Industries Ltd

UNITED KINGDOM
BSC – British Steel Corp

UNITED STATES
Armco Inc
Inland Steel Co
LTV Steel Co Inc
National Steel Corp
USS, A Division of USX Corp

YUGOSLAVIA
Metalurski Kombinat Smederevo

Carbon steel – terne-plate (lead coated) sheet/coil

ARGENTINA
Armco Argentina SA

AUSTRALIA
The Broken Hill Pty Co Ltd

FRANCE
Usinor Sacilor

WEST GERMANY
Hoesch AG
Rasselstein AG

JAPAN
Nippon Steel Corp (Shin Nippon Seitetsu)

UNITED KINGDOM
BSC – British Steel Corp

UNITED STATES
Armco Inc

UNITED STATES —*continued*
LTV Steel Co Inc
USS, A Division of USX Corp
Wheeling Pittsburgh Steel Corp

Carbon steel – colour coated sheet/coil

ARGENTINA
Comesi Saic
Aceros Revestidos SA

AUSTRALIA
The Broken Hill Pty Co Ltd

BELGIUM
Phenix Works SA

BRAZIL
Tekno SA – Construcções Industria e Comércio

CANADA
Baycoat Ltd

CHINA
Shanghai Iron & Steel Works

COSTA RICA
Metalco SA

CZECHOSLOVAKIA
East Slovak Iron & Steel Works

FINLAND
Rautaruukki Oy

FRANCE
Fabrique de Fer de Maubeuge
Usinor Sacilor

EAST GERMANY
VEB Eisenhüttenkombinat Ost

WEST GERMANY
Hoesch AG
Stahlwerke Peine-Salzgitter AG (P&S)
Thyssen Stahl AG

INDONESIA
PT Fumira, Galvanized Sheet Works
PT Industri Badja Garuda (IBG)

ITALY
Lampre Srl
La Magona d'Italia SpA
Plalam SpA
Zincor Italia SpA

Carbon steel – colour coated sheet/coil —*continued*

JAPAN
Daido Steel Sheet Corp (Daido Kohan)
Igeta Steel Sheet Co Ltd
Kawasaki Steel Corp
Kobe Steel Ltd (Kobe Seikosho)
Nisshin Steel Co Ltd
NKK – Nippon Kokan KK
Sumitomo Metal Industries Ltd
Taiyo Steel Co Ltd (Taiyo Seiko)
Toho Sheet & Frame Co Ltd
Tokai Steel Works Ltd (Tokai Kogyo)
Toyo Kohan Co Ltd
Yodogawa Steel Works Ltd (Yodogawa Seikosho)

KENYA
Steel Africa Ltd

SOUTH KOREA
Union Steel Mfg Co Ltd

NEW ZEALAND
New Zealand Steel Ltd

NIGERIA
Kolorkote Nigeria Ltd

PHILIPPINES
Philippine Steel Coating Corp

POLAND
Huta Florian

SOUTH AFRICA
Iscor Ltd

SWEDEN
Dobel AB
Gavle Verken AB
Plannja AB

SWITZERLAND
Merz & Cie Drahtwerke AG

TURKEY
Assan Demir ve Sac Sanayi AŞ

UNITED KINGDOM
BSC – British Steel Corp
Coilcolor
H.H. Robertson (UK) Ltd

UNITED STATES
American Tube Co
Inland Steel Co

UNITED STATES —*continued*
National Steel Corp
Wismarq Corp

USSR
Zaporozhstal Zavod (Zaporozhye Steel Works)

YUGOSLAVIA
Rudnici i Zelezara Skopje

Carbon steel – zinc-aluminium alloy coated sheet/coil

ARGENTINA
Comesi Saic

AUSTRALIA
The Broken Hill Pty Co Ltd

BELGIUM
Phenix Works SA

CANADA
Dofasco Inc

JAPAN
Daido Steel Sheet Corp (Daido Kohan)
Kobe Steel Ltd (Kobe Seikosho)

LUXEMBOURG
Galvalange Sarl

NETHERLANDS
Hoogovens Groep BV

SPAIN
Ensidesa – Empresa Nacional Siderurgica SA

SWEDEN
SSAB – Svenskt Stål AB

UNITED KINGDOM
BSC – British Steel Corp

UNITED STATES
Inland Steel Co
LTV Steel Co Inc
USS, A Division of USX Corp
Weirton Steel Corp

Carbon steel – coated sheet/coil

BELGIUM
Phenix Works SA

CANADA
Dofasco Inc
Stelco Inc

FRANCE
Usinor Sacilor

EAST GERMANY
VEB Eisenhüttenkombinat Ost

ITALY
Figli di E. Cavalli SpA

JAPAN
Sumitomo Metal Industries Ltd

UNITED KINGDOM
BSC – British Steel Corp

UNITED STATES
Inland Steel Co
LTV Steel Co Inc
Sharon Steel Corp

Carbon steel – electrical grain oriented silicon sheet/coil

AUSTRALIA
The Broken Hill Pty Co Ltd

BELGIUM
Cockerill Sambre SA

BRAZIL
Acesita – Cia Aços Especiais Itabira

CANADA
Dofasco Inc

FRANCE
Usinor Sacilor

WEST GERMANY
Thyssen Grillo Funke GmbH

INDIA
Sail – Steel Authority of India Ltd

JAPAN
Kawasaki Steel Corp
Nippon Kinzoku Co Ltd

JAPAN *—continued*
Nippon Steel Corp (Shin Nippon Seitetsu)

SOUTH KOREA
Pohang Iron & Steel Co Ltd

SWEDEN
Surahammars Bruks AB

UNITED KINGDOM
BSC – British Steel Corp

UNITED STATES
Armco Inc
Bethlehem Steel Corp
National Steel Corp

Carbon steel – electrical non-oriented silicon sheet/coil

AUSTRALIA
The Broken Hill Pty Co Ltd

BELGIUM
Cockerill Sambre SA

BRAZIL
Acesita – Cia Aços Especiais Itabira
Usinas – Usinas Siderúrgicas de Minas Gerais SA

BULGARIA
Kremikovtsi Iron & Steel Works

CHINA
Anyang Iron and Steel Works
Jiangxi Steel Works
Nanjing Iron & Steel Works
Taiyuan Iron & Steel Co
Wuhan Iron & Steel (Wisco)

FRANCE
Usinor Sacilor

WEST GERMANY
Stahlwerke Bochum AG (SWB)
Hoesch AG

HUNGARY
Danube Works

INDIA
Sail – Steel Authority of India Ltd
The Tata Iron & Steel Co Ltd (Tisco)

Carbon steel – electrical non-oriented silicon sheet/coil —*continued*

JAPAN
Nippon Steel Corp (Shin Nippon Seitetsu)
Sumitomo Metal Industries Ltd

SOUTH KOREA
Pohang Iron & Steel Co Ltd

NORWAY
Norsk Jernverk AS

POLAND
Huta im Lenina

RUMANIA
Tirgovişte Works

SPAIN
Grupo José María Aristrain SA
Laminaciones de Lesaca SA (LL)

SWEDEN
Surahammars Bruks AB

UNITED KINGDOM
BSC – British Steel Corp

UNITED STATES
Cyclops Corp
Inland Steel Co
LTV Steel Co Inc
USS, A Division of USX Corp

USSR
Novolipetskiy Met Zavod (Novolipetsk Iron & Steel Works)
Zaporozhstal Zavod (Zaporozhye Steel Works)

YUGOSLAVIA
Slovenske Železarne – Iron & Steel Works of Slovenia

Carbon steel – electrical non-oriented non-silicon sheet/coil

AUSTRALIA
The Broken Hill Pty Co Ltd

AUSTRIA
Voest-Alpine AG

BELGIUM
Cockerill Sambre SA

BRAZIL
Usinas – Usinas Siderúrgicas de Minas Gerais SA

CZECHOSLOVAKIA
East Slovak Iron & Steel Works
Frydek-Mistek Works
Kraluv Dvur Works

FRANCE
Usinor Sacilor

WEST GERMANY
Hoesch AG
Klöckner Stahl GmbH

HUNGARY
Borsodnádasd Sheet Metal Works
Csepel Steel & Tube Works
Salgótarján Works

INDIA
The Tinplate Co of India Ltd

JAPAN
Kawasaki Steel Corp
Kobe Steel Ltd (Kobe Seikosho)
Nippon Steel Corp (Shin Nippon Seitetsu)

POLAND
Mostostal
Huta Warszawa

SWEDEN
Surahammars Bruks AB

UNITED KINGDOM
BSC – British Steel Corp

UNITED STATES
Inland Steel Co

Carbon steel – blackplate

AUSTRALIA
The Broken Hill Pty Co Ltd

AUSTRIA
Voest-Alpine AG

BELGIUM
Cockerill Sambre SA
Sidmar NV

Carbon steel - blackplate
—continued

BRAZIL
CSN – Cia Siderúrgica Nacional
Usinas – Usinas Siderúrgicas de Minas Gerais SA

CANADA
Dofasco Inc
Stelco Inc

FRANCE
Carnaud Basse-Indre
Usinor Sacilor

WEST GERMANY
Hoesch AG
Rasselstein AG

INDIA
Sail – Steel Authority of India Ltd
The Tinplate Co of India Ltd

JAPAN
Sumitomo Metal Industries Ltd

SOUTH KOREA
Pohang Iron & Steel Co Ltd

NETHERLANDS
Hoogovens Groep BV

PHILIPPINES
National Steel Corp (NSC)

PORTUGAL
Siderurgia Nacional EP (SN)

RUMANIA
Combinat Siderurgic Galaţi (CSG)

SOUTH AFRICA
Highveld Steel & Vanadium Corp Ltd

TAIWAN
China Steel Corp

UNITED KINGDOM
BSC – British Steel Corp

UNITED STATES
LTV Steel Co Inc
National Steel Corp
USS, A Division of USX Corp
Weirton Steel Corp
Wheeling Pittsburgh Steel Corp

USSR
Kazakhskiy Met Zavod (Kazakh Iron & Steel Works)

VENEZUELA
Sidor – CVG Siderúrgica del Orinoco CA

YUGOSLAVIA
Metalurski Kombinat Smederevo

Carbon steel - electrolytic tinplate (single-reduced)

ALGERIA
SNS – Sté Nationale de Sidérurgie

ARGENTINA
Somisa – Sdad Mixta Siderurgia Argentina

AUSTRALIA
The Broken Hill Pty Co Ltd

BELGIUM
Cockerill Sambre SA

BRAZIL
CSN – Cia Siderúrgica Nacional

BULGARIA
Kremikovtsi Iron & Steel Works

CANADA
Dofasco Inc
Stelco Inc

CHILE
Cia Siderúrgica Huachipato SA

CHINA
Wuhan Iron & Steel (Wisco)

COLOMBIA
Holasa – Hojalata y Laminados SA

CZECHOSLOVAKIA
East Slovak Iron & Steel Works

FRANCE
Carnaud Basse-Indre
Usinor Sacilor

WEST GERMANY
Hoesch AG
Rasselstein AG

GREECE
Hellenic Steel Co

Carbon steel - electrolytic tinplate *—continued*

HUNGARY
Danube Works

INDIA
KR Steelunion Pvt Ltd
Sail – Steel Authority of India Ltd
The Tinplate Co of India Ltd

INDONESIA
PT Latinusa

ITALY
Industrie Cantieri Metallurgici Italiani SpA
Nuova Italsider SpA
La Magona d'Italia SpA

JAPAN
Kawasaki Steel Corp
Nippon Steel Corp (Shin Nippon Seitetsu)
NKK – Nippon Kokan KK
Toyo Kohan Co Ltd

SOUTH KOREA
Dong Yang Tinplate Industrial Co Ltd

MALAYSIA
Perstima – Perusahaan Sadur Timah Malaysia (Perstima) Sdn Bhd

MEXICO
Ahmsa – Altos Hornos de Mexico SA de CV
Hylsa SA de CV

NETHERLANDS
Hoogovens Groep BV

NORWAY
Norsk Jernverk AS

PERU
Siderperú – Empresa Siderúrgica del Perú

PHILIPPINES
National Steel Corp (NSC)

POLAND
Huta im Lenina

PORTUGAL
Siderurgia Nacional EP (SN)

SOUTH AFRICA
Iscor Ltd

SPAIN
AHV – Altos Hornes de Vizcaya SA
Ensidesa – Empresa Nacional Siderurgica SA

TAIWAN
Ton Yi Industrial Corp

THAILAND
Thai Tinplate Manufacturing Co Ltd

UNITED KINGDOM
BSC – British Steel Corp

UNITED STATES
Bethlehem Steel Corp
LTV Steel Co Inc
National Steel Corp
USS, A Division of USX Corp
USS-Posco Industries Inc
Weirton Steel Corp
Wheeling Pittsburgh Steel Corp

USSR
Karagandinskiy Met Zavod (Kazakhstanskaya Magnitka) (Karaganda Iron & Steel Works)
Kazakhskiy Met Zavod (Kazakh Iron & Steel Works)
Magnitogorskiy Metallurgischeskiy Kombinat (Magnitogorsk Iron & Steel Works Combine)
Zaporozhstal Zavod (Zaporozhye Steel Works)

VENEZUELA
Sidor – CVG Siderúrgica del Orinoco CA

YUGOSLAVIA
Zorka Chemical Works – Hemijska Industrija Zorka

Carbon steel - electrolytic tinplate (double-reduced)

AUSTRALIA
The Broken Hill Pty Co Ltd

BELGIUM
Cockerill Sambre SA
Phenix Works SA

BRAZIL
CSN – Cia Siderúrgica Nacional

CANADA
Dofasco Inc
Stelco Inc

Carbon steel – electrolytic tinplate *—continued*

COLOMBIA
Holasa – Hojalata y Laminados SA

FRANCE
Carnaud Basse-Indre
Usinor Sacilor

WEST GERMANY
Hoesch AG
Rasselstein AG

JAPAN
Kawasaki Steel Corp
Nippon Steel Corp (Shin Nippon Seitetsu)

NETHERLANDS
Hoogovens Groep BV

TURKEY
Ereğli Demir ve Çelik Fabrikalari TAS (Erdemir)

UNITED KINGDOM
BSC – British Steel Corp

UNITED STATES
Bethlehem Steel Corp
LTV Steel Co Inc
USS, A Division of USX Corp
Weirton Steel Corp
Wheeling Pittsburgh Steel Corp

Carbon steel – hot-dip tinplate

CHINA
Anshan Iron & Steel Co

CZECHOSLOVAKIA
East Slovak Iron & Steel Works

HUNGARY
Danube Works

INDIA
Sail – Steel Authority of India Ltd

POLAND
Huta im Lenina

RUMANIA
Uzina de Tevi Republica (UTR)

UNITED KINGDOM
London Tinning Co Ltd

UNITED KINGDOM *—continued*
Plated Strip (International) Ltd

Carbon steel – tin-free steel (ECCS)

BELGIUM
Cockerill Sambre SA

BRAZIL
CSN – Cia Siderúrgica Nacional

CANADA
Dofasco Inc
Stelco Inc

FRANCE
Carnaud Basse-Indre
Usinor Sacilor

WEST GERMANY
Hoesch AG
Rasselstein AG

INDIA
The Tinplate Co of India Ltd

ITALY
Nuova Italsider SpA

JAPAN
Kawasaki Steel Corp
NKK – Nippon Kokan KK

SOUTH KOREA
Dong Yang Tinplate Industrial Co Ltd

NETHERLANDS
Hoogovens Groep BV

THAILAND
Thai Tinplate Manufacturing Co Ltd

UNITED KINGDOM
BSC – British Steel Corp

UNITED STATES
Bethlehem Steel Corp
LTV Steel Co Inc
National Steel Corp
USS, A Division of USX Corp
Weirton Steel Corp

USSR
Kazakhskiy Met Zavod (Kazakh Iron & Steel Works)

Carbon steel – tin-free steel

—continued

VENEZUELA
Sidor – CVG Siderúrgica del Orinoco CA

Carbon steel – bright wire

ALGERIA
SNS – Sté Nationale de Sidérurgie

ARGENTINA
Acindar Industria Argentina de Aceros SA

AUSTRALIA
The Broken Hill Pty Co Ltd

BELGIUM
Usines Gustave Boël

BRAZIL
Siderúrgica Aconorte SA
Cia Siderúrgica Belgo-Mineira
Cosigua – Cia Siderurgia da Guanabara
Eletrometal SA – Metais Especiais
Siderúrgica Fi-El SA
Siderúrgica Mendes Junior SA (SMJ)
Cia Siderúrgica Pains
Siderúrgica Riograndense SA

CANADA
Ivaco Inc
Sidbec-Dosco Inc
Stelco Inc

CHINA
Jiangxi Steel Works

EL SALVADOR
Corinca SA – Corporación Industrial Centroamericana SA
Siderúrgica Centroamericana del Pacífico SA (Sicepasa)

EAST GERMANY
VEB Stahl- und Walzwerk Brandenburg

WEST GERMANY
Krupp Stahl AG (KS)
Saarstahl Völklingen GmbH
Westig GmbH

GUATEMALA
Hornos SA

HUNGARY
Ozdi Kohászati Uzemek (Okü)

INDIA
Agarwal Foundry & Engineering Works
Modi Industries Ltd, Steel Division

INDONESIA
PT Kratkatau Steel

ITALY
Rodacciai Industria Trafilati SpA
Metallurgica Tognetti SpA

JAPAN
Sumitomo Electric Industries Ltd (Sumitomo Denki Kogyo)

KENYA
Kenya United Steel Co Ltd

NORTH KOREA
Kangson Works

NETHERLANDS
Nedstaal BV

PAKISTAN
Metropolitan Steel Corp Ltd (MSC)

RUMANIA
Uzina de Tevi Republica (UTR)

SOUTH AFRICA
Titan Industrial Corp (Pty) Ltd

SPAIN
Industrias Bérriz
Nueva Montaño Quijano SA
Esteban Orbegozo SA

SWEDEN
AB Sandvik Steel

SWITZERLAND
Von Moos Stahl AG

TUNISIA
El Fouladh – Sté Tunisienne de Sidérurgie

UNITED KINGDOM
Allied Steel & Wire Ltd
A.E. Godrich & Son Ltd
Kiveton Park Steel & Wire Works Ltd
Unbrako Steel Co Ltd

UNITED STATES
Atlantic Steel Co
Bethlehem Steel Corp

Carbon steel – bright wire
—continued

UNITED STATES *—continued*
CF&I Steel Corp
Connecticut Steel Corp
Cuyahoga Steel & Wire Inc
Keystone Steel & Wire Co
LTV Steel Co Inc
Nelsen Steel & Wire Co Inc
Northwestern Steel & Wire Co

USSR
Met Zavod Krivorizhstal (Krivoy Rog Iron & Steel Works)

YUGOSLAVIA
Slovenske Železarne – Iron & Steel Works of Slovenia
Jadranska Zelejezara Split (Adriatic Steelworks, Split)
Unis (Associated Metal Industry in Sarajevo)

ZIMBABWE
Lancashire Steel (Pvt) Ltd

Carbon steel – black annealed wire

ARGENTINA
Acindar Industria Argentina de Aceros SA

AUSTRALIA
The Broken Hill Pty Co Ltd

BELGIUM
Usines Gustave Boël

BRAZIL
Siderúrgica Aconorte SA
Siderúrgica Barra Mansa SA
Cia Siderúrgica Belgo-Mineira
Cosigua – Cia Siderurgia da Guanabara
Siderúrgica Fi-El SA
Mannesmann SA
Siderúrgica Mendes Junior SA (SMJ)
Cia Siderúrgica Pains
Siderúrgica Riograndense SA

CANADA
Ivaco Inc
Sidbec-Dosco Inc
Stelco Inc

EGYPT
The Egyptian Copper Works

EL SALVADOR
Corinca SA – Corporación Industrial Centroamericana SA

INDIA
Agarwal Foundry & Engineering Works

JAPAN
Sumitomo Electric Industries Ltd (Sumitomo Denki Kogyo)

KENYA
Kenya United Steel Co Ltd

PAKISTAN
Metropolitan Steel Corp Ltd (MSC)

SPAIN
Nueva Montaño Quijano SA
Esteban Orbegozo SA

SWITZERLAND
Kaltband AG
Von Moos Stahl AG

TUNISIA
El Fouladh – Sté Tunisienne de Sidérurgie

UNITED KINGDOM
Allied Steel & Wire Ltd

UNITED STATES
Atlantic Steel Co
Bethlehem Steel Corp
Cuyahoga Steel & Wire Inc
Keystone Steel & Wire Co
LTV Steel Co Inc
Nelsen Steel & Wire Co Inc
Northwestern Steel & Wire Co

ZIMBABWE
Lancashire Steel (Pvt) Ltd

Carbon steel – galvanized wire (plain)

ARGENTINA
Acindar Industria Argentina de Aceros SA

AUSTRALIA
The Broken Hill Pty Co Ltd

Carbon steel - galvanized wire *—continued*

BANGLADESH
Bengal Steel Works Ltd

BELGIUM
Usines Gustave Boël

BRAZIL
Siderúrgica Aconorte SA
Cia Siderúrgica Belgo-Mineira
Cosigua - Cia Siderurgia da Guanabara
Siderúrgica Mendes Junior SA (SMJ)
Cia Siderúrgica Pains
Siderúrgica Riograndense SA

BURMA
Ywama Steel Mill

CANADA
Ivaco Inc
Sidbec-Dosco Inc
Stelco Inc

COLOMBIA
Simesa - Siderúrgica de Medellín

EL SALVADOR
Corinca SA - Corporación Industrial Centroamericana SA
Siderúrgica Centroamericana del Pacífico SA (Sicepasa)

FINLAND
Dalsbruk Oy Ab

GUATEMALA
Hornos SA

HUNGARY
Salgótarján Works

INDIA
Agarwal Foundry & Engineering Works
Modi Industries Ltd, Steel Division

INDONESIA
PT Gunung Gahapi Steel
PT Kratkatau Steel

KENYA
Kenya United Steel Co Ltd

PAKISTAN
Metropolitan Steel Corp Ltd (MSC)

SPAIN
Nueva Montaño Quijano SA
Esteban Orbegozo SA

SWITZERLAND
Von Moos Stahl AG

TUNISIA
El Fouladh - Sté Tunisienne de Sidérurgie

TURKEY
Borusan Holding AŞ

UNITED KINGDOM
Allied Steel & Wire Ltd

UNITED STATES
Atlantic Steel Co
Bethlehem Steel Corp
Keystone Steel & Wire Co
Northwestern Steel & Wire Co

ZIMBABWE
Lancashire Steel (Pvt) Ltd

Carbon steel - galvanized wire (barbed)

ARGENTINA
Acindar Industria Argentina de Aceros SA

AUSTRALIA
The Broken Hill Pty Co Ltd

BRAZIL
Siderúrgica Aconorte SA
Cia Siderúrgica Belgo-Mineira
Cosigua - Cia Siderurgia da Guanabara
Siderúrgica Fi-El SA
Siderúrgica Mendes Junior SA (SMJ)
Cia Siderúrgica Pains
Siderúrgica Riograndense SA

CANADA
Ivaco Inc
Stelco Inc

COLOMBIA
Simesa - Siderúrgica de Medellín

EL SALVADOR
Corinca SA - Corporación Industrial Centroamericana SA

GUATEMALA
Hornos SA

Carbon steel - galvanized wire —*continued*

INDIA
Agarwal Foundry & Engineering Works

KENYA
Kenya United Steel Co Ltd

PAKISTAN
Metropolitan Steel Corp Ltd (MSC)

PANAMA
Aceros Panamá SA

SPAIN
Nueva Montaño Quijano SA
Esteban Orbegozo SA

UNITED STATES
American Tube Co
CF&I Steel Corp
Keystone Steel & Wire Co
Northwestern Steel & Wire Co

ZIMBABWE
Lancashire Steel (Pvt) Ltd

Carbon steel - seamless tubes and pipes

ARGENTINA
Siderca Saic

AUSTRALIA
Email Tube Division, Email Ltd
Tubemakers of Australia Ltd

AUSTRIA
Voest-Alpine AG

BRAZIL
Cosim - Cia Siderúrgica de Mogi das Cruzes
Mannesmann SA

BULGARIA
Kremikovtsi Iron & Steel Works

CANADA
The Algoma Steel Corp Ltd

CHINA
Anshan Iron & Steel Co
Anyang Iron and Steel Works
Baotau Iron & Steel Co
Changcheng Steel Works
Changzhi Iron & Steel Works

CHINA —*continued*
Chengdu Seamless Steel Tube Plant
Guangzhou Iron & Steel Works
Guiyang Steel Mill
Hefei Iron & Steel Co
Hengyang Steel Tube Plant
Jiangxi Steel Works
Kunming Iron & Steel Works
Sanming Iron & Steel Works

CZECHOSLOVAKIA
Chomutov Tube Works
NHKG - Nová Hut Klementa Gottwalda
Vitkovice Steelworks

FRANCE
Pont-à-Mousson SA
Vallourec Industries

WEST GERMANY
Benteler AG
Röhrenwerke Bous/Saar GmbH (RBS)
Mannesmannröhren-Werke AG
Maxhütte - Eisenwerke-Gesellschaft Maximilianshütte mbH
TPS-Technitube Röhrenwerke GmbH

HUNGARY
Csepel Steel & Tube Works

INDIA
The Tata Iron & Steel Co Ltd (Tisco)

ISRAEL
Middle East Tube Co Ltd

ITALY
Dalmine SpA
Acciaierie e Ferriere Lombarde Falck SpA
Acciaierie e Tubificio Meridionali SpA
Pietra SpA - Acciaierie Ferriere e Tubifici
Seta Tubi SpA
Tubicar SpA

JAPAN
Kawasaki Steel Corp
Kobe Steel Ltd (Kobe Seikosho)
Nippon Steel Corp (Shin Nippon Seitetsu)
NKK - Nippon Kokan KK
Sumitomo Metal Industries Ltd

NORTH KOREA
Kangson Works
Kimchaek Works

Carbon steel – seamless tubes and pipes *—continued*

SOUTH KOREA
Sammi Steel Co Ltd

MEXICO
Tamsa – Tubos de Acero de Mexico SA

NETHERLANDS
Robur Buizenfabriek BV

POLAND
Huta Bierut

RUMANIA
Uzina de Tevi Republica (UTR)

SOUTH AFRICA
Tosa (Division of Dorbyl Ltd)

SPAIN
Babcock & Wilcox Española SA Tubular Products Division
Centracero SA
Transmesa – Transformaciones Metalurgicas SA
Tubacex – C.E. de Tubos por Extrusion SA
Tubos Reunidos SA

SWEDEN
Structo DOM Europe AB
Ovako Steel AB

TURKEY
MKEK Çelik Çekme Boru Fabrikasi

UNITED KINGDOM
BSC – British Steel Corp
Cameron Iron Works Ltd, Forged Products Division
Cold Drawn Tubes Ltd
English Seamless Tube Co
The Phoenix Steel Tube Co Ltd
Seamless Tubes Ltd
TI Desford Tubes Ltd
TI Hollow Extrusions Ltd

UNITED STATES
CF&I Steel Corp
Copperweld Corp
Lone Star Steel Co
North Star Steel Co (NSS)
Pacific Tube Co
Plymouth Tube Co
Quanex Corp
The Timken Co

UNITED STATES *—continued*
USS, A Division of USX Corp

USSR
Dnepropetrovskiy Met i Trubnoprokatny Savod Imeni Lenina (Lenin Iron & Steel Tube Rolling Works)
Met Zavod Imeni Karl Liebknecht (Karl Liebknecht Steel Works)
Pervouralskiy Novotrubny Zavod (Pervouralsk New Tube Works)
Volzhskiy Truboprokatny Zavod (Volga Pipe Works)
Yuzhnotrubny Zavod (Southern Tube Works)
Zakavkazskiy Met Zavod (Transcaucasian Metallurgical Works)
Zhdanovskiy Met Zavod Ilyich (Ilyich Works)

VENEZUELA
Sidor – CVG Siderúrgica del Orinoco CA

YUGOSLAVIA
Zeljezara Sisak Integrated Metallurgical Works

Carbon steel – longitudinal weld tubes and pipes

ALGERIA
SNS – Sté Nationale de Sidérurgie

ARGENTINA
Acindar Industria Argentina de Aceros SA
Armco Argentina SA
Laminfer SA
Siam SA – Division Siat

AUSTRALIA
Australian Tube Mills Pty Ltd (ATM)
Email Tube Division, Email Ltd
Hills Industries Ltd
Palmer Tube Mills Ltd
Tubemakers of Australia Ltd

AUSTRIA
Voest-Alpine AG

BANGLADESH
National Tubes Ltd

BELGIUM
SA Laminoirs de Longtain

Carbon steel – longitudinal weld tubes and pipes —*continued*

BELGIUM —*continued*
Tubel SA

BRAZIL
Apolo Produtos de Aço SA
Confab Industrial SA
Fornasa SA
Mannesmann SA
Persico Pizzamiglio SA

BULGARIA
Kremikovtsi Iron & Steel Works

CANADA
Arc Tube Inc
Associated Tube Industries Ltd (ATI)
Atlas Tube
Barton Tubes Ltd
Ipsco Inc (formerly Interprovincial Steel & Pipe Corp Ltd)
Sidbec-Dosco Inc
Standard Tube Canada, Division of TI Canada Inc
Stelco Inc
Tubular Steels Products Div of Chartcan

CHILE
Cintac – Cĭa Industrial de Tubos de Acero SA
Cia Siderúrgica Huachipato SA

CHINA
Changcheng Steel Works
Hefei Iron & Steel Co
Hengyang Steel Tube Plant
Lingyuan Iron & Steel Works
Shoudu Iron & Steel Co

COLOMBIA
Colmena – Consorcio Metalúrgico Nacional SA
Simesa – Siderúrgica de Medellín

CYPRUS
BMS Metal Pipes Industries Ltd

CZECHOSLOVAKIA
Chomutov Tube Works
NHKG – Nová Hut Klementa Gottwalda
Veselinad Moravou Works
Vitkovice Steelworks

DENMARK
Nordisk Simplex AS

DOMINICAN REPUBLIC
Metaldom – Complejo Metalúrgico Dominicano C por A

FINLAND
Rautaruukki Oy

FRANCE
Armco SA, France
Tubes de Chevillon
Exma SA
Sté Meusienne de Constructions Mécaniques
Produits d'Acier SA
Usinor Sacilor
Vallourec Industries
Sté des Tubes de Vincey

EAST GERMANY
VEB Rohrwerke Bitterfeld
VEB Walzwerk Finow
VEB Rohr- und Kaltwalzwerk Karl-Marx-Stadt

WEST GERMANY
Benteler AG
Bergrohr GmbH Herne
Bergrohr GmbH Siegen
Eisen- und Metallwerke Ferndorf GmbH
Rudolf Flender GmbH & Co KG
Hoesch AG
Eisenbau Krämer mbH
Mannesmannröhren-Werke AG
Rothrist Rohr GmbH
Saarstahl Völklingen GmbH
Schwarzwälder Röhrenwerk
TPS-Technitube Röhrenwerke GmbH

GREECE
Metallurgiki Athinon SA
Profil SA

HUNGARY
Csepel Steel & Tube Works
Danube Works

INDIA
Apollo Tubes Ltd
BST Manufacturing Ltd
Gujarat Steel Tubes Ltd (GST)
Jain Tube Co Ltd
Jotindra Steel & Tubes Ltd
Khandewal Tubes, Division of Khandelwal Ferro Alloys Ltd

Carbon steel – longitudinal weld tubes and pipes
—continued

INDIA *—continued*
Metalman Pipe Mfg Co Ltd
Sail – Steel Authority of India Ltd
Steel Tubes of India Ltd
The Tata Iron & Steel Co Ltd (Tisco)
Tube Products of India (TPI)

INDONESIA
PT Bakrie & Bros
Bakrie Pipe Industries PT
PT Inastu – Indonesian National Steel Tube Co
PT Pabrik Pipa Indonesia (The Indonesia Tube Mill Corp Ltd)
PT Radjin Steel Pipe Industry

IRAN
Navard Va Luleh Ahwaz
Kalup Corp

ISRAEL
Middle East Tube Co Ltd

ITALY
Arvedi Group
Baroni SpA
Metallurgica G. Berera SpA
Tubificio Dalmine Italsider SpA – TDI
Emiliana Tubi – Profilati Acciaio SpA (Setpa)
Acciaierie e Ferriere Lombarde Falck SpA
Gamma Serbatoi SpA
General Sider Europa SpA
General Sider Italiana SpA
Fratelli Goffi SpA
Nuova Italsider SpA
Lombarda Tubi SpA
Marcegaglia SpA
La Metallifera SpA
Neirotti Tubi SpA
Ocsa – Officine di Crocetta SpA
OMV – Off Metall. Ventura SpA
Padana Tubi & Profilati Acciaio SpA
Profilmec SpA
Profilnastro SpA
Cesare Rizzato & C SpA
Siderpo SpA
Silpa Tubi e Profilati SpA
Tanga SpA
Tubimar Ancona SpA
Tubinar SpA (formerly Maraldi SpA)
Valducci SpA

JAPAN
Daiwa Steel Tube Industries Co Ltd
Kawasaki Steel Corp
Maruichi Steel Tube Ltd
Nippon Steel Corp (Shin Nippon Seitetsu)
Nisshin Steel Co Ltd
NKK – Nippon Kokan KK
Sumitomo Metal Industries Ltd

JORDAN
The Jordan Pipes Manufacturing Co Ltd (JPMC)

SOUTH KOREA
Hyundai Pipe Co Ltd
Korea Steel Pipe Co Ltd
Pusan Steel Pipe Corp
Union Steel Mfg Co Ltd

LUXEMBOURG
Arbed SA

MALAYSIA
Amalgamated Industrial Steel Bhd
Choo Bee Metal Industries Sdn Bhd
Hume Industries Sdn Bhd
Maruichi Malaysia Steel Tube Bhd
Steel Pipe Industry of Malaysia Sdn Bhd

MEXICO
Camas y Tubos SA de CV
Hylsa SA de CV
Tubería Laguna SA de CV (Tulsa)
Productora Mexicana de Tubería SA de CV (PMT)
Protumsa – Productos Tubulares Monclova SA
Tubacero SA
Tubería Nacional SA

MOZAMBIQUE
Ima – Indústria Moçambicana do Aço Sarl

NETHERLANDS
P. van Leeuwen Jr's Buizenhandel BV
VBF Buizen BV

NEW ZEALAND
New Zealand Steel Ltd

NIGERIA
Hoesch Pipe Mills (Nigeria) Ltd

NORWAY
A/S Sønnichsen Rørvalseverket

Carbon steel - longitudinal weld tubes and pipes
—continued

POLAND
Huta Ferrum
Huta im Lenina

PORTUGAL
Oliva - Indústrias Metalúrgicas SA

RUMANIA
Uzina Metalurgică Iaşi

SINGAPORE
Bee Huat Industries Pte Ltd
Steel Tubes of Singapore (Pte) Ltd (STS)

SOUTH AFRICA
Steel Pipe Industries (Pty) Ltd (SPI)
Tosa (Division of Dorbyl Ltd)

SPAIN
Perfiles Aragón SA
Arania SA
Grupo José María Aristrain SA
Arregui SA
Talleres Balcells SA (Tabalsa)
Bornay SA
Chinchurreta SL
Condesa - Conducciones y Derivados SA
Industrias Duero SA
Industrias Garaeta
Hijos de Juan de Garay SA
Laminaciones de Lesaca SA (LL)
Pecofrisa - Perfiles Conformados en Frio SA
Perfil en Frio SA (Perfrisa)
Perfiles Rioja SA
Transmesa - Transformaciones Metalurgicas SA
Tubos de Precisión SA (Tupre)
Cooperativa Industrial Talleres Ulma
Metalúrgicas Vallisoletana SA (Meva)

SWEDEN
AB Alvenius Industrier
Wirsbo Bruks AB

SWITZERLAND
Arfa Röhrenwerke AG
Hermann Forster AG, Steel Tube Division
Jansen AG Stahlöhrenwerke
Rheintub Ltd
Rothrist Tube Ltd

SWITZERLAND *—continued*
Tubofer SA

TAIWAN
Kao Hsing Chang Iron & Steel Corp
Mayer Steel Pipe Corp

THAILAND
Thai Steel Pipe Industry Co Ltd
Thai-Asia Steel Pipe Co ltd

TURKEY
Borusan Holding AŞ
Erbosan - Erciyas Boru Sanayii ve Ticaret AŞ
Mannesmann-Sümerbank Boru Endüstrisi TAS (MSBE)
Yücel Boru ve Profil Endustrisi AŞ

UNITED KINGDOM
Armco Ltd
Armstrong Tube Co Ltd
Barton Engineering Ltd
Beam Tubes Ltd
Brasway PLC, Tubes Division
BSC - British Steel Corp
Burn Tubes Ltd
Elmtube Ltd
Fulton (TI) Ltd
Hayes Tubes Ltd
Longmore Brothers, Glynwed Steels Ltd
Monmore Tubes
Natural Gas Tubes Ltd
Newman-Tipper Tubes Ltd
The Phoenix Steel Tube Co Ltd
Premier Tubes Ltd
TI Tube Products Ltd

UNITED STATES
Acme Roll Forming Co
Allied Tube & Conduit Corp, an ATCOR Company
American Tube Co
Armco Inc
Berg Steel Pipe Corp
Central Nebraska Tubing
Copperweld Corp
Cyclops Corp
Fort Worth Pipe Co
Indiana Tube Corp
Jackson Tube Service Inc
James Steel & Tube Co
Laclede Steel Co
Lone Star Steel Co
LTV Steel Co Inc
Maruichi American Corp

Carbon steel – longitudinal weld tubes and pipes
—continued

UNITED STATES *—continued*
Newport Steel Corp
Northwest Pipe & Casing Co
Opelika Welding, Machine & Supply Inc
Pacific Tube Co
Pittsburgh Tube Co
Pixley Tube Corp
Plymouth Tube Co
Quanex Corp
Sharon Tube Co
Stupp Corp
Trent Tube
UNR-Leavitt, Div of UNR Inc
USS, A Division of USX Corp
Welded Tube Co of America
Western Tube & Conduit Corp
Wheatland Tube Co
Wheeling Pittsburgh Steel Corp
Whittymore Tube Co, Div Whittymore Enterprises Inc

USSR
Met Zavod Amurstal (Amur Steel Works)
Dnepropetrovskiy Met i Trubnoprokatny Savod Imeni Lenina (Lenin Iron & Steel Tube Rolling Works)
Novomoskovskiy Zavod (Novomoskovsk Works)
Novosibirskiy Met Zavod Kuzmin
Volzhskiy Truboprokatny Zavod (Volga Pipe Works)
Yuzhnotrubny Zavod (Southern Tube Works)

VENEZUELA
CA Conduven
Imosa – Industria Mecanica Orion SA
Univensa – Unión Industrial Venezolana SA

YUGOSLAVIA
IMK – Organizacija Za Proizvodnju Cévićno Zavarenih cevi i Profila Uroševac
11 Oktomvri Tube Works, Kumanovo
Zeljezara Sisak Integrated Metallurgical Works
Unis (Associated Metal Industry in Sarajevo)

Carbon steel – spiral-weld tubes and pipes

ALGERIA
SNS – Sté Nationale de Sidérurgie

ARGENTINA
Siam SA – Division Siat

BRAZIL
Confab Industrial SA

BULGARIA
Kremikovtsi Iron & Steel Works

CANADA
Ipsco Inc (formerly Interprovincial Steel & Pipe Corp Ltd)
Stelco Inc

CHINA
Baoji Steel Tube Works

CZECHOSLOVAKIA
NHKG – Nová Hut Klementa Gottwalda

FINLAND
Rautaruukki Oy

FRANCE
Usinor Sacilor

EAST GERMANY
VEB Rohrwerke Bitterfeld

WEST GERMANY
Eisen- und Metallwerke Ferndorf GmbH
Hoesch AG
Stahlwerke Peine-Salzgitter AG (P&S)

HUNGARY
Danube Works

INDIA
Sail – Steel Authority of India Ltd

INDONESIA
Krakatau Hoogovens International Pipe Industries Ltd PT

ISRAEL
Middle East Tube Co Ltd

ITALY
Nuova Italsider SpA

JAPAN
Kawasaki Steel Corp
Kawatetsu Steel Tube Co

Carbon steel – spiral-weld tubes and pipes —*continued*

JAPAN —*continued*
Kubota Ltd (Kubota Tekko)
Kurimoto Ltd
Nippon Steel Corp (Shin Nippon Seitetsu)
NKK – Nippon Kokan KK
Sumitomo Metal Industries Ltd
Toa Steel Co Ltd

SOUTH KOREA
Pusan Steel Pipe Corp

KUWAIT
Kuwait Metal Pipe Industries KSC

NEW ZEALAND
Humes SWP

PAKISTAN
Indus Steel Pipes Ltd
Ramna Pipe & General Mills Ltd

RUMANIA
Bucharest Welded Pipe Rolling Mill
Uzina de Tevi Republica (UTR)

SOUTH AFRICA
Steel Pipe Industries (Pty) Ltd (SPI)
Tosa (Division of Dorbyl Ltd)

SPAIN
Helisold Iberica SA

SWEDEN
AB Alvenius Industrier

SWITZERLAND
Rheintub Ltd
Romag – Röhren und Maschinen AG

TURKEY
Mannesmann-Sümerbank Boru Endüstrisi TAS (MSBE)
Ümran Spiral Kaynakli Boru Sanayii AŞ

UNITED KINGDOM
Byard Kenwest Eng Ltd
Motherwell Bridge Pipe
Natural Gas Tubes Ltd

UNITED STATES
Armco Inc
L.B. Foster Co
Lone Star Steel Co
Newport Steel Corp
Northwest Pipe & Casing Co

USSR
Novomoskovskiy Zavod (Novomoskovsk Works)
Zhdanovskiy Met Zavod Ilyich (Ilyich Works)

YUGOSLAVIA
IMK – Organizacija Za Proizvodnju Cévićno Zavarenih cevi i Profila Uroševac

Carbon steel – large-diameter tubes and pipes

ALGERIA
SNS – Sté Nationale de Sidérurgie

ARGENTINA
Siam SA – Division Siat

AUSTRALIA
Tubemakers of Australia Ltd

BRAZIL
Confab Industrial SA
Mannesmann SA

CHINA
Anshan Iron & Steel Co

CZECHOSLOVAKIA
Chomutov Tube Works

FRANCE
Usinor Sacilor

WEST GERMANY
Bergrohr GmbH Herne
Bergrohr GmbH Siegen
Hoesch AG
Eisenbau Krämer mbH
Mannesmannröhren-Werke AG
Stahlwerke Peine-Salzgitter AG (P&S)

INDONESIA
Krakatau Hoogovens International Pipe Industries Ltd PT

ITALY
ATB – Acciaieria e Tubificio di Brescia SpA
Tubificio Dalmine Italsider SpA – TDI
Nuova Italsider SpA

JAPAN
Kawasaki Steel Corp
Nippon Steel Corp (Shin Nippon Seitetsu)
NKK – Nippon Kokan KK

Carbon steel – large-diameter tubes and pipes —*continued*

JAPAN —*continued*
Sumikin Weld Pipe Co Ltd
Sumitomo Metal Industries Ltd

KUWAIT
Kuwait Metal Pipe Industries KSC

MALAYSIA
Hume Industries Sdn Bhd

MEXICO
Productora Mexicana de Tubería SA de CV (PMT)
Tubacero SA

SOUTH AFRICA
Tosa (Division of Dorbyl Ltd)

UNITED KINGDOM
BSC – British Steel Corp

UNITED STATES
Berg Steel Pipe Corp
Bethlehem Steel Corp
L.B. Foster Co
Lone Star Steel Co
LTV Steel Co Inc
Northwest Pipe & Casing Co
Plymouth Tube Co

USSR
Volzhskiy Truboprokatny Zavod (Volga Pipe Works)

VENEZUELA
Sidor – CVG Siderúrgica del Orinoco CA

YUGOSLAVIA
IMK – Organizacija Za Proizvodnju Cévićno Zavarenih cevi i Profila Uroševac

Carbon steel – oil country tubular goods

ARGENTINA
Siderca Saic

AUSTRALIA
Australian Tube Mills Pty Ltd (ATM)

AUSTRIA
Voest-Alpine AG

BRAZIL
Confab Industrial SA
Fornasa SA
Mannesmann SA
Persico Pizzamiglio SA

CANADA
The Algoma Steel Corp Ltd
Sonco Steel Tube, Division of Ferrum Inc
Stelco Inc

FRANCE
Usinor Sacilor
Vallourec Industries

WEST GERMANY
Benteler AG
Hoesch AG
Mannesmannröhren-Werke AG
TPS-Technitube Röhrenwerke GmbH

INDIA
The Tata Iron & Steel Co Ltd (Tisco)

INDONESIA
PT Bakrie & Bros

ISRAEL
Middle East Tube Co Ltd

ITALY
Dalmine SpA
General Sider Europa SpA
Seta Tubi SpA
Tubimar Ancona SpA

JAPAN
Maruichi Steel Tube Ltd
Nippon Steel Corp (Shin Nippon Seitetsu)
NKK – Nippon Kokan KK
Sumitomo Metal Industries Ltd

SOUTH KOREA
Pusan Steel Pipe Corp
Union Steel Mfg Co Ltd

MEXICO
Hylsa SA de CV
Tamsa – Tubos de Acero de Mexico SA
Tubacero SA

RUMANIA
Uzina de Tevi Republica (UTR)

SOUTH AFRICA
Tosa (Division of Dorbyl Ltd)

Carbon steel – oil country tubular goods —*continued*

SPAIN
Babcock & Wilcox Española SA
Tubular Products Division
Tubos Reunidos SA

SWEDEN
Wirsbo Bruks AB

TURKEY
Mannesmann-Sümerbank Boru Endüstrisi TAS (MSBE)
MKEK Çelik Çekme Boru Fabrikasi

UNITED KINGDOM
BSC – British Steel Corp
TI Desford Tubes Ltd

UNITED STATES
Armco Inc
CF&I Steel Corp
Cyclops Corp
Fort Worth Pipe Co
Lone Star Steel Co
LTV Steel Co Inc
Newport Steel Corp
North Star Steel Co (NSS)
Pittsburgh Tube Co
Quanex Corp
The Timken Co
UNR-Leavitt, Div of UNR Inc
USS, A Division of USX Corp

USSR
Novomoskovskiy Zavod (Novomoskovsk Works)
Taganrogskiy Metallurgicheskiy Zavod (Taganrog Iron & Steel Works, formerly Andreyevskiy Works)
Yuzhnotrubny Zavod (Southern Tube Works)

VENEZUELA
CA Conduven
Sidor – CVG Siderúrgica del Orinoco CA

YUGOSLAVIA
IMK – Organizacija Za Proizvodnju Cévićno Zavarenih cevi i Profila Uroševac
Zeljezara Sisak Integrated Metallurgical Works

Carbon steel – galvanized tubes and pipe

ALGERIA
SNS – Sté Nationale de Sidérurgie

ARGENTINA
Acindar Industria Argentina de Aceros SA
Armco Argentina SA

AUSTRALIA
Tubemakers of Australia Ltd

BELGIUM
Tubel SA

BRAZIL
Apolo Produtos de Aço SA
Fornasa SA
Mannesmann SA
Persico Pizzamiglio SA

BULGARIA
Kremikovtsi Iron & Steel Works

CANADA
Atlas Tube
Barton Tubes Ltd
Sidbec-Dosco Inc
Sonco Steel Tube, Division of Ferrum Inc
Standard Tube Canada, Division of TI Canada Inc

CHILE
Cintac – Cĭa Industrial de Tubos de Acero SA

CHINA
Hengyang Steel Tube Plant

COLOMBIA
Colmena – Consorcio Metalúrgico Nacional SA
Simesa – Siderúrgica de Medellín

CZECHOSLOVAKIA
Veselinad Moravou Works

DOMINICAN REPUBLIC
Metaldom – Complejo Metalúrgico Dominicano C por A

FINLAND
Rautaruukki Oy

FRANCE
Produits d'Acier SA
Usinor Sacilor

Carbon steel – galvanized tubes and pipe *—continued*

WEST GERMANY
Benteler AG
Mannesmannröhren-Werke AG
Saarstahl Völklingen GmbH

GREECE
Metallurgiki Athinon SA

INDIA
Apollo Tubes Ltd
BST Manufacturing Ltd
Gujarat Steel Tubes Ltd (GST)
Jain Tube Co Ltd
Jotindra Steel & Tubes Ltd
Khandewal Tubes, Division of Khandelwal Ferro Alloys Ltd
Metalman Pipe Mfg Co Ltd

INDONESIA
Bakrie Pipe Industries PT
PT Pabrik Pipa Indonesia (The Indonesia Tube Mill Corp Ltd)

ISRAEL
Middle East Tube Co Ltd

ITALY
Arvedi Group
Metallurgica G. Berera SpA
Emiliana Tubi – Profilati Acciaio SpA (Setpa)
General Sider Europa SpA
General Sider Italiana SpA
Acciaierie e Tubificio Meridionali SpA
Neirotti Tubi SpA
Pietra SpA – Acciaierie Ferriere e Tubifici
Profilnastro SpA
Tubimar Ancona SpA
Valducci SpA

JAPAN
Daiwa Steel Tube Industries Co Ltd
Maruichi Steel Tube Ltd
Nisshin Steel Co Ltd
NKK – Nippon Kokan KK
Sumitomo Metal Industries Ltd

JORDAN
The Jordan Pipes Manufacturing Co Ltd (JPMC)

SOUTH KOREA
Hyundai Pipe Co Ltd
Pusan Steel Pipe Corp

SOUTH KOREA *—continued*
Union Steel Mfg Co Ltd

KUWAIT
Kuwait Metal Pipe Industries KSC

LUXEMBOURG
Arbed SA

MALAYSIA
Maruichi Malaysia Steel Tube Bhd
Steel Pipe Industry of Malaysia Sdn Bhd

MEXICO
Camas y Tubos SA de CV
Hylsa SA de CV
Tubería Nacional SA

NETHERLANDS
VBF Buizen BV

NEW ZEALAND
New Zealand Steel Ltd

NIGERIA
Hoesch Pipe Mills (Nigeria) Ltd

NORWAY
A/S Sønnichsen Rørvalseverket

PAKISTAN
Ramna Pipe & General Mills Ltd

PORTUGAL
Oliva – Indústrias Metalúrgicas SA

RUMANIA
Uzina Metalurgică Iaşi

SINGAPORE
Bee Huat Industries Pte Ltd

SOUTH AFRICA
Tosa (Division of Dorbyl Ltd)

SPAIN
Grupo José María Aristrain SA
Talleres Balcells SA (Tabalsa)
Laminaciones de Lesaca SA (LL)

SWITZERLAND
Rheintub Ltd
Romag – Röhren und Maschinen AG

TAIWAN
Kao Hsing Chang Iron & Steel Corp
Mayer Steel Pipe Corp

TANZANIA
Aluminium Africa Ltd (Alaf)

Carbon steel – galvanized tubes and pipe *—continued*

THAILAND
Thai Steel Pipe Industry Co Ltd
Thai-Asia Steel Pipe Co ltd

TURKEY
Borusan Holding AŞ

UNITED KINGDOM
Burn Tubes Ltd
Longmore Brothers, Glynwed Steels Ltd
Monmore Tubes
Newman-Tipper Tubes Ltd
TI Tube Products Ltd

UNITED STATES
Allied Tube & Conduit Corp, an ATCOR Company
American Tube Co
Cyclops Corp
Laclede Steel Co
LTV Steel Co Inc
Northwest Pipe & Casing Co
Sharon Tube Co
UNR-Leavitt, Div of UNR Inc
Wheatland Tube Co
Whittymore Tube Co, Div Whittymore Enterprises Inc

VENEZUELA
CA Conduven
Univensa – Unión Industrial Venezolana SA

YUGOSLAVIA
Zeljezara Sisak Integrated Metallurgical Works

Carbon steel – plastic coated tubes and pipes

AUSTRALIA
Tubemakers of Australia Ltd

WEST GERMANY
Benteler AG
Bergrohr GmbH Herne
Bergrohr GmbH Siegen
Hoesch AG
Eisenbau Krämer mbH

ITALY
General Sider Europa SpA
General Sider Italiana SpA

JAPAN
NKK – Nippon Kokan KK
Sumitomo Metal Industries Ltd

SOUTH KOREA
Hyundai Pipe Co Ltd

NETHERLANDS
VBF Buizen BV

NEW ZEALAND
Humes SWP

SWEDEN
Wirsbo Bruks AB

UNITED KINGDOM
Burn Tubes Ltd
Monmore Tubes
Newman-Tipper Tubes Ltd

UNITED STATES
American Tube Co

Carbon steel – cold drawn tubes and pipes

AUSTRALIA
Email Tube Division, Email Ltd
Martin Bright Steels Ltd
Tubemakers of Australia Ltd

AUSTRIA
Voest-Alpine AG

BRAZIL
Fornasa SA
Mannesmann SA
Persico Pizzamiglio SA
Laminação Santa Maria SA – Indústria e Comércio

CANADA
Associated Tube Industries Ltd (ATI)
Standard Tube Canada, Division of TI Canada Inc

CHINA
Changzhi Iron & Steel Works
Guiyang Steel Mill
Hengyang Steel Tube Plant
Jiangxi Steel Works
Shanghai Iron & Steel Works

FRANCE
Tubes de Chevillon
Vallourec Industries

Carbon steel – cold drawn tubes and pipes *—continued*

WEST GERMANY
Benteler AG
Mannesmannröhren-Werke AG
TPS-Technitube Röhrenwerke GmbH

INDIA
Steel Tubes of India Ltd
The Tata Iron & Steel Co Ltd (Tisco)
Tube Products of India (TPI)

ITALY
Dalmine SpA
Acciaierie e Ferriere Lombarde Falck SpA
Acciaierie e Tubificio Meridionali SpA
Pietra SpA – Acciaierie Ferriere e Tubifici
Cesare Rizzato & C SpA
Tubicar SpA

JAPAN
Kobe Steel Ltd (Kobe Seikosho)
Nisshin Steel Co Ltd
NKK – Nippon Kokan KK
Sanwa Metal Industries Ltd
Sumitomo Metal Industries Ltd

SOUTH KOREA
Sammi Steel Co Ltd

MEXICO
Precitubo SA de CV
Tamsa – Tubos de Acero de Mexico SA

RUMANIA
Uzina de Tevi Republica (UTR)
Uzina Metalurgică Iaşi

SOUTH AFRICA
Tosa (Division of Dorbyl Ltd)

SPAIN
Hijos de Juan de Garay SA
Laminaciones de Lesaca SA (LL)
Transmesa – Transformaciones Metalurgicas SA

SWEDEN
Structo DOM Europe AB

SWITZERLAND
Hermann Forster AG, Steel Tube Division
Jansen AG Stahlöhrenwerke
Rothrist Tube Ltd

UNITED KINGDOM
Armstrong Tube Co Ltd
Cold Drawn Tubes Ltd
The Phoenix Steel Tube Co Ltd
Seamless Tubes Ltd
TI Tube Products Ltd

UNITED STATES
Copperweld Corp
Cyclops Corp
Lone Star Steel Co
LTV Steel Co Inc
Pacific Tube Co
Pittsburgh Tube Co
Plymouth Tube Co
Quanex Corp
Sharon Tube Co
The Timken Co
Trent Tube
Whittymore Tube Co, Div Whittymore Enterprises Inc

VENEZUELA
CA Conduven

YUGOSLAVIA
Zeljezara Sisak Integrated Metallurgical Works
Unis (Associated Metal Industry in Sarajevo)

Carbon steel – hollow bars

BRAZIL
Mannesmann SA

WEST GERMANY
Benteler AG

JAPAN
Sumitomo Metal Industries Ltd

NEW ZEALAND
Centralloy Industries – Division of Stainless Castings Ltd

SOUTH AFRICA
Tosa (Division of Dorbyl Ltd)

SPAIN
Babcock & Wilcox Española SA Tubular Products Division
Tubacex – C.E. de Tubos por Extrusion SA
Tubos Reunidos SA

UNITED KINGDOM
Bedford Steels

Carbon steel – hollow bars —*continued*

UNITED KINGDOM —*continued*
TI Desford Tubes Ltd

UNITED STATES
Lone Star Steel Co
LTV Steel Co Inc
The Timken Co

Carbon steel – precision tubes and pipes

AUSTRALIA
Palmer Tube Mills Ltd
Tubemakers of Australia Ltd

AUSTRIA
Voest-Alpine AG

BELGIUM
Tubel SA

BRAZIL
Fornasa SA
Mannesmann SA
Persico Pizzamiglio SA

CZECHOSLOVAKIA
Chomutov Tube Works

DENMARK
Nordisk Simplex AS

FRANCE
Vallourec Industries

EAST GERMANY
VEB Rohr- und Kaltwalzwerk Karl-Marx-Stadt

WEST GERMANY
Benteler AG
Mannesmannröhren-Werke AG
Mecano-Bundy GmbH
TPS-Technitube Röhrenwerke GmbH

INDIA
BST Manufacturing Ltd
Gujarat Steel Tubes Ltd (GST)
Metalman Pipe Mfg Co Ltd
The Tata Iron & Steel Co Ltd (Tisco)

JAPAN
Sumitomo Metal Industries Ltd

NETHERLANDS
Robur Buizenfabriek BV

NETHERLANDS —*continued*
VBF Buizen BV

POLAND
Huta Swierczewski

SOUTH AFRICA
Tosa (Division of Dorbyl Ltd)

SPAIN
Hijos de Juan de Garay SA
Perfil en Frio SA (Perfrisa)
Transmesa – Transformaciones Metalurgicas SA
Tubacex – C.E. de Tubos por Extrusion SA
Tubos Reunidos SA

SWITZERLAND
Jansen AG Stahlöhrenwerke
Rothrist Tube Ltd
Tubofer SA

TURKEY
Mannesmann-Sümerbank Boru Endüstrisi TAS (MSBE)

UNITED KINGDOM
Barton Engineering Ltd
Burn Tubes Ltd
Longmore Brothers, Glynwed Steels Ltd
Newman-Tipper Tubes Ltd
The Phoenix Steel Tube Co Ltd
TI Tube Products Ltd

UNITED STATES
Indiana Tube Corp
Lone Star Steel Co
Pacific Tube Co
Plymouth Tube Co
The Timken Co
Whittymore Tube Co, Div Whittymore Enterprises Inc

YUGOSLAVIA
Zeljezara Sisak Integrated Metallurgical Works

Carbon steel – hollow sections

AUSTRALIA
Australian Tube Mills Pty Ltd (ATM)
Palmer Tube Mills Ltd
Tubemakers of Australia Ltd

Carbon steel - hollow sections —*continued*

BELGIUM
SA Laminoirs de Longtain
Tubel SA

CANADA
Atlas Tube
Sonco Steel Tube, Division of Ferrum Inc
Standard Tube Canada, Division of TI Canada Inc

CHINA
Jiangxi Steel Works

COLOMBIA
Simesa – Siderúrgica de Medellín

FRANCE
Exma SA
Usinor Sacilor
Vallourec Industries

WEST GERMANY
Hoesch AG
Klöckner Stahl GmbH

INDIA
Metalman Pipe Mfg Co Ltd
The Tata Iron & Steel Co Ltd (Tisco)

INDONESIA
PT Inastu – Indonesian National Steel Tube Co

ITALY
Profilnastro SpA

JAPAN
Kubota Ltd (Kubota Tekko)
Maruichi Steel Tube Ltd
Nippon Steel Metal Products Co Ltd (Nittetsu Kenzai Kogyo KK)
Sumitomo Metal Industries Ltd

JORDAN
The Jordan Pipes Manufacturing Co Ltd (JPMC)

SOUTH KOREA
Hyundai Pipe Co Ltd
Korea Steel Pipe Co Ltd
Pusan Steel Pipe Corp

MALAYSIA
Amalgamated Industrial Steel Bhd
Maruichi Malaysia Steel Tube Bhd

MEXICO
Hylsa SA de CV

NETHERLANDS
P. van Leeuwen Jr's Buizenhandel BV
VBF Buizen BV

NEW ZEALAND
New Zealand Steel Ltd

NIGERIA
Hoesch Pipe Mills (Nigeria) Ltd

NORWAY
A/S Sønnichsen Rørvalseverket

SINGAPORE
Bee Huat Industries Pte Ltd
Steel Tubes of Singapore (Pte) Ltd (STS)

SOUTH AFRICA
Tosa (Division of Dorbyl Ltd)

SPAIN
Hijos de Juan de Garay SA
Laminaciones de Lesaca SA (LL)

THAILAND
Thai-Asia Steel Pipe Co ltd

TRINIDAD & TOBAGO
Iscott – Iron & Steel Co of Trinidad & Tobago Ltd

TURKEY
Borusan Holding AŞ

UNITED KINGDOM
Bedford Steels
BSC – British Steel Corp
TI Tube Products Ltd

UNITED STATES
Allied Tube & Conduit Corp, an ATCOR Company
Bock Ind Inc
Independence Tube Corp
James Steel & Tube Co
Lone Star Steel Co
MP Metal Products Inc
UNR-Leavitt, Div of UNR Inc
Welded Tube Co of America

VENEZUELA
CA Conduven

YUGOSLAVIA
11 Oktomvri Tube Works, Kumanovo

Stainless steel – ingots

ARGENTINA
Establecimiento Altos Hornos Zapla

AUSTRALIA
The Broken Hill Pty Co Ltd
Commonwealth Steel Co Ltd

BELGIUM
ALZ NV
SA Fabrique de Fer de Charleroi

BRAZIL
Acesita – Cia Aços Especiais Itabira
Aços Anhanguera SA (Acoansa)
Eletrometal SA – Metais Especiais

BULGARIA
Kremikovtsi Iron & Steel Works

CANADA
Atlas Specialty Steels Division of Rio Algom Ltd
Slater Steels, Sorel Forge Division

CZECHOSLOVAKIA
Vitkovice Steelworks

FRANCE
Usinor Sacilor

WEST GERMANY
Böhler AG, Edelstahlwerke
Dörrenberg Edelstahl GmbH
Krupp Stahl AG (KS)
Saarstahl Völklingen GmbH

INDIA
Bihar Alloy Steels Ltd
Ellora Steels Ltd (ESL)
Kalyani Steels Ltd
Man Industrial Corp Ltd
Punjab Con-Cast Steels Ltd
Sail – Steel Authority of India Ltd

ITALY
Acciaieria Foroni SpA

JAPAN
The Japan Steel Works Ltd
Kobe Steel Ltd (Kobe Seikosho)
Nippon Stainless Steel Co Ltd
Nisshin Steel Co Ltd
NKK – Nippon Kokan KK
Pacific Metals Co Ltd (Taiheiyo Kinzoku)
Sumitomo Metal Industries Ltd

SOUTH KOREA
Sammi Steel Co Ltd

MEXICO
Industrias CH SA (formerly Campos Hermanos SA)

POLAND
Huta Baildon

SPAIN
Acerinox SA
Forjas Alavesas SA (Fasa)
SA Echevarría (SAE)
Olarra SA (Olsa)
Roldan SA

SWEDEN
Fagersta Stainless AB

TURKEY
MKEK Celik Fabrikasi

UNITED KINGDOM
BSC – British Steel Corp
C.G. Carlisle & Co Ltd
Howmet Turbine (UK) Inc, Howmet Alloys International
Sheffield Forgemasters Ltd
Wiggin Steel & Alloys

UNITED STATES
Armco Inc
Cyclops Corp
Eastern Stainless Steel Co
Electralloy Corp
Ellwood Uddeholm Steel Corp
Jessop Steel Co
National Forge Co
Slater Steels, Fort Wayne Specialty Alloys Division
The Timken Co

Stainless steel – slabs

AUSTRALIA
The Broken Hill Pty Co Ltd

BELGIUM
ALZ NV
SA Fabrique de Fer de Charleroi

BRAZIL
Acesita – Cia Aços Especiais Itabira

BULGARIA
Kremikovtsi Iron & Steel Works

Stainless steel - slabs
—continued

CANADA
Atlas Specialty Steels Division of Rio Algom Ltd
Slater Steels, Sorel Forge Division

CZECHOSLOVAKIA
Vitkovice Steelworks

FINLAND
Outokumpu Oy

FRANCE
Usinor Sacilor

WEST GERMANY
Krupp Stahl AG (KS)
Thyssen Edelstahlwerke AG

INDIA
Mahindra Ugine Steel Co Ltd (Muscosteel)

ITALY
Terni Acciai Speciali SpA

JAPAN
Nippon Stainless Steel Co Ltd
Nippon Yakin Kogyo Co Ltd
Nisshin Steel Co Ltd
NKK – Nippon Kokan KK
Pacific Metals Co Ltd (Taiheiyo Kinzoku)
Sumitomo Metal Industries Ltd

SOUTH AFRICA
Middleburg Steel & Alloys (Pty) Ltd – Steel Division

SPAIN
Acerinox SA

SWEDEN
Avesta AB

TAIWAN
Tang Eng Iron Works Co Ltd

TURKEY
MKEK Celik Fabrikasi

UNITED KINGDOM
BSC – British Steel Corp
C.G. Carlisle & Co Ltd
Sheffield Forgemasters Ltd
UES – United Engineering Steels Ltd

UNITED STATES
Carpenter Technology Corp
Cyclops Corp
Eastern Stainless Steel Co
J & L Specialty Products Corp
Jessop Steel Co
National Forge Co
Washington Steel, Division of Blount Inc

Stainless steel - blooms

AUSTRALIA
The Broken Hill Pty Co Ltd
Commonwealth Steel Co Ltd

BRAZIL
Acesita – Cia Aços Especiais Itabira
Aços Anhanguera SA (Acoansa)
Eletrometal SA – Metais Especiais

BULGARIA
Kremikovtsi Iron & Steel Works

CANADA
Atlas Specialty Steels Division of Rio Algom Ltd
Slater Steels, Sorel Forge Division

FRANCE
Usinor Sacilor

EAST GERMANY
VEB Edelstahlwerk 8 Mai 1945

WEST GERMANY
Krupp Stahl AG (KS)
Thyssen Edelstahlwerke AG

INDIA
Bhoruka Steel Ltd
Bihar Alloy Steels Ltd
Ferro Alloys Corp Ltd (Facor) (Steel division), (Vidarbha Iron & Steel Corp Ltd)
Mahindra Ugine Steel Co Ltd (Muscosteel)
Panchmahal Steel Ltd

ITALY
DeltaCogne SpA

JAPAN
Nippon Stainless Steel Co Ltd
Nippon Yakin Kogyo Co Ltd
NKK – Nippon Kokan KK
Pacific Metals Co Ltd (Taiheiyo Kinzoku)

Stainless steel - blooms *—continued*

JAPAN *—continued*
Sumitomo Metal Industries Ltd

MEXICO
Industrias CH SA (formerly Campos Hermanos SA)

POLAND
Poldi Steelworks

SPAIN
Forjas Alavesas SA (Fasa)
Aceros de Llodio SA (Allosa)

SWEDEN
AB Sandvik Steel

TURKEY
MKEK Celik Fabrikasi

UNITED KINGDOM
C.G. Carlisle & Co Ltd
Sheffield Forgemasters Ltd
UES – United Engineering Steels Ltd

UNITED STATES
Carpenter Technology Corp
Cyclops Corp
Ellwood Uddeholm Steel Corp
National Forge Co
The Timken Co

USSR
Uzbekskiy Metallurgicheskiy Zavod (Uzbek Iron & Steel Works)

Stainless steel - billets

ARGENTINA
Establecimiento Altos Hornos Zapla

AUSTRALIA
The Broken Hill Pty Co Ltd
Commonwealth Steel Co Ltd

BRAZIL
Acesita – Cia Aços Especiais Itabira
Aços Anhanguera SA (Acoansa)
Siderúrgica Nossa Senhora Aparecida SA
Eletrometal SA – Metais Especiais
Aços Finos Piratini SA
Villares Group

BULGARIA
Kremikovtsi Iron & Steel Works

CANADA
Atlas Specialty Steels Division of Rio Algom Ltd
Slater Steels, Sorel Forge Division

CZECHOSLOVAKIA
Vitkovice Steelworks

FRANCE
Aubert et Duval
Usinor Sacilor

WEST GERMANY
Edelstahlwerke Buderus AG
Krupp Stahl AG (KS)
Schmidt & Clemens GmbH & Co
Thyssen Edelstahlwerke AG

HUNGARY
Diosgyör Works (Lenin Metallurgical Works)

INDIA
Bhoruka Steel Ltd
Bihar Alloy Steels Ltd
Kalyani Steels Ltd
Mahindra Ugine Steel Co Ltd (Muscosteel)
Mukand Iron & Steel Works Ltd
Panchmahal Steel Ltd
Sail – Steel Authority of India Ltd
Singh Electrosteel Ltd (SES)
Star Steel Pvt Ltd (Starsteel)
Usha Alloys & Steels Ltd (UASL)

ITALY
Acciaierie di Bolzano SpA
DeltaCogne SpA
Acciaieria Foroni SpA
Safas – Stà Azionaria Fonderia Acciai Speciali SpA
Seta Acciai SpA

JAPAN
Kobe Steel Ltd (Kobe Seikosho)
NKK – Nippon Kokan KK
Pacific Metals Co Ltd (Taiheiyo Kinzoku)

SOUTH KOREA
Sammi Steel Co Ltd

MEXICO
Industrias CH SA (formerly Campos Hermanos SA)

POLAND
Huta Baildon
Poldi Steelworks

Stainless steel – billets

—continued

SPAIN
Acerinox SA
Forjas Alavesas SA (Fasa)
SA Echevarría (SAE)
Patricio Echeverría SA
Aceros de Irura SA
Aceros de Llodio SA (Allosa)
Olarra SA (Olsa)
Roldan SA

SWEDEN
Avesta AB
Fagersta Stainless AB
AB Sandvik Steel

TURKEY
MKEK Celik Fabrikasi

UNITED KINGDOM
C.G. Carlisle & Co Ltd
Sheffield Forgemasters Ltd
UES – United Engineering Steels Ltd

UNITED STATES
Carpenter Technology Corp
Cyclops Corp
Electralloy Corp
Jessop Steel Co
Earle M. Jorgensen Co, Forge Division
LTV Steel Co Inc
National Forge Co
Slater Steels, Fort Wayne Specialty Alloys Division
The Timken Co

Stainless steel – semis for seamless tubemaking

CZECHOSLOVAKIA
Vitkovice Steelworks

WEST GERMANY
Thyssen Edelstahlwerke AG

ITALY
Acciaieria Foroni SpA
Seta Acciai SpA

JAPAN
NKK – Nippon Kokan KK

SWEDEN
Avesta AB
Avesta Nyby Powder AB

UNITED KINGDOM
C.G. Carlisle & Co Ltd

UNITED STATES
The Timken Co

USSR
Uzbekskiy Metallurgicheskiy Zavod (Uzbek Iron & Steel Works)

Stainless steel – hot rolled band (for re-rolling)

BELGIUM
ALZ NV

CANADA
Atlas Specialty Steels Division of Rio Algom Ltd

FINLAND
Outokumpu Oy

FRANCE
Usinor Sacilor

ITALY
Terni Acciai Speciali SpA

JAPAN
Nippon Steel Corp (Shin Nippon Seitetsu)
Nisshin Steel Co Ltd

SPAIN
Acerinox SA

SWEDEN
Fagersta Stainless AB
Surahammars Bruks AB

UNITED KINGDOM
BSC – British Steel Corp

UNITED STATES
Allegheny Ludlum Corp
Cyclops Corp
J & L Specialty Products Corp
Washington Steel, Division of Blount Inc

Stainless steel – wire rod

AUSTRIA
Joh. Pengg, Draht- & Walzwerke

Stainless steel – wire rod *—continued*

BRAZIL
Siderúrgica Nossa Senhora Aparecida SA
Eletrometal SA – Metais Especiais
Aços Finos Piratini SA
Villares Group

CHINA
Changcheng Steel Works
Guiyang Steel Mill

FRANCE
Usinor Sacilor

WEST GERMANY
Saarstahl Völklingen GmbH
Thyssen Edelstahlwerke AG

INDIA
Bihar Alloy Steels Ltd
Man Industrial Corp Ltd
Mukand Iron & Steel Works Ltd

ITALY
Acciaierie di Bolzano SpA
DeltaCogne SpA

JAPAN
Aichi Steel Works Ltd
Kobe Steel Ltd (Kobe Seikosho)
Nippon Steel Corp (Shin Nippon Seitetsu)
Pacific Metals Co Ltd (Taiheiyo Kinzoku)
Sanyo Special Steel Co Ltd
Sumitomo Electric Industries Ltd (Sumitomo Denki Kogyo)

SOUTH KOREA
Sammi Steel Co Ltd

MEXICO
Acero Solar SA de CV

SPAIN
Forjas Alavesas SA (Fasa)
SA Echevarría (SAE)
Inoxca – Inoxidables de Catalunya SA
Roldan SA

SWEDEN
Fagersta Stainless AB

UNITED KINGDOM
BSC – British Steel Corp
Sanderson Kayser Ltd

UNITED KINGDOM *—continued*
The Templeborough Rolling Mills Ltd (TRM)

UNITED STATES
Al Tech Specialty Steel Corp
Armco Inc
Cyclops Corp
Techalloy Co Inc
The Timken Co

YUGOSLAVIA
Boris Kidrič Steelworks

Stainless steel – round bars

ALBANIA
Steel of the Party Metallurgical Combine

ARGENTINA
Acindar Industria Argentina de Aceros SA

AUSTRALIA
Commonwealth Steel Co Ltd

AUSTRIA
Vereinigte Edelstahlwerke AG

BANGLADESH
Panther Steel Ltd

BRAZIL
Acesita – Cia Aços Especiais Itabira
Aços Anhanguera SA (Acoansa)
Siderúrgica Nossa Senhora Aparecida SA
Eletrometal SA – Metais Especiais
Aços Finos Piratini SA
Villares Group

CANADA
Atlas Specialty Steels Division of Rio Algom Ltd
Slater Steels, Sorel Forge Division

CHINA
Changcheng Steel Works
Guiyang Steel Mill
Shaanxi Precision Alloy Plant

FRANCE
Aubert et Duval
Aciéries de Bonpertuis
Aciéries et Laminoirs du Saut du Tarn (ALST)
Usinor Sacilor

Stainless steel – round bars
—continued

EAST GERMANY
VEB Edelstahlwerk 8 Mai 1945

WEST GERMANY
Böhler AG, Edelstahlwerke
Remscheider Walz- und Hammerwerke, Böllinghaus & Co
Edelstahlwerke Buderus AG
Dörrenberg Edelstahl GmbH
Krupp Stahl AG (KS)
Saarstahl Völklingen GmbH
Schmidt & Clemens GmbH & Co
Thyssen Edelstahlwerke AG

HUNGARY
Diosgyör Works (Lenin Metallurgical Works)

INDIA
Bihar Alloy Steels Ltd
Ellora Steels Ltd (ESL)
Ferro Alloys Corp Ltd (Facor) (Steel division), (Vidarbha Iron & Steel Corp Ltd)
Guest Keen Williams Ltd (GKW)
Kalyani Steels Ltd
Mahindra Ugine Steel Co Ltd (Muscosteel)
Mukand Iron & Steel Works Ltd
Panchmahal Steel Ltd
Punjab Con-Cast Steels Ltd
Sail – Steel Authority of India Ltd
Usha Alloys & Steels Ltd (UASL)

ITALY
Acciaierie di Bolzano SpA
DeltaCogne SpA
Acciaieria Foroni SpA
Safas – Stà Azionaria Fonderia Acciai Speciali SpA
Acciaierie Valbruna SpA

JAPAN
Aichi Steel Works Ltd
Daido Steel Co Ltd
Hitachi Metals Ltd (Hitachi Kinzoku)
Nippon Stainless Steel Co Ltd
Nippon Yakin Kogyo Co Ltd
Pacific Metals Co Ltd (Taiheiyo Kinzoku)
Sanyo Special Steel Co Ltd
Tohoku Special Steel Works Ltd (Tohoku Tokushuko)

SOUTH KOREA
Sammi Steel Co Ltd

MEXICO
Industrias CH SA (formerly Campos Hermanos SA)
Acero Solar SA de CV

SPAIN
Forjas Alavesas SA (Fasa)
SA Echevarría (SAE)
Patricio Echeverría SA
Inoxca – Inoxidables de Catalunya SA
Aceros de Irura SA
Aceros de Llodio SA (Allosa)
Olarra SA (Olsa)
Forjas y Aceros de Reinosa SA
Roldan SA

SWEDEN
Avesta AB
Avesta Nyby Powder AB
Avesta Uddeholm Stainless Bar AB
AB Sandvik Steel

TURKEY
MKEK Celik Fabrikasi

UNITED KINGDOM
Bedford Steels
C.G. Carlisle & Co Ltd
Dudley Port Rolling Mills, Glynwed Steels Ltd
George Gadd & Co, Glynwed Steels Ltd
Joseph Gillott & Sons, Glynwed Steels Ltd
Hemmings Ltd
Howmet Turbine (UK) Inc, Howmet Alloys International
Kiveton Park Steel & Wire Works Ltd
Frank Pickering & Co Ltd
Sanderson Kayser Ltd
Sheffield Forgemasters Ltd
Spartan Sheffield Ltd
Spencer Clark Metal Industries PLC
Welded Stainless Steel Tubes Ltd

UNITED STATES
Al Tech Specialty Steel Corp
Cyclops Corp
Electralloy Corp
LTV Steel Co Inc
Nelsen Steel & Wire Co Inc
Slater Steels, Fort Wayne Specialty Alloys Division
Techalloy Co Inc
The Timken Co

Stainless steel - round bars *—continued*

USSR
Met Zavod Kuybyshev (Kuibyshev Iron & Steel Works)
Uzbekskiy Metallurgicheskiy Zavod (Uzbek Iron & Steel Works)

VENEZUELA
Acerex CA

YUGOSLAVIA
Boris Kidrič Steelworks
Slovenske Železarne - Iron & Steel Works of Slovenia

Stainless steel - square bars

ALBANIA
Steel of the Party Metallurgical Combine

ARGENTINA
Acindar Industria Argentina de Aceros SA

AUSTRIA
Vereinigte Edelstahlwerke AG

BRAZIL
Acesita - Cia Aços Especiais Itabira
Aços Anhanguera SA (Acoansa)
Siderúrgica Nossa Senhora Aparecida SA
Eletrometal SA - Metais Especiais
Aços Finos Piratini SA
Villares Group

CANADA
Atlas Specialty Steels Division of Rio Algom Ltd
Slater Steels, Sorel Forge Division

CHINA
Guiyang Steel Mill

FRANCE
Aubert et Duval
Aciéries de Bonpertuis
Aciéries et Laminoirs du Saut du Tarn (ALST)
Usinor Sacilor

WEST GERMANY
Böhler AG, Edelstahlwerke

WEST GERMANY *—continued*
Remscheider Walz- und Hammerwerke, Böllinghaus & Co
Edelstahlwerke Buderus AG
Krupp Stahl AG (KS)
Saarstahl Völklingen GmbH
Schmidt & Clemens GmbH & Co
Thyssen Edelstahlwerke AG

INDIA
Bihar Alloy Steels Ltd
Ellora Steels Ltd (ESL)
Ferro Alloys Corp Ltd (Facor) (Steel division), (Vidarbha Iron & Steel Corp Ltd)
Guest Keen Williams Ltd (GKW)
Kalyani Steels Ltd
Mahindra Ugine Steel Co Ltd (Muscosteel)
Panchmahal Steel Ltd
Punjab Con-Cast Steels Ltd
Sail - Steel Authority of India Ltd

ITALY
DeltaCogne SpA
Safas - Stà Azionaria Fonderia Acciai Speciali SpA
Acciaierie Valbruna SpA

JAPAN
Daido Steel Co Ltd
Hitachi Metals Ltd (Hitachi Kinzoku)
Nippon Stainless Steel Co Ltd
Nippon Yakin Kogyo Co Ltd

SOUTH KOREA
Sammi Steel Co Ltd

MEXICO
Acero Solar SA de CV

SPAIN
SA Echevarría (SAE)
Inoxca - Inoxidables de Catalunya SA
Aceros de Llodio SA (Allosa)
Olarra SA (Olsa)
Forjas y Aceros de Reinosa SA

SWEDEN
Avesta AB
Avesta Uddeholm Stainless Bar AB

TURKEY
MKEK Celik Fabrikasi

UNITED KINGDOM
Bedford Steels
C.G. Carlisle & Co Ltd

Stainless steel - square bars *—continued*

UNITED KINGDOM *—continued*
Dudley Port Rolling Mills, Glynwed Steels Ltd
George Gadd & Co, Glynwed Steels Ltd
Hemmings Ltd
Sheffield Forgemasters Ltd
Spartan Sheffield Ltd
Spencer Clark Metal Industries PLC

UNITED STATES
Al Tech Specialty Steel Corp
Cyclops Corp
LTV Steel Co Inc
Nelsen Steel & Wire Co Inc
Slater Steels, Fort Wayne Specialty Alloys Division
Techalloy Co Inc
The Timken Co

YUGOSLAVIA
Boris Kidrič Steelworks

Stainless steel - flats

ALBANIA
Steel of the Party Metallurgical Combine

ARGENTINA
Acindar Industria Argentina de Aceros SA

AUSTRIA
Vereinigte Edelstahlwerke AG

BRAZIL
Acesita – Cia Aços Especiais Itabira
Aços Anhanguera SA (Acoansa)
Siderúrgica Nossa Senhora Aparecida SA
Aços Finos Piratini SA
Villares Group

CANADA
Atlas Specialty Steels Division of Rio Algom Ltd
Slater Steels, Sorel Forge Division

CHINA
Shaanxi Precision Alloy Plant

FRANCE
Aubert et Duval
Aciéries de Bonpertuis

FRANCE *—continued*
Aciéries et Laminoirs du Saut du Tarn (ALST)
Usinor Sacilor

WEST GERMANY
Böhler AG, Edelstahlwerke
Remscheider Walz- und Hammerwerke, Böllinghaus & Co
Edelstahlwerke Buderus AG
Krupp Stahl AG (KS)
Saarstahl Völklingen GmbH
Thyssen Edelstahlwerke AG

INDIA
Bihar Alloy Steels Ltd
Ellora Steels Ltd (ESL)
Ferro Alloys Corp Ltd (Facor) (Steel division), (Vidarbha Iron & Steel Corp Ltd)
Guest Keen Williams Ltd (GKW)
Mahindra Ugine Steel Co Ltd (Muscosteel)
Panchmahal Steel Ltd
Punjab Con-Cast Steels Ltd
Sail – Steel Authority of India Ltd

ITALY
DeltaCogne SpA
Acciaierie Valbruna SpA

JAPAN
Aichi Steel Works Ltd
Daido Steel Co Ltd
Hitachi Metals Ltd (Hitachi Kinzoku)
Nippon Kinzoku Co Ltd
Nippon Yakin Kogyo Co Ltd

SOUTH KOREA
Sammi Steel Co Ltd

MEXICO
Acero Solar SA de CV

SPAIN
Acerinox SA
Forjas Alavesas SA (Fasa)
SA Echevarría (SAE)
Inoxca – Inoxidables de Catalunya SA
Aceros de Irura SA
Aceros de Llodio SA (Allosa)
Olarra SA (Olsa)
Roldan SA

SWEDEN
Avesta AB
Avesta Uddeholm Stainless Bar AB

Stainless steel – flats

—continued

TURKEY
MKEK Celik Fabrikasi

UNITED KINGDOM
Bedford Steels
BSC – British Steel Corp
C.G. Carlisle & Co Ltd
Dudley Port Rolling Mills, Glynwed Steels Ltd
George Gadd & Co, Glynwed Steels Ltd
Sheffield Forgemasters Ltd
Spartan Sheffield Ltd
Spencer Clark Metal Industries PLC
Welded Stainless Steel Tubes Ltd

UNITED STATES
Al Tech Specialty Steel Corp
Cyclops Corp
Eastern Stainless Steel Co
LTV Steel Co Inc
McDonald Steel Corp
Nelsen Steel & Wire Co Inc
Slater Steels, Fort Wayne Specialty Alloys Division

YUGOSLAVIA
Boris Kidrič Steelworks

Stainless steel – hexagons

ARGENTINA
Acindar Industria Argentina de Aceros SA

AUSTRIA
Vereinigte Edelstahlwerke AG

BRAZIL
Villares Group

CANADA
Slater Steels, Sorel Forge Division

FRANCE
Aubert et Duval
Aciéries et Laminoirs du Saut du Tarn (ALST)
Usinor Sacilor

WEST GERMANY
Remscheider Walz- und Hammerwerke, Böllinghaus & Co
Edelstahlwerke Buderus AG
Krupp Stahl AG (KS)

WEST GERMANY *—continued*
Saarstahl Völklingen GmbH
Thyssen Edelstahlwerke AG

INDIA
Guest Keen Williams Ltd (GKW)
Mahindra Ugine Steel Co Ltd (Muscosteel)

ITALY
DeltaCogne SpA
Acciaierie Valbruna SpA

SOUTH KOREA
Sammi Steel Co Ltd

MEXICO
Acero Solar SA de CV

SPAIN
Aceros de Llodio SA (Allosa)

TURKEY
MKEK Celik Fabrikasi

UNITED KINGDOM
Bedford Steels
C.G. Carlisle & Co Ltd
Dudley Port Rolling Mills, Glynwed Steels Ltd
George Gadd & Co, Glynwed Steels Ltd
Hemmings Ltd
Sanderson Kayser Ltd
Sheffield Forgemasters Ltd
Spartan Sheffield Ltd
Spencer Clark Metal Industries PLC

UNITED STATES
Al Tech Specialty Steel Corp
Cyclops Corp
LTV Steel Co Inc
Nelsen Steel & Wire Co Inc
Slater Steels, Fort Wayne Specialty Alloys Division
Techalloy Co Inc

YUGOSLAVIA
Boris Kidrič Steelworks

Stainless steel – bright (cold finished) bars

AUSTRALIA
Commonwealth Steel Co Ltd
Martin Bright Steels Ltd

Stainless steel - bright
—continued

BRAZIL
Aços Anhanguera SA (Acoansa)
Siderúrgica Nossa Senhora Aparecida SA
Villares Group

FRANCE
Aubert et Duval

WEST GERMANY
Thyssen Edelstahlwerke AG
Westig GmbH

INDIA
BP Steel Industries Pvt Ltd
Damehra Steels & Forgings Pvt Ltd
Guest Keen Williams Ltd (GKW)

ITALY
DeltaCogne SpA
Nuova Deltasider SpA
Acciaierie Valbruna SpA

JAPAN
Aichi Steel Works Ltd
Daido Steel Co Ltd
Hitachi Metals Ltd (Hitachi Kinzoku)

SOUTH AFRICA
Flather Bright Steels (Pty) Ltd
Rand Bright Steels (Pty) Ltd

SPAIN
Forjas Alavesas SA (Fasa)
La Calibradora Mecánica SA
Itinsa - Iberica de Tubos Inoxidables SA
Aceros de Llodio SA (Allosa)
Olarra SA (Olsa)

SWITZERLAND
Vereinigte Drahtwerke AG

UNITED KINGDOM
Bedford Steels
Brasway PLC, Bright Bar Division
C.G. Carlisle & Co Ltd
K.W. Haywood & Son Ltd
Arthur Lee & Sons PLC
Frank Pickering & Co Ltd
Steel Parts, Glynwed Steels Ltd
W. Wesson, Glynwed Steels Ltd

UNITED STATES
Bliss & Laughlin Steel Co
Carpenter Technology Corp

YUGOSLAVIA
Slovenske Železarne - Iron & Steel Works of Slovenia

Stainless steel - light angles

BANGLADESH
Panther Steel Ltd

BRAZIL
Villares Group

WEST GERMANY
Remscheider Walz- und Hammerwerke, Böllinghaus & Co
Saarstahl Völklingen GmbH

ITALY
Acciaierie Valbruna SpA

JAPAN
Aichi Steel Works Ltd
Nippon Kinzoku Co Ltd
Nippon Stainless Steel Co Ltd
Nippon Yakin Kogyo Co Ltd

SOUTH KOREA
Sammi Steel Co Ltd

SPAIN
Inoxca - Inoxidables de Catalunya SA

SWEDEN
Avesta AB
Avesta Uddeholm Stainless Bar AB

UNITED KINGDOM
Bentham International Ltd
Pressfab Sections Ltd
TI Stainless Tubes Ltd

UNITED STATES
Plymouth Tube Co
Slater Steels, Fort Wayne Specialty Alloys Division

Stainless steel - light tees

WEST GERMANY
Remscheider Walz- und Hammerwerke, Böllinghaus & Co
Saarstahl Völklingen GmbH

SWEDEN
Avesta AB

Stainless steel – light tees
—continued

UNITED KINGDOM
Pressfab Sections Ltd
TI Stainless Tubes Ltd

UNITED STATES
Plymouth Tube Co

Stainless steel – light channels

BANGLADESH
Panther Steel Ltd

WEST GERMANY
Remscheider Walz- und Hammerwerke, Böllinghaus & Co
Saarstahl Völklingen GmbH

JAPAN
Aichi Steel Works Ltd

SWEDEN
Avesta AB

UNITED KINGDOM
Bentham International Ltd
Pressfab Sections Ltd
TI Stainless Tubes Ltd

UNITED STATES
Plymouth Tube Co

Stainless steel – special light sections

WEST GERMANY
Remscheider Walz- und Hammerwerke, Böllinghaus & Co
Hoesch AG
Saarstahl Völklingen GmbH
Thyssen Edelstahlwerke AG

INDIA
Guest Keen Williams Ltd (GKW)

ITALY
Metallurgica Calvi SpA
Falci SpA, Reparto Laminazione

JAPAN
Aichi Steel Works Ltd

SPAIN
Forjas Alavesas SA (Fasa)

SWEDEN
Avesta AB

UNITED KINGDOM
Bentham International Ltd
Dudley Port Rolling Mills, Glynwed Steels Ltd
Osborn Steel Extrusions Ltd
Pressfab Sections Ltd
Spencer Clark Metal Industries PLC
TI Stainless Tubes Ltd

UNITED STATES
LTV Steel Co Inc
McDonald Steel Corp
Plymouth Tube Co

Stainless steel – cold roll-formed sections (from strip)

WEST GERMANY
Hoesch AG
Tillmann Kaltprofilwalzwerk GmbH & Co KG

JAPAN
Nippon Kinzoku Co Ltd

UNITED KINGDOM
George Clark (Sheffield) Ltd
Manchester Cold Rollers Ltd
Metsec PLC
Pressfab Sections Ltd
Sections & Tubes Ltd
Warwick-Finspa Ltd, Warwick Sections Division
Welded Stainless Steel Tubes Ltd

UNITED STATES
Cyclops Corp
MP Metal Products Inc
Roll Forming Corp

Stainless steel – hot rolled hoop and strip (uncoated)

CHINA
Shaanxi Precision Alloy Plant

WEST GERMANY
Edelstahlwerke Buderus AG
Hindrichs-Auffermann AG
Krupp Stahl AG (KS)

Stainless steel – hot rolled hoop and strip *—continued*

INDIA
Sail – Steel Authority of India Ltd

ITALY
Terni Acciai Speciali SpA

JAPAN
Daido Steel Co Ltd
Nisshin Steel Co Ltd

SPAIN
Acerinox SA

SWEDEN
Avesta AB
AB Sandvik Steel

UNITED KINGDOM
BSC – British Steel Corp

UNITED STATES
Allegheny Ludlum Corp
Eastern Stainless Steel Co

Stainless steel – skelp (tube strip)

INDIA
Sail – Steel Authority of India Ltd

SPAIN
Acerinox SA

SWEDEN
Avesta AB

UNITED KINGDOM
Welded Stainless Steel Tubes Ltd

UNITED STATES
Eastern Stainless Steel Co

Stainless steel – cold rolled hoop and strip (uncoated)

BELGIUM
ALZ NV

CANADA
Atlas Specialty Steels Division of Rio Algom Ltd

CHINA
Chongqing Special Steel Works
Shaanxi Precision Alloy Plant

FRANCE
Usinor Sacilor

WEST GERMANY
Edelstahlwerke Buderus AG
Hindrichs-Auffermann AG
Krupp Stahl AG (KS)
Thyssen Edelstahlwerke AG
VDM – Vereinigte Deutsche Metallwerke AG
Westig GmbH

INDIA
Sail – Steel Authority of India Ltd

ITALY
Terni Acciai Speciali SpA

JAPAN
Daido Steel Co Ltd
Hitachi Metals Ltd (Hitachi Kinzoku)
Nippon Kinzoku Co Ltd
Nisshin Steel Co Ltd
Takasago Tekko KK
Tokushu Kinzoku Kogyo Co Ltd (Tokkin)

POLAND
Huta Baildon

SPAIN
Acerinox SA
Inoxservi SA (formerly Aceros Completos Españoles Recocidos y Aleados SA)

SWEDEN
Avesta AB
AB Sandvik Steel
Uddeholm Strip Steel AB

UNITED KINGDOM
BSC – British Steel Corp
Arthur Lee & Sons PLC

UNITED STATES
Allegheny Ludlum Corp
Cyclops Corp
Mercury Stainless Corp
Somers Thin Strip, Olin Corp
Techalloy Co Inc
Teledyne Rodney Metals
The Thinsheet Metals Co
Ulbrich Stainless Steels and Special Metals Inc
Universal Molding Co

Stainless steel – medium plates

ALBANIA
Steel of the Party Metallurgical Combine

AUSTRALIA
The Broken Hill Pty Co Ltd

BELGIUM
ALZ NV

CANADA
Atlas Specialty Steels Division of Rio Algom Ltd

CHINA
Shanghai Iron & Steel Works

FRANCE
Usinor Sacilor

WEST GERMANY
Edelstahlwerke Buderus AG
Hindrichs-Auffermann AG
Krupp Stahl AG (KS)
Thyssen Edelstahlwerke AG
VDM – Vereinigte Deutsche Metallwerke AG

INDIA
Sail – Steel Authority of India Ltd

ITALY
Terni Acciai Speciali SpA

JAPAN
The Japan Steel Works Ltd
Nippon Metal Industry Co Ltd (Nippon Kinzoku Kogyo)
Nippon Steel Corp (Shin Nippon Seitetsu)
Nippon Yakin Kogyo Co Ltd
Nisshin Steel Co Ltd
NKK – Nippon Kokan KK

NORTH KOREA
Kangson Works

POLAND
Huta Baildon

SOUTH AFRICA
Middleburg Steel & Alloys (Pty) Ltd – Steel Division

SPAIN
Acerinox SA

SWEDEN
Avesta AB

UNITED KINGDOM
BSC – British Steel Corp

UNITED STATES
Allegheny Ludlum Corp
Avesta Inc
Eastern Stainless Steel Co
J & L Specialty Products Corp
Jessop Steel Co
Mercury Stainless Corp

Stainless steel – heavy plates

ALBANIA
Steel of the Party Metallurgical Combine

AUSTRALIA
The Broken Hill Pty Co Ltd

BELGIUM
SA Fabrique de Fer de Charleroi

CHINA
Shanghai Iron & Steel Works

CZECHOSLOVAKIA
Vitkovice Steelworks

FRANCE
Usinor Sacilor

WEST GERMANY
Krupp Stahl AG (KS)
Thyssen Edelstahlwerke AG

HUNGARY
Diosgyör Works (Lenin Metallurgical Works)

INDIA
Sail – Steel Authority of India Ltd

ITALY
Terni Acciai Speciali SpA

JAPAN
The Japan Steel Works Ltd
Nippon Metal Industry Co Ltd (Nippon Kinzoku Kogyo)
Nippon Steel Corp (Shin Nippon Seitetsu)
Nisshin Steel Co Ltd
NKK – Nippon Kokan KK

Stainless steel – heavy plates —*continued*

NORTH KOREA
Kangson Works

POLAND
Huta Baildon

SPAIN
Acerinox SA

SWEDEN
Avesta AB

UNITED KINGDOM
Spartan Redheugh Ltd

UNITED STATES
Allegheny Ludlum Corp
Avesta Inc
Eastern Stainless Steel Co
Jessop Steel Co
USS, A Division of USX Corp

Stainless steel – universal plates

JAPAN
Nippon Steel Corp (Shin Nippon Seitetsu)

UNITED STATES
Eastern Stainless Steel Co
Washington Steel, Division of Blount Inc

Stainless steel – floor plates

UNITED STATES
Eastern Stainless Steel Co

Stainless steel – clad plates

FRANCE
Usinor Sacilor

JAPAN
The Japan Steel Works Ltd
Nippon Steel Corp (Shin Nippon Seitetsu)
NKK – Nippon Kokan KK

UNITED STATES
Allegheny Ludlum Corp
Lukens Steel Co

Stainless steel – hot rolled sheet/coil (uncoated)

BELGIUM
ALZ NV

BULGARIA
Kremikovtsi Iron & Steel Works

CANADA
Atlas Specialty Steels Division of Rio Algom Ltd

CHINA
Taiyuan Iron & Steel Co

FINLAND
Outokumpu Oy

FRANCE
Usinor Sacilor

WEST GERMANY
Hindrichs-Auffermann AG
Krupp Stahl AG (KS)
Thyssen Edelstahlwerke AG

INDIA
Sail – Steel Authority of India Ltd

ITALY
Terni Acciai Speciali SpA

JAPAN
Kawasaki Steel Corp
Nippon Metal Industry Co Ltd (Nippon Kinzoku Kogyo)
Nippon Stainless Steel Co Ltd
Nippon Steel Corp (Shin Nippon Seitetsu)
Nippon Yakin Kogyo Co Ltd
Nisshin Steel Co Ltd

RUMANIA
Combinat Siderurgic Galaţi (CSG)

SOUTH AFRICA
Middleburg Steel & Alloys (Pty) Ltd – Steel Division

SPAIN
Acerinox SA
Patricio Echeverría SA

SWEDEN
Avesta AB

TAIWAN
Mayer Steel Pipe Corp

Stainless steel - hot rolled sheet/coil *—continued*

UNITED KINGDOM
BSC – British Steel Corp
Spencer Clark Metal Industries PLC

UNITED STATES
Allegheny Ludlum Corp
Cyclops Corp
Eastern Stainless Steel Co
J & L Specialty Products Corp
Somers Thin Strip, Olin Corp
USS, A Division of USX Corp

USSR
Chelyabinskiy Met Zavod (Chelyabinsk Iron & Steel Works)
Zaporozhstal Zavod (Zaporozhye Steel Works)

Stainless steel - cold rolled sheet/coil (uncoated)

BELGIUM
ALZ NV

BULGARIA
Kremikovtsi Iron & Steel Works

CANADA
Atlas Specialty Steels Division of Rio Algom Ltd

CHINA
Taiyuan Iron & Steel Co

FINLAND
Outokumpu Oy

FRANCE
Aubert et Duval
Usinor Sacilor

EAST GERMANY
VEB Eisenhüttenkombinat Ost

WEST GERMANY
Hindrichs-Auffermann AG
Krupp Stahl AG (KS)
Thyssen Edelstahlwerke AG

INDIA
Sail – Steel Authority of India Ltd

ITALY
Terni Acciai Speciali SpA

JAPAN
Kawasaki Steel Corp
Nippon Metal Industry Co Ltd (Nippon Kinzoku Kogyo)
Nippon Stainless Steel Co Ltd
Nippon Steel Corp (Shin Nippon Seitetsu)
Nippon Yakin Kogyo Co Ltd
Nisshin Steel Co Ltd
Pacific Metals Co Ltd (Taiheiyo Kinzoku)
Takasago Tekko KK

SOUTH KOREA
Sammi Steel Co Ltd

MEXICO
Mexinox SA de CV

RUMANIA
Combinat Siderurgic Galaţi (CSG)
Tirgovişte Works

SOUTH AFRICA
Middleburg Steel & Alloys (Pty) Ltd – Steel Division

SPAIN
Acerinox SA

SWEDEN
Avesta AB

SWITZERLAND
Merz & Cie Drahtwerke AG

TAIWAN
Mayer Steel Pipe Corp

UNITED KINGDOM
BSC – British Steel Corp
Welded Stainless Steel Tubes Ltd

UNITED STATES
Allegheny Ludlum Corp
Cyclops Corp
Eastern Stainless Steel Co
J & L Specialty Products Corp
Mercury Stainless Corp
Teledyne Rodney Metals
Washington Steel, Division of Blount Inc

USSR
Chelyabinskiy Met Zavod (Chelyabinsk Iron & Steel Works)
Zaporozhstal Zavod (Zaporozhye Steel Works)

Stainless steel – bright annealed sheet/coil

WEST GERMANY
Krupp Stahl AG (KS)

ITALY
Terni Acciai Speciali SpA

JAPAN
Nippon Steel Corp (Shin Nippon Seitetsu)
Takasago Tekko KK

SOUTH KOREA
Sammi Steel Co Ltd

SPAIN
Acerinox SA

SWEDEN
Avesta AB

UNITED KINGDOM
BSC – British Steel Corp
Plated Strip (International) Ltd
Welded Stainless Steel Tubes Ltd

UNITED STATES
Allegheny Ludlum Corp
J & L Specialty Products Corp
Mercury Stainless Corp

Stainless steel – wire

AUSTRIA
Joh. Pengg, Draht- & Walzwerke

BRAZIL
Sandvik do Brasil SA

CHINA
Chongqing Special Steel Works
Shaanxi Precision Alloy Plant
Shoudu Iron & Steel Co

FRANCE
Usinor Sacilor

WEST GERMANY
Krupp Stahl AG (KS)
Thyssen Edelstahlwerke AG
VDM – Vereinigte Deutsche Metallwerke AG
Westig GmbH

ITALY
Nuova Deltasider SpA

JAPAN
Nippon Metal Industry Co Ltd (Nippon Kinzoku Kogyo)
Sumitomo Electric Industries Ltd (Sumitomo Denki Kogyo)

MEXICO
Acero Solar SA de CV

SPAIN
SA Echevarría (SAE)
Inoxca – Inoxidables de Catalunya SA
Roldan SA

SWEDEN
Avesta AB
Fagersta Stainless AB
AB Sandvik Steel

UNITED KINGDOM
Hemmings Ltd
Sanderson Kayser Ltd
Welded Stainless Steel Tubes Ltd

UNITED STATES
Al Tech Specialty Steel Corp
LTV Steel Co Inc
National-Standard Co
Nelsen Steel & Wire Co Inc
Techalloy Co Inc
Ulbrich Stainless Steels and Special Metals Inc

YUGOSLAVIA
Slovenske Železarne – Iron & Steel Works of Slovenia

Stainless steel – seamless tubes and pipes

AUSTRALIA
Email Tube Division, Email Ltd

AUSTRIA
Vereinigte Edelstahlwerke AG

BRAZIL
Persico Pizzamiglio SA

CHINA
Changcheng Steel Works
Chengdu Seamless Steel Tube Plant
Hengyang Steel Tube Plant
Shaanxi Precision Alloy Plant

CZECHOSLOVAKIA
Chomutov Tube Works
Vitkovice Steelworks

Stainless steel – seamless tubes and pipes —*continued*

FRANCE
Vallourec Industries

WEST GERMANY
Röhrenwerke Bous/Saar GmbH (RBS)
Mannesmann Edelstahlrohr GmbH

INDIA
Neeka Tubes Ltd

ITALY
Dalmine SpA
Sandvik Italia SpA
Seta Tubi SpA
Sitai SpA
Tubificio di Solbiate SpA (formerly Fonderie Officine Rovelli)

JAPAN
Kobe Steel Ltd (Kobe Seikosho)
Nippon Steel Corp (Shin Nippon Seitetsu)
NKK – Nippon Kokan KK
Sanyo Special Steel Co Ltd
Sumitomo Metal Industries Ltd

SOUTH KOREA
Sammi Steel Co Ltd

NEW ZEALAND
Centralloy Industries – Division of Stainless Castings Ltd

SPAIN
Centracero SA
Tubacex – C.E. de Tubos por Extrusion SA

SWEDEN
Ansab AB
Avesta AB
Avesta Nyby Powder AB
AB Sandvik Steel

UNITED KINGDOM
Accles & Pollock
Cameron Iron Works Ltd, Forged Products Division
Fine Tubes Ltd
TI Hollow Extrusions Ltd
TI Stainless Tubes Ltd
Welded Stainless Steel Tubes Ltd

UNITED STATES
Al Tech Specialty Steel Corp
Lone Star Steel Co

UNITED STATES —*continued*
Plymouth Tube Co
The Timken Co

USSR
Chelyabinskiy Met Zavod (Chelyabinsk Iron & Steel Works)

Stainless steel – longitudinal weld tubes and pipes

AUSTRALIA
Email Tube Division, Email Ltd
Sandvik Australia Pty Ltd

BRAZIL
Persico Pizzamiglio SA

CANADA
Associated Tube Industries Ltd (ATI)
Nor-Sand Metal Inc
Standard Tube Canada, Division of TI Canada Inc

FINLAND
Haato-Tuote Oy
Oy Hackman Ab
Oy Ja-Ro Ab

FRANCE
BSL Tubes et Raccords SA
Sté Meusienne de Constructions Mécaniques
Vallourec Industries

WEST GERMANY
Benteler AG
Eisenbau Krämer mbH
Mannesmann Edelstahlrohr GmbH
Poppe & Potthoff GmbH & Co
Schoeller Werk GmbH & Co KG
VDM – Vereinigte Deutsche Metallwerke AG

HUNGARY
Koebanya Tube Works

INDIA
Choksi Tube Co Pvt Ltd
Neeka Tubes Ltd

ITALY
Arvedi Group
Nuova CMC SpA
Dalmine SpA
Sitai SpA

Stainless steel – longitudinal weld tubes and pipes *—continued*

ITALY *—continued*
Tubificio di Solbiate SpA (formerly Fonderie Officine Rovelli)

JAPAN
Nippon Steel Corp (Shin Nippon Seitetsu)
Nippon Yakin Kogyo Co Ltd
Nisshin Steel Co Ltd
Sumitomo Metal Industries Ltd

SOUTH KOREA
Pusan Steel Pipe Corp

NETHERLANDS
VBF Buizen BV

NEW ZEALAND
Email Industries Ltd

PHILIPPINES
Tubemakers Philippines Inc

SOUTH AFRICA
Salmac Stainless Tube (Pty) Ltd
Sandvik (Pty) Ltd
Tosa (Division of Dorbyl Ltd)

SPAIN
Armco SA, Spain
Austinox SA
Tubos y Calibrados de Cataluña SA
Itinsa – Iberica de Tubos Inoxidables SA

SWEDEN
Avesta AB
Avesta Sandvik Tube AB (AST)

SWITZERLAND
Arfa Röhrenwerke AG
Zwahlen & Mayr

TAIWAN
Mayer Steel Pipe Corp

TURKEY
Borusan Holding AŞ

UNITED KINGDOM
Bentham International Ltd
Coventry Tubes Permatube (Glynwed Tubes & Fittings Ltd)
Energy Tubes Ltd
Fine Tubes Ltd
Stelco Hardy Ltd

UNITED KINGDOM *—continued*
Welded Stainless Steel Tubes Ltd

UNITED STATES
Allegheny Ludlum Corp
Bristol Metals Inc
Carpenter Technology Corp
Indiana Tube Corp
LTV Steel Co Inc
Northwest Pipe & Casing Co
Plymouth Tube Co
Scientific Tube Inc

Stainless steel – oil country tubular goods

BRAZIL
Persico Pizzamiglio SA

FRANCE
Vallourec Industries

ITALY
Seta Tubi SpA

JAPAN
NKK – Nippon Kokan KK
Sumitomo Metal Industries Ltd

SOUTH AFRICA
Tosa (Division of Dorbyl Ltd)

SWEDEN
Avesta Sandvik Tube AB (AST)

UNITED KINGDOM
TI Stainless Tubes Ltd

UNITED STATES
Scientific Tube Inc
The Timken Co

Stainless steel – cold drawn tubes and pipes

AUSTRALIA
Email Tube Division, Email Ltd
Martin Bright Steels Ltd

BRAZIL
Persico Pizzamiglio SA

CANADA
Associated Tube Industries Ltd (ATI)
Standard Tube Canada, Division of TI Canada Inc

Stainless steel – cold drawn tubes and pipes *—continued*

CHINA
Changcheng Steel Works
Hengyang Steel Tube Plant
Shaanxi Precision Alloy Plant

FRANCE
Vallourec Industries

WEST GERMANY
Poppe & Potthoff GmbH & Co
Schoeller Werk GmbH & Co KG

INDIA
Choksi Tube Co Pvt Ltd
Neeka Tubes Ltd

ITALY
Dalmine SpA

JAPAN
Kobe Steel Ltd (Kobe Seikosho)
Nisshin Steel Co Ltd
NKK – Nippon Kokan KK
Sanwa Metal Industries Ltd
Sumitomo Metal Industries Ltd

SOUTH KOREA
Sammi Steel Co Ltd

SOUTH AFRICA
Quadrisection (Pty) Ltd

SPAIN
Austinox SA
Transmesa – Transformaciones Metalurgicas SA
Tubacex – C.E. de Tubos por Extrusion SA

SWEDEN
Avesta Sandvik Tube AB (AST)

SWITZERLAND
Arfa Röhrenwerke AG

UNITED KINGDOM
Accles & Pollock
Durham Tube Ltd
TI Stainless Tubes Ltd

UNITED STATES
Al Tech Specialty Steel Corp
Cyclops Corp
Lone Star Steel Co
LTV Steel Co Inc
Plymouth Tube Co

UNITED STATES *—continued*
The Timken Co

Stainless steel – hollow bars

ITALY
Tubificio di Solbiate SpA (formerly Fonderie Officine Rovelli)

JAPAN
Kobe Steel Ltd (Kobe Seikosho)
Sanyo Special Steel Co Ltd

SPAIN
Tubacex – C.E. de Tubos por Extrusion SA

SWEDEN
Avesta AB
Avesta Nyby Powder AB

UNITED KINGDOM
Bedford Steels
TI Stainless Tubes Ltd

UNITED STATES
Al Tech Specialty Steel Corp
Carpenter Technology Corp
LTV Steel Co Inc
Plymouth Tube Co
The Timken Co

Stainless steel – precision tubes and pipes

AUSTRALIA
Email Tube Division, Email Ltd

BRAZIL
Persico Pizzamiglio SA

CZECHOSLOVAKIA
Chomutov Tube Works

FRANCE
Vallourec Industries

WEST GERMANY
Benteler AG
Poppe & Potthoff GmbH & Co

INDIA
Choksi Tube Co Pvt Ltd

Stainless steel – precision tubes and pipes —*continued*

ITALY
Tubificio di Solbiate SpA (formerly Fonderie Officine Rovelli)

JAPAN
Kobe Steel Ltd (Kobe Seikosho)
Nippon Kinzoku Co Ltd
Sumitomo Metal Industries Ltd

NETHERLANDS
VBF Buizen BV

SPAIN
Transmesa – Transformaciones Metalurgicas SA

SWEDEN
Avesta Sandvik Tube AB (AST)

SWITZERLAND
LN Industries SA

UNITED KINGDOM
Accles & Pollock
Fine Tubes Ltd

UNITED STATES
Cyclops Corp
Indiana Tube Corp
Lone Star Steel Co
Plymouth Tube Co
Scientific Tube Inc
The Timken Co

Alloy steel – ingots

ARGENTINA
Acindar Industria Argentina de Aceros SA
Establecimiento Altos Hornos Zapla

AUSTRALIA
The Broken Hill Pty Co Ltd
Commonwealth Steel Co Ltd

BELGIUM
SA Fabrique de Fer de Charleroi

BRAZIL
Acesita – Cia Aços Especiais Itabira
Aços Anhanguera SA (Acoansa)
Eletrometal SA – Metais Especiais

BULGARIA
Kremikovtsi Iron & Steel Works

CANADA
Atlas Specialty Steels Division of Rio Algom Ltd
Canadian Steel Wheel, Division of Hawker Siddeley Inc
Slater Steels, Sorel Forge Division
Stelco Inc

CZECHOSLOVAKIA
Vitkovice Steelworks

FRANCE
Usinor Sacilor

WEST GERMANY
Böhler AG, Edelstahlwerke
Dörrenberg Edelstahl GmbH
Klöckner Stahl GmbH
Krupp Stahl AG (KS)
Lemmerz-Werke KGaA
Saarstahl Völklingen GmbH

INDIA
Bihar Alloy Steels Ltd
Ellora Steels Ltd (ESL)
Kalyani Steels Ltd
Man Industrial Corp Ltd
Modi Industries Ltd, Steel Division
Punjab Con-Cast Steels Ltd
Raipur Wires & Steel Ltd (RWSL)
Sail – Steel Authority of India Ltd
Sanghvi Steels Ltd
Somani Iron & Steels Ltd
The Tata Iron & Steel Co Ltd (Tisco)

ITALY
ASO SpA – Acciai Speciali Ospitaletto

JAPAN
Kobe Steel Ltd (Kobe Seikosho)
Mitsubishi Steel Mfg Ltd (Mitsubishi Seiko)
Nisshin Steel Co Ltd

SOUTH KOREA
Sammi Steel Co Ltd

MEXICO
Fundiciones de Hierro y Acero SA (Fhasa)
Industrias CH SA (formerly Campos Hermanos SA)

POLAND
Huta Baildon

SPAIN
Forjas Alavesas SA (Fasa)
Grupo José María Aristrain SA

Alloy steel - ingots

—continued

SPAIN *—continued*
SA Echevarría (SAE)
Olarra SA (Olsa)
Roldan SA
Tubacex – C.E. de Tubos por Extrusion SA

TAIWAN
Taiwan Machinery Manufacturing Corp

TURKEY
MKEK Celik Fabrikasi

UNITED KINGDOM
C.G. Carlisle & Co Ltd
Howmet Turbine (UK) Inc, Howmet Alloys International
Sheffield Forgemasters Ltd
UES – United Engineering Steels Ltd

UNITED STATES
Acme Steel Co
Armco Inc
The Babcock & Wilcox Co, Tubular Products Division
Cameron Iron Works Inc
CF&I Steel Corp
The Champion Steel Co
Columbia Tool Steel Co
Cyclops Corp
Ellwood Uddeholm Steel Corp
Jessop Steel Co
LTV Steel Co Inc
Lukens Steel Co
Marathon LeTourneau Co, Longview Division
National Forge Co
Rouge Steel Co
Seattle Steel Inc
The Timken Co
USS, A Division of USX Corp

Alloy steel - slabs

ARGENTINA
Acindar Industria Argentina de Aceros SA

AUSTRALIA
The Broken Hill Pty Co Ltd

BELGIUM
SA Fabrique de Fer de Charleroi

BRAZIL
Acesita – Cia Aços Especiais Itabira

BULGARIA
Kremikovtsi Iron & Steel Works

CANADA
Atlas Specialty Steels Division of Rio Algom Ltd
Slater Steels, Sorel Forge Division
Stelco Inc

FRANCE
Usinor Sacilor

WEST GERMANY
Krupp Stahl AG (KS)
Thyssen Edelstahlwerke AG

INDIA
Mahindra Ugine Steel Co Ltd (Muscosteel)
The Tata Iron & Steel Co Ltd (Tisco)

JAPAN
Kobe Steel Ltd (Kobe Seikosho)
Nisshin Steel Co Ltd

NETHERLANDS
Hoogovens Groep BV

TURKEY
MKEK Celik Fabrikasi

UNITED KINGDOM
C.G. Carlisle & Co Ltd
Sheffield Forgemasters Ltd
UES – United Engineering Steels Ltd

UNITED STATES
Bethlehem Steel Corp
Carpenter Technology Corp
Cyclops Corp
Jessop Steel Co
LTV Steel Co Inc
Lukens Steel Co
Marathon LeTourneau Co, Longview Division
National Forge Co
Oregon Steel Mills, Division of Gilmore Steel Corp
Rouge Steel Co
Seattle Steel Inc
USS, A Division of USX Corp

USSR
Donetskiy Metallurgicheskiy Zavod (Donetsk Iron & Steel Works)

Alloy steel - blooms

ARGENTINA
Acindar Industria Argentina de Aceros SA

AUSTRALIA
The Broken Hill Pty Co Ltd
Commonwealth Steel Co Ltd

BRAZIL
Acesita – Cia Aços Especiais Itabira
Aços Anhanguera SA (Acoansa)
Eletrometal SA – Metais Especiais
Villares Group

BULGARIA
Kremikovtsi Iron & Steel Works

CANADA
Atlas Specialty Steels Division of Rio Algom Ltd
Slater Steels, Sorel Forge Division
Stelco Inc

CZECHOSLOVAKIA
Vitkovice Steelworks

FRANCE
Usinor Sacilor

EAST GERMANY
VEB Edelstahlwerk 8 Mai 1945

WEST GERMANY
Krupp Stahl AG (KS)
Maxhütte – Eisenwerke-Gesellschaft Maximilianshütte mbH
Thyssen Edelstahlwerke AG

INDIA
Bihar Alloy Steels Ltd
Ferro Alloys Corp Ltd (Facor) (Steel division), (Vidarbha Iron & Steel Corp Ltd)
Mahindra Ugine Steel Co Ltd (Muscosteel)
Panchmahal Steel Ltd
The Tata Iron & Steel Co Ltd (Tisco)

ITALY
Acciaierie di Bolzano SpA
DeltaCogne SpA
Nuova Deltasider SpA
Acciaierie e Ferriere Luigi Leali SpA

JAPAN
Kobe Steel Ltd (Kobe Seikosho)
Mitsubishi Steel Mfg Ltd (Mitsubishi Seiko)

JAPAN —*continued*
Toa Steel Co Ltd

MEXICO
Industrias CH SA (formerly Campos Hermanos SA)

NETHERLANDS
Hoogovens Groep BV

PHILIPPINES
Armco-Marsteel Alloy Corp

POLAND
Poldi Steelworks

SPAIN
Forjas Alavesas SA (Fasa)
Babcock & Wilcox Española SA Tubular Products Division
Aceros de Llodio SA (Allosa)

SWEDEN
Björneborgs Jernverks AB
AB Sandvik Steel

TURKEY
Asil Çelik Sanayi ve Ticaret AŞ (Asil Çelik)
MKEK Celik Fabrikasi

UNITED KINGDOM
C.G. Carlisle & Co Ltd
F.H. Lloyd & Co Ltd
Sheffield Forgemasters Ltd
UES – United Engineering Steels Ltd

UNITED STATES
Acme Steel Co
Bethlehem Steel Corp
Carpenter Technology Corp
Copperweld Steel Co
Cyclops Corp
Ellwood Uddeholm Steel Corp
LTV Steel Co Inc
National Forge Co
Seattle Steel Inc
The Timken Co
USS, A Division of USX Corp

USSR
Met Zavod Krivorizhstal (Krivoy Rog Iron & Steel Works)

Alloy steel - billets

ARGENTINA
Acindar Industria Argentina de Aceros SA
Establecimiento Altos Hornos Zapla

AUSTRALIA
The Broken Hill Pty Co Ltd
Commonwealth Steel Co Ltd

BRAZIL
Acesita - Cia Aços Especiais Itabira
Açopalma - Cia Industrial de Aços Várzea de Palma
Aços Anhanguera SA (Acoansa)
Siderúrgica Nossa Senhora Aparecida SA
Eletrometal SA - Metais Especiais
Aços Finos Piratini SA
Villares Group

BULGARIA
Kremikovtsi Iron & Steel Works

CANADA
Atlas Specialty Steels Division of Rio Algom Ltd
Slater Steels, Sorel Forge Division
Stelco Inc

COLOMBIA
Simesa - Siderúrgica de Medellín

CZECHOSLOVAKIA
Vitkovice Steelworks

FINLAND
Dalsbruk Oy Ab

FRANCE
Aubert et Duval
Aciéries et Laminoirs de Rives
Usinor Sacilor

EAST GERMANY
VEB Stahl- und Walzwerk Brandenburg

WEST GERMANY
Walzwerk Becker-Berlin Verwaltungsges.mbH & Co Strangguss KG
Edelstahlwerke Buderus AG
Klöckner Stahl GmbH
Krupp Stahl AG (KS)
Maxhütte - Eisenwerke-Gesellschaft Maximilianshütte mbH
Saarstahl Völklingen GmbH
Schmidt & Clemens GmbH & Co

WEST GERMANY *—continued*
Thyssen Edelstahlwerke AG

HUNGARY
Diosgyör Works (Lenin Metallurgical Works)
Ozdi Kohászati Uzemek (Okü)

INDIA
Bhoruka Steel Ltd
Bhushan Industrial Corp
Bihar Alloy Steels Ltd
Indore Steel & Iron Mills Pvt Ltd (Isim)
Kalyani Steels Ltd
Mahindra Ugine Steel Co Ltd (Muscosteel)
Modi Industries Ltd, Steel Division
Mukand Iron & Steel Works Ltd
Panchmahal Steel Ltd
Partap Steel Rolling Mills (1935) Ltd
Raipur Wires & Steel Ltd (RWSL)
Sail - Steel Authority of India Ltd
Sanghvi Steels Ltd
Singh Electrosteel Ltd (SES)
Star Steel Pvt Ltd (Starsteel)
The Tata Iron & Steel Co Ltd (Tisco)
Usha Alloys & Steels Ltd (UASL)

ITALY
Acciaierie di Bolzano SpA
Acciaieria e Ferriera del Caleotto SpA (AFC)
Acciaierie e Ferriere di Caronno SpA
DeltaCogne SpA
Nuova Deltasider SpA
Indumetal - Industrie Metallurgiche SpA
Acciaierie e Ferriere Luigi Leali SpA
Raimondi SpA

JAPAN
Kobe Steel Ltd (Kobe Seikosho)
Mitsubishi Steel Mfg Ltd (Mitsubishi Seiko)
Toa Steel Co Ltd
Yodogawa Steel Works Ltd (Yodogawa Seikosho)

SOUTH KOREA
Sammi Steel Co Ltd

MEXICO
Industrias CH SA (formerly Campos Hermanos SA)

NETHERLANDS
Hoogovens Groep BV

Alloy steel – billets
—continued

PHILIPPINES
Armco-Marsteel Alloy Corp

POLAND
Huta Baildon
Poldi Steelworks

SOUTH AFRICA
Scaw Metals Ltd
Usco – The Union Steel Corp (of South Africa) Ltd

SPAIN
Forjas Alavesas SA (Fasa)
Babcock & Wilcox Española SA Tubular Products Division
SA Echevarría (SAE)
Patricio Echeverría SA
Aceros de Irura SA
Aceros de Llodio SA (Allosa)
Olarra SA (Olsa)
Pedro Orbegozo y Cía SA (Acenor)
Roldan SA
Tubacex – C.E. de Tubos por Extrusion SA

SWEDEN
Asea Powdermet AB
Ovako Steel AB
AB Sandvik Steel

TURKEY
Asil Çelik Sanayi ve Ticaret AŞ (Asil Çelik)
MKEK Celik Fabrikasi

UNITED KINGDOM
Barworth Flockton Ltd
C.G. Carlisle & Co Ltd
F.H. Lloyd & Co Ltd
Sheffield Forgemasters Ltd
UES – United Engineering Steels Ltd
Unbrako Steel Co Ltd

UNITED STATES
Acme Steel Co
Atlantic Steel Co
Bethlehem Steel Corp
Carpenter Technology Corp
CF&I Steel Corp
The Champion Steel Co
Copperweld Steel Co
Cyclops Corp
IRI International Corp
Jessop Steel Co

UNITED STATES *—continued*
Earle M. Jorgensen Co, Forge Division
LTV Steel Co Inc
National Forge Co
Razorback Steel Corp (RSC)
Seattle Steel Inc
The Timken Co
USS, A Division of USX Corp

USSR
Donetskiy Metallurgicheskiy Zavod (Donetsk Iron & Steel Works)

YUGOSLAVIA
Boris Kidrič Steelworks

Alloy steel – semis for seamless tubemaking

FRANCE
Usinor Sacilor

WEST GERMANY
Maxhütte – Eisenwerke-Gesellschaft Maximilianshütte mbH

ITALY
Acciaierie di Bolzano SpA
Acciaierie e Ferriere Luigi Leali SpA

SPAIN
Babcock & Wilcox Española SA Tubular Products Division

UNITED KINGDOM
C.G. Carlisle & Co Ltd
F.H. Lloyd & Co Ltd

UNITED STATES
The Babcock & Wilcox Co, Tubular Products Division
Razorback Steel Corp (RSC)
The Timken Co

Alloy steel – hot rolled coil (for re-rolling)

FRANCE
Usinor Sacilor

WEST GERMANY
Klöckner Stahl GmbH

ITALY
Snar SpA

Alloy steel – hot rolled coil —*continued*

UNITED STATES
Cyclops Corp
Inland Steel Co
Rouge Steel Co
The Timken Co

Alloy steel – wire rod

AUSTRALIA
The Broken Hill Pty Co Ltd

AUSTRIA
Vereinigte Edelstahlwerke AG

BRAZIL
Siderúrgica Nossa Senhora Aparecida SA
Eletrometal SA – Metais Especiais
Mannesmann SA
Aços Finos Piratini SA
Villares Group

FINLAND
Dalsbruk Oy Ab
Ovako Steel Oy Ab

FRANCE
Commentryenne des Aciers Fins Vanadium Alloys
Usinor Sacilor

WEST GERMANY
Maxhütte – Eisenwerke-Gesellschaft Maximilianshütte mbH
Saarstahl Völklingen GmbH
Thyssen Edelstahlwerke AG

INDIA
Bhoruka Steel Ltd
Bihar Alloy Steels Ltd
Indore Steel & Iron Mills Pvt Ltd (Isim)
Man Industrial Corp Ltd
Modi Industries Ltd, Steel Division
Mukand Iron & Steel Works Ltd

ITALY
Acciaierie di Bolzano SpA
Acciaieria e Ferriera del Caleotto SpA (AFC)
DeltaCogne SpA
Nuova Deltasider SpA
Lucchini Siderurgica SpA

JAPAN
Aichi Steel Works Ltd

JAPAN —*continued*
Daido Steel Co Ltd
Sumitomo Electric Industries Ltd (Sumitomo Denki Kogyo)
Toa Steel Co Ltd

NETHERLANDS
Nedstaal BV

SPAIN
Forjas Alavesas SA (Fasa)
SA Echevarría (SAE)

SWEDEN
Ovako Steel AB

UNITED KINGDOM
BSC – British Steel Corp
The Templeborough Rolling Mills Ltd (TRM)
UES – United Engineering Steels Ltd
Unbrako Steel Co Ltd

UNITED STATES
Armco Inc
CF&I Steel Corp
Latrobe Steel Co

USSR
Beloretskiy Met Zavod (Beloretsk Iron & Steel Works)
Met Zavod Krivorizhstal (Krivoy Rog Iron & Steel Works)

Alloy steel – round bars

ALBANIA
Steel of the Party Metallurgical Combine

ARGENTINA
Acindar Industria Argentina de Aceros SA

AUSTRALIA
Commonwealth Steel Co Ltd

AUSTRIA
Vereinigte Edelstahlwerke AG

BRAZIL
Acesita – Cia Aços Especiais Itabira
Açopalma – Cia Industrial de Aços Várzea de Palma
Aços Anhanguera SA (Acoansa)
Siderúrgica Nossa Senhora Aparecida SA
Eletrometal SA – Metais Especiais
Aços Finos Piratini SA

Alloy steel - round bars
—continued

BRAZIL *—continued*
Laminação Santa Maria SA – Indústria e Comércio
Villares Group

CANADA
Lake Ontario Steel Co (Lasco Steel)
Manitoba Rolling Mills, Amca International
Slater Steels – Hamilton Speciality Bar Division
Slater Steels, Sorel Forge Division

CHINA
Benxi Iron & Steel Co
Chengdu Metallurgical Experimental Plant
Chongqing Special Steel Works
Guiyang Steel Mill
Shaanxi Precision Alloy Plant
Shanghai Iron & Steel Works
Special Steel Co of Shougang

CZECHOSLOVAKIA
Vitkovice Steelworks

FINLAND
Ovako Steel Oy Ab

FRANCE
Aciers d'Allevard
Aubert et Duval
Aciéries et Laminoirs de Rives
Aciéries et Laminoirs du Saut du Tarn (ALST)
Usinor Sacilor

EAST GERMANY
VEB Edelstahlwerk 8 Mai 1945

WEST GERMANY
Böhler AG, Edelstahlwerke
Remscheider Walz- und Hammerwerke, Böllinghaus & Co
Edelstahlwerke Buderus AG
Dörrenberg Edelstahl GmbH
Fürstlich Hohenzollernsche Hüttenverwaltung Laucherthal (FHH)
Klöckner Stahl GmbH
Krupp Stahl AG (KS)
Maxhütte – Eisenwerke-Gesellschaft Maximilianshütte mbH
Saarstahl Völklingen GmbH
Schmidt & Clemens GmbH & Co
Thyssen Edelstahlwerke AG

HUNGARY
Diosgyör Works (Lenin Metallurgical Works)
Ozdi Kohászati Uzemek (Okü)

INDIA
Bihar Alloy Steels Ltd
Ellora Steels Ltd (ESL)
Ferro Alloys Corp Ltd (Facor) (Steel division), (Vidarbha Iron & Steel Corp Ltd)
Guest Keen Williams Ltd (GKW)
Kalyani Steels Ltd
Mahindra Ugine Steel Co Ltd (Muscosteel)
Modi Industries Ltd, Steel Division
Mukand Iron & Steel Works Ltd
Panchmahal Steel Ltd
Punjab Con-Cast Steels Ltd
Sanghvi Steels Ltd
Sunflag Iron & Steel Co
Visvesvaraya Iron & Steel Ltd (formerly The Mysore Iron & Steel Ltd)

ITALY
SpA Officine Fratelli Bertoli fu Rodolfo
Acciaierie di Bolzano SpA
DeltaCogne SpA
Nuova Deltasider SpA
Las – Laminazione Acciai Speciali SpA
Acciaierie e Ferriere Luigi Leali SpA
Acciaierie Valbruna SpA

JAPAN
Aichi Steel Works Ltd
Daido Steel Co Ltd
Hitachi Metals Ltd (Hitachi Kinzoku)
Kobe Steel Ltd (Kobe Seikosho)
Nippon Koshuha Steel Co Ltd (Nippon Koshuha Kogyo)
Sanyo Special Steel Co Ltd
Toa Steel Co Ltd
Tohoku Special Steel Works Ltd (Tohoku Tokushuko)

NORTH KOREA
Songjin Works

SOUTH KOREA
Sammi Steel Co Ltd

MEXICO
Industrias CH SA (formerly Campos Hermanos SA)
Acero Solar SA de CV

Alloy steel – round bars
—continued

MEXICO *—continued*
Tamsa – Tubos de Acero de Mexico SA

PERU
Siderperú – Empresa Siderúrgica del Perú

PHILIPPINES
Armco-Marsteel Alloy Corp

POLAND
Poldi Steelworks

PORTUGAL
Fábrica de Aços Tomé Fèteira SA

SOUTH AFRICA
Scaw Metals Ltd
Usco – The Union Steel Corp (of South Africa) Ltd

SPAIN
Aforasa – Acerias y Forjas de Azcoitia SA
Forjas Alavesas SA (Fasa)
Babcock & Wilcox Española SA Tubular Products Division
SA Echevarría (SAE)
Patricio Echeverría SA
Aceros de Irura SA
Aceros de Llodio SA (Allosa)
Olarra SA (Olsa)
Esteban Orbegozo SA
Pedro Orbegozo y Cía SA (Acenor)
Roldan SA

SWEDEN
Björneborgs Jernverks AB
Kloster Speedsteel AB
Ovako Steel AB
Smedjebacken-Boxholm Stål AB
Uddeholm Tooling AB

SWITZERLAND
Von Moos Stahl AG

TURKEY
Asil Çelik Sanayi ve Ticaret AŞ (Asil Çelik)
MKEK Celik Fabrikasi

UNITED KINGDOM
Barworth Flockton Ltd
Bedford Steels

UNITED KINGDOM *—continued*
Dudley Port Rolling Mills, Glynwed Steels Ltd
George Gadd & Co, Glynwed Steels Ltd
Joseph Gillott & Sons, Glynwed Steels Ltd
Kiveton Park Steel & Wire Works Ltd
T.W. Pearson Ltd
Sanderson Kayser Ltd
Sheffield Forgemasters Ltd
Spencer Clark Metal Industries PLC
UES – United Engineering Steels Ltd
Unbrako Steel Co Ltd
Woodstone Rolling Mills Ltd

UNITED STATES
Al Tech Specialty Steel Corp
Atlantic Steel Co
The Babcock & Wilcox Co, Tubular Products Division
Bethlehem Steel Corp
Carpenter Technology Corp
CF&I Steel Corp
Chaparral Steel Co
Columbia Tool Steel Co
Copperweld Steel Co
Crucible Specialty Metals Division, Crucible Materials Corp
Cyclops Corp
Inland Steel Co
Kentucky Electric Steel Corp
Laclede Steel Co
Latrobe Steel Co
LTV Steel Co Inc
National Forge Co
Nelsen Steel & Wire Co Inc
North Star Steel Co (NSS)
Nucor Corp
Quanex Corp
Seattle Steel Inc
Slater Steels, Fort Wayne Specialty Alloys Division
Techalloy Co Inc
Teledyne Vasco
The Timken Co

USSR
Chelyabinskiy Met Zavod (Chelyabinsk Iron & Steel Works)
Met Zavod Krivorizhstal (Krivoy Rog Iron & Steel Works)

VENEZUELA
Acerex CA

Alloy steel - round bars
—continued

YUGOSLAVIA
Boris Kidrič Steelworks
Slovenske Železarne - Iron & Steel Works of Slovenia

Alloy steel - square bars

ALBANIA
Steel of the Party Metallurgical Combine

ARGENTINA
Acindar Industria Argentina de Aceros SA

AUSTRIA
Vereinigte Edelstahlwerke AG

BRAZIL
Acesita - Cia Aços Especiais Itabira
Aços Anhanguera SA (Acoansa)
Siderúrgica Nossa Senhora Aparecida SA
Eletrometal SA - Metais Especiais
Aços Finos Piratini SA
Laminação Santa Maria SA - Indústria e Comércio
Villares Group

CANADA
Slater Steels, Sorel Forge Division

CHINA
Chongqing Special Steel Works
Guiyang Steel Mill

CZECHOSLOVAKIA
Vitkovice Steelworks

FRANCE
Aubert et Duval
Aciéries et Laminoirs du Saut du Tarn (ALST)
Usinor Sacilor

EAST GERMANY
VEB Edelstahlwerk 8 Mai 1945

WEST GERMANY
Böhler AG, Edelstahlwerke
Remscheider Walz- und Hammerwerke, Böllinghaus & Co
Edelstahlwerke Buderus AG
Dörrenberg Edelstahl GmbH
Fürstlich Hohenzollernsche Hüttenverwaltung Laucherthal (FHH)

WEST GERMANY *—continued*
Klöckner Stahl GmbH
Krupp Stahl AG (KS)
Maxhütte - Eisenwerke-Gesellschaft Maximilianshütte mbH
Saarstahl Völklingen GmbH
Schmidt & Clemens GmbH & Co
Thyssen Edelstahlwerke AG

INDIA
Bihar Alloy Steels Ltd
Ellora Steels Ltd (ESL)
Ferro Alloys Corp Ltd (Facor) (Steel division), (Vidarbha Iron & Steel Corp Ltd)
Guest Keen Williams Ltd (GKW)
Kalyani Steels Ltd
Mahindra Ugine Steel Co Ltd (Muscosteel)
Modi Industries Ltd, Steel Division
Panchmahal Steel Ltd
Punjab Con-Cast Steels Ltd
Visvesvaraya Iron & Steel Ltd (formerly The Mysore Iron & Steel Ltd)

ITALY
DeltaCogne SpA
Nuova Deltasider SpA
Falci SpA, Reparto Laminazione
Las - Laminazione Acciai Speciali SpA
Acciaierie e Ferriere Luigi Leali SpA
Acciaierie Valbruna SpA

JAPAN
Daido Steel Co Ltd
Hitachi Metals Ltd (Hitachi Kinzoku)

SOUTH KOREA
Sammi Steel Co Ltd

MEXICO
Laminados Barniedo SA de CV
Industrias CH SA (formerly Campos Hermanos SA)
Acero Solar SA de CV
Tamsa - Tubos de Acero de Mexico SA

POLAND
Poldi Steelworks

PORTUGAL
Fábrica de Aços Tomé Fèteira SA

SOUTH AFRICA
Scaw Metals Ltd

Alloy steel – square bars
—continued

SPAIN
Aforasa – Acerias y Forjas de Azcoitia SA
Babcock & Wilcox Española SA Tubular Products Division
SA Echevarría (SAE)
Aceros de Llodio SA (Allosa)
Olarra SA (Olsa)
Pedro Orbegozo y Cía SA (Acenor)

SWEDEN
Björneborgs Jernverks AB
Kloster Speedsteel AB
Smedjebacken-Boxholm Stål AB
Uddeholm Tooling AB

SWITZERLAND
Von Moos Stahl AG

TURKEY
MKEK Celik Fabrikasi

UNITED KINGDOM
Barworth Flockton Ltd
Bedford Steels
Dudley Port Rolling Mills, Glynwed Steels Ltd
George Gadd & Co, Glynwed Steels Ltd
Joseph Gillott & Sons, Glynwed Steels Ltd
Kiveton Park Steel & Wire Works Ltd
T.W. Pearson Ltd
Sanderson Kayser Ltd
Sheffield Forgemasters Ltd
Spencer Clark Metal Industries PLC
UES – United Engineering Steels Ltd
Woodstone Rolling Mills Ltd

UNITED STATES
Al Tech Specialty Steel Corp
Atlantic Steel Co
Bethlehem Steel Corp
Cyclops Corp
Inland Steel Co
LTV Steel Co Inc
National Forge Co
Nelsen Steel & Wire Co Inc
Quanex Corp
Techalloy Co Inc
The Timken Co

USSR
Chelyabinskiy Met Zavod (Chelyabinsk Iron & Steel Works)

USSR *—continued*
Met Zavod Krivorizhstal (Krivoy Rog Iron & Steel Works)

VENEZUELA
Acerex CA

YUGOSLAVIA
Boris Kidrič Steelworks

Alloy steel – flats

ALBANIA
Steel of the Party Metallurgical Combine

ARGENTINA
Acindar Industria Argentina de Aceros SA

AUSTRIA
Vereinigte Edelstahlwerke AG

BRAZIL
Acesita – Cia Aços Especiais Itabira
Aços Anhanguera SA (Acoansa)
Siderúrgica Nossa Senhora Aparecida SA
Aços Finos Piratini SA
Laminação Santa Maria SA – Indústria e Comércio
Villares Group

CANADA
Manitoba Rolling Mills, Amca International
Slater Steels – Hamilton Speciality Bar Division
Slater Steels, Sorel Forge Division

CHINA
Benxi Iron & Steel Co
Chengdu Metallurgical Experimental Plant
Shaanxi Precision Alloy Plant

CZECHOSLOVAKIA
Vitkovice Steelworks

FINLAND
Ovako Steel Oy Ab

FRANCE
Aciers d'Allevard
Aciéries et Forges d'Anor
Aciéries et Laminoirs de Rives
Aciéries et Laminoirs du Saut du Tarn (ALST)
Usinor Sacilor

Alloy steel – flats *—continued*

EAST GERMANY
VEB Edelstahlwerk 8 Mai 1945

WEST GERMANY
Böhler AG, Edelstahlwerke
Remscheider Walz- und Hammerwerke, Böllinghaus & Co
Edelstahlwerke Buderus AG
Dörrenberg Edelstahl GmbH
Fürstlich Hohenzollernsche Hüttenverwaltung Laucherthal (FHH)
Klöckner Stahl GmbH
Krupp Stahl AG (KS)
Maxhütte – Eisenwerke-Gesellschaft Maximilianshütte mbH
Saarstahl Völklingen GmbH
Thyssen Edelstahlwerke AG

INDIA
Bihar Alloy Steels Ltd
Ellora Steels Ltd (ESL)
Ferro Alloys Corp Ltd (Facor) (Steel division), (Vidarbha Iron & Steel Corp Ltd)
Guest Keen Williams Ltd (GKW)
Mahindra Ugine Steel Co Ltd (Muscosteel)
Partap Steel Rolling Mills (1935) Ltd
Punjab Con-Cast Steels Ltd
Sanghvi Steels Ltd
Sunflag Iron & Steel Co
Visvesvaraya Iron & Steel Ltd (formerly The Mysore Iron & Steel Ltd)

ITALY
DeltaCogne SpA
Nuova Deltasider SpA
Falci SpA, Reparto Laminazione
Las – Laminazione Acciai Speciali SpA
Acciaierie e Ferriere Luigi Leali SpA
Acciaierie Valbruna SpA

JAPAN
Aichi Steel Works Ltd
Daido Steel Co Ltd
Hitachi Metals Ltd (Hitachi Kinzoku)

SOUTH KOREA
Sammi Steel Co Ltd

MEXICO
Acero Solar SA de CV

POLAND
Poldi Steelworks

PORTUGAL
Fábrica de Aços Tomé Fèteira SA

SOUTH AFRICA
Scaw Metals Ltd
Usco – The Union Steel Corp (of South Africa) Ltd

SPAIN
Forjas Alavesas SA (Fasa)
Babcock & Wilcox Española SA Tubular Products Division
SA Echevarría (SAE)
Aceros de Irura SA
Aceros de Llodio SA (Allosa)
Olarra SA (Olsa)
Pedro Orbegozo y Cía SA (Acenor)
Roldan SA

SWEDEN
Björneborgs Jernverks AB
Kloster Speedsteel AB
Smedjebacken-Boxholm Stål AB
Uddeholm Tooling AB

SWITZERLAND
Von Moos Stahl AG

TURKEY
Asil Çelik Sanayi ve Ticaret AŞ (Asil Çelik)
MKEK Celik Fabrikasi

UNITED KINGDOM
Barworth Flockton Ltd
Bedford Steels
BSC – British Steel Corp
Ductile Hot Mill, Glynwed Steels Ltd
Dudley Port Rolling Mills, Glynwed Steels Ltd
George Gadd & Co, Glynwed Steels Ltd
Joseph Gillott & Sons, Glynwed Steels Ltd
T.W. Pearson Ltd
Sanderson Kayser Ltd
Sheffield Forgemasters Ltd
Spencer Clark Metal Industries PLC
UES – United Engineering Steels Ltd
W. Wesson, Glynwed Steels Ltd

UNITED STATES
Al Tech Specialty Steel Corp
Atlantic Steel Co
Bethlehem Steel Corp
Cyclops Corp
Kentucky Electric Steel Corp
LTV Steel Co Inc

Alloy steel - flats *—continued*

UNITED STATES *—continued*
McDonald Steel Corp
Nelsen Steel & Wire Co Inc
Quanex Corp
Seattle Steel Inc
The Timken Co

VENEZUELA
Acerex CA

YUGOSLAVIA
Boris Kidrič Steelworks
Slovenske Železarne – Iron & Steel Works of Slovenia

Alloy steel - hexagons

ARGENTINA
Acindar Industria Argentina de Aceros SA

AUSTRIA
Vereinigte Edelstahlwerke AG

BRAZIL
Laminação Santa Maria SA – Indústria e Comércio
Villares Group

CANADA
Slater Steels, Sorel Forge Division

CHINA
Chongqing Special Steel Works

FRANCE
Aubert et Duval
Aciéries et Laminoirs du Saut du Tarn (ALST)
Usinor Sacilor

EAST GERMANY
VEB Edelstahlwerk 8 Mai 1945

WEST GERMANY
Remscheider Walz- und Hammerwerke, Böllinghaus & Co
Edelstahlwerke Buderus AG
Dörrenberg Edelstahl GmbH
Fürstlich Hohenzollernsche Hüttenverwaltung Laucherthal (FHH)
Krupp Stahl AG (KS)
Saarstahl Völklingen GmbH
Thyssen Edelstahlwerke AG

INDIA
Guest Keen Williams Ltd (GKW)

INDIA *—continued*
Mahindra Ugine Steel Co Ltd (Muscosteel)

ITALY
DeltaCogne SpA
Nuova Deltasider SpA
Falci SpA, Reparto Laminazione
Acciaierie Valbruna SpA

POLAND
Poldi Steelworks

PORTUGAL
Fábrica de Aços Tomé Fèteira SA

SOUTH AFRICA
Scaw Metals Ltd

SPAIN
Aforasa – Acerias y Forjas de Azcoitia SA
Aceros de Llodio SA (Allosa)

SWITZERLAND
Von Moos Stahl AG

TURKEY
Asil Çelik Sanayi ve Ticaret AŞ (Asil Çelik)
MKEK Celik Fabrikasi

UNITED KINGDOM
Bedford Steels
Dudley Port Rolling Mills, Glynwed Steels Ltd
George Gadd & Co, Glynwed Steels Ltd
Kiveton Park Steel & Wire Works Ltd
T.W. Pearson Ltd
Sanderson Kayser Ltd
Sheffield Forgemasters Ltd
Spencer Clark Metal Industries PLC
UES – United Engineering Steels Ltd
Woodstone Rolling Mills Ltd

UNITED STATES
Al Tech Specialty Steel Corp
Cyclops Corp
Inland Steel Co
LTV Steel Co Inc
Nelsen Steel & Wire Co Inc

YUGOSLAVIA
Boris Kidrič Steelworks

Alloy steel - bright (cold finished) bars

AUSTRALIA
Commonwealth Steel Co Ltd
Martin Bright Steels Ltd

BRAZIL
Aços Anhanguera SA (Acoansa)
Siderúrgica Nossa Senhora Aparecida SA
Villares Group

CANADA
Union Drawn Steel Co Ltd

CHINA
Changcheng Steel Works
Chengdu Metallurgical Experimental Plant
Chongqing Special Steel Works
Hefei Iron & Steel Co
Shanghai Iron & Steel Works
Special Steel Co of Shougang

FINLAND
Ovako Steel Oy Ab

FRANCE
Aubert et Duval

WEST GERMANY
Fürstlich Hohenzollernsche Hüttenverwaltung Laucherthal (FHH)
Saarstahl Völklingen GmbH
Thyssen Edelstahlwerke AG
Westig GmbH

INDIA
BP Steel Industries Pvt Ltd
Chase Bright Steel Ltd
Damehra Steels & Forgings Pvt Ltd
Guest Keen Williams Ltd (GKW)

ITALY
SpA Officine Fratelli Bertoli fu Rodolfo
DeltaCogne SpA
Nuova Deltasider SpA
Acciaierie Valbruna SpA

JAPAN
Aichi Steel Works Ltd
Daido Steel Co Ltd
Hitachi Metals Ltd (Hitachi Kinzoku)

MEXICO
Cold Rolled de México SA

MEXICO *—continued*
Industrias CH SA (formerly Campos Hermanos SA)

SOUTH AFRICA
Rand Bright Steels (Pty) Ltd

SPAIN
Forjas Alavesas SA (Fasa)
Babcock & Wilcox Española SA Tubular Products Division
Aceros de Llodio SA (Allosa)
Olarra SA (Olsa)
Pedro Orbegozo y Cía SA (Acenor)

SWEDEN
Ovako Steel AB
Smedjebacken-Boxholm Stål AB

SWITZERLAND
Von Moos Stahl AG

TURKEY
MKEK Celik Fabrikasi

UNITED KINGDOM
Bar Bright Ltd
Barlow Bright Steels, Glynwed Steels Ltd
Barworth Flockton Ltd
Bedford Steels
Brasway PLC, Bright Bar Division
Bright Steels Ltd
British Bright Bar Ltd
C.G. Carlisle & Co Ltd
Richd W. Carr & Co Ltd
Joseph Gillott & Sons, Glynwed Steels Ltd
A.E. Godrich & Son Ltd
K.W. Haywood & Son Ltd
Arthur Lee & Sons PLC
Longmore Brothers, Glynwed Steels Ltd
Midland Bright Drawn Steel Ltd
Frank Pickering & Co Ltd
Steel Parts, Glynwed Steels Ltd
UES – United Engineering Steels Ltd
Unbrako Steel Co Ltd
W. Wesson, Glynwed Steels Ltd

UNITED STATES
Baron Drawn Steel Corp
Bliss & Laughlin Steel Co
Carpenter Technology Corp
Columbia Tool Steel Co
Copperweld Steel Co
Crucible Specialty Metals Division, Crucible Materials Corp

Alloy steel – bright

—continued

UNITED STATES *—continued*
Cuyahoga Steel & Wire Inc
Greer Steel Co
Moltrup Steel Products Co
Nucor Corp
Quanex Corp
Superior Drawn Steel Co, Division of Standard Steel Specialty Co
Teledyne Columbia-Summerill
Teledyne Pittsburgh Tool Steel
Wyckoff Steel Division, Ampco-Pittsburgh Corp

YUGOSLAVIA
Slovenske Železarne – Iron & Steel Works of Slovenia

Alloy steel – light angles

CANADA
Manitoba Rolling Mills, Amca International

CHINA
Benxi Iron & Steel Co

CZECHOSLOVAKIA
Vitkovice Steelworks

FRANCE
Aciéries et Laminoirs de Rives

WEST GERMANY
Remscheider Walz- und Hammerwerke, Böllinghaus & Co
Maxhütte – Eisenwerke-Gesellschaft Maximilianshütte mbH
Saarstahl Völklingen GmbH

ITALY
Falci SpA, Reparto Laminazione

MEXICO
Laminados Barniedo SA de CV

SWEDEN
Smedjebacken-Boxholm Stål AB

USSR
Met Zavod Krivorizhstal (Krivoy Rog Iron & Steel Works)

Alloy steel – light tees

FRANCE
Aciéries et Laminoirs de Rives

WEST GERMANY
Remscheider Walz- und Hammerwerke, Böllinghaus & Co
Saarstahl Völklingen GmbH

Alloy steel – light channels

CANADA
Manitoba Rolling Mills, Amca International

CZECHOSLOVAKIA
Vitkovice Steelworks

WEST GERMANY
Maxhütte – Eisenwerke-Gesellschaft Maximilianshütte mbH
Saarstahl Völklingen GmbH

SWEDEN
Smedjebacken-Boxholm Stål AB

Alloy steel – special light sections

BRAZIL
Laminação Santa Maria SA – Indústria e Comércio

CANADA
Manitoba Rolling Mills, Amca International

FRANCE
Aciéries et Laminoirs de Rives

WEST GERMANY
Remscheider Walz- und Hammerwerke, Böllinghaus & Co
Dörrenberg Edelstahl GmbH
Klöckner Stahl GmbH
Saarstahl Völklingen GmbH
Thyssen Edelstahlwerke AG

INDIA
Guest Keen Williams Ltd (GKW)

ITALY
Metallurgica Calvi SpA
Falci SpA, Reparto Laminazione
Las – Laminazione Acciai Speciali SpA

Alloy steel – special light sections —*continued*

SPAIN
Forjas Alavesas SA (Fasa)

UNITED KINGDOM
Dudley Port Rolling Mills, Glynwed Steels Ltd
Joseph Gillott & Sons, Glynwed Steels Ltd
Osborn Steel Extrusions Ltd
Spencer Clark Metal Industries PLC
TI Stainless Tubes Ltd

UNITED STATES
LTV Steel Co Inc
McDonald Steel Corp

Alloy steel – hot rolled hoop and strip (uncoated)

BRAZIL
Acesita – Cia Aços Especiais Itabira

CHINA
Changcheng Steel Works
Shaanxi Precision Alloy Plant

FRANCE
Usinor Sacilor

WEST GERMANY
Edelstahlwerke Buderus AG
Krupp Stahl AG (KS)

ITALY
Snar SpA

JAPAN
Daido Steel Co Ltd
Kobe Steel Ltd (Kobe Seikosho)

NETHERLANDS
Hoogovens Groep BV

SWEDEN
AB Sandvik Steel

UNITED KINGDOM
BSC – British Steel Corp
Ductile Hot Mill, Glynwed Steels Ltd
W. Wesson, Glynwed Steels Ltd

UNITED STATES
Sharon Steel Corp

Alloy steel – skelp (tube strip)

FRANCE
Usinor Sacilor

Alloy steel – cold rolled hoop and strip (uncoated)

BRAZIL
Acesita – Cia Aços Especiais Itabira
Armco do Brasil SA Indústria e Comércio

CHINA
Changcheng Steel Works
Shaanxi Precision Alloy Plant

FRANCE
Usinor Sacilor

EAST GERMANY
VEB Kaltwalzwerk Oranienburg

WEST GERMANY
Edelstahlwerke Buderus AG
Krupp Stahl AG (KS)
Thyssen Edelstahlwerke AG
Westig GmbH

ITALY
Snar SpA

JAPAN
Daido Steel Co Ltd
Hitachi Metals Ltd (Hitachi Kinzoku)
Kobe Steel Ltd (Kobe Seikosho)

POLAND
Huta Baildon

SPAIN
SA Echevarría (SAE)

SWEDEN
AB Sandvik Steel

SWITZERLAND
Kaltband AG

UNITED KINGDOM
BSC – British Steel Corp
Arthur Lee & Sons PLC

UNITED STATES
Greer Steel Co
Rome Strip Steel Co Inc
Sharon Steel Corp

Alloy steel – cold rolled hoop and strip *—continued*

UNITED STATES *—continued*
Thompson Steel Co Inc

YUGOSLAVIA
Slovenske Železarne – Iron & Steel Works of Slovenia

Alloy steel – medium plates

ALBANIA
Steel of the Party Metallurgical Combine

AUSTRALIA
Bunge Industrial Steels Pty Ltd

AUSTRIA
Vereinigte Edelstahlwerke AG

CANADA
Ipsco Inc (formerly Interprovincial Steel & Pipe Corp Ltd)

CHINA
Changcheng Steel Works
Chongqing Special Steel Works

FRANCE
Aubert et Duval
Usinor Sacilor

EAST GERMANY
VEB Walzwerk Hermann Matern, Burg

WEST GERMANY
Edelstahlwerke Buderus AG
Krupp Stahl AG (KS)
Thyssen Edelstahlwerke AG

JAPAN
Chubu Steel Plate Co Ltd
Kobe Steel Ltd (Kobe Seikosho)

NORTH KOREA
Songjin Works

NETHERLANDS
Hoogovens Groep BV

POLAND
Huta Baildon

TAIWAN
China Steel Corp
Tang Eng Iron Works Co Ltd

UNITED KINGDOM
Sanderson Kayser Ltd

UNITED STATES
Avesta Inc
Inland Steel Co
Oregon Steel Mills, Division of Gilmore Steel Corp
Teledyne Vasco

USSR
Chelyabinskiy Met Zavod (Chelyabinsk Iron & Steel Works)

YUGOSLAVIA
Boris Kidrič Steelworks

Alloy steel – heavy plates

ALBANIA
Steel of the Party Metallurgical Combine

AUSTRALIA
Bunge Industrial Steels Pty Ltd

BELGIUM
SA Fabrique de Fer de Charleroi

CHINA
Changcheng Steel Works

FRANCE
Usinor Sacilor

WEST GERMANY
Krupp Stahl AG (KS)
Thyssen Edelstahlwerke AG

HUNGARY
Diosgyör Works (Lenin Metallurgical Works)

JAPAN
Kobe Steel Ltd (Kobe Seikosho)

NETHERLANDS
Hoogovens Groep BV

POLAND
Huta Baildon

SWEDEN
Björneborgs Jernverks AB

TAIWAN
China Steel Corp
Tang Eng Iron Works Co Ltd

Alloy steel – heavy plates
—continued

UNITED KINGDOM
Spartan Redheugh Ltd

UNITED STATES
Avesta Inc
Inland Steel Co
Marathon LeTourneau Co, Longview Division

USSR
Chelyabinskiy Met Zavod (Chelyabinsk Iron & Steel Works)

YUGOSLAVIA
Boris Kidrič Steelworks

Alloy steel – universal plates

WEST GERMANY
Lemmerz-Werke KGaA
Saarstahl Völklingen GmbH

UNITED KINGDOM
Ductile Hot Mill, Glynwed Steels Ltd

UNITED STATES
Seattle Steel Inc

Alloy steel – floor plates

JAPAN
Kobe Steel Ltd (Kobe Seikosho)

UNITED STATES
Inland Steel Co

Alloy steel – hot rolled sheet/coil (uncoated)

BRAZIL
Acesita – Cia Aços Especiais Itabira

CANADA
Dofasco Inc
Ipsco Inc (formerly Interprovincial Steel & Pipe Corp Ltd)

CHINA
Changcheng Steel Works
Chongqing Special Steel Works
Special Steel Co of Shougang

FRANCE
Aubert et Duval
Usinor Sacilor

WEST GERMANY
Krupp Stahl AG (KS)
Thyssen Edelstahlwerke AG

JAPAN
Kobe Steel Ltd (Kobe Seikosho)

NETHERLANDS
Hoogovens Groep BV

TAIWAN
Tang Eng Iron Works Co Ltd

UNITED KINGDOM
Spencer Clark Metal Industries PLC

UNITED STATES
Allegheny Ludlum Corp
Cyclops Corp
Inland Steel Co
Rouge Steel Co
Teledyne Vasco

USSR
Chelyabinskiy Met Zavod (Chelyabinsk Iron & Steel Works)

YUGOSLAVIA
Slovenske Železarne – Iron & Steel Works of Slovenia

Alloy steel – cold rolled sheet/coil (uncoated)

BRAZIL
Acesita – Cia Aços Especiais Itabira

WEST GERMANY
Krupp Stahl AG (KS)
Thyssen Edelstahlwerke AG

JAPAN
Kobe Steel Ltd (Kobe Seikosho)

NETHERLANDS
Hoogovens Groep BV

RUMANIA
Tirgovişte Works

TAIWAN
Tang Eng Iron Works Co Ltd

UNITED STATES
Allegheny Ludlum Corp

Alloy steel – cold rolled sheet/coil —*continued*

UNITED STATES —*continued*
Carpenter Technology Corp
Cyclops Corp
Inland Steel Co
Rouge Steel Co

USSR
Chelyabinskiy Met Zavod (Chelyabinsk Iron & Steel Works)

NETHERLANDS
Nedstaal BV

SPAIN
SA Echevarría (SAE)
Roldan SA

SWEDEN
AB Sandvik Steel

SWITZERLAND
Kaltband AG

UNITED KINGDOM
A.E. Godrich & Son Ltd
Kiveton Park Steel & Wire Works Ltd
Sanderson Kayser Ltd
Unbrako Steel Co Ltd

UNITED STATES
Bethlehem Steel Corp
Carpenter Technology Corp
Cuyahoga Steel & Wire Inc
Nelsen Steel & Wire Co Inc
Techalloy Co Inc
Teledyne Pittsburgh Tool Steel
Teledyne Vasco
The Timken Co

USSR
Beloretskiy Met Zavod (Beloretsk Iron & Steel Works)

YUGOSLAVIA
Boris Kidrič Steelworks
Slovenske Železarne – Iron & Steel Works of Slovenia

Alloy steel – wire

AUSTRIA
Joh. Pengg, Draht- & Walzwerke

BELGIUM
Cockerill Sambre SA

BRAZIL
Siderúrgica Nossa Senhora Aparecida SA
Mannesmann SA

CHINA
Changcheng Steel Works
Shaanxi Precision Alloy Plant
Shoudu Iron & Steel Co
Special Steel Co of Shougang

FINLAND
Dalsbruk Oy Ab

FRANCE
Usinor Sacilor

EAST GERMANY
VEB Edelstahlwerk 8 Mai 1945

WEST GERMANY
Krupp Stahl AG (KS)
Thyssen Edelstahlwerke AG
Westig GmbH

ITALY
Nuova Deltasider SpA

JAPAN
Kobe Steel Ltd (Kobe Seikosho)
Sanyo Special Steel Co Ltd
Sumitomo Electric Industries Ltd (Sumitomo Denki Kogyo)

MEXICO
Acero Solar SA de CV

Alloy steel – seamless tubes and pipes

AUSTRALIA
Email Tube Division, Email Ltd

BRAZIL
Mannesmann SA

CHINA
Baoshan Iron & Steel Works
Chengdu Seamless Steel Tube Plant
Guiyang Steel Mill
Shaanxi Precision Alloy Plant
Shanghai Iron & Steel Works
Special Steel Co of Shougang

CZECHOSLOVAKIA
Chomutov Tube Works

Alloy steel – seamless tubes and pipes —*continued*

EAST GERMANY
VEB Stahl- und Walzwerk Riesa

WEST GERMANY
Benteler AG
Röhrenwerke Bous/Saar GmbH (RBS)
Poppe & Potthoff GmbH & Co
Wälzlagerrohr GmbH

ITALY
Dalmine SpA
Tubicar SpA

JAPAN
Kobe Steel Ltd (Kobe Seikosho)
Sanyo Special Steel Co Ltd

SOUTH KOREA
Sammi Steel Co Ltd

MEXICO
Tamsa – Tubos de Acero de Mexico SA

SPAIN
Babcock & Wilcox Española SA Tubular Products Division
Centracero SA
Tubacex – C.E. de Tubos por Extrusion SA
Tubos Reunidos SA

SWEDEN
Avesta Nyby Powder AB
Ovako Steel AB

UNITED KINGDOM
Cameron Iron Works Ltd, Forged Products Division
Seamless Tubes Ltd
TI Desford Tubes Ltd
TI Hollow Extrusions Ltd

UNITED STATES
Al Tech Specialty Steel Corp
The Babcock & Wilcox Co, Tubular Products Division
Lone Star Steel Co
Pacific Tube Co
Plymouth Tube Co
Quanex Corp
The Timken Co

USSR
Chelyabinskiy Met Zavod (Chelyabinsk Iron & Steel Works)

USSR —*continued*
Dnepropetrovskiy Met i Trubnoprokatny Savod Imeni Lenina (Lenin Iron & Steel Tube Rolling Works)

Alloy steel – longitudinal weld tube and pipes

AUSTRALIA
Email Tube Division, Email Ltd
Palmer Tube Mills Ltd

BRAZIL
Mannesmann SA

CANADA
Associated Tube Industries Ltd (ATI)
Barton Tubes Ltd
Ipsco Inc (formerly Interprovincial Steel & Pipe Corp Ltd)
Standard Tube Canada, Division of TI Canada Inc

WEST GERMANY
Eisenbau Krämer mbH

JORDAN
The Jordan Pipes Manufacturing Co Ltd (JPMC)

SOUTH KOREA
Pusan Steel Pipe Corp

MEXICO
Galvak SA

NETHERLANDS
P. van Leeuwen Jr's Buizenhandel BV

UNITED KINGDOM
Bentham International Ltd
Manchester Cold Rollers Ltd

UNITED STATES
The Babcock & Wilcox Co, Tubular Products Division
Carpenter Technology Corp
Indiana Tube Corp
Lone Star Steel Co
LTV Steel Co Inc
Pacific Tube Co
Pixley Tube Corp
Plymouth Tube Co
Quanex Corp
Trent Tube
Whittymore Tube Co, Div Whittymore Enterprises Inc

Alloy steel – longitudinal weld tube and pipes
—continued

USSR
Elektrostal Metallurgechesky Zavod Imeni Tevosyan

Alloy steel – spiral-weld tubes and pipes

CANADA
Ipsco Inc (formerly Interprovincial Steel & Pipe Corp Ltd)

PAKISTAN
Indus Steel Pipes Ltd

UNITED STATES
Lone Star Steel Co

Alloy steel – large-diameter tubes and pipes

UNITED KINGDOM
Bentham International Ltd

UNITED STATES
Lone Star Steel Co
LTV Steel Co Inc
Plymouth Tube Co
Trent Tube

Alloy steel – oil country tubular goods

BRAZIL
Mannesmann SA

CANADA
Ipsco Inc (formerly Interprovincial Steel & Pipe Corp Ltd)

FRANCE
Aubert et Duval

WEST GERMANY
Benteler AG

ITALY
Dalmine SpA

MEXICO
Tamsa – Tubos de Acero de Mexico SA

SPAIN
Babcock & Wilcox Española SA Tubular Products Division
Tubos Reunidos SA

UNITED KINGDOM
TI Desford Tubes Ltd

UNITED STATES
The Babcock & Wilcox Co, Tubular Products Division
Fort Worth Pipe Co
Lone Star Steel Co
LTV Steel Co Inc
Quanex Corp
The Timken Co

Alloy steel – cold drawn tubes and pipes

AUSTRALIA
Email Tube Division, Email Ltd
Martin Bright Steels Ltd

BRAZIL
Mannesmann SA
Laminação Santa Maria SA – Indústria e Comércio

CANADA
Standard Tube Canada, Division of TI Canada Inc

CHINA
Changcheng Steel Works
Guiyang Steel Mill
Shaanxi Precision Alloy Plant

WEST GERMANY
Benteler AG
Poppe & Potthoff GmbH & Co

ITALY
Dalmine SpA
Tubicar SpA

JAPAN
Kobe Steel Ltd (Kobe Seikosho)
Sanwa Metal Industries Ltd

SOUTH KOREA
Sammi Steel Co Ltd

MEXICO
Tamsa – Tubos de Acero de Mexico SA

Alloy steel – cold drawn tubes and pipes —*continued*

SPAIN
Transmesa – Transformaciones Metalurgicas SA
Tubacex – C.E. de Tubos por Extrusion SA

SWITZERLAND
Rothrist Tube Ltd

UNITED KINGDOM
Seamless Tubes Ltd

UNITED STATES
Al Tech Specialty Steel Corp
Cyclops Corp
Lone Star Steel Co
LTV Steel Co Inc
Plymouth Tube Co
Quanex Corp
The Timken Co
Trent Tube
Whittymore Tube Co, Div Whittymore Enterprises Inc

Alloy steel – hollow bars

FRANCE
Aubert et Duval

WEST GERMANY
Benteler AG

SPAIN
Babcock & Wilcox Española SA Tubular Products Division
Tubacex – C.E. de Tubos por Extrusion SA

SWEDEN
Avesta Nyby Powder AB

UNITED KINGDOM
Bedford Steels
TI Desford Tubes Ltd

UNITED STATES
Al Tech Specialty Steel Corp
Carpenter Technology Corp
Lone Star Steel Co
The Timken Co

Alloy steel – precision tubes and pipes

AUSTRALIA
Email Tube Division, Email Ltd
Palmer Tube Mills Ltd

BRAZIL
Mannesmann SA

CHINA
Shanghai Iron & Steel Works

WEST GERMANY
Benteler AG
Poppe & Potthoff GmbH & Co
Wälzlagerrohr GmbH

SPAIN
Transmesa – Transformaciones Metalurgicas SA
Tubacex – C.E. de Tubos por Extrusion SA
Tubos Reunidos SA

SWITZERLAND
Rothrist Tube Ltd

UNITED KINGDOM
TI Tube Products Ltd

UNITED STATES
Lone Star Steel Co
Plymouth Tube Co
The Timken Co
Trent Tube
Whittymore Tube Co, Div Whittymore Enterprises Inc

Alloy steel – hollow sections

AUSTRALIA
Palmer Tube Mills Ltd

BRAZIL
Mannesmann SA

CANADA
Ipsco Inc (formerly Interprovincial Steel & Pipe Corp Ltd)

JORDAN
The Jordan Pipes Manufacturing Co Ltd (JPMC)

NETHERLANDS
P. van Leeuwen Jr's Buizenhandel BV

Alloy steel – hollow sections —*continued*

SWEDEN
Avesta Nyby Powder AB

UNITED STATES
Lone Star Steel Co

Cast iron – pipes

ARGENTINA
Tamet – SA Talleres Metal/urgicos San Martin

BRAZIL
Cia Metalúrgica Barbará

CHINA
Shoudu Iron & Steel Co

EGYPT
Delta Steel Mill SAE

FRANCE
Pont-à-Mousson SA

WEST GERMANY
Halbergerhütte GmbH

INDIA
Agarwal Foundry & Engineering Works

JAPAN
Kubota Ltd (Kubota Tekko)

NORTH KOREA
Hwanghai Iron Works

SPAIN
Funditubo SA

UNITED KINGDOM
The Clay Cross Co Ltd
Stanton PLC

UNITED STATES
Electralloy Corp

YUGOSLAVIA
Rudnik i Zelezara Vareš

Cast iron – pipe fittings

BRAZIL
Cia Metalúrgica Barbará

FRANCE
Pont-à-Mousson SA

INDIA
Agarwal Foundry & Engineering Works

JAPAN
Kubota Ltd (Kubota Tekko)

SPAIN
Funditubo SA

UNITED KINGDOM
The Clay Cross Co Ltd
Stanton PLC

Index to Advertisers

Index to Advertisers – cont.

Index to Advertisers – cont. Page

Index to Advertisers – cont. Page

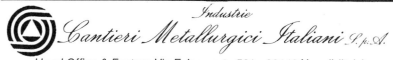

NOTES

NOTES

NOTES

NOTES

NOTES

NOTES

Printed in England by Staples Printers St Albans Limited at The Priory Press.